International Conference on Innovative Research (ICIR Euroinvent 2017)

IOP Conference Series: Materials Science and Engineering
Volume 209

Iasi, Romania
25 - 26 May 2017

Editors:

Andrei Victor Sandu
Petrica Vizureanu

ISBN: 978-1-5108-4412-4
ISSN: 1757-8981

Printed from e-media with permission by:

Curran Associates, Inc.
57 Morehouse Lane
Red Hook, NY 12571

Some format issues inherent in the e-media version may also appear in this print version.

Printed by Curran Associates, Inc. (2017)

For permission requests, please contact the Institute of Physics
at the address below.

Institute of Physics
Dirac House, Temple Back
Bristol BS1 6BE UK

Phone: 44 1 17 929 7481
Fax: 44 1 17 920 0979

techtracking@iop.org

Additional copies of this publication are available from:

Curran Associates, Inc.
57 Morehouse Lane
Red Hook, NY 12571 USA
Phone: 845-758-0400
Fax: 845-758-2633
Email: curran@proceedings.com
Web: www.proceedings.com

Table of contents

Volume 209

International Conference on Innovative Research - ICIR Euroinvent 2017
25–26 May 2017, Iasi, Romania

Accepted papers received: 7 June 2017
Published online: 27 June 2017

Preface

Synthesis and Characterization of Materials

Procedures and Technologies for Materials Engineering

Materials & Life Science

IOP Conf. Series: Materials Science and Engineering **209** (2017) 011001 doi:10.1088/1757-899X/209/1/011001

EUROINVENT
ICIR 2017

International Conference on Innovative Research

May 25th to 26th, 2017
Iasi – Romania
Palace of Culture

Organized by:

- **Romanian Inventors Forum**
- **Faculty of Materials Science and Engineering, The "Gheorghe Asachi" Technical University of Iasi, Romania**
- **ARHEOINVEST Platform, Alexandru Ioan Cuza University of Iasi**
- **Centre of Excellence Geopolymer and Green Technology CEGeoGTech), Universiti Malaysia Perlis (UniMAP)**
- **Malaysian Research & Innovation Society (MyRIS)**

With support of:

- **School of Fundamental Science, Universiti Malaysia Terengganu**
- **International Federation of Inventors' Associations - IFIA**
- **World Invention Intellectual Property Associations – WIIPA**

Editors:

Andrei Victor SANDU, Mohd Mustafa Al Bakri ABDULLAH,

Petrică VIZUREANU, Che Mohd Ruzaidi GHAZALI, Ion SANDU

IOP Conf. Series: Materials Science and Engineering **209** (2017) 011001 doi:10.1088/1757-899X/209/1/011001

EUROINVENT ICIR 2017 — International Conference on Innovative Research, Iasi, 25th–26th of May 2017

Foreword

This volume contains the information of the ICIR Euroinvent 2017 Conference and the abstracts of selected peer-reviewed papers from the 2017 International Conference on Innovative Research, which was held in Iaşi, România from 25 to 26th of May 2017.

The ICIR Conference is organized under the auspices of EUROINVENT. Euroinvent is a joint event promoting creativity in a European context, by displaying the contributions of consecrated schools from higher education and academic research and also of individual inventors and researchers.

The EUROINVENT International Conference on Innovative Research (ICIR) brings together leading researchers, engineers and scientists who will present actual research results in the field of Materials Science and Engineering.

The conference aims to provide a high level international forum for researchers, engineers and scientists to present their new advances and research results in the field of materials science and engineering.

The volume covers all the aspects of materials science, from synthesis and characterization of materials to procedures and technologies for materials engineering, as well as materials application and their involvement in the life sciences.

All the papers have been reviewed by two expert referees in their relevant topic disciplines. The papers selected for the volume depended on their quality and relevancy to the conference.

The editors hope that this volume will provide the reader a broad overview of the latest advances in the field of materials science and engineering, and that they will be a valuable references source for further research.

The editors would like to express their sincere appreciations and thanks to all the committee members of the ICIR 2017 for their tremendous efforts. Thanks also to IOP Conference Series for producing the volume with full articles.

Finally, the editors would like to thank all the authors for their contribution to this valuable volume.

IOP Conf. Series: Materials Science and Engineering **209** (2017) 011001 doi:10.1088/1757-899X/209/1/011001

EUROINVENT ICIR 2017 — International Conference on Innovative Research, Iasi, 25th–26th of May 2017

Scientific Advisory Board

Chairman

Event Coordinator

IOP Conf. Series: Materials Science and Engineering **209** (2017) 011001 doi:10.1088/1757-899X/209/1/011001

EUROINVENT ICIR 2017 International Conference on Innovative Research
Iasi, 25th–26th of May 2017

Technical and Organizing Committee

Assoc.Prof.Dr.Eng. Dragos Cristian ACHITEI	*The "Gheorghe Asachi" Technical University of Iasi, Romania*
PhD student Madalina Simona BALTATU	*The "Gheorghe Asachi" Technical University of Iasi, Romania*
PhD Bogdan ISTRATE	*The "Gheorghe Asachi" Technical University of Iasi, Romania*
Assist.Prof.Dr.Eng. Mirabela Georgiana MINCIUNA	*The "Gheorghe Asachi" Technical University of Iasi, Romania*
PhD student Marius MUNTEANU	*Alexandru Ioan Cuza University of Iasi, Romania*
PhD student Vasile PELIN	*Alexandru Ioan Cuza University of Iasi, Romania*
Assoc.Prof.Dr.Eng. Manuela Cristina PERJU	*The "Gheorghe Asachi" Technical University of Iasi, Romania*
PhD student Daniel POTOLINCA	*Alexandru Ioan Cuza University of Iasi, Romania*
Assoc.Prof.Dr.Eng. Ioan Gabriel SANDU	*The "Gheorghe Asachi" Technical University of Iasi, Romania*
Assist.Prof.Eng. Catalin Andrei TUGUI	*The "Gheorghe Asachi" Technical University of Iasi, Romania*
Dr. Viorica VASILACHE	*Alexandru Ioan Cuza University of Iasi, Romania*

IOP Conf. Series: Materials Science and Engineering **209** (2017) 011002 doi:10.1088/1757-899X/209/1/011002

Peer review statement

All papers published in this volume of *IOP Conference Series: Materials Science and Engineering* have been peer reviewed through processes administered by the proceedings Editors. Reviews were conducted by expert referees to the professional and scientific standards expected of a proceedings journal published by IOP Publishing.

IOP Conf. Series: Materials Science and Engineering **209** (2017) 012001 doi:10.1088/1757-899X/209/1/012001

Low Cost Synthesis Method of Two-Dimensional Titanium Carbide MXene

Z A M Rasid[1] , M F Omar[1*], M F M Nazeri[1], M A A A'ziz[1] and M Szota[2]
[1] Centre of Excellence Geopolymer & Green Technology (CEGeoTech), School of Materials Engineering, Universiti Malaysia Perlis, Perlis Malaysia
[2] Institute of Materials Engineering, Faculty of Production Engineering and Materials Technology, Czestochowa University of Technology, 19 Armii Krajowej Str., 42-200 Czestochowa, Poland

E-mail: firdausomar@unimap.edu.my

Abstract. A layered MAX phase of Ti_3AlC_2 was synthesized through pressureless sintering (PLS) the initial powder of TiH_2/Al/C without preliminary dehydrogenation under argon atmosphere at 1350°C. An elegant exfoliations approach was used to prepare a two-dimensional (2D) metal carbide Ti_3C_2 from layered MAX phase by removing A layer by chemical etching. The use of PLS method instead of any pressure assistance method such as hot isostatic press (HIP) and hot press (HP) lowered the cost of synthesis. Recently, some unique potential of Ti_3C_2 has been discovered leads to the proposal of potential application, mostly on electronic devices. Morphology and structural analysis was used to confirm the successful of this research.

1. Introduction

Two-dimensional (2D) materials are well known materials with unique properties in studies of materials engineering. 2D materials such as graphene defines as single atomic plane materials with 0.34 nm thickness [1]. These materials also known to have very high aspect ratio [2]. In different studies, some researcher state the range of materials to be 2D materials is 1 layer to 10 layers, and more than 10 layers is considered as 3D materials [3].

Graphene is the example of comprised 2D materials with atomically thin layers of sp^2-bonded carbon atoms connected by aromatic in-plane bonds. These 2D materials has only one-atom thickness and isolated from graphite in 2004 by Novoselov et al. [4]. Since the outstanding electronic properties of graphene has been discovered, other 2D materials such as hexagonal boron nitrides [5], transition metal dichalcogenides (TMDs) [6], metal oxides and hydroxides, have attracted much renewed attention. Graphene shows tremendous attraction to researchers from different fields and has risen as the most exciting star in materials science during the past several years. Its exceptional properties, such as half-integer quantum Hall effect, ambipolar electric field effect, extremely high carrier mobility, high thermal conductivity, high specific surface area, and the highest strength ever measured, provide a fertile ground for the possible implementation of graphene in nano-devices for a large variety of applications, and a lot of recent reviews have been directed towards its synthesis, properties, and functionalized applications [7].

In 2011, a fascinating idea [8] was proposed to prepare freestanding graphene-like carbides (and nitrides) not from their 3D parent binary phases, but from ternary layered MAX phases (known also as nanolaminates), which include various 2D-like layers of transition metal carbides (or nitrides) as building blocks [9]. These 2D materials are now known as 'MXenes'; this term denotes their genesis from MAX phases (with the loss of the A component) and their similarity to graphene [7,9].

MXenes is new 2D nanosheet materials and gaining so much attention from materials engineer because its properties and unique behaviour and has similar lamellar structure with graphene. Most of materials engineer is interested with new materials because the opportunity to become a new product with better technologies and application. Generally, 2D materials is produced by removing A-layers (figure 1) from MAX phases by chemical etching. MAX phases are ternary carbide or nitrides with the chemical formula of $M_{n+1}AX_n$, where M is an early transition metal, A is an A-group element and X is either carbon or nitrogen. The value of n can be 1, 2 or 3 [3,4].

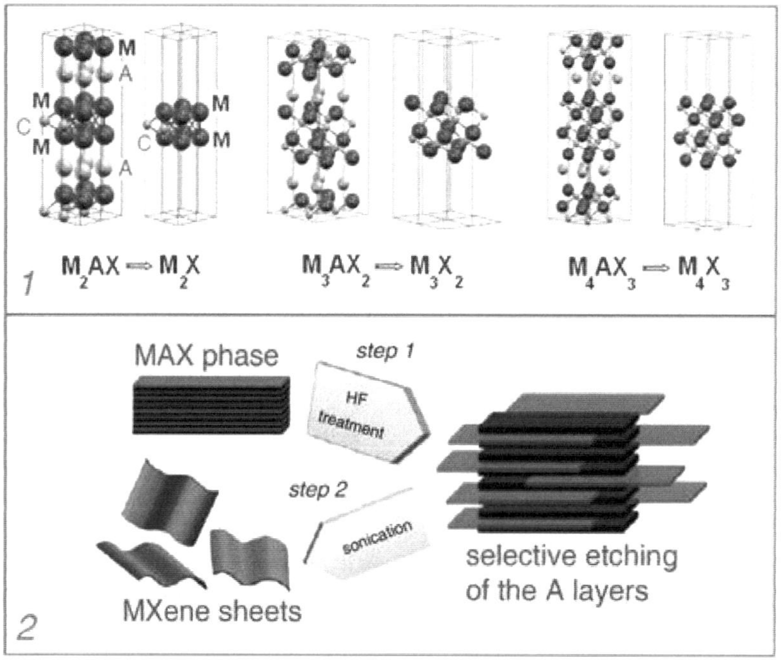

Figure 1. Atomic model of preparation of MXenes ($M_{n+1}X_n$) from MAX phases ($M_{n+1}AX_n$) via removing of A sheets [12].

Pressureless sintering (PLS) is a low-cost method to synthesized a MAX phase solid in vacuum or under the inert gas flowing. This method is used for some of researcher to commercialized MAX phase without any pressure assistance during sintering process; for example hot isostatic pressing (HIP) and hot press (HP) [13]. An elegant exfoliation approaches is used to synthesis MXene from MAX phase by selectively etching layers of A-group elements in MAX phase by using a hydrofluoric solution as etchant [14].

The aim of this research is to synthesis a Ti_3C_2 MXene form its precursor which is Ti_3AlC_2 using an elegant exfoliation approach. Morphological and structural analysis was performed to study the microstructure and element present in both Ti_3AlC_2 and Ti_3C_2.

2. Materials and methods

Starting materials of titanium hydrate (TiH_2, 99%, 325 mesh, Alfa Aesar, United State), aluminium (Al, 99%, 200 mesh, Acros Organics, Austria) and graphite (C, 99%, 550 mesh, Acros Organics, Germany) were used in this research. The powder was weighed according to the stoichiometry ratio of 3:1.1:2 based on previous research [15]. The mixture then was mixed by using a planetary ball mill for

20, 40, 60 and 80 hours respectively. The powder then was cold-pressed in stainless-steel mould with the pressure of 6000 psi to form a pellet with the diameter of 1 cm and thickness of 0.3 cm. The pellet was sintered in tube furnace by using a pressureless sintering technique. The heating rate was controlled at 10°C/min and the sintering temperature was selected at 1350°C for 2 hours in argon atmosphere. The sample then were ground to remove the surface layer in order to remove any contaminate from oxidation process during sintering.

The pellet then was pulverized, immersed and stirred in hydrofluoric solution (HF, AR grade, 49%, Qrec, New Zealand) for 20 hours in fume hood. The resulted mixture then was placed in centrifuge tube and centrifugation process was run at 5000 rpm for 7 minutes. The mixture then was washed with deionized water for several times, until the pH of solution approaching 7. The mixture then was rinsed with methanol for 3 times then placed in ultrasonic cleaner to removes contaminant from the mixture. The resulting Ti_3C_2 then was dried in vacuum at 50°C for 24 hours.

X-ray diffraction (XRD) analysis was perform using Bruker D2 Phaser at 2θ values of 20 – 60° with Cu Kα radiation to determine the corresponding peaks of the MAX phase and MXene. X'pert HighScore Plus software was used to match the corresponding peaks with the standard data from International Committee of Diffraction Data (ICDD) X-ray data file. Microstructure analysis was performed using Nova NanoSEM 450 to characterize surface morphology of both MAX phase and MXene.

3. Results and discussions

There are four phases matches for mixture of TiH_2/Al/C after ball-milled for 20, 40, 60 and 80 hours (figure 2) which is hexagonal Ti_3AlC_2 (ICDD 00-052-0875), hexagonal graphite (ICDD 00-008-0415) and tetragonal TiO_2 (ICDD 00-034-0180) and cubic TiC (ICDD 00-006-0614). This result is supported with the other Ti/Al/C researcher that uses 1:20 powder to ball ratio during ball-milling process [16]. A large amount of TiC phase was detected after ball-milled for 10 hours and the low amount desired preliminary phase of Ti_3AlC_2 also present. The intensity of Ti_3AlC_2 is decrease with the increasing milling time indicates the phase may gradually transform into amorphous phase [16].

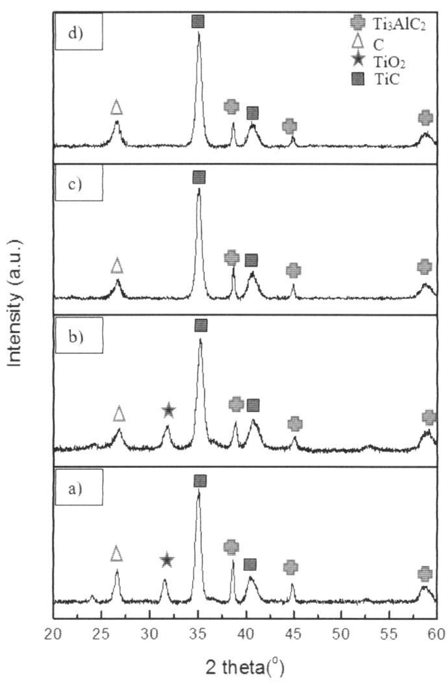

Figure 2. X-ray diffraction pattern of TiH_2/Al/C powder ball milled for a) 20 hours, b) 40 hours, c) 60 hour and d) 80 hour.

The formation of TiO_2 phase may take place during milling process since Ti is one of the most reactive metal to oxygen at elevated temperature [17]. The peak of TiO_2 decreased when the milling hour approaching 60 hours, conclude the period as the most optimize hours to ball-milling to obtain the optimum result. The phase of TiO_2 may formed during ball milling process where the reaction between O_2 and TiH_2 takes place. As to graphite, a lower peak indicates the existence of the phase and the peak does not vanish even after sinter for 80 hours. This result shows the small amount graphite powder does not distribute into the grain boundaries of Ti to form TiC and Ti_3AlC_2. In light of the ball-milling result, it is clear that the desired phase of Ti_3AlC_2 is formed after ball-milled for 20 hours and the 60 hours of milling is the most optimum period for ball-milling since the peak of TiO_2 was fully vanished.

XRD pattern shows there are four phases matches after the samples was sintered at 1350°C for 2 hours (figure 3a) which is rhombohedral Al_2O_3 (ICDD 01-075-0782), rhombohedral graphite (ICDD 01-075-2078), hexagonal Ti_3AlC_2 (ICDD 00-052-0875) and cubic TiC (ICDD 03-065-7994). Ti_3AlC_2 phase was fully vanished after chemical etched using a HF solution for 20 hours (figure 3b) and replaced by rhombohedral Ti_3C_2 phase. This result is also supported by other researchers [8,15,18-19] in this field based on the reactions of the HF solutions with Ti_3AlC_2 which include:

$$Ti_3AlC_2 + 3HF \rightarrow AlF_3 + 3/2H_2 + Ti_3C_2 \quad (1)$$

$$Ti_3C_2 + 2H_2O \rightarrow Ti_3C_2(OH)_2 + H_2 \quad (2)$$

$$Ti_3C_2 + 2HF \rightarrow Ti_3C_2F_2 + H_2 \quad (3)$$

The formation of Ti_3C_2 phase in XRD pattern (figure 3b) instead of $Ti_3C_2(OH)_2$ in reaction (2) is because the ultrasonication in methanol leads to exfoliation of MXene Ti_3C_2 [9].

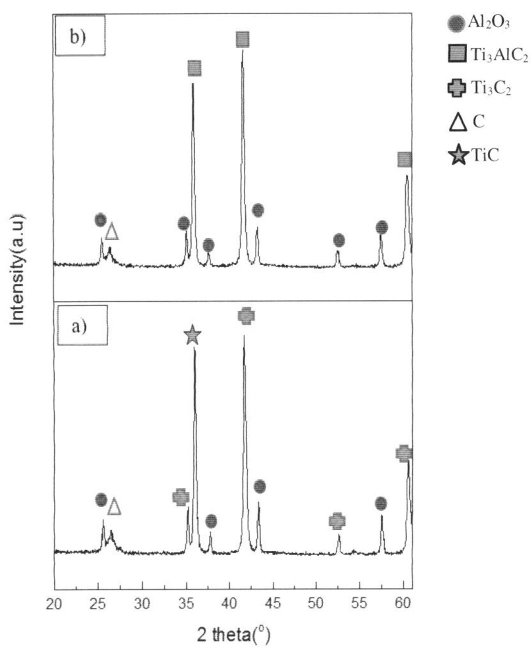

Figure 3. X-ray diffraction pattern for a) before etching and b) after etching with 49% HF solution for 20 hours.

Scanning electron image (SEM) reveals the microstructural of Ti_3AlC_2 before HF treatment which consists a number of lamellar grains (figure 4a). The product is highly porous with two types of morphology which is a layered of Ti_3AlC_2 particle with the TiC (marked as A) (figure 4c) and agglomerates TiC with the companies of Al_2O_3 and C (marked as B) (figure 4e). This result is supported by other researchers [16].

After HF treatment, SEM images shows two types of morphology which is a layer of Ti_3C_2 (marked as C) and the agglomerates Al_2O_3 with graphite, similar with B. The images of layered Ti_3C_2 with the thickness of 0.1-0.3 μm (figure 4d), confirms the successful exfoliation of Ti_3C_2 similar with reported exfoliation graphene [20]. Figure 4c and 4d is the comparison of product before and after chemical etching with the same magnification. Sonication of HF treated resulted the separation of 2D Ti_3C_2 MXene.

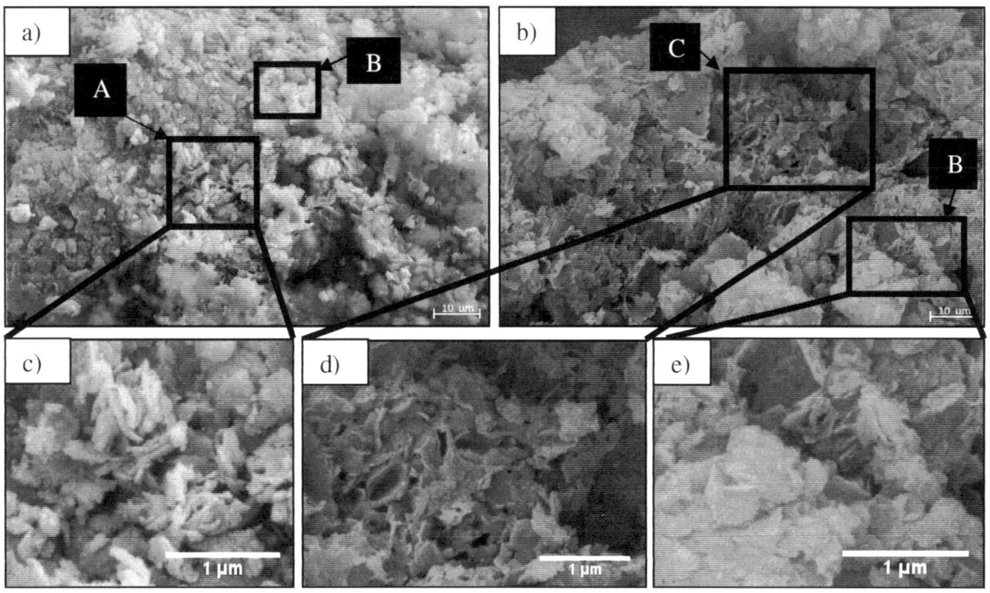

Figure 4. Scanning electron images for a) before etching and b) after etching with 49% HF solution for 20 hours c) layered of Ti_3AlC_2 particle with the TiC d) layered of Ti_3C_2 and e) the agglomerates Al_2O_3 with graphite and TiC.

4. Conclusions

A layered of titanium carbide Ti_3C_2 MXene was synthesized by low cost pressureless sintering (PLS) method using a starting powder of TiH_2, Al and graphite for two hours. A layered of Ti_3AlC_2 was successfully exfoliated and fabricate a layer of 2D Ti_3C_2 after HF treatment for 20 hours and ultrasonication. This is happening since the relatively weak bonding of Al layer compared to the Ti-C bonds. The layers of 2D Ti_3C_2 is quite thin between 0.1-0.3 μm. Two others phase was detected by XRD after the treatment which is C and Al_2O_3 and SEM image shows the agglomerates of this phase at the side of Ti_3C_2 layers. This result is helpful to confirm the successfulness of this method to produces a 2D Ti_3C_2 before the deeper study on its electrical or other properties since most of the properties was revealed by only in theoretical calculation.

References

[1] Geim A K and Novoselov K S 2007 *Nat Mater* **6**(3) 183–91
[2] Naguib M, Mashtalir O, Carle J, Presser V, Lu J, Hultman L, Gogotsi Y and Boursom M W 2012 *ACS Nano* **6**(2) 1322-31

[3] Bianco A, Cheng H M, Enoki T, Gogotsi Y, Hurt R H, Koratkar N, Kyotani T, Monthioux M, Park C R, Tascon J M D and Zhang J 2013 *Carbon N Y* **65** 1–6
[4] Novoselov K S, Geim A K, Morozov S V, Jiang D, Zhang Y, Dubonos S V, Grigorieva V and Firsov A A 2004 *Science* **306**(5696) 666–9
[5] Haubner R, Wilhelm M, Weissenbacher R and Lux B 2002 *High Perform Non-oxide Ceramic II* **102** 1–45
[6] Miro P, Ghorbani-Asl M and Heine T 2014 *Angew Chemie* **53**(11) 3015–8
[7] Tang Q and Zhou Z 2013 *Prog Mater Sci* **58**(8) 1244–315
[8] Naguib M, Kurtoglu M, Presser V, Lu J, Niu J, Heon M, Hultman L, Gogotsi Y and Barsoum M W 2011. *Adv Mater* **23**(37) 4248–53.
[9] Shein I R and Ivanovskii A L 2013 *Micro Nano Lett* **8**(2) 59–62
[10] Barsoum M W 2013 *Am Sci* **89**(4) 334–43
[11] Mironov S, Sato Y S and Kokawa H 2010 *Mater Sci Eng A* **527**(29–30) 7498–504
[12] Shein I R and Ivanovskii A L 2012 *Comput Mater Sci* **65** 104–14
[13] Low I M 2012 *Advances in Science and Technology of* $M_{n+1}ax_n$ *Phases* (Cambridge, Woodhead Publishing)
[14] Li J, Du Y, Huo C, Wang S and Cui C 2015 *Ceram Int* **41**(2) 2631–5
[15] Zhang H, Wang L, Chen Q, Li P, Zhou A, Cao X and Hu Q 2015 *Mater Des* **92**682–9
[16] Li S-B, Zhai H-X, Bei G P, Zhou Y and Zhang Z L 2006 *Mater Sci Technol* **22**(6) 667–72
[17] Park K T, Pan M, Meunier V and Plummer E W 2007 *Phys Rev B* **75**(24)
[18] Yang J, Chen B, Song H, Tang H and Li C 2014 *Cryst Res Technol* **49**(11) 926–32
[19] Li Z, Wang L, Sun D, Zhang Y, Liu B, Hu Q and Zhou A 2015 *Mater Sci Eng B* **191**(C) 33–40
[20] Lv W, Tang D M, He Y B, You C H, Shi Z Q, Chen X C, Chen M C, Hou P X, Liu C and Yang P H 2009 *ACS Nano* **3**(11) 3730–6

Acknowledgments
This present work was supported by Ministry of Higher Education of Malaysia under the Fundamental Research Grant Scheme (FRGS) number 9003-00390.

IOP Conf. Series: Materials Science and Engineering **209** (2017) 012002 doi:10.1088/1757-899X/209/1/012002

Synthesis and Characterization of TiO_2/SiO_2 Thin Film via Sol-Gel Method

D S C Halin[1,*], M M A B Abdullah[2,1], N Mahmed[1], S N A Abdul Malek[1], P Vizureanu[1,4] and A W Azhari[3]
[1] Center of Excellence Geopolymer & Green Technology (CEGeoGTech), School of Materials Engineering, Universiti Malaysia Perlis, (UniMAP), Perlis, Malaysia
[2] Faculty of Engineering Technology, Universiti Malaysia Perlis, (UniMAP), Perlis, Malaysia
[3] School of Environmental Engineering, Universiti Malaysia Perlis, (UniMAP), Perlis, Malaysia
[4] Faculty of Materials Science and Engineering, Gheorghe Asachi Technical University of Iasi, Romania

E-mail address: [*]dewisuriyani@unimap.edu.my

Abstract. TiO_2/SiO_2 thin films were prepared by sol-gel spin coating method. Structural, surface morphology and optical properties were investigated for different annealing temperatures at 300°C, 400°C and 500°C. X-ray diffraction pattern show that brookite TiO_2 crystalline phase with SiO_2 phase presence at 300°C. At higher temperatures of 400-500°C, the only phase presence was brookite. The surface morphology of film was characterized by scanning electron microscopy (SEM). The films annealed at 300°C shows an agglomeration of small flaky with crack free. When the temperature of annealing increase to 400-500°C, the films with large flaky and large cracks film were formed which was due to surface tension between the film and the air during the drying process. The UV-Vis spectroscopy shows that the film exhibits a low transmittance around 30% which was due to the substrate is inhomogeneously covered by the films. In order to improve the coverage of the film on the substrate, it has to repeatable the spin coating to ensure the substrate is fully covered by the films.

1. Introduction

TiO_2 is a material which have its chemical stability and electronic properties and well known as semiconductor.

It has been used in wide application in semiconductor electrochemistry, solar energy conversion, as gas sensor, in electronic devices, and as antireflective coatings [1]. TiO_2 becomes one of the most popular photocatalytic materials.

The TiO_2 naturally have existing in several crystalline phase such as rutile, brookite, and anatase. Among of these, anatase and rutile crystalline phase are widely used in application of solar cell. Rutile phase are suitable for protective coating on lense due to stable thermodynamically and high refractive

IOP Conf. Series: Materials Science and Engineering **209** (2017) 012002 doi:10.1088/1757-899X/209/1/012002

index. At lower temperatures, brookite and anatase phases are more stable however both will change to the rutile phase at certain temperature. Due to the photocatalytic activity, most of the researches tend to use anatase, TiO_2. The larger band gap in crystalline phase in anatase because it has better response to ultraviolet photons for photocatalysis.

Photocatalytic is the process in which the solar energy is converted into electrical energy in the solar cell. The photocatalytic performance of these compounds depends on the characteristics of the TiO_2 crystallites, such as the size and surface area [2].

Therefore, modification of its physical and chemical property is of interest to researchers [3,4]. One of the possible ways to modify the property of TiO_2 crystallites is by adding a second semiconductor into the TiO_2 matrix. Silicon dioxide (SiO_2) has been incorporated into the TiO_2 matrix to enhance the photocatalytic process [5-8].

SiO_2 has high thermal stability, excellent mechanical strength and helps to create new catalytic active sites due to interaction between TiO_2 and SiO_2 [9].

Recently, Zhou et al. [8] demonstrated that mixed metal oxides (TiO_2-SiO_2) enhance the photocatalytic performance due to improved surface adsorption and increasing surface hydroxyl group in the thin film. Also, at the same time SiO_2 acts as the carrier of TiO_2 and helps to obtain a large surface area as well as a suitable porous structure [10].

It has been proven by the researchers that by adding SiO_2 as a adding material to the TiO_2 can improve the photocatalytic activity and hydrophilicity process [11-13].

Based on previous researchers stated, additions of SiO_2 to the TiO_2 can prevent the formation of rutile phase at high temperature up to 800°C that is the undesired phase for the photocatalytic applications in the solar cell [14].

This research work is focus on synthesized and characterized the structural, morphology and optical properties of TiO_2/SiO_2 thin film with three different annealing temperatures in order to improve the properties the self-cleaning property in solar applications.

2. Methodology

The sol-gel method was used to synthesize TiO_2/SiO_2 thin films. The tetraethyl orthosilicate, TEOS, and tetra-n-butyl orthotitanate, TBOT, were used as the sources of SiO_2 and TiO_2 respectively to prepare the solutions of SiO_2 solutions and TiO_2 solutions.

Firstly, TBOT was mixed together with ethanol and deionized water by using molar ratio of mixture (TBOT: ethanol: deionized water) was 1:0.5:3. Next, hydrochloric acid was added drop by drop into the mixing solutions which give the function as a catalyst. Then, this solution was mixing together for 1 hour by using magnetic stirrer.

After 1 hour, a clear solution was obtained then TEOS was added into the solution until the ration of TBOT: TEOS was 1:1. Then, the solution was continued stirred for 30 minutes at 45°C. The final solutions was undergoes aging process for 24 hours. Figure 1 shows the flowchart of the synthesized TiO_2/SiO_2 thin film.

Thin films of TiO_2/SiO_2 were deposited by the method of spin coating in air at room temperature, on glass substrate with speed at 1000 rpm for 30s. Finally, the films were annealed at 300°C, 400°C and 500°C in air for 3 hours by using muffle furnace.

The crystalline structure was characterized by an X-ray diffractometer (XRD) in 2θ range from 20° to 80°. Scanning electron microscopy (SEM) was used to study the surface morphology and pore distribution of the produced films.

The transmission and absorbance were measure by UV/VIS Perkin-Elmer spectrophotometer.

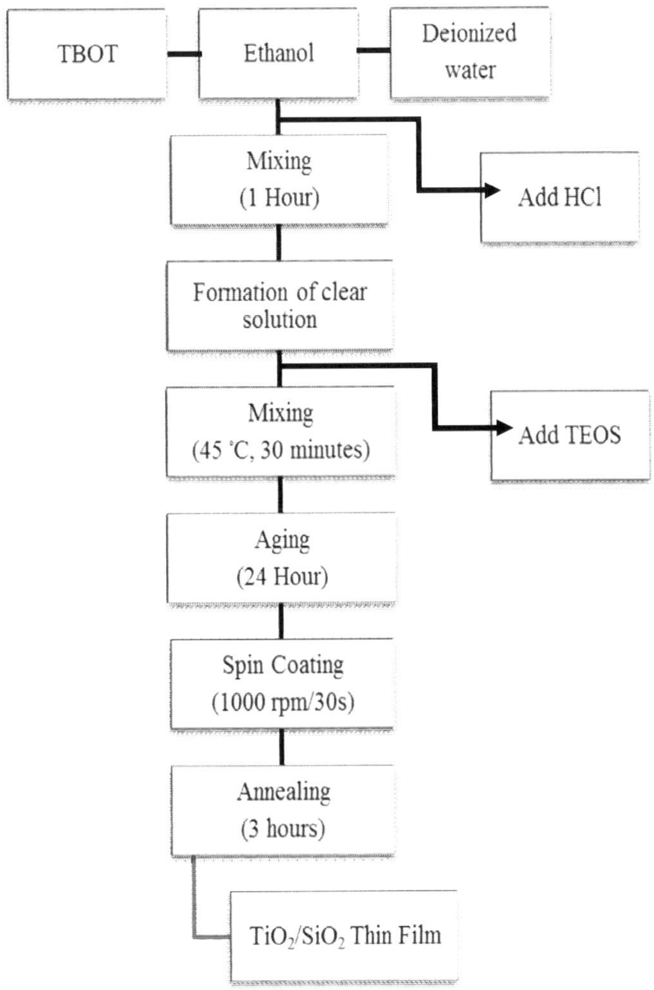

Figure 1. The flowchart of the synthesized TiO_2/SiO_2 composite thin film.

3. Results and discussion

X-ray diffraction pattern has been used to investigate the phase of the prepared TiO_2/SiO_2 films. The X-ray diffraction pattern of TiO_2/SiO_2 thin films annealed at different temperatures is shown in figure 2. The peaks have been indexed and found to be that of brookite TiO_2 and SiO_2. The diffraction peak at 2θ = 31.99° was assigned to the characteristic diffraction peak of brookite TiO_2 phase and corresponds to (1 2 1) direction. For the SiO_2 phase, the diffraction peak at 2θ = 66.54° which is corresponds to (3 0 0) direction. The diffraction pattern shows that an annealing at a temperature equal or lower than 300°C would be largely sufficient to form TiO_2/SiO_2 completely. It is well agreed with Maeda & Yamasaki [15] which found that when the films was annealed at 250°C for 180 min, the phase of titania–silica (TiO_2-SiO_2) mixed films exist. They also reported on addition of SiO_2 into the TiO_2 films can prevents the recrystallization of the TiO_2 composites. When the temperature of annealing was increase to 400°C and 500°C, the only peak reveal was brookite TiO_2 films. Mechiakh et al. [16] was reported, at higher temperatures (400°C and 450°C), in addition to anatase TiO_2 the formation of brookite TiO_2 which crystallizes with the (1 2 1) plane parallel to the surface. No peak corresponding to the SiO_2 phase is seen in the diffraction pattern of films annealed at 400°C and 500°C.

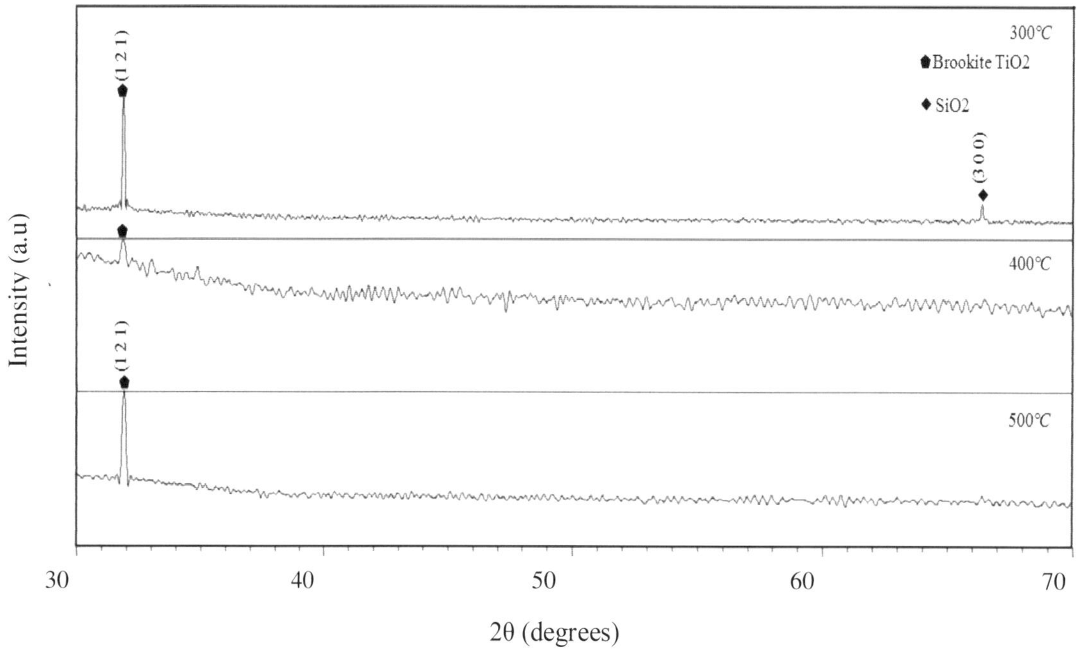

Figure 2. X-ray diffraction pattern of TiO_2/SiO_2 films annealed at different temperatures.

Many efforts have been made to modify the TiO_2 by impurities for improving the photocatalytic activity under visible light [17-19]. Modification of TiO_2 with SiO_2 has been attracted in the past decades and widely used due to photocatalysis and hydrophilicity properties [20]. The characteristic of SiO_2 itself, such as its water-trapping effect, is suggested to play an important role for surface hydrophilicity of the TiO_2/SiO_2 films. In the TiO_2/SiO_2 films, therefore, the synergistic effects of anatase polycrystalline structure of the TiO_2 and water-trapping effect of the SiO_2 are important for their strong photo-induced hydrophilicity [15].

Figure 3 shows the SEM micrographs of TiO_2/SiO_2 annealed at 300°C, 400°C and 500°C. The SEM micrographs revealed that the surface morphology of the TiO_2/SiO_2 films depend strongly upon annealing temperature. It was observed that the average grain size of the films increases significantly when annealing temperature increases. In the presence of SiO_2 for the films annealed at 300°C (figure 3(a)), it can be seen many nucleation centers are present on the substrate and agglomeration of small flaky are produced. Then, the films with small flaky size are not able to grow into bigger ones and prevent crack. Whereas for higher annealing temperatures as seen in figure 3(b) and (c) it revealed large flaky size and large cracks throughout the coatings. Temperature increasing may lead to non-uniform and cracked coating and reduce coating lucidity and transparency [21].

It also clearly seen that the substrate is inhomogeneously covered by large flaky size and it may formed during drying process due to surface tension between the film and the air [22]. In the case of sol-gel coated films, during drying process, the capillary forces might have generated which provides cracks on the surface [23].

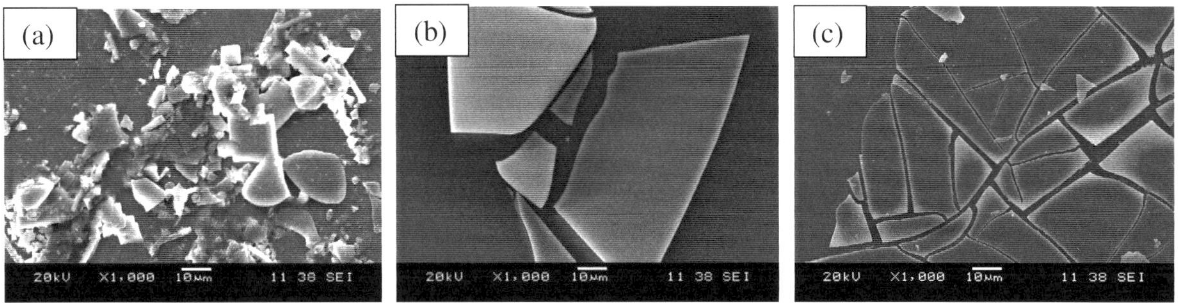

Figure 3. SEM micrographs of TiO_2/SiO_2 annealed at (a) 300°C, (b) 400°C and (c) 500°C.

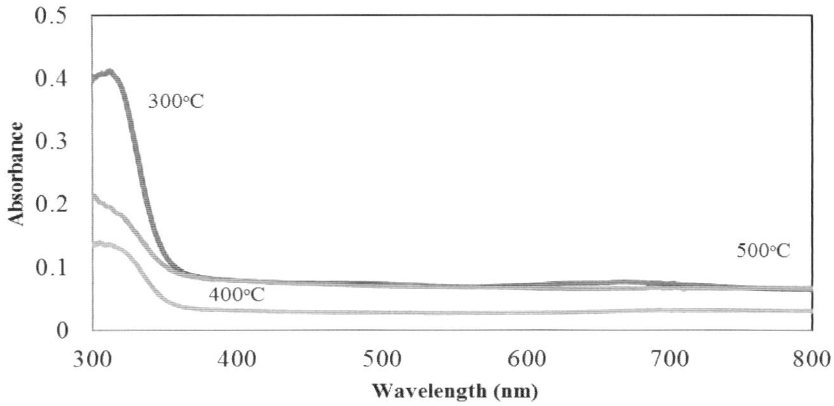

(a)

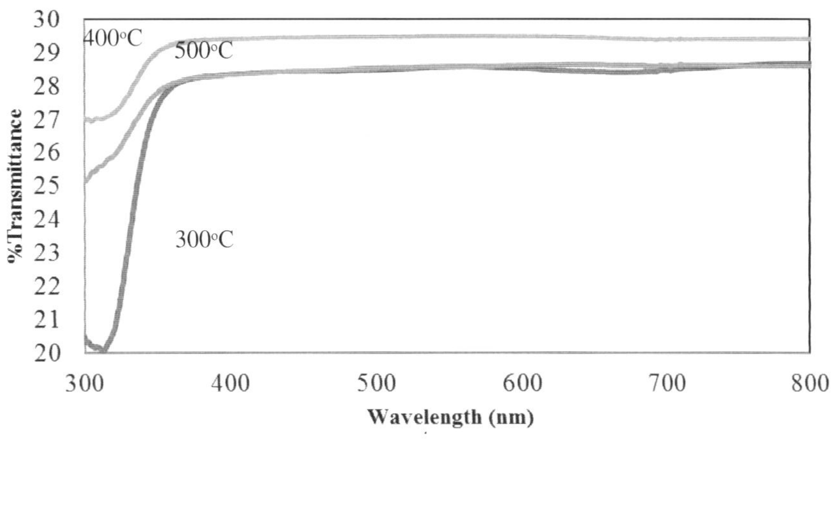

(b)

Figure 4. UV-Vis spectra of absorbance and transmittance for TiO_2/SiO_2 films annealed at different temperatures.

The optical absorbance and transmittance spectra of TiO_2/SiO_2 films deposited on glass substrates with different annealing temperatures were shown in figure 4(a) and (b). The films show sharp absorption edge in ultraviolet region at a wavelength of about 320 nm as seen in figure 4(a). The absorbance is much higher for the films which annealed at 300°C with the formation of TiO_2/SiO_2 and the films are crack free. The transmittance within the visible and near infrared region is lower than 30% including the glass substrates, which reveals the poor optical properties of TiO_2/SiO_2 produced in this study. It is well agreement with the SEM micrograph which the substrate was not fully covered with the films and the films have large cracks. When the films are having large cracks, the UV light will be scattered in all direction.

Optical transparency is prerequisite for the application of self-cleaning coating on transparent glass or plastic surface. The films annealed at 400°C and 500°C shows the same trends of spectra for absorbance and transmittance with the only brookite phase exist. The transmittance of film annealed at 400°C is much higher compared to the film annealed at 500°C is due to pure TiO_2 film at 400°C is having less crack and bigger crystal size. The increased of cracking crystal size is responsible for the slight loss in optical transmission in the visible range. The process of spin coating needs to repeat a few times, in order to ensure the substrate is fully covered by the films and it also can enhance the transmittance and the absorbance of the films [24].

4. Conclusions

This work deals with the preparation and characterization of TiO_2/SiO_2 thin films using a simple and cost effective method: sol-gel spin coating. The deposition was performed on glass substrate at room temperature and annealed at three different annealing temperatures; 300°C, 400°C and 500°C. This work presents the annealing temperature effect on the properties of TiO_2/SiO_2. X-ray diffraction pattern shows that the thin films obtained at annealing temperature 300°C, having a phase of TiO_2/SiO_2. At higher annealing temperature (400-500°C), the formation of brookite phase presence. The SEM micrograph of the film annealed at 300°C show that the film consist of agglomeration of small flaky size without any crack. The analysis of UV-Vis transmission spectra shows that TiO_2/SiO_2 thin films are transparent in the visible range with a poor transmittance due to large crack and non-uniform film. This research proved that annealing temperature is an important parameter which affects the structure; surface morphology, and optical properties of the films.

References

[1] Zhao Z, Sun J, Xing S, Liu D, Zhang G, Bai L and Jiang B 2016 *J. Alloys Compd.* **679** 88
[2] Matsuda A, Higashi Y, Tadanaga K and Tatsumisago M 2006 *J. Mater. Sci.* **41** 8101
[3] Perju M C, Vizureanu P 2014 *REV. CHIM. (Bucharest)* **65**(6) 694-696
[4] Mareci D, Bolat G, Istrate B, Munteanu C, Căilean A 2015 *Materials and Corrosion* **66**(12) 1529–1535
[5] Yu J, Yu J C and Zhao X 2002 *J. Sol-Gel Sci Technol.* **24** 95
[6] Meng X, Qian Z, Wang H, Gao X, Zhang S and Yang M 2008 *J. Sol-Gel Sci Technol.* **46** 195
[7] Machida M, Norimoto K, Watanabe T, Hashimoto K and Fujishima A 1999 *J. Mater. Sci.* **34** 2569
[8] Zhou L, Yan S, Tian B, Zhang J and Anpo M 2006 *Mater. Lett,* **60** 396
[9] Ennaoui A, Sankapal B R, Skryshevsky V and Lux-Steiner M C 2006 *Sol. Energy Mater. Sol. Cells* **90** 1533
[10] Cheng P, Zheng M, Jin Y, Huang Q and Gu M 2003 *Mater. Lett.* **57** 2989
[11] Jesus M, Neto J, Timo G, Paiva P, Dantas M and Ferreira A 2015 *Appl. Adhes. Sci*, **3**1
[12] Jafry H, Liga M, Li Q and Barron A 2011 *New J. Chem.*, **35** 400
[13] Stevens N, Priest C, Sedev R and Ralston J 2003 *Langmuir* **19** 3272

[14] Tang Y, Fu S, Zhao K, Xie G and Teng L 2015 *Ceram. Int.*, **41** 13285
[15] Maeda M and Yamasaki S 2005 *Thin Solid Films* **483** 102
[16] Mechiakh R, Meriche F, Kremer R, Bensaha R, Boudine B and Boudrioua A 2007 *Optical Materials* **30** 645
[17] Vohra M, Kim S and Choi W 2003 *J. Photochem. Photobiol.A Chem.* **160** 55
[18] Hata S, Kai Y, Yamanaka I, Oosaki H, Hirota K and Yamazaki S 2000 *JSAE Rev.* **21** 97
[19] Asahi R, Morikawa T, Ohwahi T, Aoki K and Taga Y 2001 *Science* **293** 269
[20] Guan K 2005 *Surf. Coat. Technol.* **191** 155
[21] Shokuhfar A, Alzamani M, Eghdam E, Karimi M and Mastali S 2012 *Nanoscience and Nanotechnology* **2 (1)** 16
[22] A.Elfanaoui, E. Elhamri, L. Boulkaddat, A. Ihlal, K. Bouabid, L. Laanab, A. Taleb, X. Portier, 2011 International *J. of Hydrogen Energy* **36** 4130
[23] Latthe S S, Liu S, Terashima C, Nakata K and,Fujishima A 2014 *Coatings* **4** 497
[24] Chandraboss V L, Karthikeyan B, Kamalakkannan J, Prabha S and Senthilvelan S *2013 J. of Nanoparticles* **13** 1

Acknowledgements
This work was partially supported by Ministry of Higher Education of Malaysia under Fundamentals of Research Grant Scheme (FRGS), No.: 9003-00478 and School of Materials Engineering, Universiti Malaysia Perlis (UniMAP), Malaysia.

IOP Conf. Series: Materials Science and Engineering **209** (2017) 012003 doi:10.1088/1757-899X/209/1/012003

Preparation and Characterisation of Linear Low-Density Polyethylene / Thermoplastic Starch Blends Filled with Banana Fibre

A W M Kahar[1,2*] and L Ju Ann[1]
[1] School of Materials Engineering, Universiti Malaysia Perlis,02600 Jejawi, Perlis
[2] Centre of Excellence Geopolymer and Green Technology, Universiti Malaysia Perlis

E-mail: kaharwahab@unimap.edu.my

Abstract. In this study, the influence of banana fibre (BF) loading using sodium hydroxide (NaOH) pre-treated and succinic anhydride-treated (SA) BF on the mechanical properties of linear low-density polyethylene (LLDPE)/thermoplastic starch (TPS) matrix is investigated. LLDPE/TPS/BF composites were developed under different BF conditions, with and without chemical modifications with the BF content ranging from 5% to 30% based on the total composite. The tensile strength showed an increase with an increase of fibre content up to 10%, thereby decreasing gradually beyond this level. NaOH pre-treated and SA treated BF added with LLDPE/TPS composite displays a higher tensile strength as compared to untreated BF in LLDPE/TPS composites. Thermal behaviour of the BF incorporated in LLDPE/TPS composite was characterised using differential scanning calorimetry (DSC) and thermal gravimetric analysis (TGA). This showed that SA treated BF exhibits better thermal stability, compared to other composites. This is because of the improvement in interfacial adhesion existing between both the fibre and matrix. In addition, a morphology study confirmed that pre-treated and treated BF had excellent interfacial adhesion with LLDPE/TPS matrix, leading to better mechanical properties of resultant composites.

1. Introduction

Given the current environmental issues resulting from petroleum-based plastics, many research studies have been conducted to investigate how to reduce the amount of plastic waste. Partial replacement of industrial plastics with biodegradable polymer is an effective means to solve the problem of plastic waste. It is usually produced by blending non-biodegradable plastics with biodegradable components as a dispersing filler such as a polysaccharide (cellulose and starch). The polymer combines the high-quality properties of conventional plastics and abundantly available polysaccharides as a biodegradation agent. Incorporating a high-content of a polysaccharide as a filler within the binary and ternary system is limited due to its rigid characteristics and low processing capability. In this study, starch is converted to thermoplastic to optimise the starch content in the resultant composite. The molecular structure of starch granules is destroyed and blended together with targeting polymer. This is undertaken by heating the starch under a condition of significant stress in the presence of a plasticiser, such as glycerol [1]. Although incorporating starch can accelerate the biodegradation of polymer materials, several drawbacks were observed (e.g. one drawback being lower mechanical properties). To enhance the mechanical properties of the resultant blend, BF was used as a reinforcing filler. The combination of using starch in thermoplastic form and cellulosic fibre as a reinforcing filler

IOP Conf. Series: Materials Science and Engineering **209** (2017) 012003 doi:10.1088/1757-899X/209/1/012003

could potentially be used to optimise the biodegradation agent in the synthetic polymer matrix, with acceptance mechanical properties. In addition to the increase in interface adhesion between the fibre and matrix, several surface modifications such as mercerisation, acetylation, etherification and graft copolymerization are required. This could potentially reduce the hydrophilicity of natural fibre and the tendency of the fibre to absorb moisture [2]. In this study, the blends of LLDPE/TPS filled with BF will be performed. Chemical modifications of BF are obtained by treating with sodium hydroxide (NaOH) and followed by succinic anhydride.

2. Materials and methods

LLDPE with a density of 0.92 g/cm^3 was supplied by Lotte Chemical Titan, Malaysia. Cassava starch was used in the preparation of thermoplastic starch (TPS) in the presence of glycerol as a plasticiser. Food grade cassava starch was supplied from Thye Huat Chan Sdn. Bhd, glycerol supplied from TTD Chemical, and the reagent grade succinic anhydride supplied from Sigma Aldrich.

2.1. Surface modifications

BF was dried at room temperature for three days before post-dried in an oven for 24 hours where it was then ground and sieved to 75 µm in size. The next step was to soak the BF in sodium hydroxide (NaOH) for 4 hours at 30°C to remove any lignin and hemicellulose [3]. The banana mercerised fibre was then soaked in 5% SA solution mixed with sulphuric acid (H_2SO_4) acting as a catalyst. The mixture was then heated at 80°C for 1 hour where the treated BF was washed with distilled water and dried in a heated oven at 80°C for 12 hours [4].

2.2. TPS/LLDPE filled with banana fiber preparation

The preparation of TPS was undertaken by using a heated two roll mill in the presence of 35 wt% glycerol, followed by blending with LLDPE at a temperature of 135°C for a further 10 minutes. The final stage of the process then involved the BF being charged and processed until the mixture was homogeneous. Refer table 1, showing the formulation of TPS/LLDPE filled with BF. A series (consisting of three types) was prepared to represent; untreated BF/LLDPE/TPS, pre-treated BF/LLDPE/TPS and SA treated BF/LLDPE/TPS composites with a variation of 5% to 30% BF loading.

Table 1. Formulation of the TPS/LLDPE filled with banana fiber.

Samples	LLDPE/TPS (70:30) (%)	Banana trunk fiber (%)	Pretreatment (NaOH) (%)	Treatment (SA)(%)
Untreated banana trunk fiber in LLDPE/TPS blends	100	0	-	-
	95	5	-	-
	90	10	-	-
	80	20	-	-
	70	30	-	-
NaOH pretreated banana trunk fiber in LLDPE/TPS blends	100	0	-	-
	95	5	*5	-
	90	10	*5	-
	80	20	*5	-
	70	30	*5	-
SA treated banana trunk fiber in LLDPE/TPS blends	100	0	-	-
	95	5	*5	*5
	90	10	*5	*5
	80	20	*5	*5
	70	30	*5	*5

[a]based on the % weight of BF fiber.

2.3. Testing

Tensile testing was conducted using an Instron machine based on ASTM D638. Tensile strength, Young's modulus and elongation at break data were collected from an average of five samples, for each compound. Structural analysis of the composites was determined using a Fourier Transform Spectroscopy (FTIR) with a total of 32 scans performed, at the wavelength ranging from 700 cm^{-1} to 4000 cm^{-1}. The morphology study of TPS/LLDPE filled banana fibre was conducted using a Scanning Electron Microscopic (SEM) at a magnification of 150X and 300X. The specimens were mounted on aluminium stubs and coated with palladium to avoid the possibility of any electrostatic charging occurring during the testing. Thermal analysis was conducted by utilising Differential Scanning Calorimetry (DSC) and Thermal Gravimetric Analysis (TGA). DSC was carried out using the ambient room temperature to 180°C at a scan rate of 10°C/min. The sample was then cooled to ambient room temperature, allowing the formation of crystals to occur. The amount of crystal structure (Xc) is calculated by using the following equation:

$$\frac{\Delta H^0}{\Delta H_f^0} x100\% \quad (1)$$

where: ΔH^0 = Enthalpy of fusion;

ΔH_f^0 = Enthalpy of crystallinity of a sample at 100% crystallinity.

TGA was calculated using ambient room temperature to 600^0C at a scan rate of 10°C/min to assess the thermal stability and thermal degradation of LLDPE/TPS/BF composites.

3. Results and discussions

A series of BF incorporated in LLDPE/TPS blends was characterized using a Fourier Transformation Infrared Spectroscopy (FTIR). The spectra for LLDPE/TPS blends filled with 10% untreated, pre-treated and treated BF are shown in figure 1.

A small peak was observed at 3395 cm^{-1} corresponding to -OH stretching, characterizing the TPS and BF functional group. There were two sharp peaks identified at 2917 cm^{-1} and 2849 cm^{-1} respectively, representing -CH stretching of the alkanes group of LLDPE. Compared to the untreated and pre-treated composites, a new peak was observed at 1743 cm^{-1} for treated LLDPE/TPS/BF composites, caused by -C=O stretching vibration occurring from the ester bond. This could possibly represent the esterification reaction occurring between BF and SA. Hence, the presence of SA in the sample was removed before FTIR testing by using a Soxhlet extraction technique.

Thus, the influence of -C=O vibrations occurring from the SA was eliminated. The possible esterification reaction between BF and SA is shown in figure 2. SA can act as a bridge between hydrophobic LDPE matrix and hydrophilic BF and the firm interaction between BF and SA leading to a much higher performance in the mechanical properties of the resultant composites.

The influence of BF reinforced LLDPE/TPS matrix on tensile strength is shown in figure 3. At initial BF loading, the tensile strength value for untreated composite is 4.4 MPa, being lower than in LLDPE/TPS matrix. This indicated that the incorporation of BF interfered with the alignment of LLDPE and TPS chains, therefore influencing the final properties of the composite.

The tensile strength was increased to 6.3 MPa at 10% BF loading, showing the optimum level of untreated BF loading. As the BF content increases above 10%, the tensile strength starts to decrease, attributed to the agglomeration of FB due to strong hydrogen bonding holding the fibres together [5]. The agglomeration of BF within LLDPE/TPS matrix promoted the micro-crack formation appearing when an external force was applied to the samples.

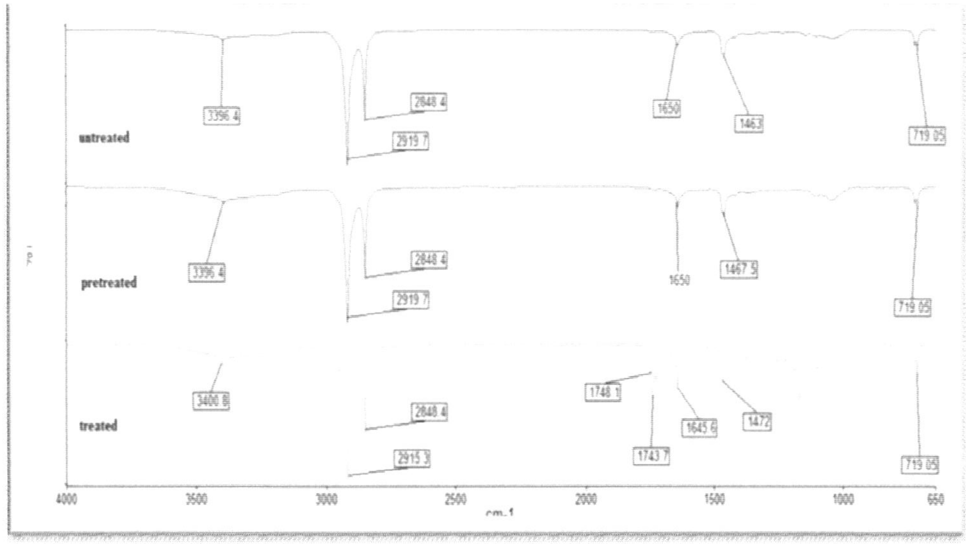

Figure 1. FTIR spectra of untreated, pretreated and treated 10% BF loading in LLDPE/TPS matrix.

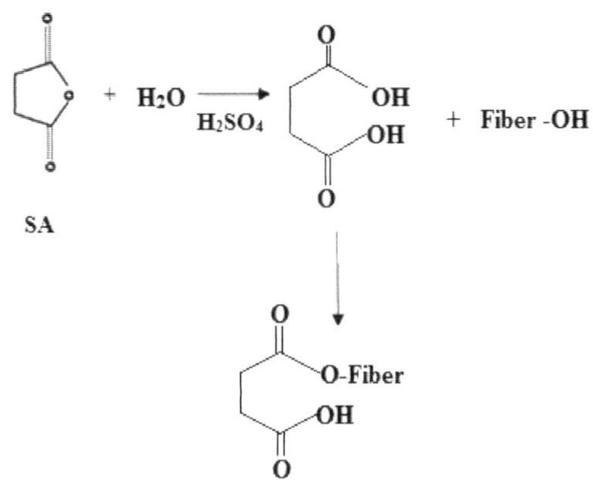

Figure 2. Mechanism of SA treated fiber.

Likewise, when comparing to other composite series, untreated BF shows an inferior value for all samples tested, whereas, the NaOH treated BF/LLDPE/TPS blends show a much higher tensile strength than untreated BF/LLDPE/TPS composite counterparts. The treatment with NaOH physically removes the non-cellulose material and lignin, leaving gaps on the surface of the BF for interacting physically with LLDPE/TPS matrix. Furthermore, NaOH treatment can also cause fibrillation due to the breaking down of the fibre bundle into smaller fibre giving a higher contact area for improved fibre-matrix interaction. In the case of SA-treated BF based composite, tensile strength is marginally lower at initial BF loading, as compared to LLDPE/TPS matrix. This is because the amount may not be adequate for proper load distributions applied to the matrix [6]. However, unlike the untreated NaOH and treated BF, SA treated BF based composites at 10% BF loading and greater; found that the tensile strength was always more than LLDPE/TPS matrix. SA treatment was found to improve the adhesion and compatibility between BF and LLDPE/TPS matrix, leading towards a significant increase in the tensile strength value. Improved interaction at the interphase level implies that the external force applied to LLDPE/TPS matrix was efficiently transferred throughout the composite system.

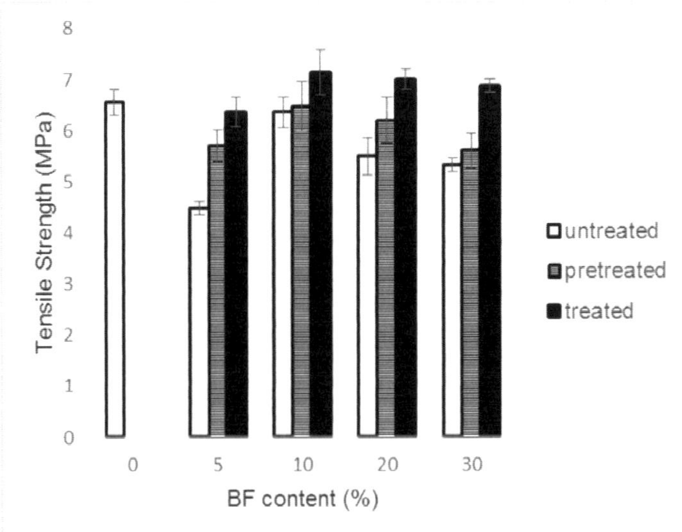

Figure 3. Tensile strength of untreated, pretreated and treated banana fiber reinforced LLDPE/TPS blends at different fiber content.

SEM micrographs of untreated and NaOH treated BF are shown in figures 4(a) and (b). Untreated BF displays a smoother surface with several impurities attached to the fibre. Upon NaOH treatment, the BF exhibited a much rougher surface due to the partial dissolution of lignin and hemicellulose and the removal of impurities on the fibre's surface. The cleaner and more irregular surface could, therefore, provide a higher contact surface area for any mechanical interaction between the BF and the polymer matrix.

The morphology of the tensile fracture surface of untreated BF, NaOH-treated BF, and SA treated BF reinforces LLDPE/TPS matrix, are illustrated in figures 5(a), (b) and (c). Yet, it is commonly known and acknowledged that synthetic polymer, LLDPE is incompatible with cellulosic fibre due to the hygroscopic nature of BF. As illustrated in figure 5(a), significant gaps between fibre and LLDPE matrix are evident, representing the higher level of interfacial tension existing between hydrophobic LLDPE matrix and hydrophilic BF. The fibre tends to agglomerate rather than disperse in LLDPE matrix.

Furthermore, various fibre pullout and voids appear in the micrograph possibly representing a lack of interfacial adhesion occurring between the fibre and LLDPE matrix. Figure 5(b) shows the SEM micrograph of the tensile fracture surface of the NaOH treated BF reinforced LLDPE/TPS blends. When compared to the untreated BF system, NaOH-treated BF showed a better level of dispersion in LLDPE/TPS matrix, leading to enhanced tensile strength of the composite. After BF undergoes NaOH treatment, fibrillation occurred by breaking down the bundle of fibre into smaller fibrils. Smaller sized BF provides a much greater contact area for mechanical interaction between fibre and LLDPE/TPS matrix. SEM micrograph of the tensile fracture surface of SA-treated BF reinforced LLDPE/TPS matrix is shown in figure 5(c).

Contrary to an untreated BF based system, the fibre was well embedded into LLDPE/TPS matrix due to the improved interfacial interaction at the interphase. The formation of ester linkages, following the surface modification, was lowering the hydrophilicity of BF and therefore, increasing its wettability and dispersion within LLDPE/TPS matrix. The matrix polymer uniformly coated the BF

IOP Conf. Series: Materials Science and Engineering **209** (2017) 012003 doi:10.1088/1757-899X/209/1/012003

resulting in gaps between the fibre-matrix being significantly reduced. The crack fracture highlights that BF was pull out together with the matrix polymer during tensile testing which implies the cohesive bond existing between SA treated BF and LLDPE matrix.

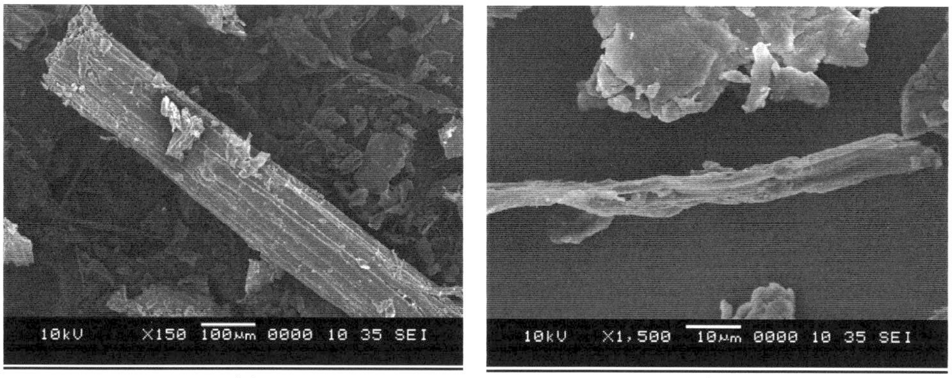

(a) (b)

Figure 4. SEM micrographs of (a) an untreated BF and (b) an alkali treated BF.

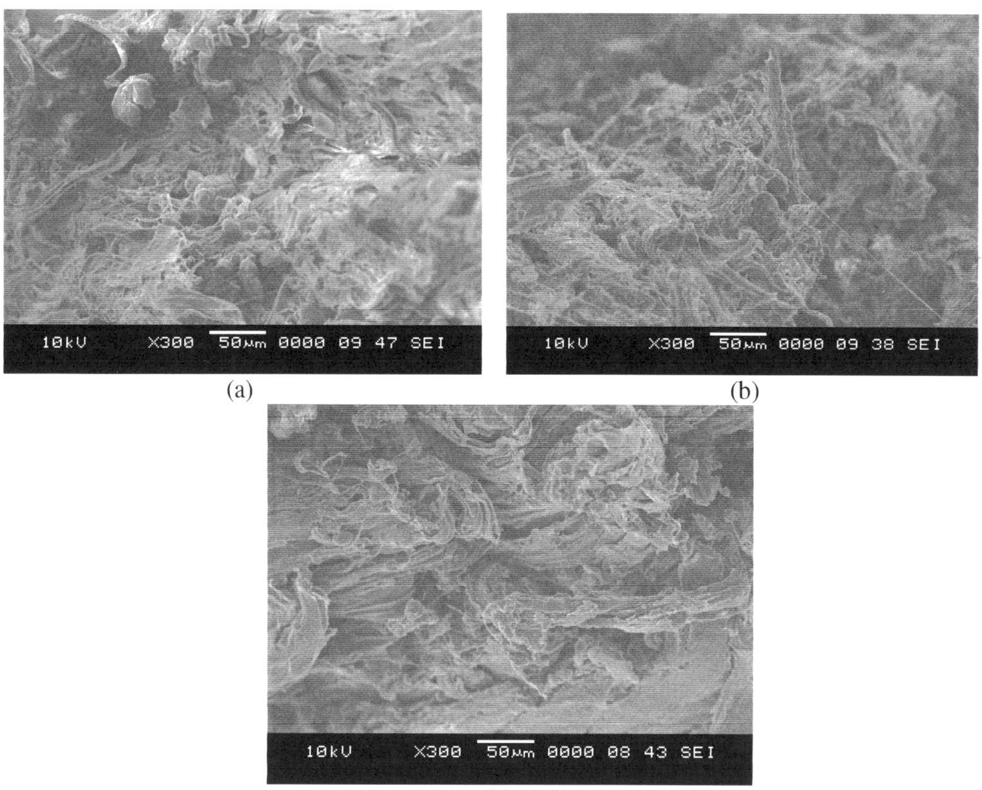

(a) (b)

(c)

Figure 5. SEM micrograph of tensile fracture surface of (a) the untreated 10% BF, (b) pretreated BF, and (c) SA treated BF reinforced LLDPE/TPS matrix.

TGA curves of LLDPE/TPS matrix filled with 10% and 30% of BF are illustrated in figures 6(a) and (b). The TGA curves highlight three significant slopes for all composite samples. The initial degradation was observed between 120°C to 180°C due to the volatilisation of glycerol used in the preparation of TPS. The second slope starts at 260°C and is correlated to the degradation of polysaccharide, (i.e. starch and cellulose) with the highest degradation recorded at a temperature of 380°C. The degradation of LLDPE occurred at 420°C, with near complete degradation observed at 500°C. The data collected from the first to the third significant degradation was then tabulated in table 2, for further analysis. The value for Td1, Td2 and Td3 was determined from the peak differential thermogravimetric (DTG). Referring to table 2, it shows that the untreated, NaOH-treated and SA treated 10% BF has a higher degradation temperature than the degradation temperature at 30% BF of composite counterparts. In comparing the untreated BF reinforced LLDPE/TPS composite, the Td1 was 123.9°C and 120.9°C for 10% and 30% BF, respectively. Similarly, all 10% BF based composites were recorded as having better thermal stability than the 30% BF based composites. On the other hand, the curves for pre-treated BF reinforced LLDPE/TPS matrix moved to a much greater temperature when compared to the untreated BF composite, indicating higher thermal stability. The weight loss at 200°C, 300°C and 400°C of the pre-treated BF reinforced LLDPE/TPS matrix was lower than the untreated BF reinforced LLDPE/TPS matrix. This indicated that the pre-treated BF reinforced LLDPE/TPS matrix had better thermal stability as compared to the untreated composite.

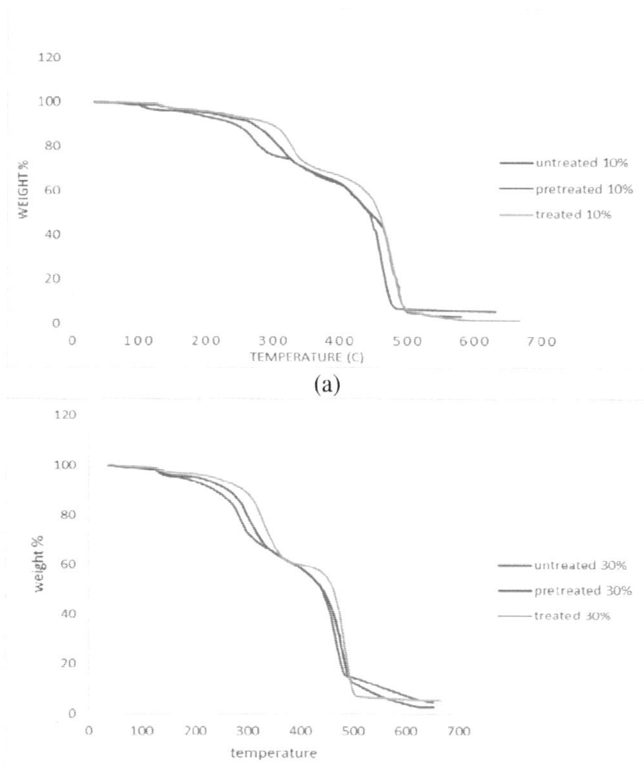

(b)

Figure 6. (a) Thermal Gravimetric Analysis (TGA) curve of 10% untreated, pretreated and treated BF in LLDPE/TPS blends; (b) Thermal Gravimetric Analysis (TGA) curve of 30% untreated, pretreated and treated BF in LLDPE/TPS blends.

Alternatively, as shown in table 2, the treated BF reinforced LLDPE/TPS matrix had the highest temperature and maximum degradation when compared to untreated and pre-treated BF counterparts, reflecting better thermal stability.

Table 2. TGA data of LLDPE/TPS matrix filled with 10% and 30% of untreated, pretreated and treated BF.

LLDPE/TPS filled with banana fiber composite		T_{dmax1} (°C)	T_{dmax2} (°C)	T_{dmax3} (°C)	Weight loss (%) 200°C	300°C	400°C
10% of banana fiber content	untreated	123.9	271.9	452.9	6.68	24.16	37.34
	NaOH treated	139.9	303.9	462.8	4.80	18.23	37.01
	SA treated	141.8	323.0	465.9	3.99	10.91	33.34
30% of banana fiber content	untreated	120.9	269.0	451.9	6.76	27.55	41.44
	NaOH treated	130.9	293.0	458.9	4.95	20.86	41.53
	SA treated	140.0	321.0	471.0	4.05	11.46	39.95

DSC examined the thermal characteristics that influenced the physical properties of LLDPE/TPS/BF composites. The melting temperature of LLDPE crystal fraction, the heat of fusion and the percentage of crystallinity (Xc%), including data for 10% and 30% BF reinforced LLDPE/TPS matrix, are shown in table 3. In table 3, the melting temperature for both 10 % and 30 % BF loading, increased after the fibre was treated with NaOH and SA. The temperature position for melting peak shifted from 115.6°C to 122.8°C, and 123.0°C for untreated, NaOH-treated and SA treated LLDPE/TPS/BF composites, respectively. A similar trend was observed for the 30% BF filled LLDPE/TPS system. The increase in the melting peak position indicates that the SA-treated BF, favoured the formation of larger crystal lamella, thereby restricting the mobility of LLDPE chains. Furthermore, as shown in table 3, the increase of filler loading also decreased the Xc of LLDPE/TPS/BF composite. LLDPE/TPS filled with 10% untreated BF had a heat of fusion value of 72.89 J/g, which corresponds to 24.9% of Xc based on the theoretical value of 293 J/g for 100% crystalline linear polyethylene [7]. This illustrates a much higher value than LLDPE/TPS filled with 30% untreated BF, which is 18.1%. The presence of excessive fibre between LLDPE chains could prevent them from being aligned closely and interfere with the crystal formation of LLDPE phase. Whereas, the pre-treated BF reinforced LLDPE/TPS matrix exhibited higher Xc than the untreated BF reinforced LLPDE/TPS blended composite. Furthermore, the 10% treated BF in LLDPE/TPS blends provided the highest Xc of the entire series. The smaller size of BF after surface modifications can act as nucleation seeds for crystal formation. It was concordance to the higher melting temperature which reflected, a greater fusion of energy required to melt the thicker crystal lamellae.

Table 3. Thermal properties of LLDPE/TPS/BF composites at different BF loadings.

Composition		Melting temperature, T_m (°C)	Heat of fusion, ΔHf (J/g)	Percentage of crystallinity, X_c (%)
10% BF reinforced LLDPE/TPS matrix	untreated	115.59	72.89	24.88
	pretreated	122.78	84.19	28.73
	treated	123.03	92.69	31.63
30% BF reinforced LLDPE/TPS matrix	untreated	123.78	52.93	18.06
	pretreated	124.51	56.81	19.39
	treated	124.85	73.33	25.03

4. Conclusions

In this study, different modified BF was used as a reinforcing agent in LLDPE/TPS matrix. Based on the tensile strength results, LLDPE/TPS with pre-treated BF showed a better tensile value as compared to untreated BF composites. This is due to the improvement in interfacial adhesion between BF and LLDPE/TPS matrix. SEM morphological confirmed that the surface roughness of BF after NaOH treatment enhanced the physical adhesion at the interphase of BF and LLDPE/TPS matrix. Further modification of BF with SA exhibited the highest tensile strength of the series studied. The chemical reaction between BF and SA has led to better interphase interactions occurring, and subsequently, increased tensile properties. SEM morphology of tensile fracture surface also showed that the proper mixing of LLDPE/TPS with treated BF exhibited less fibre pull-out. Better thermal stability and a higher degree of crystallinity also resulted from the improved interaction between SA treated BF composites as compared to other series.

References

[1] Zullo R and Iannace S 2009 *Carbohydrate Polymers* **77**(2) 376-383
[2] Li X, Tabil G and Panigrahi S 2007 *Journal of Polymers and the Environment* **15**(1) 25-33
[3] Ray D, Sarkar B K, Rana A K and Bose N R 2001 *Bulletin of materials science* **24**(2) 129-135
[4] Haque M M U, Zaman M A A, Rahaman M H and Hossain M Z 2014 *International Journals of Materials Science and Applications* **3**(3) 106-110
[5] Beckermann G W, Pickering K L, and Foreman N J 2004 *Proceedings of SPPM*, 25-27 pp. 257-265
[6] Rashed H M M A, Islam M A, and Rizvi F B 2006 *Journal of Naval Architecture and Marine Engineering* **3**(1) 1-6
[7] Thomas L C 2007 *Characterization of melting phenomena in linear low density polyethylene by modulated DSC TM. TA Instruments, TA-227*

Acknowledgments

The authors would like to thank Universiti Malaysia Perlis for the supply of raw materials and the equipment provided. The authors also gratefully acknowledge the financial support from the FRGS grant 9003-00601.

IOP Conf. Series: Materials Science and Engineering **209** (2017) 012004 doi:10.1088/1757-899X/209/1/012004

Fabrication of Porous Ceramic-Geopolymer Based Material to Improve Water Absorption and Retention in Construction Materials: A Review

N H Jamil[1,2*], W M A W Ibrahim[2], M M A B Abdullah[1,2], A V Sandu[3,2] and M F M Tahir[2]
[1]Faculty of Engineering Technology, Universiti Malaysia Perlis (UniMAP), Perlis, Malaysia
[2]Center of Excellence Geopolymer and Green Technology (CEGeoGTech), School of Materials Engineering, Universiti Malaysia Perlis (UniMAP), Perlis, Malaysia
[3]Faculty of Materials Science and Engineering, Gheorghe Asachi Technical University of Iasi, Romania

Email: noorinahidayu@unimap.edu.my

Abstract. Porous ceramic nowadays has been investigated for a variety of its application such as filters, lightweight structural component and others due to their specific properties such as high surface area, stability and permeability. Besides, it has the properties of low thermal conductivity. Various formation techniques making these porous ceramic properties can be tailored or further fine-tuned to obtain the optimum characteristic. Porous materials also one of the good candidate for absorption properties. Conventional construction materials are not design to have good water absorption and retention that lead to the poor performance on these criteria. Temperature is a major driving force for moisture movement and influences sorption characteristics of many constructions materials. The effect of elevated temperatures on the water absorption coefficient and retention remain as critical issue that need to be investigated. Therefore, this paper will review the process parameters in fabricating porous ceramic for absorption properties.

1. Introduction

Rising temperatures in large cities nowdays poses an increasing environmental threat. This phenomenan arises from the increasing amount of heat generated by human activity (e.g., vehicles and air conditioners). The increasing amount of surface areas covered by artificial materials with a high solar absorption capacity also affected this global warming issues. Some counter measures have been carried out. For example, planting on the building roof to increase the green area, and using water retentive material for construction material to cool surroundings according to the evaporation heat. These materials need- ed permanence, weathering resistance and low density. Because, in many cases, it is difficult to change constructed materials [1]. Ceramic tiles as building materials have the advantages of permanence, weather resistance and decoration. However, dense ceramics such as conventional tiles are heavy and low machinability. Porous ceramics have a lot of advantages as construction materials compare with dense ceramics. It is because of not only lightweight and high machinability but also heat insulation, sound absorption, and others. In addition, water is absorbed and retained into the pore [1].

Porous ceramic have attracted many researcher to tailor the propeties for construction industries. Porous ceramics finds it a lot of new applications in industrial areas, however, because the porosity of ceramic material can be arranged in a well-defined and homogeneous manner or heterogeneously on the other hand and offers the ceramic material many special characteristics ranging from an increased surface area, to permeability, to the control of heat transport within the structure, to the maximization of the strength and density ratio. Materials containing tailored porosity exhibit special properties and features that usually cannot be achieved by their conventional dense counterparts [2]. Water absorption and retention of porous ceramics as building materials are characterized by their permanence, heat insulation and water retention. Whilst the water retention properties of the samples depend on their pore size, larger sizes giving both higher water absorption and water release rates [4,5,6].

Recently, many researchers have reported the technique of fabricating porous ceramics made from industrial wastes such as low grade silica, glass, and waste alumina. In this technique, molding does not depend on the plasticity of the raw material. Therefore, the flow process and the solidification process can be separated and the control of the porosity is easy. Furthermore various characteristics based on the pores can be changed [1]. The development of new technologies to recycle and convert waste materials into reusable materials is important for environmental protection and sustainable improvement of our society [5].

2. Porous geopolymer

Geopolymer is an inorganic polymer composite which has a potential sustainable construction material because of its lower energy and carbon footprint as compared to Portland cement-based materials [7]. The microstructure and mechanical properties are known to depend strongly on the chemical compositions of the starting materials [2,5,6]. Geopolymers exhibit a wide variety of properties and characteristics, including high compressive strength, low shrinkage, high temperature resistance [8] and acid and fire resistance [9, 10], and seem to be a desirable alternative to ordinary Portland cement and environmentally sustainable characteristics [10,11]. Besides, the CO2 emission due to production of fly ash-based geopolymer is at least 80% less compared to manufacture of ordinary Portland cement [11,12].

Previously, many works [13-17] have been conducted on fly ash based geopolymers and have gained some achievements. However, almost all of these literatures focused on the effects of some parameters on the compressive strength of geopolymers and few researches have emphasized on adding foaming agent into the polymer pastes to produce porous geopolymer materials and taken the thermal conductivity of them into consideration. In this study, H_2O_2 was added to the polymer pastes to make porous geopolymer materials through the decomposition of H_2O_2 at alkaline environment before the pastes were concreted. Considering the fact that the organic thermal insulation materials widely used today are flammable and the inorganic thermal insulation materials need complex processing conditions and high sintering temperature, which is a big component in manufacturing cost, the porous fly ash-based geopolymer material synthesized in this study has good application potential as thermal insulation material in some situations.

2.1. Pore morphology of porous geopolymer

Based on research done by Junjie Feng, figure 1 represents the photographs of fracture surfaces of samples cured at various temperatures with different amounts of sodium water glass and H_2O_2 added [18].

It can be figured out clearly that the increase of sodium water glass amount enlarges the pore size of the samples. In addition, when adding superfluous water glass, the 5 decomposition of H_2O_2 is violent and some macropores are left inside the sample, which is unfavorable.

Comparing photographs of (d), (e) and (f), it is easy to reach a decision that increasing the dosage of H_2O_2 can also magnify the pore size because of the continuous decomposition of H_2O_2.

IOP Conf. Series: Materials Science and Engineering **209** (2017) 012004 doi:10.1088/1757-899X/209/1/012004

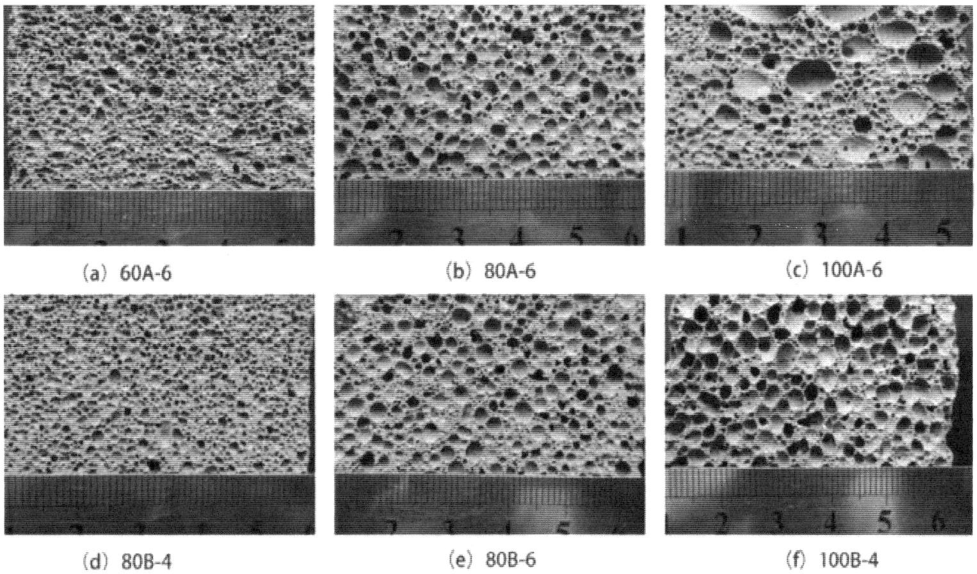

Figure 1. Photographs of fracture surfaces of samples (a) 60A-6 (b) 80A-6 (c) 100A-6 (d) 80B-4 (e) 80B-6 (f) 80B-8 [18].

Moreover, by the contrast of images (b) and (e), conclusion can be made that curing temperature has little influence on the pore morphology.

2.2. Water absorption properties porous geopolymer

The features and properties of the porous geopolymer, for example, porosity, pore size distribution, pore morphology, and pore connectivity (commonly identified as the relationship between open and closed porosity), depend strongly on the composition and processing method. Research study shows that the chemical compositions of the geopolymers (in terms of their H_2O/Al_2O_3 and Na_2O/Al_2O_3 ratios) were systematically varied, allowing the porosity and water retention properties to be adjusted [6]. Figure 2 shows that geopolymers synthesized with higher H_2O/Al_2O_3 ratios are more porous, their larger pore size and higher pore volume giving them good water absorption and water retention properties but decreasing their mechanical strength.

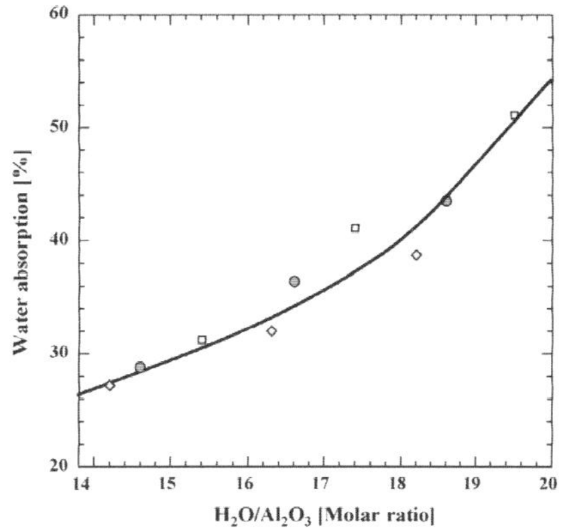

Figure 2. Water absorption of the samples as a function of their H O/Al O ratio [6].

The limited mechanical strength of this type of geopolymers may have restriction on some applications. By contrast, the geopolymers synthesized with lower H_2O/Al_2O_3 ratios are denser, with smaller pore sizes and lower pore volumes, resulting in better water retention and mechanical properties than materials with higher H_2O/Al_2O_3 ratios. Thus, these two types of geopolymers are suitable for different water retention applications [6].

3. Processing of porous waste ceramic

This study examines potential waste diatomite and coal fly ash reuse to prepare water absorption and retain porous ceramics. The coal fly ash can be converted into porous ceramics by using this method. The ability to use diatomite and coal fly ash produced as water absorption and retention of porous ceramic is also examined by studying their water retention and thermal conductivity properties. The compressive strength is the crucial index of the engineering quality of porous ceramic material. The sintering temperatures were varied to obtain the optimum temperature for fabricating the porous ceramic [2].

Table I shows the compressive strength test results for porous ceramics made from diatomite and coal fly ash mixtures. The compressive strength of the porous ceramics increased when the sintering temperature increased from 1000°C to 1270°C. The results showed that the optimal sintering temperature that maximized the compressive strength was 1270°C. The compressive strength of the mixed porous ceramic samples of diatomite that contained the coal fly ash decreased slightly when the sintering temperature increased above 1200°C. The porous diatomite ceramic strength decreased when up to 20% of the coal fly ash was added to the porous ceramics that were heated to 1200°C [2].

Table 1. Mechanical characteristic of porous ceramics [2].

TABLE II: MECHANICAL CHARACTERISTICS OF POROUS CERAMICS

Mechanic Characteristic	Heating Temperature (℃)	Coal Fly Ash Replacement Level (%)				
		0	5	10	15	20
Porosity (%)	1000	64.02	62.14	61.86	61.63	60.23
	1100	63.41	61.29	60.49	59.69	58.95
	1200	62.59	59.97	57.51	56.01	54.61
	1270	61.27	57.23	55.07	53.83	51.71
Water absorption (%)	1000	93.39	88.08	84.72	81.65	75.91
	1100	91.13	83.83	80.59	77.79	74.12
	1200	87.85	80.60	75.04	71.36	67.34
	1270	85.23	75.83	70.92	67.43	63.44
Compressive strength (MPa)	1000	2.48	3.29	3.89	4.39	4.89
	1100	4.36	5.55	6.48	7.20	8.17
	1200	5.92	8.10	11.25	13.65	16.31
	1270	6.10	9.67	12.77	15.38	17.04

Consequently, the coal fly ash can be blended with diatomite to produce porous ceramics. The porous ceramic samples containing the coal fly ash exhibited excellent slow water-releasing properties, which may be attributed to the smaller pores, compared to those in the foamed glass. The large amounts of coal fly ash in the porous ceramic samples also facilitated a slow water release, which yielded acceptable water-retention properties. These properties make the porous ceramic samples containing coal fly ash samples promising for use as water-retaining materials to combat this rising temperatures effects [2]. Li L et al also reported about the use of vermiculite as a cementitious material in agricultural field which also known as 'planting concrete'. It was found that vermiculite has a high water absorption and water retention capacity.

The water absorption increases according to open porosity. The water absorption is calculated from sample weight. Because the density of the high porosity sample is low, the water absorption is estimated high. In fact, it turned out that the amount of the water contained in the sample decreased. Porous ceramics fabricated from waste material shows better water absorption and retention. Porous ceramics can also being fabricated from waste glass and fly ash using the milling method and annealing procedures. This technique is effective in changing waste materials into valuable porous ceramics by mechanically applying a milling method and low sintering process [5].

The mechanical properties such as hardness or bending strength for porous ceramics are related to the pore size or generation of pores of porous ceramics. In other words, when a number of variables such as sintering temperature or soaking time, it will caused the pore size to grow bigger or the pores to grow in number. Hence the mechanical properties of porous ceramics become such that ceramics having a small pore sizes or small pore volumes yield a better result than ceramics having large pore sizes or greater pore volumes [5]. The bending strength of the porous ceramics prepared indicated that with an increase in the heat treatment temperature from 600°C to 700°C, the bending strength improved from 9.25 to 26.81 MPa. The porous ceramic specimens exhibited a decreasing tendency of strength with increasing pore size and pore volume, which may be attributed to defect formation [5].

The porous ceramic specimens at all sintering temperature range used are strong enough for practical usage. The water absorption and retention were depending on the pore structures. These properties were deteriorated with the large pore size, or the too high porosity. As a result, in the design of water retentive build- ing materials, the control of pore structure is important to control of water absorption and retention. Ideally, the small contact area of each pore is effective to improve in a water retention property at high porosity [6].

4. Conclusion

The water absorption and capillary lift properties of geopolymers synthesized with higher H_2O/Al_2O_3 ratios could be enhanced by the introduction of pores generated by thermal and/or chemical treatment of organic pore-forming fibers. The porous ceramics fabricated from waste material exhibited enhanced water absorption and retention capabilities. Moreover, the results of this study also indicated that water absorption and retention are dependent on the pore structures. In summary, porous ceramic samples containing the coal fly ash have excellent mechanical properties, making them feasible for use in water absorption and retention of porous ceramic applications. The formation of porous ceramic can be tailored from the raw material's properties and processing method for it's absorption properties but the mechanical strength must be major consideration as to make it acceptable in construction industry.

References

[1] Kato T, Ohashi K, Fuji M and Takahashi M 2008 *Journal of the Ceramic Society of Japan* **212-215**
[2] Lin K L and Lan J Y 2013 *Journal of Clean Energy Technologies* **1**
[3] Lin K L, Lee T C, Chang J C and Lana J Y *Water Absorption and Retention of Porous Ceramics Cosintered from Waste Diatomite and Catalyst*
[4] Thokchom S, Ghosh P and Ghosh S 2009 Effect of water absorption, porosity and sorptivity on durability of geopolymer mortars, *Journal of Engineering and Applied Sciences*
[5] Yoona S D, Leeb J U, Yunb Y H and Kimb M J 2013 *Journal of Ceramic Processing Research* **14**
[6] Okada K, Ooyama A, Isobe T, Kameshima Y, Nakajima A and MacKenzie K J D 2009 *Journal of the European Ceramic Society* **29** 1917–1923
[7] Li Z, Ding Z, Zhang Y 2004 *Proceedings of the International Workshop on Sustainable Development and Concrete Technology*

[8] Barbosa V F F, Mackenzie K J D 2003 *Mater Res Bull* **38**(2) 319-331
[9] Wu H C, Sun P 2007 *Constr Build Mater* **21**(1) 211-217
[10] Duxson P F J A, Provis J, Lukey G, Palomo A, Van Deventer J 2007 *J Mater Sci* **42**(9) 2917-2933
[11] Nematollahi B, Sanjayan J 2014 *Mater Des* **57** 667-672
[12] Duxson P, Provis J L, Lukey G C, Van Deventer J S 2007 *Cem Concr Res* **37**(12) 1590-1597
[13] Swanepoel J C, Strydom C A 2002 *Appl Geochem* **17**(8) 1143-1148
[14] Görhan G, Kürklü G 2014 *Composites Part B: Engineering* **58** 371-377
[15] Somna K, Jaturapitakkul C, Kajitvichyanukul P, Chindaprasirt P 2011 *Fuel* **90**(6) 2118-2124
[16] Chindaprasirt P, Jaturapitakkul C, Chalee W, Rattanasak U 2009 *Waste Manag* **29**(2) 539-43
[17] Bakharev T 2005 *Cem Concr Res* **35**(6) 1224-1232
[18] Feng J, Zhang R, Gong L, Li Y, Cao W, Cheng X 2014 Development of porous fly ashbased geopolymer with low thermal conductivity, *Materials and Design*

Acknowledgements
This work was supported by the Centre of Excellence Geopolymer and Green Technology (CeGEOGTech), Universiti Malaysia Perlis.

IOP Conf. Series: Materials Science and Engineering **209** (2017) 012005 doi:10.1088/1757-899X/209/1/012005

Electromagnetic Nondestructive Evaluation of Tubes using Data Mining Procedure

A Savin[1*], N Iftimie[1], P Vizureanu[2], R Steigmann[1] and G S Dobrescu[1]
[1] National Institute of Research and Development for Technical Physics, Nondestructive Testing Department, Iasi, Romania
[2] Gheorghe Asachi Technical University, Faculty of Materials Science and Engineering, Iasi, Romania

E-mail: asavin@phys-iasi.ro

Abstract. The fundamental issues in nondestructive evaluation consists in the identification of events corresponding to the flaws which can appear in the examined object and their extraction from noises. This is usually done by comparison with pre-established thresholds, experimentally determined by using standard samples or in the basis of the solution of the forward problem and simulations. This paper presents the features extraction using data mining procedure in the case of tubes from steam generators having different flaws. The data mining is carried on using simulated models in CIVA 9 and experimental data gathered using an inner differential sensor developed in this purpose.

1. Introduction

Electromagnetic (EM) nondestructive evaluation (NDE) is an interdisciplinary domain, which plays an essential role in the assuring components and systems reliability. NDE tests are periodically effectuated, without affecting the further use of object or material, in order to characterize the materials and locate the flaws, preventing the apparition of events that can alter the functioning of the product. NDE offers a balance between quality/price and efficiency and can be applied especially to industrial inspections. Many papers have combined ultrasonic tests with electromagnetic tests to increase the reliability and researchers continue to obtain new ways of applying physics and other disciplines to develop better NDE methods.

In NDE domain, impressive amount of data are collected and stored, with specific details and relevance for each applied inspection method and also the nature of inspected product. The amount of data and their complexity exceed the ability of human selection and analysis of data, automatic techniques being required to adapt the corresponding work sequence. For NDE methods, the data extraction from data basis implies the processing of data and identification of patterns and tendencies from these information, in order to adopt a decision for selection.

The data mining principles have been known for a long time, but with the emergence of large data bases they have gained a broad spreading.

In order to streamline NDE operations (costs/time), nowadays, the NDE procedures store large amount of data at relatively low costs (data which often do not deliver relevant information) and then are post processed by special departments. The system "Online analytical processing" (OLAP) [1] is part of the broader category of business intelligence and comprises also reports writing and data mining. Adopting new technologies to obtain data basis, which integrates data from multiple sources into a given location, allows for quick answers. In NDE, detailed data are obtained, allowing the establishing of a general trend of the evolution of the inspected product. Their visualization is difficult, and therefore the data storage concept is split into components for easier understanding and use.

The paper is based on the use of database data processing methodology and some possibilities of applying this procedure in NDE of steam generators using differential inner sensors.

2. Principles of temporal data mining

Data mining, known as knowledge discovery in databases [2,3], represents a domain which grown rapidly, having multiple applications due the necessity of computer processing of information. According to [4] data mining is "the process of discovering interesting knowledge, such as patterns, associations, changes, anomalies and significant structures from large amounts of data stored in databases, data warehouses, or other information repositories". The simplified steps taken in process are: collect data; filtering by eliminating errors; selecting properties that interest the analysis; application of filter and detection/analysis of new knowledge; validation/visualization and evaluation of results.

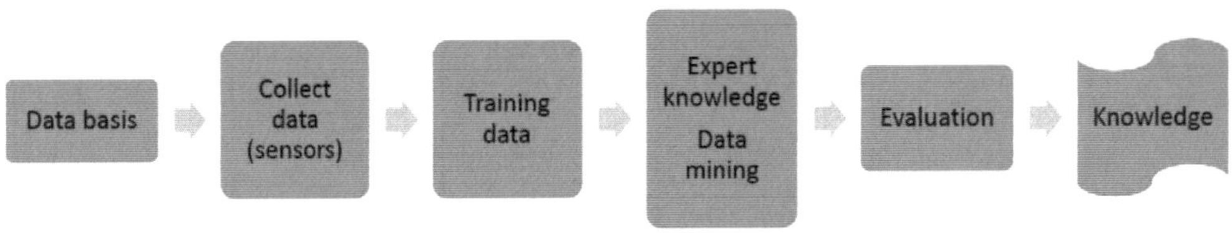

Figure 1. Data mining principle.

The data provided by eddy current equipment is sampled and digitized and the results of an inspection can be presented as a time series

$$X = \{x_t, t = 1,2,...N\} \tag{1}$$

is a collection of length N where x_t is the value of a time series x at time t. It is assured that the length of each time interval $[t, t+1]$ is equal for all $1 \le t \le N$. Each series contain data from a large number of samples, each determination being affected by noises due to the equipment or displacement system, or can contain information that can be considered as noises and which are in fact local modifications of electrical permittivity or magnetic permeability, variations of lift-off, temperature fluctuations, etc. prefiltering is required to avoid high frequency noises and the common method use filtering with the average oven a slide window [5]. After prefiltering, few time series transformations are applied: shifting; scaling and rank-space.

Then, the possibility to exist an event for all starting points and for different window widths is calculated, an event in time series corresponds to region where the probability is maxim. The common method uses the filtering with the average over a sliding window [5].

$$x_t = x_t - \frac{1}{L}\sum_{k=-\frac{L}{2}}^{\frac{L}{2}} x_{t+k} \tag{2}$$

where L is the window length.

The shifting transform on a time series is to get a new time series by adding some real number to each item in the old time series [5]. A shifting of δ on a time series X is

$$x + \delta = \{x_t + \delta, t = 1,2,...N\} \tag{3}$$

The scaling transform of a on a time series X is [6].

$$X = \{ax, t = 1, 2, ...N\} \quad (4)$$

such transform on a time series in to a new time series by multiplying same real number to each item in the old time series. A simple way to make a similarity measure invariant to shifting and scaling is to first normalize the time series. The normal form $\hat{X}$ of a time series is transformed from X by shifting the time series by its standard deviation

$$\hat{X} = \frac{X - mean(X)}{std(X)} \quad (5)$$

The Pearson's correlation coefficient between two times series is defined as follows

$$corr(X,Y) = \frac{mean(X * Y) - mean(X)mean(Y)}{std(X)std(Y)} = \frac{1}{N}\sum_{i=1}^{N}\hat{X}_i\hat{Y}_i \quad (6)$$

This shifting and scaling is performed on the amplitude axis of the time series. A widely used transform in data mining is rank-space transformation. We rank each point in a time series T_R from 1 to the number of total point N, where the lowest value is assigned a value of 1 and the largest is assigned a value of N. This will allow us to consider a time series with a uniform distribution of noise, regardless of the noise distribution before the transformation [7].

Using the sum of the values inside any arbitrary sliding window of *TR*, we can obtaining its p-value (the probability that the sum occurred by chance). For a starting poin s and windows size *w*, *S(s,w)* the sum of the ranked values inside that window, with $1 \leq s \leq N$,$1 \leq s + w - 1 \leq N$. The sum is defined as $S(s,w) = r_s + r_{s+1} + ... + r_{s+w-1}$, for $r_s, ..., r_{s+w-1} \in X_R$. According to [7] for each pair of *(w,N)* exist a distinctly probability distribution, because each sum is dependent only on the number of values being added together *(w)* and their bounds $1 \leq s \leq N$.

Dividing the formed sums by the number of possible combinations for *s* and *w*, and respecting the above method, the probability density defined as $\begin{pmatrix} N-w \\ w \end{pmatrix}$ can be evaluated [8]. Two problems appear: identification of event in time series and the evaluation of shape and severity of discontinuities which had generated this event. The identification of the event is made knowing that an event in time series corresponds in *(s,w)* plane where the probability is maximum, being characterized by two values – starting point and window width (corresponding to its end point).

The NDE of discontinuities can be carried out by clustering procedure (based on the appurtenance of the event to a cluster determining Mahalanobis distance in spatial feature associated to the possible flaws), either using a neuro-fuzzy system (trained with possible data answers and tested in time series in the region of event apparition).

The procedure of clustering requires the construction of clusters on the basis of simulated models, for a large variety of flaws, 6-10th order Fourier harmonics amplitudes being chose as features. For the second procedures, the database must be obtained by simulation for the training of the network.

3. Methodology

Steam generator tubes from Pressurized Heavy Water Reactors, CANDU 600 type, made from Incoloy 800 having the inner diameter 18.2 mm and wall thickness 1.80 mm were taken into consideration. On these samples, artificial discontinuities were practiced using electro discharge machine, table 1.

Table 1. Dimensions and location of EDM slots.

No	Defect type	Dimension [mm]	Depth [mm]
1	Hole	0.1	Through hole
2	Hole	0.3	Through hole
3	Hole	0.3	Through hole+fretting
4	Hole	0.4	Through hole
5	Inner circumferential slot	1*	0.1
6	Inner circumferential slot	0.5*	0.05
7	Inner circumferential slot	1*	0.2
8	Inner circumferential slot	0.5*	0.1
9	Inner circumferential slot	0.5*	0.2

Obs * mark the opening flaw

The artificial discontinuities presented above were distributed on the tube sample having 800 mm length, similarly with those used in steam generator construction figure 2. The precision of slots execution was 0.01 mm, the distance between them being 30 mm.

In order to obtain minimum noise, the sensor developed in this purpose is inner differential type, each coil has 50 turns, from CuEm wire 0.1 mm diameter and the distance between coils is 3mm. During the measurements the sensor is maintained fixed meanwhile the tube is displaced with a motorized XY stage- Newmark type, coupled with the PC through RS 232 interface. The sensor is connected to Network/Impedance/Spectrum Analyzer 4395A Agilent USA, connected with PC through IEEE 488.2 Keithley interface. The scheme and the experimental set-up are presented in figure 3. Once the level of noise was estimated, the optimal frequency was established at 240 kHz. The measurements were effectuated with 0.01 mm displacement step.

Figure 2. Steam generator mock-up.

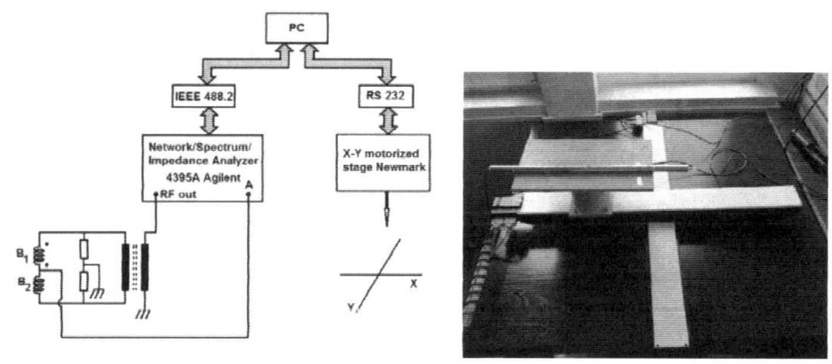

Figure 3. The measurement system.

4. Results and discussions

According to [9] the results obtained by simulation have supported a database with the simulated responses of the sensor at the main type of flaws, grouped on categories (through holes, inner circumferential slots, etc.). In order to locate the position of artificial discontinuities practiced on the tubes, for obtaining simulated results of response of the differential sensor described above in presence of discontinuities presented in table 1, software CIVA 9.2 has been employed. For Incolloy 800, the electrical conductivity is $1.1x10^6$ S/m and for flaws electrical conductivity is 0 S/m. The relative magnetic permeability is 1. The database obtained is used for data mining procedure. The obtained data were processed in order to correspond to the set-up prescribed by [10]. These types of discontinuities form clusters, their appurtenance at a cluster being given by Mahalanobis distance.

The real and imaginary components of the signal obtained at the scanning over the discontinuities through holes #1 – 0.1 mm diameter and #3 – 0.3 mm respectively, as well as inner circumferential slot #5 with1x0.1 mm dimensions are presented in Figure 4a. Due to the high level of signal sampling (over 20100 for a scanning step of 0.01 mm), the quantization noise appears, so the signal must be filtered, the best choice being the use of sliding window [9]. Figure 4b present the filtered signal. The representations show the good signal to noise ratio, the discontinuities being well emphasized.

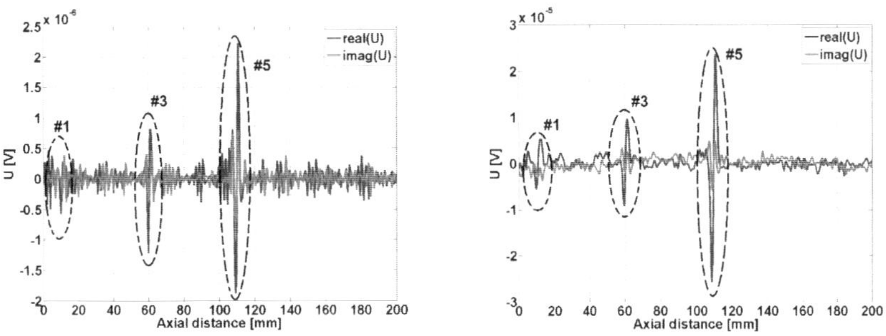

(a) (b)

Figure 4. Response of sensor, real and imaginary components of signal: (a) raw signal; (b) filtered signal.

It can be shown that the quantization noise was reduced. Down sampling until signals reach 100 samples can be performed to decrease the sampling rate without deformation of signal shape. The signals invariant to shifting and scaling can be obtained by normalization (eq. (5)), figure 5a. Applying rank-space transformation to signals presented above, the uniform distribution of the noise is obtained (figure 5b).

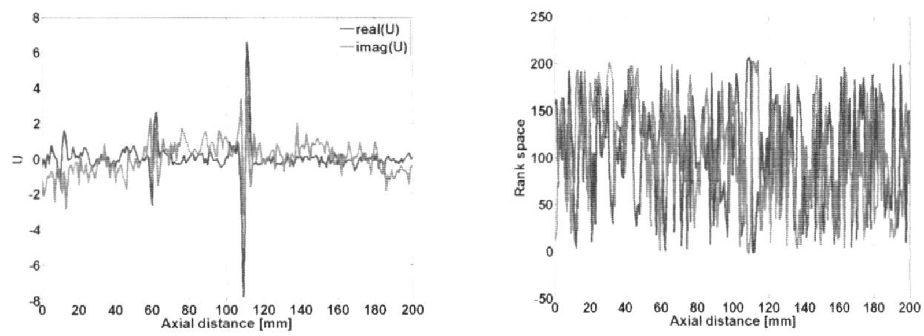

(a) (b)

Figure 5. Post processing of the signals: (a) Normalized signal; (b) Rank space transformation of the signal.

Figure 6 presents the calculated distribution probability of event apparition in the data series, in (S,w) plane. S is the starting point and w is the window width, distinctive for two channels, real and imaginary, respectively.

The start point of the events and their width can be evaluated from the high density of probability zones, the flaw being defined by the starting point and its length by the window width.

The nature of flaws and its appurtenance to a cluster is carried on following the clustering procedure described in [5], function of Mahalanobis distances between the point representing the emphasized flaw ad the weighting centers of the clusters formed on the basis of simulation and experimental data.

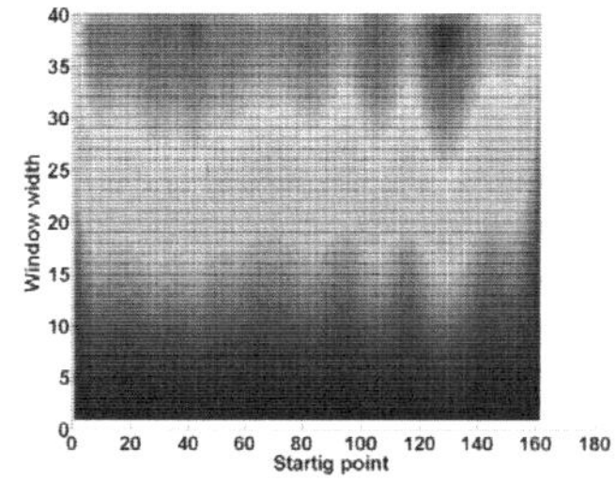

(a)

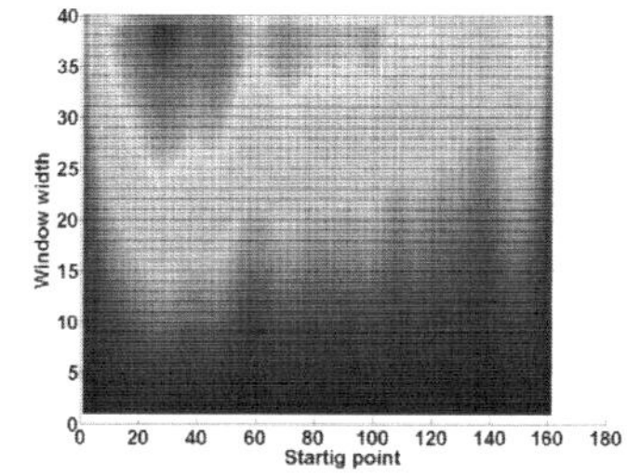

(b)

Figure 6. The density of probability of event apparition: (a) real; (b) imaginary.

Figure 7 shows the clusters of vector signal in amplitude plane (a) and in 3D space (b), the axes being the coefficient amplitudes of the 6th, 7th and 8th harmonics. The results of classification are presented in table 2, only for three events visible in figure 7.

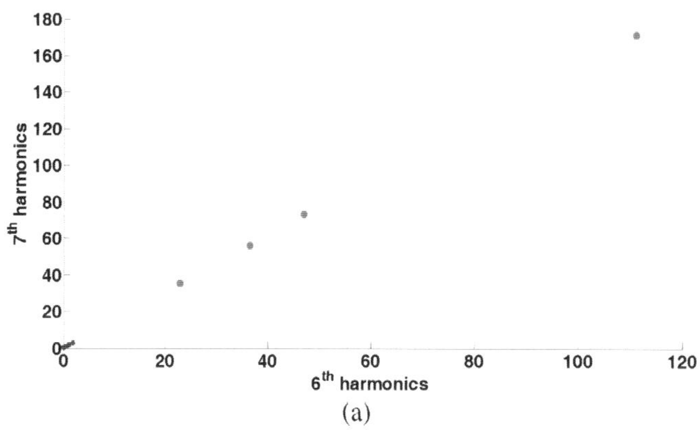

(a)

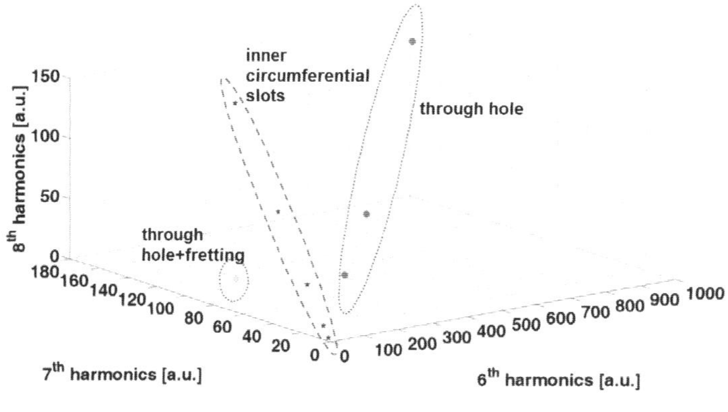

(b)

Figure 7. Classification on flaws categories: (a) amplitude-phase plane; (b) 3D in 6th, 7th and 8th harmonics of Fourier spectrum.

Table 2. Probability of characterization of events I, II and III.

Event	Probability								
	Through wall hole diameter [mm]				Inner circumferential slots - w×d [mm]				
	0.1	0.3	0.3*	0.4	1x0.1	0.5x0.05	1x0.2	0.5x0.1	0.5x0.2
I	0.91	0.51	0.42	0.35	0.43	0.21	0.32	0.17	0.21
II	0.65	0.89	0.69	0.61	0.31	0.19	0.37	0.41	0.52
III	0.15	0.31	0.41	0.27	0.94	0.41	0.72	0.31	0.37

It can be shown that clusters containing different categories of the flaws can be constructed, even if one flaw have been detected into a certain category. The cluster formed by the signals provided by the through hole with different diameters generate smaller Mahalanobis distance than the cluster consisting in inner circumferential slots. If the number of features is increased, the classification can be carried on with more refinement, decreasing the probability of incorrect classification.

5. Conclusions

The use of an automatic classification system allows the classification of the flaws into classes or even into indications with estimated geometrical dimensions. Eddy current examination operations have two temporal series as output (one for real component of induced voltage and the second for the imaginary component). The identification and estimation of artificial flaws practiced on steam generator tubes has been performed by specific data mining procedures allowing a correct identification, classification and estimation of flaws dimension. The clusters for various flaws classes were obtained using synthetic data obtained by simulations and real data obtained by eddy current testing of samples.

References

[1] Thomsen E 2002 *OLAP solutions: building multidimensional information systems. John Wiley & Sons*
[2] Chen M S, Han J and Yu P S 1996 *IEEE T Knowl Data En*, **8** (6) 866-883
[3] Fayyad U and Piatetsky-Shapiro G, Smyth P 1996 *From Data Mining to Knowledge Discovery: An Overview* (Massachusetts: The MIT Press).
[4] Hui S C and Jha G 2000 Inform Manage **38**(1) 1-13
[5] Cosarinsky G, Ruch M, Rugirello G, Domizzi G, Grimberg R, Steigmann R, Savin A 2012 *Insight* **54**(6) 316-321
[6] Ryan D ed., 2013 *High performance discovery in time series: techniques and case studies. Springer Science & Business Media*
[7] Preston D, Protopapas P, Brodley C 2009 Stat Anal Data Min **2**(5-6) 396-411
[8] Andrews G E 1984 *The Theory of Partition*, Cambridge University Press, Cambridge
[9] Cosarinsky G, Ruch M, Rugirello G, Domizzi G, Savin A, Steigmann R, Grimberg R 2011 *Eddy current testing reliability improvement; application to nuclear fuel cladding tubes*, VIII Congreso Regional de Ensayos No Destructivos y Estructurales – VIII CORENDE , Campana , Buenos Aires, Argentina
[10] Code ASME, Section V, art.8, app.1 and 2., 2002

Acknowledgments

This paper is partially supported by Romanian Ministry of Research and Innovation under project PN-II-PCCA-2013-4-0656 and Nucleus Program PN 16-37 01 01.

IOP Conf. Series: Materials Science and Engineering **209** (2017) 012006 doi:10.1088/1757-899X/209/1/012006

Porous AlMg-SiC Composites Structure Modeling By Means of Fractal Analysis

O Rusu[1*] and I Rusu[1]
[1] Technical University "Gheorghe Asachi" of Iasi-Romania, Department of Materials Science and Engineering, Blvd. Mangeron, No. 59A, 700050, Iasi, Romania

E-mail: oana84rou@yahoo.com

Abstract. This work is a continuation of the authors research in the field of ultralight metallic composite materials, based on AlMg10 alloy and SiC particles and obtained by salt dissolution method. We used for the fractal analysis the fractal geometry modeling by means of fractal dimension types of composites obtained from performed experiments. We achieved the following fractal dimensions for the samples: 1.37 (for 5% SiC sample), 1.41 (for 10% SiC sample) and 1.45 (for 15% SiC sample). Fractal analysis indicated that all the obtained samples have cells with a statistically regular form. We conclude that this kind of composite materials can be included in ultralights porous metal composite materials, with a tendency to a metal foam structure.

1. Introduction

This paper describes, on the basis of a quantitative analysis, the configurations and geometrical characteristics of the cellular composites structure (size, shape and distribution of the cells), using for this purpose the fractal geometric modeling, which will lead to the determination of the obtained porous composite fractal dimension [1].

Analyzed materials are ultra light composite materials based on AlMg10 alloy with various percent of silicon carbide (5, 10 and 15%), obtained by dissolution of salts method. This method is based on the use of particles of various salts, in the form of powders, which are soluble in suitable solvents. The method adopted by me used as soluble salt particles of sodium chloride and water as solvent and is based on the introduction in the molten metal alloy of a mixture of powders of sodium chloride and silicon carbide, followed, after cooling, by the dissolution of the sodium chloride in water [2]. Their microscopic analysis showed that these materials are part of cellular metallic materials, tending to metal foams.

2. Application of fractal analysis to the study of AlMg10-SiC porous composite materials

The analysis method, based on the properties of fractals (fragmentation to infinity, self-similarity, the shift-invariance and fractal dimension was described in [3]. So, we obtain a direct method for the study of different porous type composites, function of added SiC amount to the AlMg10 alloy [4].

In order to quantify such changes, we follow these steps:

(i) a black-white digitalization from each image was used in order to reproduce topographical features on the analyzed surface;

(ii) using MATLAB software, based on an existing procedural model which uses numerical optimized calculation routines, we obtain fractal dimension dependencies on cell uniformity degree for each for each investigated material.

Therefore, we select three images for structural representation of the three types of obtained composites (figure 1) and we translate them to a grayscale image and then, for simplicity, we will use only a black and white picture (figure 2).

a.

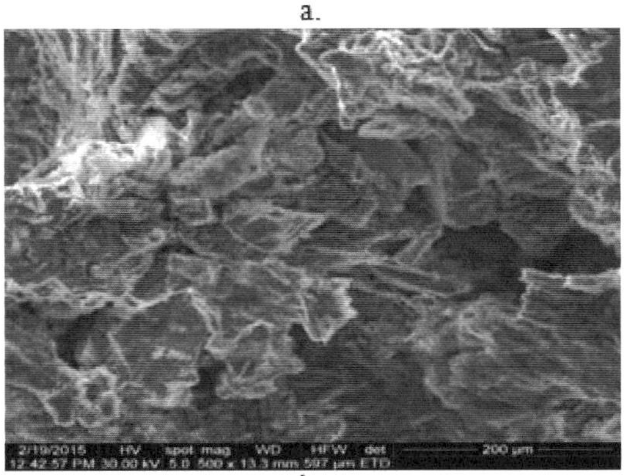

b.

c.

Figure 1. Electron microscopy image: a – AlMg10 - 5% SiC, b – AlMg10 - 10% SiC; c – AlMg10 - 15% SiC. X500.

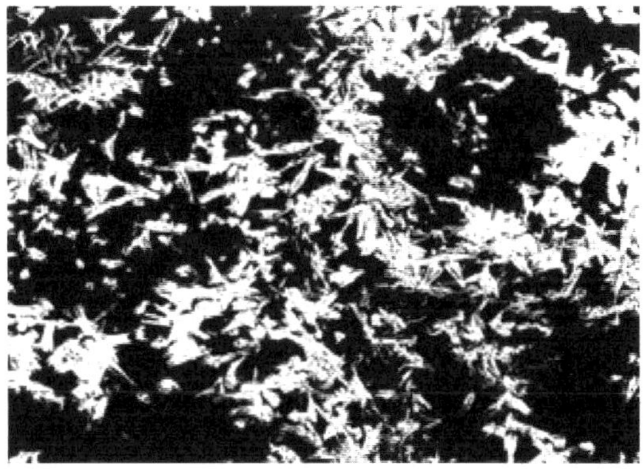

a.

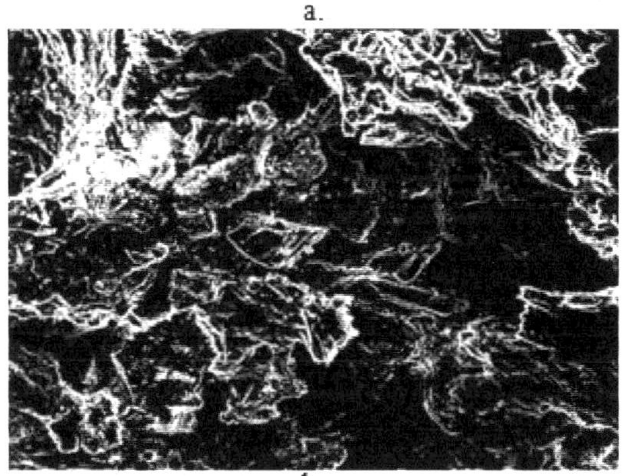

b.

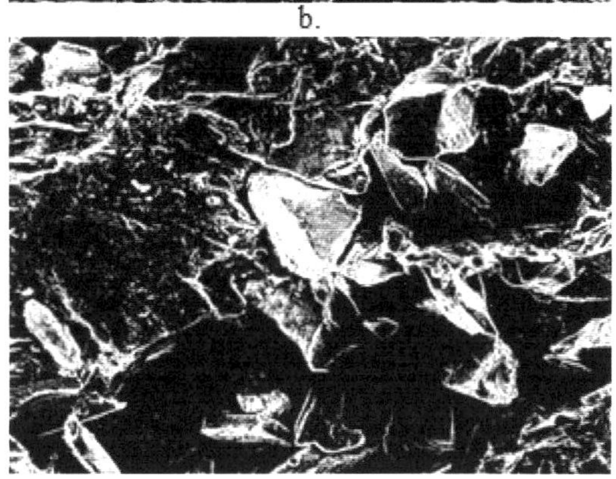

c.

Figure 2. Electron microscopy image in black and white of AlMg10-SiC alloy with 5% SiC (a), 10% SiC (b) and 15% SiC (c).

The dependence of fractal dimension on the cell uniformity degree for the analyzed three composites types are presented in tables 1 ÷ 3 and 3 ÷ 5 figures.

Table 1. Values of β and Df for composite sample AlMg10 - 5% SiC.

Analyzed cell	Fractal dimension D_f	Cell uniformity degree β
1	1,2745	0,8891
2	1,2222	0,7488
3	1,4438	0,4735
4	1,1849	0,6006
5	1,0842	0,8375
6	1,1734	0,8934
7	1,3499	0,3738
8	1,2858	0,7881
9	1,3728	0,9542
10	1,1759	0,8054

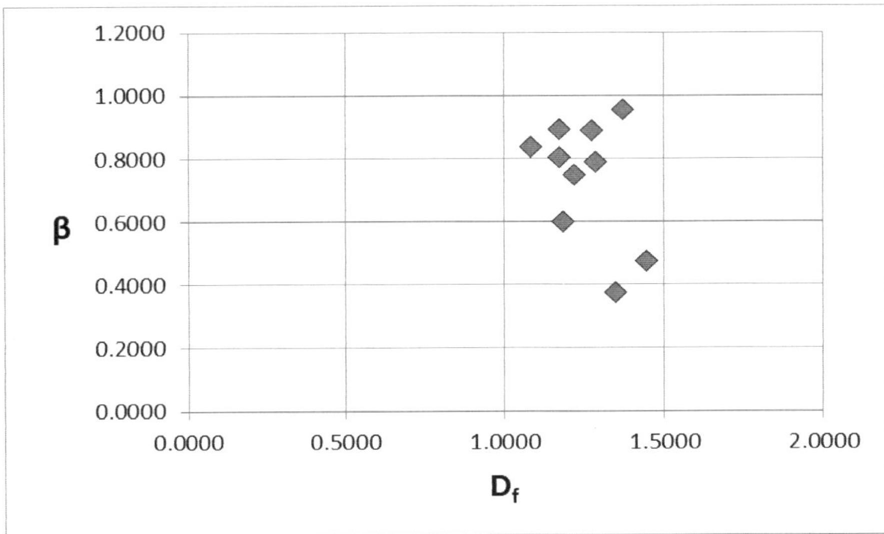

Figure 3. Logarithmic correlation D_f – β for composite sample AlMg10 - 5% SiC.

Table 2. Values of β and Df for composite sample AlMg10 - 10% SiC.

Analyzed cell	Fractal dimension D_f	Cell uniformity degree β
1	1,2933	0,4748
2	1,1939	0,5723
3	1,5033	0,8234
4	1,3703	0,9883
5	1,2943	0,9383
6	1,1657	0,8494
7	1,2228	0,6308
8	1,5771	0,7569
9	1,4144	0,9897
10	1,3557	0,9799

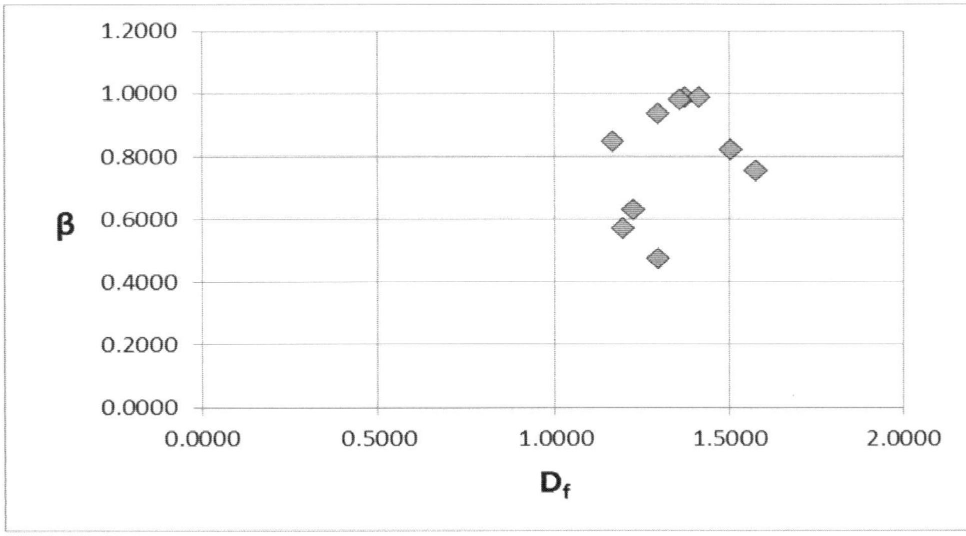

Figure 4. Logarithmic correlation $D_f - \beta$ for composite sample AlMg10 - 10% SiC.

Table 3. Values of β and Df for composite sample AlMg10 - 15% SiC.

Analyzed cell	Fractal dimension D_f	Cell uniformity degree β
1	1,0099	0,8782
2	1,1461	0,6480
3	1,2520	0,9684
4	1,1955	0,9808
5	1,2778	0,4255
6	1,4601	0,8479
7	1,3002	0,9553
8	1,2299	0,9835
9	1,2889	0,9606
10	1,4546	0,9898

It can be observed that, in the case of composite sample AlMg10 - 5% SiC (table 1 and figure 3), the earliest cell form exhibiting a tendency for levelling (with a degree of cell uniformity β closest to value 1) appears in a total of three cases, namely for positions 1, 6 and 9. The optimal fractal dimension is that for position 9 for which the value of fractal dimension Df = 1.3728 is closest to the theoretical fractal dimension Df = 1.66.

In the case of AlMg10 - 10% SiC sample the nearest types of cells with a tendency for levelling (table 2 and figure 4) appears in the positions 4, 5, 6, 9 and 10, in a greater number of cases as compared to previous sample. In this case, the fractal dimension closest to the theoretical value is that in the position 9, for which Df = 1.4144.

For the third sample AlMg10 - 15% SiC we can observe from table 3 and figure 5 that the tendency for levelling grows – positions 3, 4, 7, 8, 9 and 10. The fractal dimension closest to the theoretical value is that in the position 10, for which $D_f = 1.4546$.

So, it can be concluded that the composite materials obtained by this method can be included in ultralight porous metallic composite materials, with a high tendency to metal foam structure (morphology).

Also, we can conclude that the AlMg10 - 15% SiC composite sample has the highest cell uniformity and the nearest fractal dimension to theoretical value versus all analyzed samples, what issues him to be closest to the concept of porous composite material.

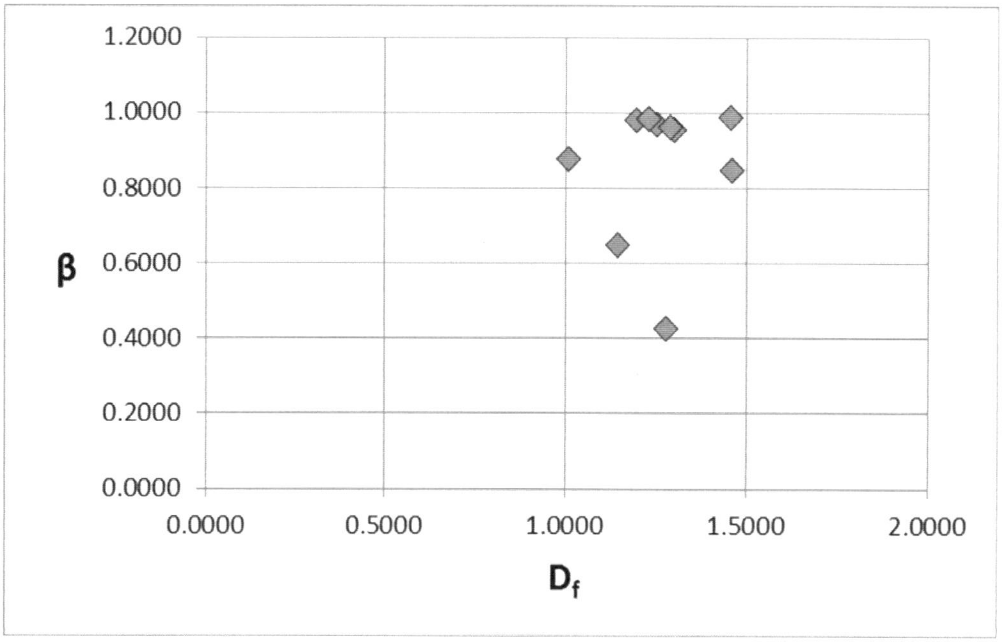

Figure 5. Logarithmic correlation D_f – β for composite sample AlMg10 - 15% SiC.

This issue can be explained by increased stability of the cell walls and connection bridges between cells due to the growth of SiC particles percentage added in AlMg10 alloy groundmass.

3. Conclusions

Fractal analysis confirms that the AlMg10 - SiC composite materials obtained by salt dissolution method can be included in ultralight porous metallic composite materials, with a high tendency to metal foam structure. The closest to this structure is the composite with 15% SiC, fact explained by the increased stability of the cell walls and connection bridges between cells due to SiC particles.

We can recommended fractal analysis for the study of various complementary reaction mechanisms associated to fractal behavior of cellular composites, such as, for example, deformation or impact behavior, vibrations damping, which represents new research directions associated with the fractal character of these types of composites.

References

[1] Lu S Z and Hellawell A 1995 *Fractal Analysis of Complex Microstrutures* in Materials, Proc. MC95 International Metallographic Conference, Colmar, France, ASM, 119–126

[2] Bălţătescu O, Axinte M, Barbu G, Manole V 2015 *Materials Science and Engineering* **95**, 012018

[3] Bălţătescu O, Florea R M, Carcea I, Rusu I 2015 *Journal of Optoelectronics and Advanced Materials* **17**(11–12) 1862–1867

[4] Bălţătescu O 2016 *Cercetări asupra materialelor compozite superuşoare cu matrice metalică*

[5] Durowoju M O 2007 *Journal of Engineering and Applied Sciences* **2**(10) 1489–1492

IOP Conf. Series: Materials Science and Engineering **209** (2017) 012007 doi:10.1088/1757-899X/209/1/012007

The Assessing of the Failure Behavior of Glass/Polyester Composites Subject to Quasi Static Stresses

M D Stanciu [1], A Savin[2*] and H Teodorescu-Drăghicescu[1]
[1] Transylvania University of Brasov, Faculty of Mechanical Engineering, 29 Eroilor Blvd, 500036, Romania
[2]National Institute of Research and Development for Technical Physics, Nondestructive Testing Department, 47 D. Mangeron Blvd, Iasi, 700050, Romania

E-mail: asavin@phys-iasi.ro

Abstract. Using glass fabric reinforced composites for structure of wind turbine blades requires high mechanical strengths especially to cyclic stresses. Studies have shown that approximately 50% of composite material failure occurs because of fatigue. Composites behavior to cyclic stresses involves three stages regarding to stiffness variation: the first stage is characterized by the accelerated decline of stiffness with micro-cracks, the second stage - a slight decrease of stiffness characterized by the occurrence of delamination and third stage characterized by higher decreases of resistance and occurrence of fracture thereof. The aim of the paper is to analyzed the behavior of composites reinforced with glass fibers fabric type RT500 and polyester resin subjected to tensile cyclic loading with pulsating quasi-static regime with asymmetry coefficient $R = 0$. The samples were tested with the universal tensile machine LS100 Lloyd Instruments Plus, with a load capacity of 100 kN. The load was applied with different speeds of 1 mm/min, 10 mm/min and 20 mm/min. After tests, it was observed that the greatest permanent strains were recorded in the first load cycles when the total energy storage by material was lost due to internal friction. With increasing number of cycles, the glass/polyester composites ability to store energy of deformation decreases, the flow phenomenon characterized by large displacements to smaller loading forces appearing.

1. Introduction

Numerous studies on the fatigue behavior of glass-reinforced composites (GFRC) have mainly addressed the effect of environmental conditions (temperature variation, glass transition temperature, humidity variation, artificial aging).

The results of these studies have shown that there is a close link between the dimensions and shape of the fibers, their orientation and distribution, and the viscous-elastic behavior of the matrix [1-3]. The behavior of materials in fatigue is very important in all engineering applications [4,5]. Since the composites have a polymer as matrix, their behavior is viscous-elastic or viscous-plastic, being grounded in the theory of cooperative [6,7].

Mechanical and/or temperature loading cause changes in matrix-fibers linkages or in chemical-physical rearrangements or in van der Waals forces distribution. If these changes exceed the internal cohesion forces or in some areas the chains are weak, the material has viscous-plastic behavior [2,8]. Even though the fibers determine the mechanical properties like elastic modulus or strength, the matrix provides the physical and chemical durability by protecting the fibers from environmental influences [2]. Although the viscous-elastic behavior of glass fibers composites is well-known [1-3], however, there is a knowledge gap about the dependence of mechanical properties evolution on cyclic loading and speed of loading. The high tensile strength of the composite is due to reinforcement and type of fibers used, while flexural strength is due to the elastic characteristics close of the two components - matrix and reinforcement, which makes both components to work together [9-11]. Experimental tests have shown that the failure in a laminated composite is very often progressive in nature, occurring by a process of damage accumulation [12,13].

Fatigue life predictions for service load histories are generally based on fatigue data performed with constant amplitude loading, which can be substantially influenced by cyclic frequency in thermoplastics [14]. The paper aims to determine the speed of deterioration of the mechanical properties under the influence of the velocity of the stress and the pulsating cycles.

2. The experimental set-up

2.1. Materials

The specimens were realized from four layers of unidirectional RT500 fabric and polyester matrix with specific geometry of tensile test according to SR-EN ISO 527-2, having the geometrical characteristics in table 1.

Table 1. Dimensions of the specimens before testing.

Input Data / Samples Code	RT1	RT2	RT3	RT4	RT5
Sample Length [*mm*]	150	150	150	150	150
The length of the calibrated part [*mm*]	70	70	70	70	70
Witdh [*mm*]	10	10.3	10	10.2	10
Thickness [*mm*]	2.5	2.5	2,2	2.5	2.5
Area [mm^2]	25	25.75	22	25.5	25
Load [*kN*]	2.5	2.5	2.5	3.0	3.0
Loading speed [*mm/min*]	1	10	20	10	20
Number of loading cycles			1-10-15		

2.2. Experimental method

The samples were subjected to tensile loading in an algorithm test consisting of three series of pulsating cycles (1, 10 and 15 cycles) at 1, 10 and 20 mm/min and maximum load of 2.5 kN and 3.0 kN respectively. An important consideration in axial fatigue testing is the uniformity of stress and strains in the specimen gage section.

The tests were made with the LR5K Plus Lloyd's Instruments Universal Machine and the results were processed using the NEXYGEN Plus software. Several tests were performed to analyze the influence of the loading cycles on the integrity of the composite material in terms of modulus of elasticity, stiffness and storage capacity of the deformation energy. The elastic-viscous-plastic behavior of the composites was also evaluated.

3. Results and discussion

The changes in cyclic deformation behavior are more pronounced at the beginning of cyclic loading (transient behavior) when the plastic deformation reached almost 50% from the total plastic strain at the final of cyclic loading (figure 1).

The area within a hysteresis loop is energy dissipated during a cycle; with increasing the number of cycles, the material usually gradually stabilizes (steady-state).

The yield strength in tension or compression was reduced after applying a load of the opposite sign that caused inelastic deformation (figure 2, a).

An appreciable progressive change is observed in stress-strain behavior during inelastic cycling (figure 2, b).

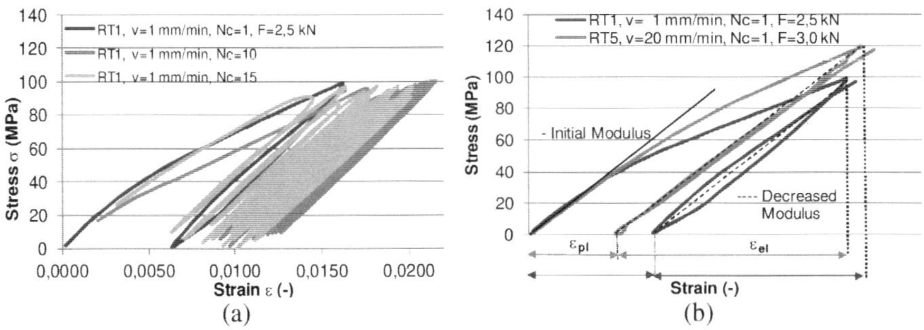

Figure 1. The hysteresis loops: a) comparison of hysteresis loops obtained at 1, 10 and 15 cycles; b) elastic and plastic strains with increasing the loading for monotonic stress.

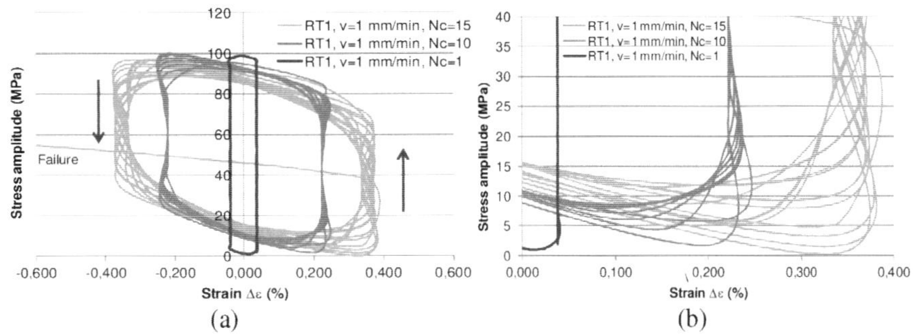

Figure 2. Comparison between stress-strains curve obtained for different cycles: (a) the cyclic stress-strain response for load speed of 1 mm/min; (b) detail of inelastic deformation during softening cycles.

In figure 3b, the values of the percentage specific deformation of GFRC for different loading rates are observed. Thus, the highest deformation values (0.30%) are recorded at 20 mm/min during the 10 cycles. For monotonic loading (figure 3a), the most stable behavior is recorded for load speed of 1 mm/min.

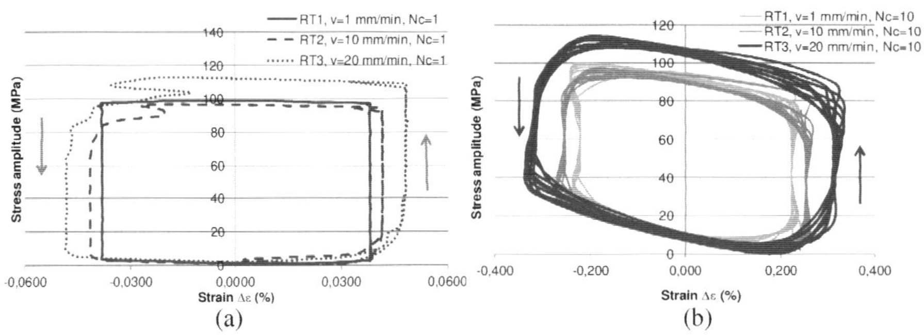

Figure 3. Strain/stress for different loading rated: (a) The monotonic stress-strain response at different loading speed; (b) the increasing of strain with increasing of loading rate.

Because of crosslink degradation in the matrix structure or plastic deformation of interfaces between matrix and glass fibers with continued cycling, the mechanical properties in terms of Young's modulus, stiffness and strength decreased (figure 4).

In the same time, a decreased resistance at deformation, irrespective a softening of materials is recorded. With increasing the loading rate with 100%, the Young's modulus value increased with 2% for monotonic stress-strain, which lead to appearance of a hardening zone which is a precedent of fatigue cracks and failure.

The intensity of load play influences the elastic modulus value: with increasing of applied force with 20%, the rigidity of materials decreased with almost 5%. The rate is useful for determining the life time of composites under fatigue test.

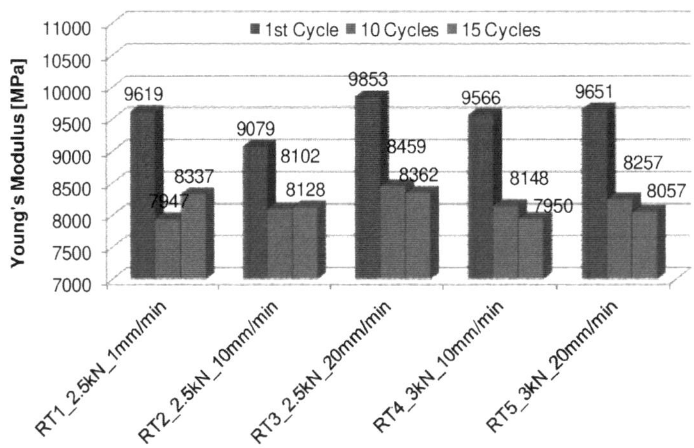

(a)

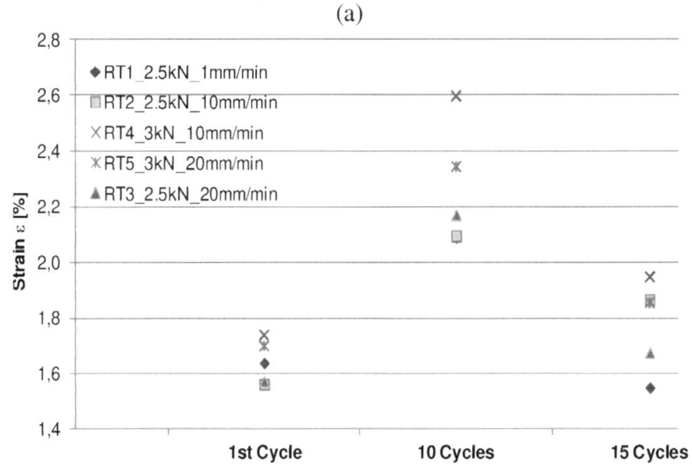

(b)

Figure 4. Variation of mechanical properties: (a) variation of elasticity modulus; (b) strain variation with respect the number of cycles.

In the case of the GFRC, it has been observed that the matrix had broken first, the inner efforts being distributed to longitudinal fibers of the glass fabric which continuous the deformation until reaching the breakage at the maximum stress (figure 5).

At the monotonic stress-strain (figure 5, a) there is no surface damage; with increasing the number of loading cycles with asymmetry coefficient $R = 0$, it can be seen the matrix failure and buckling of fibers (figure 5, b). Finally, at 15th cycles, the GFRC samples failure, both matrix and fabric, the fracture region being characterized by cracking, debonding, crushing. In that stage, the material has reached the plasticity limit and damage.

(a) (b) (c)

Figure 5. Failure area for the tested specimens: (a) after first cycle; (b) after 10 cycles; (c) after 15 cycles.

During cyclic stresses, beside the axial stresses which are developed in matrix and fibers, at the interfaces between both components appear tangential stresses due to different stiffness of each component. So, these stresses lead to sliding between layers because the elastic limits of matrix and fibers are not overlap. Than appears the first fractures from composite. The majority of the criteria proposed in literature identify the following failure modes: fiber fracture; transverse matrix cracking; shear matrix cracking [12-14].

4. Conclusions

The analyses the elastic-plastic behavior of glass fabric reinforced composite under cyclic loading compared to behavior under static loading have been carried on. The parameters which were varied were: the number of cycle, the speed of loading, the intensity of the applied force.

The stress-strain response of the GFRC samples subjected to cyclic loading is quite different from that under monotonic loading:

- the material have an linear and homogeneous response when is subjected to one loading cycle;
- with increasing the number of cycles, the material responded with viscous and plastic deformation with losing the capacity to recover the strain;
- cyclic deformation behavior indicated progressive modulus reduction and unsymmetrical straining in tension and compression;
- the plastic strain clearly depends on the applied load level.

References

[1] Amiri M, Khonsari MM 2010 *Int J Fatigue* **32**(2) 382–389
[2] Santa J F, Vanegas-Jaramillo J D, Patiño I D 2016 *Latin American Journal of Solids and Structures* **13** 344-364
[3] Zhang Y, Vassilopoulos A P, Keller T 2008 *Int J Fatigue* **30** 1813-1820
[4] Khan R, Alderliesten R, Badshah S, Benedictus R 2015 *Compos Struct* **124** 214-227
[5] Liang S, Gning P B, Guillaumat L 2012 *Compos Sci. Technol.* **72**(5) 535-543
[6] Samborsky D D, Mandell J F, Cairns D S 2002 *Sandia Contractors Report – Fatigue of Composite Materials and Substructures for Wind Turbine Blades*, Montana State University
[7] Flore D, Wegener K, Mayer H, Karr U, Oetting C C 2017 *Comp. Sci. Tech*; **141** 130-136
[8] Jiang Y and Zhang J 2008 *International Journal of Plasticity* **24** 1481-1515
[9] Stanciu M D, Harapu A, Teodorescu Drăghicescu H, Curtu I, Savin A 2016 Comparison between viscous elastic plastic behavior of the composites reinforced with plain glass fabric

and chopped strand mat. *7th International Conference on Advanced Concepts in Mechanical Engineering, Materials Science and Engineering* 147

[10] Steigmann R, Iftimie N, Dobrescu G S, Barsanescu P D, Curtu I, Stanciu M D, Savin A 2016 Nondestructive evaluation and characterization of GFRP using non-contact ultrasound and complementary method, *ModTech International Conference - Modern Technologies in Industrial Engineering IV, IOP Conf. Series: Materials Science and Engineering* 145 032010 doi:10.1088/1757-899X/145/3/032010

[11] Teodorescu-Draghicescu H, Sorin V, Stanciu M D, Curtu I, Mihalcica M 2015 *Materiale Plastice* **52**(1) 62-64

[12] Camanho P P, Bowron S, Matthews F L 1998 *Journal of Reinforced Plastics and Composites* **17** 205-33

[13] Mortazavian S, Fatemi A 2017 Fatigue of short fiber thermoplastic composites: A review of recent experimental results and analysis. *Int J Fatigue* http://dx.doi.org/10.1016/j.ijfatigue.2017.01.037

[14] Eftekhari M, Fatemi A 2016 *Int J Fatigue* **87** 153–66.

Acknowledgments

This work was supported by the Ministry of National Education, CNCS – UEFISCDI, project PN- PN-II-PCCA-2013-4-0656. The paper is in the memory of Professor Ioan CURTU, we are grateful for all scientific knowledge and skills he taught us.

IOP Conf. Series: Materials Science and Engineering **209** (2017) 012008 doi:10.1088/1757-899X/209/1/012008

Influence of Melt State on the Properties of Nickel-based Powders for Wear-Resistant Coat

E Baryshev, G Tyagunov, T Kostina and K Shmakova
Ural Federal University, Mira str., 19, Russia

E-mail: e.e.baryshev@urfu.ru

Abstract. The influence of time-temperature treatment on properties of the PG-SR-type alloys in cast condition, in produced powders and coatings is studied in this work. The temperature dependence of physical properties were obtained. These results and the comparative metallographic analysis have allowed offering the mode for time-temperature treatment of the liquid PG-SR-type alloys. Use of the time-temperature treatment when producing metal powders and coatings are as follows: the dispersibility of powders increases; the yield of powder fractions used for applying coatings increases; the powder porosity and oxidation decreases; oxidation resistance in various media increases; corrosion-, heat-, and wear resistance increases.

1. Introduction

The PG-SR-type nickel-based alloys containing chromium, boron and carbon for applying wear- and corrosion-resistant coatings, thanks to their chemical composition, have a great variety of simple phases and complex structures, which are usually far from equilibrium and micro-homogeneous state. [1,2].

Attempts to stabilize or improve service properties are normally associated only with additional charging of such expensive materials as tungsten, titanium and others to the PG-SR-type matrix alloys [3-5]. Such an approach allows sometimes solving private tasks, but can not be scientifically justified and is of low commercial effectiveness.

Numerous experimental works on physical properties and structure of liquid steel and alloys show that in most cases after melting metal is in non-equilibrium state [6-8]. And depending on the choice of charge materials, heating conditions, time of dwelling metal in liquid state the deviation degree from equilibrium may be different.

One of the methods of forming equilibrium and homogenous structure of the melt is its time-temperature treatment in smelting process based on a study of the melt structure formation processes and an analysis of the relationship of the structure in liquid and solid states [7].

In this regard, the influence of time-temperature treatment on properties of the PG-SR-type alloys in cast condition, in produced powders and coatings is studied in this work.

2. Materials and methods

Optimal temperature and time parameters for melting process were searched for and proved on the basis of the study results of temperature dependence of melts physical properties. The kinematic viscosity (ν) was measured by a method of torsional oscillations of a crucible with a melt, the surface tension (σ) and density (d) were measured by a sessile drop method, the specific electrical resistivity (ρ) by a non-contact method in a twisted magnetic field, the magnetic susceptibility (χ) was measured by the Faraday method. All studies were performed in an inert atmosphere. Before conducting the experimental test the system, the main elements of which are a metal chamber with an electric resistance furnace and a measuring cell with a test sample, was preliminary vacuumed to reach the pressure of $1.33 \cdot 10^{-3}$ Pa, was flushed out and filled with purified helium to reach the pressure of $1.11 \cdot 10^{5}$ Pa. The temperature in all plants was measured with the tungsten-rhenium thermocouple BP 5/20, hot junction of which was in the vicinity of the test sample.

The PG-SR-type nickel-based alloys used for applying wear- and corrosion-resistant coatings were selected to conduct the study. The chemical composition of the studied alloys is shown in the table 1.

Table 1. Steel composition of PG-SR-type alloys.

Grade	Content of elements in %wt					
	Ni	C	Cr	Si	B	Fe
PG-SR1	Basic	0.2-0.4	12-14	2.0-2.8	1.2-1.8	<5
PG-SR2	Basic	0.35-0.60	14-16	2.8-3.5	1.8-2.3	<5
PG-SR3	Basic	0.60-0.90	15-17	2.7-3.7	2.3-3.0	<5
PG-SR4	Basic	0.80-1.20	16-18	3.8-4.5	3.1-4.0	<5

3. Results and discussion

The study results of temperature dependence of physical properties of the PG-SR1 alloy are given in figure 1.

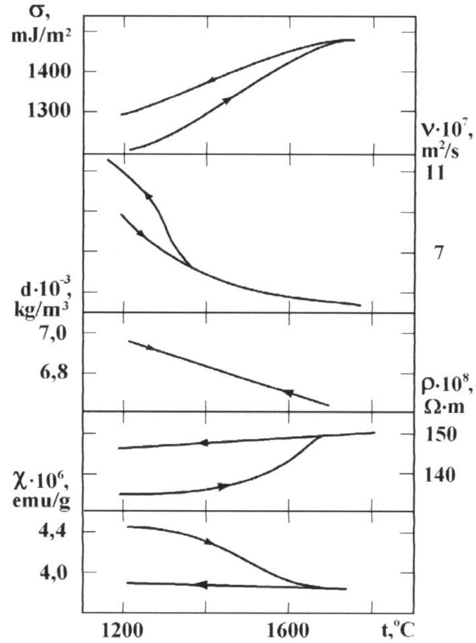

Figure 1. Temperature dependences of physical properties of the PG-SR1 alloy during heating and cooling.

The obtained dependencies ν(t) of all the studied alloys of the PG-SR group allowed to mark out two structural regions of a melt. Low temperature area is characterized by a sharp change in viscosity from extremely high to low inherent in metallic liquids, i.e. this is a region of a heterophase state of the melt. Temperature values separating these regions depend on the chemical composition of the alloy. Improving an alloying degree increases the marginal temperature from 1230 to 1470°C.

In high-temperature regions, i.e. above the marginal temperature, there are changes in the nature of heating and cooling polytherms ν and their hysteresis (Figure 1), which can be achieved only when heating is performed above the critical temperature (t_f).

According to the nature of the polytherm property change the main parameters of the time-temperature treatment (TTT) were defined. It involves heating of the liquid metal above the critical temperature, dwelling over a certain period of time and cooling up to the casting temperature or up to the dispersing temperature at a given speed.

The structure of cast samples proved to be dendritic. However, in samples, that were cast according to the technology existing at the plant, strong development of first-order dendrites was observed. This was eliminated by introducing the TTT.

A quantitative metallographic analysis of the PG-SR2 alloy has showed that using the TTT can reduce the overall number of strengthening phases (table 2), eliminate the visible appearance of primary needle shaped carbides and reduce the number of eutectic carbides almost 1.8 -2 times. The structure of strengthening phases also changes. It forms a continuous and rigid lattice in samples cast according to the plant technology. The time-temperature treatment allowed for the lattice transformation, its division into individual fragments and the emergence of regions spherulitic eutectic. The latter shows that the forms of primary crystallization centers have changed in alloys.

Table 2. Structure characteristics of cast samples of the PG-SR2 alloy.

Smelting technology	Dendritic Cell parameters, μm	Number of strengthening phases, volume %	Micro-hardness of the γ-phase, hPa	Number of spherulite-like eutectics, volume %	Wear-resistance, times
Plant	12.4	32	84	0	1
Under the TTT of the alloy	11.9	23	101	1.80	1.8

Discontinuity of the silicate-boron-carbide lattice and appearance of spherulite-like formations in alloys of the eutectic type contributes to enhancing their plasticity [6-8]. Introduction of the technology using the TTT of the melt has led to a reduction in the size and number of slip lines, formation of units concentrated in the fields of spherulite eutectic formation inherent in ductile fracture [7]. In general, the outlined structural changes enhance plasticity of the PG-SR-type alloys.

The TTT of the melt contributes to improving the components solubility in a nickel matrix and the uniform elements distribution by metal volume and between phases. Also the alloying degree of γ-solid solution increases, as evidenced by the increase in micro-hardness of the alloy matrix (table 2).

Based on the study results of the temperature dependence of physical properties of a melt and interrelationships of liquid and solid states the melting conditions before dispersing have been offered, experimental melt heats have been carried out and a comparative study of properties of the manufactured powder and coatings has been performed.

The following service properties were studied: dispersion, oxygen content, porosity, micro-hardness, heat resistance and structure of powders manufactured according to the plant technology and experimental technology.

The grain-size analysis showed that the TTT increased dispersion of powders. With this the particle yield with the size of less than 150 μm, that is the most suitable for applying coatings, has increased 3-6 times, and the average particle size of powders decreased 1.8 -2.0 times.

Preparation of the melt for dispersing under TTT modes significantly reduces the oxygen content in powders as well as lowers their porosity.

Powders with particle sizes from +0.1 to 0.2 mm were used for applying coatings by thermal spraying on cold steel backplate followed by fusion of the working surface in acetylene-oxygen flame. After fusion the coatings made of PG-SR-type self-fluxing alloys were polished with silicon-carbide grinding tools.

If powders were manufactured under the TTT of the melt, the non-fused powder coatings have had lower porosity. After fusion this trend picks up. With this, the porosity of powder coatings, manufactured according to the existing plant technology, equals to 0.41% and according to the experimental technology equals to 0.18%.

Coatings made of experimental powders enjoy hardness increased by 15% that is connected not only with the reduction in their porosity, but also with great content of fine boron carbide particles in their structure.

Coatings manufacturing technology allows for materials heating in the air in the temperature range from 75°C to 1300°C. Hence the study of the kinetics of the samples oxidation was carried out in this temperature range. Experimental data allowed to make the conclusion that the PG-SR1 cast alloy that was obtained from the equilibrium and homogenous state of the melt before hardening is marked by the 35-55% lower speed of oxidation. In terms of oxidation resistance the experimental technology contributes to approximation of this alloy to the samples of more doped PG-SR4 alloy manufactured in accordance with the plant smelting technology (figure 2).

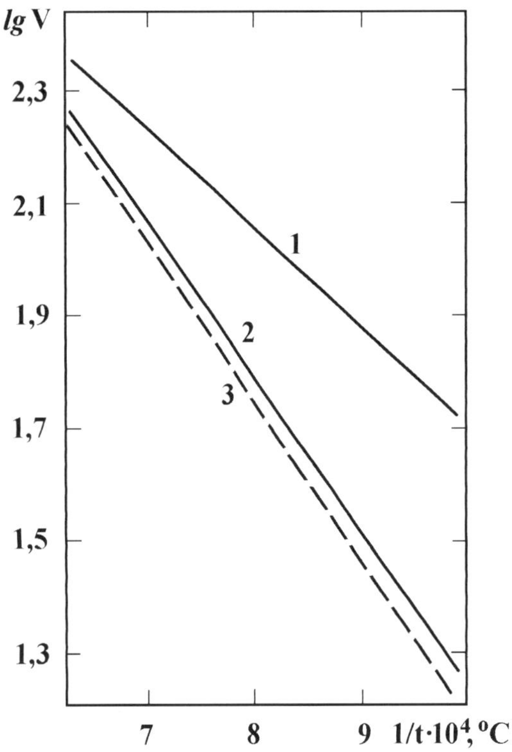

Figure 2. The dependence of the logarithm of the air oxidation speed of PG-SR1 (1 and 2) alloy samples and PG-SR4 (3) alloy on the return heating temperature. Temperature range of the experiment is from 750 to 1300°C; 1,3 is a smelting technology existing at the plant; 2 is a technology using the TTT of the melt.

The most important service property of PG-SR- type alloys is wear-resistance. Changes in weight of a reference standard and a sample after 10 test cycles were compared to quantify the wear resistance index. The time-temperature treatment of the melt of the constant chemical composition helped to increase the wear resistance of the PG-SR1 alloy 1.8 times and PG-SR2 - 1.6 times.

4. Conclusion

The PG-SR-type alloys were found out to be in a non-equilibrium and micro-inhomogeneous state after melting. Measurements of the temperature dependence of physical properties allowed to determine the temperature range when the melt progresses to the equilibrium state. These results and the comparative metallographic analysis have allowed offering the mode for time-temperature treatment of the liquid PG-SR-type alloys.

Experimental data on the high efficiency of the smelting technology using the time-temperature treatment when producing metal powders and coatings are as follows: the dispersibility of powders increases; the yield of powder fractions used for applying coatings increases; the powder porosity and oxidation decreases; oxidation resistance in various media increases; corrosion-, heat-, and wear resistance increases.

References

[1] Zhang X C, Xu B S, Xuan F Z, Tu S T, Wang H D, Wu Y X 2009 *Appl. Surf. Sci.* **255** 4362-4371
[2] Tam K F, Cheng F T, Man H C 2002 *Surf. Coat. Technol.* **149** 36-44
[3] Wang K L, Zhang Q B, Sun M .L, Wei X G, Zhu Y M 2001 *Appl. Surf. Sci.* **174** 191–200
[4] Kawahara Y 1997 *Mater. High Temp.* **14** 261-268
[5] Skulev H, Malinov S, Sha W, Basheer PAM 2005 *Surf Coat Technol.* **197** 177–184
[6] Elanskiy G N 2006 *The structure and properties of metallic melts* (Moscow: Moscow State Evening Metallurgical Institute Press) p 228
[7] Tyagunov G V, Baryshev E E, Tsepelev V S 2014 *Liquid metal. Powders* (Yekaterinburg: UMC-UPI Press) p 192
[8] Baryshev E E, Tyagunov G V, Baum B A, Batanin AP 2008 *J. Phys.: Conference Series* **98** 1- 7

Acknowledgments

This work has been supported by the Scientific researchers of higher education institutions within the State task of the Russian Federation No.4.5533.2017.

Investigation of Electrical Resistivity of Ni-Cr-Al Alloys in Liquid and Solid State

A Tyagunov, E Baryshev, G Tyagunov, V Vuykhin, V Mushnikov and K Shmakova
Ural Federal University, Mira str., 19, Russia

E-mail: e.e.baryshev@urfu.ru

Abstract. The subject of the study was temperature dependence of electrical resistivity of alloys based on Ni-Cr-Al system in liquid and solid state. Objects of research were alloy compositions contained 10 and 15 % of chromium and alloyed by 5 and 10 % of aluminium each. The temperature dependences of electrical resistivity was a characteristic to commercial superalloys. It has been founded that the higher the amount of alloyed elements the lower critical temperature heating up to which leads to the formation of equilibrity and microhomogeneouty of the melt structure and the lower the temperature of hysteresis of electrical resistivity polyterms. Melt state before crystallization has had a significant influence on solidification process and structure of alloys in solid state.

1. Introduction

The Ni-Cr-Al system occupies a special place in the development of dispersion-hardening nickel-based alloys [1]. The distortion of the crystal lattice of solid solutions based on nickel accrues continuously and smoothly with increasing number of alloying components and their concentrations. Apparently, this is due to the appearance of additional polar bonds in such alloys and the determining factor of phase stability is the electron concentration but not the difference in the dimensions of the atoms [2].

The alloying of nickel with chromium has a strengthening effect but mainly during short-term tests. The increase in the heat resistance of Ni, Cr-alloys is achieved by alloy age in particular by aluminum. Alloys are strengthened by creating a dispersive heterogeneous structure consisting of a γ-solid solution and a strengthening phase of Ni_3Al that is variable in solubility; therefore, dispersion hardening is possible because of heat treatment. Intermetallide Ni_3Al has a small area of homogeneity. Chromium is dissolved in small amounts in this phase.

In the ternary alloys of the Ni-Cr-Al system, the solubility of aluminum in the nickel-chromium solid solution is lower than in the alloys of the binary Ni-Al system. So, the isolation of the hardening intermetallic γ'-phase occurs at lower aluminum concentrations in comparison with the binary system. Consequently, substantial hardening can be obtained with significant amounts of aluminum and this dramatically reduces the ability of alloys to deform.

A general increase in the amount for intermetallic phase leads to an increase in the working temperature of the product. However, in this case it is necessary to take into account the technically feasible reduction in the plasticity of the alloy that is observed with an increase in the amount for intermetallic compounds and the possibility of changing in morphology of the strengthening phase as well.

The diagram approach used in the structural-phase analysis of alloys is developed for binary systems including Ni-Cr and Ni-Al systems [3,4]. As for ternary systems, the Ni-Cr-Al system in particular, the interest in research in this area is insignificant and the published results are few in number but they come into notice [5-7].

For example, L. Kaufman and H. Nesor [5] consider that Ni_3Al, NiCr, CrAl and CrNi compounds can be a part of the hard alloys of the Ni-Cr-Al system and Ni_2Al and Cr_2Al_3 are present in two structural modifications differing in the type of crystalline lattice. The sections at 1500 and 1800 K are presented as heterophase systems consisting of two condensed phases- liquid, i.e. melt and solid metal.

It is known that iron atoms are able to replace nickel and aluminum atoms in Ni_3Al intermetallide [6]. The electronic structure of chromium and iron atoms differs insignificantly. Therefore, despite the low solubility of chromium in the intermetallic compound, its presence is possible by the principle of substitution. In the $Ni_{80}Cr_9Al_9W_2$ alloy the quantitative ratios of γ and γ'-phases can be indeed represented as 0.05 and 0.95 respectively. The alloy $Ni_{64}Cr_{27}Al_9W_2$ containing less nickel but more chromium has a phase ratio close to the previous composition [6].

The results of research [7] indicate the presents of alloys of more complex composition. The authors of this paper indicate the possibility of existence in the Ni-Cr-Al system of atomic compositions of the type $Al_{81}Ni_3Cr_{16}$, $Al_{76,5}Ni_3Cr_{20,5}$, $Al_{71}Ni_9Cr_{19,5}$ etc along with binary alloys known as phases γ ', α, β, μ, σ, θ.

Thus, combinational manifold of atomic clusters of the Ni-Cr-Al system indicates their considerable disequilibrium and inhomogeneity. The characteristic features of short-range order as a certain regularity in the arrangement of neighboring atoms are preserved during melting process and have a hereditary effect on the structure and properties of the melt [1,8-10]. Therefore, the information on duration of the period or the effective relaxation temperature of the structure of melt is needed during the design process of the melting modes of industrial high-temperature alloys. Data on alloys synthesized in the laboratory can serve as a guide.

2. Materials and methods

Modern methods of density studies by analyzing the intensity of penetrating gamma radiation, magnetic susceptibility and electrical resistivity are successfully used for research of physical properties of metals and alloys in a wide temperature range and solid-liquid phase transition. The latter method is the most informative, since it allows to determine contactless the conductivity type and the beginning of the structural ordering of intermetallides.

The electrical resistivity of the studied alloys has been investigated in a rotating magnetic field by the contactless method. The measurement procedure is described in the research [8] in sufficient detail.

Taking into account the results of studies of binary alloys [3,4], the compositions Ni + 10 and + 15 wt. % Cr each is aluminum alloyed in an amount of 5 and 10% by weight.

The samples used for electrical resistance studies have been melted in a vacuum induction furnace in an argon atmosphere at a temperature of 1500°C. The reliability of the compositions is confirmed by atomic absorption analysis. The concentration of gases in the various samples has differed insignificantly and amounted to ~ 0.003 wt. % [O] and ~ 0, 001% by weight [N].

The duration of structural relaxation has not exceeded 7 minutes. Proceeding from this, during the measurement of ρ the exposure period at each temperature has been established within 10-12 minutes.

3. Experiment and discussion

As results of the experiments it has been shown that, the form of the temperature dependence of the resistivity ρ (t) is a characteristic of high-temperature alloys [1,9].

A typical form of the ρ (t) dependence is shown in figure 1. The general patterns inherent in the studied high-temperature nickel alloys are as follows:

- sigmoid character of the location of polytherms during heating and cooling, branching or hysteresis of polytherms;

- the interval of thermal stability (Δtts) of the primary structure of the melt located between the liquidus temperature (tl) and the abnormal temperature (tan), i.e. Δtts = tan - tl and fluctuated within the limits of 50-260°C;
- abnormal temperature (tan) - the beginning of intensive restructuring of the melt structure;
- critical temperature (tc) – the temperature, heating to which leads to irreversible restructuring of the structure of the melt;
- the temperature range of the intensive reorganization of the melt structure – Δtirs is located between the temperature of the anomaly and the hysteresis temperature of the polytherm, i.e. Δtirs = th - tan (th -temperature of hysteresis);
- coefficient of intensive restructuring of the melt structure Cirs = Δρh / Δtirs, where Δρh is the hysteresis value of the polytherm ρ for th; the index of Cirs is used as an analog of the activation energy for express analysis of the structural rearrangement process.

The phenomenon of hysteresis of polytherm ρ (t) is observed only when a certain melt temperature is reached usually corresponded with the initial stage of a sharp increase in the electrical resistivity (tan) and stabilized at a temperature (th) (figure 1).

4. Results and discussion

The study results of temperature dependence of physical properties of the PG-SR1 alloy are given in figure 1.

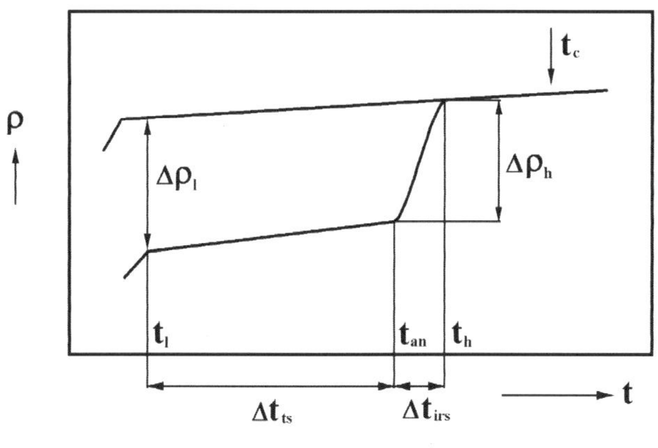

Figure 1. Typical form of the temperature dependence of the electrical resistivity (ρ) of high-temperature nickel alloys.

The temperature dependences of the electrical resistivity of Ni-Cr-Al alloy system are shown in figure 2. All of them also have a form of characteristic of high-temperature nickel alloys.

It is found that the most thermally stable is the structure of the alloy containing Ni + 10% Cr + 5% Al as evidenced by the lowest value of the coefficient dρ / dt.

Aggregate of data indicates that the initial state of the melt is microinhomogeneous, the atoms of the doping components form independent structural groups that inherit the features of the solid state. After heating the melt to ta1 its structure becomes more homogeneous which is apparently due to the destruction of nonequilibrium associations and the formation of a melt based on liquid nickel.

The well-known information on the dependence of the critical temperature (t_c) and hysteresis of polytherms (t_h) from the degree of alloying of the samples has been also confirmed. It has been established that the greater the alloy doping the lower the critical temperature heating to which increases the equilibrium and microhomogeneity of the melt structure and the lower the hysteresis temperature of the polytherm ρ (figure 2).

IOP Conf. Series: Materials Science and Engineering **209** (2017) 012009 doi:10.1088/1757-899X/209/1/012009

Upon repeated examination of noncrystallized samples, the polytherms ρ of the second measurement cycle has coincided with the cooling polytherm of the first cycle. However, cooling to room temperature has passed through the stage of γ'-phase separation at temperatures of 1100-1050°C which has had an effect on the formation of the solid metal structure. Upon repeated examination of such samples, the polythermal heating-cooling hysteresis has been observed but the hysteresis value has not exceeded 20% of the Δρ of the first study cycle.

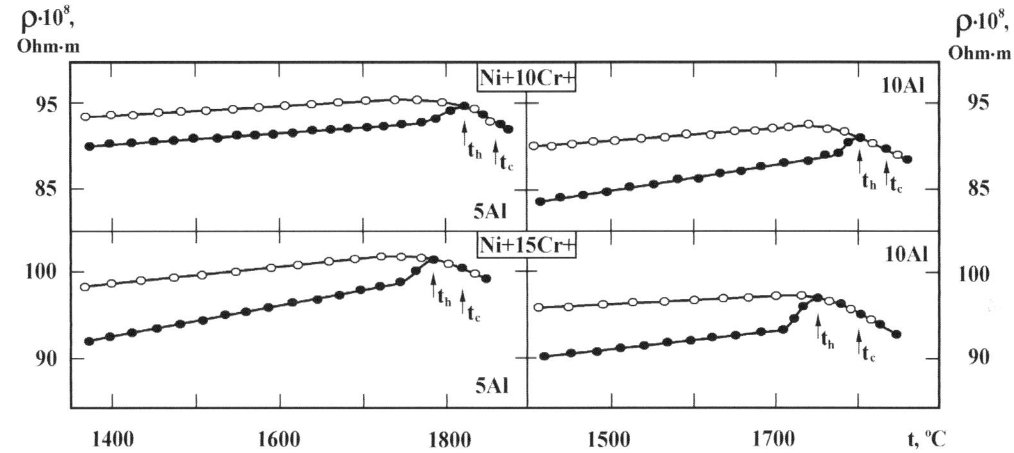

Figure 2. Temperature dependences of the electrical resistivity of Ni-Cr-Al alloy system.
• - heating; o - cooling.

The influence of the heating temperature of the Ni + 15Cr + 10Al alloy on the supercooling value $\Delta t = t_l - t_c$, the parameters of the dendritic cell of the cast metal d_{II} and the ductility for hot deformation δ are shown in figure 3 and indicates the essential role of the state of the structure of the melt before solidification.

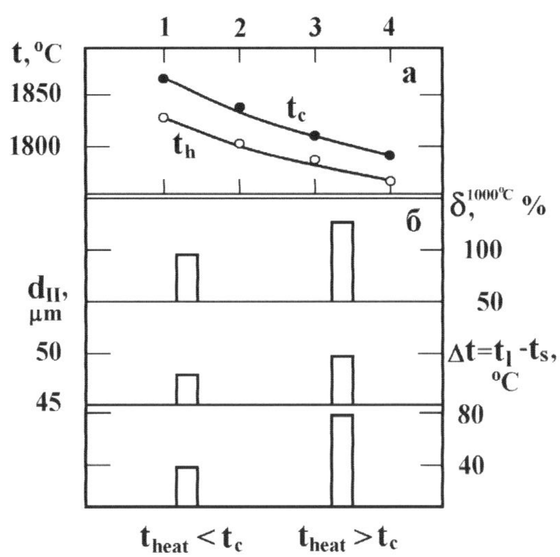

Figure 3. The effect of the alloy composition on the critical temperature (t_c) and the hysteresis temperature of the polytherm (t_h) of the electrical resistivity (a) and the effect of the melting temperature on the supercooling interval Δt, the parameter of the dendritic cell of the cast metal d_{II} and the relative elongation of δ at 1000°C temperature of the Ni + 15% Cr + 10% Al alloy (b).
1 - Ni +% Cr + 5% Al;
2 - Ni + 10% Cr + 10% Al;
3 - Ni + 15% Cr + 5% Al;
4 - Ni + 15% Cr + 10% Al.

5. Conclusion

The results of the temperature dependences research of the electrical resistivity has shown that after melting the initial state of the Ni-Cr-Al melt has been microinhomogeneous, the atoms of the doping components have formed independent structural groups inheriting the solid-state signs. After heating the melt above the hysteresis temperature, its liquid structure has become more homogeneous which is apparently due to the destruction of nonequilibrium associations and the formation of a solution based on liquid nickel.

It has been established that the higher alloy doping the lower the critical temperature, the heating up to which increases the equilibrium and microhomogeneity of the melt structure and the lower the hysteresis temperature of the polytherm of electrical resistivity.

References

[1] Tyagunov G, Baryshev E et al 2016 *Metallic liquids. Steels and alloys* (Yekaterinburg: Ural Federal University Press) p 268
[2] Beskorovayny N, Kalinin B and Platonov P 1995 *Structural materials of nuclear reactors* (Moscow: Energoatomizdat) p 704
[3] Tyagunov A, Vyukhin V, Baryshev E, Tyagunov G and Akshentsev Y 2016 *Steel Trans.* **46** 558-562
[4] Tyagunov A, Vyuhin V, Baryshev E 2015 *Bull. South Ural State University Metallurgy* **15** 51-56
[5] Kaufman L, Nesor H 1974 *Metall. Trans.* **6** 1623-29
[6] Buršik J, Brož P, Picha R 2003 *Intermetallics* **11** 483-90
[7] Grushko B, Kowalski W, Pavlychkov D , Surowiec M 2008 *J. Alloys Compd* **460** 299-304
[8] Tyagunov A, Baryshev E, Tyagunov G, Shmakova K and Tsepelev V 2017 *Acta Metall. Slovaca* **23** 79-86
[9] Tyagunov A, Baryshev E, Tsepelev V, Kostina T, Baum B , Savin O 1996 *Rasplavi* **6** 26-28
[10] Tyagunov A, Vyukhin V, Baryshev E, Tyagunov G , Kostina T Russ. *Metall.y* **12** 998-1001

Acknowledgments

This work has been supported by the Scientific researchers of higher education institutions within the State task of the Russian Federation No.4.5533.2017БЧ.

IOP Conf. Series: Materials Science and Engineering **209** (2017) 012010 doi:10.1088/1757-899X/209/1/012010

Micro-Encapsulated Porphyrins and Phthalocyanines – New Formulations in Photodynamic Therapy

R M Ion[1,2*]

[1] Nanomedicine Department, ICECHIM, 202 Splaiul Independentei, Buicharest-060021, Romania

[2] Department of Engineering Doctoral Studies, Valahia University, 13 Aleea Sinaia, Targoviste, Romania

E-mail: rodica_ion2000@yahoo.co.uk

Abstract. Photodynamic therapy (PDT), as an innovative method for cancer tretament is based on a concerted action of some drugs, called sensitizers, which generate reactive oxygen species via a photochemical mechanism, leading to cellular necrosis or apoptosis. The present work aims at loading some sensitizers, as porphyrins (P) and phthalocyanines (Pc) into alginate particles. Particles were prepared by dropping alginate into an aqueous solution containing P or Pc and $CaCl_2$, which allows the formation of particles through ionic crosslinking. It was obtained P or Pc loaded alginate beads with an average diameter of about 100 μm. For these systems, this paper analyses the spectroscopic properties, encapsulation into microcapsules, controlled releasing action and their photosensitizer capacity (singlet oxygen generation).

1. Introduction

In photodynamic therapy (PDT), as one of the most efficient methods for cancer treatment, the photosensitizer (PS) used as drug, should be administered and subsequently activated by light of a specific wavelength, leading to selective damage of the tumoral tissue [1].

Due to their essential role in PDT, and to their reduced stability, the photosensitizers should be incorporated in different delivery systems, such as hydrogels and liposomes, as some of the potential strategies to improve their efficiency in various cancer type [2,3].

Porphyrins are the most important sensitizers for PDT, due to its strongest absorption band in the region 600-800 nm, which is recognized as "therapeutic window", which enhances their selectivity [4]. In addition, the phthalocyanines as the second - generation PSs similar to porphyrins, have superior photophysical and photochemical properties [5], could be easily synthesized, have good photo- and chemical-stability, long-wavelength absorption in the 650–750 nm region and high singlet oxygen quantum yields [6].

Polymers offer several advantages in PDT due to their ability to deliver a large amount of PS to the target area, flexibility toward surface modification for better efficiency, and the possibility of being loaded in specific position [7,8]. Polymers are playing an important role in the preparation of nanoparticles and being used widely, and natural polymers composed of polysaccharides such as chitosan and alginate represent an important issue [9,10]. Normally a particle with 10-500 nm size is considered nanoparticle and the classification of nanoparticles in nanocapsules and nanospheres has been evolved [11].

Alginate is a naturally occurring linear unbranched polysaccharide extracted from brown seaweed, such as *Laminaria hyperborea, Ascophyllum nodosum* and *Macrocystis pyrifera*. Alginates are composed of (1-4)-linked β-R-mannuronic acid (M units) and α-Lguluronic acid (G units) monomers [12]. Sodium alginate is a sodium salt of alginic acid, a naturally occurring polysaccharide obtained from marine brown algae, figure 1.

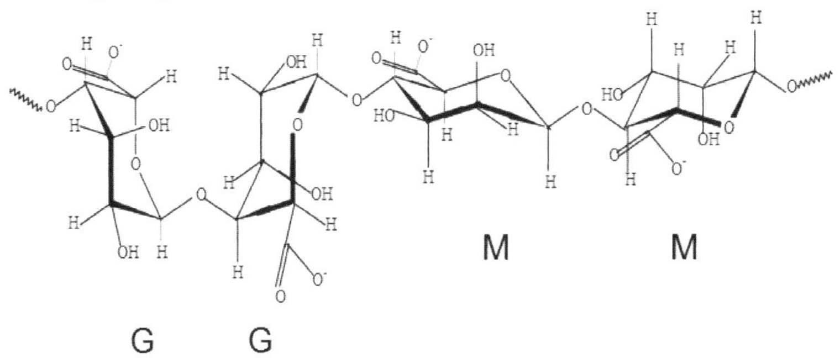

Figure 1. The structure of alginate.

Alginate can easily be gelled with different cations by the stacking of guluronic acid (G) blocks with the formation of "egg-box" calcium linked junctions (figure 2). The alginate particles are formed through external gelation and typically have a diameter of 100 –5000 μm [13].

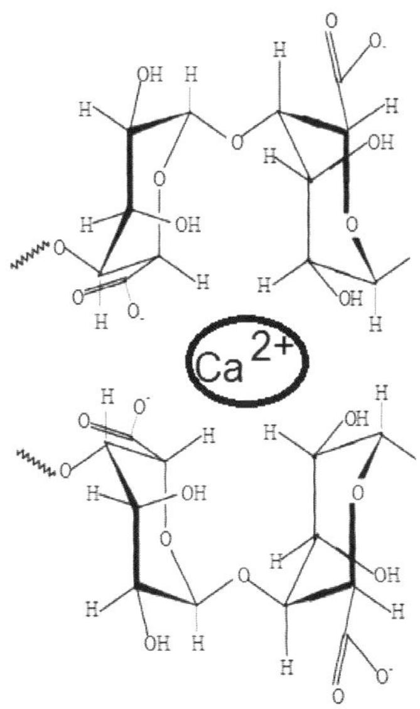

Figure 2. The strcuture of the complex "egg-box" calcium linked junctions.

The use of alginate based drug delivery systems presents many advantages mainly related to the encapsulation process which takes place at room temperature and without the use of organic solvents [14].

In this perspective, the present work is aimed at the preparation and characterization of alginate based polymeric particles loaded with different porphyrins, a free base (TSPP) and its metallic complex (ZnTSPP); and a free base phthalocyanine (TSPc) and its metallic complex (ZnTSPc).

2. Materials and methods

2.1. Materials
Sodium Alginate (250 cps at 25°C, 2% solution), were purchased from Sigma Aldrich. Calcium Chloride was purchased from Carlo Erba. Porphyrins: TSPP (5,10,15,20-p-tetra-sulphonated-porphyrin) and its metallic complex (ZnTSPP) and phthalocyanine TSPc (2,9,16,23-tetra-sulphonated-phthalocyanine) and its metallic complex ZnTSPc (zinc-3,4',4'',4'''- tetrasulfonated-phthalocyanine) have been synthesized in our laboratory, after literature receipts [15-17].

2.2. Instrumentation
UV-Vis absorption spectra measurements and the degree of porphyrins/phthalocyanines sorption has been monitored in the solution with a SPECORD M400 spectophotometer with monochromator and double beam. Optical microscopy has been achieved with a NOVEX 100 microscope by using proper magnitudes.

2.3. P/Pc Loaded Alginate Particles
P/Pc loaded alginate particles (beads and microparticles) were obtained by means of ionotropic gelation. Alginate was dissolved in physiological solution and mixed with P/Pc in distilled water. The resulting solution was added dropwise, using a syringe equipped with a 22G needle, to a water solution containing calcium chloride, under magnetic stirring.

2.4. Evaluation of P/Pc Encapsulation
The encapsulation efficiency of porphyrin or phthalocyanine was calculated by using formula:

$$EE\ (\%) = (P_{loading} - P_{filtration}) / P_{loading} \times 100$$

where: EE = encapsulation efficiency, %; $P_{loading}$ = total amount of P or Pc loading; $P_{filtration}$ = amount of P or Pc in ultrafiltrate (calculated after filtration with CHROMAFIL O-45/15 MS filters (Machinery-Nagel GmbH, Germany).

2.5. In Vitro Release of Porphyrin or Phthalocyanine
The nanoparticles in the filter were diluted with water solution of pH=5.5 to 2 mL and incubated at 37°C. The amount of extract released, was estimated from the ratio between the amount of porphyrin or phthalocyanine determined in the release medium and the initial amount of porphyrin or phthalocyanine in the nanoparticles.

2.6. Measurements of singlet oxygen ($^1O_2$) generation
Porphyrin or phthalocyanine (10 μM) were mixed with 9, 10-anthracenediyl-bis(methylene) dimalonic acid (ABDA) (20 μM) in water bi-distilled and placed in a quartz cuvette. The sample solutions were irradiated with polychromatic light from a Hg medium-pressure lamp (400 nm-800 nm) for 2 min and then the fluorescent emission of ABDA was measured at 431 nm (excitation wavelength = 380 nm). The destruction of ABDA is an indicator for the generation of singlet oxygen [18].

3. Results and discussion
The absorption spectra of porphyrins are the most useful technique for characterizing porphyrins. The presence of the conjugated double aromatic aromatic system in the porphyrin structure translates into a characteristic absorption spectrum in the visible field, porphyrins being highly colored substances [19]. In neutral solvents, the free base porphyrins, H_2P, exhibit a very high band (Soret band) around 400 nm in the absorption spectrum, and yet another series of four bands, marked IV, III, II, I in the 500- 700 nm [20].

Metalloporphyrins have an absorption spectrum characterized by a very intense Soret band between

380 and 420 nm (known as B (0,0) and by a two-band system in the 500-600 nm area, known to be composed of the bands β [Q (0,0)] and the α band [Q (1, 0)], which are 10-20 times less intense than the Soret band, depending of the central metal ion.

Tetrakis-4-sulfonato-phenyl-porphyrin (TSPP) is an anionic porphyrin, which possesses four negative charges the sulfonate groups from the four corners. In aqueous solutions, at neutral pH, the electronic absorption spectrum of TSPP is characterized by an intense Soret band at arround 420 nm and four Q bands in the 500-700 nm range. In acidic medium, new absorption bands (from 490, 707 nm) appear: J-aggregate (edge-to-edge interaction) at 490 nm and H-aggregate (face-to-face interaction) at 422 nm [21].

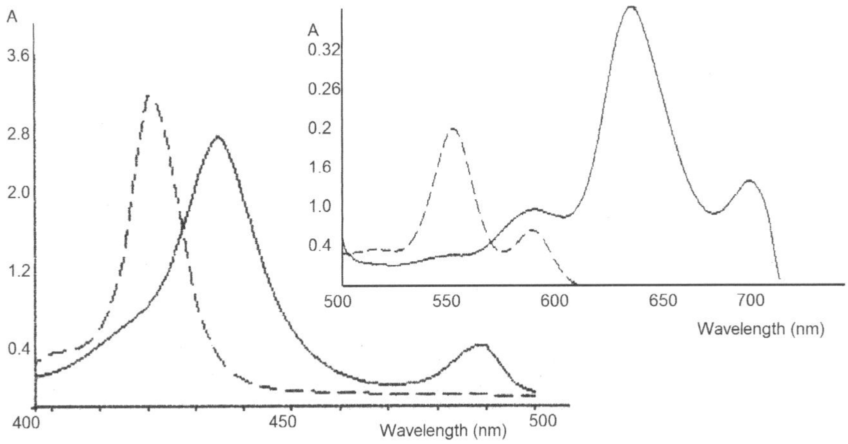

Figure 3. The absorption spectra of TSPP (full line) and ZnTSPP (dotted line).

The phthalocyanines show electronic spectra substantially different from those of the porphyrins. Phthalocyanines as expected from the extensively conjugated aromatic chromophore, exhibit UV-VIS absorption spectra with intense π–π* transitions, usually referred to as Q bands in the range 660-700 nm ($\varepsilon > 10^5 M^{-1} cm^{-1}$) [22].

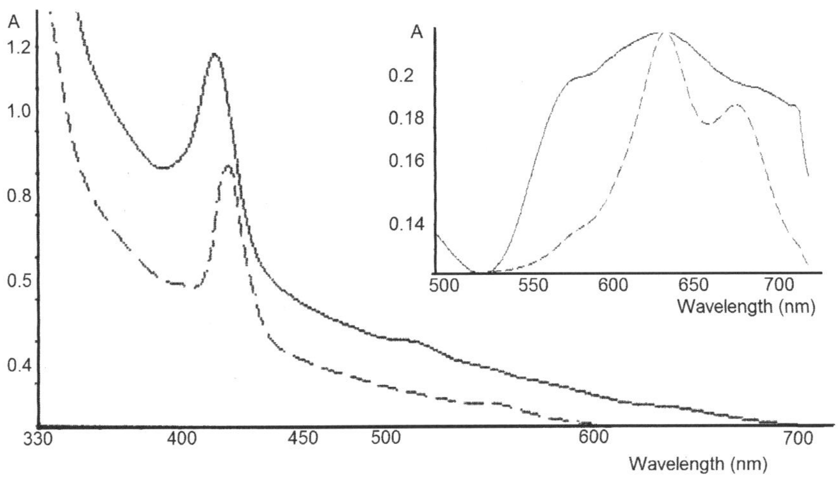

Figure 4. The absorption spectra of TSPc (full line) and ZnTSPc (dotted line).

By analyzing the obtained experimental results, some remarks could be formulated:

- The encapsulation efficiencies are completely different at free-base compounds (TSPP and TSPc): 1.11% for TSPP and 10.16% for TSPc;
- for metallo-complexes, the encapsulation efficiencies are higher than for free-base compounds: 17.32% for ZnTSPP, and 36.15% for ZnTSPc.

All the tested compounds have different releasing capacity, figure 5. Microparticles were obtained by ionic gelation lowering alginate concentration to 2% (w/v) and Calcium chloride to 2.5% (w/v) in agreement with literature data [23].

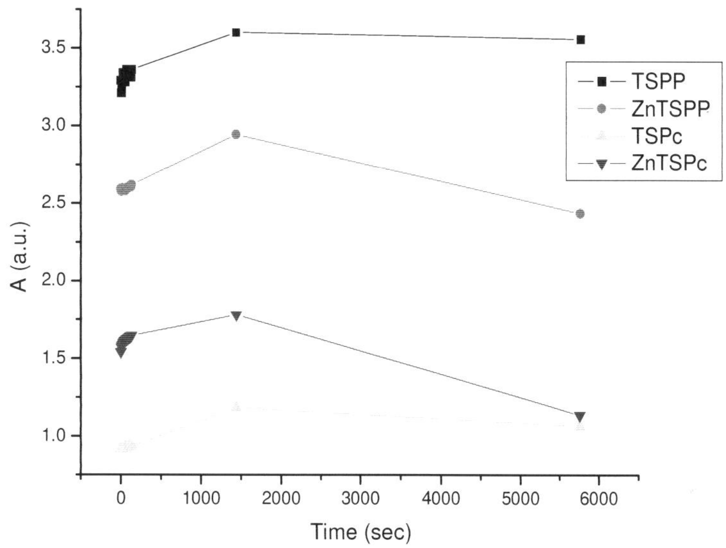

Figure 5. The controlled release of P and Pc from alginate micro-capsules.

These results could be very well correlated with their sizes, as follows:

- An initial burst release of the encapsulated P or Pc was observed at 2.5 h;
- The values of controlled release capacity are: 97.85% for TSPP, 99.23% for ZnTSPP, 97.75% for TSPc and 93.62% for ZnTSPc;
- Alginate alone is generating spherical micro-capsules with the size of 170.78 μm;
- With TSPP is generating larger sphere, because this porphyrins is able to aggregate, increasing its size from 108.07 μm at initial micro-capsules to 141.68 μm at the micro-capsules after 5 hours (assigned to a dimerisation process) and to 177.12 μm after 24 hours (assigned to aggregation process), and 55.45 μm after 96 hours (assigned to monomerisation process).
- After drying process, the micro-capsules became with unclear and damaged edges (106.44 μm), and release the monomers, which clear fitted visible (30 μm). TSPP is able to disaggregate and to slowly release from the capsules, generating monomers.
- The forms of TSPP could be identified by specific absorption maxima, too, figure 6. monomer form: Soret band at 415-416 nm; dimeric - aggregated form: 417.5 nm - 423.75 nm, in good agreement with literature reports [24];
- Similar behaviour has been observed at TSPc, but due to smaller size of this compound, it is able to encapsulate it more easily into capsules.

IOP Conf. Series: Materials Science and Engineering **209** (2017) 012010 doi:10.1088/1757-899X/209/1/012010

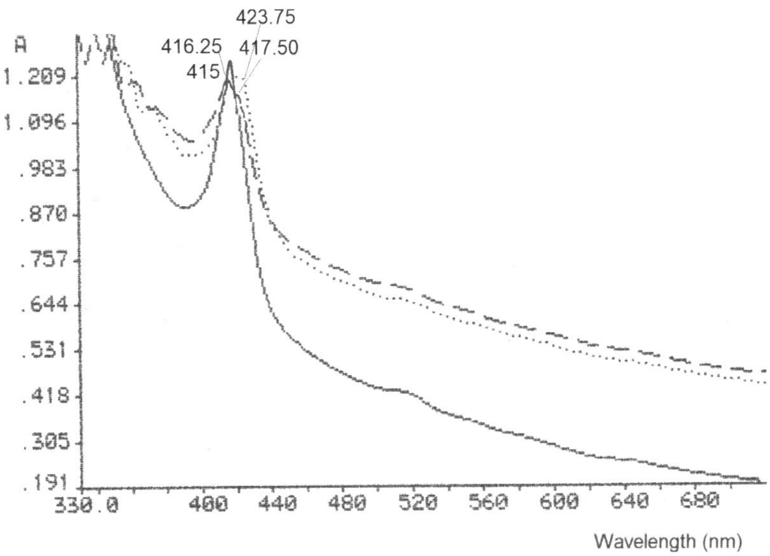

Figure 6. The changes of spectra for TSPP as micro-capsule: initial (full line), after 24 h (dotted line), and after 96 h (dried micro-capsule) (brushed line).

The aspects of TSPP encapsulated into alginate micro-capsules in different experimental conditions are visible in figure 7.

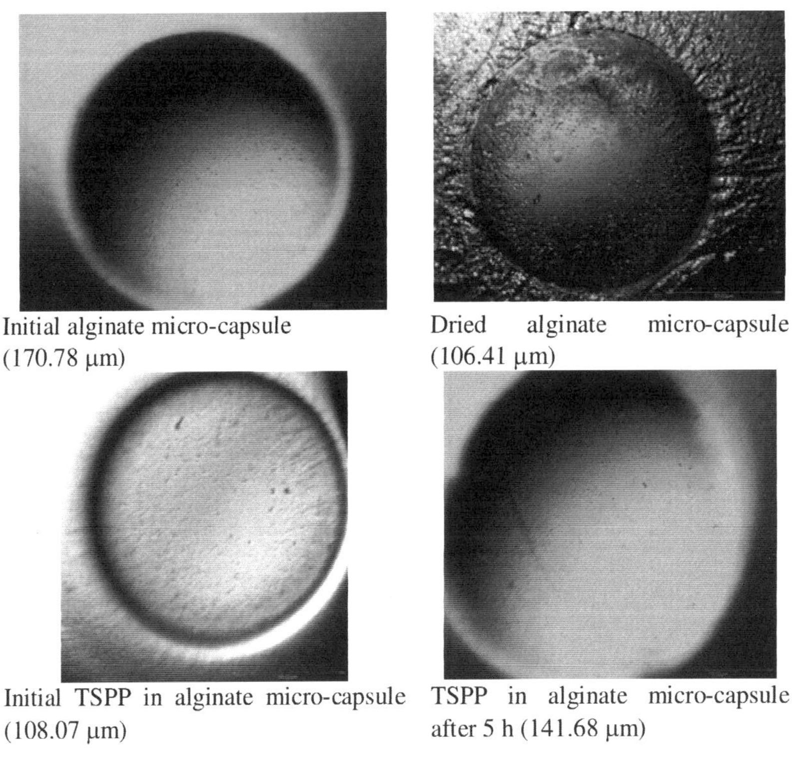

Initial alginate micro-capsule (170.78 μm)

Dried alginate micro-capsule (106.41 μm)

Initial TSPP in alginate micro-capsule (108.07 μm)

TSPP in alginate micro-capsule after 5 h (141.68 μm)

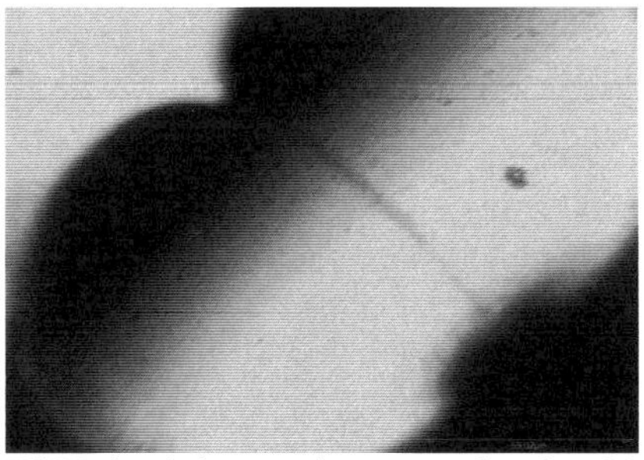

TSPP in alginate micro-capsule after 24 h (177.12 μm)

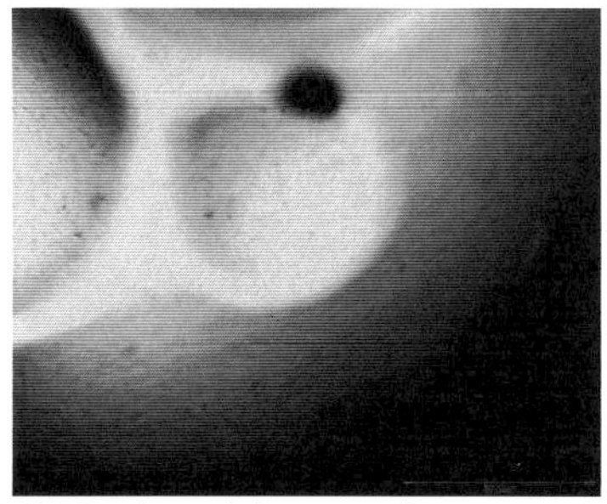

TSPP in alginate micro-capsule after 96 h (55.45 μm)

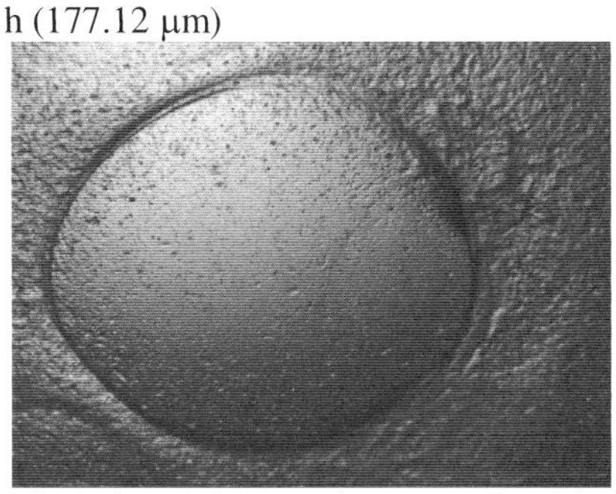

TSPP in alginate micro-capsule dried after 24 h (106.44 μm)

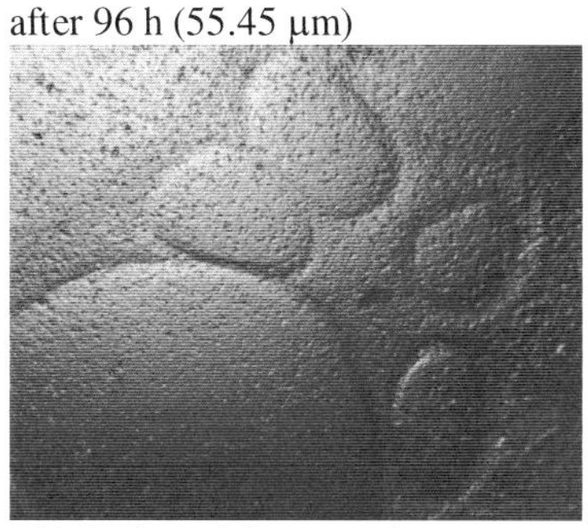

TSPP in alginate micro-capsule dried after 96 h (30.44 μm)

Figure 7. Optical microscopy for microcapsules with encapsulated P or Pc compounds.

To check whether P or Pc maintained its own functional properties after the loading into alginate based beads, singlet oxygen generation were performed on the studied compounds, table 1.

Table 1. Singlet oxygen generation in solution and in microcapsules.

Compound	φ [$^1O_2$]	φ [$^1O_2$]
	Water solution	microcapsules
TSPP	0.67	0.45
ZnTSPP	0.48	0.25
TSPc	0.11	NA
ZnTSPc	0.68	0.25

By light irradiation, has been observed the ability of the porphyrins or phthalocyanines to induce the formation of singlet oxygen and the subsequent controlled release, followed by the disruption of the microcapsules. The obtained results are partially similar with other literature reports for porphyrins as TSPP [25], but not for Pcs. In the present study, Pcs showed not very promising results, they are too light sensitive and unstable. The same has been observed for ZnTSPP, which is very unstable.

4. Conclusions

The use of alginate for the production of P or Pc loaded particles showed interesting results. The studied formulation clearly demonstrated that the proposed system posses good properties of alginate particles for therapeutic applications. The porphyrins or phthalocyanines are able to be encapsulated into microcapsules and to induce the formation of singlet oxygen, and the subsequent controlled release, followed by the subsequent disruption of the microcapsules.

References

[1] Ion R M 2012, *Photosensitizers in Medicine, Environment, and Security,* T.Nyokong; Ahsen, Vefa (Eds.), 1st Edition., Springer Ed., pp.315-349

[2] Varga S, Patachia S, and Ion R M 2007 *Pollack Periodica* **2**(2) 131-140

[3] Galateanu B, Hudita A, Negrei C, Ion R M, Costache M, Stan M, Nikitovic D, Hayes A W, Spandidos D A, Tsatsakis A M, and Ginghina O 2016 *International journal of oncology* **48** (6), 2295-2302

[4] Baldea I, Olteanu D E, Bolfa P, Tabaran F, Ion R M, Filip G A 2016, *Journal of Photochemistry and Photobiology B: Biology* **161** 402-410

[5] Allen C M, Sharman W M, and van Lier J E 2001 *J. Porphyrins Phthalocyanines* **5** 161–169.

[6] Tudor D, Nenu I, Filip G A, Olteanu D, Cenariu M, Tabaran F, Ion R M 2017 *PloS one* **12**(3), 0173241

[7] Petrisor G, Ion R M, Brachais C H, Boni G, Plasseraud L, Couvercelle J P, Chambin O 2012 *Journal of Macromolecular Science, Part A: Pure and Applied Chemistry* **49**(9) 699-705

[8] Petrisor G, Ion R M, Brachais C H., Couvercelle J P, Chambin O 2012 *Journal of Macromolecular Science, Part A: Pure and Applied Chemistry* **49**(5) 439-444

[9] Stoica R, Pop S F, Ion R M, 2013 *J.Optoelec. Adv.Mat.,* **15**(7-8) 893-898

[10] Stoica R, Somoghi R, Doncea S M, Ion R M 2017 *J.Optoelec. Adv.Mat.,* **11**(1-2) 113-118

[11] Susan M, Baldea I, Senila S, Macovei V, Dreve S, Ion R M 2011 *International Journal of Dermatolog,* **50**(3) 280-286

[12] Sachan N K, Pushkar S, Jha A, Bhattcharya A 2009 *J Pharm Res* **8**:1191-1199

[13] Dessy A, Piras A M, Chiellini F 2013 *J Biomim Biomater Tissue Eng* **18** 102-110

[14] Reis C P, Neufeld R J, Vilela S, Ribeiro A J, Veiga F 2006, *J Microencapsule* **23** 245-257

[15] Clichici S, Filip A, Daicoviciu D, Ion R M, Mocan T, Tatomir C 2010 *Acta Physiologica Hungarica* **97**(1) 41-51

[16] Siejak A, Wróbel D, Siejak P, Olejarz B, Ion R M 2009 *Dyes and Pigments* **83**(3) 281-290

[17] Neagu M, Manda G, Constantin C, Radu E, Ion R M 2007, *Journal of Porphyrins and Phthalocyanines* **11**(1) 58-65

[18] Qian H S, Guo H C, Ho P C L, Mahendran R, Zhang Y 2009 *Small* **5** 2285-2290

[19] Falk J E 1964 *Porphyrins and Metalloporphyrins*, Amsterdam, pp.129-141

[20] Hambright P 1975 *Porphyrins and Metalloporphyrins,* Elsevier, Amsterdam, pp.248

[21] Ion R M, Planner A, Wiktorowicz K, Frackowiak D 1998 *Acta Biochimica Polonica* **45**(30) 833-845

[22] Lukyanets E A 1999 *J.Porphyrins Phthalocyanines* **3** 424-432

[23] Rajaonarivony M, Vauthier C, Couarraze G, Puisieux F, Couvreur P 1993 *J Pharm Sci* **82** 912-917

[24] Ion R M 1999 *J. Biomed. Optics* **4**(3) 319-326

[25] Bédard M F, Sadasivan S, Sukhorukov G B and Skirtach A 2009 *J. Mater. Chem.* **19** 2226-2233

Acknowledgments

This paper received the financial support from MEN-UEFISCDI, through the projects: PNII 185/2014, PNIII 120 BG/2016 and PN 16.31.02.04.03.

IOP Conf. Series: Materials Science and Engineering **209** (2017) 012011 doi:10.1088/1757-899X/209/1/012011

Understanding AlN Obtaining Through Computational Thermodynamics Combined with Experimental Investigation

R M Florea[1]

[1]"Gheorghe Asachi" Technical University of Iasi, Department of Materials Science and Engineering, Prof. Dr. D-tru Mangeron Street, Iasi, Romania, ZIP 700050

E-mail: raluca.m.florea@gmail.com

Abstract. Basic material concept, technology and some results of studies on aluminum matrix composite with dispersive aluminum nitride reinforcement was shown. Studied composites were manufactured by „in situ" technique. Aluminum nitride (AlN) has attracted large interest recently, because of its high thermal conductivity, good dielectric properties, high flexural strength, thermal expansion coefficient matches that of Si and its non-toxic nature, as a suitable material for hybrid integrated circuit substrates. AlMg alloys are the best matrix for AlN obtaining. Al_2O_3–AlMg, AlN–Al_2O_3, and AlN–AlMg binary diagrams were thermodynamically modelled. The obtained Gibbs free energies of components, solution parameters and stoichiometric phases were used to build a thermodynamic database of AlN–Al_2O_3–AlMg system. Obtaining of AlN with Liquid-phase of AlMg as matrix has been studied and compared with the thermodynamic results. The secondary phase microstructure has a significant effect on the final thermal conductivity of the obtained AlN. Thermodynamic modelling of AlN–Al_2O_3–AlMg system provided an important basis for understanding the obtaining behavior and interpreting the experimental results.

1. Introduction

There are more than 50,000 materials available to engineers for the design and manufacturing of products for various applications. These materials range from ordinary materials (e.g., copper, cast iron, brass), which have been available for several hundred years, to the more recently developed, advanced materials (e.g., composites, ceramics, and high-performance steels). Due to the wide choice of materials, today's engineers are posed with a big challenge for the right selection of a material and the right selection of a manufacturing process for an application [1].

The development of composites is given by the possibility of taking advantage of particular properties of the constituent materials to meet specific demands. It can say that a composite is a material made with several different constituents intimately bonded. Composites definition is very large, and includes a lot of materials such as the Roman ways (constituted of different layers of stones, chalk and sand), wood, human body etc. Industries and materials scientists use a more restrictive definition: a composite material consists of elements produced via a physical combination of pre-existing ingredient materials to obtain a new material with unique properties when compared to the monolithic material properties [2]. This definition distinguishes a composite from other multiphase materials which are produced by bulk processes where one or more phases result from phase transformation ("in situ" composites).

Composite materials have been utilized to solve technological problems for a long time but only in the 1960s did these materials start capturing the attention of industries with the introduction of metallic-based composites. Since then, composite materials have become common engineering materials and are designed and manufactured for various applications including automotive components, sporting goods, aerospace parts, consumer goods, and in the marine and oil industries [3]. The growth in composite usage also came about because of increased awareness regarding product performance and increased competition in the global market for lightweight components.

Among all materials, composite materials have the potential to replace widely used steel, and many times with better performance. Replacing steel components with composite components can save 60 to 80% in component weight. Today, it appears that composites are the materials of choice for many engineering applications.

2. Materials and methods

Fabrication techniques of metal matrix composites are depending on whether they are aimed at continuously or discontinuously reinforced MMC production. According to whether they are primarily based on treating the metal matrix in a liquid or a solid form, the techniques can further be subdivided. An important influence on the type of component to be produced, on the micro-structures, on the cost and the application of the MMC is given by the production factors. Special classes that will be discussed in this paper are the "in situ" composites.

2.1 Materials Selection

The most popular matrix for the metal matrix composites (MMCs) is aluminum. The Al alloys are quite attractive due some characteristics: low density, capability to be strengthened by precipitation, good corrosion resistance, high thermal and electrical conductivity, and high damping capacity. Aluminum matrix composites (AMCs) have been widely studied since the 1920s and are now used in automotive industries, electronic packaging, armors and sporting goods. They offer a large variety of mechanical properties depending on the chemical composition of the Al-matrix. They are usually reinforced by Al_2O_3, SiC, SiO_2, B, BN, B_4C, AlN. The aluminum matrices are in general Al-Si, Al-Cu, Al-Mg, 2xxx or 6xxx alloys [4].

The aluminum matrix composites may offer specific advantages (and disadvantages) compared to unreinforced Al alloys, to ceramic matrix composites and to polymer matrix composites. An overview is given in table 1.

Table 1. Aluminum Matrix Composites comparison of Advantages and Disadvantages.

ADVANTAGES	*DISADVANTAGES*
Compared to un-reinforced aluminum alloys:	
- higher specific strength - higher specific stiffness - improved high temperature creep resistance - improved wear resistance	- lower toughness and ductility - more complicated and expensive production method
Compared to ceramic matrix composites:	
- higher toughness and ductility - ease of fabrication - lower cost	- inferior high temperature capability
Compared to polymer matrix composites:	
- higher transverse strength - higher toughness - better damage tolerance - improved environmental resistance - higher thermal and electrical conductivity - higher temperature capability	- less developed technology - smaller data base of properties - higher cost

IOP Conf. Series: Materials Science and Engineering **209** (2017) 012011 doi:10.1088/1757-899X/209/1/012011

This article is concerned with metal matrix composites and more specifically on the aluminum alloys matrix composites (AMCs). In AMCs one of the constituent is aluminum alloy (AlMg), which forms percolating network and is termed as matrix phase. The other constituent is embedded in this aluminum alloy matrix and serves as reinforcement, which is usually non-metallic and commonly ceramic such as AlN and Al_2O_3.

Aluminum Nitride, formula AlN, is a newer material in the technical ceramics family. While its discovery occurred over 100 years ago, it has been developed into a commercially viable product with controlled and reproducible properties within the last 20 years. Aluminum nitride has a hexagonal crystal structure and is a covalent bonded material. The use of sintering aids and hot pressing is required to produce a dense technical grade material [5]. The material is stable to very high temperatures in inert atmospheres. In air, surface oxidation begins above 700°C. A layer of aluminum oxide forms which protects the material up to 1370°C. Above this temperature bulk oxidation occurs. Aluminum nitride is stable in hydrogen and carbon dioxide atmospheres up to 980°C. The material dissolves slowly in mineral acids through grain boundary attack, and in strong alkalis through attack on the aluminum nitride grains. The material hydrolyzes slowly in water. Most current applications are in the electronics area where heat removal is important. This material is of interest as a non-toxic alternative to beryllium. Metallization methods are available to allow AlN to be used in place of alumina and BeO for many electronic applications [5]. Aluminum nitride engineering properties are given in table 2.

Table 2. AlN Engineering Properties.

Properties	*Units of Measure*	*SI/Metric*
Mechanical		
Density	gm/cc	3.26
Porosity	%	0
Elastic Modulus	GPa	330
Compressive Strength	MPa	2100
Hardness	Kg/mm^2	1100
Thermal		
Thermal Conductivity	W/m•°K	140–180
Coefficient of Thermal Expansion	10^{-6}/°C	4.5
Specific Heat	J/Kg•°K	740
Electrical		
Volume Resistivity	ohm•cm	$>10^{14}$

Table 3. 99.5% Al_2O_3 Engineering Properties.

Properties	*Units of Measure*	*SI/Metric*
Mechanical		
Density	gm/cc	3.89
Porosity	%	0
Elastic Modulus	GPa	375
Compressive Strength	MPa	2600
Hardness	Kg/mm^2	1440
Thermal		
Thermal Conductivity	W/m•°K	35
Coefficient of Thermal Expansion	10^{-6}/°C	8.4
Specific Heat	J/Kg•°K	880
Electrical		
Volume Resistivity	ohm•cm	$>10^{14}$

Alumina is one of the most cost effective and widely used materials in the family of engineering ceramics. The raw materials from which this high performance technical grade ceramic is made are readily available and reasonably priced, resulting in good value for the cost in fabricated alumina shapes. With an excellent combination of properties and an attractive price, it is no surprise that fine grain technical grade alumina has a very wide range of applications [6]. Aluminum oxide, commonly referred to as alumina, possesses strong ionic interatomic bonding giving rise to its desirable material characteristics. It can exist in several crystalline phases which all revert to the most stable hexagonal alpha phase at elevated temperatures. Alpha phase alumina is the strongest and stiffest of the oxide ceramics. Its high hardness, excellent dielectric properties, refractoriness and good thermal properties make it the material of choice for a wide range of applications.

Aluminum Oxide, Al_2O_3 engineering properties are given in table 3.

IOP Conf. Series: Materials Science and Engineering **209** (2017) 012011 doi:10.1088/1757-899X/209/1/012011

2.2. "In situ" processing of AlMg-AlN composites

There are several different processes that would fall under this category including liquid-gas, liquid-solid, liquid-liquid and mixed salt reactions. In these processes refractory reinforcement are created in the aluminum alloy matrix. One of the examples is directional oxidation of aluminum. In this process the alloy of Al–Mg is placed on the top of ceramic preform in a crucible. The entire assembly is heated to a suitable temperature (1000°C) in the atmosphere of free flowing nitrogen bearing gas mixture. Al–Mg alloy soon after melting infiltrates into the preform and composite is formed.

Gas–liquid reaction (eq. 1 and eq. 2) is also utilized to produce AlN reinforced aluminum matrix composites:

$$2N_2 + 4Al \rightarrow 4AlN \tag{1}$$

$$2N_2 + 6Mg \rightarrow 2Mg_3N \tag{2}$$

A major limitation of "in situ" technique is related to the thermodynamic restrictions on the composition and nature of the reinforcement phase that can form in a given system, and the kinetic restrictions on the shape, size and volume fraction of the reinforcement that can be achieved through chemical reactions under a given set of test conditions.

3. Experimental procedure

The matrix (AlMg15) was prepared from aluminum and magnesium of high purity (>99.9%) by conventional casting technique. Metals were melted in graphite crucibles at 1000°C under argon gas to reduce any contamination. The experiments where carried on in a sealed reaction chamber which was placed in a vertical electrical furnace. After closing the cover, argon was forced into the chamber. Afterwards, the gas was pumped out from the chamber using a vacuum pump, thus lowering the pressure below atmospheric. In order to create a protective atmosphere in the chamber this process was repeated twice. After the matrix had been melted, argon was pumped out at a temperature of 850°C and nitrogen was introduced into the chamber. The matrix alloy was heated at 1000°C and held for 1 hour.

The growth of AlN is predominantly based on capillary mass-transfer phenomena, which occur through a network of micro-channels developed in the empty space among the AlN (columnar crystals) as well as the hollows inside these crystals [7]. N_2 diffusion in the liquid matrix is the rate-determining steps of the process. Mg and Mg_3N_2 resulting after nitridation maintain a very low partial pressure of oxygen during the entire process and favor the AlN formation.

The shape, the orientation of AlN crystals with the liquid metallic droplets at the top of the crystals indicate that mass transfer of liquid matrix through the network of micro-channels is the key point of the entire process, as schematically depicted in figure 1.

According to figure 1, after the formation of the first AlN crystals, capillary forces govern the transfer of liquid matrix to the surface through the micro-channels. In the opposite direction, nitrogen diffuses into the liquid matrix. The Al mass transfer may also occur via surface diffusion mechanism on the AlN walls, which at the investigated temperature, may also result in a significant Al mass transfer towards the reaction front. However, the occurrence of capillary phenomena should be beyond any doubt because at the investigated temperature the viscosity of the matrix should be very low while the size of the micro-channels fits well the size of capillaries [6]. This claim is supported in conjunction with the fact of the good wetting regime observed between the molten alloy and the ceramic crystals of AlN.

AlN crystals should be precipitated immediately after the concentration of the dissolved N_2 in the AlMg matrix reaches the saturation point, since thermodynamics (at 1000°C, $\Delta G^{\circ}_{f,AlN}$ = 179,84 kJ/mol [8]) and experimental results indicate high chemical affinity between Al and N and fast reaction kinetics, respectively.

The formation of AlN evidently enriches the liquid matrix in the doping metals. The experimental results showed, however, that Mg is rapidly and intensively evaporated.

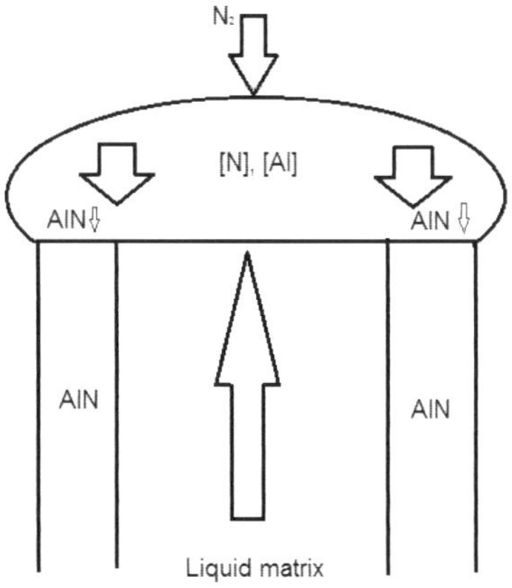

Figure 1. Schematic representation of the mechanism of the AlN crystals production and growth during "in situ" process of AlMg/AlN composite.

4. Results and discussion

4.1. Thermodynamic modeling of AlMg/AlN system

The development of thermodynamic databases presents an increasing demand. A simple model is based on speculation about ionic structure of the phases [9]. The thermodynamic data storage, retrieval and manipulation require accurate analytical representation of thermodynamic properties of solutions. Values of the standard Gibbs energies, G°, of each component are entered and stored in the solution database along with parameters which define the Gibbs energy of mixing according to Kohler/Toop polynomial model. The ternary excess Gibbs energy was calculated from a binary data interpolation. Binary terms of the polynomial expansion for G^E were entered as Redlich–Kister or Legendre polynomials due to their simplicity and the fact that their functional form is consistent with the empirical observations of solutions behavior [10].

The effect of doping elements from Al alloy matrix during its melting and nitridation was clarified through many investigations [11]. The presence of magnesium lowers the melting point of Al from 660°C to 635°C. This temperature decrease is reflected by the formation of solid solution between Mg and Al. During the melting of this material, the molten alloy was covered with an oxide film. Magnesium vaporized and its vapor caused rupture of the protecting oxide film [12]. Accordingly interruption of the oxide film would expose molten metals at the surface to N_2 atmosphere and nitridation took place successfully. Also, [13] indicated that the presence of Mg deoxidized the particles surfaces above 400°C to allow Al surfaces to react with N_2. The presence of doping elements (Mg) enhanced the nitridation reaction of Al in the matrix alloy. Also the presence Mg can reduce the surface tension of molten aluminum on the gas-liquid surface. The enhancement depends on the formation of nonprotective catalytic nitride (Mg_3N_2) which donates its nitrogen to aluminum in molten state to form AlN. Thermodynamically, nitridation of either Al or Mg is very favorable and strongly exothermic [14]. Also, AlN has higher thermal stability than magnesium nitride over the entire temperature range and thus molten aluminum reacts with Mg_3N_2 to form the more stable phase AlN. The reaction mechanism it is described through the following reactions:

$$Al(Mg) \rightarrow Al(l) + Mg(g) \quad (3)$$

$$2Mg + 2[N] \rightarrow Mg_3N_2 \quad (4)$$

$$2Al + Mg_3N_2 \rightarrow 2AlN + 3Mg \quad (5)$$

IOP Conf. Series: Materials Science and Engineering **209** (2017) 012011 doi:10.1088/1757-899X/209/1/012011

4.2. X-Ray Diffraction of AlMg-AlN composites

Samples of the ingot were taken for X-Ray examinations. Phase analysis was performed using the JDX-7S diffractometric from JEOL. The source of radiation was a lamp with a copper anode, powered by a constant voltage of 40 kV and current of 20 mA. Monochromatisation of the beam was performed on graphite monochromatic. Phase identification was supported by PCSIWIN computer program, based upon the JCPDS-International Centre for Diffraction Data 2000 file.

The X-Ray diffraction pattern (figure 2) confirmed the presence of Al-α solid solution and AlN. The formations of oxide phases were due to the reaction between the metals and residual O_2 gas present in the reaction atmosphere. However no Mg_3N_2 phase was detected in the final product.

Also the X-Ray analysis indicates that AlN forms from the very beginning of the process but the overall crystalline state of the alloy changes over increasing reaction time. In particular, the peaks of Al registered at short reaction times (1 h) were totally vanished in the prolonged experiments (10 h).

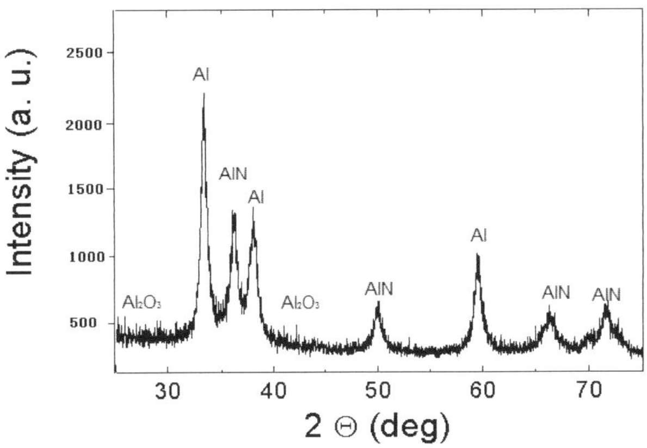

Figure 2. XRD for AlMg15/AlN after reaction with N_2 at 1000°C temperature.

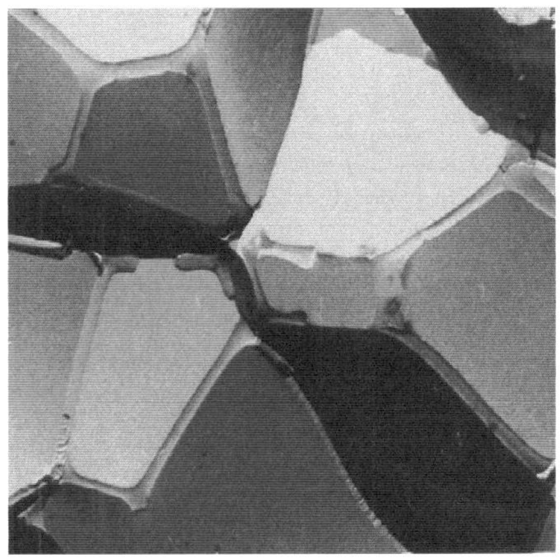

Figure 3. TEM replica of AlN reinforcing element from AlMg/AlN composite.

4.3. SEM Microstructure of AlMg15/AlN composite

The microstructure of the AlMg15/AlN composite has been investigated using a scanning electron microscope. The typical accelerating voltage used was 10–15 kV. The surface micro topography of samples was also observed by a high resolution TEM using Pt/C replicas. The replicas consisted of a thin Pt/C metal film (1–2 nm thick, consisting of 95% Pt and 5% C) and a 15–20 nm thick supporting C film. Replicas were produced from the fractured surfaces of the samples.

Nitrating for one hour at 1000°C, the ceramic reinforcing element, AlN, shows a density of 99%. A TEM micrograph of this sample is shown in Figure 3. This high density is due to the formation of an aluminum oxynitride (Al–O–N) liquid by reaction between Al_2O_3 and AlN. The addition of Mg in matrix composition promotes more extensive liquid formation due to reaction to form a quaternary liquid (Mg–Al–O–N) at lower temperature.

The SEM microstructure of a typical AlMg/AlN composite is shown in figure 4. The grain boundary distribution of the secondary phase is quite different from that of figure 3.

The secondary phase (white) shows up at the grain junctions only and not along grain edges.

The addition of Mg appears to have a further effect of changing the wetting behavior of the liquid phase with respect to the AlN grains. Indeed the dihedral angle of the secondary phase at the grain junction is high (>70°). This has a significant effect on the thermal conductivity (k). [15] suggest that thermal conductivity rises to a maximum and then decreases due to the increase in volume fraction of aluminates with further Mg addition.

Figure 4. SEM microstructures of AlMg/AlN composite: (a) at higher magnification, 6000 X; (b) at higher magnification, 4000 X; (c) at lower magnification, 500 X.

5. Conclusions

AlN crystals were successfully grown via "in situ" process in an AlMg matrix at 1000°C. From T*hermodynamic modelling of AlMg/AlN system* discussion, it could be said that presence of doping element like Mg during the Al nitridation served as a recyclable catalytic nitrogen carrier. In other words the presence of Mg_3N_2 and/or Mg may serve as an effective nucleation site for the formation of AlN. Mg plays a key role on reducing the partial pressure of oxygen in the furnace chamber during the entire process favouring the formation of AlN and suppressing the formation of Al_2O_3. The experiments made have corroborated that the reaction between liquid aluminum and nitrogen in the conditions of an increased gas pressure may lead to the formation of AlN phases, and the presence of Mg in the matrix speeds up the reaction intensity. In the initial stage, the reaction proceeds on the surface. High pressure of nitrogen inhibits vaporization of Mg and reaction between Al and N_2 proceed in limited scope. Since the reaction is of an exothermic nature, it is difficult to control. Using the in situ reaction in order to form AlN dispersion reinforcement in the aluminium alloy matrix seems possible with limited intensity of the process, for example, through lowering the Mg fraction in the matrix or setting an appropriate reaction time.

References

[1] Surappa M K 2003 *Aluminium matrix composites: Challenges and opportunities* **28** Parts 1 & 2 Sadhana Printed in India 319–334
[2] Mazumdar S K 2002 *Composites manufacturing materials, product, and process engineering*, CRC Press LLC, ISBN 0-8493-0585-3, United States of America
[3] Medraj M, Drew R A L, Thompson W T 2003 *Can. Metall. Q.* **42**(4) 495–506
[4] Medraj M, Hammond R, Thompson W T, Drew R A L 2003 *J. Am. Ceram. Soc.* **86**(4) 717–726
[5] Accuratus Corporation 35 Howard Street Phillipsburg NJ 08865, www.accuratus.com
[6] Jackson T B, Virkar A V, More K L, Dinwidie R, Cutler B 1997 *J. Am. Ceram. Soc.* **80**(6) 1421–1435
[7] Burton B, Chart T B, Lukas H L, Pelton A D, Seifert H S P 1995 *Calphad* **4**(19) 537–553
[8] Chase Jr. M W, Davies C A, Downey Jr. J R, Frurip D J, McDonald R A, Syverud A N *JANAF Thermochemical Tables* **3** Part I American Chemical Society Michigan 131
[9] Pelton A D, Bale C W, Thompson W T 2006 *F*A*C*T (Facility for the Analysis of Chemical Thermodynamics)*,McGill University/Ecole Polytechnique Montreal
[10] Farghaly F E 2001 Ph.D. Thesis Ain-shams University Cairo Egypt
[11] Kondoh K, Takeda Y J 1999.*Jpn Soc. Powder/Powder Metall.* **46**(8) 801
[12] Liu J, Gao X, Chen J, Wang Q, Shen Y, Gao Q 2005 *Thermochim. Acta* 253–265
[13] Hou Q H, Mutharasan R, Koczak M 2005 *Mater. Sci. Eng. A,* 195–121
[14] Florea R M, Peter I, Rosso M, Mitrica M, Carcea I 2013 *JOAM* **15** (7- 8) 833 – 840
[15] Haibo J, Chen K, Heping Z, Agathopoulos S, Fabrichnaya O, Ferreira J M F 2005 *J. Cr. Growth* **281** 639–645

IOP Conf. Series: Materials Science and Engineering **209** (2017) 012012 doi:10.1088/1757-899X/209/1/012012

Development and Investigation of Tungsten Copper Sintered Parts for Using in Medium and High Voltage Switching Devices

M V Lungu[1,3,*], M Lucaci[1], V Tsakiris[1], A Brătulescu[1], C D Cîrstea[1], M Marin[1], D Pătroi[1], S Mitrea[1], V Marinescu[1], F Grigore[1], D Tălpeanu[1], N Stancu[1], P Godeanu[2] and C Melnic[2]

[1] Advanced Materials Department, National Institute for Research and Development in Electrical Engineering ICPE-CA (INCDIE ICPE-CA), 313 Splaiul Unirii Street, 030138 Bucharest, Romania

[2] MAIRA MONTAJ SRL, 166 Stirbei Voda Street, 010121 Bucharest, Romania

E-mail: magdalena.lungu@icpe-ca.ro

Abstract. Tungsten-copper (W–Cu) sintered parts with 75 wt.% W, 24 wt.% Cu and 1 wt.% Ni for using as arcing contacts in medium and high voltage switching devices were developed successfully by powder metallurgy (PM) techniques. Sintered parts with diameter of 50±0.5 mm and height of 6±0.5 mm were manufactured by pressing–sintering–infiltration (P–S–I) and spark plasma sintering (SPS) at sintering temperature of 1150°C, and 1050°C, respectively. Physical, chemical, electrical, thermal and mechanical properties of the samples were investigated. Microstructure was analyzed by optical microscopy and scanning electron microscopy. Material properties were influenced by the consolidation processes. The best results were achieved by SPS process. The relative density was more than 95 %, Vickers hardness HV1/15 was over 227, elastic modulus was over 143 GPa, and homogeneous microstructure was revealed. These good properties can contribute to higher lifetime of arcing contacts under severe working conditions.

1. Introduction

W–Cu materials are of a major interest for thermal–electrical applications like electrical contacts used in low voltage (LV), medium voltage (MV) and high voltage (HV) switching devices working in vacuum, sulfur hexafluoride (SF_6) gas or oil [1-14].

Protecting and controlling the transmission networks and the distribution systems of electric power for industrial and public consumption require technically reliable switching devices such as circuit breakers (CBs) with long service life.

A CB suffers arcing contacts wear during normal operation, as well as when breaking short–circuit currents. During arcing in the switching chamber, arcing contacts are exposed to high mechanical and thermal stresses. Mechanical wear and lubrication of electrical contacts often affect the performance of the CBs. That is why there is a continuous need to be developed high performance electrical contacts.

W–Cu materials are commonly used for arcing contacts in SF_6 gas or oil CBs for MV and HV applications [5-14]. Functional properties of W–Cu materials are influenced by their microstructure and physical, electrical, thermal and mechanical properties. These properties are influenced in turn by W and Cu content and their particle size and shape, as well as by obtaining methods for W–Cu composite powder mixtures and their processing techniques. The final electrical contact pieces are manufactured at the desired compositions, sizes and various simple geometrical shapes or complex non-geometrical shapes depending on their applications. Proprietary methods for manufacture W–Cu materials vary significantly among producers, leading to different properties of the contact materials.

W–Cu materials are heterogeneous compounds made of a refractory component (W) and a soft, malleable and ductile component (Cu). W–Cu composites are considered as pseudo-alloys and advanced engineering materials due to their specific physical, electrical, thermal and mechanical properties. W–Cu powder mixtures are produced by various methods, from which the common methods are mechanical mixing, mechanical alloying and high energy ball milling of elemental powders in dry or wet conditions [2-14].

W–Cu composites with a W content of over 50 wt.%, balance being Cu are manufactured by liquid phase sintering of a pressed W–Cu powder mixture or by pressing–sintering–infiltration (P–S–I) that implies molten Cu infiltration into single pressed and sintered W or W–Cu skeleton with low Cu content [5,8-15]. The porous skeleton is filled up with liquid Cu through capillary force since Cu acts as a filling agent of pores [15]. Amounts of up to 1 wt.% of additives can be used in W–Cu composites to improve the sinterability. Sintering additives are chose from transition metals (Ni, Fe, Co, Zn, Ag, a.o.) having the role of wetting agents between W and Cu particles [2,8,11-15].

It is known that it is very difficult to achieve full densification through the above mentioned techniques due to the absence or negligible solubility between W and Cu under the equilibrium condition. High sintering temperatures or long holding times help to improve the densification but Cu may leach out from the skeleton which leads to Cu segregation and results in non-homogeneous microstructure and poor product performance [7,8]. Therefore it is necessary the achievement of new W–Cu composites by alternative techniques than conventional ones like spark plasma sintering (SPS) to minimize microstructure coarsening and to improve homogeneity of contact materials [2,3,16-19]. By SPS process, consolidation of powders is performed rapidly with uniaxial force, high heating/cooling rates and a pulsed (on-off) direct electrical current [18,19].

In this study we present the research results on the development and investigation of W–Cu contact materials with 75 wt.% W, 24 wt.% Cu and 1 wt.% Ni for using as arcing contacts in MV and HV switching devices. The W–Cu composites were manufactured in form of large size sintered parts with diameter of 50±0.5 mm and height of 6±0.5 mm by P–S–I and SPS techniques. The sintered parts were mechanically processed to achieve final products in form of protection rings with non-geometrical shapes. The obtained results recommend using of the realized sintered parts in practical applications as protection rings in MV and HV circuit breakers.

2. Experimental

2.1. Materials

As elemental powders we used W powder (purity of 99.95%, Fisher diameter of 4.37 μm, Scott density of 4.22 g/cm^3, tap density of 7.57 g/cm^3, Eurotungstene Metal Powder, France), electrolytic Cu powder (purity of 99.9%, particle size up to 63 μm, apparent density of 1.6 g/cm^3, GGP Metalpowder AG, Germany) and Ni powder (purity of 99.9%, average particle size of 5 μm, Merck, Germany).

2.2. Methods for manufacture of W–Cu sintered pieces

W–Cu powder mixtures with 75 wt.% W, 24 wt.% Cu and 1 wt.% Ni were prepared by mechanical homogenization for 8 h of the starting powders weighted in the desired amounts using a Turbula mixer and 5 mm stainless steel balls to powder ratio of 1:1.

The W–Cu sintered parts were manufactured in cylindrical shape and large size with diameter of 50±0.5 mm and height of 6±0.5 mm by PM techniques like P–S–I and SPS.

The sintered parts obtained by P–S–I technique were manufactured in a protective hydrogen and nitrogen atmosphere in two sintering steps. W skeletons containing Ni additive were compacted with a hydraulic press (Meyer, Switzerland) and sintered at 950°C for 30 min in a conveyor belt furnace

(Safed, Switzerland) to form a porous network. After that the skeletons were densified by infiltration with molten Cu for 30 min at a temperature of 1150°C that is above the melting point of Cu (1083°C).

The SPS–ed parts were manufactured in vacuum by processing of W–Cu powder mixtures in graphite molds using a SPS installation of HP D25 type (FCT Systeme GmbH, Germany) equipped with a power source supplying DC pulses. The pressing pressure of the W–Cu powder mixtures increased linearly from 2.5 MPa to 50 MPa being maintained at 50 MPa for 30 min at sintering temperature of 1050°C with heating/cooling rate of 100°C/min.

2.3. Characterization of W–Cu sintered parts

The investigation of physical, electrical and mechanical properties of the samples was carried out at room temperature (RT) on as obtained parts with diameter of 50±0.5 mm and height of 6±0.5 mm. For microstructure analysis and measurements of thermal properties, small size samples with diameter of 12.65±0.5 mm and height of 3±0.1 mm were cut from the core of the initial large parts.

The microstructure was analyzed by optical microscopy (OM) and scanning electron microscopy (SEM) using Carl Zeiss (Germany) optical microscope and an Auriga CrossBeam workstation equipped with backscattered electron (BSE) detector and energy dispersive X-ray spectroscopy (EDS) detector used for compositional (elemental) analysis. Previously the samples were put in resin, polished with fine alumina slurry and etched with Murakami's solution for 15 s.

The density was determined at 16°C by Archimedes' principle, in ethyl alcohol with density of 0.7928 g/cm^3, using an analytical balance (Kern AEJ, Switzerland) equipped with a set for density determination of liquids and solids. The measurements were carried out in triplicate, and their mean values are presented. Relative density was determined as percentage of the theoretical density (TD) of the W–Cu–Ni 75–24–1 composite (14,977 g/cm^3).

The electrical conductivity, σ (m/Ω·mm^2) was determined at 22°C with the phase-sensitive eddy current method using a conductivimeter of EX8 type (Sigmascope, Germany). Three readings were performed on both flat circular surfaces of the samples and their mean values are presented. The electrical resistivity, ρ (μΩ·cm) was computed as the inverse of the electrical conductivity. The resulted resistivity values were converted to electrical conductivity (σ) in percentages of International Annealing Copper Standard (% IACS) with the equation (1) [10]:

$$\sigma = 172.41/\rho \tag{1}$$

The thermal diffusivity and specific heat of the W–Cu small size samples were measured at 25°C by "flash" method using a LFA 447 NanoFlash apparatus (Netzsch, Germany) and a reference of Inconel 600 having specific heat at 25°C of 0.444 J/(g·K). Previously the samples were coated by spraying on all the surfaces with a thin film of graphite. A xenon lamp was used as radiation energy, and the irradiation time on the front surface of each sample was 0.18 ms. Temperature increase on the back surface of the sample was measured with an infrared (IR) detector of InSb type cooled with liquid nitrogen. The samples were measured three times and their mean values are presented.

The thermal diffusivity, α (mm^2/s) was calculated with the equation (2) [20], where l (mm) is sample thickness and $t_{1/2}$ (s) is the half-rise time:

$$\alpha = 0.1388 \cdot l^2/t_{1/2} \tag{2}$$

The specific heat of the W–Cu samples, C_p (J/(g·K)) was determined from the equation (3) [20], where Q (J) is absorbed energy, m (g) is mass and ΔT (K) is change in temperature, G is detector amplifier gain that is a constant equal to 50020, ΔV (V) is output voltage of the IR detector:

$$C_{p_{sample}} = \frac{Q_{sample}}{(m \cdot \Delta T)_{sample}} = \frac{(m \cdot C_p \cdot \Delta T)_{reference}}{(m \cdot \Delta T)_{sample}} = \frac{(m \cdot C_p \cdot \Delta V)_{reference} \cdot G_{sample}}{(m \cdot \Delta V)_{sample} \cdot G_{reference}} \tag{3}$$

The thermal conductivity, λ (W/(m·K)) of the W–Cu samples was computed with the equation (4) [20], where α (m^2/s) is thermal diffusivity, C_p (J/(kg·K)) is specific heat and d (kg/m^3) is density:

$$\lambda = \alpha \cdot C_p \cdot d \tag{4}$$

The mechanical properties were determined at RT by instrumented indentation technique using an Open Compact Platform (CSM Instruments, Switzerland) equipped with Micro Hardness Tester (MHT) and Vickers diamond indenter. The measurements were performed by microindentation with linear loading from 0.01 N to 10 N, holding time of 15 s at F_{max}, where approach speed of the indenter was 4 μm/min, loading/unloading rate was 20 N/min and duration of loading from contact to F_{max} and unloading from holding time end to last contact point was 30 s. Ten measurements were performed on both flat circular surfaces of the samples and their mean values are presented. The indentation imprints were visualized by optical microscopy using 50X objective attached to the Open Compact Platform.

The values of the mechanical properties were calculated by the power law method (Oliver & Pharr) [21] shown in the equation (5) to describe the upper part of the unloading data, where F_{max} (N) is the maximum force, h (μm) is the indentation depth under applied test force, h_p (μm) is the remnant depth, h_{max} (μm) is the maximum depth at F_{max} and m is a constant ranging 1.15...1.3 that depends of indenter shape. The values of m and h_p were determined by a least squares fitting procedure.

$$F = F_{max} \cdot [(h - h_p)/(h_{max} - h_p)]^m \tag{5}$$

The contact stiffness, S (N/μm) at F_{max} was calculated from the equation (6) [21]:

$$S = (dF/dh)_{h=h_{max}} = m \cdot F_{max} \cdot (h_{max} - h_p)^{-1} \tag{6}$$

The contact depth of the indenter, h_c (μm) with the sample at F_{max} was determined from the equation (7), where ε is the geometric constant ranging 0.77...0.81 that depends on indenter shape:

$$h_c = h_{max} - \varepsilon \cdot F_{max}/S \tag{7}$$

The projected contact area, A_p (μm^2) was determined from a fit method based on the equation (8), where C_0 and C_1 are constants determined on fused silica reference:

$$A_p = C_0 \cdot h_c^2 + C_1 \cdot h_c \tag{8}$$

The indentation hardness, H_{IT} (N/mm^2) was determined from the equation (9) [21]:

$$H_{IT} = F_{max}/A_p \tag{9}$$

The Vickers hardness HV_{IT} was determined from the equation (10) [21], where F_{max} (N) is the maximum force, and A_c (μm^2) is the developed contact area:

$$HV_{IT} = F_{max}/(9.81 \cdot A_c) \cong H_{IT}/10.58 \tag{10}$$

The reduced modulus, E_r (GPa) of the indentation contact was calculated from the equation (11) [21], where β is the geometric factor (β = 1.012 for Vickers indenter):

$$E_r = (\pi/A_p)^{1/2} \cdot S/(2 \cdot \beta) \tag{11}$$

The plane strain modulus, E^* (GPa) was calculated from the equation (12) [21], where E_r is the reduced modulus, E_i is the elastic modulus of the indenter (E_i = 1141 GPa for diamond) and υ_i is the Poisson's ratio of the indenter (υ_i = 0.07 for diamond):

$$E^* = [1/E_r - (1 - \upsilon_i^2)/E_i]^{-1} \tag{12}$$

The indentation modulus, E_{IT} (GPa) that represents the elastic modulus of the sample was calculated from the equation (13) [21] using an estimated Poisson's ratio of W–Cu samples (υ_s) of 0.29:

$$E_{IT} = E^* \cdot (1 - \upsilon_s^2) \tag{13}$$

The elastic part of the indentation work, η_{IT} (%) was calculated from the equation (14) [21], where W_{total} (N·m) is the total mechanical work of indentation, $W_{elastic}$ (N·m) is the elastic reverse deformation work of indentation and $W_{plastic}$ (N·m) is the plastic deformation work of indentation:

$$\eta_{IT} = (W_{elastic}/W_{total}) \cdot 100 = [W_{elastic}/(W_{elastic}+W_{plastic})] \cdot 100 \tag{14}$$

The difference between 100 % and η_{IT} [%] represents the plastic part of the indentation work.

The creep indentation, C_{IT} (%) was determined from the equation (15) [21], where h_1 is the indentation depth at time t_1 of reaching the test force that is kept constant and h_2 is the indentation depth at time t_2 of holding the constant test force:

$$C_{IT} = [(h_2 - h_1)/h_1] \cdot 100 \tag{15}$$

3. Results and discussions

OM and SEM micrographs along with EDAX results of the W–Cu–Ni 75–24–1 composites manufactured by P–S–I and SPS techniques are depicted in figure 1 to figure 6.

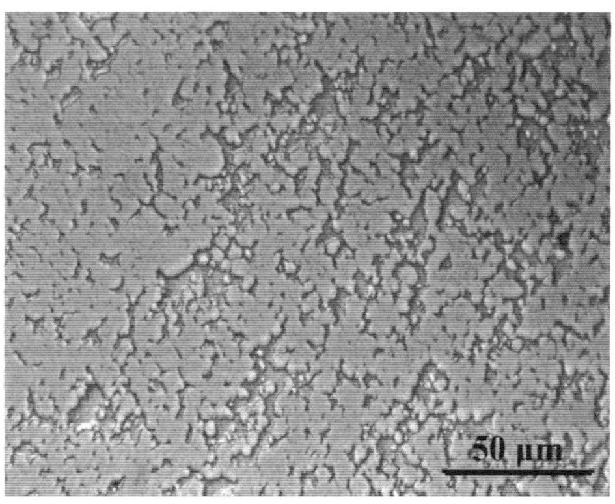

Figure 1. OM image of the infiltrated composite (500X).

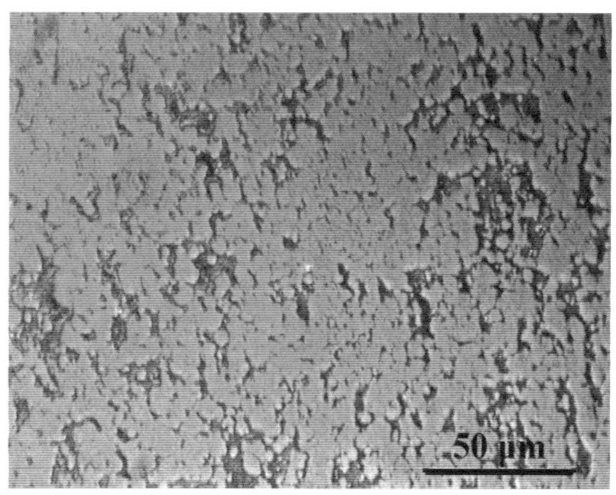

Figure 2. OM image of the SPS-ed composite (500X).

Microstructure investigation revealed the finer structure in the SPS-ed composite compared to the sintered part obtained by P–S–I technique.

The microstructure appears to be dense for both W–Cu–Ni 75–24–1 composites with lower porosity in case of the SPS-ed composite. It confirms the existence of three phases corresponding to W, Cu and Ni elements.

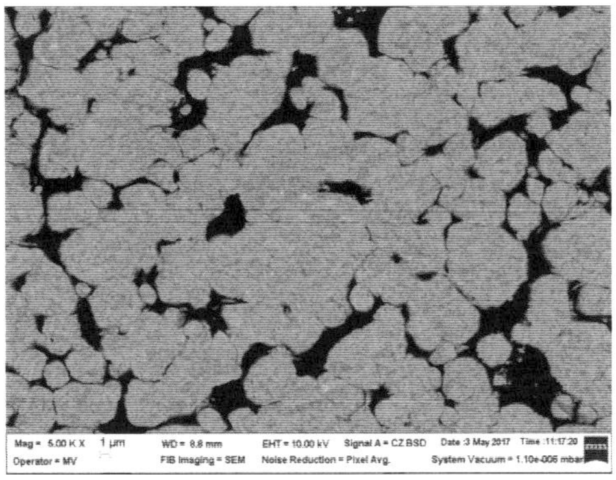

Figure 3. SEM backscattered electron image of the infiltrated composite (5000X).

Figure 4. SEM backscattered electron image of the SPS-ed composite (5000X).

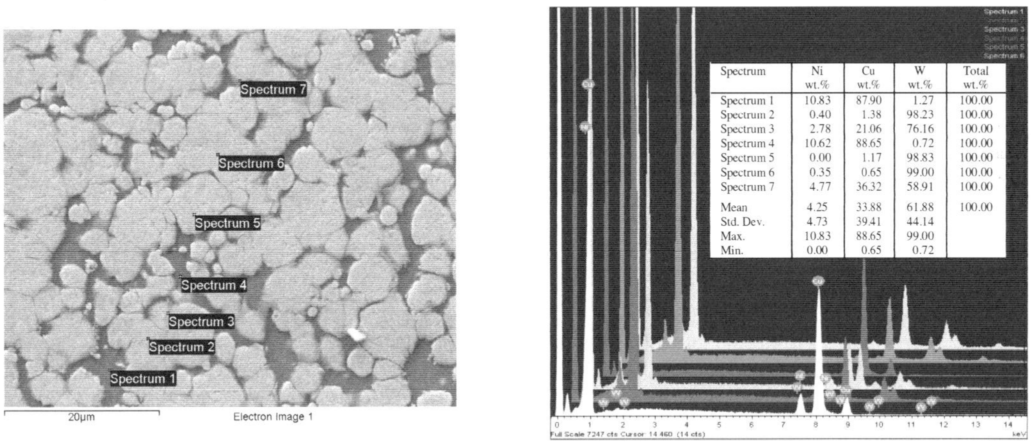

Spectrum	Ni wt.%	Cu wt.%	W wt.%	Total wt.%
Spectrum 1	10.83	87.90	1.27	100.00
Spectrum 2	0.40	1.38	98.23	100.00
Spectrum 3	2.78	21.06	76.16	100.00
Spectrum 4	10.62	88.65	0.72	100.00
Spectrum 5	0.00	1.17	98.83	100.00
Spectrum 6	0.35	0.65	99.00	100.00
Spectrum 7	4.77	36.32	58.91	100.00
Mean	4.25	33.88	61.88	100.00
Std. Dev.	4.73	39.41	44.14	
Max.	10.83	88.65	99.00	
Min.	0.00	0.65	0.72	

Figure 5. EDAX results of the W–Cu–Ni 75–24–1 composite manufactured by P–S–I technique.

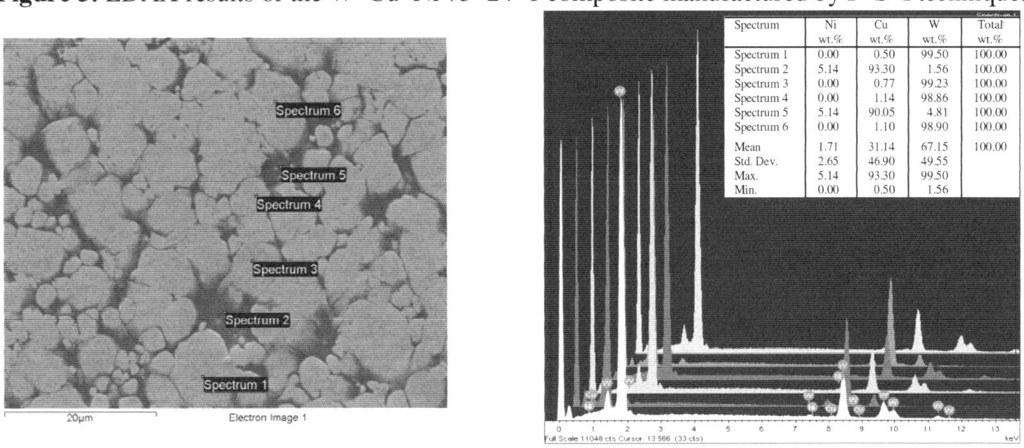

Spectrum	Ni wt.%	Cu wt.%	W wt.%	Total wt.%
Spectrum 1	0.00	0.50	99.50	100.00
Spectrum 2	5.14	93.30	1.56	100.00
Spectrum 3	0.00	0.77	99.23	100.00
Spectrum 4	0.00	1.14	98.86	100.00
Spectrum 5	5.14	90.05	4.81	100.00
Spectrum 6	0.00	1.10	98.90	100.00
Mean	1.71	31.14	67.15	100.00
Std. Dev.	2.65	46.90	49.55	
Max.	5.14	93.30	99.50	
Min.	0.00	0.50	1.56	

Figure 6. EDAX results of the W–Cu–Ni 75–24–1 composite manufactured by SPS technique.

Since these three metals can be differentiated based on contrast and their atomic number (Z) in backscattered electron (BSE) images, the heaviest element, W (Z of 74) can be identified by the bright gray colour, and the transition elements Cu (Z of 29) and Ni (Z of 28) can be noticed by the dark gray and black colour, respectively (figure 3 and figure 4). These observations are in agreement with other reports from the literature [10], [13].

EDAX spectra (figure 5 and figure 6) confirmed also the presence of W, Cu and Ni elements although compositions of both samples differ from nominal one probably due to the lack of mutual solubility and sufficient bonding between W–Cu and W–W particles that contributed to the removal of some particles from the surface of the composites during polishing with alumina particles [17].

Physical, electrical and thermal properties (mean values) of the developed W–Cu–Ni 75–24–1 sintered parts are summarized in table 1.

Table 1. Physical, electrical and thermal properties (mean values) of the W–Cu–Ni 75–24–1 sintered parts manufactured by P–S–I and SPS techniques.

Composites and fabrication method	Density (g/cm^3)	Relative density (% of TD)	Electrical conductivity ($m/(\Omega \cdot mm^2)$)	Electrical resistivity ($\mu\Omega cm$)	Electrical conductivity (% IACS)	Thermal diffusivity (mm^2/s)	Specific heat ($J/(g \cdot K)$)	Thermal conductivity ($W/(m \cdot K)$)
W–Cu–Ni 75–24–1 P–S–I	14.26	95.21	12–14.5	6.90–8.33	21–25	35±0.5	0.197± 0.002	98±2
W–Cu–Ni 75–24–1 SPS	14.45	96.48	14–14.5	6.90–7.14	24–25	46±0.4	0.188± 0.005	124±2

The measurements of the densities in ethyl alcohol show that the P–S–I composite achieved a density of 14.26 g/cm^3 (95.21% of TD) after infiltration of the skeleton with melted Cu. A better result (14.45 g/cm^3 meaning 96.48% of TD) was obtained in case of the SPS-ed composite, indicating a higher densification of the composite. Accordingly, the SPS-ed composite showed lower porosity (3.52%) than the infiltrated composite (4.79%) and better electrical and thermal properties.

Better consolidation of the W–Cu–Ni 75–24–1 powder mixture and lower porosity of the SPS-ed composite can be explained by the mechanism of SPS. It relays on rapid consolidation of powders without significant grain grow due to very rapid heating rates under pulsed (on-off) direct electrical current that generates spark plasma, spark impact pressure, Joule heating and an electrical field diffusion effect through powder particles [18]. The granular boundaries of the initial powder particles start to partially heat up because the pulse current passes through the graphite die, leading to the occurrence of an electrical field having a plasma effect [19]. Thus a short process cycle of about an hour was achieved in our study in comparison with P–S–I process that lasts over ten times longer with supplementary costs for protective gases and energy consumption at higher sintering temperature and duration.

The variation of electrical conductivity and resistivity on both surfaces of the SPS-ed part is in a narrow range (~3.4%) indicating a good material homogeneity. On contrary, the infiltrated composite showed five time higher variation (~17.2%) in electrical properties on both surfaces. The bottom surface that contained an excess of Cu resulted after infiltration exhibited similar values for electrical properties as SPS-ed part whereas on the opposite surface electrical conductivity has smaller values due to lower Cu content. This happened since during Cu infiltration the capillary force acts on the W and Ni particles from the skeleton to pull the melted Cu between W and Ni hard particles and to fill in the pores but Cu viscosity decreases with temperature [10]. Besides capillary force there are some other forces like gravity, drag and buoyancy forces that influence sintering and infiltration process [10]. Our results are in agreement with other studies in which it was revealed that Ni addition lowers significantly electrical conductivity of W–Cu composites but contributes to the improvement of hardness [8].

The measured thermal diffusivity of the SPS-ed part (46±0.4 mm^2/s) was 1.3 times greater than in case of the sample manufactured by P–S–I technique (35±0.5 mm^2/s). Both values are lower than the values obtained in other reports (73.2…77.6 mm^2/s) where Ni free W–Cu composite contained 22.33…26 wt.% Cu and balance W [13]. However, specific heat values obtained by "flash" method for both composites ranging 0.183...0.199 $J/(g \cdot K)$ conform with other reported data [13]. A possible explanation arises from the fact that specific heat of W–Cu–Ni composites can be estimated using the

IOP Conf. Series: Materials Science and Engineering **209** (2017) 012012 doi:10.1088/1757-899X/209/1/012012

mixture rule [13], [22]. Thus, the computed specific heat of W–Cu–Ni composites was defined from the equation (16) adapted from [13], [22], where w means weight fraction of each component:

$$C_p = w_W \cdot C_p^W + w_{Cu} \cdot C_p^{Cu} + w_{Ni} \cdot C_p^{Ni} \tag{16}$$

The computed specific heat of the W–Cu–Ni 75–24–1 composite is 0.196 J/(gK) since the specific heat at 25°C of W (C_p^W) is 0.132 J/(g·K), Cu (C_p^{Cu}) is 0.385 J/(gK) and Ni (C_p^{Ni}) is 0.444 J/(gK) [23].

Thermal conductivity of the SPS-ed composite was with 26% higher than in case of the infiltrated composite. However, the values of thermal conductivity of both developed W–Cu–Ni composites are lower than other reported values for Ni free W–Cu composites. Also, the existence of 3.52...4.79% porosity in the sintered parts may contribute to hinder the heat transfer to some extent [24].

The mean force-displacement curves for all the sintered parts on each flat circular surface, together with optical images of some residual imprints are shown in figure 7 to figure 10.

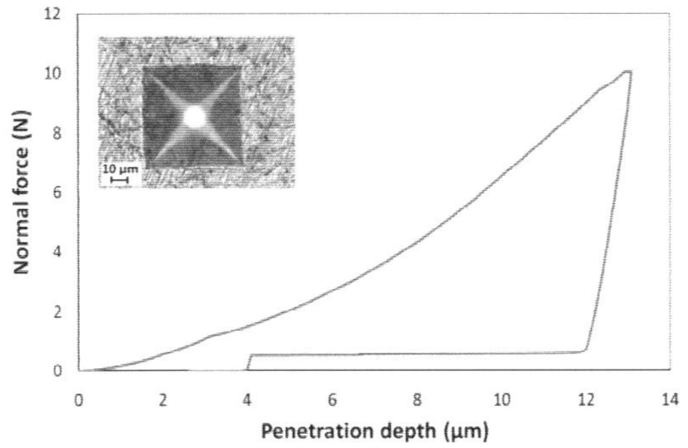

Figure 7. Mean force-displacement curve for W–Cu–Ni 75–24–1 SPS part (top surface). The inset represents an optical micrograph (50X) of a Vickers indentation into the sample (imprint diagonal, d = 87.12 μm, HV = 249).

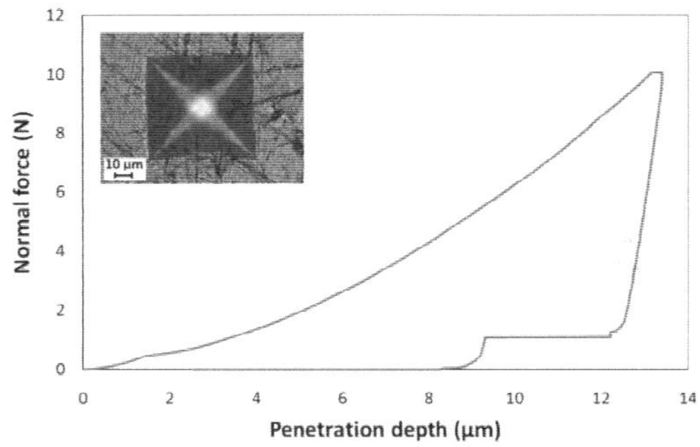

Figure 8. Mean force-displacement curve for W–Cu–Ni 75-24-1 P-S-I part (bottom surface). The inset represents an optical micrograph (50X) of a Vickers indentation into the sample (imprint diagonal, d = 89.76 μm, HV = 235).

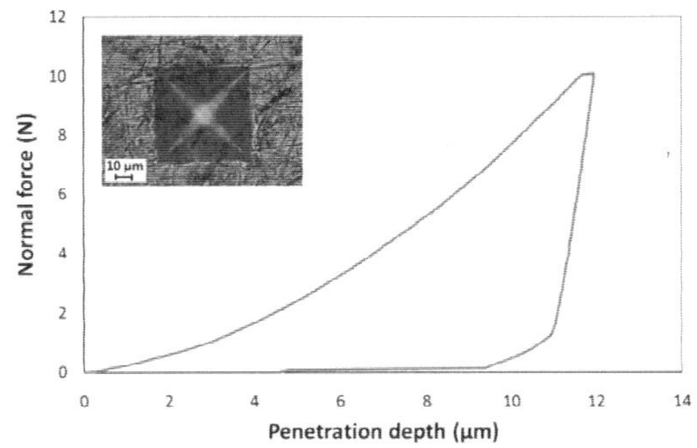

Figure 9. Mean force-displacement curve for W–Cu–Ni 75–24–1 SPS part (top surface). The inset represents an optical micrograph (50X) of a Vickers indentation into the sample (imprint diagonal, d = 81.53 μm, HV = 285).

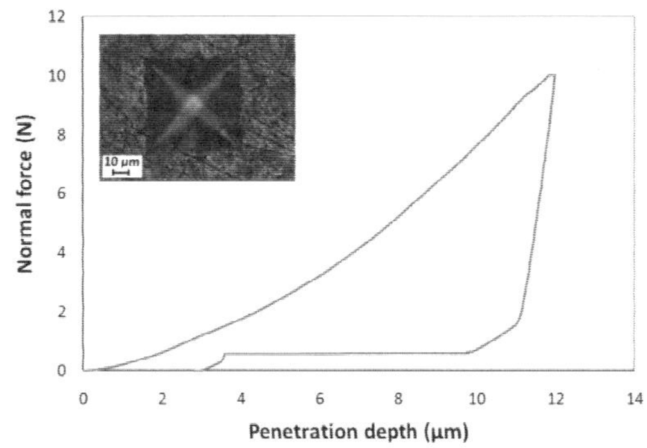

Figure 10. Mean force-displacement curve for W–Cu–Ni 75-24-1 SPS part (bottom surface). The inset represents an optical micrograph (50X) of a Vickers indentation into the sample (imprint diagonal, d = 81.65 μm, HV = 284).

The samples obtained by P–S–I and SPS processes show no significant difference in imprint shape. Only the diagonal sizes of the imprints are various indicating different values for Vickers hardness. All the imprints exhibit convex edges, which suggest a rigid plastic response to indentation.

The mean results obtained by instrumented indentation with Vickers indenter and Oliver & Pharr method for the W–Cu sintered parts are presented in table 2 and table 3.

Table 2. Main mean results obtained by instrumented indentation with Vickers indenter and Oliver & Pharr method for the W–Cu sintered parts investigated on both flat circular surfaces.

Composites and fabrication method (flat circular surface)	H_{IT} 10/30/15/30 (MPa)	HV_{IT} 10/30/15/30	E^* 10/30/15/30 (GPa)	E_r 10/30/15/30 (GPa)	E_{IT} 10/30/15/30 (GPa)	C_{IT} 10/30/15 (%)
W–Cu–Ni 75–24–1 P–S–I (top surface)	2690±109	254±10	162±5	142±4	148±5	2.51±0.16
W–Cu–Ni 75–24–1 P–S–I (bottom surface)	2535±125	239±12	167±7	146±5	153±6	2.47±0.17
W–Cu–Ni 75–24–1 SPS (top surface)	3255±105	307±10	193±4	165±2	176±3	2.58±0.15
W–Cu–Ni 75–24–1 SPS (bottom surface)	3211±233	303±12	190±7	163±5	174±6	2.46±0.10

Table 3. Additional mean results obtained by instrumented indentation with Vickers indenter and Oliver & Pharr method for the W–Cu sintered parts investigated on both flat circular surfaces.

Composites and fabrication method (flat circular surface)	F_{max} (N)	h_{max} (μm)	S (N/μm)	h_c (μm)	h_r (μm)	h_p (μm)	A_p (μm²)	W_{elast} (μJ)	W_{plast} (μJ)	η_{IT} (%)
W–Cu–Ni 75–24–1 P–S–I (top surface)	10.04± 0.01	13.10± 0.25	9.88± 0.08	12.29± 0.25	12.08± 0.25	11.93± 0.25	3734± 148	5.32± 0.12	44.20± 2.01	10.75± 0.64
W–Cu–Ni 75–24–1 P–S–I (bottom surface)	10.03± 0.01	13.42± 0.32	10.47± 0.22	12.66± 0.31	12.47± 0.32	12.33± 0.34	3963± 194	5.18± 0.05	45.94 2.05	10.15± 0.46
W–Cu–Ni 75–24–1 SPS (top surface)	10.07± 0.04	11.93± 0.17	10.47± 0.36	11.18± 0.20	10.97± 0.20	10.78± 0.15	3094± 113	5.97± 0.08	40.20± 0.24	12.93± 0.22
W–Cu–Ni 75–24–1 SPS (bottom surface)	10.02± 0.01	12.00± 0.41	10.40± 0.08	11.25± 0.41	11.04± 0.41	10.85± 0.42	3133± 227	5.96± 0.04	40.36± 0.70	12.87± 0.25

Table 2 clearly shows that the indentation hardness (H_{IT}), Vickers hardness (HV) and elastic modulus (E^*, E_r and E_{IT}) depend on the consolidation process. The SPS-ed composite with a finer and more homogeneous microstructure yielded increased values compared to the composite manufactured by P–S–I technique for maximum penetration depths in the order of about 12 μm. Thus, H_{IT} that is a measure of the resistance to permanent deformation or damage showed an increase of 6...43% for the SPS-ed composite along with an increase of 10...40% for HV_{IT} hardness and an increase of 6...26% for E_{IT}. The homogeneity of the SPS-ed composite was proved by the small variation between results concerning mechanical properties determined by microindentation on both surfaces of the sample.

The contact stiffness (S) at F_{max} that corresponds graphically on the load-displacement curve to the slope of the tangent to the unloading curve yielded close values for all the samples, ranging 10.11...10.83 N/μm for the SPS-ed composite and 9.80...10.69 N/μm for the infiltrated composite.

The creep behaviour is similar for both composites that yielded C_{IT} values in a narrow range, namely 2.36...2.73% for the SPS-ed composite and 2.30...2.67% for the infiltrated composite. The report between Vickers hardness (HV_{IT}) and elastic modulus (E_{IT}) was included in the range of 1.72...1.77 GPa^{-1} for the SPS-ed composite and 1.54...1.73 GPa^{-1} for the infiltrated composite.

The SPS-ed composite revealed better elastic behaviour in contrast to the infiltrated composite, with higher values for the elastic part of the indentation work (η_{IT}) ranging 12.62...13.15%, whereas the infiltrated composite showed η_{IT} values ranging 9.69...11.39%. As a result, the plastic part of the indentation work was higher for the infiltrated composite.

The enhancement of the mechanical properties of W–Cu–Ni composites due to the addition of Ni transition element and different processing techniques was reported in other studies, too [2], [24]. The very short processing time of about one hour used in SPS process was beneficial in controlling the grain growth and thus obtaining finer microstructure. Moreover, the finer microstructure of the SPS-ed composite has a positive effect in improving its mechanical properties.

The sintered parts produced by both processes had a good behaviour at mechanical processing in form of protection ring. In figure 11 it is exemplified the macrographic image of the W–Cu–Ni 75–24–1 SPS part, whereas in figure 12 it is shown the final aspect of the protection ring.

Figure 11. Macrographic image of W–Cu–Ni 75–24–1 SPS part.

Figure 12. Macrographic images of protection ring (left image – top surface and right image – bottom surface).

The obtained results recommend the developed W–Cu sintered parts to manufacture protection rings for use in MV and HV switching devices. However, further investigation concerning functional tests of the protection rings as arcing contacts in MV and HV circuit breakers and their certification are needed in order to introduce these products into the market.

4. Conclusions

W–Cu composites with 75 wt.% W, 24 wt.% Cu and 1 wt.% Ni for using as arcing contacts in MV and HV switching devices were developed successfully in form of sintered parts having diameter of 50±0.5 mm and height of 6±0.5 mm by P–S–I and SPS techniques.

The infiltrated composite showed a relative density of about 95%, Vickers hardness HV1/15 of 227...264, elastic modulus of 143...159 GPa, contact stiffness of 9.8...10.69 N/µm, electrical conductivity of 12...14.5 m/(Ω·mm^2) and thermal conductivity of 96...100 W/(m K).

After SPS processing, the W–Cu composite exhibited a relative density of about 96.5%, Vickers hardness HV1/15 of 291...317, elastic modulus of 168...180 GPa, contact stiffness of 10.11...10.83 N/µm, electrical conductivity of 14...14.5 m/(Ω·mm^2) and thermal conductivity of 122...126 W/m K.

The investigations showed that material properties were influenced by the consolidation processes. Preliminary results regarding mainly microstructure, density, Vickers hardness and elastic modulus were very encouraging for the sintered part achieved by SPS technique.

The sintered parts were mechanically processed to achieve final products in form of protection rings having non-geometrical shapes. The obtained results recommend using of the realized sintered parts in practical applications as protection rings in MV and HV circuit breakers.

References

[1] Wei X, Yu D, Sun Z, Yang Z, Song X and Ding B 2014 *Vacuum* **107** 83–9
[2] Lungu M, Tsakiris V, Enescu E, Pătroi D, Marinescu V, Tălpeanu D, Pavelescu D, Dumitrescu Gh and Radulian A 2014 *Acta Phys. Pol. A* **125** 327–330

[3] Tsakiris V, Enescu E, Lungu M, Lucaci M, Radulian A, Tălpeanu D, Sbarcea G, Caramitu A, Marinescu V and Ion I 2015 *9th Int. Symp. on Advanced Topics in Electrical Engineering (ATEE) 7–9 May 2015 Bucharest (IEEE)* pp 490–495
[4] Tsakiris V, Lungu M, Enescu E, Pavelescu D, Dumitrescu Gh, Radulian A and Mocioi N 2014 *Acta Phys. Pol. A* **125** 348–52
[5] Meng Y, Shen Y, Chen C, Li Y and Feng X 2014 *J. Alloy Compd.* **585** 368–75
[6] Zhang Q, Liang S and Zhuo L 2017 *J. Alloy Compd.* **708** 796–803
[7] Mondal A, Upadhyaya A and Agrawal D 2010 *Mater. Res. Innov.* **14** 355–60
[8] Ştefan I, Nicolicescu C and Gavrila M 2011 *Mater. Sci. Forum* **672** 276–80
[9] Maneshian M H and Simchi A 2008 *J. Alloy Compd.* **463** 153–9
[10] Ibrahim H, Aziz A and Rahmat A 2014 *Int. J. Refract. Met. Hard Mater.* **43** 222–6
[11] Chen P, Luo G, Li M, Shen Q and Zhang L 2012 *Mater. Design* **36** 108–12
[12] Azar G T P, Rezaie H R and Razavizadeh H 2012 *Int. J. Refract. Met. Hard Mater.* **31** 157–63
[13] Wang C P, Lin L C, Xu L S, Xu W W, Song J P, Liu X J and Yu Y 2013 *Int. J. Refract. Met. Hard Mater.* **41** 236–40
[14] Boonyongmaneerat Y 2009 *J. Mater Process. Technol.* **209** 4084-7
[15] German R M Suri P and Park S J 2009 *J. Mater. Sci.* **44** 1–39
[16] Elsayed A, Li W, El Kady O A, Daoush W M, Olevsky E A and German R M 2015 *J. Alloy Compd.* **639** 373–80
[17] Yusefi A, Parvin N 2017 *Fusion Eng. Des.* **114** 196–202
[18] Orrù R, Licheri R, Locci A, Cincotti A. and Cao G. 2009 *Mater. Sci. Eng. R.* **63** 127–287
[19] Suárez M, Fernández A, Menéndez J L, Torrecillas R, Kessel H U, Hennicke J, Kirchner R and Kessel T 2013 Challenges and opportunities for spark plasma sintering: a key technology for a new generation of materials *Sintering Applications* chapter 13 ed (Rijeka, Croatia: InTech) pp 319–42
[20] Parker W J, Jenkins R J, Butler C P and Abbott G L 1961 *J. Appl. Phys.* **32** 1679–1984
[21] Oliver W C and Pharr G M 1992 *J. Mater. Res.* **7** 1564–83
[22] Zhou Y, Sun Q X, Liu R, Wang X P, Liu C S and Fang Q F 2013 *J. Alloy Compd.* **547** 18–22
[23] D R Lide (ed) 2003 Properties of the elements and inorganic compounds; heat capacity of the elements at 25°C *CRC Handbook of Chemistry and Physics* (84th ed) section 4 (Boca Raton, Florida: CRC Press) p 4-133
[24] Chen P, Shen Q, Luo G, Li M and Zhang L 2013 *Int. J. Refract. Met. Hard Mater.* **36** 220–4

Acknowledgments
This work was supported by a grant of the National Romanian Authority for Scientific Research and Innovation, CNCS/CCCDI – UEFISCDI, project number PN–III–P2–2.1–PED–2016–1987, contract number 118 PED/2017, within PNCDI III.

IOP Conf. Series: Materials Science and Engineering **209** (2017) 012013 doi:10.1088/1757-899X/209/1/012013

Characterization of Polysulfone Membranes Prepared with Thermally Induced Phase Separation Technique

L G Tiron*, Ş C Pintilie, M Vlad, I G Birsan and Ş Baltă
"Dunarea de Jos" University of Galati, str. Domnească nr. 111, 800201, Galaţi, Romania

E-mail: geanina.tiron@ugal.ro

Abstract. Membrane technology is one of the most used water treatment technology because of its high removal efficiency and cost effectiveness. Preparation techniques for polymer membranes show an important aspect of membrane properties. Generally, polysulfone (PSf) and polyethersulfone (PES) are used for the preparation of ultrafiltration (UF) membranes. Polysulfone (PSf) membranes have been widely used for separation and purification of different solutions because of their excellent chemical and thermal stability. Polymeric membranes were obtained by phase inversion method. The polymer solution introduced in the nonsolvent bath (distilled water) initiate the evaporation of the solvent from the solution, this phenomenon has a strong influence on the transport properties. The effect of the coagulation bath temperature on the membrane properties is of interest for this study. Membranes are characterized by pure water flux, permeability, porosity and retention of methylene blue. The low temperature of coagulation bath improve the membrane's rejection and its influence was most notable.

1. Introduction

Technologies based on membrane separation are often applied in several industries: food, medicine, pharmacy and wastewater treatment [1].

The membrane can be fabricated from organic or inorganic materials. Nowadays, the most studied membrane materials are polymers because of their properties (mechanical, chemical and flexibility).

Polysulfone is an organic polymer widely used due to the chemical and thermal stability provided to the membrane [2].

Polymeric membranes can be obtained by applying phase separation process. There are two techniques according to the driving force to use phase separation: non-solvent induced phase separation (NIPS) and thermally induced phase separation (TIPS). To obtain membranes used in microfiltration, ultrafiltration, nanofiltration and reverse osmosis, the non-solvent induced phase separation (NIPS) is most used [3].

By this method, the thin film is immersed into the coagulation bath and the solvent from the casting solution is replaced by the nonsolvent (pure water) and phase inversion takes place. This technique is used to obtain asymmetric membranes [4–7].

Obtaining membranes with the phase inversion technique, their properties can be influenced by same parameters like: the polymer and solvent type and their concentration, casting solution temperature, the nonsolvent bath coagulation properties (temperature and composition), evaporation conditions and the process variables (casting speed) [8].

In this study, the polymer used is polysulfone (PSf) because of its chemical and mechanical properties and is one of the most used material in ultrafiltration process [9]. Although the membranes obtained from PSf are restricted due to its hydrophobic properties [10], [11], resulting in a low water flux and serious membrane fouling [12].

A thermally induced phase separation (TIPS) technique is a method for preparing membranes which was implemented in the 1980s. Membranes prepared by the TIPS method usually have high strength and flux [13], [14].

Temperature reduction and double diffusion between solvent and water would occur at the same time by using water soluble solvents to prepare the membrane when the incipient membrane is immersed in a water coagulation bath. At the same time, a denser skin layer forms during the membrane formation process. Pores from membrane surface skin layer prevent activated sludge and other granular substances from entering inside the membrane pores. The antifouling property of membrane is improved [15].

2. Materials and methods

The support layer (Viledon FO2471) used for manufacturing the membrane was obtained from Freudenberg (Winheim, Germany). The membrane was obtained from polysulfone (PSf) polymer. The solvent used was 1-methyl-2-pyrrolidone (NMP, 99%). Both were purchased from Sigma-Aldrich.

The membranes were obtained using the immersion precipitation method, by phase inversion. The casting solution was prepared by adding the polysulfone polymer of 23 weight percentage in the NMP solvent under continuous stirring at 1200 rotations per minute for 24 hours.

The obtained solution was cast on the support layer (non-woven polyester). The support layer was wetted with the solvent to prevent the polymer solution from blocking the pores of the support layer. The homogeneous solutions were cast with 250 μm thickness on the support layer using a film applicator (Automatic Film Applicator PA-2101, BYC-Gardner GmbH). The casting film was immersed into the coagulation bath and kept in the bath for 30 min. The temperature of the coagulation bath has been set at 5, 20 and 40°C.

The polymer concentrations and the coagulation bath temperatures used to obtain the membranes are shown in Table 1.

Table 1. Membrane number by coagulation bath temperature.

Membrane no.	PSf concentration [wt%]	Coagulation bath temperature [^{0} C]
PSf-1	23	5
PSf-2	23	20
PSf-3	23	40

2.1. Pure water flux

The membrane obtained were characterized by flux experiments with pure water in a dead-end filtration cell (Sterlitech HP4750 Stirred Cell) at an operating pressure of 10 bar using an inert gas (nitrogen) at room temperature (aprox.25°C). The pure water flux (J_w) was calculated as follows

$$Jw = \frac{V}{A \cdot t} Jw = \frac{V}{A \cdot t} \quad (1)$$

Where V is the volume of permeate collected during the time interval t through an effective membrane area A. The permeability tests were performed at four different pressures 6, 8, 10 and 12 bar.

The pure water permeability (L/m^2 · h · bar) was calculated using the following expression:

$$Pure\ water\ permeability = \frac{Pure\ water\ flux}{\Delta p} \quad (2)$$

Where Δp is the operating pressure (bar).

2.2. Retention tests

The retention of solids from a solution is the quantity of sludge that remains on the surface of the membrane through which the feed solution flows.

In this study, it was observed the retention properties of the membrane samples with Methylene Blue dye with a molecular mass of 319.85 $g \cdot mol^{-1}$. The dye concentration is determined spectroscopically using a UV-Vis Spectrophotometer HACH DR 5000 (Hach Lange GmbH, Germany).

The rejection ratio was calculated by the following equation:

$$Retention\ [\%] = \left(1 - \frac{C_f}{C_0}\right) \cdot 100 \quad (3)$$

Where C_0 represents dye concentrations in feed solution (100 ppm) and C_f is the permeate concentration.

2.3. Hydrophilicity

The water affinity of the membranes was determined using contact angle measurements. The surface hydrophilicity of the prepared PSf samples was evaluated by using a contact angle goniometer (OCA 15EC, DataPhysics) to measure the contact angle between the polymer surface and water droplet.

Before contact angle measurements, membrane samples were dried at 5°C for 24 hours and stored in a vacuum desiccator. The dropping was repeated for several times for each sample.

2.4. Porosity

Porosity of a membrane has a very important role in permeability and retention. The porosity of the membrane is determined by the equation:

$$Porosity = \frac{W_w - W_d}{\rho_w \times V} \quad (4)$$

Where ρ_w is pure water density at room temperature (kg/m^3) and V is membrane volume in wet case (m^3). W_w and W_d are the masses of a membrane, in wetted and dry state.

3. Results and discussions

In figure 1 it was observed that the flux of a membrane depends on the temperature of the coagulation bath. Increasing the temperature of the coagulation bath at 40°C leads to an increase in water flux.

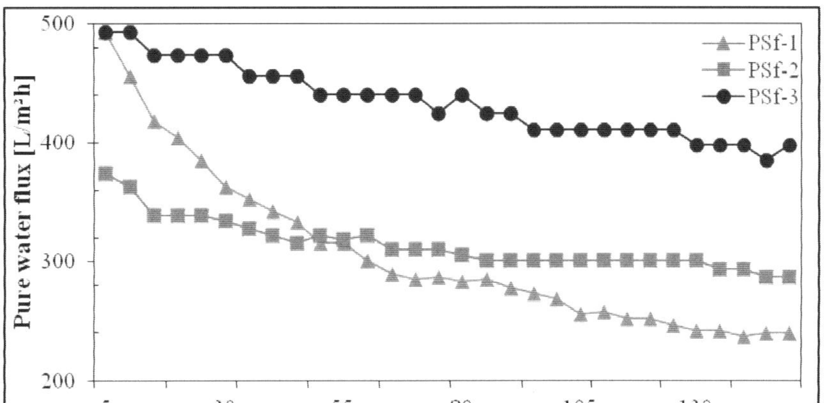

Figure 1. Comparative flux Figure.

According to the Brownian motion, in a liquid-liquid system, the irregular motion of the molecules from the two liquids is directly proportional to temperature. Temperature increase will create a more chaotic movement of molecules [16]. In the phase inversion method, temperature is of great interest in the coagulation bath. Higher temperature of the coagulation bath will lead to a faster withdraw of the solvent from the polymer solution, creating larger and irregular pores, approved by the flux of PSf-3 membrane in figure 1. In the same time, lower temepratures will make the solvent exit the polymer solution in a more controlled manner, producing smaller pores, as show in figure 1 and 2.

The rejection of methylene blue with 100 ppm concentration was investigated for prepared membranes by measuring the retention flux as it can be seen in figure 2.

Figure 2 shows that the flux of PSf-3 membrane, obtained at 40°C, has a high decline comparative with the other two. When the coagulation bath temperature is low, the retention flux of membranes is more stable.

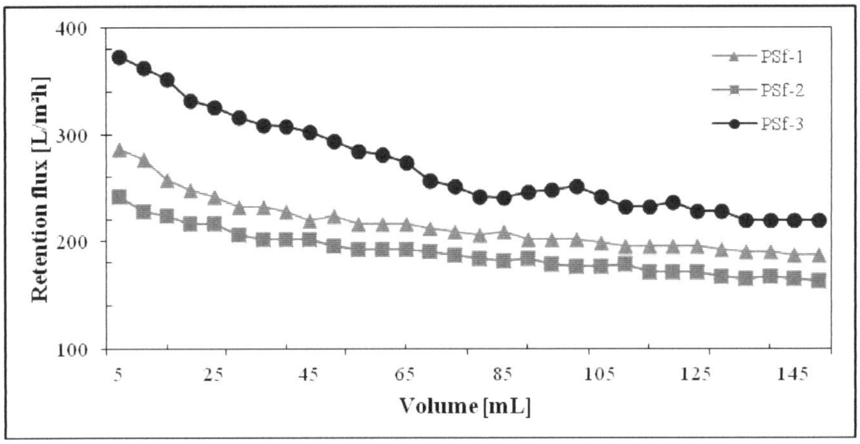

Figure 2. Comparative retention flux.

When analysing the average flux values, both in pure water and dye, PSf-1 has the lowest and unstable pure water flux. When we put the PSf-1 membrane in retention tests, we notice that the dye flux is higher than the membrane made at room temperature (PSf-2).

The retention coefficient tested is showed in figure 3. There is a tendency to increase the separation effect with decreasing the coagulation bath temperature. The permeability of the membrane is inversely proportional with the retention of methylene blue. The maximum performance of membrane retention is obtained at 5°C temperature of coagulation bath.

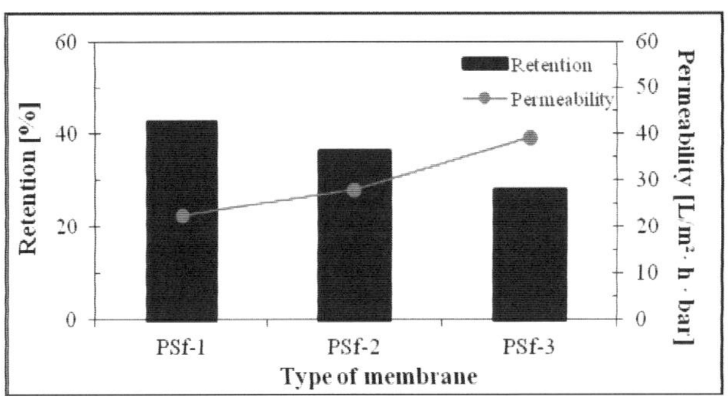

Figure 3. Methylene blue retention and the permeability of the membranes.

Water flux of the membrane obtained in coagulation bath with 40°C (PSf-3) is higher than water flux of the membrane obtained in coagulation bath with 5°C (PSf-1) due to the improvement of membrane hydrophilicity.

As can be seen in figure 4, the contact angle decreases as the coagulation bath temperature increases, meaning that the membrane surface is more hydrophilic and the porosity is higher.

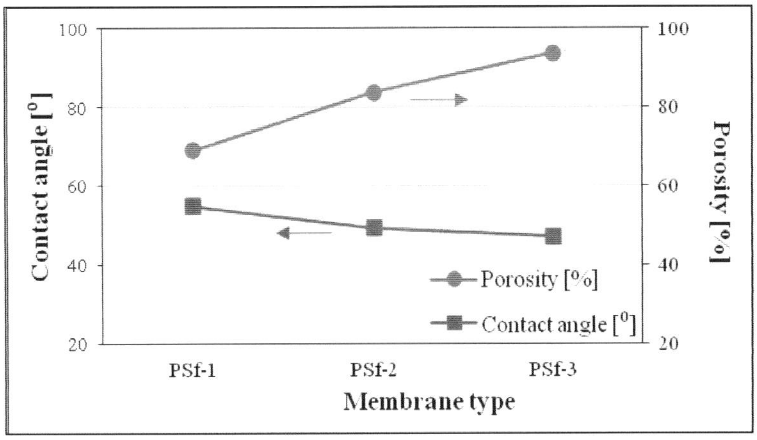

Figure 4. Hydrophilicity and porosity.

Membrane porosity is higher by increasing the coagulation temperature because the heat transmission rate decreased. Increasing the porosity leads to the increase of pure water flux and the decrease of rejection process, also have an impact on the decrease of the membrane resistance.

4. Conclusions

The properties of the membrane are significantly influenced by the coagulation bath temperatures. As observed, the pure water flux increase when the coagulation temperature rises from 5°C to 40°C, but the retention decreases. The increasing temperature has also an impact on the increase of porosity because the solvent and the water diffusion is accelerated and forms a large pore volume. On the formation process of the thin layer, the higher temperature increases the dissolving rate of the polymer (PSf).

At 40°C of coagulation bath, membrane properties are improved, less retention of the methylene blue. The membranes obtained at low temperatures have a good stability and retention.

References

[1] Jainesh H Jhaveri and Z V P Murthy 2016 *A comprehensive review on anti-fouling nanocomposite membranes for pressure driven membrane separation processes* (Surat, India: Desalination)

[2] P C Wankat 2006 *Separation Process Engineering* (New Jersey, Statele Unite: Pearson Education)

[3] Yuelian Peng, Hongwei Fan, Yajun Dong, Yanna Song, Hua Han 2012 *Effects of exposure time on variations in the structure and hydrophobicity of polyvinylidene fluoride membranes prepared via vapor-induced phase separation* (China: Applied Surface Science) p 7872–7881

[4] Sourirajan S and Loeb S 1963 *Sea water demineralization by means of an osmotic membrane* (Washington DC: Advance Chemistry*)* p 117

[5] M Mulder 1998 *Basic Principles of Membrane Technology* (London: Kluwer Academic Publichers) p 564

[6] M Cheryan 1998 *Ultrafiltration and Microfiltration Handbook* (USA: *Publisher CRC Press)*

[7] C K Kima, J H Kim, I J Roh, J J Kim 2000 *The changes of membrane performance with*

polyamide molecular structure in the reverse osmosis process (South Korea: Journal of Membrane Science) p 189–199

[8] N Pezeshk, D Rana, R M Narbaitz, and T Matsuura 2012 *Novel modified PVDF ultrafiltration flat-sheet membranes* (Ontario, Canada: Journal of Membrane Science) p 280–286

[9] H L Richards, P G L Baker and E Iwuoha 2012 *Metal Nanoparticle Modified Polysulfone Membranes for Use in Wastewater Treatment: A Critical Review* (Bellville, South Africa: Journal of Surface Engineered Materials and Advanced Technology) p 183–193

[10] H F Mark, N M Bikales and C G Over-berger 1988 *Encyclopedia of Polymer Science and Engineering*, vol. 13 (New York: Publisher John Wiley & Sons)

[11] J E Yoo, J H Kim, Y Kim and C K Kim 2003 *Novel ultrafiltration membranes prepared from the new miscible blends of polysulfone with poly (1-vinylpyrrolidone-co-styrene) copolymers* (South Korea: Journal of Membrane Science) p 95

[12] J H Kim, M S Kang and C K Kim 2005 *Fabrication of membranes for the liquid separation: Part 1. Ultrafiltration membranes prepared from novel miscible blends of polysulfone and poly (1-vinylpyrrolidone-co-acrylonitrile) copolymers* (South Korea: Journal of Membrane Science) p 167–175

[13] Rinku Thomas, Elena Guillen-Burrieza and Hassan A Arafat 2014 *Pore structure control of PVDF membranes using a 2-stage coagulation bath phase inversion process for application in membrane distillation (MD)* (United Arab Emirates: Journal of Membrane Science) p 470–480

[14] Ziyi Wang, Linquan Sun, Qin Wang, Baoan Li and Shichang Wang 2014 *A novel approach to fabricate interconnected sponge-like and highly permeable polyvinylidene fluoride hollow fiber membranes for direct contact membrane distillation* (Tianjin, China: European Polymer Journal) p 262–272

[15] Lishun Wu and Junfen Sun 2015 *An improved process for polyvinylidene fluoride membrane preparation by using a water soluble diluent via thermally induced phase separation technique* (PR China: Materials & Design) p 204–214

[16] D Braun, H Cherdron, H Ritter 2001 *Polymer Synthesis: Theory and practice, Fundamentals, Methods, Experiments - Third edition* (Berlin, Germany: Publisher Springer)

Development of Electrodeposited Zn/nano-TiO_2 Composite Coatings with Enhanced Corrosion Performance

L Benea[1] and E Dănăilă
Competences Center: Interfaces-Tribocorrosion-Electrochemical Systems (CC-ITES), Faculty of Engineering, Dunarea de Jos University of Galati.
47 Domneasca Street, RO-800008, Galati Romania

E-mail: Lidia.Benea@ugal.ro, lidiabd@yahoo.com

Abstract. Pure zinc coatings have been found ineffective when are used in aggressive environments such as those which contain chlorides or industrial pollutants [1]. In this paper, Zn/nano-TiO_2 composite coatings with various contents of TiO_2 nanoparticles (diameter size of 10 nm) were prepared on low-carbon steel by electro-codeposition technique. The deposition was carried out at different cathodic potentials ranging from −1600 mV to −2100 mV for different deposition times between 5-15 min. Pure Zn coatings were also produced under the same experimental conditions for comparison. Present work aims to investigate the effects of selected electrodeposition parameters (cathodic potential, TiO_2 nanoparticle concentration in the plating bath and electrodeposition time) on the corrosion behavior of electrodeposited Zn/nano-TiO_2 composite obtained. The corrosion experiments were performed in natural seawater, using electrochemical methods such as open circuit potential, potentiodynamic polarization and linear polarization resistance. The results showed that the inclusion of TiO_2 nanoparticles into zinc matrix lead to an improved corrosion resistance comparatively with pure zinc coatings obtained under similar conditions.

1. Introduction

The composite layers obtained by the process of codeposition the various dispersed phases with the metals in the process of electro-crystallization. There are still little information due to the limited conditions of production and the properties of the heterogeneous surfaces obtained [1-3]. A method of manufacturing these composite layers, which began to be investigated in 1962, is based on electrolysis of the dispersed phase suspension in a metallic plating bath [4].

Pure and composite zinc coatings have been and are widely used since ancient times, as the most advantageous methods of protection against the corrosion of ferrous and other materials [1-3].

Electrodeposition is one of the most common methods in the industry as the electrochemical coating method, presenting some advantages, requiring simple apparatus, requiring no expensive reagents, and last but not least, it shows great efficacity in the sense that it can be produced a large quantity, high quality and a variety of applicability of the resulting products. The development of these electrodeposition processes has led to a substantial increase in empirical information not only on electrodeposition processes in electrolyte suspensions, but also on the conditions required to obtain a satisfactory composition of composite layer with controlled composition [5-9].

Electrodeposition of composite layers offers the possibility of obtaining new materials by incorporating different dispersion particles into the metal matrix during the electro reduction of the metal in the electrolysis process of its salts.

One of the additives studied and most often used in the electrodeposition process is gelatin, since it reduces the adsorption of hydrogen ions used as an additive along with the basic chloride and, of course, increases cathodic efficiency [10].

A significant current and future interest is to use ceramic nanoparticles in the electrodeposition process in order to obtain nanocomposite materials as layers on top of other materials [11,12]. These types of ceramic particles can be: Al_2O_3, ZrO_2, TiO_2, SiO_2, which have a better corrosion resistance than pure zinc. In this case, where particulate matter is used in the electrolyte solution, the process of agglomerating the ceramic particles could occur and therefore a stirring method is needed [1-13].

The method of preparing composite coatings is the most important but least controllable variable that leads and explains sometimes contradictory data about the heterogeneous properties of composite layers [1-10]. This problem of particle formation and sedimentation can be solved by introducing some additive into the electrolyte solution [14,15]. The purpose of the paper is to improve the surface in the sense of lowering the roughness of the electro-codeposited layer and to increase the mechanical properties, especially the hardness. In practice, solutions containing between 150-300 g / L $ZnSO_4$ x $7H_2O$, i.e. about 1-2 equivalents-gram, are usually used. Solutions with high concentration of zinc sulphate at medium current density give macrocrystalline deposits. Other salts, in addition to the zinc sulphate electrolyte, are usually alkali metal or alkaline earth metal sulphates or chlorides, salts soluble in the environment, for example: Na_2SO_4, $(NH_4)_2SO_4$, NaCl, NH_4Cl, Al_2SO_4, and so on. Sulphates and chlorides of alkaline metals are introduced into the zinc electrolyte in particular to increase electrical conductivity, because the solutions of zinc sulphate and sulphates of bivalent metals (cadmium, nickel, copper, etc.) do not have good electrical conductivity [10-16].

Addition of sodium sulfate between 10 g / L and 100 g / L in $ZnSO_4$ solution has almost no influence on cathodic polarization. In turn, the electrical conductivity of the solution greatly increases, which has a favorable influence on the dispersion capacity of the electrolyte. Electrolysis of zinc in acidic solutions, the value of anodic and cathodic polarization also depends on the nature of the anions to which the zinc or other cations introduced into the solution are bound. Zinc matrix composites are typically obtained from sulphate solutions, but sometimes alkaline electrolytes with cyanide and non-cyanide alkali electrolytes can be used. Particles that can be codeposited in the zinc matrix are: silicon carbide, silicon oxide, aluminum oxide, titanium dioxide, cerium oxide, vanadium oxide, graphite. [17,18,19].

Our research work aims to investigate the effects of selected electrodeposition parameters (cathodic potential, TiO_2 nanoparticle concentration in the plating bath and electrodeposition time) on the corrosion behaviour of electrodeposited Zn/nano-TiO_2 composite obtained. The corrosion experiments were performed in natural seawater, using electrochemical methods such as open circuit potential, potentiodynamic polarization and linear polarization resistance. The results showed that the inclusion of TiO_2 nanoparticles into zinc matrix lead to an improved corrosion resistance comparatively with pure zinc coatings obtained under similar conditions.

2. Experimental set-up

2.1. Preparation of nanocomposite layers and kinetic studies

Experiments to obtain Zn /TiO_2 nanocomposite coatings in zinc matrix, with nanoparticles of titanium oxide (TiO_2), having a mean diameter of 10 nm, were performed at room temperature in an electrolyte with the composition specified in Table 1.

The electrodeposition experiments were performed in a three electrode electrochemical cell, where the cathode (the support of the layers) was made of carbon steel of dimensions 25 mm x 25 mm x 0.5 mm. The anode made from the zinc bar positioned at a distance of 30 mm from the cathode, having the same size with the cathode to maintain the zinc ions content constantly in the electrolyte solution.

A reference electrode (Ag / AgCl with saturated KCl solution having - E = -199 mV vs. the normal NHE hydrogen electrode) was used. All three electrodes are positioned vertically and parallel to each other. In this configuration, the particles are held in suspension in solution by means of a magnetic stirrer. A potentiostat type VoltaLab PGZ 100 connected to a computer was used for electrochemical measurements.

Table 1. Composition and parameters of Zn plating bath.

Compound	Concentration, g / L
$ZnSO_4 7H_2O$	200 g/L
Na_2SO_4	15 g/L
$ZnCl_2$	35 g/L
TiO_2 Nanoparticles	5, 10, 20 g/L
pH	4,85
Temperature	25^0C
Cathodic potentials	-1600, -1900, -2100 mV vs. Ag/AgCl
Electrodeposition time	5, 10, 15 min.
Magnetic stirring	300 rpm

For the electro-codeposition kinetics study, a VoltaLab PGZ 100 potentiostat connected to a computer was used. In this step a circular zinc plate representing the anode (pure Zn) and the copper substrate in cylindrical form representing the cathode were used. The same reference electrode was used for kinetics studies. In this configuration, the TiO_2 nanoparticles are also kept suspended in the solution by means of a magnetic stirrer.

The following layers were obtained:

Pure Zinc defined layers as: Zn / nano TiO_2 (0 g / L);

Composite layers with a concentration of 5 g / L TiO_2 nanoparticles in the defined solution as: Zn / nano TiO_2 (5 g / L);

Composite layers with a concentration of 10 g / L TiO_2 nanoparticles in the defined solution as: Zn / nano TiO_2 (10 g / L);

Composite layers with a concentration of 20 g / L TiO_2 nanoparticles in the defined solution as: Zn / nano TiO_2 (20 g / L).

2.2. Corrosion study of obtained layers

The electrochemical cell used in the electrochemical corrosion tests was also a three electrode one. The cell contains a reference electrode (Ag / AgCl with saturated KCl solution having - E = -199 mV vs. normal NHE hydrogen electrode), a Pt-Rh network representing the counter electrode and pure Zn or Nanocomposite layers representing the cathode. All three electrodes are positioned vertically and parallel to each other. The corrosion experiments were carried out in sea water, taken from the Romanian seaside area, namely Mangalia. This solution has a pH = 7.95, salinity = 11.9‰, conductivity = 19.8 mS, and the volume of solution used in the corrosion tests was 210 mL.

3. Results and discussion

3.1. The effect of TiO_2 nanoparticles on electrochemical measurements during electrodeposition

The electro-codeposition mechanism is considered to occur in the following steps [1]:

- At the time of introduction into the electrolyte, a layer of adsorbed ionic species is formed at the surface of the particles forming the diffuse double electric layer around each particle.

- The particles are conveyed by stirring and under the influence of electric current, they will spread to the surface of the cathode.
- When the adsorbed ionic species on the particles are reduced, the particles are embedded in the electrodeposited cathodic layer.

These ionic species cannot directly pass through a charge transfer stage, so that the first step in depositing a metal must be its chemical transformation into an electroactive specie. This stage takes place during the diffusion transport of the ionic species to the surface of the electrode which takes place inside the hydrodynamic limit layer or in some cases in the double layer [1].

Cyclic voltammetry is one of the most widely used electrochemical methods of investigation, providing information about the electroactive species involved in electrochemical processes (intermediates, products) about the coupling of chemical steps with electrochemical steps, about electro sorption phenomena, as well as kinetic data on the studied processes [2,3].

In this method of investigation, the controlled parameter is the electrode potential that varies over time, and the measured parameter is the current flowing through the electrode.

Using the potentiostat, a controlled potential is applied to the working electrode and the current flowing is measured against a reference electrode. In voltammetry, the two electrodes between which the current flows are: the working electrode and the auxiliary electrode (this electrode must have a surface much larger than the working electrode, which makes it not polarized) [2].

For the cyclic voltammetric study of electro-codeposition process, a voltage starting from a potential of -100 mV (Ag / AgCl) electrode potential was applied to the circuit terminals, reaching the value of -500 mV (vs Ag / AgCl) and then Returned to the potential value of -1500 mV (vs Ag / AgCl). The required potential variation was made for a scanning rate of 1 mV / sec.

Figure 1 (a) and (b) shows the cyclic voltammetry curves plotted at a cathodic potential with and without dispersion of TiO_2 nanoparticles, following the variation of the current density with the reduction potential of the metal.

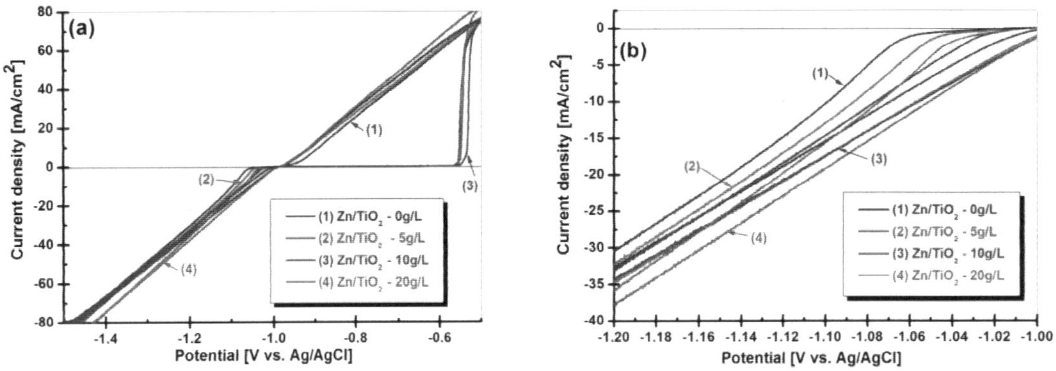

Figure 1. Comparative cyclic voltammetric diagrams performed for electrodeposition of: (1)-Zn/TiO_2 (0 g/L), (2)-Zn/TiO_2 (5 g/L), (3)-Zn/TiO_2 (10 g/L), (4)-Zn/TiO_2 (20 g/L), (a) corresponding to all scanned potential domain and (b) zoom in the cathodic potential range.

According to figure 1 (a) - (b), the presence of TiO_2 nanoparticles in the solution has different effects. Compared to the pure zinc diagram, the TiO_2 nanoparticle curves at a concentration of 5 g / L, 10 g / L and 20 g / L in the zinc plating electrolyte show a shift to more positive potentials.

Considering a constant value of the -1.4 V (vs Ag / AgCl) as cathodic potential in figure 1 (a), it is noted that the current density reveals the lowest value for the pure Zn layer, the corresponding value being of -66.34 mA / cm^2 (t = 100 sec). For the Zn / nano-TiO_2 layer (5 g / L), the current density value increases to -68.68 mA / cm^2 (t = 100 sec) and increases even more for the Zn / nano-TiO_2 layer (20 gL), the current density value reaching -74.86 mA / cm^2 (t = 100 sec).

The presented results confirm the activation of the Zn electroreduction process in the presence of TiO_2 dispersed nanoparticles. This behavior demonstrates that one of the intermediate steps of particle codeposition is the adsorption of metal ions on their surface and their migration to the surface of the cathode, the effect of which is the activation of the entire zinc electro-crystallization process.
Other authors concluded also that the addition of dispersed nanoparticles moves the nickel reduction curves to more positive potentials. The change in the reduction potential is attributed to the increase in the active surface due to cathodic adsorbed nanoparticles and a possible increase in ionic transport through nanoparticles with adsorbed ionic layers.

3.2. The dependence of the thickness of the practical layer obtained versus the codeposition time
Following electrodeposition, a first examination was visual, and the conclusion drawn from the examination is that at elevated cathodic potential between -1600 mV and -2100 mV, the layer is uniformly deposited having a bright silver color.

Determination of the (practical) thickness of the obtained layers was performed by weighing the samples before and after deposition using the electronic balance of the KERN type, and applying the following formulas:

$$G = \frac{m_d}{S} \text{ and } \Delta G = \frac{G}{\rho_{Zn}}$$

Where: G - layer thickness (g / cm^2);
m_d - mass deposited at cathode (g);
S - active area (cm^2);
ΔG - layer thickness (µm);
ρ_{Zn} - zinc density (g / cm^3).

The investigations made on these samples confirm that at higher electrodeposition times, the deposited layer or electro-codeposited is more consistent in terms of the practical thickness obtained. The variation of the comparative values of the practical thicknesses corresponding to the layers deposited at the cathodic potential E = -1600mV depending on the time of the electrodeposition process (5 min, 10 min, 15 min) is shown in figure 2.

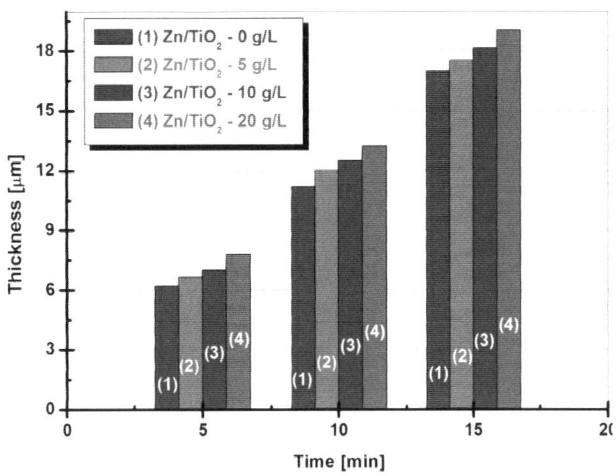

Figure 2. Layer thicknesses of electrodeposited layers obtained at cathodic potential of E=-1600mV, vs. time of electrodeposition (5 min, 10 min, 15 min): (1) - Pure Zn pur; (2) - Zn/nano-TiO_2 (5 g/L), (3) - Zn/nano-TiO_2 (10 g/L and (4) - Zn/nano-TiO2 (20 g/L).

From figure 2 it can be noticed that the thickness of all Zn / nano-TiO_2 nanocomposites, with all the concentration of TiO_2 added in the electrolyte, is higher than the thickness of pure zinc layers, considering the same electrodeposition parameters (cathodic potential and time). From figure 2 it can also be noticed that the increase of the concentration of dispersed TiO_2 dispersed phase (5 g / L, 10 g / L and 20 g / L) in the electrolyte also increases the thickness of Zn / nano-TiO_2 nanocomposites.

Thus, at the cathodic potential of E = -1600mV and the time of 5 minute, the thickness of pure Zn layer has a value of 6.2097 µm. With the addition of the first TiO_2 concentration (5 g / L), the layer reaches a thickness value equal to 6.6484 µm and increasing even further to 7.0196 µm (at TiO_2 concentrations of 10 g / L) and 7.7958 µm (for TiO_2 concentrations of 20 g / L).

3.3. Evaluation of the corrosion resistance for obtained nanocomposite layers

The evolution of open circuit electrode potential (OCP) during time of immersion in corrosive environment is an electrochemical, corrosion test method, indicating the electrochemical oxidation tendency of a material in a corrosive environment. After a period of immersion it stabilizes around a steady state value. This potential may vary over time as changes occur at the surface of the electrode (oxidation, passive layer formation or immunity). The open circuit potential is used as a corrosion behaviourr criterion.

Passivation is an electrochemical and / or chemical process which consists in lowering the corrosion rate to very low values due to the qualitative modification of the metal interface / corrosive agent. This change may be due to the formation of a poorly soluble oxide or salt film or a chemisorbed, adherent oxygen layer lacking discontinuities acting as a kinetic barrier and preventing oxidation and solvation of the formed ions. In the state of passivity, metals are virtually immune to corrosion and, from an electrochemical point of view, behave like noble metals.

Figure 3 shows the time variation of the open circuit electrode potential for the uncoated support material (1), the pure Zn layers (2), Zn / TiO_2 nanocomposites - 5 g / L (3), Zn / TiO_2 - 10 g / L (4) and Zn / TiO_2 - 20 g / L (5), obtained at the imposed cathode potential of E = -2100mV and the electrodeposition time of 15 minutes.

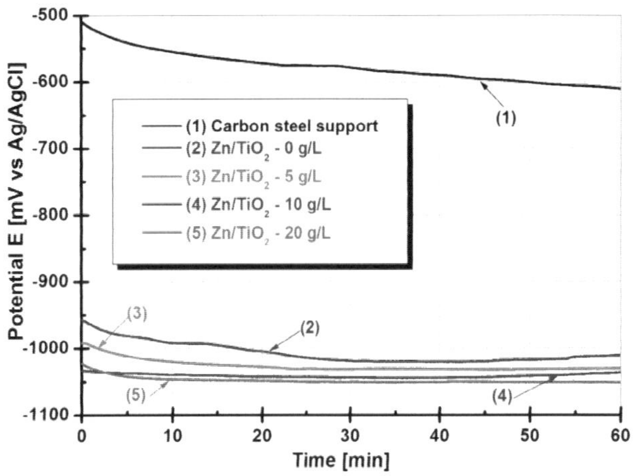

Figure 3. Time variation of the open circuit electrode potential corresponding to: (1) - the uncoated support material; (2) - Pure Zn layer; (3) - Zn / TiO_2 nanocomposite (5 g / L); (4) - Zn / TiO_2 nanocomposite (10 g / L); and (5) - Zn / TiO_2 nanocomposite (20 g / L), obtained at the imposed cathodic potential of E = -2100 mV vs. Ag/AgCl and the electrodeposition time of 15 minutes.

It can be seen from figure 3 that the uncovered support material does not reach the passivation state, its potential decreasing continuously during the immersion time and the end of the first monitoring period. The time evolution of the free potential of pure Zn and Zn / TiO_2 nanocomposites reveals an almost instantaneous passive state since immersion. These observations confirm the beneficial contribution of the studied Zn / TiO_2 coatings deposited on carbon steel, from the point of view of corrosion immunity. These Zn / TiO_2 nanocomposite coatings act as a kinetic barrier and prevent oxidation and solvation of the formed corrosion products.

From figure 3 it can also be noticed that with the increase of the concentration of TiO_2 nanoparticles in the electrolyte, the potential is more electronegative. This behavior can be attributed to the ceramic nanoparticles embedded in the zinc matrix.

Linear polarization is an electrochemical corrosion monitoring technique, being the only method to determine corrosion rate directly in the real time. Response time and data quality of this technique make this method clearly superior to all other forms of corrosion monitoring.

Figure 4 shows the polarization resistance obtained by several linear polarization around open circuit potential with a small perturbation of ± 40 mV. The diagrams correspond to pure Zn layers and Zn / TiO_2 nanocomposites obtained with 5 g / L to 20 g / L of dispersed TiO_2 nanoparticles. All the electrodeposited layers were obtained at the imposed cathodic potential of E = -2100mV and the time of the electrodeposition process equal to 15 minutes.

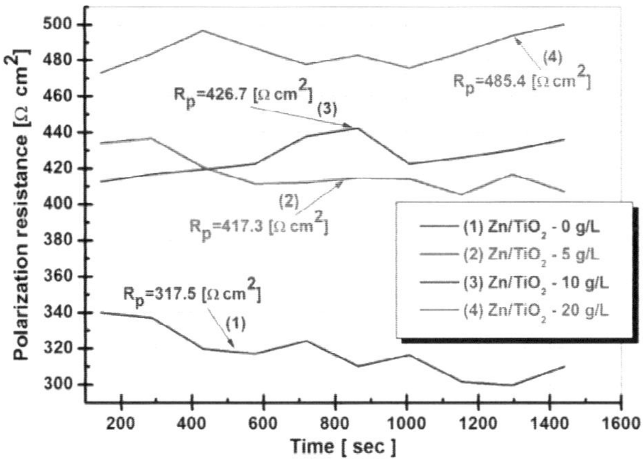

Figure 4. Polarization resistance corresponding to electrodeposited layers: (1) - Pure Zn; (2) - Zn/TiO_2 – 5 g/L; (3) - Zn/TiO_2 (10 g/L) (3); (4) - Zn/TiO_2 (20 g/L), obtained at an imposed cathodic potential of E=-2100mV and electrodeposition time of 15 minutes.

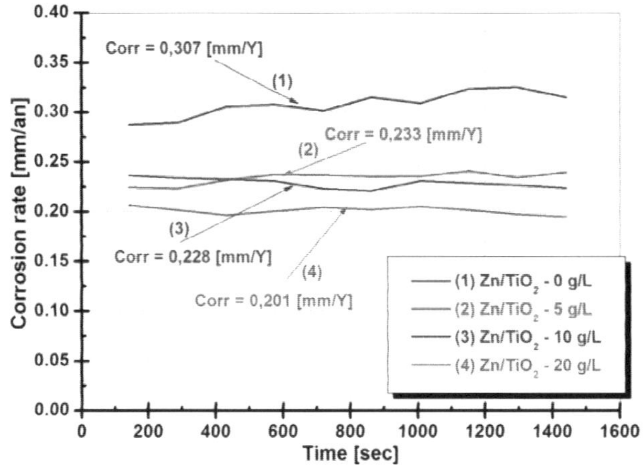

Figure 5. The corrosion rate expressed as penetration rate of: : (1) - Pure Zn; (2) - Zn/TiO_2 – 5 g/L; (3) - Zn/TiO_2 (10 g/L) (3); (4) - Zn/TiO_2 (20 g/L), obtained at an imposed cathodic potential of E=-2100mV and electrodeposition time of 15 minutes.

It should be noted that the polarization resistance is directly proportional to corrosion resistance, better if the polarization resistance is high then the studied material exhibits increased corrosion resistance. From figure 4, it can be clearly seen that the smallest polarization resistance belongs to the pure Zn layer, while that corresponding to Zn / TiO_2 nanocomposite layers increases with increasing the dispersed phase concentration added to the electrolyte. Linear polarization measurements have demonstrated that the incorporation of TiO_2 nanoparticles into the zinc matrix leads to improved polarization resistance and therefore to the improved corrosion resistance in tested corrosive environment.

Figure 5 shows the corrosion rate obtained by linear polarization, corresponding to tested electrodeposited layers, expressed as penetration rate.

From figure 5, it can be clearly seen that the highest corrosion rate belongs to the pure Zn layer, whereas that corresponding to Zn / TiO_2 nanocomposite layers decreases with the increasing of the dispersed phase concentration added to the electrolyte.

Linear polarization measurements have once again demonstrated that the incorporation of TiO_2 nanoparticles into the zinc metallic matrix leads to improved corrosion resistance.

4. Conclusions

The paper aimed to obtain nanocomposite coatings by electrochemical method with dispersed phase of nanometric TiO_2 particles (10 nm) in zinc matrix with the application of improving the anti-corrosion properties. The properties of the Zn / TiO_2 nanocomposite layers obtained are compared with the pure zinc layers obtained at the same time and cathodic potential parameters.

Following the analysis of the experimental results of this research paper led to the following conclusions:

The important factors that influence the quality of nanocomposite layers are: electrolyte composition, electrolysis conditions (cathodic potential, deposition time, temperature, stirring mode).

The nano dispersed TiO_2 particles in the zinc plating solution are contributing to the electrolysis process, which is confirmed by their effect on the the cyclic voltammetry at the cathodic potential traced in the electrolyte

Both in the case of pure zinc layers and nanocomposite layers, the thickness of the layer increases with the increase of the required cathodic potential. For nanocomposite coatings, layer thickness is slightly higher comparatively with that of pure zinc at the same deposition time, thus confirming the inclusion of TiO_2 nanoparticles in the zinc matrix.

The thickness of the electrodeposited layers increases with the increase of the electrodeposition time, a higher increase is observed in the thickness of the nanocomposite layers compared to the thickness of the pure zinc layers obtained at the same imposed cathodic potential.

The comparative study of corrosion resistance shows an improvement by including TiO_2 nanoparticles in the zinc matrix. Thus, the polarization resistance, R_p, increases from 317 $\Omega \cdot cm^2$ for pure zinc layers to 485 $\Omega \cdot cm^2$ for Zn / nano-TiO_2 nanocomposites (20 g / L).

The corrosion rate, expressed as penetration rate, decreases from 0.307 mm / year for pure zinc layers to 0.201 mm / year for Zn / nano-TiO_2 nanocomposite layers.

References

[1] Benea L 1998 *Composite Electrodeposition - In Theory and Practice*, Ed. Porto Franco Galati, Romania
[2] Benea L, Bonora P L, Borello A, Martelli S, Wenger F, Ponthiaux P and Galland J 2001 *J Electrochem Soc* **148** C 461-C 465
[3] Benea L and Danaila E 2016 *J Electrochem Soc* **163 (13)** D655 - D662
[4] Ciubotariu A C, Benea L, Mitoşeriu O, Ponthiaux P and Wenger F 2009 *J Optoelectron Adv M*, **11(6)** 892-897
[5] Prabhu R A, Shanbhag A V and Venkatesha T V 2007 *J. Appl. Electrochem.* **37** 491-497
[6] Li M C, Jiang L L, Zhang W Q, Qian Y H, Luo S Z and Shen J N 2007 *J. Solid State Electrochem.* **11** 549–553
[7] Sachin H P, Achary G, Naik A and Venkatesha T V 2007 *Bull. Mater. Sci.* **30** 57-63
[8] Abdel Aal A, Barakat M A and Mohamed R M 2008 *Appl. Surf. Sci.* **254** 4577-4583
[9] Baik D S and Fray D J 2001 *J. Appl. Electrochem.* **31** 1141–1147
[10] Alfantazi A M and Dreisinger D B 2001 *J. Appl. Electrochem.* **31** 641–646

[11] Grincevichene L, Vishomirskis R, Jakobson S, Williams B and Stasiuk T 1995 *Met. Finish.* **93** 10–14
[12] Kokate M, Garadkar K and Gole A 2016 *J Colloid Interface Sci* **483** 249–260
[13] Afifi S E, Ebaid A R, Hegazy M M and Barakat, A K 1992 *JOM* **44** 32–34
[14] Li Q, Lu H, Cui J, An M and Li D 2016 *Surf. Coat. Technol.* **304** 567-573
[15] Otani T, Nagata M, Fukunaka Y and Homma T 2016 *Electrochim. Acta* **206** 366-373
[16] de Carvalho M F and Carlos I A 2013 *Electrochim. Acta* **113** 229-239
[17] Abbott A P, Barron J C, Frisch G, Ryder K S and Silva A F 2011 *Electrochim. Acta* **56** 5272-5279
[18] Xia X, Zhitomirsky I and McDermid J R 2009 *J. Mater. Process. Tech.* **209** 2632–2640
[19] Celis J P, Roos J R and Buelens C 1987 *J. Electrochem. Soc.* **134** 1402–1408

The Effect of Normal Force on Tribocorrosion Behaviour of Ti-10Zr Alloy and Porous TiO_2–ZrO_2 Thin Film Electrochemical Formed

E Dănăilă[1] and L Benea[1*]
[1] Competences Center: Interfaces-Tribocorrosion-Electrochemical Systems (CC-ITES), Faculty of Engineering, Dunarea de Jos University of Galati, 47 Domnească Street, RO-800008, Galați Romania

E-mail: Lidia.Benea@ugal.ro

Abstract. The tribocorrosion behaviour of Ti-10Zr alloy and porous TiO_2–ZrO_2 thin film electrochemical formed on Ti-10Zr alloy was evaluated in Fusayama-Mayer artificial saliva solution. Tribocorrosion experiments were performed using a unidirectional pin-on-disc experimental set-up which was mechanically and electrochemically instrumented, under various solicitation conditions. The effect of applied normal force on tribocorrosion performance of the tested materials was determined. Open circuit potential (OCP) measurements performed before, during and after sliding tests were applied in order to determine the tribocorrosion degradation. The applied normal force was found to greatly affect the potential during tribocorrosion experiments, an increase in the normal force inducing a decrease in potential accelerating the depassivation of the materials studied. The results show a decrease in friction coefficient with gradually increasing the normal load. It was proved that the porous TiO_2–ZrO_2 thin film electrochemical formed on Ti-10Zr alloy lead to an improvement of tribocorrosion resistance compared to non-anodized Ti-10Zr alloy intended for biomedical applications.

1. Introduction

Commercially pure Ti and its alloys are mainly used for biomedical applications as implantable materials due to their good properties such as high strength, low density, moderate elastic modulus and excellent corrosion resistance, compared to other metallic biomaterials. The highly corrosion resistance of Ti based materials is due to the stable and adhesive passive oxide film (TiO_2) formation on their surface, in the presence of oxygen. This TiO_2 film protects material and provides chemical inertness in different media and is generally around 5 nm thick [1-2]. Nevertheless, the native TiO_2 film is not entirely stable under certain conditions such as mechanical stress, when this one can be easily damaged leading to an accelerated corrosion. Consequently, a strong destructive attack may occur on the surface of Ti based material due to its reaction with hostile environment. This instability of TiO_2 film will affect the biocompatibility of metallic implant, leading to the implant failure [3].

To overcome this problem, surface modification methods are often performed in order to enhance the mechanical, biological and chemical properties of Ti and its alloys. Anodic oxidation treatment is one of the simplest among the different techniques employed to form rough, porous and uniform films across which oxygen ions can diffuse and oxidize the substrate, increasing the thickness of the film.

By anodic oxidation process the ion release is decreases and the corrosion and tribocorrosion resistance of Ti and its alloys is enhances.

The tribocorrosion process has applications in biomedical fields, since it is known that the material implants are often exposed to simultaneous chemical/electrochemical and mechanical stresses [3-4]. The previous studies from the literature focused on the importance of Ti-xZr binary alloy as implant material [5-6] only in terms of corrosion performance of this alloy. Consequently, the main goal of this paper was to bring in a great contribution to the applicability of Ti-10Zr binary alloy as biomaterial. Therefore in this work was study the tribocorrosion performance of treated and untreated Ti-10Zr alloy in Fusayama-Mayer artificial saliva, under various solicitation conditions.

2. Materials and Methods

2.1. Anodic oxidation treatment

The fabrication method of porous TiO_2–ZrO_2 thin film on Ti-10Zr alloy consisted in anodic oxidation. The chemical composition of Ti-10Zr alloy is shown in table 1. Before anodic oxidation, samples of the Ti-10Zr alloy in the form of rods were cut into cylinders, embedded in epoxy resin, exposing a geometrical area of about 2.5 cm^2. Then the samples were ground using successively finer SiC abrasive papers (320 - 4000 μm), diamond paste (3 and 1 μm) and a suspension of SiO_2 solution (0.04 μm size of particles), finally achieving a mirror finish. Subsequently, the polished samples were cleaned with ethanol in an ultrasonic bath for 5 min and dried using dry, cold air.

Table 1. Chemical composition of Ti-10Zr alloy.

Element	Ti	Zr	Fe	Ni	Cu
Chemical composition [wt %]	88.23	10.89	0.39	0.28	0.21

The electrochemical setup for the growth of the porous oxide films by controlled anodic oxidation method consisted of a two-electrode configuration, with Ti–10Zr specimens previously prepared worked as anode and a Ti–6Al–4V alloy sample (in a size of 5.7 cm^2) used as counter electrode. The anodization was performed in a glass cell containing a 1M H_2SO_4 solution, by applying a DC voltage of 200 V, for 2 min. under magnetic stirring. During the anodic oxidation, 1M H_2SO_4 solution was stirred (300 rpm) in order to maintain the homogeneity of electrolyte and to accelerate the escape of gas produced in the electrochemical reaction from the surface of the Ti–10Zr substrate.

2.2. Materials characterization

The porous surface morphology of Ti–10Zr alloy specimens after anodic oxidation treatment was examined by scanning electron microscopy (SEM–EDX, FEI Quanta 200 FEG) and the chemical composition was evaluated with an energy dispersive X-ray spectroscopy (EDX) device attached to the SEM.

Tribocorrosion tests were performed using a unidirectional pin–on–disc tribometer permitting to carry out experiments in corrosive aqueous media under controlled electrochemical conditions. For electrochemical measurements was used a three–electrode arrangement which was controlled by a 1287 potentiostat and a 1255 frequency response analyser Solartron. The working electrode (WE) consisted of untreated Ti–10Zr alloy surface and porous TiO_2–ZrO_2 film formed by anodic oxidation. A circular Pt-Rh grid was used as a counter electrode (CE) whereas the reference one an Ag/AgCl electrode with KCl saturated solution (E = 200 mV vs. standard hydrogen electrode – SHE). The schematic set–up of electrochemical cell used in tribocorrosion experiment is shown in figure 1.

The electrolyte utilized in tribocorrosion tests was the Fusayama - Mayer artificial saliva [3], with the pH equal to 4.8. Alumina pin cylinder (7 mm diameter with 100 mm radius spherical tip) was used as counterpart in the sliding tests due to its high hardness, high wear resistance, chemical inertness and electrical insulating properties. The alumina pin was rotated under continuous sliding conditions drawing a circular wear track on the test sample. Prior to any sliding test, the counterbodies were degreased with ethanol. The sliding tests were carried out at a constant rotation speed of 120 rotation min^{-1}, by applying two normal force of 2N and 5 N for 5000 cycles.

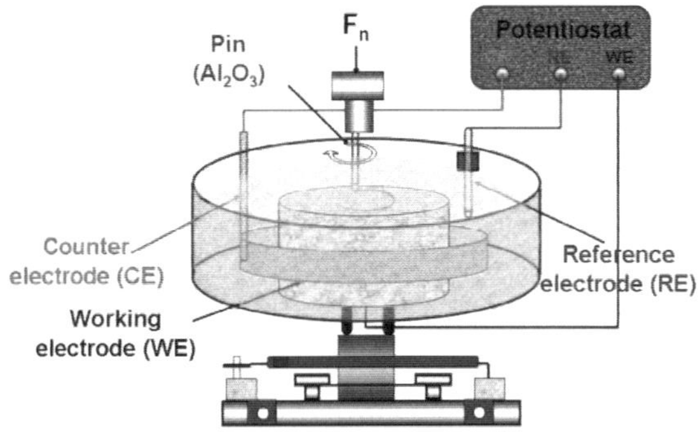

Figure 1. Schematic view of the experimental set up used at tribocorrosion tests.

The protocol used for the tribocorrosion experiments consisted of three steps: (1) measuring the open circuit potential - OCP (E^1_{oc}), before the sliding for a period of 30 min till a stable E^1_{oc} value of the test sample is achieved; (2) measuring the OCP (E^2_{oc}) of the test sample under sliding conditions; and (3) measuring the OCP (E^3_{oc}) of the test sample after sliding test, for 30 min.

3. Results and discussions

3.1. Materials characterization

Figure 2 (a) shows the surface morphology obtained after anodic oxidation treatment of Ti-10Zr alloy and figure 2 (b) the corresponding elemental composition.

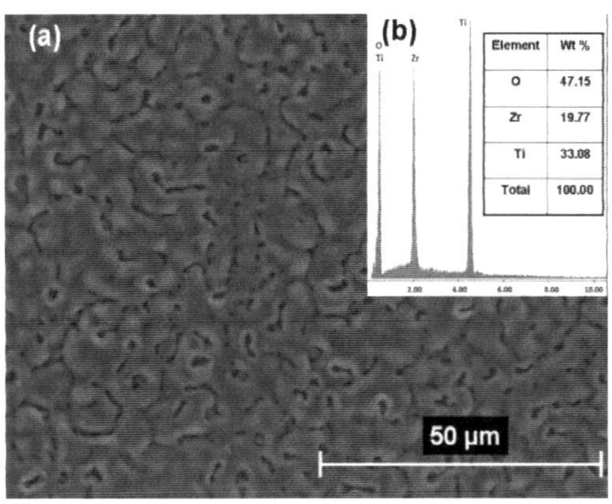

Figure 2. (a) SEM surface morphology and (b) EDX spectrum of anodic porous TiO_2-ZrO_2 film formed.

Figure 2 (a) shows that the oxide coating formed by anodic oxidation exhibits rough and porous surface with pores distributed randomly over the surface. Also it can be seen that the anodic film reveal irregular and in some regions large pores. This porous structure is able to form biological adhesive interface between bone and implant.

From figure 2 (b) which shows the wt. % of anodic porous TiO_2–ZrO_2 surface it can be seen that the wt. % corresponding to Zr element almost doubles compared to those of untreated Ti–10Zr alloy (Table 1). This can be explained by the fact that Zr element which is a better reducing agent, competes with Ti during anodization, and will oxidize first under the same electrochemical conditions.

3.2. Tribocorrosion behaviour

In figure 3 (a) and (b) are shown the evolutions of open circuit potential (OCP) as a function of the normal force applied, corresponding to treated and untreated Ti-10Zr alloy immersed in artificial saliva before (30 min.), during (5000 sliding cycles) and after (30 min.) sliding tests, at a constant rotation speed of 120 rotation min^{-1}.

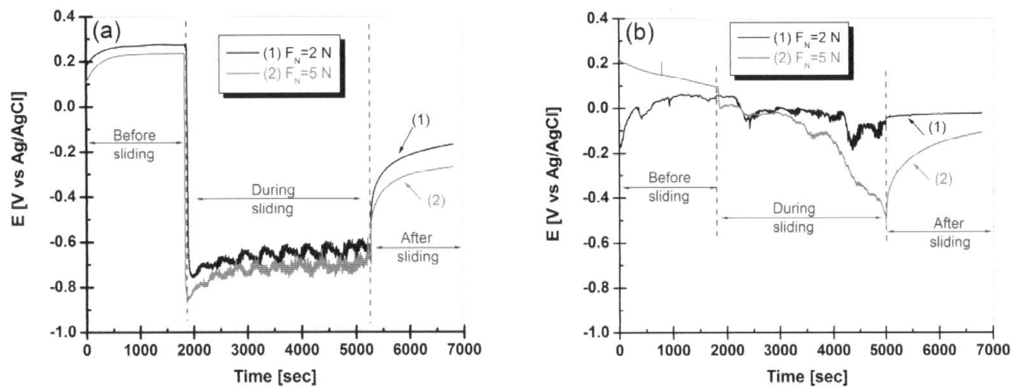

Figure 3. Variation of the open circuit potential of: (a) untreated Ti–10Zr alloy and (b) anodic porous TiO_2–ZrO_2 surface immersed in Fusayama–Mayer saliva before, during and after sliding tests by applying (1) - F_n = 2 N and (2) - F_n = 5 N, for 5000 cycles, at a constant rotation speed of 120 rotation min^{-1}.

From figure 3 (a) it is observed that the OCP recorded before sliding (E^1_{oc}) of untreated Ti–10Zr alloy showed a shift of potential in the active direction and also that by the end of 30 min. a steady state potential is attained for this surfaces. From the same figure 3 (a) it can be clearly seen that with the onset of sliding a sudden decrease of the OCP occurs for both normal forces applied. The cathodic shift potential increases with increasing the normal load. This sudden drop in potential is associated with the damage of native passive film by its partial or complete removal [7]. The sudden drop in OCP observed for untreated Ti–10Zr alloy surface induces its susceptibility for corrosion. In the period of sliding motion some oscillations in the OCP for were reported, these being attributed to the periodic removal caused by the contact with counter body (depassivation) and growth caused by contact with the electrolyte (repassivation) of the passive film in the sliding zone [8-9]. After unloading, the OCP of untreated Ti–10Zr alloy reveal an anodic shift caused by repassivation of the substrate forming immediately a protective thin film. Both tested surfaces fail to reach the initial steady state value in the given duration of time after the sliding is stopped.

From figure 3 (b) it is observed that the OCP recorded before sliding (E^1_{oc}) of anodic porous TiO_2–ZrO_2 surface remained nearly constant, confirming that the samples are in the passivated state. By applying the normal force of 2 N, the anodic porous TiO_2–ZrO_2 surface did not exhibit a sudden cathodic drop in OCP, being rather gradual until the sliding motion is stopped. With the onset of sliding under the normal force of 5 N, the anodic porous TiO_2–ZrO_2 surface presents a very small drop of the OCP in the cathodic domain. Also for this surface it can be said that the cathodic shift potential increases with increasing the normal load. The lack or the insignificant of such a sudden drop in OCP for anodic porous TiO_2–ZrO_2 layer surface suggest that the hard and thick oxide film is not totally removed with the onset of sliding. The potential fluctuations observed during sliding are more pronounced for untreated Ti–10Zr alloy surface compared to anodic porous TiO_2–ZrO_2 film surface. This behaviour of anodic porous TiO_2–ZrO_2 film reveals its enhanced tribocorrosion properties.

The effect of normal loads (2 and 5 N) on the evolution of friction coefficient for treated and untreated Ti–10Zr alloy recorded for 5000 sliding cycles, at a constant rotation speed of 120 rotation min-1 in artificial saliva is observed in figure 4.

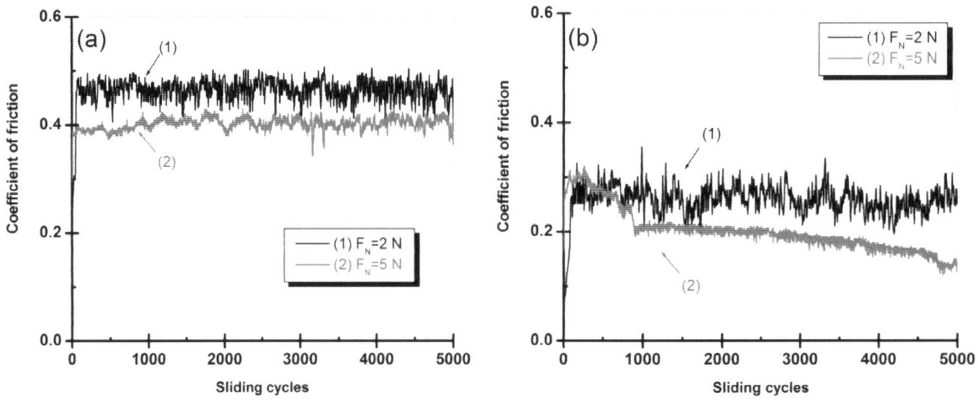

Figure 4. Friction coefficients versus sliding cycles obtained in Fusayama–Mayer saliva for: (a) untreated Ti–10Zr alloy and (b) anodic porous TiO_2–ZrO_2 surface by applying (1) - $F_n = 2$ N and (2) - $F_n = 5$ N, for 5000 cycles, at a constant rotation speed of 120 rotation min^{-1}.

Figure 4 (a) and (b) reveals a decrease in the coefficient of friction with increasing normal force at constant rotation speed and constant number of cycles, on the same type of sample. This behaviour can be explained by the smoothening of the surface under high load and diminishing the shear strength of the contacting interfaces.

Comparing the coefficients of friction of the untreated Ti–10Zr alloy surface (figure 4 (a)) with those of anodic porous TiO_2–ZrO_2 film surface (figure 4 (b)), it can be noticed that the lowest coefficients of friction are obtained for the last type of surface. These results highlight once again that anodic oxidation treatment extend the lifetime of a biomedical implant made of Ti–10Zr alloy.

4. Conclusions

The tribocorrosion behaviour of untreated Ti–10Zr alloy surface and anodic porous TiO_2–ZrO_2 surface sliding against Al_2O_3 pin in an artificial saliva solution was investigated in a pin–on–disc contact configuration combined with in situ electrochemical measurements (OCP).

Applied normal forces were found to greatly affect the potential of untreated Ti–10Zr alloy during tribocorrosion experiments. An increase in the normal force induces a decrease in potential accelerating the depassivation process, due to the removal of the native passive oxide film induced by sliding, confirming its increase in susceptibility for corrosion.

The lack or the insignificant shift in OCP for anodic porous TiO_2–ZrO_2 layer surface with the onset of sliding suggests that the hard and thick oxide film is not totally removed. This behaviour of anodic porous TiO_2–ZrO_2 film reveals its enhanced tribocorrosion properties.

Applied normal forces greatly affect also the coefficients of friction. The coefficient of friction decreases with increasing normal force under constant rotation speed and constant number of sliding cycles, on the same type of tested material. Lower friction coefficients were obtained by anodic porous TiO_2–ZrO_2 surface compared with untreated Ti–10Zr alloy surface.

All experimental results have proved that porous TiO_2–ZrO_2 film surface formed by anodic oxidation reveal higher tribocorrosion performances as compared with untreated Ti–10Zr alloy, under lubricated conditions.

References

[1] Bahrami Nasab M and Hassan M R 2010 *Trends Biomater. Artif. Organs* **24** 69-82
[2] Mohammed M T, Khan Z A and Siddiquee A N 2014 *Procedia Materials Science* **6** 1610-1618
[3] Benea L, Danaila E and Ponthiaux P 2015 *Corrosion Science* **91** 262–271
[4] Mathew M T, Srinivasa Pai P, Pourza R, Fischer A and Wimmer M A 2009 *Advances in Tribology* **2009** 1–12
[5] Han M K, Hwang M J, Yang M S, Yang H S, Song H J and Park Y J 2014 *Materials Science & Engineering A* **616** 268–274
[6] Ferreira E A, Rocha-Filho R C, Biaggio S R and Bocchi N 2010 *Corrosion Science* **52**, 4058–4063.
[7] Ponthiaux P, Wenger F, Drees D and Celis J P 2004 *Wear* **256** 459–468
[8] Kumar S, Narayanan T S N S, Raman S G S and Seshadri S K 2010 *Tribology International*, **43** 1245-1252
[9] Berradja A, Bratu F, Benea L, Willems G and Celis J P 2006 *Wear* **261** 987-993

Corrosion Behavior of Aluminum Oxide Film Growth by Controlled Anodic Oxidation

V Dumitrascu[1], L Benea[1,*] and E Danaila[1]
[1]Competences Center: Interfaces-Tribocorrosion-Electrochemical Systems (CC-ITES), Faculty of Engineering, Dunarea de Jos University of Galati, 47 Domneasca Street, RO-800008, Galati, Romania.

E-mail: Lidia.Benea@ugal.ro

Abstract. Due to the light weight and good corrosion resistance, nowadays aluminum and its alloys are used in different industries in order to decrease the maintenance costs and also to increase the equipments lifetime cycle. When aluminum and its alloys are exposed to the extreme environments, the native aluminum oxide film lose the anticorrosive properties that lead to the damage of equipments and increasing the costs. In order to improve the anticorrosive and mechanical performances of aluminum and its alloys, different techniques are used: organic coatings, the growth of a thick aluminum oxide film through different methods, etc.. The most used method for aluminum oxide growth is anodic oxidation. Anodic oxidation is an electrochemical method that allows to growth an aluminum oxide film with controllable characteristics. The aim of present paper was to growth on 1050 aluminum alloy surface nanoporous aluminum oxide films with improved anticorrosive properties. The obtained nanoporous aluminum oxide films were characterized morphological and structural by scanning electron microscopy coupled with X-ray energy dispersive analyzer. The anticorrosive properties were evaluated by electrochemical methods such as: open circuit potential, electrochemical impedance spectroscopy and cyclic voltammetry. The results showed that anodic oxidation treatment improve the anticorrosive performances of 1050 aluminum alloy.

1. Introduction

The problems of costs received significant attention from economists, researchers and engineers in the last years. New materials were developed by the researchers, to increase the life services of equipments and also to decrease the fabrication prices, in order to reduce the costs.

Aluminum and its alloys received significant attention from the researchers because of its high technological value and its wide range of industrial applications especially in aerospace and household industries [1, 2].

The researchers tried to improve the mechanical properties of aluminum by alloying with different elements [3] but they encountered a new problem: the decreasing of corrosion resistance. The decreasing of corrosion resistance in many instances reduces the applications of aluminum alloys.

Developing successful protective coatings on aluminum and its alloys is still a challenging task even if different methods are used. Different techniques such as: conversion coatings, sol-gel deposition or anodic oxidation [1,2,4-6] are used to improve the corrosion resistance of aluminum and its alloys, but everyone method present its limitations.

Anodic oxidation is an electrochemical technique used to increase the thickness of native aluminum oxide layer and to obtain a protective coating on aluminum and its alloys, with adhesive bonding to the substrate, improved wear resistance and anti-corrosive properties [7,8].

The type of anodic oxide film that can be produced upon aluminum surface depend upon several parameters and the most important it is the nature of the electrolyte. The formed anodic aluminum oxide layer could be a barrier-type layer if the aluminum oxide is insoluble in chosen electrolyte or a porous-type layer if the aluminum oxide is slightly soluble in the electrolyte [9].

Beside the electrolyte type, the involved parameters in anodic oxidation process affect the structure and the morphology of obtained aluminum oxide layer. Controlling the anodic oxidation parameters, the obtained aluminum oxide layer could be tailored for particular applications.

In order to obtain a protective aluminum oxide layer on 1050 aluminum alloy (1050AA), the anodic oxidation processes were conducted in aged sulphuric acid, at different applied potentials. In this paper we reported the effect of anodic oxide process on pores formation and improving the corrosion resistance of 1050AA. The morphological and structural analysis were investigated using a scanning electron microscope (SEM) coupled with X-ray energy dispersive analyzer. The corrosion resistance of obtained coatings were measured by electrochemical methods and compared.

2. Materials and methods

An aluminum sheet (1050AA, min 99.5% Al, 2 mm thickness) was cut in coupons with dimensions 35 mm x 30 mm. The coupons were mechanical wet-polished with SiC grinder paper (#1500 and #2000), etched in 5 M NaOH for 30 seconds at room temperature and finally washed in deionized water and alcohol under ultrasonic stirring. The samples were electropolished at 2 V, 80°C for 30 minutes in a mixture of 5% Na_3PO_4 and 15% Na_2CO_3 [10]. The electropolished aluminum pieces were placed in a two electrode cell with a parallel electrode arrangement using a large 1050AA sheet as cathode. For anodic oxidation process the electrolyte was 1 M H_2SO_4 in which it was added 1 gL^{-1} $Al_2(SO_4)x18H_2O$, in order to simulate an aged electrolyte. The anodic oxidation process was performed in potentiostatic mode using a high voltage power source (TDK LAMBDA GEN 300-8) by applying constant voltages of 15 V, 18 V and 21 V at room temperature for 35 minutes.

The electrochemical assays were done in 3.5 % NaCl solution using a standard electrochemical cell. The prepared samples with an exposed surface of 1.7 cm^2, were served as working electrode, Pt-Rh grid as counter electrode and Ag/AgCl (saturated KCl solution, E= 200mV vs. NHE – normal hydrogen electrode) as reference electrode were used. The electrochemical cell was connected to a Voltalab PGZ100 potentiostat/galvanostat during the electrochemical corrosion measurements.

In order to reach the steady-state, the open circuit potential (OCP) was performed for 17 hours. The electrochemical impedance spectroscopy (EIS) tests were conducted over a frequency range of 10^5 down to 10 Hz, with 10 points per decade, using 10 mV sin waves. The EIS results were analyzed using the fit program Zview 3.4. Before the cyclic voltammetry measurements, the analyzed samples were cathodically treated at -2100 mV for 180 seconds in 3.5% NaCl. The cyclic voltammetry measurements were initiated from -1.45 V vs. Ag/AgCl to -0.4V vs. Ag/AgCl with a potential scan rate of 1 mV/s and then reversed with the same scan rate until one complete cycle is formed.

To observe morphology and chemical composition of the 1050AA before and after anodic oxidation treatment, scanning electron microscopy (SEM, FEI QUANTA 200) coupled with X-ray energy dispersive unit was used.

3. Results and discussion

3.1. SEM and EDX analysis

In figure 1 are presented the SEM image of the top surface of 1050AA alloy taken before the anodic oxidation process. The electropolished surface of 1050AA present a cellular structure composed of randomly located depressions, bumps and small etch pits randomly distributed, which could be seeds for pore nucleation [6].

The influence of anodizing potential on the surfaces morphology formed in aged 1 M H_2SO_4 from 15 V up to 21 V in 3 V steps is shown in figure 2(a-c). It could be seen that the anodic oxidation process change the surface morphology of 1050AA samples and the nanopores are developed on anodic aluminum oxide layer. At the lowest potential (figure 2(a)), the 1050AA surface presents a disorganized structure with numerous irregularities and only a few nanopores randomly distributed could be observed. The mean diameter of nanopores is around 13 nm.

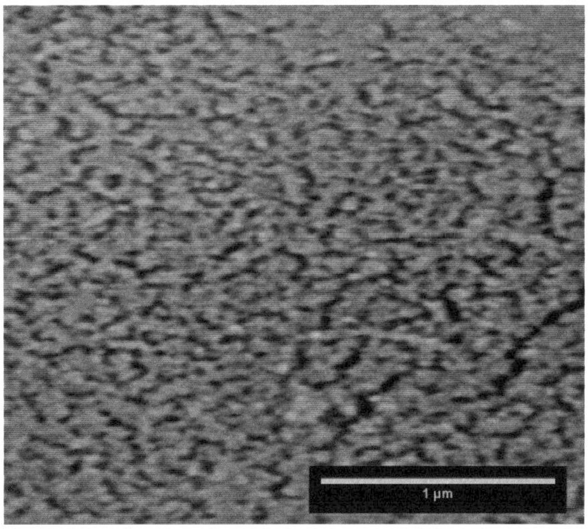

Figure 1. SEM images of electropolished 1050AA alloy surface.

Increasing the anodizing potential from 15 V to 18 V and 21 V, the surfaces morphology maintains the same irregular structure. The pattern distribution of nanopores still remains irregular, the nanopores density, as well as the nanopores diameters have increased and also, the numbers of defects have decreased. The mean diameter of nanopores increases up to 22 nm for anodizing potential at 18 V and 29 nm for anodizing potential at 21 V respectively.

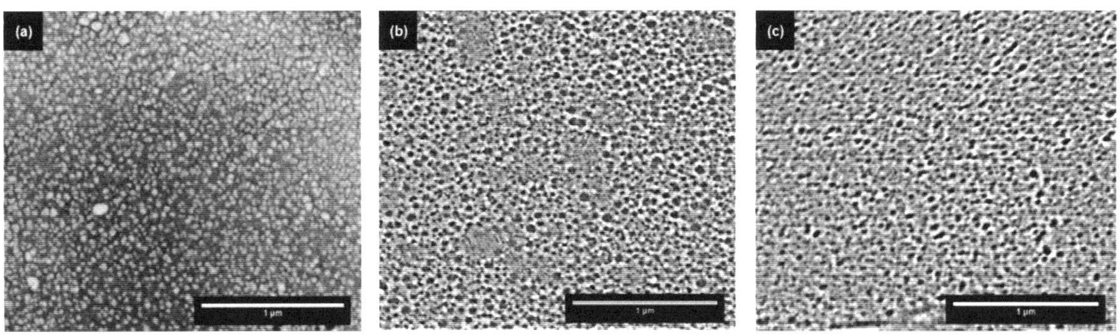

Figure 2. SEM images of 1050AA alloy anodized at different potentials: (a) 15V, (b) 18V and (c) 21V.

The changing of surface morphology involves changes in chemical composition of surfaces as can be seen from table 1, where are presented the average contents of principal elements (Al, O, C and S) of analyzed samples.

From the dates presented in table 1 it could be observed that the increasing of anodizing potential lead to the decreasing of Al content and increasing of oxygen content. Also, from the EDX spectrum could be observed the presence of C and S peaks. The C presence is due to the fact that the samples were cleaned with alcohol before SEM-EDX analyzes, and the presence of S could be attributed to SO_4^{2-} ions entrapped into nanoporous aluminum oxide layer during anodic oxidation process [6, 12].

Table 1. The average contents of aluminum, oxygen, carbon and sulphur (wt %) as function of applied potential.

Compound content (wt %)	Al	O	C	S
0 V	76.75	10.65	12.6	0
15 V	46.74	46.27	2.19	4.8
18 V	45.53	46.93	2.09	5.45
21 V	44.33	47.42	2.69	5.56

3.2. Corrosion tests

3.2.1. Open circuit potential (OCP). In figure 3 are presented the evolution of open circuit potential with immersion time for electropolished 1050AA and anodized 1050AA samples at different applied potentials in 3.5% NaCl solution. After exposing to solution of electropolished sample, the open circuit potential oscillates around -700 mV vs. Ag/AgCl and has a slightly decreasing trend in evaluated interval. The potential fluctuations could be associated with the anodic reaction kinetics due to formation of natural oxide film and it dissolution under chloride ions presence [13].

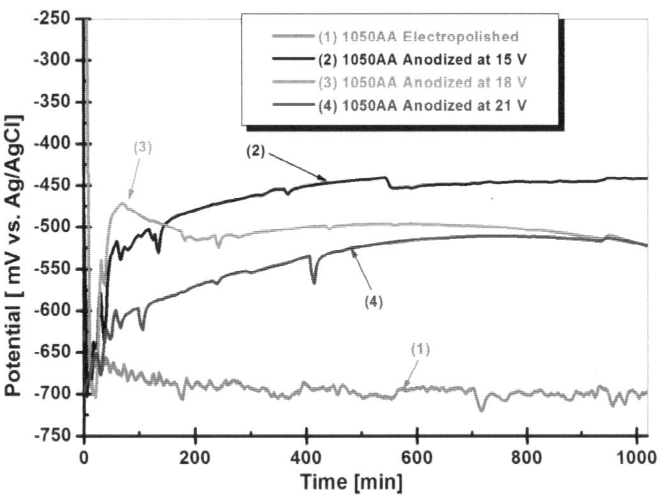

Figure 3. The evolution of open circuit potential with immersion time for 1050AA: (1) Electropolished, (2) Anodized at 15V, (3) Anodized at 18V and (4) Anodized at 21V.

For anodized sample, the stationary steady state is attained after 9 hours of immersion in 3.5% NaCl solution. The anodized 1050AA sample at 15 V present the highest value of free potential, around -440 mV vs. Ag/AgCl, due to the small diameters and decreased number of pores. By increasing the anodization potential the potential steady state decreases to -550 mV vs. Ag/AgCl for anodized samples at 18 V and 21 V respectively due to the increases of numbers and nanopores diameters. For the both anodized at 18 V and 21 V samples, the free potential present a decreasing trend that could be associated with the dissolution of the barrier layers from the bottom of anodic oxide layers under chloride ions actions.

3.2.2. Potentiodynamic polarization (PD). In order to evaluate the electrochemical stability of prepared samples in 3.5% NaCl, the potentiodynamic polarization curves were determined and are presented in figure 4(a, b) for electropolished and anodized 1050AA samples. In order to assess the polarization domains for analyzed samples, the potentiodynamic polarization measurements were carried out.

The linear forms of potentiodynamic polarization curves showed in figure 4(a, b) present all three potential domains, namely: cathodic domain, the passive domain and the transpassive domain. The larger passive domain and the smaller value of current density in the transpassive domain show a higher corrosion resistance for the analyzed sample [14].

From figure 4(a) for electropolished 1050AA and anodized 1050AA at the lower potential (15 V) could be observed all three domains, in comparison with the anodized samples at 18 V and 21 V respectively. In order to observe the passive domain for all anodized samples it was necessary to present in figure 4(b) a zoom of diagrams from figure 4(a).

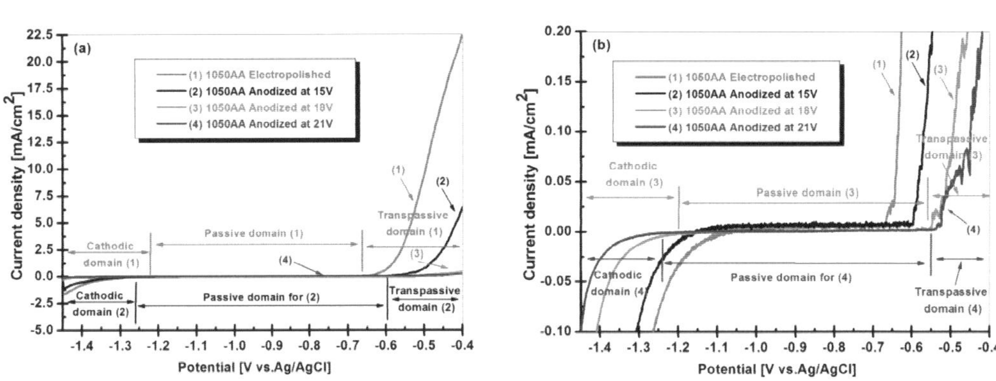

Figure 4. (a) Potentiodynamic polarization curves of 1050AA with different treatments in 3.5% NaCl. (b) Zoom of potentiodynamic polarization curves presented in figure 4(a).

For the electropolished 1050AA (figure 2(a)) it can be seen that the passive domain is more limited compared with anodized surfaces, ranging from -1.2 V to -0.66 V vs. Ag/AgCl. The lower passive domain means a poor corrosion resistance against chloride ions. Also, for electropolished 1050AA sample, the current density increase rapidly in the transpassive region which cause the dissolution of passive film and allow the chloride ions to reach metal base. Also, from figure 2(a), for anodized samples at 15 V it can be observed that the passive domain value is larger, ranging from -1.25 V to -0.59 V vs. Ag/AgCl. The larger passive domain for anodized 1050AA at 15 V in comparison with electropolished 1050AA shows that the anodic oxidation process improves the corrosion resistance of 1050AA alloy surface. The current density increase in the transpassive region and it is caused by the chloride ions penetration of nanoporous layer and dissolution of the aluminum oxide layer.

Increasing the anodizing potential, the passive domains for anodized samples at 18 V and 21 V increase to a more larger potential domains, ranging from -1.23 V to over -0.55 V vs. Ag/AgCl and -1.25 V to -0.53 V vs. Ag/AgCl, respectively. In the transpassive region for anodized samples at 18V and 21V, the current densities present lower values which mean that the electrolyte penetrate very slow the nanoporous aluminum oxide layer and the barrier layer from the bottom of aluminum oxide layer is thick enough in order to protect the metal base against the chloride ions.

3.2.3. Cyclic voltammetry (CV). The best way to analyze the pitting corrosion susceptibility for aluminum and its alloys in the presence of chloride ions is the cyclic voltammetry. The presence of the hysteresis in the transpassivation region of the recorded voltammogram curves shows that the pitting corrosion phenomena appear. The existence of a hysteresis loop in cyclic voltammograms indicates a delay in repassivation of an existing pit when the potential is scanned toward negative direction [15].

The cyclic voltammograms of analyzed samples were recorded from -1.45 V to -0.4 V vs. Ag/AgCl, with a scan rate of 1 mV/s, and after that reversed maintained the same scan rate until a complete cycle is done.

The cyclic voltammograms (*E* vs. *i*) in linear form for electropolished and anodized 1050AA samples are presented in figure 5(a, b). The electropolished 1050AA alloy surface presents the largest hysteresis that could be associated with the pitting susceptibility and the poor repassivation performance of aluminum oxide layer.

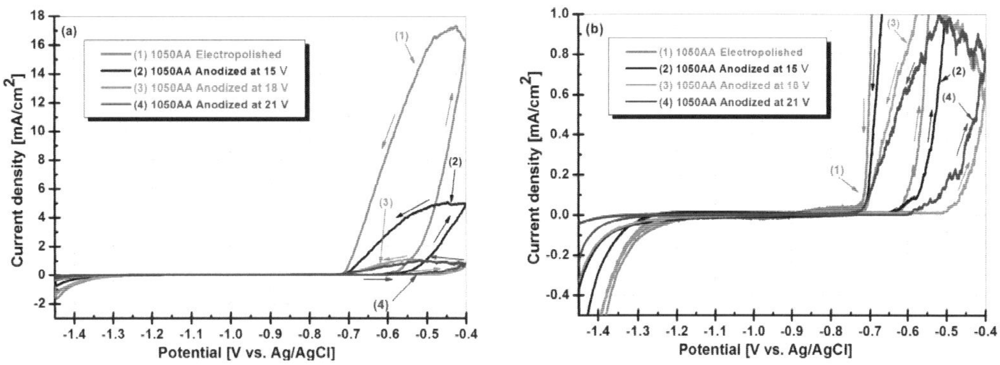

Figure 5. (a) Cyclic voltammograms of 1050AA alloy: (1) Electropolished, (2) Anodized at 15 V, (3) Anodized at 18 V and (4) Anodized at 21 V. (b) Zoom of cyclic voltammograms presented in figure 5(a).

Anodizing the electropolished samples at 15 V, the obtained nanoporous aluminum oxide layer show a decreased surface hysteresis in comparison with electropolished sample which mean a rapid repassivation and also a better corrosion resistance. In order to obtain a better view of pitting susceptibility hysteresis for anodized samples at 18V and 21V, respectively, it was necessary to present in figure 5(b) a zoom of diagrams form figure 5(a). Increasing the anodizing potential, the pitting susceptibility decrease and the repassivation process occurred more rapidly confirming that the anodic oxidation improve the corrosion resistance of 1050AA.

The nanoporous aluminum oxide layer formed on electropolished 1050AA sample at 15 V show a decreased surface of hysteresis in comparison with the hysteresis recorded for electropolished sample and also a better corrosion resistance. By increasing of the anodizing potential the corrosion resistance of 1050AA samples is increased and also the pitting susceptibility is decreased.

4. Conclusion

The surface morphology and the anti-corrosive performance of electropolished and anodized 1050AA samples were studied.

The anodic oxidation process in aged sulphuric acid at different potentials form on electropolished 1050AA samples a thick nanoporous aluminum oxide layer with improved anti-corrosive performances in comparison with electropolished samples.

By increasing the anodizing potential, the nanopores diameters have been increased and they become more regulate and also the thickness of aluminum oxide layer increase.

The electrochemical measurements show that in comparison with electropolished sample the anodic oxidation increase the passive domain and the anti-corrosive performances of 1050AA.

By increasing the anodizing potential, the obtained nanoporous aluminum oxide layers show a decreasing pitting susceptibility and an improved corrosion resistance.

The electrochemical assays confirm that the anodic oxidation improve significantly the anti-corrosive performances of 1050AA alloy, being a cheap method to increase the lifecycle of aluminum alloy parts and pieces.

References

[1] Raj V and Mumjitha M 2014 *Mat. Sci. Eng. B* **179** 25
[2] Mohammadi I, Ahmadi S and Afshar A 2016 *Mater. Chem. Phys.* **183** 490
[3] Shin J, Kim T, Kim D, Kim D and Kim K 2017 *J. Alloy. Compd.* **698** 577
[4] Whelan M, Cassidy J and Duffy B 2013 *Surf. Coat. Tech.* **235** 86
[5] Sharifi Golru S, Attar M M and Ramezanzadeh B 2015 *J. Ind. Eng. Chem.* **24** 233
[6] Michalska-Domanska M, Norek M Stepniowski W J and Budner B 2013 *Electrochim. Acta.* **105** 424
[7] Dasquet J P, Caillard D, Conforto E, Bonino J P and Bes R 2000 *Thin Solid Films* **371** 183
[8] Zechner J, Mohanty G, Frantz C, Cebeci H, Philippe L and Michler J 2013 *Surf. Coat. Tech.* **260** 246
[9] Diggle J W, Downie T C and Goulding C W 1969 *Chem. Rev.* **69** 365
[10] Rauf A, Mehmood M, Rasheed M A and Aslam M 2009 *J. Solid State Electrochem.* **13** 321
[11] Raj V and Mumjitha M 2014 Mat. Sci. Eng. B **179** 25
[12] Stojadinovic S, Nedic Z, Belca I, Vasilic R, Kasalica B, Petkovic M and Zekovic L J 2009 *Appl. Surf. Sci.* **256** 763
[13] Dumitrascu V M and Benea L 2017 *Rev. Chim. (Bucuresti)* **68** 77
[14] Mardare L, Benea L, Danaila E and Dumitrascu V 2016 *Key Eng. Mater.* **699** 71
[15] Amin M A, El Rehin S S A and El-Lithy A S 2010 *Corros. Sci.* **52** 3099

IOP Conf. Series: Materials Science and Engineering **209** (2017) 012017 doi:10.1088/1757-899X/209/1/012017

Technique Incorporating Cooling & Contraction / Expansion Analysis to Illustrate Shrinkage Tendency in Cast Irons

S Stan, M Chisamera, I Riposan*, L Neacsu, A M Cojocaru and I Stan
Materials Processing & Ecometallurgy, Politehnica University of Bucharest, 313, Independentei Spl., 6th District, 060042 Bucharest, Romania

E-mail: iulian.riposan@upb.ro

Abstract. With the more widespread adoption of thermal analysis testing, thermal analysis data have become an indicator of cast iron quality. The cooling curve and its first derivative display patterns that can be used to predict the characteristics of a cast iron. An experimental device was developed with a technique to simultaneously evaluate cooling curves and expansion or contraction of cast metals during solidification. Its application is illustrated with results on shrinkage tendency of ductile iron treated with FeSiMgRECa master alloy and inoculated with FeSi based alloys, as affected by mould rigidity (green sand and resin sand moulds). Undercooling at the end of solidification relative to the metastable (carbidic) equilibrium temperature and the expansion within the solidification sequence appear to have a strong influence on the susceptibility to macro - and micro - shrinkage in ductile iron castings. Green sand moulds, as less rigid moulds, encourage the formation of contraction defects, not only because of high initial expansion values, but also because of a higher cooling rate during solidification, and consequently, increased undercooling below the metastable equilibrium temperature up to the end of solidification.

1. Introduction

Nodular [spheroidal] graphite cast iron [Ductile iron] is a ferrous material, subjected to casting process, typically including [wt.%] 3.4 – 3.8 C, 1.8 – 2.8 Si, 0.1 – 0.6 Mn, < 0.05 P, < 0.02 S, 0.03 – 0.06 Mg_{res}, with or without rare earth elements [RE] contribution, with nodular [spheroidal] graphite precipitated during solidification. Without spheroidising treatment application [Mg or/and RE], lamellar [flake] graphite morphologies will result. Including nodular graphite particles, at higher compactness degree and consequently at reduced cutting effect of stress concentration in the metal matrix, the mechanical properties of ductile iron are significantly superior to compact graphite cast irons and especially to grey [lamellar graphite] cast irons. Despite of the eutectic – hypereutectic composition, spheroidising treatment applied to prior solidification leads to a higher solidification undercooling, favouring free carbides formation. Due to the peculiar solidification pattern of ductile iron, it is much easier for intercellular cavities to form, comparing to grey cast iron solidification. Consequently, macro and micro shrinkage will usually characterize ductile iron castings. According to Z. Jiyang [1] these contraction defects (figure 1) occur first of all because solidification will be in 'mush" mode and the resulted weak solidified shell cannot withstand the expansion which occurs during eutectic freezing, due to graphite precipitation; the casting expands outwards but does not compact inwards, thus cannot form self-feeding. On the other hand, a strong segregation of carbides-promoting elements in the eutectic inter-cells regions is typical for ductile iron solidification, favouring micro-shrinkage formation in these regions.

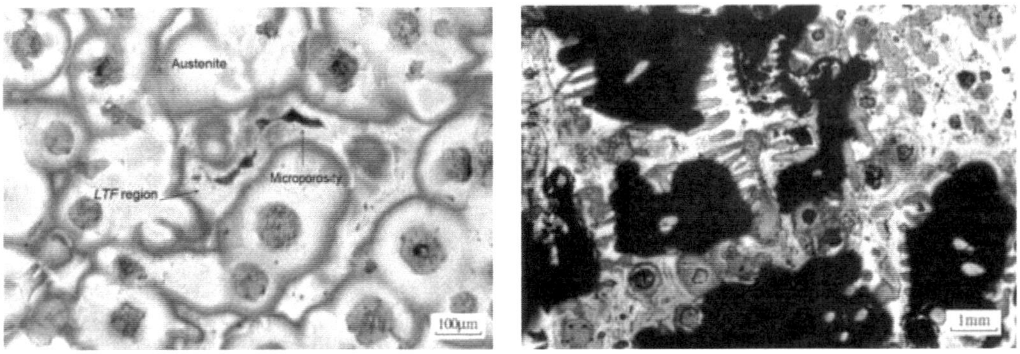

(a) Micro porosity (b) Macro porosity

Figure 1. Shrinkage porosity in ductile iron casting: a) micro-shrinkage, in inter eutectic cells regions; b) macro-shrinkage.

As the most important factors influencing the contraction defects [shrinkage and micro-shrinkage] formation, mould and metal variable act by different ways, such as mould deformation and enlargement tendency, volume-to-surface area ratio (cooling modulus) of casting, chemical composition of cast iron [especially expressed by carbon equivalent], graphite phase [nodularity, nodule count, size distribution], carbides [amount and distribution] and eutectic cells [size, count, distribution] characteristics, pouring temperature, metallurgical treatments [spheroidising and inoculation], difference as expansion in the surface region than in the inner region of a casting due to temperature differences [2-11]. With the more widespread adoption of thermal analysis testing, thermal analysis data have become an indicator of cast iron quality. The cooling curve and its first derivative display patterns that can be used to predict the characteristics of a cast iron. On the other hand, different techniques were developed to evaluate the succession of contraction and expansion stages during cast iron solidification, which are involved in macro- and micro-shrinkage formation.

The main objective of the present work was to introduce a specific experimental instrument and technique for simultaneously evaluating cooling curves and expansion or contraction of cast metals during solidification. Previous papers [12-14] presented information on the development of such specific equipment and a technique. Some recently obtained data are now added, to identify the most representative parameters on the cooling/contraction/expansion curves which are connected with contraction defects formation, focused on mould rigidity influence in inoculated ductile cast irons.

2. Materials and methods

Figure 2 illustrates the recorded equipment, using specialized software created in a LABVIEW program, which is able to simultaneously display both cooling and contraction / expansion curves and their respective source data. The equipment stand includes two parallel moulds (sample size 200 mm and 30 mm square cross section bar, at 0.72 cm cooling modulus –CM, and 1250 g weight). A high speed type AT-MIO10 interface records simultaneous temperature and linear displacement data. The program was capable of 5,000 readings/s and a display rate of 10 samples/s, with each sample represented by an average of 500 readings. The assembly for linear displacement measurement was checked, and an opto-mechanical transducer having a precision of 0.001 mm was used. The thermal analysis system was calibrated by a commercial (Chekmate III/Quik-Cup) instrument, which claims a resolution of 0.1°C according to IEC584/IPTS68, for K type thermocouples.

The calibration instrument accuracy is ± 0.05% + 1°C. The accuracy of the thermal analysis system was evaluated taking into account the full measurements chain from the cup (typical error: 0…+2°C, the compensation wire (error ±1°C), the contact block (error ±1°C) and the hardware (error ±1°C). As a combined result for these chain components, the calculated error was 2.65°C, without calibration. After calibration it was reduced to less than 1°C.

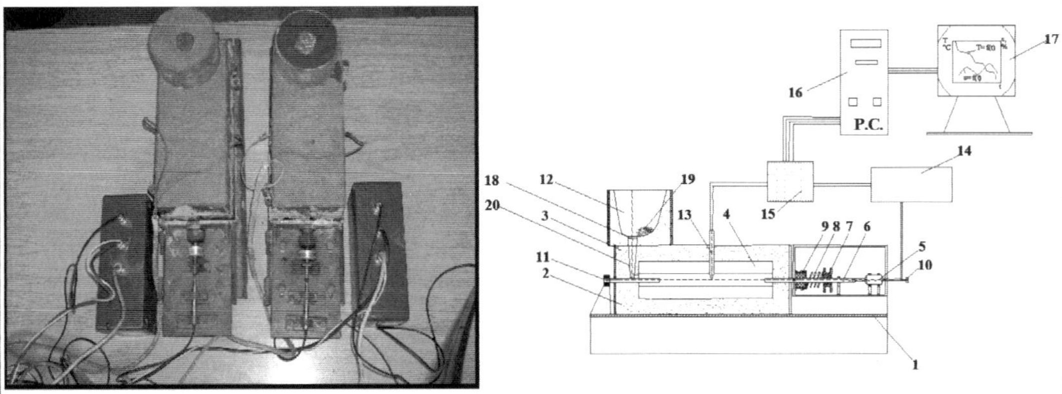

1. Stand; 2. Lower mold; 3. Upper mold; 4. Sample (bar); 5. Displacement transducer; 6. Mobile rod; 7. Thermal radiant; 8. Pretensioning spring; 9. Spring tensioning screw; 10. Fine regulation screw; 11. Rigid rod; 12. Pouring basin; 13. K- type thermocouple; 14. L.V.D.T. amplifier; 15. Connector box; 16. Computer (AT- MIO 10 interface); 17. Display monitor; 18. Fusible steel plate; 19. Inoculant; 20. Sprue.

Figure 2. Arrangement of twin mould cooling and contraction / expansion curves analyser (CCCA) and representative components.

A base iron was melted in a graphite crucible induction furnace (10 kg, 8000 Hz), with 5-7 min superheat at 1550-1570°C. A tundish cover ladle technique was used to treat the iron at 1500-1520^{0}C with a 2.5 wt.% addition of Mg-FeSi alloy with a 3-10 mm grain size (45.4% Si, 6.1% Mg, 0.6% RE, 0.86% Ca, 0.69% Al). Treated irons were poured at 1350-1380°C in the specially prepared moulds included in the Cooling / Contraction Curves Analyser-CCCA (figure 2). Typical Cooling and Contraction / Expansion Curves of ductile test irons and their first derivatives are shown in figure 3, including the significant parameters. To obtain consistency in iron treatments, late inoculation (inoculant addition in the pouring basin) was preferred (0.2 wt.% inoculant, 0.2-0.7 mm grain size), from several other inoculation techniques tested. The CCCA equipment includes two work stations (moulds), which allow simultaneous testing of two inoculants with the same conditions. Different FeSi-based alloys were tested as inoculants. The time of mould filling was 3-4 sec, while the total time of iron processing (tapping, Mg-treatment, deslagging, sample pouring) was maintained within a 3.0-3.5 min range.

Green Sand Mould (GSM) with constant compactness (rigidity) (60-70 Dieter Hardness on the mould separation plane and 125 N/cm^2 as compressive strength on standard cylindrical samples) was compared to Resin (furan) Sand Mould (RSM) (about 89 Dieter Hardness and 1400 N/cm^2 tensile strength on standard samples).

The shrinkage tendency of these experimental ductile irons using the CCCA-test bars (30 mm square cross-section bar) was evaluated by three methods. (a) An X-Ray investigation was made in the front longitudinal plane of the samples. The negative films were scanned and then automatic image analysis was applied to both concentrated and macro-dispersed shrinkage (as absolute and relative surface areas); (b) A water displacement method was used to measure both open and closed [TM - maximum pouring temperature, °C; TAL/TGL/TL - temperature of austenitic / graphitic liquidus, °C; TSEF/TEN - temperature of the start of eutectic freezing (nucleation), °C; TEU – lowest temperature of eutectic undercooling, °C; TER - temperature of graphitic recalescence, °C; TES-temperature of the end of solidification (end of solidus), °C; TEM - maximum recalescence rate, °C/s; Tst – stable (graphitic) eutectic equilibrium temperature, °C; Tmst – metastable (carbidic) eutectic equilibrium temperature, °C; ΔTs - range of equilibrium eutectic temperature (ΔTs = Tst - Tmst), °C; ΔTm – maximum degree of undercooling (ΔTm = Tst - TEU), °C; ΔTr - recalescence (ΔTr = TER - TEU), °C; τ_{ES} - duration of eutectic solidification, sec; τ_{ts} - duration of total solidification, sec; τ_{dM} - maximum dilation position, sec; τ_{dr} – time of dilation recalescence (contraction curve), sec; τ_{er} - time of eutectic recalescence (cooling curve), sec; τ_t - time between mould filling and the end of solidification, sec; $(\Delta\tau)_{ES}$ - the disparity between the moments of the end of solidification indicated by first derivatives of cooling and contraction curves; FDES - minimum values of the first derivative of cooling curves at the end of eutectic solidification, °C/s; FDESC - minimum values of the first derivative of contraction curves at the end of eutectic solidification, %/s; $(\varepsilon_{di})_{max}$ - maximum initial expansion value, %; $(\varepsilon_{di})_{TEN}$ - expansion at the initial start of eutectic freezing (nucleation), %; $(\varepsilon_{di})_{TES}$ –expansion in the final moment of eutectic freezing, %; $(\varepsilon_{di})_{ES}$ - expansion value at the end of eutectic solidification established by the first contraction curve derivative, %; ε_{ap} - contraction prior to pearlite transformation, %; ε_{dp} – the pearlitic dilation, %; TEM(e) – maximum rate of eutectic dilation, %/s; MP - the main derivative peak position relative to TL (before TL, it is positive), sec; Ir(e) - first derivative integral of the contraction curve during eutectic dilation; It(s) – total integral of the first contraction curve derivative (prior to pearlitic contraction) concentrated shrinkage, as a volume.

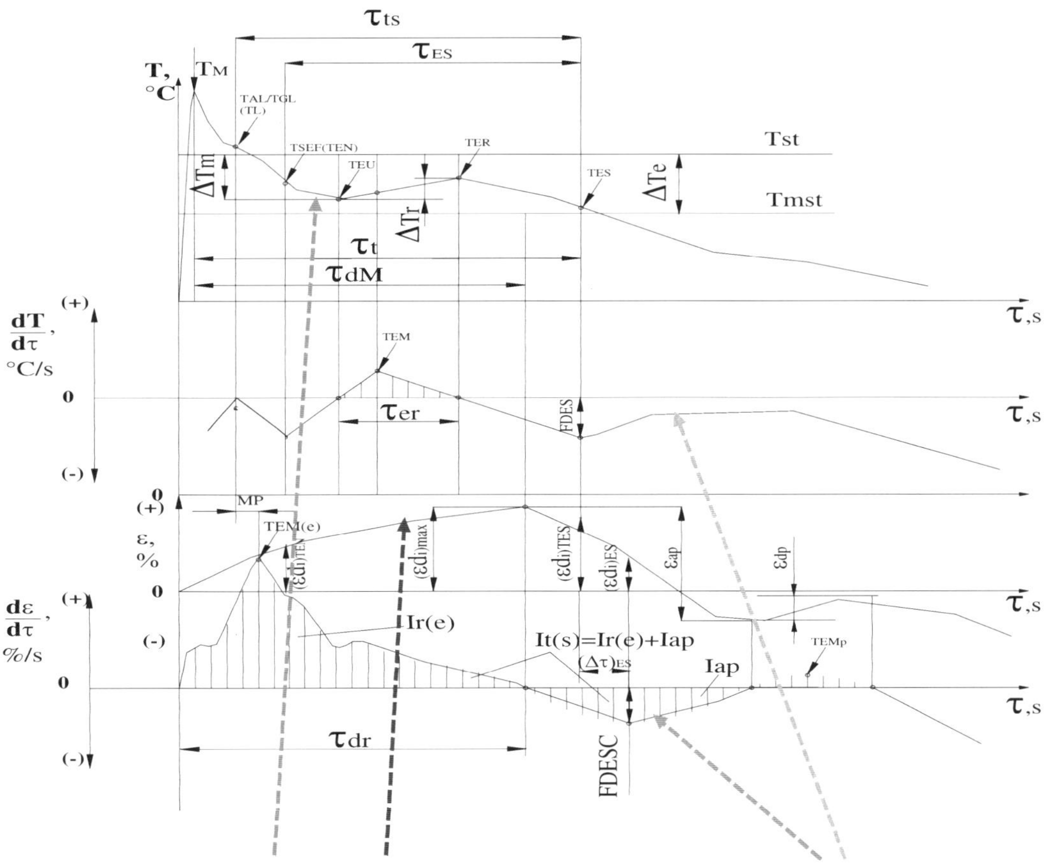

Figure 3. Scheme of cooling / contraction curves of ductile iron and their first derivatives.

The volume of cast samples, with covered and uncovered open concentrated shrinkage was measured. The apparent density as a result of open concentrated shrinkage and total shrinkage was calculated, after weighing the samples. (c) A sample weighing method in both air and water was used to evaluate the apparent density due to the influence by total shrinkage in the region where all shrinkage types are present (ρ_1); also the apparent density as a result of only micro-shrinkage effect (ρ_2) in the region where only micro-shrinkage could be present in apparently sound samples. Apparent density (ρ_1) was evaluated by weighing the section of CCCA bar where both concentrated and macro-dispersed shrinkage were identified by X-Ray examination (it was assumed that a sample would also contain micro-dispersed shrinkage in this region). Apparent density (ρ_2) was also determined by weighing the section of CCCA bar where no concentrated or macro-dispersed shrinkage were found by X-Ray examination.

3. Results and comments

The final chemical composition of all inoculated ductile irons was within a very narrow range (wt.%): 3.71-3.76 C, 2.57-2.63 Si, 0.08-0.11 Mn, 0.022-0.023 P, 0.010-0.012 S, < 0.005 Ce, CE=4.52-4.57%. All cast irons had relatively high final magnesium levels (Mg_{res}=0.065-0.075 wt.%) in order to emphasize the shrinkage sensitivity of the cast iron. A low content of other residual elements, typically (wt.%) 0.05 Cr, 0.03 Ni; 0.07 Cu; 0.067 Mo; 0.006 Ti; < 0.001 V; < 0.002 W; 0.003 Co; 0.010 Al; 0.007 As; 0.005 Sn; < 0.003 Sb; 0.0042 B; 0.0005 Bi; < 0.001 Nb; 0.001 Pb; Te < 0.002 excludes them as possible factors related to shrinkage.

Two major effects of minor elements were considered on the cast iron structure, mainly in ductile iron, in conjunction with base chemistry: pearlite promotion and anti-nodularizing effect. According to Thielman [15], pearlite promotion factor (P_x) and antinodularising factor (K) were calculated:

$$K= 4.4\ (\%Ti) + 2.0\ (\%As) + 2.4\ (\%Sn) + 5.0\ (\%Sb) + 290\ (\%Pb) + 370\ (\%Bi) + 1.6\ (\%Al) \quad (1)$$

$$P_x = 3.0\ (\%Mn) - 2.65\ (\%Si - 2.0) + 7.75\ (\%Cu) + 90\ (\%Sn) + 357\ (\%Pb) + 333(\%Bi) + 20.1\ (\%As) + 9.60\ (\%Cr) + 71.7\ (\%Sb) \quad (2)$$

Typically, these irons have a low anti-nodularising influencing factor K=0.55 and a medium level of pearlite promoting factor P_x= 4.15. It could be concluded that the content of anti-nodularising elements in both Mg-treated irons is low enough, so the role of rare earths to counteract these elements is limited. The pearlite factor P_x indicates a pearlite forming sensitiveness, for conventional solidification conditions. Structure characteristics are in according to the final chemical compositions of Mg-treated and inoculated ductile irons and the specific solidification conditions in the test bars of cooling/contraction curve analysis equipment: more than 80% graphite nodularity, 100 – 175 nodule count [1/mm^2], 25 – 30% pearlite, 70 - 75% ferrite, no free carbides presence.

The thermal analysis evaluation (see figure 3) mainly considered some important temperatures on the cooling curves of cast irons, such as TGL - temperature of the liquidus (zero point on the first derivative), TSEF-temperature of the start of eutectic freezing (minimum negative on the first derivative at the beginning of solidification), TEU - the lowest eutectic temperature (zero point on the first derivative), TER - temperature of graphitic recalescence and TES - temperature of the end of solidification (minimum on the first derivative at the end of solidification). The level of these temperatures is usually compared to the equilibrium temperature, in stable / graphitic [Tst] and in metastable / carbidic [Tmst] systems. There are many elements which individually have favourable or un-favourable influence on the equilibrium temperatures. Silicon appears to be the most important influencing element in un-alloyed irons especially at very low content of trace elements [Tst = 1153 + 6.7 (%Si); Tmst = 1147 – 12 (%Si)] [16]. In the present program, ΔTs = Tst – Tmst = 54 – 56°C, according to high silicon content in final ductile irons.

Conventionally, undercooling is defined with reference to the graphitic equilibrium eutectic temperature (Tst), as ΔTm = Tst - TEU. Free carbides occurrence is typically for ΔTm > ΔTs or TEU

IOP Conf. Series: Materials Science and Engineering **209** (2017) 012017 doi:10.1088/1757-899X/209/1/012017

< Tmst condition, respectively. The importance of the position of the start of eutectic reaction (TEU) comparing to metastable (white) eutectic temperature (Tmst) is revealed by ΔT_1 = TEU – Tmst. For the end of eutectic reaction temperature, ΔT_2 = TER - Tmst parameter was introduced. Free carbides formation is initiated by $\Delta T_1 < 0$, while graphite presence by $\Delta T_2 > 0$. Carbidic formation structure is characterized by $\Delta T_1 < 0$ and $\Delta T_2 < 0$, mottled [graphite + carbide] structure is formed for $\Delta T_1 < 0$ and $\Delta T_2 > 0$ conditions, while a graphitic structure [without free carbides] could be formed only for $\Delta T_1 > 0$ and $\Delta T_2 > 0$ conditions.

Thermal analysis results confirm the real obtained structural parameters. ΔTm = 15 – 19°C < ΔTs, ΔT_1 = 35.8 – 40.1°C > 0 and ΔT_2 = 37 – 42°C > 0 illustrate a structure without free carbides content and a graphitic structure, respectively.

Intercellular carbides or/and inverse chill formation is dependent on the position of the temperature of the end of solidification (TES), compared to the metastable (white) eutectic temperature (Tmst), expressed by ΔT_3 = TES - Tmst parameter. Beneficial end of solidification means high solidus temperature and low level of the ΔT_3 parameter (usually at low negative value, as TES < Tmst in the most of cases). In actual experiments ΔT_3 = -2…-15°C was obtained.

As cooling curve representative parameters, undercooling at the end of solidification [ΔT_3] appears to have a visible influence on the contraction defects of ductile iron castings, especially as shrinkage formation and affected apparent density of castings (figure 4).

More negative ΔT_3, higher shrinkage sensitiveness and lower casting density, respectively, for both tested mould media [soft green sand mould and rigid resin sand mould]. The relationship between the concentrated shrinkage volume level and undercooling at the end of solidification is strong [$R^2 > 0.97$], in both cases.

For the same undercooling at the end of solidification, shrinkage formation sensitiveness is visible influenced by the rigidity of the mould [higher for soft green sand mould]. The apparent density of ductile iron castings is more depending on the rigidity on the mould, comparing to the undercooling of the end of solidification [R^2 = 0.34 – 0.49].

As normally, apparent density in the casting regions where all of shrinkage types were present is lower than for only micro-shrinkage presence, but at higher dependence on undercooling at the end of solidification.

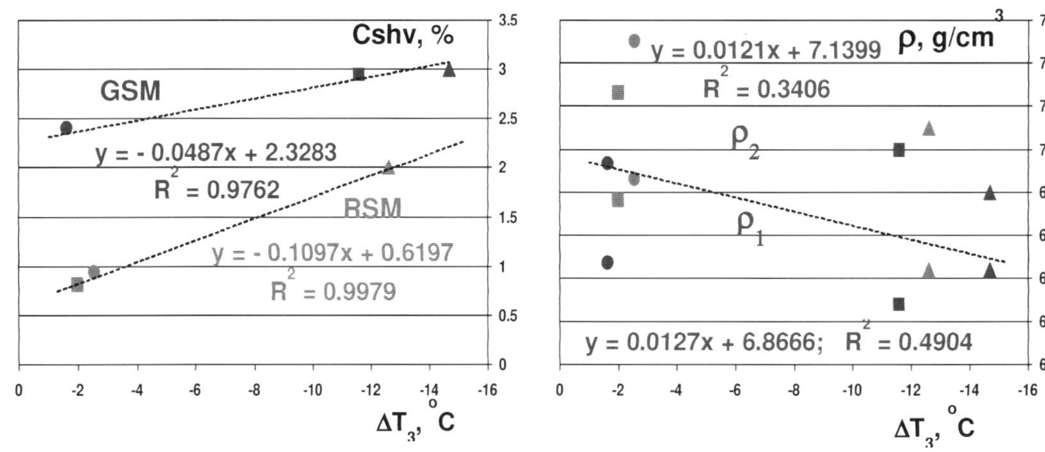

(a) (b)

Figure 4. Influence of the undercooling of the end of solidification [ΔT_3, °C] on concentrated shrinkage volume Cshv (a) and apparent density ρ (b) of inoculated ductile iron [GSM–green sand mould; RSM – resin sand mould; ρ_1 - all shrinkage type; ρ_2 – only micro-shrinkage presence].

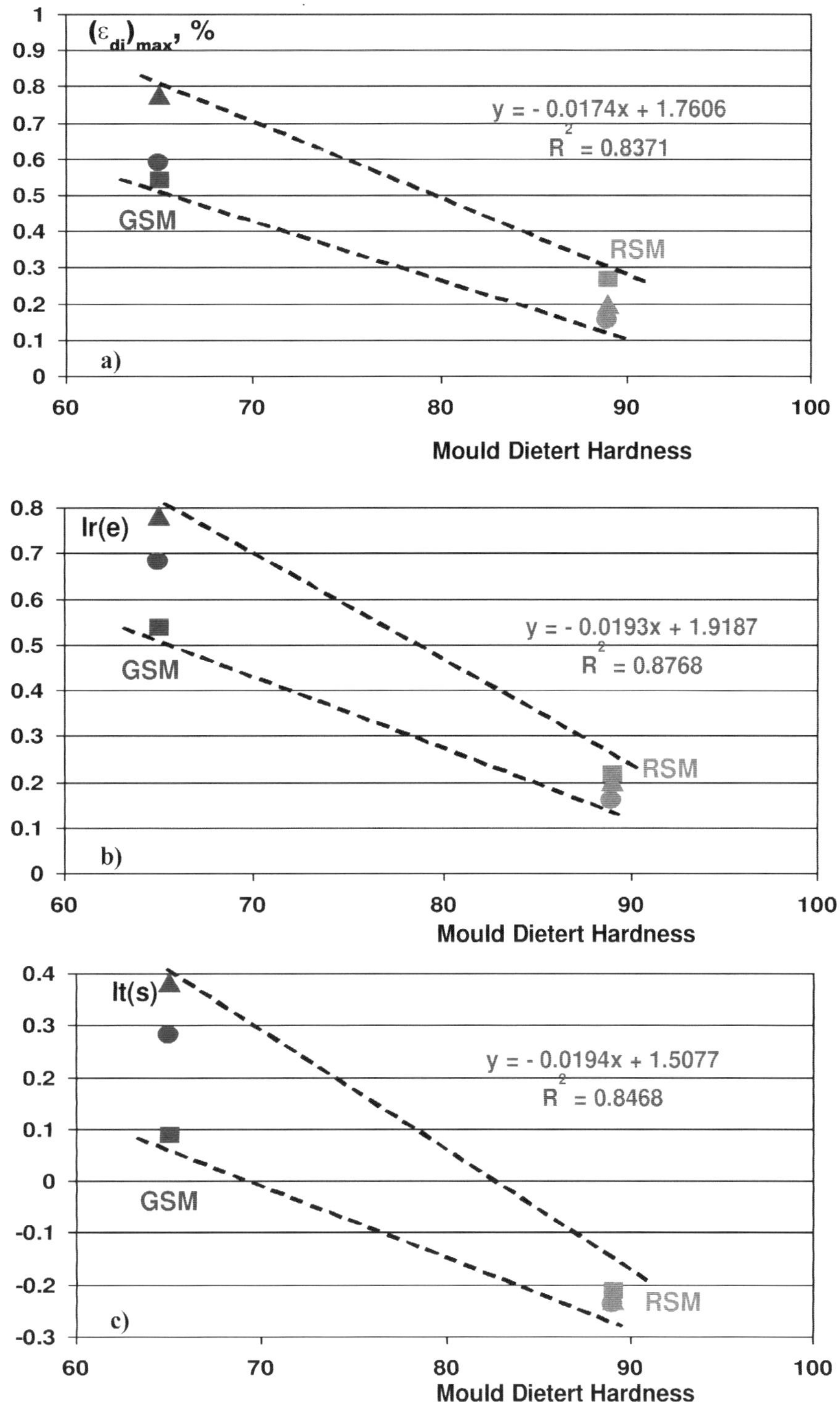

Figure 5. Influence of mould hardness [rigidity] on the $({}^{-}{}_{di})_{max}$ [a], Ir(e) [b] and It(s) [c] contraction curves parameters of inoculated irons [GSM – green sand mould; RSM – resin sand mould].

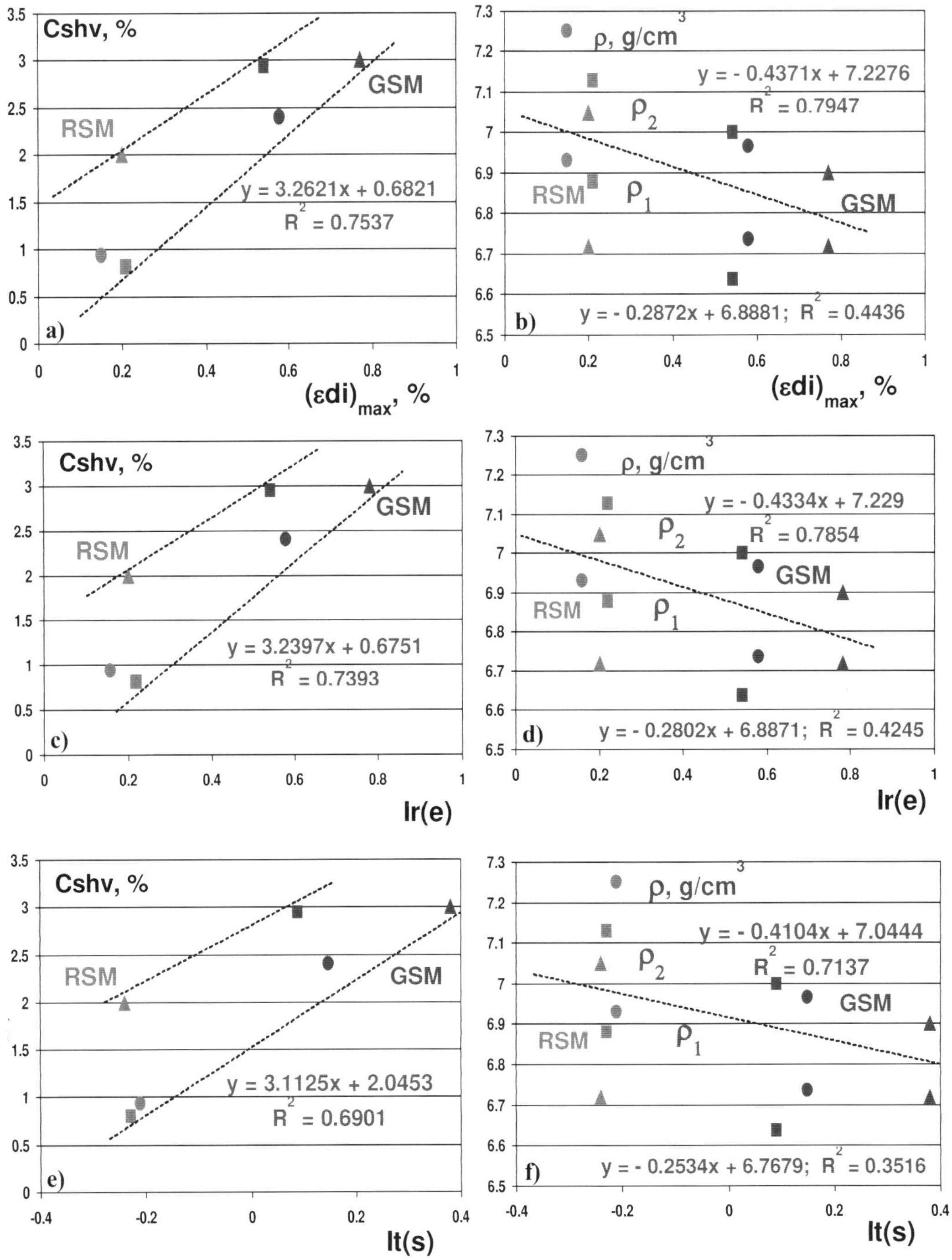

Figure 6. Influence of the maximum initial expansion (a, b), the first derivative integral of the Contraction curve during eutectic dilation (c, d) and the total integral of the first contraction curve derivative, prior to pearlitic contraction (e, f) on concentrated shrinkage volume Cshv (a, c, d) and apparent density ρ (b, d, f) of inoculated ductile iron [GSM – green sand mould; RSM – resin sand mould; ρ_1 - all shrinkage type; ρ_1 – only micro-shrinkage presence].

Three parameters on the contraction / expansion curves appear to be influenced by mould media rigidity (figure 5), with visible effects on the contraction defects formation (figure 6) in ductile iron castings, such as maximum initial expansion [$(\varepsilon_{di})_{max}$], the first derivative integral of the contraction curve during eutectic dilation [Ir(e)] and the total integral of the first contraction curve derivative up to the end of contraction prior to the start of pearlite formation (eutectoid) [It(s)] as previous papers have pointed out [12–15].

Lower mould rigidity [Dietert hardness], higher maximum initial expansion level (a short time before the end of eutectic solidification), 0.5 – 0.8% for green sand mould [GSM] and 0.15 – 0.30% for resin sand mould [RSM], as figure 5a shows, for a high relationship rule [$R^2 > 0.8$]. Similarly the first derivative integral of the contraction curve during eutectic dilation [Ir(e)] was influenced [$R^2 > 0.85$] (figure 5b). The total integral of the first contraction curve derivative (up to the end of ante-pearlitic contraction), $It_{(s)}$ is very strongly affected by mould type (figure 5c). This parameter has positive values on GSM and negative values on FRM pouring. So, by increasing of mould rigidity, $It_{(s)}$ sign is changed from positive to negative values. On a constant value of ante-pearlitic contraction (ε_{ap}), the $It_{(s)}$ value is in direct connection with $(\varepsilon_{di})_{max}$, and by this way, a good correlation between $It_{(s)}$ and ductile iron shrinkage tendency could be found.

As a representative parameter, the maximum initial expansion [$(\varepsilon_{di})max$, %] was found to have a strong influence on shrinkage and micro-shrinkage formation in ductile iron castings, as Figures 6a and 6b show: higher $(\varepsilon_{di})_{max}$, higher shrinkage sensitiveness and lower apparent density of the obtained ductile iron castings. Generally, the mould rigidity has an obvious effect on ductile iron shrinkage tendency. Both concentrated and total shrinkage were greatest in green sand mould (GSM) solidification, where there was much higher initial eutectic expansion $(\varepsilon_{di})max$ than in resin sand mould (RSM). Consequently, the apparent density of ductile iron samples was higher in RSM versus GSM. The GSM and RSM mould media are differentiated more by the contraction / expansion curve analysis results, such as the $(\varepsilon_{di})max$ parameter, in contrast to cooling curve parameters.

As also the first derivative integral of the contraction curve during eutectic dilation [Ir(e)] and the total integral of the first contraction curve derivative (up to the end of ante-pearlitic contraction), It(s) were influenced by mould rigidity, a good relationship between contraction defects sensitiveness and the level of these two important parameters on the contraction / expansion curves were observed (figures 6c,d and 6e, f). Higher Ir(e) and It(s) levels, typically for green sand mould, higher concentrated shrinkage volume Cshv [$R^2 \geq 0.7$] and lower apparent density of castings, for both evaluated situations: the lowest in the casting regions where all of shrinkage types were present [$R^2 = 0.35 – 0.42$] comparing to the regions were present than for only micro-shrinkage presence [$R^2 > 0.7$].

4. Conclusions

The obtained technique for simultaneous cooling / contraction / expansion analysis is useful to evaluate the sensitiveness of iron castings to form contraction defects, such as macro-shrinkage and micro-shrinkage.

This technique, incorporating two moulds, was used to test different inoculants, different inoculant additions, different mould media, and different iron chemistries, in laboratory experiments.

The mould rigidity has an obvious effect on the ductile iron shrinkage tendency. The highest level of both concentrated and total shrinkage was recorded in green sand mould solidification where a much higher level of the initial eutectic expansion $(\varepsilon_{di})_{max}$, as compared to furan resin mould was recorded.

With higher maximum initial expansion (shortly before the end of eutectic solidification) and / or greater undercooling (more negative) at the end of solidification, there is an increase in concentrated and dispersed shrinkage volumes, which lowers the apparent density of ductile iron castings.

The dual purpose technique is easy to apply in the melt shop, similar to current thermal analysis systems. Using specialized software, both cooling and contraction / expansion curves and their pertinent readings are able to be displayed simultaneously.

References

[1] Jiyang Z 2010 *China Foundry* **7**(2) 183-198
[2] Wallace J F, Samal P K and Voss J D 1984 *AFS Trans.* **92** 765-784
[3] Skaland T 2003 *Int. J. Cast Metals Res.* **16** (1-3) 11-16
[4] Skaland T 2003 *Proc. 2003 Keith Millis Symposium on Ductile Cast Iron* (Hilton Head Island, USA, Oct. 2003) 61-69; *Proc. 66th World Foundry Congress* (Istanbul, Turkey, Sept. 2004) 975-987
[5] White D 2011 *Proc. 115th AFS Casting Congress*, (Schaumburg, Il, USA) pp 11-141
[6] Li W Z, Liu B C and Li J R 2001 *J. Mater. Sci. Techn.* 2001 **17**(6) 610-614.
[7] Han Z Q, Su J Y and Cai K K 2000 *J. Univ. Sci. Technol. Beijing*, **17** (1) 24-29.
[8] Ohnaka I, Iwane J, Yasuda H and Zhu J D 2003 *Int. J. Cast Met. Res.*, **16** (1-3) 293-299.
[9] Stefanescu D M 2005 *Int. J. Cast Met. Res.*, **18** (3) 129-143.
[10] Winter B P, Ostrom T R, Hartman D J, Trojan P K and Pehlke R D 1984 *AFS Trans.* **92** 551-560
[11] Bates C E, Oliver G L and McSwain R H 1977 *AFS Trans.* **85** 289-298
[12] Riposan I, Chisamera M, Stan, S, Gadarauteanu C and Skaland T 2003 *Proc. of 2003 Keith Millis Symposium on Ductile Cast Iron*, (Hilton Head Island, USA) 125-135
[13] Chisamera M, Riposan I, Stan S, Toboc P, Skaland T and White D 2006 *Proc. of 8th Int. Symp. & Processing of Cast Iron,* (Beijing, China) 116-121; 2011 *Int. J. Cast Met. Res.* **24** (1) 28-36.
[14] Chisamera M, Riposan I, Stan S and Barstow M 2012 *AFS Trans.* **120** 375-388.
[15] Thielman T 1970 *Giessereitechnik* **1** 16-24.
[16] Sillen R V 1998 *Proc. AFSInt. Inoculation Conf.*, Rosemont, Illinois (USA), April 6-8.

Acknowledgments

This work was partially financed by a grant of the Romanian National Authority for Scientific Research and Innovation, CNCS/CCCDI - UEFISCDI, project number PN-III-P2-2.1-PED-2016-1793, within PNCDI III.

IOP Conf. Series: Materials Science and Engineering **209** (2017) 012018 doi:10.1088/1757-899X/209/1/012018

Microstructural Investigations on Alloy Mg-2Ca-0.2Mn-0.5Zr-1Y

S Lupescu[1], C Munteanu[1], B Istrate[1], S Stanciu[2], N Cimpoesu[2] and B Oprisan[3]*
[1] "Gheorghe Asachi" Technical University of Iasi, Faculty of Mechanical Engineering, 43 "D. Mangeron" Street, 700050, Iasi, Romania
[2] "Gheorghe Asachi" Technical University of Iasi, Faculty of Material Science and Engineering, 61-63 "D. Mangeron" Street, 700050, Iasi, Romania
[3] "Gr.T. Popa Medicine" University of Iasi, Faculty of General Medicine, University street nr. 16, Iasi, Romania

E-mail: bogdano_22@yahoo.ro

Abstract. Authors elaborated a five component material using specific magnesium master alloys with several elements resulting a potential biocompatible and biodegradable alloy.The main goal of the present paper is to investigate the properties of some master Mg-2Ca-0.2Mn-0.5Zr-1Y alloy. The surface morphology was characterized using X-ray diffraction (XRD), optical microscopy and scanning electron microscopy (SEM). After the XRD analysis, it have been identified the following compounds, such as Mg, Mg_2Ca, $Mg_{24}Y_5$, respectively MgY. These compounds have the hexagonal crystallographic structure for Mg and Mg_2Ca type, respectively cubic form for Mg-Y and $Mg_{24}Y_5$. The microstructure presents a uniform morphology and an undisclosed zirconium. Also, manganese is embedded in magnesium and Ca forms a lamellar eutectic mixture of Mg_2Ca type. In conclusion, Mg-2Ca-0.2Mn-0.5Zr-1Y alloy shows similar characteristics from the microstructure point of view with other biodegradable materials, these alloy could be used as biodegradable implant.

1. Introduction

Magnesium alloys are widely used in the automotive industry, in the medical and aerospace industries due to their low density, high specific strength, good pouring capacity and machining [1-5]. By comparing with other commonly used biomaterials [18,19], magnesium and its alloys have excellent biocompatibility and biodegradability as well as a more elastic modulus near that of human bones and a possibility of gradual dissolution and absorption after implantation in a physiological environment. Although they are so used, their corrosion and wear resistance limits the field of application of Mg to other industries. Moreover, the rate of degradation of these alloys in the corrosive environment, especially in the Cl-containing physiological environment, is so rapid that implants cannot maintain their mechanical integrity before restoring human tissue [11-13]. Therefore, Mg align with other elements to increase corrosion resistance. Mg alloys have attracted particular attention with regard to the degradability of the implanted materials due to their high biocompatibility and satisfactory mechanical properties. As a result, Mg alloys have the advantage that they do not exhibit the stress-shielding effect normally occurring in conventional metal implants made of stainless steel or titanium choices that are widely used for bone implants [6,8,16]. Various Mg alloys have been investigated as biodegradable materials and some have shown good biocompatibility, such as Mg-Ca, Mg-Zn, Mg-Mn, Mg-Zr, Mg-Y [17]. In the paper presented Mg-2Ca-0,2Mn-0,5Zr-1Y showed specific magnesium microstrusture and good mechanical properties. Mn is mainly used to increase ductility. In Mg-Ca binary alloys, the Mg_2Ca

phase is formed, which can improve creep resistance due to solid solution consolidation, precipitation consolidation and grain boundary fixation. Higher amounts of Ca (> 1% by weight) can lead to problems such as hot breakage or sticking during casting [14]. The RE elements can stop the grain boundaries at high temperatures and can help increase resistance, especially by strengthening the resistance. Mg-Y alloys are known to form three different intermetallic phases at different temperature intervals with the increase in Y: $Mg_{24}Y_5$, Mg_2Y and MgY [15]. This mechanism increases the operating temperature of Mg alloys in the transport industry and improves creep resistance and corrosion resistance [13]. However, the rate of rapid degradation of Mg choices in human bio-medical limits their clinical applications [9,10]. It is therefore important to improve the corrosion resistance of Mg alloys in order to broaden the field of application. The results will allow us to provide useful data for understanding these intermetallic and designing alloys based on Mg-2Ca-0,2Mn-0,5Zr-1Y high performance.

2. Materials and methods

The samples used in this experiment were prepared by blending the corresponding proportions of commercial Mg ingots (& gt; 99.95%) and 4 master alloys Mg-25% Y, Mg-15% Ca, Mg-3% Mn and Mg-25% Zr. Experimental compositions of Mg-2Ca-0.2Mn-0.5Zr-1Y were prepared. These compositions were prepared using an induction furnace at 750°C under the protection of an inert atmosphere (Ar), the graphite crucible was allowed to cool to room temperature after which it was broken to obtain a 50 mm ingot and Diameter of 20mm this ingot was cut into pills with a diameter of 20mm and a thickness of 5 mm. The samples were sanded with abrasive paper with grain size up to 2000 SiC to obtain a uniform smooth surface. The microstructures were investigated by scanning electron microscopy (SEM Quanta 200 3D) equipped with an EDX detector. The analyzed area of each sample was milled and polished to a final finishing with a alumina suspension of 0.05 lm. Prior to observation, the samples were etched using 4% w / w nitric acid in ethanol for 5-8 s, washed thoroughly with water and alcohol and then dried with hot air. To highlight microstructures, use the Leica 5000DMI optical microscope. We used an XRD Expert Pro MPD diffractometer for diffraction. The microhardness was done using a tribometer with indentation and scratch. The micro scratch test was carried out as follows: a constant load of 5N was applied over a distance of 4mm with a displacement velocity of 0.1mm / s.

3. Results and discussions

3.1. Structural analysis

3.1.1.Optical images. The images shown below represent the morphology of the analyzed sample area, captured at the Leica 5000DMI microscope at different magnification powers, and show high material compaction.

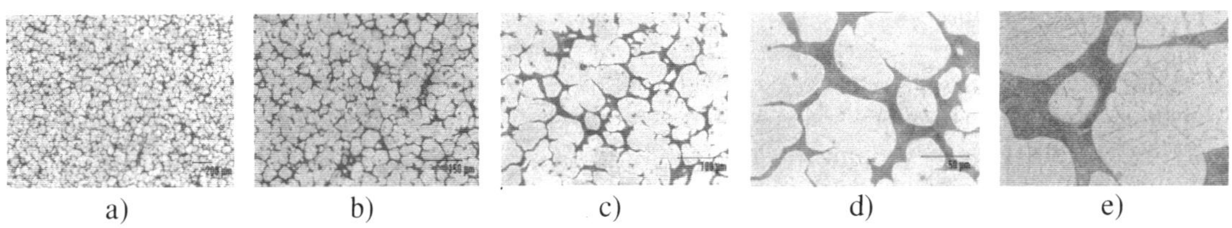

a) b) c) d) e)

Figure 1. Optical analysis of the material.

3.1.2. Scanning electron microscopy. The aspects of morphology and microstructure are presented in figure 2 (a, b, c, d, e). The SEM images taken at up to 5000x highlight the phases of Mg alloys after casting and the compactness and homogeneity of the material and surface morphology. The grains formed after casting the four alloys have different shapes depending on the alloy that predominates in a certain area.

IOP Conf. Series: Materials Science and Engineering **209** (2017) 012018 doi:10.1088/1757-899X/209/1/012018

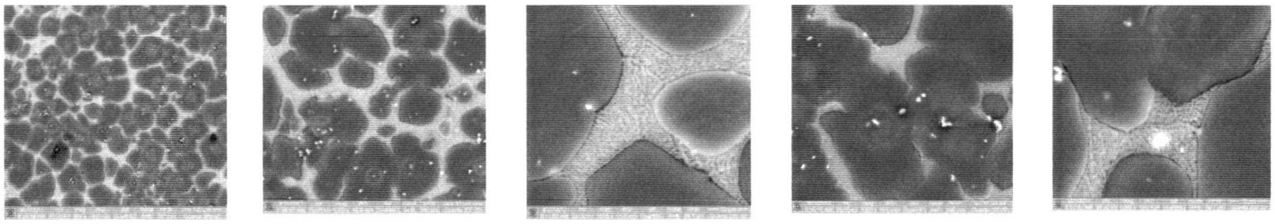

a) b) c) d) e)

Figure 2. Surface SEM images for Mg-2Ca-0,2Mn-0,5Zr-1Y.

3.1.3. EDX analysis. To see the estimative composition of Mg choices after SEM, an EDX analysis followed and the results obtained after the test are shown in figure 3. After the EDX test we found that MgY has a Y-concentration of 24.790%, Mg-Zr 19.636%, Mg-Mn 1.537% and Mg-Ca 16.835.

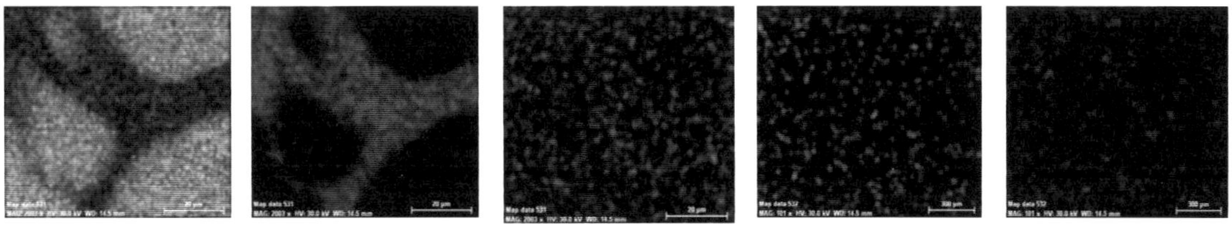

a)Mg b)Mg-15Ca c)Mg-3Mn d)Mg-25Zr e)Mg-30Y

Figure 3. EDX images of Mg alloys.

3.2. XRD analysis

The Xrd analysis was performed on a Panalytical XPERT Pro MPD diffractometer, CuKα X-ray.

Table 1. Lattice parameters of Mg-2Ca-0,2Mn-0,5Zr-1Y.

Compound	Space Group	Crystal system	a (Å)	b (Å)	c (Å)	α (°)	β (°)	γ (°)	Cell volume (10^6 pm^3)
CaMg2	P63/mmc	Hexagonal	6,2300	6,2300	10,1200	90	90	120	340,16
Mg	P63/mmc	Hexagonal	3,2200	3,2200	5,2300	90	90	120	46,96
Mg24Y5	I-43m	Cubic	11,2500	11,250	11,2500	90	90	90	1423,83
MgY	Pm-3m	Cubic	3,8000	3,8000	3,8000	90	90	90	54,87

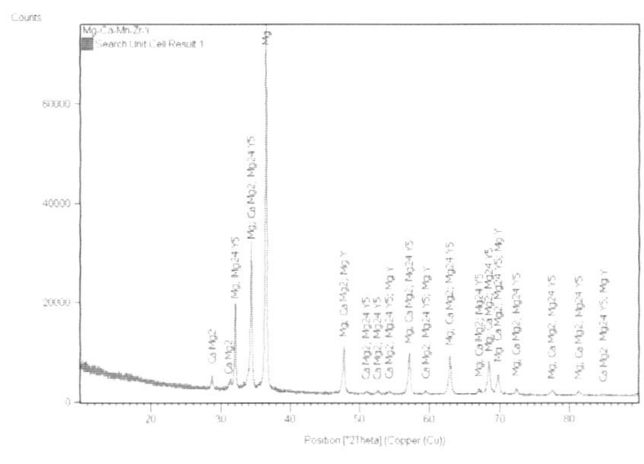

Figure 4. Graph of Xrd analyzes for Mg-2Ca-0,2Mn-0,5Zr-1Y.

The XRD models are shown in figure 4, Pure Magnesium is highlighted at 33.46° (2 Theta - as the highest peak). There has been highlighted the Mg_2Ca and $Mg_{24}Y_5$ and MgY compounds first having the hexagonal crystallographic structure and the other two crystallographic cubic structures. Mg-Mn and αZr or less revealed they have the same hexagonal crystallographic structure as the Mg. Parameters of all compounds are shown in table 1.

3.3. Scratch and micro-indentation analysis

Figure 5 shows a scratch test of the Mg-2Ca-0.2Mn-0.5Zr-1Y sample. Three attempts were made in 3 different areas, resulting in small differences between them, which show the influence of alloying elements in different areas. Table 2 shows the values for young modulus, hardness, scratch, coefficient of friction and stiffness for all three tests. The coefficient of friction is rather low about 0.747. The results of the microindentation test showed a depth between 6,372 µm and 6,682 without fracture.

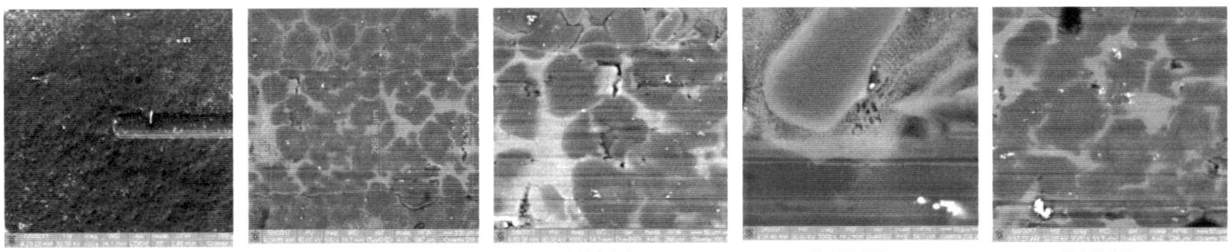

a)100x b)500x c)1000x d)5000x e)1000x

Figure 5. Scratching on diffrent magnifying powers.

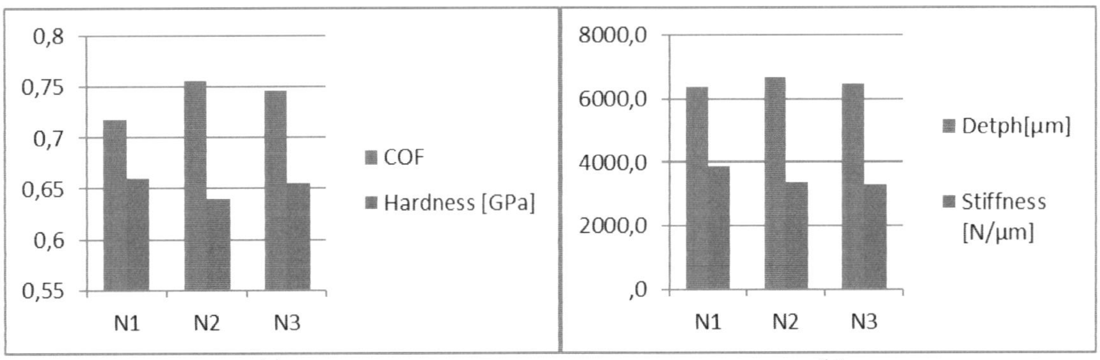

(a) (b)

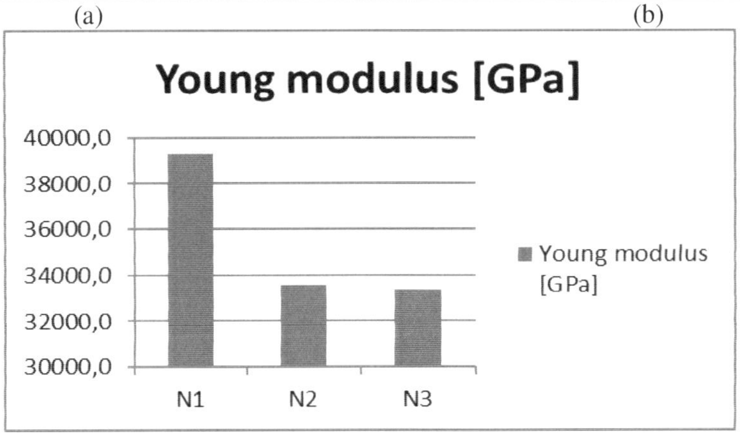

(c)

Figure 6. Young modules.

The penetration trace of the scratch test is not uniform and has different widths as observed in figure 5b, widths between 294.32μm and 269.84μm. This test demonstrates that the material obtained after casting does not have a uniform but well-structured and fairly compact composition.

Figure 5c shows that during scratching, the material breaks to the limit of Mg_2Ca grains and grains of Mg.Because of all the master alloys Mg-Ca has the lowest breaking resistance.

Table 2. Some mechanical properties of Mg-2Ca-0,2Mn-0,5Zr-1Y.

Mg-2Ca-0,2Mn-0,5Zr-1Y	N1	N2	N3
Detph [μm]	6.372	6.682	6.503
Stiffness [N/μm]	3.877	3.383	3.308
Young modulus [GPa]	39.292	33.567	33.371
COF	0.718	0.756	0.747
Hardness [GPa]	0.663	0.647	0.652

The young modulus has a fairly high value of 39,292, but in the following tests it decreases gradually 33,567GPa and then 33,371GPa, which shows us that the casting has a satisfying elasticity and is close to that of the human bone. The hardness of the alloys is rather small but higher than that of pure magnesium.

4. Conclusions

The purpose of the paper was to form a new biodegradable biomaterial formed in the base of Mg and alloyed with several elements such as Ca, Mn, Zr and Y for a high biocompatibility, an increased resistance to corrosion and degradation in time as long as possible. A possible application could be using these materials in medical and not only. Influence of Y shows us greater resistance and at the same time other biodegradable elements in small quantities bring benefits to the human body. Researching morphology, X-ray diffraction, micro scratch and micro indentation, and even the EDX test helps us to improve the material through future studies for the best biocompatibility with the human body. X-ray analysis showed the presence of crystallized compounds: Mg_2Ca, $Mg_{24}Y_5$, MgY, MgMn and αZr having the hexagonal crystalline structure for Mg_2Ca, MgMn, αZr as well as the crystalline cubic structure for $Mg_{24}Y_5$ and MgY. The modulus of elasticity is quite high, which shows us that it is a good thing for Mg-based biodegradable implants. The research of the material will continue with further experiments in order to achieve biodegradable prostheses and implants successfully replacing the classic stainless steel Co-Cr and Ti alloys.

References

[1] Shrestha S 2016 *Surf. Eng.* **26** 313–316
[2] Cao X, Jahazi M, Immarigeon JP, Wallace W 2006 *J. Mater. Proc. Technol.* **171** 188–204
[3] Mareci D, Bolat G, Izquierdo J, Crimu C, Munteanu C, Antoniac I and Souto R M 2016 *Materials Science and Engineering* C **60** 402-410
[4] Gray JE, Luan B 2002 *J. Alloys Compd.* **336** 88–113
[5] Mordike B L, Ebert T 2001 *Mater.Sci. Eng.* A **302** 37–45
[6] Zhang E L, Yin D S, Xu L P, Yang L, Yang K 2009 *Mater Sci Eng* C **29** 987–93
[7] Zhang S X, Li J N, Song Y, Zhao C L, Zhang X N, Xie C Y 2009 *Mater Sci Eng* C **29** 1907–12
[8] Reifenrath J, Krause A, Bormann D, von Rechenberg B, Windhagen H, Meyer- Lindenberg A 2010 *Mat-wiss u Werkstofftech* **41**(12) 1054–61
[9] Rad H R B, Idris M H, Kadir M R A, Farahany S 2012 *Mater Des* **33** 88–97
[10] Zhang E L, Yang L 2008 *Mater Sci Eng* A **497** 111–8
[11] Witte F, Kaese V, Haferkamp H, Switzer E, Meyer-Lindenberg A, Wirth C J, Windhagen H 2005 *Biomaterials* **26**(17) 3557

[12] Staiger M P, Pietak A M, Huadmai J, Dias G 2006 *Biomaterials* **27**(9) 1728
[13] Mueller W D, Nascimento M L, de Mele M F L 2010 *Acta Biomater* **6**(5) 1749
[14] Li Z J, Gu X N, Lou S Q, Zheng Y F 2008 *Biomaterials* **29** 1329
[15] Liu M, Schmutz P, Uggowitzer PJ, Song G, Atrens A 2010 *Corrosion Science* **52** 3687–3701
[16] Minciuna, MG, Vizureanu, P, Achiţei DC, Abdullah MMAB, Sandu AV 2016 *Key Engineering Materials* **700** 86-92
[17] Nicoara M, Raduta A, Locovei C, Buzdugan D, Stoica M 2015 *Journal of Thermal Analysis and Calorimetry* **124** DOI 10.1007/s10973-016-5532-5
[18] Costan A, Dima A, Ionita I, Forna N, Perju M C, Agop M 2011 *Optoelectronics and Advanced Materials-Rapid Communications* **5** 1-2
[19] Baltatu M S, Vizureanu P, Cimpoesu R, Abdullah M M A B, Sandu A V 2016 *Revista de Chimie*, **67**(10) 2100-2102

Obtaining and Mechanical Properties of Ti-Mo-Zr-Ta Alloys

M S Bălţatu[1], P Vizureanu[1,2*], V Geantă[3], C Nejneru[1], C A Ţugui[1] and S C Focşăneanu[1]
[1]"Gheorghe Asachi" Technical University of Iasi, Faculty of Materials Science and Engineering, 41 "D. Mangeron" Street, 700050, Iasi, Romania
[2] Centre of Excellence Geopolymer & Green Technology School of Materials Engineering, University Malaysia Perlis, Kompleks Pengajian Jejawi 2, 02600 Arau, Perlis
[3] University Politehnica of Bucharest, Romania 313, Splaiul Independentei, sector 6, 060042, Bucharest, Romania

E-mail: peviz2002@yahoo.com

Abstract. Ti-based alloys are successfully used in the area of orthopedic biomaterials for their enhanced biocompatibility, good corrosion and mechanical properties. The most suitable metals as an alloying element for orthopedic biomaterials are zirconium, molybdenum and tantalum because are non toxic and have good properties. The paper purpose development of two alloys of Ti-Mo-Zr-Ta (TMZT) prepared by arc-melting with several mechanical properties determined by microindentation. The mechanical properties analyzed was Vickers hardness and dynamic elasticity modulus. The investigated alloys presents a low Young's modulus, an important condition of biomaterials for preventing stress shielding phenomenon.

1. Introduction

Use of metals as biomaterials started from the ancient time used to replace a part of the human body or to function directly with tissue. Despite the large number of metals and alloys able to be produced in industry, only a few are biocompatible and capable of long-term success as an implant material. Typical materials currently found in medical applications are classical metal alloys: titanium alloys (Ti6Al4V), cobalt alloys (CoCrMo) and stainless steels in many important medical applications such as orthopedic implants, dental materials, cardiovascular devices and others [1-4].

The number of titanium alloy biomaterials is higher than cobalt alloys and stainless steels because of low elasticity (close to bone), high corrosion resistance, high biocompatibility, high strength and density low [5-7].

Research in the field of biomaterials over the last decade has helped to improve properties by introducing non-toxic elements, for improving surface characteristics, mechanical properties, corrosion resistance, biocompatibility etc., thus aiming at replacing existing ones [8-10].

Designing and selecting biomaterials consists in choosing the correct material compositional limits so that unwanted reactions do not occur in the human tissue-implant interaction.

The literature has provided information on the negative influence of some components on classical biomaterials, such as Al, which show acute toxicity at very high doses, favoring an increased risk of breast cancer and other neurological conditions such as Alzheimer's disease. While elements like Mo,

Zr and Ta do not produce toxic corrosion products in human tissue and have minimal cytotoxicity, highlighting excellent biocompatibility, thus favoring vascularization in the tissue [11-14].

Despise this facts, I selected elements non-toxic like Mo, Zr and Ta to development a a new Ti-based alloys to decreased the elastic modulus to avoid stress shielding effect for designing a new β-type of Ti-based alloy with lower elastic modulus, corrosion resistance and good biocompatibility.

This paper is proposing elaboration of two alloys and obtain mechanical properties of TiMoZrTa (TMZT) samples.

2. Experimental procedures

For the elaboration of TMZT alloys, were used as raw materials with a high purity, elements such as: Ti - 99.8%, Mo - 99.7%, Zr -99.2% and Ta - 99.5%. In order to obtain TMZT alloys, it has been chosen to use a vacuum arc remelting MRF ABJ 900 [15].

The preparation of TMZT alloys in the vacuum arc remelting furnace comprised a succession of operations as described in figure 1.

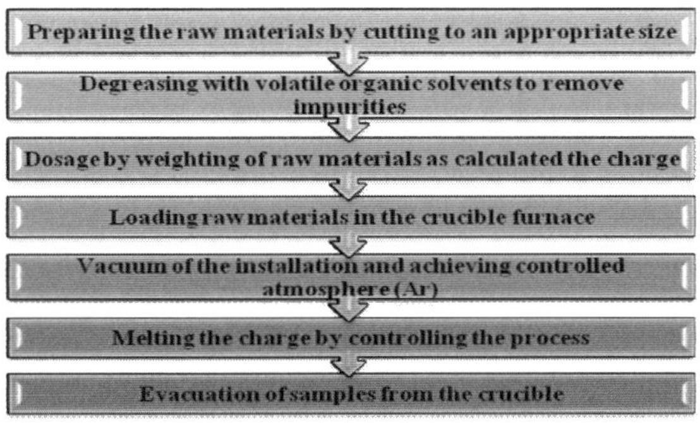

Figure 1. Technological flow of TMZT alloys.

At weighting the load, account was taken of the characteristics of the different alloying elements and their physicochemical properties. Accordingly, account must be taken of the method of elaboration, the construction of the melting furnace used because this can influences in many parts the losses of metal, at melting [16].

In figure 2 are presented all stages of TMZT alloying, which included the weighing of the raw material, the loading of the alloying elements and the final semi-finished products obtained.

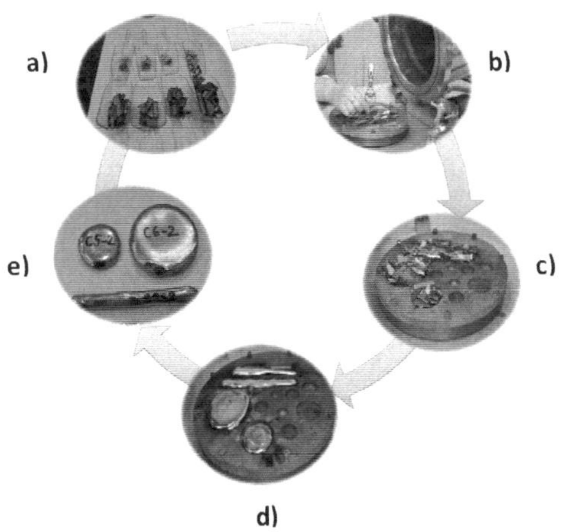

Figure 2. Stages of development of TMZT alloys: a) weighing of raw materials and gravimetric dosing; b), c) loading of the raw material; D), e) TMZT semi-products obtained after solidification.

Determination of the chemical composition was performed by EDX chemical analysis. The chemical composition of the alloy was analysis with EM VEGA II LSH scanning electron microscope manufactured by the TESCAN Co., the Czech Republic, coupled with an EDX QUANTAX QX2 detector manufactured by the BRUKER/ROENTEC Co., Germany.

The Vickers hardness measurements were measured using a Wilson Wolpert universal hardness tester, 751N model with a load of 9,807 N and 12 s.

Aparrent coefficient of friction and elastic modulus were measured using CETR UMT-2 Tribometer. For micro-scratch analysis it was used a constant load method with a load of 5N on a distance of 4 mm, for a single determination.

3. Results and discussions

The chemical composition are shown in table 1. The chemical analyses (EDX) of the alloys studied were performed in many different points for a precise determination, with an uniform distribution of elements and a homogeneous alloy.

Table 1. Chemical composition of the TiMoZrTa (TMZT) sample after melting.

Element	Mo	Zr	Ta	Ti
wt.(%)	12.56	6.99	14.45	Balance
	18.45	6.48	11.7	Balance

The variation-curve of the micro-indentation test is presented in figure 3. This represents the response of the material during a load–displacement curve. For this measurement were carried out three determinations and their average was calculated.

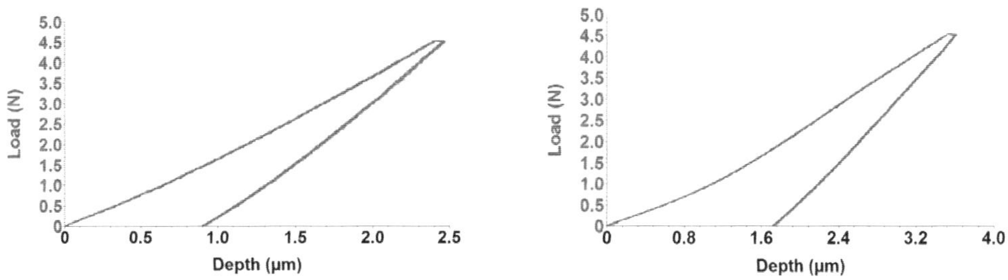

Ti15Mo7Zr15Ta **Ti20Mo7Zr15Ta**

Figure 3. The curve of micro-indentation of TMZT samples.

Few mechanical properties of TMZT alloys are presented in table 2 (hardness, young modulus, stiffness).

Table 2. Some mechanical properties of TMZT samples.

Alloy	Hardness (HV)	Young Modulus (GPa)	Stiffness (N/µm)
Ti15Mo7Zr15Ta	390.88	51.93	3.01
Ti20Mo7Zr15Ta	397.56	43.57	3.20

The hardness values of alloys were close, between of 390.88 – 397.56 HV. Compared to other biomaterials, TMZT alloys have a higher hardness than stainless steels (155 HV), less than CoCrMo (601 HV), but close to the Ti6Al4V alloy (349 HV), which is the most used in implantology [1,17].

The modulus of elasticity is a very important criterion underlying the choice of metallic materials used in orthopedics and should be as close as possible to that of the human bone (17-30 GPa). TMZT alloys exhibit values of the elastic modulus close to those of human bone, values between 51.93-43.57 GPa. Compared to other metallic biomaterials: CoCrMo alloys (210-253GPa) and stainless steels (190-210 GPa), the investigated TMZT alloys exhibit much lower values and the closest values to the human bone (17-30 GPa). TMZT alloys show a significant improvement in mechanical properties even for the titanium alloys: Ti6Al4V (100-114 GPa) and C.P. Ti (102-104 GPa) [14,18,19].Due to Mo and Ta stabilizing elements, these alloys have the advantage of increasing mechanical strength and a modulus of elasticity close to that of the biological bone.It can be seen that with the increase of the molybdenum percentage, the properties of the experimental TMZT alloys improve.

4. Conclusions

The paper presents the steps of elaborating two new titanium alloys with non-toxic elements such as Mo, Zr and Ta. For these alloys were studied mechanical properties like hardness, young modulus and stiffness.

These alloys have the advantage of increasing mechanical strength and a modulus of elasticity close to the biological bone due to alloying with elements like Mo, Zr and Ta.

In order to use these alloys in medical applications it will be necessary to continue with future studies to establish the cytotoxic behavior and corrosion resistance.

References

[1] Geetha M, Singh A K, Asokamani R, Gogia A K 2009 *Mater. Sci.* **54** 397-425

[2] Bălţatu M S, Vizureanu P, Ţierean M H, Minciună M G, Achiţei D C 2015 *Advanced Materials Research* **1128** 105-111

[3] Antoniac I 2007 Biomateriale metalice utilizate la executia componentelor endoprotezelor totale de sold, Ed. Printech, Bucureşti

[4] Istrate B, Munteanu C, Matei M, Oprisan B, Chicet D, Earar K 2016 *IOP Conf Ser Mater Sci Eng* **133** 12010

[5] Minciuna M G, Vizureanu P, Achitei D C, Ghiban N, Sandu A V, Forna N C 2014 Structural Characterization of Some CoCrMo Alloys with Medical Applications (3) 335-338

[6] Niinomi M 1998 *Mater. Sci. Eng.* **243** 231-236

[7] Elias C N, Lima J H C, Valiev R, Meyers M A 2008 *Biological Materials Science* pp 46- 49

[8] Leyens C, Peters M 2003 Titanium and Titanium alloys. Fundamentals and Applications, Ed. Wiley-VCH pp. 423-431

[9] Baier R E 2006 *Journal of material science: Materials in medicine* **17**(11) 1057-1062

[10] Bunea D, Nocivin A 1998 Materiale biocompatibile, Ed. şi Atelierele Tipografice Bren, Bucureşti

[11] Chelariu R, Bujoreanu G, Roman C 2006 Materiale metalice biocompatibile cu baza titan, Ed. Politehnium, Iaşi

[12] Long M, Rack H J 1998 *Biomaterials* **19** 1621–1639

[13] Song Y H, Kim M K, Park E J, Song H J, Anusavice K J, Park Y J 2014 *Dent. Mater.* **30**(9) 977–983

[14] Chen Q, Thouas G A 2015 *Materials Science and Engineering R* **87** 1–57

[15] ***www.eramet.ro

[16] Geantă V, Tripşa I, Ştefănoiu R 2007 Tratat de Ştiinţa si Ingineria Materialelor, Vol. 2., Bazele teoretice şi ingineria obţinerii materialelor metalice, Cap. 7 - Elaborarea oţelurilor. Academia de Ştiinţe Tehnice din Romania, Editura AGIR, Bucureşti

[17] Bombac D M, Brojan M, Fajfar P, Kosel F, Turk R 2007 *Materials and Geoenvironment* **54**(4) 471-499

[18] Gordin D M, Gloriant T, Nemtoi G, Chelariu R, Aelenei N, Guillou A, Ansel D 2005 *Materials Letters* **59** 2936 – 2941

[19] Costan A, Dima A, Ionita I, Forna N, Perju M C, Agop M 2011 *Optoelectronics and Advanced Materials-Rapid Communications* **5** 1-2

IOP Conf. Series: Materials Science and Engineering **209** (2017) 012020 doi:10.1088/1757-899X/209/1/012020

Morpho-Structural Characterization of WC20Co Deposited Layers

C A Tugui[1,2] and P Vizureanu[1]
[1] Gheorghe Asachi Technical University, Faculty of Materials Science and Engineering, Iasi, Romania
[2] National Institute of Research and Development for Technical Physics, Nondestructive Testing Department, Iasi, Romania

E-mail: tzugui.andrei@yahoo.com

Abstract. Hydroelectric power plants use the power of water to produce electricity. In this paper we propose a solution that will increase the efficiency of turbine operation by implementing new innovative technologies to increase the working characteristics by depositing hard thin films of tungsten carbide. For this purpose hard tough deposits with WC20Co and Jet Plasma Jet on X3CrNiMo13-4 stainless steel were used for the realization of the Francis turbine with vertical shaft.

1. Introduction

The new technical requirements imposed by rapid technological development must find solutions that meet these functional requirements of new equipment, assemblies and subassemblies.

Materials used in the construction of Francis turbine blades with a vertical shaft have good corrosion resistance [1-2], but have low abrasion resistance properties [3,4], which lead to a lower operating reliability.

The paper tried to solve these problems by depositing hard thin layers [5-8] to improve wear resistance by increasing the surface hardness of the material used in the construction of the blades of Francis turbines with vertical shaft [9]. In this purpose, wear-resistant deposits were made by the plasma jet method with WC20Co based electrodes on stainless steel X3CrNiMo13-4 [10] used for the development of the Francis vertical shaft guide blades.

By this method, different layer thicknesses can be obtained depending on the number of plasma jet passes over the sample. Layers based on tungsten carbide have been deposited since this material increases the hardness of the substrate layer without affecting the corrosion resistance of the blades from Francis turbine with vertical shaft.

Analyzes performed by the electron scanning microscope will highlight the structure of the deposited layer, suport layer and their interface. Chemical analyzes were performed to emphasize the structure of the deposited layer and the substrate as well as the distribution of the deposition elements on the surface and in cross section [11,12].

Chemical analyzes were performed using the EDX probe mounted on the VEGA II LMH I (SEM). Also, cross section microhardness tests were carried out to determine the microhardness of the base material and the deposited layer, these tests were carried out with a Vickers penetrator, and to determine the roughness of the deposited layer, tests were carried out using the equipment Mitutoyo SJ 301.

2. Materials and deposition method

The material on which the thin thin layers were deposited was cut from a blade of a Francis vertical shaft turbine, and then the specimens of the specific dimensions for each type of test and test to which they would be subjected were made from the obtained piece. The sample was cropped on a cooling cutting machine to avoid thermal influence on the base material.

After the cut, some of the samples were left blank and some of them were deposited with thin hard layers to make a comparison between the properties of the material from the blade of Francis turbine with vertical shaft (material not deposited) and the material obtained by plasma jet deposition. Samples obtained will be subjected to SEM and EDX investigations and to microhardness and adhesion tests of the deposited material.

2.1. Basis material

As a substrate, a stainless steel from a Francis vertical-axis turbine was used. The chemical composition of the material from the blade was made using the Foundry Master Spectrometer. The chemical composition of the stainless steel on which thin layer will be deposited is: Fe - 80.9%; C - 0.10%; Cr - 12.7%; Ni - 4.12%; Mn - 0.54%; Others - 1.64%. This material is a stainless steel with a high nickel and chromium content and is part of the stainless steel grade X3CrNiMo13-4 (1.6982) EN 10213-2007.

Optical microscopy analyzes were performed on Olympus optical microscope BX52, at Non-Destructive Testing Department of the National Institute of Research and Development for Technical Physics, Iasi. These tests were performed to analyze the surface, microstructure of the base material, grains shape, and possible material flaws (holes and cracks).

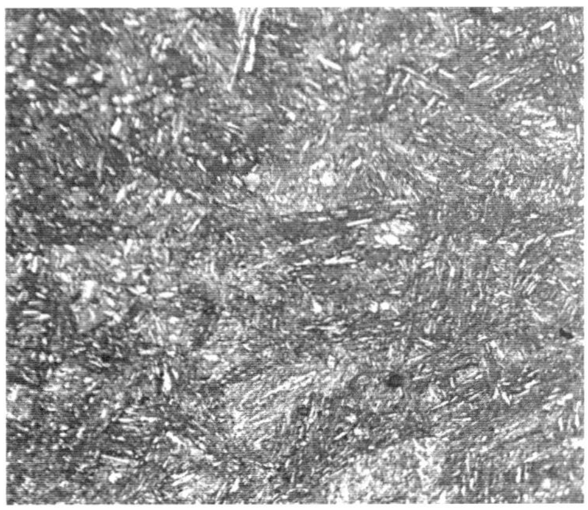
(a)

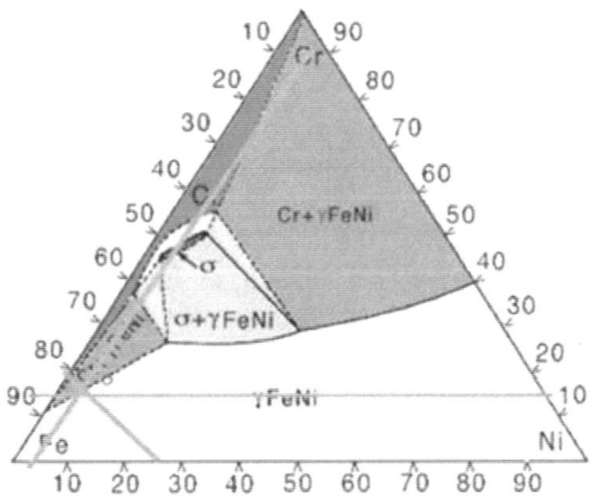

(b)

Figure 1. Optical microscope analysis of base material at 400X magnitude (a) and ternary diagram of Fe-Cr-Ni alloy (b).

The analysis of the structure of the GX3CrNi13-4 base material is shown in Figure 1a, it presents a martensitic acicular microstructure and elongated grains but also some material defects such as some microholes. Figure 1b shows sections of the ternary diagram for Fe-Cr-Ni alloys.

With the green line the zones representing the percentages of the alloying elements of the studied steel (Cr - 13%, Ni - 4.5%, Fe – in balance) are plotted. The diagram shows both the solid state transformation lines and the presence of solid solutions and intermetallic compounds that influence the physico-chemical and mechanical properties of the studied alloy.

2.2. Deposition material

Powders were also studied on Olympus optical microscope, but also in the SEM to highlight the shapeand sizes of the grains. The deposition material for thin layer deposition is WC20Co powder comprising tungsten carbide with cobalt, its chemical composition is 82.8% Fe, 13% Co, 4.2%. Coatings with this powder have good corrosion and abrasion resistance [13].

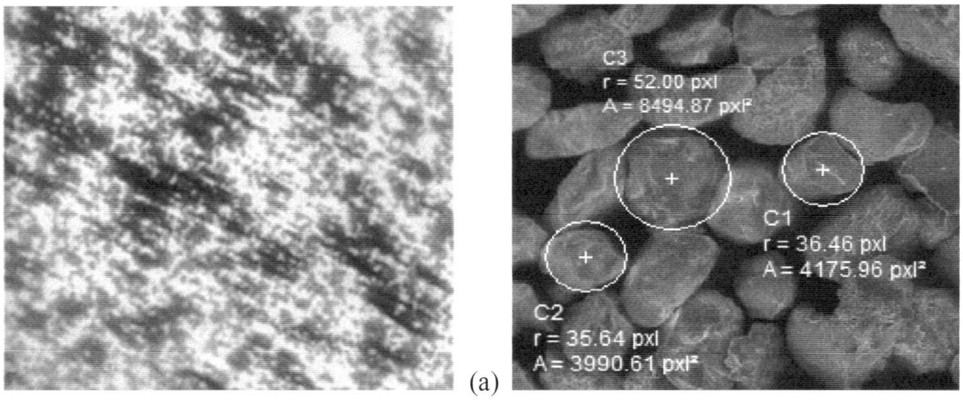

Figure 2. Analysis of WC20Co powders: a) microscopic image at 200X magnitude; b) SEM analysis and determination of grains sizes at 500X magnitude.

It can be observed that WC20Co have very thin gains, irregular, mainly ovoid.

2.3. Deposition method

The Sulzer Metco 9MCE facility of the Faculty of Mechanical Engineering was used for the deposition of thin layers by the plasma jet method (figure 3). The spraying of the melted material by plasma jet was carried out at a temperature of 16000°C [14].

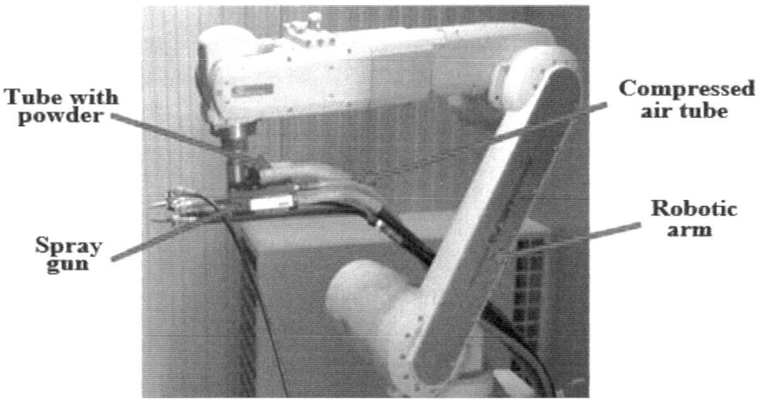

Figure 3. Sulzer Metco 9MCE plasma jet deposition installation.

This method consists in melting and spraying metal or non-metal powders by means of plasmatrons on the base material (in our case a stainless steel type GX3CrNi13-4). Due to the high velocity and pressure of the plasma, the powder is sprayed onto the surface of the base material at a speed of 450 to 650 m / s and the spray distance was about 100 mm.

3. Results nd discussions

Analyzing images obtained at the SEM microscope on the surface of WC20Co powder deposition by the PTJP method, a rugged appearance of the surface with many microdents, microcracks and partially fused powders is observed as presented in figure 4.

The surface does not exhibit visible oxides and layer combustion but exhibits dents and layered areas characteristic of WC20Co powder which melts at very high temperatures.

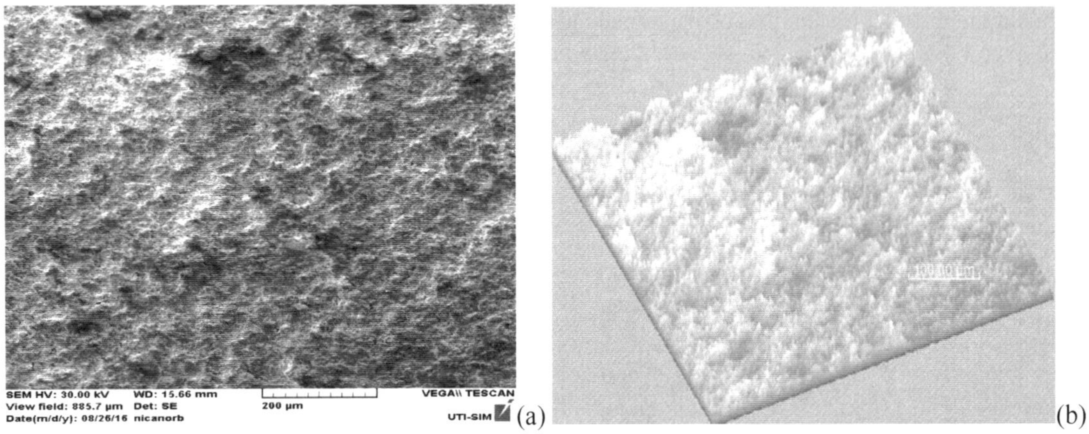

Figure 4. Analysis of the deposited layer with WC20Co powders by the plasma jet method: a) SEM analysis; b) 3D surface analysis.

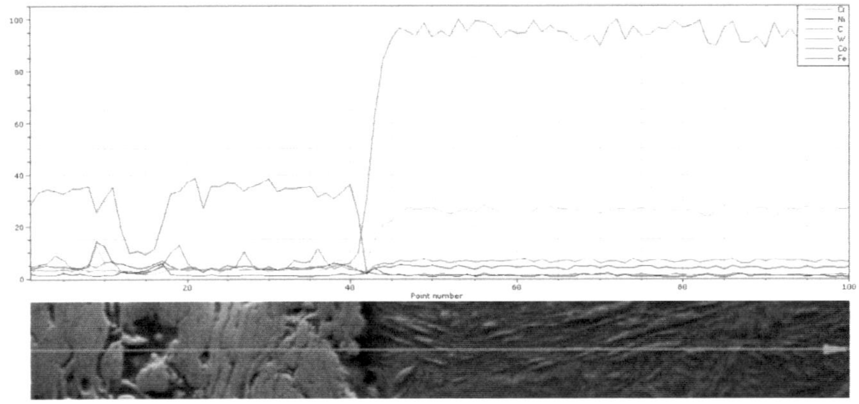

Figure 5. EDX line analysis of the sample with WC20Co based powder by the PTJP method.

Line EDX analysis in figure 5 shows that there is tungsten in the layer area and is missing in the contact area between the base material and the substrate. The deposited layer shows a slight non-uniformity along the interface between the layer and the substrate, observing the presence of a thin layer of oxides on the separation line.

This phenomenon is due to the plasma jet deposition method that does not create transient structures of melted material but only to the structure formed by flattened droplets due to the spraying dynamics when are sent to the surface.

Figure 6 shows the section through the deposited WC20Co powder layer by the plasma jet method. It is seen in the image obtained by the scanning microscope, that the deposited layer structure is uneven with gaps and oxides at the substrate layer interface, which affects the adhesion of the deposition on the base material. The deposited layer is relatively thin, with a thickness of about 73 µm.

The deposited layer shows a slight non-uniformity along the interface and a strong delimitation between the layer and the substrate, even observing the presence of a thin layer of oxides on the separation line.

The adhesion tests were carried out using a tribological and mechanical test equipment Universal Micro-Tribometer (CETR-UMT-2) at the Tribology Laboratory at the Faculty of Mechanical Engineering, Iasi. The following parameters were calculated and plotted with the aid of the equipment program: variation of the response force Fx (N), variation of the normal load force, Fz (N), variation of the friction force, Ff (N) and variation of the coefficient of Friction, COF.

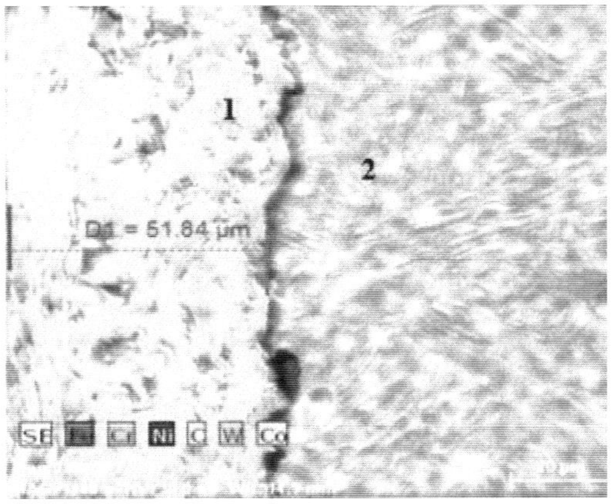

Figure 6. EDX analysis of the thin layer deposited with WC20Co powders by the plasma jet method, 1 - layer, 2 - substrate.

The method used to test WC20Co-based powder samples is that of progressive loading. The test parameters are: penetration velocity that is equal to 1 mm/s, normal load force that varies linearly from 0 ÷ 19 N. The load time was 10 seconds and the length at which the adhesion test was performed is 10 mm.

From the analysis of the data on the loading diagram presented in figure 7 it can be shown that the deposited layer primes very fine cracks at an Fz force of approximately 4.8 N, ie 6.3 to 7.1 mm from the start of the test.

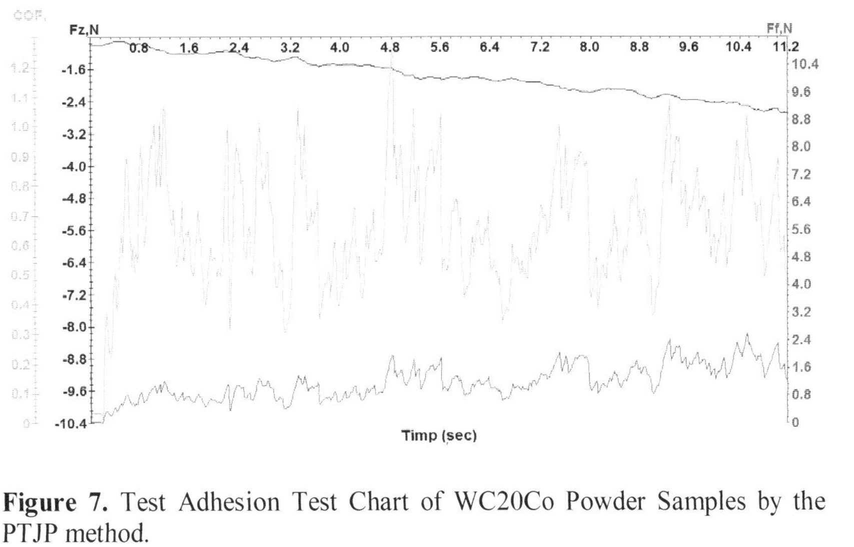

Figure 7. Test Adhesion Test Chart of WC20Co Powder Samples by the PTJP method.

Very fine cracks occur at a distance of 6.2 ÷ 6.4 mm from the start of the adhesion test. Upon priming the cracks in the deposited WC floor, COF increases from 0.05 to 1.2 and the frictional force Ff increases from 0.025 N to over 2.4 N. These values indicate a very good adhesion of the deposited WC20Co powder layer by the jet plasma method to the substrate.

4. Conclusions

WC20Co powder deposition can be successfully carried out by the plasma jet method since the plasma jet temperature is over 16000°C which results in a good melting of tungsten carbide having a melting temperature of 2830°C.

The plasma jet method can produce layers of different thicknesses, which can vary depending on the number of passes of the plasma jet over the deposited material in our case over a stainless steel from the blade of Francis turbine with shaft vertical.

In deposition by the plasma jet method, the appearance of superposition deposited is influenced by particle granulation. When depositing with WC20Co powders, fine surfaces are obtained because the powder has a smaller granulation than other types of deposition obtained with other WC-based powders.

Line EDX analysis shows that WC is present in the layer area and is missing in the base and substrate area. The deposited layer has a non-uniformity along the layer-substrate interface, also observing the presence of oxides at the interface. This phenomenon is due to the plasma jet deposition method that does not create transient structures of substrate melt material but only to the structure formed by flattened droplets due to the splashing dynamics that are glued to the surface of the sample.

References

[1] Chang J T, Yeh C H, He J L, Chen K C 2003 *Wear* 162–169
[2] Nejneru C, Perju M C, Sandu A V, Axinte M., Quaranta M., Sandu I., Costea M., Abdullah M M A B 2016 *Revista de Chimie* **67** 1191–1194
[3] Yahya Z, Abdullah M M A B, Hussin K, Ismail K N, Sandu A V, Vizureanu P, Razak R A 2013 *Revista de Chimie* **64**(12) 1408- 1412
[4] Mustafa Al Bakri A M, Kamarudin H, Bnhussain M, Rafiza A R, Zarina Y 2012 *ACI Materials Journal* **109**(5) 503-508
[5] Junaid H Masoodi, Harmain G A 2017 *Energy* **118**, 644 – 649
[6] Perju M C, Nejneru C, Vizureanu P, Axinte M, (2013) *Metalurgia International* **18** 174 – 177
[7] Bull S J, Rickerb D S, 1990, *Surface and Coatings Technology* 149 – 164
[8] Chen Z, Zhou Y, 2016 *Surface and Coatings Technology* 2419 – 2430
[9] Ţugui C A, Vizureanu P, Savin A, Iftimie N, Perju M C, Cimpoesu N, Nejneru C 2016 *Advanced Materials Research* **1138** 62 – 68
[10] Fedorov A, Rymkevich A, Bazhenov V, Zubchenko A S, Davydova N V 2015 *J. Welding International*, 894 –900
[11] Mustafa Al Bakri A M, Kamarudin H, Bnhussain M, Nizar K, Rafiza A R, Zarina Y 2012 *Reviews on Advanced Materials Science* **30**(1) 90-97
[12] Sandu A V, Bejinariu C, Nemtoi G, Sandu I G, Vizureanu P, Ionita I, Baciu C 2013 *Revista de Chimie* **64**(8) 825-827
[13] Neopane H P, Dahlhaug O G, Cervantes M, 2011 *Global Journal of researches in engineering Mechanical and mechanics engineering* 131– 149
[14] Istrate B, Mareci D, Munteanu C, Stanciu S, Crimu C I, Trinca L C, Kamel E 2016 Environmental Engineering and Management Journal **15**(5) 955-963

Acknowledgments

This paper is partially supported by Romanian Ministry of Research and Innovation under project PN-II-PCCA-2013-4-0656 and Nucleus Program PN 16-37 01 01.

IOP Conf. Series: Materials Science and Engineering **209** (2017) 012021 doi:10.1088/1757-899X/209/1/012021

Dissimilar Brazed Joints Between Steel and Tungsten Carbide

I Voiculescu[1], V Geanta[1], H Binchiciu[2], D Iovanas[3] and R Stefanoiu[1]
[1]Politehnica University of Bucharest, 313 Splaiul Independenţei, 060042, Bucharest, Romania
[2]S.C. Sudotim AS S.R.L., 30 Mihai Viteazu Boulevard, Timişoara, Romania
[3]Transilvania University of Brasov, 29 Eroilor street, 500036, Braşov, Romania

e-mail: ioneliav@yahoo.co.uk

Abstract. Brazing is a joining process used to obtain heterogeneous assemblies between different materials, such as steels, irons, non-ferrous metals, ceramics etc. Some application, like asphalt cutters, require quick solutions to obtain dissimilar joints at acceptable costs, given the very short period of operation of these parts. This paper presents some results obtained during the brazing of dissimilar joints between steel and tungsten carbide by using different types of Ag-Cu system filler materials alloyed with P and Sn. The brazing techniques used were oxygen-gas flame and induction joining. The brazing behaviour was analysed in cross sections by optical and electron microscopy. The metallographic analysis enhanced the adhesion features and the length of penetration in the joining gap. The melting range of the filler materials was measured using thermal analysis.

1. Introduction

Joints between different materials, such as steel and tungsten carbide, are often made by brazing. During brazing, filler materials having a liquidus temperature above 450°C and below the solidus temperature of the base metal are melted and then enter the joining gap, by means of the capillarity action [1].

The main problem encountered during the brazing of heterogeneous materials (ceramics and steels) is the fact that most common filler materials cannot cover, at the same time, the entire ceramic surface, due to their strong ionic and covalent bonds [2].

Ceramics are used in a multitude of industrial fields and the possibilities to use of them in combination with other materials (metallic or ceramic) have great economic importance. The evolution and development of more advanced types of ceramics was favoured by the challenge to work in difficult operating conditions, such as abrasive and erosive environments. To be able to work in such severe conditions, substantial high values of mechanical properties (hardness and wear resistance) must be obtained.

One of these applications is the asphalt cutter, which requires an assembly of numerous small cutters made of tungsten carbide placed on the large steel rollers. The main issue in this case is to obtain a good assembly between the ceramic parts and the steel support, which allows long periods of operation until the replacement of the worn parts [3]. The cooling of a dissimilar joint after brazing (i.e. steel brazed with tungsten carbide) is very important due to the high expansion coefficient of the steel, which tends to contract faster than the part of tungsten carbide to which it is brazed.

This mismatch develops shear stresses within the interfaces of the joint. If high strength brazing filler material is used, these stresses will be directly transmitted to the brittle tungsten carbide and could cause its cracking (figure 1). For this reason, the filler metal must have a low yield point, to be able to deform and to allow the dissipation of the shear stresses [4-9].

Figure 1. Crack formation in tungsten carbide brazed with steel components (500x, 100x) [3].

To solve this problem, direct brazing of the ceramic cutters on the steel support can be applied by using active filler metals that contain special alloying elements. The addition to regular silver-based brazing alloys of metals having strong affinity for the elements constituting the ceramic can improve wetting and adhesion effects. Thus, metals having strong affinity for oxygen, such as titanium, aluminium, zirconium, hafnium, lithium, silicon or manganese, help the conventional brazing alloys in the wetting of oxide ceramics, without special preparation. Metals that react with silicon, carbon or nitrogen help the wetting of silicon carbide or silicon nitride. The tungsten carbide inserts have high hardness, wear resistance, long life and high cutting performance, but they are prone to breakage during brazing. For this reason, brazing requires a special sandwich filler metal which includes a copper layer to reduce residual stresses.

Another issue that can occur during heating is the oxidation of the tungsten carbide component. The oxidation can occur on both surfaces causing superficial damage and the emergence of micro-cracks. The brazing surface must be completely protected against oxidation by the coating provided by paste or flux. When brazing tungsten carbide with steel components, the heat flow is more effective on the sides and less at the bottom of the joint, as the complete penetration of the brazing alloy in these zones is prevented. Therefore, to balance the heat flow, a few pieces of solder wire can be placed under the tungsten carbide to contribute to the better warming and heat dispersion into the gap, by wetting the entire surface of the parts, from the bottom up [6].

Recent filler materials used for the brazing of tungsten carbide are those alloyed with Ag and Zn (Cd-free), whose drawback is the high potential of evaporation during soldering. For this reason, in the most recent filler materials, Zn is replaced by Sn, Cu, Ni, P and Mn, sometimes accompanied by small amounts of indium or tin, to decrease the melting temperature [10-11].

This paper presents some results obtained during the brazing of dissimilar joints between tungsten carbide and steel, by using different types of Ag-Cu system filler materials alloyed with P and Sn. Phosphorus added to the alloy improves the flowability of the molten alloy, deoxidizing the surfaces (by forming of P_2O_5) and improves mechanical properties by cleaning grain boundaries. Excess phosphorus can decrease the electrical conductivity and it is detrimental for the induction brazing process. Copper combines with tin to form solid solution for concentrations below 15.8% Sn in the temperature range (520-586°C).

For high concentrations of tin (40% Sn), the μ phase occurs together with the α phase. For about 11% Sn, at temperatures in the range of 200 - 350°C, precipitates unstable ε and ε' phases occur, as shown by the Cu-Sn binary diagram [12, 13]. The brazing techniques used were oxygen-gas flame and induction joining. The brazing behavior was analysed in cross sections by optical and electron microscopy. The metallographic analysis enhanced the adhesion features, the length of penetration in the joining gap and the wetting angle. The melting range of the filler materials was measured using thermal analysis.

2. Experimental procedure

In order to establish the optimal regime for the brazing process, two different filler materials and two brazing processes were studied. For each brazing process, 4 tungsten carbide and carbon steel assemblies were prepared for the destructive test (cross sections obtained using the conventional EDM process) in order to study the adherence between the steel parts, tungsten carbide and filler material, the spreading mode into the brazing gap and any possible imperfections.

2.1. Oxy-flame brazing

For the manual oxy-flame brazing process, the experimental filler material used was Ag15CuP6 (cadmium free) (max. 15%wt Ag, max. 82% Cu, max. 3%wt P), 2 mm in diameter and 450 mm long (figure 2). The FH10 flux used in the experimental program is generally used for brazing and soldering, according to EN 1045. The operating temperature range of this flux is 550-80°C, coupled with filler materials that have a melting point below 800°C (typically below 75°C). The condition required is for the selected flux to be active at minimum 50°C, over the liquidus temperature of the brazing alloy. The flux is recommended for homogenous or heterogeneous joints, for components made of carbon or stainless steel, copper, brass, ceramic materials but not aluminium. It is recommended especially for the brazing of aluminium bronzes, steels, tungsten or molybdenum carbide, or whenever protection is needed for components during heating [14-18].

(a) (b)

Figure 2. Manual oxy-flame brazing process (a) and filler materials (coated rods) (b).

Before brazing, the wetting test was performed in order to choose the appropriate working regime, the filler material and the cleaning procedure. The water droplet method was applied to test the surface preparation mode. For this purpose, a water droplet is applied on the completely clean surface using a pipette and then it is observed to detect if it has a tendency to agglomerate or to scatter on a larger surface of metal or ceramic. Figure 3 shows clearly that, if a cleaning solution is used, the water droplet spreads on the higher surface both on the surface of the steel part and on the flat surface of the tungsten carbide (figure 3b and 3d).

(a) (b) (c) (d)

Figure 3. The influence of the cleaning process on the spreading of the water droplet: a) unclean steel surface; b) degreased steel surface; c) tungsten carbide contaminated surface; d) polished and degreased tungsten carbide surface.

The solderability testing of the tungsten carbide is essential before starting the brazing process. During the experimental program, the analysis of the wettability of the filler material was conducted by making small deposits on the surfaces ready to be brazed. It was found that, in the case of tungsten carbide parts, by modifying the conditions of cleaning, the brazed surfaces behave completely different in the case of the 4 analysed samples. In the case of the unclean surface of tungsten carbide, the spreading of the filler material is poor and the droplet of molten metal tends to detach from the surface, the wetting angle being over 90 degrees (figure 4a). The best scattering of the liquid metal is achieved if complete preparation of the ceramic surface was carried out by polishing, scouring and pickling (figure 4d). Based on these assertions, the following succession of phases was introduced into the surface preparation procedure: grinding with 400 grit sandpaper SiC until obtaining a smooth surface, DR-60 solvent degreasing and pickling using brazing flux (FB 10).

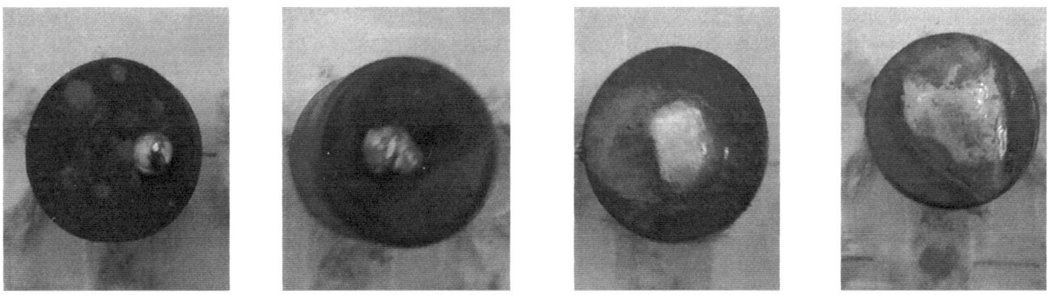

(a) (b) (c) (d)

Figure 4. Filler material spreading on the tungsten carbide surface: a) unclean; b) grinded surface; c) degreased surface; d) fully degreased and pickled surface.

The stages of the brazing process were as follows:

- machining the brazing gap (figure 5a);
- application of flux into the brazing gap;
- positioning of the tungsten carbide tip into the brazing gap (figure 5b);
- protecting the free surface of the ceramic body using refractory slurry (dip coating, leak and drying in furnace at 150°C for 1 hour);
- heating the assembly with flame by moving the flame around the steel body until reaching a prescribed preheating temperature, without exceeding the melting temperature of the brazing alloy;
- continuing the heating by positioning the flame on the ceramic tip and adding filler material until a slow refluxing in the brazing gap can be seen;
- slow cooling of the assembly in air;

- removal of the flow excess using hot water (40°C) and alcohol, followed by drying with hot air.

(a) (b) (c)

Figure 5. Brazing gap prepared for assembly (a), assembly prepared by oxy-flame brazing (b) and brazed piece (c) [3].

2.2. Induction brazing

A second brazing method used in this study was induction brazing (figure 6), which allows the assembly with the lowest values of actual work time. Brazing is recommended to be performed in protective atmospheres by blowing an inert gas (argon) because tungsten carbide is overheated and exposed to oxidation (figure 6).

Figure 6. Installation for brazing in protective atmosphere and the brazed assembly.

The steps of the induction brazing process are as follows:

- brazing gap processing;
- mechanical and chemical cleansing of assembly surfaces;
- application of flux on the inner surface;
- positioning of filler material at the bottom of the brazing zone;
- positioning of tungsten carbide tip into the brazing gap;
- positioning of the assembly under the inductor coil;
- pre-heating the assembly, without exceeding the melting temperature of the brazing alloy;
- brazing (by increasing progressively the electrical values for the melting of the filler material);
- slow cooling of the assembly in argon atmosphere;

IOP Conf. Series: Materials Science and Engineering **209** (2017) 012021 doi:10.1088/1757-899X/209/1/012021

- removing excess flow by polishing with wire brush followed by washing with warm water (40°C) and alcohol, followed by drying in hot air;
- visual inspection of the brazed assembly.

The parameter values for induction brazing are presented in table 1.

Table 1. Brazing parameters.

Sample	Gap dimension, [mm]	Voltage, [V]		Current, [A]		Frequency, [Hz]		Time, [sec]	
		P*	B**	P	B	P	B	P	B
P1	0.1	175	240	7	8	30	30	60	90
P2	0.05	160	240	8	10	30	30	60	120
P3	0.15	160	217	8	8	30	30	60	120

P* - Pre-heating; B** - Brazing

3. Results

After brazing, the assembly was sectioned as well: Cutting transversely to the axis of the assembly, at a distance of 3 mm from the brazing gap, using a cutting precision machine Isomet 4000 and a metallographic disk for hard materials, 0.76 mm thick; Cutting in the cross section of the assembly, through the tungsten carbide, filler material and steel part, using the spark erosion cutting procedure (figure 7).

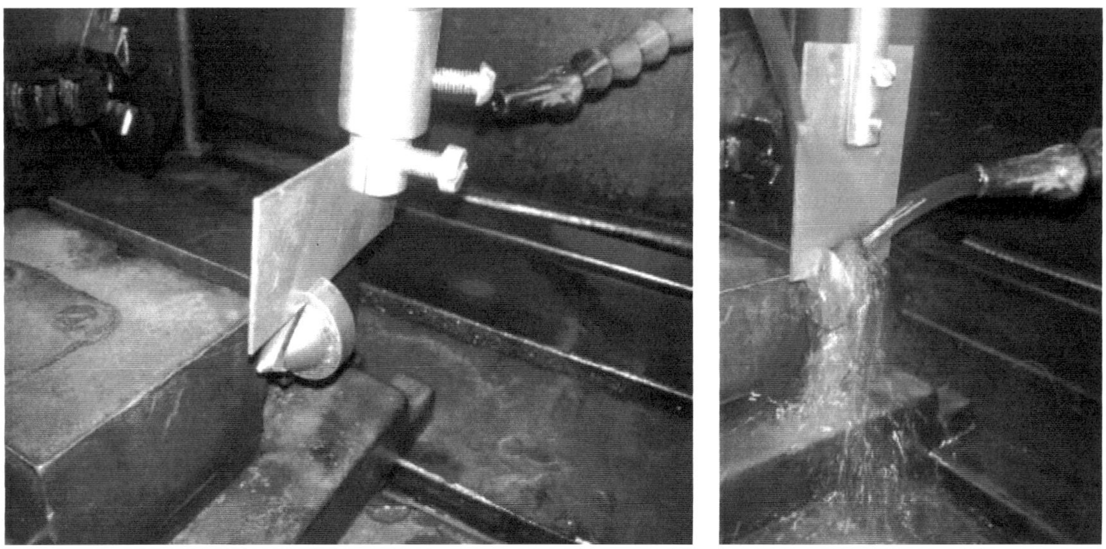

Figure 7. Spark erosion cutting of the brazed assembly.

The cross sections were then polished using sandpaper SiC, alumina and diamond paste, in order to analyse the joining zone. The analysis of the joining zones was carried out by optical (OM) and scanning electron microscopy (SEM) (figure 8).

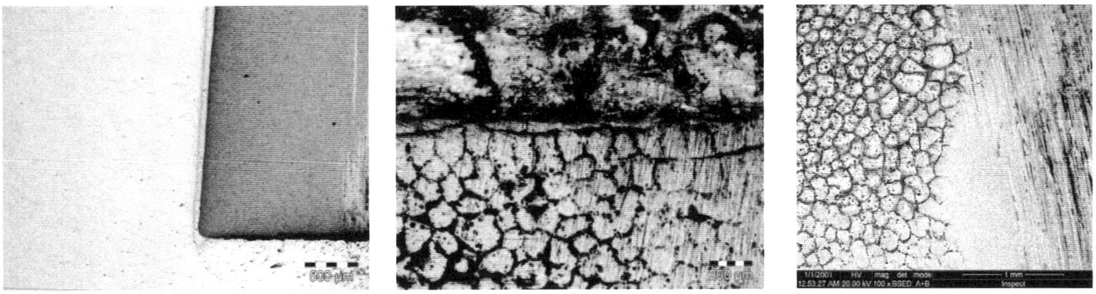

(a) (b) (c)

Figure 8. Cross section microstructure of the brazed assembly: a) global picture of the assembly zone, 50x (OM); b) details on transition between the tungsten carbide and steel (200x); c) bonding zone with very good adherence (SEM), 100x.

4. Conclusions

During the brazing of tungsten carbide with steel parts, there must be ensured a good penetration of the molten filler material into the brazing gap, given that the wettability of the two types of material is different and problems related to the bond strength can occur.

Oxy-flame brazing allows getting good quality brazed joints between steel and tungsten carbide, using a Ag-Cu alloy filler material. As silver is in greater proportion (over 30% wt Ag), the brazing behaviour is better and the carbide surface wetting is achieved more easily. The main drawback in this case is the lower productivity.

During induction, brazing requires a protective atmosphere to mitigate the oxidation effects of tungsten carbide. At the same time, excessive heating of tungsten carbide can promote hardness reduction. Recent filler materials used for the brazing of tungsten carbide are those alloyed with Ag and Zn (Cd-free), whose drawback is the high potential of evaporation during soldering. For this reason, in the most recent filler materials, Zn is replaced by Cu, Ni and Mn, sometimes accompanied by small amounts of indium or tin, to decrease the melting temperature.

References

[1] Schwartz M 1990 *Ceramic Joining* (ASM International, Materials Park OH) 166
[2] Nicholas M G 1989 *Ceramic-Metal Interfaces. In Surfaces and Interfaces of Ceramic Materials* (Kluwer Academic Publishers, United Kingdom) 393
[3] Voiculescu I, Binchiciu H, Geanta V and Iovanas DM 2014 *Research project „Materials and performant technologies designed to realize knives for milling the asphalt"no. PCCA 188/2012 – MATFREZ*;
[4] Huhne T, Hofmann E and Fussel U 2015 *Soldering and brazing of copper and copper alloys* (Deutches Kupferinstitut, Deutchland) 4
[5] Binchiciu E, Fleser T, Voiculescu I, Tihanov Tanasache D and Dragomir A 2016 *Advanced Materials Research* **1138** 119
[6] Willingham J 2005 *Filler metals and fluxes for brazing - Tungsten Carbide* (Jhonson Matthey Metal joining) 1 http://www.jm-metaljoining.com
[7] Klomp T 1990 *Interface Chemistry and Structure of Metal-Ceramic Interfaces, Joining of Ceramics* (M. G. Nicholas, Chapman and Hall, London)
[8] Voiculescu I, Geanta V, Daisa D D and Alexandrescu E2007 *Research project „Nanomaterials filler materials for joining ceramics – NANOCERAD*
[9] Voiculescu I, Geanta V, Stefanoiu R, Binchiciu H and Daisa D D 2012 *Alumina brazed using metallic alloys* (International Conference on Innovative Technologies, IN-TECH 2012, Rijeka) 523
[10] Suganuma K, Okamoto T, Koizumi M, Fujita T and Niihara K 1990 *J. Am. Ceram. Soc.* **73**, 1221

[11] Binchiciu A, Voiculescu I, Geantă V, Binchiciu H, Ştefănoiu R and Binchiciu E 2010 *Patent Number: RO125836/18.03.2010, Coated silver alloy rods of reduced hygroscopicity for brazing, comprises silver and tin deposited by extrusion deoxidizing flux made of synthetic vitreous precursors and metal powder additions.* (Derwent Innovation Index ISI-Web of Science/Science Citation Index Expanded)
[12] Smith W F 1993 *Structure and properties of engineering alloys, second edition* (McGraw-Hill)
[13] Davis J R 2001 *Phase diagrams* (ASM Specialty Handbook: Copper and Copper Alloys , ASM International) 369
[14] Binchiciu E, Fleser T and Voiculescu I 2015 *Advanced Materials Research* **1111** 25
[15] Binchiciu E, Fleser T, Voiculescu I and Binchiciu A 2015 *Advanced Materials Research* **1128** 224
[16] Matthey J *Metal joining, Silver Brazing Alloys and Fluxes* 4 http://www.jm-metaljoining.com
[17] Sim R F 1987 *Copper Phosphorus Based (Self –fluxing) Brazing Alloys used for Joining Copper and its Alloys* (FWP Journal) 33
[18] Binchiciu A, Voiculescu I, Geantă V, Binchiciu H, Ştefănoiu R, Negriu R M and Binchiciu E 2012 *Patent Number: RO125835/30.07.2012. Ecological coated core wires of silver alloys for brazing, is made by extrusion coating with deoxidizing flux with metal powder addition* (Derwent Innovation Index ISI-Web of Science/Science Citation Index Expanded).

Acknowledgements
This work was carried out through the Partnerships in Priority Areas Program - PN II, implemented with the support of MEN – UEFISCDI, Project MATFREZ – PCCA 188/2012 and PTE 14/2016 - BIOLEV.

IOP Conf. Series: Materials Science and Engineering **209** (2017) 012022 doi:10.1088/1757-899X/209/1/012022

Surface Characterization of New Biomaterials

M G Minciuna[1,3], P Vizureanu[1,3,a], M M B Abdullah[3], D C Achitei[1,3], B Istrate[2], R Cimpoesu[1] and S C Focsaneanu[1]

[1]Faculty of Materials Science and Engineering, „Gheorghe Asachi" Technical University from Iaşi, Blvd. D. Mangeron 41, Iaşi, 700050, România

[2]Faculty of Mechanical Engineering, „Gheorghe Asachi" Technical University from Iaşi, Blvd. D. Mangeron 41, Iaşi, 700050, România

[3]Center of Excellence Geopolymer & Green Technology (CEGeoGTech), School of Materials Engineering, Universiti Malaysia, Perlis, 01000 Kangar, Perlis, Malaysia

E-mail : peviz2002@yahoo.com

Abstract: This paper presents the characterization of new alloys CoCrMoSi6, CoCrMoSi7, CoCrMoSi10, in terms of hardness determinations, fractographic analysis and surface analysis. The original version of the alloy was obtained by casting process in a vacuum arc furnace. Experimental results obtained from this study confirms that by increasing content of silicon, the mechanical properties are superior and the positive results obtained at surface studies favoring the formation of compounds, that lead to the reduction of alloying grade for α solid solution and the plasticity of the alloys.

1. Introduction

The worldwide researches aiming to improve the traditional manufacture technologies of both implants and of the biomaterials of which they are made has the final goal to promote a new generation of multifunctional implants with long-term performances. In regard to the implant material, the current trends in both medical practice and in researches focus on the use of materials with enhanced biologic and biomechanical characteristics, which provide advantages both in regard to biocompatibility with the human tissue and in avoiding the risk of infection or rejection after implanting.

The CoCrMo alloys were highlights by their advantageous properties. In composition of these alloys it is found non-noble metals, which improve the properties by increase the resistance and protect the alloys from oxidation, by passivation [7].

Due to low cost of alloying elements, the obtaining techniques less demanding, but also the advantageous properties, it can specify that CoCrMo alloys have a large utilization, being followed by titanium alloys [3,4,15].

The opportunity of experimental researches is the result of numerous observations, which highlights that the general properties and surface properties (roughness, fractografy) of alloys affect the quality of implants [4].

The experimental researches, present in this study, have like principal objective to improve the properties of cobalt alloys, used in medical applications [1,10,11]. To fulfill this objective, were elaborated new variants of biomaterials, which have a long lifetime, as well as replacement of current materials based on stainless steels and titanium alloys.

2. Material and methods

2.1. Obtaining technologies of new alloys by CoCrMo system

The establish of utilizations domain of CoCrMo alloys is based on knowledge of their physical, chemical, biological, technological properties, as well as their mutual influences between these.

On base the CoCrMo commercial alloy, with known properties, was obtained new alloys, by CoCrMo system, with improved properties, by addition of silicon [9,11].

The materials used (CoCrMo commercial alloy and silicon), for production of these alloys are with advanced purity and was properly selected, in the terms of chemical composition and grain size, in the order to be placed in alveoli made on cooper plate, cooled with water, by installation.

Table 1. Medium contents of elements by CoCrMo commercial alloy.

Alloy	Chemical composition [%]					
	Co	Cr	Mo	C	Si	Mn
"C" commercial alloy	65	29	5	0.4	0.35	0.25

In table 2 are present the characteristics of CoCrMo commercial alloy, according to quality certificate from producer.

Table 2. Characteristics of CoCrMo commercial alloy, according technical sheet of product.

Melting domain	1330-1400°C
Hardness	402 HV
Tensile strength (R_m)	760N/mm^2
Density	8.26 g/cm^3
Modulus of elasticity	219000 MPa
Elongation	3%

To improve the properties of alloys by CoCrMo system is adding silicon, in varied content, in the elaboration time.

The increase of silicon content, for the alloys of CoCrMo system, determines the properties improvement by: hardness increase, realization of micro-crystalline structure and low values of roughness after cast process [2,13,14].

Table 3. New alloys obtained after cast process.

Silicon content added	Resulted alloy	Notation	Origin
	CoCrMo		commercial
6%	CoCrMoSi6	CSi6	original
7%	CoCrMoSi7	CSi7	
10%	CoCrMoSi10	CSi10	

To highlight the properties improvement, realized by content modification of silicon, was considered that the important laboratory investigations, for characterization of new alloys by CoCrMo

IOP Conf. Series: Materials Science and Engineering **209** (2017) 012022 doi:10.1088/1757-899X/209/1/012022

system, are: electronic microscopy, structural analysis by X-ray diffraction, hardness tests, fractographic analysis and roughness determinations [5,8,12].

2.2. Micro-structural characterization of cobalt based alloys

For identification of structure, phases and constituents, it is used a SEM, with BSE detector, at 500X magnitude.

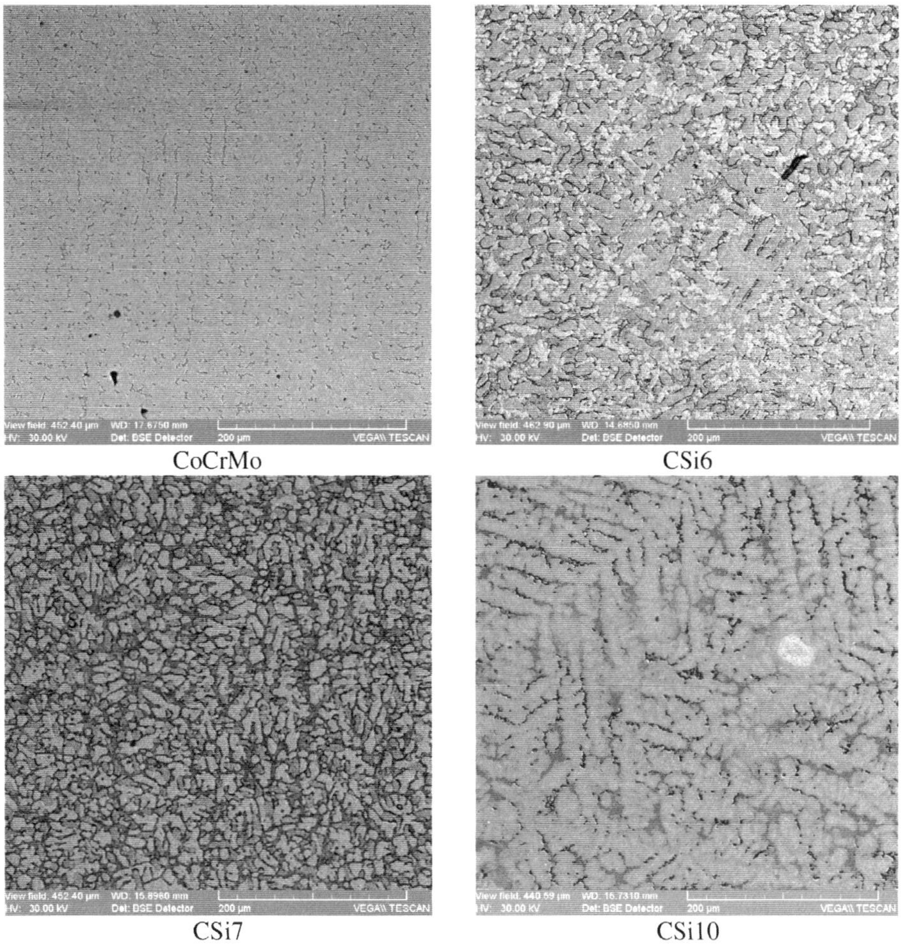

Figure 1. Microstructure of investigated alloys at 500X BSE magnitude.

SEM microstructure for commercial alloy and original variants allied with silicon in 6, 7 and 10% contents, at 500X BSE magnitude, present a dendrite structure, specific to cobalt based alloys, in cast state, with existence of α and β phases, in varied proportions, modified once with the increase of silicon content, in the time of elaboration and cast processes.

2.3. Fractographic analysis

This analysis offer, with help of SEM, a complete characterization of mechanical properties, about the behavior at different mechanical actions (elasticity, fragility, tensile strength, creep).

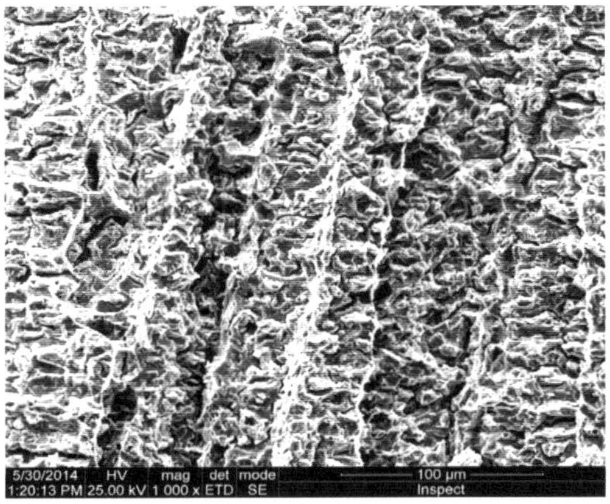

Figure 2. The aspect of surface for a sample made by CoCrMo commercial alloy.

In the figure 2, it can observe the aspect of breaking plans, with SEM, for CoCrMo commercial alloy. The aspect of surface for a sample made by CoCrMo commercial alloy, present a mixed breaking ductile-fragile, with micro-cracks and hard compounds.

Figure 3. The aspect of surface for a sample made from CSi6 original variant of alloy.

In the figure 3 CSi6 original variant of alloy, present a break by cleavage (fragile) with cracks, slides and nests of hard compounds.

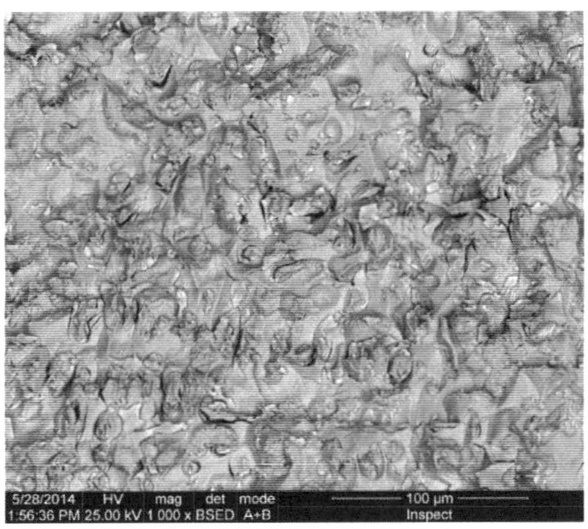

Figure 4. The aspect of surface for a sample made by CSi7 original variant of alloy.

In the figure 4 CSi7 original variant of alloy present a break by cleavage and cracks developed around the nests of hard compounds.

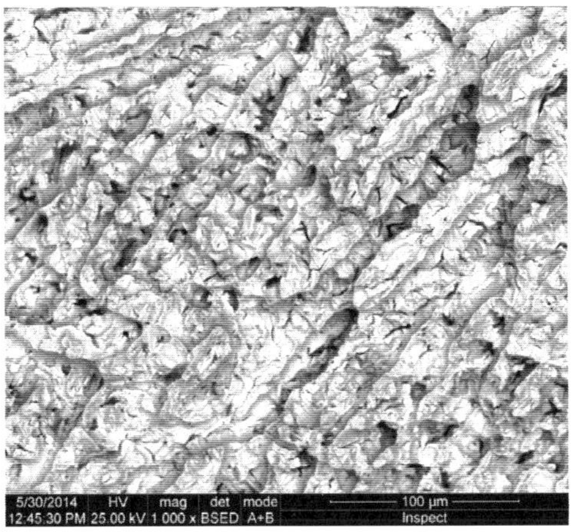

Figure 5. The aspect of surface for a sample made by CSi10 original variant of alloy.

In the figure 5 CSi10 original variant presents a break by semi-cleavage, oriented on dendrite formations and highlights many micro-cracks.

2.4. Investigations using X-ray diffraction

The establishing of compositional phases was made by quantitative analysis, with X-ray diffraction, using X-ray difractometer, model Panalytical X'Pert PRO MPD.

It is used an X-ray fascicule, characteristic CuKα, monochrome with Ni filter. The 2θ analysis domain is between 20…1000, step time is 3 second/step and step value is 0.0010.

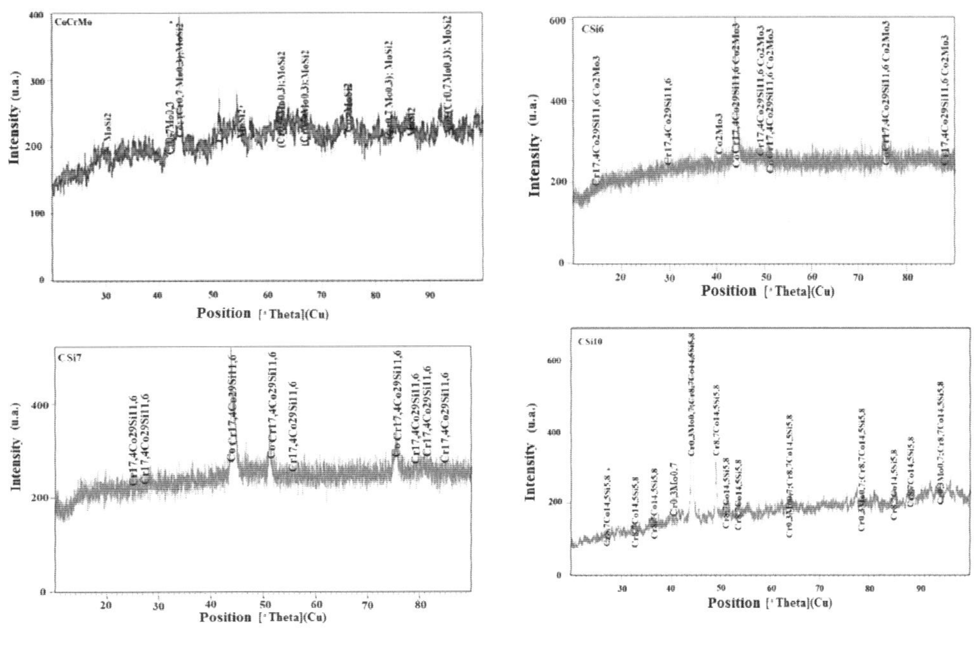

Figure 6. X-ray diffractograms obtained for new alloys by CoCrMo system.

Table 4. Phases identified at alloys by CoCrMo system.

Alloy	Phase	2θ angle
CoCrMo	Co with cubic crystalline network	44.19°
	$Cr_{0.7}Mo_{0.3}$ with cubic crystalline network	43.56°
	$MoSi_2$ crystallized in tetragonal network	44.48°
CSi6	Co with cubic crystalline network	43.91°
	Co_2Mo_3 crystallized in tetragonal network	44.59°
	$Cr_{17.4}Co_{29}Si_{11.6}$ with cubic crystalline network	44.10°
CSi7	Co with cubic crystalline network	44.10°
	$Cr_{29}Co_{17.4}Si_{11.6}$ with cubic crystalline network	44.10°
CSi10	$Cr_{8.7}Co_{14.5}Si_{5.8}$ with cubic crystalline network	44.10°
	$Cr_{0.3}Mo_{0.7}$ with cubic crystalline network	41.39°

2.5. Surface analysis of alloys by CoCrMo systems

It was made roughness measurements, on samples whose surfaces are finished by polishing and it was use a specialize apparatus, type Tester, Mitutoyo SJ-301 mark. The average values of parameters specific to roughness for CoCrMo commercial alloy and CSiK (K = 6, 7, 10) original variants, are present in table 6. In the table 5 is present the investigation realized on surface quality, have like principal objective the highlighting of roughness variations, corresponding to samples of CoCrMo and CsiK (K = 6, 7, 10) alloys.

Table 5. The average values of all parameters specific to roughness.

Alloys	Average value measured for R_z, R_a, R_q[μm]		
	R_z	R_a	R_q
CoCrMo	0.63	0.09	0.24
Csi6	0.24	0.03	0.04
CSi7	0.53	0.08	0.10
CSi10	1.17	0.19	0.25

After the study of roughness, the alloy which presents the lowest values (high quality) is CSi6 original variant alloy.

2.6. Hardness determinations of cobalt based alloys

To obtain the optimal result was made three determinations for each alloy, with same measurement conditions. Hardness measurements was made on Wilson Wolpert universal hardnessmeter, model 751N, using a press force by 9.807 N and a measurement time by 12 seconds. Hardness measurements made on new alloys by CoCrMo system, offer information about mechanical resistance, imposed or not the utility of heat treatments.

Table 6. Hardness of cobalt based alloys.

Alloys	CoCrMo	CSi6	CSi7	CSi10
Method	HV			
Punctual measured value	423	633	720	800
	446	633	746	832
	446	653	772	772
Medium value	438.3	639.6	746.0	801.3

3. Conclusions and discussions

By increase the silicon content (6, 7 and 10%), decrease the contents of cobalt, molybdenum and chrome, which are found on samples surface.

For cobalt alloys, the structure become homogenous and all the alloying element are found finely dispersed in structure.

The fractografic analyses were highlight the fact that new alloys by CoCrMo system, present a linear behavior until break, being part by fragile materials category. In the case of fragile break, micro-cracks initiated in breaking surface suddenly propagates, without a global deformation of materials, but only in micro-volumes localized on breaking surface, like: $Co_xCr_yMo_z$ with cubic crystalline network, CrSi with cubic crystalline network and Cr_yMo_zSi with cubic crystalline network. The fragile breaks produce sections in a plan approximately perpendicular on request plan and have a crystalline aspect (the break is initiated on grain limits with appearance of cleavage plans). The phase's composition of structure, as well as the characteristics of elemental cell for each identified phases were established by X-ray diffraction, but the principal objective was determination of compositional phases, microstructure and micro-composition of commercial alloy and for new alloys by CoCrMo system.

The values obtained at surface analysis confirm the fact that CSi6 alloy is the most indicate to be use in medical applications, because present a high corrosion resistance. Roughness with values slightly upward is specific to CSi10 alloy and implied a probability by high absorption, but and physical – chemical interactions of elements by biologic liquid with surface materials, more powerful.

The hardness tests have demonstrated that the silicon addition at the commercial variant of alloy by CoCrMo system, improve the mechanical characteristics, by formation of solid solution and chemical compounds like Cr_3Si and Mo_3Si, favoring also a structure with fine grains. After 60%Si, the hardness increase very much (639.6 HV), sometimes making impossible the mechanical processing. In this context, it is considered a possible processing, with a special device (super-hard mills).

References

[1] Achiţei D C, Minciună M G, Vizureanu P, Sandu A V, Cimpoeşu R and Istrate B 2016, *IOP Conference Series: Materials Science and Engineering* **133** 120
[2] Băitanu D D, Găluşcă D G, Achiţei D C and Minciună M G 2016 *Key Engineering Materials*, **700** 70–77
[3] Bose S, Banerjee D and Bandyopadhyay A 2017, Chapter 1 – Introduction to Biomaterials and Devices for Bone Disorders, in *Materials for Bone Disorders*, pp. 1–27
[4] Canto C E, Andrade E, de Lucio O, Cruz J, Solís C, Rocha M F, Alemón B, Flores M and Huegel J C 2016 *Nuclear Instruments and Methods in Physics Research Section B: Beam Interactions with Materials and Atoms* **371** 258–262
[5] Chen Q and Thouas G A 2015 *Materials Science and Engineering: R: Reports* **87** 1–57
[6] Mahapatro A 2015 *Materials Science and Engineering: C* **55** 227–251
[7] Minciuna M G, Vizureanu P, Hanganu C, Achitei D C, Popescu D C and Focsaneanu S C 2016 *IOP Conference Series: Materials Science and Engineering* **133** 19
[8] Minciună M G, Vizureanu P, Achiţei D C and Cimpoeşu R 2015 *Key Engineering Materials*, **660** 68–74
[9] Nouri A and Wen C 2015, 1 – Introduction to surface coating and modification for metallic biomaterials, in *Surface Coating and Modification of Metallic Biomaterials*, pp. 3–60
[10] Raghavendra G M, Varaprasad K and Jayaramudu T 2015, Chapter 2 – Biomaterials: Design, Development and Biomedical Applications, *Nanotechnology Applications for Tissue Engineering*, pp. 21–44.
[11] Vajpai S K, Sawangrat C, Yamaguchi O, Ciuca O P and Ameyama K 2016, *Materials Science and Engineering: C* **58** 1008–1015
[12] Vizureanu P, Sandu A V, Berbecaru A, and Sandu I G 2016 *Technology* **18**(1) 174–178
[13] Wang Y, Yan Y, Su Y and Qiao L 2017 *Journal of the Mechanical Behavior of Biomedical Materials* **68** 124–133
[14] Zavaglia C A C and Prado da Silva M H 2016 'Feature Article: Biomaterials, in *Reference Module in Materials Science and Materials Engineering*
[15] Costan A, Dima A, Ionita I, Forna N, Perju M C, Agop M 2011 *Optoelectronics and Advanced Materials-Rapid Communications* **5** 1-2

IOP Conf. Series: Materials Science and Engineering **209** (2017) 012023 doi:10.1088/1757-899X/209/1/012023

The Structure and Properties of Rapid Cooled Iron Based Alloy

B Jeż[1*], M Nabialek[1], P Pietrusiewicz[1], K Gruszka[1], K Błoch[1], J Gondro[1], J Rzącki[1], M M A B Abdullah[2], A V Sandu[3], M Szota[4], K Jeż[5] and A Sałagacki[6]

[1] Institute of Physics, Faculty of Production Engineering and Materials Technology, Czestochowa University of Technology, 19 Armii Krajowej Str., 42-200 Czestochowa, Poland

[2] Gheorghe Asachi Technical University of Iasi, Faculty of Materials Science and Engineering, 41 D. Mangeron Blvd., 70050, Iasi, Romania

[3] Geopolymer & Green Technology (CeGeoGTech), School of Material Engineering, Universiti Malaysia Perlis (UniMAP), 01000 Kangar, Perlis Malaysia

[4] Institute of Materials Engineering, Faculty of Production Engineering and Materials Technology, Czestochowa University of Technology, 19 Armii Krajowej Str., 42-200 Czestochowa, Poland

[5]Majkowice 25, 97 – 510 Ręczno\

[6]Olszowiec 53, 97-217 Lubochnia

E-mail: bartek199.91@o2.pl

Abstract. In this paper we studied the structure of rapidly cooled alloy which composition was based on iron. Samples were prepared using arc melting under protective atmosphere of argon and then casted in the process of rapid cooling into water cooled copper mold. Samples of the same composition were also made using the liquid melt casting method on a spinning copper roll. The high purity samples in the form of rods and ribbons were obtained. As expected, the obtained samples were characterized by an amorphous structure as confirmed by Mössbauer spectroscopy and X-ray diffraction studies.

1. Introduction

Rapidly cooled alloys have been the focus of science and industry for decades. Rapidly cooled alloys are those which are produced in such a way as to ensure a fast cooling of the melted ingredients resulting in significant changes in the structure of the resulting material. Rapidly cooled alloys differ significantly from their classic crystalline counterparts, with identical chemical composition. Despite the identical chemical composition, the alloy obtained by the rapid cooling method is significantly different in comparison to the conventional alloy [5-8].

Rapidly cooled materials are usually obtained by cooling the molten material. Suitable high cooling rate results in changes in the structure of the resulting material. Depending on the chemical composition of alloy, cooling rate and alloy manufacturing methods, different structures can be obtained. When talking about fast-cooled alloys, amorphous alloys and alloys with only partial crystallization of the structure are considered. Such alloys depending on the size of the crystallized grains may be called nanocrystalline. Nanocrystalline alloys may also be in the form of an amorphous matrix containing crystalline phase grains of nanometer size. Amorphous and nanocrystalline materials are extremely interesting due to their superior magnetic properties compared to classical counterparts [9-15].

The unique properties of fast-cooled alloys are directly derived from their structure. The high cooling speed prevents crystallization of the material, creating a structure characterized by the absence long range order of atoms. Production of alloys by rapid cooling causes disturbances in the structure of the material, resulting in structural defects in the form of free volumes and pseudodislocational dipoles [16-23].

The purpose of rapid cooling is to obtain an amorphous or only partially crystallized structure. The high speed of occurrence of the crystallization process greatly complicates the preparation of such materials. The need for a very fast cooling of the material reduces the size of the alloys produced, which is one of the biggest problems with rapidly cooled alloys. Rapidly cooled materials are often obtained on the base of iron. This type of material is characterized by very good magnetic properties such as low loss for demagnetization and high magnetization saturation. Rapidly quenched alloys are produced at various rates of cooling. The needed speed depends primarily on the chemical composition of the material. pending on the alloy produced, cooling rates of the order 10^5 K/s for classical amorphous materials and about 10^2 K/s for the production of massive amorphous alloys [24-28].

In this paper the results of structural studies of fast-cooled iron based alloys are presented. Samples were made of multi-component alloy$Fe_{61}Co_{10}Hf_2.5Zr_{2.5}Ti_2W_2B_{20}$.

2. Materials and experiments

The study used elements with a purity of 99.99%. The alloy components were weighed on the analytical balance. The accuracy of the weight and the purity of the elements ensure the desired alloy composition. The weighed alloy components were mixed together and then melt with a plasma arc on a water-cooled copper plate. Because of the high reactivity of oxygen with alloying elements, the batch melting process is carried out in a protective atmosphere of argon. Several times of ingot melting resulted in high homogeneity of the produced material, thus ensuring high reliability of the conducted investigations of the properties of the produced material. A fast-cooled alloy was obtained in the form of rod and ribbons. The used method of casting allows typical cooling rate of 10^0 – 10^2K/s.

The melted material was injected under pressure into a water-cooled copper mold. Ribbons were made by fast cooling of liquid melt on a rotating copper roll. The structure of the obtained alloy was examined using X-ray diffractometer and structural studies using Mössbauer spectroscopy. X-ray diffraction was performed using Brucker's Advanced D8 automated x-ray equipment. The diffractometer was equipped with a semiconductor counter and CuKa X-ray tube. Measurements were made for powdered samples for 2Θ angle ranging from 30° to 120°. Samples were exposed for 7 seconds per measurement step at the resolution 0.02°. Mössbauer research was performed using the Polon Mössbauer spectrometer. The Mössbauer spectrometer was equipped with the source ^{57}Co(Rh) with 50 mCi activity. Mössbauer spectra were analyzed using NORMOS software developed by R.A. Brand gave the possibility of obtaining a distribution of hyperfine induction fields of on the ^{57}Fe nucleus [17]. Structural studies were performed for samples in the post-solidified and post-heat-treated state. Samples were heated in a vacuum oven at 850K for 30 minutes. X-ray and Mössbauer studies were performed for low-energy powdered samples. The average grain size present in the sample volume in the form of a rod after the isothermal heating process at 850K / 30minutes was estimated based on the Scherrer's formula:

$$D = \frac{\lambda \cdot K}{2B_0 \cos\Theta}$$

K – Scherrer's coefficient (set to K = 0.91)
λ – characteristic wavelength
B_0 – peak half width (also background was considered)
Θ – Bragg angle

3. Results

X-ray diffraction patterns from the analyzed samples are shown in figure 1.

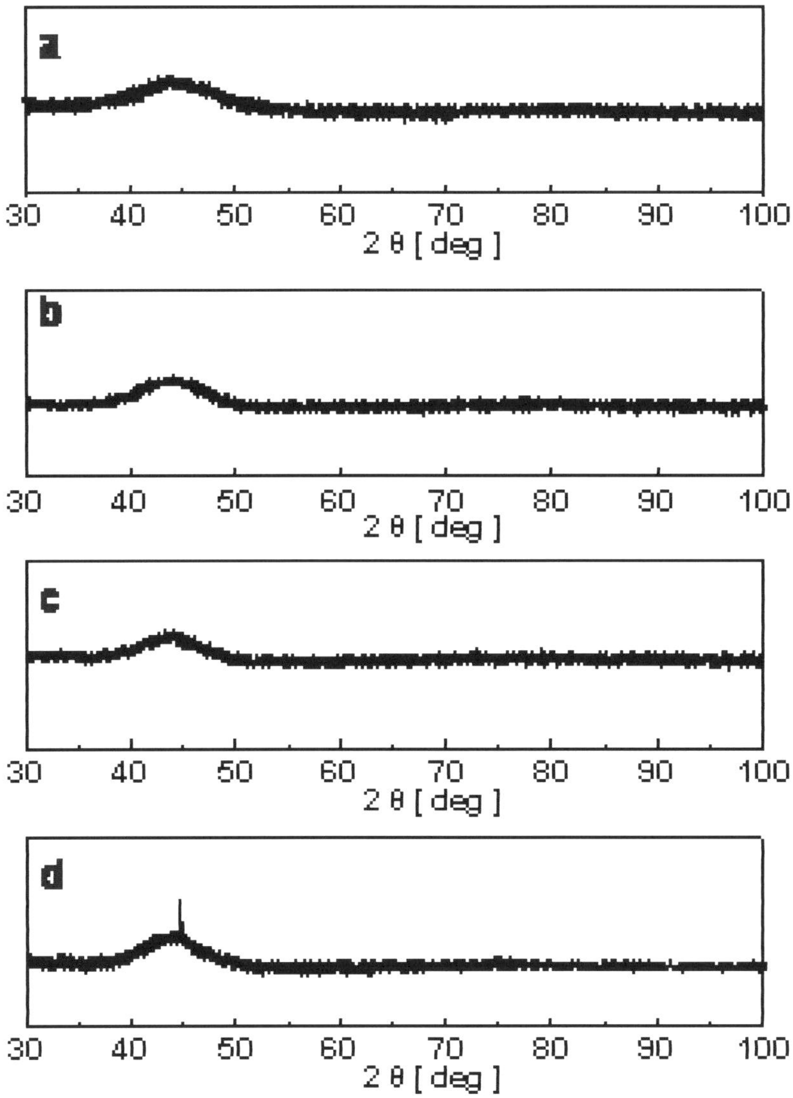

Figure 1. The XRD images for sample $Fe_{61}Co_{10}Hf_2.5Zr_{2.5}Ti_2W_2B_{20}$ in as cast state (a, c) and after isothermal annealing at 850K/30 minutes (b, d): a, b – ribbon 35 μm, c, d – rod of 1 mm diameter.

Shown in figure 1 (a, b) diffraction images for the samples after solidification state are typical of materials with an amorphous structure. They consist only of a single broad maximum known as the amorphous halo. Thermal treatment temperature for the alloy samples $Fe_{61}Co_{10}Hf_2.5Zr_{2.5}Ti_2W_2B_{20}$ in the form of ribbons and rods, was determined on the basis of differential scanning calorimetry (DSC). This temperature was determined slightly below the crystallization temperature for both two sample shapes. Therefore, it should be expected that the pretreatment of both rods and strips should not lead to the formation of privileged systems in the volume of alloys that meet the laws of spatial spacing and angular translations between atoms. In the case of a ribbon sample (figure 1b), the conducted isothermal treatment did not result in a crystalline structure in the volume of the material. On the other hand, performing the same thermal treatment (850 K) for a rod sample (figure 1d) resulted in a decrease in the energy state of the material due to topological and chemical changes which resulted in the emergence of fine nuclei seed of the crystalline phase. X-rays penetrate into the sample just for a few interplanar distances. Therefore, to obtain reliable X-ray results, they should be done for powdered samples. Using the Scherrer equation, the mean crystal particle size present in the rod sample after the isothermal heating process was calculated.

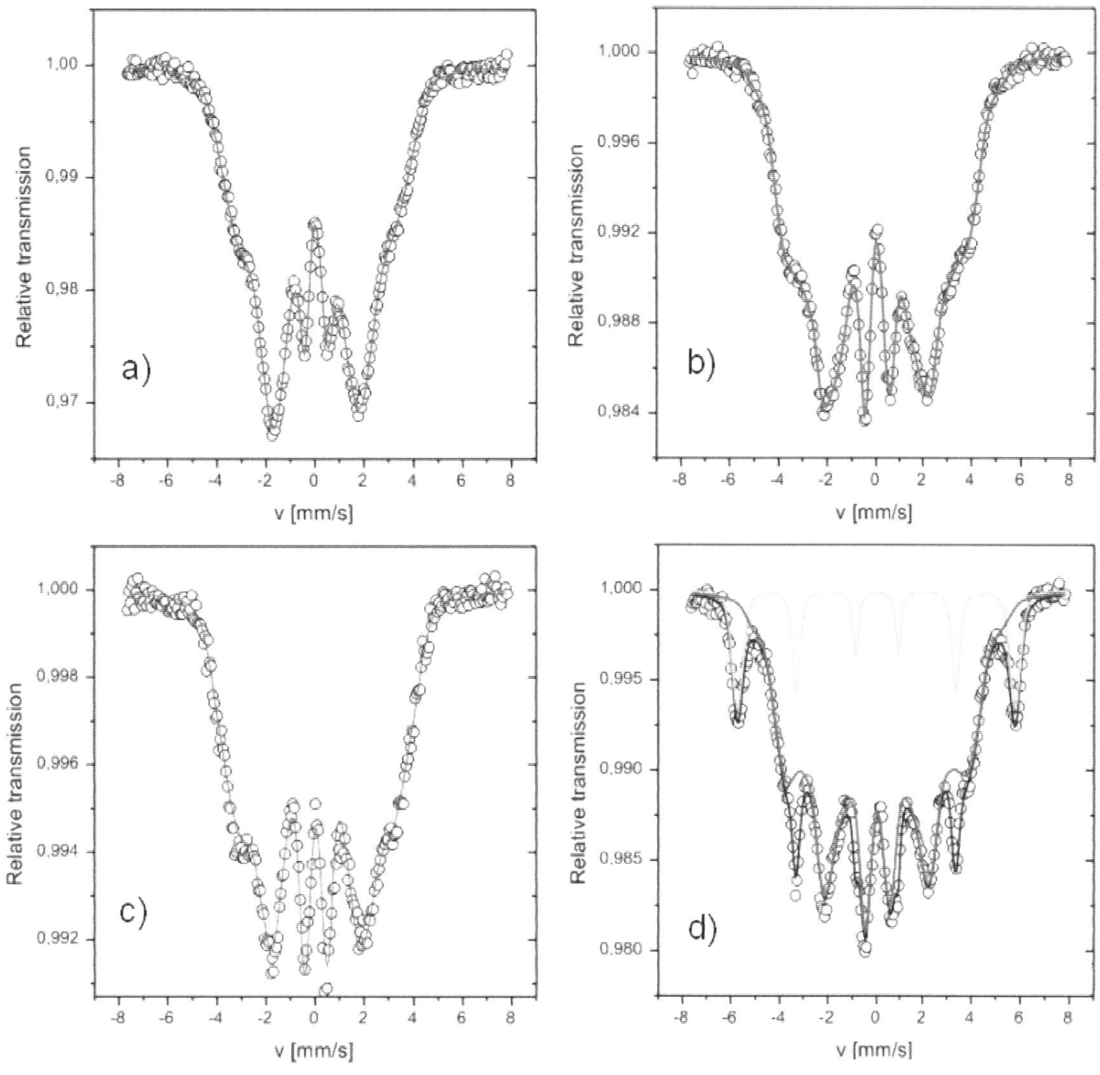

Figure 2. Mössbauer transmission spectrum recorded for the investigated alloy $Fe_{61}Co_{10}Hf_2.5Zr_{2.5}Ti_2W_2B_{20}$ in as cast state (a, c) and after isothermal annealing at 850K/30 minutes (b, d): a, b – ribbon 35μm, c, d – rod with 1 mm diameter.

The estimated grain size is 7 nm. Obtaining a finely granular structure at nanometric grains may be a factor in improving the properties of a given material. At present, metallic nanomaterials are heavily displacing their crystalline equivalents with the same chemical composition. The analysis of the sample structure was performed using a more sensitive test method, which is Mössbauer's spectroscopy. It is assumed that in the case of X-ray diffraction, samples can be analyzed with a minimum volume of crystalline phase of 7% relative to the rest of the material. Mössbauer studies, on the other hand, provide the possibility to determine the volume fraction of the crystalline phase present in the sample even at less share.

Figure 2 shows the Mössbaeuer spectra recorded for the samples of the alloy $Fe_{61}Co_{10}Hf_{2}.5Zr_{2.5}Ti_{2}W_{2}B_{20}$ In the state after solidification and after thermal treatment at 850K for 30 minutes. Mössbauer spectra obtained for samples in the as cast state in the form of a strip and rod are similar. Their shape is typical for samples with an amorphous structure. They consist of a broad asymmetric overlapping lines.

A similar shape of the Mössbauer spectra was observed for the strip sample after the annealing process. This means that the soaking process itself did not result in the formation in the sample volume of lower free energy areas corresponding to the crystalline phase.

It is known that raising the temperature promotes diffusion processes that take place in the sample volume. Amorphous systems are characterized by many states with different free energy values making them metastable. It follows that the amorphous state itself has many configurations of energy states which, with each successive jump through the potential barriers, reduce their free energy and eventually reach crystalline state. The initial phase of crystallization of amorphous alloys is manifested in the formation in the volume of alloys of clusters with similar arrangement of atoms from which crystalline seed is formed and then grains. It is assumed that such systems are evenly distributed in the volume of the material. This is the case with biphasic materials of an amorphous and crystalline structure. In the case of a rod sample subjected to isothermal heating at temperature 850K/30 min.

There is an additional line on the Mössbauer transmission spectrum associated with the presence of the crystalline phase in the sample (which is consistent with results of the X-ray examinations). Based on the Mössbauer spectra on ^{57}Fe nuclei an analysis was made.

Figure 3 shows the distributions of the superfine field induction for the investigated samples. For the sample in the shape of ribbon in the as cast state and after isothermal annealing process the distributions of hyperfine induction fields are similar. There are clearly visible two lines low- and high-field components. The mean value of the low-field component for the ribbon sample in the as cast state is about 7 T whereas the value of this component for the same sample after heating is 10 T. The similar tendency is observed for second well seen high field component, where for sample in as cast state its value is about 16 T and after isothermal annealing is 20 T. In addition, it has to be stated that for the sample after heating in the distribution of induction of the hyperfine induction fields appeared the third component with the least probability occurring in the fields range from 25 to 35 T. Configuration changes in the environment of ^{57}Fe are connected with nearest atoms. It is evident that the process of isothermal annealing influences the magnetic properties of the ribbon of $Fe_{61}Co_{10}Hf_{2}.5Zr_{2.5}Ti_{2}W_{2}B_{20}$ alloy.

Based on a visual analysis of the distribution of the hyperfine induction field for a sample annealed at 850 K, it can be concluded that clusters of atoms begin to form in the volume of the melt. This is evidenced by the presence of the third field component and is consistent with the thermodynamics of amorphous systems.

A completely different shape of hyperfine induction is observed for rod shape sample. For the rod in as cast state in the distribution of hyperfine field there are three components, in which the low-field component is well separated. The mean values of induction of hyperfine fields are about 3 T, while for the components in the higher fields the values are 11 and 19 T.

After heating of rod samples, the shape of the induction of the hyperfine fields has changed considerably. It was possible to distinguish at least 5 components of which 4 were related to amorphous structure and one to crystalline structure. The mean values of induction of superfine fields associated with the amorphous structure were respectively: 5, 14, 23 and 30 T. The field component located at the highest magnetic induction is associated with the appearance of clusters with similar atomic configurations. The average value for this field was about 36 T.

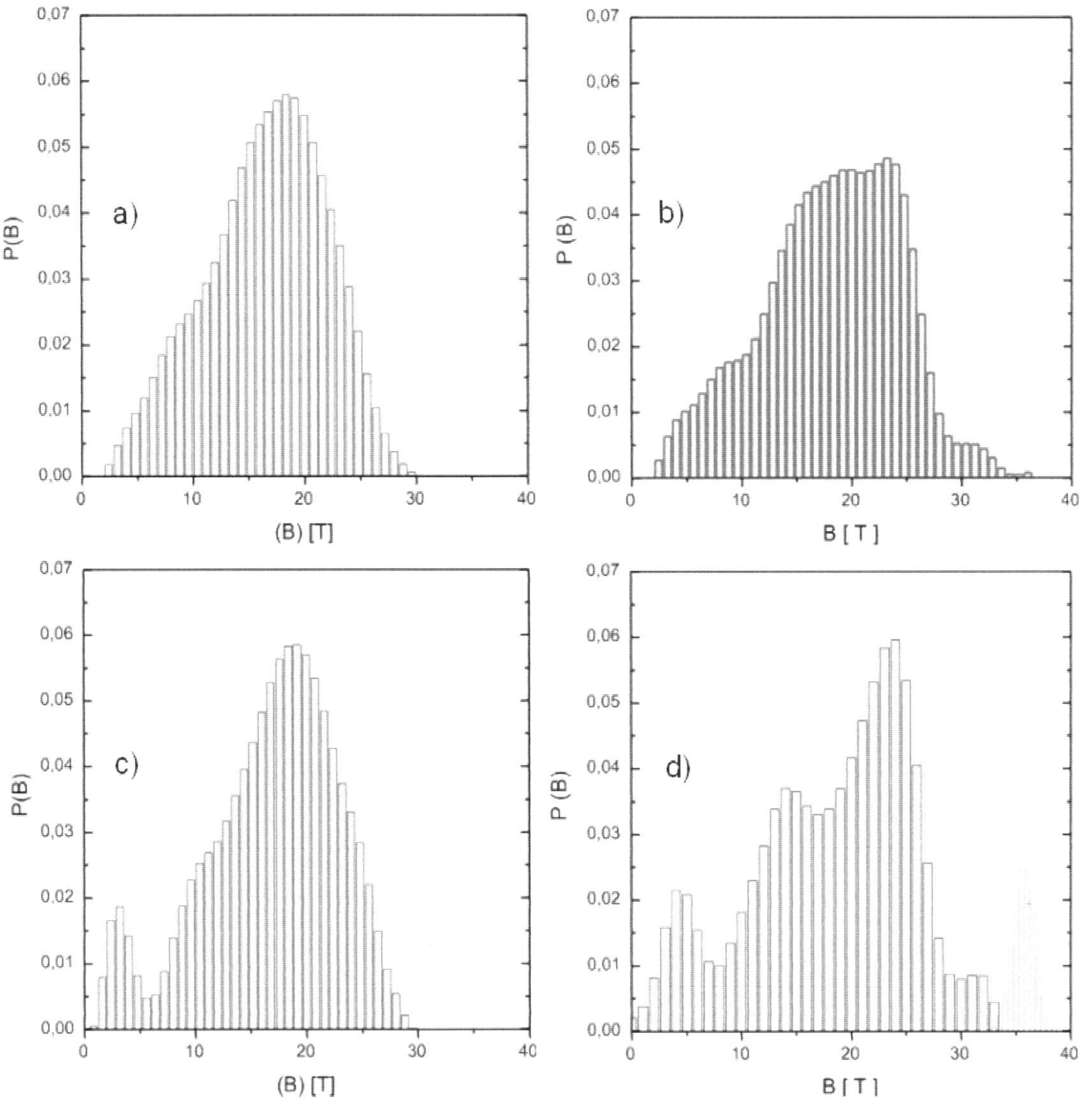

Figure 3. Mössbauer transmission spectrum recorded for the investigated alloy $Fe_{61}Co_{10}Hf_{2}.5Zr_{2.5}Ti_{2}W_{2}B_{20}$ in the as cast state (a, c) and after isothermal annealing at 850K for 30 min. (b, d): a, b – strip 35 μ m, c, d – rod of 1 mm diameter.

4. Conclusions

Using suitable laboratory equipment an amorphous $Fe_{61}Co_{10}Hf_{2}.5Zr_{2.5}Ti_{2}W_{2}B_{20}$ alloy samples be obtained in the form of strips of 35 μm width, and rods of 1 mm diameter. Determining the right temperature and time for the isothermal annealing process is a decisive factor for the structure of the materials under study. s is known, the structure determines the properties of alloys. It has been shown in the present paper that the temperature of the heat treatment did not caused crystallization of the alloy in the form of a ribbon. However, as the Mössbauer studies indicate, changes in the topological and chemical arrangement of atoms in the amorphous state occurred in the volume of the sample. This is related to the reconfiguration to lower energy metastable states. Amorphous ribbons are produced at much higher cooling rates (10^4 – 10^6 K/s) than massive amorphous rods (10^0 – 10^3 K/s). The shape of the Mössbauer spectra obtained for rod samples was quite different from that of ribbon shaped samples. This means that the manufacturing process and the cooling speed of the liquid alloy itself have the greatest influence on the forming of the structure. For larger specimens where the solidification time is longer, volume diffusion occurs at further distances, which promotes the formation of energy-prone systems such as clusters of atoms or nuclei.

References:

[1] McHenry M E, Willard M A and Laughlin D E 1999 *Prog. Mater. Sci.* **44** 291
[2] Lesz S, Szewczyk R, Szewieczek D and Bienkowski A 2004 *J. Mater. Process. Tech.* **743** 157-158
[3] Inoue A, Shen B L and Chang C T 2006 *Intermetallics* **14** 936–944
[4] Shen B and Inoue A 2004 *Applied Physics Letters* **85** (21) 4911–4913
[5] Hasiak M, Sobczyk K, Zbroszczyk J, Ciurzyńska W, Olszewski J, Nabiałek M, Kaleta J, Świerczek J and Łukiewska A 2008 *IEEE Trans. Magn.* **11** 3879-3882
[6] Sobczyk K, Świerczek J, Gondro J, Zbroszczyk J, Ciurzyńska W, Olszewski J, Brągiel P, Łukiewska A, Rzącki J and Nabiałek M 2012 *J.Magn. Magn. Mater.* **324** 540-549
[7] Pawlik P, Nabiałek M, Żak E, Zbroszczyk J, Wysłocki J J, Olszewski J and Pawlik K 2004 *Arch. Mat. Sci.* **177** 25/3
[8] Inoue A, Yano N and Masumoto T 1984 *J. Mater. Science* **19** 3786-3795
[9] Wang W H, Dong C and Shek CH 2004 *Materials Science and Engineering* R **44** 45-89
[10] Li W, Yang Y Z and Xu J 2017 *Journal of Non-Crystalline Solids* **461** 93–97
[11] Willard M A and Daniil M 2013 in: *Handbook of Magnetic Materials*, Vol. 21, ed Buschow K H J (London: North Holland) p 173
[12] Hasiak M, Sobczyk K, Zbroszczyk J, Ciurzyńska W, Olszewski J, Nabiałek M, Kaleta K, Świerczek J and Łukiewska A 2008 *IEEE Trans. Magn.* **44** 3879
[13] Gondro J, Świerczek J, Rzącki J, Ciurzyńska W, Olszewski J, Zbroszczyk J, Bloch K, Osyra M and Łukiewska A 2013 *J. Magn. Magn. Mater.* **341** 100-107
[14] Chiriac H and Lupu N 2001 *Physica B: Condensed Matter* **299** (3–4) 293–301
[15] Sun H J, Man Q, Dong Y Q, Shen B, Kimura H, Makino A and Inoue A 2010 *J. Alloys Comp.* **504** 31
[16] Gruszka K, Nabiałek M, Szota M, Bloch K, Gondro J, Pietrusiewicz P, Sandu A V, Mustafa Al Bakri A M, Walters S, Walters K, Garus S, Dośpiał M and Mizera J 2016 *Arch. Metall. Mater.* **61** 641–644
[17] Makarov V A, Belenkii A Ya and Kozlova O S 1993 *Phys. Stat. Sol.* (A) **139** 173–179
[18] Nabiałek M, Pietrusiewicz P, Błoch K and Szota M 2015, *Int. J. Mater. Res.* **106** 682-688
[19] Pietrusiewicz P, Błoch K, Nabiałek M and Walters S 2015 *Acta Phys. Pol.* A **127** 397-399
[20] Gondro J, Błoch K, Brągiel P, Nabiałek M and Szota M 2016 *Arch. Metall. Mater.* **61** 451–456
[21] Kronmüller H 1981 *Journal of Applied Physics* **52** (3) 1859-1864
[22] Lange H and Kronmüller H 1986 *Phys. Stat. sol.* (a) **95** 621-633
[23] Kronmüller H 1984 *J. Magn. Magn. Mater.* **41** 366
[24] Nabiałek M 2016 *Arch. Metall. Mater.* **61** 439–444
[25] Inoue A, Shen B and Ohsuna T 2002 *Mater. Trans.* **43** (9) 2337–2341
[26] Inoue A 1998 *Mater. Sci. Fundations* 6, Trans Tech Publications, 1-116
[27] Sanchez Ll J L, Bustamante S R, Barthem V M T S and de Miranda P E V 2005 *Journal of Magnetism and Magnetic Materials* **294** 131–135
[28] Inoue A 1997 *Materials Science and Engineering* A **226-228** 357-363.
[29] Brand R A 1987 *Nuclear Instruments and Methods in Physics Research* B **28** 398-416
[30] Gu B X, Xue D S, Shen B G and Li F S 1997 *J. Magn. Magn. Mater.* **167** 105
[31] Szewieczek D and Lesz S 2005 *J. Mater. Process. Tech.* **162- 163** 254

Preliminary Microstructural and Microscratch Results of Ni-Cr-Fe and Cr_3C_2-NiCr Coatings on Magnesium Substrate

B Istrate[1], C Munteanu[1*], S Lupescu[1], M Benchea[1] and P Vizureanu[2]
[1] "Gheorghe Asachi" Technical University of Iasi, Faculty of Mechanical Engineering, 43 "D. Mangeron" Street, 700050, Iasi, Romania
[2] "Gheorghe Asachi" Technical University of Iasi, Faculty of Material Science and Engineering, 61-63 "D. Mangeron" Street, 700050, Iasi, Romania

E-mail: cornelmun@gmail.com

Abstract. Thermal coatings have a large scale application in aerospace and automotive field, as barriers improving wear mechanical characteristics and corrosion resistance. In present research, there have been used two types of coatings, Ni-Cr-Fe, respectively Cr_3C_2-NiCr which were deposited on magnesium based alloys (pure magnesium and Mg-30Y master alloy). There have been investigated the microstructural aspects through scanning electronic microscopy and XRD analysis and also a series of mechanical characteristics through microscratch and indentation determinations. The results revealed the formation of some adherent layers resistant to the penetration of the metallic indenter, the coatings did not suffer major damages. Microstructural analysis highlighted the formation of Cr_3C_2, Cr_7C_3, Cr_3Ni_2, Cr_7Ni_3, $FeNi_3$, Cr-Ni phases. Also, the apparent coefficient of friction for Ni-Cr-Fe coatings presents superior values than Cr_3C_2-NiCr coatings.

1. Introduction

Thermal coatings represent some special coatings processes, which has seen an accelerated development in the last decades, especially in automotive, naval, aerospace domains, like: turbine blades, cylinders, pistons, valves, in order to improve wear resistance, corrosion resistance and lifetime of the machine part [1–3]. The most common thermal deposition techniques are APS and HVOF, but there are also other coatings methods like: flame spraying, wire arc spraying, PLD etc [4–9]. Coatings obtained by plasma jet deposition (APS) with formation of Ni, Cr ceramic layers are among used materials in mechanical applications [10]. The main disadvantage of these coatings could represent the isolated inhomogeneity, micro-cracks and residual stresses. These stresses occur during cooling process of thermal spraying [11].

Mayrhofer et al. highlighted the wear resistance improvement of steel substrates by using Cr_3C_2 stabilized with NiCr layers in comparison with WC coatings [12]. It has been also shown that Cr_3C_2-NiCr coatings present comparable wear with tungsten carbide, but it has superior oxidation resistance [2]. Varies et al. studied the fatigue resistance improvement Cr_3C_2 based coatings by performing some kinetic energy layers and special treatments [13]. The researchers showed that the coatings which contain iron, improve the layers ability to make diffusion with base material being also cheaper and less pollutant with the environment than Co and Ni based coatings [14,15].

In the present paper, authors deposited two types of plasma jet coatings (Cr_3C_2-NiCr and Ni-Cr-Fe) on magnesium alloys substrate (Pure Mg and Mg-30Y master alloy). It have been used these based materials due their applicability in automotive and aerospace domains. Kula et al. highlighted that the presence of Yttrium in magnesium alloys lead to hardness and ductility improvement [16].

There have been performed morphological analyses of deposited layers, phase identification using XRD determination and also identification of some mechanical characteristics using scratch and indentation method.

2. Materials and methods

Ni-Cr-Fe, respectively Cr_3C_2-NiCr coatings were obtained using atmospheric plasma spray facility (SPRAYWIZARD-9MCE, Sultzer-Metco, USA/ 9MB spraying gun) with typically particle size between 10-90 microns. The substrates of these coatings used in this present research were pure magnesium and a Mg-30Y master alloy (HB New Material, Changsta China). In table 1 are highlighted the technological parameters used for the coating process.

Morphological and microstructure have been determined using a LFD (large field detector), respectively dual BSD (backscattered detector) on a scanning electron microscope SEM Quanta 200 3D DUAL BEAM. For phase analysis, XRD analysis was carried out using an X'Pert Pro MPD diffractometer, with a copper X-ray tube – Kα: 1.54 A, in 2Theta scanning range of 20°-90°. Apparent coefficient of friction, hardness and elastic modulus were measured using CETR UMT-2 Tribometer. Micro-scratch analysis parameters consist in: a constant load of 5N for a distance of 4 mm, for a single determination, on a flat sample of 20mm x10mm sample. Also, for the indentation determination, there were performed three determinations for each sample with a constant load of 5 N and using a metallic indenter.

Table 1. Technological parameters used for coating process.

Powder	Gun	Ar		H_2		Electric		Powder feeder 9MP			Spray distance (inch)
		Pressure (psig)	Gas flow (SCFH)	Pressure (psig)	Gas flow (SCFH)	DC (A)	DC (V)	Carrier gas flow (SCFH)	Air Preasure (psig)	Rate (lb/h)	
Cr3C2-NiCr coating	9MB	75	111	50	10	500	60-70	13.5	20	5.5	2.25
Ni-Cr-Fe coating		75	90	50	20	500	70-80	13.5	20	4	9

3. Results and discussions

3.1. Structural analysis

3.1.1. Scanning electron microscopy. Microstructure and morphology aspects of the ceramic coatings are presented in figures 1-4. In figure 1 (a-f) is presented the specific deposition aspects of APS method of a Cr_3C_2-NiCr, respectively Ni-Cr-Fe based coatings.

These ceramic layers have on the surface an average roughness with the presence of some small unmolded particle, few micropores and some microcracks, due the recovery process of thermal deposition. Different scale images present that in both cases an adherent coating with the substrate.

In figures 2-4 are shown cross-section images (a-f) for Cr_3C_2-NiCr / Ni-Cr-Fe layers. Ceramic coatings have the appearance of columnar grains called "splats", due the layer by layer spraying technique and having an average thickness of the splat of 3 microns (figure 4).

The average thickness of coatings are presented in figure 2(a, e), respectively in figure 3 (a, d) which is aproximatively 200 μm for Cr_3C_2-NiCr coating and respectively 150 μm for Ni-Cr-Fe coating.

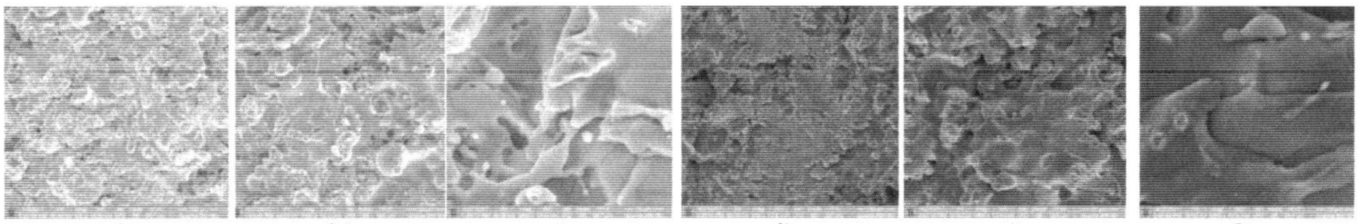

a) b) c) d) e) f)

Figure 1. Surface SEM images for Cr_3C_2-NiCr coating: a), b), c); Ni-Cr-Fe coating: d), e), f).

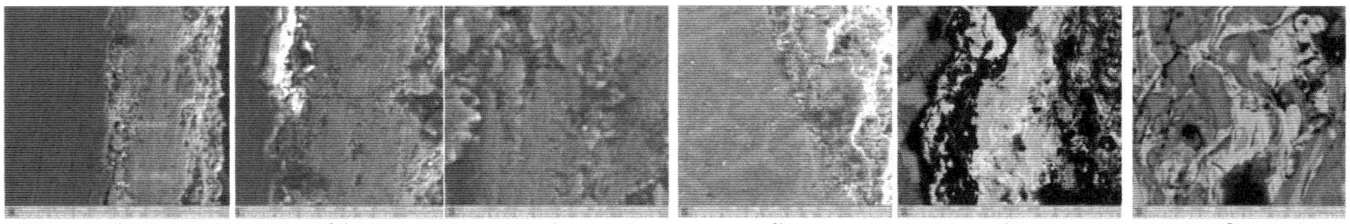

a) b) c) d) e) f)

Figure 2. Cross-section section SEM images for Cr_3C_2-NiCr on pure Mg substrate: a), b), c); Mg-30Y substrate: d), e), f).

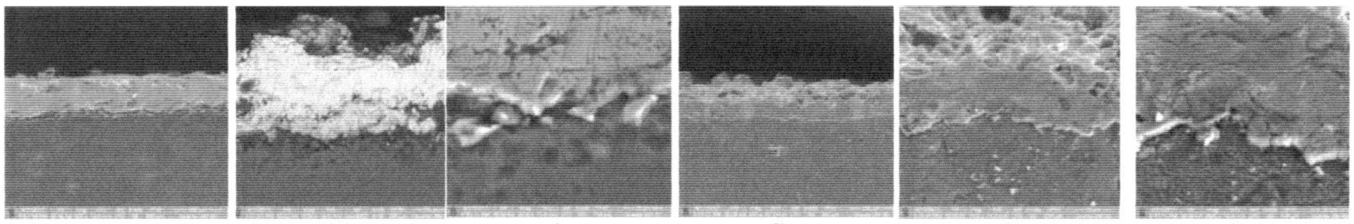

a) b) c) d) e) f)

Figure 3. Cross-section section SEM images for Ni-Cr-Fe on pure Mg substrate: a), b), c); Mg-30Y substrate: d), e), f).

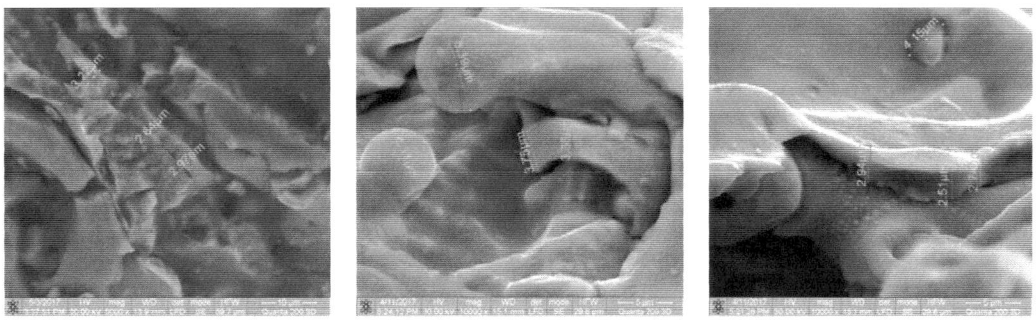

a) b) c)

Figure 4. Aspects of spraying results: a) Cr_3C_2-NiCr; b),c) Ni-Cr-Fe coating.

3.1.2. XRD analysis. XRD diffraction analyses are shown in Figure 5. The diffraction analysis highlighted Cr-based compounds: Cr_3C_2, Cr_7C_3, Cr_3Ni_2, $Cr_{0.1}Ni_{0.9}$, Cr_7Ni_3 having both orthorhombic and tetragonal crystal structure.

For the coatings, Cr_3C_2 and Cr_7C_3 are the predominant phases around at 44.22° 2θ angle. Cr_3Ni_2 and $FeNi_3$ are the minority diffraction peaks of the coatings around 51.67° 2θ angle and 76.24° 2θ angle, with a tetragonal, respectively cubic structure. Base materials consist three types of compounds: α-Mg (hexagonal structure) and MgY, respectively Mg24Y5 (cubic structure). Lattice parameters of all identified compounds are shown in table 2.

Table 2. Lattice parameters of Cr_3C_2-NiCr / Ni-Cr-Fe coatings / Base materials (Pure Mg / Mg-30Y).

	Compound	Space Group	Crystal system	a (Å)	b (Å)	c (Å)	α (°)	β (°)	γ (°)	Cell volume (10^6 pm^3)
Cr3C2-NiCr coating	Cr_3C_2	Pnam	Orthorhombic	5,5250	11,4680	2,8266	90	90	90	179,10
	Cr_7C_3	Pnma	Orthorhombic	4,5260	7,0100	12,1420	90	90	90	385,23
	Cr_3Ni_2	P42/mnm	Tetragonal	8,8200	8,8200	4,5800	90	90	90	356,29
	Cr_7Ni_3	P42/mnm	Tetragonal	8,7100	8,7100	4,4900	90	90	90	340,63
	$Cr_{0.1}Ni_{0.9}$	P-3m1E	Cubic	3,5350	3,5350	3,5350	90	90	90	44,17
Ni-Cr-Fe coating	$Cr_{1.12}Ni_{2.88}$	Fm-3m	Cubic	3,5400	3,5400	3,5400	90	90	90	44,36
	$FeNi_3$	Fm-3m	Cubic	3,5556	3,5556	3,5556	90	90	90	44,95
	$Cr_{0.1}Ni_{0.9}$	Fm-3m	Cubic	3,5350	3,5350	3,5350	90	90	90	44,17
Pure Mg / Mg-30Y	Mg	P63/mmc	Hexagonal	3,2088	3,2088	5,2099	90	90	120	46,46
	MgY	Pm-3m	Cubic	3,7900	3,7900	3,7900	90	90	90	54,44
	$Mg_{24}Y_5$	I-43m	Cubic	11,2500	11,2500	11,2500	90	90	90	1423,83

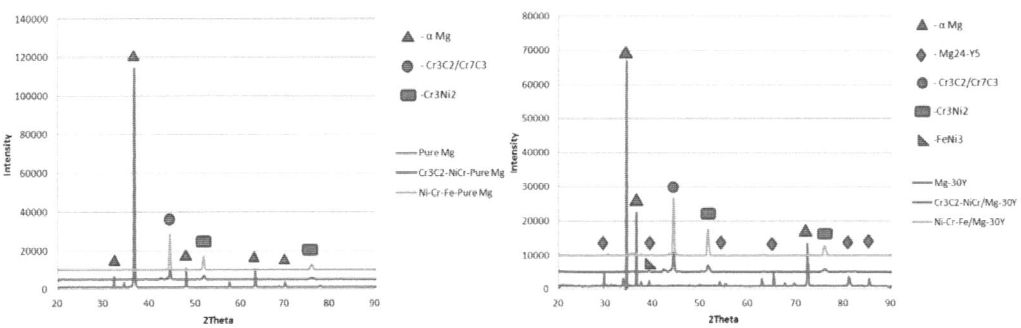

Figure 5. XRD patterns of Cr_3C_2-NiCr / Ni-Cr-Fe coatings.

3.1.3. Scratch and micro-indentation analysis. Figure 6 presents SEM images of scratch test of Cr_3C_2-NiCr coatings (a, b), respectively Ni-Cr-Fe coatings. The apparent coefficient of friction for the Cr_3C_2-NiCr layer is in the range of 0.55. Same analyses were carried out for the second layer, respectively Ni-Cr-Fe, which presented an apparent coefficient of friction around 1.25.

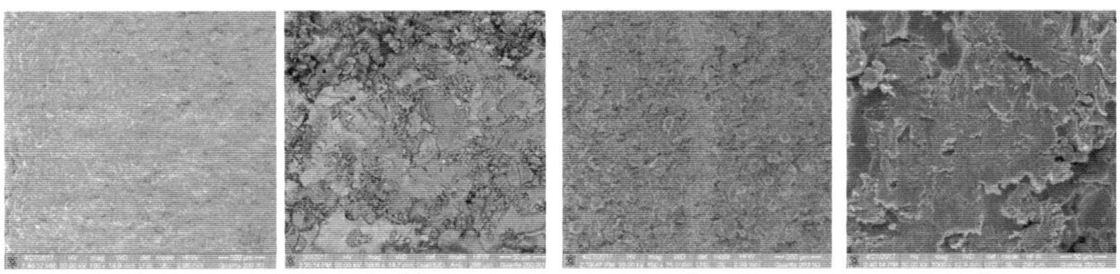

a) b) c) d)

Figure 6. SEM images of scratch test for Cr_3C_2-NiCr coatings (a,b), Ni-Cr-Fe coatings (c,d).

SEM images showed adherent layers without removing to the base material. It have been observe a small number of microcracks in both of the coatings, resulting less fragile layers. Coatings present similar values of the stiffness between 3.589 and 3.953 N/μm. The results of microindetation test have shown a depth of between 6.15 μm and 16.33 μm for Cr3C2-NiCr layer, respectively 8.46 μm and 18.91 μm for the second, without fracturing. It can be concluded that both coatings have high resistance to fracturing, a fact confirmed by hardness values and coefficient of friction apparently values presented in table 3 and figure 7. Also the hardness is influenced by the substrate, Ytrrium has superior behavior in magnesium alloys [14].

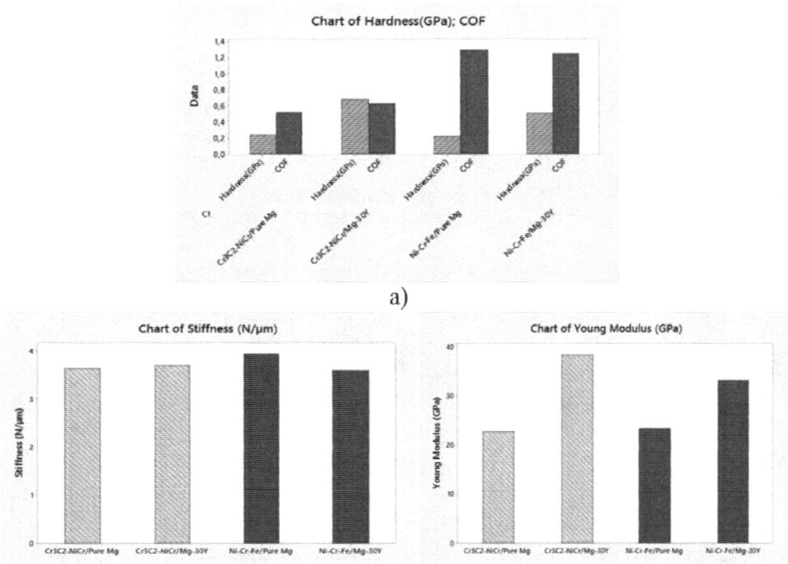

a)

b) c)

Figure 7. Some mechanical characteristics obtained by scratch and microindentation analysis: a) Hardness/COF; b) Stiffness; c) Young modulus.

Young modulus is influenced by the substrate characteristics. The coatings performed on pure magnesium substrate show similar values between 22.79 GPa and 23.37 GPa and also the coatings deposited on Mg-30Y revealed higher values of elastic modulus between 33.17 GPa and 38.34 Gpa.

Table 3. Some mechanical properties of Cr3C2-NiCr / Ni-Cr-Fe coatings on Pure Mg, respectively Mg-30Y substrate.

	COF	Hardness (GPa)	Young Modulus (GPa)	Stifness(N/µm)
Cr3C2-NiCr / Pure Mg	0.515	0.239	22.795	3.630
Cr3C2-NiCr / Mg-30Y	0.629	0.683	38.343	3.701
Ni-Cr-Fe / Pure Mg	1.298	0.217	23.375	3.935
Ni-Cr-Fe / Mg-30Y	1.253	0.513	33.172	3.589

4. Conclusions

The aim of this research was to investigate the morphological appearance, X-ray diffraction, micro-scratch and micro-indentation aspects on Cr3C2-NiCr / Ni-Cr-Fe coatings on Pure Mg, respectively Mg-30Y substrate. These coatings confirm the presence of "splats" particles with an average size of 3 microns. Also, on the surface of the coatings it has been observed typically plasma deposition morphology like, few microcracks, unmolded particles etc. X-Ray analysis showed the presence Cr-based compounds: Cr_3C_2, Cr_7C_3, Cr_3Ni_2, $Cr_{0.1}Ni_{0.9}$, Cr_7Ni_3 having both orthorhombic and tetragonal crystal structure as predominant phase. Also was identified as secondary phase Cr_3Ni_2 and $FeNi_3$ with a crystal structure of tetragonal and cubic type. Young modulus is influenced by the substrate characteristics. The coatings performed on pure magnesium, respectively Mg-30Y substrate show similar values. Also, the apparent COF of Cr_3C_2-NiCr is lower than COF of Ni-Cr-Fe coatings. Research will continue with future experiments in order to evaluate the coatings characteristics for different types of applications.

References

[1] Lu Z, Myoung S W, Kim E H, Lee J H and Jung Y G 2014 *Mater. Today Proc.* **1** 35–43
[2] Matikainen V, Bolelli G, Koivuluoto H, Sassatelli P, Lusvarghi L and Vuoristo P 2017 Sliding wear behaviour of HVOF and HVAF sprayed Cr3C2-based coatings *Wear* 0–1
[3] Suarez M, Bellayer S, Traisnel M, Gonzalez W, Chicot D, Lesage J, Puchi-Cabrera E S and Staia M H 2008 *Surf. Coatings Technol.* **202** 4566–71
[4] Avram P, Imbrea M S, Istrate B, Strugaru S I, Benchea M and Munteanu C 2014 *Indian J. Eng. Mater. Sci.* **21**
[5] Sadeghimeresht E, Markocsan N and Nylén P 2016 Surface & Coatings Technology Microstructural characteristics and corrosion behavior of HVAF- and HVOF-sprayed Fe-based coatings *Sct*
[6] Berger L M 2015 *Int. J. Refract. Met. Hard Mater.* **49** 350–64
[7] Raţoi M, Dascəlu G, Stanciu T, Gurlui S O, Stanciu S, Istrate B, Cimpoesu N and Cimpoesu R 2014 *Preliminary results of FeMnSi+Si(PLD) alloy degradation* **638**
[8] Perju M C, Vizureanu P 2014 REV. CHIM. (Bucharest) 65(6) 694-696
[9] Cimpoesu N, Stanciu S, Vizureanu P, Cimpoesu R, Achitei D C, Ionita I 2014 *Journal of Mining and Metallurgy* **50**(1) 69-76
[10] Gariboldi E, Rovatti L, Lecis N, Mondora L and Mondora G A 2016 *Surf. Coatings Technol.* **305** 83–92
[11] Tsui Y C and Clyne T W 1997 *Thin Solid Films* **306** 23–33
[12] Mayrhofer E, Janka L, Mayr W P, Norpoth J, Rodriguez Ripoll M and Gröschl M 2015 *Surf. Coatings Technol.* **281** 169–75
[13] Varis T, Suhonen T, Calonius O, ??uban J and Pietola M 2016 *Surf. Coatings Technol.* **305** 123–31
[14] Liu W-H, Shieu F-S and Hsiao W-T 2014 *Surf. Coatings Technol.* **249** 24–41
[15] Wang S, Cheng J, YI S-H and KE L 2014 *Trans. Nonferrous Met. Soc. China* **24** 146–51
[16] Kula A, Jia X, Mishra R K and Niewczas M 2017 *Int. J. Plast.* **92** 96–121

IOP Conf. Series: Materials Science and Engineering **209** (2017) 012025 doi:10.1088/1757-899X/209/1/012025

Corrosion Evaluation of Some Phosphated Thin Layers on Reinforcing Steel

P Lazar[1], C Bejinariu[1], A V Sandu[1,2*], A M Cazac[1], O Corbu[2,3], I G Sandu[1,4] and M C Perju[1,2]
[1] Gheorghe Asachi Technical University of Iasi, Faculty of Materials Science and Engineering, Blvd.
[2] Center of Excellence Geopolymer & Green Technology (CEGeoGTech), School of Materials Engineering, Universiti Malaysia Perlis, 01000 Kangar, Perlis, Malaysia
[3] Faculty of Civil Engineering, Technical University of Cluj-Napoca, 28 Memorandumului Street,400114 Cluj-Napoca, Romania
[4] Romanian Inventors Forum, Str. Sf.P.Movila 3, L11, III/3, 700089, Iasi, Romania

E-mail: sav@tuiasi.ro (*corresponding author)

Abstract. In this work we studied the evaluation of corrosion of chemical deposited phosphate layers on the OB37 steel used in construction. The experimental determinations were performed for short term immersion, and the samples were characterized by scanning electron microscopy. By using electro-chemical impedance spectroscopy, the corrosion speed of the alloys was investigated. The corrosive resistance of the reinforcing steel is much improved.

1. Introduction

These materials are commercial alloys used commonly in civil and industrial buildings.

The prediction of the properties and performance of a metallic material performance is not easy to evaluate, being necessary numerous circumstances [1-6].

The corrosion behavior depends on numerous factors: the metal or alloy and its properties, the corrosive medium and its concentration, temperature, and other environmental factors [6-9].

The known anticorrosion procedures of iron object surfaces by chemical phosphatation consist in precipitation of low soluble pyrophosphate, as Zn(II), Ni(II), Mn(II), Ca(II) and, very rarely, Fe(II) in order to deposit thin layers [10-12].

These procedures have the disadvantage that involves a large number of cations in the composition of the in-situ obtained layers.

These are applied after degreasing and pickling [11-15].

Electrochemical methods (electro-chemical impedance spectroscopy) and surface imaging (SEM) were used to evaluate the average corrosion of coated reinforcing steel.

The electrochemical methods were used to determine the corrosion rate and microscopy was used to evaluate the surface [8, 10].

2. Materials and methods

The layers were obtained by chemical deposition, after immersion of the samples in a NaOH based solution for degreasing, respectively in a HCl based solution for pickling, in a solution made of H_3PO_4

and HNO_3 in which are added Zn chips and $NaNO_2$ and $Na_3P_3O_{10}$.

The samples are kept in solution for 30 mins at 80°C.

The morphology was performed using scanning electron microscopy coupled with energy dispersive X-ray spectroscopy (SEM-EDX) model Tescan Vega II LSH with a Bruker Quantax QX2 detector, being computer-controlled and was used to obtain the micrographs at 100x, 500x, 1000x, and 5000x for secondary (SE) and backscattered (BSE) electron detectors.

All the corrosion tests were done at room temperature (20°C), using a electrochemical cell with three electrodes connected to an Autolab PG STAT 302 (Metrohm Autolab) Potentiostat.

The reference electrode is saturated calomel (ECS), the auxiliary electrode is made of platinum and the third one was the tested sample.

The corrosion potential was calculated with the Nova 1.6 software.

3. Results and discussions

The surface imaging shows a dendritic structure (figure 1) with compact layers with high roughness, which provide good adesion to the concrete.

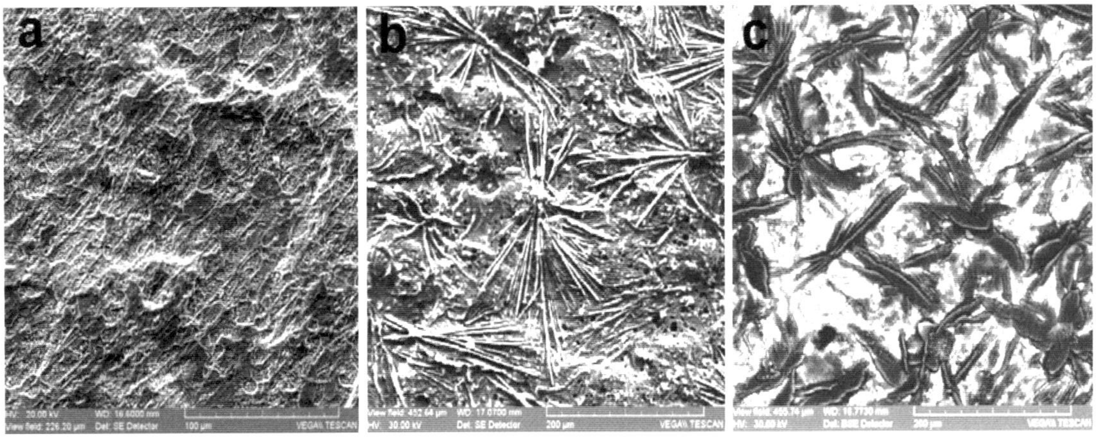

Figure 1. SEM images of the samples: a – the support (1000X), b – SE image of deposited layers (500X); c – BSE image of deposited layers (500X).

The chemical deposition allows obtaining of thin phosphate layers obtained from thin acicular dendritic cristallites, with visible germination centers.

Distribution of the atoms on surface highlight two processes of structural reformation by sequential co-precipitation, first substractive – the formation of iron phosphate and second the formation of zinc phosphate [12].

The corrosion speed can be correlated with the intensity of the corrosion current or with the current density based on the general law of electrolysis.

The superficial corrosion speed is defined as the weight loss on time unit on surface unit [10].

Figure 2 presents the Evans diagram of the standard sample – not deposited and respectively the phosphated reinforcing steel.

Corrosion speed of the samples in 0.1M NaCl solution was indicated as following: untreated sample: $45 \cdot 10^{-3}$ mm/year compared to coated sample with $1.6 \cdot 10^{-3}$ mm/year.

It is known in literature that on the coprecipitation process a thin ceramic layer of iron and zinc phosphate is formed ($Zn_2Fe(PO_4)_2 \cdot 4H_2O$), which asures a good corrosion resistance and as a suport layer for concrete [9-13].

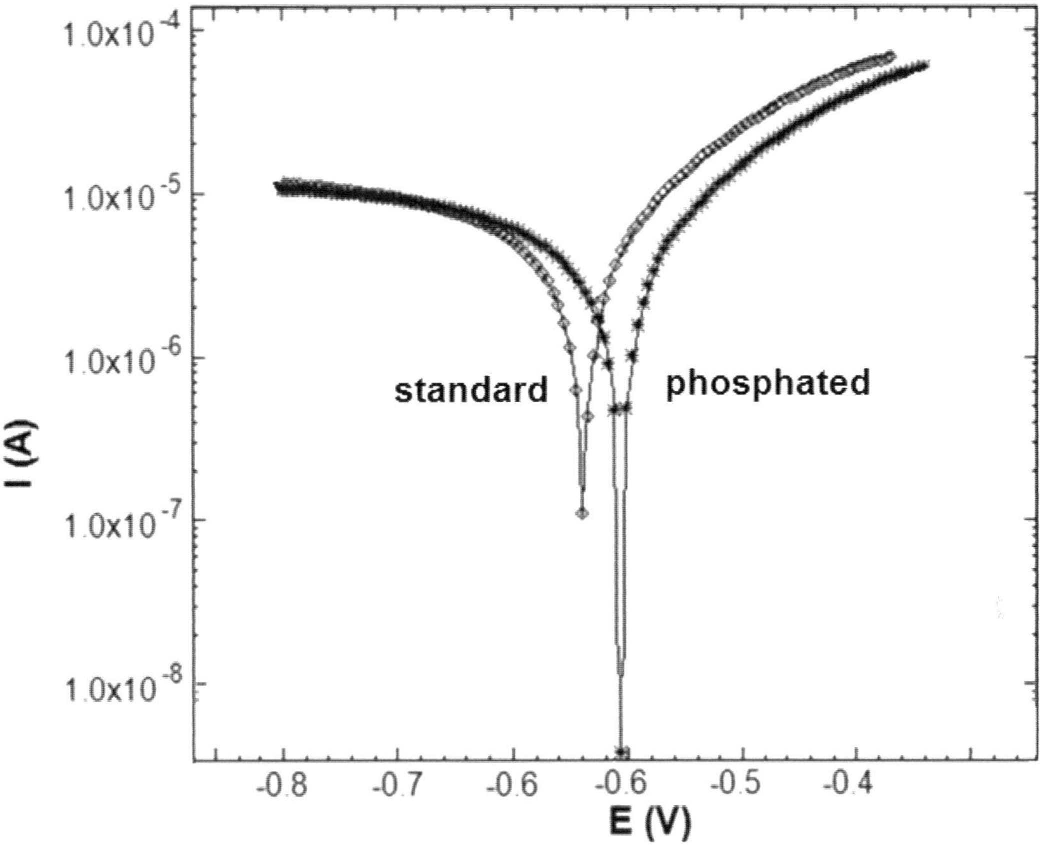

Figure 2. Evans diagram of the samples: in 0.1m NaCl.

4. Conclusions

The result obtained from present work indicate that phosphate layers can be used for enhancing the corrosion resistance of reinforcing steel.

Assessed with previous studies we can say that the obtained structures present a compact layer, results confirmed by the corrosion evaluation.

The phosphate coated reinforcing steel bars can be successfully used on concrete for enhancing the life of buildings.

References

[1] Amrudin A, Thierry D 1996 *Progr Org Coat* **28** 59
[2] Zimmermann D, Munoz AG, Schultze JW 2005 *Surf Coat Technol* **197** 260
[3] ASTM A767/A767M: Specification for Zinc-Coated (Galvanized) Steel Bars for Concrete Reinforcement
[4] Creus J, Mazille H, Idrissi H 2000 *Surface and Coatings Technology* **130** 224
[5] Bârcă E S, Plăiașu A G, Abrudeanu M, Istrate B, Luca D, Munteanu C 2015 Journal *of Optoelectronics and Advanced Materials* **17**(9-10) 1522-1527
[6] Tugui C A, Vizureanu P, Nejneru C, Perju M C, Achitei D C, Axinte M 2014 *Advanced Materials Research* **1036** 201-206
[7] Minciuna M G, Vizureanu P, Achitei D C, Ghiban B, Sandu A V, Mareci D, Balan A 2014 *Revista de Chimie* **65**(10) 1138-1141

[8] Fang J., Jiang JH, Tan SY, Ma AB, Jiang JQ 2010 *Surface & Coatings Technology* **204** 2381
[9] Ningshen S, Mudali UK, Amarendra G, Rai B 2009, *Corrosion Science*, **51**, 322
[10] Bockris J. O'M., Reddz AKN 1970, 'Modern Electrochemistry', 2nd edn, New York, Plenum Press
[11] Lazar P, Bejinariu C, Sandu AV, Cazac AM, Sandu IG 2015 *Key Engineering Materials* **660** 213
[12] Sandu AV, Ciomaga A, Nemtoi G, Bejinariu C, Sandu I 2012 *Journal of Optoelectronics and Advanced Materials* **14** 704
[13] Sandu AV, Ciomaga A, Nemtoi G, Bejinariu C, Sandu I 2012 *Microscopy Research and Technique* **75** 1711
[14] Sandu AV, Bejinariu C, Nemtoi G, Sandu IG, Vizureanu P, Ionita I, Baciu C 2013 *Revista de Chimie (Bucharest)*, **64** 825
[15] Bejinariu C, Lazar P, Sandu AV, Cazac AM, Sandu IG, Corbu O 2015 *Applied Mechanics and Materials* **754-755** 310

IOP Conf. Series: Materials Science and Engineering **209** (2017) 012026 doi:10.1088/1757-899X/209/1/012026

Corrosion Studies of Fly Ash and Fly Ash-Slag Based Geopolymer

F F Zainal[1*], S F M Amli[1], K Hussin[1,2], A Rahmat[1] and M M A B Abdullah[1,2]
[1] Center of Excellence Geopolymer and Green Technology (CEGeoGTech), School of Materials Engineering, Universiti Malaysia Perlis (UniMAP), Perlis, Malaysia
[2] Faculty of Engineering Technology, Universiti Malaysia Perlis (UniMAP), Perlis, Malaysia

E-mail: farahfarhana@unimap.edu.my

Abstract. This paper presents the results of corrosion studies between Fly Ash Geopolymer (FG) paste and Fly Ash-Slag Geopolymer (FSG) paste. Geopolymer was made from aluminosilicate inorganic polymers mixed with the alkaline activator in order to reduce the carbon dioxide (CO_2) to the ecosystem. Samples then were cured at 60°C for 24 hours in the oven. Reinforcement bar is placed at the center of the paste. The samples were examined after 7, 14 and 28 days in terms of Open Circuit Potential (OCP) test, phase analysis and morphology analysis. The potential values regarding OCP test for FSG paste from 7 days until 28 days are 0.464 V, 0.474 V and 0.498 V more positive than FG paste which the potential values are 0.087 V, 0.133 V and 0.206 V respectively. From the Pourbaix diagram, all the potential values for FG paste and FSG paste were located in the same Fe_2O_3, passivity region. Passive layer which is the oxide form exists in this region to protect the reinforcement bar from corrosion agents. It can be proved from phase analysis results which iron oxide hydroxide (FeOOH), hematite (Fe_2O_3) and magnetite (Fe_3O_4) peaks exist. The differences of morphological structures of these pastes were observed by Scanning Electron Microscope (SEM). It shows that FSG paste had good corrosion resistance and low corrosion rate compared to FG paste.

1. Introduction

An early degradation of reinforced concrete structures is caused by corrosion of steel in aggressive environments. Steel is passive due to the alkalinity of concrete, which the oxide formed to protect the steel concrete. However, the passive film could damage if the reinforcement bar exposed to the active environments such as carbon, chlorine, and acid conditions [1].

A few studies have been conducted to prevent the corrosion of reinforcement bar by improving the quality of the concrete. The uses of polymer to improve quality of concrete have attracted and obtained great attention nowadays.

Thus, geopolymer concrete is used as an alternative way to replace Ordinary Portland Cement (OPC) concrete. Geopolymer have a few advantages such as light weight, good tensile strength, durability and high corrosion resistance [1]. Geopolymer is inorganic materials produced by geosynthesis from alumina-silica materials with an alkaline activators and their chains or networks of inorganic molecules linked by covalent bonds. The precursors that always used as main raw material in geopolymer are volcanic ash, zeolite, kaolin, fly ash and granulated blast furnace slag [2].

2. Materials and Method

2.1. Materials and Sample Preparations

Fly ash geopolymer (FG) paste in this study was made from a mixture of class F fly ash, sodium hydroxide (NaOH) solution and sodium silicate (Na_2SiO_3) solution. While fly ash-slag geopolymer (FSG) paste was made from a mixture of 70% fly ash, 30% slag, NaOH solution and Na_2SiO_3 solution. The American Society for Testing and Materials (ASTM) defines fly ash as a residue resulting from the combustion of coal in power plants [3].

Fly ash has the advantage of reducing concrete permeability and mitigates against corrosion of reinforcement, Class F fly ash is highly pozzolanic material and contains almost 70% pozzolanic compounds such as silica oxide (SiO_2), alumina oxide (Al_2O_3) and iron (III) oxide (Fe_2O_3) [4].

NaOH solution was prepared by dissolving the NaOH pellets in distilled water. Then, the NaOH solution was mixed with Na_2SiO_3 solution. An alkaline activator was prepared 24 hrs prior to use with the ratio of the mixture of Na_2SiO_3/NaOH is 2.5.

The ratio is very important in order to obtain a homogeneous solution. The alkaline activator was later mixed with the fly ash for about 30 minutes.

The prepared mixtures were placed in moulds and compacted. After that, the reinforcement bar was embedded in the geopolymer paste. The samples were kept at ambient temperature in the moulds until it becomes hardened. All the samples were taken out from the moulds after 24 hrs. Then, the samples were cured at 60°C in the oven for 24 hrs [5-7].

2.2. Experimental Methodology

2.2.1. Open Circuit Potential (OCP). In this research, the OCP test set up as shown in figure 1. The Sanwa digital multimeter CD 771 was used to measure the potential difference between the steel in the concrete and the metal in the reference electrode.

The positive side was connected to the reinforcement bar embedded in the geopolymer paste while the negative side was connected to the reference electrode. Reference electrode that was used in this research is Saturated Calomel Electrode (SCE) because this electrode is suitable to measure potential value in laboratory scale [8]. The potential value readings were taken accordingly with the time set.

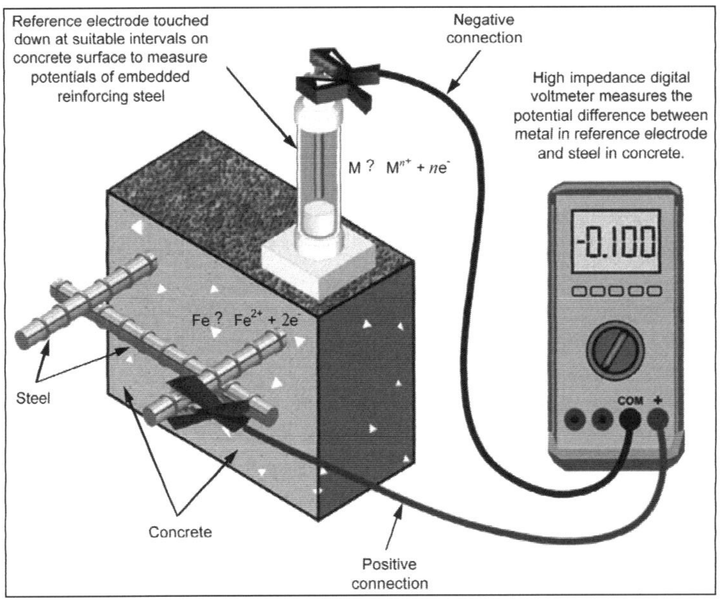

Figure 1. Reference Electrode Circuitry [8].

2.2.2. Phase Analysis. This test was set up to know the phase exist between reinforcement bar and geopolymer paste. The test was conducted by X-Ray Diffraction (XRD) method.

2.2.3. Morphological Analysis. This test was run to observe the fly ash geopolymer (FG) and fly ash-slag geopolymer (FSG) paste surface morphology and pore. Scanning Electron Microscope (SEM) was conducted by using JEOL machine JSM6460-LA.

3. Results and Discussions

3.1. Open Circuit Potential (OCP)

Figure 2 shows graph of time (day) against redox potential (volts) for open circuit potential (OCP) of reinforcement bar embedded in fly ash geopolymer (FG) and fly ash-slag geopolymer (FSG) paste at ambient temperature. From the graph, the FSG paste samples have higher potential values compared to the FG paste samples from day 7 until day 28.

The potential values of FSG paste samples were 0.464 V, 0.474 V and 0.498 V from day 7 until day 28 while FG paste samples potential values were 0.087 V, 0.133 V and 0.206 V respectively.

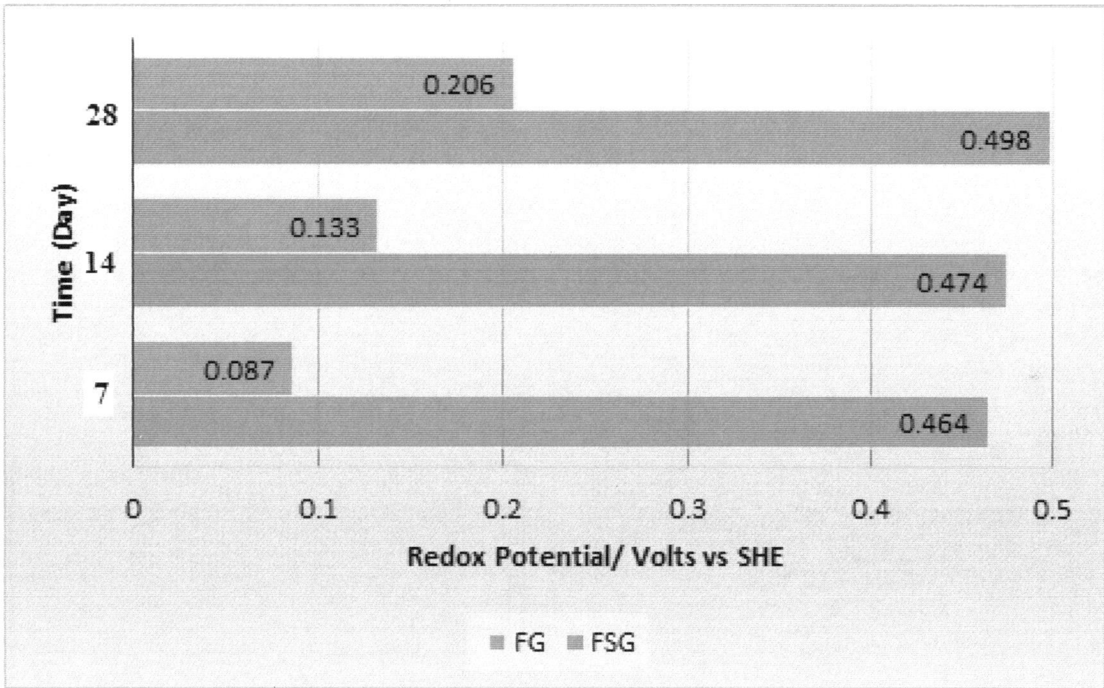

Figure 2. Graph of time (day) against redox potential (volts) for OCP of reinforcement bar embedded in FG and FSG paste at ambient temperature.

The potential values obtained from OCP test were interpreted in Pourbaix diagram to determine which region these value located. The pH value for these FG and FSG paste were observed by using the pH paper.

The pH value for these paste is 12 as the concrete were always being alkaline in nature [9].

Figure 3 shows potential-pH diagram or Pourbaix diagram for Fe-H_2O at 25°C. When referring the Pourbaix diagram at pH 12, all samples of FG and FSG paste were in the passivity region as labelled. The passive layer was formed to protect the reinforcement bar from corrosion in this region.

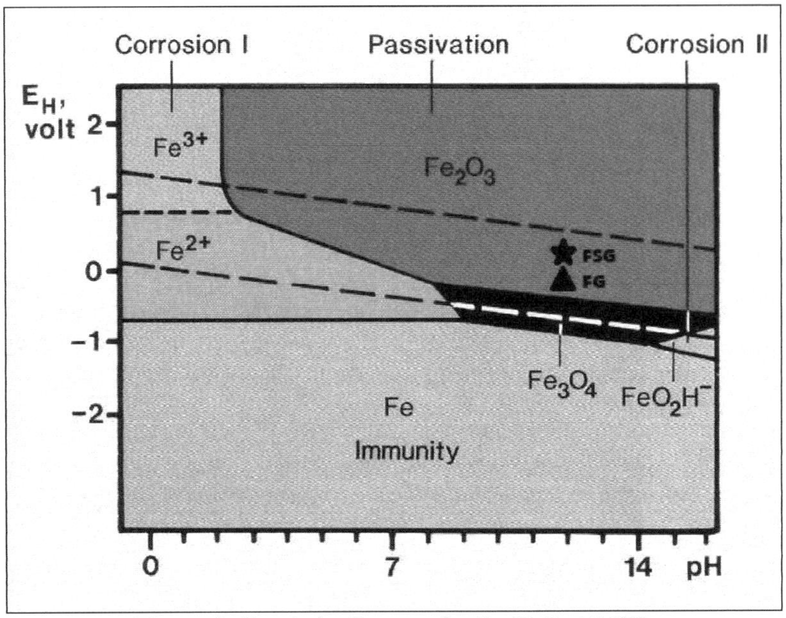

Figure 3. Pourbaix diagram for Fe-H_2O at 25°C.

3.2. Phase Analysis

Figure 4 (a) shows the X-Ray Diffraction (XRD) pattern for reinforcement bar and reinforcement bar embedded in fly ash geopolymer (FG) paste and figure 4 (b) was XRD pattern for fly ash-slag geopolymer (FSG) paste. Both figures shows that peak iron oxide hydroxide (FeOOH), hematite (Fe_2O_3) and magnetite (Fe_3O_4) were exist as stated in OCP analysis which mean in passive region.

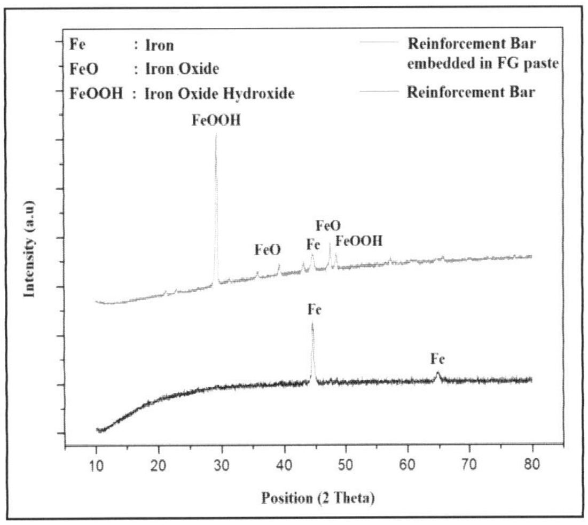

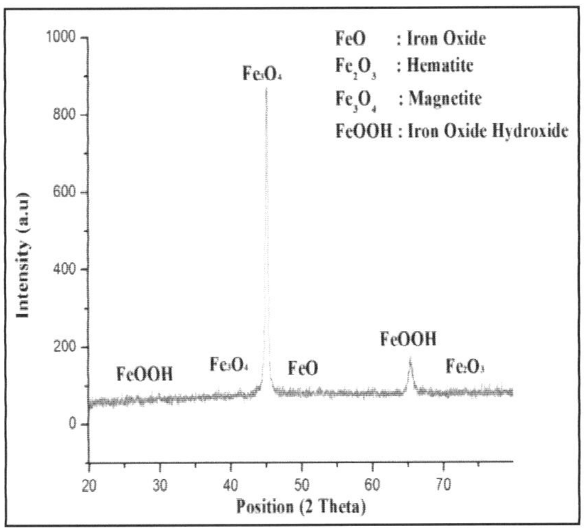

(a) (b)

Figure 4. XRD pattern for (a) Reinforcement bar and reinforcement bar embedded in FG paste; (b) Reinforcement bar in FSG paste.

3.3. Morphology Analysis

Table 1 shows micrographics Scanning Electron Microscope (SEM) of fly ash geopolymer (FG) paste and fly ash-slag geopolymer (FSG) paste for day 7 until 28. For the FG paste in day 7, it shows a lot of

IOP Conf. Series: Materials Science and Engineering **209** (2017) 012026 doi:10.1088/1757-899X/209/1/012026

pore, unreacted fly ash and micro crack. It was determined that the existence of unreacted fly ash and microcracks had caused the increasing of pores and reduced strength [10]. The corrosive agents will easily penetrate to the reinforcement bar with these pores. When comparing with FSG paste for day 7, most of the fly ash and slag has already reacted with alkaline activators and only have a few pore. For day 14, the reaction at FSG paste was almost done, only have slightly unreacted slag. Different with FG, there were still many pores and unreacted fly ash. There were few minor cracks on FG and FSG paste on day 28. The reactions of FSG paste are already done while FG paste microstructure is still changing. It shows that the geopolimerization process occured in FSG paste was better than FG paste.

Table 1. Micrographics SEM of FG and FSG paste for 7, 14 and 28 days.

Days	Fly Ash Geopolymer (FG) Paste	Fly Ash-Slag Geopolymer (FSG) Paste
7 Days	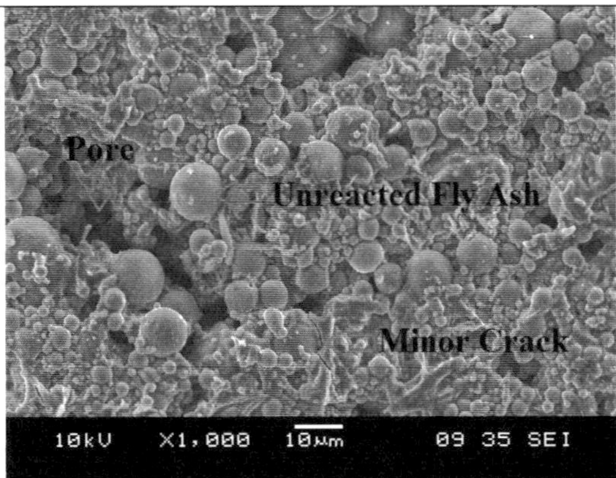	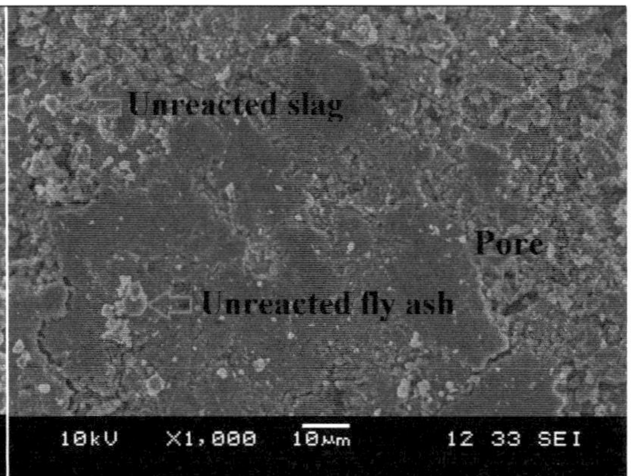
14 Days	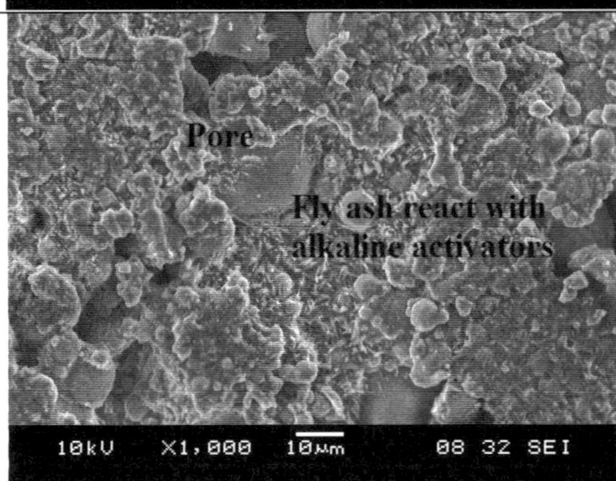	
28 Days	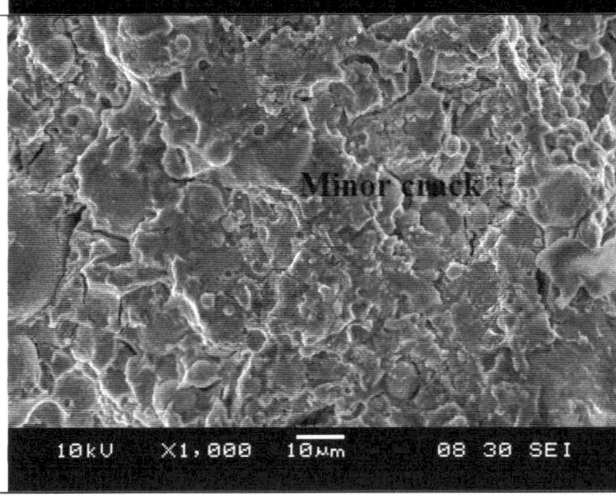	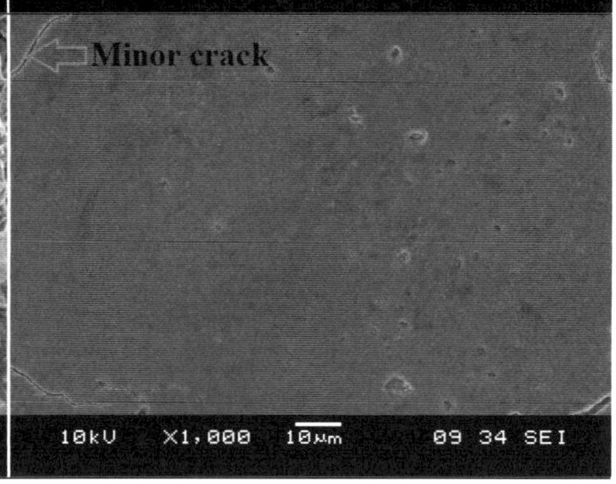

4. Conclusions

Based on this research, the following conclusions were drawn:

- Fly ash-slag geopolymer (FSG) paste had good corrosion resistance and low corrosion rate compared to fly ash geopolymer (FG) paste.
- From the Pourbaix diagram, all the potential values for FG paste and FSG paste were located in the same Fe_2O_3, passivity region. Passive layer which is the oxide form exists in this region to protect the reinforcement bar from corrosion agents.

References

[1] Asmara Y P, Siregar J P, Tezara C, Nurlisa W and Jamiluddin J 2016 *International Journal of Corrosion* 3853045, 5 pages.
[2] Nikolov A, Rostovsky I and Nugteren H 2017 *Case Studies in Construction Materials, Elsevier* **6** 198-205
[3] ASTM C618-12a, 2012 *Standard Specification for Coal Fly Ash and Raw or Calcined Natural Pozzolan for use in Concrete.*
[4] Boral F A 2002 *Boral Material Technologies,* San Antonio, Texas.
[5] Montes C, Gomez S A, Khadka N and Allouche E N 2013 *World of Coal Ash (WOCA) Conference.*
[6] Mastura W I W, Kamarudin H, Nizar I K and A M Mustafa Al Bakri 2013 *Advanced Science Letters* **19** 186-189
[7] Tennakoon C, Shayan A, Sagoe-Crentsil K and Sanjayan J G 2014 *Austroads Bridge Conference,* pp.12
[8] ASTM C876-09, 2009 *Standard Test Method for Corrosion Potentials of Uncoated Reinforcing Steel in Concrete*
[9] Farhana Z F, Kamarudin H, Rahmat A and A M Mustafa Al Bakri 2013 *Australian Journal of Basic and Applied Sciences,* **7**(5) 230-235
[10] Abdullah M M A B, Hussin K, Bnhussain M, Nizar I K, Yahya Z and Razak R A 2012 *Int. J. Mol. Sci.* **13**(6) 7186-7198

Acknowledgments

The authors would like to thank the staffs of School of Materials Engineering, Universiti Malaysia Perlis (UniMAP) for their involvement in the research. This work was supported and funded by the Center of Excellence Geopolymer & Green Technology (CeGeoGTech), UniMAP.

IOP Conf. Series: Materials Science and Engineering **209** (2017) 012027 doi:10.1088/1757-899X/209/1/012027

The Physical and Mechanical Properties of Gellan Gum Films Incorporated Manuka Honey as Wound Dressing Materials

N A N Mohd Azam[1] **and K A M Amin**[1,2*]

[1]School of Fundamental Science, Universiti Malaysia Terengganu (UMT), 21030 Kuala Nerus, Terengganu, Malaysia.

[2]Centre of Excellence Geopolymer System Research, School of Materials Engineering, Universiti Malaysia Perlis (UniMAP), P.O. Box 77, D/A Pejabat Pos Besar, 01000 Kangar, Perlis, Malaysia

E-mail: kerol@umt.edu.my

Abstract. In this study, the physical and mechanical properties of gellan gum (GG) films incorporated manuka honey were investigated. The result shown that by increasing the honey content in GG film, the swelling, gel fraction and mechanical properties were decreased. Gellan gum films incorporated with highest concentration of honey at 10% (GEL-H10) has lowest tensile strength and Yong's modulus at 0.91 ± 0.2 MPa and 38.9 ± 1.7 kPa, respectively compared to GG films at lower concentration of honey. The water vapour transmission rate of GEL-H10 was recorded at 1145 ± 175 g $m^{-2}d^{-1}$ which comparable with the commercial wound dressing product. This film has shown promising results to be used as wound dressing application.

1. Introduction

Biopolymers are receiving greater attention than synthetic petrochemical-based polymers due to the environmental concerns. A variety of renewable biopolymer such as a polysaccharide; i.e. gellan gum (GG) derived from *Pseudomonas elodea*, has been studied in the development of wound dressings materials. GG is approved by the United States Food and Drug Administration (US FDA) and the European Union (EU) for use in the food industry and is an emerging scaffold material for tissue engineering application. For instance, GG has been studied for biomedical application as it has the potential to be used as matrices to repair and regenerate a wide variety of tissue and organ [1]. The development of GG hydrogels receives a lot of attentions due to the exceptional properties which meet the quintessential prerequisites of ideal wound dressings. A few studies have observed the good biocompatibility of GG against human skin fibroblasts cells (CRL-2522) [2], human fetal osteoblasts (hFOBs 1.19) [3] and rat bone marrow cells (rBMC) [4]. In wound dressing application, GG has shown promising results to proliferate the cell growth [5]. Although GG has been reported to biocompatible on live cells, researchers keep working to improve the proliferation rate of GG, for example by forming a complex gel with carboxymethyl chitosan [6] and incorporated with branched polyethyleimine nanoparticles [7]. Honey is a well-known substance to threat the wounds and burns throughout the ages. Commonly, the healing or other properties of honey is studied in its pure state, i.e. in liquid form. Numerous studies have been reported the advantages of honey such as the anti-bacterial, anti-inflammatory, anti-viral and anti-oxidant effects [8-10]. Besides numerous studies of using pure honey has been used to treat wound and burns, limited studies have been reported to incorporated honey as a composite with bio-polymer, i.e. transforming honey with biopolymers to a film or hydrogels. A few studies have been reported to used chitosan and gelatin incorporated with honey as a sheet [11], and some using polyvinylpyrrolidone (PVP) entrapped with honey to produce hydrogels [12]. Other study using alginate, chitosan and gelatin to produce hydrogel with an addition of honey [13]. Up to our knowledge, no single study has been conducted to produce gellan gum hydrogels incorporated with honey. This study fabricated and characterized the gellan gum hydrogels incorporating manuka honey. The mechanical characteristics, water vapour transmission rates, and swelling of the hydrogels were examined.

2. Materials and method

2.1. Materials
High-acyl gellan gum (KELCOGEL® Gellan Gum, lot number 5C1574A) was obtained from CP KELCO, USA. Glycerine (product number-G2289, lot number SHBC2650V), and anhydrous calcium chloride, $CaCl_2$ (product number-C5670, lot number SHBC2650V) were obtained from Sigma Aldrich, St Louis, MO, USA and Manuka honey (product number FL522990/1) was obtained from Nature's Way, New Zealand. All materials were used as received without any further purification.

2.2. Preparation of gellan gum films incorporated Manuka honey (GEL-H)
The gellan gum (GG) films were prepared via casting method. The GG solution was prepared by dissolving 1% (w/v) of GG in 70 mL deionized water and with 55% (w/w) glycerine at continuous stirring for 1 hour and 45 minutes at 70°C. After GG was fully dissolved, 5 mM $CaCl_2$ was added into the mixture and stirred again. Lastly, Manuka honey was added and the solution was stirred for another 15 minutes. These gellan gum solutions containing 2%, 4%, 6%, 8% and 10% (w/v) Manuka honey were known as GEL-H2, GEL-H4, GEL-H6, GEL-H8 and GEL-H10 respectively. The solution was poured into petri dish (90 mm x 5 mm) and dried in the oven at 50°C for 24 hours.

2.3. Characterization of GEL-H films

2.3.1. FTIR Characterization. ATR-FTIR spectra were determined by using Perkin Elmer Spectrum 100 FTIR spectrophotometer with PIKE Miracle ATR accessory with single-bounce beam path, 45° incident angle, and 4 cm^{-1} resolution. Then all spectra were corrected by using Perkin Elmer Spectrum 100 software. The wave number region studied was between 400-400 cm^{-1}. The resulting spectra and peak of functional groups were recorded.

2.3.2. Swelling Percentage. The swelling percentage test was carried out by immersing GEL-H (20 mm x 20 mm) in phosphate buffer solution at pH 7.2 in water bath (37 ± 0.5 °C). The samples were removed after 24 hours and lightly wipe with wet filter paper to expel surface solution. Finally, the swelling degree was determined from equilibrium swelling ratio as follow:

$$\text{Swelling percentage } (\%) = (M_f - M_i) / M_i$$

where, M_f = final weight of sample and M_i = initial weight of sample

2.3.3. Gel fraction. The film sample was cut into 20 mm x 20 mm and let dried for 24 hours. Then the film was weighed (W_i) and the film was immersed in 10 mL deionised water at room temperature for 24 hours. After removing film from solution, the film was dried in oven at 50 °C for 24 hours and weighed again (W_f) and gel content was calculated as below:

$$\text{Gel fraction } (\%) = (W_f - W_i) / 100$$

where, W_i = initial weight, W_f = final weight.

2.3.4. Water Vapour Transmission Rates . The film was cut into 30 mm x 30 mm and put as a cap on a glass vial with diameter of 16 mm that contain 10 ml deionised water. Then it was weight (W_i) and left in the desiccator containing silica gel for 24 hours. After 24 hours, it was weighed again (W_f). The value of WVTR (g $m^{-2} d^{-1}$) was calculated as below:

$$WVTR = (W_i - W_f) / A$$

where, W_i= initial weight, W_f= final weight and A= area of vial opening

2.3.5. Tensile Strength. The tensile strength was obtained by using Instron universal testing machine, model 3366 with cross-speed set at 10 mm/ min. All films were cut into 20 mm x 60 mm for tensile stress-strain measurement. The tensile stress and strain at break were calculated from the slope of linear part of stress-strain curve. The elastic modulus was also recorded. The tests were repeated triplicates per sample for defined ratio.

3. Results and discussion

3.1. FTIR Spectrum
Figure 1 shows the FTIR spectra of GEL film incorporated manuka honey. There are shifting and disappearance of few functional peaks in spectrum of GEL film compared to a representative of GEL-H10 film. Shifting or disappearance of the frequency in the functional groups of peaks indicates that there is an interaction between polymer and drug [14].

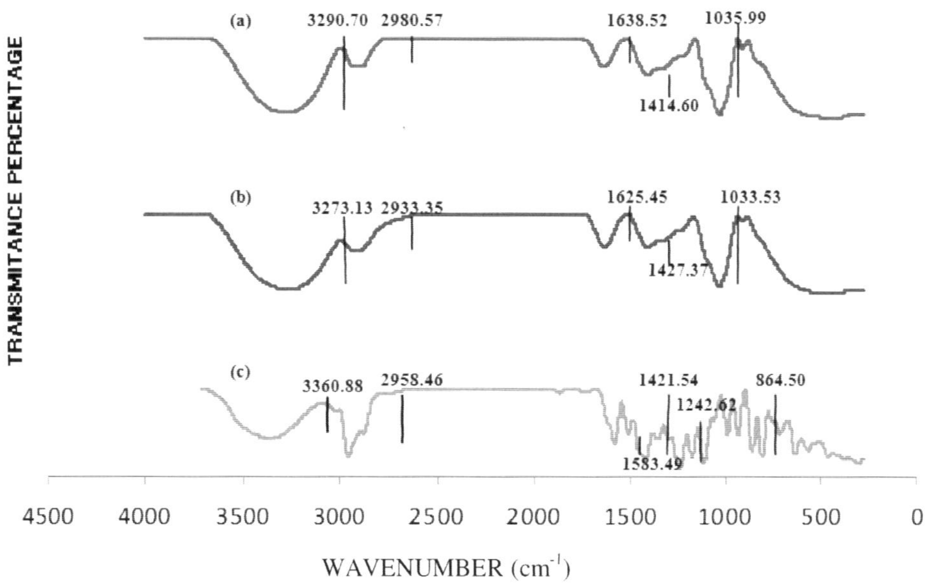

Figure 1. ATR-FTIR spectra of (a) GEL-H10 film, (b) GEL film and (c) Manuka honey.

GEL film shows a few main peaks at 3273 cm^{-1} due to the stretching vibration belong to O-H group. A peak was shown at wavenumber of 2935 cm^{-1} which due to sp3 hybridization of C-H stretching branched alkane, and a few other peaks at 1625 cm^{-1} and at 1427 cm^{-1} due to C=C bond and C-H bond saturated sp^3 hybridization, respectively. There was also C=O stretching band form at1033 cm^{-1} and a peak 919 cm^{-1} due to alkane C-H bending. The incorporation of manuka honey in gellan gum film shows a shifting of stretching vibration in gellan gum film, i.e. GEL film at 3273 cm^{-1} to 3290 cm^{-1} of GEL-H10 which indicate the hydroxyl O-H vibration of H-bonded between gellan gum and Manuka honey. The shifting of carbonyl group, C=O group was observed in GEL-H10 at 1035 cm^{-1} compared to GEL film at 1033 cm^{-1}. The band of C-H stretching branched alkane of GEL film at 16255 cm^{-1} was shifted to 1638 cm^{-1} of GEL-H10 which suggested that the interaction between gellan gum and Manuka honey.

3.2. Swelling Percentage

Swelling behaviour is an important characteristic to indicate the ability of film to absorb any exudates or wound liquids from wound area. The ideal values for swelling behaviour depends on the wound condition, in which wound with huge amounts of exudates needed a dressing with high swelling behaviour and vice versa. The swelling percentages of all GEL-H films were summarized in table 1. The swelling behaviour of GEL-H films was decreased upon addition of Manuka honey. For example, the GEL film absorbed 1321 ± 46% in PBS solution (pH 7.2) and decreased to 1174 ± 24% for GEL-H2 and further reduced to 302 ± 18% for GEL-H10 film. The decreased of swelling of GEL film with high content of honey (GEL-H10) could be due to lesser hydroxyl functional (OH) groups of GEL film available to bind with hydroxyl group from water, or in another word the addition of honey limits the hydrogen bond between GEL film with water [15]. The hydrogen bond is expected to occur between GEL and honey and thus decreased the swelling percentage of GEL-H films.

Table 1. Swelling, gel fraction and water vapour transmission rates (WVTRs) value of GEL-H films.

SAMPLES	SWELLING (%)	GEL FRACTION (%)	WVTR ($g\ m^{-2}d^{-1}$)
GEL	1321 ± 46	29 ± 5	1841 ± 214
GEL-H2	1174 ± 24	23 ± 4	1734 ± 184
GEL-H4	786 ± 25	18 ± 3	1668 ± 146
GEL-H6	588 ± 14	15 ± 4	1442 ± 167
GEL-H8	437 ± 21	13 ± 4	1369 ± 153
GEL-H10	302 ± 18	11 ± 2	1145 ± 175

3.3. Gel Fraction and Water Vapour Transmission Rates

Gel fraction is a parameter that was utilized to indicate the crosslinking behaviour of the samples. From the result obtained, GEL film has the highest gel fraction recorded at 29 ± 5% (Table 1). The inclusion of honey decreased the gel fraction values up to 11 ± 2% of GEL-H10. It can be deduced that the addition of honey disrupted the crosslinking properties of the film.

Water vapour transmission rates (WVTRs) is important to control the loss of body liquid due to the exudation and evaporation. The loss of body fluid may affect the decrease in body temperature thus accelerate metabolism rate. However, the WVTR value less than 76 $g\ m^{-2}\ d^{-1}$ can cause accumulation of exudates and lead to growth of bacteria [16]. This problem can be improved by controlling the transmission of the body liquid in order to maintain high humidity area of wound itself. For that reason, WVTR test was necessary to determine the potential of film to allow the transmission of body fluid. Table 1 showed the result of WVTR values of gellan gum hydrogel with different concentration of manuka honey. WVTRs value of GEL-H10 film was significantly decreased to 1145 ± 175 $g\ m^{-2}\ d^{-1}$ compared to GEL films at 1841 ± 214 $g\ m^{-2}\ d^{-1}$. Lowest WVTR value for GEL-10 is expected due to huge amount of honey incorporated into the gellan gum film. The present of huge amount of honey (GEL-H10) disturbs the diffusion of water through the film and reflecting to the lowest WVTR obtained. In contrast, GEL-H2 film with lower amount of honey exhibited higher WVTR values.

Nevertheless, all the hydrogel samples show the WVTR values are suitable for wound dressing product because it is in the range of commercial wound dressing product (90-2893 g $m^{-2} d^{-1}$) [16].

3.4. Tensile test

Tensile test was performed to determine the strength of the fabricated film. It is a versatile tool for the mechanical characterization of pharmaceutical film formulations including film dressings. Figure 2 shows the stress-strain curves of the GEL-H films and table 2 summarized the tensile strength, Young's Modulus and tensile strain of the films. GEL-H2 film exhibited the highest tensile strength and tensile strain at 2.1 ± 0.9 MPa and 12 ± 5% respectively. The addition of higher amount of honey decreased the tensile strength and tensile strain of the film. The GEL-H10 shows the tensile strength at 0.9 ± 0.37 MPa and tensile strain at 3 ± 1%. This is a result of a decreased in hydrogen bonding interactions between gellan gum-gellan gum (GG-GG) due formation of hydrogen bonding between gellan gum-honey (GG-Honey) [17]. The GEL-H10 film physically soft which reflected to the low tensile performance compared to free-standing GEL films.

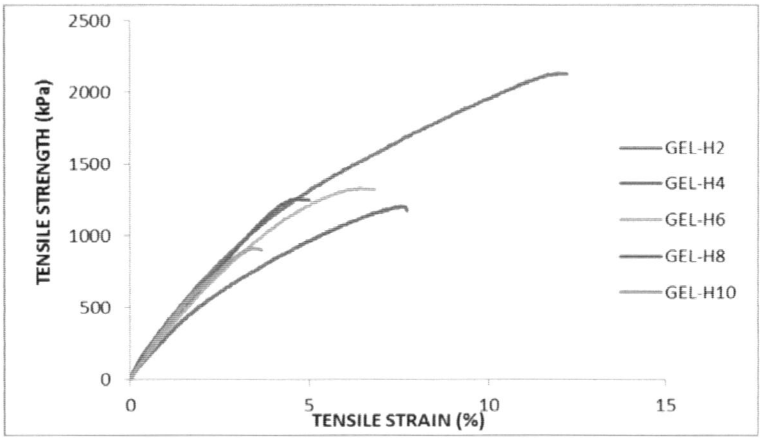

Figure 2. Tensile stress-strain curve of GEL-H films.

Table 2. Tensile strength (σ), Young's Modulus (YM) and tensile strain (ε) of gellan gum films incorporated with Manuka honey.

SAMPLES	σ (MPa)	YM (MPa)	ε (%)
GEL-H2	2.1 ± 0.96	32.2 ± 3	12 ± 5
GEL-H4	1.2 ± 0.72	27.0 ± 3	8 ± 3
GEL-H6	1.3 ± 0.53	31.5 ± 3	6 ± 3
GEL-H8	1.2 ± 0.54	37.1 ± 2	4 ± 2
GEL-H10	0.9 ± 0.37	38. 9 ± 2	3 ± 1

4. Conclusion

Gellan gum films incorporated with Manuka honey (GEL-H) were successfully prepared. Inclusion of manuka honey into GEL films decreased the swelling and gel fraction of the films. Water vapour transmission rates of GEL-H films were in the range of commercial dressing products and the tensile performance shows all the films were flexible with higher strain values by lowering the concentration of manuka honey. The film has shown promising properties to be used in dressing application.

References

[1] Oliveira J T, Santos T C, Martins L, Picciochi R, Marques A P, Castro A G, Neves N M *Tissue Engineering Part A* **16** 343-53
[2] Amin M, Anuar K, Gilmore K J, Matic J, Poon S, Walker M J and Wilson M R 2012 *Macromolecular bioscience* **12** 374-82
[3] Wang C, Gong Y, Lin Y, Shen J and Wang D-A 2008 *Acta Biomaterialia* **4** 1226-34
[4] Smith A M, Shelton R M, Perrie Y and Harris J J 2007 *J Biomater Appl* **22** 241-54
[5] Mohd S S, Abdullah M A A and Mat Amin K A 2016 *Journal of Bioactive and Compatible Polymers: Biomedical Applications* **31** 648-66
[6] Tang Y, Sun J, Fan H and Zhang X 2012 *Carbohydrate Polymers* **88** 46-53
[7] Goyal R, Tripathi S K, Tyagi S, Ravi Ram K, Ansari K M, Shukla Y, Kar Chowdhuri D, Kumar P and Gupta K C 2011 *European Journal of Pharmaceutics and Biopharmaceutics* **79** 3-14
[8] Lee H, Churey J J and Worobo R W 2008 *Int. J. Food Microbiol.* **126** 240-4
[9] Tan H T, Rahman R A, Gan S H, Halim A S, Hassan S A, Sulaiman S A and Kirnpal-Kaur B 2009 *BMC complementary and alternative medicine* **9** 34
[10] Yaghoobi R and Kazerouni A 2013 *Jundishapur journal of natural pharmaceutical products* **8** 100-4
[11] Wang T, Zhu X-K, Xue X-T and Wu D-Y 2012 *Carbohydrate polymers* **88** 75-83
[12] Yusof N, Hafiza A A, Zohdi R M and Bakar M Z A 2007 *Radiation Physics and Chemistry* **76** 1767-70
[13] Kurhade S T, Momin M, Khanekar P and Mhatre S 2013 *International Journal of Drug Delivery* **5** 353
[14] Prakash S J, Santhiagu A and Jasemine S 2014 *Journal of Pharmaceutical and BioSciences* **2** 63-71
[15] Yang L and Paulson A 2000 *Food Research International* **33** 563-70
[16] P. Wu, A. C. Fisher, P. P. Foo, D. Queen and Gaylor J D S 1995 *Biomaterials* **16** 171
[17] Silva C L, Pereira J C, Ramalho A, Pais A A C C and Sousa J J S 2008 *Journal of Membrane Science* **320** 268-79

Acknowledgements

The authors wish to thanks the Ministry of Higher Education, Malaysia for financial assistance under the Fundamental Research Grand Scheme (FRGS– vote no. 59449) and University Malaysia Terengganu especially School of Fundamental Science and Central Laboratory for providing the facilities for undertaking this work.

IOP Conf. Series: Materials Science and Engineering **209** (2017) 012028 doi:10.1088/1757-899X/209/1/012028

ZnO Photoanode Effect on the Efficiency Performance of Organic Based Dye Sensitized Solar Cell

I S Mohamad[1,2,3*], S S Ismail [1], M N Norizan[1,2,3], S A Z Murad[1] and M M A Abdullah[2,4]

[1] School of Microelectronic Engineering, Universiti Malaysia Perlis, Pauh Putra Campus, 02600 Arau, Perlis, Malaysia.

[2] Centre of Excellence Geopolymer and Green Technology (CeGeoGTech), Universiti Malaysia Perlis, Perlis, Malaysia.

[3] Advanced Multi-Disciplinary MEMS-Based Integrated NCER Centre of Excellence (AMBIENCE), Universiti Malaysia Perlis, Perlis, Malaysia

[4] Faculty of Engineering Technology, Universiti Malaysia Perlis (UniMAP), P.O Box 77, D/A Pejabat Pos Besar, Kangar, Perlis, Malaysia

E-mail: *ilisalwani@unimap.edu.my

Abstract. Dye sensitized solar cell has been emerged as one of the most promising candidates for photovoltaics applications in good quality of their low manufacturing cost and impressive conversion energy. Titanium dioxide (TiO_2) which is used as photoanode in the market has the advantage of wide bandgap energy but low in electron mobility (~10 $cm^2/(V.s)$). Ruthenium in the other hand, as the dye sensitizer is a rare noble metal and harmful to human health. Thus, this article reveals the performance of photo-to-electric conversion efficiency with the usage of Zinc Oxide as photoanode with higher electron mobility (155 $cm^2/(V.s)$) compared to TiO_2 utilizing three natural fruit dyes of Prunus domestica, Magnifera indica and Citrus limon. ZnO and TiO_2 photoanodes were fabricated using sol gel and dr blade method respectively. The morphology of the photoanodes were characterized using Scanning Electron Microscope and the efficiency of the complete DSSC with all different fruit dyes were characterized using Semiconductor Parametric Analyzer. The different property of electron mobility photoanodes effect in DSSC proved to give better performance with the photoconversion efficiency of 3.082% using ZnO with Prunus domestica dye. This article also reveals that pH indicator does not affect the selection and the performance of DSSC.

1. Introduction

Dye sensitized nanocrystalline solar cell (DSSC) is the third generation of solar cell invented by O'Reagan and Gràtzel [1-3] is a cheap and versatile alternative to silicon based solar cells and has already achieved solar-to-electricity conversion efficiency as high as 11.5% [4,5]. Major advantages of DSSCs are the large flexibility in shape, colour, transparency, and performance also under diffuse light [6,7]. DSSCs can be integrated into large varieties of products, e.g. handbags or clothing, indoor applications, and building-integrated photovoltaics such as roll-able devices for walls of buildings or windows [8,9]. The main stream of the research has been focusing on the development of materials which would enhance the conversion efficiency, simplify the production of DSSC and assure their long-lifetime [2].

Figure 1 shows the structure of DSSC which consists of titanium dioxide (TiO_2) photoanode semiconductor, synthetic dye (Ruthenium N79) as a photosensitizer, carbon counter electrode and iodine as the redox electrolyte mediator. The monolayer of the photosensitizer is attached to TiO_2 inside the cell and electrocatalyst material is separated from photoanode by the electrolyte redox mediator. The sensitizer becomes excited after absorbing sunlight and injects electrons to the conduction band of the semiconductor. The electrolyte then regenerates the sensitizer by donating electrons and diffuses towards the counter electrode, where the reduction reaction occurs. Instead of these forward charges transport processes, some back-electron transport processes also occur which drastically reduce the power conversion efficiency of solar devices [10].

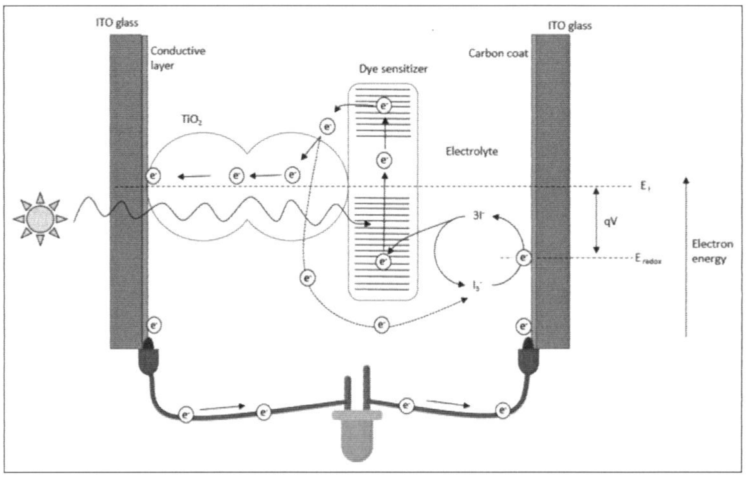

Figure 1. The DSSC structure and graphical working operation.

Among these components, the photoanode directly determines the photo-current density because it not only transports photo-induced electrons but also adsorbs dyes [11]. Accordingly, the semiconductor material used in photoanode play an important role in the working process of DSSCs. Among various semiconductors, nanocrystalline anatase TiO_2 is a promising photoanode material in DSSCs and has been widely investigated in recent years [12-19]. However, the large band gap of TiO_2 (3.2 eV) limits its utilization of visible light and many dyes failed to inject electrons from excited dyes into the conduction band of TiO_2 owing to insufficient electron injection driving force [20]. In this study, Zinc Oxide of having a band gap similar to the TiO_2 with 3.37 eV is proposed to enhance the performance of DSSC due to its advantage of higher excitonic electron mobility ~115 $cm^2/(V.s)$ compared to anatase TiO_2 which is reported to be only ~10 $cm^2/(V.s)$ [21-24].

Ruthenium in the other hand is widely used in the industry as the synthetic dye sensitizer which functions to absorb energy from the sunlight and excite its electron from the lower band to higher band energy for the electrical conversion. However, according to chemical properties report from Lenntech, besides of being one of the rarest material on earth, Ruthenium is a harmful chemical to human health even though it is not harmful to the environment. All Ruthenium compounds are regarded as highly toxic and as carcinogenic which may lead to cancer [25,26]. Thus, many researches have been done in exploring the possible alternative solutions for the sensitizer in DSSC using natural pigment dyes such as cyanin [27-38], carotene [39-40], tannin [41] and cholorophyll [42]. Calogero and Marco reported that a conversion efficiency of 0.66% was obtained using red Sicilian orange juice dye as sensitizer [34]. Wongchareeet al. employed rosella as a sensitizer in their DSC, which achieved a conversion efficiency of 0.70% [29]. Thus, in this study, three different fruit dyes of prunus domestica, magnifera indica and citrus limon is used in DSSC fabrication and the photo-to-electric conversion efficiency is compared.

2. Materials and methods

2.1. Materials

The conductive glass substrates (ITO, 25 mm x 25 mm, 10 Ω/sq, thickness 180 nm), ZnO powder (Sigma Aldrich, 205532), TiO_2 (Sigma Aldrich, 14027), prunus domestica (dark prune) fruit (pH 4.8), Magnifera indica (mango) fruit (pH 5.9), Citrus limon (lemon) fruit (pH 2.3), iodine, I_3 (Sigma Aldrich) and candle.

2.2. Fabrication of photoanodes

The process started with cleaning the ITO glasses by sonicating the glasses inside acetone solution for 10 minutes in the first stage. Next, the glasses were rinsed with DI water before sonicating it again in the acetone for 5 minutes. The glasses were blown dry using air dryer. Two photoanodes were prepared in this research which were ZnO and TiO_2. ZnO photoanode was prepared by mixing 16.93 g ZnO powder in 15.49 ml diethanolamine (DEA). This solution was stirred on a hot plate until the solution was homogeneous at 60°C for one hour. The solution was applied on the conducting side of the ITO glass using the sol-gel method. The process was done on a spin coater by dropping 3 drops of ZnO solution on the ITO at 250 RPM for 30 seconds.

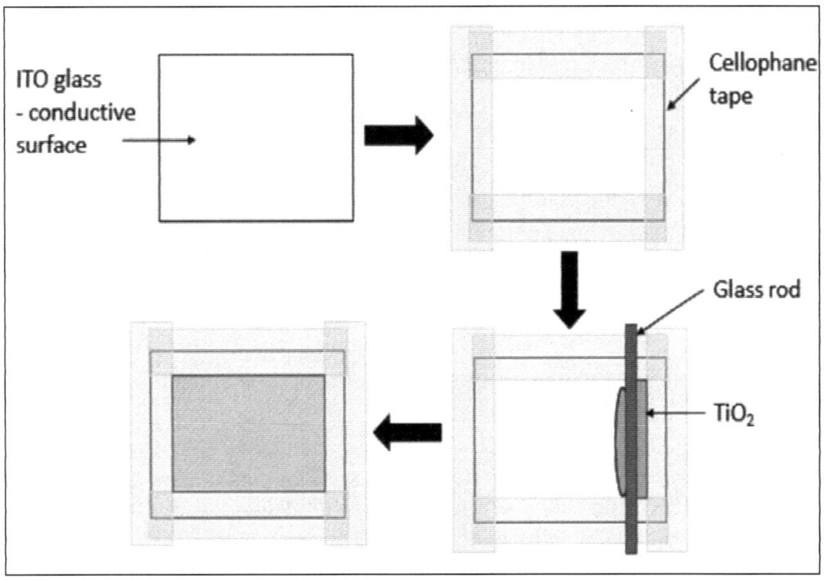

Figure 2. Doctor Blade Technique.

TiO_2 photoanode was prepared by mixing 6 g of TiO_2 powder with 9 ml acetic acid solution. This combination was grinded using mortar and pestle. A drop of Triton X was added to the mixture which act as a surfactant to facilitate spherical TiO_2 nanoparticles. Using Dr Blade method as in figure 2, the conductive side of the ITO glass was taped with cellophane tape as to reserve a conductive part for conduction and to secure the thickness of the TiO_2 on the surface of the photoanode. Both ZnO and TiO_2 photoanodes were placed on a hot plate for soft bake at 150°C for 10 minutes. Next the samples were going through hard bake in the furnace at 450°C for 1 hour. The active cell area for both photoanodes are 4 cm^2.

2.3. Synthesis of organic dyes

Prunus domestica (dark prune) fruit (pH 4.8), Magnifera indica (mango) fruit (pH 5.9), Citrus limon (lemon) fruit (pH 2.3) chosen as the organic dyes in this research. The fruits are placed in the mortar with a ratio of 10–20 g and 3–5 ml of DI water. Then by using a pestle, the fruits were crushed until it became a soupy mixture as shown in figure 3 (a). Next, the ZnO and TiO_2 photoanodes were immersed in the dye solution as shown in figure 3 (b) for 24 hours to allow the dye molecules attached to the photoanodes. After this process, the dye coated photoanodes were rinsed with DI water and were dried.

Figure 3. (a) Dye preparation (b) Coating process of the dye to the photoanodes and (c) Deposition of soot of candle carbon.

2.4. Fabrication of counter electrode
The counter electrode is prepared using the deposition of soot from a burning candle. A candle was lit and then the conductive side of ITO glass was placed facing down about 10 cm above the candle as shown in figure 3 (c). The carbon from combustion of wax from the candle will produce smoke and cause the black carbon deposition on the conductive side of the glass.

2.5. DSSC assembly
Coated ZnO and TiO_2 photoanodes were combined with the carbon coated counter electrode using the clip. Iodine was injected in between the two electrodes.

2.6. Characterization
The morphology of the ZnO and TiO_2 surfaces on the ITO glass was characterized by Scanning Electron Microscopy (SEM) (JOEL, JSM-6460LA). The current density (I-V) data are collected using Semiconductor Parametric Analyzer (SPA) (Keithly, 4200-SCS).

The incident light intensity was 100 W from the bulb used as the source power. Based on the I-V curve, the fill factor (FF) is defined as: $FF = (J_{max} \times V_{max})/(J_{sc} \times V_{oc})$, where the J_{max} and V_{max} are the photocurrent density and photovoltage for maximum power output; J_{sc} and V_{oc} are the short circuit photocurrent density and open circuit photovoltage respectively. The overall energy conversion efficiency is defined as $\eta = (FF \times J_{sc} \times V_{oc})/P_{in}$ where P_{in} is the power of the incident light.

3. Results and discussions

3.1. SEM Results
Figure 4 shows the SEM cluster images for TiO_2 and ZnO respectively. The clusters composed of these particles tended to fuse together to form large aggregates.

The ZnO spherical cluster has a uniform size and consist of well-defined, bigger surface area compared to TiO_2 cluster. Higher surface area of ZnO particle has the advantage of having more dye attached to it as compared to TiO_2.

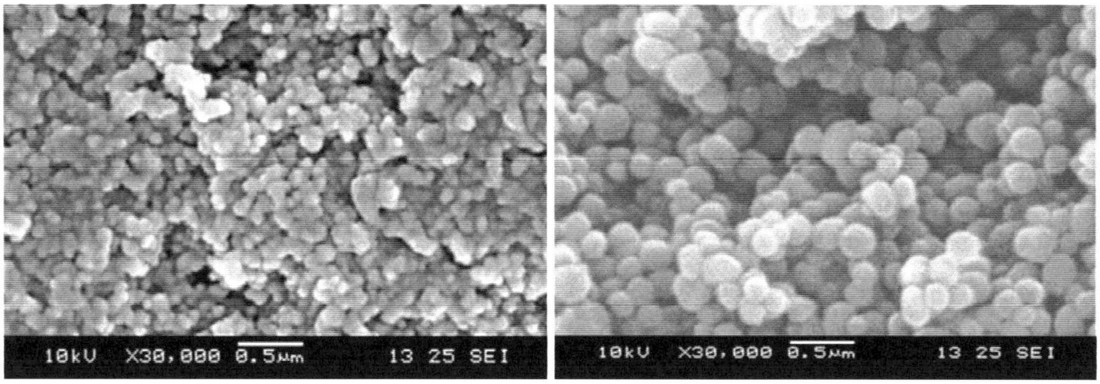

Figure 4. SEM images for (a) TiO2 (b) ZnO.

3.2. J-V Characteristics

The performance of the DSSC was evaluated through the efficiency calculation of the cells when it is illuminated. J-V characteristics of the ZnO and TiO_2 cells with all 3 dyes (Prunus domestica, Magnifera indica and Citrus limon) were evaluated using SPA tool and the parameters of J_{sc}, V_{oc}, FF and efficiency were calculated as in table 1. In this project, P_{in} is equal to 100 W and the surface area is 4 cm^2 for all samples.

Table 1. Efficiency for all evaluated DSSCs.

Sample	J_{sc}(mA/cm^2)	V_{oc}(mV)	FF	η(%)
ZnO with Prunus domestica dye	0.006	124	3.95	3.08
TiO_2 with Prunus domestica dye	0.008	145	2.48	2.88
ZnO with Magnifera indica dye	0.009	175	1.26	1.99
TiO_2 with Magnifera indica dye	0.012	185	0.76	1.72
ZnO with Citrus limon dye	0.013	201	0.51	1.34
TiO_2 with Citrus limon dye	0.014	224	0.31	0.96

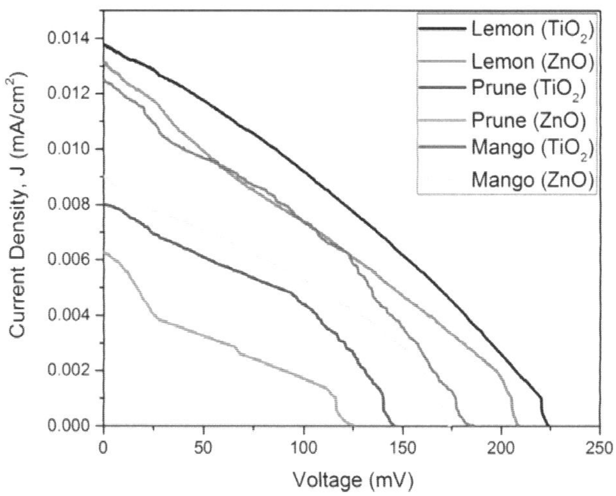

Figure 5. J-V curves for ZnO and TiO2 photoanodes with 3 different dyes.

Based on table 1, the ZnO photoanode trend of efficiency was higher compared to TiO_2 for all 3 different dyes tested. As for the dye factor comparison, Prunus domestica (pH 4.8) contributed the highest efficiency followed by Magnifera indica (pH 5.9) and Citrus limon (pH 2.3). The best performance of DSSC was given by the combination of ZnO photoanode and prunus domestica dye compared to other samples with 3.08% photoconversion efficiency. Based on these efficiency results, it is shown that pH indicator does not affect the performance of the DSSC. Figure 5 shows the J-V curve for all the samples of DSSCs tested, which clearly shows that the short circuit current density, open circuit voltage and fill factor of each sample.

4. Conclusions

Spherical clusters of ZnO photanode was obtained via SEM evaluation and it shows that it has a bigger surface area which helps in increasing the performance of DSSC as the more dye can be attached to the ZnO particle as compared to TiO_2. Prunus domestica dye with the darkest color of dye gave the highest efficiency compared to magnifera indica and citrus limon with the efficiency of 3.08% when fabricated with ZnO photoanode. This reseach also prove that pH indicator does not affect the efficiency of the DSSC.

References

[1] O'Reagan B and Gràtzel M 1991 *Nature* **353**(6346) 737-740
[2] Al-Omair N A, Reda S M, Al-Hajri F M 2014 *Scientific Research of Advances in Material Physics and Chemistry* **4** 29-32
[3] Mehmood U 2017 *Journal of Organic Electronics* **42** 187-193
[4] Chen C Y, Wang M, Li J Y, Pootrakulchote N, Alibabaei L, Ngoc-le C H, Decoppet J D, Tsai J H, Gratzel C, Wu C G, Zakeeruddin S M and Gratzel M 2009 *ACS NANO* **3**(10) 3103-3109
[5] Premalala E V A, Dematagea N, Kumarab G R A, Rajapakseb R M G, Murakamib K and Konno A 2012 *Electrochimica Acta* **63** 375-380
[6] Hagfeldt A, Boschloo G, Sun L, Kloo L 2010 *Chem Rev* **110** 6595–663
[7] Weerasinghe H C, Huang F, Cheng Y 2013 *Nano Energy* **2** 174–89
[8] Jelle B P, Breivik C, Røkenes H D 2012 *Solar Energy Mater Solar Cells* **100** 69–96
[9] Heiniger L P, O'Brien P G, Soheilnia N, Yang Y, Kherani N P, Gratzel M et al. 2013, *Adv Mater*
[10] Guo X, Lu G, Chen J 2015, *Front. Energy Res.* **3** 50
[11] Dong C K, Xiang W C, Huang F Z, Fu D C, Huang W C, Bach U, Cheng Y B, Li X 2014 *Spiccia Nanoscale* **6** 3704
[12] Tang X, Liu X, Zhang L, Xing Y, Tian Y 2014 *Chem. Phys.* **441** 121
[13] Huang Q, Zhou G, Fang L, Hu L 2011 *Wang Energy Environ. Sci.* **4** 2145
[14] Edvinsson T, Pschirer N, Schoneboom J, Eickemeyer F, Boschloo G, Hagfeldt A 2009 *Chem. Phys.* **357** 124
[15] Li Z Q, Que Y P, Mo L E, Chen W C, Ding Y, Ma Y M, Jiang L, Hu L H, Dai S Y 2015 *ACS Appl, Mater. Interfaces* **7** 10928
[16] Li Z Q, Ding Y, Mo L E, Hu L H, Wu J H, Dai S Y 2015 *ACS Appl, Mater. Interfaces* **7** 22277
[17] Li Z Q, Chen W C, Guo F L, Mo L E, Hu L H, Dai S Y 2015 *Sci. Rep.* **5** 14178
[18] Ding Y, Zhou L, Mo L E, Jiang L, Hu L H, Li Z Q, Chen S H, Dai S Y 2015 *Adv. Funct. Mater.* **25** 5946
[19] Vizureanu, P 2009 *Metalurgia International* **14**(5) 5-9
[20] Wang Z S, Cui Y, Dan-oh Y, Kasada C, Shinpo A, Hara K 2008 *J. Phys. Chem. C* **112** 17011
[21] Gra¨tzel M 2001*Nature* **414** 338
[22] Kaidashev E M, Lorenz M, von Wenckstern H, Rahm A, Semmelhack H C, Han K H, Benndorf G, Bundesmann C, Hochmuth H, Grundmann M 2003 *Appl. Phys. Lett.* **82** 3901
[23] Curri M L, Comparelli R, Cozzoli P D, Mascolo G, Agostiano A 2003 *Mater. Sci. Eng. C* **23** 285
[24] Mohamad I S, Norizan M N, Hanifiah M K F M, Amin I A M, Shahimin M M 2015 *AIP*

Conference Proceeding **1660** 0700049
[25] Lenntech B V 2017 *Report on Chemical Property of Ruthenium*
[26] Sirimanne P M, Senevirathna M K I, Premalal E V A, Pitigala P K D D P, Sivakumar V, Tennakone K 2006 *J. Photochem. Photobiol. A* **177** 324–327
[27] Hao S, Wu J, Huang Y, Lin J 2006 *Sol. Energy* **80** 209–214
[28] Polo A S, Murakami Iha N Y 2006 *Sol. Energ. Mat. Sol. Cell* **90** 1936–1944
[29] Wongcharee K, Meeyoo V, Chavadej S 2007 *Sol. Energ. Mat. Sol. C* **91** 566–571
[30] Tennakone K, Kumarasinghe A R, Kumara G R R A, Wijayantha K G U, Sirimanne P M 1997 *J. Photochem. Photobiol. A* **108** 193–195
[31] Zhang D, Lanier S M, Downing J A, Avent J L, Lumc J, McHale J L 2008 *J. Photochem. Photobiol. A* **195** 72–80
[32] Roy M S, Balraju P, Kumar M, Sharma G D 2008 *Sol. Energ. Mat. Sol. C* **92** 909–913
[33] Fernando J M R C, Senadeera Natural G K R 2008 *Curr. Sci.* **95** 663–666
[34] Calogero G, Di. Marco G 2008 *Sol. Energ. Mat. Sol. C* **92** 1341–1346
[35] Dai Q, Rabani J 2002 *J. Photochem. Photobiol. A* **148** 17–24
[36] Cherepy N J, Smestad G P, Grätzel M, Zhang J Z 1997 *J. Phys. Chem. B* **101** 9342–9351
[37] Luo P, Niu H, Zheng G, Bai X, Zhang M, Wang W 2009 *Acta Part A* **74** 936–942
[38] Furukawa S, Iino H, Iwamoto T, Kukita K, Yamauchi S 2009 *Thin Solid Films* **518** 526–529
[39] Gómez-Ortíz N M, Vázquez-Maldonado I A, Pérez-Espadas A R, Mena-Rejón G J, Azamar-Barrios J A, Oskam G 2010 *Sol. Energ. Mat. Sol. C* **94** 40–44
[40] Yamazaki E, Murayama M, Nishikawa N, Hashimoto N, Shoyama M, Kurita O 2007 *Sol. Energy* **81** 512–516
[41] Espinosa R, Zumeta I, Santana J L, Martı'nez-Luzardo F, Gonza'lez B, Docteur S, Vigil E 2005 *Sol. Energ. Mat. Sol. C* **85** 359–369
[42] Kumara G R A, Kaneko S, Okuya M, Onwona-Agyeman B, Konno A, Tennakone K 2006 *Sol. Energ. Mat. Sol. C* **90** 1220–122

Acknowledgments
The authors wish to thank School of Microelectronic Engineering, Center of Excellence Geopolymer and Green Technology (CeGeoGTech), and Universiti Malaysia Perlis (UniMAP) for the support while doing this research. This research and paper is financially supported by Research Acculturation Grant Scheme 9018-00092 from Ministry of Higher Education (MOHE) Malaysia.

IOP Conf. Series: Materials Science and Engineering **209** (2017) 012029 doi:10.1088/1757-899X/209/1/012029

Indium (In) Effects to The Efficiency Performance of $Ga_{1-X}In_xP$/GaAs Based Solar Cell Using Silvaco Software Modelling & Simulation

M N Norizan[1,2,3]*, S M Zahari[1], I S Mohamad[1,2,3], R A M Osman[1], M M Shahimin[4] and S A Z Murad[1]

[1] School of Microelectronic Engineering, Universiti Malaysia Perlis, Pauh Putra Campus, 02600 Arau, Perlis, Malaysia.

[2] Centre of Excellence Geopolymer and Green Technology (CeGeoGTech), Universiti Malaysia Perlis, Perlis, Malaysia.

[3] Advanced Multi-Disciplinary MEMS-Based Integrated NCER Centre of Excellence (AMBIENCE), Universiti Malaysia Perlis, Perlis, Malaysia.

[4] Department of Electrical and Electronic Engineering, Faculty of Engineering, National Defence University of Malaysia (UPNM), Kem Sungai Besi, 57000 Kuala Lumpur, Malaysia.

E-mail: *mohdnatashah@unimap.edu.my

Abstract. $Ga_{1-x}In_xP$ composition has been applied to the top cell of multi-junction GaInP/GaAs based solar cell and currently have achieving a conversion efficiency of more than 46%, however its capability is unclear. We performed an analysis using Silvaco simulation method to evaluate the effect of In and the substitution was made to the $Ga_{1-x}In_xP$ for the range of x from 0 to 1. We found that the highest efficiency recorded was 17.66% when the composition of Indium was x=1. The efficiency has been increasing about 11.71% from x=0 to x=1 In content. As the composition of In raised, the value of efficiency and short circuit current density, Jsc also become higher (13.60 mA/cm^2) by having a greater photon absorption in a wider band gap energy. In addition to that, V_{oc}, P_{max}, V_{max}, I_{max} and fill factor was measured to be 2.15 V, 2.44 mW/cm^2, 2.0 V, 1.22 mA/cm^2 and 83.34 respectively. In conclusion, this study confirms that the existence of In in $Ga_{1-x}In_xP$ improves the solar cell efficiency by gaining a higher energy gap and producing more electrons for best achievement in multilayer solar cell applications.

1. Introduction

Multi-junction solar cells is a natural extension of single-junction cells and has been one of the most promising choices to capably convert the light into electrical energy [1–4]. GaAs based compound multi-junction devices are low band gap components of the most efficient solar cells [5]. It is required to limit the losses and improve the efficiency of photovoltaic (PV) cells in order to obtain a high performance PV cell [6,7]. In order to control the amount of current produced at a time, the selection of materials with certain energy gap for each layer is very important [8].

Recent papers have reported that the efficiency around 30% for GaInP on a Ge substrate have been obtained [9]. After that, Si-based has replaced Ge-based bottom cells because it is cheaper than other material and most widely used in semiconductor device and solar cell industry [10]. The growth of compound semiconductor-based cells on Si-based bottom cell has been reported [11]. It is difficult to develop a multilayer cell with a large number of subcells on Si substrates because of a different thermal coefficients and lattice constant of semiconductor materials for the respective subcells. Hence, to prevent a lattice mismatch among materials, a two-layer cells or tandem cell is preferable to be investigated. GaInP/Si solar cell previously reported to achieve the experimental efficiency about 9.9–10.4% [12] and our previous research reported that the simulation efficiency of $Ga_{0.5-x}In_{0.5}P$/Si with the contribution of x=0.5 In content was only 10.12% [13]. Presently, compound semiconductor using GaAs has been developed and used as a space solar cell material because of more superior radiation resistance and high conversion compared to Si cells [14].

National Renewable Energy Laboratory (NREL) has demonstrated a high efficiency (~27% terrestrial) GaInP/GaAs n-on-p tandems [15]. GaInP/GaAs solar cells have the potential for achieving a conversion efficiency of more than 30% and 35%, respectively, and monolithic tandem cell with an efficiency of 29.5% has already been reported from previous research [10]. The wider gap of 1.9 eV top cell design is an intermediate step towards developing a 2.0 eV $In_{0.5}Al_{0.1}Ga_{0.5}P$ top cell which also gives better current matching to the GaAs substrate. AlGaAs can also be an alternative to GaInP. However, due to its high sensitivity to water and oxygen contamination has made it undesirable choice [16]. Therefore, In ($\geq$0.5) is found to be promising material composition of solar cell [17]. GaInP material is a wide range of band gap semiconductor nearly 1.9 eV that can be grown lattice-matched to Ge and GaAs and was selected as a proper material to be used as a top cell for multi-band gap tandem solar cells [18–20].

The objective of this research work is to study and to explore the effect of In contribution in tandem cell $Ga_{1-x}In_xP$/GaAs substrate using simulation model. $Ga_{1-x}In_xP$ cell within the range of x from 0 to 1 was expected to absorb the photon energy differently since their energy gap will be changed accordingly depending on the In content. Introducing In into the GaP was expected to increase the efficiency of solar cell to gain high power density. In this work, Silvaco TCAD tools were selected to accomplish this research objective. The Atlas simulator from Silvaco International has the feature to model a wide variety of physical device characteristics. This model generates the electrical characteristics of a solar cell based on the virtual fabrication of its physical structure, allowing for direct manipulation of materials, doping, and dimensions. The usage of Atlas is already investigated for solar cell application in previous research. This ability shows remarkable promise towards the multilayer solar cell design [21-23].

2. Details of The Simulated Solar Cell

The experiment was conducted by using a simulation software application, Silvaco Atlas by having the composition of x in the $Ga_{1-x}In_xP$/GaAs. Silvaco TCAD was chosen as a tool because of its feature to make the structure of semiconductor devices process physically and able to generate the electrical characteristics for solar cell simulation. Figure 1 shows the modelling design of $Ga_{1-x}In_xP$/GaAs for ($0\leq x\leq 1$) layer using Silvaco.

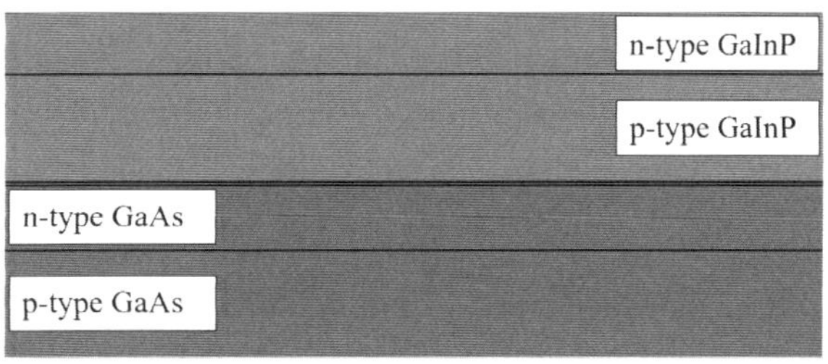

Figure 1. $Ga_{1-x}In_xP$/GaAs Solar Cell layer.

The simulation was done by changing the x composition for the $Ga_{1-x}In_xP$/GaAs tandem cell within the range of x from 0 to 1. In this simulation, efficiency and material electrical properties were calculated. The current matching condition was determined by n-type of 0.05 µm thickness with $2x10^{18}$ cm^{-3} of doping concentration while for p-type, 0.55 µm thickness with $7x10^{16}$ cm^{-3} concentration was used. The composition of Indium (In) was varied to determine the effectiveness and improve the cell structure. The range of $0 \leq x \leq 1$ for In was used and has been analysed in terms of its performance.

3. Results and discussions

Table 1 shows the output of the simulation which is the short current density (J_{sc}), open circuit voltage (V_{oc}), maximum power (P_{max}), maximum current (I_{max}), maximum voltage (V_{max}), fill factor (FF) and efficiency.

Based on table 1, J_{sc} was only 11.23 mA/cm^2 without the content of In. J_{sc} is the current through the solar cell due to the generation and collection of light-generated carriers, which is the involvement of electron and hole in the cell. It is the largest current from the solar cell and depends on the area, the number of photons, the spectrum of the incident light and optical properties. When the composition of In increase, the area of cell also increase and the energy gap become wider so that, it can capture more photons at the area. This is one of the reasons of thin layer should be made thinner than a bottom layer cell as we had discussed formerly. At x=1.0, the J_{sc} reaches 13.60 mA/cm^2, which is the highest value of J_{sc}.

The energy gap also enlarges in this work, proves by the increment number of open circuit voltage, V_{oc}. This occurs as the recombination current falls on the solar cell. At a certain case, there is a drop off to V_{oc} at a very high band gap due to the very low short circuit current, I_{sc}.

Table 1. $Ga_{1-x}In_xP$ /GaAs solar cell efficiency.

Range ($0 \leq x \leq 1$)	Short Circuit Current Density, J_{sc} (mA/cm^2)	Open Circuit Voltage, V_{oc} (V)	Maximum Power, P_{max} (mW/cm^2)	Maximum Voltage, V_{max} (V)	Maximum Current, I_{max} (mA/cm^2)	Fill Factor, FF	Efficiency (%)
0	11.23	0.86	8.23	0.8	1.03	84.86	5.95
0.3	13.17	0.25	2.64	0.20	1.30	79.38	7.47
0.5	13.24	1.31	1.46	1.20	1.21	83.66	10.54
0.7	13.37	1.63	1.87	1.50	1.21	83.75	13.23
1.0	13.60	2.15	2.44	2.00	1.22	83.34	17.66

Maximum power, P_{max} can be calculated by having $P_{max} = V_{oc}I_{sc}FF$, which is the product of short-circuit current (I_{sc}), open-circuit voltage (V_{oc}) and fill factor (FF). The P_{max} irregularly fluctuate as the In content enlarges in the region between 0.1 to 0.5.

The current-voltage (I-V) characteristic shifts as the cell begins to produce power when the light hit on the cell. The amount of the shifts in term of the I-V curve becomes greater as the light intensity increase. Identically, the input of the photo-current to the total current through the solar cell becomes less significant when the voltage is increased after the maximum point is reached and the total current changes sign, is said to be drop off because of the very low I_{sc}. P_{max} was displayed to be slightly differ from the range of x from 0 to 0.5 depends on the material properties.

By introducing GaInP as a top cell on GaAs substrate, the GaInP itself must be able to absorb the amount of light with maximum conversion efficiency. To attain this necessity, the technique is by thinning the top cell layer. Figure 2 (a, b, c) shows the current-voltage (I-V) curve and (d) the efficiency graph respectively.

At the beginning of the simulation, the efficiency was about 5.95% without In content (In_x; where x=0) known as GaP/GaAs. The trend increase to 7.47 % (x=0.3), gaining about 1.52%.

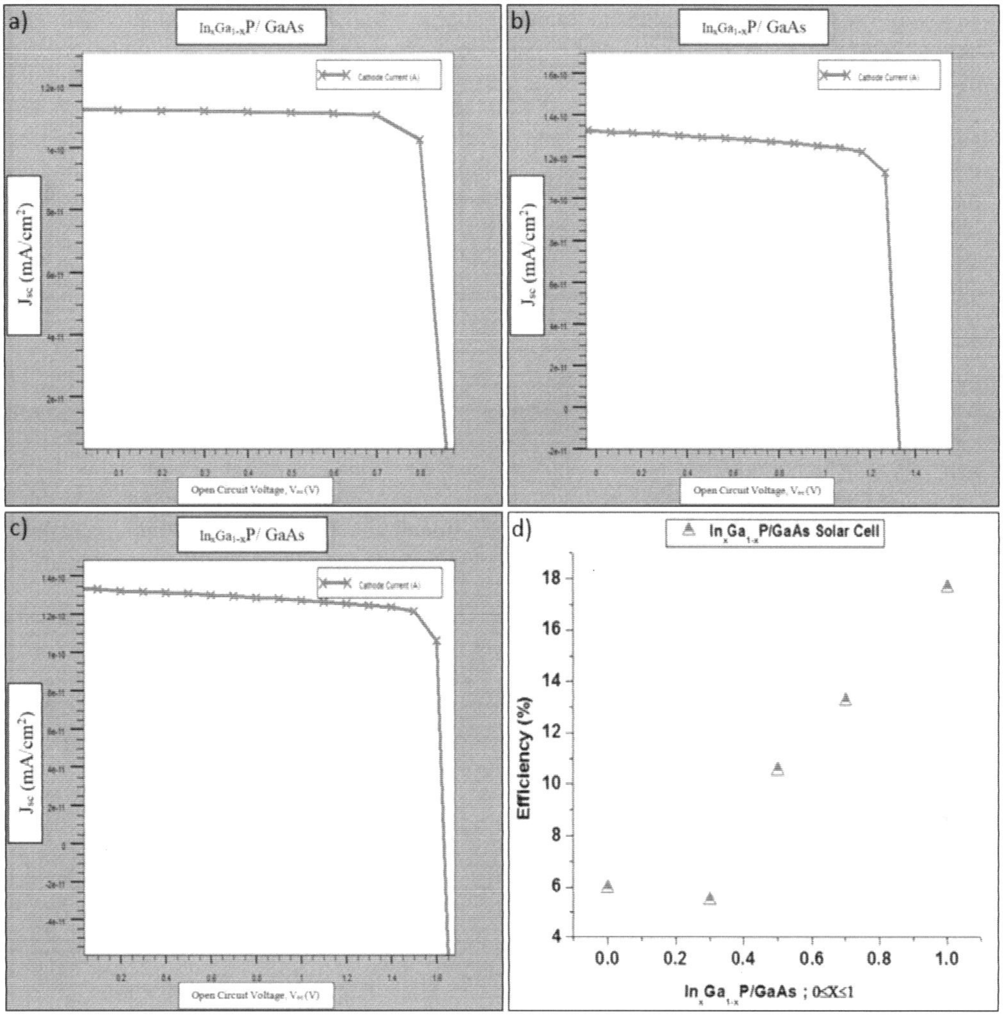

Figure 2. Current-voltage (I-V) curve $Ga_{1-x}In_xP$/ GaAs when a) x=0, b) x=0.5 c) x=0.7, and d) efficiency graph from the range of (0≤x≤1).

The efficiency resumes to rise up to 10.54% to 13.23% (x=0.5~0.7). Finally, with the In content of x=1, the efficiency was of the highest value, which is 17.66% after having about 7.12% increment from In content (x≥0.5).

Table 2. Material Properties of GaAs, GaP, InP and GaInP.

Material	Band gap, E_g	Lattice Constant, Å	Electron Mobility, μ_n (cm^2 V^{-1}s^{-1})	Hole Mobility, μ_p (cm^2 V^{-1}s^{-1})
GaAs	1.43	5.65	8500	400
GaP	2.26	5.45	250	150
InP	1.27	5.86	4600	150
InGaP	1.90	5.65	1945	141

Table 2 shows the III-V compound semiconductor material properties consist of the band gap, lattice constant, electron mobility and hole mobility. These properties can be related to the simulation output in Table 1. Without the substitution of In composition, $Ga_{1-x}In_xP/GaAs$ (x=0), the cell is known as GaP/GaAs. The GaP material took part in the elementary simulation with a high band gap (2.26 eV) and if the photon has lower energy than band gap material ($eV \leq E_g$), the loss mechanism occurs when the solar cell is unable to extract power in the photon. This case refers to the fluctuation of the maximum power, P_{max} and also the maximum of current and voltage (I_{max}, V_{max}) in table 1 above. The lower electron mobility for GaP which is 250 $cm^2\ V^{-1}s^{-1}$ also affects the solar cell system when it slows down the movement of electron and hole to be paired off. As a movement of carrier becomes slow, the recombination took place. This is another loss mechanism in solar cell where the electron holes left behind by previous excitations meet the electrons created by the photoelectric effect.

It is important to stack the layer using suitable material, having a greater mobility such as GaAs material. When the photon absorbed to the bottom layer of solar cell, the electron in GaAs material with high mobility promotes the acceleration of electron-hole in the conduction band area so that the electron is quickly getting excited. This excitation will then produce current and voltage. Stacking more layers which consists of different material can eliminate the loss, but the selection of material must have almost equal lattice constant. The lattice constants are prescribed as physical dimensions of unit cells in the crystal lattice. Choosing the proportionate value or in the other words, matching the lattice structures between two different semiconductor materials, allows a region of band gap change to be formed in a material without introducing a change in crystal structure. Recombination centres can occur when there are mismatched in the crystal lattice constants which creates defects or dislocation in the lattice.

4. Conclusions

This work demonstrates the existence of In in $Ga_{1-x}In_xP/GaAs$ based solar cell which improves the solar cell efficiency by gaining a higher energy gap and producing more electrons for best achievement in multilayer solar cell applications. In contribution with x=1, lead to a higher efficiency linearly from the range of x from 0 to 1. GaP structure was tuned to gain higher energy gap which led to high absorption of photon energy and enhanced the number of electrons released by increasing the In content. Greater efficiency was obtained by eliminating loss mechanisms such as recombination in a cell and power losses which due to the photon energy gap less than the energy gap of the material chosen ($E_v \leq E_g$). Nominating multilayer structure using III-V compound semiconductor also improves the solar cell applications instead of varying the composition of metal into a top layer of multilayer structure.

References

[1] Yang W, Dai P, Ji L, Tan M, Wu Y, Uchida S, Lu S and Yang H 2016 *Appl. Surf. Sci.* **389** 673–8
[2] Jiang C S, Friedman D J, Geisz J F, Moutinho H R, Romero M J and Al-Jassim M M 2003 *Appl. Phys. Lett.* **83** 1572–4
[3] Friedman D J, Kurtz S R, Bertness K A, Kibbler A E, Kramer C, Olson J M, King D L, Hansen B R and Snyder J K 1995 *Prog. Photovoltaics Res. Appl.* **3** 47–50
[4] Barrera M, Rubinelli F, Rey-Stolle I and Plá J 2012 *Phys. B Condens. Matter* **407** 3282–4
[5] Green M A, Emery K, Hishikawa Y, Warta W and Dunlop E D 2015 *Prog. Photovoltaics Res. Appl.* **23** 805–12
[6] Verma J, Dey P, Prajapati A and Das T D 2016 *International Conference on Microelectronics, Computing and Communications (MicroCom)* (IEEE) pp 1–4
[7] Chang S T, Liao M H and Lin W-K 2011 *Thin Solid Films* **519** 5022–5
[8] Zahari S M, Norizan M N, Mohamad I S, Osman R A M and Taking S 2015 *AIP Conf. Proc.* **1660**
[9] Yamaguchi M, Takamoto T, Araki K and Ekins-Daukes N 2005 *Sol. Energy* **79** 78–85

[10] Takamoto T, Ikeda E, Kurita H and Ohmori M 1994 High efficiency InGaP solar cells for InGaP/GaAs tandem cell application *Proceedings of 1994 IEEE 1st World Conference on Photovoltaic Energy Conversion - WCPEC (A Joint Conference of PVSC, PVSEC and PSEC)* vol 2(IEEE) pp 1729–32
[11] Law D C, King R R, Yoon H, Archer M J, Boca A, Fetzer C M, Mesropian S, Isshiki T, Haddad M and Edmondson K M 2010 *Sol. Energy Mater. Sol. Cells* **94** 1314–8
[12] Shigekawa N, Morimoto M, Nishida S and Liang J 2014 Surface-activated-bonding-based InGaP-on-Si double-junction cells *Jpn. J. Appl. Phys.*
[13] Zahari S M, Osman R A M, Norizan M N, Mohamad I S, Malini A and Taking S 2015 *Appl. Mech. Mater.* **754–755** 1182–6
[14] Yamaguchi M 2001 *Sol. Energy Mater. Sol. Cells* **68** 31–53
[15] Wojtczuk S J, Vernon S M and Sanfacon M M 1993 Comparison of windows for P-on-N InGaP solar cells *Conference Record of the Twenty Third IEEE Photovoltaic Specialists Conference - 1993 (Cat. No.93CH3283-9)* (IEEE) pp 655–8
[16] Singh K J and Sarkar S K 2012 *Opt. Quantum Electron.* **43** 1–21
[17] Yamaguchi M 2012 *Energy Procedia* **15** 265–74
[18] Novák J, Hasenöhrl S, Kúdela R and Kučera M 2005 *J. Cryst. Growth* **275** e1281–6
[19] King R R, Law D C, Edmondson K M, Fetzer C M, Kinsey G S, Yoon H, Sherif R A and Karam N H 2007 *Appl. Phys. Lett.* **90** 183516
[20] Takamoto T, Ikeda E, Kurita H and Ohmori M 1997 *Appl. Phys. Lett.* **70** 381–3
[21] Michael S 2005 A Novel Technique for Modeling Radiation Effects in Solar Cells Utilizing SILVACO® Virtual Wafer Fabrication Software *2005 8th European Conference on Radiation and Its Effects on Components and Systems* (IEEE) p LN9-1-LN9-6
[22] Jahanshah F, Sopian K, Othman Y and Fallah H R 2009 *Sol. Energy* **83** 1629–33
[23] Vizureanu, P 2009 *Metalurgia International* **14**(5) 5-9

Acknowledgments

This research work was supported by the Fundamental Research Grant Scheme (FRGS) Grant No:9003-00339 and Research Acculturation Grant Scheme (RAGS) No:9018-00092, Universiti Malaysia Perlis (UniMAP) and Ministry of Higher Education (MoHE) Malaysia. The authors wish to thank School of Microelectronic Engineering, Center of Excellence Geopolymer and Green Technology (CeGeoGTech), and Universiti Malaysia Perlis (UniMAP) for the support while doing this research.

IOP Conf. Series: Materials Science and Engineering **209** (2017) 012030 doi:10.1088/1757-899X/209/1/012030

Structural and Morphological Investigation for Water-Processed Graphene Oxide/Single-Walled Carbon Nanotubes Hybrids

M R Muda[1], M M Ramli[1, 2*], S S Mat Isa[1, 2], D S C Halin[2], L F A Talip[1,2], N S Mazelan[1, 2], N A M Anhar[1] and N A Danial[1, 2]

[1] School of Microelectronic Engineering, University Malaysia Perlis, 01000 Malaysia

[2]Center of Excellence Geopolymer and Green Technology, School of Materials Engineering, Universiti Malaysia Perlis (UniMAP), Perlis, Malaysia

E-mail: mmahyiddin@unimap.edu.my

Abstract. New group of materials derived from hybridization of single walled carbon nanotubes (SWCNTs) and graphene oxide (GO) which resulting novel three dimensional (3D) materials generates an outstanding properties compared to corresponding SWCNTs and GO/Graphene. In this paper, we describe a simple approach using water processing method to develop integrated rGO/GO-SWCNT hybrids with different hybrid ratios. The hybrid ratios were varied into three divided ratio and the results were compared between pristine SWCNTs and GO in order to investigate the structural density and morphology of these carbonaceous materials. With an optimized ratio of rGO/GO-SWCNT, the hybrid shows a well-organized hybrid film structures with less defects density sites. The optimized mixture ratio emphasized the important of both rGO and SWCNTs in the hybrid structures. Morphological structural and defects density degrees were examined by Field Emission Scanning Electron Microscopy (FESEM) and Raman spectroscopy.

1. Introduction

Carbon is the one of the most abundant elements found in nature which exists in various allotropic forms.

Recently, CNTs and graphene which representative of 1D and 2D carbon nanomaterials has attracting worldwide attentions during last decades due to their remarkable physical, electrical, and chemical properties [1,2].

Besides that, integrating CNTs and graphene into hybrid structures provide a novel approach to new 3D materials with advantageous properties.

Therefore, a unique and intelligent graphene-CNTs hybrid films could overcome some of the limitations and shortcomings of individual components [3–5].

Owing to the its vast properties, these materials had shown the great potential to be applied in various areas of nano-technology research such as super capacitors, batteries, biosensors, and especially for optical devices.

In this study, we reported a simple approach for preparing hybrid G-SWCNT materials using water-processed method and the carbonaceous materials were characterized using FESEM and Raman Spectroscopy.

2. Materials and methods

2.1. Materials
SWCNTs was purchased from NanoLab, US (initial purity > 95%) and functionalized using acid treatment. Fine graphite powder purchased from NE Scientific (initial purity > 90%) was pre-treated before being oxidized as reported by Muda et. al [6]. Sulphuric acid 98%, nitric acid 70%, $KMnO_4$ powders, HCl acid 30% and H_2O_2 30% were purchased from Sigma–Aldrich.

2.2. GO-SWCNT mixture preparation
First, GO was synthesized by a modified hummers method as described in our previous study [6]. Pre-oxidized graphite flakes and 15 g of potassium permanganate ($KMnO_4$) were stirred in a solution containing concentrated sulphuric (H_2SO_4) acid for several hours. The solution reaction was then terminated by the addition of hydrogen peroxide (H_2O_2) and subsequently washed with hydrochloric (HCl) acid in order to remove the sulfate ions. The final product was washed with deionized (DI) water and dried in vacuum. The acid treatment of SWCNTs was performed using a mixture of H_2SO_4 and nitric (HNO_3) in the volume ratio (3:1). The mixture was refluxed for several hours, after which time the acid mixture was decanted. At the end of this process, the mixture was repeated washed with DI water until the pH of the solution become neutral. In order to obtain the GO/rGO-SWCNT mixture, GO powder and acid treated SWCNT were mixed thoroughly with different ratios in DI water before subjected to sonication process for well dispersion.

2.3. Film Preparation and Thermal Reduction Process
Carbonaceous films were prepared by spray pyrolysis method using a specific volume of centrifuged solution on a silicon dioxide (SiO_2) substrate. After the coating process, the samples were immediately annealed at 300°C for 3 hours in a vacuum furnace. Then the samples were characterized by field emission scanning microscopy (FESEM) and Raman spectroscopy.

3. Results and discussions

3.1. FESEM Analysis

3.1.1. GO. Figure 1 shows the surface morphology of GO sheets which produced by modified Hummer's method with an accelerating voltage of 3kV and 150kX magnification. Based on FESEM micrograph, it can be seen obviously that the aggregated GO sheets are closely associated with another sheets.

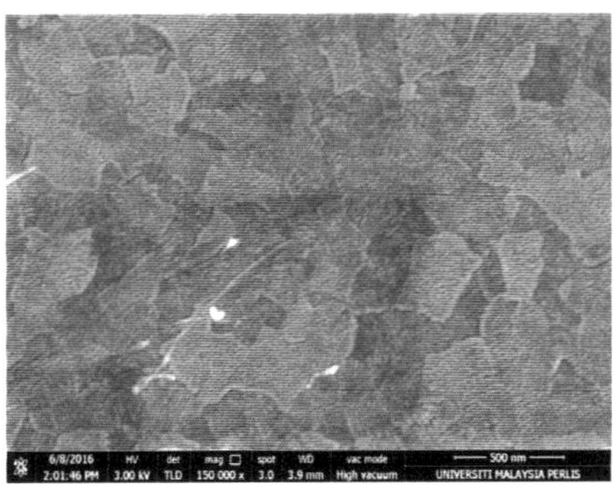

Figure 1. Micrograph of FESEM of the GO sheets.

This means that Van der Waals forces between GO sheets stack them tightly during spray deposition process. On the other hand, the GO sheets exhibit a rather wavy and crumpled structure on the basal plane. As reported by Schniepp [7], crumpled and wrinkles formation might be attributed to hydroxyl and epoxy group that located on basal plane which tend to form chains on the graphitic surface, leading to the generation of topological defects along GO sheets.

3.1.2. SWCNT. Figure 2 shows the morphology of raw single-walled carbon nanotubes (SWCNTs) and acid functionalized SWCNT (f-SWCNTs).

Generally, the raw SWCNTs contain a large quantity of impurities such as amorphous carbon and catalytic particles. Besides, the hydrophobic nature makes SWCNTs are tends to agglomerates which hindering it to be dispersed in various solution. Thus, acid functionalization SWCNTs is the one of the established method that was carried out to increase solubility and purity of SWCNTs.

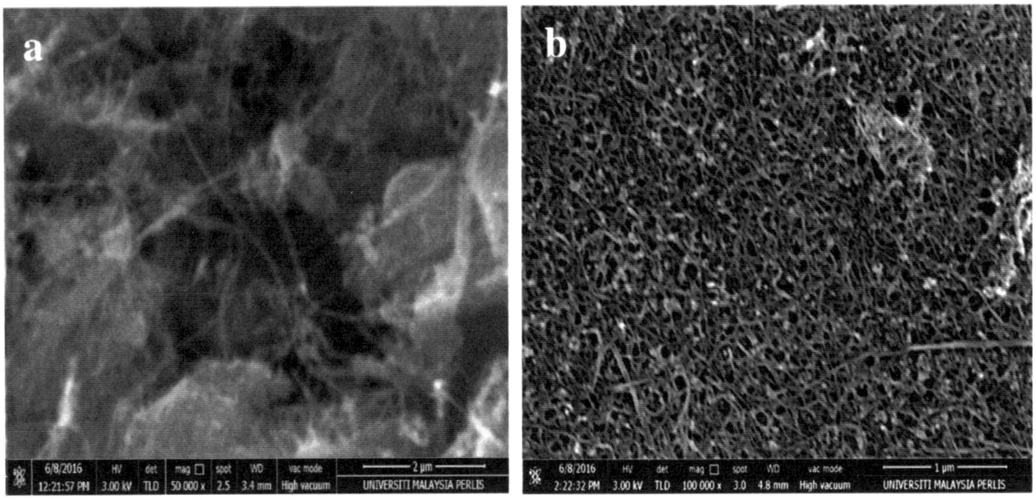

Figure 2. The micrograph of FESEM of (a) raw SWCNT and (b) acid treated SWCNT.

According to the figure 2, it can be clearly seen the differences of before and after acid functionalization process where, the raw SWCNTs are agglomerated with the metal catalyst particles. However, after functionalization process, the metal particles were totally removed.

As been discussed by Goyanes et. al [8], during acid functionalization process, the raw SWCNTs were purified and a part of catalyst particles were possibly eliminated. They also reported that, the functional groups were attached on the sidewall tubes during oxidation which resulting improvement in its hydrophilicity and dispersibility.

3.1.3. GO-SWCNT Hybrid. The microstructures of the as-prepared rGO-SWCNT hybrid films with different ratios were shown in figure 3. The nano-composite of hybrid rGO-SWCNT is a well-defined 3D structure which contains fragmented rGO flakes connected with entangled of SWCNTs.

As seen in figure 3, it can be clearly observed that, the rGO-SWCNT hybrid films with different ratio provides a better mixture for rGO and SWCNT, in which SWCNTs randomly bridge between graphene sheets and tend to form a well-organized hybrid structure.

From figure 3(a), the ratio 9:1 shows that the individual SWCNTs appeared to be separated on the surface of rGO sheets. In the meantime, this result proved that the interaction between rGO and SWCNTs tubes caused by the π-π conjugated aromatic domain [9].

In contrast, the SWCNTs that attached to the rGO sheets in ratio 5:5 showed a high degree of dispersion as shown in figure 3 (b).

It can be deduced that the dispersion between these two materials resulted homogeneously and well-uniformed hybrid thin films. However, by increased the ratio SWCNTs as seen in figure 3(c), the

IOP Conf. Series: Materials Science and Engineering **209** (2017) 012030 doi:10.1088/1757-899X/209/1/012030

SWCNTs tubes are not individually stacked on rGO surface, but rather formed a clusters or bundles with minor aggregated morphology.

Based on observation with all these results, it can be conclude that rGO sheets were covered and connected with SWCNTs, and form a network structure.

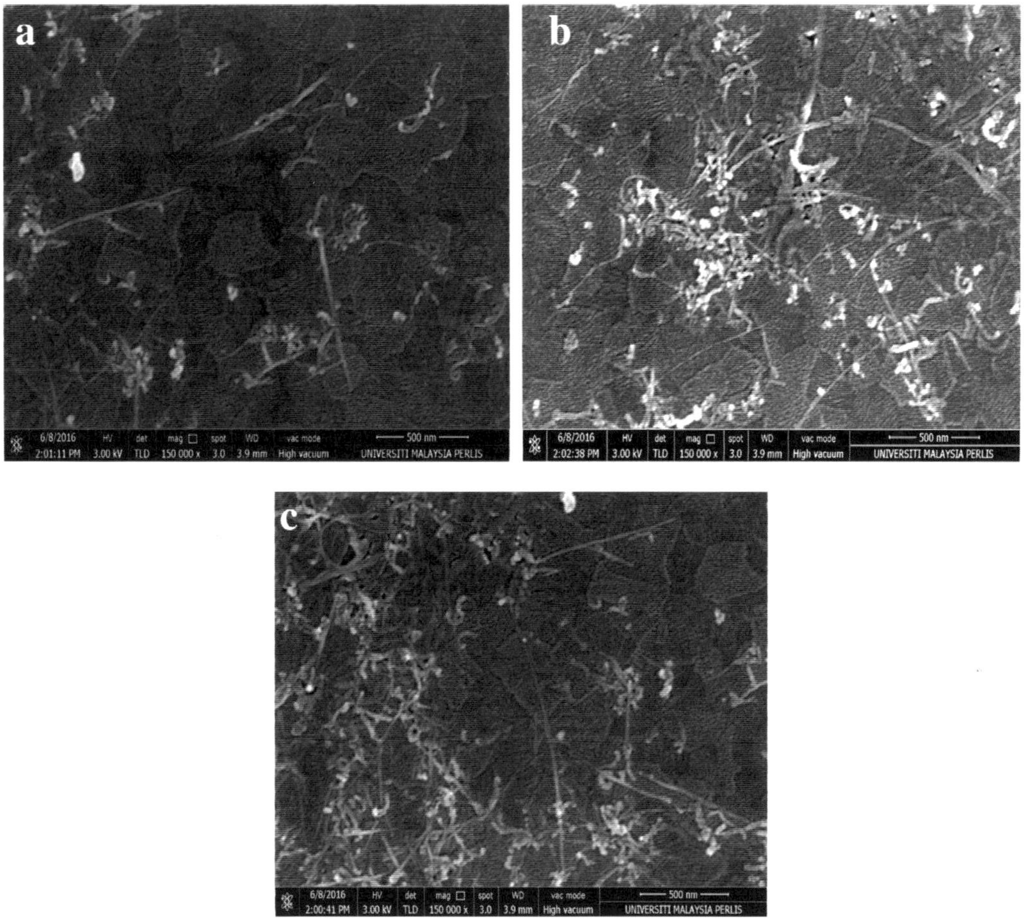

Figure 3. The micrograph of FESEM of G-SWCNT with different ratios of (a) 9:1, (b) 5:5, and (c) 1:9 ratios.

These results proved that SWCNTs were well intercalated onto rGO layer, which prevented the rGO from self-assembly restacking during thermal reduction process. On the other hand, these results also meet the agreement with previous research that was done by Chang et. al [9], which state that, the rGO-SWCNTs hybrid composite might constitute a more efficient conductive path for electrons, thereby improving the electrons transfer rate [9]. In addition, the presence of charging during the FESEM imaging suggested that the rGO-SWCNTs hybrid films were electrically conductive [10].

3.2. Raman Analysis

3.2.1. GO. Figure 4 shows the Raman spectra for GO. The formation of GO was further characterized by Raman spectroscopy in order to determine the well-ordered and disordered GO structures by evaluating the quantitative index of intensity D to G (I_D/I_G) ratio.

In general the spectra of GO consist of three peaks, namely, D mode ≈ 1350 cm^{-1}, G mode ≈ 1590 cm^{-1}, and 2D mode ≈ 2698 cm^{-1}. In particular, D mode corresponding to the defects in hexagonal carbon network and G mode arises due to the vibration of carbon-carbon bonds in basal plane [11].

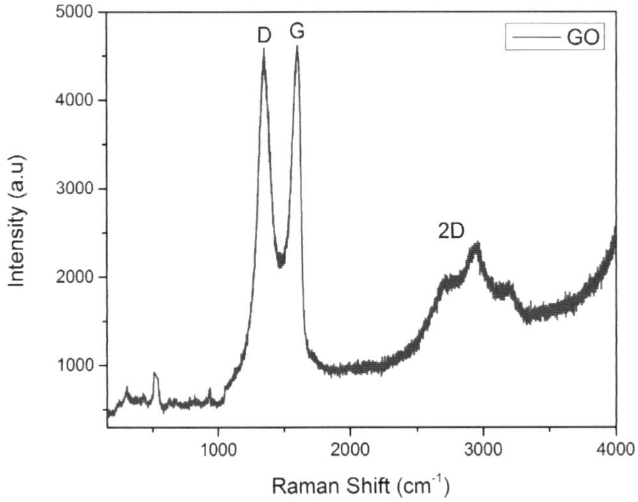

Figure 4. Raman spectra for GO.

Based on figure 4, it can be observed that D and G peak have a sharp peak, where indicates the presence of functional groups at the basal plane and edge of graphene sheets produced during the oxidation of graphite. However, the 2D peak was not reflected with the micrographs of FESEM since, the FESEM imaging and Raman spectroscopy does not directly correlate Raman peaks with monolayer GO topography. Thus, 2D peak resulted substantially lower and wider than G band, suggesting that the presence of multilayered graphene derivatives.

The intensity of I_D/I_G of GO is 0.994, which regarding to the defects site graphitic structural and distortion of carbon bonds due to reduction in the in-plane sp^2 carbon atoms caused by oxidation. These result also consistent with previous research by Feng et. al [12] which reported that the carbon lattice in GO has developed some degree of amorphous character due to oxidation process itself.

3.2.2. SWCNT. The Raman spectra of SWCNTs exhibits a few characteristic modes that can be used to determine the diameter of nanotubes, structural defects of SWCNTs and also vibration characteristics which called as radial breathing mode (RBM). The RBM band is originated from the out-of-plane tangential acoustic modes of monolayer graphene sheet and all carbon atoms vibrate in phase along the radial direction [13]. Figure 5 shows the Raman spectra for SWCNT-COOH. On the other hand, the Radial Breathing Mode (RBM) also gives precise information about the nanotube diameter which can be measure by equation:

$$d = \frac{223.75}{\omega_{RBM}} \quad (1)$$

Figure 5 shows the Raman spectra of SWCNT-COOH. The RBM band located at low Raman shifts in range 158 cm^{-1} ~ 269 cm^{-1} show the calculated diameter of SWCNT-COOH range from 0.87 nm to 1.47 nm. These results give significant evidence regarding to the theoretical and actual diameter of SWCNT. In addition, 2D band with sharp intensity peak located at 2689 cm^{-1} show the presence of single layer graphene sheet which roll and folded into a 1D structure. The G band had shown a highest intensity peak which located at range 1496 cm^{-1} ~ 1665 cm^{-1} while, the D band was appeared as small intensity peak at range 1227 cm^{-1} ~ 1425 cm^{-1}. The quality of the nanotubes was established by determining the ratio of I_D/I_G.

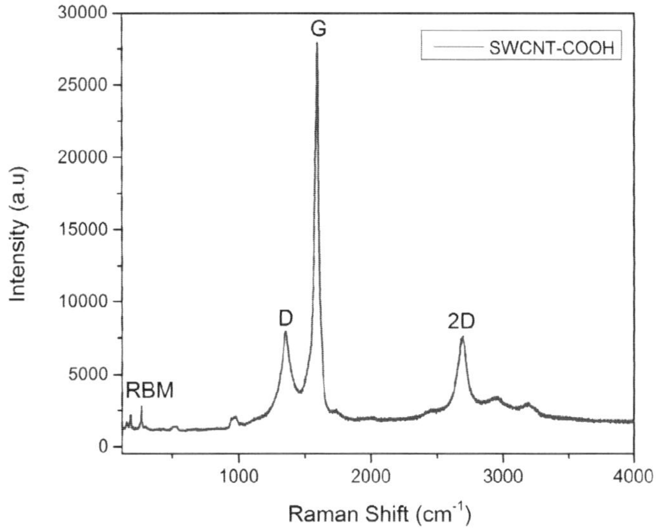

Figure 5. Raman spectra for SWCNT-COOH.

The average ratio of I_D/I_G is 0.284, which indicate that, the nanotubes had low ratio structural defects emanated from mild oxidation process. As revealed by these results, it can conclude that, D band appeared in Raman spectrum related to the indication of the covalent sidewall functionalization, as it reflects the conversion of the hybridization of some carbon atoms from sp^2 to sp^3 [14].

3.2.3. GO/rGO-SWCNT Hybrid. The intensity of I_D/I_G ratio also was measured in order to determine the density of structural defects hybrid material.

Figure 6 shows the Raman spectra of (a) GO-SWCNT and (b) rGO-SWCNT with different hybrid ratio which obtained at 514 nm laser excitation wavelength. According to figure 6(a) and figure 6(b), the intense band of D, G, and 2D peak can be observed at range 1183-1452 cm^{-1}, 1515-1667 cm^{-1}, and 2700-2900 cm^{1}, respectively.

Based on calculated I_D/I_G ratio from both spectra, it can be clearly observed that the intensity of I_D/I_G ratio shows the same pattern, which decreased from 0.88 to 0.77, from 0.76 to 0.75, and from 0.68 to 0.64 for hybrid material with ratio of 9:1, 5:5, and 1:9, respectively. As been discussed previously, oxidation from synthesizing and functionalization process will introduce functional groups where the disrupting of sp^2 bonding network as well as break the π-conjugated structure in the original hexagonal carbon structure.

However, after thermal reduction process, the functional groups that attached on the hybrid materials were majority removed thus, restoration from sp^3 to sp^2 bonding networks resulting from less defective site on carbon structure. According to both spectra, the ratio of 1:9 peak show the lowest I_D/I_G ratio compared to other peaks, which attributed from minimal defects at high SWCNTs loading.

The low structural defects and cluster forms of SWCNTs tubes on rGO sheets contribute to the uniform connection paths between SWCNTs and rGO sheets.

These results were consistent and well agreed with Shahriary et.al [15], which reported that the hybrid materials composed to the low D to G ratio emanated from influence of the incorporation SWCNTs on the rGO sheets.

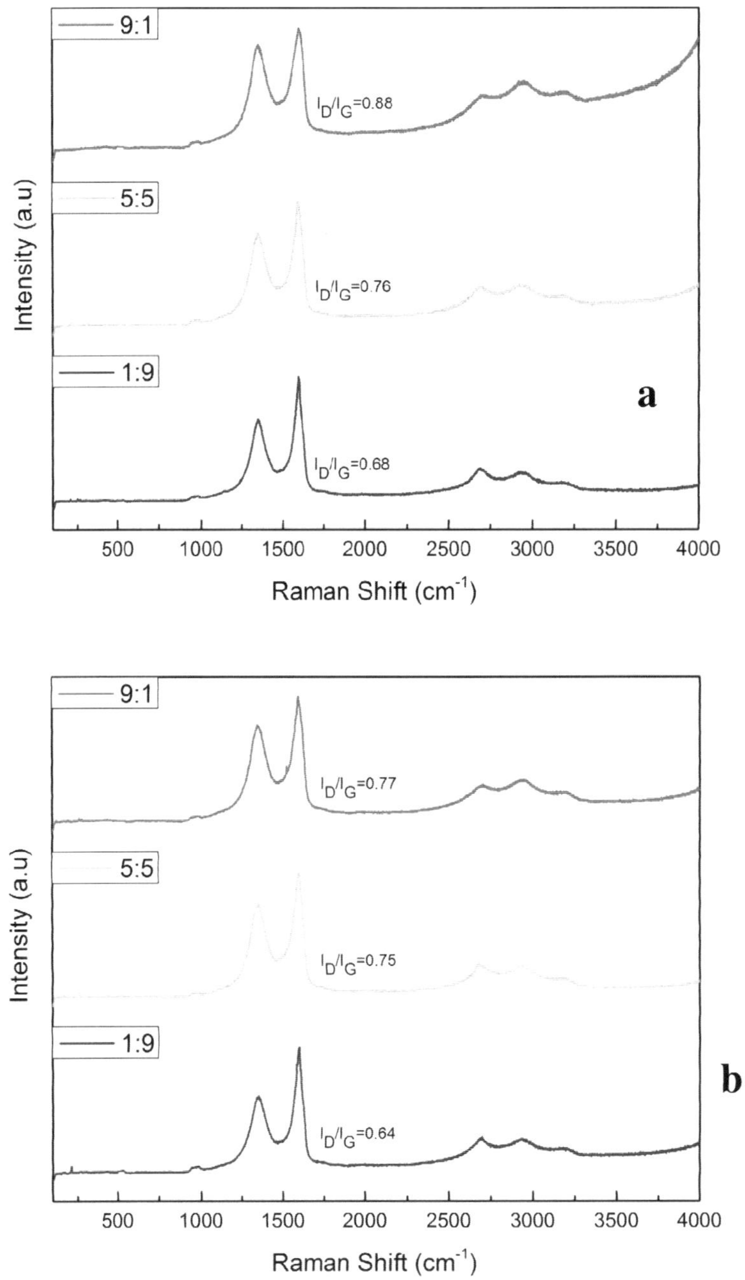

Figure 6. Raman spectra of (a) GO-SWCNT and (b) rGO-SWCNT with different hybrid ratio.

Due to connection between the rGO and SWCNT, the isolated aromatic domain on the surface rGO might cause a higher extension of the delocalized π-electrons.

4. Conclusions

In summary, graphene oxide nanosheets were successfully hybridized with functionalized single walled carbon nanotubes in deionized (DI) water dispersion and then thermally reduced to form the rGO-SWCNT composite. It was found that, the incorporation SWCNT into rGO sheets not only increased the electrochemically effective surface area, but also forms a better network structure with minimal agglomeration and restacking effects. On the other hands, rGO-SWCNT with mass ratio 1:9 shows the lowest defect rates which attributed from high loading of low defects density SWCNT. This finding also proved that, the rGO-SWCNT hybrid materials have a better structural morphology and less structural defects density compared to pristine rGO and SWCNT.

References

[1] Baughman R H, Zakhidov A A and De Heer W A 2002 *Science (80)* **297** 787–92
[2] Geim A K and Novoselov K S 2007 *Nat. Mater.* **6** 183–91
[3] Yen M-Y, Hsiao M-C, Liao S-H, Liu P-I, Tsai H-M, Ma C-C M, Pu N-W and Ger M-D 2011 *Carbon N. Y.* **49** 3597–606
[4] Fan Z, Yan J, Zhi L, Zhang Q, Wei T, Feng J, Zhang M, Qian W and Wei F 2010 *Adv. Mater.* **22** 3723–8
[5] Kim Y-S, Kumar K, Fisher F T and Yang E-H 2011 *Nanotechnology* **23** 15301
[6] Muda M R, Hanim K N, Isa S S M, Ramli M M and Jamlos M F 2015 High Throughput Graphene Oxide in Modified Hummers Method and Annealing Effect on Different Deposition Method *Applied Mechanics and Materials* vol 815(Trans Tech Publications)pp 141–7
[7] Schniepp H C, Li J-L, McAllister M J, Sai H, Herrera-Alonso M, Adamson D H, Prud'homme R K, Car R, Saville D A and Aksay I A 2006 *J. Phys. Chem. B* **110** 8535–9
[8] Goyanes S, Rubiolo G R, Salazar A, Jimeno A, Corcuera M A and Mondragon I 2007 *Diam. Relat. Mater.* **16** 412–7
[9] Chang L-H, Hsieh C-K, Hsiao M-C, Chiang J-C, Liu P-I, Ho K-K, Ma C-C M, Yen M-Y, Tsai M-C and Tsai C-H 2013 *J. Power Sources* **222** 518–25
[10] Wimalasiri Y and Zou L 2013 *Carbon N. Y.* **59** 464–71
[11] Kondratowicz I, Żelechowska K, Majdecka D and Bilewicz R 2015 *Mater. Sci.* **33** 292–300
[12] Feng H, Wang X and Wu D 2013 *Ind. Eng. Chem. Res.* **52** 10160–71
[13] Liu Z, Zhang J and Gao B 2009 *Chem. Commun.* **6902** 18
[14] Kim U J, Furtado C A, Liu X, Chen G and Eklund P C 2005 *J. Am. Chem. Soc.* **127** 15437–45
[15] Shahriary L and Athawale A A 2014 *Int J Renew Energy Env. Eng* **2** 58–63

Acknowledgments

This work was partially supported by Fundamental Research Grant (FRGS) and Research Acculturation Grant Scheme (RAGS) funded by Ministry of Education Malaysia. Authors would also want to thank School of Bioprocess, UNIMAP and Nanotechnology Research Centre, Faculty of Science and Mathematics, UPSI.

Surface Topographical Modification of Coronary Stent: A Review

C H Tan[1,2,a], **N Muhamad**[1,2,b] **and M M A B Abdullah**[3,2,c]
[1] School of Manufacturing Engineering, Pauh Putra Main Campus, Universiti Malaysia Perlis, Perlis, Malaysia
[2] Center of Excellence Geopolymer & Green Technology, School of Materials Engineering, Universiti Malaysia Perlis, Perlis, Malaysia
[3] Faculty of Engineering Technology, Uniciti Alam Campus, Universiti Malaysia Perlis, Perlis, Malaysia

E-mail: [a] jimmyhao91@gmail.com, [b] noorhafiza@unimap.edu.my

Abstract. Driven by the urge of mediating the inflammatory response from coronary stent implant to improve patency rates of the current coronary stent, concern has been focusing on reducing the risk of in-stent restenosis and thrombosis for long-term safety. Surface modification approach has been found to carry great potential due to the surface is the vital parts that act as a buffer layer between the biomaterial and the organic material like blood and vessel tissues. Nevertheless, manipulating cell response in situ using physical patterning is very complex as the exact mechanism were yet elucidated. Thus, the aim of this review is to summarise the recent efforts on modifying the surface topography of coronary stent at the micro- and nanometer scale with the purpose of inducing rapid in situ endothelialization to regenerate a healthy endothelium layer on biomaterial surface. In particular, a discussion on the surface patterns that have been investigated on cell selective behaviour together with the methods used to generate them are presented. Furthermore, the probable future work involving the surface modification of coronary stent were indicated.

1. Introduction

Cardiovascular diseases (CVD) are the leading cause of death globally, in 56 million deaths occurred worldwide during 2012, 17.5 million deaths was caused by cardiovascular diseases which are 31% of total deaths [1]. In other words, this statistic shows approximately one in every four deaths is caused by CVD worldwide. Therefore, currently, this issue is a global active issue as everyone is aiming on reducing the overall mortality from CVD. Generally, CVD is caused by the deposition of fatty material called plaque beneath the endothelium that comprises the innermost layer of a vessel, this condition called atherosclerosis [2,3]. This caused arterial narrowing and an inhibition of blood flow called stenosis. Currently, for stenosis cases a preferable treatment is angioplasty with stent placement; because this approach is the least invasive treatment after all compare to bypass surgery [4].

Stent is a scaffolding device that is introduced to cooperate with plain balloon angioplasty, in order to re-establish vascular patency and prevent elastic recoil of the vessel occur after plain balloon angioplasty that leads to restenosis [5]. However, as time passed it become apparent that bare metal stents too are restenotic due to intimal proliferation, thus ever since it has evolved to incorporate biological functionality in order to mitigate the risk of restenosis by inhibiting the Neointimal hyperplasia and decrease immune responses initiated by injury caused by stent implantation [4,6].

The invention of coronary stenting has transformed the treatment of coronary artery disease to be safer and ultimately save a lot of life or at least extent the patient life successfully. However, there were still a lot of room for improvement in order to overcome the current limitation. For example, although the introduction of Drug Eluting Stent (DES) has successfully reduced the risk of restenosis but due to the delayed healing that causes by the anti-proliferation drug, it also increases the risk of late stent thrombosis [7]. Therefore, this has led to rapid innovation over the past decade in order to optimise the stent performance.

This review aims to provide valuable insight into the recent work on enhancing the attachment and proliferation of endothelial cells by modifying the surface topography of coronary stent. In particular, a discussion on the techniques that has been explored to develop the surface topography and those surface topographies that has been investigated are presented to aid the future research directions. Prior to this, an overview of the rationale of physical surface modification to attain in situ endothelialization will be presented.

2. Rationale of surface topographical modification

Restenosis and late-stent thrombosis remain as the most challenging limitation after stent placement. Restenosis is the event of re-narrowing of blood vessel which occurs when the excessive proliferation of smooth muscle cells in the response to acute vessel wall injury induced by angioplasty and stenting [8]. Thrombosis is the event of blot clot on the surface of implanted medical device due to the surface electrical charge differences between stent surface and the blood elements, and other surface properties might contribute to this such as surface topology that determine the binding properties between the stent surface and blood cells [9]. Drug-eluting stent has been invented in order to overcome the restenosis issue by locally releasing anti-proliferation agent to control and inhibit the proliferation of smooth muscle cells [10]. While this has successfully reduced the risk of restenosis, the use of DES has caused the late-stent thrombosis risk increases due to the delayed healing of the vessel wall as the anti-proliferation drug also inhibit the proliferation of endothelial cells which is a desire for healing purposes [5,11–13]. Moreover, all patient with DES placement needs to undergo dual antiplatelet therapy, which has unwanted complication such as the risk of bleeding [14]. Thus, the current research direction has shifted from inhibiting the proliferation of smooth muscle cells as shown by DES to stimulate healing by enhancing the adhesion and proliferation of endothelial cells. A lot of research have proven the value of functional and intact endothelium in the prevention of restenosis and thrombosis [15–17]. Thus, the current priority was to modify the stent in term of function that will induce the adhesion and proliferation of healthy endothelial cells.

In conjunction, surface topographical modification attempt to alter the surface properties and equip the surface with specific functions while not disturb or damage the bulk material properties which normally is the valuable mechanical properties of the material. The specific functions aforementioned are referred to the surface that has better biocompatibility which should show selective cell growth on those favourable cells and simultaneously prevent the rejection response by the immune system on a foreign object. In other words, the surface topographical modification is capable of minimising the thrombogenicity of the material while stimulating vessel wall healing by enhancing the adhesion, proliferation and migration of endothelial cells.

Therefore, the current active area of research on improving the performance of stent was on manipulate the surface topography such as surface roughness, surface pattern and feature size in order to optimise the biological responses. This has led to the efforts on developing non-thrombogenic surface modifications that facilitate vascular wall healing in a manner that leads to the generation of an intact and functional endothelium layer. Despite the fact that various biomaterial like 316L stainless steel, AISI 310, nitinol or biodegradable material is publicly regarded to be biocompatible, it is still needed to modify its surface properties such as surface roughness and feature size in order to improve their performance as vascular stent materials [18,19].

3. Methods of surface topographical modification

Surface topographical modification should just intent on minimal alteration to the topmost layer of the surface without affecting the bulk material properties. There are a vast variety of methods to perform surface topographical modification with the ultimate aim of achieving better blood compatibility and rapid endothelialization can be grouped into 2 categories. The categories are surface roughening approach and surface patterning approach, which are illustrated in figure 1 below. In general, surface roughening means to roughen the stent surface by creating uneven structure across the surface structure randomly and unspecific as illustrated in figure 1(a). On the other hand, Surface patterning approach is a modify the surface with specific tailor patterns that are more organised and almost identical throughout the surface as illustrated in figure 1(b) and 1(c) which is the cross-section view of surface patterning with arrays of spikes and grooves respectively. Table 1 demonstrated those common surface topographical modification methods.

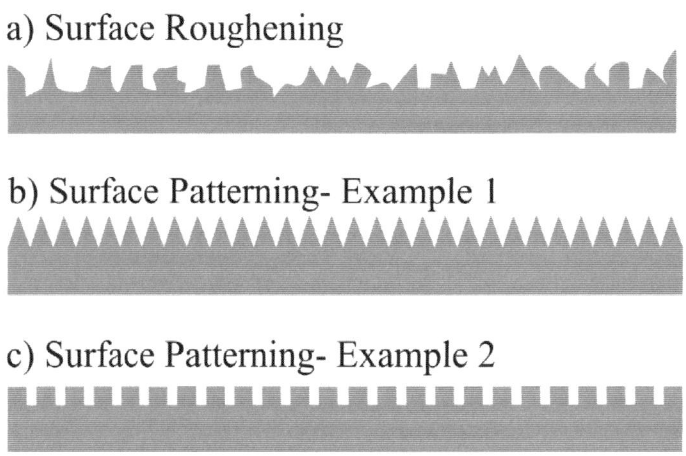

Figure 1. Illustration of the surface modification approaches.

4. Recent development of surface topographical modification

A vast variety of surface structures and methods have been explored to enhance the biocompatibility of coronary stent by developing nano/microstructure on the stent structure. Thus, this paper summarise the methods and the surface topographies that has been investigated and highlight the important and more recent studies. A summary table of all studies that have been reviewed in this paper is provided in 1.

In 2008, Lu et al. [20] speculated that rationally design surface structure with an elongated and aligned morphology which mimic the natural structure of a healthy vessel wall would enhance rapid endothelialization. Thus, they has introduced a novel plasma-based dry etching technique to modify surface structure into periodic arrays of grooves with high precision and reproducibility on titanium, and had compare the endothelial cell response on surfaces with the width of grooves equal to the ridge that is ranging from 750 nm to 100 µm [20].

In vitro study showed endothelial cell proliferated faster on the surface with the smallest feature size that is the grooves with 750 nm. Moreover, the cell guidance effect can be clearly noticed when the groove width is below 10 µm that show cellular alignment and elongated along the grooves. They also suggested that there is a potential for even smaller feature size on enhancing endothelial cell proliferation [20].

However, the effects of competitive adhesion and proliferation of smooth muscle cells should be assessed in order to draw a more conclusive finding. Besides, the surface modification of plasma-based dry etching process has altered not only the surface topology but also other surface properties such as the surface wettability and surface chemistry. Thus, it is hard to conclude that the smaller features have obvious effects on cell guidance from above studies.

Table 1. Fabrication methods employed to develop surface topographical features.

Approaches	Methods	Description
Surface Roughening - This method is to create a random surface structure to roughen surface that has low reproducibility and cannot form a specific surface pattern.	Chemical pickling [21]	Chemical pickling is an acid etching process, which can be used to roughen the surface and produce micropores on the surface when the acid solution is selected properly.
	Anodization [22]	An electrochemical process that used to coat the surface with an anodic oxide. With proper parameter control, nanotube surface structure can be developed.
	Sandpaper one-directional grinding [21]	Sandpaper grinding is being used to roughen surface by creating micro scratch on the surface.
	Chemical vapour deposition [23]	A chemical process often used in the semiconductor industry to produce high-quality thin films. With a proper parameter control, the surface structure generated could be in nanoscale or microscale.
Surface Patterning – develop well define surface structure with specific patterns. Normally have high reproducibility and precision.	Photo-lithography	Utilise the ultraviolet light to expose light sensitive photoresist and form patterns on the following the mask patterns then follow by direct etching to remove the substrate material then lastly the removal of the resist. Normally only at microscale level due to the limit of light diffraction.
	Electron-beam lithography and hot embossing [24]	Extent version of photolithography that utilises the high energy electrons to expose the electron sensitive resist. Widely used to create nanoscale surface features.
	Plasma dry etching [20]	Advanced micromachining techniques with high precision and reproducibility that can be used to create define and uniform patterns surface, popular in microelectronic fabrication. A Periodic array of microgrooves is one of the examples that can be produced using this method.
	Reactive ion etching [25]	A dry etching technology uses chemically reactive plasma to remove material. Widely used in microfabrication. Can be used to form predefine patterns with high precision.
	Laser Surface Engineering [19,25]	Nanosecond laser and femtosecond laser has been widely used to develop micro and nanostructure on the surface. Either direct write surface patterning or laser induces periodic surface structuring.

Therefore, in 2009, Shen et al. [21] had tried to study the solo effect of surface feature sizes on endothelialization by controlling the surface wettability and surface chemistry by using nano-coating approach. They developed micropores and microgrooves surface features on biomedical nitinol alloy by using chemical pickling approach and sandpaper one-directional grinding approach respectively and the combination of both approaches was used to create the surface structure that composes of both micropores and microgrooves [21]. They also found out that micro-pores surface showed better cell adhesion than the surface patterned with micro-grooves, while surface patterned with both micropores and micro-grooves showed even better cell adhesion among all [21]. Moreover, the plasma nano-coating that had enhanced the surface hydrophilicity also showed better cell adhesion results compared with the surface without nano-coating. In contrary, Aktas et al. [27] found that there was no correlation between the wetting behaviour of substrate and cell adhesion behaviour.

Likewise, in 2012, Aktas et al. [27] highlighted the need to investigate the effects of each individual surface properties separately on endothelialization, this is because normally the surface modification technique not only alters the surface topography but also made changes to the surface chemistry and/or surface free energy at the same time. Thus, they has developed Al_2O_3 nanowires surface structure of different density by using chemical vapour deposition technique and investigated the selecting cell behaviour of both endothelial cells and smooth muscle cells [27]. They claimed that the chemical vapour deposition technique have the ability to develop surface topography of different roughness with the identical surface chemistry and wetting behaviour [27]. The nanowires having diameters in the range of 20 to 30 nm, the two surfaces density developed were having an average roughness (R_a) and maximum roughness (R_{max}) of 21 and 187 nm; and 29 and 272 nm respectively. Through the in-vitro studies, Aktas et al. [27] found that microstructured Al_2O_3 plates have better cell growth compare to nanostructured Al_2O_3 for both endothelial cells and smooth muscle cells. However, the low-density nanowires structured Al_2O_3 demonstrated the desired cell selective behaviour, which is favouring the adhesion and proliferation of endothelial cells while inhibiting the growth of smooth muscle cells. Since the low-density nanowires structure is showing lower aspect ratio when compare with high-density nanowires structure. Thus, Aktas et al. [27] suspected that the endothelial cells are favouring lower aspect ratio surface structure while smooth muscle cells are favouring high aspect ratio surface structure. This finding suggesting that surface topography plays an important role in manipulating the cell responses.

On the other hand, since the stent will be contacted with blood inside the vessel thus it is important to investigate the effect of the surface feature's changes onto blood cells too and compare its adhesion rate with the healing endothelial cells. Thus, in 2010, Csaderova et al. [24] has demonstrated that the nano-pillars surface structure on poly-Ɛ-caprolactone (PCL) has a strong cell-selective effect on two different cell lines, which inhibiting the proliferation of fibroblast while enhancing the proliferation of the endothelial cell. They suggested that the cell-selective effect is caused by the reduction of available surface area generated by nano-pillars [24]. The nano-pillars with a diameter of around 105 nm, a depth around 116 nm and a pitch of 300nm were fabricated on silicon master by utilising the technique of electron-beam lithography and transferred to PCL substrate by a hot embossing technique.

Furthermore, in 2014, Lee et al. [12] has investigated the effects of surface morphology of nanotubular structure on cell selecting behaviour which include the cell spreading, proliferation and migration. These surface features were created through anodization onto Nitinol biomaterial as a coating. This nanotubular surface structure has shown it could enhance re-endothelialization by improving the spreading and migration of endothelial cells and diminish the neointimal hyperplasia by decreasing the proliferation, ECM production, and migration of smooth muscle cells. In addition, the nanotubular surface did not induce inflammation by observing the ICAM-1 expression in smooth muscle cells as compare to a flat surface. However, there was no proliferation of endothelial cells shown on the nanotubular structure by analysing the cell numbers. Thus, Lee et al. [22] suggested that further studies on optimising the diameter of the nanotube can be conducted in the future in order to improve the endothelial cells proliferation. Although the surface structure generated in this studies is not exactly the same with the surface structure developed by Csaderova [24] as mention above, but it clearly showing the same cell response which can be deduced that the nanotube-like structure does improve endothelialization. However, more studies have to be done on comparing this surface features with those with an array of grooves patterns in order to elucidate more conclusive results.

Moreover, at the same year, Vandrangi et al. [25] intended to fill the knowledge gap of how endothelial cells response towards with micrometre to sub-micrometer groove-based feature on titanium and silicon substrate. Thus, they employed the Ti Deep Reactive Ion Etching method and conventional Silicon micromachining techniques to develop periodic groove feature on titanium and silicon substrate respectively [25]. A highly uniform periodic groove array were precisely formed with nearly rectangular groove profiles that have groove width ranging from 0.5 to 50 μm with a groove depth of 1.3 μm and groove pitch equal to twice the groove width. In the thorough in-vitro studies, Vandrangi et al. [25] found that endothelial cells have shown favourable response on cell proliferation, adhesion, function, and morphology with decreasing feature size on both patterned titanium and silicon substrate just that the titanium substrate is showing better cell response over silicon substrate at comparable feature size. Although this study has shown the most comprehensive in-vitro study on the effects of patterned substrate produced so far, it does suffer from a major drawback that it does not take account of smooth muscle cells nor does examine the cell response of smooth muscle cell compare to endothelial cells. This is very important in order to show the cell selective capability of the patterned surface features as the desired cell response is to inhibit the growth of smooth muscle cells will enhancing the proliferation of endothelial cells.

On the other hand, laser surface modification approach had been widely employed to develop surface topography with cell selective behaviour. Generally, those studies employed laser surface modification approach can be divided into two categories that are a direct-write modification and self-organizing effects modification. Direct-write modification approach is almost like laser milling which the surface patterns output is greatly depends on path of laser movement the generate defined pattern, while the self-organizing effects modification approach is an implicit process that the surface structure is being altered by using ultrashort pulse laser that normally develops induced surface structure where the dimension of the patterns is independent of the focused laser spot size such as cone-like protrusions and laser surface induced periodic structure.

The study of the endothelial cells' behaviour on surface features created by using nanosecond laser was first carried out by Li et al. [26] in 2013. They demonstrated that there was a significant increase in adhesion and proliferation of endothelial cells on laser treated surface with defined surface pattern and surface chemistry [26]. Besides, they showed that surface chemistry is one of the factors that have great influence on cell behaviour as from the in-vitro study showed when nitrogen assisted gas is employed during the laser surface structuring process, there were obvious changes in the surface chemistry and which ultimately contribute to the significant effects on cell response [26]. Similarly at the same year, Oberringer et al. [28] reported the use of femtosecond laser on creating direct write patterns of laser pulse spots array on 316L stainless steel which composed of micro and nano features and investigated the cell-surface interaction of endothelial cells and fibroblast with the ultimate aim of reducing the risk of thrombosis. They demonstrated that the surface with more nanofeatures had better performance on decreasing the proliferation of myofibroblast proliferation while not affecting the proliferation of the endothelial cell [28]. This study has led the priority focus of future study to focus on nanoscale features.

While for laser-induced surface structures, in 2015, Nozaki et al. [29] demonstrated that the nitinol surface composed of both periodic micro and nanostructures that were induced by femtosecond laser was capable of improving the morphology of endothelial cells and inhibit the adhesion of platelets cells. They showed that there was a contradiction on the surface wettability between the periodic microstructured surface and periodic nanostructured surface that induced by femtosecond laser [29]. Where the Nanostructures surface was hydrophilic and have successfully improved the morphology of endothelial cells, while the micro/nanostructures surface was hydrophobic and successfully prevent the adhesion of platelets cells. Similarly, in 2016, Liang et al. [19] has investigated on how to enhance rapid in-situ re-endothelialization of 316L stainless steel coronary stent by altering the surface features using femtosecond laser with the goal of mimicking the native micro and nano structure of vascular smooth muscle cell (SMC) wall. The biomimetic stent has been tested in-vivo in the iliac artery of the rabbit and the results showed that the bionic surface patterns do regulate the growth orientation and enhance the adhesion, proliferation, and migration of endothelial cells through cell guidance effect. Liang et al. [19] have demonstrated that the microgrooves of around 24 μm do enhance the stability of HUVECs on the surfaces by controlling the cell growth orientation, and the nanometer surface fibrillary structure of about 700 nm have guides the migration of the HUVECs. In the other words, this patterned surface has the potential for enhancing the re-endothelialization process and ultimately reduce the incident of restenosis and thrombosis. As this result has also been supported by McDaniel et al. [30], where the laser-induced surface structure by femtosecond laser was able to reduce the adhesion of monocytes cells.

Table 2. Summary of previous works on developing surface topographical features.

Methods	Substrate	Pattern & Dimension	Description	Author (year) [ref.]
Chemical vapour deposition (CVD) with nanosecond laser treatment	Al_2O_3	Nanostructure with micro-craters. Micro-craters diameter: 45-50 µm with 100 nm depth and the spacing between craters is 150 µm.	Chemical vapour form nanostructures and nanosecond laser form micro-craters. Both surface structure managed to promote the adhesion and proliferation of endothelial cells and successfully reduced the adhesion of smooth muscle cells.	K Kiefer et al. (2016) [23]
Ti deep reactive ion etching (Ti DRIE)	Silicon and Titanium	Groove-based topography (micron to submicron). Groove widths: 0.5-50 µm, depth: 1.3 µm, pitch: twice the groove width	Successfully demonstrated the significant improvement in cellular adhesion, proliferation and morphology with decreasing feature size on patterned Ti substrates	P Vandrangi et al. (2014) [25]
Anodization	Nitinol	Upright nanotubular structure. Outer Diameter ≈ 110 nm	Successfully demonstrated the increased of endothelial cell spreading and migration and decreased of smooth muscle cell proliferation.	P P Lee et al. (2014) [22]
Chemical Vapour Deposition	Al_2O_3	Nanostructured surface with nanowires feature Nanowire diameter: 20-30 nm	Low-density nanowires structured surface have cell selective behaviour of endothelial cells over smooth muscle cells.	C Aktas et al. (2012) [27]
Electron-beam lithography and hot embossing	Poly-Ɛ-caprolactone	Nano-pillars Diameter: 105±5 nm, depth: 116±5 nm, pitch: 300 nm	The nano-pillars structures have shown cell selective behaviour of enhancing endothelial cells proliferation and inhibiting the proliferation of fibroblast.	L Csaderova et al. (2010) [24]
Chemical Pickling, Sandpaper grinding and plasma nano-coating	Nitinol	Micro-pores, micro-grooves, Surface roughness, Ra: 10 to 600 nm	Better cell adhesion on feature with both structure and with better hydrophilicity.	Y Shen et al. (2009) [21]
Plasma-based dry etching	Titanium	Mimic native vessel endothelial cell wall Periodic arrays of grooves. Groove and ridge: 750 nm to 100 µm	Nanometer- scale grooves structure has shown better performance on endothelialisation compares to micrometer-scale grooves structure as well as the random nanostructure control.	J Lu et al. (2008) [20]
Femtosecond laser	316LSS	Mimic native SMC structure: Microgrooves (≈24 µm) and nanometer fibrillary structure (≈700 nm).	Successfully promote adhesion, proliferation and migration of endothelial cells on the bionic surface with patterns of vascular smooth muscle cells wall.	C Liang et al. (2016) [19]
	Platinum stainless steel (Pt: SS) and 316LSS	Laser-induced periodic surface structure (LIPSS) with roughness ranging from 29 nm to 50 nm.	Alter surface topology with chemical composition together. LIPSS successfully decreased the adhesion of monocytes that shows it's capable on reducing thrombosis occurring.	C McDaniel et al. (2015) [30]
	Nitinol	Laser-induced periodic surface structure (LIPSS), Hierarchical periodic nanostructures and micro/nanostructures surface.	Nanostructures surface was hydrophilic and has successfully improved the morphology of endothelial cells. While the micro/nanostructures surface was hydrophobic and successfully prevent the adhesion of platelets cells.	K Nozaki et al. (2015) [29]
	316LSS	Array of pulsed laser scan with different hatch distance, composed of micro and nanostructured surface.	Successfully decrease the proliferation of myofibroblast on laser treated surface while there are no significant effects on endothelial cells proliferation.	M Oberringer et al. (2013) [28]
	Stainless steel 304	Laser-induced periodic structure with linear and circular polarisation.	Successfully demonstrated that the laser treated surfaces have better biocompatibility than untreated surface.	C Y Lin et al. (2012) [31]
Nanosecond Laser	316LSS	Array of pulsed laser scan with different hatch distance, composed of micro and nanostructured surface.	Manage to alter the surface topography and surface chemistry in a single step with the help of nitrogen assist gas that has shown improvement in endothelialisation	L Li et al. (2013) [26]

5. Conclusions and future prospects

Overall, these reviews highlight the viability of manipulating cells-substrate response by surface topographical modification. Despite intensive research, the ideal stent surface topography with optimal cell response remain to be achieved. There were an obvious contradiction in the literature regarding the surface features dimension as some suggesting that nanoscale features were better than microscale features on modulating the selective cellular behaviour while others suggesting the other way round, thus further investigation is needed to refine the optimum features dimension and structure. From the literature, those bionic surface that mimics the native vascular wall is the quite promising approach, but either mimicking the endothelial cell wall or the smooth muscle cell wall has greater performance has to investigate further. Besides, in term of surface topography, many studies have investigated the width of groove and ridge, however, no study has been done on investigating the effects of the groove depth and the groove profile. In addition, the effect of surface topography on cell selective response have to be studied further on other widely used biomaterials of stent to proof its viability.

Besides, although plenty of surface topographical modification methods has been assessed on developing a better stent surface, the ideal surface topography method yet to be identified. This is because, almost all modification methods that alter the surface features is always accompanied with other surface properties changes such as surface chemistry, surface wettability, surface charge distribution etc. Thus, it is very difficult to investigate and elucidate the independent contribution of surface topography effects on collective cell growth. Some literature suggesting that applying a coating on top of different surface topography were able to isolate the surface topography effects by homogenised others surface properties. Thus, further research is needed in this particular area. Moreover, regarding the cell culture and assays, other than the responses of endothelial cells and smooth muscle cells, it is a must to investigate the thrombogenicity of the surface topography in order to demonstrate the cells-substrate response in details. Lastly, in vivo study is crucial as the complex in vivo environment may induce different result as in vitro study.

References

[1] Mendis S, Armstrong T, Bettcher D, Branca F, Lauer J, Mace C, Poznyak V, Riley L, Silva VDCE, Stevens G and Kwok CT 2014 *World Health Organization*
[2] Tomkin GH and Owens D 2012 *Atheroscler Thromb J* **5** 13–21
[3] Ţugui C A, Nejneru C, Găluşcă D G, Perju M C, Axinte M, Cimpoeşu N, Vizureanu P 2015 *Journal of Optoelectronics and Advanced Materials* **17**(11-12) 1855 – 1861
[4] Tammareddi S, Sun G and Li Q 2016 *Mater Des* **90** 682–92
[5] Jeewandara T, Wise S and Ng M 2014 *Materials (Basel)* **7** 769–86
[6] Zhang K, Liu T, Li J-A, Chen J-Y, Wang J and Huang N 2014 *J Biomed Mater Res Part A* **102** 588–609
[7] Chong DST, Lindsey B, Dalby MJ, Gadegaard N, Seifalian AM and Hamilton G 2014 *Eur J Vasc Endovasc Surg* **47** 566–76
[8] Bhatia V and Kaul U 2008 In: Medicine update 2008 p. 70–5.
[9] Kereiakes D J, Yeh R W, Massaro J M, Driscoll-Shempp P, Cutlip D E, Steg P G, Gershlick A H, Darius H, Meredith I T, Ormiston J, Tanguay J-F, Windecker S, Garratt K N, Kandzari D E, Lee D P, Simon D I, Iancu A C, Trebacz J and Mauri L 2015 *JACC Cardiovasc Interv* **8** 1552–62
[10] Iqbal J, Gunn J and Serruys PW 2013 *Br Med Bull* **106** 193–211
[11] Giacoppo D, Gargiulo G, Aruta P, Capranzano P, Tamburino C and Capodanno D 2015 *BMJ* **351** h5392
[12] Grabow N, Martin DP, Schmitz K-P and Sternberg K 2009 *J Chem Technol Biotechnol* **85** 744–51
[13] Lüscher T F, Steffel J, Eberli F R, Joner M, Nakazawa G, Tanner F C and Virmani R 2007 *Circulation* **115** 1051–8
[14] Mohammad R A, Goldberg T, Dorsch M P and Cheng J W M 2010 *Clin Ther* **32** 2265–81

[15] Ren X, Feng Y, Guo J, Wang H, Li Q, Yang J, Hao X, Lv J, Ma N and Li W 2015 *Chem Soc Rev* **44** 5680–742
[16] Pang J H, Farhatnia Y, Godarzi F, Tan A, Rajadas J, Cousins B G and Seifalian A M 2015 *Small* **11** 6248–64
[17] Liang C, Hu Y, Wang H, Xia D, Li Q, Zhang J, Yang J, Li B, Li H, Han D and Dong M 2016 *Biomaterials* **103** 170–82
[18] Istrate B, Mareci D, Munteanu C, Stanciu S, Crimu C I, Trinca L C, Kamel E 2016 *Environmental Engineering and Management Journal* **15**(5) 955-963
[19] Minciuna, M G, Vizureanu P, Hanganu C, Achitei D C, Popescu D C, Focsaneanu S C 2016 *Materials Science and Engineering* **133**(1) 28
[20] Lu J, Rao M P, MacDonald N C, Khang D and Webster T J 2008 *Acta Biomater* **4** 192–201
[21] Shen Y, Wang G, Chen L, Li H, Yu P, Bai M, Zhang Q, Lee J and Yu Q 2009 *Acta Biomater* **5** 3593–604
[22] Lee P P, Cerchiari A and Desai T a 2014 *Nano Lett* **14** 5021–8
[23] Kiefer K, Akpınar G, Haidar A, Ikier T, Akkan C K, Akman E, Lee J, Miró M M, Kaçar E, Demir A, Veith M, Ural D, Kasap M, Kesmez M, Abdul-Khaliq H and Aktas C 2016 *RSC Adv* **6** 17460–9
[24] Csaderova L, Martines E, Seunarine K, Gadegaard N, Wilkinson C D W and Riehle M O 2010 *Small* **6** 2755–61
[25] Vandrangi P, Gott S C, Kozaka R, Rodgers V G J and Rao M P 2014 Pisignano D, editor. *PLoS One* **9** e111465
[26] Li L, Mirhosseini N, Michael A, Liu Z and Wang T 2013 *Lasers Surg Med* **45** 608–16
[27] Aktas C, Dörrschuck E, Schuh C, Miró M M, Lee J, Pütz N, Wennemuth G, Metzger W, Oberringer M, Veith M and Abdul-Khaliq H 2012 *Mater Sci Eng C* **32** 1017–24
[28] Oberringer M, Akman E, Lee J, Metzger W, Akkan C K, Kacar E, Demir A, Abdul-Khaliq H, Pütz N, Wennemuth G, Pohlemann T, Veith M and Aktas C 2013 *Mater Sci Eng C* **33** 901–8
[29] Nozaki K, Shinonaga T, Ebe N, Horiuchi N, Nakamura M, Tsutsumi Y, Hanawa T, Tsukamoto M, Yamashita K and Nagai A 2015 *Mater Sci Eng C* **57** 1–6
[30] McDaniel C, Gladkovskaya O, Flanagan A, Rochev Y and O'Connor G M 2015 *RSC Adv* **5** 42548–58
[31] Lin C Y, Cheng C W and Ou K L 2012 *Phys Procedia* **39** 661–8

Acknowledgments
The authors gratefully acknowledge School of Manufacturing Engineering, UniMAP for their expertise and support. The authors would also like to thank for the funding support from the Fundamental Research Grant Scheme (FRGS-9003-00584) under Ministry of Education Malaysia (MOE).

IOP Conf. Series: Materials Science and Engineering **209** (2017) 012032 doi:10.1088/1757-899X/209/1/012032

Morphological Analysis (SEM) of the Surface of a Non-Noble Dental Alloy Subjected to Electrocorrosion

E R Baciu[1], I Grădinaru[1], M Baciu[2], R I Vasluianu[1], R Cimpoeşu[2], C Baciu[2]* and C Bejinariu[2]

[1]Department of Dental Materials, Faculty of Dental Medicine, "Gr.T.Popa" University of Medicine and Pharmacy of Iaşi

[2] Faculty of Materials Science Engineering, "Gheorghe Asachi" Technical University of Iaşi

E-mail: constantin_baciu@yahoo.com

Abstract. Corrosion consists in the degradation of a material under the chemical or electrochemical action of the environment where it is placed. The investigations carried out aimed to show the structural modifications produced in Co-Cr-Mo alloy, Robur 400 (Eisenbacher Dental – Waren ED GmbH, Germany) subjected to electrocorrosion in Fusayama-Mayer artificial saliva. The specimens prepared by mechanical polishing were analysed structurally by using a scanning electron microscope. During the tests run we could notice a general corrosion of the surfaces of the specimens made from Robur alloy. Through 2D and 3D microscopy and qualitative determinations of the luminous variation we could notice the effects of electrocorrosion tests on the surface of the metal material.

1. Introduction

In the1970s, as the price of gold was going up significantly, a cheaper alternative, namely the cobalt-chrome alloys, was introduced in the market of dental materials. Thanks to the modern technology of vesting and polishing, cobalt-chrome-based dental alloys are widely used today in dental prosthetics to make skeletonized removable partial dentures. These alloys also have indications for the making of fixed restorations [1].

Co-Cr-based alloys represent an advantageous alternative as compared to the Ni-Cr-based alloys due to their biocompatibility and the lack of allergic reactions that the latter may develop. At the same time, Co-Cr-based alloys have an increased hardness as they are very resistant to corrosion due to chrome which forms a protective oxide coating on the alloy surface [2,3,4,5].

Since metal biomaterials are subjected to an intense corrosive attack after they have been introduced in the environment of the human body, the resistance to corrosion represents an important quality of them.

2. Goal

The goal of this study consists in determining the structural modifications taking place in Co-Cr-Mo alloy, Robur 400 (Eisenbacher Dental – Waren ED GmbH, Germany) used in the metal components of the skeletonized removable dentures following electrocorrosion in Fusayama-Mayer artificial saliva.

3. Materials and method

The semi products from non-noble dental alloy under study were purchased from the manufacturing companies and they have values of the chemical composition, physical and mechanical properties in accordance with the product's technical sheet.

As for the chemical composition, we may notice that the material exhibits the features of a complex alloy as it has a high number of alloying elements (table 1).

Table 1. Chemical composition of the non-noble dental alloy under study [6].

Alloy trade name	Chemical composition, [%]											Technical norm
	Co	Ni	Cr	Mo	Ti	Nb	Al	W	Si	Mn	C	
Robur 400	~62	-	29.10	5.85	-	-	-	0.72	0.48	0.57	0.52	DIN EN ISO 22674:2007

For the experimental researches we made specimens from the alloy mentioned above in the shape of cylinders that were mounted in Teflon holders (figure 1).

Figure 1. Robur alloy.

The specimens prepared this way were analysed structurally. For this we used an electron scanning microspore. Then they were subjected to electrocorrosion tests in Fusayama-Mayer artificial saliva (table 2). To identify the structural modifications taking place in the alloy following the corrosion process, we repeated the microscopy tests.

Table 2. Electrolytic solution of artificial saliva for electrocorrosion tests.

Artificial saliva Compounds	NaCl g/l	KCl g/l	$NaHCO_3$ g/L	Urea g/L	$CaCl_2$ g/L	NaH_2PO_4 H_2O g/L	Na_2S g/L
Fusayama	0.4	0.4	0.69	1	0.65	0.69	0.05

The pH of the solution was measured by a multi-parameter analyser of CONSORT 831C type, and the pH value of Fusayama Mayer saliva was 7.

4. Results and discussions

Robur experimental alloy was polished before the corrosion resistance test by means of metallographic paper having granulation between 200 and 2000. The surface of the material is given in figure 2.

We may notice a poor highlighting of the microstructure, figure 2 a), and we mention that no chemical attack was performed on the surface of the metal alloy.

In figure 2 b) w present the 3D state of the alloy surface after polishing with a slight difference, below 3-4 μm, between the dendritic phase, which is a level higher and it is characteristic to the alloy, and the rest of the metal matrix. The difference occurred due to the different response the two phases had to the mechanical polishing stress.

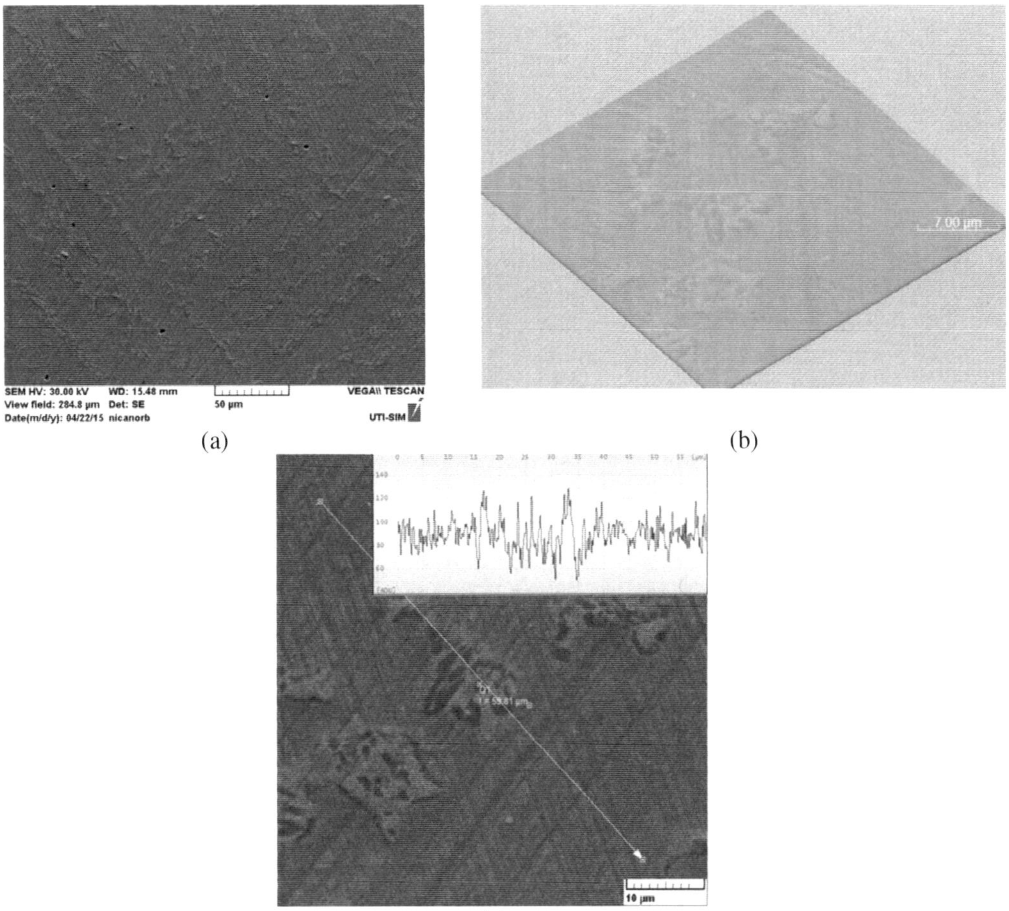

Figure 2. SEM microscopies of Robur alloy before the electrocorrosion test a) 2D, b) 3D and c) variation of luminous intensity on surface.

In figure 2 c) w present a variation of the luminous intensity on the surface of the metal material given by the difference of composing phases of the alloy and obtained by microscopy based on VegaTescan analysis software and the registered spectrum of secondary electrons. We may notice small variations of the luminous intensity on the surface ranging between 50 and 70 ADU (adimensional units).

After having run the electrocorrosion test in the experimental conditions specified above, the surface of the corroded material is presented in fig.3. We may notice a serious and generalized corrosion of the material, figure 3 a), with the formation of cracked films of material, figure 3 b). The dendritic compounds remained stable, figure 3 b), and the cracked pellicular corrosion product formed only on the rest of the metal matrix. The cracking of the pellicular coating is general and the coating is relatively stable when rinsed under water.

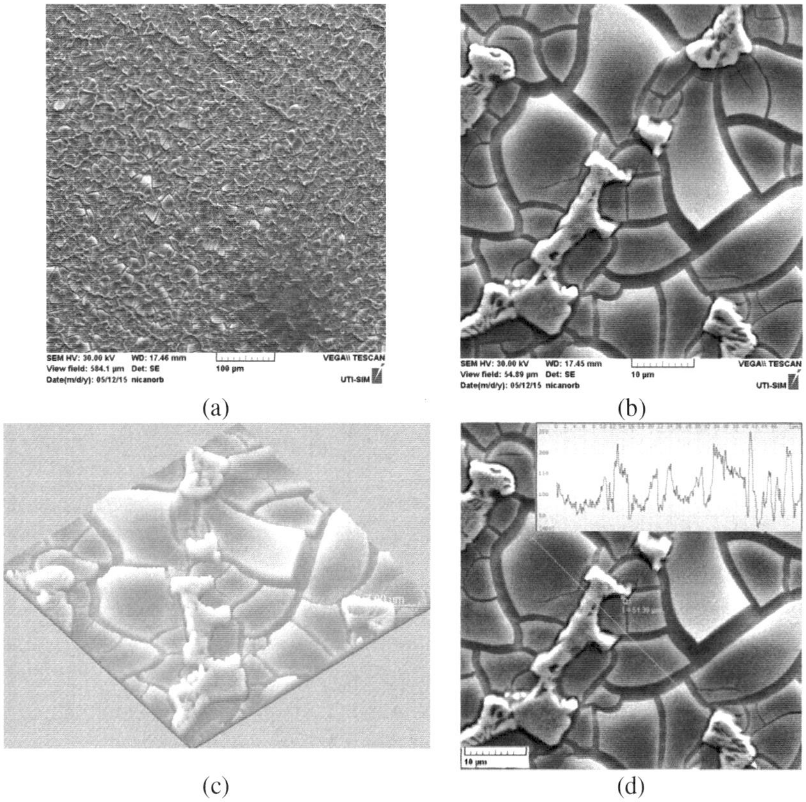

Figure 3. Images of Robur alloy surface after the corrosion test, a) 2D – 500x, b) 2D – 5000x, c) 3D and d) variation of the luminous intensity.

The state of the surface after the formation of the corrosion compounds is presented also through 3D micrography in figure 3 c), where we may see a layer 2-3 μm thick and its tendency of exfoliation from the surface of the basic metal material. From the variation of the luminous intensity in figure 3 d), also confirmed by the 3D image of the surface, we may notice that the layer formed on the surface of the metal matrix is smaller than the remained dendritic phases highlighted while preparing the specimen by polishing. In this case, variation was about 200 ADU (adimensional units).

5. Conclusions

We analysed a Co-Cr-Mo-based dental alloy with applications in the field of dental prosthetics, before and after an electrocorrosion test run in electrolytic solution of Fusayama artificial saliva.

The specimens from the selected alloy prepared by mechanical polishing are characterized by two chemically different phases, one being the metal matrix, and the other of dendritic type.

After the running of tests, we noticed a general corrosion of the surface of the metal material, Robur alloy presenting a stable layer on the surface with numerous fissures and exfoliations mainly made of oxides.

Through 2D and 3D microscopy and qualitative determinations of the luminous variation, we could see the effects of the electrocorrosion tests on the surface of the metal material.

References

[1] Rincic N, Baucic I, Miko S, Papic M and Prohic E 2003 *Coll. Antropol.* **27** 99
[2] Giacomelli F C, Giacomelli C, Spinelli A 2004 *J. Braz. Chem. Soc.* **15** 541

[3] Cimpoeşu N, Trincă L C, Dascălu G, Stanciu S, Gurlui S O, Mareci D 2016 *Journal of Chemistry* Article ID 9520972, http://dx.doi.org/10.1155/2016/9520972
[4] Hsu R W, Yang C, Huang C, Chen Y 2005 *Mater. Chem. Phys.* **93** 531
[5] Metikoš-Huković M, Babić R 2007 *Corros. Sci.* **49** 3570
[6] Roberts H W, Berzins D W, Moore B K 2009 *J. Prosthodont.* **18** 188
[7] Izquierdo J, Bolat G, Cimpoesu N, Trinca L C, Mareci D, Souto R M 2016 *Applied Surface Science* **385** 368
[8] Baciu E R, Grădinaru I, Baciu M, Forna N C 2012 Considerations regarding the balance structure of Co-Cr-Mo alloys for removable partial denture, *Buletinul Institutului Politehnic din Iaşi, Tomul* LVIII (LXII) Sectia Matematică. Mecanică Teoretică. Fizică, 93
[9] Chelariu R, Suditu G D, Mareci D, Bolat G, Cimpoesu N, Leon F, Curteanu S 2015 *JOM*, DOI: 10.1007/s11837-015-1362-3
[10] Mareci D, Cimpoesu N, Popa M I 2012 *Materials and Corrosion* **63** 176

IOP Conf. Series: Materials Science and Engineering **209** (2017) 012033 doi:10.1088/1757-899X/209/1/012033

Microstructural Analysis of Ti-Based Shape Memory Alloys Following the Electrochemical Corrosion in Artificial Saliva

C Baciu[1], E R Baciu[2]*, R Cimpoeşu[1], C G Levente[3], D G Bosinceanu[4], M Baciu[1] and C Bejinariu[1]

[1]Faculty of Materials Science Engineering, "Gheorghe Asachi" Technical University of Iaşi

[2]Department of Dental Materials, Faculty of Dental Medicine, "Gr.T.Popa" University of Medicine and Pharmacy of Iaşi

[3]Department of Endodontics, Faculty of Dental Medicine, "Gr.T.Popa" University of Medicine and Pharmacy of Iaşi

[4]Department of Total prosthesis, Faculty of Dental Medicine, "Gr.T.Popa" University of Medicine and Pharmacy of Iaşi

E-mail: elena.baciu@umfiasi.ro

Abstract. The investigations carried out aimed to highlight the structural modifications occurred in the Ti-based shape memory alloys subject to electrocorrosion in Afnor artificial saliva. The behavior to corrosion was highlighted by fast electrochemical tests, mainly by dynamic potentiometry. From the microstructural analysis we noticed that the specimens of the two Ti-based shape memory alloys show traces of "pitting" corrosion on their surface of diverse sizes, a fact that will raise issues in terms of cytotoxicity due to the corrosion products released.

1. Introduction

NiTi (Nitinol) alloy shows the SME (shape memory effect) where its shape and rigidity is thermally controlled as it is a material capable to sense the external temperature modifications and to be transformed into a pre-programmed shape in elastic and plastic deformation energies [1].

The superelastic NiTi orthodontic wires allow the increase, transmission and control of light and continuous forces necessary for dental shifts [1-3]. Due to the specificity of the orthodontic fixed appliances many authors suggest the inclusion of some constraint elements on the model assemblies so as to reproduce the clinical reality [4-6] as opposed to ISO 15841: 2006 methodology [7].

The beta-titanium (β-Ti) wires are titanium molybdenum alloys. Goldberg and Burstone envisioned this alloy for orthodontic use after recognizing such advantages as elastic modulus below stainless steel and near to nickel-titanium (NiTi) conventional alloy, excellent formability, weld-ability, and low potential for hypersensitivity [8].

Corrosion consists in the degradation of a material under the chemical or electrochemical action of the environment where it is located.

Due to the high content of nickel of the NiTi alloy, there is the possibility that the Ni in the alloy might dissolve due to corrosion with a carcinogenic potential on the body [9].

The specialized studies have shown that the corrosion products of the dental alloys solubilized in the gum tissues depend on the alloy composition which influences the resistance to corrosion by the structure formed during casting and the subsequent processing protocols.

2. Paper goal

The goal of this study consists in the determination of the structural modifications occurred in two shape memory alloys intended for orthodontic fixed appliances following electrocorrosion in Afnor artificial saliva.

3. Materials and methods

We used two types of orthodontic wires from Ti-based alloys supplied by the same manufacturer, namely Nitinol Clasic-*3M* and Beta III Titanium-*3M*.

For the electrocorrosion tests we used a solution of Afnor artificial saliva having pH=8.2 and a chemical composition very close to the natural saliva.

The behavior to corrosion was studied by fast electrochemical tests, mainly through dynamic potentiometry. The potential measurements within open circuit and potentiodynamic polarizations were carried out on a VoltaLab 21 (PGP201 - Radiometer Copenhagen) potentiostat equipped with VoltaMaster 4 software for data acquisition and processing. The linear polarization curves were registered for a scanning speed of the potential of 1mV/s, within a potential interval close to the corrosion potential. To verify the susceptibility to corrosion in points and the main parameters of the corrosion process in case of some high overpotentials, the cyclical polarization curves were registered for the speed of 100 mV/s, for the -500 ... +1500 mV potential range. All measurements were performed at 25°C in naturally aired solution. For I=0 corrosion current we determined, by means of Volta Master 4 software, the shape of branches (b_a and b_c) of Tafel diagram, the polarization resistance R_p, the density of the corrosion current J_{cor} and the corrosion speed V_{cor}.

The microstructural analysis was performed by electronic microscopy for different powers of amplification of the image depending on the structural details identified on the surface. We studied the main characteristics of the surface before and after the electrocorrosion resistance tests.

4. Results and discussions

Through the corrosion tests we obtained Tafel diagram (figure 1a) and the cyclical diagram (figure 1b) characterizing the electrocorrosion behavior of the Nitinol Clasic-*3M* alloy.

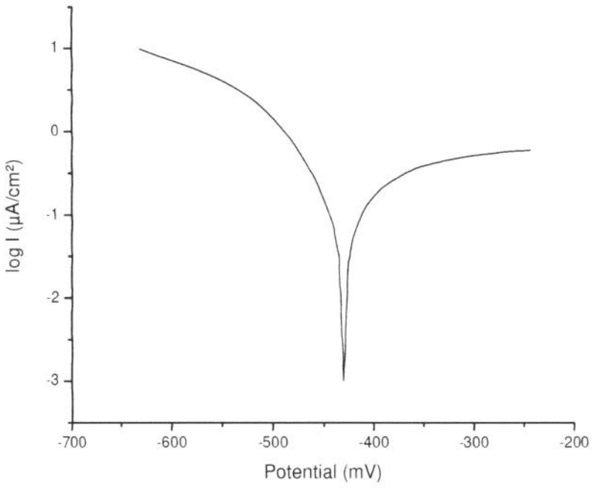

a)

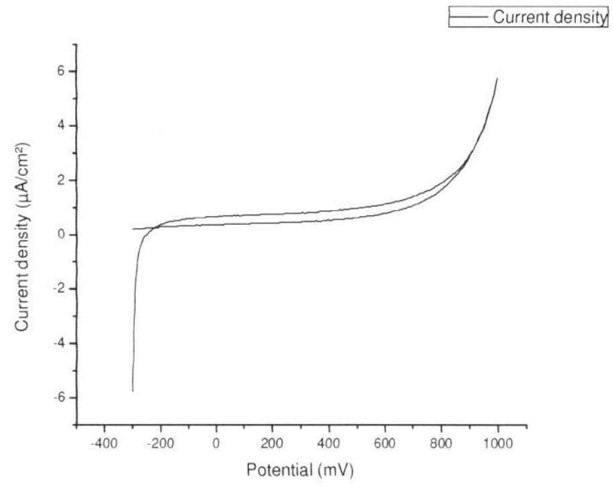

b)

Figure 1. Polarization curves of Nitinol Clasic-3M alloy in Afnor artificial solution a- linear and b-cyclical.

The results registered during the electrocorrosion process in solution of Afnor artificial saliva indicate an adequate behavior of Nitinol Clasic-*3M* material characterized by a polarization resistance of 152.09 kohm.cm and a corrosion speed per year of 15.04 µm/Y (table 1).

Table1. Parameters of electrocorrosion process in Afnor artificial solution for Nitinol Clasic-3M alloy.

Parameter	Value
E(I=0)	-426.8 mV
J_{cor}	1.2865 µA/cm
Rp	152.09 kohm.cm
Corrosion rate	15.04 µm/Y

The analysis of the cyclical diagram of corrosion highlights the "pitting" effect through corrosion, as it is of hysteresis type with the return branch on the exterior. The "pitting" type of corrosion represents a deficient way of behavior of the material at the interaction with the exterior environment due to the quantity of material transferred to the electrolyte environment (in our case Afnor artificial saliva).

In figure 2 we presented Tafel diagram (figure 2a) and the cyclical diagram (figure 2b) characterizing the electrocorrosion behavior of Beta III Titanium-*3M* alloy.

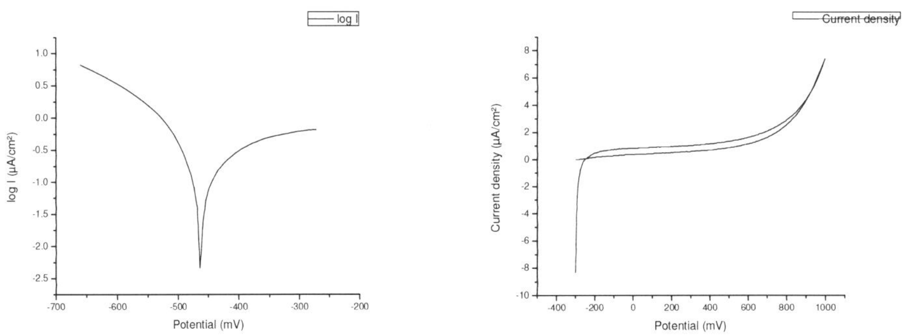

a) b)

Figure 2. Polarization curves of Beta III Titanium-3M alloy in Afnor artificial solution a- linear and b-cyclical.

The results registered for the corrosion in solution of Afnor artificial saliva are given in table 2. We may notice a superior behavior of *Beta III Titanium-3M* material characterized by a high polarization resistance of 160.77kohm.cm and a corrosion rate per year of 6.114 µm/Y.

Table 2. Parameters of electrocorrosion process in Afnor artificial solution for Beta III Titanium-3M alloy.

Parameter	Value
E (I=0)	-459.8 mV
J_{cor}	0.5228 µA/cm
Rp	160.77 kohm.cm
Corrosion rate	6.114 µm/Y

In figure 3, we presented the electronic microscopy images for the surface of Nitinol Clasic-*3M* alloy at different amplification scales. The surface of the material shows "pitting" traces characterized by pits appearing on the surface of the specimens.

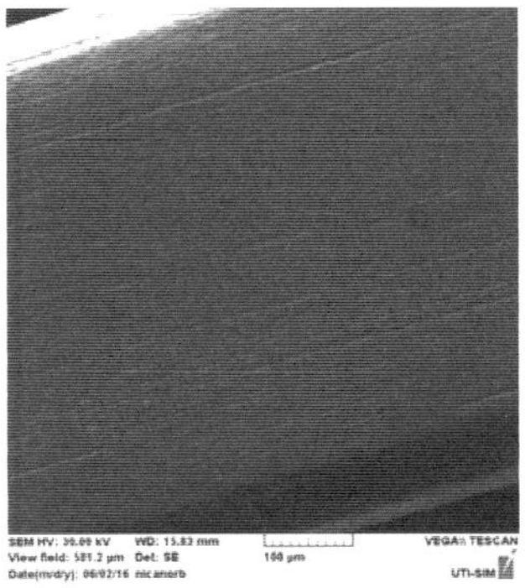
a)

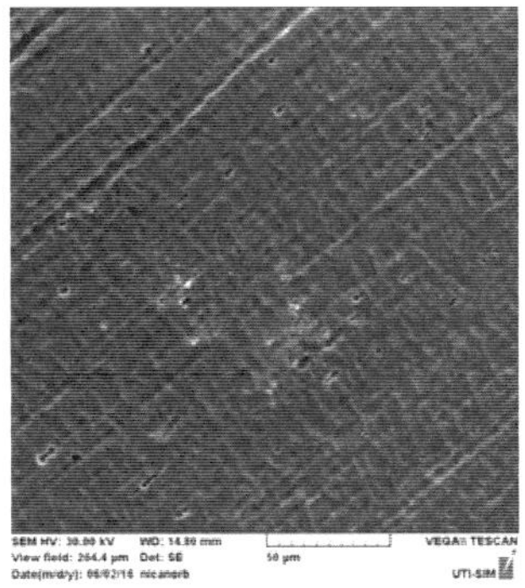
b)

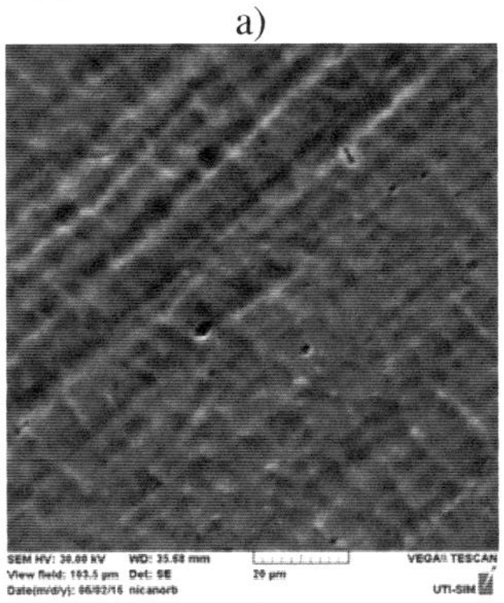
c)

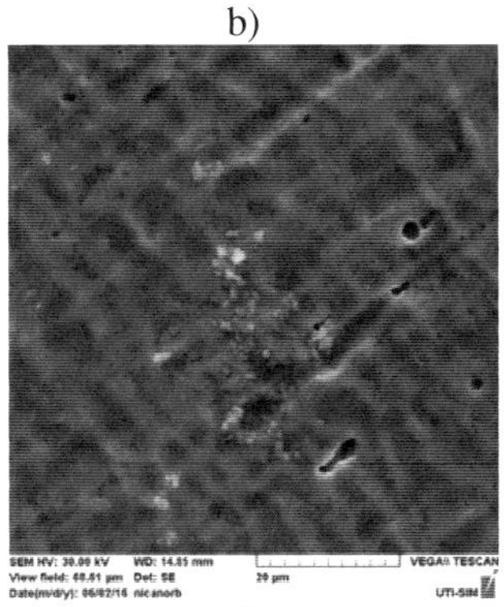
d)

Figure 3. Nitinol Clasic-3M at different amplification scales: a) 100:1 amplification scale, b) 1000:1 amplification scale, c) 2000:1 amplification scale and d) 3000:1 amplification scale.

At low amplification powers (figure 3a) the traces of the electrocorrosion test are not visible.

Figures 3b-3d clearly highlight the traces of the corrosion process, material dislocations measuring between 1 and 10 μm and having an orderly character. In figure 3d, we may also notice the formation of some corrosion compounds on the surface of material grouped around the corrosion points.

The connection between the manner of grouping of the reaction compounds and the orderly arrangement of the craters on the surface of the material is determined by the interaction between compounds and the phasic constitution of NiTi alloy. Being still separate, the craters do not affect at this moment the functionality of nitinol smart alloy.

Beta III Titanium-*3M* alloy corroded on the direction of mechanical processing (figure 4a-4c) through the formation of some crevices on the surface of the material with widths of 1-3 μm and lengths up to 40 μm.

In figure 4c, we may notice the existence of corrosion compounds at the level of the surface of the smart metal material. Compounds are 5-15 μm wide and 20-30 μm long.

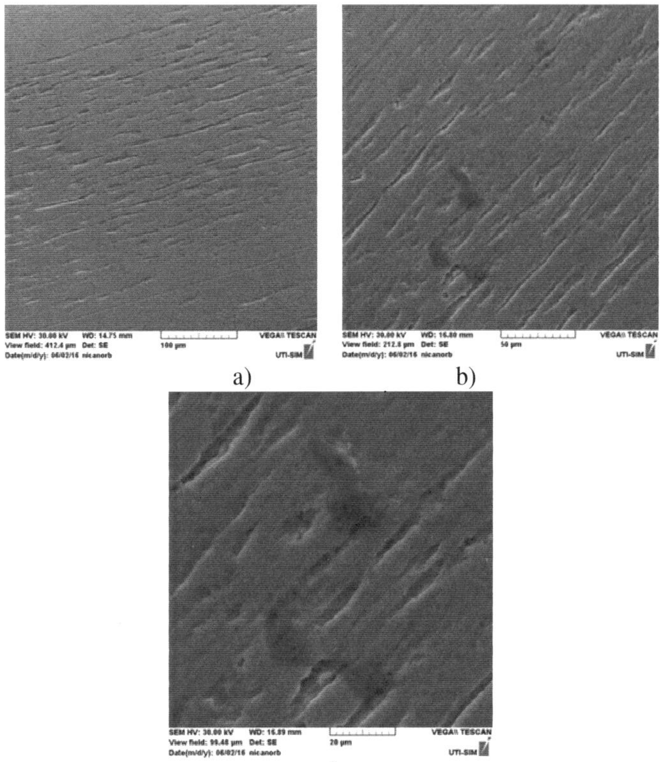

a) b) c)

Figure 4. Electronic microscopies of the surface of Beta III Titanium-3M alloy at different amplification scales: a) 500:1 amplification scale, b) 1000:1 amplification scale, c) 2000:1 amplification scale.

The surface of Beta III Titanium-*3M* alloy is totally corroded, the corrosion formations being unified and making a long line of corrosion. These "crevices" will affect both the mechanical stability of the smart material and the special properties characteristic to the shape memory alloys.

5. Conclusions

To define the behavior to corrosion of the dental alloys we used the linear polarization method and the cyclical polarization method.

By the linear polarization method we obtained corrosion potentials providing information about the tendency towards corrosion of the alloy in the corrosion environment (for example Afnor artificial saliva).

The results registered during the electrocorrosion process indicate a superior behavior of *Beta III Titanium-3M* material characterized by a high polarization resistance and a lower corrosion speed per year.

The microstructural analysis of the surface of the two nitinol alloys after the electrocorrosion resistance test in Afnor artificial solution was performed through SEM electronic microscopy by using a SE secondary electron detector. Both alloys show "pitting" traces of diverse sizes. Although both materials corroded through the dislocation of material from the superficial layer, they had different behaviors through the formation of distinct craters having an orderly arrangement (Nitinol Clasic-*3M*alloy), and by the unification of craters and the formation of crevices (Beta III Titanium-*3M*), respectively. Both forms of manifestation of electrocorrosion effects in Afnor artificial saliva will negatively influence both the mechanical properties and the special properties of the two shape memory alloys.

References

[1] Chen M F, Yang X J, Hu R X, Cui Z D and Man H C 2004 *Materials Science and Engineering C* **24** 497
[2] Rodriguez C, Brown L C 1975 The Mechanical Properties of SME Alloys in "Shape Memory Effects in Alloys" *J., Plenum Press* 29-58
[3] Humbeeck V 1990 *Trait. Them.* **234** 369
[4] Henderson E, Nash D H, Dempster W M 2011 *Journal of the Mechanical Behavior of Biomedical Materials* **4** 261
[5] Sun L, Huang W M, Ding Z, Zhao Y, Wang C C, Purnawali H and Tang C 2011 Stimulus-responsive shape memory materials: a review *Materials and Design* DOI: 10.1016/j.matdes.2011.04.065
[6] Revie R W and Uhlig H H 2008 Corrosion and corrosion control: an introduction to corrosion science and engineering *John Wiley & Sons* 205
[7] Duffo G S and Quezada Castillo E 2004 *Corrosion* **60** 594
[8] Goldberg A J and Burstone C J 1979 *J Dent Res.* **58** 593
[9] Paraschiv C, Ştirbu I, Cimpoeşu R, Bernevig M, Nejneru C, Manole V, Cimpoesu N and Zegan G 2016 *Optoelectronics and advanced materials – rapid communications* **10** 87
[10] Huang T H, Ding S J, Min Y and Kao C T 2004 *Eur. J. Orthodont.* **26** 171
[11] Mareci D, Cimpoesu N and Popa M I 2012 *Materials and Corrosion* **63** 176

IOP Conf. Series: Materials Science and Engineering **209** (2017) 012034 doi:10.1088/1757-899X/209/1/012034

A Recovery Process of Active Cathode Paste from Spent Li-Ion Batteries

C M Toma[1], G V Ghica[1], M Buzatu[1], M I Petrescu[1], E Vasile[1] and G Iacob[1,*]
University POLITEHNICA of Bucharest, 313 Splaiul Independentei Blvd, 060042, Bucharest, Romania

E-mail: iacob_gh@yahoo.com (*corresponding author)

Abstract. In this work, the depleted active paste from spent lithium–ion batteries was separated from cathode by means of ultrasonic vibration. First the unit cells were discharged in brine at room temperature, for safety reasons. Then anode, separator, electrolyte and cathode were separated. Spent Li-Ion batteries were introduced into a washing container to separate electrode materials from their support substrate: active paste (lithium cobalt oxide – $LiCoO_2$) from cathode (Al foil) and graphite from anode (Cu foil). The Al foil and Cu foil were also recovered. A cleaning efficiency of 91% was achieved using a solution of 1.5 M acetic acid after a 6 minute time of exposure into an ultrasonic washing container with a frequency and electric power of 50 kHz and 50 W, respectively. The XRD patterns and the morphology of $LiCoO_2$ powder were presented.

1. Introduction

Li-ion batteries have a wider variety of materials in each cell. All Li-ion batteries pose a potential environmental hazard. The first lithium batteries were using lithium metal as the cathode and not contained toxic metals, but there is the possibility of fire if metallic lithium is exposed to moisture while the cell are corroded [1]. Lithium-ion secondary rechargeable batteries contain no metallic lithium but uses as cathode a material such as $LiXMA_2$ ($LiCoO_2$, $LiNiO_2$, $LiNiMnCo_2$) and graphite at anode [2]. These contain toxic and flammable electrolyte. Lithium-ion batteries used, partially spent and stored in dumps, can self-ignite and extinguishing these fires is particularly difficult.

On the other hand, Cobalt is a rare and precious metal, and is a relatively expensive material compared with the other constituents of lithium ion batteries. Cobalt is used in chemical production, super alloys and steels. There are few countries that possess exploitable deposits of cobalt (Congo, Canada, China and Russia). The price of cobalt is about 4.7 times more expensive than nickel, 6.6 times more expensive than titanium, 7 times more expensive than lithium [3].

A lithium-ion rechargeable battery comprises:

- A cathode consists of Al foil covered by a fine layer of powdered $LiCoO_2$ linked (usually) together with aluminum foil through adhesive agent such as polyvinylidene fluoride (PVDF - a highly non-reactive thermoplastic fluoropolymer) binder. PVDF could not react with most of strong acid and base, most of strong oxidant and halogen, and could not be dissolved in such organic reagents as fatty hydrocarbon, aromatic hydrocarbon or aldehyde alcohol at room temperature and could only partly be dissolved in oleum, some special ketenes and ethers [4]; NMP (N-methylpyrrolidone) is the better solvent system for PVDF, but can be tested and but can be tested and DMAC (Dimethylacetamine), DMF (Dimethylformamide), DMSO (Dimethylsulfoxide, THF (Tetrahydrofuran);

- An anode made from Cu foil covered by a line layer of powdered graphitic carbon linked together by the same adhesive agent used in the cathode;
- A toxic and flammable electrolyte, consists of an organic liquid solvent which dissolve Li salt (substances like $LiClO_4$, $LiBF_4$ and $LiPF_4$ [5-8]);
- Separator (polyethylene, polypropylene);
- Casing (aluminum, plastics, stainless steel).

The typical chemical composition of lithium-ion rechargeable batteries with $LiCoO_2$ as active cathode paste is shown in table 1.

Table 1. Chemical composition of a typical lithium-ion rechargeable battery [1].

Component	wt.%
$LiCoO_2$	27.5
Steel/Ni	24.5
Cu/Al	14.5
Carbon	16
Electrolyte	3.5
Polymer	14

Spent lithium-ion batteries can be subjected to different recycling processes, physical process, mechanical separation, mechanochemical process, chemical process, electrochemical process etc.
$LiCoO_2$, which is used as the cathode material for almost all lithium-ion batteries, is considered to be the most important target of this type of batteries. It is possible to recycling cobalt and other metals from the cathode, or to separate and reuse the $LiCoO_2$.

Contestabile et al., [9] treats the battery rolls with N-methylpyrrolidone (as a solvent to dissolve PVDF) at 100°C for 1 h; $LiCoO_2$ was separated from their support substrate. The recycling of $LiCoO_2$ cathode material by ultrasonic technology was used for the first time by Jinhui Li et al., [10] develop a combined process that used among other methods, the ultrasonic washing and agitation to reduce costs and pollution. They have determined that by using both agitation and ultrasonic washing, a good percentage of electrode materials have been separated from Al foils. Jinhui et al. also studied the influence of temperature on recovering efficiency.

Li et al. [11] used an ultrasonic-assisted leaching process to remove the metals from spent Li-ion batteries. They investigate the effect of ultrasonic power, leaching time and temperature on active materials in order to obtain the optimal parameters of this process. In order to achieve the best result, they use different acidic reagents and came to the conclusion that citric acid was more efficient and eco-friendly than the others two acids that were analyzed. They also have shown that ultrasonic agitation plays an important role in leaching efficiency of the process.

The ultrasonic cleaning was investigate in detail by Li-Po He et al. [12], who also proposed an environmentally process for the recycling of cathode materials from spent Li-ion batteries.

In this paper, we propose a recycling method of recover cathode material based on ultrasonic leaching with acetic acid, at room temperature.

2. Materials and methods

Spent Li-ion batteries Li-ion batteries from telephones were used in this study. The recovery process of active paste, graphite, and films of Al and Cu is shown in figure 1.

For the first, the batteries were put into a 10 wt. % NaCl solution to discharge for 30 minutes. Then, Li-ion batteries were manually dismantled and physically separated into their different parts:

cathode, anode, steel, separators, and current collectors. Cathodes and anodes were treated in an ultrasonic cleaning machine (Emmi 12-HC).

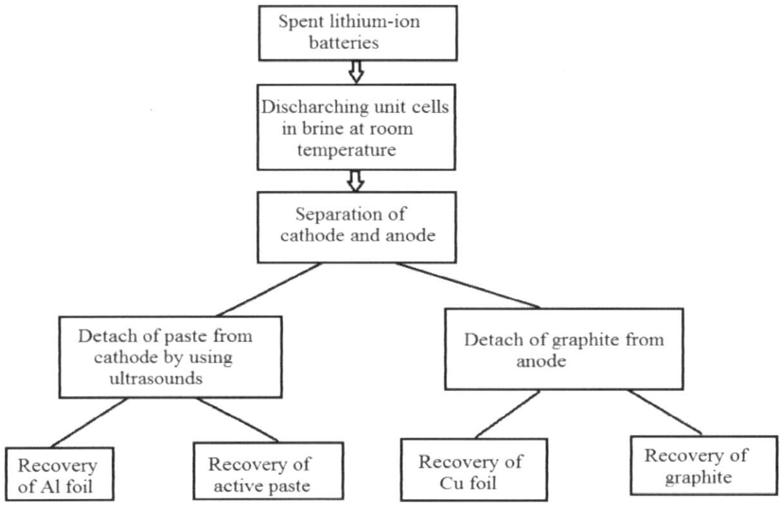

Figure 1. Flow sheet of the metal recovery process.

The technical specifications of the cleaning machine were as follows: housing – stainless steel, cleaning frequency = 45 kHz; cleaning time = 1 - 60 min; volume = 1.2 l; heating temperature = 20 - 80°C; bath dimension 200x100x65 mm; maxim power = 100 W; ultrasonic power = 50/75/100 W. The position of cathode foils in the basket of ultrasonic bath was maintained at 15 mm from the ultrasonic generator.

This process was aimed at separating $LiCoO_2$ from the Al foils, and separating carbon powder from Cu foil. Figure 2 shows the mechanism by which the ultrasonic help dislodge active paste (or graphite powder) of the carrier foil.

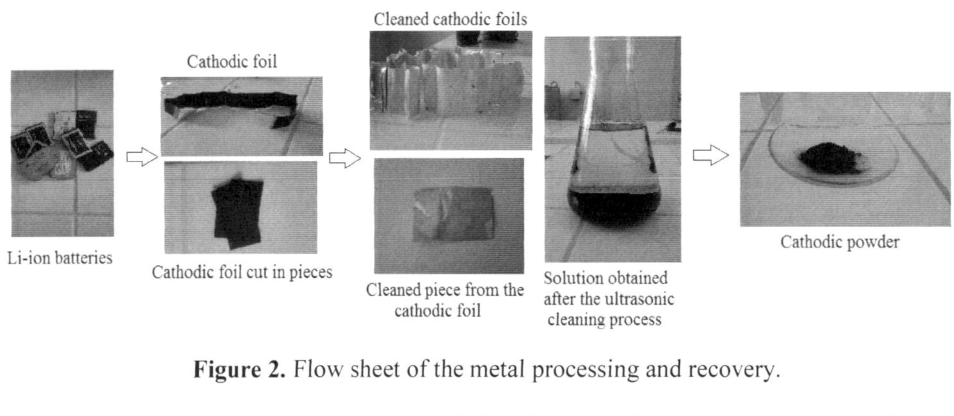

Figure 2. Flow sheet of the metal processing and recovery.

Ultrasonic cleaning uses cavitations bubbles induced by high frequency pressure (sound) waves to agitate a liquid. The cavitations bubbles can increase in dimension, till a critic dimension, when are taking place the bubbles implosion. Stirring produce large forces that help the detachment of active paste from the metallic foil substrate.

IOP Conf. Series: Materials Science and Engineering **209** (2017) 012034 doi:10.1088/1757-899X/209/1/012034

The ultrasound can be used with just water, but using a solution of acetic acid as solvent the cleaning efficiency grows.

We have conducted the experimental research in order to identify the molar optimal concentration of acetic acid in the recovery of active quasi-full slurry of the cathode occurs.

Weight of pieces from cathode, before and after separation of active paste was determined using a balance with three decimal.

All experiments were performed at room temperature (no need additional heat sources). Purified water and acetic acid (technical grade) was use as the cavitations medium (figure 3) to investigate the effect of acetic acid concentration on the recovery time.

Cleaning efficiency was calculated with the following equation:

$$\eta = \frac{m_i - m_f}{m_i} \cdot 100 \tag{1}$$

η - Cleaning efficiency;

m_i - initial weight of cathode piece, g;

m_f - final weight of cathode piece, g.

The cathode of spent Li-ion batteries was characterized by X-ray diffraction, SEM and EDAX. Also, active paste containing $LiCoO_2$ (collected after decantation, filtered and dried) was investigated at the end of the process by X-ray diffraction, SEM and EDAX. Processing solutions is the subject of subsequent works.

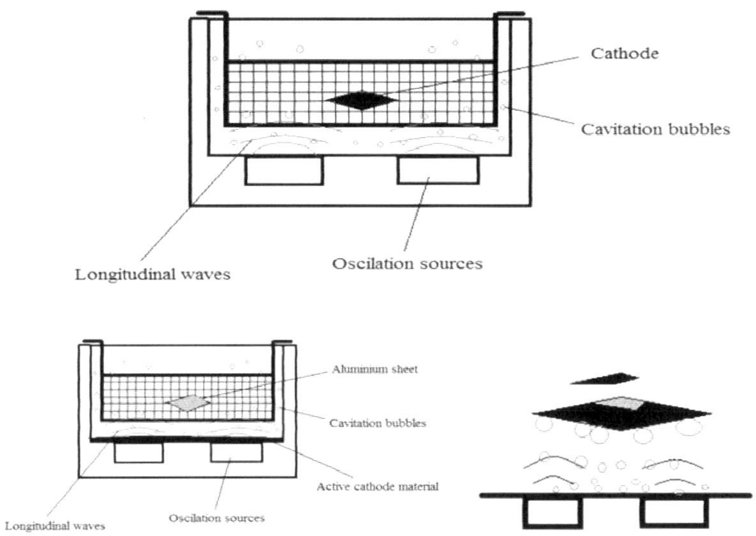

Figure 3. Visualization of cavitations effect on cathodic foils.

3. Results and discussions

All the experiments were carried out in the Laboratory of Hydrometallurgy and Metal Recycling of the Department of Engineering and Management of Obtaining Metallic Materials from Faculty of Materials Science and Engineering, University Politehnica of Bucharest.

The results have been summarized below:

Discharging waste batteries before manually dismantling and physically separated into their different parts, it has an important effect in the recovery flow of the slurry active; cleaning time in acetic acid solution is halved (figure 4).

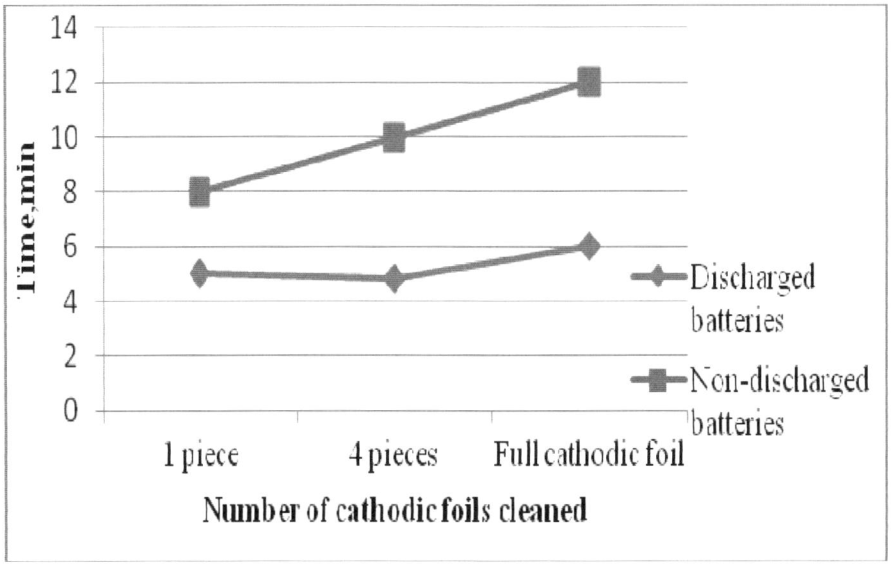

Figure 4. Discharge treatment effect on the recovery time for a different number of pieces.

Table 2. Cleaning efficiency for cathode paste from discharged batteries.

Cathode	m_i (initial weight), g	m_f (final weight), g	$\Delta m = m_i - m_f$	η^*, %	*Time*, min.
2 pieces	1.030	0.913	0.913	88.64	5
3 pieces	1.696	1.549	1.549	91.33	4.8
4 pieces	2.350	2.141	2.141	91.10	4.9
16 pieces (entire cathode)	7.756	7.064	7.064	91.07	6

η^*, [%] - it was calculated without taking into account the amount of solubilized paste.

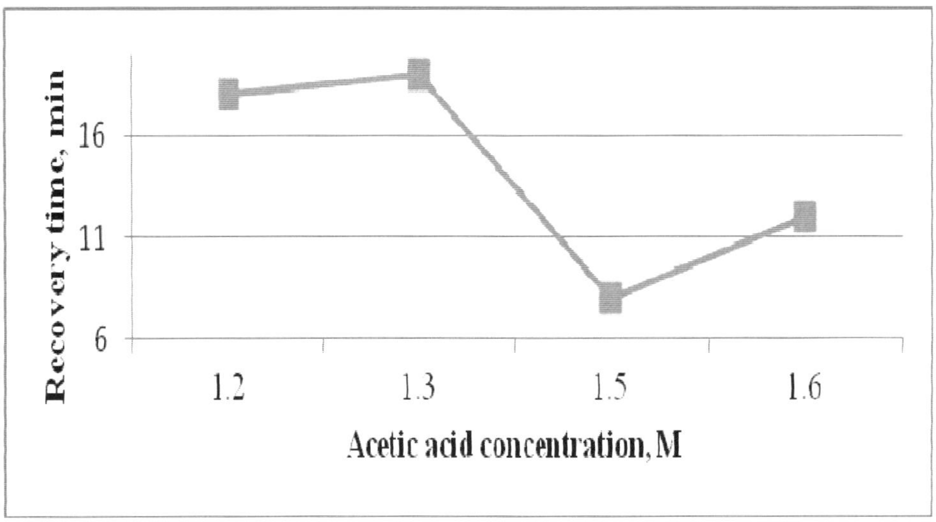

Figure 5. Recovery time of active paste at different acetic acid concentration.

The shortest time to remove the active pulp from cathode was achieved using a solution of 1.5M acetic acid, after exposure to ultrasounds in the following conditions: 50 W ultrasonic power; 45 kHz frequency; 20°C temperature; 8 minutes time (figure 5).

The samples were analyzed by Scanning Electron Microscopy (figure 6) using a Quanta Inspect F50, with a field emission gun (FEG) with 1.2 nm resolution and an Energy Dispersive X-ray Spectrometer (EDXS) having 133 eV resolution at MnK_{α}.

XRD data (figure 7) were obtained using a Panalytical X'PERT MPD X - ray diffractometer with a Cu K - a radiation source (λ = 1.5418 Å) in the range 2θ = 10–90°.

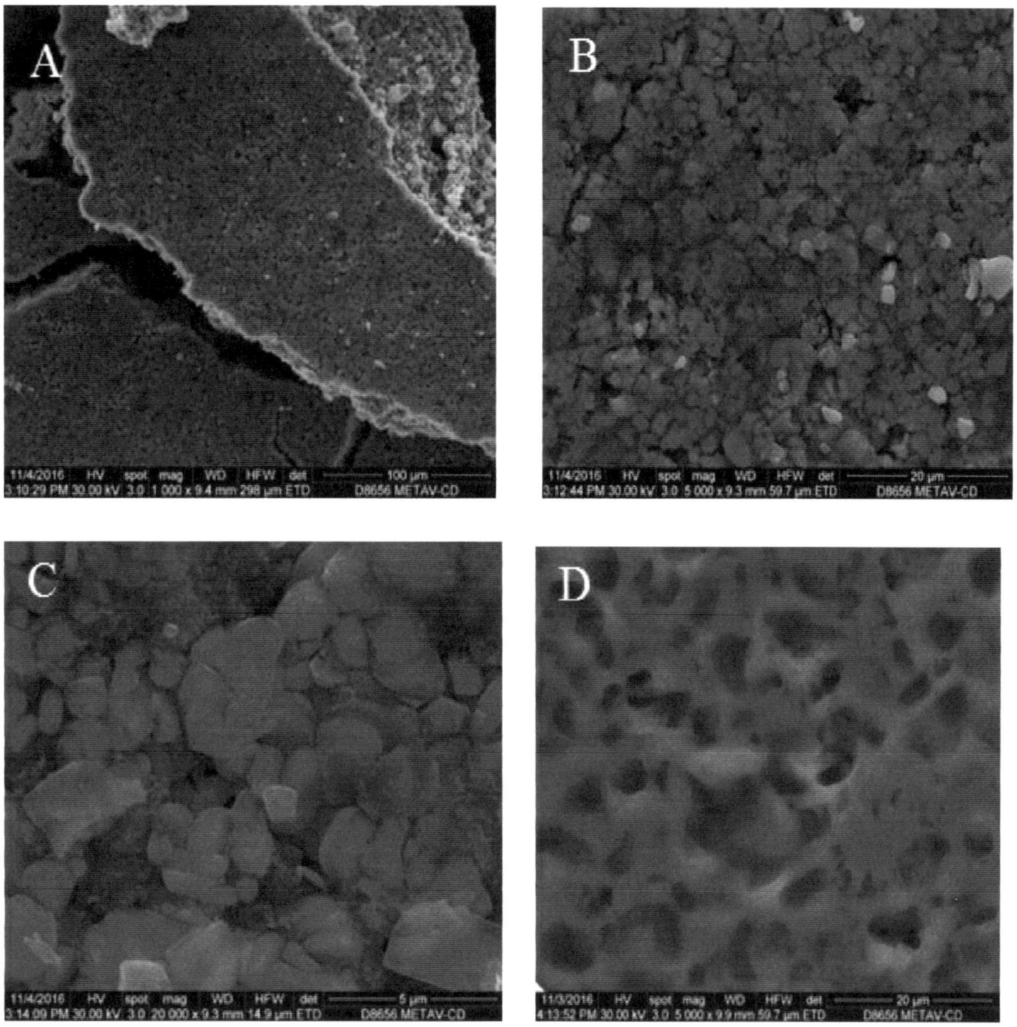

Figure 6. SEM image of cathode materials after treated in an ultrasonic cleaning machine (a, b, c) and Al cathode surface after removing active paste (d).

The appearance of the pulp (identified as $LiCoO_2$) is caught in figure 6 (a, b, c) after the ultrasonic treatment in a solution of acetic acid. $LiCoO_2$ particles are bonded through adhesive agent (polyvinylidene fluoride –PVDF probably).

Figure 6 (d) presents the surface topography of the aluminum foil after ultrasonic attack.

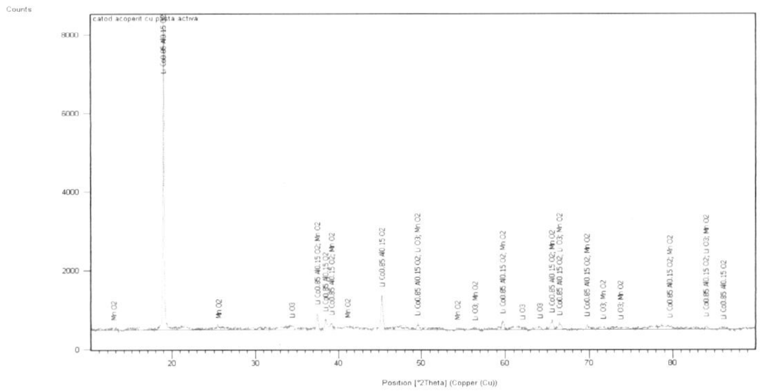

a

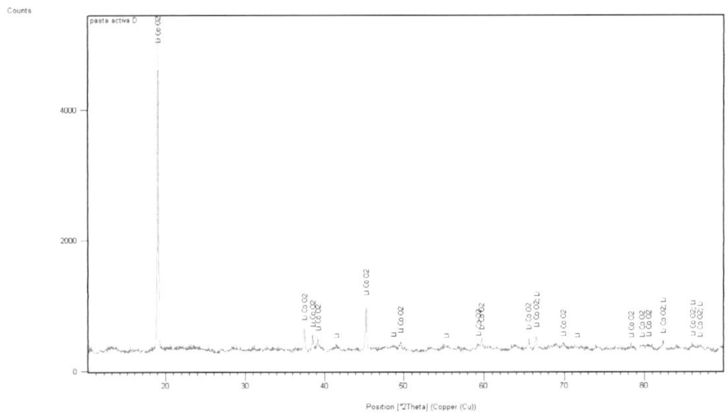

b

Figure 7. XRD patterns of cathode foil with active paste (a) and after separation from Al foil (b).

Identification of $LiCoO_2$ compound was realized by XRD.

4. Conclusions

The result obtained from present work can be summarized in the following points.

1. Our experiments indicate that this method allows to recovery of active paste from cathode by ultrasonic in acetic acid medium.
2. Even if we had obtained for the moment cathode/anode foils through manual dismantling, in the next future, when electric vehicle will dominate the automotive market, standardization and design will require a slight separation of parts.
3. Cathode active paste containing $LiCoO_2$ has been detached entirely from Al foils.
4. Optimal value of molar concentration of acetic acid for the quasi-full recovery of active paste from the cathode was situated at 1.5 M.
5. $LiCoO_2$ compound it has been identified by X-ray diffractometer.
6. The proposed method is cheap (acetic acid is cheap and biodegradable).
7. Analysis of chemical elements solubilized by sonication process of cathodes in acetic acid solution (1.5 M) will be subject to further work.

References

[1] Xu J, Thomas H R, Francis R W, Lum K R, Wang J and Liang B 2008 *J. Power Sources* **177** 512
[2] Battery University, http://batteryuniversity.com
[3] London Metal Exchange, https://www.lme.com/metals/minor-metals/cobalt/
[4] Qin Y and Qi S 2006 *Nonferr. Meetal (Part of Metallurgy)* **1** 13
[5] Bernardes A M, Espinoza D C R and Tenorio J A S 2004 *J. Power Sources* **130** 291
[6] Wu Q, Lu W and Prakash J 2000 *J. Power Sources* **88** (2) 237
[7] Iwakura C, Fukumoto Y, Inoue H, Ohashi S, Kobayashi S, Tada H and Abe M 1997 *J. Power Sources* **68** (2) 301
[8] Chen J, Yao C, Sheu S, Chiou Y C and Shin H C 1997 *J. Power Sources* **68** (2) 242
[9] Contestabile M, Panero S and Scrosati B 2001 *J. Power Sources* **92** 65
[10] Jinhui L, Pixing S, Zefeng W, Yao C and Chein-Chi C 2009 *Chemosphere* **77** 1132
[11] Li L, Zhai L, Zhang X, Lu J, Chen R, Wu F and Amine K 2014 *J. Power Sources* **262** 380
[12] Li-Po H, Shu-Ying S, Xing-Fu S and Jiang-Guo Y 2015 *Waste Management* **46** 523

IOP Conf. Series: Materials Science and Engineering **209** (2017) 012035 doi:10.1088/1757-899X/209/1/012035

Research on Lessening of Bonding Effects Between the Metallic and Non-Metallic Surfaces Through the Graphite Films Deposited with Pulsed Electrical Discharges Process

L Marin[1] and P Topala[2]
[1] Polymers Department, National Institute for Research and Development in Chemistry and Petrochemistry, Splaiul Independentei 202 sector 6, Bucharest, Romania
[2] Engineering, State University Alecu Russo, Balti, Rep Moldova

E-mail: andree_marr@yahoo.co.uk

Abstract. The paper presents the results of experimental research on the physics of natural graphite film formation, the establishment of chemical composition and functional properties of the graphite films, formed on metal surfaces, as a result of the action of plasma in the air environment, at a normal pressure, under the electrical discharge in impulse conditions (EDI). The researchings were performed in the frame of doctoral thesis "Research on lessening of the bonding effects between the metallic and nonmetallic surfaces through the graphite films" and aimed to identify the phenomena that occur at the interface metal/ film of graphite, and to identify also the technological applications that it may have the surface treatment for submitting the films of graphite on metallic surfaces achieved through an innovative process of electrical pulsed discharges. After the research works from the PhD theme above mentioned, a number of interesting properties of graphite pellicle have been identified ie reducing of metal surface polarity. This led to drastic decreases for the values of adhesion when bonding of metal surfaces was performed using a structural polyurethane adhesive designed by ICECHIM. Following the thermo-gravimetric analysis, performed of the graphite film obtained by process of electrical pulsed discharges, have been also discovered other interesting properties for this, ie reversible mass additions at specific values of the working temperature Chemical and scanning electron microscopy analysis have revealed that on the metallic surface subjected to electrical pulsed discharges process, outside the graphite film, it is also obtained a series of spatial formation composed of carbon atoms fullerenes type which are responsible for the phenomenon of addition of mass.

1. Introduction

The formation of graphite films at micrometric and nanometric scale on the surfaces of alloys-made parts causes their diffusion in the surface layer accompanied by the formation of high hardness carbides, and, as a result, the wear resistance of this layer increases [1-5].

The process of formation of graphite films, in all cases, leads to a decrease in surface roughness of the processed surface. According to the results obtained by the authors [4, 6-9], the application of films on the surfaces of the components that work in cinematic couples, leads to a decrease of the friction coefficient of at least 3 times. Experimental research and industrial tests are aimed at demonstrating that the deposits are formed more efficiently if the processed piece is included in the discharge contour of the current pulse generator used as anode. The formed films can reach up to 7 micrometers in thickness, increasing, at least twice, the operating durability of the casting molds components, due to their solid lubricant and anti-refractory properties [5, 10, 11].

2.Technology of graphite film forming

Experimental tests of forming graphite deposits on metal surfaces were performed in normal working environment – air. To achieve this technology a special installation was applied and it was further thoroughly described as a construction and as an operating system in the paper [6]. To achieve this, electrical discharges in impulse were applied, which interacted with the electrodes surfaces in electrode patches maintenance regime "cold" in order to avoid melting, vaporization and removal of material from them. Bars, made of technical graphite, were used as electrode tools, having a cylindrical form and the cross-sectional area of 5-7 mm^2. This ensures the formation of current impulses with duration ranging between 10^{-6}-10^{-7} s, which corresponds to life duration of "cold" electrode spots. Generator offers the formation of current impulses with the following parameters: gap released energy W_S=0 - 4,8J, accumulated energy on the capacitor W_c= 0-12J, at the voltage applied to the charging capacitor U_c=0 - 250V, for its capacity comprised within C=100-600 μF with 100 μF pace. This ensures the priming of electrical discharges in impulse at the gap values of S=0,01-2,5 mm, with the discharge frequency f=0-50Hz.

The morphology of the treated surfaces was studied by means of SEM method, and the EDX and XPS phase methods were used to study the its chemical composition.

The adherence of the formed film to the treated surface was made by attempts of shearing on a HECKERT FPZ 100 dynamometer type. To identify the emerged changes in adhesion properties, as a result of applying graphite films, a comparative measurement was performed, related to the detachment forces of the assemblies, made with the help of a strong adhesive - polyurethane - between a set of samples treated with graphite and non-treated samples. A set of 3 graphite-treated samples and a set of 3 non-treated samples are glued; the treated samples with graphite film at the end, as it was shown in the paper [3]. After bonding, the samples are overlapped on top of an area of 25x25 mm, and afterwards, when the adhesive hardens, they are subjected to traction.

3. Physical model of graphite film formation by applying electrical discharges in impulse

The formation efficiency of the graphite deposits on the processed surface is also determined by the fact that, while subduing the plasma to electric discharges in impulse according to [6], the temperature in the plasma jet reaches values up to 10^4 K, which is more than sufficient for graphite vaporization and its polar transfer to the piece surface.

The physical model, offered below, refers to the transfer of electrode tool material made of graphite from the processed piece surface, through impulse electrical discharges under a sub-excitement regime. It is worth mentioning that the electro-erosion of the graphite is a special one, with regard to metallic and semiconductor materials.

The concept of physical model is based on the analysis of experimental results, previously obtained by the authors [11, 12], where, it was established that, a more evident erosion of the graphite occurs when connecting the electrode to the discharge circuit of electric impulse generator, functioning as cathode. While concerning the general case of processing electrically conductive materials in work [12] it was established that the amount of material collected from the surface of the electrode subdued to electric discharges in impulse can be determined by the following relationship:

$$m = k\rho U_e \int_0^{\tau} i(t)dt \tag{1}$$

where: k – is the proportionality coefficient; ρ- is the density of the electrode material; U_e – is the voltage drop at the electrode surface; i – is the actual value of the electric discharge in impulse; τ- is the duration of the electric discharge in impulse. With the increasing of the voltage drop at the electrode-cathode surface, the amount of electric power, released from its surface, increases as well. It is equal to:

$$W_c = U_c \int i(t)dt \tag{2}$$

Considering that the process of electro-erosion is, actually, an electrochemical one, arising at high temperatures, we can assume that recombining and dissociative processes occur at the anode-electrode surface, at the electrode-cathode surface and that of the plasma channel.

Summarizing the above mentioned facts in work [8, 13] and considering the distributed processes, we suggest the following physical model of the formation of graphite deposits under the action of the plasma subdued to electric discharges in impulse: in the initial phase (a) (see figure 1) the priming of electrical discharge occurs according to Townsend mechanism and to the formation of conductive channel; in phase (b) there take place a whole range of dissociation of atmospheric components in separate atoms, ions of oxygen, hydrogen and nitrogen, and, definitely, the electrons from the previous phase are present in the plasma; in phase (c) they interact one with each other and with electrode surfaces, producing surface activation and causing their erosion; both in phase (c) and in phase (d) there are many series of intensive oxidation reactions of the surface of the graphite electrode, as well as there is a dissociation of the product in oxygen and carbon, with a further transfer on the treated surface, both under the action of the electric field gap and under the action of the plasma component atoms (e); the process ends with film synthesis, which contains carbon of various crystallization forms and phases formed from component elements of plasmagene gas.

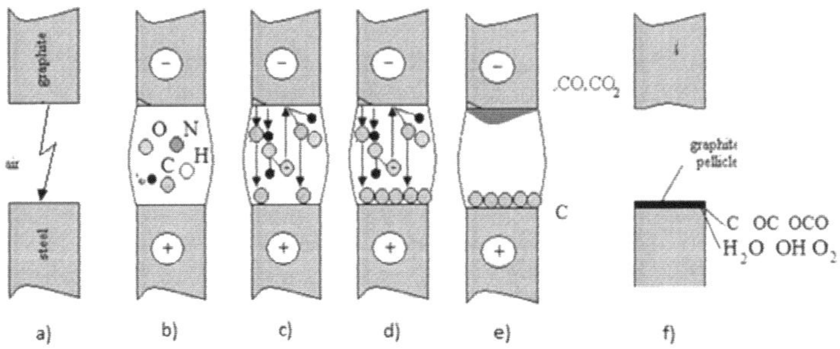

Figure 1. Physical model of forming the graphite film on the piece surface under the influence of electric discharges in impulse.

Proceeding from concrete conditions (air working environment, at the atmospheric pressure), we might assume the followings:

Due to the fact that the oxygen, in the plasma channel, interacts more intensively with the cathode-electrode surface, the oxidation reactions take place, accompanied by the release of carbon oxide CO, complemented by additional discharge of heat Q on the cathode surface and accompanied probably by the formation of carbon dioxide CO_2. As a confirmation of the processes of graphite oxidation at the cathode surface can serve the obtained results by work authors [14] in the process of formation of oxides film on metal parts surfaces, with the application of electric discharges in impulse.

4. SEM electron microscopy analysis of the graphite films

A set of silicon samples were submitted on an electron microscopy analysis and a set of images have been obtained with magnification in the range (1000x – 20 000 x) for a large number of samples.

From these images is clearly evidenced, that besides graphite film, was also obtained a series of globular formations with spherical shape. These spherical formations appear randomly on the sample surface and they due to the electrical discharges in pulse – EDI treatment. It is very clear that these formations it can not have another composition than carbon. The resulting shape and the dimensions of these formations, from the images taken by electron microscopy, they are comparable to those of fullerenes, molecules spatial type with C 60 – C 80 [15]. These images bring a new argument in favor of a priori assertions regarding the occurrence of these chemical species - fulerene type - to application of the procedure of electrical discharges pulsed EDI.

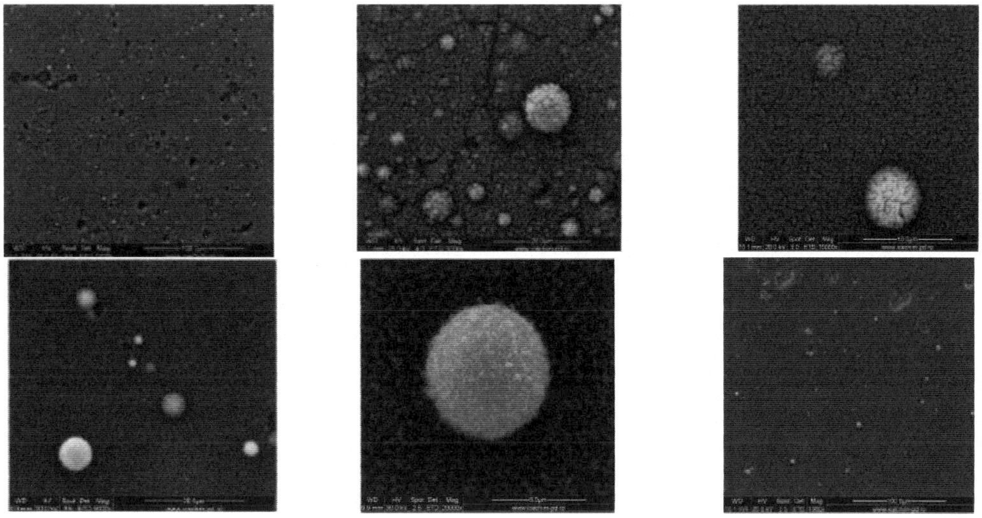

Figure 2. Samples of the graphite film on the metal surface magnification 2000 - 20 000x.

5. Thermal Gravimetric Analysis (TGA) of graphite film

TGA tests on graphite films were conducted on a Du Pont Instruments 951 device. In order to avoid errors – due to the oxidative process that takes place at high temperature and results in a weight increase – TGA tests on graphite films were carried out in nitrogen atmosphere. Graphite powder was collected from three conditioned samples and, further, introduced in the platform unit. Afterwards, it is placed on the platform and the sample is inserted into the heating chamber.

The following parameters were set for the reference sample and for the graphite deposit material on the surface of the work piece: 20-800°C temperature range; 10°C/min heating rate; medium N_2 analysis at a working pressure of 760 mm Hg.

5.1. TGA tests on reference samples

First of all, TGA tests were done on a pure graphite sample, which was thought to be the reference sample.

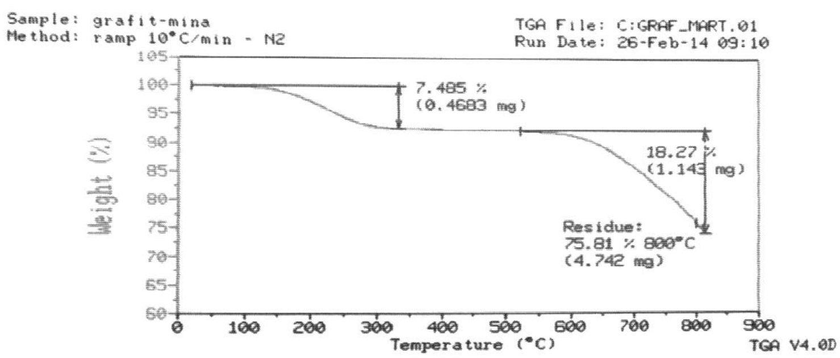

Figure 3. TGA curve corresponding to pure graphite reference sample.

About 5 mg of powdered graphite reference sample was inserted into the device platform. TGA adsorption curve of reference sample is given in figure 3.

The following features are observed from figure 5 up to 100°C – the curve shape does not reveal any unusual thermal behaviour for graphite; within the temperature range of 100-280°C a weight loss of 7.485% (0.4683 mg) occurs as a result of volatiles and water evaporation; between 280-600°C no weight loss occurs which shows a high thermal stability of graphite; some non-significant decomposition (of about 18% of baseline) occurs within the temperature range of 650-800°C, caused by decomposition of tars and heavy hydrocarbons that are found in most varieties of graphite and, finally, considerable amount of graphite is found in the residue – more than 75%, at the end of determination 800°C. These aspects are characteristic for graphite behavior, a compound well-known for its particularly high chemical and thermal stability.

5.2. TGA tests on experimental sample deposition

A similar procedure was used to analyse the graphite films, deposited on the experimental samples. Since the available amount was extremely low, the graphite powder was collected from 3 of the 4 samples of each alternative. The collection process was done by the mechanical graphite scraping, without contacting the sample. The registration of the tested sample weight change versus the temperature is presented in figure 4.

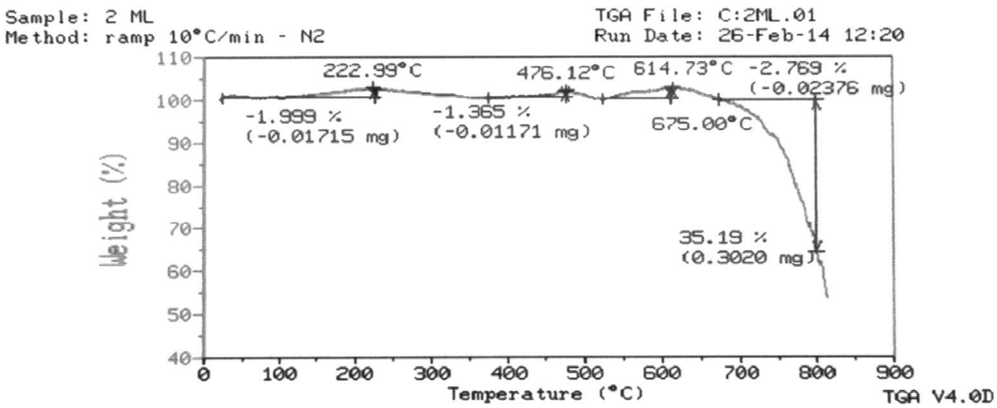

Figure 4. TGA curve corresponding to 10/1,5/600/250 sample.

TGA curve of 10/1.5/600/250 sample (figure 6) shows a completely different allure from that characteristic of pure graphite. The graph shows a number of very interesting aspects suggesting that the graphite film, deposited by electrical discharge in impulse, has a completely different structure from that of pure graphite or that, besides graphite, other chemical compounds of carbon are formed. From figure 5, one may notice that within 200-300°C temperature range (at 222.99°C) a significant weight increase (of 1.999%) occurs, showing that the graphite sample gains weight, instead of losing it due to decomposition or loss in volatiles – showing a material adsorption in graphite film structure (as the adsorbed material cannot be anything but nitrogen from the atmosphere in which the tests are run). The phenomenon is reversible – at about 300°C the weight does not increase anymore, which may be the result of substance volatilization at this evaporation temperature.

Within the temperature range of 450-550°C (476.12°C) a new significant increase in weight (of 1.365%) may be observed, suggesting that, as a sample of graphite increases its weight instead of losing weight due to decomposition or loss in volatiles, a material adsorption occurs in the graphite film structure. Predominantly, a high temperature denies the assumption of a device error; the phenomenon of weight increases, being concrete and real. The phenomenon becomes reversible again at about 550°C and a weight increase does not occur anymore.

Within the temperature range of 600-750°C (614.73°C), a new weight increase occurs in a significant percentage of 2.769%, showing that the graphite sample weight increases instead of decreasing due to volatile decomposition or loss of weight gain – showing a material adsorption within the graphite film structure. Particularly, high temperature denies the assumption of a device error, the phenomenon of weight increase being concrete and real.

Reversible weight increase occurs within temperature ranges of 200-300°C, 450-550°C, 600-700°C having enough increased values, that cannot be assigned to device errors (permissible error being of max. 0.1%). The phenomenon of adsorption/desorption (reversibility) is very similar to the one occurring in zeolites, which presents spatial molecular structures made of atoms of Si, O, Al, able to close smaller molecules within them and to release them later. A similar phenomenon occurs also in graphite film, showing that besides graphite, spatial structures are also synthesized, in this example consisting of carbon atoms, which are known as fullerenes. These structures are made up of spatial molecules of 40-70 carbon atoms and have 20-40 nm in size. Because TGA tests were carried out under N_2 atmosphere, the chemical oxidation process, that could eventually bring mass intake, is excluded. The only scientific explanation consists of the presence of spatial structures able to store small molecules (N_2) and release them later. The emergence of the adsorption/desorption phenomenon at different temperatures and different percentages also shows that the potential spatial structures of carbon atoms have different sizes.

It is noteworthy that within the 20-750°C temperature range graphite film does not suffer any changes arising from thermal degradation.

In literature it is mentioned that spatial structures, containing fullerenes-type carbon atoms, could be found in carbon particles, resulting from incomplete combustion of hydrocarbons and, generally, in everything that involves violent combustion – combustion in internal combustion engines. It is also stated that spatial structures of these molecules, consisting of carbon atoms, can embed – without reacting to them – small molecules (H_2, N_2, monatomic inert gas molecules, HOH) inside their spatial structure (figure 5).

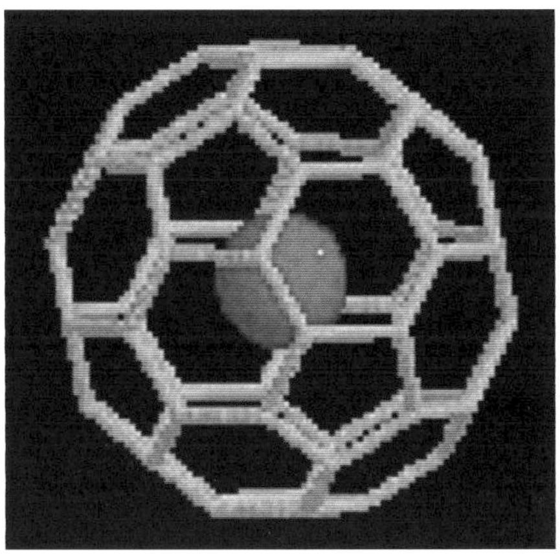

Figure 5. Graphical representation of the properties of spatial structures formed of carbon atoms.

These scientific literature [28-30] explanations provide a plausible elucidation on weight increase phenomena, which were observed in TGA tests on graphite coating sample and nearly double the proportion in which the sample absorbs water – compared to other samples under investigation.

6. Chemical analysis of the graphite films

Globular formations present on the surface of graphite film, obtained by the procedure of electrical discharges in pulsed EDI, have led to the preliminary conclusion that besides graphite film itself mostly, of which is composed the deposited film on the metallic surface of the target, was also obtained a number of other carbon chemical species [16,17] Starting from the idea that these globular formations are compounds of carbon, with spatial molecules fullerenes and / or nanotubes type, with relatively high number of carbon atoms, (60 – 80), the samples presented in figure 2, previously subjected to a non-destructive SEM microscopy analysis, have been subjected subsequently to chemical analysis. The chemical analysis of samples, consisted of their treatment with many organic solvents [18] for 48 hours, under continuous stirring. The best results were obtained in case of using α-chloronaphthalene, in which the carbon spatial formations have the highest solubility [19-21]. The most representative graphite film sample, after the action of α-chloronaphthalene for 48 hours, it was subject to a new electron microscopy SEM analyse. Following this further analysis in electron microscopy SEM it has been found that a large part of globular formations have been removed. In figure 6 they are shown the initial SEM image (left) and the final image (right) after the action of α-chloronaphthalene for 48 hours.

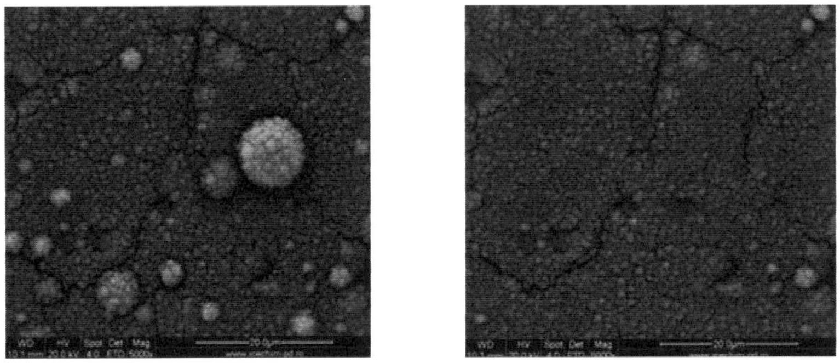

Figure 6. Sample before and after the action of α-chloronaphthalene for 48 hours.

Table 1. Mass variation of a graphite film sample as a result of the action of α-chloronaphthalene for 48 hours.

Initial Mass of graphite film sample, g	*6*	*12*	*18*	*24*	*30*	*36*	*42*	*48*
				hours				
5.7998	*5.7997*	5.7997	5.7997	5.7996	5.7996	5.7995	5.7995	5.7995

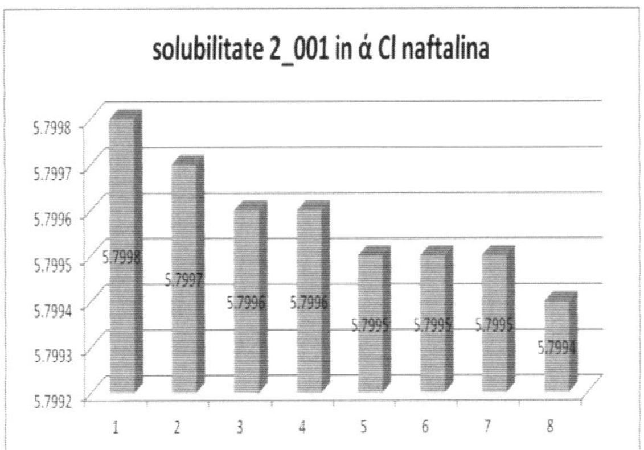

Figure 7. The loss of mass for the sample of graphite film coded 2_001 as the result of the action of α-chloronaphthalene.

The result of the action of α-chloronaphthalene, is that the most significant sample of graphite film coded 2_001 suffers a significant loss of mass – 0.4 mg - perceptible for measuring instruments (from

5.7998g to 5.7994g). This fact indicates that some component of graphite film it is removed by solubilisation.

7. Functional properties of graphite films

Carbon and its allotropic forms in which it crystallizes, shows a particular interest for researchers both as material and in particular as functional properties for technical applications in contemporary technologies. Following are presented a number of functional properties of graphite films deposited on active surfaces of the various types of pieces of machinery and equipment industry.

7.1. Anti socket

In order to broaden the applicability of the practical domain related to the scientific elaborations and, particularly, to graphite films deposited on metal surfaces, with the help of EDI method, a number of attempts, on the timing of screw driving - unscrewing of threaded joints (bolt-nut), were made by the authors [22, 23]. Measurements were made concerning the value of screwing moment in bolt couplers - screw nut with the formation of deposits on its front surface and the second solution - on the surface of the thread bolt (see table 2).

If we compare the results shown in table 2, we can see that the moment of screwing is increasing, being variable and dependent on the number of tracers (surface processing with EDI). It would seem that the deposit formation is not beneficial and leads to an increase of the screwing moment value.

This unimportant increase of the screwing moment is also caused by increasing the size of the pieces that form the couple. In contrast to the reverse - the unscrewing was not registered, together with the outlet effect for parts of couplings with graphite film on the active surface (also in case of keeping them at high temperatures, ad in case of placing mounted joints in solutions presenting active chemical environments).

Table 2. Values of the unscrewing moment of screw nuts, covered with a graphite film, M_{des}, Nm.

Number of passes	*Screwing force, kN*			
	0,5	1	1,5	2
2	2	6	9,9	13
3	2,2	5,8	9,5	12,5
4	2,5	4,5	8,2	10,15
5	3	6,1	10,2	14
6	2,9	8	12	16
8	3,5	6,2	11	14,1
10	3,5	5,9	11	13,2

7.2. Antiblocking properties of graphite films

Graphite films deposited by pulsed electrical discharges with 7-8 μm thickness, react physico-chemical as theoretical model, with metallic surface which leads to a high adhesion of graphite film [51-53].

A first method of verifying the theory on the creation of an anti-adhesive film on the metal surface as a result of treatment with graphite, achieved by electrical discharges, it consisted in achieving of a set of specimens made of two metal plates bonded by an structural adhesive to verify the shear forces.

According to the valid standardization, structural adhesives are the adhesives used to create structures- assemblies of components- which can not be removed only with the destruction of the structure. It follows that adherence achieved with this type of adhesive is stronger than the resistance of the glued components. Structural adhesives are used to achieve structures that no longer can be sold in parts. Therefore, a permanent structure is produced.

In these conditions the adhesives that made the structure should have excellent bonding characteristics and excellent resistance. Assemblies made with structural adhesive obtained were subjected to shear stress test. As the result, the values presented in table 3 were obtained for the shear test.

Table 3. Shear stress values for different ways of gluing the specimens.

both specimens untreated daN/cm^2			Mixed treated specimens daN/cm^2			both specimens treated with graphite daN/cm^2		
86.4	85.80	83.7	62.1	58.6	62.0	49.6	51.2	50.4

From the shear stress test results table it can be observed that the untreated specimens have the higher values about 85 daN/cm^2. Mixed treated specimens are averages of shear stress values about 60 daN/cm^2, which proves that the graphite film, deposited on one of specimens, reduces the adherence between the adhesive and the metal surface.

The last version where both specimens are treated with graphite tension the shear stress values diminishes up to the approximate value about 50 daN/cm^2, which represents a decrease of approximately 40% compared to untreated specimens [26, 27] (version 1) (figure 8).

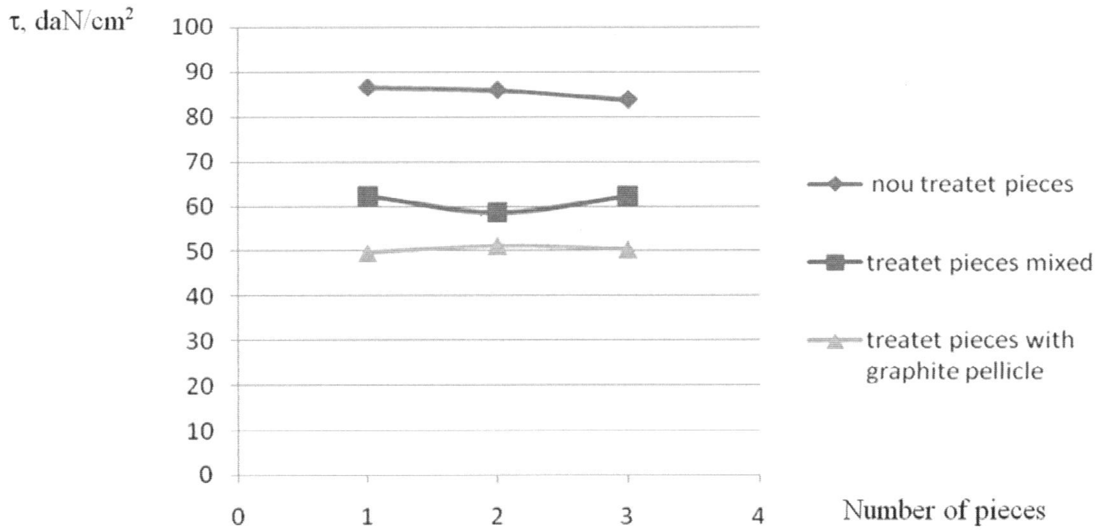

Figure 8. The dependence of the shearing stress on the way of conditioning the test pieces.

In this context, it is also observed that the shear occurs in two ways as it is shown in figure 9. In the first version breakage occurs in the adhesive mass figure 9 b) which is characteristic of specimens that were not subjected to the treatment of electrical discharges with graphite tool-electrode.

In the second version breakage occurs at interface graphite pellicle and adhesive surface which is characteristic of specimens that were subjected to the treatment of electrical discharges with graphite electrode too.

As a confirmation of this goal comes the surface morphology research for EDI treated and untreated specimens after the shear.

In the case of the specimens which do not have been subjected to electrodeposition treatment, all assemblies cleavage occurred in the adhesive mass, which is demonstrated by the pictures shown in figure 10. This shows strong adhesion to metal substrates of the adhesive.

It can be seen from figure 10 that breaking assembly occurred in adhesive mass. This is noticed by the presence of adhesive on whole active surface of the specimen (dark coloured zone).

IOP Conf. Series: Materials Science and Engineering **209** (2017) 012035 doi:10.1088/1757-899X/209/1/012035

In this case the higher shear forces have been observed (over 80 daN/cm^2). In the case of the specimens treated with graphite by electrical discharges pulse treatment EDI breaking structure if they applied a tangential force occurs at the adhesive support interface (figure 11). So the whole mass of the adhesive or much of it will remain on one of the supports. It says in this case that the inner resistance of the adhesive is greater than adherence to the metal support, the structure is not destroyed as a result of unbundling in parts, and i.e. the adhesive cannot works as structural adhesive. Determined share forces in this case are only to half (50 daN/cm^2) compared with the superficial untreated specimens.

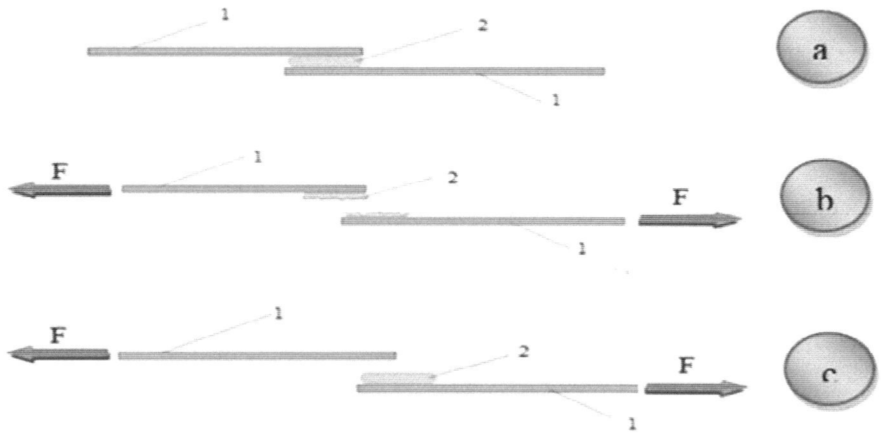

Figure 9. Main scheme shear rupture assemblies: a - initial state; b - final state for untreated specimens; c-final state for EDI treated specimens; 1- metallic sheets, 2- structural adhesive.

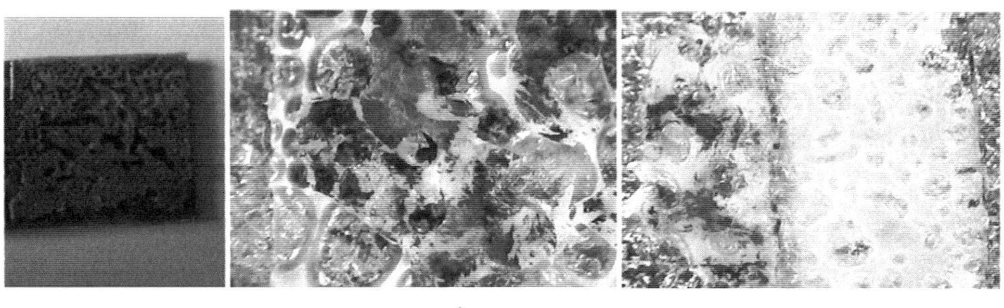

a b c

Figure 10. Component of the assembly without surface treatment after shearing: a-general view; b, c - microscope image, ×80.

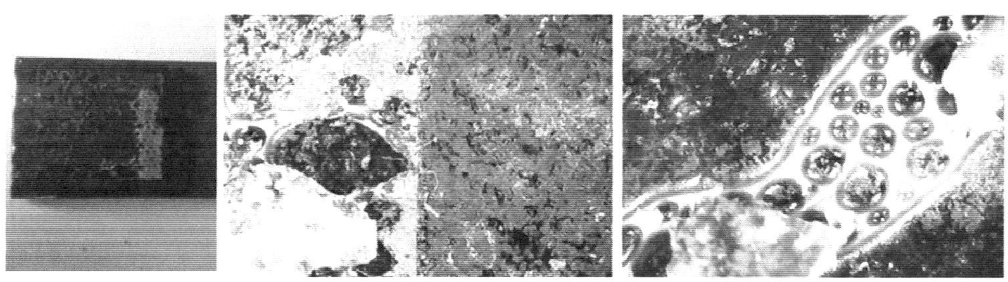

a b c

Figure 11. Component of the assembly with EDI surface treatment after shearing: A - general view; b, c - microscope image, ×80.

It can be seen from figure 11, that the existence of large stretching dark area, means that breaking assembly occurred at adhesive graphite film interface.

In figure 11, a may also be better observed, an area where film deposition of graphite treatment was not performed properly with a continuity 100% - area with a lighter colour - where there was a breakage in the adhesive mass, and in other areas where electro-deposition treatment of graphite was performed properly with a continuity 100% assembly rupture occurred at the graphite adhesive interface - area with a dark colour.

These images taken under the microscope demonstrate these statements. In figure 11, c is observed really well the area where the adhesive remains – area with a lighter colour - in contrast with the dark area, where the cleavage occurred at the interface adhesive - graphite film from the metallic surface.

All these observations allow asserting that the graphite films deposited by pulsed electrical discharges - EDI- determine non-adherence properties, which confirm the assumptions made in previous papers [28].

8. Conclusions

From the above mentioned are drawn the following conclusions:

The erosion of the graphite and its transfer on the metal surfaces with deposition film formation by the action of electrical discharges in impulse is more effective when the tool electrode made of pyrolytic graphite is connected in the discharge circuit of the current impulse generator as cathode. The erosion process in ordinary conditions (air at atmospheric pressure and room temperature) is one complex, of electro-physical and electro-chemical nature that devolves at high temperatures.

The presence of the sample made of steel containing an important amount of the iron stimulates the formation on its surface of 3D type formations (fullerenes or nanotubes). These goals are confirmed by the behaviour of depositions at thermo-gravimetric actions (stating mass increase under the influence of nitrogen at the temperature of 222.99°C, 476,12°C and 614.73°C) and by solubility tests in different environments: it varies from 51 for chloronaphthalene down to 0,006 for tetrahydrofuran).

The graphite films formed by applying electrical discharges in impulse possess a set of beneficial functional properties, such as: decrease the surface adherence by 1.4 times, decrease the wear coefficient from -0.1 down to 4, increase the wear resistance of piece components of glass molding forms, condition the increase of corrosion resistance in chemically aggressive media by 1.5 times.

Functional properties of the graphite depositions, their chemical composition and morphology indicate the fact that they presents 3D structures and in particular carbon nanotubes. The presence of iron in the sample, graphite purity of the tool-electrode, as well as energetic parameters of electrical discharges in impulse provide sufficient and necessary conditions for carbon nanotubes synthesis.

References

[1] Luneva V P, Verhoturov A D, Coziri A V, Glabet T V, Brui V N 2005 Using the Cr-Ni alloys for the electric discharge deposition forming of. EPI, Nr 4, pp.11-18

[2] Burumculov F H, Lezin P P, Senin P V, Ivanov V I 2003 The electric discharge tehnology aimed at retreading and hardening of the machining pieces and tools. USM "Ogareva", Saransk

[3] Topala P 2007 The transfer of mass and diffusion processes on surface layers of engoine parts during electrosparkle processing. The materials of international conference: the technologies of maintaining, retreading and hardening of engine parts. Saint-Petersburg. **2** 234-242

[4] Sidorenco S I, Ivaschenco E V, Mazanko V F, Lobachva G F, Mironov D V, Khranovskaia E N 2005 Forming of ferrous alloys surface layer by combination of nitriding and electrospark alloying by chrome and titanium and certain properties of the afore-mentioned layer. 6-th International Conference "Interaction of Radiation with solids", Minsc, Belarus, **28-30** 430-432

[5] Beshliu V 2008 *Technologies in machine building* **1** 24-30

[6] Topala P, Stoicev P 2008 Technologies of conductible material processing by applying pulsed electrical discharge machining. Chisinau: Edition Tehnica-Info, pp 265

[7] Topala P, Beşliu V, Stoicev P, Ojegov A 2013 *International Journal of Modern Manufacturing Technologie* **2** 97-102
[8] Topala P, Marin L, Beşliu V 2013 Applying graphite micropellicles to decrease the coefficient of superficial adhesion. Advanced Manufacturing Technologies 2013, 7th international seminar Advanced Manufacturing Technologies, Sozopol, Bulgaria, pp. 97-104
[9] Kurochkin Yu V and Demin Yu N 2001 *Chemical and Petroleum Engineering* **37** 404-408
[10] Mihaliuc A 2003 The roughness reduction of electrosparkle Co rings during the following processing with graphite electrode. EPI. **3** 21-23
[11] Topala P 2007 *Nonconventional technologies review* **1** 129-132
[12] Topala P, Beshliu V 2008 Graphite deposits formation on innards surface on adhibition of electric discharges in impulses". Bulletin of the Polytehnic Institute of Iassy, T.LIV, pp.105-111
[13] Beşliu V 2008 Structure and Properties of Surface Layers of Pieces Cemented when Interacting with the Plasma Channel of Electric Discharges in Pulse. The annals of „Dunărea de Jos" University of Galaţi, Fascicle V, Technologies in machine bulding, Vol.1, Year XXIV (XXIX), pp. 75-82
[14] Topala P, Stoicev P, Epureanu A, Beshliu V 2006 The hardening of steel surfaces on the sections for electrosparkle alloyage. International Scientific and Technical conference Machinebulding and technosphere of the XXI century. Donetk, pp. 262-266
[15] Buckminsterfullerene and Buckyballs – Definition, Discovery, Structure, Production, Properties and A. AZoM.com. July 15, 2006
[16] Howard J B, McKinnon J T, Makarovsky Y, Lafleur A L, Johnson M E 1991 *Nature* **352** (6331): 139–141
[17] Zhao Y, Kim Y H, Dillon A C, Heben M J, Zhang S B 2005 *Physical Review Letters* **94** (15): 155504
[18] Beck M T, Mándi G 1997 *Fullerenes, Nanotubes and Carbon Nanostructures* **5**(2): 291
[19] Bezmel'nitsyn V N, Eletskii A V, Okun' M V 1998 *Physics-Uspekhi* **41**(11):1091
[20] Ruoff R S *et al.* 1993 *Journal of Physical Chemistry* **97**(13): 3379
[21] Sivaraman N, Dhamodaran R, Kaliappan I, Srinivasan T G, Vasudeva Rao P R P, Mathews C K C 1994 *Fullerene Science*
[22] Topala P, Mazuru S, Besliu V, Cosovschii P, Ojegov A 2011 Application of EDI in increasing durability of glass moulding forms poansons. *Proceedings of The 15th International Conference, Modern Technologies, Quality and Innovation. Vadul-lui-Vodă, Chişinău,* V.II, 1093-1096. ISSN 2069-6736
[23] Topala P, Besliu V, Marin L 2014 *Advanced Materials Research* **1036** 172-177
[24] Topala P, Besliu V, Ojegov A 2011 Application of pulsed electrical discharges with graphite tool-electrode. Machine building and techno-sphere of XXI century *Proceedings of XVIII international scientific-technical conference,* Donetsk, pp. 240-245
[25] Topala P, Mazuru S, Besliu V, Cosovschii P, Stoicev P 2010 Increasing the durabilitz of glass moulding forms applying graphite pillicles. *Proceedings of The 14th International Conference, Modern Technologies, Quality and Innovation.* Slănic-Moldova, pp. 635-638. ISSN 2066-3919
[26] Marin L, Topala P, Stoicev P, Besliu V, Ojegov A 2014 *Engineering meridian* **1** 39-44
[27] Besliu V, Topala P, Stoicev P, Ojegov A, Hîrbu A, Guzgan D 2015 *Engineering meridian* **3** 40-46
[28] Li Y, Guo M, Zhou Z, Hu M 2002 *Journal of the International Societies for Precision Engineering and Nanotechnology* **26** 7–14

IOP Conf. Series: Materials Science and Engineering **209** (2017) 012036 doi:10.1088/1757-899X/209/1/012036

A Versatile Method for Nanostructuring Metals, Alloys and Metal Based Composites

G Gurau[1], C Gurau[1], L G Bujoreanu[2] and V Sampath[3]
[1]"Dunărea de Jos" University of Galati, Faculty of Engineering, Domnească Street, 47, RO-800008, Galati, Romania
[2]Faculty of Materials Science and Engineering, "Gheorghe Asachi" Technical University of Iaşi, Blvd. Dimitrie Mangeron 61A, 700050 Iaşi, Romania
[3]Department of Metallurgical and Materials Engineering, Indian Institute of Technology Madras, Chennai-600 036, India

E-mail: carmela.gurau@ugal.ro

Abstract. A new severe plastic deformation method based on High Pressure Torsion is described. The method patented as High Speed High Pressure Torsion (HSHPT) shows a wide scope and excellent adaptability assuring large plastic deformation degree on metals, alloys even on hard to deform or brittle alloys. The paper present results obtained on aluminium, magnesium, titan, iron and coper alloys. In addition capability of HSHPT to process metallic composites is described. OM SEM, TEM, DSC, RDX and HV investigation methods were employed to confirm fine and ultrafine structure.

1. Introduction

Ultrafine grained (UFG) and nanoscale metals and alloys have been of significant interest to research community due to their exhibited physical and mechanical properties accurately enhanced, not achievable in their coarse-grained counterparts [1-3]. Since the discovery of the production of UFG microstructures, in the early 1990s, using the ''top-down'' approach several publications dealing with pure metals and alloys have been appeared [4]. These techniques refer to imposing severe mechanical strains to a bulk solid with a relatively coarse grain size and processing the solid to produce a UFG or nanoscale microstructure [4]. Severe plastic deformation (SPD), among every production technique of bulk nanostructured materials (BNMs) proved to be a powerful tool for manufacture of nanostructures of different metallic materials down to ultrafine grain range, leading to remarkable combination of properties [1,5]. As much as, through variation of SPD processing parameters materials manifest various properties resulting from a wide range of grain sizes (below 1 μm), phase composition grain boundaries and interfaces evolve [6,7]. In this context, many methods have been developed to produce bulk fine-grained materials have been using very diverse SPD techniques such as: accumulative roll bonding (ARB) [8, 9], equal channel angular extrusion (ECAE) [1-13], high pressure torsion (HPT) [6,14,15,16], local canning compression [17,18], pure shear extrusion (PSE) [19] and so on. These procedures are capable of introducing large plastic straining (in excess of Von Misses strains of 1) and cause significant microstructural changes in bulk crystalline solids [20].

One of the most common and widely used methods of severe plastic deformation, used to obtain BNMs is high pressure torsion. High pressure torsion is a severe plastic deformation process, able to drastically reduce grain size, down to nanostructure or amorphous level, in bulk materials [21]. HPT, known ever since 1943, provided a direct demonstration of the ability to employ heavy plastic straining in the production of bulk materials having microstructures with grain sizes in the submicrometer range (under 500μm until amorphous) and with a high fraction of high-angle grain boundaries [21,22,23]. Among methods designed based on the HPT approach, stands High Speed High Pressure Torsion (HSHPT) original technique. Basically this technology is a combination of friction stir technology and High Pressure Torsion. In this article, we present new successfully produced UFG on different types of alloys and comprise ours previous recent published results by means HSHPT processing route. In addition capability of HSHPT to process metallic composites is described.

2. Materials and methods

2.1. HSHPT processing route

HSHPT is a patented, reliable method able to produce ultrafine grained and BNMs, for appreciable types of alloys [23-25]. As HPT, refers to the processing of metals whereby samples are simultaneously subjected to a compressive force and torsion strains, efficient in grain-refinement, contributing to a marked increase of mechanical properties [24-26]. Unlike classic HPT, to which rotations of the superior anvil performs at the low speed, of the order of 10-1 rotation per minute (rpm), in HSHPT technology the torsional stress is performed at high speed of the order of hundreds rpm up to 2000 rpm. The technology involves heat generation by friction between the anvils (a and b figure 1) and the sample (c) due to high pressure (h) on inferior anvil (around 1GPa) somewhat lower than commonly used in HPT and also the shear of the crystalline grains due to rotation of superior anvil at high speed. The HSHPT process is very fast, lasting about 10 s. Because entire HSHPT process is developing in a few seconds, a fast metal flow and thermal transfer takes place on the cold surfaces of the anvils, acquiring fine structure [27]. The HSHPT method has been designed to severe plastic deformation of brittle alloys. This technology is able to perform much larger, with a diameter as large as 45 mm and more complex-shaped specimens [25]. The extent of deformation/plastic strain was estimated using the following relationship: $\varepsilon=\ln(h_i/h_f)$; where h_i and h_f indicate the initial and final thickness of the sample, respectively [23].

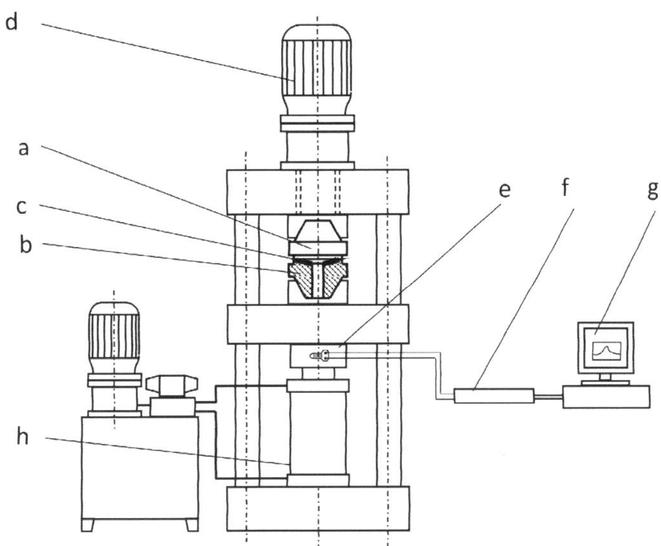

Figure 1. Principle scheme of HSHPT machine.

The HSHPT machine is powered by a 15 kW AC electric motor (figure 1d) operated by an EATON SVX024A1-4A1B1 frequency converter via PLC XC 200. The force variation is recorded with Hottinger Spider 8 system (figure 1e, 1f and 1g). Also the process temperature and displacement of lower anvil are monitored with a temperature sensor CT laser radiation pyrometer T2 MHCF OPTC and 5LMO LiU5X3 displacement magnetic-inductive sensor.

2.2. Materials

In order to identify the potential for severe plastic deformation of the HSHPT technology, various samples of metals and alloys were manufactured in the flat and coned discs or rings. Technical grade metals such as cooper, aluminium, magnesium were selected in the tests. Based on the results of previous works shape memory alloys (SMAs) group were systematically investigated, since offer a myriad of possibilities for different applications in the industry, based on unique properties: shape memory effect (SME) and the superelasticity (SE). Cu-Al-Ni SMA, Ti-Ni with slightly different composition and, consequently, different transformation temperatures were studied, Fe-Mn-Si-Cr SMAs and Ni-Co-Fe-Ga magnetic SMA have been the main subject of research. The new research direction consists of producing composite multilayers by the HSHPT method. HSHPT process can be used to produce multilayers coherent and well-bonded inasmuch as is similar to stir friction method. In this context we perform Cu/Al and NiTi (SME) /NiTi (SE) composites.

Microstructural observations were realized on the as-cast, extruded or rolled state, as well as the HSHPT-processed alloys specimens, described above, and subjected to different degrees of deformation. Samples were prepared for primary microstructural examination by optical microscopy (OM) as well as scanning electron microscopy (SEM). The samples were cut, enclosed in cold mounting resin, and polished using standard metallographic techniques; the samples were then etched using a typical solution for each alloy system.

The OM examination was carried out using an OLYMPUS BX51 microscope equipped with a video camera and the QCapture software package. A Zeiss microscope was used for the SEM studies. The microstructural analysis of the samples by TEM was carried out using a Tecnai 20G2 transmission electron microscope operating at a voltage of 200 kV and equipped with an EDX spectrometer. Thin foils from HSHPT-processed specimens were prepared using an electrolyte with 90vol% of methanol and 10vol% of perchloric acid. The phase identification was carried out using an X-ray diffractometer (Bruker AXS D8 Discover) equipped with a Cu Kα radiation source. The transformation temperatures of the specimens processed were determined by differential scanning calorimetry (DSC TA Q20). A heating/cooling rate of 50°C /min in the range of 0 to 250°C was adopted for the experiments. Room temperature Vickers microhardness values were determined using an indenter under a load of 0.98067N (HV0.1) applied for 10s.

3. Results and discussion

3.1. Cu-Al-Ni SMA

The HSHPT technology was developed for plastic deformation of severe brittle polycrystalline Cu-Al-Ni shape memory alloys, attempted to improve the ductility by grain refinement. The presence of coarse grains of extruded billets, larger than 200μm are shown in Figure 2a does not recommend it for severe deformation. The maximum pressure applied was 0.7GPa. The strain levels applied was between 0.7 and 3.62. The HSHPT discs were achieved up to 29mm in diameter. The SPD occurs in about 20s. Figure 2b present elongated grains in case of 1.7 deformation degree. As shown in Figure 2c, subjecting the alloy to HSHPT processing to a logarithmic strain of 3.5 the coarse austenitic grains have completely been replaced by flowlines.

The room temperature TEM micrographs show a heavily deformed microstructure of the HSHPT processed alloy incorporating a logarithmic strain of 1.7. In figure 3 martensite plates and stress fields are clearly visible for extruded initial sample.

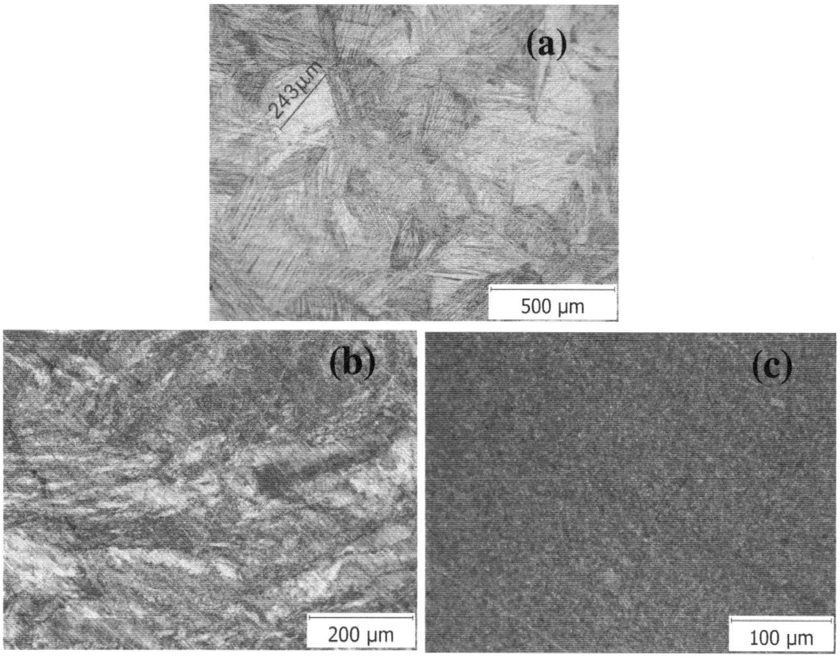

Figure 2. Optical micrographs of Cu-13Al-4Ni: a) extruded; severely plastic deformed by HSHPT logarithmic strain of: b) 0.7; c) 2.5.

Grain refinement is not accompanied by a high density of lattice defects as can be seen in figure 3 b. but finer twins and needles are accompanied by numerous stacking faults. Parallel-banded of two martensite plate variants that spatial intersecting are notable. These morphologies are referred to as twin type martensite: γ1′ (2H) orthorhombic predominantly after SPD and β1′ (18R) monoclinic, confirmed by XRD investigations [27].

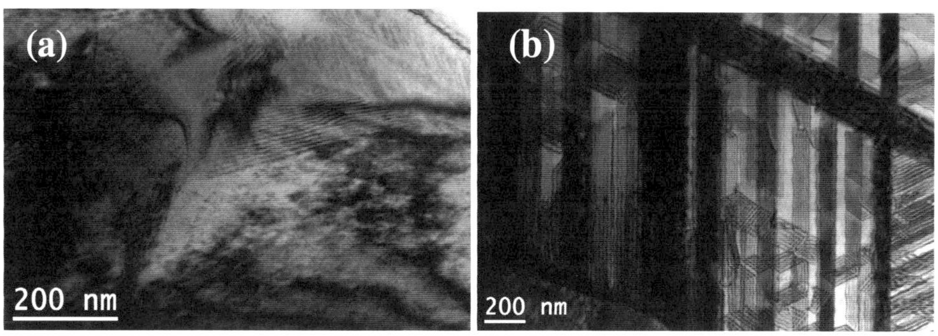

Figure 3. TEM images of Cu-13Al-4Ni alloy a) extruded b) martensite morphology.

The alloy processed by HSHPT shows reversible martensitic transformation without the necessity for post-deformation annealing and phase transformation stability after 10 thermal cycles. Due to microstructural refinement the microhardness increases by nearly 50%.

3.2. Ti-Ni SMA

All selected Nickel–Titanium (Ni–Ti) alloys of two different near-equiatomic chemical compositions (Ni-rich and Ti-rich) subjected to severe plastic deformation by means of HSHPT exhibit unique shape memory effect (SME) or superelasticity (SE) quickly after deformation without any post-deformation heat treatment. The DSC and in situ XRD studies confirm reversible martensitic transformation in modules severe plastic deformed [28]. The rotation speed applied was 1795 rpm and the maximum force was about 120kN. The torsion time for Ni-Ti samples varied between 7 to 19 seconds determining different degrees of refining of the crystalline grains. The strain levels applied was between 0.92 and 3. The HSHPT modules were achieved up to 45mm in diameter as against those manufactured by means HPT usually about 10 mm in diameter.

For Ni rich Ni-Ti, in the case of the deformation degree of 0.92 (figure 4a) the austenitic grains are finer than few micrometres, equiaxed and distributed almost uniformly. The microstructure of Ni-Ti alloy after HSHPT processing with 1.52 deformation degree is characterized by the obviously elongated grains with waved grain boundaries (figure 4b).

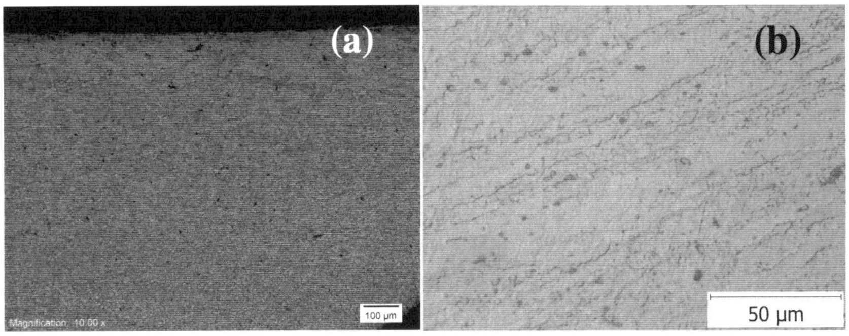

Figure 4. Optical micrographs of Ni-Ti severely plastic deformed by HSHPT logarithmic strain of: a) 0.92; b) 1.52.

After subjecting the alloy to the HSHPT processing to a logarithmic strain of 0.92, the grains exhibit apparent grinding. Inside the B2 austenite grains appear especially randomly oriented twins (figure 5a). The refinement of the grain size is more evident after increasing degree of deformation as shown in figure 5b. In discs subjected to a logarithmic strain level of 2.33 the individual grains or grain boundaries were no longer observed being outside the range of detection by standard SEM observation. The sample reveals only large precipitates and lines of one martensite plates variant.

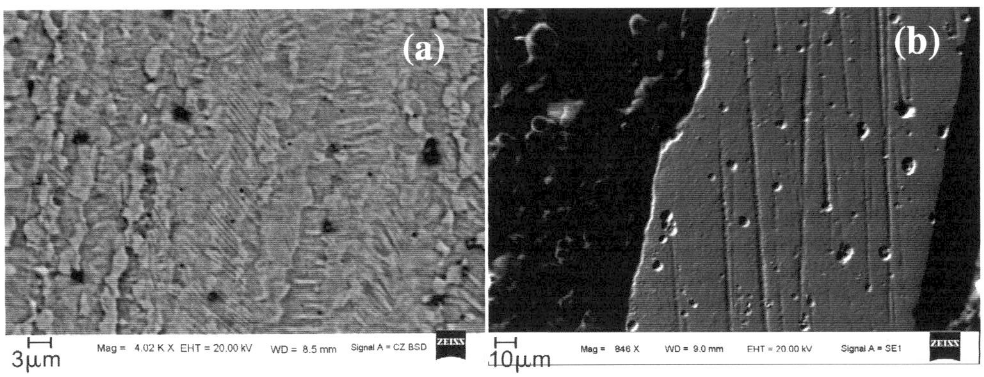

Figure 5. SEM images of Ni-Ti severely plastic deformed by HSHPT logarithmic strain of: a) 0.92; b) 2.33.

TEM micrographs of the Ni-Ti sample that was subjected to a deformation level of 1.71. Figures 6a and b highlight the heavily deformed microstructure showing a complex strain contrast, having no preferred orientation, lattice defects such as dislocation accumulated in clusters, dislocation tangles and high density of dislocations, also residual stress field can be find. Figure 6c shows multiple fine twins and variants with parallel orientations of b19' martensite with a monoclinic crystal structure.

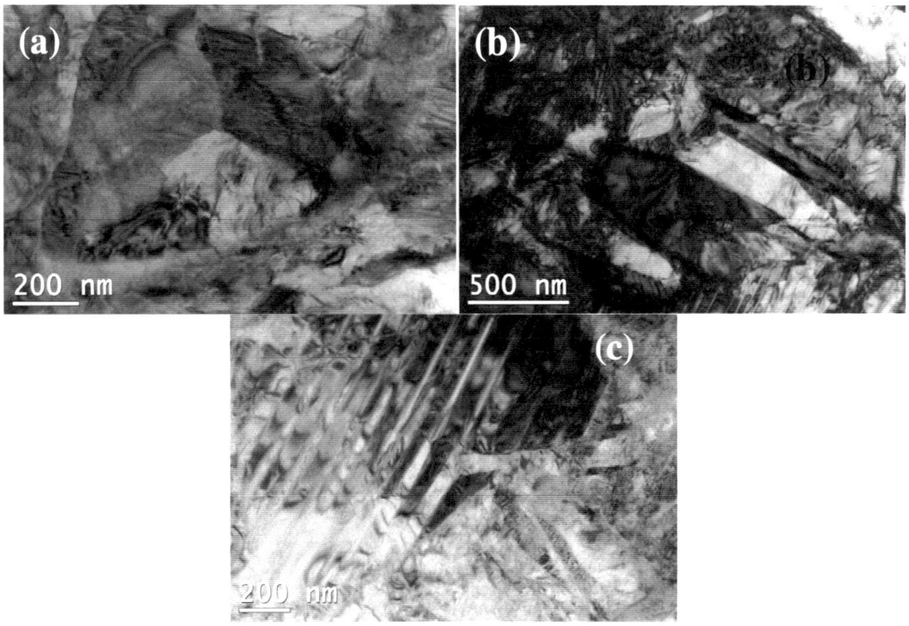

Figure 6. TEM images of Ni-Ti severely deformed with logarithmic strain 1.72 highlighting areas of different structural morphologies.

The hardness of HSHPT processed specimens s increased with increasing of the deformation degree up to 180%, being essentially homogeneous along diameter [28].

3.3. Fe-Mn-Si-Cr SMA

HSHPT Fe-Mn-Si-Cr modules can behave as actuators, being able to generate compensation strokes or constrained recovery forces.

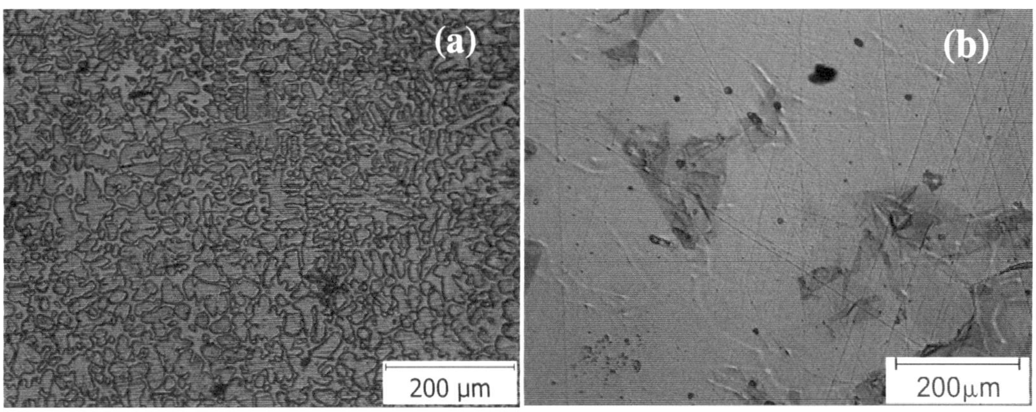

Figure 7. Optical micrographs of Fe–Mn–Si–Cr alloy (a) cast state and (b) after severe plastic deformation by HSHPT of 2.3.

A superelastic-like response was obtained during compression loading-unloading to coned-disk spring shape modules, obtained from as cast state alloy (figure 7a) by means of HSHPT procedure [29]. The presence of ε (hcp) stress-induced in samples severe plastic deformed was confirmed by OM (figure 7b), SEM (figure 8), TEM (figure 9) micrographs and XRD patterns, too [23, 25]. The distorted SPD alloy microstructure can be seen in Figure 7b, revealing only the dark contrast of the triangular morphology of ε-HCP martensite.

The micrograph shown in figure 8a reveals two variants of fine martensite plates on surface relief of samples with 2.21 logharthmic strain. As evident in Figure 8b, the individual grains or grain boundaries in discs subjected to a logarithmic strain level of 3.0 were outside the range of detection by standard SEM observation.

The TEM analysis of the alloy after HSHPT subjected to a logarithmic strain level of 0.9 and 1.53 highlighted grain refinement down to 50 nm with two predominant variants of ε-HCP martensite and numerous stacking fault (figure 9a and b). Also ascertain the presence of a high density of dislocations characteristic of cold severe plastic deformation, were not found in this alloy processed by HSHPT. No post-deformation annealing was required after HSHPT processing for actuator applications.

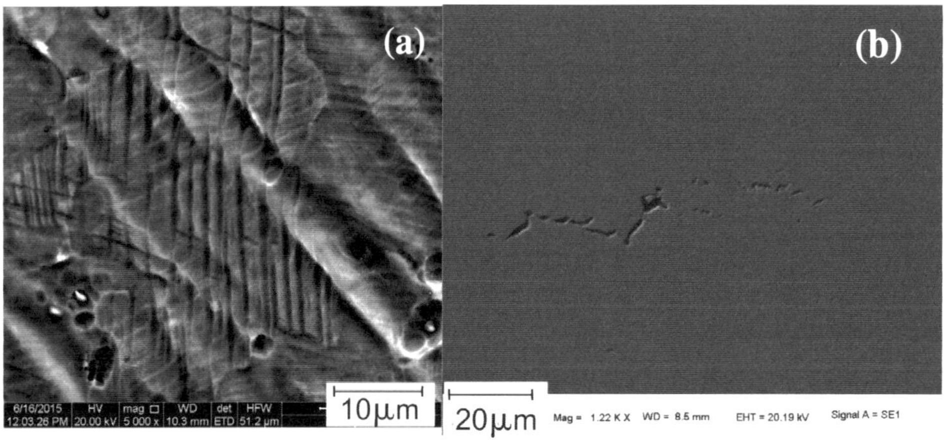

Figure 8. SEM micrographs of Fe-Mn-Si–Cr alloy after severe plastic deformation by HSHPT to a longitudinal logarithmic strain of: (a) 2.21 and (b) 3 (transversal cross section).

The maximum force inflicted was approximately 120kN. The processing time varied between 2 and 11 s for the Fe- based alloy. The strain levels applied was between 0.15 and 3. The HSHPT modules exhibited maximum 35 mm diameter having thickness of 0.18 mm.

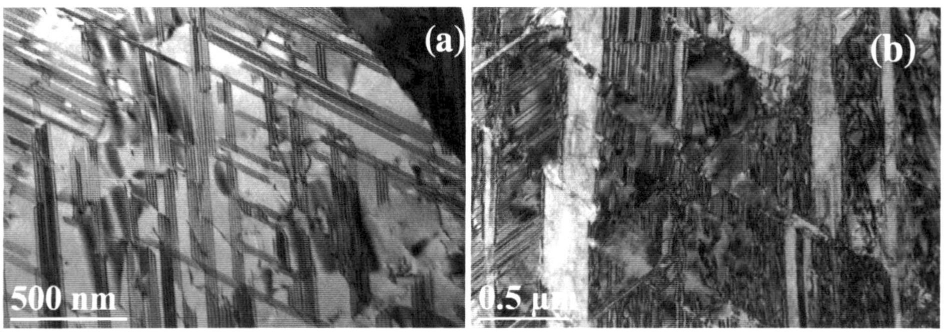

Figure 9. I TEM images of the Fe–Mn–Si–Cr after severe plastic deformation by HSHPT to logarithmic strain: a) 0.73 and b) 1.53.

3.4. Ni-Co-Fe- Ga SMA

An ongoing rcscarch on magnetic Ni–Fe–Ga alloys points out that by using HSHPT severe plastic deformation method the magnetic properties are improved on the strength of promoting the formation of the β phase martensitic. There is effect of the deformation on decrease of γ phase which obstructs the formation of the β phase. The γ phase were highly deformed elongated long the flow lines (Figure 10 a) and waved (figure 10 b).

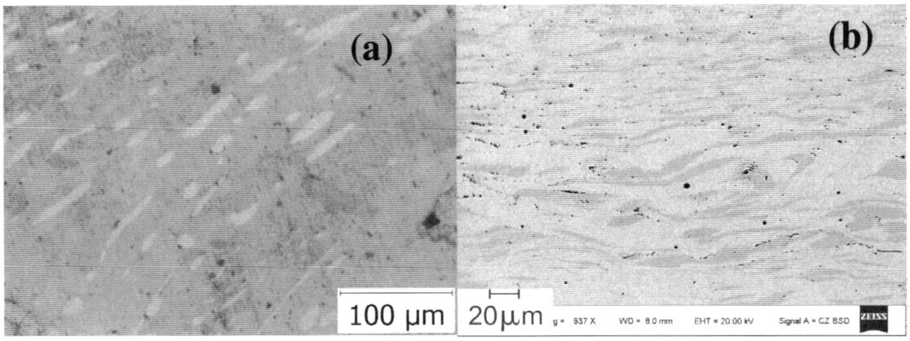

Figure 10. HSHPT 'ed Ni-Co-Fe-Ga magnetic SMA a) Optical and b) SEM micrographs.

3.5. Cu/Al multilayers composite

The initial copper specimens possessed average grains size of 159μm, and the average grains size of aluminum 56 μm (figure 11a and 11b) respectively. HSHPT is able to bring about Cu/Al composite by 2 to 11 layers, showing coherent, efficient bonding with fine structure (figure 11c and 11d)).

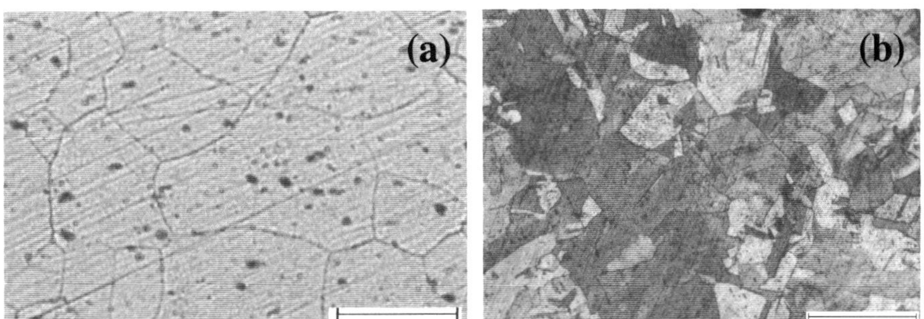

Figure 11. I OM micrograph of: a) Al; b) Cu; c) HS HPT processed multilayers Cu/Al composite cross sections (c) seven layers and d) highlighting Al layers by metallographic etching

During SPD, large cooper clusters (average size of about 10μm) moved deeply into the aluminum matrix, markedly for all Cu/Al composites, showing a homogeneous distribution, with clear boundaries, slightly elongated along the radial direction. A remarkable increase in the Vickers microhardness (up to 16 time compared to the initial state) was noticed for all the multilayers composites, regarding of the numbers of layers as a consequence of the microstructural refinement achieved by HS HPT method. SPD occurs quickly within 2-5s relying on number of layers. The discs diameter was up to 45 mm.

3.6. Ni-Ti (SME) /Ni-Ti (SE) multilayers composite

SEM and OM micrographs of Ni-Ti (SME) /Ni-Ti (SE) multilayer after HSHPT processing is shown in Figure 12 a and b. It is observed, even if the number of layers increase at 24, the interface between the layers is straight and adherent. This multilayer composite has potential for use in MEMS.

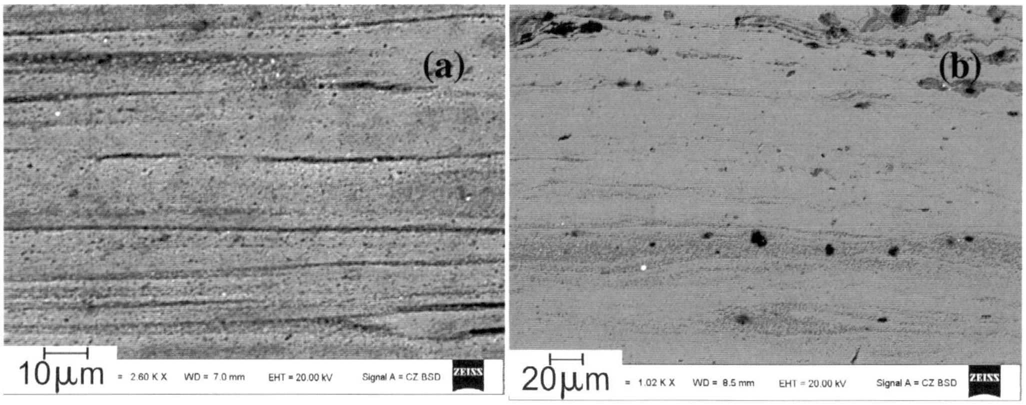

Figure 12. HS HPT processed multilayers NiTi (SME) /NiTi (SE) composite cross sections (24 layers): a) SEM and b) OM.

4. Conclusion

1. SPD processing by HSHPT technique, leads to the occurrence of significant grain refinement down to ultrafine range until the nanocrystalline range, in various metals and alloys like: aluminium, magnesium, titan, iron and coper. The process can be applied even for brittle and hard deformable alloys. SMAs group benefit of this technique based on improved properties: shape memory effect (SME) and the superelasticity (SE).
2. Among methods designed based on the HPT approach, stands High Speed High Pressure Torsion (HSHPT) original technique being a combination of friction stir technology and High Pressure Torsion.
3. HSHPT is a promising and reliable method for achieving semi-products with rotational symmetry, whose dimensions exceed those of the HPT samples by four times depending on system alloy.
4. The capability of HSHPT to process multilayers metallic composites is described.

References

[1] Abramova M M, Enikeev N A, Kim J G, Valiev R Z, Karavaeva M V, Kim H S 2016 *Mater. Lett.* **166** 321–324
[2] Agra K, Bohn F, Mori T J A, Callegari G L, Dorneles L S, Correa M A, Magn J 2016 *Mater.* **420** 81–87
[3] Bujoreanu L, Goanță V, Cimpoeșu N, Gurău C, Suru M 2015 *MATEC Web Conf.* **1** 3–8
[4] Valiev R Z, Langdon T G 2006 *Prog. Mater. Sci.* **51** 881–981

[5] Valiev R Z, Enikeev N A, Murashkin M Y, Kazykhanov V U, Sauvage X 2010 *Scr. Mater.* **63** 949–952
[6] Shahmir H, Langdon T G 2016 *Mater. Sci. Eng.* A **667** 293–299
[7] Tohidi A A, Ketabchi M, Hasannia A 2013 *Mater. Sci. Eng.* A **577** 43–47
[8] Karimi M, Toroghinejad M R, Dutkiewicz J 2016 *Mater. Charact.* **122** 98–103
[9] Ghalandari L, Mahdavian M M, Reihanian M, Mahmoudiniya M 2016 *Mater. Sci. Eng.* A **661** 179–186
[10] Druker A V, Baruj A, Isola L, Fuster V, Malarría J, Bolmaro R 2016 *Mater. Des.* **107** 153–162
[11] Kim S M, Arockiakumar R, Park J K 2012 *Mater. Sci. Eng.* A **546** 53–58
[12] Ma J, Karaman I, Kockar B, Maier H J, Chumlyakov Y I 2011 *Mater. Sci. Eng.* A **528** 7628–7635
[13] Gurău G, Gurău C, Braz-Fernandes F M, Bujoreanu L G 2014 *Indian J. Eng. Mater. Sci.* **21**
[14] Panigrahi A, Sulkowski B, Waitz T, Ozaltin K, Chrominski W, Pukenas A, Horky J, Lewandowska M, Skrotzki W, Zehetbauer M, Mech J 2016 *Behav. Biomed. Mater.* **62** 93–105
[15] Qiao X, Li X, Zhang X, Chen Y, Zheng M, Golovin I S, Gao N, Starink M J 2016 Intermetallics Formed at Interface of Ultrafine Grained Al/Mg Bi-Layered Disks Processed by High Pressure Torsion at Room Temperature
[16] Kuramoto S, Nagasako N, Furuta T, Horita Z 2013 *J. Alloys Compd.* **577** S147–S150
[17] Jiang S, Zhang Y, Zhao L, Zheng Y 2013 *Intermetallics* **32** 344–351
[18] Jiang S, Hu L, Zhao Y, Zhang Y, Liang Y 2013, *Mater. Sci. Eng.* A **569** 117–123
[19] Rahimi F, Eivani A R 2015 *Mater. Sci. Eng.* A **626** 423–431
[20] Mara N A, Beyerlein I J 2015 *Curr. Opin. Solid State Mater. Sci.* **19** 265–276
[21] Alizadeh R, Mahmudi R, Ngan A H W, Huang Y, Langdon T G 2016 *Mater. Sci. Eng.* A **651** 786–794
[22] Rentenberger C, Waitz T, Karnthaler H P 2007 *Mater. Sci. Eng.* A **462** 283–288
[23] Gurău G, Gurău C, Potecaşu O, Alexandru P, Bujoreanu L G 2014 *J. Mater. Eng. Perform.* **23** 2396–2402
[24] Resnina N, Belyaev S, Zeldovich V, Pilyugin V, Frolova N, Glazova D 2016 *Thermochim. Acta* **627–629** 20–30
[25] Gurau G, Gurau C, Sampath V, Bujoreanu L G 2016 **23** 1–8
[26] Kang J Y, Kim J G, Kim S K, Chin K G, Lee S, Kim H S 2016 *Scr. Mater.* **123** 122–125
[27] Gurau G, Gurau C, Bujoreanu L G, Fernandes F M B 2014 in: Int. Multidiscip. Sci. GeoConference Surv. Geol. Min. Ecol. Manag. SGEM
[28] Gurau G, Gurau C, Sampath V 2015 *MATEC Web Conf.* **33** 3003
[29] Gurău G, Bujoreanu L G, Gurău C, Comăneci R I, Lohan N M, Pricop B, Suru M G 2015 *Int. J. Mod. Manuf. Technol.* **7** 23–27

IOP Conf. Series: Materials Science and Engineering **209** (2017) 012037 doi:10.1088/1757-899X/209/1/012037

Study on Fired Clay Bricks by Replacing Clay with Palm Oil Waste: Effects on Physical and Mechanical Properties

A A Kadir[1,2], N A Sarani[1], M M A B Abdullah[3,2], M C Perju[2,4] and A V Sandu[2,4]
[1]Department of Water and Environmental Engineering, Faculty of Civil and Environmental Engineering, Universiti Tun Hussein Onn Malaysia, Johor, Malaysia
[2] Center of Excellence Geopolymer and Green Technology (CEGeoGTech), School of Materials Engineering, Universiti Malaysia Perlis (UniMAP), Perlis, Malaysia
[3]Faculty of Engineering Technology, Universiti Malaysia Perlis (UniMAP), Perlis, Malaysia
[4]Faculty of Materials Science and Engineering, Gheorghe Asachi Technical University of Iasi, Romania

E-mail: aeslina@uthm.edu.my[1], nramira1987@gmail.com[2]

Abstract. Palm oil is one of the major agricultural industries in Malaysia. Due to the poor management system, the discarded palm oil waste has always been linked to the environment issues. During processing of palm oil, a considerable amount of solid waste by-products in the form of fibres, shells, empty fruit bunches and fly ashes are produce rapidly. Therefore, this study was conducted to incorporate 1%, 5% and 10% of palm oil waste into fired clay brick. Samples of brick were fired at 1050°C temperature with heating rates of 1°C/min. Manufactured bricks were tested with physical and mechanical properties including firing shrinkage, dry density, water absorption and compressive strength. The results demonstrated that the replacement of 1% up to 5% of palm oil waste had improved several properties, although, a decrease of performance in certain aspects has also been observed. As a result, palm oil waste can be utilized in an environmentally safe way into fired clay brick thus providing adequate properties of fired clay brick.

1. Introduction

Oil palm plantation is one of the successful stories of agricultural industry in Malaysia. From a modest beginning in the early 1920s, the palm-oil industry developed rapidly where Malaysia is the second largest palm oil production after Indonesia [1]. From figure 1, production of crude palm oil has recorded a marginal decrease until December 2016 with 15% to 17.3 million tonnes compared to 19.9 million tonnes in year 2015 [2]. The decrease was due to lingering effects of the El Nino, rejection of additional palm oil levy in France and low stockpiles [3]. However, it is projected to increase in year 2017 since world palm oil production increase by 5.65 million tonnes or 9.6% compared to last year with 64.5 million tonnes [1]. This is due to higher fresh fruit bunch processed from newly matured area especially in Sabah and Sarawak and also a better management of the organization.

Meanwhile, oil planted area in Malaysia as the year of 2015 has reached 5.64 million hectares with increasing of 4.6% compared to previous year. According to Figure 2, Sabah is still the largest planted area with 27.4%, followed by Sarawak with 25.5%. Peninsular Malaysia accounted with 47% of total planted area [4].

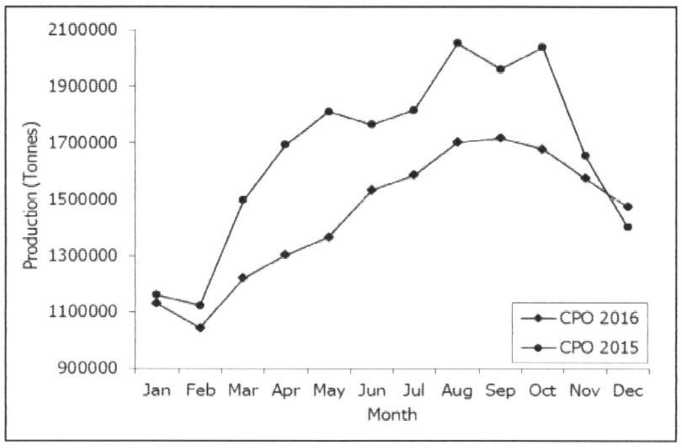

Figure 1. Monthly production of crude palm oil for the year of 2016 [2].

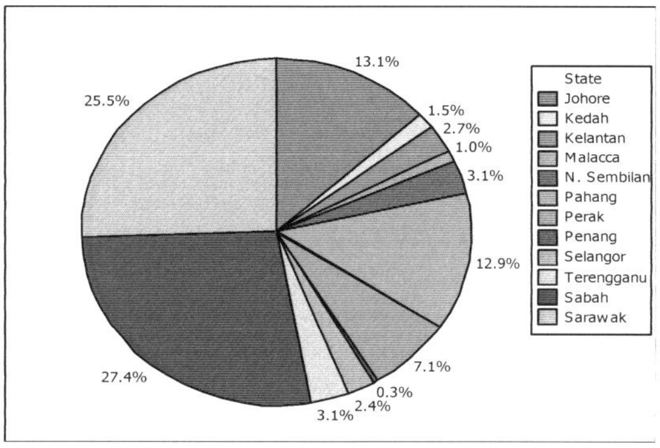

Figure 2. Oil planted area by state as at December 2015 [4].

The nation's pollution dilemma caused by the oil palm sector is increased by an annual crop of 4 million tons of solid waste [5]. A large area is required for the disposal of these palm oil waste materials. According to [6], one of the problems associated with the production of palm oil is large quantities of processed residues that have no economic value and it only disposed of by illegal and uncontrolled open burning which is discouraged by Department of Environment Malaysia.

Table 1. Waste generated during palm oil process [8].

Wastes	Quantity of waste generated
Empty fruit bunch	22%
Palm fibre	13.5%
Palm kernel shell	5.5%
Palm oil mill effluent	67%

In the processing of crude palm oil and palm kernel from fresh fruit bunch, the palm-oil industry produces substantial amount of biomass wastes in the form of fibres, shells and empty bunches which is higher in fibre content [7]. From the extraction of fresh fruit bunch, palm oil consists only 10% of the total biomass and the rest 90% are discarded as waste [8]. Currently, shell and fibre are used extensively as fuel for the production of steam in the palm-oil mills as a means of waste disposal and energy recovery. Table 1 shows the waste generated during palm oil processing. Palm oil fiber (PF) is natural fiber extracted from palm oil vascular bundle the empty fruit bunch (EFB) through the retting process Palm fiber is stable and non-hazardous biodegradable material which can use or processed for various purposes such as ceramic, soil stabilization, compaction, brick manufacturing and mattress erosion control. Palm kernel shell is hard to decompose and has been reused as mulch. According to [9], approximately 5 tons of palm kernel shell is obtained from 66 tons of fresh fruit bunches. Palm kernel shell has high grade solid, low ash and low sulphur content.

Meanwhile, empty fruit bunch is the by-product from crude palm oil mill which having low economic value before more application being developed. Currently, the major application of empty fruit bunch has extracted the fibre for others industry. By realizing empty fruit bunch has potential as energy sources, it has been used in clay brick to save energy during firing stage [6], [10].

Another concern is palm oil fuel ash (POFA). POFA is the ash produced from the husk fibres and shells during palm oil burning. Currently, shells and fibre wastes are utilized widely as fuel for the production of steam in palm oil mills. However, about 5% of ash weight is engendered after combustion process. This ash is amassed from the boiler and directly dumped onto open fields close to the mill. Hence, disposal of this waste becomes equipollent compulsory.

Agreeing by [11], 998 million tonnes of agricultural waste is produced every year and in Malaysia itself, 1.2 million tonnes of agricultural waste is disposed of in landfills annually. Waste material from agricultural activities would cause environmental pollution and this waste must be managed properly to eliminate the effect towards the environment. Other than that, health implication of poor waste management can be very damaging to the people.

1.1. Incorporation Waste into Fired Clay Bricks

Reduction of dumped waste and towards environment sustainability can be ensured by appropriate consumption or recycling of these materials. Incrementing world population and life demand are persistently raising the price of crude materials and reducing the natural resources. Along with this line, researches have been concentrated to utilise waste materials as a potential alternative in the construction industry.

Brick is the man-made building materials that have been used since the early human civilisation. Bricks are broadly used for construction, civil engineering work and landscape design since it has an attractive appearance, strength and durability, fire and weather resistance, thermal and sound insulation. The quality of brick usually depends on the composition of raw materials, production method, firing method and firing temperature [12]. Because of the versatility of clay, when it mixed with water, clay mineral give a plastic mass that can be formed by pressure and fired in a furnace [13]. This firing process transforms the clay into a building component in high performance of properties.

Fired clay brick can describe as common, facing or engineering brick. Common brick has no visual finish and ordinarily used for general building work particularly when brickwork is to be rendered, plastered or unseen in the finished work. It is also widely used as interior wall surface or exterior walls that are below the ground level. Facing brick are manufactured to give an attractive finish. It was intended to be used in both structural and non-structural masonry, where appearance required. The colour which may be uniform or multi coloured resulted from the mixes of clay used and the firing condition. Facing brick is available in a wide range of colours, textures, smooth, light or sandy. It often applies to the top or the surface of the wall for ecstatic factor. Engineering bricks usually are dense, strong, used to support heavy loads, and also where the effects of impact damage, water absorption or chemical attack that needs to be minimised. Moreover, they are further subdivided into facing and common according to their properties.

2. Materials and method

2.1. Materials

Clay soil was supplied by a brick manufacturer located at Yong Peng, Johor. Upon delivery, clay soil was oven dried at 105°C for 24 hours. Dried clay soil was crushed and ground to yield homogenous size pass through a 500 μm sieve.

Meanwhile, palm oil waste was collected from palm oil mill plantation at Kluang, Johor. Palm oil wastes used in this study consists of palm kernel shell (PKS), palm oil fuel ash (POFA), palm fibre (PF) and empty fruit bunch (EFB). Palm oil wastes also were oven dried to keep constant mixing water content during brick manufacturing. The samples of PKS, PF and EFB need to be crushed and ground pass through 4.75 mm and retained at 2.36 mm nominal sieve. Meanwhile, POFA was sieved to remove the leftover of nutshells and fibres which are not completely burned during burning process in boiler. Figure 3 to figure 6 shows different type of palm oil waste used in this study. The chemical composition of clay soil and palm oil waste were analysed using X-Ray Fluorescence (XRF).

Figure 3. Palm oil fuel ash.

Figure 4. Palm kernel shell.

Figure 5. Palm fibre.

Figure 6. Empty fruit bunch.

2.2. Methods

The geotechnical properties of clay soil and palm oil waste such as soil classification and specific gravity test were determined according to British Standard [12]. Soil classification is necessary to analysed because it define or classify type of soil or predict soil performance when combining with waste. This test covers the determination of plastic limit, liquid limit and plasticity index. Meanwhile, specific gravity test was performed using small pyknometer method. Distilled water was used as a standard for density bottle fluid. However, samples contain salts such as palm oil waste, kerosene was used as fluid. Before conducting a trial mix of control brick and clay-POW brick, necessary amount of water need to determine in order to achieve adequate plasticity of the mixture. Adequate water can minimize absence of cracking during drying and firing process. Optimum moisture content can be performed by Standard Proctor Test according to British Standard [13].

2.3. Brick manufacturing

The trial mix was prepared to determine suitable percentage of palm oil waste content for brick manufacturing. The ratio of palm oil waste was calculated based on specific gravity of the samples. The sample was started by mixing clay soil and clay-waste sample with predetermined water. The mixture was then pressed into mould for the following size; 215 mm x 102.5 mm x 65 mm with pressure of 3000 psi. The wet brick sample was kept for 24 hours at room temperature followed by oven dried at 105°C for another 24 hours. The dried bricks finally were fired in a furnace at 1050°C with heating rates of 1°C/min. Control brick with zero waste also was prepared for control purpose.

The manufactured bricks were designed as control brick (CB) for brick without waste, palm kernel shell brick (PKSB), palm oil fuel ash brick (POFAB), palm fibre brick (PFB) and empty fruit bunch brick (EFBB). The manufactured clay bricks then went through a series of test including physical and mechanical properties such as firing shrinkage, dry density, water absorption [14] and compressive strength [15].

3. Results and discussion

3.1. Characterization of raw materials

The clay soil presents a typical composition and mainly consists of Silica (SiO_2), Alumina (Al_2O_3) and Ferric Oxide (Fe_2O_3) with 55.7%, 24.4% and 4.46% respectively. Minor contents in clay soil consist of Sodium Oxide (Na_2O) and Calcium Oxide (CaO) with 0.3% and 0.25% respectively. Palm Oil Fuel Ash (POFA) was found high in Silica (SiO_2), Ferric Oxide and Calcium Oxide (CaO) content with 32.7%, 13.36% and 11.1% respectively. This significant amount of Silica in POFA is greatly contributing to the pozzolanic reaction thus increased the binding between clay soil particles [16], [17]. Meanwhile, chemical compositions in Palm Kernel Shell (PKS) were high in Ferric Oxide (Fe_2O_3), (SiO_2) and Calcium Oxide (CaO) with 34.37%, 22.58% and 14.3% respectively. Chemical composition test also found that Silica (SiO_2), Ferric Oxide (Fe_2O_3), Calcium Oxide (CaO) and Sulphur Trioxide (TiO_2) are the major element present in both Palm Fibre (PF) and Empty Fruit Bunch (EFB). The concentrations are 22.58%, 10.13%, 15.00% and 7.38% for PF and 25.4%, 15.84%, 13.33% and 7.56% for EFB. It can be concluded that major chemical composition of palm oil wastes are extremely similar to clay soil and high silica in raw materials is beneficial to increase strength of bricks [18].

Table 2. Chemical composition of raw materials.

Oxides	Clay soil	PKS	POFA	PF	EFB
SiO_2	55.7	22.58	32.7	26.9	25.4
Al_2O_3	24.4	9.08	3.32	9.8	8.27
Na_2O	0.3	0.37	NA	NA	0.31
K_2O	2.24	2.14	15.9	8.46	7.41
Fe_2O_3	4.46	34.37	13.36	10.13	15.84
CaO	0.25	14.3	11.1	15.0	13.33
MgO	1.2	1.88	2.34	3.19	4.25
P_2O_5	NA	1.65	NA	6.07	0.87
TiO_2	0.94	0.79	NA	1.28	1.59
MnO	0.94	0.16	NA	0.19	0.51
SO_3	NA	1.56	3.65	7.38	7.56
CuO	NA	NA	NA	0.2	0.15
Cr_2O_3	NA	0.2	NA	0.15	0.4
ZnO	NA	0.3	NA	NA	NA

3.2. Properties of manufactured brick

3.2.1. Firing shrinkage. The addition of palm oil waste into fired clay brick has changed the shrinkage of the clay bricks where total shrinkage was increased with increasing of palm oil waste. The results in figure 7 have shown that the addition of 10% of palm oil waste might result in various pore sizes after firing thus increased the shrinkage values of bricks. By comparing different types of palm oil waste, the presence of EFB demonstrates highest shrinkage compared to PKS, POFA and PF into the mixture. It was found that cellulose fibre in EFB absorbed the water the within the existing particles and became the leading cause of the increased need for water thus resulted in the increased firing shrinkage of brick. This data obtained was in good agreement with the results reported by [19], controlling shrinkage during firing process could minimize deforming of sample due to excessive shrinkage.

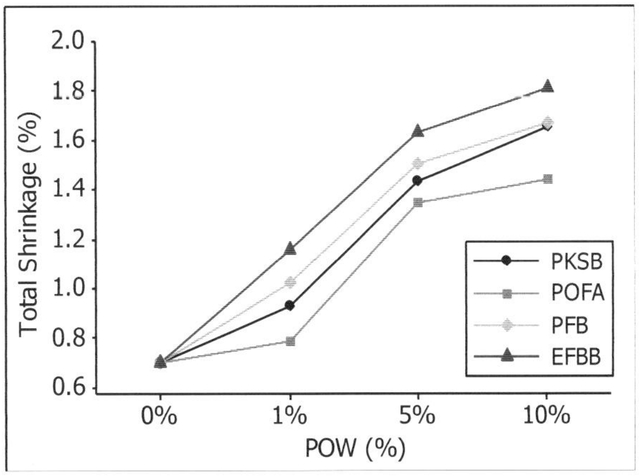

Figure 7. Firing shrinkage of POW brick.

3.2.2. Dry density. The result in figure 8 shows that the density of palm oil waste brick was inversely proportional to the percentages of palm oil waste added in the mixture. As a result, there was a linear decreased in dry density.

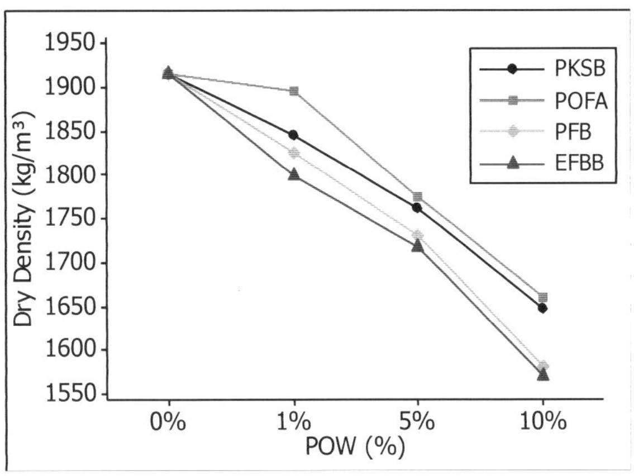

Figure 8. Dry density of POW brick.

It was found that the dry density was decreased with an increasing amount of palm oil waste ranging from 1% to 10%. In the meantime, different types of palm oil waste show different pattern of dry density where EFBB is lighter compared to other palm oil waste brick (with addition 1% to 10% of EFB). This is because of the characteristic of EFB and PF itself which is high in organic matter, could be eliminated during sintering process. This is similar to [20] that weight loss could be due to elimination of the organic matter from the clay and residue using combustion and to the removal of water content from clay mineral due to dehydroxylation reactions in the clay. Meanwhile, POFAB is likely high compared to others palm oil waste brick due to the fineness of the POFA particles which is eventually filling the voids between clay particles.

3.2.3. Water absorption. Figure 9 shows a correlation between the increasing percentages of palm oil waste which tend to increased water absorption. Increasing the percentages of POWs from 1% to 10% into fired clay brick resulted in an increment of water absorption. The same pattern of water absorption for different types of palm oil waste is quite similar with the result in firing shrinkage where POFAB absorb low water during 5 hours boiling test compared to PKSB, PFB and EFBB. Meanwhile, it can be seen in the case of incorporation 5% and 10% of PF and EFB, there are narrow differences in the water absorption; however both bricks still absorb high water content. This may happen because the elimination of the organic matter from the brick bodies during sintering process created numerous pores inside the brick.

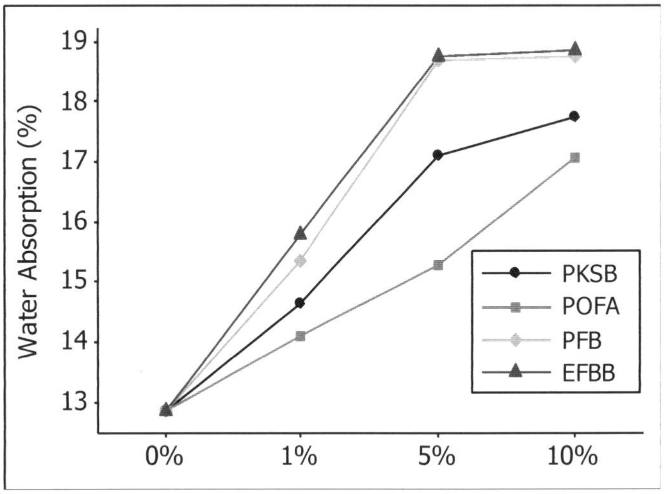

Figure 9. Water absorption of POW brick.

3.2.4. Compressive strength. Figure 10 shows that brick with higher percentage of waste obtained lower strength. It is expected that PFB and EFBB contain high organic matter thus leave pores inside brick during sintering process. By comparing different of POWs, waste contains lower organic content such as POFA achieved higher compressive strength compared to waste with high organic content.

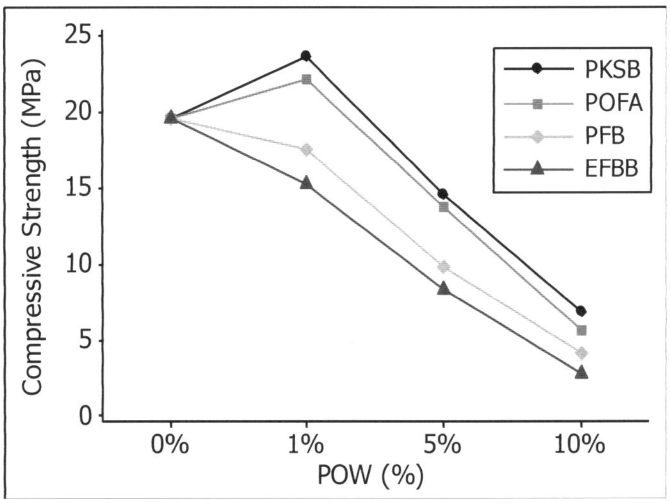

Figure 10. Compressive strength of POW brick.

This might be due to a significant amount of silica in POFA is believed to contribute towards the pozzolanic reaction during hydration thus improving binder with clay soil.

4. Conclusion

The brick in this study was manufactured from clay and potential palm oil waste (PKS, POFA, PF and EFB) by controlling the optimum ratio/percentages of waste into fired clay brick. Initial examination on the brick appearance found that there are minimum defects such as minor crack and no bloating were observed after firing. Regarding physical and mechanical properties, incorporation of palm oil waste to the clay body increases the required water content for maintained the plasticity of the mixture. Shrinkage of the brick is firmly increased to the expected stabilization effect of cellulose-fibre which is mostly due to very high water content. The study has found that all the manufactured POWs brick were under allowable limit of shrinkage and size variation of the brick samples were virtually uniformed. The results indicated that is possible to incorporate 1% up to 10% of palm oil waste into fired clay brick, which fulfills several limit standards and possesses mechanical properties similar to those of control brick. However, incorporation of 20% and 30% of palm oil waste has found to be failed due to major disintegration effect during sintering stage. It is noteworthy that 20% palm oil waste content and above has a significant effect on the poor structure which leads to black coring effect and causing poor surface texture of brick. However, this could be improved by increasing mixing times or prolongs firing time in the next stage.

The recycling of these agricultural wastes may present clear advantages from economical and technological aspect which are the reduction of costs related to use of alternative raw materials, reduction of consume of virgin raw materials, the obtainment of porous bricks with insulating ability when organic waste is used and the obtainment of bricks with excellent mechanical properties of using palm oil waste.

References

[1] World Palm Oil Production, https://www.worldpalmoilproduction.com/default.asp
[2] Malaysian Palm Oil Board, http://bepi.mpob.gov.my/index.php/en/statistics/production/168-production-2016/746
[3] Kong S, http://www.theborneopost.com/2016/11/14/sarawaks-palm-oil-sector-sees-decline-as-el-nino-effect-lingers/
[4] Malaysian Palm Oil Board, http://bepi.mpob.gov.my/images/area/2015/Area_summary.pdf
[5] Ismail M, Ismail M A, Lau S K, Muhammad B and Majid Z 2010 *Concr. Res. Lett.* **1** 60-66
[6] Abdullah N and Sulaiman F 2013 *Biomass Now - Sustainable Growth and Use* (Malaysia, InTech)
[7] Awalludin M F, Sulaiman O, Hashim R and Nadhari W N A W 2015 *Renew. Sustain. Energy Rev.* **50** 1469-84
[8] Mahlia T M, Abdulmuin M, Alamsyah T M and Mukhlishien 2011 *Energy Convers. Manag.* **42** 2109-18
[9] Embrandiri A, Singh R P, Ibrahim H M and Ramli A A 2012 *Environmentalist* **32** 111-117
[10] Yuhazri M Y, Sihombing H, Yahaya, Said S H, Nirmal M R U, Lau S and Tom P P 2009 *Glob. Eng. Technol. Rev.* **2** 26-32
[11] Agamuthu P 2013 *Waste Manag. Res.* **31** 1-2
[12] BS 1377-1 1990 *British Standard (BS). Methods of Test for Soils for Civil Engineering Purposes - Part 1: General Requirements and Sample Preparation*
[13] BS 1377-4 1990 *Methods of Test for Soils for civil engineering purposes - Part 4: Compaction-Related Tests*
[14] BS EN 772-21 2011 *Methods of test for Masonry Units Part 21 : Determination of Water Absorption of Clay and Calcium Silicate Masonry Units by Cold Water Absorption*
[15] BS EN 772-1 2011 *Methods of Test for Masonry Units Part 1 : Determination of Compressive Strength*
[16] Pourakbar S, Asadi A, Huat B B K and Fasihnikoutalab M H 2015 *Transp. Geotech.* **3** 24-35

[17] Oyeleke R B, Yusof M B, Salim M R and Ahmad K 2011 *Int. J. of Renewable Energy Resources* **1** 1-8
[18] Hegazy B E E, Fouad H A and Hassanain A M 2011 *J. Am. Sci* 569-576
[19] Baccour H, Medhioub M, Jamaoussi F and Mhiri T 2009 *J. Mater. Process. Technol.* **209** 2812-17
[20] Eliche-Quesada D, Martínez-Martínez S, Pérez-Villarejo L, Iglesias-Godino F J, Martínez-García C and Corpas-Iglesias F A 2012 *Fuel Process. Technol.* **103** 166-173

Acknowledgment
The authors gratefully acknowledge the Center for Graduate Studies UTHM for research support.

IOP Conf. Series: Materials Science and Engineering **209** (2017) 012038 doi:10.1088/1757-899X/209/1/012038

Performance and Characterization of Geopolymer Concrete Reinforced with Short Steel Fiber

M M A B Abdullah[1,2*], M A Faris[1,2], M F M Tahir[2], A A Kadir[3,2] , A V Sandu[2,4], N A A Mat Isa[1] and O Corbu[2,5]

[1]Faculty of Engineering Technology, Universiti Malaysia Perlis (UniMAP), Perlis, Malaysia

[2]Center of Excellence Geopolymer and Green Technology (CEGeoGTech), School of Materials Engineering, Universiti Malaysia Perlis (UniMAP), Perlis, Malaysia.

[3]Department of Water and Environmental Engineering, Faculty of Civil and Environmental Engineering, Universiti Tun Hussein Onn Malaysia, Johor, Malaysia

[4]Faculty of Materials Science and Engineering, Gheorghe Asachi Technical University of Iasi, Romania

[5] Central Laboratory of the Faculty of Civil Engineering, Technical University of Cluj Napoca, Romania

E-mail: mustafa_albakri@unimap.edu.my

Abstract. In the recent years, geopolymer concrete are reporting as the greener construction technology compared to conventional concrete that made up of ordinary Portland cement. Geopolymer concrete is an innovative construction material that utilized fly ash as one of waste material in coal combustion industry as a replacement for ordinary Portland cement in concrete. The uses of fly ash could reduce the carbon dioxide emission to the atmosphere, redundant of fly ash waste and costs compared to ordinary Portland cement concrete. However, the plain geopolymer concrete suffers from numerous drawbacks such as brittleness and low durability. Thus, in this study the addition of steel fiber is introduced in plain geopolymer concrete to improve its mechanical properties especially in compressive and flexural strength. Characterization of raw materials also determined by using chemical composition analysis. Short type of steel fiber is added to the mix in weight percent of 1 wt%, 3 wt%, 5 wt% and 7 wt% with fixed molarity of sodium hydroxide of 12M and solid to liquid ratio as 2.0. The addition of steel fiber showed the excellent improvement in the mechanical properties of geopolymer concrete that are determined by various methods available in the literature and compared with each other.

1. Introduction

Nowadays, the urge of using geopolymer concrete as green construction materials has increased parallel with infrastructure development besides increasing in awareness as a result of global warming. Therefore, the utilization of fly ash as waste material also improved and overcome a major problem for disposal [1]. However, plain cement concrete faced various imperfection such as brittleness, fracture resistance, and etc. [2]. Hence, the addition of steel fibres is an alternative to promote these problems such as toughness, flexural strength, and energy absorption improved.

Production of conventional concrete customarily used ordinary Portland cement (OPC) as the binder and the usage of OPC is on the increase to meet infrastructure developments [3]. However, the calcination of limestone and combustion of fossil fuels during the manufacturing of OPC resulting the huge amount of carbon dioxide, (CO_2) released to the environment [4]. Geopolymer is said to be a new alternative material which the greenness of conventional concrete is improved besides having a better performance compared to concrete with OPC. Plus, the production of geopolymer does not emit carbon dioxide to the environment thus overcoming the global warming issue [5].

According to findings reported by other researcher, the term geo-polymer referred covering class of synthetic alumino-silicate materials with potential use in a number of areas, essentially as replacement for Portland cement and for advanced high-tech composites, ceramic applications or as a form of cast stone [4]. Besides, as the chemical compositions of geopolymer materials are similar to natural zeolitic materials (but it has amorphous microstructure) geopolymer is considered as members of the family of inorganic polymers. The composition of zeolites is based on an aluminosilicate framework and a three-dimensional network of inorganic polymers that are made up of SiO_4 and AlO_4 tetrahedral that are linked by shared oxygen atoms into rings and cages [6].

In the present study, an experimental investigation on mechanical and physical performance of geopolymer concrete with addition of steel fiber has been carried out. Geopolymer concrete mixes were prepared with alkaline solution to fly ash ratio of 2.0 and short types of steel fiber is used with varying amount.

2. Materials and methods

Experimental work is designed to study the effect of addition short steel fibers on mechanical properties on geopolymer concrete. The main material used for making fly ash based geopolymer concrete composite specimen is fly ash along with other material such as coarse and fine aggregates, alkaline activator solution, steel fibers, and water.

2.1. Fly Ash

Low-calcium ASTM Class F dry fly ash obtained from local power station is used as the source material. In this experimental work, the fly ash used also from low-calcium ASTM Class F fly ash that is obtained from Manjung Power Plant, Lumut, Perak. With contains less than 10% of calcium oxide, Class F fly ash is suitable for geopolymer concrete production as it will makes substantial contributions to the workability, chemical resistance, and reduction in thermal cracking [2].

2.2. Alkaline Solution

The laboratory grade for sodium hydroxide (NaOH) is in the flakes form and sodium silicate (Na_2SiO_3) solution is used as alkaline activators. With about 98% purity, NaOH flakes is dissolved in water before it mixed together with the activator solution. The concentration of sodium hydroxide is fixed at 12M also the ratio of $NaOH/Na_2SiO_3$ at 2.5 as suggested by previous research [7].

2.3. Aggregates

The formulation of geopolymer concrete also need aggregates that consists of fine and coarse aggregates to occupy the largest volume about 75% to 80 % by mass. Available fine aggregates from river sand are used and the coarse aggregates came from crushed stones with nominal size of 5 mm and the maximum size is 20 mm. The ratio of fine and coarse aggregates are also included with fly ash ratio because they are considered solid constituent in the geopolymer concrete as fly ash/fine aggregates/coarse aggregates is 1:1:1.5 in ratio respectively.

2.4. Steel fibers

The steel fiber used in this study is short type of steel and with range of 0.03 mm to 0.06 mm length, which can make it sufficient to be randomly dispersed in geopolymer mixture. Short steel fibers is used with aspect weight ratio of 0 wt%, 1 wt%, 3 wt%, 5 wt%, and 7 wt%. The geopolymer concrete with no addition of steel fiber is used as a reference in this study.

2.5. Mix Proportion of Geopolymer Concrete

The mixing procedure used for this study is similar to conventional concrete and the process is done in the laboratory at room temperature. With ratio of 2.0 for solid to liquid ratio, fly ash and fine aggregates together with respective percentage of short steel fiber are initially mixed together [7]. Then the liquid component of the mixture is added to prepare wet mix until it gives homogenous mix. After that, the coarse aggregate is finally added to the mixture to ensure the homogeneity of mixture as well as to ease the mixing process. It is noted that, the mixture is quite faster to harden. Then, the mix proportion with respective fiber quantity is casted following the respective mold testing and finally cylindrical concrete molds were filled with geopolymer concrete mixture and the cured at ambient temperature.

Table 1. Quantity of fibers for various mixes.

Steel fiber (wt%)	GPC	GPC1	GPC3	GPC5	GPC7
Fiber quantity (kg/m^3)	0.0	26.0	78.0	130.0	182.0

3. Results and discussions

3.1. Chemical Composition Analysis

Table 2 presented the chemical composition of fly ash. Fly ash was obtained from Manjung Power Plant, Lumut, Perak and it was found to be corresponding to ASTM Class F with CaO content is less than 20%. The fly ash was used as the main constituent material for the production of geopolymer concrete.

Table 2. Chemical composition analysis of fly ash.

Component	SiO_2	Al_2O_3	Fe_2O_3	TiO_2	CaO	MgO	SO_3	K_2O	MnO	BaO	SrO
Percentage (%)	38.80	14.70	19.48	1.02	18.10	3.30	1.50	1.79	0.16	0.27	0.11

Based on result, it is important to have higher percentage of SiO_2 in fly ash composition because silica is the main element of the structural skeleton for the reaction of product formed in the alkaline activation of fly ash. As reported by Feng, chemical composition of fly ash especially SiO_2, Al_2O_3 and Fe_2O_3 are important in order to establish a long-term pozzolanic activity in which fly ash reactivity with the lime (calcium hydroxide) in the chemical reaction [8]. Besides, these acidic oxide content (SiO_2, Al_2O_3 and Fe_2O_3) will resulting in widely range of strength activity index between 64 to 100%.

In addition, there are limits that should be acknowledge for the fly ash to be a leading construction material with optimal binding properties which are; the percentage of unburned material (LOI) should be lower than 5%, the content of Fe_2O_3 must not be higher than 10%, low CaO content and the amount of reactive silicate must be between 40% to 50% [9].

3.2. Analysis of Fresh Geopolymer

3.2.1. Workability. The trend graph shows the decreasing value of workability with increasing fiber content. The maximum slump value recorded is at 0 wt% of fiber content which is 100 mm and the lowest slump value recorded is 32 mm with the highest fiber content. Desirable workable flow is indeed important for workable and easily placed into the mold [10]. Besides, the workability of geopolymer mixture can be improved with addition of extra water and superplasticiser. However, superplasticiser contribute to negative impact on strength of geopolymer thus, it is preferable to have extra water rather than addition of superplasticiser.

There are several factors that can be affect the slump value of fresh geopolymer which are moisture content of aggregates, variation of ambient temperature, mixing time and degree of condensation reaction between binder and alkaline solution [11].

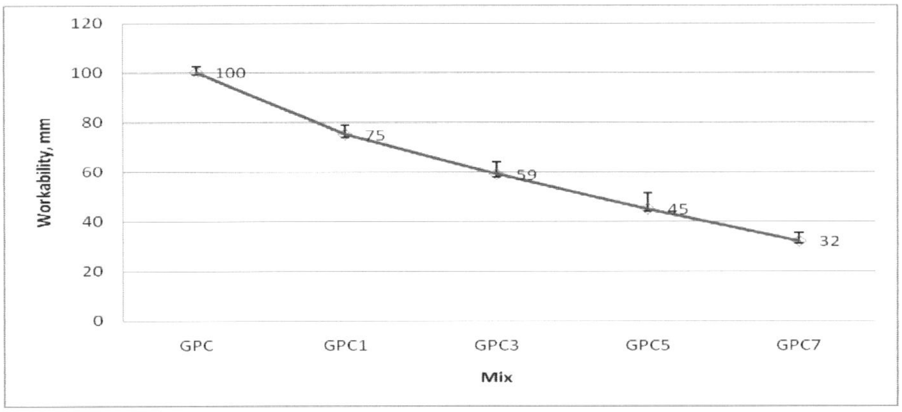

Figure 1. Workability of fresh geopolymer.

For this experimental work, if the fiber content is increases more than 7 wt% the compressive strength and flexural strength of geopolymer concrete is impossible to increases as the slump value continue to decrease. This will be difficulties for casting of geopolymer samples. Besides, the best workability of geopolymer according to The range of 100 mm to 175 mm and the lowest is below than 25 mm [12]. The workability of fresh geopolymer with addition of fiber which is the lowest workability is at the highest percentage of fiber content that is only 12 mm of slump value [13].

3.3. Physical Analysis of Geopolymer Concrete

3.3.1. Density. Figure 2 shows an increasing trend density of geopolymer concrete parallel to the increasing in volume of steel fiber in the mixture. The maximum density of the geopolymer concrete can reach up to 2500 kg/m^3 when the percentage of addition steel fiber is at 7 wt%.

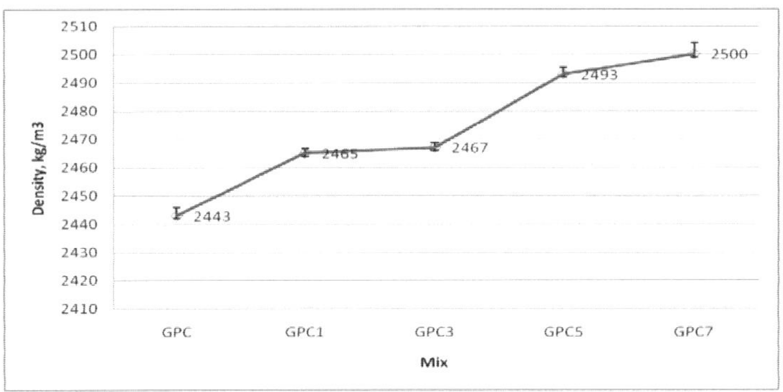

Figure 2. Density of geopolymer with addition of steel fiber.

However, GPC mix is recorded the lowest density of geopolymer concrete which is 2443 kg/m^3 when there is no steel fiber is added and considered to be reference point in this study. Besides, density of GPC1 with the lowest percentage inclusion of steel fiber is recorded higher than the reference mix which is 2465 kg/m^3.

3.3.2. Water Absorption. According to Luhar S., water absorption characteristic of geopolymer concrete contribute a large influence towards the durability of the structure [14]. Penetration of water into the geopolymer concrete will resulting the structure to become rubble thus, spalling concrete occurs. In the case of steel fiber reinforced concrete, this causes the steel bar embedded to corrode and finally reduces the life span of the concrete structure. The result in figure 3 also reveals that the water absorption of geopolymer concrete with addition of steel fiber is lower than the 0 wt% (control concrete) which is the lowest reading is at 7 wt% (2.7%). This shows that the concrete become less permeability to fluids (water) because of the pore is been occupied by addition of short steel fiber in the design of geopolymer concrete.

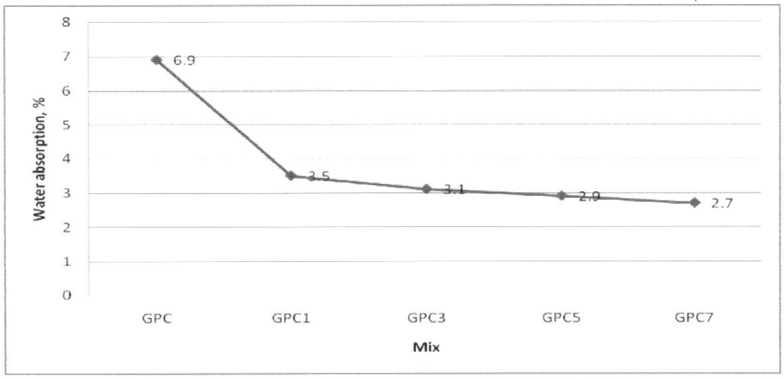

Figure 3. Water absorption of geopolymer concrete with various percentage of fiber.

Besides, if the permeability of geopolymer concrete tends to be high thus, the porosity also high due to interconnected in pores. Farhana stated that, as the concrete is a porous material it tends to have contact with surrounding environment and causing any movement of water through concrete structure hence gives an impact to the durability of mortar and concrete [15].

3.4. Mechanical Performance of Geopolymer Concrete

3.4.1. Compressive Strength. The results shown in the figure 4 are the average values of three specimens tested on the test age. Compressive strength of geopolymer concrete increases with respect to short steel fiber content in the mixture. The maximum increase in compressive strength is recorded when the percentage of short steel fiber is at the highest which is 69.80 MPa. Meanwhile, the lowest compressive strength is recorded at 1 wt% of addition steel fiber which is 45.40 MPa.

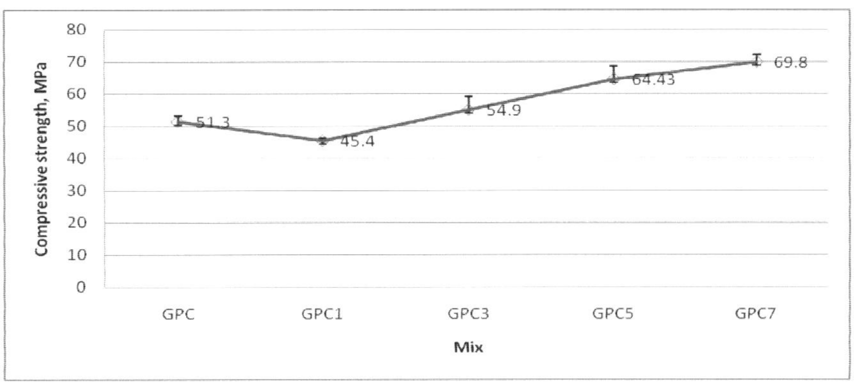

Figure 4. Compressive strength of geopolymer concrete with respect to various addition fiber content.

Based on the results, there are several reasons that may contribute to the value of compressive strength of all samples which can be divided into parameters to the compressive strength of geopolymer concrete; Firstly, the ratio of steel fiber content is very large, which exactly increases the compressive strength value of the geopolymer concrete as early as 14 days plus the value is considered higher than ordinary Portland cement concrete. Randomly and homogenous oriented steel fibers will arrest cracks formation and propagation when the concrete cracks thus, promote both ductility and strength of the geopolymer concrete [2]. Further mechanical performance of plain concrete have been enhance just through the expansion of fibers [16].

The bridging action is controlled by debonding, sliding and pulling out of fibers thus; the demand of energy for the crack to propagate is expanded. Hence, the optimum percentage of short steel fiber used in the production of geopolymer concrete is determined which is 7 wt% along with good mechanical performance of geopolymer concrete at its highest compressive strength of 69.80 MPa.

Next, as it can be seen in Figure 5, density of geopolymer concrete gives a direct relation towards the compressive strength of geopolymer concrete. This trend also similar with conventional concrete as increasing in curing time, the density of concrete also increased.

The increased in percentage of addition short steel fiber also one of the factors that increasing the density value the concrete.

The more compacted the concrete, resulting higher in compressive strength thus, to reduce the porosity of concrete, filler materials such as silica fume is added into the mix design of the concrete [17]. Meanwhile, in this case porosity of geopolymer concrete has been filled with aggregates and fly ash plus the inclusion of short steel fiber into the mixture.

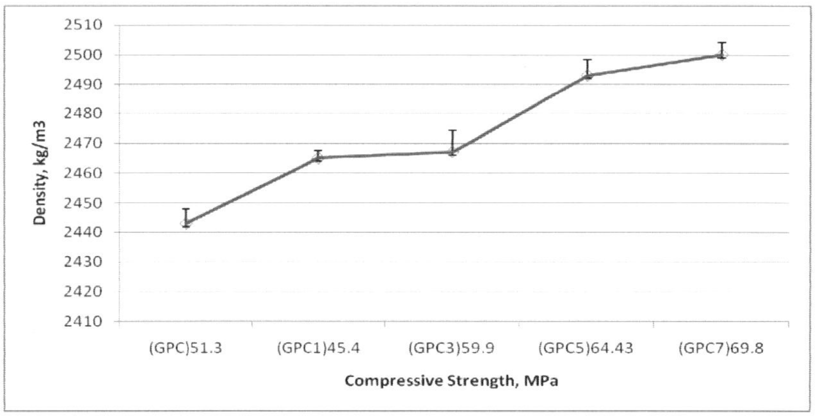

Figure 5. Density of geopolymer concrete against compressive strength.

Furthermore, to study the effect of curing age to compressive strength of geopolymer concrete reinforced with short steel fiber, the highest addition of steel fiber is chosen which is 7 wt% for the reason that it possessed the highest value of compressive strength which is up to 69.80 MPa.

As shown in the figure 6, the development compressive strength of geopolymer concrete is rising with longer curing age in which the value of compressive strength obtained is 4.1 MPa as early as 3 day. Meanwhile, the value of compressive strength continue to increase to 9.3 MPa, 13.7 MPa at the curing age of 5 day and 7 day respectively.

In addition, compressive strength of geopolymer concrete enhance by longer the curing time and curing at elevated temperature thus, promote geopolymerization process that resulting in higher compressive strength [18].

Despite the fact that reaction accelerate because of the heat, one must concern to minimize the loss of water [3]. However, geopolymer concrete experienced cracking as the negative response for curing at too high temperature thus, curing in temperature range of 60 °C to 90 °C within time in 24 hours to 72 hours is recommended by [6].

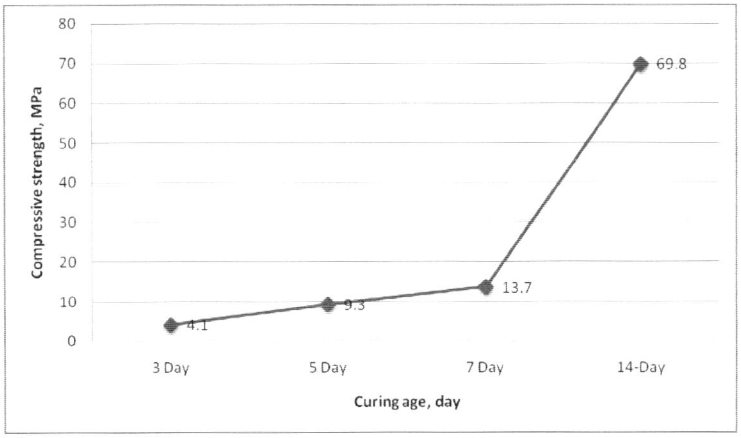

Figure 6. Compressive strength of geopolymer concrete for 7 wt% addition of short steel fiber with varying curing age.

3.4.2. Flexural Strength . Figure 7 shows the flexural strength results were plotted after 14 days of curing and the curing also takes place in ambient temperature same with compressive test samples. The increasing trending flexural strength for all samples is observed and the highest flexural strength value is up to 5.94 MPa at maximum fiber content which is 7 wt%. Meanwhile, the lowest fiber content in geopolymer concrete mixture has 1.45 MPa of flexural strength value.

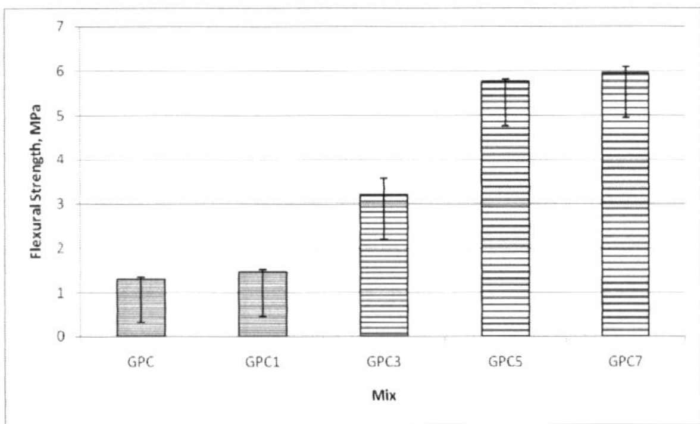

Figure 7. Flexural strength value for various percentage of steel fiber content.

Flexural strength value of geopolymer concrete indicates the durability characteristics of concrete structure hence determined the application of the concrete in many ways such as for railway sleepers, sewer pipes and also other precast concrete products. According to Llyod, by having an excellent early strength gain as a novel characteristic of geopolymer concrete can further exploited in precast industry where steam curing is used in order to increase the rate production of elements although curing in ambient temperature also possible [19]. Flexural strength of geopolymer concrete mix with addition of short steel fiber can decreases the crack propagation in concrete and attain higher peak value in terms of strength (include compressive strength). All the factors that contribute in higher value of flexural strength is the same as discussed in compressive strength results.

3.5. Relationship between Compressive Strength and Flexural Strength of Geopolymer Concrete

As discussed in compressive strength results there are several factors that contribute in order to obtained maximum value of strength included flexural strength which are; Firstly, the chemical composition of fly ash as raw materials itself that has less than 20% of CaO content by ASTM C 618 that can provide strength of geopolymer concrete. Besides, large amount of SiO_2, Al_2O_3 and Fe_2O_3 content contribute in the reactivity with lime element (Ca) in the chemical reaction to produce polymeric Si-O-Si bonds together with alkaline activator solution.

Next, the increasing trend shows in figure 5 density of geopolymer against compressive strength. It can be seen in that as density of geopolymer concrete continue to increase because of percentage of fiber content increase, therefore it also resulting the value of compressive strength increases. High amount of fiber content makes concrete become denser as it cover up large amount of pore in the concrete structure whereas strength of the concrete also maximize.

Besides, curing age of geopolymer concrete samples for both compressive and flexural strength also same (14 days) in ambient temperature. Therefore, the rate for polymerization process takes place is relatively low compare to when geopolymer concrete is cured in elevated temperature of range 60°C to 90°C as recommended by [6].

3.6. Morphology Analysis

All geopolymer concrete samples undergoes morphology analysis after mechanical testing (compressive test) is done. There are a few data that needed to be observed by doing morphology analysis such are to inspected cracks pattern on the concrete samples after been subjected to high load then to determine the effectiveness of fiber reinforced in geopolymer concrete structure.

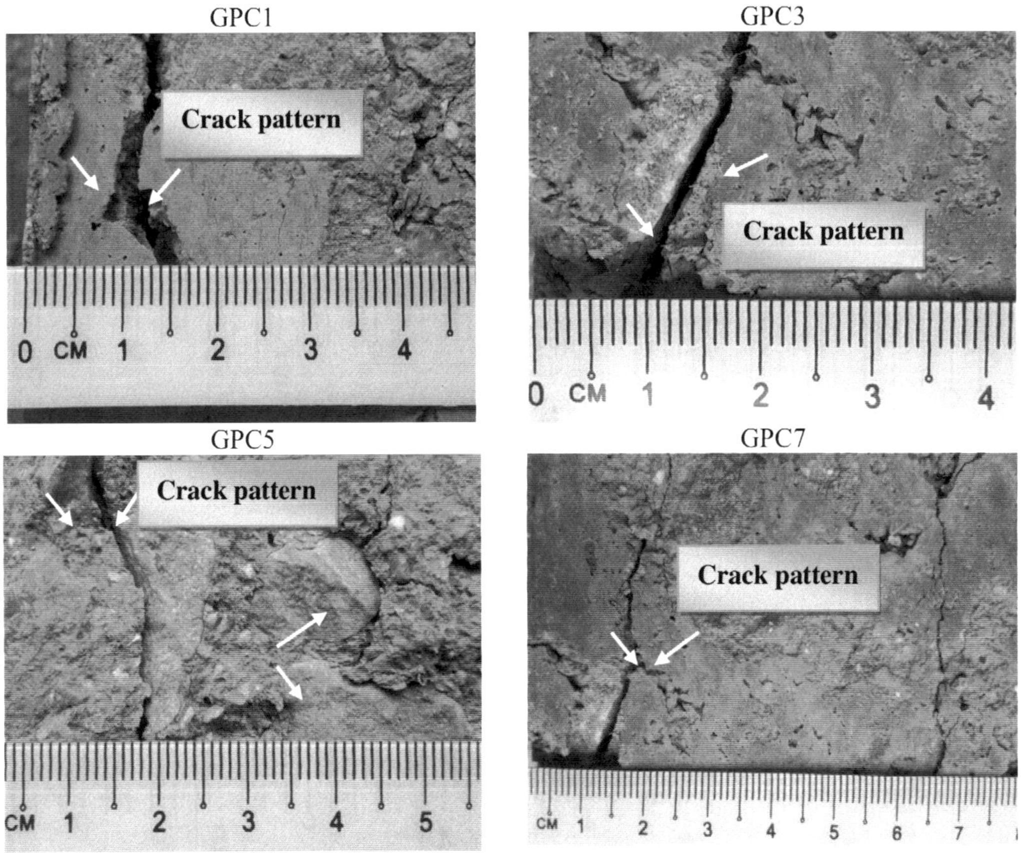

Figure 8. Crack patterns for geopolymer concrete structure after compressive strength testing.

As it can be seen in figure 8, as increasing in fiber content in geopolymer concrete smooth cracks surfaces are obtained. This indicates that maximum compressive strength of concrete is achieved as expected. Besides, the direction for cracks to propagates at all specimens is the same in which the cracks tends to avoid from bumping towards coarse aggregates in the concrete structure. Mix GPC specimen that have the lowest compressive strength value reveals that crack arresting did not happened as effective than other mix GPC1, GPC3, GPC5 and GPC7 because of there are no steel fiber content in the mixture. As mentioned before, plain concrete suffers major drawbacks such as brittleness and low elastic properties.

4. Conclusions

Based on the experimental results it is concluded that the density of geopolymer concrete increased with increase in fiber content, whereas both workability and water absorption results of geopolymer concrete reduced with increased in fiber content. The objective of this research is achieved when the optimum percentage of fiber content in geopolymer concrete structure is discovered. From the results obtained 7 wt% of short steel fiber content in geopolymer concrete structure shows maximum value of various strength which are compressive strength and flexural strength. The maximum value in compressive strength and flexural strength can achieved up to 69.80 MPa and 5.94 MPa respectively.

Eventually, these result display an excellent mechanical performance of geopolymer concrete thus accomplished another objective in this research which are the inclusion of steel fiber in mix design indeed contribute to good strength of the concrete structure. Besides, in order to prove geopolymer concrete to have better performance with addition of steel fiber as reinforced all samples are compared with control concrete (no steel fiber content). Therefore, both of research objectives are achieved.

References

[1] Suriya A P and Senthil K G 2015 *International Journal of Engineering Trends and Technology (IJETT)* **21** 396-399
[2] Sayyad A S and Patankar S V 2013 *Indian Journal of Materials Science* 1-8
[3] Rangan B V, Hardjito D, Wallah S E, and Sumajouw D M J 2013 *Geopolymer: green chemistry and sustainable development solutions* 133-138
[4] Motorwala A, Vineet S, Ravishankar K, Praveena N, and Rajiwala D B 2013 *International Journal of Emerging Technology and Advanced Engineering* **3** 159-166
[5] Meor A F, Mustafa Al Bakri A M, Khairul N I, Ratnasamy M, Aeslina A K, Kamaruddin H, and Mien V T 2016 *Materials Science Forum* **857** 387-394
[6] Mustafa Al Bakri A M, Kamarudin H, Norazian M N, Ruzaidi C M, Zarina Y & Rafiza 2011 Microstructure studies on the effect of the alkaline activators ratio in preparation of fly ash-based geopolymer. *2011 International Conference on Chemistry and Chemical Process* **10** 13-17
[7] Mustafa Al Bakri A M, Kamarudin H, Bnhussain M, Khairul Nizar I, Rafiza A R, and Zarina I 2012 *Reviews on Advanced Material Science* **30** 90-97
[8] Feng X, and Clark B 2011 Evaluation of the physical and chemical properties of fly ash products for use in Portland cement concrete. *World of Coal Ash (WOCA) Conference* **2011** 1-8
[9] Fernandez-Jimenez A and Palomo A 2003 *Fuel* **18** 2259-2265
[10] Chindaprasirt P, Chareerat T, and Sirivivatnanon 2007 *Cement and Concrete Composites* **29** 224-229
[11] Nath P and Sarker P K 2012 Geopolymer concrete for ambient curing condition. *Australasian Structural Engineering Conference 2012: The past, present and future of structural engineering* **7** 225-232
[12] Patankar S V, Ghugal Y M, and Jamkar S S 2015 *Advances in Structural Engineering* **205** 1619-1634
[13] Ranjbar N, Mehrali M, Behnia A, Pordsari A J, Johnson U, and Jumaat M Z 2016 *Journal Pone* **1** 1-5
[14] Luhar S, and Khandelwal U 2015 *Seventh Sense Research Group (SSRG) International Journal of Civil Engineering* **2** 1-10

[15] Farhana Z F, Kamarudin H, Rahmat A, Al Bakri A M M 2015 *Materials Science Forum* **803** 166-172
[16] Ramkumar G, Sundarkumar S, and Sivakumar A 2015 Development of steel fibre reinforced geopolymer concrete. *International Journal of Advanced Research In Science and Engineering (IJARSE)* **4** 1717-1725
[17] Rasa E, Ketabachi H, and Afshar M H 2009 *Asian Journal of Civil Engineering (Building and Housing)* **16** 33-42
[18] Memon F A, Nuruddin M F, Demie S, and Shafiq N 2011 *International Journal of Civil, Environmental, Structural, Construction and Architectural Engineering* **5** 342-345
[19] Llyod N A, and Rangan B V 2010 Geopolymer concrete with fly ash. *Second International Conference on Sustainable Construction Materials and Technologies* **2010** 1-13

Acknowledgments
The author would like to express his gratitude to the Center of Excellence Geopolymer & Green Technology (CEGeoGTech) for funding the research project.

The Influence of Thermoplastic Starch and Banana Fibre Contents on Physical and Thermal Properties of LLDPE

A W M Kahar[1,2*] and A E Abduati Salem[1]
[1] School of Materials Engineering, Universiti Malaysia Perlis,02600 Jejawi, Perlis.
[2] Centre of Excellence Geopolymer and Green Technology, Universiti Malaysia Perlis.

E-mail: kaharwahab@unimap.edu.my

Abstract. Blending of linear low density polyethylene (LLDPE), thermoplastic starch (TPS) and banana fiber (BF) have been studied. Two types of systems were prepared; the matrix having different ratio of LLDPE/TPS and, the LLDPE/TPS composites having 5 – 30 wt% BF. Morphological changes using scanning electron microscope (SEM) were observed and its showed that TPS particle are homogenously dispersed in LLDPE matrix. On the other hand BF was found to be well embedded in TPS phase, showing the good interaction between BF and TPS phases. This observation show an agreement with the Young's modulus value which is increased with the BF contents. The increment in Young's modulus value was also attributed to the difficulties in LLDPE/TPS chains movement with the presence of BF.

1. Introduction

Linear low density polyethylene (LLDPE) has interesting mechanical properties which provide good tensile properties, dimensional stability and high thermal and chemical resistance. Currently there are many intensive study on the biodegradable plastic based on the natural resources due to the many environmental downsides caused by conventional plastics such as LLDPE, starting from the production of plastics until the problem of waste disposal [1-2]. As a response for the environmental issues, blending synthetic polymer with a cheap natural biopolymer from the natural bio-based plant provides new attention to manufacture biodegradable or partial biodegradable products. Most commonly used are cellulose and starch. Incorporation of cellulose and starch into synthetic polymer, which enhances the environmental fragmentation and degradation have been investigated by many researchers [3-5]. This new generation of bio-based plastic can compete in the market which is currently being dominant by the products based on petroleum feedstock.

It is not necessary for producing 100% of biodegradable materials as substitution of petroleum based products immediately, but a partial replacement also could help to save our environment. Economical problem associated with expensive process for producing biodegradable polylactic acid (PLA), polyhydroxyalkanoates (PHAs) polycaprolactone (PCL), polybutylene succinate (PBS) and low mechanical properties of inexpensive thermoplastic starch (TPS) makes the production of fully biodegradable materials difficult. A viable solution is to combine different features and benefits from both petroleum and bio-resources based materials to produce useful products having the requisite combination of cost-performance properties for real world applications. Most of the synthetic polymers are incompatible at molecular level with TPS and leads to poor interfacial adhesion strength. Due to strong intermolecular hydrogen bonding, TPS tend to agglomerate and do not easily disperse in LLDPE matrix. These drawbacks cause the properties of the blends to reduce significantly with increasing TPS content. In order to overcome incompatibility of these blends, the area of study was concentrated on adding banana fiber that can be compensated the properties deterioration caused by the incorporation of TPS. Therefore, this study proposed a preparation and characterization of linear low density polyethylene LLDPE/thermoplastic starch (TPS) / banana fiber composites for partial biodegradable composites production.

2. Materials and methods

Linear low density polyethylene (LLDPE) granulates with density 0.92 g/cm3 was supplied by Titan Chemicals, Malaysia. Tapioca starch used was a food grade and obtained from Thye Huat Chan Sdn Bhd., Malaysia. Reagent grade glycerol was obtained from Merck, Darmstadt, Germany and used as received.

2.1. Preparation of Thermoplastic Tapioca Starch (TPS)

Tapioca starch was dried at 80°C in vacuum oven for 24 hours. Tapioca starch was first pre-mixed with 35wt% of glycerol using a kitchen blender (3000 rpm, 2 min) until a homogenous mixture was obtained. Then, the mixture was stored overnight in a dry place. Later the mixture was processed using heated two-roll mill at the temperature of 150°C and the mixing time was 10 min.

2.2. Preparations of LLDPE/TPS/Banana Fiber Composites

The blends were prepared by melt blending LLDPE/TPS/banana fiber in a Haake Rheomix 600 mixer equipped with roller rotors. The mixing was carried out at temperature 150°C and rotor speed was fixed at 55rpm. The LLDPE was first charged into the mixing chamber for 3 min and followed by TPS. Then, the mixing time was continued for 4 minutes before mixed with banana fiber and the process continued until plateau torque was reached. The whole process proceeded for 12 min. The blend was converted into 1mm sheet using a hydraulic hot press Gotech Testing Machine. The hot press procedures involved by preheating at 150°C for 6 min and followed by 2 min compressing at the same temperature, subsequent cooling under pressure for 4 min.

2.3 Tensile Test

Measurements of the tensile properties were performed by using an Instron Universal Testing Machine (model 3366), with crosshead speed 50 mm/min. Dumbbell sample (1mm thick) with 50 mm gauge length were tested according to ASTM D 638. The mean value, out of five samples was reported with standard deviation to show the error range.

2.4. Scanning Electron Microscopy (SEM)

Fracture surface of the samples were performed with a Leo Supra – 3SVP Field Emission Scanning Electron Microscopy at an acceleration rate of 20kV. The fracture surfaces of the specimen were mounted on aluminium stubs and sputter coated with thin layer of gold to avoid electrostatic charge during examination.

2.5. Thermogravimetric Analyses (TGA)

The thermogravimetric analyses of the blends were operated using a Perkin Elmer Pyris 6 machine. Samples were tested at heating rate 10°C/min from ambient temperature to 600°C. Two types of samples were studied: LLDPE/TPS blends and LLDPE//TPS/banana fiber composites. Some text.

3. Results and discussions

In this TPS based LLDPE blends, LLDPE as a rigid phases poses the excellent tensile properties; whereas, the TPS plays a function as a biodegradation agent of resultant blends. TPS based LLDPE blends offer the combinations between easy processing of thermoplastic with added biodegradable functional of tapioca starch. Besides that, the starch conversion from granule to plasticized form was also considered as an impact modifier to absorb external force applied to the blends and preventing the crack propagation. The tensile properties of LLDPE/TPS blends were represented in order to understand the mechanical characteristics of the blends at different LLDPE and TPS ratio (figure 1 (a) – (c)). The blends were varied from 100% of LLDPE to a ratio of LLDPE/TPS 60/40 in order to observe the mechanical behavior of the blends with incorporation of TPS. At low TPS contents, TPS phase existed as a dispersed phase, and thus the deformation behavior of the blends depends on the LLDPE phase. Up to 20% of TPS contents, the tensile strength of the blends almost identical to the virgin LLDPE value. As increasing TPS contents, TPS phase changed from the small size dispersed phase into agglomerate TPS phase, and thus LLDPE plastic response decreased as indicated by decrease in tensile strength and Young's modulus values. Based on these tensile properties results, LLDPE/TPS with a blend ratio of 80/20 was chosen for the further work.

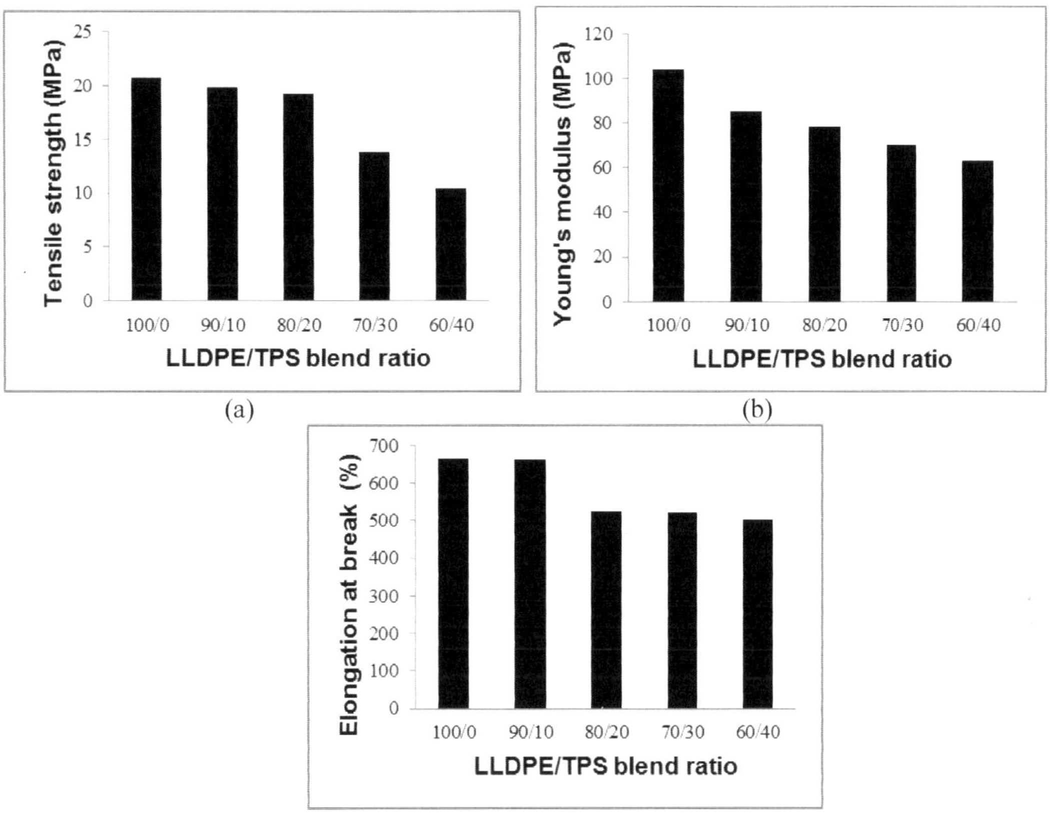

Figure 1. (a) – (c) Variation of tensile properties in LLDPE/TPS blends at different blend ratio.

The effect of BF incorporation into tensile strength of LLDPE/TPS (80/20) was shown in figure 2(a). The graph exhibited that the tensile strength decrease sharply at initial addition of BF component and the value remains unchanged up to 10% BF. The possible explanation for this observation was due to the dilution effect as hygroscopic BF component was generally incompatible to the LLDPE phase. Further addition of BF up to 30% showed mark decrease in tensile strength, indicated that the optimum BF loading was achieved. On the other hand, the Young's modulus values were strongly affected by the increment of BF loading in the LLDPE/TPS blends figure 2 (b). The values increased linearly with the BF weight fraction. Its can be explained by the interlocking mechanism between BF surface with LLDPE chains. As expected that the increment in Young's modulus values was proportionally inverse to that elongation at break values figure 2 (c). Insufficient interaction at the BF-LLDPE interphase and high amount of BF contents (30%) might result in early sample failure. It can be conclude that, the acceptable mechanical strength of LLDPE/TPS-BF composites is only feasible at lower BF content, and from graphs, it can be approximated at around 10-20% of BF portion.

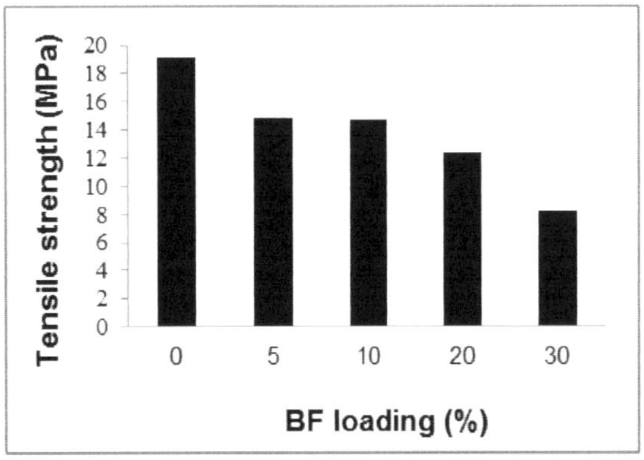

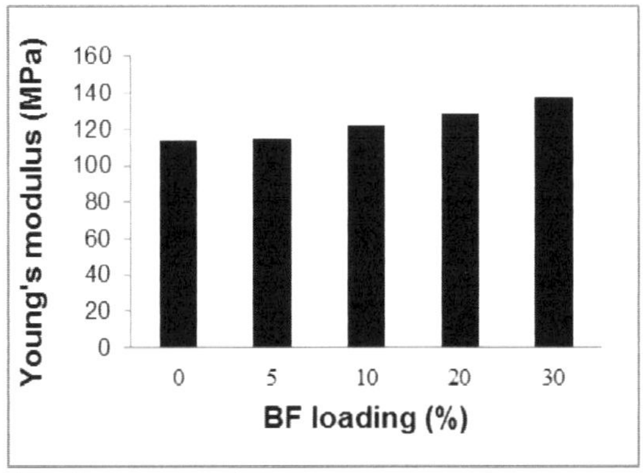

(a) (b)

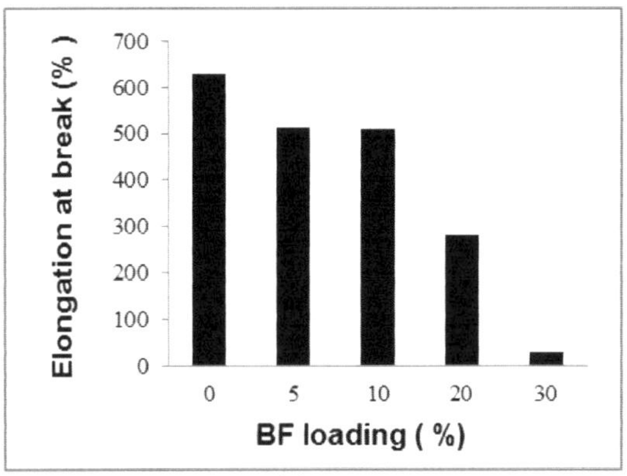

(c)

Figures 2. (a) – (c) Variation tensile properties in LLDPE/TPS-BF blends at different BF contents.

SEM images of LLDPE/TPS blends containing 10% and 30% of BF were given in figures 3 (a) and (b). The micrographs at 300 X magnification revealed that BF was well embedded in TPS phase. This observation was expected due to the BF and TPS have almost similar characteristics in nature. At the higher BF loading, the fracture surface of the LLDPE/TPS blends become coarser indicating better resistance to the force applied to the blends. This observation can be correlated with Young's modulus value where it increases when BF was introduced to the LLDPE/TPS blends.

The TGA and DTG experimental results of LLDPE/TPS blends and LLDPE/TPS/banana fiber are shown in figure 4 (a) and figure 4 (b). It can be observed in the TGA curve that the degradation of pure LLDPE starts at temperature of 480°C and for TPS, the degradation start as low as 150°C. This observation indicates the addition of TPS has promoted the degradation of the blends and this might due to volatilization of moisture and glycerol in TPS. Besides that, one significant slope occurred at about 290°C-350°C and it can be attributed to the pyrolitic process in starch chains.

The similar result was reported by Kahar et al. [6] for pyrolitic temperature of starch polymer. For the DTG thermograms of LLDPE/TPS/banana fiber, it's showed that the decomposition point for both 10% and 30% banana loading shifting to higher decomposition temperature. With increasing the amount of banana, the first degradation peak of the TPS shifted from 324°C to 330°C and 335°C with respect to 10% and 30% TPS loading.

(a) (b)

Figure 3. (a) – (b) SEM micrograph of LLDPE/TPS at 10% and 30% BF loadings.

Another peak of the blends also show a similar trend, which is 10 – 15°C higher with the increased in banana fiber loading, showing the significant increase in thermal stability.

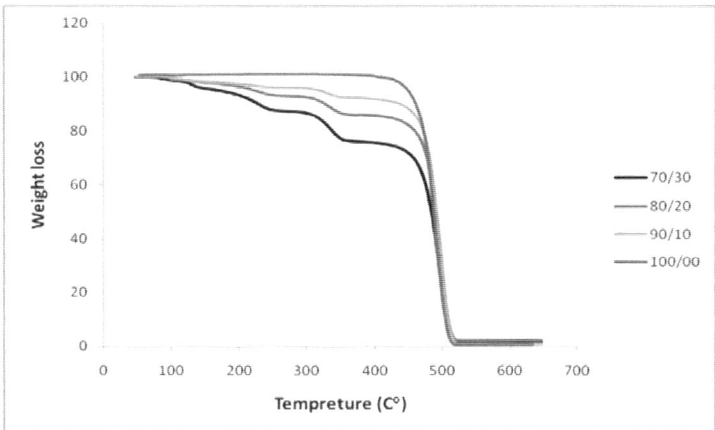

(a)

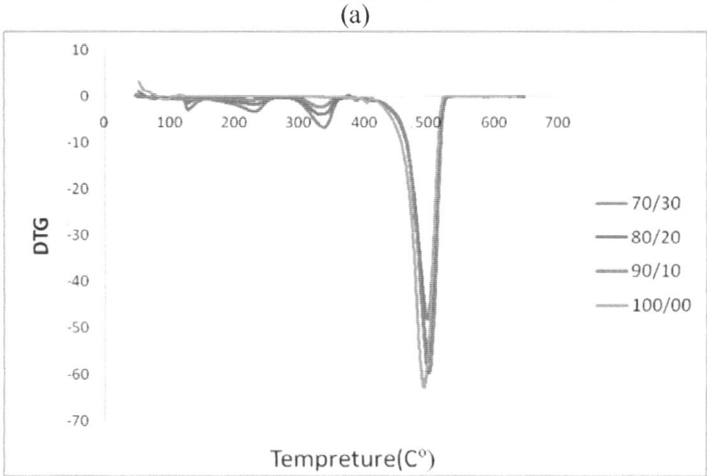

(b)

Figure 4. (a) and (b) Themogravimetric curves and DTG of LLDPE/TPS blends.

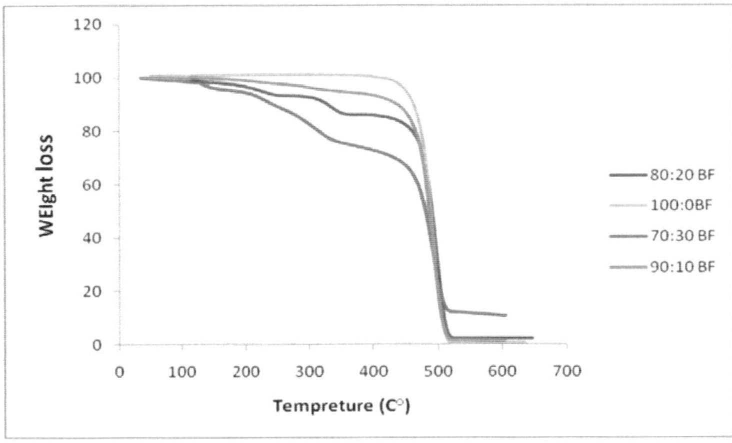

(a)

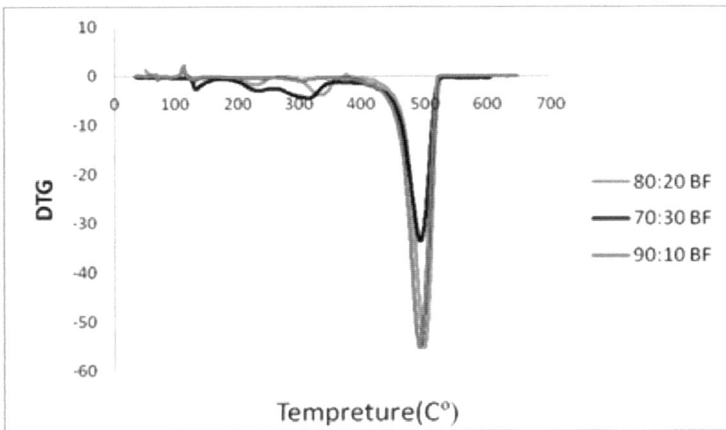

(b)

Figures 5. (a) and (b). Themogravimetric curves and DTG of LLDPE/TPS/banana fiber.

4. Conclusions

The purpose of this study was aimed to develop partial biodegradable plastics by utilizing TPS and BF. The optimum contents of TPS and BF in LLDPE matrix, and with acceptable mechanical properties were assessed in terms of mechanical properties such as tensile strength and Young's modulus. The mechanical data was also correlated with the blends morphologies in order to understand the interaction behaviour between LLDPE, TPS and BF. In general, the tensile strength was decreased as increased in TPS and BF contents, but the Young's modulus values showed good increment as BF contents was increased. SEM micrographs showed that BF has good adhesion with TPS and also well embedded in TPS phase. The good interaction between BF and TPS resulted in better failure resistance when load was applied to the blends. Banana fiber was introduced into LLDPE/TPS matrix as a reinforcing agent in order to promote better thermal stability of the LLDPE/TPS blends. In the presence of the banana fiber, LLDPE chains are difficult to form crystal due to the interfering chains alignment from the neighboring components. TGA and DTG thermograms displayed the effect of banana fiber in LLDPE/TPS blends. The shift of thermal stability for compatibilized blends towards higher temperature also can be explained by the increase in interfacial interaction between banana fiber and TPS. In conclusion, the composites showed better thermal properties than the LLDPE/TPS blend, as a result from the stronger interfacial interaction between banana fiber and TPS and the influence of banana fiber on the LLDPE chains alignment.

References

[1] Saeoui P, Sirisinha C, Sanguanthammarong P and Thaptong P 2010 *Polymer Testing.* **29** 346-351
[2] Kahar A W M, Ismail H and Nadras O 2012 *Polymer-Plastics Technology and Engineering* **51** 298–303
[3] Ning W, Jiugao Y, Xiaofei M and Ying 2007 *Carbohydrate Polymer.* **67** 446–453
[4] George R N and Carol D S 2008 *Polymer.* **49** 1-173
[5] Feng S Y, Lauke B and Mai Y W 2008 *Composites Part B: Engineering.* **39** 933-961
[6] Kahar A W M, Ismail H and Nadras O 2012 *Journal of Applied Polymer Science* **128**(4) 2479–2488

Acknowledgments

The authors would like to thank Universiti Malaysia Perlis for the supply of raw materials and the equipment provided. The authors also gratefully acknowledge the financial support from the FRGS grant 9003-00601.

IOP Conf. Series: Materials Science and Engineering **209** (2017) 012040 doi:10.1088/1757-899X/209/1/012040

Correlation of the Processing Parameters in the Formation of Granulated Ground Blast Furnace Slag Geopolymer

I H Aziz[1], M M A B Abdullah[2,1], H C Yong[1], L Y Ming[1], D Panias[3] and K Sakkas[3]
[1]Center of Excellence Geopolymer and Green Technology (CEGeoGTech), School of Materials Engineering, Universiti Malaysia Perlis (UniMAP), Perlis, Malaysia
[2]Faculty of Engineering Technology, Universiti Malaysia Perlis (UniMAP), Perlis, Malaysia
[3]School of Mining and Metallurgical Engineering, National Technical University of Athens, Athens, Greece

Email: ikmalhakem@gmail.com & mustafa_albakri@unimap.edu.my

Abstract: Geopolymers are inorganic materials with huge potential applications including building material, fire resistant materials, and agricultural construction materials. Various parameters influenced the final properties of these geopolymer concretes. This study developed the effects of several factors such as solid-to-liquid ratio, NaOH concentration, and Na2SiO3/NaOH ratio on the compressive strength of granulated ground blast furnace slag (GGBFS) by statistical design of experiment (DOE) approach. Analysis of the experimental results through ANOVA exhibited that the specimen with NaOH concentration of 10M, Na2SiO3/NaOH ratio equals to 2.5, and solid-to-liquid ratio of 3.0 curing at room temperatures for 28 days was potential of highest strength (168.705 MPa) in the considered procedure. Besides, the relationship between compressive strength and influential factors could be suitably by fraction factorial design method.

1. Introduction

Cement concrete is the main material used worldwide in building construction in the world due to the durability in various environment and appropriate mechanical properties. Contrary, the Portland cement clinker production is an energy- intensive process. Thus, the development of geopolymer materials is addressed by researchers as an alternative replacement to Portland cement [1-3]. The term 'Geopolymer' has been coined by Davidovits in 1970s, symbolizes the inorganic aluminosilicate based on geological materials which reacted with an alkaline solution to form binder through polycondensation reaction [4]. Geopolymer is an amorphous aluminosilicate structure consisting three-dimensional Si-O-Al polymeric network produced by the poly-condensation of the individual $(SiO)^{4-}$ and $(AlO_4)^{5-}$ tetrathedrals.

Geopolymer system can be established into two types of binding system which are silica-aluminum (Si+Al) with medium to high alkaline solution and silica-calcium (Si+Ca) with a mild alkaline solution [5]. Class F fly ash and metakaolin were the source material used in the (Si+Al) binding system due to silica and alumina content as the main composition. Meanwhile, ground granulated blast furnace slag (GGBFS) was used in the (Si+Ca) system due to the higher content of silica and calcium in the composition. There are many potential benefits of using GGBFS as aluminosilicate sources material in geopolymer process include excellent mechanical strength [6], fast setting time [7], great thermal behavior [8], and fire resistant materials [9]. Practically, the mixture of sodium hydroxide (NaOH) and sodium silicate (Na_2SiO_3) are often used as an alkaline solution. The sodium silicate was acted as alkali activator, plasticizer, and binder [10]. Meanwhile, the alkali hydroxide appropriate for the dissolution of aluminosilicates sources [11]. In addition, alkaline solutions also promoted a certain quantity Si and Al atoms to dissolve the aluminosilicates sources, form monomers in solutions, and then polycondense to form rigid network [12]. Sodium silicate much preferable activating solutions owing to its soluble content, which leads to enhance the rate of polymerization reaction [13]. The properties of geopolymer are affected by the composition of raw materials, the type and the relative amount of alkali activator during the initial period of the geopolymerization reaction [13]. The solid-to-liquid (S/L) ratio and Na_2SiO_3/NaOH ratio have exceptional effects on mechanical properties [14]. Besides, these parameters affect the workability of the geopolymer slurry [15].

Design of Experiment (DOE) is a systematic approach to investigate of a system or process. A series of structured tests are designed in which planned changes are made to the input variables of a system or process. The effects of these changes on a pre-defined output are then assessed statistically. Subsequently, one of the outcomes of this statistical analysis is the ability to predict the optimum combination and level of variables in an experiment by examination of the response of a relatively small number of samples. Considering all the parameters in a single study may not be achievable. Thus, an appropriate experiment design can help in understanding the effect of various parameters at the same time by the minimum required works is performed in this study. The DOE studies were carried out to determine the effect of NaOH concentration, solid-to-liquid ratio, and Na_2SiO_3/NaOH ratio on the compressive strength of GGBFS. The factors were varied according to the fraction factorial design, devised using MINITAB. In addition, the results were analyzed by ANOVA (Analysis of Variance) to evaluate the compressive strength of GGBFS as a function of the above mentioned factors. The P-value was used to determine whether the results are statistically significant.

2. Experimental

2.1. Material

In this study, the fineness of particle size of GGBFS is below 150 μm. The raw materials of GGBFS are drying in an oven at 100°C in order to evaporate the moisture content. The GGBFS are supplied by Ann Joo Integrated Steel Sdn. Bhd, Penang, Malaysia. Table 1 shows the chemical composition of the GGBFS as determined by X-Ray Fluorescence (XRF) analysis.

Table 1. Chemical composition of GGBFS (wt.%).

Compound	CaO	SiO_2	Al_2O_3	MgO	Fe_2O_3	TiO_2	ZrO_2	MnO_2
Wt %	50.37	30.4	10.5	3.2	0.53	0.98	0.05	0.71

NaOH in pellet form with 99% purity, and Na_2SiO_3 consists of Na_2O = 9.4%, SiO_2 = 30.1% and H_2O = 60.5% with weight ratio SiO_2/Na_2O = 3.20-3.30 and specific gravity at 20°C = 1.4 gm/cc, were used as a mixture of alkaline activator.

2.2. Sample preparation and testing

Experiment design and analysis was approach by Minitab 17. A factorial design was used to obtain required experimental procedures. The acceptance ranges for parameters are listed in table 2 [16-18] and 25 suggested experiments in table 3. To study the effect of the NaOH concentration, solid-to-liquid ratio, Na_2SiO_3/NaOH ratio on the compressive strength of the GGBFS geopolymer samples, sodium silicate, and sodium hydroxide were introduced in various amounts into GGBFS. The design included 24 model points and 1 additional center points. Total of 25 series of GGBFS geopolymer samples with mixture was prepared. The mixture of geopolymer was cast in acrylic plastic cube molds (50mm x 50mm x 50mm). The molds sealed with plastic wrap and all samples cured in room temperature for 24 hours before undergo analysis.

Table 2. Range for each factor in factorial design (low level indicates the lowest range factor in this studies and high level for the optimum strength achieved).

Factors	Low	High
NaOH concentration (M)	6	10
Solid-to-liquid ratio	1.0	3.0
Na_2SiO_3/NaOH ratio	1.5	2.5

Compressive test was performed according to ASTM C109. Three samples of each formulation were tested and the average data were obtained. A universal Instron machine series 5569 was used for the measurement. The loading was displacement controlled at a constant rate of 0.5mm/min for all the tests. Then, to generate the model with the optimum statistical significance, ANOVA and prediction and were used.

Table 3. Full factorial design method for three factors (NaOH ratio, solid-to-liquid ratio, and Na_2SiO_3/NaOH ratio) affecting compressive strength of GGBFS.

Run Order	NaOH	S/L	Na_2SiO_3/NaOH (SS/NaOH)	Compressive strength (MPa)
1	6	1	1.5	47.9
2	10	1	1.5	31.8
3	6	3	1.5	109.3
4	10	3	1.5	80.4
5	6	1	2.5	60.9
6	10	1	2.5	53.1
7	6	3	2.5	109.5
8	10	3	2.5	161.7
9	6	1	1.5	34.8
10	10	1	1.5	31.6
11	6	3	1.5	115.4
12	10	3	1.5	97.5
13	6	1	2.5	61.4
14	10	1	2.5	65.3
15	6	3	2.5	115.4
16	10	3	2.5	168.7
17	6	1	1.5	60.9
18	10	1	1.5	37.1
19	6	3	1.5	103.3
20	10	3	1.5	76.8
21	6	1	2.5	62.3
22	10	1	2.5	62.2
23	6	3	2.5	103.3
24	10	3	2.5	155.3
25	8	2	2.0	101.9

3. Result and Discussion

Table 4 showed the overall main affects for GGBFS compressive strength found to be statistically significant as indicated by the F-value of 141.31 where the corresponding P-value of 0.00 as the result obtained from Minitab. The solid-to-liquid ratio exhibited highest positive effect on compressive strength, respectively. Contrary, NaOH concentration gives the lowest effect on compressive strength. Moreover, as the P-value for NaOH concentration is 0.391, it will give less effect (3.08) on the compressive strength of GGBFS as response in these studies. Hence, with the P-value unequal to 0.00 gives no significant effects on the compressive strength of GGBFS.

Statistically it can be observed that all interaction (three way and two-way) was found to be significant effect on the compressive strength of GGBFS geopolymer due to the P-value equal to 0.00 and the effect value is higher (1533.3 and 4259.7) compared to others one-way interaction factors. Meanwhile, parametric study of the combination factors affecting the GGBFS geopolymer compressive strength are strongly recommended first before conducting further experimental related to mechanical properties and other application of the geopolymer concrete [19, 20]. Furthermore, this statistical study generated the information which enhanced the manufacturing of GGBFS properties those benefits for further researches.

Table 4. ANOVA values of the models for compressive strength.

Source	Degree of freedom	Effect	Adjusted mean square	F-value	P-value	Significance
Main effect	3	31051.4	10350.5	141.31	0.000	Yes
NaOH	1	3.08	56.8	0.78	0.391	No
S/L	1	65.61	25826.9	352.6	0.000	Yes
SS/NaOH	1	29.35	5167.7	70.55	0.000	Yes
2-way interaction	3	4259.7	1419.9	19.39	0.000	Yes
NaOH*S/L	1	10.96	720.7	9.84	0.006	Yes
NaOH*SS/NaOH	1	22.48	3031.4	41.39	0.000	Yes
S/L*SS/NaOH	1	9.20	507.6	6.93	0.017	Yes
3-way interaction	1	1533.3	1533.3	20.93	0.000	Yes
NaOH*S/L*SS/NaOH	1	15.99	1533.3	20.93	0.000	Yes
Residual error	17	1245.2	73.2	-	-	-
Curvature	1	321.1	321.1	5.56	0.031	No
Pure error	16	924.1	57.8	-	-	-

3.1. The effect of various parameters

Main effects of the three factors include: NaOH concentration, solid-to-liquid ratio, Na_2SiO_3/NaOH ratio are shown in Figure 1. The solid-to-liquid ratio (S/L) has a huge effect than NaOH concentration and Na_2SiO_3/NaOH ratio. The length of solid-to-liquid ratio from the low level (1.0) to the high level (3.0) was immense compared to NaOH concentration and Na_2SiO_3/NaOH ratio effects. Thus, this proved that the solid-to-liquid ratio had a significant effect on the compressive strength of GGBFS.

In figure 2, the Pareto chart exhibited all factors; solid-to-liquid (S/L) ratio and Na_2SiO_3/NaOH (SS/NaOH) ratio except the concentration of NaOH factors alone have significant effect on the compressive strength of GGBFS geopolymer. All factors except NaOH concentration extends beyond the reference lines (2.11) demonstrated potentially important to the compressive strength of GGBFS as shown in figure 2 [21]. Following are the interpretation of the main effect plot for GGBFS compressive strength:

IOP Conf. Series: Materials Science and Engineering **209** (2017) 012040 doi:10.1088/1757-899X/209/1/012040

i. Solid to Liquid ratio (S/L)	:	Runs with 3.0 S/L ratio had higher compressive strength than runs at 1.0 S/L ratio
ii. Concentration of NaOH (NaOH)	:	Runs at higher and lower concentration of NaOH had comparable compressive strength of GGBS geopolymer.
iii. Na_2SiO_3/NaOH ratio (SS/NaOH)	:	Runs at higher and lower SS/NaOH ratio had comparable compressive strength of GGBS geopolymer.

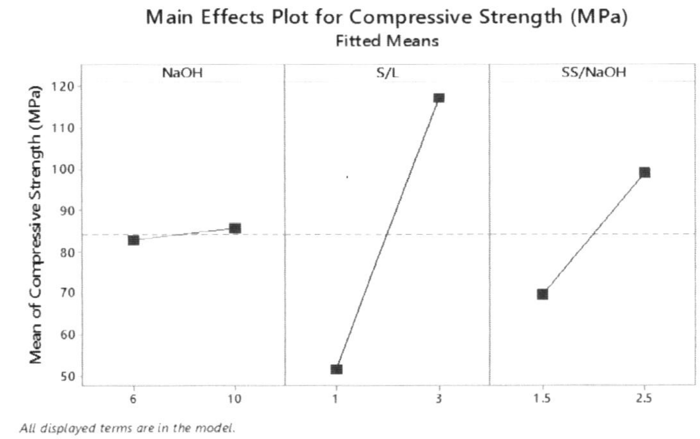

Figure 1. Main effect plots for compressive strength of GGBS geopolymer.

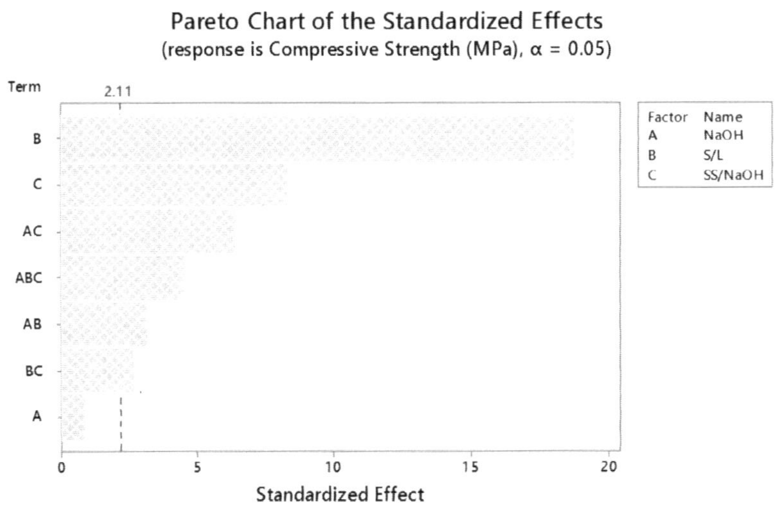

Figure 2. The effects of concentration of NaOH, solid to liquid ratio, activator ratio, on compressive strength of GGBFS geopolymer.

The interaction plot exhibited the relationship between one categorical factor and continuous response depends on the value of the second categorical factors.

The interactions between the combinations factors on compressive strength for GGBFS geopolymer are illustrated in figure 3. The line in the interaction plot represent the interaction affected the relationship between the factors and the response. In addition, the more nonparallel the lines are, the greater the strength of the interaction [22].

This interaction effect indicated that the relationship between NaOH concentration and compressive strength of GGBFS depends on the value of SS/NaOH ratio. The SS/NaOH ratio is equal to 1.5 and the NaOH concentration at 6M is associated with the highest mean compressive strength of GGBFS. However, when the SS/NaOH ratio is equal to 2.5 and the NaOH concentration at 10M is associated with the highest mean compressive strength. Thus, the combination (SS/NaOH ratio*NaOH) has a significant effect compared to other combination of factors due to nonparallel lines in the interaction plots as obtained in figure 3.

Additionally, increasing the sodium silicate to alkaline solution ratio has enhanced the compressive strength of the geopolymer [23]. Sodium silicate promotes the compressive strength of GGBFS based geopolymer paste in two ways; (1) the sodium silicate solution improves the dissolution rate of Si and Al; (2) then, the Al-O bonds are weaker than Si-O bonds in raw material, Al dissolve rapidly in alkali solution [2]. Thus, if Si ion is provided prior to being available through dissolution of raw material, it can improve the degree of geopolymerization and enhanced the mechanical properties [24]. Otherwise, the dissolution and formation of Si-O-Si oligomers as a strong bond can be enhance due to a simultaneous increase in Na and Si ions [25]. Besides, the hydration degree of GGBFS increase due to ability of OH- ions from NaOH solution then create intermediary reactants which last precipitate into C-S-H gel and other final product [26].

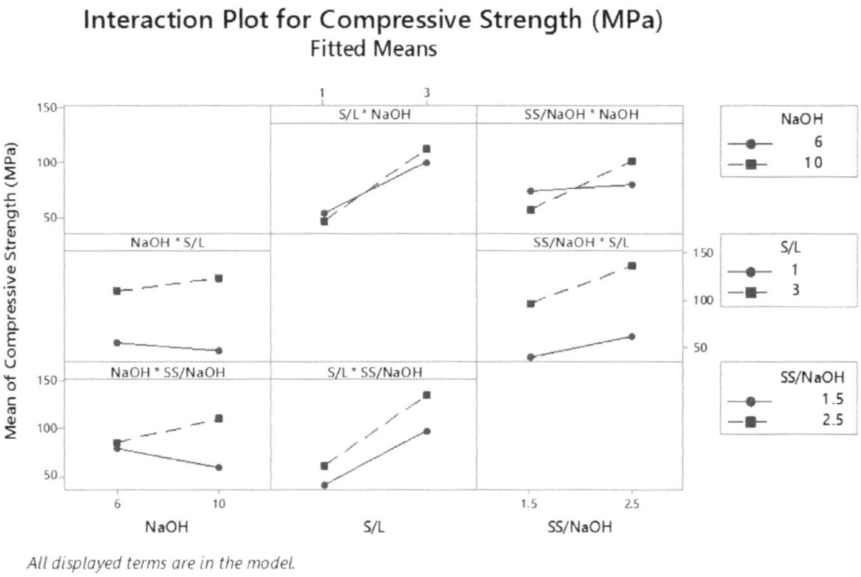

Figure 3. Interaction plots for compressive strength of GGBS geopolymer.

3.2. Prediction of compressive strength

The model equation for the GGBFS geopolymer can be used to mathematically estimate the compressive strength as the response for any combination of factor levels within the design space. The equation predicts the compressive strength of GGBFS based on the concentration of NaOH, solid-to-liquid ratio, alkaline activator ratio and other combination factors (NaOH*S/L, NaOH*SS/NaOH, S/L*NaOH, S/L*NaOH). Thus, in order the get the optimum compressive strength of GGBFS by using the fixed S/L ratio and alkaline activator ratio, the concentration of NaOH can be predict. The model represents the data of the present study are written as follows:

Compressive strength = -41.5 + 4.78 (NaOH) + 120.4 (S/L) + 48.9 (SS/NaOH) -13.25 (NaOH*S/L) – 4.75 (NaOH*SS/NaOH) - 54.7 (S/L*NaOH) + 7.99 (c)

Where; NaOH = concentration of NaOH, S/L = solid-to-liquid ratio, SS/NaOH = alkaline activator ratio.

The equation gives the benefits to evaluate the manufacturing of GGBFS compressive strength in order required minimum experiment. Also, by using this prediction equation, the researchers can focus on the further application such as fire resistant material by using the GGBFS as the raw precursor material

4. Conclusions

In this study, the effect of three parameters, namely, NaOH concentration, solid-to-liquid (S/L) ratio, and Na_2SiO_3/NaOH ratio on compressive strength was investigated. In summary, the following summarized the most important outputs derived statistically from DOE analysis on GGBFS geopolymer.

i. Increasing the concentration of NaOH from 6M to 10M leads to an increase in compressive strength of GGBFS geopolymer.
ii. The concentration of NaOH at 10M, solid-to-liquid ratio (S/L) of 3.0, and Na_2SiO_3/NaOH ratio equal to 2.5 exhibited an optimum compressive strength of GGBFS geopolymer of 168.7 MPa.
iii. ANOVA data analysis obtained that ratio of the solid-to-liquid (S/L) was the highest influenced than NaOH concentration and Na_2SiO_3/NaOH ratio.

References

[1] Provis, J L 2014 *Materials and Structures* **47** 11-25
[2] Li C, Sun H, and Li L 2010 *Cement and Concrete Research* **40** 1341-1349
[3] Shi C, Jiménez A F, and Palomo A 2011 *Cement and Concrete Research* **41** 750-763
[4] Yun-Ming L, Cheng-Yong H, Al Bakri M M, and Hussin K 2016 *Progress in Materials Science* **83** 595-629
[5] Palomo A, Grutzeck M, and Blanco M 1999 *Cement and Concrete Research* **29** 1323-1329
[6] Bernal S A, Rodríguez E D, de Gutiérrez R M, Gordillo M, and Provis J L 2011 *Journal of Materials Science* **46** 5477
[7] Puligilla S 2011
[8] Puligilla S and Mondal P 2013 *Cement and Concrete Research* **43** 70-80
[9] Cheng T and Chiu J 2003 *Minerals Engineering* **16** 205-210
[10] Liew Y, Kamarudin H, Al Bakri A M, Binhussain M, Luqman M, Nizar I K, Ruzaidi C and Heah C 2011 *Physics Procedia* **22** 312-317
[11] Lloyd N and Rangan B 2010
[12] Singh P S, Trigg M, Burgar I and Bastow T 2005 *Materials Science and Engineering: A* **396**, 392-402
[13] Tempest B, Sanusi O, Gergely J, Ogunro V, and Weggel D, 2009
[14] Rovnaník P 2010 *Construction and Building Materials* **24** 1176-1183
[15] Chindaprasirt P, Chareerat T and Sirivivatnanon V, 2007 *Cement and Concrete Composites* **29**, 224-229
[16] Oh J E, Monteiro P J, Jun S S, Choi S and Clark S M 2010 *Cement and Concrete Research* **40** 189-196
[17] Sakulich A R 2009
[18] Villa C, Pecina E, Torres R and Gómez L 2010 *Construction and Building Materials* **24** 2084-2090
[19] Vora P R and Dave U V 2013 *Procedia Engineering* **51** 210-219

[20] Ghanbari M, Hadian A and Nourbakhsh A 2015 *Procedia Materials Science* **11** 711-716
[21] Richard N M 2014
[22] Levine D M, Ramsey P P and Smidt R K 2001 *Pearson*
[23] Nadoushan M J and Ramezanianpour A A 2016 *Construction and Building Materials* **111** 337-347
[24] Fernández Carrasco L, Fernández-Giménez A and Palomo A 2007
[25] Mustafa Al Bakri A M, Kamarudin H, Abdulkareem O A, Ruzaidi C, Rafiza A and Norazian M 2012: *Trans Tech Publ.*
[26] Sakulich A R, Anderson E, Schauer C and Barsoum M W 2009 *Construction and Building Materials* **23** 2951-2957

Acknowledgement

The authors gratefully acknowledge Center of Excellence Geopolymer and Green Technology (CEGeoGTech) and School of Materials Engineering , UniMAP for their expertise and support. The authors would also like to thank for the funding support from the Fundamental Research Grant Scheme (FRGS-9003-00540) under Ministry of Education Malaysia (MOE) and support from "Partnership for Research in Geopolymer Concrete" (PRIGeoC-689857) sponsored by the European Union.

IOP Conf. Series: Materials Science and Engineering **209** (2017) 012041 doi:10.1088/1757-899X/209/1/012041

Obtaining and Characterization of Polyolefin-Filled Calcium Carbonate Composites Modified with Stearic Acid

C Croitoru*, A Pascu, I C Roata and E M Stanciu
Materials Engineering and Welding Department, Transilvania University of Brasov, Eroilor 29 Str., 500039, Brasov, Romania

E-mail: c.croitoru@unitbv.ro

Abstract. In order to obtain high performance calcium carbonate-reinforced HDPE and PP composites, the dispersibility and compatibility of the inorganic phase in the polymer has been achieved through surface treatment of the amorphous calcium carbonate filler with stearic acid. The surface coating of the inorganic phase has been proved by XRD and FTIR spectroscopy, through forming of an intermediate layer of calcium stearate which acts as a surfactant, efficient in providing an optimum compatibility with the dominatingly hydrophobic polymer matrix, as determined from the structural information obtained through samples cross-sections analysing.

1. Introduction

Calcium carbonate represents one of the most frequently used fillers for polyolefins, due to its low cost, good chemical stability under plastic processing conditions, fire retardancy, color neutrality and so forth [1,2]. It is a rather ubiquitous material, found in nature as ground mineral (marble, limestone, chalk) or in precipitated form (calcite, aragonite, amorphous calcium carbonate). The latter is usually the most frequently used, accounting for about 60% of all the carbonaceous filler types used in conjunction with thermoplastic matrices [3]. When incorporated into the polyolefin matrix, it can efficiently reduce the overall cost of the material, while providing a series of useful properties, such as high impact resistance, stiffness, good colorability, resistance to environmental stress-cracking and low mold shrinkage [4].

Calcium carbonate loadings, reported to the thermoplastic resin is usually in the range of 20 to 40% wt. range. The hardness impacted by this type of filler to polyethylene is reported in the 70-98 HB unit range. Calcium carbonate filled composites with polyolefin matrix are commonly used as appliances, electrical components, houseware, toys, furniture for indoor and outdoor use, automotive industry, and several utility products [5].

Recently, a lot of research has been conducted in the field of toughening of semicrystalline polymers with the addition of inorganic fillers. It has been reported that both crystalline as well as amorphous calcium carbonate is able to efficiently increase the stiffness of amorphous polypropylene, isotactic polypropylene, or high-density polyethylene due to stress concentration, debonding at the polymer/filler interface and the formation of shear bands, which lead to more efficient energy absorption during deformation [6]. The main condition to achieve an optimum performance for the polyolefin/calcium carbonate composites is to efficiently achieve a good dispersion of the calcium carbonate particles.

Due to the dissimilar nature of polyolefins and calcium carbonate in terms of polarity and surface energy, implicitly, a strong tendency of agglomeration for the inorganic phase has been reported. Many strategies have been employed to address this shortcoming, among which the most frequently employed are the particles surface treatment to reduce their surface energy and in some cases to increase the compatibility with the dominatingly hydrophobic matrix.

Among the surface functionalization agents, γ-methacryloxypropyl trimethoxy silane or trivinyl ethoxysilane, long alkyl chain organophosphorus coupling agents, non-ionic surfactants and so forth. These types of additives are often expensive, carcinogenic and tend to dissolve only in non-polar volatile solvents, such as toluene, aliphatic hydrocarbons [7, 8], etc.

One of the most economically-efficient functionalization agent is stearic acid (SA). Its loading on the calcium carbonate particles surface usually ranges from 1 to 14 wt.%, and the advantage is that it could be used also in aqueous media (at high temperatures) or in ethanol, a milder volatile solvent [8].

In this study, amorphous calcium carbonate powder has been surface treated with stearic acid by treatment of the particles with different amounts of stearic acid dissolved in ethanol at 60°C. The structure and properties of the obtained surface modified carbonate has been determined by SEM microscopy, XRD and FTIR spectroscopy. The functionalized calcium carbonate has been introduced and blended in high density polyethylene (HDPE) and polypropylene (PP) matrices through thermoforming at 150°C for HDPE and 190°C for PP. The composites cross-section has been examined by electron microscopy, evidencing a good wetting of the functionalized inorganic phase by the polymer.

2. Materials and methods

2.1. Materials

Amorphous calcium carbonate powder (50-90 μm particle diameter) has been purchased from Sigma-Aldrich. Stearic acid has been purchased from the same company. Both filler and functionalizing agent were used as such.

High density polyethylene (melt flow rate of 3.15 g/10 min) and polypropylene (melt flow rate of 2.65 g/10 min) powder (~0.5 mm sieve) have been used for the forming of the thermoplastic matrix.

2.2. Methods

2.2.1. Calcium carbonate powder SA surface treatment. A concentration of 10% stearic acid in 98% ethanol has been prepared. A determined amount of the calcium carbonate powder (10g) has been immersed in 50 mL of the SA ethanol solution and magnetically stirred at 60°C under reflux for 1 hour. After the surface functionalization, the solution has been filtered, washed several times with distilled water and dried in a vacuum drying stove at 40°C and 20 mbar.

2.2.2. Composites obtaining through thermoforming. PP and HDPE powders were mixed with the surface treated stearic acid calcium carbonate (10% wt. reported to the polymer), placed in a stainless-steel matrix, and pressed at 100 atm for 30 s. After pressing, the metal matrix has been introduced in an oven at 150°C for HDPE and 190°C for PP for 1 hour, followed by cooling to room temperature. The nominal dimensions of the obtained composites are 100 mm x 10 mm x 4 mm (length x width x thickness).

2.2.3. Characterization methods. To evidence the surface functionalization of the amorphous calcium carbonate with SA, XRD spectra have been obtained, with a Bruker Discover diffractometer ($K_{\alpha 1}$ =0.154 nm, scan speed 5 degrees/min). To characterize the surface chemistry of the calcite particles and the composites, FTIR spectra have been acquired in the 4000-600 cm^{-1} interval (Bruker VERTEX 70 spectrophotometer working in reflectance mode). A QUANTA 200 scanning electron microscope was used to study the morphological appearance of the functionalized powder and composites cross-sections.

3. Results and discussion

As it could be observed from figure 1, the amorphous calcium carbonate powder presents a spherical morphology, with a mean particle diameter of 70 μm.

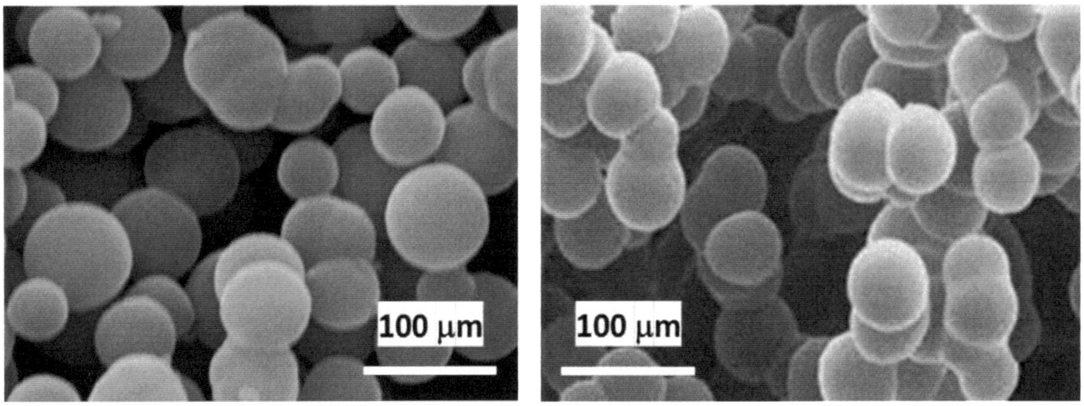

Figure 1. Initial $CaCO_3$ powder (left) and SA-treated $CaCO_3$ powder (right).

Stearic acid functionalization preserves the initial morphology of the carbonate particles. No significant modifications in mean diameter occurs, probably due to the adsorption of the organic molecule in a monolayer on the particles surface, as the reference literature suggests [8].

Stearic acid can efficiently coat the amorphous calcium carbonate powder, as determined from the infrared spectra of the functionalized and neat powders (figure 2). The coated calcium carbonate presents several structural features, common of both components, namely stearic acid, and calcium carbonate.

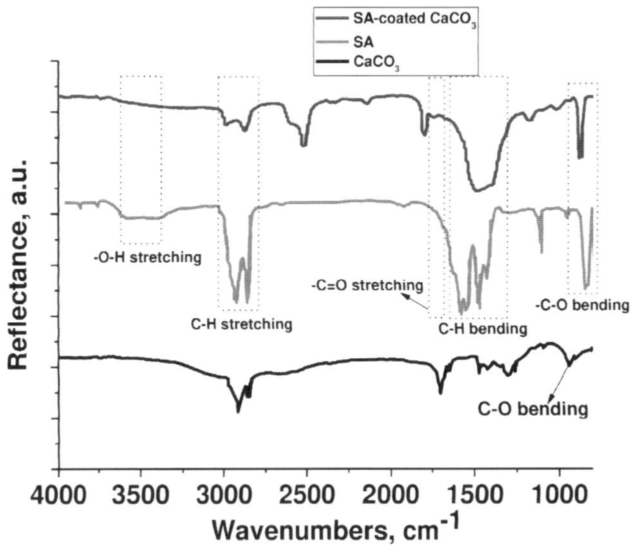

Figure 2. FTIR spectra of the neat and SA-coated calcium carbonate particles.

The band centred at 3400 cm^{-1} could be ascribed to in plane -O-H stretching vibration from the carboxyl groups of the stearic acid or from the weakly adsorbed water on the surface of the carbonate particles [9]. Also, alkyl groups vibration (hydrocarbonate chain of stearic acid) could be noted on the

FTIR spectra of the functionalized particles, as well as C-O bending vibration in carbonates (946 cm^{-1}) [10]. The mechanism responsible for carbonate particles functionalization employs in a first step, the monolayer adsorption of SA molecules on $CaCO_3$ particles surface (figure 3), followed by their reaction with the carbonate, ultimately forming a layer of calcium stearate, which acts as a surfactant, improving the dispersibility of the inorganic phase in the selected polymer matrices and lowering their surface energy.

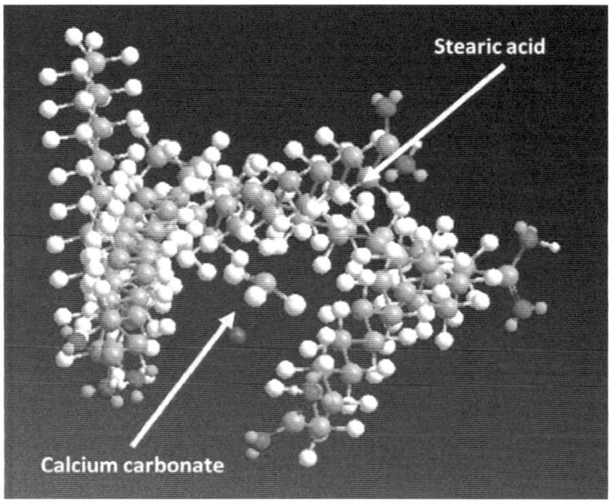

Figure 3. Mechanism of SA interaction with the calcium carbonate particles.

The particles functionalization has been also evidenced form the XRD diffractograms (figure 4), albeit, considering the amorphous character of the inorganic phase and the low crystallinity of SA, the diffraction peaks are broad and less characteristic.

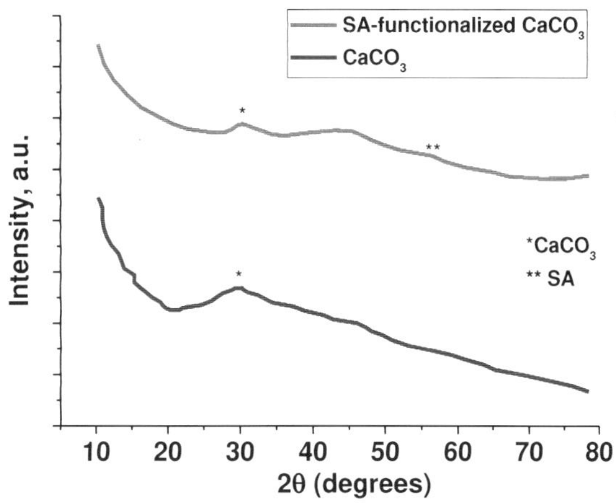

Figure 4. XRD diffractograms of neat calcium carbonate powder and SA-treated calcium carbonate powder.

The cross-sections of the composites obtained through functionalized carbonate particles inclusion in HDPE and respectively PP reveal a satisfactory dispersibility of the inorganic phase throughout the observed area (figure 5).

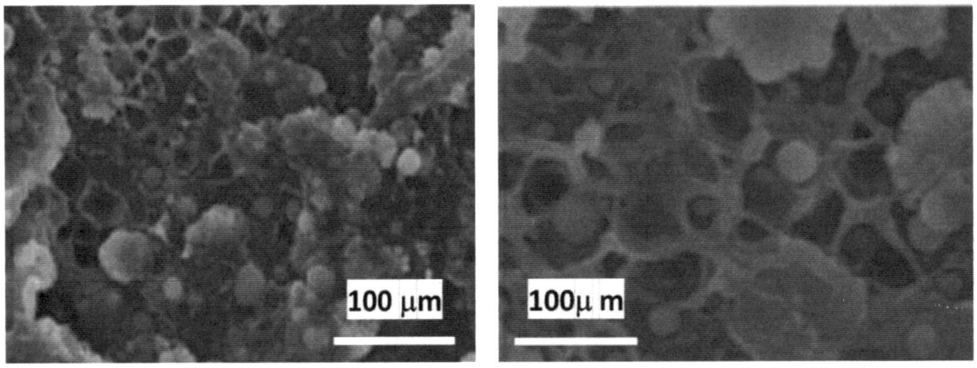

Figure 5. FTIR spectra of neat PP, HDPE and stearic acid coated calcium carbonate filled PP and HDPE.

Also, a good "wetting" of the inorganic phase by the thermoplastic matrix has been achieved, as determined from the continuity of the material in the interphase zone, determined by SEM micrographs image analysis.

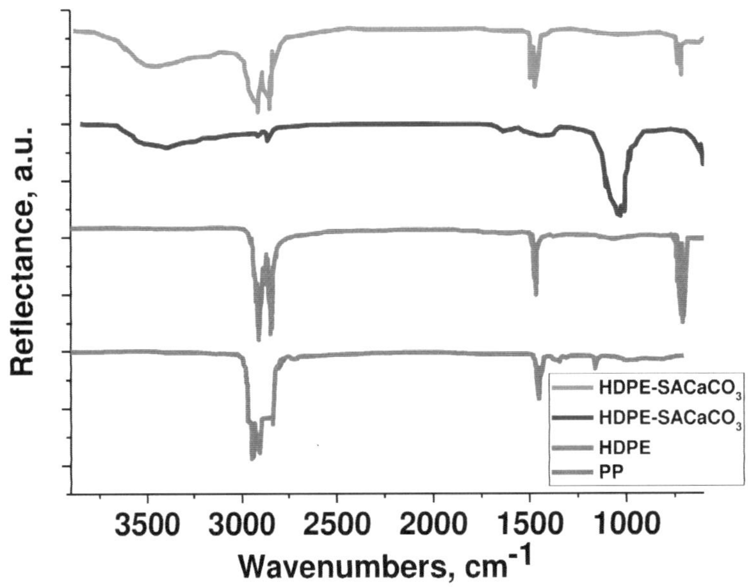

Figure 6. FTIR spectra of neat PP, HDPE and stearic acid coated calcium carbonate filled PP and HDPE.

The FTIR spectra of neat polymers (HDPE and PP) reveal the presence of the usual absorption bands, such as methyl group -CH_3 centered at ~2970 cm^{-1}, vibration of methylene group - CH_2 at ~2870 cm^{-1} (methylene rocking vibration ~1460 cm^{-1}, methyl out-of-plane vibration ~1370 cm^{-1}). Supplementary from PP, HDPE presents a strong absorption at ~1483 and 1493 cm^{-1}, ascribed to collective bending vibration overtones of the alkyl groups. It could be seen that the obtained composites present several features common to stearic acid, such as the -OH groups stretching vibration [11].

4. Conclusion

Amorphous calcium carbonate with spherical particle morphology has been coated by stearic acid to improve its dispersibility and compatibility with several hydrophobic thermoplastic polymer matrices, such as high-density polyethylene and polypropylene. An efficient hydrophobization of the powder has been achieved, determined from the scanning electron micrographs of the obtained hybrid organic/inorganic composites.

References

[1] Essabir H, Ouadi Bensalah M, Rodrigue D, Bouhfid R, Qaiss A 2017 *Construction&Building Materials* **134** 549
[2] Croitoru C, Giubega A, Patachia S, Baltes L, Pascu A, Roata I, Tierean M 2016 *Bulletin of the Transilvania University of Braşov* **9** 58
[3] El-Sherbiny S, El-Sheikh S, Barhoum A 2015 *Powder Technol* **279** 290
[4] Xiong Z, Li Y, Pan L, Yu J, Lu S 2017 *Composite Structures* **171** 370
[5] Chan C, Wu J Li J Cheung Y 2002 *Polymer* **43** 2981
[6] Pradittham A, Charitngam N, Puttajan S, Atong D, Pechyen C 2006 *Energy Procedia* **56** 264
[7] Rungruang P, Grady BP, Supaphol P 2006 *Coll and Surf A* **275** 114
[8] Cao Z, Daly M, Clemence L, Geever LM, Major I, Higginbotham CL, Devine DM 2016 *Appl Surf Sci* **378** 320
[9] Jiang X, Li S, Xiang G, Li Q, Fan L, He L, Gu K 2016 *Food Chem* **212** 585
[10] Konopacka-Łyskawa D, Kościelska B, Karczewski J, Gołąbiewska A 2017 *Mat Chem Phys* **193** 13
[11] Misra RDK., Nerikar P, Bertrand K, Murphy D 2004 *Mat Sci Eng: A* **384** 284
[12] A, Daik R, Zainudin E S, Tahir P M. 2014 *J Reinf Plastic Comp* **33** 440
[13] Livi J, Duchet-Rumeau P., Thi Nhan P., Gerard J F 2011 *J. Colloid Interface Sci.* **354** 797
[14] Sheng Y, Zhou B, Zhao J, Tao N, Yu K, Tian Y, Wang Z 2004 *J. Colloid Interface Sci.* **272** 326
[15] Bosch Reig F, Gimeno Adelantado J V, Moya Moreno M C M 2002 *Talanta* **58** 4
[16] Vitucci F M, Trequattrini F, Palumbo O, Brubach J B, Roy P, Paolone A 2014 *Vibr Spectrosc* **74** 81

Acknowledgements

This work was supported by the Romanian National Authority for Scientific Research and Innovation, CNCS-UEFISCDI, project number PN-II-RU-TE-2014-4-0173.

Thermal Spraying of CuAlFe Powder on Cu_5Sn Alloy

I C Roata*, A Pascu, C Croitoru[1], E M Stanciu[1] and M A Pop[2]
[1]Materials Engineering and Welding Department, Transilvania University of Brasov, Romania
[2]Materials Science Department, Transilvania University of Brasov, Romania

E-mail: ionut.roata@unitbv.ro

Abstract. To improve the corrosion and wear resistance of copper and its alloys, flame spraying has been employed to obtain a relatively homogenous Cu/Al/Fe-based coating. To minimize the defects that usually occur by using this method, a post-coating annealing step has been employed, by using concentrated solar energy as means of thermal surface treatment. Scanning electron micrographs have indicated a reduction in the cracks/pores density and accelerated corrosion testing have indicated a higher performance of the solar-annealed sample, in comparison with the initial reference material. The coating approach mentioned in this paper could be successfully applied to restore several worn tools and instruments, and could also be of use in the renewable energy field (IR-absorbent coatings) or in advanced oxidation processes, such as photocatalysis.

1. Introduction

Thermal spraying comprises in a group of processes through which various metallic or non-metallic feedstock materials (in powder, rod, or wire form) are brought in melted or semi-melted state using a heat source, after which they are accelerated and propelled on a substrate. The propelling of the coating material is responsible for spreading and bonding to the desired substrate [1].

Among the conventional thermal spraying methods, flame spraying is by far the most economically efficient approach to deposit a large variety of coatings, using as fuel sources hydrogen or several saturated and unsaturated hydrocarbons (propane, acetylene and so forth) which in reaction with a comburant (air, oxygen) produce an appropriate heat source, used for melting or plastic state achieving, as well as in creating a compressed gas steam that atomizes the molten particles and accelerates them onto the substrate [2-4].

Usually, all the thermal processes generate coatings with lamellar grain structure, because of rapid solidification of small globule particles, which are flattened by striking the cold substrate surface at high velocities. The thermal sprayed coatings usually present a large density of defects (cracks, pores, foreign particles inclusions), coating material and substrate oxidation due to the high temperatures which generates delamination, de-bonding, and therefore poor mechanical and corrosion-resistance performance. Currently, approaches designed to mitigate these shortcomings are limited to applying a post-deposition thermal heat treatment that reduces the number of defects through splats re-melting. This has detrimental effects on coating hardness and bonding to the substrate [5-10].

In this study, the flame spraying method has been used to obtain a cooper-aluminium coating deposited on a Cu_5Sn alloy, with potential use in manufacturing of electrical contacts, in the renewable energy field (IR absorbent coatings) or as photocatalysis substrate. To obtain an improved homogeneity of the coating, both in compositional as well as in morphological terms, a concentrated solar post-deposition thermal treatment has been employed. The benefit of using solar energy for thermal treatment resides in its accessibility, economic effectiveness and "green" character, thus avoiding the use of other usual and more environmentally-demanding heating processes, such as oven heat treatments, induction heating or molten salts thermal treatment [11].

2. Experimental

2.1. Materials

Rectangular samples (l=20 mm, L=20 mm, h=25 mm) of Cu5Sn alloy have been used as substrate for the flame spraying process, in conjunction with a Cu-Al-Fe powder, acquired from Castolin Eutectic, Germany. The elemental composition of the powder is presented in table 1.

Table 1. The chemical composition of Cu-Al-Fe coating powder.

Element	Composition (wt. %)
Cu	Balance
Al	8.5 – 10.75
Fe	0.5 - 2

The powder presents a core-shell type of morphology, with a copper core and an aluminum-iron coating. This configuration has been proven to aid interphase elemental diffusion and thus the formation of hard phases.

2.2. Methods

2.2.1. Thermal spraying and solar heat treatment of obtained coatings. The coatings obtaining has been realised in two steps. Firstly, the powder has been applied on the copper substrate by using the flame spraying method and secondly, a surface thermal treatment has been applied to improve splat adhesion, minimize the number of defects and to promote the formation of hard-phases in the material and refinement of the microstructure.

The coating was obtained by using a CastoDyn DS8000 flame and powder metallizing unit. C_2H_2 (0.7 bar pressure) has been employed as carburant gas and oxygen (4 bar pressure) as comburant, by utilizing a spraying distance of 150 mm, as shown in table 2. This distance has been selected by taking into consideration that at spraying distances lower than 100 mm, superficial layer exfoliation associated with material oxidation takes place because of overheating, and at large spraying distances (> 200 mm) an enhanced level/degree of porosity and poor adhesion could occur owing to the embedding a high quantity of cold particles in the coated material.

Because thermally sprayed coatings are inhomogeneous, all prepared samples have been thermally treated at the surface by solar heat fusion. Solar heat treatment has been performed with the help of a SF5 furnace (1 – 5 kW) having a 2.5 cm solar spot diameter, at the Plataforma Solar de Almería (PSA), a dependency of the Centro de Investigaciones Energéticas, Medioambientales y Tecnológicas (CIEMAT), the biggest concentrating solar technology inquiry, improvement, and test centre in Europe [11].

The solar thermal treatment plant scheme is described in figure 1.

A thorough control of solar thermal treatment has been carried out by using a thermocouple inserted in a drilled hole in the middle of the samples. Several samples were fully melted to define the operational temperature domain (maximum of 970°C).

Thermal treatment of the samples has been achieved in four distinct steps, illustrated in figure 2: I) sample preheating at the temperature of 600°C (T1); II) maintaining the preheating temperature for 5 min (T_1-T_2); III) heating at 950°C with 50°C/min (T3), followed by IV) cooling the samples at room temperature (T4).

Table 2. Operational parameters of the thermal spraying process.

Parameters	Sample 1 S1	Sample 2 S2
Thermal spraying distance	150 mm	150 mm
Powder feed rate	38 g/min	38 g/min
Preheating	120 °C	120 °C
Oxygen pressure	4 bar	4 bar
Acetylene pressure	0.7 bar	0.7 bar

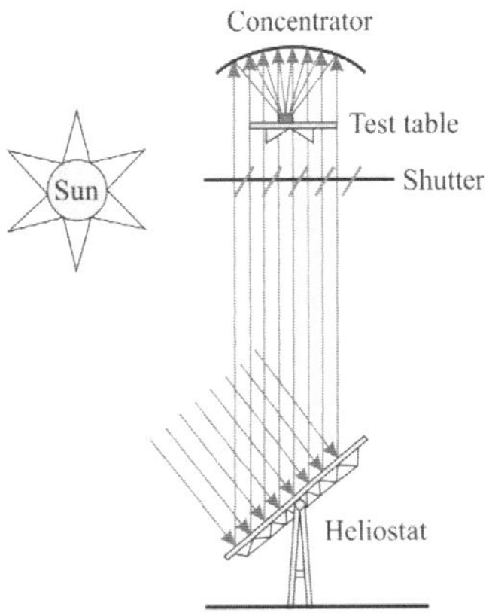

Figure 1. Functional diagram of the SF05 solar furnace.

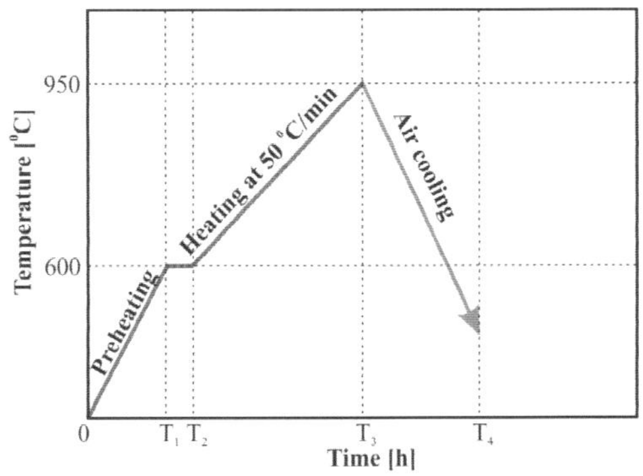

Figure 2. Solar heat treatment diagram.

2.2.2. Scanning electron microscopy (SEM) and microhardness tests The cross-section electron micrographs of the flame-sprayed samples have been achieved by utilizing a scanning electron microscope (SEM) Quanta 3D 200i model at various magnifications, mentioned on each micrograph. The elemental distribution of Cu, Al and Fe within the coating as well as in the boundary region was outlined with the help of energy-dispersive X-ray spectroscopy (EDS).

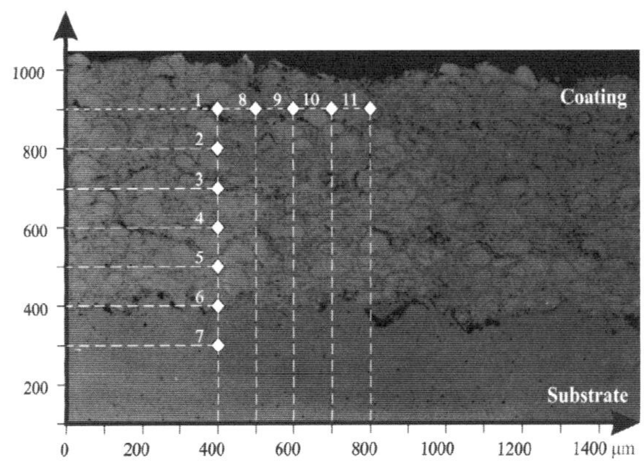

Figure 3. Positioning of microhardness indentations.

The microhardness tests were obtained on a FM 700 durometer, by employing 11 tests of HV_{01} in distinct zones, both on the coating as well as in the substrate material, as indicated in figure 3. A force of 1 N with a 15 seconds penetrator holding time has been used.

2.2.3. Corrosion testing. For the accelerated corrosion tests, an aqueous solution with global concentrations of 25% wt. NaCl; 1% wt. HNO_3 and 3% wt. H_2O_2 has been used. This solution is especially used in conjunction with aluminum and copper alloys to test for inter-grain corrosion or surface pitting. The samples have been weighed initially (m_0) and then immersed in 10 ml of the corrosion test solution for 1 hour, with the coating facing downward. The volume of the corrosion solution and the diameter of the recipients in which the corrosion testing has been performed were chosen such as only the coating is in contact with the corrosive environment. After 1 hour, the samples have been washed several times with distilled water, dried and reweighted (m_1). The relative mass loss due to corrosion (Δm_{corr}) has been determined with eq. 1:

$$\Delta m_{corr} = \frac{m_0 - m_1}{m_0} \cdot 100(\%) \qquad (1)$$

Copper ions leaching from the coating to the aqueous environment has been visually assessed.

3. Results and discussion

3.1. Structure of the Cu-Al-Fe coatings

Figure 4 (S1) presents the cross-section electron micrographs of the Cu-Al-Fe intermetallic coatings before heat treatment and (S2) after heat treatment at 950°C.

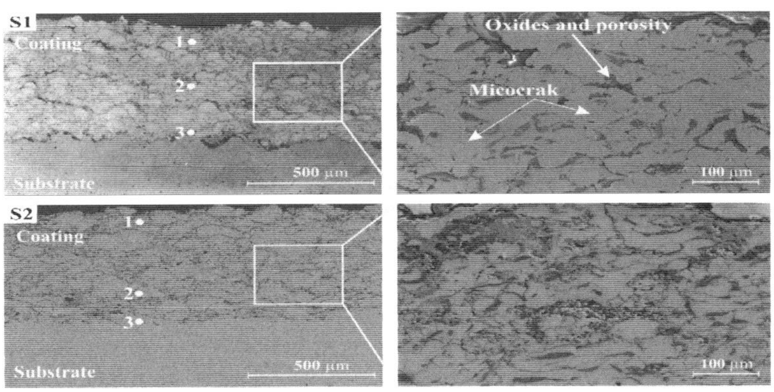

S1

Element	1 Wt%	2 Wt%	3 Wt%
CuK	86.50	88.32	96.27
AlK	9.81	10.21	6.97
FeK	1.35	1.47	0.98

S2

Element	1 Wt%	2 Wt%	3 Wt%
CuK	84.47	93.30	93.59
AlK	14.21	5.66	5.32
FeK	1.32	1.04	1.09

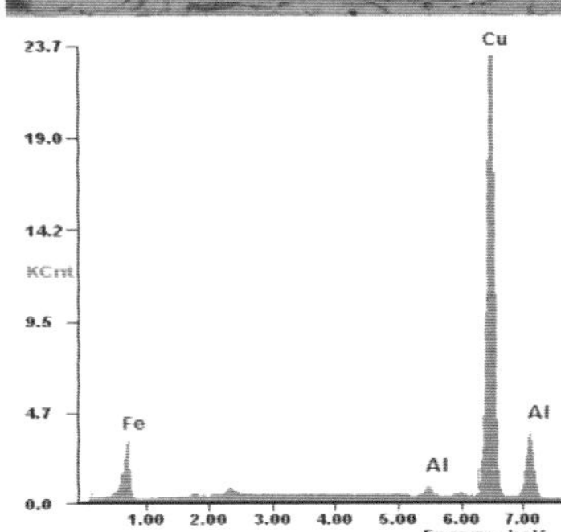

Figure 4. Cross-sectional SEM micrographs of S1 (initial sample) and S2 (heat treated at 950°C).

IOP Conf. Series: Materials Science and Engineering **209** (2017) 012042 doi:10.1088/1757-899X/209/1/012042

The S1 sample presents a typical lamellar (multi-layered) aspect and in some regions, partly unmelted splats could be evidenced as result of the thermal spraying process. Both dense splat and close-grained structures can be observed in the SEM micrographs as outcome of consecutive coating processes and resolidification of molten or semi-molten droplets.
Moreover, the interface with the substrate displays pores, partial melting, and high density of oxides.

An improvement in the coating bonding to the substrate has been achieved by employing solar energy, as it could be observed from Figure 4. A clear reduction in the cracks and pores density of the coating has been obtained, which aids in an improved wear and corrosion resistance performance of the material. The chemical composition variation along the coating (from the top to the substrate) indicates an improved inter-mixing of elements in the case of S2 sample, possibly with the formation of different hardphases, such as $CuAl_2$ mentioned also in other research performed on the Cu-Al coating system [12-14], that may be responsible for the improved corrosion resistance of the material.

3.2. Micro-hardness of the coatings
The minimum and maximum micro-hardness values of the Cu-Al-Fe coating marked along the cross-section of the coating are 149.78 and 158.09 HV_{01}. Along the coating a ± 1 % variation occurred, indicating a relatively high compositional homogeneity of the coating material.

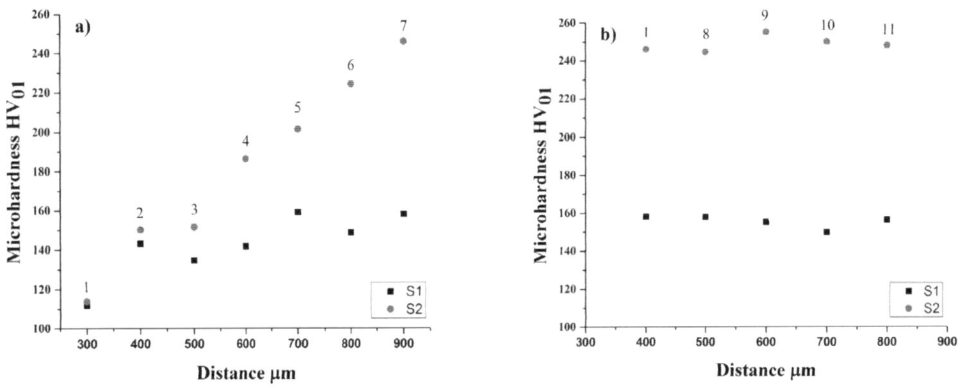

Figure 5 The influence of the temperature on the hardness values at different distances measured from the coating to the substrate (a) and the influence of temperature on the hardness along the coating (b).

The usage of solar heat treatment leads to a coating hardness increase from 158.09 $HV_{0.1}$ (S1) to 255.18 (S2) (figure 5). Figure 5b reveals that the micro-hardness longitudinally measured on the cross section for sample S2 is around 249 HV_{01} compared with 155 HV_{01} in the case of reference sample S1. This hardness increase in the solar annealed sample could be related to hardphases formation and subsequently could be linked with an improved corrosion performance.

3.3. Corrosion performance
Figure 6 illustrates the surface morphology before and after accelerated corrosion testing of the reference and solar-annealed samples.

An advanced corrosion of both aluminium and copper has been registered in the case of the non-annealed S1 sample, also by evidencing the colour differences of the solutions after the corrosion experiments. The mass loss value for S1 was 0.4996 %, in contrast with only 0.1919% for S2, indicating approximatively a threefold increase in the corrosion resistance after solar annealing.

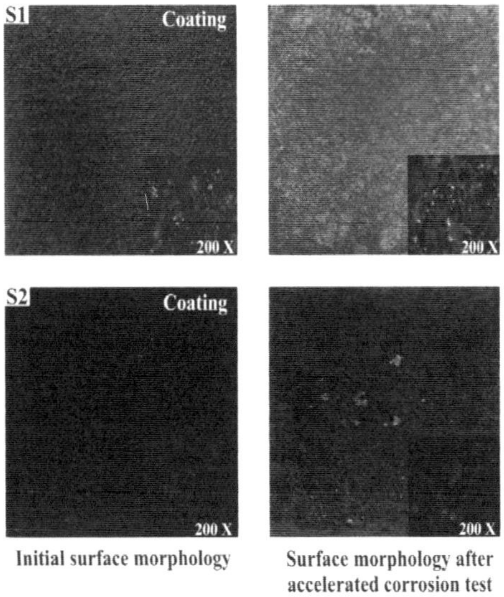

Figure 6. Corrosion testing of the flame-sprayed samples.

4. Conclusion

Ternary copper-aluminum-iron composite coatings have been obtained through flame spraying of a special core/shell powder. The characteristic defects, such as pores and cracks density has been reduced through applying a short post-deposition concentrated solar energy thermal treatment. Through this treatment, intermetallic hard-phases are formed which contribute to microhardness and corrosion resistance increase.

References

[1] Zhang P, Lau YY, 2010 *Journal of Applied Physics* **108** 44914
[2] Rahmana A, Jayaganthana R, Prakasha S, 2009 *J Alloys Compd.* **478** 472
[3] Hefei Li, Shufeng Tao, Zhaohui Zhou, Lidong Sun, A. Hesnawi, Shengkai Gong, 2007 *Surf. Coat. Techno* **201** 6589
[4] Mulero M.A., Zapata J, Vilar R, Martinez V, Gadow R, 2015 *Surf. Coat. Techno.* **278** 1
[5] Davis J.R. et al. 2004 *Handbook of Thermal Spray Technology,* ASM Int. 3-54
[6] Picas J.A, Forn A, Rilla R, Martin E, 2005 *Surf. Coat. Technol.* **200** 1178
[7] Fanicchia F, Axinte D.A, Kell J et al., 2017 *Surf. Coat. Technol.* **315** 546
[8] Houdková Š, Smazalová E, Vostřák M, Schuber J, 2014 *Surface & Coatings Technology* **253** 14
[9] Suutala J, Tuominen J, Vuoristo P, 2006 *Surface & Coatings Technology* **201** 1981
[10] Li C.J, Yang G.J, 2013 *Journal of Refractory Metals and Hard Materials* **39** 2
[11] Rodríguez J, Cañadas I, Zarza E, 2014 *Energy Procedia,* **49** 1511
[12] Yang J, Zhang Y, Zhao X, An Y, Zhou H, Hou J.G, 2015 *Tribology International* **90** 96.
[13] Zhao X, An Y, Hou G, Zhou H, Chen J, 2016 *Tribology International* **101** 255
[14] Zhang J, Wang B, Chen G, Wang R, Miao C, Zheng Z, Tang W, 2016 *Trans. Nonferrous Met. Soc. China* **26** 3283

Acknowledgements

Financial support by the Access to Research Infrastructures activity in the 7th Framework Programme of the EU (SFERA 2 Grant Agreement n. 312643) is gratefully acknowledged. The ICIR Euroinvent 2017 Conference mobility expenses have been supported by a grant of the Romanian National Authority for Scientific Research and Innovation, CNCS - UEFISCDI, project number PN-II-RU-TE-2014-4-0173.

IOP Conf. Series: Materials Science and Engineering **209** (2017) 012043 doi:10.1088/1757-899X/209/1/012043

Simulation of Impact Phenomena on the Composite Structures Containing Ceramic Plates and High Entropy Alloys

V Geantă[1], T Chereche₈[2], P Lixandru[2], I Voiculescu[1], R Ştefănoiu[1], D Dragnea[2], T Zecheru[3] and L Matache[3]
[1]Politehnica University of Bucharest, 313 Splaiul Independenţei, 060042, Bucharest, Romania
[2]UPS PILOT ARM, 2 Laminorului Street, 137210, Târgovişte, Romania
[3]Scientific Research Center for CRBN Defence and Ecology Bucharest, 225 Olteniţei Road 041309, Bucharest, Romania

e-mail: victorgeanta@yahoo.com

Abstract. Due to excellent mechanical properties, high entropy alloys from the system AlxCrFeCoNi can be used successfully to create composite structures containing both metallic and ceramic plates, which resists at dynamic load during high speeds impact (like projectiles, explosion). The paper presents four different composite structures made from a combination of metallic materials and ceramics plates: duralumin-ceramics, duralumin-ceramics-HEA, HEA-ceramics-HEA, HEA-ceramics-duralumin. Numerical simulation of impact behavior of the composite structures was performed by virtual methods, taking into account the mechanical properties of both materials. The best results were obtained using composite structures HEA-ceramics-HEA, HEA-ceramics-duralumin.

1. Introduction

The specifications regarding the security of the collective and individual defense equipment and structures in the military field have imposed increased requirements for the strength of the panels/flooring/protection elements to the penetration of different types of projectiles, due to the diversification of the types of interventions in the military activities. In order to ensure protection against impacts with high-speed projectiles, the materials must have both mechanical strength and impact energy absorption capacity. This can be accomplished by making materials that combine both requirements or by making composite materials that separately have the two characteristics [1].

The research directions in the field of ballistic protection have resulted in:

- Design and obtainment of modern and efficient devices for personal protection;
- Design and obtainment of protection devices for important facilities;
- Obtainment of new materials with special qualities compared to those considered conventional;
- Design and obtainment of adequate technical devices for intervention and neutralization of DEI;
- Establishing the conditions for testing materials and equipment so as they are fully useful, at a low cost price and with accessible technological possibilities [2 - 6].

Ballistic protection materials are used to obtain armored fighting machines - collective protection - and to manufacture individual protection systems. The main characteristic of the materials used for ballistic protection is the high resistance to the impact with kinetic and explosive ammunition. The provision of equipment for the protection of the armed forces, the police and the intervention structures has demonstrated their full utility during conflicts and events. In addition to the real protection it offers, this equipment also provides moral support for its use as a result of more advanced ergonomics and aesthetics.

Polyamide and polyethylene type polymeric materials, woven or non-woven, offer the best solutions for the production of soft armours, resistant to ballistic fragments and soft-core, sharp-nosed bullets. In order to stop steel-core, sharp-nosed bullets, metal armours or armours made of polymeric composite materials are required. In this area, a Kevlar composite armor with a mass of 28 kg / m^2 provides the same protection as a steel armour with a mass of 46 kg / m^2 [7].

To ensure protection against piercing bullets with a mass higher than 10 g and speeds exceeding 850 m/s, none of the solutions above is acceptable. The steel armour facing this threat is 14 mm thick and weighs 130 kg/m^2 and it cannot be used in the composition of bulletproof vests. The bullet protection problem has been satisfactorily solved by the use of structures that combine the high hardness properties of ceramic materials with the mechanical properties of the polymers used in ballistic protection [8]. Such a structure is composed mainly of a ceramic layer and a polymer composite layer. The two layers of strength are embedded with a polymeric fabric. The ceramics used are characterized by a hardness that exceeds 9 on the Moss scale. The Vickers microhardness of these materials is 2000 - 3000 VHN, compared to 750 VHN in the case of steel.

Currently, the best ballistic protection structures are made entirely of composite materials or incorporate, in part, material with outstanding features of impact resistance. The design and implementation of new structures for ballistic protection is based on the knowledge of the required properties of materials and the creation of composite structures for specific conditions of use.

Composite structures for ballistic protection with the use of ceramic and metallic materials are proposed as a solution for effective ballistic protection. One of the classes of metal alloys that can be used for this purpose is the high entropy alloys class (HEA) [9-11]. By definition, high entropy alloys contain at least five main metallic elements with concentrations ranging from 5 to 35% atomic. High entropy alloys (HEA) are composed of n major alloying elements with $n \geq 5$, introduced in equimolar or nearly equimolar ratios, which easily lead to the formation of simple solid phase solutions with BCC or FCC, nano-structures or even amorphous states as cast [12, 13, 14]. Therefore, the high entropy alloys are solid solutions with high strength, good thermal stability and hardening capacity above classical alloys, combined with superior strength characteristics under various environmental conditions [15, 16]. High entropy alloys, due to their high mechanical strength and high energy absorption capacity, provide multiple possibilities for use in individual protection applications or military or non-military armoured vehicles.

The paper presents four composite structures made from a combination of metallic materials and ceramics plates. The numerical simulation of impact behaviour of the composite structures was performed by virtual methods in order to assess the performance of the proposed structures.

In order to limit the material costs, the analysis of the ballistic protection performances of the constructive solutions of the composite structures, aimed at choosing the optimum solution, is done by virtual numerical simulation methods using a methodology for dynamic stress behaviour with high speeds of deformation of the composite structures [9]. The proposed methodology uses the Finite Element Method (FEM) for which there is a wide range of software packages [17]. The methodology is formulated in general terms, common to all finite element programs, and thus it can be applied to any program that has the analysis mode. The accuracy of the solutions given by numerical simulations is guaranteed by the correctness of the physical models used. Nowadays, the mathematical support of numerical simulation methods - the mathematical model - is well established. If used correctly, the mathematical support does not adversely affect the solutions [18 – 20]. In other words, the use for

IOP Conf. Series: Materials Science and Engineering **209** (2017) 012043 doi:10.1088/1757-899X/209/1/012043

numerical simulations of the current program packages: ABAQUS, ANSYS, AUTODYN, COSMOS M, LS-DYNA, NASTRAN etc. guarantees the correctness of solutions [17,19].

2. Composite structures made of metallic material – ceramic material used for the numerical simulation

For the virtual testing of the composite structures, there were analysed two types of metallic materials, i.e. high entropy alloys from the AlCrFeCoNi system and duralumin as well as ceramic materials. Thus, the composite structures for ballistic protection were designed in the following constructive variants:

- VARIANT 1 - Composite layered structure consisting of a 2 mm thick duralumin layer and a 10 mm thick ceramic layer, joined with adhesive, as shown in figure 1;
- VARIANT 2 - Composite layered structure consisting of a 2 mm thick duralumin layer, a 10 mm thick ceramic layer and a 6 mm thick layer of HEA, joined with adhesive, as shown in figure 2;
- VARIANT 3 - Composite layered structure consisting of a 6 mm thick HEA layer, a 10 mm thick ceramic layer and a 6 mm thick HEA layer, joined with adhesive, as shown in Figure 3;
- VARIANT 4 - Composite layered structure consisting of a 6 mm thick HEA layer, a 10 mm thick ceramic layer and a 10 mm thick duralumin layer, joined with adhesive, as shown in figure 4.

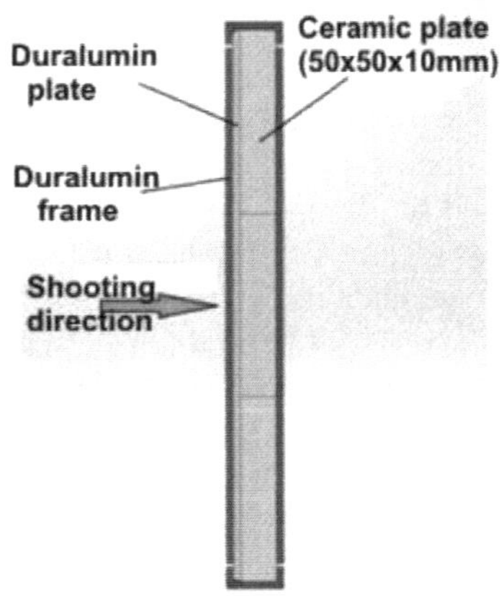

Figure 1. Composite structure – VARIANT 1.

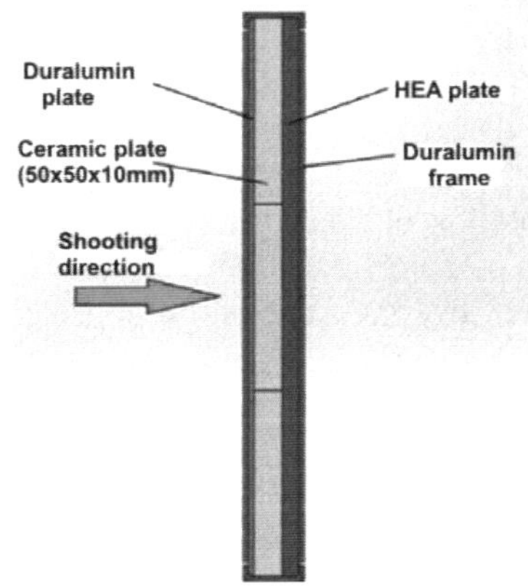

Figure 2. Composite structure – VARIANT 2.

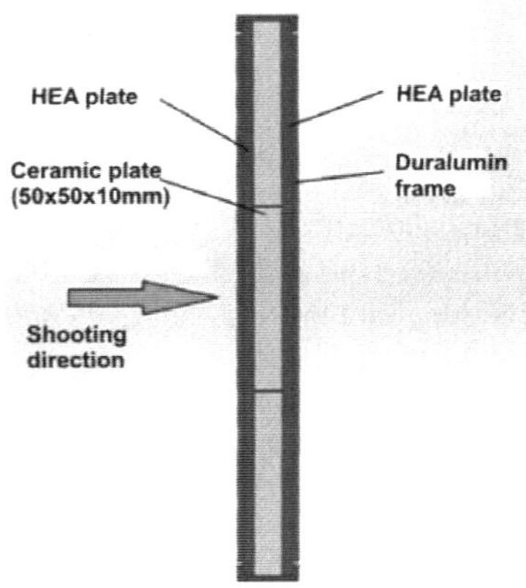

Figure 3. Composite structure – VARIANT 3.

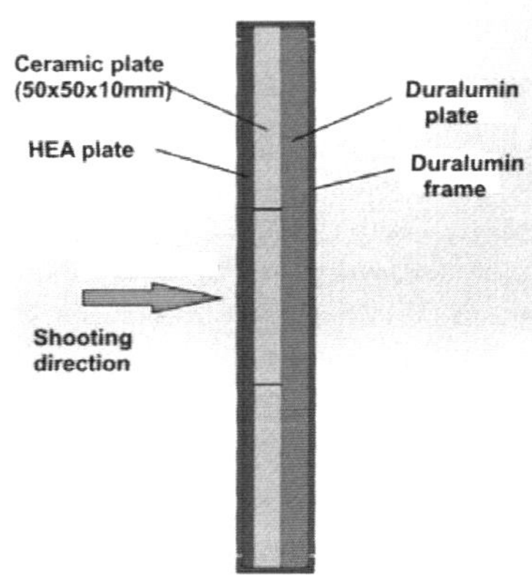

Figure 4. Composite structure – VARIANT 4.

3. Input data for the virtual testing of the composite structures

As mentioned above, the analysis of the performance of the proposed composite structures variants was done using a method of numerical simulation. For each analysis variant there were made three virtual shootings at different speeds, using 7.62 x 39 mm caliber incendiary, armor-piercing bullets.

The virtual testing was made using the same kind of projectile, with the following values of the specific parameters:

- Bullet calibre of 7.62 x 39 mm;
- Shooting angle of 0°;
- Incendiary piercing bullet weight of 7.67g;
- Steel core weight of 4g.

The setting of the speed bullets was done so that the effects on targets to be at the limit of perforating. The speeds of projectiles were adjusted during simulations to fall within these limits, different for each of the four types of structures analysed.

4. Results and discussions

The results of the numerical simulation are presented in Figures 5, 7, 9, 11 and in the graphics from Figures 6, 8, 10, 12, according to the testing speeds. The analysis of these representations highlights the role of correctly positioning the materials relative to the direction of impact.

4.1. Variant 1 (duralumin – ceramics)

In the case of the constructive variant 1 (duralumin - ceramics), the shooting through numerical simulation highlighted the following aspects (figure 5): after 7.4 e-3 ms the aluminium plate was perforated and the bullet reached the ceramic layer. It is noted that after 14.6 e-3 ms cracks appeared and the material of the ceramic plate began to be displaced and after 27 e-3 ms the perforation was complete, the bullet and armour fragments still have high velocity (450 m/s) and can cause lesions or may be lethal. Following the simulation, the speed variation vs. time was plotted for two nodes, from the core of the bullet (node 20966) and of the ceramic plate (node 107237) that was dislocated (figure 6). After 5 μs from the impact between the bullet and the composite structure, the dislocation and acceleration of the considered node from the ceramic plate begins, finally increasing its speed up to 450 m/s while the bullet is slowed down from the impact speed of 1250 m/s at 400 m/s. Subsequently, the speeds of the two nodes reach the same value, establishing a balance between the remaining core and the element dislodged from the ceramic plate.

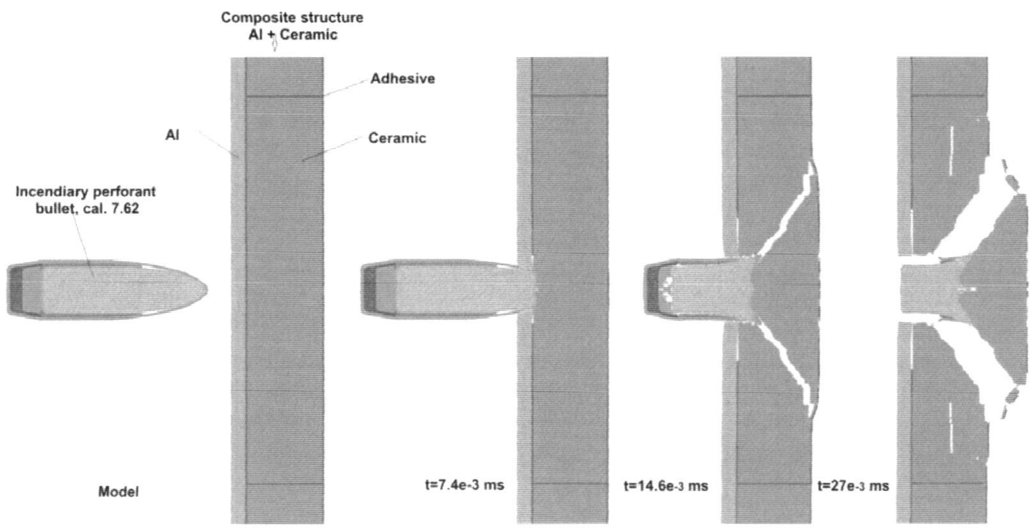

Figure 5. The behavior of the composite structure VARIANT 1 at impact with an incendiary armor piercing bullet (7.62 mm caliber) with initial speed of 1250 m/s.

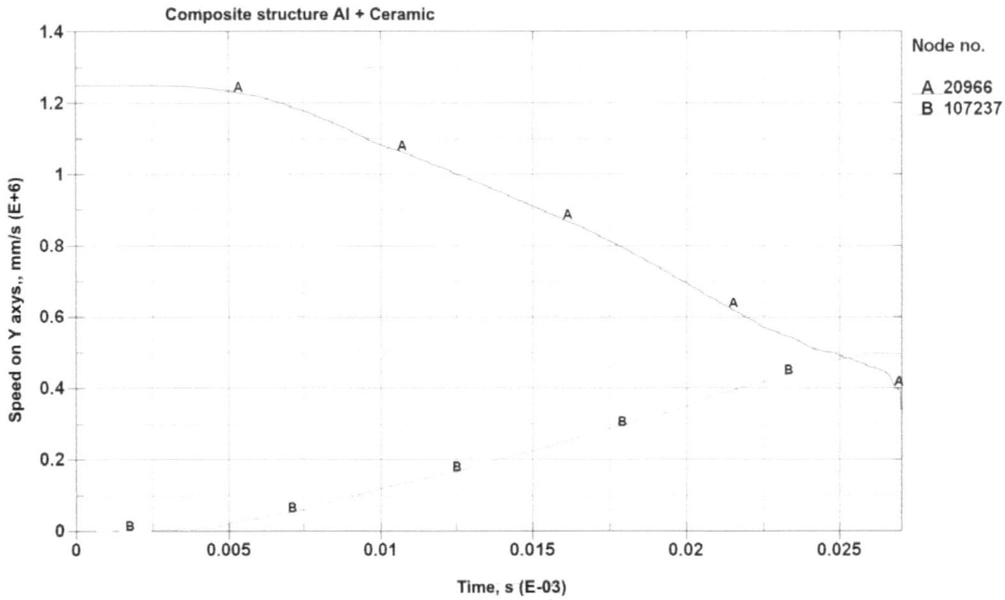

Figure 6. Variation of the bullet speed vs. time at the impact with the composite layered structure – VARIANT 1.

This constructive solution does not provide adequate protection for personnel and fighting techniques.

4.2. Variant 2 (duralumin – ceramics – HEA)
Considering the situation in the previous case, the composite structure was modified by adding a layer of HEA to the previous structure. In this case, after 7.8 e-3 ms the duralumin plate was perforated and the bullet touched the ceramic layer (figure 7).

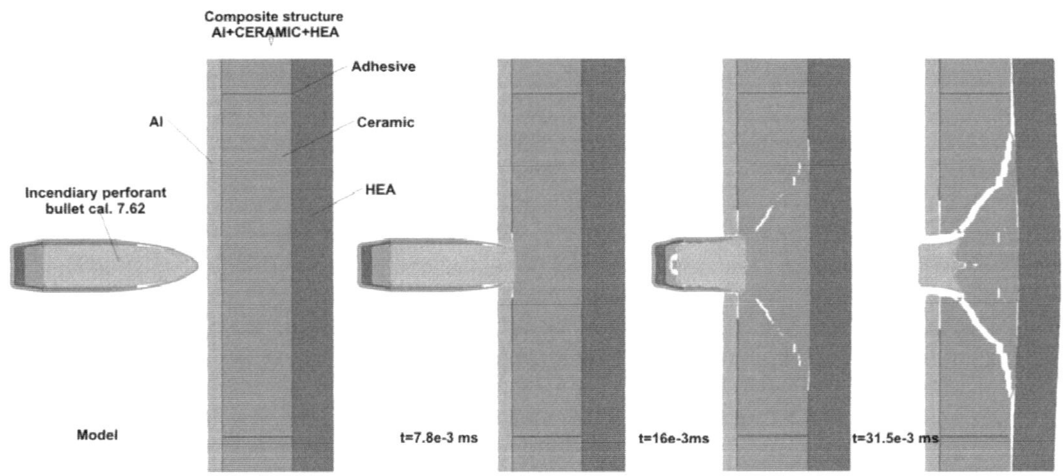

Figure 7. The behavior of the composite structure VARIANT 2 at impact with an incendiary armor piercing bullet (7.62 mm caliber) with initial speed of 1250 m/s.

It is noted that after 16 e-3 ms cracks appeared and the process of dislocation of the material from the ceramic plate began and after 31.5 e-3 ms the movement of the bullet and of the dislodged ceramic element stopped, without cracks in the HEA layer. Following the simulation, the speed variation vs. time was plotted for the node in the bullet core (node 25572 - figure 8). After 30 μs from the impact between the bullet and the structure, its velocity is null and then it has a slight return motion.

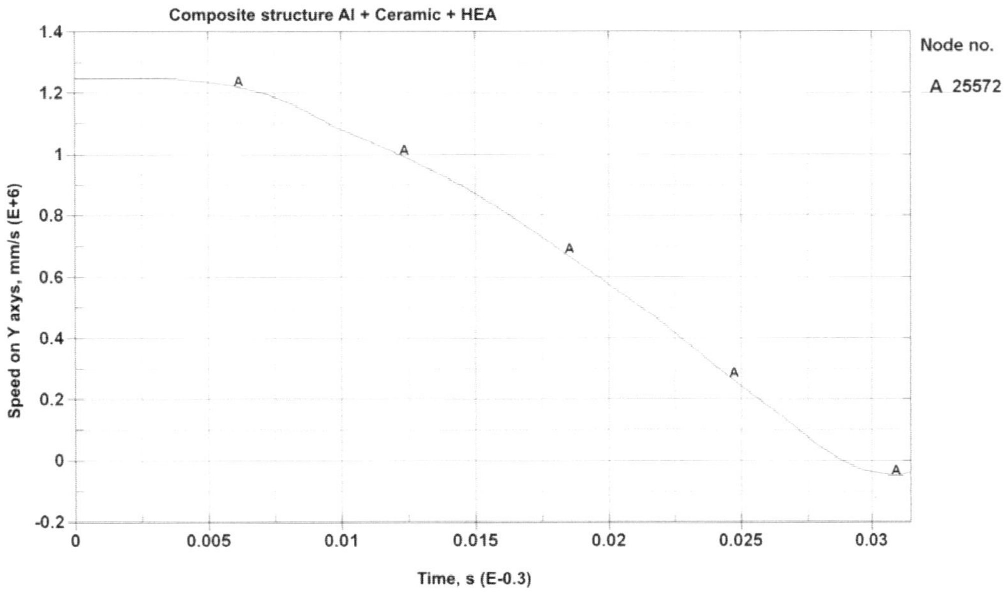

Figure 8. Variation of the bullet speed vs. time at the impact with the composite layered structure – VARIANT 2.

This constructive solution can be considered for the obtainment of composite structures used for effective ballistic protection.

4.3. Variant 3 (HEA – ceramics – HEA)

After 6.2 e^{-3} ms the HEA plate was perforated and the bullet reached the ceramic layer (figure 9).

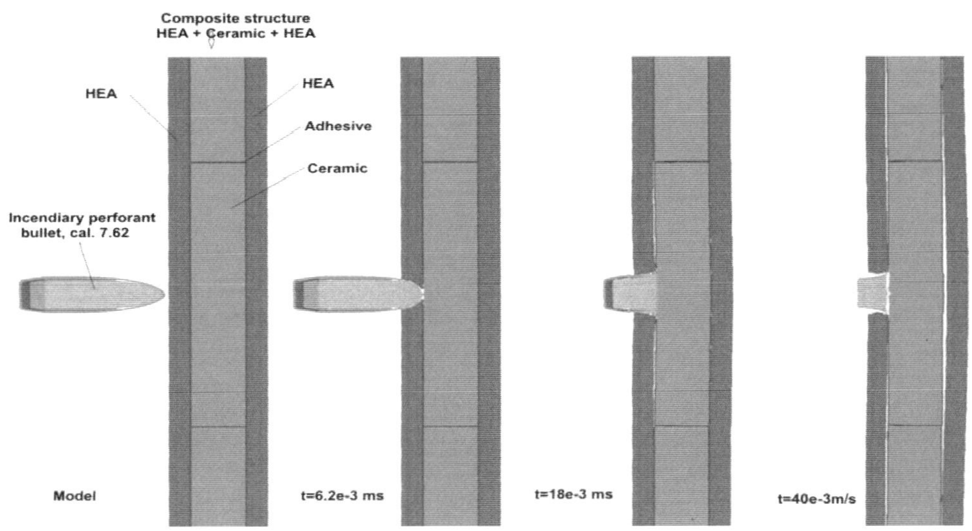

Figure 9. The behavior of the composite structure VARIANT 3 at impact with an incendiary armor piercing bullet (7.62 mm caliber) with initial speed of 1250 m/s.

It is noted that after 18 e^{-3} ms the core is consumed more than half without perforating the ceramic layer. It can be observed that the second HEA layer resisted and there is no perforation. After 26 μs from the impact between the bullet and the composite structure, its velocity is null and it has a slight return motion. The variation of the bullet speed versus time at impact with the composite structure 3 is shown in figure 10. This constructive solution can be considered for the obtainment of ballistic protection structures.

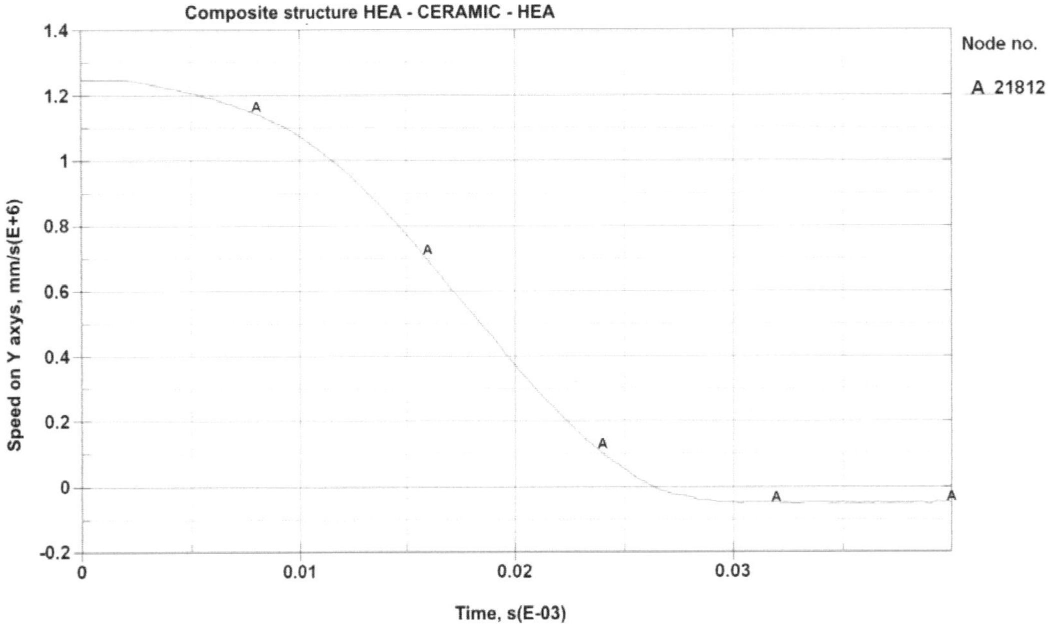

Figure 10. Variation of the bullet speed vs. time at the impact with the composite layered structure – VARIANT 3.

4.4. Variant 4 (HEA – ceramics – duralumin)

Another configuration used for virtual testing is made of HEA, ceramic and duralumin (figure 11).

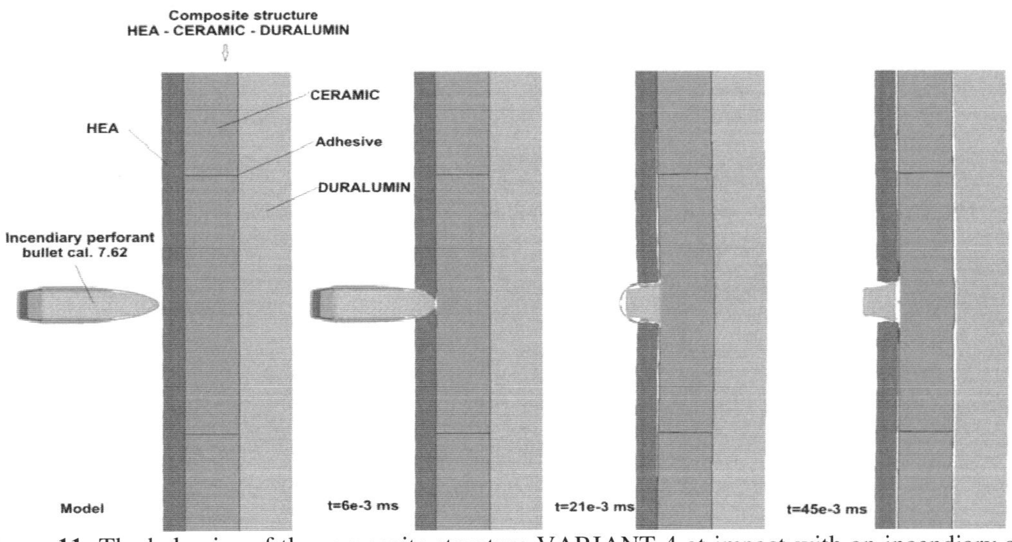

Figure 11. The behavior of the composite structure VARIANT 4 at impact with an incendiary armor piercing bullet (7.62 mm caliber) with initial speed of 1250 m/s.

After 6 e-3 ms the HEA plate was perforated and the bullet touched the ceramic layer. It is noted that after 21 e-3 ms the core is consumed more than half without perforating the ceramic layer. At the same time, the other components of the bullet are consumed almost totally without the ceramic layer being cracked and dislodged. It can be observed that the duralumin layer resisted and there is no perforation. After 26 μs from the impact between the bullet and the composite structure, its velocity is null and then it has a slight return motion (figure 12).

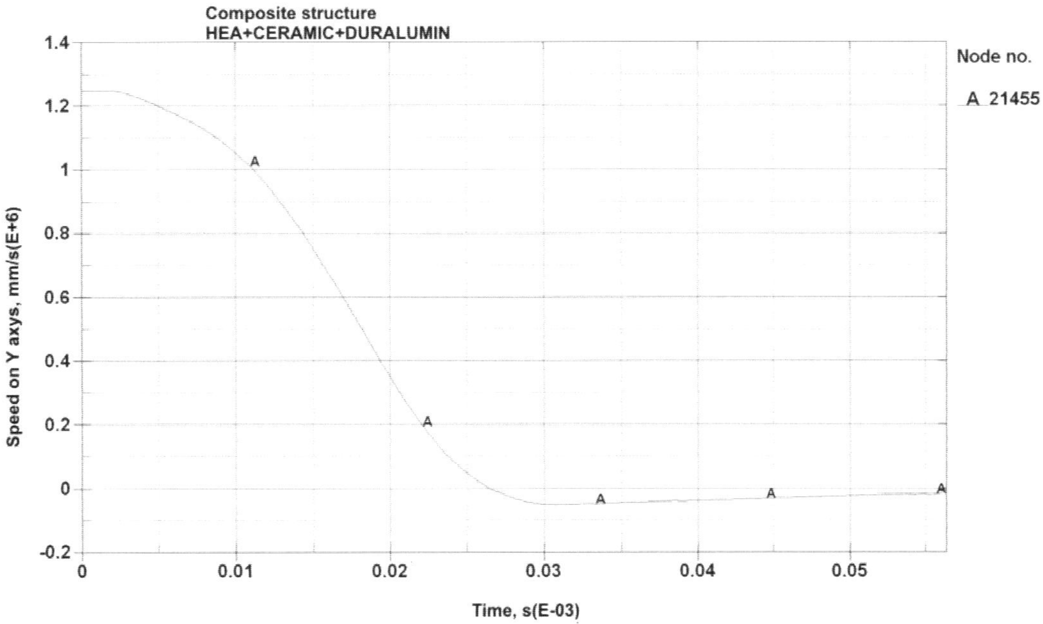

Figure 12. Variation of the bullet speed vs. time at the impact with the composite layered structure – VARIANT 4.

This constructive solution can be considered for the obtainment of composite structures for adequate protection.

5. Conclusions

The composite structure of duralumin-ceramics does not provide adequate protection for personnel and fighting techniques.

The numerical simulations showed that the composite structure of duralumin-ceramics-HEA can be considered for building ballistic protection structures.

The research shows that the composite structure of HEA-ceramic-HEA behaves well during simulations.

Increased attention should be paid to the behaviour of the second HEA layer at the interference between the direct shock wave and the reflected one (it is possible to produce ballistic fragments from the second layer of HEA).

The composite structure of HEA-ceramics-duralumin behaves well at bullet impact and can be used for the obtainment of ballistic protection structures.

References

[1] Cherecheş T 1996 *Muniţii* (Bucureşti, Editura Academiei Tehnice Militare)
[2] Ben – Dor G, Dubinsky A and Elperin T 2002 *Compos. Struct.* **56** 243
[3] Billon H H and Robinson D J 2001 *Int. J. of Impact Engng* **25** 411
[4] Briscoe B J and Motamedi F 1992 *Wear* **158** 229

[5] Duan Y, Keefe M, Bogetti T A and Cheeseman B A 2005 *Int. J. of Impact Engng* **31** 996
[6] Hamouda A M S and Hashmi M S J 1997 *J. Mater. Process. Tech.* **64** 189
[7] Johnson G R, Beissel S R and Cunniff P M 1999 *Proceedings of the 18th International Symposium on Ballistics*, San Antonio, Texas, 962 – 69
[8] Loi F 2005 *Quantitative Risk-based System Safety Assessment for Ammunition Process Facility* (Defence Science & Technology Agency, Singapore)
[9] www.uefiscdi.ro *Composite structures resistant at dynamic loadings applied with high deformation speeds used in the field of collective protection – HEAMIL*. Research Project – PCCA 209/2012.
[10] Geantă V, Voiculescu I, Ştefănoiu R, Savastru D, Csaki I, Patroi D and Leonat L 2013 *Optoelectron. Adv. Mat.* **7** 874
[11] Voiculescu I, Geantă V, Ştefănoiu R, Pătroi D and Binchiciu H 2013 *Rev. Chim-Bucharest* **64** 1441
[12] Zhou Y J, Zhang Y, Wang Y L and Chen G L 2007 *Mat. Sci. Eng. A-Struct.* **454-455** 260
[13] Hsua U S, Hunga U D, Yeh J W, Chen S K, Huanga Y S and Yang C C 2007 *Mat. Sci. Eng. A-Struct.* **457** 77
[14] Lee C P, Chang C C, Chen Y Y, Yeh J W and Shih H C 2008 *Corros. Sci.* **50** 2053
[15] Lin C M and Tsai H L 2009 *J. Alloy. Compd.* **489** 30
[16] Chen Y Y, Duval T, Hung U D, Yeh J W and Shih H C 2005 *Corros. Sci.* **47** 2257
[17] Century Dynamics Inc. 1997 *AUTODYN. Interactive Non – linear Dynamic Analysis Software*
[18] Lim C T, Shim V P W and Ng Y H 2003 *Int. J. Impact Eng.* **28** 13
[19] Mahfuz H, Zhu Y, Haque A, Abutalib A, Vahadilla U, Jeelani S, Gama B, Gillespie J and Fink B 2000 *Int. J. Impact Eng.* **24** 203
[20] Moaveni S 2003 *Finite Element Analysis: Theory and Application with ANSYS 2-nd EDN* (Prentice – Hall)

Acknowledgments
The research work was financially supported by the Romanian National Program for Research within the framework of the Project No. PCCA 209/2012 “Composite structures resistant to dynamic loadings applied at high deformation speeds used in the field of collective protection – HEAMIL”.

The Simulation of Point Contact Stress State for APS Coatings

D Chicet[1], A Tufescu[1*], C Paulin[1], M Panţuru[1] and C Munteanu[1]
[1]Gheorghe Asachi Technical University of Iasi-Romania, Department of Materials Science and Engineering, Blvd. Mangeron, No. 41, 700050, Iasi, Romania
[2]Gheorghe Asachi Technical University of Iasi-Romania, Department of Mechanical Engineering, Blvd. Mangeron, No. 43, 700050, Iasi, Romania

E-mail: ana.tufescu@yahoo.com

Abstract. Surface engineering has been conquered in recent decades by the versatility of the layers produced by thermal spraying, both in terms of spraying methods, of the materials types and their applications. In some cases, the coatings can be subjected during operation to rolling contact fatigue, with the main wear factors: thermal spray coatings structure and state of stress and strain in the contact area. In this paper was studied how three types of coatings deposited by APS (Atmospheric Plasma Spray) behaved at the contact fatigue tests. Subsequently they were carried out simulation of pressures and von Mises stresses distribution. It has been observed that the presence of asperities on the surface causes the development of local micro-contacts and therefore high values of pressure and local stresses in the vicinity of the surface.

1. Introduction

Plasma spray coatings have begun to be increasingly used in recent years to improve the life span of key components operating in hostile environments. These coatings act as the "first line of defence" against the usual surfaces degradation mechanisms, such as: wear of any type, corrosion, oxidation, defects caused by overheating and so on [1].

Obtaining a high performance coating on a particular alloy / metal piece with high mechanical strength provides an attractive way to combine surface and track requirements for just about any imaginable application. This concept of surface modification plays an important role in reducing the destruction frequency of relatively costly components [1,2,3].

One of the most common wear-related cases occurs in rolling parts with point-to-point contact - rolling bearings, in which wear is caused by contact fatigue.

It has the following main damage modes in the case of superficial coatings: abrasion wear, spalling wear, inlayer and interfacial flaking [4,5].

In this case, the surfaces of the two tribo-elements (the ball and the bearing ring surface - which represent a class I tribological coupling) are in direct contact with the development of normal and tangential stresses on the surface and in the depth, as well as with local deformation of the surfaces in the contact area.

This tension and deformation state in the contact area can be characterized by applying the theory of elastic contact, established by Hertz, in 1880, according to which, if there is a point or a contact line between two elastic bodies under the influence of the external forces, which are normal on the contact area, on this one appear normal stresses and the conjugated bodies elastically deform.

The hypotheses admitted by Hertz make this theory to maintain its validity today [6,7,8]:
1. the first hypothesis is relative to the shape of the contact field, which is assumed to be elliptical;
2. each body was considered to be an elastic semi-space;
3. the surfaces are considered as ideal, without roughness, between the bodies being transmitted only a normal pressure distribution;
4. although the real contact surface may not be flat, the body idealization by semi-spaces considers that the normal load is transmitted in a direction parallel to the Oz axis, and the tangential border stresses act in the xOy plane.

2. Methods and materials

2.1. Materials

In this paper, the research was carried out on three types of plasma sprayed coatings on a bearing steel substrate with the axial bearing ring geometry (with a 9MBM spray gun), realised using the following powders:

- a tungsten carbide powder with the nominal composition WC10Co4Cr, especially used for applications where both wear resistance (abrasion, erosion, fretting) and corrosion is required at a working temperature of up to 500°C (Test 1 - P1);
- a powder of Ni-based materials with nominal composition Ni 5Al 5Mo, used for high density coatings, high oxidation and corrosion resistance, good wear and erosion resistance, and a low contraction coefficient (Sample 2 - P2);
- a ceramic grade powder with the nominal composition Al2O3 - 13 TiO2, used for abrasion and slipping coatings with good oxidation resistance at working temperatures of approximately 550°C and good behaviour in alkaline or acidic environments (Test 3 - P3).

There were produced three types of axial bearing ring samples coated with the above mentioned layers, with a maximum thickness of 100μm, which were then tested for contact fatigue wear on a specially designed device.

2.2. Methods

In order to evaluate how the coatings underlying this study had behaved in contact fatigue tests and to compare the real and theoretical aspects, a modelling of the pressure and von Mises stress distribution was performed. For this purpose, a program developed in C ++ and Matlab was used [6,8].

In order to increase the efficiency of the numerical algorithm, a computational routine dedicated to the discrete convolution technique and rapid techniques (DC-FFT) for the convolution products specific to three-dimensional elastic contact problems has been developed. The algorithm used admits an elastic behaviour of the material, without imposing the flow limit of the material according to an plastic-elastic behaviour.

To evaluate the samples after the contact fatigue test on ball groove surfaces, they were transversally cut and metallographic prepared by sanding, polishing and reagent attack with a 3% nital solution to see their microstructure.

This was done using an electron microscope Quanta 200 3D, obtaining secondary electron topography images. There were used both the LowVacuum mode at a pressure of 60 Pa with Large Field Detector (LFD) and High Vacuum mode with ETD detector at working distance WD ≈ 14-16 mm and at different magnification (1500x ÷ 16000x).

3. Results and discussions

The distribution graphs were made for each of the three samples, both for the smooth and real contact, using the roughness values of each.

The variables according to which these models were made are: the modulus of elasticity and the roughness represented by the standard deviation, the input and output data being synthesized in table 1 for each of the three samples.

IOP Conf. Series: Materials Science and Engineering **209** (2017) 012044 doi:10.1088/1757-899X/209/1/012044

Table 1. The input and output data for each of the three samples.

Sample		P1	P2	P3
Input data	Normal load per ball, Q [N]		72,6153	
	Plane, R1 [mm]		10000	
	Ball radius, R2 [mm]		8	
	Modulus of elasticity, E1, E2 [MPa]	$1{,}798\cdot10^5$; $2{,}1\cdot10^5$	$1{,}329\cdot10^5$; $2{,}1\cdot10^5$	$1{,}46\cdot10^5$; $2{,}1\cdot10^5$
	Poisson coefficient, ν1, ν2		0,3; 0,3	
	Roughness Ra [μm]	0,05	0,11	0,7
Output data	Maximum stress, σ0 [MPa]	1356	1207	1254
	Contact ellipse major semi-axis b0 [mm]	0,159920	0,169472	0,16631

3.1. Sample 1- P1

In case of sample P1, the distribution graphs are presented in figures 1 - 3. Figure 1 shows the pressures and von Mises stresses distribution in the case of hertzian contact, and it can be observed that von Mises stresses have maximum values at a depth between 0.05 - 0.1 mm (about 850 MPa) and the maximum stress calculated in the Ox plane is σ0 = 1356 MPa.

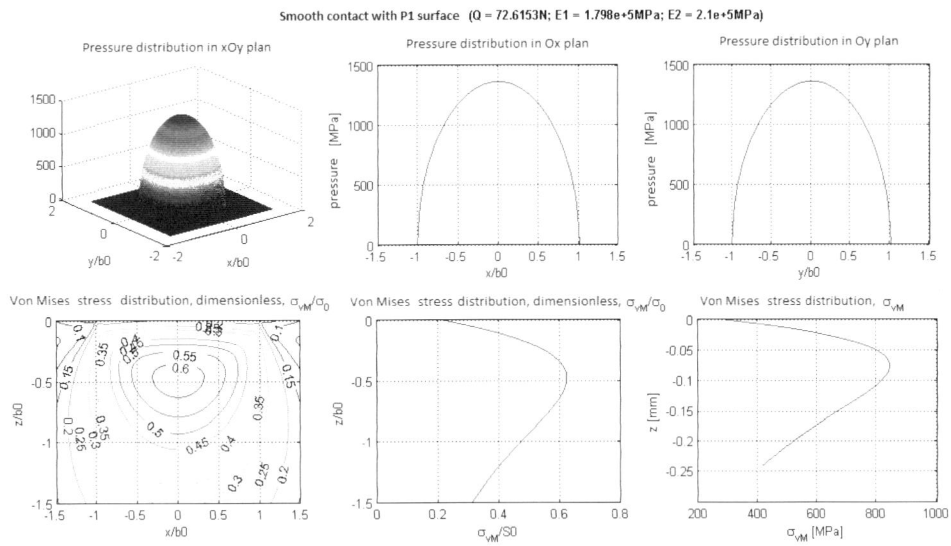

Figure 1. The distribution of pressures and von Mises stresses in the case of smooth contact with P1 surface.

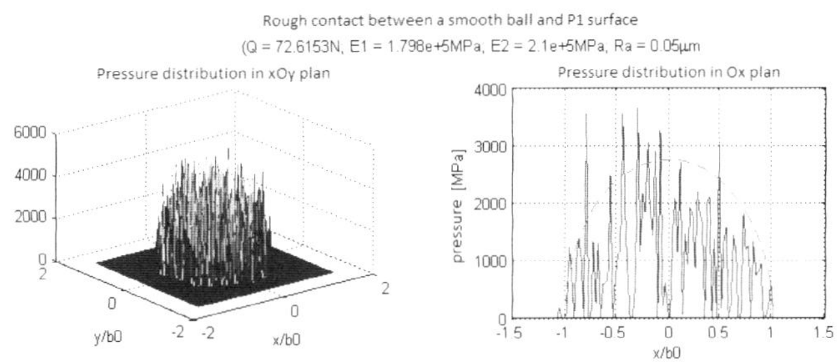

Figure 2. The pressures distributions in the case of the contact smooth ball – rough P1 surface.

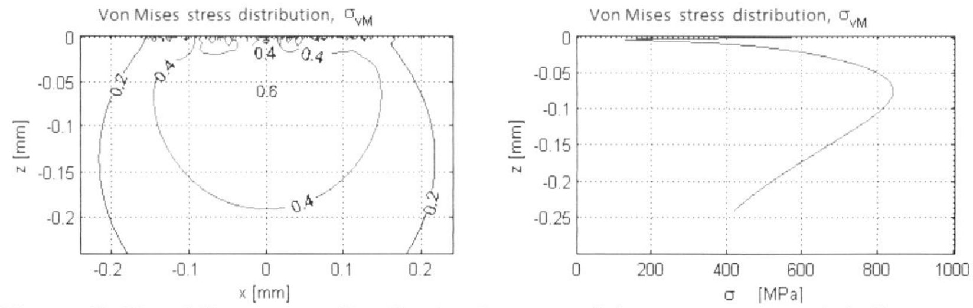

Figure 3. Von Mises stress distribution in case of the contact smooth ball – rough P1 surface.

In the real case of the contact smooth ball - rough surface shown in figures 2, 3, it is observed that the presence of surface asperities determines the realization of local micro-contacts with contact micro-areas and consequently with high pressure values (as can be seen in figure 2, where the maximum stress in the Ox plan is $\sigma_0 \approx 3700$ MPa). Thus, local high values are determined, as can be seen in figure 3 in the immediate vicinity of the surface, at a depth of 0 to 6 μm with a von Mises stress caused by roughness at a depth of 2μm of about 600 MPa.

It should be noted, however, that at depths corresponding to the maximum value of von Mises stress for smooth contact, the distribution of tensions is practically not influenced by the presence of roughness. This can be seen in figures 2 and 3, and von Mises stresses have maximum values at a depth of 0.05-0.1 mm, about 850 MPa.

Figure 4 shows a detail of the coating microstructure in the superficial area (at a depth of 16 μm), which is composed of successive layers of sprayed particles, which crystallize under polygonal geometries of different sizes. By correlating with the von Mises stresses distribution graph in figure 3, we can see that in the area where these stresses reach the maximum value (at 2 μm below the surface) no changes in particle geometry have occurred, but the degree of cohesion between them, which led to material separation (abrasion and spalling wear).

Figure 4. Sample P1. Secondary electron images in the superficial area of the (16 μm): a) coating, in the bearing race area, 12000x; b) coating, outside the bearing race area 12000x.

3.2. Sample 2 – P2

In case of sample P2, the distribution graphs are shown in figures 5-7. For sample P2, the pressure and von Mises stresses distribution for the smooth contact is shown in figure 5, were it can be observed that they have maximum values at a depth between 50-100 μm (about 750 MPa) and the maximum stress determined in the Ox plan is $\sigma_0 = 1207$ MPa.

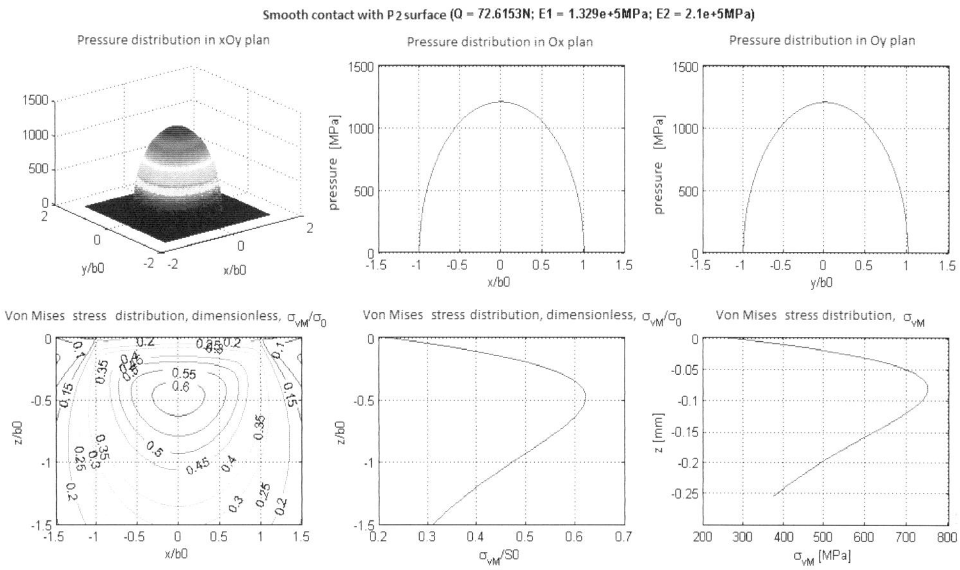

Figure 5. The distribution of pressures and von Mises stresses in the case of smooth contact with P2 surface.

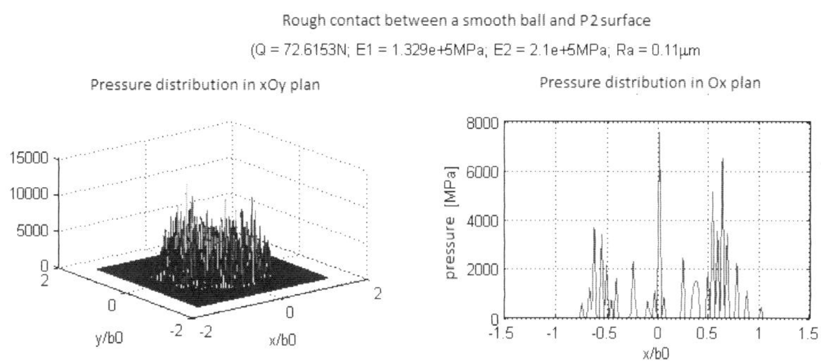

Figure 6. The pressures distributions in the case of the contact smooth ball – rough P2 surface.

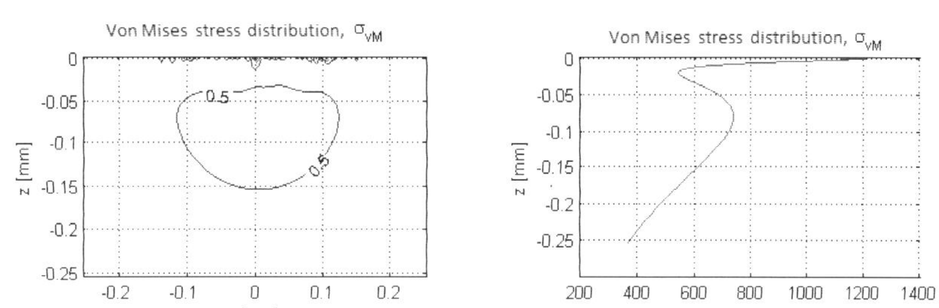

Figure 7. Von Mises stress distribution in case of the contact smooth ball – rough P2 surface.

It is obvious the effect of the surface roughness on the pressure distribution (see figure 6) and the maximum values that can be recorded: $\sigma_0 \approx 8000$ MPa, but also on the von Mises stresses distribution, as can be seen in figure 7. In this figure one can see a maximum of stress recorded near the surface at a depth of about 2µm with a value of 1200 MPa. In the case of sample P2, there was a coating delamination in 4 areas of the bearing race. Figure 8a presents the aspect of the coating - substrate interface in the area outside the bearing race, being obvious the influence of the spraying process on the substrate through the yielded temperature at the moment of impact and the high kinetic energy of the sprayed particles. Figure 8b shows the marginal delamination area of the coating with the exposure of the bearing steel substrate (at the bottom of the image).

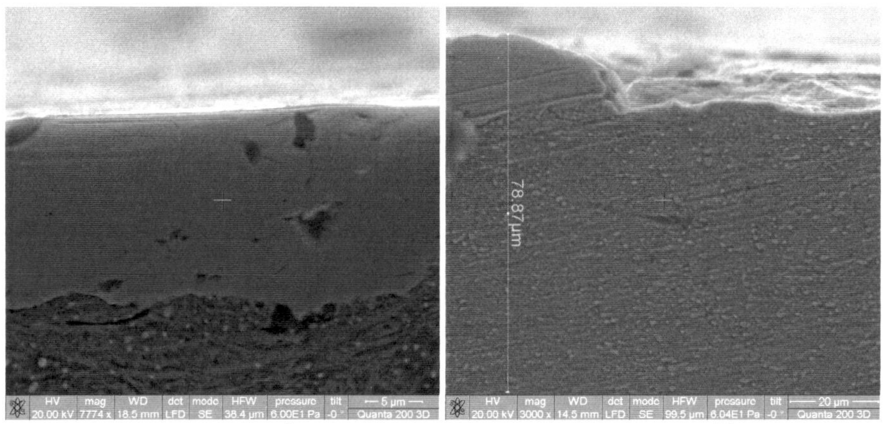

a) b)

Figure 8. Sample P2. Secondary electron images of the: a) coating – substrate contact area (outside the bearing race area, 7700x); b) marginal delamination area of the bearing race, 3000x.

3.3. Sample 3- P3

In case of sample P3, the distribution graphs are presented in Figures 9-11. For the sample P3, as in the case of the other two samples, it can be seen from figure 9 that in the case of herzian contact, the von Mises stresses have maximum values at a depth of 0.05 - 0.1 mm (about 780 MPa), and on the pressure distribution graphs in the Ox and Oy plans it is observed that they reach a maximum value of σ_0 = 1254 MPa.

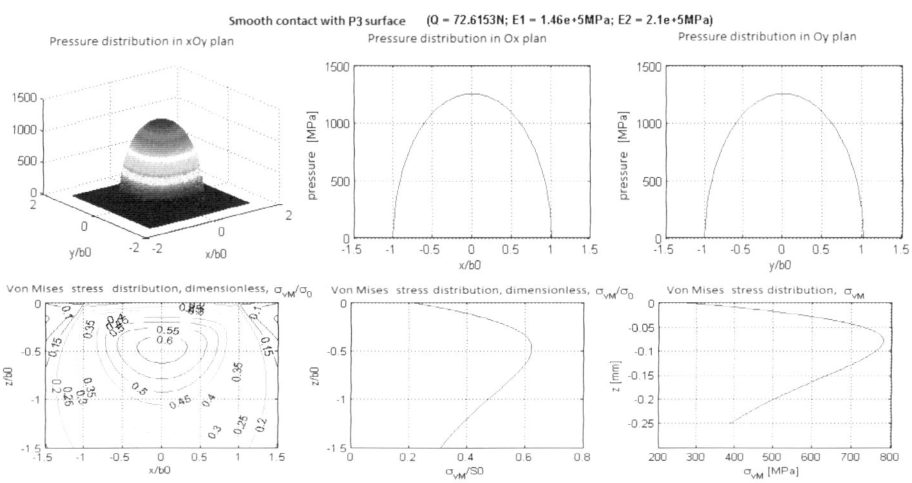

Figure 9. The distribution of pressures and von Mises stresses in the case of smooth contact with P3 surface.

In the case of the contact smooth ball - rough surface, the roughness effects on the pressure distribution are obvious, with a maximum stress value of $\sigma_0 \approx 4 \cdot 10^4$ MPa, which occurs in the immediate vicinity of the surface, as can be seen in Figure 12, at a depth of 0 to 8 μm, the maximum of these stresses being at a depth of 6 μm with a value of 1000 MPa. It is obvious that we are dealing with an area where the maximum values of the von Mises stresses are much higher than the maximum values for smooth contact, the latter being at much greater depths.

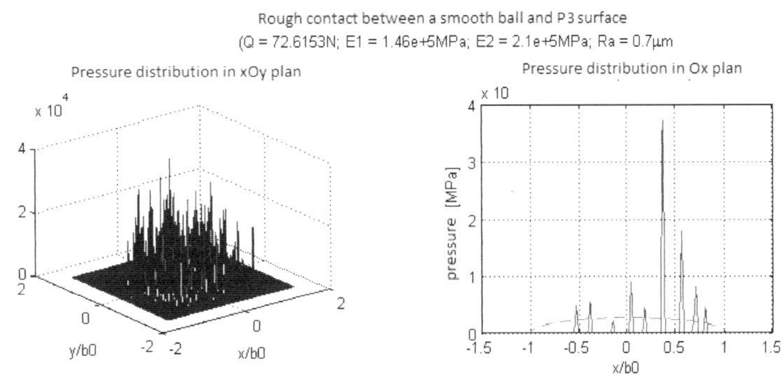

Figure 10. The pressures distributions in the case of the contact smooth ball – rough P3 surface.

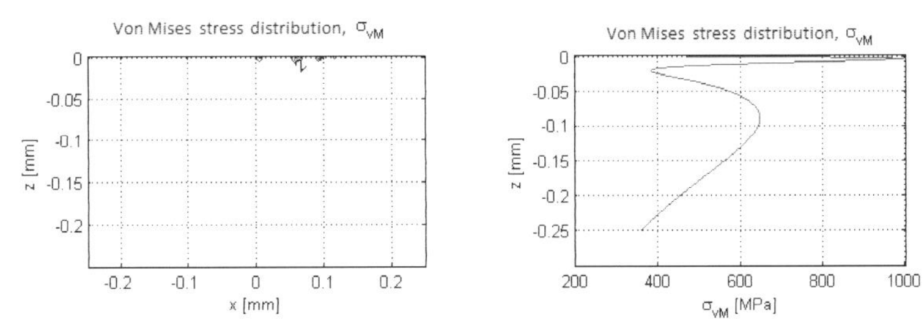

Figure 11. Von Mises stress distribution in case of the contact smooth ball – rough P3 surface.

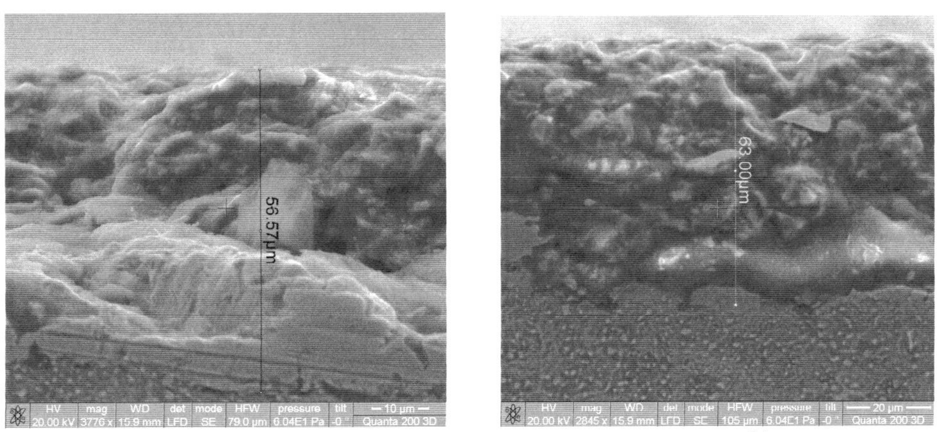

a) b)

Figure 12. Sample P3. Secondary electron images in the superficial area of the: a) coating, in the bearing race area, 3700x; b) coating, outside the bearing race area 2800x.

In figure 12 there are presented details of the coating both in the bearing race area (figure 12 a) and in the outside of the bearing race area (figure 12 b). It is possible to observe the lamellar aspect of the layer and the good cohesion of the splats as resulted after the spraying process. In these images it can also be seen that the von Mises stresses peak (located at a depth of approximately 6 μm according to the graph in figure 11) did not significantly affect the microstructure of the coating.

4. Conclusions

After examining the results obtained by modelling the distribution of pressures and von Mises stresses in the case of hertzian contact and after correlating them with the microstructural images of the samples subjected physically to rolling contact tests, the following observations can be made:

- the lesser the height of the roughness, the more the pressure distribution obtained in the rough contact is closer to the Hertzian distribution;
- at depths corresponding to the maximum value of von Mises stresses for smooth contact, the stress distribution is not influenced by the presence of roughness;
- the presence of asperities on the contact surface determines the realization of local microcontacts with contact micro-areas and consequently with high pressure values, which also cause local high values of stresses;
- the depth distribution of the von Mises equivalent stress highlights the concentrating effect of the stresses on the layer in the immediate vicinity of the surface as a result of the roughness presence;
- in the immediate vicinity of the surface the maximum values of the von Mises stresses are much higher than the maximum values corresponding to the smooth contact, the latter being also located at much greater depths.

References

[1] Suryanarayanan R 1993 *Plasma spraying: theory and applications* (Singapore; World Scientific Publishnig)
[2] Avram P, Imbrea M S, Istrate B, Strugaru S, Benchea S I and Munteanu C 2014 *Indian Journal of Engineering and Materials Sciences* **21** 315-321
[3] Zhong-yu P, Bin-shi X, Hai-dou W and Chun-Huan P 2010 *Surface & Coatings Technology* **204** 1405–1411
[4] Fujii M, Yoshida A, Ma J, Shigemura S and Tani K 2006 *Tribology Inter.* **39** 856–862
[5] Cretu S 2009 *Contactul concentrat elastic-plastic* (Iași; Ed. "Politehnium")
[6] Creţu S and Antaluca E 2003 *Analele Universitatii Dunarea de Jos Galati* I f VIII (Tribology) 39-47
[7] Creţu S 2006 The Influence of the Correlation Length on Pressure Distribution and Stress State in Concentrated Rough Contacts, *Proc. of ASME/ASLEIJTC* (track 15, USA)
[8] Urzică A, Bălan R and Creţu S 2009 *AMM* **52** vol. III 341-346

IOP Conf. Series: Materials Science and Engineering **209** (2017) 012045 doi:10.1088/1757-899X/209/1/012045

New Bond Coat Materials for Thermal Barrier Coating Systems Processed Via Different Routes

A Soare[1], I Csaki[2], M Sohaciu[2*], C Oprea[3], S Soare[1], I Costina[4] and M I Petrescu[1]
[1] National R&D Institute for Cryogenics and Isotopic Tech., Rm. Vâlcea, Romania
[2] University Politehnica Bucharest, Romania
[3] SC TENSOR SRL Bucharest, Romania

E-mail: * msohaciu@yahoo.com

Abstract. This paper aims at describing the development of new Ru-based Bond Coats (BC) as part of Thermal Barrier Coatings. The challenge of this research was to obtain an adherent and uniform layer of alumina protective layer after-high temperature exposure. We have prepared a RuAl 50/50 at% alloy in an induction furnace which was subsequently subjected to oxidation in an electric furnace, in air, at 1100C, for 10h and 100h. Mechanical alloying of Ru and Al powders was another processing route used in an attempt to obtain a stoichiometric RuAl. The alloy was sintered by Spark Plasma Sintering (SPS) and then oxidized at 1100C for 1 and10h. The alloys obtained as such were analysed before and after oxidation using advanced microscopy techniques (SEM and TEM). The encouraging results in case of RuAl alloys prepared by induction melting reveal that we obtained an adherent and uniform layer of alumina, free of delta-Ru. The results for the samples processed by powder metallurgy were positive but need to be further investigated. We should note here the novelty of this method for this particular type of application – as a BC part of a TBC system.

1. Introduction

A Thermal Barrier Coating (TBC) system is presented in figure 1 and consists of: superalloy, bod coat (BC), thermally grown oxide (TGO) and thermal barrier coating (TBC).

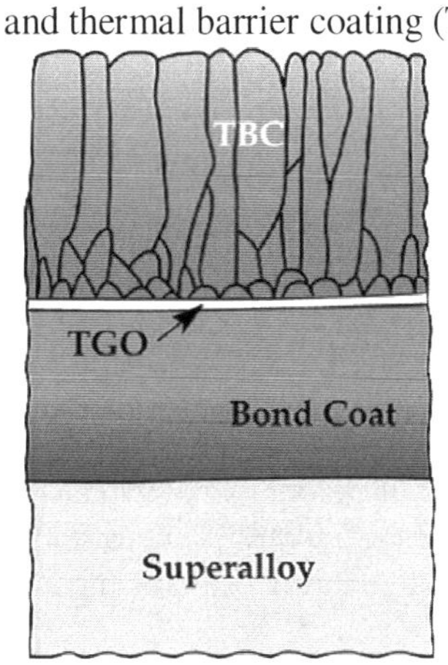

Figure 1. Thermal Barrier Coating System [1].

The main component of the system is the Ni-based superalloy representing the base for the TBC. The role of the TBC is to provide protection against corrosion and high temperature developed in the gas turbine working environment.

The upper layer, the thermal barrier coating, has a low thermal conductivity [2,3] meaning a low heat flux from the combustion gases to the main component of the system (the superalloy) resulting in a decrease of temperature of over 100°C [3,4].

This layer (TBC) is usually yttria-stabilised zirconia (YSZ – 7wt% Yttria), has a certain porosity degree and has a thickness ranging between 100 and 500μm. Microcracks and aligned porosity are induced in these layers during the coating process, to accommodate the of thermal expansion effects.

The main function of the bond coat (BC), the layer between the superalloy and the TBC, is to protect the superalloy from oxidation and to act as a bond between the TBC and the superalloy. During high temperature exposures, a thermally grown oxide (TGO) layer forms between the BC and the TBC. BC plays a very important role as being the promoter of the thermally grown oxide and the layer provided should be a uniform, dense, stable. α- Al2O3 is the preferred oxide as TGO because it is thermodynamically stable, it is compatible with YSZ, has an oxygen low diffusivity and it is a very good thermal insulator [5, 6].

The bond coat processing techniques including noble metals with favourable properties have been already described in the literature. Pd layers have a very good oxidation resistance at high temperature [7] and Ir layers form in fact two layers: one containing Ni and a lower Ir-rich layer [8,9].

Over the years, the most used BC alloys have been platinum-modified nickel aluminide (Ni-Pt)Al (37,44) and McrAlY (M, metal) [11,12]

Systems containing Ru have also been the subject of many studies but their mechanism of oxidation was not fully understood. The main goal of this research was to obtain a RuAl alloy which forms a uniform and adherent alumina layer after high temperature exposures.

The work focused on preparation and microstructural characterisation of RuAl alloys subjected to oxidation.

2. Materials and methods

Two processing routes were used for producing RuAl50/50 at%. First route was processing in liquid state, in a levitation induction furnace at 2400°C for 25 minutes. The second route was processing in solid state, using powder metallurgy technique.

The elemental powders were milled together in a planetary ball mill PM 400 RETSCH, in stainless steel vials and balls, at 300rpm for 25h. The process control agent (PCA) used was stearic acid in percentage of 1at%. The PCA was used to prevent the welding of the particles during mechanical alloying process and improving the alloying behavior of RuAl. For sintering we used Spark Plasma Sintering (SPS) technique in the following conditions: 1060°C, 80MP and 9kN.

The consolidated and as cast samples were subjected to oxidation at 1100°C for 10 hours, in an electric furnace. The samples were introduced in the furnace when the heating started.

The samples were investigated before and after oxidation. The SEM images were obtained with a Field Emission Gun (EG) Variable Pressure (VP) SEM equipped with WDX and EDX spectrometer.

3. Results and discussions

The SEM images obtained for RuAl 50/50 before oxidation are presented in the figure 2 a and b. Figure 2 reveals the microstructure of the RuAl50/50 at% presenting the backscattered image revealing three different phases in the alloy (a) and in figure 2b a detail of the Ru-RuAl eutectic is presented.

To prepare a stoichiometric RuAl alloy is the main challenge of this work due to the high difference between Ru and Al melting points: Ru has the melting point at 2334°C and Al at 660°C. In the phase diagram presented in figure 3 the narrow stoichiometry interval proves that processing RuAl 50/50 at% could be difficult. The narrow stoichiometry gap is also a matter to be taken into consideration when processing this alloy. The alloy obtained indicates the presence of more phases of RuAl compound due to the Al leaving the melt resulting in secondary phase formation.

IOP Conf. Series: Materials Science and Engineering **209** (2017) 012045 doi:10.1088/1757-899X/209/1/012045

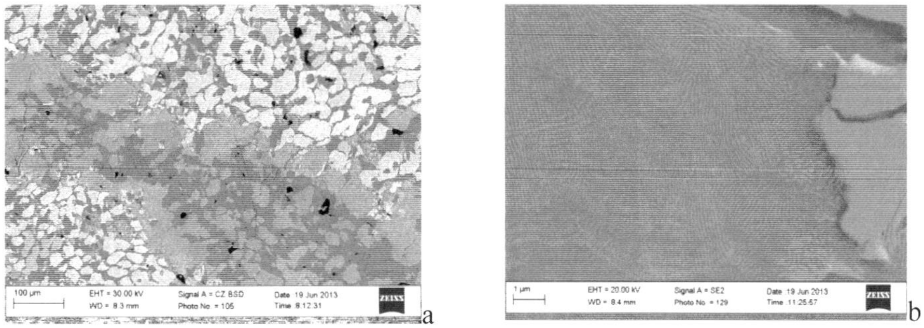

Figure 2. Microstructure of the RuAl processed in liquid state before oxidation: a) three phases present in RuAl alloy and b)detailed image of Ru-RuAl eutectic.

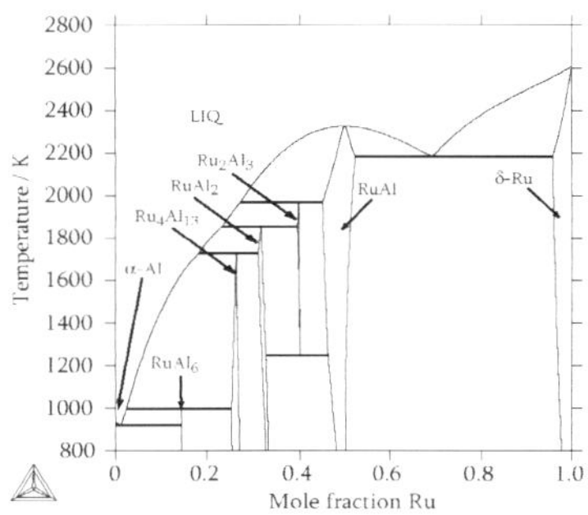

Figure 3. Ru-Al phase diagram.

In figure 4 a SEM image of the solid state processed samples is shown.

Figure 4. The SEM image of SPS sample (processed in solid state).

Figure 5 and 6 present the EDX result analyses for the as cast and SPS consolidated samples.

The oxidation process is important for assessing the new alloy capabilities. The ability of the bond coat to resist formation of the thermally grown oxide depends on the method of surface preparation and the thermal cycles in the corrosive environment where the chemical composition is constantly changing due to interdiffusion and oxidation. The most important part of the TGO is to act as a diffusion barrier between the oxidant atmosphere and by oxygen diffusion inhibiting in the sublayers and rapid diffusions towards the oxidant atmosphere. This layer is essential for the chemical and mechanical compatibility of the super alloy with the thermal barrier system. The oxidation time is important in order to evaluate the behavior of Ru phase. The chosen time for this set of experiments was 10h.

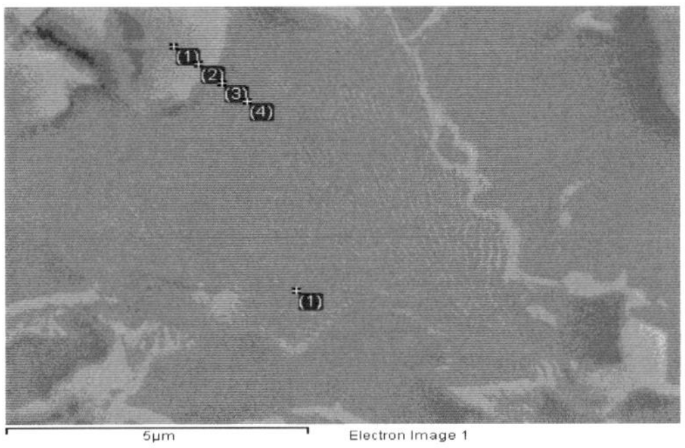

Spectrum	Al	Ru
	64.37	35.63

Figure 5. The EDX analyses results for as cast sample.

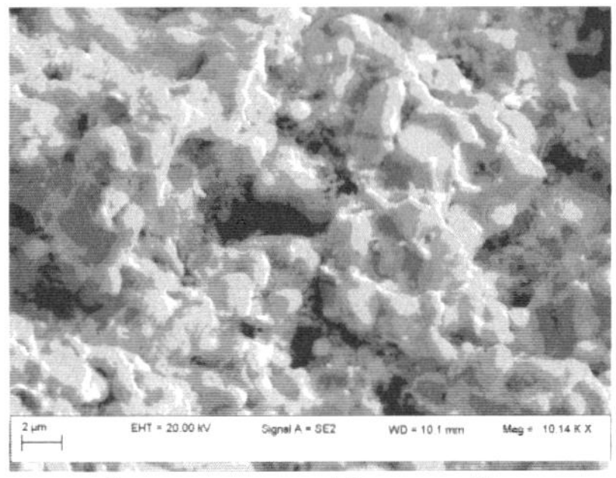

Spectrum	Al	Ru
	66.34	34.56

Figure 6. The EDX analyses results for SPS consolidated sample.

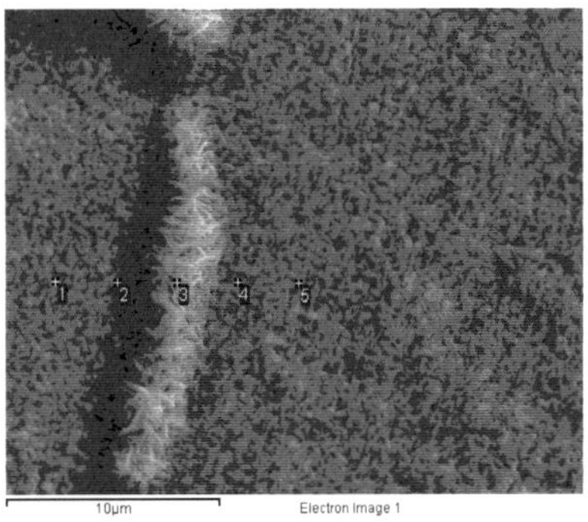

Figure 7. Acicular growth of the Al2O3 oxide on the RuAl surface.

Figure 8. SEM image of the oxide grown on the RuAl SPS consolidated sample.

In figure 7 and 8 the microstructures of the as cast and SPS consolidated samples after 10 h oxidation in the furnace are shown.

The thermally grown oxide obtained after 10h oxidation time seem to be dens and uniform and the most important is the absence of the delta – Ru phase which is very good indicator of the successful oxidation of RuAl 50/50at%.

4. Conclusions

RuAl 50/50at% was processed using two different routes: liquid state processing and solid state processing. The RuAl alloy was not reported yet to be processed by powder metallurgy and the results obtained after oxidation process encourage further investigation for this processing route. The alloy processed in liquid state grew an oxide layer dense and uniform, without delta-Ru phase indicating a very good behavior as a bond coat.

References

[1] Tryon B, Feng Q, Pollock T M, et al. 2006 *Metall and Mat Trans A* **37** 33 - 47
[2] Wright P K, Evans A G 1999 *Current Opinion in Solid State and Materials Science* **4** 255
[3] DeMasi-Marcin J T, Gupta D K 2000, *Surface and Coatings Technology* **68/69** 1 - 7
[4] Nicholls J R 1994, *Journal of Metals* **28**
[5] Clarke D R 2003 *Acta Materialia* **51** 1393 - 6
[6] Levi C G 2004 *Emerging materials and processes for thermal barrier systems. Curr. Opin. Sol.St&Mat.Sci.* **8** 77
[7] Angenete J, Stiller K, Langer V 2003, *Oxidation of Metals* **60** 47

[8] Li M J, Sun X F, Guan H R, Jiang X X, Hu Z Q 2003, *Surface and Coatings Technology* **167** 106
[9] Datta P K, Burnell-Gray J S, Natesan K 2002 *Intermetallic Compounds, Westbrook, J.H. and Fleischer, R.L. New York: John Wiley & Sons, Ltd.* pp. 561.
[10] Fisher G, Datta P K, Burnell-Gray J S 1999*bSurface and Coatings Technology* **113** 259
[11] Leyens C, Pint B A, Wright I G 2000, *Surface and Coatings Technology* **133-134** 15

Acknowledgments
This research is supported by the National Program PN II contract number 182/2012.

Obtaining of High Cr Content Cast Iron Materials

C Florea, C Bejinariu[1], I Carcea, N Cimpoesu, D L Chicet and C Savin
„Gheorghe Asachi" Technical University of Iasi, Strada Prof.dr.doc. D. Mangeron nr. 41, Iasi, 700050, Romania

Email: nicanornick@yahoo.com

Abstract. We have obtained, through the classic casting process, 3 highly chromium-based experimental alloys proposed for replacing the FC 250 classical cast iron in braking applications. Casting was carried out in an induction furnace and cast into moulds made of KALHARTZ 8500 resin casting mixture and HARTER hardener at SC RanCon SRL Iasi. It is known that the microstructure of the cast iron is a combination of martensite with a small amount of residual austenite after the heat treatment of the ingot. In the case of high-alloy chromium alloys, the performance of the material is due to the presence of M7C3 carbides distributed in the iron matrix Resistance to machining and deformation is based on alloy composition and microstructure, while abrasion resistance will depend on properties and wear conditions.

1. Introduction

Friction between two dry surfaces in contact and moving relative to each other is known as Coulomb friction and even if it is a phenomenon that occurs several times a day the frictional force can only be estimated from previous experience and experimental results [1].

In the case of braking systems (for any type of vehicle using braking systems) the contact surfaces are often covered with different materials to result in a sliding process and are not just two metallic materials sliding over each other. The metallic surface will be affected by abrasion, adhesion of compounds other than the base material, or deformation of the material during sliding, which will create metallic fragments of material as well as other particles [2].

Any metal friction surface, even if it appears geometrically fine at a macroscopic level, is quite rugged on a microscopic scale, with a large distribution of roughness along it. An explanation of friction formation is the interaction of microscopic asperities of two surfaces, but when two such surfaces are forced to slide one on the other, it is very difficult to specify where and how contact occurs between them.

Even if, from the point of view of an engineer, it can be assumed that a coefficient of friction between two surfaces, especially metallic, sliding one at the other, can be considered constant in the applicative work with the braking systems, it is very important to understand that this coefficient The friction is variable and it is very useful to understand why [4,5]. There are several reasons why the braking performance of a vehicle can change over time and any damage can be recognized by the user through a reduction in braking parameters such as distance or vehicle stopping time [6-8].

Alloy cast iron with a high percentage of chromium are recognized as having a high wear resistance and have been widely used for wear-affected parts operating under extreme conditions. Exceptional wear resistance of Cr-rich castings results primarily from the high percentage of M7C3 carbides formed in the ferrous matrix with M = Fe, Cr and which have the role of preventing graphite formation and generally stabilize carbides [14].

The chromium content is depending on the destination of the cast iron, being 11 ... 30% for the wear resistant cast iron, 15 ... 25% and 29 ... 35% for the high temperature cast iron and 20 ... 35% for the cast iron corrosion. Typically, molybdenum is present in these cast irons at a maximum of 30%. If the chromium content is less than 9.5%, the presence of Carbide (Fe, Cr) 3C, with orthorhombic network, as part of eutectic colonies, is found in the structure. In this paper we present some preliminary results, chemical and microstructural insights, obtain on three experimental cast-iron materials.

2. Experimental details

Experimental samples with varying percent chromium (≈11, 14, 22 wt%) were obtained in the form of further heat treated ingots and experimental samples. For the present study, in addition to a class 250 FC cast iron, three types of chromium alloy cast iron were developed at SC Rancon S.R.L. Iaşi in the induction furnace and cast into moulds made from Kalhartz 8500 resin binder and Harter hardener. The chemical composition of the three materials was determined on the Foundry-Master 01J0013 Spectrometer.

Their hardness was measured on the Wilson Wolpert Model 751h hardness meter (equipped with the SIM Iasi faculty) with a diamond-shaped cone penetrator with a load force of 150kgf and a standard pushdown time of 12 seconds. At micro-scale chemical composition was realized using an EDS detector (Bruker XFlash – EDAX type) in order to establish some chemical details from the cast iron materials. For microstructure, after mechanical grinding and polish plus chemical etching (with nital at room temperature, 4% alcohol solution of nitric acid (HNO_3)) the structure was observed by scanning electron microscopy (VegaTescan LMH II SEM with SE detector at 30 kV).

3. Experimental results

The three experimental materials obtained by the classical method of industrial casting are referred, in the present work, as the alloy A, B and C, these having the possible applications for producing different automotive parts such as brake discs. The chemical composition of the three materials is shown in table 1. At the same time we also analysed a standard material used for industrial automotive brake discs, a classical cast iron, FC 250 which we named the sample D or the initial sample. The chemical composition is presented as a percentage by weight (wt%) and was performed after 5 measurements by optical spectroscopy on a 4 mm^2 area for each determination.

Chromium, in quantitative terms, varies in the casting of the oligo-element (less than 0.3%) to the alloy element quality (over 0.3% - the bibliographic source [15] indicates cast iron with a maximum of 36% - 45% Cr [16] materials were also experimentally analysed. Chromium is an anti-graphitize on a scale, in an ascending order of the anti-graphitizing (whitening) effect, the order being: W, Mn, Mo, Sn, Cr, V, B etc. As an anti-graphitize chromium increases the number of eutectic cells and the proportion of pearlite in the case of grey cast iron.

Table 1. Chemical composition of experimental alloys.

Alloy	C wt%	Si wt%	Mn wt%	P wt%	S wt%	Cr wt%	Mo wt%	Ni wt%	Cu wt%
A	2,76	0,64	0,65	0,03	0,067	11,2	0,27	0,3	0,16
B	2,48	0,65	0,53	0,03	0,078	14,7	0,38	0,38	0,18
C	2,64	1,11	0,66	0,03	0,056	20,9	0,11	0,16	0,1
D	3,1	1,2	1,0	0,03	0,04	-	-	-	-

For a better evaluation and characterization of the alloys tested, their hardness was measured on the Wilson Wolpert Model 751h hardness meter (equipped with the SIM faculty) with a diamond-shaped cone penetrator with a loading force of 150 Kgf and a standard push time of 12 Seconds, the results are shown in table 2.

Table 2. Hardness values of the experimental materials.

Alloy	Hardness in 3 points (HRC)			Mean value of hardness (HRC)
A	56,7	55,4	56,3	56,1
B	58,9	57,5	58,8	58,4
C	43,2	42,2	45,3	43,6
D	57,3	57,8	56,7	57,3

Experimental alloy micro-hardness values show close average values for samples A, B and D (i.e., the one alloyed with 11 and 14 wt% Cr respectively the initial sample) and a lower value for sample C, the sample containing a higher percentage of chromium (21 wt%).

The chemical analysis of the constituents of material C was performed on the Bruker EDAX equipment by analyzing at various points on the micrograph. Three areas for chemical analysis were selected, showing microstructural differences due mainly to their chemical differences, figure 1.2 a).

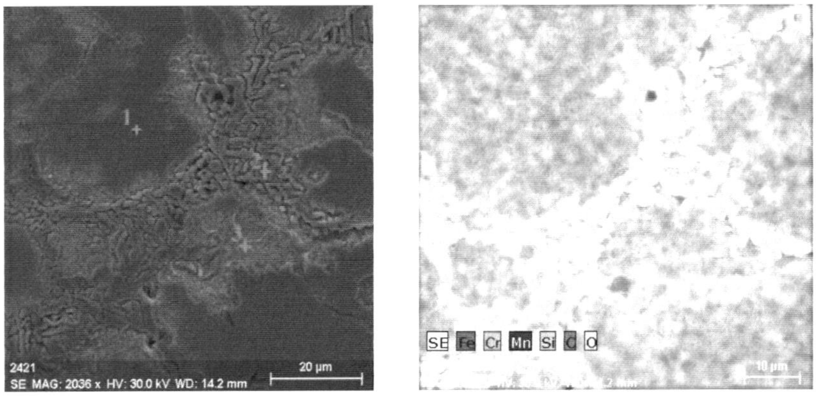

(a) (b)

Figure 1. Alloy C (a) Surface micrograph of experimental material C with selection of micro-zones (1-3) for chemical analysis and (b) distribution of Fe, Cr, Mn, Si, C and O chemical elements on a specific surface.

Figure 1 b) shows the distribution of Fe, Cr, Mn, Si, C and O on the surface of the metallic material. table 3 shows the chemical compositions characteristic of the 3 points selected in figure 1 a). At point 1, the iron composition of iron-based matrix composed of Fe, Cr, Si and C. The selected area of point 2 shows the Cr-based compounds in dendrite like forms mostly Cr-based carbides and Fe. The area selected by point 3 is a transition zone between the dendrites with a higher percentage of chromium and the matrix of the iron-based material. There is a slight oxidation of the compounds with more chromium and especially of the interface between the carbides and the ferrous matrix (the main cause being due to the chemical attack made to highlight the metal alloy microstructure).

Table 3. Chemical analysis of alloy C by weight percent (wt%) and atomic percent (%).

Area (Fig. 1 a))	Fe		Cr		C		Si		O	
	wt%	at%	wt%	at%	wt%	at%	wt%	at%	wt%	at%
Point 1	76,9	63,5	16,9	14,8	5,1	19,6	1,2	1,9	-	-
Point 2	39,7	32,7	55,3	48,9	4,03	15,4	-	-	1	3
Point 3	74,1	64,9	21,5	20,3	2,5	10	0,9	1,6	1,1	3,2
EDAX Error	1,8		0,5		0,8		0,1		0,6	

IOP Conf. Series: Materials Science and Engineering **209** (2017) 012046 doi:10.1088/1757-899X/209/1/012046

The chemical composition obtained in point 3 is very close to the general chemical composition shown in Table 1 representing, at least in terms of carbon, a mediation between the two main component phases. It is known that the microstructure of the cast iron is a combination of martensite with a small amount of residual austenite after the heat treatment of the ingot. In the case of chromium-high alloy castings, the performance of the material is due to the presence of carbons of the M7C3 type distributed in the iron matrix [10]. Resistance to machining and deformation is based on alloy composition and microstructure, while abrasion resistance will depend on properties and wear conditions.

The microstructural material is characterized in figure 2 at two amplification powers showing a homogeneous dendrite tree structure distributed in the iron matrix.

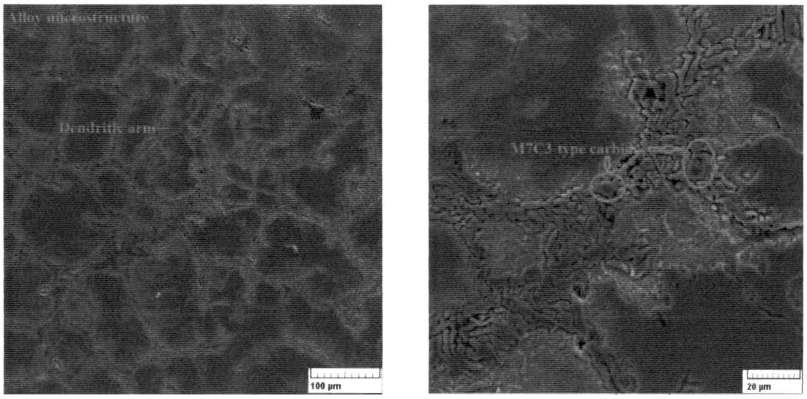

(a) (b)

Figure 2. SEM Microscopes (a) Arboreal spacing of homogenous dendrites distributed and (b) distribution of carbides in the ferrous matrix.

The distribution of M7C3 carbides and secondary carbides, in a much lower percentage, demonstrates the significant effects of Si and Cr on dendrite sparing. Since the cast iron chromium ratio is greater than 9.5%, carbon (Cr, Fe) 7C3 is also present in the structure, along with carbon (Fe, Cr) 3C, the latter determining cylindrical and conical eutectic colonies, figure 2b). Carbon (Fe, Cr) 7C3 has a triangular network and develops in the direction of the longitudinal axis of the cylindrical or tapered column of the eutectic colony.

4. Conclusions

We have obtained, through the classic casting process, 3 highly chromium-plated experimental alloys to replace the FC 250 classic cast iron in braking applications. Casting was carried out in an induction furnace and cast into moulds made of KALHARTZ 8500 resin casting mixture and HARTER hardener at SC RanCon SRL Iasi. The experimental results show a very good chemical and microstructural homogeneity of the made materials.

References

[1] Sun T, Song R, Wang X, Deng P and WU C 2015 *Journal of Iron and Steel Research, International* **22** 84-90
[2] Fei L, Yehua J, Han X and Jun T 2015 *Journal of Alloys and Compounds* **618** 380-385
[3] Toma S L, Bejinariu C, Baciu R and Radu S 2013 *Surf & Coat. Technol.* **220** 266–270
[4] Mihai D and Cimpoesu N 2010 *Metalurgia Internaţional* **XV** 30-35 ISSN 1582 – 2214
[5] Mareci D, Cimpoesu N and Popa M I 2012 Electrochemical and SEM characterization of NiTi alloy coated with chitosan by PLD technique *Materials and Corrosion* DOI: 10.1002/maco.201206501 **63** 176-180

[6] Harun M A, Abd Rahim S Z, Mat Saad M N and Ghazali M F 2016 *European Journal of Materials Science and Engineering* **1** 9-18
[7] Aelenei N, Lungu M, Mareci D and Cimpoeşu N 2011 *Environmental Engineering and Management Journal* **10** 1951-1958
[8] Sandu A V, Ciomaga A, Nemtoi G, Bejinariu C and Sandu I 2012 SEM-EDX and micro-FTIR studies on evaluation of protection capacity of some thin phosphate layers *MICROSCOPY RESEARCH AND TECHNIQUE* **75** 1711-1716
[9] Predescu C, Pantilimon C, Sohaciu M, Matei E, Savastru D, Berbecaru A, Predescu A, Anton M G and Coman G 2016 *Journal of optoelectronics and advanced materials* **18** 873-877
[10] Predescu A M, Matei E, Savastru D, Coman G, Predescu C, Vlad G and Favier L 2014 *Digest journal of nanomaterials and biostructures* **9** 987-995
[11] Pevec M, Oder G, Potrč I and Šram M 2014 *Engineering Failure Analysis* **42** 221-230
[12] Irimiciuc Ş A, Agop M, Nica P, Gurlui S, Mihaileanu D, Toma St and Focsa C 2014 *Jpn. J. Appl. Phys.* **2014** 5
[13] Abbasi S, Teimourimanesh S, Vernersson T, Sellgren U, Olofsson U and Lundén R 2014 *Wear* **314** 171-180
[14] Zarina Y, M A Bakri, A M, Kamarudin H, Nizar I K, Rafiza A R 2013 *Reviews on Advanced Materials Science* **34** (1) 37-43
[15] Minciuna M G, Vizureanu P, Achitei D C, Sandu A V, Mareci D, Balan, A 2014 *Revista de Chimie* **65** (10) 1138-1141

IOP Conf. Series: Materials Science and Engineering **209** (2017) 012047 doi:10.1088/1757-899X/209/1/012047

Experimental Equipment for Damping Capacity Analyze of High or Low Internal Friction Metallic Materials

I A Gârnet, S Stanciu*, I Hopulele, M G Zaharia, N Cimpoesu, D L Chicet and R C Crăciun
„GheorgheAsachi" Technical University of Iasi, Strada Prof.dr.doc. D. Mangeron nr. 41, Iasi, 700050, Romania

*sergiustanciu2003@yahoo.com

Abstract. An experimental equipment, type torsion pendulum was made in laboratory in order to analyze the damping capacity of metallic materials. The scheme of the equipment is presented, 2D and 3D visions at real scale. The equipment functioning (mechanical and electrical part) and principles are presented. In this article we present some preliminary experimental results obtained on different materials (aluminium, steel etc.) using two different methods for registration the outputs (one based on optoelectronic device with Arduino acquisition board and second on video analyze (cinematic review: video to jpeg) of the damped motion of the lead pendulum). Steel materials were with shoot penning surface modification with and without heat treatment in order to establish the heat treatment influence on the damping capacity property.

1. Introduction

Beside damping capacity materials (vibration amortization, earthquake etc.) in other industrial applications, specifically for electronic equipments used for guiding, materials (preferably metallic for very good mechanical properties) with a very low damping capacity value are required [1-4]. An example is the resonator element used in the gyroscope equipment. In this case, the metallic material vibrations are used to detect rotation and to be maintained as long time is possible without a new excitation (usually up to 30-60 minutes) [5,6]. For crystalline materials (metallic/alloys), damping capacity of mechanical external solicitation is mainly based on crystal defects (vacancies, interstitial atoms, dislocations, grain boundaries or plates presence) [7-10].

In case of amorphous materials (like metallic glasses, concrete [9]), the idea of defect is not so clear and, therefore, the damping level is not easy to predict [11]. In amorphous materials case the damping capacity behave differently based on the material. Amorphous polymers are often absorbing materials due to their high intrinsic molecular mobility, while amorphous oxides are often characterized by low damping. Silica (SiO_2) is a typical example: the damping coefficient can be lower than 10^{-6}; i.e., the quality factor Q can exceed 10^6 and represent a good option for low damping materials [12-14]. However, this amorphous material encounters two problems: it can be processed only at very high temperature, because its glass transition is higher than 1200 °C; and due its isolating nature, a conducting layer has to be deposited on its surface to enable electromagnetic excitation, limiting its damping performance [15,16]. Based on very good mechanical properties many industrial areas prefer metallic materials in crystalline state. A good compromise solution, taking in account the

IOP Conf. Series: Materials Science and Engineering **209** (2017) 012047 doi:10.1088/1757-899X/209/1/012047

working facilities of crystalline materials, will be a crystalline material with very low damping capacity and mechanical properties like cast-irons [17, 18].

This article presents an experimental laboratory pendulum to determine low or high damping capacity materials. It is present the main principle and two modalities to establish the damping capacity, both with electronic devices. Preliminary results are made on cast iron material after a mechanical operation of shot-penning.

2. Experimental details

In figure 1 a) is presented the full 2D and 3D schematic of the experimental laboratory pendulum and real scale dimensions. The equipment is formed from two parts: the mechanical part, figure 1 a), and the electrical part, figure 1 b). The electrical part is used to start the oscillation using an electromagnet. The mechanical part has a pre-strain possibility at the griping system so it can be analyze pre-strain sample, a function that cannot be used on a usual damping analyzer (DMA).

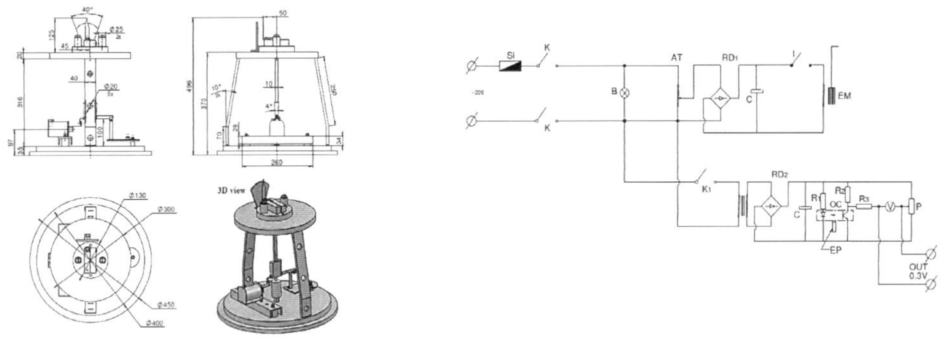

a) b)

Figure 1. a) Equipment schematic and b) electric scheme with Si: fuse, K: general switch, B: voltage signalling, At: autotransformer for adjustment, RD1: regulation cell, C: electro-condenser, I: magnet switch, EM: electromagnet, K1: measure switch, RD2: recovery cell, R1-R3: resistances, OC: optocoupler, EP: screen closure, V: voltmeter, P: potentiometer and OUT: signal output.

Cast iron material, plates with 5x20x1 mm were made using wire cutting procedure from experimental plates in order to fit the pendulum striping elements. The material surface, after shot-penning was analyzed using SEM equipment by 2D and 3D images. For the first type of results interpretation we use a video camera and a transition from video to image software, we register the main amplitudes and use the formula: $\delta=(1/tf)\ln[(A_0-a_o)/(A_n-a_n)]$ (1) to calculated the damping capacity of the material. In the second case we use an Arduino board to translate the signal from the pendulum in digital values for transfer and storage on a PC. The program that runs on Arduino board is:

```
// the setup routine runs once when you press reset:
void setup() {
  // initialize serial communication at 9600 bits per second:
  Serial.begin(9600);
}
// the loop routine runs over and over again forever:
void loop() {
  // read the input on analog pin 0:
  int sensorValue = analogRead(A0);
  // Convert the analog reading (which goes from 0 - 1023) to a voltage (0 - 5V):
  float voltage = sensorValue * (5.0 / 1023.0);
```

```
 // print out the value you read:
 Serial.println(voltage);
}
```

After the experiment data acquisition all the value transmitted from the board it is saved on an excel document and the data processed in order to determine the parameters required for equation (1). The advantage of this type of processing data is given by the long time possibility to record data from the pendulum so low damping capacity materials can be analyzed from 30-60 minutes or more.

3. Experimental results

The experimental material surface, after shot-penning operation, is presented in figure 2 by SEM analyze. In figure 2 a) and b) are presented 2D insights for two different amplification scales and highlight the surface state, with modifications of the surface around 50 μm. In figure 2 c) by 3D SEM analyze the surface modification present different area, formed after shot-penning operation, with variations of height less than 10 μm.

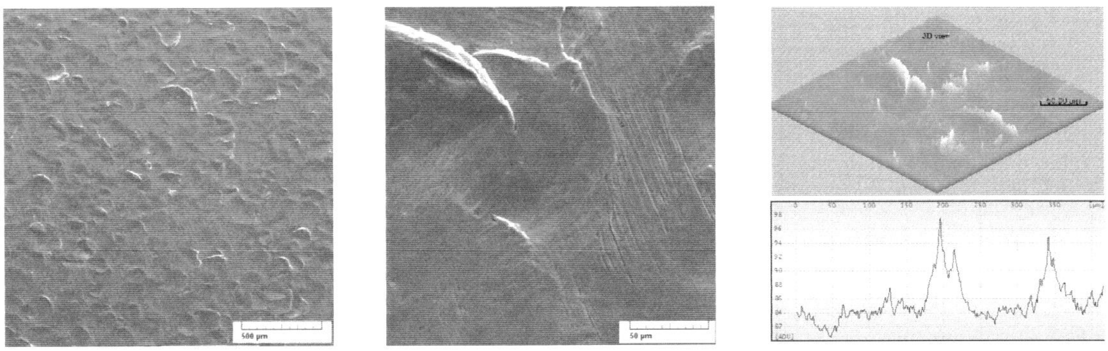

a) b) c)

Figure 2. Surface SEM analyze of shot-penning cast iron sample.

The experimental sample movement during the pendulum oscillation was filmed using a high performance camera and the movement film divided in 100 pictures. In figure 3 is presented the cinematic movement snapshots of the experimental pendulum I (initial state) and HT (heat treated). Using a picture analyze software the movement can be registered with a very high precision and the displacement can be easily determined. These preliminary results analyze the movement of the pendulum for 60 seconds. In case of high damping materials this period is sufficient to determine the amplitude variation but for low damping capacity materials no variation was observed in this period.

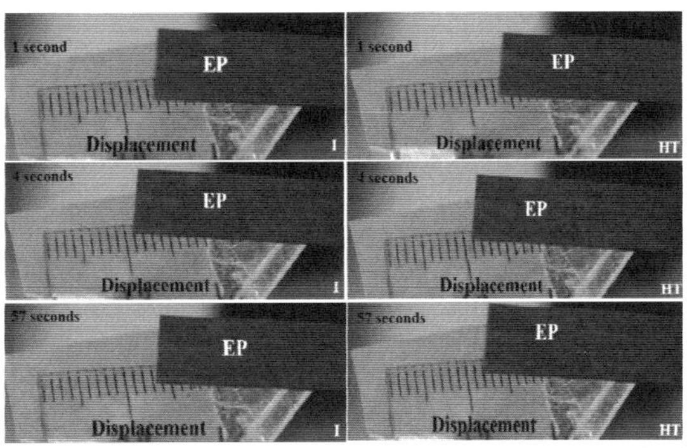

Figure 3. Cinematic movement snapshots of the experimental pendulum I (initial state) and HT (heat treated)

We analyze de movement of pendulum and determine the values of A_0 (first biggest amplitude above the zero line), A_n (last biggest amplitude used for calculus), a_0 (first biggest amplitude under the zero line), a_n (last biggest amplitude under the zero line), register the analyze time, seconds, calculate the frequency and apply equation (1) we obtain a value of $6.75x10^{-6}$ for Q^{-1} and $0.15x10^{6}$ quality factor which represent a very good value for low damping materials. The reduced video registration time can influence the final calculation value of the Q^{-1} so further investigations with bigger registration periods is required.

In figure 4 is present the schematic of the acquisition board used to transform the analog signal of the pendulum in digital signal in order to be registered and after that transferred to a PC as digital results (tensions given by the optocoupler).

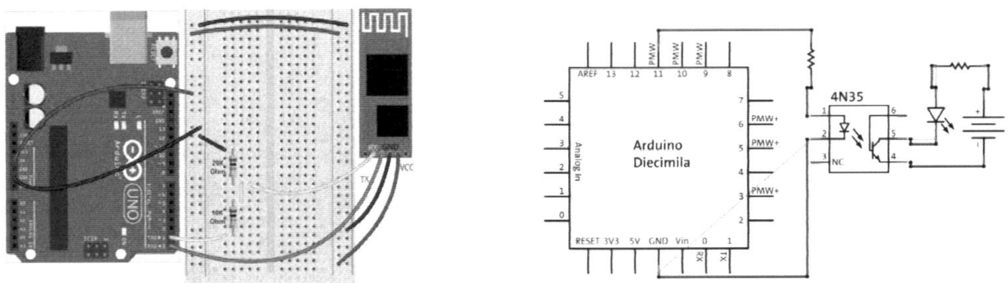

Figure 4. Schematic of the acquisition part of the signal from the pendulum using and optocoupler sensor.

The results, after registration and interpretation, based on the diagrams presented in figure 5, show very small variation of the damping capacity in the first minute and few modifications appear after 5 minutes. Even the damping capacity can be evaluated after this period, taking care of the time analyze, frequency and oscillation amplitudes, at the beginning of the experiment and after a while, further analyze with time extended (30-60 minutes) are required to increase the quality of the analyze.

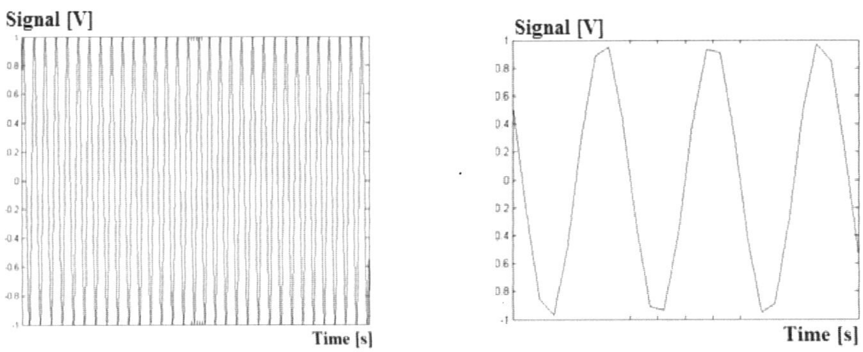

a) b)

Figure 5. Analog to digital signal analyze a) first part of the experiment (first 60 seconds) and b) last registered part of the experiment (last 10 seconds after 300 seconds).

In the initial condition using a machined and shot-penning sample, the damping coefficient Q^{-1} is about $3.75x10^{-5}$; thus, the quality factor ($1/Q^{-1}$) is about $0.27x10^{5}$. In comparison, for pure silica, this coefficient is similar [8]. Residual stresses are induced by the shot penning process itself, especially by the surface deformation and impact energy. A tempering heat treatment will induce a decrease of all the defects (residual stresses not previously eliminated or induced by shot penning operation and intrinsic material defects), and thus, an enhancement of the quality factor Q should be observed. At room temperature, the quality factor of such an tempered sample was measured and found equal to Q $= 0.77x10^6$ the result can be explain through the elimination of different defects from the material, surface and structure, that will decrease the damping capacity of the material.

As further work we propose to improve the equipment experimental conditions: a vacuum system will be applied and a heating system of the sample (using an electrical resistance) in order to establish the damping capacity at various temperatures for different application fields. Both registrations of movement systems will be re-analyzed and improved for a higher quality and larger time periods. Also the damping capacity of the material can be observed using a thermal camera to register thermal variation during oscillation, especially between the beginning and the end of the movement. High damping materials will present a high temperature variation in time that the low damping materials smaller temperature variations.

4. Conclusions

An experimental pendulum was obtained in laboratory conditions in order to analyze the damping capacity of metallic materials with possibility of a pre-stress apply. Preliminary results of Q^{-1} values determination for a cast-iron sample were obtained using two different acquisition systems. The results present a future of cast-irons in low damping capacity materials with different application possibility.

A tempering heat treatment applied will contribute to a decrease of the defects number and normally an enhancement of the quality factor Q should be observed. At room temperature, the quality factor of such an tempered sample was measured and found equal to $Q = 0.77x10^6$.

References

[1] Wong S J, Fox C H, McWilliam S, Fell C P, Eley R J 2004 *Micromech. Microeng* **14** S108–S112
[2] Treviso A, GenechtenVan B, Mundo D and Tournour M 2015 *Composites Part B: Engineering* **78** 144-152
[3] Simion P, Istrate B, Munteanu C, Dia V, Avram P, Imbrea M S 2014 *Advanced Materials Research* **1036** 134-139
[4] Achitei D, Galusca, D G, Vizureanu P, Carabet R, Cimpoesu N 2009 Metalurgia International **14** 45-48
[5] Cimpoesu N, Stanciu S, Tesloianu D, Cimpoesu R, Popa R F and Moraru E 2017 *Metal Science and Heat Treatment* **58** 11 – 12
[6] Delpueyo D, Balandraud X, Grediac M, Stanciu S and Cimpoesu N 2017 Measurement of Mechanical Dissipation in SMAs by Infrared Thermography *Residual stress, thermomechanics& infrared imaging, hybrid techniques and inverse problems* **9**, *Book Series: Conference Proceedings of the Society for Experimental Mechanics Series* 9-14
[7] Paun V P, Cimpoesu N, Hanu Cimpoesu R, Munceleanu G V, Forna N and Agop M 2010 *MaterialePlastice* **47** 158-163
[8] Cimpoeşu N, Stanciu S, Meyer M, Ioniţă I and Cimpoeşu R 2010 *Journal of Optoelectronics and Advanced Materials* **12** 386-391
[9] Heah C Y, Kamarudin H, Mustafa Al Bakri, A M, Luqman M, Khairul Nizar I, Liew Y M 2011 *Australian Journal of Basic and Applied Sciences* **5** (7) 1026- 1035
[10] Cimpoesu N, Axinte M, Cimpoesu Hanu R, Nejneru C, Achitei D C and Stanciu S 2010 *Journal of Optoelectronics and Advanced Materials* **12** 1772-1776
[11] Haon C, Camel D, Drevet B And Pelletier J M 2008 *Metallurgical and materials transactions A* **39A** 1791.
[12] Duwel A, Gorman J, Weinstein M, Borenstein J, Ward P 2003 *Sens. Actuat.* **103** 70–74
[13] Can Aydiner C, Ustundag E, Prime M B, Peker A J 2003 *Non-Crystal. Solids* **316** 82–99
[14] Ţugui C A, Nejneru C, Găluşcă D G, Perju M C, Axinte M, Cimpoeşu N, Vizureanu P 2015 *J. of Optoel. and Advanc.Mater.* **17** 1855-1861

[15] Harun M A, Rahim Z S, Mat Saad M N, Fathullah M, Ghazali, 2016 *Europ. J.of Mater. Scienc. and Engineer.* **1** 9-18
[16] Paleu V, Georgescu S, Baciu C, Istrate B, Baciu E R, 2016 *Book Series: IOP Conference Series-Materials Science and Engineering* **147**
[17] Karnaukhova T. V, *J. of Mathematical Sciences* 2010 **4** 168
[18] Craciun R C, Stanciu S, Cimpoesu R, Ursanu A I, Manole V, Paraschiv P; Chicet D L, 2016 *Book Series: IOP Conference Series-Materials Science and Engineering* **147**

IOP Conf. Series: Materials Science and Engineering **209** (2017) 012048 doi:10.1088/1757-899X/209/1/012048

Preliminary Results from Duplex Procedure for Obtain of Fe Based Materials for Automotive Applications

R C Crăciun[1], S Stanciu[1], V Geantă[2], I Voiculescu[2], V Manole[1], I A Gârneţ[1], A Alexandru[1], N Cimpoesu*[1] and F Săndulache[1]
[1]„GheorgheAsachi" Technical University of Iasi, Strada Prof.dr.doc. D. Mangeronnr. 41, Iasi, 700050, Romania
[2]University Politehnica Bucharest, Materials Science and Engineering Faculty, SplaiulIndependentei 313, Sector 6, 060042, Bucharest, Romania

E-mail: nicanornick@yahoo.com

Abstract. Iron based materials still represent a high percentage from metallic materials used in industry, in general, and in automotive industry, in particular. In this case we used a duplex process in order to obtain the FeMnSiAl experimental alloy for a more efficient use of various units. In the first stage iron, manganese, silicon and aluminum were melted and mixed together using arc melting technology and for the second stage the alloy was re-melt for homogeneity in an induction furnace. Chemical composition, after each melting step, was analyzed using EDS Bruker detector for various areas and microstructural characterization using SEM, VegaTescan LMH II with SE detector, equipment. This alloy is proposed as a metallic approach of mechanical dumpers used in automotive industry for low and medium impact contacts.

1. Introduction

The duplex process, which was first used in the second half of the 19th century, makes possible more efficient use of various units. For example, the duplex process using a basic open hearth furnace and an arc furnace or a converter and an arc furnace is used to increase the output of the electric furnace (that is, to increase the output of high-quality steel) and to reduce the specific consumption of electric power. This decrease is accomplished by removing from the electric furnace such operations as the melting of the charge and the partial refining of the metal; in the duplex process, only the final refining and deoxidation of the steel are carried out in the arc furnace.

The application of the duplex process is limited [1-5]. Shape memory alloys present a high damping capacity especially in the transformation domain from austenite to martensite and reverse based on dislocations movement, plates re-orientation and other phenomena already discussed [6-9]. From shape memory alloys categorizes most used is NiTi but this one have a quite expensive production procedure and cannot be suitable for automotive applications at industrial scale.

By economical point of view is more suitable to obtain a Fe-based shape memory alloy and the system FeMnSi present very good results in damping capacity as well as shape memory effect [10-13]. Aluminium element can be used as additional element to system FeMnSi in order to improve the mechanical damping capacity.

In this paper is presented few experimental results from the obtaining part of this alloy using two melting methods by arc melting and induction melting for chemical homogenization.

2. Experimental setup

An experimental alloy, FeMnSiAl , was proposed and obtained from high purity materials. Vacuum Arc Melting Facility MRF ABJ 900 (University Politehnica Bucharest) ensures the melting of metallic materials in argon-controlled atmosphere after pre-emptive working chamber up to 10-5 mbar by means of a non-consumable throttle tungsten mobile electrode. The RAV MRF ABJ 900 has the following technical features: The working chamber (vacuum container) is made of 304L stainless steel and is provided with double water cooled walls (temperature below 500°C). The base of the working chamber is made of aluminium with a thickness of 20 mm. It is placed above the electrical generator to minimize the surface and volume of the plant. The installation loading is done at the bottom of the work stand by opening the articulated bell and placing the load in the alveoli configured in the copper base plate. The loading of the materials for elaboration and refining, as well as the unloading of the samples, is done simply by lifting the lid that will be hung with a hinge from the rest of the room. Visualization of the crucible and electrode can be performed through the 4 "diameter window located at the front of the plant. The work area illumination during the operation of the installation is through a 1 "diameter window, placed in the rear of the container with a halogen lamp included. The viewing window is equipped with a folding welding glass to protect the view during melting and refining of the metallic load. The vacuum pump system is connected to the bottom of the working chamber. The working container is provided with clamps for closing and isolating it from the atmosphere. The rewinding part, made at Gheorghe Asachi Technical University in Iasi, took place in an induction furnace and the materials were heated to a temperature of 1600°C.

The chemical analysis of the materials was performed with an EDS - Bruker EDAX spot detector or by line distribution of the main elements Fe, Mn, Si and Al. The microstructural alloy was analyzed by SEM VegaTescan LMH II electronic scanning microscopy after the samples were mechanically polished with SiC metallographic maps and etching with nital solution followed by rinsing with distilled water.

3. Experimental results

The experimental alloy was first obtained using arch melting technique and secondly induction furnace for chemical homogenization. The experimental composition proposed for obtaining was Fe15Mn3Si3Al. In figure 1 a) is presented the image of the surface and the chemical analyze points proposed for analyze on the inferior part of the ingot and b) XRD spectrum of sample FeMnSiAl in three points (1, 3 and 5 from figure a)).

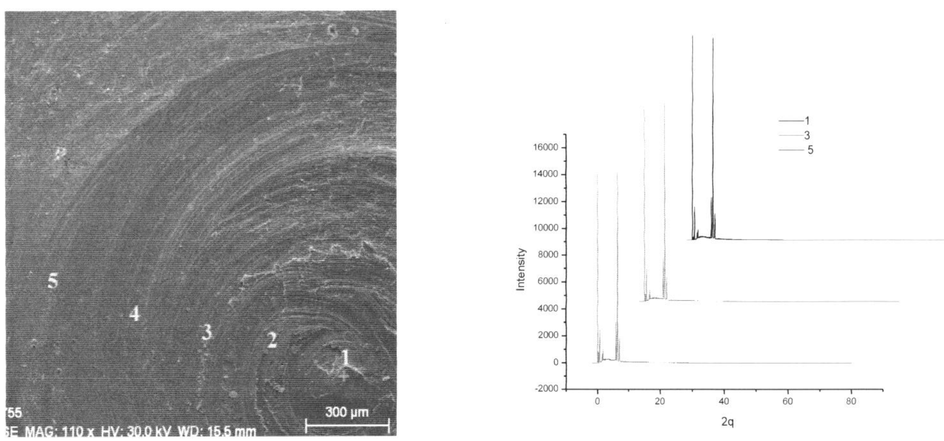

a) b)

Figure 1. a) Image of the surface and the chemical analyze points and b) XRD spectrum of sample FeMnSiAl in three points (1, 3 and 5).

The material have a mixed microstructure which consists mainly of γ-austenite (11 and 200) but also some reduced peaks of ε-martensite (100), figure 1 b) at room temperature.

The necessity of the second melting stage appear after we analyze the ingot on the length and we observe high variations between the ingot ends, not only for chemical composition but also for magnetic properties.

We propose a chemical composition analyze from the centre of the sample, point 1 to exterior, point 5 and determine the homogeneity of the ingot on the transversal way. The XRD results, figure 1 b), present no variation between the selected points (1, 2 and 3) so no chemical variation are observed on this part of the ingot. The results are presented in table 1, in wt% and at%, also with the equipment error. Small variations between the same element chemical composition and beside the average value and with some low differences for Mn, Si and Al against the proposed chemical composition (15, 3 and 3 wt%).

Table 1. FeMnSiAl alloy after arch melting.

Elements/	Fe		Mn		Si		Al	
points	wt[%]	at[%]	wt[%]	at[%]	wt[%]	at[%]	wt[%]	at[%]
Point 1	74,85	71,51	20,92	20,31	2,52	4,79	1,17	3,39
Point 2	76,14	70,88	20,18	19,08	2,23	4,12	1,53	2,94
Point 3	75,36	70,33	20,74	19,66	2,49	4,61	1,69	3,25
Point 4	76,08	71,80	20,85	19,99	1,98	3,72	1,20	3,04
Point 5	75,87	73,30	21,02	20,64	2,01	3,84	1,11	2,22
Average	75,66	71.56	20.74	19.94	2.25	4.216	1.34	2.97
Erorr %	1,5		0,35		0,15		0,10	

By chemical analyze on the bottom part of the ingot we observe a higher percentage of Mn, as against top part (11,5%) so during the preparation of the materials for arch melting and the process itself appear this variation.

In table 2 are presented the results of chemical analyze made on six points. For induction melting we wire cut, the initial ingot and put all the pieces in a ceramic crucible. Melting starts around 1600°C temperature and during 30-60 seconds for homogenization, under Ar atmosphere and poured in a metallic pre-heated form. The ingot obtain was more 60% weight from the starting mass. It was lost some material from the first quantity that remain stuck on the crucible.

Table 2. FeMnSiAl (S2) alloy re-melted in inductionfurnace.

Elements/	Fe		Mn		Si		Al	
points	wt[%]	at[%]	wt[%]	at[%]	wt[%]	at[%]	wt[%]	at[%]
Point 1	81,43	77,96	14,44	14,06	2,37	4,51	1,75	3,48
Point 2	81,48	78,52	15,07	14,76	2,09	4,0	1,36	2,72
Point 3	80,37	76,33	14,69	14,18	2,88	5,44	2,06	4,06
Point 4	79,97	76,01	15,18	14,66	2,88	5,45	1,97	3,88
Point 5	80,07	75,71	14,54	13,98	2,95	5,55	2,44	4,77
Point 6	79,12	74,32	14,80	14,14	3,32	6,20	2,75	5,35
Average	80.41	76.48	14.79	14.29	2.75	5.19	2.10	4.04
Erorr %	1,5		0,35		0,15		0,10	

In figure 2 the distribution of main elements Fe, Mn, Si and Al are presented in the selected image from 2 a). The distribution of the elements is homogeneous for all the elements including Al. For alloys with shape memory effect and damping capacityit is important for the alloy to present a high chemical homogeneity, even if the damping capacity usually decrease with elimination of defects, chemical variations etc.

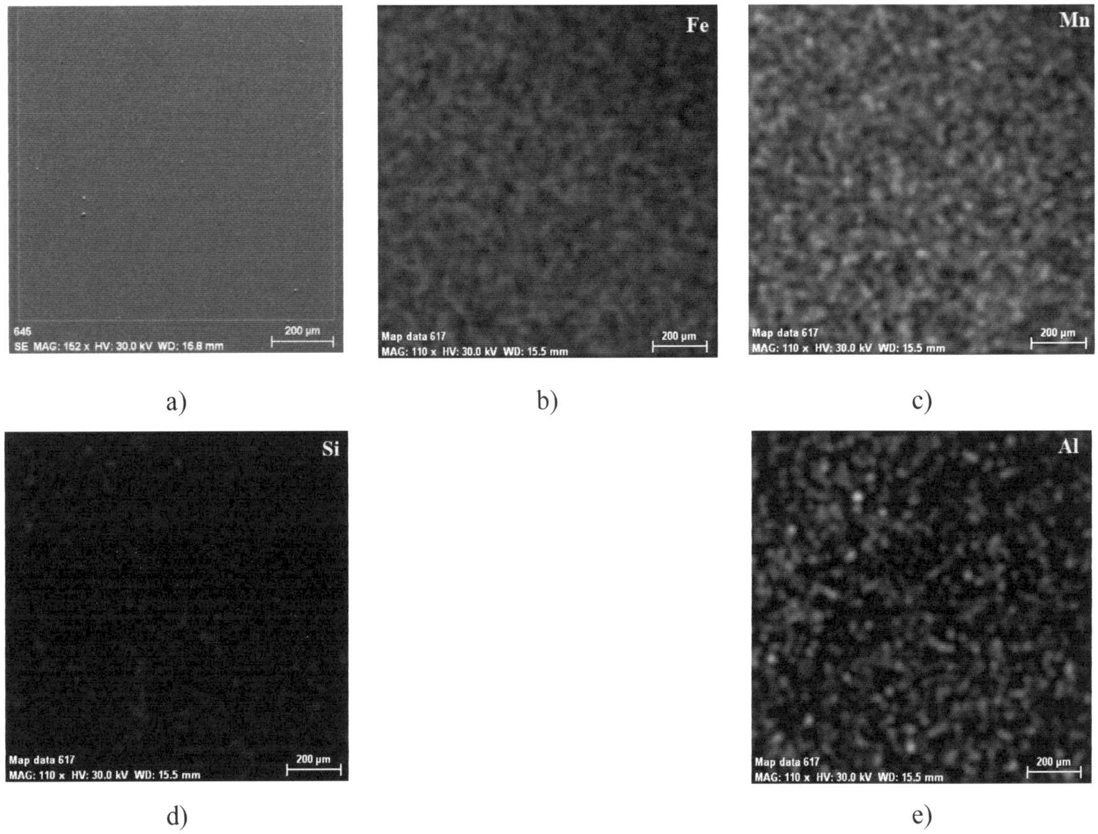

Figure 2. Chemical elements Fe, Mn, Si and Al after the re-melting operation a) area selected for distribution analyze, b), iron distribution, c) manganese distribution, d) silicon distribution and e) aluminium distribution.

The alloy microstructure is presented in figure 3 using SEM technology. The sample is after heat treatment of solution hardening with 0% deformation so the structure suppose to be austenite.

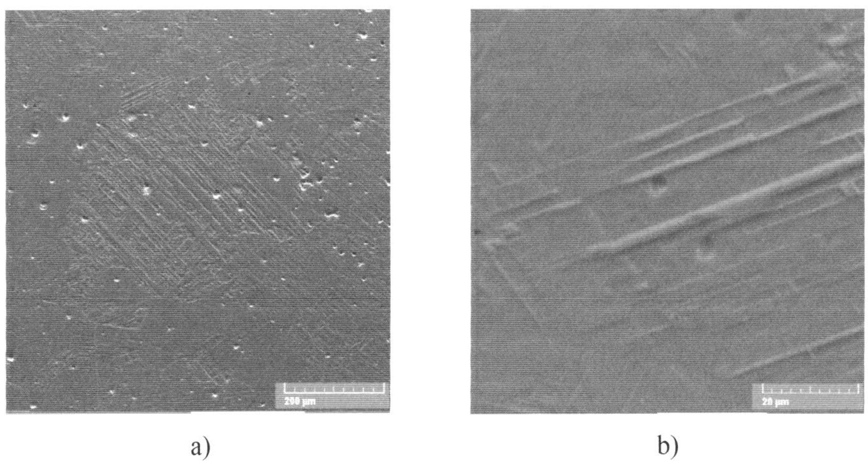

Figure 3. SEM images of FeMnSiAl experimental alloy a) 500x and b) 1000x.

The present of martensite plates, figure 3 b), can be put on the contraction forces that appear at cooling stage after the heat treatment at 1100°C and chill in oil. The appearance of a relief on the austenite structure was reported in CuMnAl superelastic alloy [14]. The alloy presents at room temperature a stable austenitic phase (A, γ, FCC) [15-17]. The stacking fault energy of austenite phase is an important factor in controlling the deformation behaviour and the occurrence of deformation in this kind of materials.

Fe–Mn–Si-based alloys for specific compositions are expected to have a shape memory effect associated with the deformation-induced martensitic transformation from face-centred cubic austenite to hexagonal martensite on loading.

As further work we will obtain the hot-rolled samples and analyze the damping behaviour of the materials using Dynamic Mechanical Analyzer (DMA) in order to obtain the coefficient tanδ values with temperature and frequency variation. Also the materials will be analyzed using an laboratory pendulum to establish the internal friction with and without a pre-strain value.

4. Conclusions

We obtain a Fe-based alloy with possible shape memory effect (based on the chemical composition) and present some preliminary results about the chemical composition evolution in the duplex melting process. At room temperature the material have a microstructure from especially from γ-austenite but also with reduced quantities of ε-martensite. The experimental sample, FeMnSiAl, after the heat treatment of solution hardening with, have no deformation degree applied, so the structure suppose to be austenite but from the scanning electron microscopy images it seems to present martensite plates that can be put on the contraction forces phenomena that appear at cooling stage.

References

[1] Dia V, Bujoreanu L G, Stanciu S and Munteanu C 2008 *Materials science and engineering a-structural materials properties microstructure and processing* **481** 697-701

[2] Cimpoeşu N, Stanciu S, Meyer M, Ioniţă I and Cimpoeşu R 2010 *Journal of Optoelectronics and Advanced Materials* **12** 386-391.

[3] Paun V P, Cimpoesu N, Hanu Cimpoesu R, Munceleanu G V, Forna N and Agop M 2010 *Materiale Plastice* **47** 158-163

[4] Delpueyo D, Balandraud X, Grediac M, Stanciu S and Cimpoesu N 2017 Measurement of Mechanical Dissipation in SMAs by Infrared Thermography *Residual stress, thermomechanics& infrared imaging, hybrid techniques and inverse problems Book Series: Conference Proceedings of the Society for Experimental Mechanics Series* 99-14

[5] Takahiro S, Ilya N, Kazuyuki O, Kaoru S, Susumu T, Tadakatsu M, Yuya C, Atsumichi K, Yasuhiko I and Kaneaki T 2015 *ScriptaMaterialia* **99** 49-52

[6] Suru M G, Lohan N M, Mihalache E, Pricop B, Mocanu M and Bujoreanu L G 2017 *Journal of testing and evaluation* **45** 419-427

[7] Yang H K, Tian Y Z, Zhang Z J and Zhang Z F 2016 *Materials Science and Engineering: A* **655** 251-255

[8] Gurau G, Gurau C; Potecasu O, Alexandru P and Bujoreanu L G 2014 *Journal of Materials Engineering and Performance* **23** 2396-2402

[9] Mahato B, Shee S K, Sahu T, GhoshChowdhury S, Sahu P, Porter D A and Karjalainen L P 2015 *ActaMaterialia* **86** 69-79

[10] Mahato B, Sahu T, Shee S K, Sahu P, Sawaguchi T, Kömi J and Karjalainen L P 2017 Simultaneous twinning nucleation mechanisms in an Fe–Mn–Si–Al twinning induced plasticity steel *ActaMaterialia* **132** 264-275

[11] Yang H K, Zhang Z J, Tian Y Z and Zhang Z F 2017 *Materials Science and Engineering: A***690** 146-157

[12] Motomichi K, Takahiro S, Kazuyuki O, Takehiko K and Masato M 2008 *Materials Science and Engineering: A* **497** 353-357

[13] Pricop B, Soyler U, Lohan N, Ozkal M, Bujoreanu L G, Chicet D, Munteanu C 2012 *Journal of Materials Engineering and Performance* **21**(11) 2407-2416
[14] Cimpoesu N, Stanciu S, Tesloianu D, Cimpoesu R, Popa R F and Moraru E 2017 *Metal Science and Heat Treatment* **58** 11 – 12
[15] Motomichi K, Yusuke Y, Rinqing C, Takahiro S, Kaneaki T and Hiroshi N 2017 *International Journal of Fatigue* **94** 1-5
[16] Kowalska J, Ratuszek W, Witkowska M, Zielińska-Lipiec A and Tokarski T 2015 *Journal of Alloys and Compounds* **643** S39-S45
[17] Ţugui C A, Nejneru C, Găluşcă D G, Perju M C, Axinte M, Cimpoeşu N, Vizureanu P 2015 *J. of Optoelectr. and Advanc. Mater.* **17** 1855-1861

Acknowledgement
This work was carried out through the Partnerships in priority areas program - PN II, implemented with the support MEN - UEFISCDI, project no. PTE 48/2016.

IOP Conf. Series: Materials Science and Engineering **209** (2017) 012049 doi:10.1088/1757-899X/209/1/012049

Preliminary Results on the Surface of a New Fe-Based Metallic Material after "In Vivo" Maintaining

F Săndulache[1*], S Stanciu[1], N Cimpoeşu[1], T Stanciu[2], R Cimpoeşu[1], A Enache[3] and R Baciu[3]

[1] „Gheorghe Asachi" Technical University of Iasi, Str. Prof.dr.doc. D. Mangeron nr. 41, Iaşi, 700050, Romania

[2] "Ion Ionescu de la Brad" University of Agricultural Sciences and Veterinary Medicine Iaşi, Romania

[3] University of Medicine and Pharmacy „Grigore T. Popa" of Iaşi, Romania

E-mail: * florinsandulache@yahoo.com

Abstract. A new Fe-based alloy was obtained using UltraCast melting equipment. The alloy, after mechanical processing, was implanted in five rabbit specimens (with respect for the "in-bone" procedure). After 30 days of implantation the samples were recovered and analyzed by weight and surface state meanings. Scanning electron microscopy technique was used to determine the new compounds morphology from the metallic surface and X-ray dispersive energy spectroscopy for chemical analyze results. A bond between the metallic material and biological material of the bone was observed through increasing of sample weight and by SEM images. After the first set of tests, as the samples were extracted and biologically cleaned, the samples were ultrasonically cleaned and re-analyzed in order to establish the stability of the chemical compounds.

1. Introduction

Progress in materials science and engineering, particularly in the areas of processing and thermo-mechanical treatments of metallic materials allow them structural adjustment [1-6]. Increasingly growth in higher expectations of society for a higher level of life rushed forward biomaterials and especially new technologies for producing implants with very good clinical performance. Implants paradigm to be inert and corrosion resistant in these moments is caused by the emergence and development of new classes of materials, that of degradable biomaterials [7-10].

Some special clinical problems (illness / injury) only require temporary support for healing. This support can be provided only temporary degradable materials that allow the implant to degrade gradually when fully function is fulfill. The concept of biodegradable medical applications is known as the "biodegradable sutures". However, degradable implants, especially those made of metallic materials, can be considered a new concept effectively responding paradigm that "metallic biomaterials must be corrosion resistant" [11-14]. The alloys based on Fe or Mg is two classes of metals which have been proposed for applications, such as biodegradable materials. Some of the Mg based alloys have been dealt with Mg-Al, Mg-Ca and M- RE. In the case of alloys based on Fe were considered pure Fe and Fe-Mn. They are mostly proposed for cardiovascular applications [15-18]. The general concept is simple degradable biomaterials as their name, some implants require their presence only temporarily damaged bone to support the healing process of diseased tissue. This temporary intervention (implant) is considered in the application only in certain cases such as cardiovascular, orthopedic and pediatric [19-23].

The article present some experimental results from the analyze of the surface of FeMnSi alloy after a 30 days implantation period in a biological living environment.

2. Experimental

The experimental alloys were obtained using UltraCast furnace (Ar atmosphere) from high purity metals and master alloys.

The in vivo experiments were conducted according to the Faculty of Veterinary Medicine of Iași Ethics Committee guidelines, and ISO 10993-2: 2006 animal welfare requirements, on lots of each five experimental rabbits, males, 12 weeks age and 1000 g average weight, in Agronomic and Veterinary Medicine University (USAMV) Biobase. The experimental rabbits were hosted in individual cages with free water and food access.

Scanning electron microscopy technique was used to determine the new compounds morphology from the metallic surface and EDS technique for chemical analyze results. A SEM VegaTescan LMHII model was used at 30 kV filament power supply. For chemical determinations the preliminary results were obtained using Quantax Bruker EDAX detector and Esprit software. The metallic samples were cleaned in an ultrasound bath with technical alcohol for 60 minutes and re-analyzed by surface point of view. Digital balance was used in order to be weighted the samples and information till milligrams values were modifications appear in 30 days.

3. Results

Two of five FeMnSi implanted samples were analyzed after 30 days of implantation in a rabbit bone with respect to the implantation protocol. The samples were weighed initially before the implantation, after the implantation and after a cleaning session through sonication for 60 minutes in an alcohol bath. Metallic implant information, weight and dimensions, before and after a 30 days implantation period are presented in table 1. By weight point of view, table 1, we observed two situations, first – sample 1 – when the sample is corroded and lose weight before and after sonication and secondly – sample 2 – for similar conditions the sample is corroded but win weight based on the compounds formed on the surface and additional bone tissue from the implantation area.

Table 1. Metallic implant information before and after a 30 days implantation period.

Sample 1 before and after implantation (30 days in a rabbit bone)					
Length [mm]	Initial width [mm]	Thickness [mm]	Initial weight [g]	After implantation weight [g]	After sonication of the implant [g]
25.3	2.45	0.45	0.2284	0,2277	0.2177
Sample 2 before and after implantation (30 days in a rabbit bone)					
Length [mm]	Initial width [mm]	Thickness [mm]	Initial weight [g]	After implantation weight [g]	After sonication of the implant [g]
25.2	2.5	0.5	0.2195	0.2305	0.2276

The weight lost by first sample is 0,7 mg and after the sonication process 10 mg of the material is removed from the implant surface as compounds with low stability and almost detached from the surface after the implantation time.

In figure 1 are SEM micrograph of the material surface after implantation period and sonication of the material a) along the implant surface and b) detail of the surface. The images present a corroded surface, at macro and micro scale, with different types of compounds form.

Chemical insights were taken from the implant surface before and after sonication process. Results on distribution of elements Fe, Mn, Si, Ca, P, O and Na are presented in figure 2.

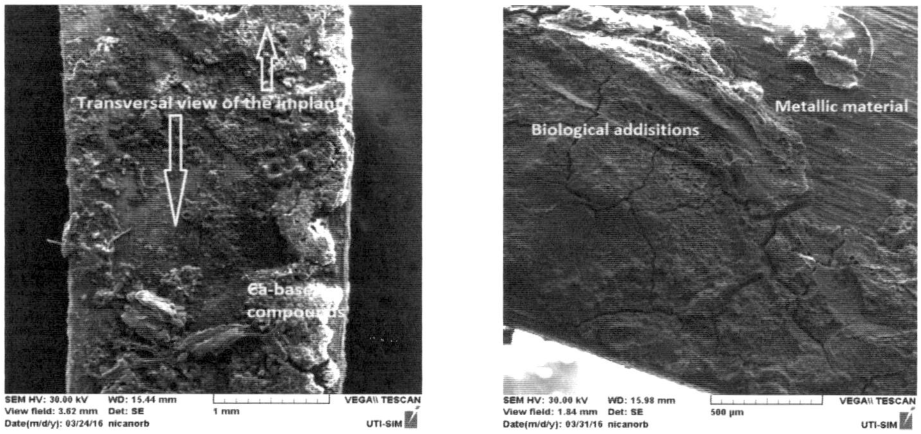

a) b)

Figure 1. SEM micrograph of the material surface after implantation period and sonication of the material a) along the implant surface and b) detail of the surface.

The elements distributions present the appearance of Ca-P compounds, oxides and before sonication reduced traces of Na, figure 2 a). After sonication a stabile oxide layer, figure 2 b), is present on the right side and on the left a metal corroded substrate is observed.

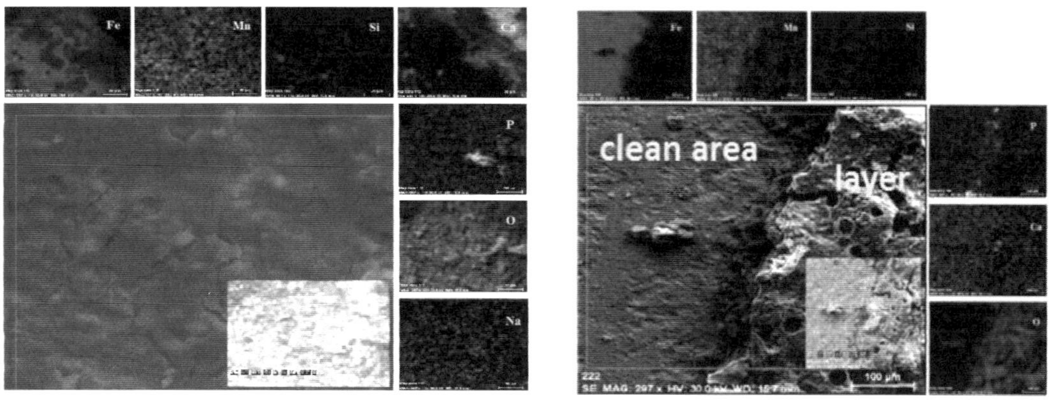

a) b)

Figure 2. SEM image in centre of the surface after implantation and before sonication stage in a) and after in b) and distributions of elements Fe, Mn, Si, Ca, P, O and Na.

In table 2 chemical compositions from different area are presented. We present the results for chemical composition on a large 4 mm2 area, a reduce 0,25 μm2 area point1 – from the SEM detail presented in table 2, on another reduced are of 0,25 μm2 area point 2, from the SEM detail presented in table 2, after sonication process on a clean area and on deposited layer, figure 2 b).

All the chemical composition results present corroded areas with formation of new compounds, especially oxides but also compounds with calcium, phosphorus, magnesium and sodium. On macro-scale we observed a high quantity of oxides, C-based and Ca-based compounds (new compounds formed on the metallic surface or biological tissue traces). In point 1 the oxides layer is thinner so a bigger signal for substrate is registered. Manganese element is smaller in all areas and means that Mn compounds from the implanted alloy are prone to corrosion more than Si compounds which keep a higher percentage.

Table 2. EDS results after implantation period before and after sonication process.

	Fe		Mn		Si		O		C		Ca		Na		P		Mg	
	[wt.%]	[at.%]	[wt.%]	[at.%]	[wt.%]	[at.%]	[wt.%]	[at.%]	[wt.%]	[at.%]	[wt.%]	[at.%]	[wt.%]	[at.%]	[wt.%]	[at.%]	[wt.%]	[at.%]
Chemical composition on 4 mm^2 area	43,92	18,01	0,25	0,1	0,51	0,41	39,18	56,1	11,43	21,81	2,26	1,29	1,75	1,74	0,70	0,52	-	-
Chemical composition on 0,25 μm^2 area point1	75,03	46,83	0,29	0,19	0,82	1,02	19,13	41,68	2,79	8,11	0,77	0,67	-	-	0,52	0,58	0,65	0,93
Chemical composition on 0,25 μm^2 area point 2	35,6	16,7	-	-	-	-	36,71	60,09	3,3	7,2	24,01	15,69	-	-	0,38	0,32	-	-
After sonication	93,03	78,16	0,33	0,28	0,8	1,34	1,88	5,51	3,64	14,23	-	-	-	-	0,31	0,48	-	-
Chemical composition on the clean area	78,25	47,78	-	-	0,58	0,71	14,73	31,59	3,71	10,61	1,07	0,913	-	-	1,363	1,51	-	-
Chemical composition on deposited layer	38,24	15,3	-	-	0,64	0,51	48,7	68,01	6,44	11,97	1,84	1,02	-	-	2,64	1,9	1,01	0,92
Error	1,27		0,05		0,06		5,57		1,67		0,11		0,19		0,06		0,09	

In point 2 large amounts of oxygen and calcium are observed. After sonication compounds are detached from the surface so we can conclude that in time the degradation of the material continues. After sonication are still stable, in small areas, oxides and small quantities of Ca-P material. After sonication more than 85% of the compounds detach from the surface.

4. Conclusions

Preliminary results after the implantation of a FeMnAl alloy for 30 days in a biological environment are presented using scanning electron microscopy, X-ray energy dispersive analyze and weight measurements. The results present the metallic material surface after implantation period and before and after a cleaning process by sonication. All the surface present corrosion traces and formation of unstable with low adhesion compounds especially oxides and Ca-P compounds. Traces of Na and Mg were also observed on different parts of the surface.

References

[1] Aelenei N, Lungu M, Mareci D and Cimpoeşu N 2011 *Environmental Engineering and Management Journal* **10** 1951-58
[2] Allori A C, Sailon A M and Warren S M 2008 *Tissue Engineering* B **14** 259-73
[3] Cimpoeşu N, Stanciu S, Vizureanu P, Cimpoeşu R, Achiţei D C and Ioniţă I 2014 *Journal of Mining and Metallurgy* B **50** 69-76
[4] Dorozhkin S V 2014 *Acta Biomaterialia* **10** 2919-34
[5] Farraro K F, Kim K E, Woo S L-Y, Flowers J R and McCullough M B 2014 *Journal of Biomechanics* **47** 1979-86
[6] Baltatu S, Vizureanu P, Mareci D, Burtan LC, Chiruta C, Trinca LC 2016 *Materials and Corrosion-Werkstoffe und Korrosion* **67** (12) 1314-1320
[7] Răţoi M, Dascălu G, Stanciu T, Gurlui S O, Stanciu S, Istrate B, Cimpoesu N and Cimpoeşu R

2015 *Key Engineering Materials* **638** 117-22
[8] Cimpoeşu N, Trincă L C, Dascălu G, Stanciu S, Gurlui S O, and Mareci D 2016 *Journal of Chemistry* **2016** 9
[9] Henderson S E, Verdelis K, Maiti S, Pal S, Chung W L, Chou D-T, Kumta P N, Almarza Al J 2014 *Acta Biomaterialia* **10** 2232–2332
[10] Kini U and Nandeesh B N 2012 Physiology of Bone Formation, Remodeling, and Metabolism, In: Radionuclide and hybrid bone imaging, Fogelman I., Gnanasegaran G., Wall H. (Eds.) *Springer* Berlin 29–57
[11] Liu B, Zheng Y F and Ruan L 2011 *Materials Letters* **65** 540-43
[12] Cimpoeşu R H, Pompilian G O, Baciu C, Cimpoeşu N, Nejneru C, Agop M, Gurlui S and Focşa C 2010 *Optoelectronics and Advanced Materials-Rapid Communications* **4** 2148 – 53
[13] Peuster M, Hesse C, Schloo T, Fink C, Beerbaum P and von Schnakenburg C 2006 *Biomaterials* **27** 4955-62
[14] Istrate B, Mareci D, Munteanu C, Stanciu S, Crimu C I, Trinca L C, Kamel E 2016 *Environmental Engineering and Management Journal* **15** (5) 955-963
[15] Quadbeck P, Stephani G, Kümmel K, Adler J and Standke G 2007 *Materials Science Forum* **534/536** 1005-08
[16] Izquierdo J, Bolat G, Cimpoesu N, Trinca L C, Mareci D and Souto R M 2016 *Applied Surface Science* **385** 368-78
[17] Ryan G, Pandit A and Apatsidis D P 2006 *Biomaterials* **27** 2651-70
[18] Schinhammer M, Steiger P, Moszner F, Jörg F Löffler and Peter J Uggowitzer 2013 *Materials Science and Engineering* **C 33** 1882–93
[19] Staiger M P, Pietak A M, Huadmai J and Dias G 2006 *Biomaterials* **27** 1728-34
[20] Tan L, Wang Q, Lin X, Wan P, Zhang G, Zhang Q and Yang K 2014 *Acta Biomaterialia* **10** 2333–40
[21] Tran P and Webster T J 2008 *International Journal of Nanomedicine* **3** 391-96
[22] Wadley H N G 2002 Cellular Metals Manufacturing *Advanced Engineering Materials* **4** 726-33
[23] Stanciu S, Ursanu A, Trincă L C, Trofin A E, Solcan C, Munteanu C, Cimpoesu N, Acatrinei D, Sindilar E V, Stanciu T, Fântanariu M and Topliceanu L 2016 *Environmental Engineering and Management Journal* **15** 973-80

Acknowledgements
We would like to express our sincere thanks to Professor Lucia Carmen Trinca for the help given in implantation at the University of Agricultural Sciences and Veterinary Medicine, Iaşi.

IOP Conf. Series: Materials Science and Engineering **209** (2017) 012050 doi:10.1088/1757-899X/209/1/012050

Electro-Chemical Behavior of Low Carbon Steel Under H_2S Influence

M G Zaharia[1], S Stanciu[1], R Cimpoesu*[1], C Nejneru[1], C Savin[1], V Manole[1] and N Cimpoeşu[1]
Technical University "Gheorghe Asachi" of Iasi, Faculty of Material Science and Engineering, Prof. Dr. Doc. Dimitrie Mangeron Street, number 67, Iaşi, 700050, Romania

E-mail: ramonahanu@yahoo.com

Abstract. A commercial low carbon steel material (P265GH) with application at industrial scale for natural gas delivery and transportation systems was analyzed in H_2S atmosphere. The article proposed a new experimental cell in order to establish the behavior of the material in sulfur contaminated environment. In most of the industrial processes for gas purification the corrosion rate is speed up by the presence of S (sulfur) especially as ions or species like H_2S. The H_2S (hydrogen sulfide) is, beside a very toxic compound, a very active element in the acceleration of metallic materials deterioration especially in complex solicitations like pressure and temperature in the same time. For experiments we used a three electrodes cell with Na_2SO_4 + Na_2S solution at pH 3 at room temperature (~ 25°C) to realize EIS (electrochemical impedance spectroscopy) and potentio-dynamic polarization experiments. Scanning electron microscopy and X-ray dispersive energy spectroscopy were used to characterize the metallic material surface exposed to experimental environment.

1. Introduction

Human race demand energy in an increasing way which will reach with a plus of 55% by 2030 manly because of the industry consume and the population growth [1,2]. The main important energy source remain the oil and natural gas for at least 30-50 years even progresses in the renewable energy were recorded most of all in strong industrialized and very good research programs countries. Most of the natural gas or oil production and transport domains encountered the appearance of various quantities of carbon dioxide (CO_2), hydrogen sulfide (H_2S), water and chemical components [3]. At the moment when these gaseous pass through dissolution into water compose a weak acid that is very corrosive.

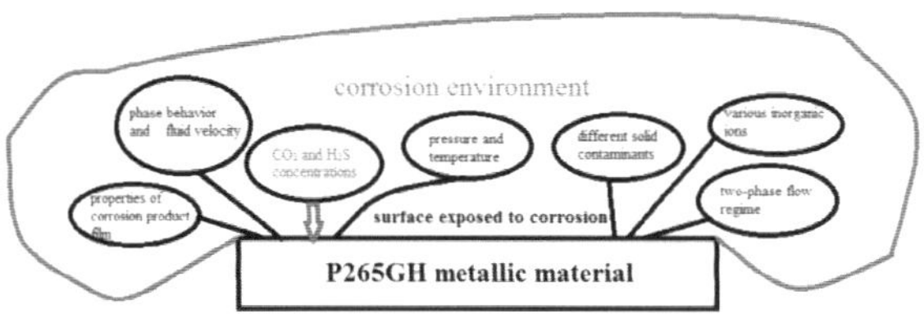

Figure 1. Factors that influence the metallic material corrosion resistance exposed to an environment with CO_2+H_2S.

This corrosive environment will cause high intensity corrosion (generalized, pitting, cracking or combinations). Having CO_2, H_2S and water products the corrosion process in gas and oil production operations is conditional on different factors that act separately or together on the metallic parts, figure 1. The influence of these parameters and interactions among them is a very important aspect in the corrosion research on oil and gas production systems.

In this article we proposed an experimental cell to analyze the effects of an environment with CO_2+H_2S on the commercial steel P256GH for different concentrations of H_2S (0, 10, 20 and 30 ppm/mol).

2. Experimental setup

Analytical grade Na_2SO_4, Na_2S, and H_2SO_4 reagents were used. To prepare the solution, 1 M Na_2SO_4 solution was buffered by H_2SO_4 to a pH of 3. Usually, if is necessary to analyze the H_2S corrosion mechanism, experimentally we have to purge an inert gas into the solution to remove dissolved oxygen [4] but in our experimental case we analyze the behavior of the material in an atmosphere with oxygen, this case being recognized in few industrial applications. In this case few cathodic chemical reactions will interfere (like O_2 and HS^- reduction). We proposed for the article four analyze cases which depend on the quantity of produced Na_2S, initial case with 0 and producing 10, 20 and 30 ppm/mole of H_2S, through addition to the solution (concentrations are in ppm/mole). In this case next reaction will occur based on the dissociation of Na_2S to H_2S in a solution with pH smaller than 7 [5]: $Na_2S + 2H^+ \rightarrow 2Na^+ + H_2S$ now a reduce amount of H_2S is locally generated at the vicinity of alloy surface. An isolated electro-chemical cell (600 cm^3 volume) was used for adding adding Na_2S. During the experiment the Na_2S was continuously added in order to produce H_2S also an inert gas (argon) was purged above the experimental cell with a very low pressure control and prevent the decreasing in H_2S concentration during the experiment. The sample, P256 was mounted cylindrical specimens with 1 cm (diameter) have the role of working element.

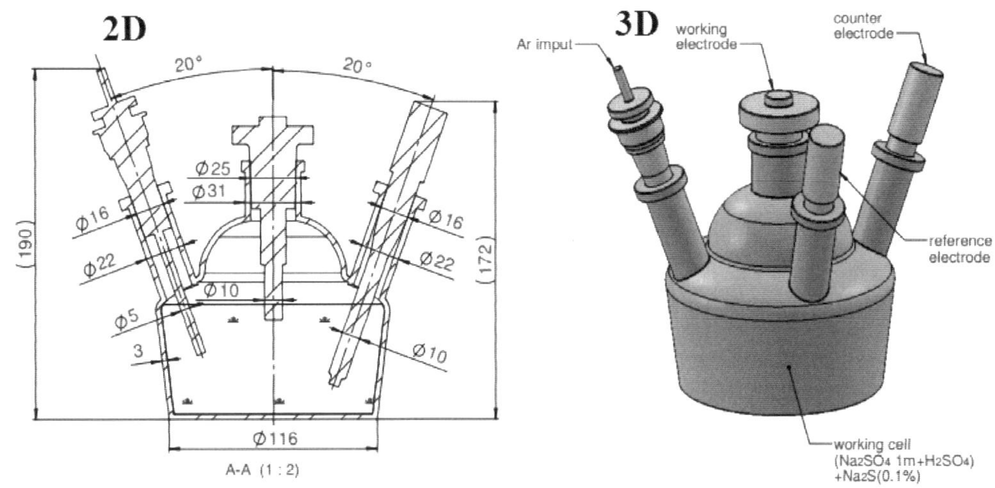

Figure 2. Experimental cell for study the H_2S influence.

We perform more corrosion experiments keeping the order: (I) 300 s OCP recording, (II) LPR test with potential domain of ± 100 mV and the rate of 1 mV/s, (III) potentiostatic EIS test at OCP with amplitude of 10 mV and normally a frequency range of 10^2 to 10^5 Hz, and (IV) linear and cyclic tests with a rate of 1 mV/s and potential range from 200 to 1500 mV with respect to OCP. All measurements were repeated 3 times for results reproducibility. The corrosion experiments were realized by EG201 Potentiostat equipment, using saturated calomel and Pt wire as reference and counter electrodes, respectively. The equipment software was used to extract the EIS equivalent circuit components. Simulated data was fitted by less than 2% error in calculation.

The experimental sample was, after the electro-corrosion tests, analyzed on the surface by scanning electron microscopy (SEM VegaTescan LMH II, SE detector, 30 kV) and X-ray dispersive energy spectroscopy (EDAX Bruker, XFlash).

3. Experimental results

The diffused H atoms can react to form H_2 inside the steel, causing hydrogen bubbles on the surface of the material [6].

$$H_2S(aq) = H_2S(ad) \quad (1)$$

$$H_2S(ad) + e- = H(ad) + HS^-(ad) \quad (2)$$

$$H_2S(ad) + H(ad) + e- = H_2(g) + HS^-(ad) \quad (3)$$

At anode the main reaction is the oxidation of iron. A reaction mechanism for solid materials in case of reaction between H_2S and Fe is presented in equation (4). During the reaction, characterized mainly by loss of iron contribute at the metallic material destroying and to decrease the strength.

$$Fe(s) + H2S(aq) = FeS(s) + 2\ H+(aq) + 2\ e^- \quad (4)$$

The results of OCP variations are presented in figure 1 a). The increase in H_2S concentration decreases the OCP from -590 mV (0 ppm/mole) to -630 mV (36 ppm/mole). Decrease in OCP with the increase of H_2S indicate that the alloy suffer a deterioration and the activity on the surface is accelerated. In figure 3 b) is represented the influence of H_2S concentration on LPR (polarisation resistance, RP) values.

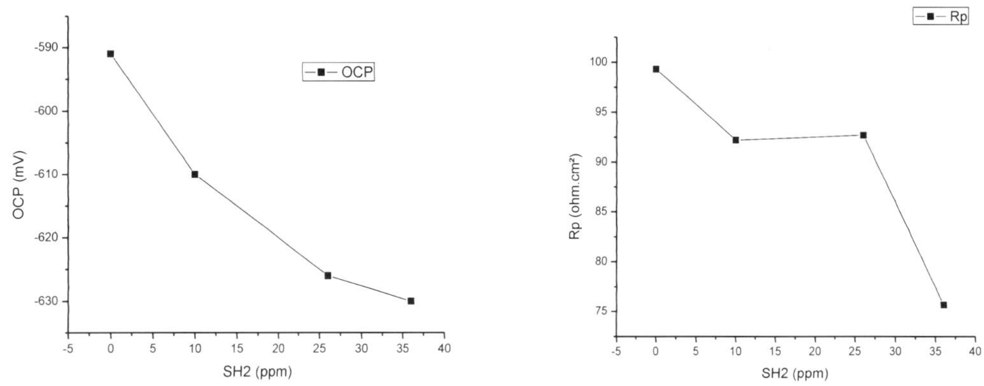

Figure 3. a) OCP variations of P265GH alloy vs. H_2S concentration (ppm by mole) in Na_2SO_4 + Na_2S (pH 3). b) LPR (Rp) variations of P265GH H_2S concentration (ppm by mole) in Na_2SO_4 + Na_2S (pH 3).

During the increase of the H_2S concentration the RP values decrease, fact that confirm the results of OCP measurements are follow a similar trend. Between initial sample and the one expose to 36 ppm/mole H_2S the RP suffer a high decrease which is relative smaller for 10 and 26ppm/mole samples. Based on the experimental results the corrosion rate increase. The increase in corrosion rate by introducing H_2S in contact with P256GH can collaborate with the increase in proton reduction, equations (2)–(4) and automatically the anodic dissolution. In the same time the elements based on sulfide ions can have an important role as surface catalysts and to increase the anodic dissolution presented in both in LPR and OCP results. In figure 4 are presented in a) linear polarization curves of P265GH steel in Na_2SO_4 + Na_2S, pH 3, at various H_2S concentrations (ppm by mole), T ≈ 25°C and b) cyclic polarization curve of P265GH steel in Na_2SO_4 + Na_2S, pH 3, at various H_2S concentrations (ppm/ mole), T ≈ 25°C. Linear polarization, figure 4 a), present the increasing character of the corrosion rate with the percentage of H_2S with near values for 0 and 10 ppm/mole samples and bigger values of corrosion for 20 and 30 ppm/mole samples.

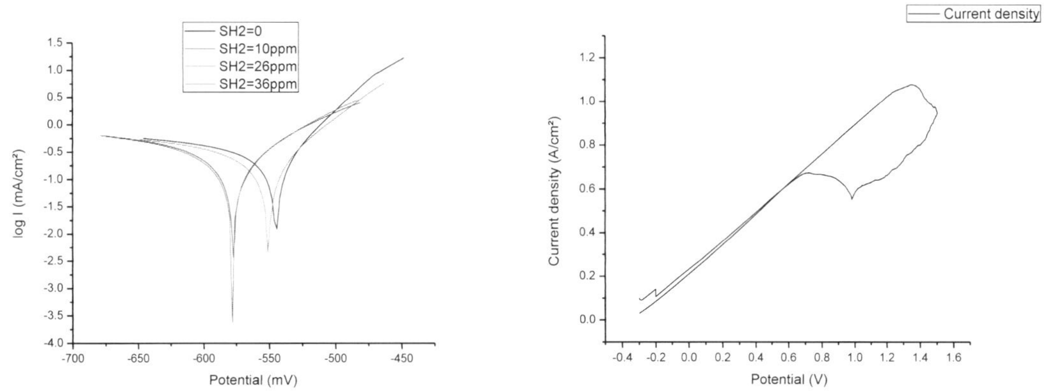

a) b)

Figure 4. a) Linear polarization curves of P265GH steel in Na_2SO_4 + Na_2S, pH 3, at various H2S concentrations (ppm by mole), T ≈ 25°C b) Cyclic polarization curve of P265GH steel in Na2SO4 + Na_2S, pH 3, at various H_2S concentrations (ppm/ mole), T ≈ 25°C.

The cyclic curve present definitely a pitting corrosion type with a trend to generalization of the pitting on the entire material surface.

In figure 5 are presented in (a) Nyquist and (b) Bode plots of P265GH in Na2SO4 + Na2S, pH 3, at various H_2S concentrations (ppm by mole), T ≈ 25°C. In all conditions, the Nyquist curves indicate a single capacitance loop having a time constant in Bode plots. Therefore, the alloy-electrolyte interface can be simulated by a Randles equivalent circuit [7-17].

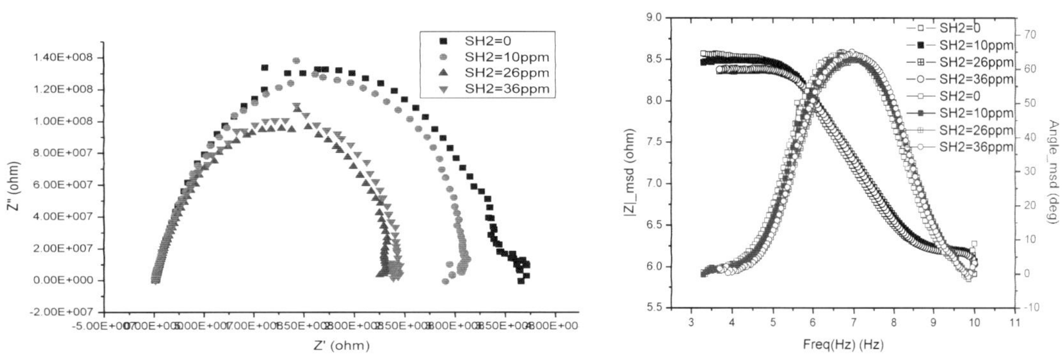

a) b)

Figure 5. (a) Nyquist and (b) Bode plots of P265GH in Na_2SO_4 + Na_2S, pH 3, at various H_2S concentrations (ppm by mole), T ≈ 25°C.

During the increase in the H_2S concentration, RP of the experimental alloy reduce more or less with the same trend. The participation of H_2S didn't induce any other time constants on the graphs presented in figure 5, in the same time a decrease in the capacitance loop diameter and charge transfer

IOP Conf. Series: Materials Science and Engineering **209** (2017) 012050 doi:10.1088/1757-899X/209/1/012050

resistance can be observed by an increase in the H_2S concentration. The EIS results are proper to be used to identify the corrosion and electro-corrosion mechanisms by the presence of H_2S.

The reactions between H_2S and P256GH has an determining role on the initiation and propagation of cracking and failure. It can be observed from figure 6 a) and b) the formation of corrosion pits on the surface of material and the tendency to connect and form a crack. In conditions of a higher pressure (for example in transport of oil and gas products) these cracks can produce the material failure. As cathodic reaction can be considered the H_2 evolution reaction.

Hydrogen could appear as a result of the reduction reaction of H_2S on the steel surface. The produced hydrogen atom can penetrate into the metallic lattice and diminish the mechanical properties of carbon steels, causing Hydrogen Induced Cracking (hydrogen embrittlement). The corrosion pits, after an entire corrosion cycle, describe in experimental part, have diameters between 1-3 μm.

a) b)

Figure 6. SEM images of the P265GH surface after the electro-chemical tests a) 250x and b) 5000x.

Analyzing the electrochemical corrosion reactions between H_2S and Fe will improve the understanding of the initiation and propagation of cracking. It is important in this applications field where the corrosion mechanism act in time or long time like decades to analyze if we can correlate a higher corrosion rate of metallic materials tends to provoke a higher susceptibility of C-steel to cracking.

In table 1 is presented the chemical composition of the metallic material surface after the electro-chemical test made on a 4 mm^2 area. Comparing to the C-steel used for tests we observe a decrease of iron and manganese chemical elements mass percentages which probably pass to oxides and detach from the material. On the surface remain oxides which are still attached to the surface, indicated by the presence of oxygen with 1.47 wt%. In the same time the lost percentages of iron and manganese are connected also with the presence of sulfur on the surface. The compound CO_2 was adsorbed on the steel surface prior to H_2S and the H_2S adsorption reduced.

Based on the chemical composition result, table 1, and the literature [18] the compound CO_2 corrosion was the main process at the beginning of corrosion process and the main product is probably $FeCO_3$. Based on cyclic curve and EIS experiments the structure form at the beginning, probably $FeCO_3$ and precipitated on the metal surface was not compact at the start of the corrosion. In this case the specie Fe^{2+} has the ability to penetrate through the scales by diffusion.

After that Fe^{2+} reached the surface film / liquid interface, the compound iron sulfide form when combining with the S^{2-} anion and then precipitated on $FeCO_3$.

Table 1. Chemical composition after the experimental test.

Element	wt.[%]	at.[%]	Error in %
Iron	96.43	88.98	0.95
Oxygen	1.47	4.72	0.41
Carbon	1.27	5.45	0.35
Manganese	0.75	0.69	0.16
Sulfur	0.1	0.16	0.04

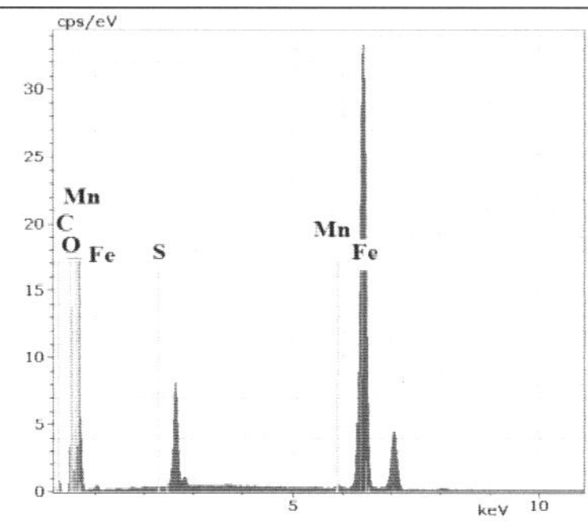

4. Conclusions

OCP, LPR and EIS results showed almost similar trends in corrosion behavior of the material. By an increase in H_2S concentration from 0, 10, 20 and then 30 ppm, the corrosion resistance of the material decrease, substantially for sample in 30 ppm/mole solution. The compound H_2S have the effect to retard the formation of passive layer from the substrate. Linear potentiometry confirm the increase of corrosion rate with the increase of H_2S percentage. Cyclic potentiometry present a clearly pitting corrosion of the surface at the beginning and with a general character at the end. The pitting corrosion is confirmed by the SEM images.

References

[1] Holditch S A and Chianelli R R 2008 *MRS Bull.* **33** 317
[2] Liew Y M, Kamarudin H, Mustafa Al Bakri A M, Luqman M, Khairul Nizar I, Heah C Y 2011 *Australian Journal of Basic and Applied Sciences* **5** (9) 441-449
[3] Perju M C, Vizureanu P 2014 *REV. CHIM. (Bucharest)* 65(6) 694-696
[4] Ma H, Cheng X, Li G, Chen S, Quan Z, Zhao S, Niu L 2000 *Corros. Sci.* **42**, 1669
[5] Betova I, Bojinov M, Hyökyvirta O, Saario T 2010 *Corros. Sci.* **52** 1499
[6] Lee K L J, Nesic S 2005 *NACE proceedings* 05630
[7] Izquierdo J, Bolat G, Cimpoesu N, Trinca L C, Mareci D, Souto R M 2016 *Applied Surface Science* **385** 368
[8] Chelariu R, Mareci D, Bolat G, Peptu C A, Cailean D 2015 *Materials And Corrosion-Werkstoffe Und Korrosion* **66** 573
[9] Cimpoeşu N, Trincă L C, Dascălu G, Stanciu S, Gurlui S O, Mareci D 2016 *Journal of Chemistry* Article ID 9520972
[10] Chelariu R, Bolat G, Izquierdo J, Mareci D, Gordin D M, Gloriant T, Souto R M 2014 *Electrochimica Acta* **137** 280
[11] Bolat G, Mareci D, Iacoban S, Cimpoeşu N, Munteanu C 2013 *Journal of Spectroscopy* ID 714920
[12] Zhenguang L, Xiuhua G, Linxiu D, Jianping L, Ping L, Chi Y, Misra R D K, Yuxin W 2017 *Electrochimica Acta* **232** 528
[13] Wang P, Wang J, Zheng S, Qi Y, Xiong M, Zheng Y 2015 *International Journal of Hydrogen Energy* **40** 11925
[14] Davoodi A et al. 2011 *Corrosion Science* **53** 399
[15] Zepon G et al., 2015 *Materials & Design* **83** 214
[16] Ziomek-Moroz M, Hawk J A, Thodla R, Gui F 2013 *ECS Trans.* **50**, 79
[17] Holditch S A and Chianelli R R 2008 *MRS Bull.* **33** 317
[18] Wu H B, Liu L, Wang L, Liu Y 2014 *Journal of iron and steel research international* **21** 76

CMT Welding of Low Carbon Steel Thin Sheets

E M Stanciu [1], A Pascu [1] and I Gheorghiu [2]
[1] Department of Materials Engineering and Welding, Transilvania University of Brasov, 29 Eroilor st. 500036, Brasov, Romania
[2]Product Design, Mechatronics and Environment Department, Transilvania University of Brasov, 29 Eroilor st. 500036, Brasov, Romania

E-mail: elena-manuela.stanciu@unitbv.ro

Abstract. This paper addresses to the cold metal transfer MAG welding of low carbon steel thin sheets. The paper highlights the advantages of using CMT process for but joining of S235 carbon sheets by comparing the CMT with the conventional synergic pulse MAG welding. A lower weld bead area and heat affected zone is obtained by the continuous movement of the wire, digitally synchronised with the short-circuit of the arc. CMT welding can produce good welds even in unfavorable conditions, such as using thin plates and high diameter wire.

1. Introduction

Nowadays, after almost 80 years of industrial implementing, GMAW welding remains the most used welding technology up to date. Lighter materials are used to reduce the manufacturing costs and the overall weight of the welded structures (eg. automotive industry). Although numerous advantages can emerge from using thinner and lighter materials, welding of these structures is challenging, considering that a thinner material will be quickly overheated, thus resulting in major drawbacks residing in the welded strength and on the dimensional characteristics of the assembly. Laser technology is the best available technology for welding of similar and dissimilar materials [1, 2], especially as thin sheets, but the equipment is expensive and special preparation of materials are required (no gaps).

Conventional MIG welding, standard, synergetic or pulsed synergic, is used for joining thin sheets of low carbon steel in common industrial applications. An improved method for the MIG/MAG welding is obtained by adding a continuous movement of the wire synchronised with the arc short circuit. The process control commands the retraction of the wire (up to 90 time/sec) and allows a clear and precise detaching of the melted droplet [3,4]. Using this process control, Chen et al. [5] investigates the process stability and the deposition rate in the case of CMT welding of 3 mm mild steel workpiece. He reveals that welding frequency chance very little with welding parameters and has values between 80.1 and 90.1 Hz. The cold metal transfer welding – CMT is frequently used for welding of low carbon thin sheets, aluminium, or dissimilar joints [6, 7, 8]. Pickin et al. [9] have investigated the CMT cladding of thin aluminium alloy and determine that CMT process exhibits better control of the dilution and lower cracking susceptibility when welding with Al-2319 filler. In a complex study, R. Cao et al [10], successfully welded pure titanium with copper using ERCuNiAl copper wire as filler material. By using the CMT, a satisfying dual welding-brazing joint between Ti and Cu with sufficient strength and good mechanical behaviour could be obtained. In the recent years, the CMT process became increasingly attractive in research concerning welding and brazing of materials with low melting point or of dissimilar materials with high differences of physical and mechanical proprieties.

The objective of this study is to investigate the capabilities and advantages of the cold metal transfer process for welding of thin plates of carbon steel. The welds are analyzed from the geometrical and microstructural point of view and the mechanical behavior is also discussed.

2. Materials and methods

S235JR low carbon steel was used as base material for the experimental tests were with the nominal dimension of 100x100x1.2 mm and A5.18 wire of 1.2 mm diameter as filler material. S235JR steel possess good cold forming capabilities and is the most used structural steel for the construction industry. The 1.2 mm thick wire was chosen in order to highlight the advantages and versatility of the motion control system of the filler material.

The nominal composition of both materials is presented in table1.

Table1. Chemical composition of the base and filler material.

Specification	*C* %	*Mn* %	*P* %	*S* %	*Si* %	*Cu*	*Cr*
S235JR	0,17	1,28	0,03	0,04	0,04	0,45	-
AWS A5.18/A5.18M - ER70S-6	0,07	1,7	0,016	0,011	0,85	-	0,013

The test specimens were clamped in a butt weld geometry using a special designed system. The specimens were positioned with a 0.4 mm gap. The experimental set-up is illustrated in figure 1.

The welding experiments were carried out using a CMT MIG-MAG power source made by FRONIUS (*TRANS PULSE SYNERGIC 5000*) equipped with a CMT PullMig welding torch manipulated by a Fronius programmable moving control unit as illustrated in figure 1.

The test plates were mechanically cleaned by steel wire brush, sand paper and degreased with isopropyl alcohol prior to welding.

Figure 1. Experimental set-up used for the cold metal transfer MAG welding of low carbon steel thin sheets.

All the welding tests were performed using CORGON (82% argon and 18% carbon dioxide) as shielding gas, with a constant flow rate of 15 L/min. The tests were conducted in synergic mode and therefore the parameters interact with each other. From this reason, the tests were made by keeping a fixed value of welding parameters for pulse mode and respectively for CMT. The influence of the welding speed on the weld bead profile and the mechanical behavior of the joints has been assessed by gradually modifying the welding speed.

As summarized in table 1, the welding current was 60 A for conventional pulsed mode and 85 A for the CMT with very low variation during the welding process (+-1 %, indicated by the digital display of the power source).

Table 2. MAG welding process parameters and response factors.

Sample		Current [A]	Tension [V]	Welding speed [mm/min]	Weld area [mm^2]	Weld bead width [mm]	HAZ [mm]
Pulse Synergic	1	60	18.3	600	6.92	4.5	1.55
	2	60	18.3	800	4.65	3.44	1.03
	3	60	18.3	1000	2.92	3.34	0.47
CMT	4	85	10.8	600	3.24	3.34	0.47
	5	85	10.8	800	2.85	3.00	0.77
	6	85	10.8	1000	1.50	2.20	-

3. Results and discussions

After the welding process the samples were prepared by extracting specimens for microstructure investigation and tensile testing. The cross-section specimen has been grinded, polished, and chemically etched in Nital reagent (3% HNO_3in ethanol). A Leica DM ILM inversed microscope was used for the macro and microstructural analyses of the welds.

In the figure 2 is illustrated the upper surface of the joints and the heat affected zone associated with sample 2 and sample 5, different magnitude of the HAZ being clearly visible.

CMT

Conventional pulsed welding

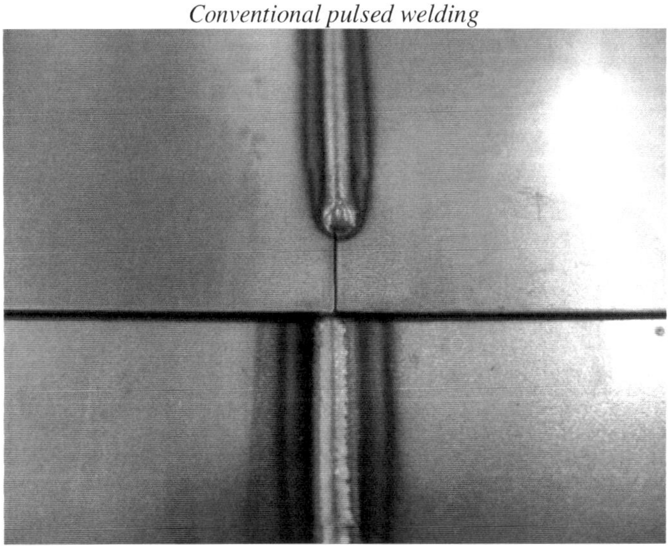

Figure 2. Upper side of the sample welded by synergic mode - pulse and CMT at 800 mm/min welding speed.

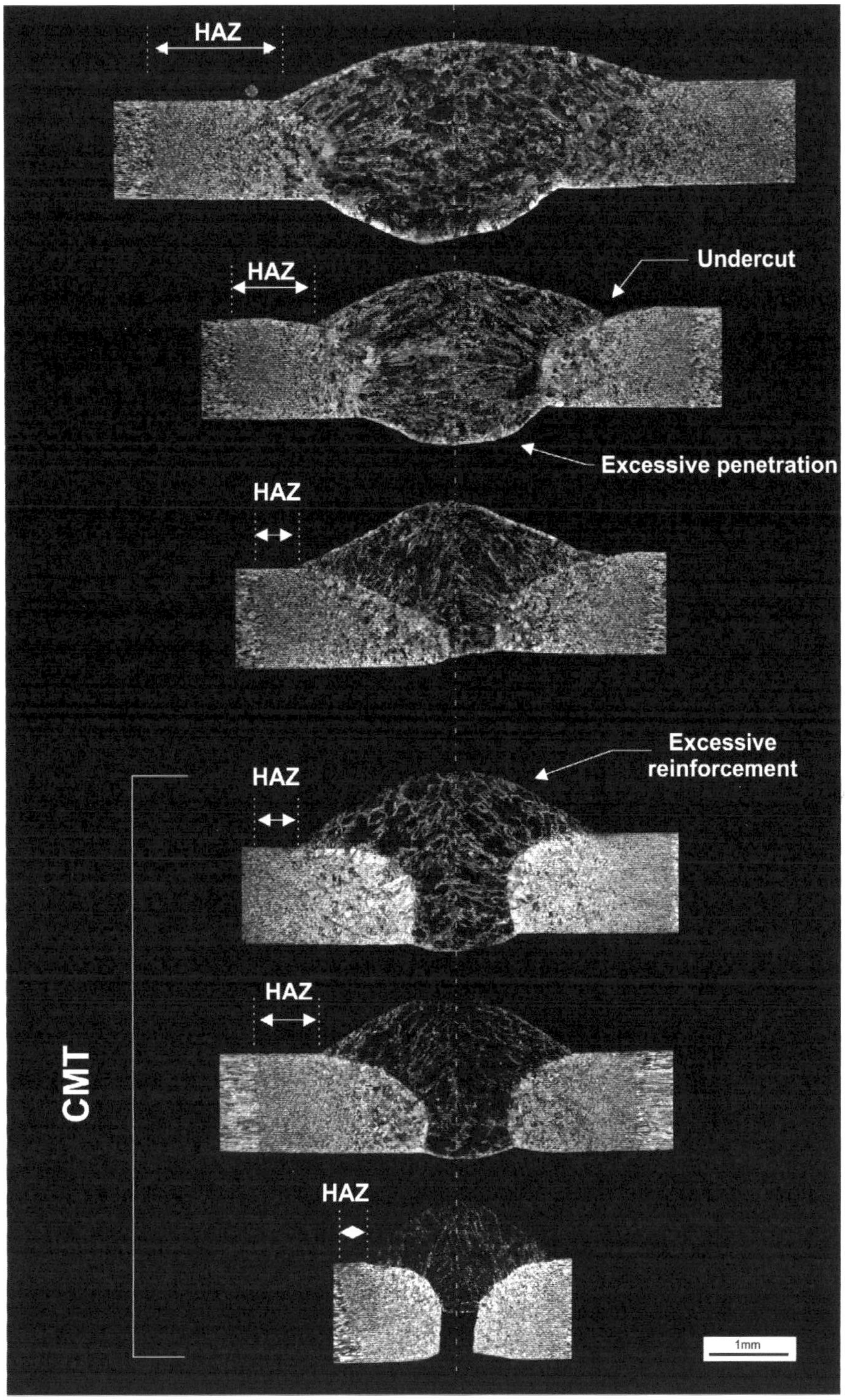

Figure 3. Cross-section profile of the samples 1 to 6 (A stitching picture software was used for composing the images).

Figure 3 shows the weld bead profile of samples welded by conventional synergic pulse (samples 1 to 3) and by cold metal transfer welding samples 4 to 6. In both welding mode, the current was kept as low as possible but enough to achieve welded joints at speeds between 600 and 1000 mm/min.

Excessive penetration, improper contour or undercut characterize the samples realized by synergic pulse and excess reinforcement or lack of penetration marks the samples realized by CMT. Although the defects are present, the results show that welding speed has a direct influence on the weld bead profile and on the heat affected zone and that CMT set-up can produce a better weld bead geometry in case of thin plates assembly. By synchronizing the wire feed motion with the short circuit of the arc (CMT) a lower amount of temperature in the welding process is induced, being obvious that a lower amount of filler material is used for construction of the welded joint.

An increasing of weld bead high is produced due to the lower amount of heat induced in the process that can allow a lower dilution and therefore a better formation of the reinforcement. The weld bead area is almost half in case of using the CMT as can be observed from figure 4.

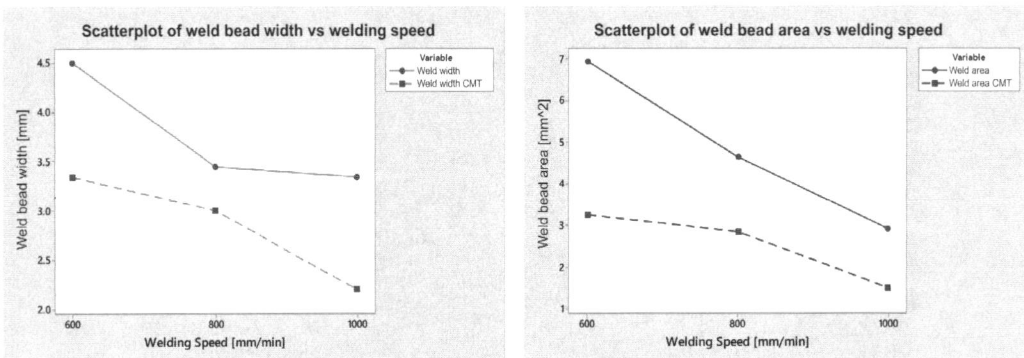

Figure 4. Influence of the welding speed on the weld bead area and width.

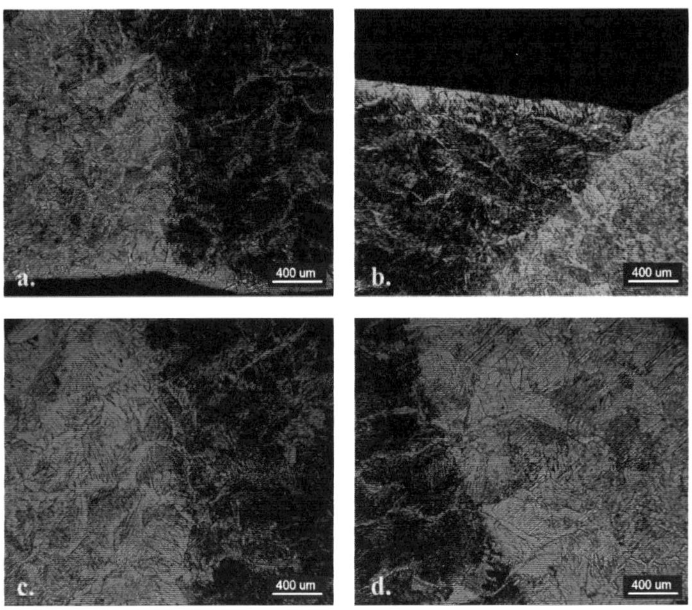

Figure 5. Microstructure of the sample 5.

The base material is characterized by a ferito-pearlitic microstructure with specific elongated grains formed due to the material rolling. The morphology of the base material changes near the welding bead to an oversized grain structures formed from pearlite and acicular ferrite (figure 5). The same microstructure characterizes the weld bead and the fusion zone.

A partial grain refinement occurs at the interface of the weld bead with the base material and continues with a grain refinement with small grains of ferrite and pearlite formed due to the recrystallization process as presented in figure 6.

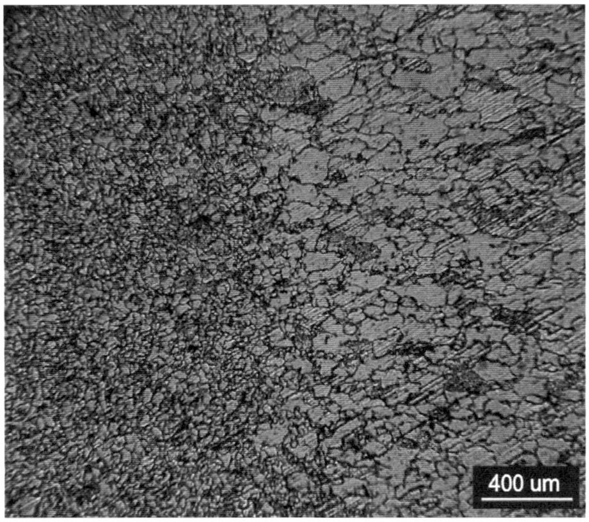

Figure 6. Microstructural change at the limit of the heat affected zone associated with sample 5.

Table 3. Results of the tensile tests of sample 5.

No.	W to Fmax Nmm	W to break Nmm	ReH M/mm^2	Rm N/mm^2	Break	ε-F max %	ε-Break %
1	33651,87	48907,07	-	492,50	1,00	31,07	41,84

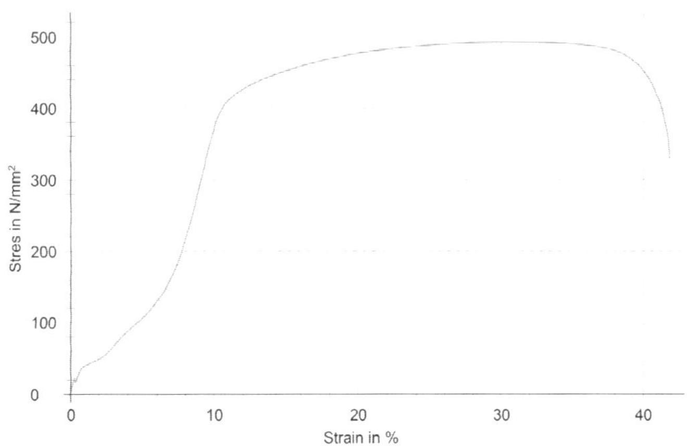

Figure 7. Tensile strength of sample 5 welded with CMT process.

Except the sample 6, all the welded samples present a good mechanical behavior as failure during the tensile test occurs far away from the weld bead or the heat affected zone. In the figure 7 is presented the strength – elongation curve acquired in the case of sample 5.

4. Conclusions

In this study, the CMT welding was used for welding of S235JR low carbon steel in butt geometry with a 0.4 mm gap. Compared with conventional synergic pulse welding the CMT technology can produce better results in term of heat affected zone and weld bead profile.

It was determined that an increased reinforcement of welds is favorized by the cold metal transfer process and must be controlled by reducing the wire speed. Increasing the welding speed will not reduce the reinforcement but will provide an improper penetration. Although the differences of weld bead geometry are major, the mechanical behavior of the analyzed specimens are almost similar. All the welding joints presents a superior tensile strength compared with the base material.

References

[1] Alhamed M and Abdullah W 2013 *Journal of Electron Devices* **18** 1563-1567
[2] Ohya Y, Ito S, Ban T, and Takahashi Y 2000 *Key Engineering Materials* **181** 113-116
[3] Feng J, Zhang H, He P 2009 *Materials & Design* **30** 1850–1852
[4] Roată I C, Pascu A, Stanciu E M, Pop M A 2014 *Advanced Materials Research* **1029** pp. 140-145
[5] Chen M, Zhang D, Wu C 2017 *Journal of Materials Processing Technology* **243** 395–404
[6] Shu F Y, Tian Z, Lü Y H, He W X, Lüjian-Jun Lin F Y, Zhao H Y, Xu B S 2015 *Trans. Nonferrous Met. Soc. China* **25** 2701–2707
[7] Kadoi K, Murakami A, Shinozaki K, Yamamoto M, Matsumura H 2016 *Materials Science and Engineering: A* **666** 11–18
[8] Gungor B, Kaluc E, Taban E, Sik A 2014 *Materials & Design (1980-2015)* **54** 207–211
[9] Pickin C G, Williams S W, Lunt M 2011 *Journal of Materials Processing Technology* **211** (3) 496–502
[10] Cao R, Feng Z, Lin Q, Chen J H 2014 *Materials & Design (1980-2015)* **56** 165–173

Acknowledgments

This work was supported by a grant of the Romanian National Authority for Scientific Research and Innovation, CNCS/CCCDI – UEFISCDI, project number PN-III-P2-2.1-BG-2016-0349, within PNCDI III.

IOP Conf. Series: Materials Science and Engineering **209** (2017) 012052 doi:10.1088/1757-899X/209/1/012052

Ionic Liquid Surface Treatment of Calcite for Improved Compatibility with Polyolefin Matrix

C Croitoru* I C Roata, A Pascu and E M Stanciu
Materials Engineering and Welding Department, Transilvania University of Brasov, Eroilor 29 Str., 500039, Brasov, Romania

E-mail: c.croitoru@unitbv.ro

Abstract. Preliminary surface treatment of crystalline marble powder filler with a dominatingly hydrophobic ionic liquid, namely triethylsulfonium-bis(trifluoromethylsulfonyl) imide has been employed to design a filler with improved compatibility towards non-polar polymer matrices, such as polyolefins. The maximum uptake of ionic liquid per 1g of marble powder was 132 mg. The surface coating of the crystalline phases has been proved by XRD and FTIR spectroscopy, making our method useful in designing fillers with improved dispersibility/ compatibility in/with polyolefins. The floating test method has proved the efficiency of the alkylsulfonium ionic liquid in hydrophobization of the crystalline calcite-rich phase, thus proving useful in obtaining composites with improved properties.

1. Introduction

Calcium carbonate is the most widely used filler for particulate-reinforced polyolefin composites. Its low price, coupled with good thermal stability, workability, and neutral color [1-3]. It has been reported to efficiently increase the impact resistance, fire-resistance and water stability in various combinations with high density polyethylene (HDPE) and polypropylene (PP) matrices [3].

However, due to its high surface polarity and hydrophilicity it is mainly incompatible with hydrophobic polymer matrices such as polyolefins [4].

To improve its dispersibility in polymer media, calcium carbonate is often surface treated with a variety of modifiers/dispersion agents such as silanes, phosphates, titanates, fatty acids (mainly stearic acid), etc [5,6]. Because of this treatment, a layer of hydrophobic organic molecules is attached to the mineral surface, via physical bonding. The structure and properties of such layers, being the interface between the inorganic filler and the polymer matrix bears a strong influence on the final properties of the composites [5].

Among the most frequently used calcite coating techniques at laboratory scale, include dry coating (through mixing with the additive at different temperatures, dependent on its nature) and the solution method. In the solution method, the calcite surface is usually treated with the solution of the acid in a nonpolar solvent (n-alkanes, toluene) [7].

Ionic liquids (ILs) have been scarcely used as dispersing agents for inorganic fillers so far. Their use as surface hydrophily-enhancing agents could impart also some useful properties to the composites, such as antimicrobial character, improved UV-resistance, fire retardancy and so forth [8,9].

In this study, crystalline marble powder has been surface treated with triethylsulfonium-bis(trifluoromethylsulfonyl) imide by treatment of the particles with different amounts of ionic liquid dissolved in ethanol. The structure and properties of the obtained products were determined by SEM microscopy, XRD and FTIR spectroscopy. The effect of surface modification has been evaluated by the floating test, confirming the efficiency of the studied ionic liquid use as surface functionalizing agent.

2. Materials and methods

2.1. Materials

Marble dust powder (sieved, average particle diameter < 0.01mm) has been provided by TITAN MAR Co., Chiajna, Romania.

The ionic liquid used in this study (figure 1), namely triethylsulfonium-bis(trifluoromethylsulfonyl) imide (TesIm) has been purchased from IoLiTec-Ionic Liquids Technologies GmbH, Germany (99.5% purity) and was used without further purification.

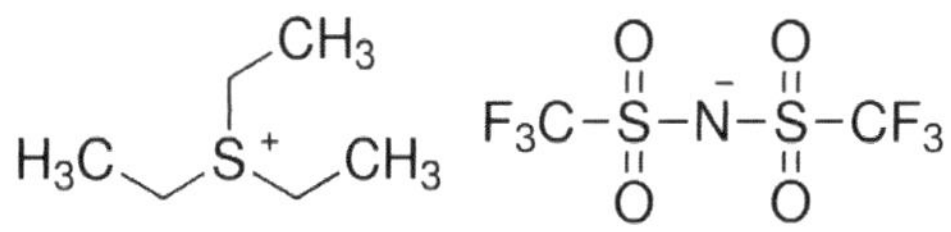

Figure 1. Chemical structure of used ionic liquid.

2.2. Methods

2.2.1. Marble powder IL surface treatment. Due to the high viscosity of the pure ionic liquid, to obtain a uniform coating of the marble powder with the ionic liquid, three solutions with concentrations of 30%, 50% and 50% vol. TesIm in 98% ethanol have been prepared. A determined amount of marble powder (10g) have been immersed in 50 mL of the TesIm ethanol solutions and magnetically stirred for determined time periods at 25°C. For the IL uptake kinetic determination, aliquots of the solutions have been extracted at specified time intervals and analysed by UV spectrophotometry at 235 nm, based on the absorption maximum of the IL in ethanol (determined previously).

2.2.2. Coated marble powder characterization. For structural characterization, the marble powder has been removed from the ethanol ionic liquid solution and dried in vacuum (300 mbar, 50°C). To determine the modification in crystalline phases that occur in marble after ionic liquid treatment, XRD spectra have been obtained, with a Bruker Discover diffractometer ($K_{\alpha 1}$ =0.154 nm, scan speed 5 degrees/min). To characterize the surface of the calcite particles, FTIR spectra have been acquired in the 4000-600 cm^{-1} interval (Bruker VERTEX 70 spectrophotometer working in reflectance mode). A QUANTA 200 scanning electron microscope was used to study the morphological appearance of the IL-treated marble powder and reference.

The effect of surface modification has been evaluated by the floating test, initially reported by Sheng et al. [10], which represents the ratio of the free-floating functionalized powder to the overall weight of the powder after being magnetically stirred in water at 1000 rpm and room temperature (the active ratio). To assess the modification in surface energy of the coated marble powder, surface energy measurements have been performed. The powder has been dried at 105°C for 5 hours then it was pelleted by means of a hydraulic press at 100 atm for 10 s. The surface energy of the pellet (related to the sample polarity) has been determined through contact angle measurements. As a general methodology, three drops (5 µL each) of distilled water, respectively octanol, with different surface tensions have been placed on the sample's surface, and the contact angles have been assessed by ImageJ software (contact angle plugin). With the help of SCA software (DataPhysics GmbH), the surface energy has been assessed, based on the Owens, Wendt, Rabel and Kaelble (OWRK) method.

Untreated marble powder and neat PP and HDPE powders have been pelleted and analysed in a similar manner.

3. Results and discussion

The ionic liquid uptake kinetic of the calcite powder is presented in figure 2. It can be observed that equilibrium uptake of IL is reached in approx. 150 min of contact with the IL ethanol solution.

A pseudo-equilibrium zone could be registered at the 40-80 min interval, which could be due to the concurrent elimination of several crystalline phases from the inorganic material because of the magnetic stirring process, as well as of different hydrophobic IL monolayers stacking. The active ratio of the functionalized powder is 32.45% for the marble immersed in the 30% vol. TesIm ethanol solution; 57.23% for the marble immersed in the 50% vol. TesIm ethanol solution and 65.23% for the 70% vol. solution, respectively.

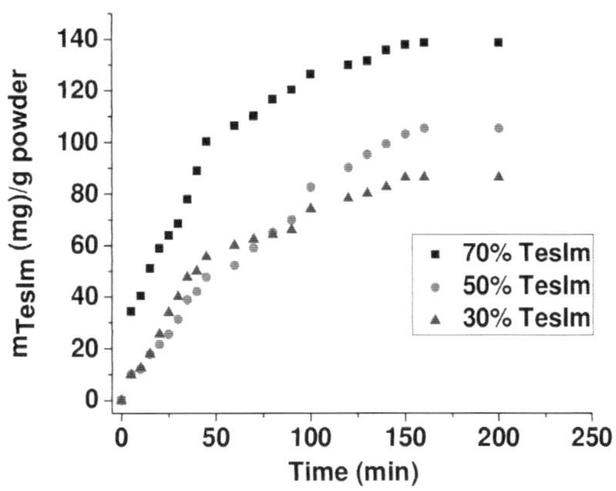

Figure 2. TesIm IL sorption kinetic onto the marble powder.

The maximum loading amount of ionic liquid is 132 mg/g of marble powder, registered for the solution with the highest IL content (70% vol.).

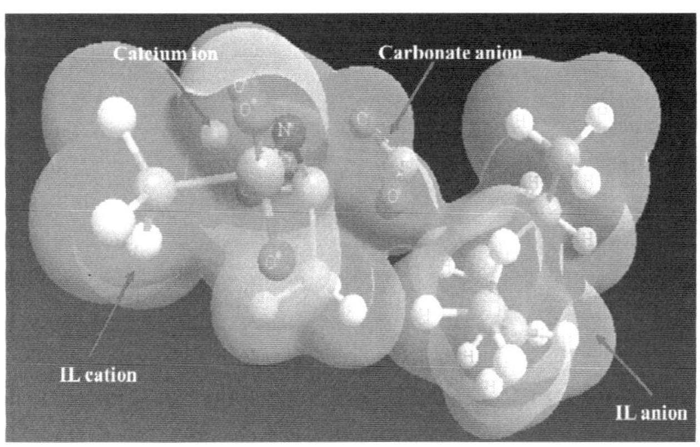

Figure 3. Calcium carbonate interactions with the TesIm IL.

It is expected that the calcium and carbonate ions in the marble filler interact with the anion/cation of the ionic liquid. Molecular dynamics simulations (MMF94 interface, 5000 iterations, minimum

RMS gradient of 0.1) have indicated that calcium carbonate molecule is placed within the "solvation surface" of the ionic liquid. The Ca^{2+} anion surface charge is efficiently "masked" by the $-CF_3$ groups of the bis(trifluoromethylsulfonyl) imide anion (effective charge of +0.041) while CO_3^{2-} anion is attracted by charge-dipole interactions by the S^+ atom in the ionic liquid cation (effective charge of the two O^- ions being -0.64 and -0.71).

The highest IL uptake speed has been registered for the solution with the highest concentration gradient. The irregularities present on the powder surface (figure 4) favour the sorption of the IL, which probably is oriented with the polar end towards the calcite phase and with the hydrophobic alkyl chains towards the exterior, possibly favouring the interaction with dominatingly hydrophobic polymer matrices, such as polyolefins.

Figure 4. Initial marble powder (left) and TesIm-treated marble powder (right).

Ionic liquid treatment of the marble powder seems to increase the surface roughness of the particles, probably due to the leaching of several weakly bonded crystalline phases, aided by the magnetic stirring process.

The marble dust powder presents a dominatingly crystalline structure, as determined from the XRD diffractogram (figure 5a). The as-received marble powder comprises mainly of trigonal calcite, rhombohedral dolomite, and several aluminosilicate minerals, such as aegirine, hypersthene, diopside, alongside quartz.

Ionic liquid adsorption on the surface of the marble particles determines a decrease in the overall intensity of the spectrum, probably due to the coating of the crystalline phases with the dominatingly amorphous ionic liquid.

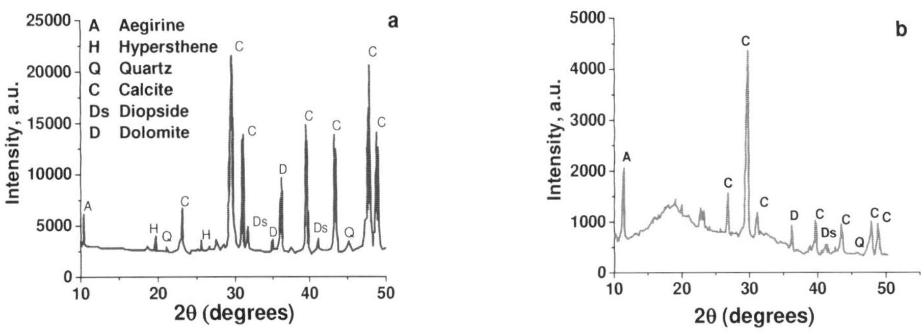

Figure 5. XRD diffractograms of: (a): neat marble powder; (b): TesIm-treated marble powder (70% vol. IL solution).

This hypothesis could be confirmed by the presence of the amorphous peak, centered at ~19.7° (figure 5b). Also, several peaks, ascribed to hypersthene, diopside and dolomite disappear from the diffractogram, being leached into the IL solution. As the calcite phase has the higher hardness among the minerals present in marble, leaching of compounds with lower mechanical resistance could prove useful for the hardness and impact resistance of particulate-reinforced polyolefin composites.

The surface sorption of ionic liquid is also confirmed by the FTIR spectra presented in figure 6.

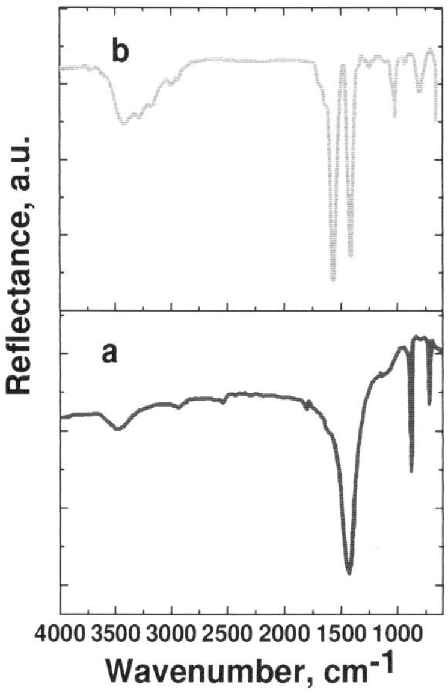

Figure 6. FTIR spectra of (a): reference marble powder and (b): IL-treated marble powder.

The FTIR spectra of neat marble powder from Fig. 5a shows the presence of weakly bonded water (-OH group, band centered at ~ 3520 cm^{-1} and 1620 cm^{-1}). Several intense bands are present, belonging to C-O bending vibration (~1470 cm^{-1}) and C-O rotation (~870 cm^{-1}) in the carbonate anion [11].

Ionic liquid-treated marble powder presents the spectral features of neat IL, namely, the presence of S-O and C-F vibrations (3000-4000 cm^{-1} domain), as well as of alkyl stretching vibration band (~2920 cm^{-1}) and the band associated with the C-F vibration in the trifluoromethanesulfonate anion (~ 1230 cm^{-1}) [12].

The values presented in table 1 for the IL-treated marble pellet indicate a decrease with up to 54% in the values of the surface energy comparing with the untreated marble powder indicating an optimum functionalization.

Table 1. Surface energy values of the analysed samples.

Sample	γ (mN/mm)	Contact angle (°)
Marble	75.4	23.5
Functionalised marble	34.1	68.7
PP	30.4	56.1
HDPE	29.9	50.9

Through functionalization it could be seen that the values for the surface energy of the functionalized marble resemble those of PP and HDPE, indicating a good potential compatibilization between the two phases, beneficial to the obtaining of composites, either through injection or thermoforming. Even if the surface energy is also affected by roughness and porosity (HDPE, PP and marble powder having rather different compressibility and particle dimensions), the determinations have clearly indicated a hydrophobization of the highly polar marble powder.

4. Conclusion

Crystalline calcite-rich marble powder has been surface treated with ethanol solutions of different concentrations comprising the triethylsulfonium-bis(trifluoromethylsulfonyl) imide ionic liquid. An efficient hydrophobization of the marble powder has been achieved, determined by an active ratio of over 60%, for the most concentrated ionic liquid solution.

The XRD and the FTIR spectra confirm the presence of the ionic liquid on the surface of the inorganic phase, proving the potential usefulness of our method in designing particle-reinforced composites with polyolefin matrices.

References

[1] Rungruang P, Grady B P, Supaphol P 2006 *Colloids Surf. A.* **275** 114
[2] Keler D S, Luner P 2000 *Colloids Surf. A.* **161** 401
[3] Croitoru C, Giubega A, Patachia S, Baltes L, Pascu A, Roata I, Tierean M 2016 *Bulletin of the Transilvania University of Braşov* **9** 58
[4] Osman M A, Suter U W 2002 *Chem. Mater.* **14** 4408
[5] Baoqing S, Jipeng L, Zhixu W 2006 *Chinese J. Chem. Eng.* **14** 814
[6] Bellayer S, Tavard E, Duquesne S, Piechaczyk A, Bourbigot S 2009 *Polym Degrad Stabil* **94** 797
[7] Rezaei Gomari K A, Hamouda A A 2006 *J. Pet. Sci. Eng.* **50** 140
[8] Shamsuri A A, Daik R, Zainudin E S, Tahir P M. 2014 *J Reinf Plastic Comp* **33** 440
[9] Livi J, Duchet-Rumeau P., Thi Nhan P., Gerard J F 2011 *J. Colloid Interface Sci.* **354** 797
[10] Sheng Y, Zhou B, Zhao J, Tao N, Yu K, Tian Y, Wang Z 2004 *J. Colloid Interface Sci.* **272** 326
[11] Bosch Reig F, Gimeno Adelantado J V, Moya Moreno M C M 2002 *Talanta* **58** 4
[12] Vitucci F M, Trequattrini F, Palumbo O, Brubach J B, Roy P, Paolone A 2014 *Vibr Spectrosc* **74** 81

Acknowledgements
This work was supported by the Romanian National Authority for Scientific Research and Innovation, CNCS-UEFISCDI, project number PN-II-RU-TE-2014-4-0173.

IOP Conf. Series: Materials Science and Engineering **209** (2017) 012053 doi:10.1088/1757-899X/209/1/012053

Structural Modifications of Superficial Layer of C45 Steel Samples Through WT20 and WZr8 Depositions

D C Achiţei [1,3], M M A B Abdullah[1,3], A V Sandu [1,3], A C Ţugui [1] and M Benchea[2]

[1]Faculty of Materials Science and Engineering, „Gheorghe Asachi" Technical University from Iaşi, Blvd. D. Mangeron 41, Iaşi, 700050, România

[2]Faculty of Mechanical Engineering, „Gheorghe Asachi" Technical University from Iaşi, Blvd. D. Mangeron 41, Iaşi, 700050, România

[3]Center of Excellence Geopolymer & Green Technology, Universiti Malaysia Perlis

E-mail: dragos_adc@yahoo.com

Abstract: The paper presents technological aspects regarding the modification of mechanical chracteristics in the superficial layer of C45 steel samples, through thin layers deposition using WT20 and WZr8 electrodes. Deposition of thin layers was made through electrical discharge method in impulse. The obtained samples were microstructural analyzed, at various magnitudes, on an VegaTescan electronic microscope. Also, measurements of mechanical chracteristics were made through indentation, highlighting the improved values after layers deposition with the 2 electrods.

1. Introduction

The steel presents the posibility of allloying with the most various chemical elements, resulting highly improved properties [1,2].

The carbon steels contain beside iron and carbon, other alloying elements. Other chemical elements are considered impurities. The exception from this rule are some elements, which are imposed by the technological process of elaboration and casting (manganese, silicon, aluminium).

The regular steel carbon is manufactured curently and used without heat treatments at civil constructions, metallic constructions and machines facturing.

The quality carbon steel is used in machine manufacturing, usually heat treated and has guaranteed chemical composition and mechanical characteristics.

The superior carbon steel is a quality steel in which the content of sulfur and phosphorus is limited at below 0.035% each and conditions are imposed regarding the structure (size of austenitic grain, depth of hardening) and the maximum content in non-metallic inclusions.

Low alloyed steels are steels which contain alloying elements willingly introduced in mininum quantities which influence the physical-chemical properties and mechanical characteristics.

The alloyed steels are the medium and high alloyed and are used only heat treated. The alloyed steels have the sum of alloying elements concentrations higher than 10% or one of this concentration higher than the following limits: Si 6%; Mn 6%; Cr 6%; Ni 4.5%; Mo 1%; W 4%; Co 1%; V 1%.

The steels are delivered to the beneficiaries in the form of cast parts or laminated: blooms, slabs, wires, billets, sinkers, round steel, square, hexagonal, flat, profiles, thin sheets, plats, strips [8].

Table 1. Influence of alloying elements on steel properties [8].

Silicon	Increases the corrosion resistance Has deoxidizing effect
Manganese	Increases hardenability Increases hardness Increases tensile strength Increases the resistance at abrasive usage
Chrome	Increases the mechanical resistance Increases thermal stability Increases the corrosion resistance Increases the resistance at abrasive usage
Nickelum	Increases the static and dynamic tenacity Increases the corrosion resistance and refractarity
Molibdenum	Increases hardenability Increases mechanical resitance Increases the stability at temperature Increases the dynamic resilience Increases the corrosion and rupture resistance
Vanadium	Increases the mechanical reistance Increases the plastic deformability Forms carbides, nitrides, oxides Favorizes the fineness of structure Increases hardenability
Tungsten	Increases hardness Increases refractarity Increases the mechanical resistance and hardenability Forms hard carbides, resistant at abrasive usage
Cobalt	Increases hardness Decreases the adherence at rapid machining
Aluminium	Deoxidyzing Eases the hardening of the grain at nitriding
Titanum	Deoxidyzing Reduces the sulphurs Forms hard and refractary carbides
Bor	Increases the hardening of low carbon steels
Copper	Increases the corrosion resistance Increases the mechanical resistance
Tin	Increases the machining

2. Deposition of thin layers through electrical discharge in impuse method

The deposition of layers through electrical discharge on Fe-C alloys are based on electroerosion phenomena and polar transfer of electrode material to the metallic parts. Through the proximity of the electrode to the part, at the critical distance of puncture, electrical discharge through impulse is triggered [3,4,7]. Due to polar effect, the transfer of electrode material to the part assures the forming of the superficial layer, with highly determined physical-chemical properties. As a result of the material transfer and thermal modification from the discharge area, the superficial layer of the part is modifying the structure and the chemical composition. The characteristics of this layer can variate in high limits according to the electrode material, the environment composition between electrod and part, the parameters of impulse discharging. The superficial alloying, realized through layers deposition with electrical discharge method, can have the following purposes [5,6,7]:

- Increasing the resistance properties
- Increasing the thermostability
- Increasing the corrosion resistance
- Increasing the mechanical and electrical properties
- Increasing the steels hardenability
- Changing of electrical and magnetical properties

3. Experiments

3.1. Determination of chemical composition through spectral quantitative analysis

The analyze of chemical composition was made on prepared samples through polishing, using an Foundry Masters optical spectrometer, type 01J0013. With WasLab software and calibration programs, has been obtained a analysis bulletin, which present the values determined by the apparatus.

Table 2. Chemical composition determinate through spectral quantitative analysis.

Fe	C	Si	Mn	P	S	Cr	Ni	Mo	Al	Cu
97.8	0.62	0.25	0.59	0.013	0.072	0.12	0.16	0.026	0.07	0.17

The analized sample is part of steel class, mark C45.

3.2. Superficial processing through electric discharge with compact electrodes

As hardness measured values after the improving treatment are low, on the normalized samples thin layers were applied using Elitron 22A apparatus, through electrical discharge method. For the experimental attempts, 2 tungsten electrodes were used, having different compositions (table 3).

Table 3. Composition of electrodes used in the experimental attempts.

Designation	Composition			
	Dioxide	Dioxide in %	Impurities %	Tungsten
WT20	ThO_2	1.70 – 2.20	< 0.20	Remainder
WZr8	ZrO_2	0.70 – 0.90	< 0.20	Remainder

3.3. The structure analysis of superficial layers

After the WT20 and WZr8 layers deposition, were made SEM photos at 100x, 200x, 500x, 1000x magnitudes (figures 1, 2), [9].

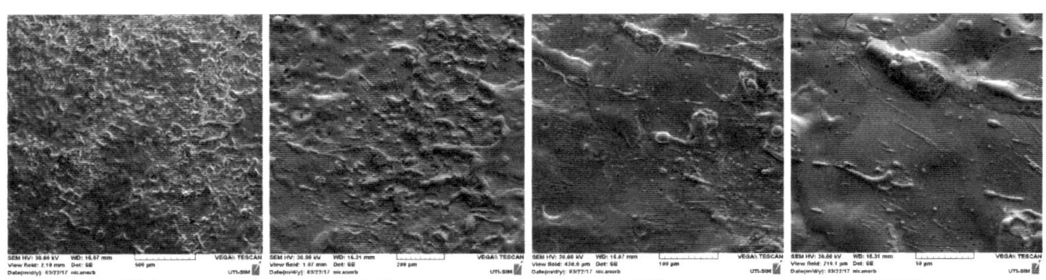

100x 200x 500x 1000x

Figure 1. The aspect of WT20 deposited layer on C45 steel.

From metallographic photos achieved through electronic microscopy, at different magnitudes, it can be observed that in the case of WZr8 deposition, the layer is more uniform, and in case of WT20, the surface formed structure are smoother.

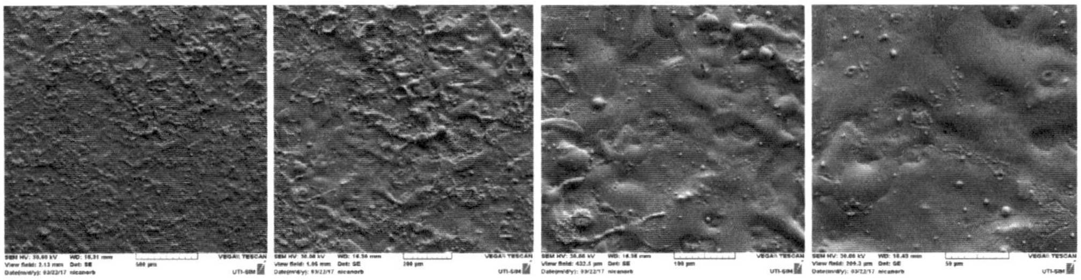

Figure 2. The aspect of WZr8 deposited layer on C45 steel.

3.4. Measurements of mechanical characteristics at the superficial layer level

The determination of mechanical characteristics was made on CETR-UMT2 microtribometer. Hardeness and Young modulus modifications were analized.

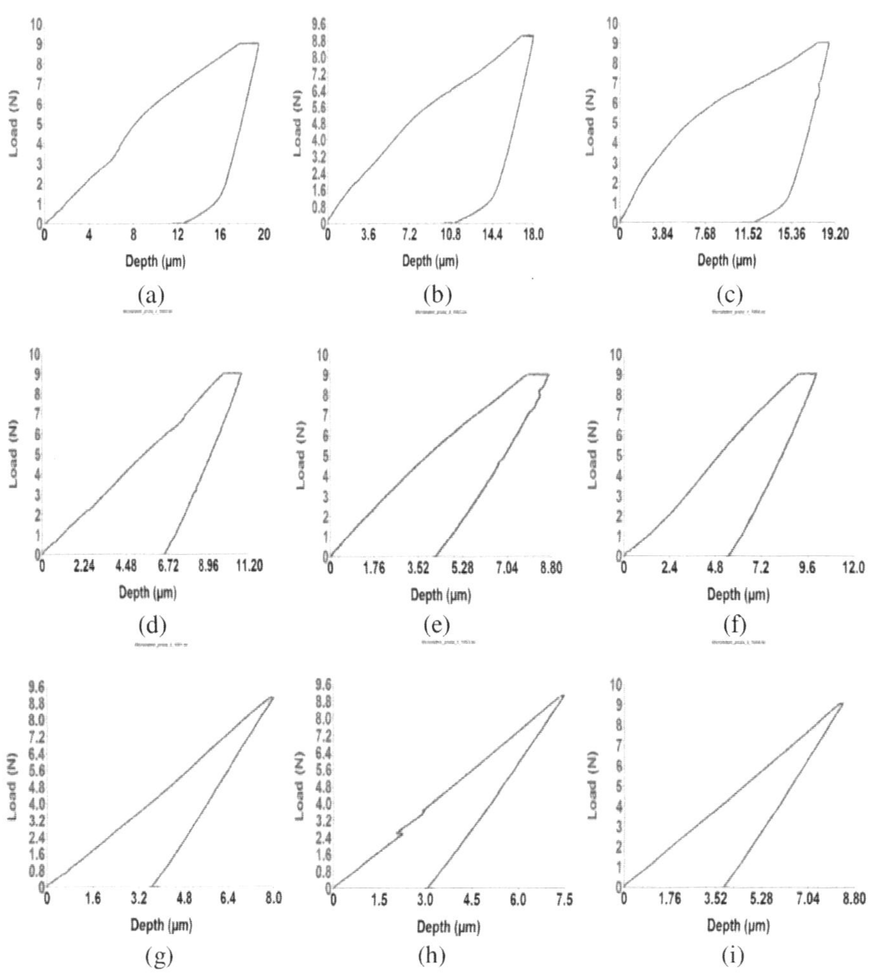

Figure 3. Aspects of microindentation curves for base material (figures a, b, c) and WT20 (figures d, e, f) and WZr8 (figures g, h, i) deposited layers.

The analized samples had small dimensions (according with the requirements of apparatus manufacturer) and were prepared in advance (without oxides or other substances resulting from previous machining). The achieved layers through depositions with WT20 and WZr8 electrodes, were presented with different structures and their hardness average is improved against the obtained values for the basic material (C45 steel).

Table 4. Measured values of mechanical characteristics.

State	Nr.	HV (Gpa)	Average	Young (GPa)	Media
Based material C45	1	0,427483	0,451694	18,125	17,918
	2	0,471523		18,273	
	3	0,456076		17,357	
Deposition with electrode WT20	1	1,317432	1,203496	22,300	21,475
	2	1,271933		21,955	
	3	1,021125		20,170	
Deposition with electrode WZr8	1	1,451079	1,508652	23,673	22,863
	2	1,662807		24,185	
	3	1,412071		20,732	

In the case of Young modulus there is a 25% increase in measured values for deposited layers (WT20 and WZr8) compared to the base material (steel C45).

4. Conclusions

The mechanical characteristics of the superficial layers can be improved by superficial alloying, using the electrical discharge method and electrodes that have chemical elements in the composition with a modifying role (according to table 1). The samples used in the experimental test are part of C45 steels (table 2 – quantitative spectral analysis bulletin), and the two electrodes were based on tungsten, with different compositions (WT20 and WZr8 – table 3). Superficial layer structures were analyzed by electronic microscopy at different magnification powers and it can be seen that in case of deposition of WZr8 the layer is more uniform (figure 1), and in the case of WT20 deposits, the structures formed on the surface are finer (figure 2). The HV hardness measured on the deposited layers has values superior to the base material (steel C45). The increase is in the range 250 ... 300% (according to table 4). Young modulus shows an increase of approximately 25% of the measured values for layers deposited (WT20 and WZr8) compared to the base material (steel C45).

References

[1] Muralidharan B, Chelladurai H, Praveen S et al. 2016 *Materials And Manufacturing Processes* **31** 1853-1864
[2] Moscicki T, Radziejewska J, Hoffman J et al. 2015 *Ceramics International* **41** 8273-8281
[3] Salih D, Salim Levent A, Kemal K 2015 *Acta Physica Polonica A* **127** 1410-1413
[4] Alexandru A, Carabet R G, Carcea I 2008 *Metalurgia International* **13** 18-22
[5] Ţugui C A, Nejneru C, Galuşcă D G, Perju M C, Axinte M, Cimpoeşu N, Vizureanu P 2015 *Journal Of Optoelectronics And Advanced Materials* **17** 1855-1861
[6] Vizureanu P, Perju M C, Galuşcă D G, Nejneru C, Agop M 2010 *Metalurgia International* **15** 59-64
[7] Găluşcă D G ş.a. 2011 *Tratamente termice – îndrumar de laborator*, Editura Tehnopress
[8] Alexandru I, ş.a. 1995 *Alegerea şi utilizarea materialelor metalice*, Editura D.P, Bucureşti
[9] Perju M C, Tugui C A, Nejneru C, Axinte M, Vizureanu P 2016 *IOP Conference Series-Materials Science and Engineering* **133** 012025

IOP Conf. Series: Materials Science and Engineering **209** (2017) 012054 doi:10.1088/1757-899X/209/1/012054

Scientific Ground of a New Optical Device for Contactless Measurement of the Small Spatial Displacements of Control Object Surfaces

I P Miroshnichenko[1*] and I A Parinov[2]
[1] Don State Technical University, 1 Gagarin Square, Rostov-on-Don, Russia
[2] Southern Federal University, 105/42 Bolshaya Sadovaya street, Rostov-on-Don, Russia

E-mail: ipmir2011@yandex.ru

Abstract. It is proposed the computational-experimental ground of newly developed optical device for contactless measurement of small spatial displacements of control object surfaces based on the use of new methods of laser interferometry. The proposed device allows one to register linear and angular components of the small displacements of control object surfaces during the diagnosis of the condition of structural materials for forced elements of goods under exploring by using acoustic non-destructive testing methods. The described results are the most suitable for application in the process of high-precision measurements of small linear and angular displacements of control object surfaces during experimental research, the evaluation and diagnosis of the state of construction materials for forced elements of goods, the study of fast wave propagation in layered constructions of complex shape, manufactured of anisotropic composite materials, the study of damage processes in modern construction materials in mechanical engineering, shipbuilding, aviation, instrumentation, power engineering, etc.

1. Introduction

At present, the development and use of high-precision contactless measuring devices based on modern laser technologies and new methods of optical interferometry are very actual for solution different scientific and applied problems. It allows one to significantly improve the accuracy, quality and informative of the measurement results of the small displacements of control object surfaces [1–9].

Optical devices for measuring small linear and angular displacements of control surfaces are known and already used, representing various modifications of a two-way laser interferometer with combined branches [1-11].

The functional capabilities of the known devices can be significantly expanded for account of providing simultaneous measurement of linear and all angular components of the small displacements of control objects surfaces. The need for this is caused by that during the testing of real control objects the small displacements of their surfaces have both linear and angular components. At the same time, the impossibility of their simultaneous registration reduces the informative and accuracy of the measurement results, requires the installation of additional measuring devices, complicating the design and operation of the experimental-measuring setup.

The aim of the study was the development and calculation-experimental ground of an optical device for the contactless measurement of small spatial displacements of control object surfaces based on the use of new methods of laser interferometry.

IOP Conf. Series: Materials Science and Engineering **209** (2017) 012054 doi:10.1088/1757-899X/209/1/012054

2. Numerical study and analysis of its results

To prepare the calculation-experimental ground, a numerical study was performed of the intensity distributions in the interference patterns produced by a two-way laser interferometer with aligned branches, using new mathematical models and original software. The effect of different types of beam splitters and various cases of polarization on the optical fields of the intensity of interference patterns was investigated.

The obtained results are shown in illustrations. Figure 1 shows the intensity distribution $I(\Theta)$ in the horizontal section of the interference pattern along the coordinate Θ in the observation region of the interference pattern for the case of parallel polarization, when in the considered interferometer the amplitude sinusoidal grating is used as the beam splitter.

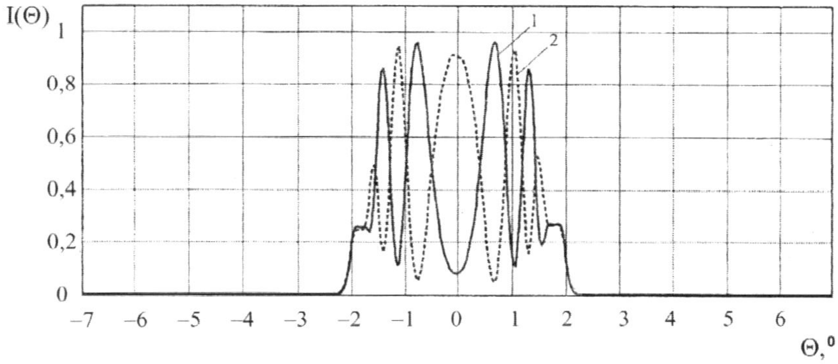

Figure 1. Intensity distribution in horizontal section of interference patterns.

Figure 2 presents a similar distribution for the case when the beam splitter is an amplitude sinusoidal lattice with perpendicular polarization.

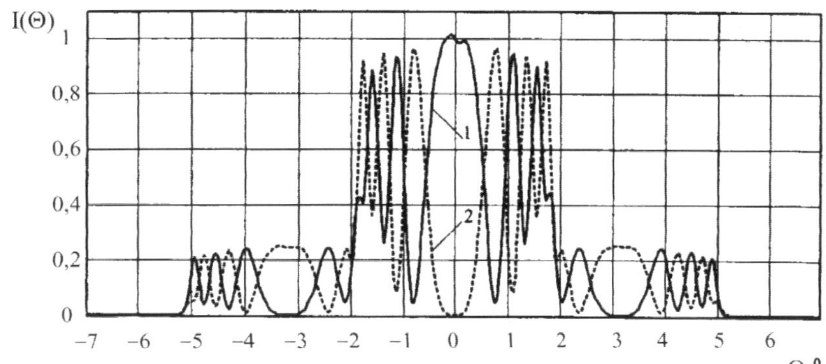

Figure 2. Intensity distribution in horizontal section of interference patterns.

Figure 3 shows an intensity distribution for the case when the beam splitter is a phase sinusoidal lattice with perpendicular polarization.

Figures 1-3 present the following notations: 1 - the dependence is obtained for linear displacement $\Delta h = 0$ and angular displacement $\Delta\alpha = 0°$; 2 - the dependence is obtained for $\Delta h = \lambda/4$ and $\Delta\alpha = 0°$ (where λ is the wavelength of the optical radiation of the source).

An analysis of the results of numerical studies, obtained for different types of beam splitters, makes it possible to note the following. If we compare the parallel and perpendicular polarization, then in the latter case, when the electric field strength vector is in a plane perpendicular to the plane of incidence, it is possible in greater degree to study the effect of the diffraction properties of beam splitters on the parameters of the interference pattern formed.

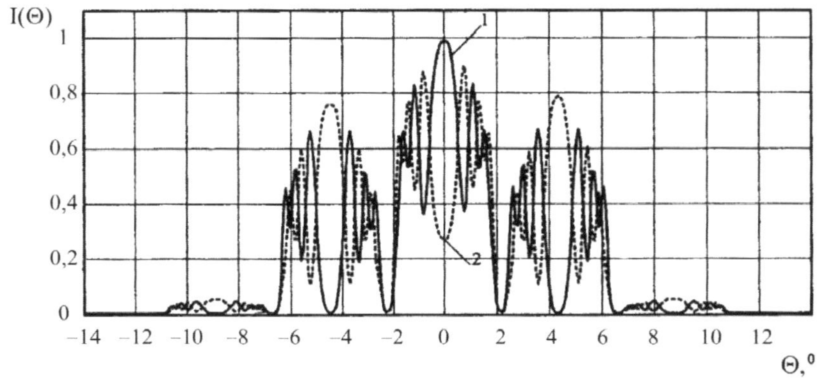

Figure 3. Intensity distribution in horizontal section of interference patterns.

Analysis of the results also showed that when using the amplitude gratings as the beam splitters, the diffraction peaks of −1 and +1 orders are formed (see Figure 2), and when using phase gratings, higher-order peaks are present in addition to them (see Figure 3).

The use of phase gratings in solving measurement problems is quite promising, since it allows ones to obtain approximately the same amplitudes of intensity variation at peaks of −1, 0 and +1 orders. In this case, the intensity amplitude at the peaks of −1 and +1 orders is higher than in the corresponding peaks, formed by using an amplitude grating.

The results of the numerical study described above made it possible to develop a method for measuring the linear Δh and angular $\Delta\alpha$ components of the small displacements of control object surfaces. This method consists in analyzing the intensity distributions at the peaks of −1 and +1 orders of the interference pattern.

In the case of the carrying out the matching condition, the intensities I^{-1} and I^{+1}, respectively, at the peaks of −1 and +1 orders of the interference pattern varies identically, otherwise the nature of the intensity variation at the peaks of the pointed orders is different.

Measurement of the intensities I^{-1} and I^{+1} at the peaks of −1 and +1 orders of the interference pattern allows us to obtain, instead of an equation with two unknowns, a set of equations for Δh and $\Delta\alpha$ (respectively, the linear and angular components of the small displacement of control object surface):

$$\begin{cases} I^{-1} = f_1(\Delta h, \Delta\alpha) \\ I^{+1} = f_2(\Delta h, \Delta\alpha) \end{cases} \quad (1)$$

where $f_1(\Delta h, \Delta\alpha)$, $f_2(\Delta h, \Delta\alpha)$ are the known dependences for the peaks of −1 and +1 orders, respectively, linking the corresponding intensities with linear Δh and angular $\Delta\alpha$ components of the small displacement of the control object surface.

The values of Δh and $\Delta\alpha$, satisfying Equation set (1), will correspond to the actual displacement of the object.

3. Description of the proposed optical device and the principle of its operation

The proposed technical solution is a further development of the method described above. The scheme of the proposed optical device for measuring small spatial displacements is shown in figure 4.

The proposed device consists of the following elements: a source of coherent optical radiation (laser) (1); an optical system (2) that converts the radiation of source (1) into a divergent beam; a beam splitter (3) made in the form of a phase grating; a reflector (4) rigidly fixed on the surface (5) of the control object; screen (6), in the plane of which photodetector devices (7) (for example, photodiodes) are installed, divided into three groups (8, 9, 10), each of which is located in the corresponding regions of peaks: of –1 order (11), of 0 order (12) and of +1 order (13) of the interference pattern (14) (see figure 5).

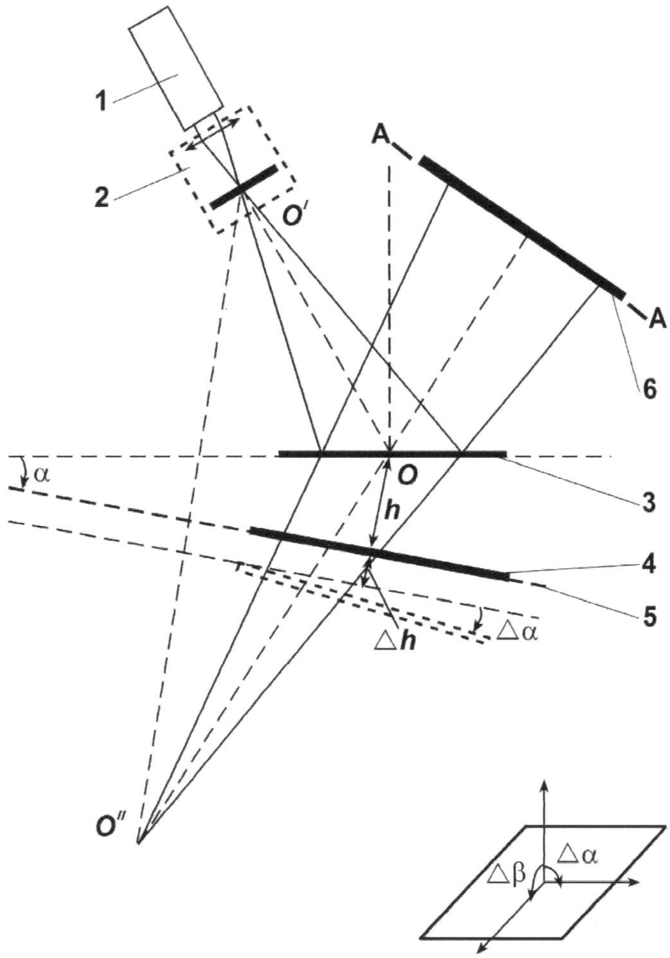

Figure 4. Scheme of the proposed optical device for measuring small spatial displacements of control object surfaces.

The photodetectors (7) are connected to a system for recording, processing and displaying measurement results. The reflector (4) and beam splitter (3) are disposed to each other at an angle α, while the reflector 4 is removed from the inner surface of the beam splitter (3) by a distance h. The reflector (4) can absent, in this case, its functions are performed by the reflecting surface (5) of the control object. The peaks of –1 order (11), of 0 order (12) and of +1 order (13) of the interference pattern (14) are projected onto the screen (6).

The principle of operating the device is as follows. In the process of testing, when linear Δh and (or) angular components $\Delta\alpha$ and $\Delta\beta$ of a small displacement of the control object surface (5) arise, it is occurred a change in the intensity of the optical field at the peaks of –1 order (11), of 0 order (12) and of +1 order (13) of the interference pattern (14). This change takes place also in their corresponding areas at the installation sites of the allocated groups (8), (9) and (10) of the photodetector devices (7).

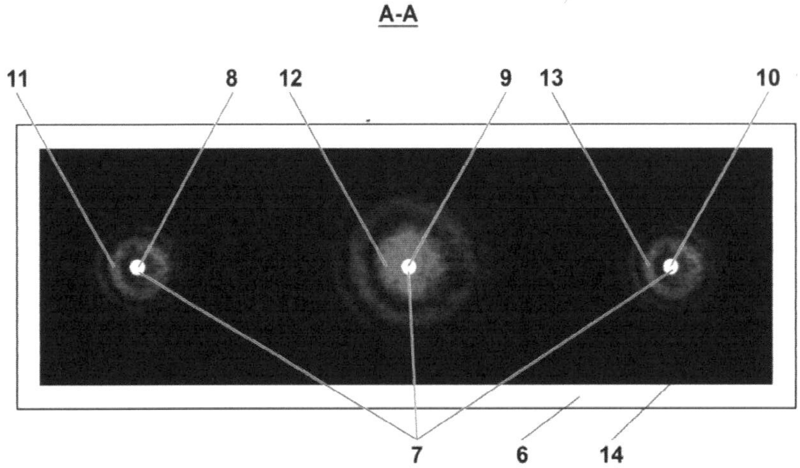

Figure 5. Appearance of the interference pattern in the screen area.

The photodetectors (7) detect the intensity of the optical field, and the intensities, obtained from each of the groups (8), (9), and (10) of the photodetectors (7), are the measurement results. The system for recording, processing and displaying the measurement results provides the recording of the intensity values from each of the groups (8), (9) and (10) of the photoreceptors (7) and performs their processing.

One of possible options of the processing is, for example, the solution of the following equations set:

$$\begin{cases} I^{-1} = f_1(\Delta h, \Delta\alpha, \Delta\beta) \\ I^{0} = f_2(\Delta h, \Delta\alpha, \Delta\beta) \\ I^{+1} = f_3(\Delta h, \Delta\alpha, \Delta\beta) \end{cases} \tag{2}$$

where $f_1(\Delta h, \Delta\alpha, \Delta\beta)$, $f_2(\Delta h, \Delta\alpha, \Delta\beta)$, $f_3(\Delta h, \Delta\alpha, \Delta\beta)$ are the known dependences for the peaks of –1 order (11), of 0 order (12) and of +1 order (13), respectively; they link the intensities with linear Δh and angular $\Delta\alpha$ and $\Delta\beta$ components of the displacement of the control object surface (5), respectively; I^{-1}, I^{0} and I^{+1} are the intensities, measured by each of the groups (8), (9) and (10) of the photodetectors (7), installed in the respective regions of the peaks of –1 order (11), of 0 order (12) and of +1 order (13) of the interference pattern (14).

The results of the processing are the values of the linear Δh and angular $\Delta\alpha$ and $\Delta\beta$ components of the displacement of the control object surface (5) that simultaneously satisfy the intensity values I^{-1}, I^{0} and I^{+1}.

Thus, the linear and all angular components of the small displacement of control object surface can be determined by using the intensity values, measured by groups of photodetectors at the peaks of –1, 0 and +1 orders, knowing for each peak the dependences of intensity *versus* linear and angular displacements of the control object surface. In this case, the measurement results are those values of the linear and angular components of displacement, which simultaneously satisfy the measured values of intensity at the peaks of –1, 0 and +1 orders of the interference pattern.

The described optical device is protected by the Russian Federation patent for an invention [12].

4. Experimental research and analysis of its results

The present study was carried out for the experimental confirmation of the method described above, and also for the preparation of an experimental ground for the proposed technical solution. During the research, an experimental setup was used, constructed according to the scheme depicted in figure 4 and allowing to provide separate and simultaneous reproduction of the given linear Δh and angular $\Delta\alpha$ components of small displacements. The source of coherent optical radiation was He-Ne laser (wavelength $\lambda \approx 0.63$ μm). By creating the simulated linear Δh and angular $\Delta\alpha$ components of the small displacements, the disposition of the reflector 4 (see figure 4) was changed by means of piezoelectric transducers linked to it, which were connected electrically with voltage sources. During the investigation, the photodetector devices recorded intensities in the specified regions of the peaks of –1 and +1 orders of the interference pattern.

Figure 6 shows an image of a typical interference pattern, where the notations: –1, 0, +1 correspond to the peaks of –1, 0 and +1 orders of the interference pattern, the mutual position of which depends on the grating period and on the distance between the beam splitter (4) and screen (6) (see figure 4).

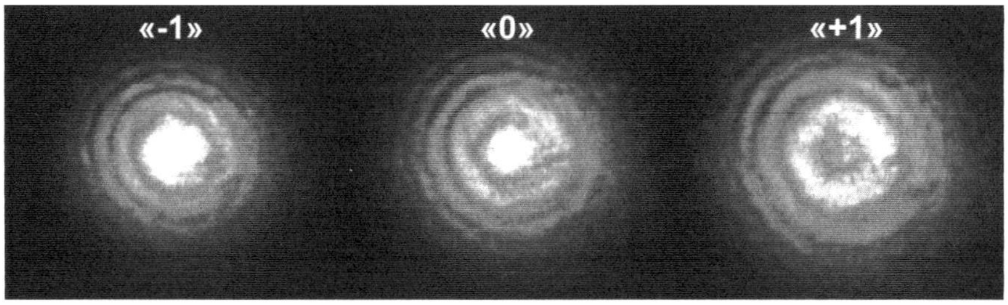

Figure 6. Image of typical interference pattern.

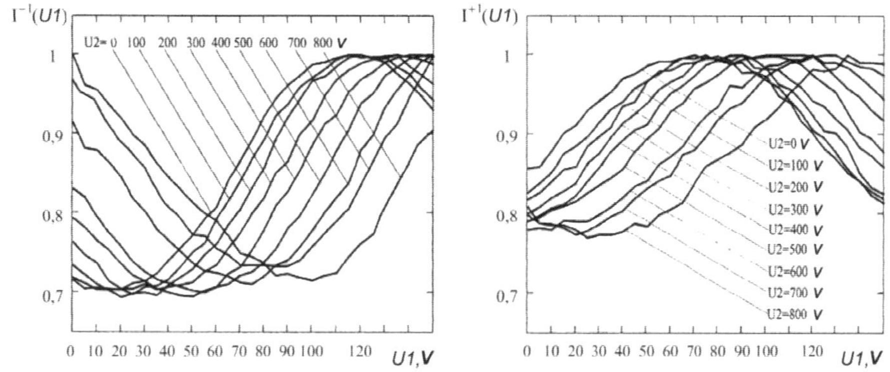

(a) (b)

Figure 7. Results of experimental study (explanation into text).

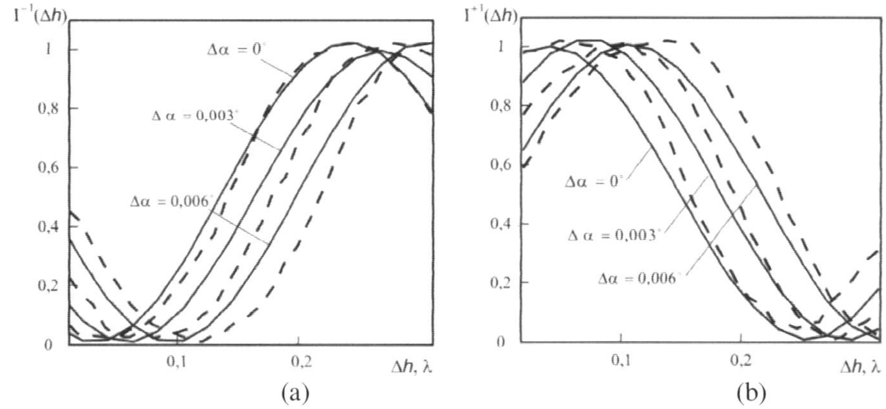

(a) (b)

Figure 8. Comparison of numerical and experimental results (explanation into text).

The experimental procedure consisted of stepwise assignment of the linear component Δh of the small displacement of reflector at fixed value of the angular component $\Delta\alpha$, while the recording of the signals of the photodetectors was carried out at each step.

During numerical study, the main parameters of the mathematical model of the proposed optical device were brought into line with the parameters of experimental setup. It allowed one to compare the results of numerical and experimental studies, and to confirm the feasibility of the developed method for measuring small displacements and to ground the proposed technical solution.

The intensities I^{-1} and I^{+1} are obtained at the peaks of –1 and +1 orders in dependence on the linear component Δh of the small displacement of reflector (presented by the voltage $U1$ on the piezoelectric transducer, modeling the linear component Δh) (see figures 7a and 7b, respectively) at various values of angular component $\Delta\alpha$ of the small displacement of reflector (presented by the voltage $U2$ on the piezoelectric transducer, modeling the angular component $\Delta\alpha$).

Figure 8 shows the dependences of the intensities I^{-1} and I^{+1} obtained by comparing the results of numerical experimental studies at the peaks of –1 order (Figure 8a) and peaks of +1 order (Figure 8b). The solid line corresponds to the results of a numerical study and the dotted line presents the results of an experimental study.

Analysis of the numerical and experimental results allowed one to clearly confirm the described method and ground the proposed technical solution.

At present, the experimental investigation continues in the direction of investigating the features of simultaneous detection of linear (Δh) and angular ($\Delta\alpha$ and $\Delta\beta$) components of the small displacement of the control object surface.

5. Conclusions

The obtained in the paper results allow us to make the next conclusions:

1. On the basis of the use of new methods of laser interferometry, a new optical device for contactless measurement of small spatial displacements of control object surfaces has been developed. It allows simultaneous registering the linear and all angular components of small displacements without installing additional measuring means.
2. The proposed device is scientifically grounded by carrying out numerical and experimental studies of its functional properties.
3. The described device allows one to expand the functionality of known domestic and foreign small displacement meters, to increase the information content and significantly up to 30% the accuracy of measurement results. Moreover, it also provides the possibility of correction of measurement results, excluding the influence of internal and external destabilizing effects without the use of additional measuring means, which improves the quality of measurement results up to 40%.
4. The obtained results are most expedient for using in the process of high-precision measurements of small linear and angular displacements of the control object surfaces during experimental studies, evaluation and diagnostics of the state of construction materials for the power elements of operating goods, investigation of fast-moving wave processes in layered structures of complex shape made of anisotropic composite materials, the study of the processes of defect formation in new construction materials in engineering, shipbuilding, aircraft building, instrument making, power engineering, etc.

References

[1] Miroshnichenko I P, Parinov I A, Rozhkov E V, Serkin A G, Sizov V P 2012 Novel Test Means and Techniques Based on Optical Interferometry and Acoustic Emission to Study Displacements of Object Surfaces and Damage of High-Temperature Superconductive Tapes and Related Composites *Piezoelectrics and Related Materials: Investigations and Applications* (New York: Nova Science Publishers) 239-282

[2] Miroshnichenko I P, Parinov I A, Rozhkov E V, Sizov V P, Shevtsov V A 2013 Novel Optical Means for Measurement of Displacements of Surfaces of the Control Objects at Diagnostics

of Materials and Goods *Physics and Mechanics of New Materials and their Applications* (New York: Nova Science Publishers) 145-154

[3] Miroshnichenko I P, Parinov I A, Rozhkov E V, Serkin A G 2006 *Metallurgist* **50**(7-8) 408-409

[4] Miroshnichenko I P, Serkin A G 2010 Improved Experimental Setup for Study of Defect Formation Processes in Samples of Construction Materials *Metallurgist* **54(3-4)**189-190.

[5] Alekhin V E, Miroshnichenko I P, Sizov V P 2007 *Russian Journal of Nondestructive Testing* **43**(2) 112

[6] Miroshnichenko I P, Serkin A G 2007 *Russian Journal of Nondestructive Testing* **43**(4) 234

[7] Alekhin V E, Miroshnichenko I P, Nesterov V A, Sizov V P 2007 *Russian J. of Nondestructive Testing* **43**(9) 592

[8] Miroshnichenko I P, Serkin A G, Sizov V P 2008 *Journal of Optical Technology* **75**(7) 437

[9] Alekhin V E, Miroshnichenko I P, Serkin A G 2008 *Measurement Techniques* **51**(10) 26-30

[10] Miroshnichenko I P, Parinov I A, Chang S H 2015 Novel Optic Devices for Measurement of Displacements Based on Method of Control Object Highlighting by Using Laser Interferometer *Advanced Materials. Studies and Applications* (New York: Nova Science Publishers) 437-451

[11] Miroshnichenko I P, Parinov I A, Rozhkov E V, Chang S-H 2016 *Advanced Materials Manufacturing, Physics, Mechanics and Applications. Springer Proceedings in Physics* **175** 341-356

[12] Miroshnichenko I P, Parinov I A, Sizov V P, *Measurement Method of Linear and Angular Components of the Small Displacements of the Control Object Surfaces* Patent RU No. 2606245/ 10.01.2017.

Acknowledgements

This study has been partially supported by grant of the Russian Foundation for Basic Research No. 1608-00740.

IOP Conf. Series: Materials Science and Engineering **209** (2017) 012055 doi:10.1088/1757-899X/209/1/012055

Wireless Sensors for Wind Turbine Blades Monitoring

N Iftimie[1*], R Steigmann[1], N A Danila[1], D Rosu[2], P D Barsanescu[3] and A Savin[1]
[1]National Institute of R&D for Technical Physics, Nondestructive Testing Department, Iasi, Romania
[2]S.C. Compozite S.A, Brașov, Romania
[3]Gh.Asachi Technical University, Faculty of Mechanical Engineering, Iasi, Romania

E-mail: niftimie@phys-iasi.ro

Abstract. The most common defects in turbine blades may be faulty microscopic and mesoscopic appeared in matrix, no detected by classical nondestructive testing (i.e. using phased array sensors), broken fibers can also appear and develop under moderated loads, or cracks and delaminations due to low energy impacts, etc. The paper propose to present the results obtained from testing of glass fiber reinforced plastic used in the construction of the wind turbine blades as well as the monitoring of the entire scalable blade using wireless sensors placed on critical location on blade. In order to monitories the strain/stress during the tests, the determination of the location and the nature of defects have been simulated using FEM.

1. Introduction

In accordance with to growing concern on global warming, it is expected to use more clean energy including wind energy. The designs of large wind turbines (WT) are rapidly increasing in 21st century. So then, wind induced damages of WT are increasing particularly in sever wind storm dominant region [1]. EU studies the possibility to update air pollution standards for thermal power station based on coal, in the conditions in which this type of installation represent the biggest source of sulfur dioxide and mercury, nitrogen oxides, arsenic, lead and cadmium emissions. Thus, the market for electrical energy obtained by harvesting wind power is expanding. In these conditions, in order to increase the power of turbine, wind turbine producers use glass fiber reinforced plastics in the construction of wind turbine blades (WTB) because these allow due to low weight, the increase of blade's dimensions [2,3]. Nondestructive testing (NDT) techniques for in-service inspection and determination of high risk degradation regions of blades are growing, function of the type and blade dimension, being based on numerous studies and researches, function on the type and the size of the WT [2]. The NDT method must be carried out both during the fabrication process as well as during the functioning of the blades, first for the decreasing of fabrication and maintenance costs and for reduced downtime. The blades are usually subject to random and complex mechanical stresses. Condition monitoring can be carried up using thermography [4] and fatigue crack detection performance can be done [5]. Implementing a process of fault detection and characterization of structures is refers as Structural Health Monitoring (SHM) [6]. The SHM (e.g. vibration monitoring and damage detection of wind blade) process involves the observation of a system over time using periodically sampled dynamic response measurements from an arrangement of sensors, the extraction of damage-sensitive features from these features to determine the current state of system health [7-11]. In order to decrease the fabrication and maintenance costs, as well as for avoiding unproductive time, NDT is required both during fabrication and in-service. Starting from the scheme presented in figure 1, flaws mentioned above must be detected and evaluated by NDT methods with good probability of detection (POD) under high reliability coefficient.

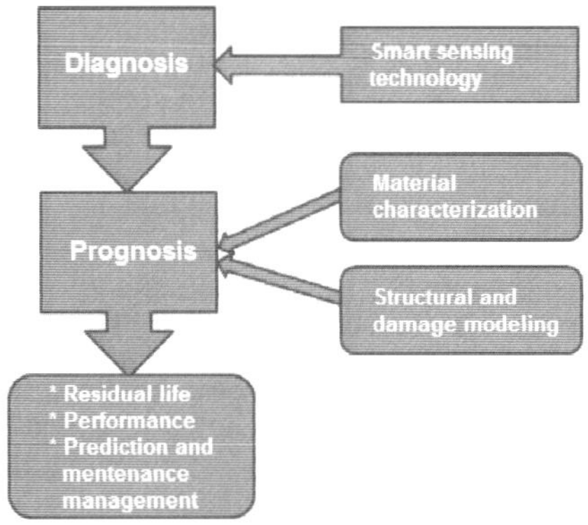

Figure 1. Structural Health Monitoring (SHM) principle.

Wireless sensors (WRS) are very attractive for SHM applications [12,13]. The last years have brought different wireless technologies that can be used for structural damage detection. This includes radio-frequency identification (RFID) technology that has been proposed for a wireless damage detection platform in civil engineering application [14-16]. RFID tag sensors provide identification data and monitor of physical parameters of tagged objects without having an active sensor in the tag circuitry [17,18]. The RFID tag sensor has distinguished impact in research publication field, economic and social areas. To this day, a passive, printable, mass deployable, low cost, environmentally RFID tag sensor has not been reported for monitoring the sensing parameters.

Passive RFID systems consist of an active read out device (reader) and one or several passive tags. They are used for the wireless identification of objects with a tag attached to them. Within this tag there is a microcontroller with memory where identifiers and supplementary information can be provided. The fundamental characteristic of passive RFID sensor is that the involved tags is full passive, gathering all their energy from a magnetic or electromagnetic field emitted by the reader [19-21]. The exclusiveness of RFID sensor compared reported studies are as follows:

I. data ID and sensing information are in both magnitude and phase spectrum;
II. a single tag has multiple parameter sensing capabilities and
III. RF sensing is incorporated using metamaterials rather than external sensors or lumped components in the tag circuitry.

The paper presents the results of testing of a WTB, in static conditions, using WRS embedded within in order to monitor possible damages that further can be transformed into flaws. To determine the righteous of solutions, reliability of correct diagnosis probability, prognosis, and evaluation of residual lifetime and maintenance management, a scalable WTB has been constructed, that had the scaled dynamics of a full scale blade as well as enough sensors for measurements as well as monitoring. The maximum stress zones [22] and damage evolution and remnant stress [23] have been determined using Finite Element Modelling (FEM). The results of complex mechanical tests, performed on scalable WTB models (in our case a blade of 1750 mm length) are used to give efficiency to monitoring strategy.

2. SHM based on transducers. Experimental set-up

The passive WRS designed to monitor stress/deformation status have as sensitive element a special type of MM, that follows the relative displacement (compression or tension) of its components due to an impact and can develop slows or either fast. The ratio of relative displacement to the structure

IOP Conf. Series: Materials Science and Engineering **209** (2017) 012055 doi:10.1088/1757-899X/209/1/012055

length allows the calculation of materials deformation and the ratio of stress at deformation provides the elastic modulus which is an intrinsic property of the structures materials. Thus, the measurement of relative displacement provides critical information on the force presented on material. In function of type and geometrical dimensions, the sensor has the resonance frequency in the range of RF and microwaves.

In order to obtain the best sensor with MM for monitoring the GFRP structure, a sensor which allows the determination of resonance frequency displacement at the changing of dielectric properties from the sample has been developed. The system consist in a RFID reader and a RFID tag (figure 2), the tag including the sensor and the integrate circuit. When the effort is small, the resonance frequency modifies almost linear in function of effort, meaning that the effort can be determined if the resonance frequency is measured. The RFID tag antenna assures that the interrogation frequency shall be equal with the one of RFID tag to obtain perfect matching of impedance between antenna tag and IC chip. The interrogation range is one the important factors for LC passive WRS. The smallest amount of energy must be transmitted toward reader to activate RFID tag, meaning that the transmitted power threshold reach minimum value at resonance frequency. The RFID tag is bonded to the WTB structure subdued to loadings.

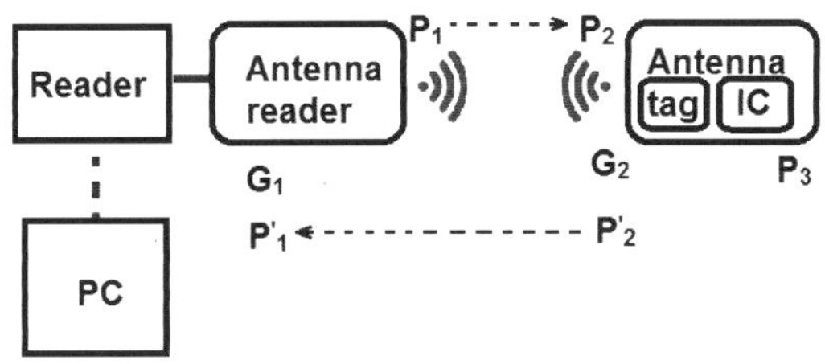

Figure 2. Wireless sensor system principle.

The SHM of WTB is characterized by sensors network around the blade to detect and locate any occurring damage. Taking into account the efficacy of the method and *a priori* knowledge about WTB, the sensors are distributed according to the most expected damage area, aiming the minimizing of the sensors number (figure 3).

Figure 3. Testing stand and experimental set-up for WRS measurements.

The blade has been realized from E-glass/epoxy EPIKOTE Resin MGS LR 385 composite. The leading edge is straight and the trailing edge conical for an easy construction. The profile follows NACA airfoil. For increased structural strength and stiffness at 0.286R (R- the total length of distance between the center of rotor and the tip of blade), the same NACA has been applied both for the upper surface and lower surface, keeping the aerodynamically performances of the blade's tip. The profiles

IOP Conf. Series: Materials Science and Engineering **209** (2017) 012055 doi:10.1088/1757-899X/209/1/012055

between 0.268R and tip were linear interpolated. A compromise has to be found between high-resolution and long propagation distance. The results are obtained at preliminarily testing of a scalable WTB, in laboratory, using passive RFID - WRS as uni-dimensional displacements sensor.

3. Wireless sensors (WRS) and FEM simulation

The passive WRS designed to monitory stress/strain status have as sensitive element a special type of metamaterial (MM), that follows the relative displacement (compression or tension) of its components due to an impact and can develop slows or either fast. The measurement of strain provides critical information on the force presented on material. In specific construction as 2D geometry on flexible support, copper split ring resonators can be used as stress/strain sensors [24,25]. In function of type and geometrical dimensions, the sensor has the resonance frequency in the range of RF and microwaves. The quality factor of the equivalent circuit has value in the range of tens or hundreds. MM presents resonant properties and around resonance frequency displays unusual properties. The inductance and the capacitance are given by [24].

In order to obtain the best sensor with MM for monitoring the glass fiber composite structure, a sensor which allows the determination of resonance frequency displacement at the changing of dielectric properties from the sample has been developed. The system consist in a RFID reader and a RFID tag [25], the tag including the sensor and the integrate circuit. When the sensor detect a modification of the strain ε, the frequency is

$$f_r' = \frac{c}{4(1+\varepsilon)(L+\Delta L)\sqrt{\varepsilon_r}} = \frac{f_r}{1+\varepsilon} \approx f_r(1-\varepsilon) \quad (1)$$

Thus, when the load is small, the resonance frequency modifies almost linear with load, the load can be determined if the resonance frequency is measured. The RFID tag antenna assures that the interrogation frequency f shall be equal with the one of RFID tag to obtain perfect matching of impedance between antenna tag and IC chip. The smallest amount of energy must be transmitted toward reader to activate RFID tag, the transmitted power threshold (measured through the reader) reach minimum value at resonance frequency.

The reader antenna (Figure 3) is fixed on the upper part of the stand, at 30 cm from the middle area of WRS placement locus. The antenna is connected to a logger reader coupled to PC by USB. The loading force is configured to have an increment of 243με at each step. The mean threshold of interrogation power (dBm) is read by logger reader at each loading step.

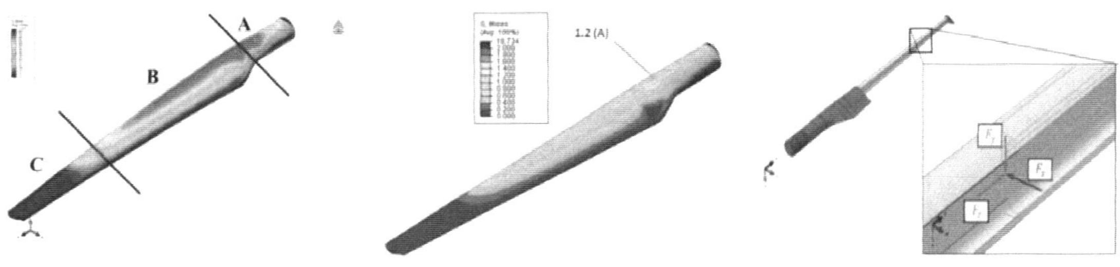

(a) (b) (c)

Figure 4. FEM simulation of WTB:(a) division of blade in critical regions; (b) Von Mises stress; (c) loading forces disposition.

WRS can assure and enhance the functioning of WTB due to the detection/monitoring properties, small dimensions and reduced weight, allowing the embedding them into the base material of the blade (being embedded at intersection of longeron with NACA profile, inside the WTB). In this structure, 3 WRS were placed along the central longitude of the blade, into critical points determined by FEM of blade under vibration. Figure 4a presents the division of the blade in critical regions, in Figure 4b is presented the result of simulation under a force F_x = 200N, producing the displacement of the tip with 1.59 mm.

The FEM model shown below takes into account the presence of the reinforcement structure, who's mass cannot be neglected, especially if one considers that the distance from the axis of rotation increase the inertia and can reduce the frequency associated with the first mode of vibration of WTB. Fortunately, the glass fibers employed makes the structure very rigid.

Three WRS sensors were placed in region with maximum critical points established by FEM, in region B at 307 mm, 362 mm and 406 mm from the hub fixing in stand. All the loadings are static.

4. Experimental results

The dependency strain-load for WTB in a loading-unloading cycle for three sets of experimental measurement is presented in figure 5. It can be observed the same linear dependency strain-load, even the existence of a remnant stress at force removal, indicating an accumulation of energy in WTB composite structure, preponderant in the resin. These stresses are relieving in relatively short time once with the material relaxation, so that the new tests are not influenced. The reader antenna (Figure 2) is fixed on the upper part of the stand, at 30cm from the middle area of WRS placement locus. The antenna is connected to a logger reader coupled to PC by USB. The loading force is configured to have an increment of 243με at each step. The mean threshold of interrogation power, in dBm, is read by logger reader at each loading step. Figure 6 plots the experimentally dependency strain-load for WTB into a cycle loading for the three WRS placed in different critical regions.

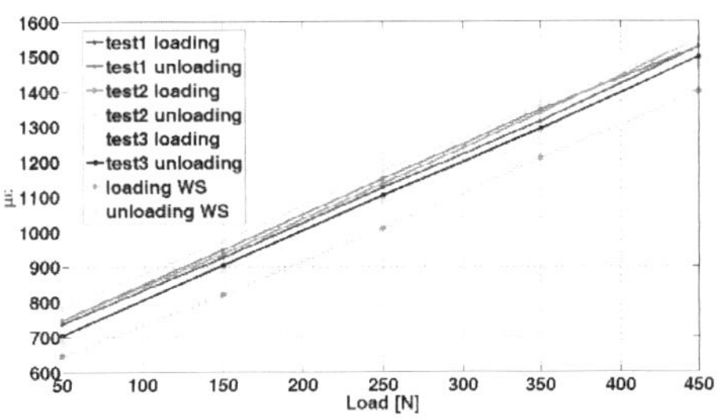

Figure 5. The dependence of strain vs. loadings.

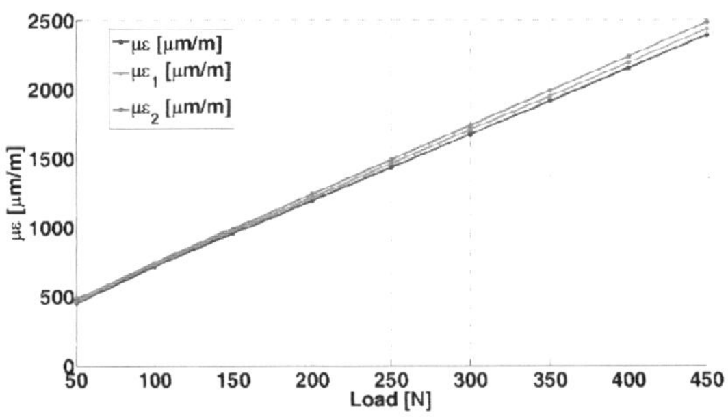

Figure 6. Signal delivered by the 3 wireless sensors for loading of WTB.

It can be observed that the strain has a practically linear dependency on to the applied load on all WRS with small difference on slopes. Although the sensors are not located at exactly the same location, a great resemblance has been found. The wireless sensors offer many advantages over traditional resistive strain gauges. These include remote sensing, easy installation, non-corrosive and lower maintenance cost. This shows that complementary technologies are a good alternative for structural strain monitoring.

5. Conclusions

A sensor wireless RFID for WTB monitoring based sensing platform has been development. The platform is maintenance free and can be used for the damage detection/monitoring in SHM applications. The sensing capability and the damage detection/monitoring performance of the system have been illustrated using detection in a GFRP composite. The results from the former experiment have been validated using numerical simulations. Practical applications demonstrates that, in order to avoid environmental disasters, it is necessary to establish diagnosis and prognosis methods, based on using information obtained from sensors constructed on known physical principles. The monitoring is close related with nondestructive evaluation and the trend is to obtain real time information. Scalable

WTB have been constructed and tested to loadings using WRS embedded in the maximum concentration stress zones. The sensors responses obtained in static conditions were analyzed following a linear tendency; even they were not placed in the same regions. The measurements show advantages as remote sensing, simple embedding, and easy interpretation. The tests were carried on scalable models because in the frame of the project that sustains the paper, the blades will be employed into a demonstrator to show the righteous of solutions, reliability of correct diagnosis probability, prognosis, and evaluation of residual lifetime and maintenance management.

References

[1] Kawai H, K. Michishita and A. Deguchi 2008 *BBAA VI International Colloquium on: Bluff Bodies Aerodynamics & Applications* Milano Italy July 20-24
[2] Adams D, White J, Ramsey M and Farror C 2011 *Wind Energy* **14**(4) 603-623
[3] Yang B and Sun D 2013 *Renew. Sust. Energ. Rev.* **22** 148-167
[4] Gao B, Woo W L, He Y and Tian G Y 2016 *IEEE T Ind. Inform.* **12**(1) 371-383
[5] Taylor S G, Farinholt K, Choi M, Jeong H, Jang J, Park G, Lee J R and Todd M D 2014 *J. Intel. Mat. Syst. Str.* **25**(5)
[6] Balageas D, Fritzen C P and Güemes A 2010 *John Wiley & Sons* **90**
[7] Farrar C R and Worden K 2007 *Philos. T. Roy. Soc. A.* **365**(1851) 303-315
[8] Worden K, Farrar C R, Manson G and Park G 2007 *Proc. Roy. Soc. A* **463**(2082) 2961-2981
[9] Wind turbine accident compilation, http://www.caithnesswindfarms.co.uk/fullaccidents.pdf
[10] Wind Energy Targets for 2020 – 2030, EWEA (2011) http://www.ewea.org/fileadmin/files/library/publications/reports/Pure_Power_III.pdf
[11] www.gerenewableenergy.com/wind-energy/turbines.html
[12] Tanner N A, Wait J R, Farrar C R et al. 2003 *J. Intel. Mat. Syst. Str.* **14** 43–56
[13] Lynch J P, Sundararajan A, Law K H et al. 2004 *In: Proceedings of SPIE, San Diego USA* **5394**. SPIE
[14] Lynch J P, Wang Y, Swartz R A et al. 2008 *Structural Control and Health Monitoring* **15** 518–539
[15] Lehpamer H 2012 *RFID Design Principles (2nd edn)*, (Artech House: Norwood)
[16] Bolic M, Simplot-Ryl D and Stojmenovic I 2010 *RFID Systems: Research Trends and Challenges.* (Wiley: Chichester)
[17] Chen T, Li S and Sun H 2012 *Sensors* **12** 2742-2765
[18] Schueler M, Mandel C, Puentes M and Jakoby R 2012 *IEEE Microwave Magazine* **13** 57-68
[19] Smith J R 2013 *Springer: New York* **13** 13-29
[20] Smith J R, Sample A P, Powledge P S et al. 2006 *In UbiComp 2006: Ubiquitous Computing, Dourish P* (Friday A (eds). Springer: Berlin) 495–506
[21] Ikemoto Y, Suzuki S, Okamoto H et al. 2009 *Sensor Review* **29** 127–136
[22] Grimberg R, Tian G Y, Savin A, Steigmann R and Dobrescu G S 2014 *Stud. Appl. Electromag.* **39** 25-27
[23] Antaya S and Ahmed Z 2013 *Eng. Fail. Anal.* **35(2)** 147-158
[24] Savin A, Steigmann R, Iftimie N, Stanciu M D, Danila N A and Barsanescu P D 2017 *Innovative manufacturing engineering & energy international conference, The 21th edition of IManEE 2017 International Conference* May 25-26 2017 Iasi Romania
[25] Savin A, Steigmann R, Dobrescu G D 2014 *12th Biennial Conference on Engineering Systems Design and Analysis (ESDA2014)* V002T07A027

Acknowledgments

This paper is partially supported by Romanian Ministry of Research and Innovation under project PN-II-PCCA-2013-4-0656 and Nucleus Program PN 16-37 01 01.

Development of Anticorrosive Polymer Nanocomposite Coating for Corrosion Protection in Marine Environment

L Mardare[1] and L Benea[1,*]
[1]Competences Center: Interfaces-Tribocorrosion-Electrochemical Systems, Faculty of Engineering, Dunarea de Jos University of Galati, Domnească Street, 47, RO-800008, Galaţi Romania.

*e-mail: Lidia.Benea@ugal.ro , lidiabd@yahoo.com

Abstract. The marine environment is considered to be a highly aggressive environment for metal materials. Steels are the most common materials being used for shipbuilding. Corrosion is a major cause of structural deterioration in marine and offshore structures. Corrosion of carbon steel in marine environment becomes serious due to the highly corrosive nature of seawater with high salinity and microorganism. To protect metallic materials particularly steel against corrosion occurrence various organic and inorganic coatings are used. The most used are the polymeric protective coatings. The nanostructured TiO_2 polymer coating is able to offer higher protection to steel against corrosion, and performed relatively better than other polymer coatings.

1. Introduction

Marine atmosphere with its high level of salinity and humidity is very corrosive, and the effects of corrosion are allegedly responsible for 30% of failures on ships and other marine equipment. The effects of corrosion are expensive costs to vessel, structural and equipment operations, compounded by the financial penalties on breakdowns, outages, and repairs [2].

The Port of Constanta is located on the Western Coast of the Black Sea, through which the Danube flows into the sea. The two breakwaters located northwards and southwards shelter the port creating the safest conditions for port activities. Constanta harbour is a major port on the western coast of the semi-enclosed Black Sea. Its brackish waters and low species richness make it vulnerable to invasions. The intensive maritime traffic through Constanta harbour facilitates the arrival of alien species. In marine and brackish water environments, shipping is considered the main vector of alien species introductions [2].

Inside Constanta harbour at the end of July 2009, average values of salinity, temperature and dissolved oxygen at the surface were 12.1 PSU, 25°C and 14 mg/L, respectively. At the end of August 2010, average values of the parameters registered in the same locations inside the harbour were 13.8 PSU, 27.3°C and 9.4 mg/L [2].

The majority of the pH results was found in the domain of most natural waters that falls in the 6 to 8.5 range [3,4].

The unique hydrological and hydro chemical characteristics of the Black Sea confer the status of the largest anoxic basin in the world. The Black Sea is subject to environmental variations due to the natural influences (e.g., climate change, global warming, sea level rising, coastal erosion) [3].

The corrosion of steels in fresh and brackish waters is known to be a function not so much of salinity itself but rather of the calcium–calcium carbonate balance (alkalinity and hardness) of the water and how this is influenced by its pH [5].

One of the most widely used methods for protection against metallic corrosion is the application of organic polymeric coatings, often described as paintings. Among the main organic coatings that improve the corrosion resistance of steel are paintings with an epoxy resin, in which inorganic particles or passivating compounds may be incorporated, in order to delay the corrosion phenomenon.

The resin exerts a barrier effect, separating the metal from the corrosive aqueous medium. However, the barrier properties are often insufficient, since aggressive ions can penetrate through the pores of the coating, initiating the corrosive process [6-7].

Polymeric coatings have been widely used for corrosion protection of metallic substrates because they are easy to apply with a reasonable cost. Among possible choices of polymeric coatings, have been extensively used due to its excellent gloss, good durability as well as relatively low cost [6].

2. Experimental set-up

2.1. Materials

Three different coatings were painted (table 1) on 3 mm thickness *E 32* naval steel plates. The chemical composition of the steel substrate is shown in table 2 and immersed in seawater harvested from the port of Constanta. Seawater parameters are shown in table 3.

Table 1. Tested sample types.

Coat code	Description
E 32	Uncoated naval steel
E32G	Primer: Two-component polyamide adduct cured, high solids, high build epoxy paint.
E32GNP	Primer: Two-component polyamide adduct cured, high solids, high build epoxy paint, blended with TiO_2 nanoparticles.

The corrosion experiments were performed on naval steel EN32, uncoated or coated with a two-component polymeric primer or two-component polymeric primer with epoxy polyurethane paint layer. For corrosion measurements, all the samples were cut to dimensions of 25 x 25 x 2 mm.

After that the surface was covered with a polymeric primer two-component polyamide adduct cured, high solids, high build epoxy paint. (having a dry thickness (DFT) equal to 40 μm). Finally this polymeric primer was coated with two-component glossy acrylic polyurethane with DFT equal with 80 μm.

Table 2. The chemical composition of the E32 naval steel.

	C [%]	Cr [%]	Mo [%]	Mn [%]	Ni [%]	Si [%]	P [%]	S [%]	N [%]	Al [%]	Cu [%]	As [%]	Fe [%]
E32	0.18	0.2	0.08	1. 6	0.4	0.5	0.035	0.035	0.008	0.06	0.35	0.08	Balance

Samples used in the experimental protocol, have been bonded with copper wire and insulated with epoxy resin to obtain a measurable area. The naval steel samples were degreased with alcohol and acetone. The samples covered with polymers have been degreased with water and detergent.

In order to perform the corrosion experiments it was used a standard three-electrode cell consisting of the tested samples as working electrode, a Pt-Rh grid used as auxiliary electrode and Ag/AgCl electrode (saturated KCl solution, E = 200 mV vs. normal hydrogen electrode (NHE)) as a reference electrode.

Table 3. Black Sea water characteristics.

pH	Salinity [ppt]	Conductivity [mS]
7.93	11.9	20.9

The experiments were done using a Potentiostat/Galvanostat PGZ 100 and the data were recorded with VoltaMaster 4 software.

3. Results and discussion

3.1. Open Circuit Potential

Corrosion tests were initiated by monitoring the evolution of open circuit potential, after immersion in the tested solution, until they reached a stationary steady state. The potential-time measurements of the four different surfaces studied in sea water are shown in figure 1. The evolution of OCP was monitored for 12 hours for each sample.

The OCP trend of uncoated E 32 naval steel decreases slight from the value of - 488 mV to -705 mV vs. Ag/AgCl during the immersion. The potential shift towards negative direction indicates that this material is unable to form a protective oxide layer on its surface after exposure to marine environmental conditions [8-9].

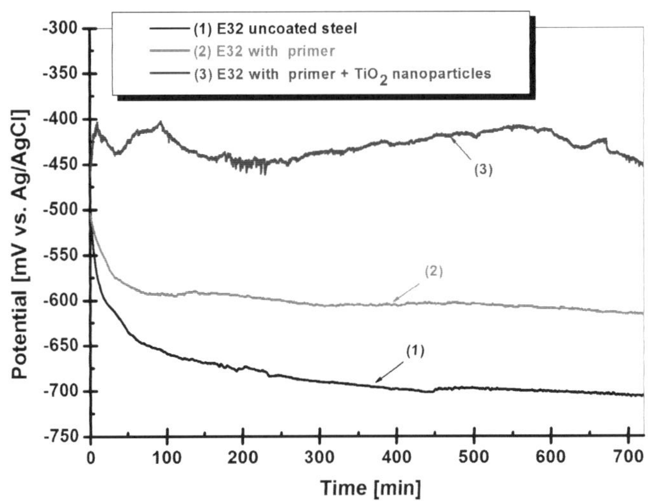

Figure 1. Variation of open circuit potential with time immersed in sea water for: (1) uncoated E 32, (2) E 32 coated with primer, (3) E 32 coated with primer and blended with TiO_2 nanoparticles.

For E 32 steel coated with polymeric primer, the OCP is almost constant in all over the immersion period and the steady state potential is already achieved. This almost constant potential of polymeric primer coating at a value around -599 mV vs. Ag/AgCl, is associated with an improved corrosion resistance.

For E32 coated with polymeric primer with TiO_2 nanoparticles the value growing slowly from –483mV to -442 mV vs. Ag/AgCl.

Generally a steady potential indicates that the film remains intact and protective. Comparing the steady state potential of the three different surfaces studied in sea water, it is clearly revealed that the uncoated E 32 naval steel does not reach a noble steady state potential, while the two types of polymeric coatings applied on E 32 naval steel offers improvement in its corrosion performances.

3.2. Evolution of Polarization Resistance during immersion time

The evolution of polarization resistance it was performed by linear polarization method around free potential value with a very small potential difference (± 40 mV) in order to keep the steady state surface. Linear polarization resistance is the only corrosion monitoring method that allows corrosion rates to be measured directly, in real time. From figure 2 (a), it can be seen that the lowest polarization resistance (R_p) value is attained by the uncoated E 32 naval steel being equal to 0.195 kohm·cm^2.

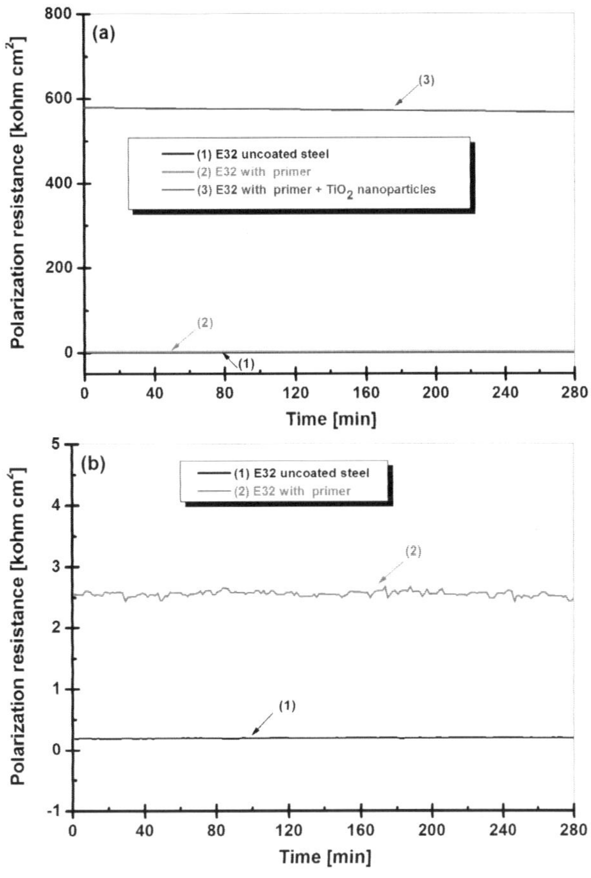

Figure 2 (a). The evolution of R_p values during immersion time of: (1) uncoated E32, (2) E32 coated with primer,(3) E32 coated with primer and blended with TiO_2 nanoparticles, (b). Zoom of polarization resistance diagrams from Fig. 2 (a) for: (1) uncoated E32, (2) E32 coated with primer.

This R_p value corresponding to uncoated EN32 naval steel remains constant during the immersion time in sea water. By applying the polymeric primer onto the E 32 naval steel the R_p increases to a mean value around 2.557 kohm·cm^2, being higher than that of uncoated E 32 naval steel. For E32 coated steel with polymeric primer blended with TiO_2 nanoparticles the R_p value is around 573.391 kohm·cm^2. An increased value of polarization resistance it means a lower corrosion current density and therefore a lower corrosion rate, as it can be seen from figure 2 (b). The higher corrosion current density value corresponds to E32 coated steel with polymeric primer and primer blended with TiO_2 nanoparticles.

According with the data presented in figure 3 it can be seen that the higher corrosion rate, as penetration rate, correspond to uncoated E32 naval steel compared with two types of polymeric primer coatings. From corrosion rate evolution, as penetration rate showed in figure 3 it can be seen that coated steel with primer and primer plus TiO_2 nanoparticles have very low values of 29.922 μm/year and 0.141 μm/year respectively as compared with uncoated steel which reveal a corrosion rate value of 386.79 μm/year. These results suggest that the two types of polymeric primer coatings increase the corrosion resistance of E32 naval steel and are in good agreement with the polarization resistance data results.

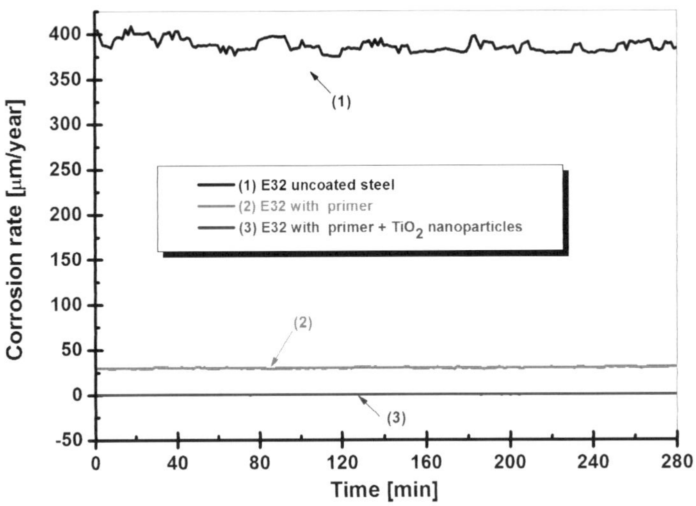

Figure 3. The corrosion rates, as penetration rate versus time corresponding to all three analysed surfaces immersed in seawater.

3.3. Cyclic voltammetry

In figure 4 (a, b) are presented the cyclic voltammograms of all three studied surfaces in seawater. These diagrams were recorded in the potential range of -1500 mV vs. Ag/AgCl to +1500 mV vs. Ag/AgCl at a scan rate of 5 mV/s and then reversed with the same scan rate till the starting potential to form one complete cycle.

Highlighting the localized corrosion susceptibility in the presence of chloride ions in the seawater of uncoated naval steel E32 can be seen very well due to the specific hysteresis aspect which presents the anodic transpassivation part of the curve (1) from figure 4 (a). The specific hysteresis aspect indicates that uncoated naval steel E32 is susceptible to pitting and crevice corrosion in marine environment. The steel coated only with polymeric primer show a smaller passive domain as compared with primer mixed with nanoparticles. Also some small hysteresis are revealed on cyclic voltametry, probably do to some porosity of coating. The steel coated with primer mixed with TiO_2 nanoparticles don't show the transpassive domain (anodic current did not increase) and reveal the absence of specific hysteresis for localized corrosion.

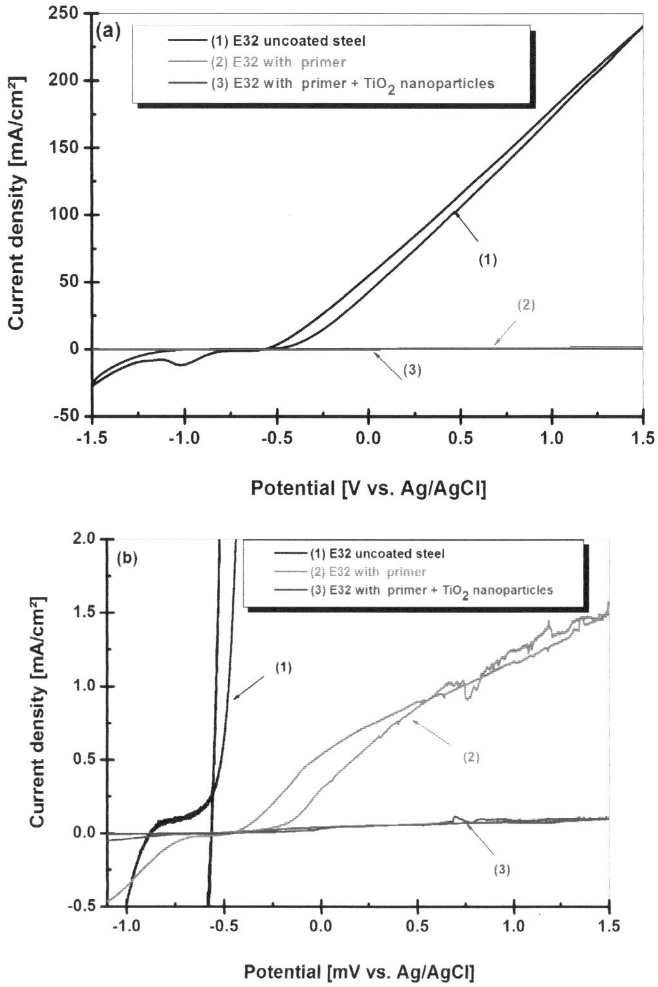

Figure 4 (a). Cyclic voltammograms of: (1) uncoated E32, (2) E32 coated with primer, (3) E32 coated with primer and blended with TiO_2 nanoparticles, (b). Zoom of Cyclic voltammograms in the lower range of current densities for: (1) uncoated E32, (2) E32 coated with primer, (3) E32 coated with primer and blended with TiO_2 nanoparticles.

From figure 4 it can be observed that upon the onset of the transpassive region the current still raises up until the potential is reversed in the case of uncoated naval steel E32 immersed in seawater indicating a susceptibility to pitting corrosion. On the other hand, the current density is almost constant for the naval steel coated with polymeric coatings on entire studied range of potential.

The recorded polarization curves of uncoated naval steel E32 are characterized by narrow passive regions, as shown in the figure 4 (a). The end of this narrow passive region denotes the pitting potential, E_{pit}, above which the passive film breaks down and pits initiate on the uncoated steel surface. This is indicated by the rapid increase in the passive current density without any sign for oxygen evolution, as a consequence of the passivity breakdown. These narrow passive regions of uncoated naval steel E32 reflects without any doubt the severity nature of the pitting attack induced by

Cl^- anions, compared with polymeric primer coatings.

The polymeric primer coatings have revealed the extended passive domains indicating once again a much higher corrosion resistance compared with uncoated naval steel E32.

4. Conclusion

From the present experimental research work and the obtained results, the following can be concluded:

By coating the naval steel with polymeric primer, the open circuit potential is shifted to more positive values from E=-681 mV to E=-599 mV. Adding to polymeric primer TiO_2 nanoparticles, the open circuit potential is even more shifted to positive values confirming the beneficial effect of TiO_2 nanoparticles.

From 200 point linear polarization, it was calculated the evolution of polarization resistance (Rp), resulting the highest value for modified polymeric primer with TiO_2 nanoparticles. Thus, Rp of steel coated with polymeric primer plus TiO_2 is about two hundred times higher than Rp of primer without naoparticles and about 1000 time higher than uncoated steel.

The good effect of TiO_2 nanoparticles addition to polymeric primes is also revealed by the corrosion rate, as penetration rate, which is smallest for coated steel with primer modified with nanoparticles, having the value of 0.141 μm/year. The uncoated steel show a higher corrosion rate of 386.79 μm/year, which was diminished only to 29.92 μm by coating with simply polymeric primer.

The cyclic voltametry measurements proved the better behavior of steel coated with primer plus TiO_2 nanoparticles, showing the highest passive domain in the scanned potential range. Also from cyclic voltametry it was revealed that uncoated steel is sensitive to localized corrosion, by showing the specific hysteresis. The steel coated with primer mixed with TiO2 nanoparticles don't show the transpassive domain (anodic current did not increase) and reveal the absence of specific hysteresis for localized corrosion.

The polymeric coatings increases the lifetime for naval steel and this is materialized in lowering maintenance costs.

References

[1] Soares G, Garbatov Y, Zayed A and Wang G 2009 *Corros Sci* **51** 2014–2026
[2] Upadhyay V and Battocchi D 2016 *Prog Org Coat* **99** 365–377
[3] Samad A A, Soud Y and Zaki M 2014 Applications *Journal of Surface Engineered Materials and Advanced Technology* **4** 189
[4] Catianis I, Ungureanu , Magagnini L, Ulazzi E, Campisi and Stanica A 2016 *Open Geosci* **8** 174–194
[5] Moreno J D, Bonilla M, Adam J M, Borrachero M V and Soriano L 2015 *Constr Build Mater* **100** 11–21
[6] Corvo F, Perez T, Dzib LR, Martin Y, Castaneda A, Gonzalez E and Perez J 2008 *Corros Sci* **50** 220–230
[7] Kim I T, Dao D K, Jeong Y S, Huh J and Ahn J H 2017 *J Constr Steel Res* **133** 256–268
[8] Jin F L, Li X and Park S J 2015 *J Ind Eng Chem* **29** 1–11
[9] Soares C G, Garbatov Y, Zayed A and Wang G 2009 *Corros Sci* **51** 2014–2026
[10] Touzain S, Le T Q and Bonnet G 2005 *Prog Org Coat* **52** 311–319
[11] Gurrappa I and Malakondaiah G 2005 *Mater Sci Eng* **A 391** 235–242

Thermokinetic Simulation of Precipitation in NiTi Shape Memory Alloys

C D Cirstea[1], E Karadeniz-Povoden, E Kozeschnik, M Lungu, P Lang, A Balagurov and V Cirstea
[1] National Institute for Research and Development in Electrical Engineering ICPE-CA, 313 Splaiul Unirii, 030138, Bucharest, Romania

E-mail: [1] diana_cirstea@yahoo.com, diana.cirstea@icpe-ca.ro

Abstract. Considering classical nucleation theory and evolution equations for the growth and composition change of precipitates, we simulate the evolution of the precipitates structure in the classical stages of nucleation, growth and coarsening using the solid-state transformation Matcalc software. The formation of Ni_3Ti, Ni_4Ti_3 or Ni_3Ti_2 precipitate is the key to hardening phenomenon of the alloys, which depends on the nickel solubility in the bulk alloys. The microstructural evolution of metastable Ni_4Ti_3 and Ni_3Ti_2 precipitates in Ni-rich TiNi alloys is simulated by computational thermokinetics, based on thermodynamic and diffusion databases. The simulated precipitate phase fractions are compared with experimental data.

1. Introduction

During the last years thermokinetic simulation have the best techniques to predict the evolution of the precipitates, allowing us to obtain materials with desired properties. NiTi shape memory materials show a reversible and diffusion less martensitic transformation between the austenite cubic phase with B2 structure and martensite monoclinic or orthorhombic phases with B19' or B19 structure [1-3]. In NiTi near-equiatomic alloys trained for practical (or industrial) applications, the properties like shape memory and superelasticity are controlled by the presence of the precipitates. These precipitates can be induced in the B2 matrix of NiTi materials by thermo-mechanical treatment or by aging/annealing supersaturated Ni-rich from the matrix composition. Thus, by balancing the number and size of the precipitates with the aging/annealing time and temperature, the precipitates disturb the simple B2-B19' transformation to B2-R-B19' multi-stage transformation due to intermediate R-phase, in the case of Ni_4Ti_3 [4,5]. These transformations are extremely important, providing a Ni-depletion in the matrix and strain field induced in the surrounding Ni_4Ti_3 precipitates [1,6-8] with direct influence on martensitic transformation, opposite to Ni_3Ti_2, Ni_3Ti or $NiTi_2$, precipitates that don't have any influence on martensitic transformation. Ageing or annealing of the Ni-rich NiTi alloys can lead to four types of precipitates and among them Ni_4Ti_3 and Ni_3Ti_2 are metastable phases, while Ni_3Ti and $NiTi_2$ are stable phases [1,2,5,6].

In order to design NiTi advanced materials we must understand the thermodynamics and kinetics processing design which often starts with a parametric design for chemical composition. For this reason it was shown in literature [1,2,5,9-11] that the change of composition (until 0.1wt% Ni in composition), temperature (between 325-650°C for aging/annealing), pressure (protective or not protective atmosphere in the processed of those materials) and external force field (cold or hot working of these materials for training) or every small adjustment lead to significantly different evolution of microstructure driven by the thermodynamics and kinetics of the NiTi system. It is known that thermodynamics is the core of the precipitation simulation and generally it is used the Calphad (calculation of phase diagram) method. The Calphad method is correlated with predictive models for kinetic process of particle size evolution (by modeling techniques quantitatively) [12]. The most kinetic models used for coarsening is represented by the Lifshitz-Slyozov-Wagner (LSW) theory [13-17]. Due to the relations between processes-structure-property of materials, were developed some kinetic models for particle size, or coarsening rate (Morral-Purdy model), or concomitant nucleation, growth and coarsening of precipitates (Kampmann and Wagner numerical (KWN) model) extended to multicomponent alloys systems [15,16,18]. Another design model from Calphad thermodynamic and kinetic to predict phase transformation kinetic is the Kolmogorov-Johnson-Mehl-Avrami (KJMA) model, which is used to study precipitation processes and to generate Precipitation- Temperature – Time (PTT) diagrams [12]. At present, there are a lot of commercial software package available on the Calphad technique like Thermo-Calc, Pandat, MTDATA, FactSage. A commercial precipitation-kinetics software package, MatCalc, that has been developed based on thermodynamic extremum principles [19-20] will be used in this paper to correlate the process-structure modeling of solid-state transformation with practical applications for NiTi materials obtained by spark plasma sintering (SPS) and aging at 400°C.

2. Experimental data

The thermokinetic simulations were performed with the MatCalc software, version 6.00 release 0.100, and the thermodynamic database mc_sma_v1.000, using optimized Gibbs energy functions of metastable phases from Povoden-Karadeniz et al. [21] and kinetic database mc_ni_v2.001.ddb [22]. The selected phases for simulations were B2, Ni_3Ti_2, Ni_4Ti_3 and Ni_3Ti, considering the kinetic behavior from literature [2,5,9] and corresponding to the SPS, AG treatments and to different conditions presented above.

For all simulations, the interfacial energy was set to γ = 0.12 Jm^{-2} for incoherent Ni_3Ti phase. For Ni_4Ti_3 and Ni_3Ti_2 phases we used two correction files. The nucleation model used in the simulations was classical nucleation theory (CNT) and the growth model was Svoboda–Fischer–Fratzl–Kozeschnik (SFKK) [19,20].

For the present simulation, the nucleation site for each type of precipitate phase was defined according to experimental observations presented in papers [1-5,10,11,23]: nucleation was allowed in the bulk for Ni_4Ti_3, on dislocations for Ni_3Ti_2 and respectively, on grain boundary for Ni_3Ti phases. All elements were kept in supersaturated solid solution at time $t = 0$. The results of coherent precipitation of Ni_4Ti_3 were compared with data observed experimentally for the sintering treatment that was made by spark plasma sintering (SPS) by applying a stress of 50MPa for 5 minute, heating at 900°C and 850°C for 2 minute and followed by aging (AG) at 400°C and water quenched. The aging treatment was carried out at 400°C, for 3 h, in a tubular furnace under argon atmosphere. The entire heat treatments take about 5 hours.

In order to test the applicability of the thermokinetic simulations, the NiTi materials must be characterized from the microstructural point of view by X-ray diffraction (XRD) and scanning electron microscopy (SEM). In addition, it must be done some nanoindentation tests for two compositions, respectively 50.8% Ni-Ti and 51.5% Ni-Ti. By the thermokinetic simulations we test the capability to reproduce the experimental microstructures of NiTi samples obtained by spark plasma sintering (SPS) for two different temperatures (900°C and 8500°C) and aging (AG) at 400°C in the oven in argon atmosphere. All samples were embedded, grinded, polished and afterwards etched. The conditions of XRD and SEM and nanoindentation measurements are described in the papers [24]. In order to estimate the phase fraction of each crystalline phase, the XRD experimental spectrum was simulated using Rietveld method by Maud software, where the values of reliability factors were below 5%. The chemical compositions of precipitates were determined by EDS.

The NiTi samples with 51.5at% Ni-Ti composition, obtained in SPS, were coded E1 for 900°C, respectively E2 for 850°C. Similarly, the NiTi samples with 50.8at%Ni-Ti composition, were coded E3 for 900°C, respectively E4 for 850°C. The same code is used for thermokinetic simulation.

3. Results and discussions

In this section are presented the results of thermokinetic precipitation simulations in the Ni-Ti system for two different compositions and two different technological fluxes. Only for SPS we have different temperatures, namely 900°C and 850°C, while the conditions for aging (AG) applied on NiTi materials are the same (400°C). The SPS and AG treatment results from simulations conducted for NiTi materials with the atomic composition of $Ni_{50.8}Ti$ and $Ni_{51.5}Ti$ are presented in the figure 1.

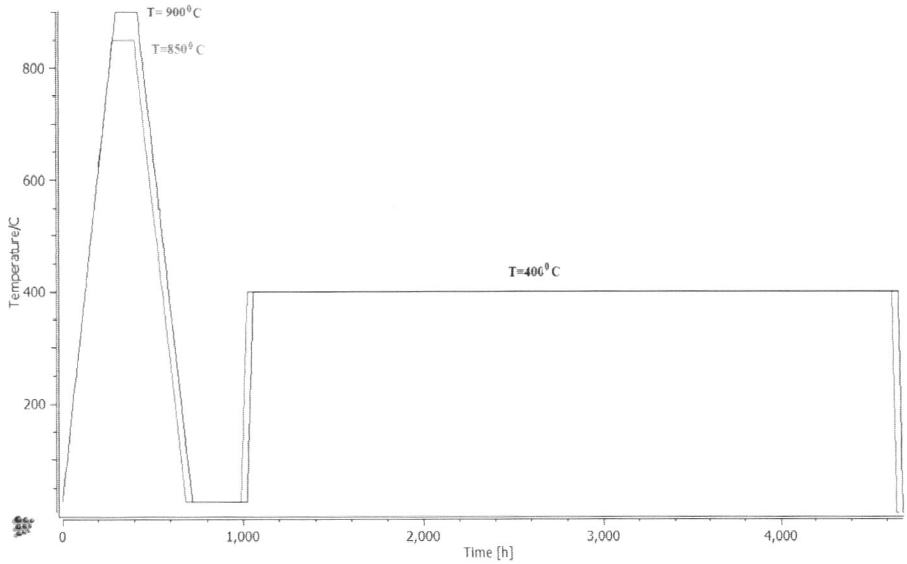

Figure 1. Simulations results for $Ni_{50.8}Ti$ and $Ni_{51.5}Ti$ obtained by SPS at 900°C or 850°C and AG at 400°C (MatCalc software).

In the first step, the precipitation kinetic was simulated for $Ni_{51.5}Ti$ composition taking into account the different SPS temperatures and AG. The results of this simulation are summarized in figure 2 and figure 3. The phase fractions of Ni_3Ti and Ni_4Ti_3 precipitates starts to increase for both heat treatments temperatures, being higher in E1 and E2 (more Ni in composition) than E3 and E4. Moreover, it was found that, during AG treatment, in E2 there is a tendency to decrease the Ni_3Ti phase fraction, suggesting a coarsening process which is in accordance with table 1. The calculated precipitate radius at this point (at 400°C) is about 100 nm for E1 and 80 nm for E2, whereas the number density is in the range of $3\cdot10^7 m^{-3}$ (E1) or $1\cdot10^{18} m^{-3}$ (E2). The values of precipitate radius calculated for E1 are higher than E2, while the number precipitate of E1 is smaller than E2. In addition, the Ni_3Ti phase fraction decrease during AG in E2 and increase slowly in E1, while the Ni_4Ti_3 phase fraction decrease very slow in E2. The calculated mean radius for Ni_4Ti_3 precipitate is about 5-10 nm in E1 case and is much higher than in E2. The number density for Ni_4Ti_3 precipitate is high and this can be correlated with the values of the material hardness presented in table 2.

In the second step ($Ni_{50.8}Ti$), the phase fraction of Ni_3Ti and Ni_4Ti_3 precipitates start to grow for both E3 and E4, but the maximum values are lower than the E1 and E2. Similar to the results from the first step, it was found that during AG treatment, in E4 there is a tendency to decrease the Ni_3Ti phase fraction, while in the E3 is growing. The calculated mean radius is about 140 nm for E3 and 60-80 nm for E4. In case of Ni_4Ti_3 precipitate, the mean radius decreases after AG. The number density for

Ni_4Ti_3 precipitate is very low according to simulations. The results obtained from simulations for the three parameters under consideration are comparable to [10,11,23,25] for a composition of 50.7%atNi.

Results of thermokinetic simulations of precipitate evolution for Ni_4Ti_3, Ni_3Ti_2 and Ni_3Ti precipitates, particle size versus number density and the histograms of Ni_4Ti_3 and Ni_3Ti_2 precipitates are shown in figures 4 and 5 and are in accordance with table 1.

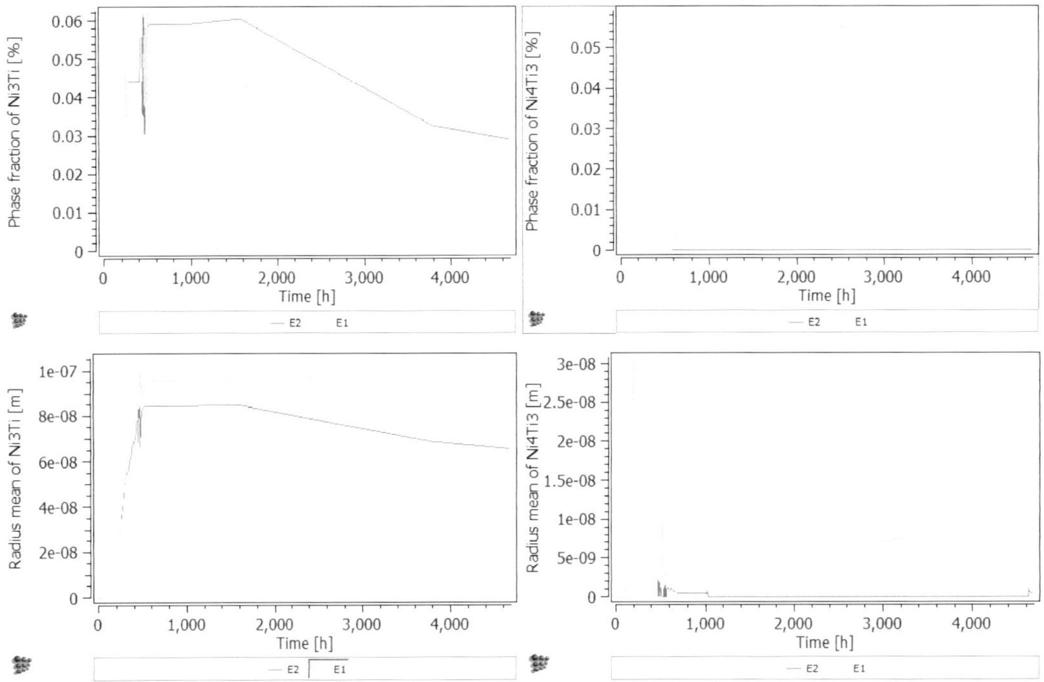

Figure 2. Results of the thermokinetic simulation of the precipitate evolution (phase fraction [%] and mean precipitation radius [m]) for E1 and E2 of $Ni_{51.5}Ti$.

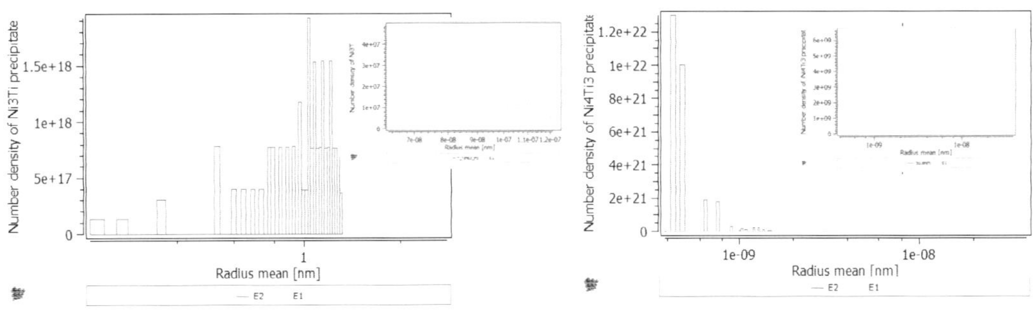

Figure 3. Histogram plot for Ni_3Ti and Ni_4Ti_3 precipitates related to figure 2.

Since no thermal events were identified for Ni_3Ti_2 precipitates during the heat treatments, they were considered zero and, therefore, has not been taken into account.

The results obtained from XRD, SEM and nanoindentation characterizations performed on the manufactured materials can be correlated with the results calculated by simulations. Thus, the phase fraction calculated by the Rietveld method approaches the values obtained by simulation. Also these results usually occurs in NiTi alloys with concentrations of Ni between 50.8 and 51 at.% [2].

The experimental results obtained by XRD shown the existence of a structure corresponding to B2, B19', Ni_4Ti_3, and Ni_3Ti precipitates with B19' major phase.

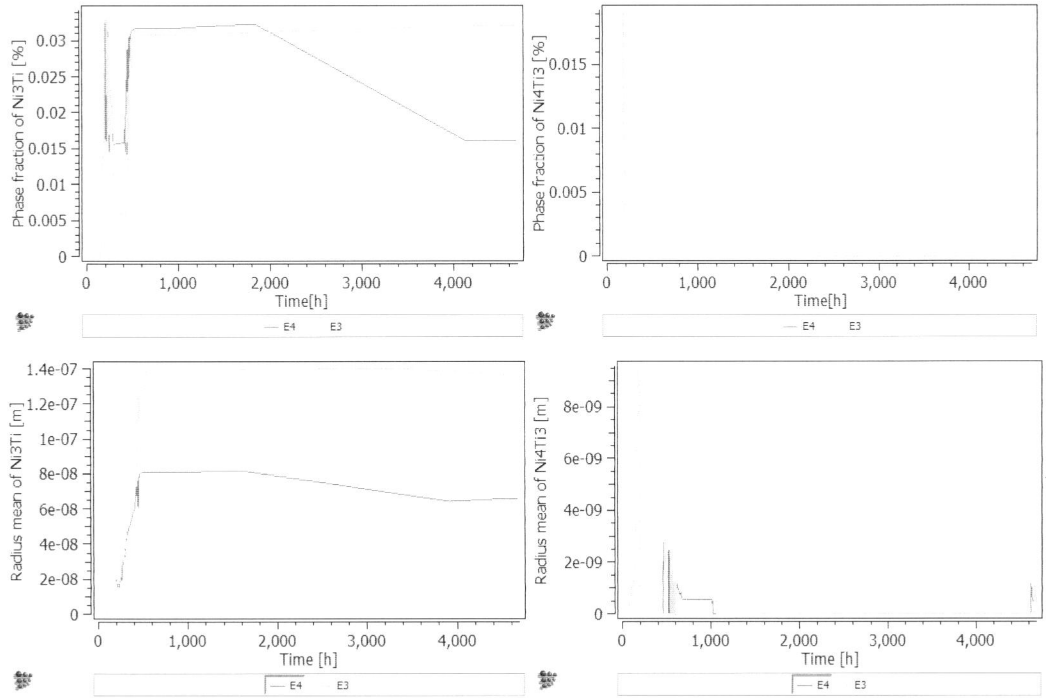

Figure 4. Results of the thermokinetic simulation of the precipitate evolution (phase fraction [%] and mean precipitation radius [m]) for E3 and E4 of $Ni_{50.8}Ti$.

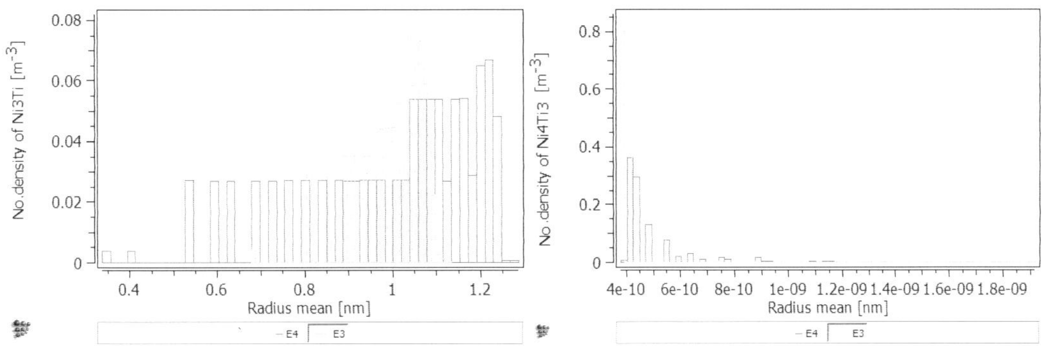

Figure 5. Histogram plot for Ni_3Ti and Ni_4Ti_3 precipitates related to figure 4.

This suggests that the materials obtained by SPS are martensite type, fact confirmed by Young's module values ranging from 42 to 62 GPa according to [26,27]. Also, SEM micrographs (figure 7) show a fully recrystallized microstructure containing few dispersed precipitates at grains and twins boundaries.

Table 1. Relative percentage of the phases calculated by Rietveld method for NiTi materials.

Sample code	Phase [wt%] of B19'	Phase [wt %] of B2	Phase [wt %] of Ni_3Ti	Phase [wt %] of Ni_4Ti_3
E1	81.75±4.85	9.181±0.52	6.49±0.81	2.56±0.31
E2	81.49±4.79	10.43±0.67	5.33±0.95	2.76±0.36
E3	95.42±4.17	2.17±0.91	1.18±0.32	1.21±0.21
E5	94.19±0.0	2.29±0.29	1.63±0.4	1.86±0.42

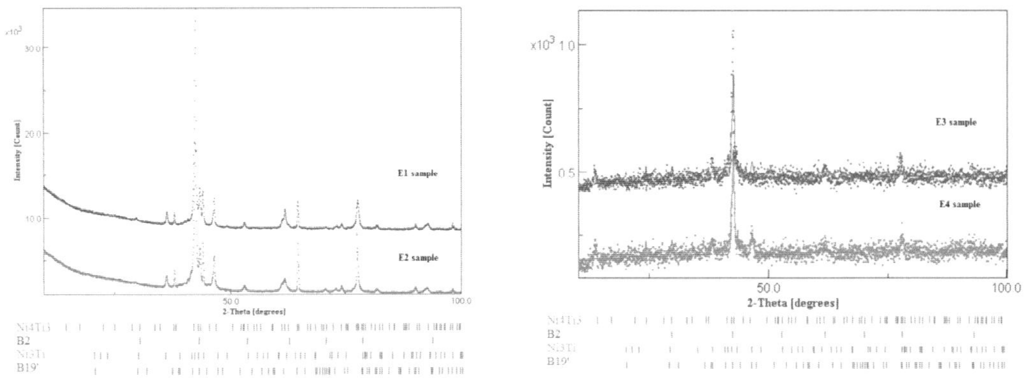

Figure 6. Simulated XRD results of NiTi sample processed by SPS and AG.

In figure 7 are presented several SEM microstructures representative of different scales.

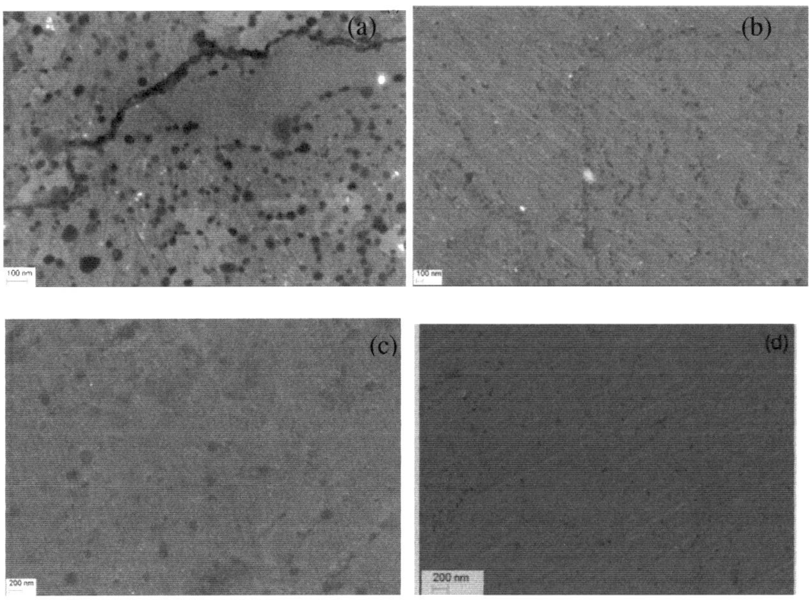

Figure 7. SEM Microscopy for E1(a), E2(b), E3(c) and E4(d).

Table 2. Results of nanoindentation measurements.

	E1	E2	E3	E4
HV	478±39	393 ±38.34	336±7	285±38.66
E (GPa)	64.59±2.59	44.59±1.06	46±8.99	42±5.5

4. Conclusions

We demonstrated that thermodynamics and mobility databases, combined with nucleation theory and the SFFK growth model for the precipitate evolution are successfully used to evaluate experimental microstructures in SMA related to stable Ni_3Ti and metastable Ni_4Ti_3 precipitates. The simulations deliver a prediction of precipitation trends as a function of aging time and temperature, which is important information for the realization of optimized NiTi materials.

The heat treatments applied to shape memory alloys lead to significant structural changes like variations of Young's modulus values characteristic to the martensite phase, depending on time of mechanical alloying and on the applied thermal cycles. Due to structural changes that appear during the heat treatment processes, the mechanical characteristics also change.

References

[1] Beyer J, Brakel R A and Lloyd J R T 1986 *ICOMAT* **86** 703-8
[2] Nishida M and Wayman C M 1986 *Metallurgical Transactions* **A 17** 1505-15
[3] Chapman and Hall Ltd. 1972 The Strengthening of Ti-55at.%Ni by formation of Ti_2Ni_3 Intermetallic Phase *Journal of Material Science* 7
[4] Paryab M, Nasr A, Bayat O, Abouei V, Eshraghi A 2010 *Metalurgija* **16,** 123-131
[5] Otsuka K, *, X. Ren Physical metallurgy of Ti–Ni-based
[6] Nishida M, Hara T, Ohba T, Yamaguchi K, Tanaka K and Yamauchi K *Mater. Trans.* **44** 2631-2636
[7] Michutta J, Somsen Ch, Yawny A, Dlouhy A and Eggeler G 2006 *Acta Mater.* **54** 3525
[8] Michutta J, Carroll M C, Yawny A, Somsen Ch, Neuking K and Eggeler G 2004 *Mater. Sci. Eng.* **A 378** 152
[9] Sinha A, Datta S, Chakraborti P C and Chattopadhyay P P 2013 Understanding the Shape-Memory Behavior in Ti-(~49 at%)Ni Alloy by Nanoindentation Measurement *Metallurgical and materials transactions* **44A**
[10] Allafi J K, Ren X and Eggeler G 2002 *Acta Materialia* **50** 793-803
[11] Bujoreanu L G, Yg M L, Gollerthan S, Somsen C and Eggeler G 2010 *Int. J. Mat. Res* **101** 623-30
[12] Xiong W and Olson G B 2016 Accelerating cybermaterials innovation, *Computational Materials*
[13] Lifshitz I M and Slyozov V V 1961 *J. Phys. Chem. Solids* **19** 35-50
[14] Wagner C 1961 Theorie der Alterung von Niederschlägen durch Umlösen (OstwaldReifung) *Zeitschrift für Elektrochemie Berichte der Bunsengesellschaft für Phys Chemie* **65** 581–591
[15] Morral J E and Purdy G R 1994 *Scr. Metall. Mater.* **30** 905-8
[16] Morral J E and Purdy G R 1995 *J. Alloys Compd.* **220** 132-5
[17] Kuehmann C J and Voorhees P W 1996 *Metall. Mater. Trans.* **A 27** 937-43
[18] Wagner R, Kampmann R and Voorhees P W 2005 Phase Transformations in Material (ed. Kostorz G.) 309–407 (Wiley-VCH Verlag GmbH and Co. KGaA)
[19] Svoboda J, Fischer F D, Fratzl P and Kozeschnik E 2004 *Theory, Mater. Sci. Eng.* **A 385** 166-174
[20] Kozeschnik E, Svoboda J, Fratzl P and Fischer F D 2004 *Mater. Sci. Eng.* **A 385** 157-165
[21] Povoden-Karadeniz E, Cirstea C D, Lang P, Wojcik T and Kozeschnik E 2013 Thermodynamic of Ti-Ni shape memory alloys *Calphad Journal* 128-139
[22] www.matcalc.tu
[23] Panchenko E Yu, Chumlyakov Yu I, Kireeva I V, Ovsyannikov A V, Sehitoglu H, Karaman I and Maier yHj 2008 *The Physics of Metals and Metallography* **106** 577-89
[24] Cirstea C D, Lungu M, Balagurov A M, Marinescu V, Culicov O, Sbarcea G and Cirstea V 2015 *Adv. Eng. Forum* **13** 83-90
[25] Fan G, Zhou Y, Chen W, Yang S, Ren X and Otsuka K 2004 *Acta Mater* **52** 4351-4362
[26] Neves F, Martins I, Correia J B, Oliveira M and Gaffet E 2007 *Intermetallics* **15**(12) 1623–31
[27] Mahesh K K, Braz Fernandes F M and Gurau G 2012 *Journal of materials science* **47** 6005-14

IOP Conf. Series: Materials Science and Engineering **209** (2017) 012058 doi:10.1088/1757-899X/209/1/012058

Pulsed Laser Cladding of Ni Based Powder

A Pascu*, E M Stanciu, C Croitoru, I C Roata and M H Tierean
Department of Materials Engineering and Welding, Transilvania University of Brasov, 29 Eroilor str, 500036, Brasov, Romania

E-mail: alexandru.pascu@unitbv.ro

Abstract. The aim of this paper is to optimize the operational parameters and quality of one step Metco Inconel 718 atomized powder laser cladded tracks, deposited on AISI 316 stainless steel substrate by means of a 1064 nm high power pulsed laser, together with a Precitec cladding head manipulated by a CLOOS 7 axes robot. The optimization of parameters and cladding quality has been assessed through Taguchi interaction matrix and graphical output. The study demonstrates that very good cladded layers with low dilution and increased mechanical proprieties could be fabricated using low laser energy density by involving a pulsed laser.

1. Introduction

Up to date, there are many technologies available to fabricate coatings layers on different types of steel substrates, among which laser cladding is one of the most used technique for improving the properties of new surfaces or to recondition the worn components [1,2]. A higher compactness of the coatings high precision and superior mechanical proprieties are only several advantages of laser cladding over the conventional coatings methods (plasma coating, flame coating or welding) The main goal in the laser cladding technology is to not overheat the base material and to obtain a low dilution, typically (< 5%) and in the same time to have an increased adhesion between the coating and base material [3,4].

Usually for laser cladding a continuous wave laser is used that can provide a high and constant power for the process [5,6,7]. CO_2, Nd:YAG and diode are the most common continuous lasers involved in laser cladding processes. In contrast, a pulsed laser is able to provide the laser energy in plusses meaning that the power density can be tuned by setting the pulse duration and the number of pulses per time unit. In a complex study, Sun at el. [8] investigated the influence of pulse energy, frequency and powder feed rate on the Stellite 6 cladding on stainless steel and concluded that dilution has a major influence on the hardness of the cladded layer and that an 89% pulses overlap is the optimal values in order to prevent the occurrence of small cracks in the cladded layer. A similar study was carried out by Farnia et al. [10] which analyse and optimize the melting ratio as the key factor for the process design. It was determined that pulse duration and overlapping factor have different effects at low and high values. Moreover, the melting ratio increases at low values of pulse duration and decrease at longer pulse duration that mean there is more interaction time between the laser pulse and the coating in case of short pulses.

A comparison between continuous and pulsed laser was made by Zhang et al [9] using preplaced powder method to coat titanium–vanadium carbides reinforced in a Fe-based matrix. They report advantages like nanoscale carbides formation, finer grain structure and improved hardness in case of using the pulsed laser.

Pulsed laser cladding of WC carbide in overlapped geometry was successfully realized by C.P. Paul [11]. Full dense and crack free coating were fabricated using different parameters for pulse width, frequency and laser power.

Muwala et al. [12] uses a modulated Yb-fibre laser to investigate the preplaced cladding of Inconel 718 powder on AISI 304 steel substrate. Using PVA as binder to deposit the powder was determined that stacks of columnar dendrites are formed predominately in case of using the pulsed laser due to the repeated melting solidifications cycles.

The aim of this study is to further investigate the influence of laser pulse width and frequency on the geometry profile and hardness of Ni based powders. Using the design of experiments method is determined what parameters have the most influence on the cladded tracks and what is the interaction between the process parameters.

2. Materials and methods

2.1. Materials

Stainless steel was used as substrate for the laser cladding tests, respectively AISI 304 grade plates with nominal dimensions of 60x60x6 mm. Currently 304 stainless steel in used for manufacturing of a large number of components in marine equipment, automotive industry, petro-chemical industry and for applications that require moderate to high corrosion resistance,

Nickel based Inconel 718 atomized powder has been used for laser cladding through different process parameters. This type of powder is characterized by a very good behavior for laser cladding and can be used for various reconditioning applications of worn stainless steel components.

The Inconel coatings are dense, pore free and exhibit excellent creep and stress rupture coupled with a good corrosion resistance at high temperatures (up to 700°C).

Table 1. Chemical composition of the powder.

Material	Element (%)											Powder dimension
	Ni	Cr	Fe	Mo	Cu	Nb	Ti	Si	Mn	C	B	
MetcoClad 718	Bal	19	18	3	-	5	1	0.2	0.08	0.05	0.005	44…90 μm
AISI 304*	9.6	19.2	Bal	-	-	-	-	0.80	1.61	0.053	-	

* analysed by SPECTROMAXx M spectrometer

2.2. Cladding obtaining

The experimental tests have been carried out using a TRUMPH Trupulse 552 pulsed laser with an average power of 552 W and a Precitech YC50 coaxial cladding head manipulated by a CLOOS 7 axes welding robot (figure 1). The powder was dosed using a AT-1200HPHV Termach (Thermach Inc. USA) powder feeder and argon with 99.99% purity was used as carrier gas.

The cladding module consists in a conical powder injection system having the laser beam positioned coaxially with the powder jet.

A stand-off distance of 10 mm was used between the cladding head and the substrate and a 5° tilting angle in the cladding direction was necessary to protect optical system of the laser. To investigate the effects of the cladding parameters on the clad profile and mechanical proprieties, experiments consisting in coaxial laser cladding of a single track was performed.

Altogether nine individual track of 40 mm length were fabricated with a specific set of parameters and the optimum ones have been determined through the DOE analysis method.

Figure 1. The experimental frame used for the laser cladding of Inconel 718.

Table 2 summarizes the process parameters and the main geometrical and mechanical characteristics of the laser cladded tracks. All the claddings have been performed using a 2 mm pulsed laser spot diameter with rectangular shaped profile and a powder feed rate of 4 g/min.

Table 2. Laser cladding parameters.

Sample	Power [W]	Pulse duration [ms]	Frequency [Hz]	Width [mm]	Hight [mm]	Clad Area [mm²]	Melt Area [mm²]	Melt depth [mm]	Wet angle [°]	Dilution [%]	Micro Hardness [HV_{02}]
1.1	2000	0.8	130	1.35	0.20	1.49	0	0	152	0	182
1.2	2300	1	130	1.83	0.33	0.39	0	0.03	150	0	200
1.3	2600	1.2	130	2.19	0.45	0.60	0.15	0.22	146	20	214
2.1	2000	1	150	1.73	0.36	0.42	0	0.04	145	0	215
2.2	2300	1.2	150	1.93	0.45	0.57	0.18	0.23	142	24	223
2.3	2600	0.8	150	1.66	0.35	0.40	0	0	144	0	203
3.1	2000	1.2	170	1.65	0.44	0.52	0.09	0.12	132	14.75	228
3.2	2300	0.8	170	1.66	0.32	0.38	0	0	150	0	208
3.3	2600	1	170	2.07	0.52	0.72	0.12	0.20	125	14.28	220

The laser cladded tracks have been cut and prepared by grinding, polishing and electrochemical etching in 10% wt. oxalic acid solution using continuous current (5 V and 400 mA). An inversed Leica DM IL optical microscope was used for the cross-section analysis of the single tracks. The micro-hardness analyses were realised using a Shimadzu HMV 2T micro-hardness tester. Five HV_{02} were made on each sample using the following set-up: load 200gf and dwell time of 15 sec

3. Results and discussions

Figure 2 illustrates the geometrical appearance of the obtained samples cross-section.

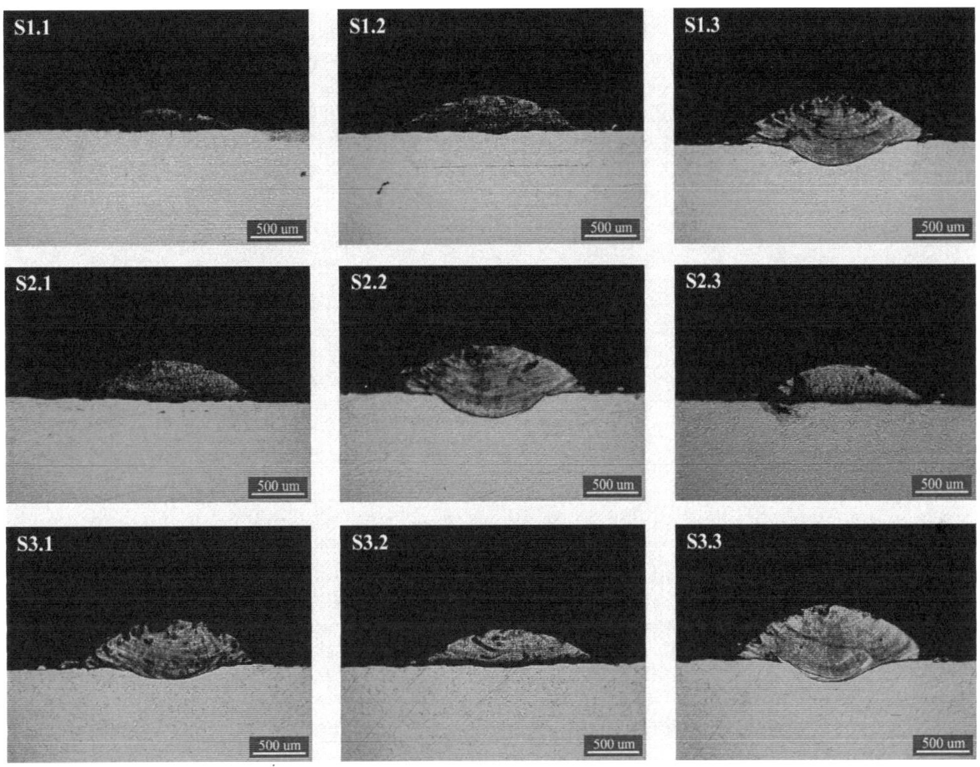

Figure 2. Low magnification microscopy of the single tracks laser claddings.

Compared with the continuous wave laser cladding presented in our previous works [13,14,15], the pulsed laser deposition bears more specific parameters regarding pulse frequency, duration, energy, and so forth which have a pronounced influence on the quality of the coatings. It can be clearly observed from figure 1 that the process parameters have a major influence on the geometrical profile of the laser cladded tracks. The pulse duration and pulse frequency of the laser have a direct influence on the amount of energy that is transferred into the cladding process per unit of time and area. As the general aim of the laser cladding process is to use low amounts of energy to avoid overheating the substrate. In the same time, it is also necessary to have enough power density to melt the powder and to create a strong bonding with the substrate. This fine tuning of the laser energy could be obtained by using a pulsed laser that allows controlling the pulse length and the number of pulses employed per unit area. Using the Taguchi design of experiments matrix, the influence of the power, pulse duration and frequency on the properties of the coatings (clad area, melt depth, dilution, micro-hardness) has been assessed. The nine tracks presented in the figure 2 shows that all process parameters have a significant influence on the tracks profile, as determined from the interaction plots depicted in figures 3 a and b. As a general rule for interpreting this type of graphs, the highest mean value (clad area, hardness) or the lowest (dilution, melt depth) represents the optimum value for the respective parameter. Also, the value of the slope dictates the intensity of the influence for the chosen variables. A higher variation slope means a significant interaction/dependency.

The most obvious influence of the operational parameters is on the clad area that represent the area of the material coated on the surface of the base material. Using the Taguchi three step optimization design approach, the dependence between the process parameters (power, pulse and frequency) and the clad area has been assessed (figure 2).

The cladded area is mainly influenced by the first two parameters variation, respectively from 2000 to 2300 W for laser power, 0.8 to 1 ms for laser pulse width and from 130 to 150 Hz for laser frequency. The optimal values for laser power is 2000W, 0.8 ms pulse width and 130 Hz frequency.

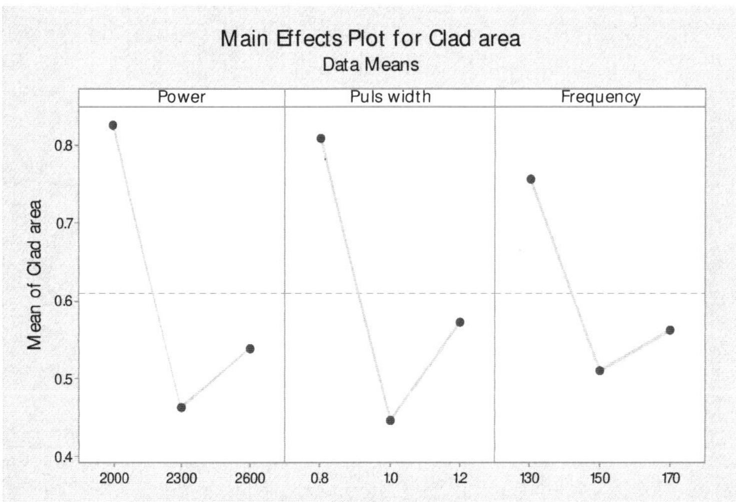

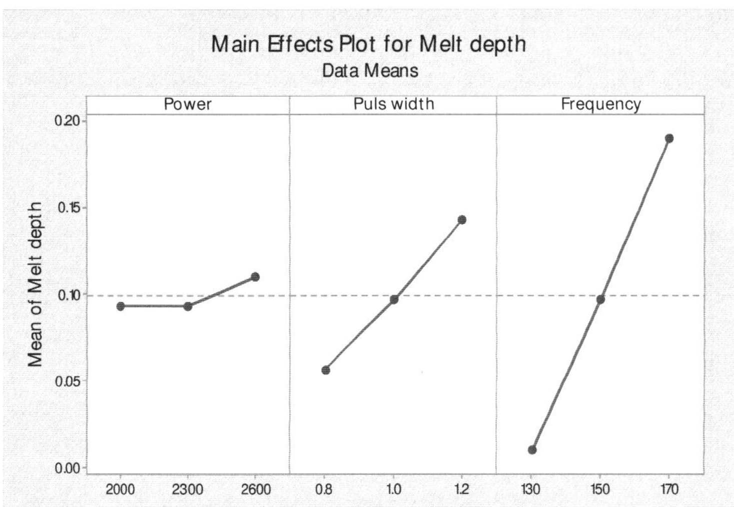

Figure 3. Main effect plot of clad area and melt depth as function of laser power, pulse width and frequency.

The process parameters have a different effect on the melt depth of the cladding tracks. Figure 1 suggests that laser power has no significant influence on the melt depth compared with the laser frequency that can drastically modify the melt depth form 0 up to 0.20 mm with direct consequence on the dilution and mechanical proprieties of the coating. It results that even if the laser energy is influenced by the power and pulse duration, the increasing of the number of pulses on the same unit of area will produce an increase of the melt depth of the substrate.

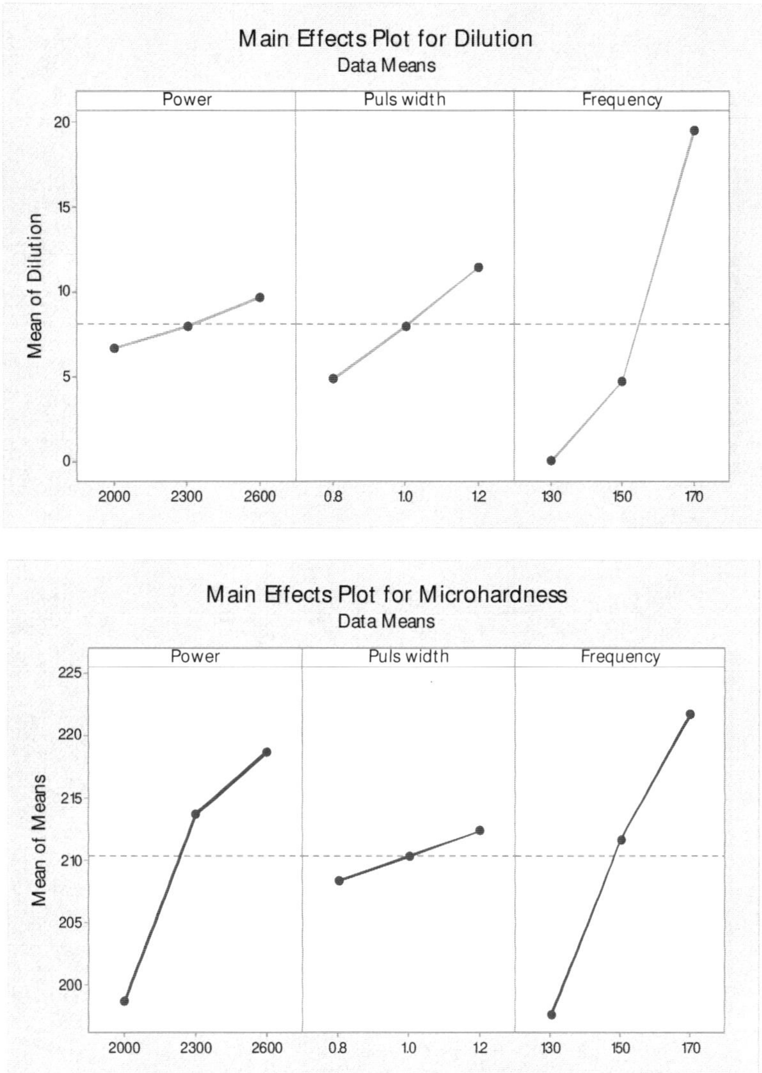

Figure 4. Main effect plot of dilution and microhardness as function of laser power, pulse width and frequency.

By analyzing the main effect plot for dilution, it has been determined that dilution is influenced by the melt depth of the cladded tracks the similarity between the two plots being obvious (figure 4 a and b). The frequency can produce the most visible effect on the melt depth and microhardness as resulted from figure 3b.

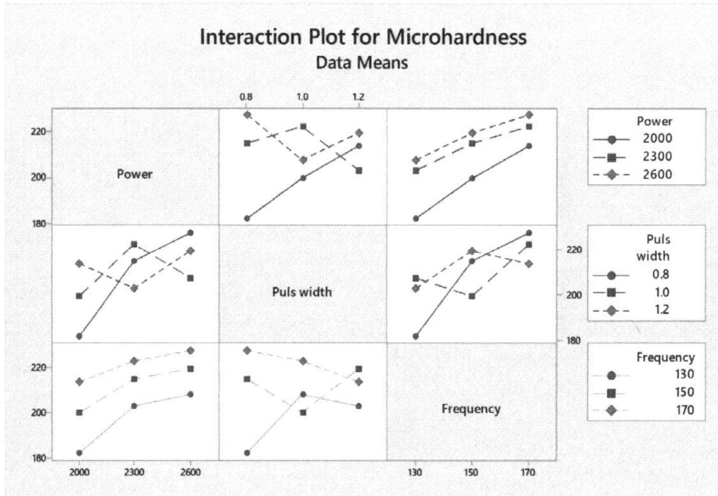

Figure 5. Interaction plot between power, pulse width and laser frequency.

Finally, all the effect plot for clad area, melt depth, dilution and microhardness must be considered as been related and interacted with-other.

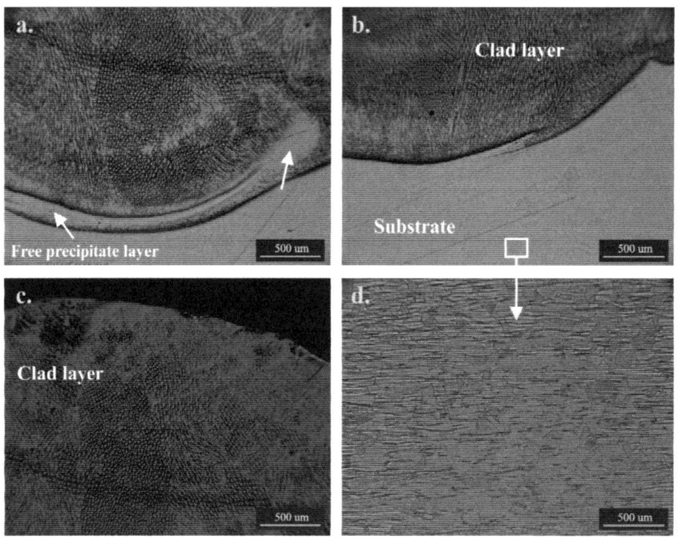

Figure 6. Microstructure of the Inconel 718 cladded tracks.

The process parameters can interact with-other and create a particular mean effect. The interaction plot presented in figure 5 highlights the interaction between the power, pulse width and frequency related to the clad micro-hardness. By analyzing the interaction slopes is clearly that only in case of power and frequency combination there is no significant interaction so the main effect, respectively the hardness increasing can be predictable. The interaction is severe in case of pulse with combination settings with power or laser frequency. It results that for a predictable result in case of laser cladding using a pulsed laser the best way to obtain the desired clad geometry profile is to set the power and frequency and to maintain the pulse with constant. The cladding microstructure is composed from coarse dendrite near the interface with the base material and more fine dendrite structure in the upper area of the coating. A noninterference line or a free precipitate layer as it could be seen in figure 6 a is the interface zone between the materials. No visible or microstructural modifications have been noted in the heat affected zone due to the low heat input of the pulsed laser. The cross-section profile and microstructure are modified depending on experimental parameters set-up. Therefore, a threshold temperature must be ensured to promote the nucleation of the dendrites and the formation of a dense and compact structure. This threshold temperature was not reached on samples 1.1 2.3 and 32 where large islands of unmelted powder are present in the coating surface.

4. Conclusions

In this study, a nickel based Inconel 718 atomized powder has been cladded on AISI 304 steel substrates by means of a pulsed laser. In order to find the optimal parameters of the laser which have an increased influence on the cladding quality, namely laser power, pulse width and frequency, a DOE approach has been used, by using a Taguchi design. The results, which express the interrelationship between laser cladding parameters and the characteristics of the clad produced, can be used to find optimum laser parameters, to predict the responses, and to contribute to a better understanding of the laser cladding process.

References

[1] Zhang P and Liu Z 2016 *Materials & Design* **100** 254–262
[2] Hemmati I, Ocelı´k V and Hosson J Th M 2015 *Electronic and Optical Materials* pp.137-162
[3] Hemmati I, Ocelı´k n V and DeHosson J Th M 2012 *Materials Letters* **84** 69–72
[4] Song T B, Hussain K and Voisey T 2016 *Physics Procedia* **83** 706-715
[5] Luo X, Li J and Li G J 2015 *Journal of Alloys and Compounds* **626** 102–111
[6] Damian Janicki 2017 *Optics & Laser Technology* **94** 6–14
[7] Hemmati I, Ocelı´k V and De Hosson J.Th.M 2013 *Physics Procedia* **41** pp. 302 – 311
[8] Sun S, Durandet Y and Brandt M 2005 *Surface and Coatings Technology* **194** pp. 225–231
[9] Zhang H, Zoun Y, Zou Z and Zhao W 2015 *Materials Letters* **139** pp. 255 – 257
[10] Amirreza F, Farshid M G and Sabbaghzadeh J 2013 *Optics and Lasers in Engineering* **51** 69–76
[11] Paul C P, Alemohammad H, Toyserkani E, Khajepour A and Corbin S 2007 *Materials Science and Engineering* A **464** 170–176
[12] Muvvala G, Karmakar D P and Nath A K 2017 *Optics and Lasers in Engineering* **88** 139–152
[13] Stanciu E M, Pascu A, Tierean M H, Voiculescu I, Roata I C, Croitoru C and Hulka I 2016 *Materials and Manufacturing Processes* **31** 1556-1564
[14] Iosif H, Serban V A, Utu D, Duteanu N M, Pascu A, Roată I C and Maior I 2016 *Romanian Journal of Materials* **46**(1) 49 – 54
[15] Pascu A, Stanciu E M, Roata I, Croitoru C, Baltes L and Tierean M 2016 *Bulletin of the Transilvania University of Brasov* **9**(58) 9-16

Acknowledgments

We hereby acknowledge the structural funds project PRO-DD (POSCCE,0.2.2.1., ID 123, SMIS 2637, ctr. No 11/2009) for providing the infrastructure at the Research& Development Institute ICDT-PRO-DD of the Transilvania University of Brasov used in this work.

The ICIR Euroinvent 2017 Conference mobility expenses have been supported by a grant of the Romanian National Authority for Scientific Research and Innovation, CNCS - UEFISCDI, project number PN-II-RU-TE-2014-4-0173.

IOP Conf. Series: Materials Science and Engineering **209** (2017) 012059 doi:10.1088/1757-899X/209/1/012059

Aeronautical Industry Requirements for Titanium Alloys

D T Bran, C F Elefterie and B Ghiban
University Politechnica of Bucharest, Splaiul Independenţei nr. 313, sector 6, Bucureşti, Romania

E-mail: dragos.bran@yahoo.com

Abstract. The project presents the requirements imposed for aviation components made from Titanium based alloys. A significant portion of the aircraft pylons are manufactured from Titanium alloys. Strength, weight, and reliability are the primary factors to consider in aircraft structures. These factors determine the requirements to be met by any material used to construct or repair the aircraft. Many forces and structural stresses act on an aircraft when it is flying and when it is static and this thesis describes environmental factors, conditions of external aggression, mechanical characteristics and loadings that must be satisfied simultaneously by a Ti-based alloy, compared to other classes of aviation alloys (as egg. Inconel super alloys, Aluminum alloys).For this alloy class, the requirements are regarding strength to weight ratio, reliability, corrosion resistance, thermal expansion and so on. These characteristics additionally continue to provide new opportunities for advanced manufacturing methods.

1. Introduction
To properly construct and maintain any equipment, especially airframes is vital to understand the uses, strengths, limitations, and other characteristics of this structural metals [1,2]. In aircraft may result in the loss of equipment and lives if is not taken into consideration the maintenance and repair, the substitution of inferior materials or even a slight deviation from design specification [3]. The finest craftsmanship can be readily erased by the use of unsuitable materials. For a specific repair job the selection of the correct material demands familiarity with the physical properties of various metals.

2. Properties of Metals
In the aircraft maintenance the general properties of metals and their alloys are the primary concern [5]. These terms are explained in this thesis to be able to establish a basis for further discussion of structural metals.

2.1. Brittleness
When a metal permits only a small amount of deformation without shattering we say it is a brittle material. Such a metal can break or crack without modifications in the shape. Structural materials are many times exposed to shock loads and this is why this characteristic is not advantageous. Examples of brittle materials are the cast iron and aluminum and sometimes some very hard grades of steels.

2.2. Density
The weight of a unit volume of a material is named density. To determining the weight of a part before actual manufacture can be used the specified weight of a material per cubic inch, in aircraft work. In

order to maintain the proper weight and balance of the aircraft when choosing a material to be used in the design of a part is important to have into consideration the density.

2.3. Ductility
The property of a metal that can be permanently drawn, twisted or bent into different shapes without breaking is named ductility. For metals used in making wire and tubing this property is essential.

In aircraft industry, it is better to use ductile metals due to an ease of forming and also they are more resistant to shock loads and other types of failure. This is why the aluminum alloys are preferred for the fuselage and wing skin, for the cowl rings and for the extruded components such as the spars, bulkheads and also the ribs. Another material with a high ductility used in aircrafts is chrome molybdenum because it can be formed into shapes. Another property similar to ductility is malleability.

2.4. Elasticity
The property that allows a material to come back to the original shape and dimension after a tension which caused the modification has been removed is called elasticity. Elasticity is one of the most important properties because knowing it, components that don't have a permanent deformation when the load is removed can be designed. Each material has an elastic limit. When the tension is greater, the deformations will be permanent. In the aeronautical industry, the components are conceived in such a way that that the maximum tolerances at which they will be exposed will not stress these components further than their elastic limit. An example of component where this property can be observed is a spring.

2.5. Electric and heat Conductivity
The ability of a material to conduct heat or electricity is called conductivity. Metals are generally good conductors of heat and electricity. Heat conductivity at metals is an important characteristic since during the welding it is valuable to know the amount of heat necessary for good fusion. In order to eliminate the radio interference, aeronautical engineers must consider the electrical conductivity of a metal in conjunction with bonding.

2.6. Fusibility
The property of a metal to turn into liquid phase after heat is applied is called fusibility. Welding is an example of process where the metals are fused together. The temperature of fusion for steels is about 1426°C and the temperature of fusion for most aluminium alloys is around 590°C.

2.7. Hardness
The ability of a material to resist abrasion, penetration, cutting action, or permanent distortion is named hardness. We can increase this ability in the case of steel and certain aluminum alloys, by cold working the metal and by heat treatment[6,7].

Usually the parts with structural character are formed while they are soft and afterwards thermal-treated to obtain hardness properties in such a way that the finished shape could be retained. The metals properties of hardness and the strength are usually directly related.

2.8. Malleability
A metal can be called malleable when is pressed into various shapes without cracking, is hammered or rolled. The metal that is worked into curved shapes, such as cowlings, fairings, or wingtips this property is necessary. An example of a malleable metal is copper.

2.9. Strength
Strength is an important material property. Is the ability of a material to resist deformation and to resist stress without breaking. The strength of the material is affected by the type of load or stress applied.

2.10. Thermal Expansion
The contraction and expansion of a metal as reaction to changes in temperature is the property of thermal expansion. After a metal has been heated, it will grow its dimensions. After cooling it down, the dimensions will shrink. The tolerances generated by this phenomenon must be taken into account while designing aircraft parts.

2.11. Toughness
The property of a material to be resistant to tearing or shearing and also to stretching and other deformation without damaging is called toughness. It is another very important property for a material used in aeronautics.

3. Titanium and Titanium Alloys
Titanium was first discovered by an English priest named Gregot. In 1825, a rudimentary separation process for the titanium was first performed. Enough quantity of pure titanium was obtained until 1906 for a study to begin. After the first study, the Kroll process was developed in 1932. Titanium sponge was first made by the United States Bureau of Mines as early as 1946. The melt process was developed four years after this and today the usage of this metal is very spread. Titanium has applications in many industries and there are high commercial orders for this material and its alloys for all kinds of components such as tools, fixtures, pumps and screens, mostly due to a high resistance to corrosion of titanium. In the aeronautical industry, titanium is used for the skin of the fuselage, for the engine components, for firewalls, for frames and other parts such as ducts, fasteners and fittings [4].

Other parts made from titanium are compressor disks, blades and vanes, spacer rings, bolts, housings for turbines and many others [7].

The visual aspect of titanium is similar to the stainless steel. To quickly distinguish titanium the spark test can be performed. During the burst, the titanium will generate a white brilliant trace. Another quick way to identify titanium is by wetting it and drawing a line on glass with it. The effect will be a dark line similar to the one traced by a pencil. When it comes to some of the most important properties, such as elasticity, strength to high temperatures and density, the titanium is between steel and aluminum. Titanium has a low coefficient of thermal expansion and also a low conductivity but a relatively high melting point, ranging from 1500°C to 1730°C. The most important physical aspects regarding titanium is that it is a light, corrosion resistant and strong material, with a density 60% bigger than aluminums but half of the stainless steel density. The properties for high temperatures are not great due to a high melding point and thus, for example the yield strength will rapidly decrease at temperatures higher than 430°C.

The avidity of titanium for nitrogen and oxygen increase once the temperature is above 500°C and thus making the metal brittle if a long contact is maintained , the metal becomes brittle and loses its value.

At high temperatures of up to 1650°C, titanium has its credit in applications where the strength is not a primary concerning aspect. Such requirements are for the firewall of the aircraft.

The magnetic character of titanium is nonmagnetic and in comparison with stainless steels, the electrical conductivity of titanium is similar. The harness values for most titanium alloys are high.

By the heat treatment, the titanium and its alloys don't achieve the high level of hardness as some steel heat treated alloys.

In the recent period, a new type of titanium alloy has been developed with triple treatable ability. Before this, the only method of forming that could be achieved was first heating and afterwards rolling the material. There is still the possibility that before treating the titanium for hardness to form the new obtained alloy in a soft condition.

The stabilizing elements which are used to produce titanium alloys are Iron, Molybdenum and Chromium. These alloys will be quenched and age hardened. The increase of ductility is also a consequence of adding these elements.

Steel and Aluminum have a lower fatigue tolerance than titanium does.

The hardness of Titanium is inversely proportional to the purity of the alloy. By chemical analysis, distinguishing between the unalloyed titanium or diverse grades of commercially pure titanium is not an efficient method. In conclusion, the mechanical characteristics of the material are the ones which are used at determining the grades.

3.1. The Designations of Titanium

In order to provide a advantageous and transparent way for characterizing all the titanium alloys, a classification was established and thus, three categories were proposed: A-B-C because there are three kinds of crystals which are determined by titanium and its alloys: A (alpha), B (beta), and C (combined alpha and beta) with the following properties:

• The alpha alloys — all around characteristics; good weldability; in both hot and cold conditions, they are strong and tough. Good corrosion resistance.

• The beta alloys —very good ductility and ability to bend; in hot and cold conditions the alloys are strong, but the contamination is problematic because these alloys are susceptible.

• Alpha + beta alloys— these alloys are strong in cold and warm condition, but, nevertheless, weak at high temperatures; the bendabilityis good; moderate resistance to corrosion; the forging properties are very good but these alloys represent a compromise for performances.

There are two ways of manufacturing the titanium in order to use it commercially: it can have commercial purity or it can be alloyed. The yield strength of the titanium and its alloys can varry from 380 MPa to 760 MPa. This material has a generic scope, and is manufactured in processes where moderate to severe forming apply. Occasionally, titanium and its alloys are used in nonstructural parts for aircraft design for multiple applications which require a high corrosion resistance. One of this applications is tubing with its fittings. Other applications for the titanium are due to the fact that it has a good weldability and where a high stress resistance is required in high temperature environments.

3.2. Resistance to Corrosion for Titanium

Titanium has a very good resistance to corrosion. A shielding surface layer is generating the corrosion resistant behavior, being made of stable oxide or chemically absorbed oxygen. Due to the existence of the oxygen and its agents, the layer is created. The character of the titaniums corrosion is uniform. A tiny proof of pitting exist and also forms of localized attack. The titanium and its alloys are not usually subjected to corrosion fatigue, stress corrosion galvanic or intergranular corrosion. These materials have an equal or larger resistance to corrosion than stainless steel grades.

Experimentally, it was determined that the titanium and its alloys tend to polarize readily when attacked with acids or saline solutions. This leads to a lower current flow in corrosion and galvanic cells. At the surface of titanium, there are corrosion currents and metallic couples which are restricted in a natural way. It is an effect which offers a good protection to many chemicals and also, the titanium and its alloys can be used in combination with other elements without the worry of the galvanic effect on eachother.

3.3. The Heat Treating of Titanium and its alloys

The reasons why titanium and its alloys need heat treatment are:

• The elimination of the tensions which have accumulated after forging

• To anneal the material subsequently to forging or to offer a high level of ductility for the plastic deformation processes.

• Improving the strength by thermal hardening

3.3.1. Tension Relieving. Usually, the tension relieving is used to eliminate stress concentrations consequential as of forming of titanium forged products.

This heat treatment has a range of temperatures which start at 343°C and go up to 538°C. The exposure in time is different from a rather thin plate which takes a few minutes to more than an hour for larger parts. One representative cure of stress relieving is at 480°C for half an hour, being subsequent to an atmospheric cooling. Pickling in solution with acid character is an easy solution to remove the discoloration or scales formed on the exterior of the treated metal. The percent of nitric acid which is recommended is aprox. 10 – 20% and up to 30% hydrofluoric acid. Room temperature is indicated for the operation.

3.3.2. Complete Annealing. The toughness and ductility are properties which can be obtained at room temperature for titanium and its alloys by annealing. Other obtained properties are the structural and dimensional stability at a high range of temperatures and also enhanced machinability properties.

A complete anneal is necessary to prepare the material for supplementary operations. This step of the treatment is carried to completion at a range of 650°C – 900°C. The exposure time at the high temperature can start from 1/4h and up to multiple hours, build upon parameters such as the dimensions and also the magnitude of the forging which is to be done. Generally, such a treatment is performed at 700°C for one hour, with subsequent atmospheric cooling. After the complete annealing treatment a caustic descaling process is mandatory because the treatment produces usually enough scale and so a sodium hydride salt bath must be provided.

3.3.3. Temperature Hardening. Pure titanium is not possible to be thermal treated, unlike the alloyed state, where the aeronautical industry alloys benefit by an increase of strength by thermal treating but this property comes most of the times with a compromise for the ductility. To have the highest properties, the recommendation is to perform a water quenching process at a temperature of 800°C and afterwards another heating up to 480°C with a time of 8h.

3.3.4. Casehardening. Special applications indicate the usage of titanium and its alloys due to the elements fast assimilation of other gases such as oxygen and nitrogen but also carbon. These can be absorbed at relatively low temperatures. The processes of carburizing, nitriding and carbonitriding are being used at the production of water-resistant case of 3 μm to 5 μm in depth.

4. Mechanical properties of Titanium Alloys

4.1. Tensile properties

Profitable (over 99% purity) grades of titanium poses a ultimate tensile strength of about 434 MPa, which is equal to that of common, low-grade steel alloys, but with a much lower density. Titanium has a density 60% higher than aluminums, but with properties which make it more than twice as strong as the most general used 6061-T6 aluminium alloy. Some titanium alloys achieve tensile strengths of over 1400 MPa. Nevertheless, when heated above 430°C titanium loses its strength. The tensile properties of the most used titanium alloys in aeronautics is presented in figure 1.

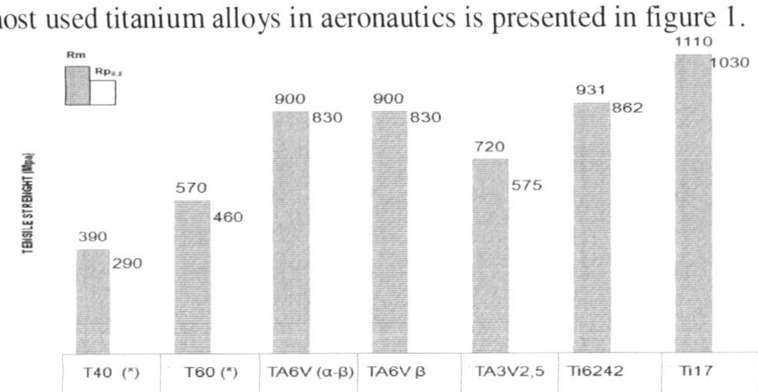

Figure 1. Titanium alloys ultimate and yield tensile strength.

4.2. Fracture toughness properties

In materials science, fracture toughness is a property which describes the ability of a material containing a crack to resist fracture, and is one of the most important properties of any material for many design applications, especially in the field of aeronautics. Figure 2 describes the fracture toughness for titanium alloys. A large scatter band is noticeable.

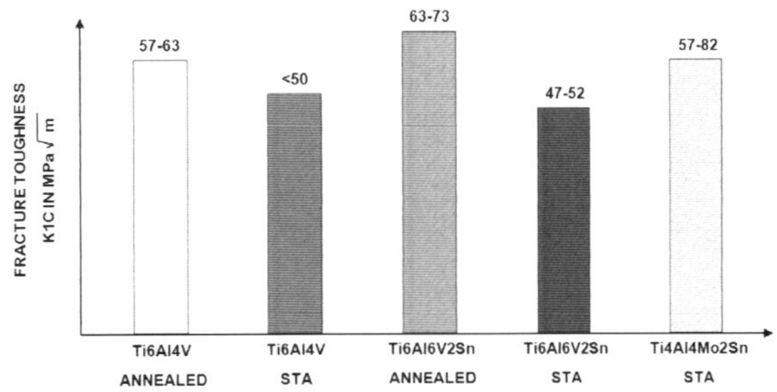

Figure 2. Fracture Toughness for Titanium Alloys.

4.3. Fatigue properties

Titanium isn't as hard as some grades of heat-treated steel, it is non-magnetic and has a low coefficient of electrical and heat conductivity. Precautions must be taken in order to machine titanium, because the material might gall unless sharp tools and proper cooling methods are used. The titanium structures have a fatigue limit that guarantees longevity in some applications, similar to steel structures. The fatigue requirements in aeronautics for the Ti-6Al-4V alloy is described in table 1 for the annealed forged state, for the Alpha-Beta

Hot Isostatically Pressed and Annealed state and for the β Annealed Forging state.

Table 1. Fatigue requirements.

State of the alloy	The curve of the test results, calculated in accordance with EN 6072, shall not lay below the reference curve defined by the following data:	
	σ max (MPa)	Number of cycles
annealed forged state	410	$1x10^4$
	270	$1x10^5$
	185	$1x10^6$
Alpha-Beta Hot Isostatically Pressed and Annealed state	400	$3x10^4$
	285	$1x10^5$
	230	$1x10^6$
β Annealed Forging state	410	$1x10^4$
	270	$1x10^5$
	185	$1x10^6$

4.4. Crack propagation properties

A occurrence in fatigue loading where the crack remains in a closed position is crack closure , even if on the material some external forces are present. A crack can be opened during this process, beyond a stress with a certain limit.

Phase transformation during the propagation of the crack and the plastic deformation of the material are some of the factors which contribute to this phenomenon and thus, the surface of the crack can present traces of corrosion, due to a fluid presence inside the crack which is amplified by a rough quality of the surface inside the crack. By diminishing the speed of the crack rate, a longer life for the material is gained.

A wide range of fatigue data can be explained by the effect of the crack closure. This is the standard analysis for the effects of load ratio. Almost all the fatigue prediction models use it. Nevertheless, predicting such effects of crack closure in laboratory conditions is not quite possible.

The crack propagation requirements in aeronautics for the Ti-6Al-4V alloy is described in table 2 for the annealed forged state, for the Alpha-Beta Hot Isostatically Pressed and Annealed state and for the β Annealed Forging state.

Table 2. Crack propagation requirements.

State of the alloy	The crack propagation rates shall not be greater than the following data:	
	ΔK (MPa√m)	da/dN (mm/cycle)
annealed forged state	15	$1,5x10^{-4}$
	20	$3,5x10^{-4}$
	30	$8x10^{-4}$
Alpha-Beta Hot Isostatically Pressed and Annealed state	15	$1x10^{-4}$
	20	$3x10^{-4}$
	30	$1,3x10^{-3}$
	40	$3x10^{-3}$
β Annealed Forging state	15	$4x10^{-5}$
	20	$1,5x10^{-4}$
	30	$7x10^{-4}$
	40	$1,5x10^{-3}$

5. Industry specifications for Titanium alloys

A list of specifications which state requirements for titanium alloys is shown in table 3.

Table 3. Specifications for titanium alloys.

Material	ASTM		AMS	German WL	British BS	French AIR
C.P.Ti	ASTM 265	Grade 1	AMS 4902	3.7024	BS TA1	AIR 9182 T-35
		Grade 2	AMS 4900	3.7034	BS TA2	AIR 9182 T-40
		Grade 3	AMS 4901	3.7064	BS TA6	AIR 9182 T-65
Ti3Al2.5V	ASTM 338 Grade 8		AMS 4944	3.7194		
Ti6Al4V	ASTM 265 Grade 5		AMS 4911	3.7164	BS TA10	AIR 9182 T-A6V
Ti4Al4Mo2Sn				3.7184	BS TA51 (IMI 550)	AIR 9183 T-A4DE
Ti6Al6V2Sn	MIL-T-9047 (Ti6-6-2)		AMS 4978 AMS 4979	3.7174		

IOP Conf. Series: Materials Science and Engineering **209** (2017) 012059 doi:10.1088/1757-899X/209/1/012059

6. Conclusions

Having a relatively low density, titanium is a low weight material, with good resistance to corrosion which indicate it in structural applications. Titaniums mechanical properties can be increased by allowing with other elements and through heat treatment processes. The titaniums ratio for strength to weight is first-class, it also has a low thermal expansion coefficient. Titanium has good oxidation resistance at intermediary temperatures. The weldability and toughness of titanium are good.

Through many application of titanium and its alloy, one with the most importance is the aerospace industry. With a large number of customers from this industry, it is directly linked with the economy of manufacturing this material. Due to a high demand, progress is made and better aeroplanes are manufactured. The higher demand of titanium is on the rise. In the aeronautical industry, many parts are manufactured from titanium, such as frames, engine components and others. The titanium performances provides the manufactures with what they desire in order to have the best designs. Since the jet propulsion engines and also the parts of the frame need to resist a wide range of temperatures from belo 0°C and up to 600°C, the titaniums high temperature performances are ideal. Parts of the engine which are manufactured of titanium are the blades , discs, caings and shafts. Due to such characteristics and applications, and the inter-dependence between the titanium and the aerospace industry, the market for this material should be strong in the years to come.

References

[1] ASM International Handbook Committee 1992 *ASM Handbook Volume 2 Properties and Selection Nonferrous Alloys and Special Purpouse Materials*
[2] ASM Handbook Committee 1985 *ASM Handbook Volume 9 Metallography And Microstructures*
[3] FAA Aviation *Maintenance Technician Handbook*
[4] Robert Baboian 2005 *Corrosion Tests and Standards Application and Interpretation - Second Edition*
[5] S. Banerjee and P. Mukhopadhyay 2007 *Phase Transformations - Examples from Titanium and Zirconium Alloys*
[6] Meyers M A and Chawla K K 2009 *Mechanical Behavior of Materials*
[7] William D and Callister Jr 2007 *Materials Science and Engineering An Introduction*

Aeronautical requirements for Inconel 718 alloy

C F Elefterie[1*], C Guragata[1], D Bran[1], B Ghiban[1]
[1] University Politechnica of Bucharest, Splaiul Independenţei no. 313, district 6, Bucharest, Romania

*E-mail: elefterie_florina@yahoo.com

Abstract. The project goal is to present the requirements imposed by aviation components made from super alloys based on Nickel. A significant portion of fasteners, locking lugs, blade retainers and inserts are manufactured from Alloy 718.
The thesis describes environmental factors (corrosion), conditions of external aggression (salt air, intense heat, heavy industrial pollution, high condensation, high pressure), mechanical characteristics (tensile strength, yield strength, and fatigue resistance) and loadings (tensions, compression loads) that must be satisfied simultaneously by Ni-based super alloy, compared to other classes of aviation alloys (as egg. Titanium alloys, Aluminum alloys).
For this alloy the requirements are strength, durability, damage tolerance, fail safety and so on. The corrosion can be an issue, but the fatigue under high-magnitude cyclic tensile loading it's what limits the lifetime of the airframe.
Also, the excellent malleability and weldability characteristics of the 718 system make the material physical properties tolerant of manufacturing processes. These characteristics additionally continue to provide new opportunities for advanced manufacturing methods.

1. Introduction

In the recent time the use of the Alloy 718 increased both from economical and technical point of view. Subsequently it becomes the most widely used superalloy for aerospace applications. With applications in critical rotating parts, supporting structure, airfoils and pressure vessels within an aircraft engine, makes it the most used alloy.

Improving the Alloy 718's quality and cost are the goals of all manufacturers, while personalize the requirements for melting, forging, heat treatment process, forging and keep the balance for creep and crack growth behaviour.

The intermetallic phases known to precipitate in Inconel718 are the metastable y' and y" and the 8 equilibrium phase. Current applications are limited due to the metastability of the y'' strengthening system. The efforts to increase the temperature stability of Alloy 718 with the normal approaches are not successful. In spite of wide ranges of applications, the number of investigations is limited for the mechanisms associated with tensile and cyclic deformation behaviour for this particular alloy.

In this paper, are reported the conditions designed to give the optimum properties for this alloy. The substructural changes published in the literature in Inconel 718 under cycling loading and monotonic condition are characterized and compared with other important alloys for having a better view of the importance of Inconel 718.

This subject is a matter of interest for other researchers in material science. Robert E. Schafrik, Douglas D. Ward and Jon R. Groh who studied the past, present and future applicability of the Inconel 718 and during 35 years discovered that the evolution of this alloy was a success and they believe that will be interesting to look back in another 35 years more to see the evolution of 718 alloy that set its bar high.[3]

Sreeramesh Kalh, K. Bhanu Sankara Rao, Gary R. Halford and Michael A. McGaw" [4] explained the deformation and damage mechanisms in Inconel 718 tested under monotonic tensile strains and fully-reversed fatigue.[4]

2. Environmental factors

Many variables influence the performance of a specific material in a specific environment. These include concentration, temperature, aeration, liquid or gaseous flow rates, impurities, abrasives, and cycling process conditions.

The atmospheric factors (e.g. relative humidity, temperature, sulfur dioxide content, hydrogen sulfide content, chloride content, amount of rainfall, dust) and even orientation of the exposed metal, all can effect marked influences in corrosion behavior. In an arid atmosphere, free of contaminants, only negligible corrosion would be expected.

Due to the presence of chromium and aluminum and formation of their respective stable oxides, the passivation of Inconel 718 surfaces is achieved, conferring its characteristic corrosion resistance.

Metals that form passive layers are susceptible to localized corrosion (pitting), which consists in the formation of short extension and deep cavities. This kind of corrosion initiates due to the break of passivation film, usually in defects such as inclusions, dislocations, grain boundaries or other interfaces.

Corrosion problems can be found in many forms. Metallic corrosion under aqueous conditions can take place by many mechanisms with quite varied results.

3. Experimental procedures

3.1. Materials and process parameters

The Inconel 718 will be heat treated according to the AMS2750 requirement with the follow parameters:

- solution annealed - 980°C – 60-80 minutes soaking time, followed by polymer quench;
- precipitation hardened - 720°C – 8 hours soaking time, cooled down to 620°C – 8 hours.

After this step will be finished, the desire of this experiment is to change the parameters, as for example the quench used, temperature, etc.

By changing the parameters slowly and compare, it will provide us important information regarding the behavior of this alloy and its mechanical properties changes.

The working process will be strictly observed knowing that for excellent stress and creeping resistance, the application temperature should be kept lower that 700°C as gamma double prime is meta stable. Beyond this limit in a prolonged use it quickly ages.

3.2. Characterization methods of mechanical properties testing

The mechanical properties of the Inconel 718 samples will be investigated first by the following methods: Tensile and Fatigue tests.

- Tensile testing

The properties that are usually determined during a tensile testing are the following:

Name of determined property	Unit	Symbol
Ultimate Tensile Strength	MPa	UTS/ R_m
Yield Strength	MPa	YS/ $Rp_{0.2}$
Elongation (Calculated)	%	A
Reduction of Area	%	Z/ RA

IOP Conf. Series: Materials Science and Engineering **209** (2017) 012060 doi:10.1088/1757-899X/209/1/012060

The UTS is usually determined by performing a tensile test and recording the engineering stress versus strain. The highest point of the stress–strain curve is the UTS. It is an intensive property; therefore its value does not depend on the size of the test specimen. However, it is dependent on other factors, such as the preparation of the specimen, the presence of surface defects, and the temperature of the test environment and material.

The Yield Strength is the material property defined as the stress at which a material begins to deform plastically. Prior to the yield point the material will deform elastically and will return to its original shape when the applied stress is removed.

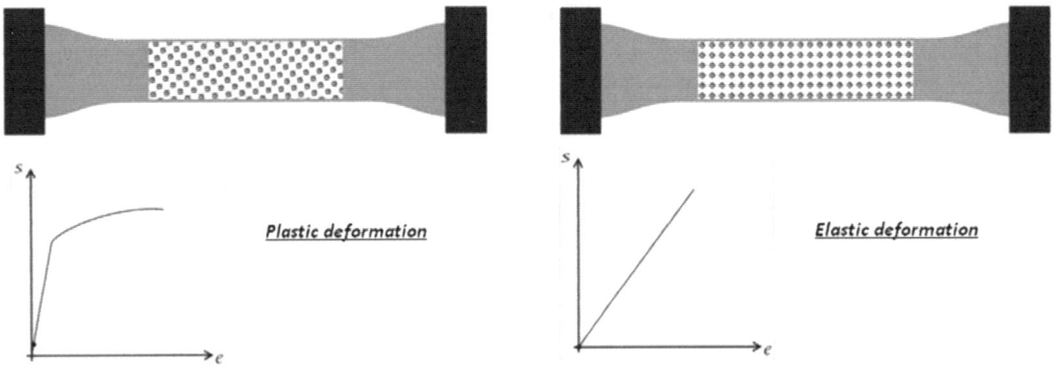

The reduction of area and the elongation are reported as additional information on the deformational characteristics of the material. The two are used as indicators of ductility.

Test process

The test process involves placing the test specimen in the testing machine and slowly extending it until it fractures. During this process, the elongation of the gauge section is recorded against the applied force. The data is processed so that the result is not dependent on the geometry of the test sample. The elongation measurement is used to calculate the engineering strain, ε, using the following equation:

$$\varepsilon = \frac{\Delta L}{L_0} = \frac{L - L_0}{L_0}$$

Where ΔL is the change in gauge length, L_0 is the initial gauge length, and L is the final length. The force measurement is used to calculate the engineering stress, σ, using the following equation:

$$\sigma = \frac{F_n}{A}$$

Where F is the tensile force and A is the nominal cross-section of the specimen. The machine does these calculations as the force increases, so that the data points can be graphed into a stress–strain curve.

The most common testing machine used in tensile testing is the universal testing machine. This type of machine has two crossheads; one is adjusted for the length of the specimen and the other is driven to apply tension to the test specimen. There are two types: hydraulic powered and electromagnetically powered machines.

- Fatigue test

In materials science, fatigue is the weakening of a material caused by repeatedly applied loads. It is the progressive and localised structural damage that occurs when a material is subjected to cyclic loading. The nominal maximum stress values that cause such damage *may be much less* than the strength of the material typically quoted ***as the ultimate tensile stress limit.***

Fatigue strength is determined by applying different levels of cyclic stress to individual test specimens and measuring the number of cycles to failure. Standard laboratory test use various methods for applying the cyclic load, e.g. rotating bend, cantilever bend, axial push-pull and torsion.

The data are plotted in the form of a stress-number of cycles to ***failure (S-N) curve***. S-N test data are usually displayed on a log-log plot, with the actual S-N line representing the mean of the data from several tests.

As a rough guide, the fatigue limit is usually ***about 40% of the tensile strength***. In principle, components designed so that the applied stresses do not exceed this level should not fail in service. The difficulty is a localized stress concentration may be present or introduced during service which leads to initiation, despite the design stress being normally below the 'safe' limit.

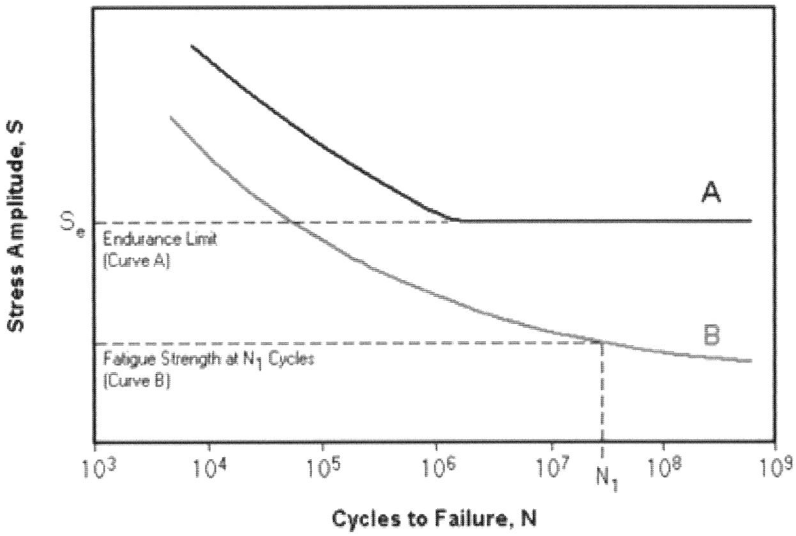

Fig.1 – fatigue limit[9]

Most materials, however, exhibit *a* ***continually falling curve*** as in (b) and the usual indicator of fatigue strength is to quote the stress below which failure will not be expected in less than a given number of cycles which is referred to as the endurance limit.

4. Results expected

4.1. Macrostructure and Microstructure

The macrostructure of Inconel 718 needs to be without defect, no abnormal grain growth and with the macro grain size homogeneous through thickness. No evidence of non-concentric tree rings, freckles or white spots observed.

The samples will be compared with the requirements of EN2950 or ASTM E340.
The microstructure needs to consists with a fine equiaxed grain structure with grain size below the requirements of ASTM E112 – G=5 and ALA – G=3.

Twin boundaries, fine grain boundary and intergranular precipitation should be observed.
Also on the microstructure, no laves phase have to be seen.

IOP Conf. Series: Materials Science and Engineering **209** (2017) 012060 doi:10.1088/1757-899X/209/1/012060

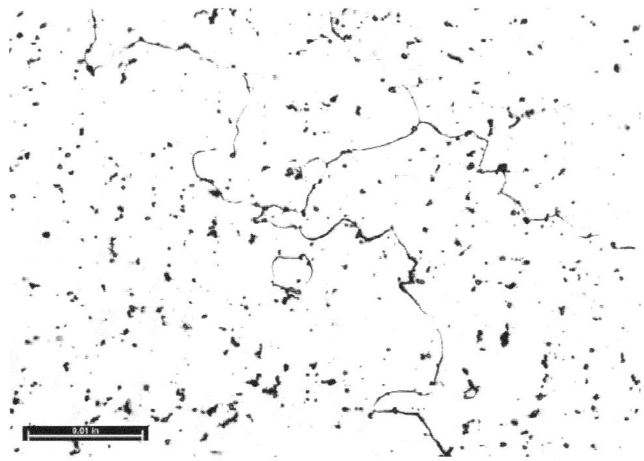

Fig.2 example of Inconel 718 microstructure from 100x[8]

4.2. Tensile test

Tensile test will be performed for room temperature and elevated temperature (<650°C).
The expectations of the tensile test results are to be well above the minimum values of the ASTM E 10 requirement.

For room temperature			
UTS (MPa)	**$YS_{0.2}$(MPa)**	**A%**	**Z%**
1350 ≤ UTS ≤ 1550	**≥ 1100**	**≥ 12**	**≥ 15**
		≥ 8	**≥ 12**

For elevated temperature <650°C			
UTS (MPa)	**$YS_{0.2}$ (MPa)**	**A%**	**Z%**
≥ 1000	**≥ 860**	**≥ 12**	**≥ 15**
		≥ 8	**≥ 12**

The behaviour of all the test sample results will be preferred as representative of the whole population. Having the results in the middle of the cloud it will be unlikely to lead to any production problems.

4.3. Fatigue test

Fatigue test will be performed at room temperature and represented thru a Wöhler regression curve.
It will be applied a strain range of 2% during fatigue and increase until the damage of the material.

5. Conclusions

An affordable Alloy 718 derivative with greater temperature capability is a long-standing desire.

This study looks to revisit the potential for an Alloy 718 derivative that would have processing characteristics similar to 718.

The new alloy is targeted to have the malleability and fusion weldability characteristics of Alloy 718, with stability at higher temperature range.

As we look to the future, the versatility of Alloy 718 ensures its continued usage, albeit for a smaller portion of future engine components due to increasing engine operating temperatures. Finding a suitable higher temperature counterpart to Alloy 718 has proven to be a daunting task.

The substructural changes in Inconel 718 will be characterized and compared with those published in the literature.

IOP Conf. Series: Materials Science and Engineering **209** (2017) 012060 doi:10.1088/1757-899X/209/1/012060

6. References

[1] N.Geru – Metalurgie fizică, Bucureşti, E.D.P. 1981

[2] E.A. Ott, J.R. Groh, A. Banik, I. Dempster, T.P. Gabb, R. Helmink, X. Liu, A. Mitchell, G.P. Sjöberg andA. Wusatowska-Sarnek, *The Microstructure and Mechanical Properties of Inconel 718 Fine Grain Ring Forging*

[3] Robert E. Schafrik, Douglas D. Ward, Jon R. Groh, *Application of Alloy 718 in GE Aircraft Engines: Past, Present and Next Five Years*, Cincinnati. Ohio 452 15

[4] Sreeramesh Kalh, K. Bhanu Sankara Rao, Gary R. Halford and Michael A. McGaw, *Deformation and damage mechanisms in inconel 718 superalloy*, Cleveland, Ohio 44135

[5] V.N. Shukla, R. Jayaganthan and V.K. Tewari, Oxidation and Hot Corrosion Behaviour of Ni-based Superalloy Inconel 718 in Na2 So4 -75%V2O5 Environment at Elevated Temperature, India

[6] E. Andrieu, R. Cozar and A. Pineau, Effect of environment and microstructure on the high temperature behavior of alloy 718, France

[7] High-Performance Alloys for Resistance to Aqueous Corrosion – Special Metals

[8] http://egr.uri.edu/materialslab/current-research/

[9] http://www.totalmateria.com/page.aspx?ID=CheckArticle&site=kts&NM=282

Preventing Bacterial Infections using Metal Oxides Nanocoatings on Bone Implant

L D Duceac[1], S Straticiuc[1], E Hanganu[2*], L Stafie[1], G Calin[1] and S L Gavrilescu[2]
[1] Faculty of Medical Dentistry, „Apollonia" University of Iasi, Romania
[2] Faculty of Medicine, "G. T. Popa" University of Medicine and Pharmacy, Iasi, Romania

E-mail: dr.elenahanganu@gmail.com

Abstract. Nowadays bone implant removal is caused by infection that occurs around it possibly acquired after surgery or during hospitalization. The purpose of this study was to reveal some metal oxides applied as coatings on bone implant thus limiting the usual antibiotics-resistant bacteria colonization. Therefore ZnO, TiO_2 and CuO were synthesized and structurally and morphologically analized in order to use them as an alternative antimicrobial agents deposited on bone implant. XRD, SEM, and FTIR characterization techniques were used to identify structure and texture of these nanoscaled metal oxides. These metal oxides nanocoatings on implant surface play a big role in preventing bacterial infection and reducing surgical complications.

1. Introduction

It is well known that from decades nosocomial infections are present in hospitals affecting seriously most of patients. This ancient problem regarding microbial infections is caused by high drug resistance of pathogenic bacteria (gram positive and gram negative) to antibiotics. Many researches focused on finding different methods and materials for fighting infections caused by resistant microorganisms. There are several factors involving microorganism load, surgical technique and type of implant that influence infection appearance on bone implant surface inducing sometimes septic complications. This risk implies patients presenting co-morbidities that could increase the probability of infection.

Nanotechnology offers the opportunity to develop a wide variety of nanomaterials with specific use in medical area especially in limiting damages caused by pathogen microbes.

Moreover, structural and textural features of implant such as material, size, shape and intended use have an important role in this field of medicine. Researcher's attention in focused on finding implant coatings acting as antimicrobial agents and able to reduce bacterial adhesion and proliferation of the surrounding tissue [1,2].

It is well known that prophylactic antibiotics are administrated to patients undergoing surgery in order to prevent postoperative infection, being well aware that antibiotics has some disadvantages including compliance time of drug administration and the reduced ability to kill pathogenic germs present at the implant-surrounding tissue interface.

Structural and surface characteristics of an implant such as electrostatic charge, porosity and hydrophobicity play a major role knowing that microorganisms have an interesting strategy to adhere and colonize all natural and synthetic surfaces [3,4]. Therefore, researchers should look for solution to avoid bacteria colonization on the implant surface and to create a barrier on its surface in contact with living tissue especially applied to the patients with compromise immunity and patients whose risk of infection is greatly increased [5-7].

Bone implants are constantly subject to improvement in bone interaction to ensure the success regarding orthopedic surgery. Textural and structural surface characteristics of metal oxides thin films used as nanocoatings such as morphology, topography and chemistry can represent important tailoring tools to enhance biological action at the bone implant – tissue interface. It is well known that surrounding biological area is affected by the coated titanium implant wettability requiring a long term osseointegration and a good biological environment including the prevention of bacteria multiplication on hard and soft surrounding tissue.

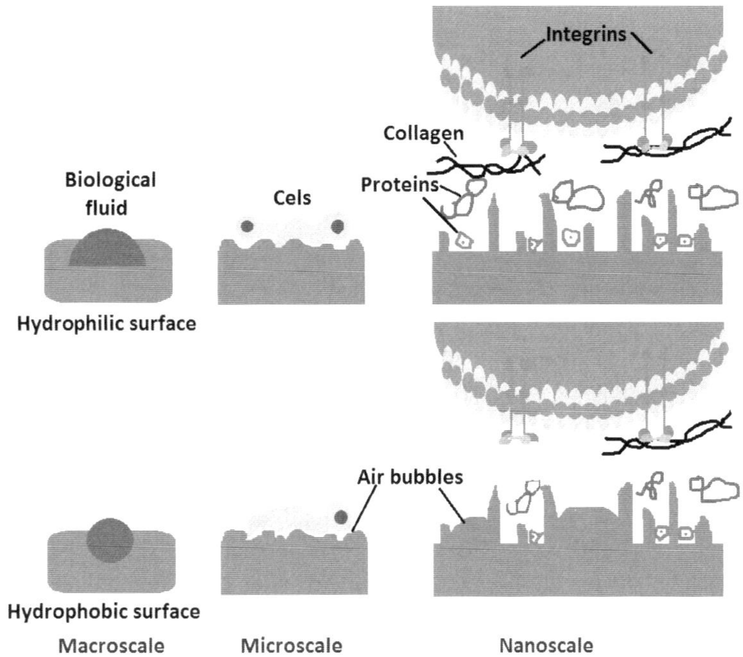

Figure 1. Representation on implant-tissue interactions at macro-, micro- and nanometric level.

Bacterial adhesion and multiplication on bone implant may cause a severe disease that leads to bone re-sorption and in special cases the need to remove the implant. Several studies take into account a major interest in designing implants with antimicrobial nanocoatings acting independently or synergistically with the antibiotic administered to limit bacterial infection [8].

A wide range of nanomaterials and technologies were developed and tested as antibacterial nanocoatings. For their manufacturing were been considered certain aspects such as biocompatibility, anti-infective efficiency, durability and resistance to mechanical stresses during surgery and postoperatively respectively [9,10].

Recent researches propose suitable implant coatings possessing multifunctional surfaces and acting as antimicrobial agents with specific properties [11]. Figure 2 reveals most important properties of an antimicrobial coated bone implant.

Nowadays, implant nanostructured coatings are of great interest especially those inhibiting bacteria adhesion and colonization. Among them, silver nanoparticles and some metal oxides used as nanocoatings proved to be suitable to prevent infection of medical device type bone implant [12-16].

When the content of metal oxides is adequate, the antibacterial effect of implant coating can be carried out. The amount of ZnO, TiO_2 and CuO metal oxides can be controlled in order to check the bactericidal degree of these thin films used as bone implant coatings.

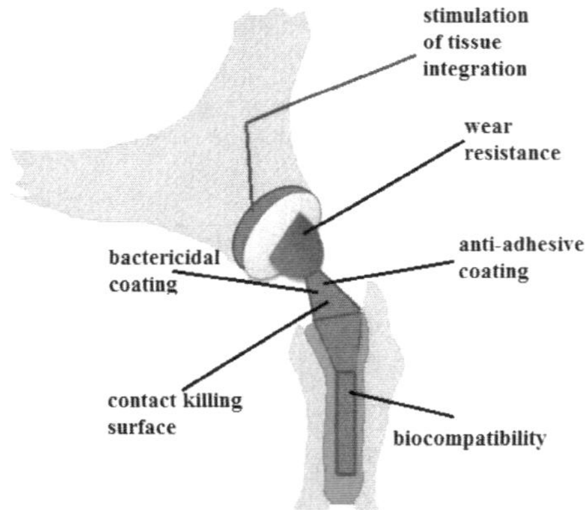

Figure 2. Multifunctional surfaces of an implant.

Different methods were used for preparing CuO, ZnO and TiO_2 metal oxides used as nanocoatings on bone implant in order to prevent bacterial infection.

2. Experimental procedures

2.1. Synthesis of CuO thin film on bioglass

Sol-gel method was used to obtain thin films of bioglass dopped with copper oxides with different percentages of oxides as follows: 6% P_2O_5, 26% CaO, 48% SiO_2 and 20% CuO. The first step consist in the hydrolysis of tetraethyl orthosilicate (19 mL of bioglass containing 20% of copper oxide) catalyzed with 30 mL HNO_3 of 0.1M and stirred for 1 h.

The other reagents including tetraethyl phosphate, calcium nitrate tetrahydrate and copper (II) acetate monohydrate (1.7 g) were added to obtain the sol which completed hydrolysis at room temperature for 2h. The resulting sol was kept at room temperature for 10 days to get the gel using ultrasonic procedure for 30 minutes. After this step, the final gel was dried at 60°C for 25 h and the applied a heat treatment at 600°C for 6 hours. Finally, the obtained sample was ground to obtain the nanopowders for further characterization.

2.2. Synthesis of TiO_2 thin layers on bioglass

Spray pyrolysis method was used for TiO_2 thin layers obtaining. Great advantages of the method are that thin layers are rough, can be obtained on large surfaces and most important, at low costs. 0,5 mL of $TiCl_4$ used as precursor was dissolved in a 5 mL solution of ethanol and spryed on a heated substrate of glass were the pyrolytic rection take place, thus obtaining nanocrystalline metal oxide. The deposition was performed at atmospheric pressure to obtain thermodynamically stable phases of TiO_2 then thin layers were calcinated at 500°C for 5 hours.

2.3. Synthesis of ZnO thin films

Sol-gel synthesis was used to prepare zinc oxide thin films. Zinc acetate dihydrate used as precursor was dissolved in a 0,75 mol/L mixed solution of ethanol and monoethanolamine. The obtained film was thermally treated at 600°C for 2 hours thus obtaining crystallized ZnO ready to undergo physicochemical characterization.

For further characterization of the obtained metal oxides thin films were used XRD, SEM and FTIR techniques.

3. Results and discussions

XRD patterns (figure 3a) of ZnO correspond to the diffraction peaks (100), (101) and (102) exhibiting growth crystallites of zinc oxide. For CuO sample it can be observed in figure 3c oriented peaks at 2Θ equal to 35.6° and 38.6° corresponding to the reflections (-111) and (111) respectively. Crystalline nature of CuO increased under reduced atmosphere in oxidation process. In case of TiO_2 the diffraction peaks shown in figure 3b are well defined at 2Θ corresponding to the planes (101) and (111).

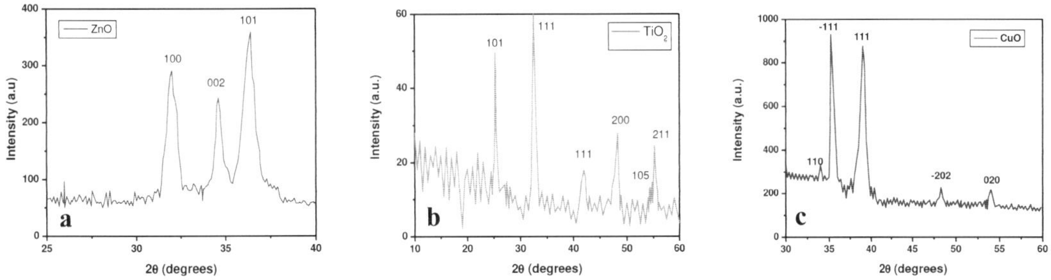

Figure 3. XRD pattern of as synthesized samples: a) ZnO; b) TiO_2; c) CuO.

Scanning electron microscopy (SEM) was used to provide details regarding textural characteristics of metal oxides thin films al nanomatric level. This technique is very important for revealing metal oxides thin films morphology. Figure 4a shows morphological characteristics of CuO multilayer nanoclusters hierarchically arranged having 15-25 nm in width that covers the glass surface.

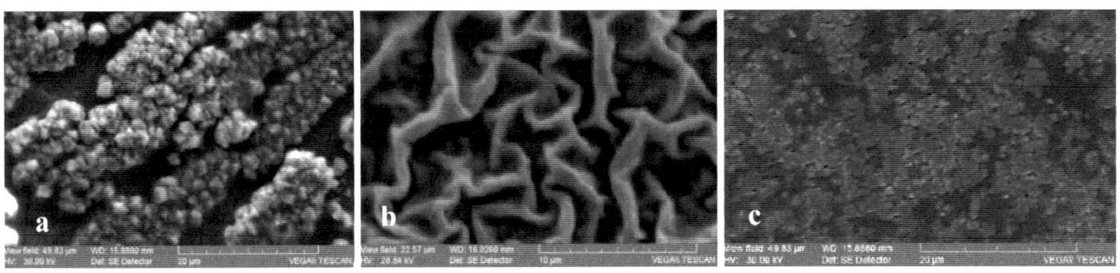

Figure 4. SEM images of metal oxides thin films.

ZnO images (figure 4b) shows irregular ZnO rods with high surface area and small rod shaped crystals. SEM features of TiO_2 thin films (figure 4c) reveals a smooth surface covered with a granular and uniform surface.

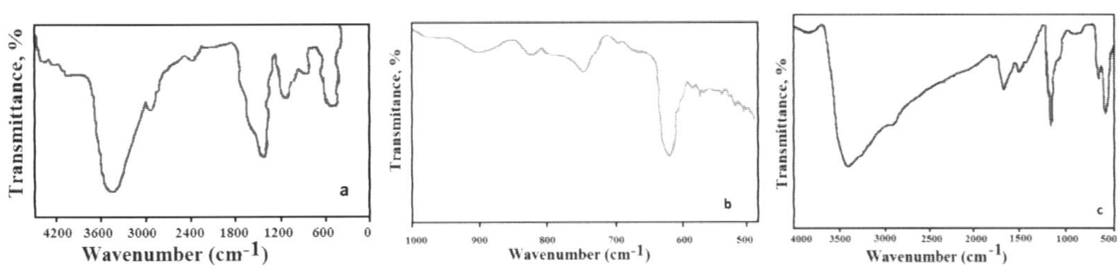

Figure 5. FTIR spectra of thin films: a) ZnO; b) TiO_2 and c) CuO.

FTIR spectrum of CuO sample (figure 5a) indicates the presence of – OH groups at 3363 cm^{-1}. O-H bending vibration associated with Cu atoms can be observed at around 1624 cm^{-1} – 1102 cm^{-1} absorption peaks. Broad absorption peaks at 600 cm^{-1} and 520 cm^{-1} are attributed to Cu-O stretching suggesting the presence of copper oxide and incorporated water in this structure. Absorption peaks of zinc oxide sample are shown in figure 5b. ZnO stretching vibration can be observed at 483.5 cm^{-1} absorption band. Absorption peak at 1424.5 cm^{-1} is associated with C-O stretching and around 3443 cm^{-1} is attributed to O-H groups present in the film. The absorption band at around 1222.5 cm^{-1} may correspond to acetic acid presence. For TiO_2 sample, IR spectra presented in figure 5c show the presence of TiO_2 at 610 cm^{-1} and 740 cm^{-1} absorption peaks.

FTIR spectra of synthesized CuO, ZnO and TiO_2 thin films suggest the presence of metal oxides bonds.

4. Conclusions

Worldwide healthcare people are aware that prevention is the best solution for bone implant infection. A few factors can influence the osseointegration and infection appearance regarding bone implant. Particularly, modification of bone implant surface by coating with metal oxides thin films results in improving clinical aspects by enhancing osseointegration and preventing infections at the bone implant-surrounding tissue interface.

Presently, antibiotics treatment for infection caused by bone implant may require device removal and elaborated procedures to restore damaged surrounding tissue. The use of nanoscaled metal oxides thin films to coat bone implants is a good option for osseoconduction and bactericidal effect to prevent this stringent problem. The major purpose for this study was to synthesize and characterize metal oxides thin films using advanced techniques for their subsequent applications as nanocoatings on bone implant.

Researches on antibacterial nanocoatings have proved the efficiency of metal or metal oxides deposited on implant surfaces. As ZnO, TiO_2 and CuO thin films used as nanocoatings on titanium implant already demonstrated their biocompatibility they can be suitable for deposition on bone implant in order to improve osseointegration in same time with limiting the adhesion and multiplication of pathogenic bacteria.

CuO, TiO_2, ZnO metal oxides plays an important role due to their biocompatibility and well adherence on implant improving the performance of this medical device and making it more efficient. Metal oxides nanocoatings on bone implant improve osseointegration enhancing its safety and antibacterial efficacy.

References

[1] Zhao L and Chu P K 2009 *J. Biomed. Mater. Res. B Appl. Biomater.* **91** 470-480
[2] Kargupta R and Bok S 2014 Wiley *Interdiscip. Rev. Nanomed. Nanobiotechnol* **6** 475-496
[3] Costerton W and Veeh R 2003 *J Clin Investig.* **112** 1466–1477
[4] Chen Y and Busscher H J 2011 *Appl Environ Microbiol.* **77** 5065–5070
[5] Zimmerli W and Lew P D 1984 *J Clin Investig.* **73** 1191–1200
[6] Higgins D M and Basaraba R J 2009 *Am J Pathol* **175** 161–170
[7] Zimmerli W and Sendi P 2011 *Semin Immunopathol.* **33** 295–306
[8] Engelsman A F and Saldarriaga-Fernandez I C 2010 *Biofouling.* **26** 761–777
[9] Corvec S and Portillo M E 2012 *Int. J. Artif. Organs* **35** 923–934
[10] Dale H and Fenstad A M 2012 *Acta Orthop.* **83** 449–458
[11] Gallo J and Holinka M 2014 *Int. J. Mol. Sci.* **15** 13849-13880
[12] Panacek A and Kvitek L 2006 *J. Phys. Chem. B* **110** 16248–16253
[13] Della Valle C and Visai L 2012 *Int. J. Artif. Organs* **35** 864–875
[14] Campoccia D and Montanaro L 2013 *Biomaterials* **34** 8018–8029
[15] Vishwakarma V and Josephine J 2009 *Biofouling* **25** 705–710
[16] Jan T and Iqbal J 2013 *Int. J. Nanomed.* **8** 3679–3687

IOP Conf. Series: Materials Science and Engineering **209** (2017) 012062 doi:10.1088/1757-899X/209/1/012062

Fundamental Aspects on Conductive Textiles Implemented in Intelligent System

L R Manea[1*], L Hristian[1], D Ene[1], N Amariei[1] and A Popa[2]
[1] "Gheorghe Asachi" Technical University of Iasi, Faculty of Textile, Leather & Industrial Management, Department of Engineering and Design of Textile Products, Blvd. Mangeron, No.28, Iasi, Romania
[2] "Aurel Vlaicu" University of Arad, Romania, Engineering Faculty, Textile Department, Romania

E-mail: lili191065@yahoo.com

Abstract. Conductive fibers, which are electrically conductive elements having the structure of a fiber, have a fairly long history and have been used for applications in electronic textiles as well as for aesthetics, anti-static and shielding purposes. Electrically conducting textile fibers, such as gold-coated threads, were produced in antiquity for aesthetic purposes, before the discovery of electricity, using various manufacturing methods. The textile intelligent systems, which comprise conducting textile structures (electroconducting wires or structures), present a dynamic behavior which favors the self regulation of the thermal insulation and vapor permeability with the purpose to maintain the thermo-physiological balance; the clothing assembly aims at monitoring the biologic potential, used only in critical situation (ex. accidents, falling down in a precipice etc.).

1. Introduction

In the eighteenth century, after the discovery of electricity, copper, iron, steel, brass, platinum, silver electrically conducting wires have been used in non-textile applications [1,2]. Fibers explicitly designed for electrical conductivity appeared in the late nineteenth century, and they were even used in textile structures, such as the electric corset which enabled the ladies to recover from different illnesses [3]. In the early twentieth century, more useful blends of electrical functionality and textiles appeared, such as the electrically heated glove for drivers, and, after the discovery and development of conducting polymers in 1977, illuminated clothing and heated clothing [4]. Modern methods to produce electrically conducting filaments comprise conductive substrates, metal wires, metallized yarns and inherently conductive polymers [1].

2. Main sorts of conductive fibers

In There are many types of conductive fibers:

- Substrates and conductive elements – mainly used in anti-static applications. Textile materials that conduct electricity can be produced by the polymerization of pyrrole on an appropriate substrate that provides the required characteristics of flexibility, strength and processability. Besides applications in anti-static fabrics, such materials are used in radar-absorbing products for camouflage [5]. Other approaches are to embed conducting elements strait within a fabric or yarn, or to apply conductive inks to a fabric [6] using screen printing [7].

- Metal fibers – which have very low electrical resistance, but also low elasticity and bending resistance. Such metal conductive fibers were woven into a fabric, producing a bus to link conventional rigid printed circuit board (PCB) electronics [8] or used in sports bra to form pulse-sensing electrodes [9].
- Metallized fibers – obtained from silk wound with thin gold strips [10], or adding a metallized coating directly to a core yarn. Many metallized fibers were obtained from silver-coated polyamide yarns or from polyester fibers with copper sulphide suffused to their surface [11].

Electrically conducting strips – strips of 2mm wide, containing the metal bond pads, are connected to link components and are woven in the weft direction into the textile in the place of standard yarns [12], or conducting lines are printed onto insulating plastic strips [13]. Inherently conductive polymers - conductive polymers such as polyaniline (PANi) and polypyrrole (Ppy) are used in smart clothing as sensors and actuators [14]. The electrically conducting yarns are obtained by melt spinning or a coating process [15]. PANI is one of the most promising conductive polymers for several reasons: its conductive form has excellent chemical stability combined with a relatively high degree of electrical conductivity; it can be processed in melt or solution; the monomer (aniline) is relatively cheap and the chain polymerization reaction is highly efficient.

The conductive states of polyaniline appears upon protonic doping of the emeraldine form of polyaniline. Emeraldine, which is one of three idealized oxidation states of PANi, consists of equal numbers of reduced and oxidized repeat units. One of the methods for processing PANI, without affecting the polymer structure, consists in mixing it with conventional polymers.

These blends combine the desirable properties of the two components, such as high electrical conductivity of the PANI with the physical and mechanical properties of the matrix polymer. The morphology of such mixtures has a dominant effect on their properties. The level of interaction between the PANi and matrix controls the morphology of the polymer mixture and thus, its electric conductivity.

Mixtures of PANI-based electrically conductive polymer can be obtained using conventional solution or melt processing techniques. Common polymers used for producing electrically conductive polymer nanocomposites are: polyethylene, polypropylene, polystyrene, polyvinyl chloride, polymethylmethacrylate resins, melaminformaldehide phenol resins, epoxy and thermoplastic elastomers (polyethylene terephthalate). There are different methods for obtaining the conductive yarns:

- the SOL–GEL method, which consists in the chemical coating of different yarns (for example, polyethylene terephthalate (PET) or polypropylene (PP)) with the conductive polyaniline polymer (PANi) [16 -21];
- the melt-spinning of the polymeric composites obtained by mixing the polymers (PET or PP) with the conductive PANi polymer [22-32].
- fibers with conducting elements introduced during extrusion - metal particles and tiny particles of carbon have been used, but the method has the drawback that the physical properties of the fiber can be compromised [16].
- conductive carbon nanofibers - manufactured using a wet spinning technique based on short carbon nanotubes [33].

Less conductive fibers and textiles are mainly used in static dissipation, electromagnetic shielding [34], signal and power transfer, while higher resistance conductive fibers are used as a heating element. Their flexibility and ability to be used in existing textile represent significant advantages over the solid or stranded metal.

Conductive fibers have been used in aeronautics for shielding aircraft and light-weight, high strength specialty cables, and in medical equipment industry, for electroencephalogram electrodes. Another medical application is the strain measurement. PPy-coated Lycra can be used as strain gauge material [35]. In electronic textiles, the conducting fibers can deliver power, transport input and output signals or perform as a transducer [1].

3. Implementation of electroconductive textile structures in obtaining communicating clothes

3.1. Smart materials

The term"smart" textiles covers a very broad range. Thus, one can talk about smart materials and smart systems, respectively. The *smart materials* are the unconventional structures capable of taking over and processing the information, then reacting specifically to external stimuli. These materials include specific characteristics of multi-functionality and adaptability, having the capacity to data processing by exclusively using the intrinsic characteristics of that respective material. The assembly of smart materials integrated at a nanomeric scale is called a *smart structure*, a structure which can unitary respond to the action of some external stimuli [36-38]. The explosion of micro/nano/biotechnologies, electronic devices, computers, communications under the form of sensors and incorporable systems has supplied the natural need of man living in permanent interaction with the environment, to increase its claims to clothes adaptable and synchronized with the individual. The new information processing systems for the choosy persons need not only to deliver information about certain conditions or interactions with environment conditions, but also to be able to feel, think and act. In the textile area, the concept of artificial intelligence can be attached to the new generation of textile fibers, of non-conventional materials (knitting, woven and unwoven fabrics) within a unitary product structure that can change some of its properties under the influence of an external stimulus, generating optimal answers adapted to each situation. A smart system is the result of assembling some active elements (attachment or integration of active particles) within a unitary product structure, to facilitate the mobile and personalized information processing, in order to offer comfort, safety or individual protection. At the same time, textiles offer large surfaces that can "host" artificial intelligence. Therefore textiles are another dimension of the artificial intelligence that can serve as a real information-processing superstructure, with the capability to feel, think and act, based on the wearer and/or environment stimuli. In other words, one considers the intelligent textile systems as products that have in their structure intelligent materials assembled within a complex structure, in order to obtain some special effects concerning their functionality under the action of stimuli. Their structure is different, depending on the intelligence degree of the considered textile system (passive or active intelligence, very intelligent) [39].

Referring to the articles of clothing, their aesthetic function was considered as the most important until recently, but its place has been taken by the functions of comfort (thermo-physiologic, psycho-sensorial), protection and availability. In a few words, the functionality got the front rank. Therefore, the user's requirements have directed specialists' concerns toward the elaboration and utilization of intelligent materials. These confer the human body good heat insulation, but the basic requirement is to ensure a body constant temperature, which is one of the conditions of preserving the wearer's disposition and moving capability [40]. Besides temperature, the characteristics of the materials and of the clothing products have also been chosen according to the weather conditions (air currents, wind, moisture etc.).

For accomplishing the integrated apparel, there have been taken into account the requirements imposed by the usage environment. The thermo-physiological comfort is provided when the inner temperature of the body is maintained at a constant value (37°C), under various environment conditions and different intensities of the effort made by the wearer. The human body is subjected, below 36°C and above 39°C, to the hypo- or hyper-thermic risk which can lead to death when the variations exceed 5' up to 12°C. The providing of the thermo-physiological comfort mainly depends on the balance between the heat produced and lost by the body and, respectively, on the equilibrium between the thermogenesis (the heat production) and the thermolysis (the heat loss). The heat exchange between the human body and the environment is accomplished through different ways, namely, conduction, convection, irradiation (thermal radiation), evaporation. The heat lost from the skin surface implies a heat transfer from the inner part of the body towards the surface, a heat transfer accomplished by the *circulatory system* in two ways, out of four, namely, convection and conduction [41]. The loss is done by evaporation (the saturation of the expired air with vapours) and by convection (the heating of the inspired air) through the agency of the *circulatory system*. About 85% of the heat quantity is eliminated by skin through evaporation, thermal radiation, convection.

3.2. Technical and constructive characteristics of communicant products

The achieved communicating clothing is addressed to sportsmen eager to perform sports at extreme temperatures, and to alpinists. In order that the equipment should finally accomplish the conditions of maximum efficiency with respect to the wearer's protection, it is necessary that all the materials and technologies implied in equipment realization correspond to the final goal. Generally, the humidity is mostly due to the organism's reaction meant to loose the additional heat produced through physical effort, as well as to the warm and wet environment created within the under-cloth microclimate, through the utilization of ensembles that "do not breath", as they are not sufficiently permeable to vapors.

The electronic systems for monitoring the user, accompanying the conductive structure, can include the following parts:

1. sensors (temperature, humidity, conductibility, acceleration) of reduced dimensions (nanometrical) on various types of carriers (flexible, ceramics/polymer-type) with the purpose to ensure the adjustment of temperature and humidity of the body;
2. stabilized source board (with adjustable specific parameters, namely, voltage); power generator – specialized circuits for the electrical supply of the conductive structures to the sensors network;
3. control board of the intensity of the supplied power, with a role in the reduction of the power consumption and the prevention of overheating the integrated structure, which will be able to control separately each electroconductive structure, with the possibility to being individually supplied;
4. mixed interface electronical unit (for the user) with the following roles: 1. on/off heating (hand-driven); 2. on/off temperature and humidity measurements, with the possibility to change the values set for in order to be adjusted according to the environment parameters; 3. the ensuring of the information transmission towards the hardware unit, which processes the received information with the purpose to adjust/monitor the state parameters of the users.

The operation of the *resistive temperature factors* is based on the variation of the electric resistance R of the resistive element, which constitutes the sensor according to temperature. The *resistive temperature factors* convert a physical value (heat quantity / temperature) in another physical value of an electric nature (electric resistance / electric voltage), which can be measured or used for connecting some electrotechnical components (the electric supply system) onto certain functions (e. g. heating by means of conductive textile strutures).The resistive temperature factors can be accomplished by thermoresistive materials (in which the electric resistance varies at the same time with the temperature) under the form of yarns, bands, foils and thin layers. For monitoring the temperature of the environment respectively, of the user, between the smart garment and the body, the most suitable form of material for the resistive temperature sensor is the thin layer obtained by the deposition on a carrier. The thermoresistive materials under the form of thin layers have a series of advantages, as compared with the thermoresistive materials of other geometrical shapes (e.g. yarns); these advantages are the following:

- high resistivity, this being the reason why the resistive route for obtaining high values of the electric resistance (e.g. of $1k\Omega$ order) is much shorter;
- the high contact surface, a fact which leads to the improvement of the heat transfer from the inner environment to the sensor, and therefore, to the increase of the estimation precision of the environment / human body temperature;
- the material can be geometrically structured as a resistive route, under the form of a meander to increase the value of the sensor electric resistance;
- reduced material consumption;
- low thermal inertia.

The software of the system for electric monitoring, temperature control and humidity detection permits memory intervals from 1 second to 24 hours, data transfer compatible with Windows

98/2000/XPE, USB interface to connect with the computer. Figure 1 presents examples of temperature and humidity records, taken from performed experiments [37-41].

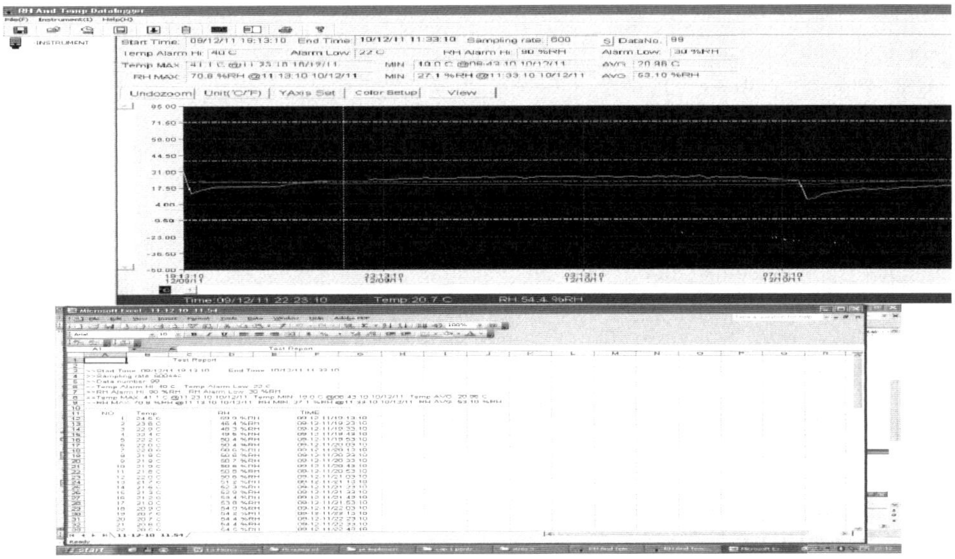

Figure 1. Records of the parameters temperature and humidity.

3.3. Embedded wireless networked device for internet/GSM network control

A communicating clothing article provides the following functions: a. communication and navigation functions performed through functional architectures implemented by GPS utilization; b. positioning (location and posture); c. user's monitoring (temperature, humidity, pulse); d. environment parameters monitoring. Intelligent clothing also has a computerized module (embedded wireless networked device – EWND) for control through the internet/GMS networks. This module can includes: a. local PC provided with software for the management of the physiological monitoring process and for ensuring an user interface to set the specific working conditions; b. embedded-type computer for controlling the physiological monitoring process; c. sensors of position and motion, humidity, temperature, electric conductibility and impact detection; d. GSM modems and proper software for M2M-type remote control. Monitoring the environment parameters (temperature, humidity) provides the user with concrete data necessary in difficult surviving situations. The researches on the development of smart clothing require novelty elements from the following domains: textile materials (composition, structure, technology, manufacturing process, non-textile insertion; clothing comfort (ergonomics, modular design, integration of the data processing modules and communication modules); microelectronics (electronic devices miniaturization, sensors integration, textile aerials integration, aerials flexibility); computation technique (specific processing systems, communication proceedings); service (maintenance, reconfiguration).

The integrated apparel ensemble has been structured as follows: a. conductive structures accomplished by electroconductive yarns; b. physiological monitoring electronic device. c. basic apparel. The realized communicating clothing has in its composition the following elements: a. the accomplishment of conductive textile structures by nanometric processing (superficially/in the mass) through the insertion of organic materials with high levels of conductibility; b. the integration of electronic systems (for physiological monitoring) through which the interface with the external environment is achieved on the basis of the conductive properties of materials; c. the putting together of the effects (a and b) for accomplishing a perfectly compatible and functional product structure, which constitutes a technical solution at hand for any user; d. the accomplished apparel ensemble ensures the temperature of the human body under low conditions to provide the wearer's

comfort; e. the maintaining of the temperature is provided by some textile structures with conductive yarns, which operate on the basis of the Joule effect, having a reduced mass which does not inconvenience the wearer; f. the garment can be used under the climatic conditions during winter.
Smart clothing allows monitoring on a permanent base, without affecting the comfort of the person wearing them.

The textile intelligent systems, which comprises conducting textile structures (wires and electroconducting structures) present a dynamic behavior which favors the self regulation of the thermal insulation and vapor permeability with the purpose to maintain the thermo-physiological balance; the integrated clothing assembly has as an option the module for monitoring the biologic potential, used only in critical situation (ex. accidents, falling down in a precipice etc.).

4. Conclusions

Smart clothing is a reality of our times. Even if they are still the prerogative of some special social categories such as sportsmen, informatics specialists, businessmen or researchers, the explosive expansion of the mobile communications and of the computation technique, will bring them into everybody's wardrobe. The textile smart systems represent a challenge that one has to be able to cope with everyday.

References

[1] Cork C R 2015 *Woodhead Publishing* Elsevier Ltd. Cambridge UK
[2] Csiszar G, Ungar T and Jaro M 2013 *Appl. Phys. A* **111** 897
[3] Fishlock D, Doctor volts. 2001 *IEE Rev.* **47** 23
[4] Ghosh T K, Dhawan A and Muth J F 2006 *Intelligent Textiles and Clothing. Woodhead Publishing* 239
[5] Kuhn H H, Child A D and Kimbrell W C 1995 *Synth. Met.* **71** 2139
[6] Parashkov R, Becker E, Riedl T, Johannes H and Kowalsky 2005 W *Proc. IEEE* **93** 1321
[7] Paul G, Torah R, Beeby S and Tudor J 2014 *Sens. Actuators A: Phys.* **206** 35
[8] Park S, Mackenzie K and Jayaraman S 2002 *PMIP, Proceedings of the Thirty-Ninth Annual Design Automation Conference* 170
[9] Adidas 2015 *http://www.numetrex.com/wearable-technology/*
[10] Post E R 1996 E-broidery: An Infrastructure for Washable Computing. MSc Thesis *MIT* http://cba.mit.edu/docs/theses/99.02.post.pdf.
[11] R-STAT 2016 *http://www.r-stat.fr/*
[12] Zysset C, Kinkeldei T W, Munzenrieder N, Cherenack K and Troster G 2012 *IEEE Trans* **2** 1107
[13] Beeby S, Cork C, Dias T, Grabham N, Torah R, Tudor J and Yang K 2014 *Advanced manufacturing of Smart and Intelligent Textiles* SMIT
[14] Cho G 2010 *Crc Press* Boca Raton
[15] Kim B, Koncar V, Devaux E, Dufour C and Viallier P 2004 *Synth. Met.* **146** 167
[16] Gibbs P and Asada H H 2004 *IEEE International Conference on Robotics and Automation. Proceedings* ICRA'04 **5** p 4753
[17] Manea L, Buhu L, Buhu A and Aniculaesei G 2007 *Intelligent Textiles and Mass Customisation* (Casablanca, Morocco) **1** p 439
[18] Manea L R and Pieptanariu M 2006 *Proceedings of the International Scientific Conference, Unitech* Technical University Gabrovo (Gabrovo, Bulgaria) **3** p 330
[19] Buhu A, Buhu L, Manea L and Amariei N 2007 *Intelligent Textiles and Mass Customisation* (Casablanca, Morocco) **1** p 350
[20] Amariei N, Pieptanariu M and Manea L R 2006 *International Scientific Conference Unitech* Technical University Gabrovo (Gabrovo, Bulgaria) **3** p 265
[21] Pieptanariu M, Amariei N, Manea L R 2006 *International Scientific Conference Unitech* Technical University Gabrovo (Gabrovo, Bulgaria) **2** p 313

[22] Manea L R, Amariei N and Buhu A 2007 *Intelligent Textiles and Mass Customisation* (Casablanca, Morocco) **1** p 356
[23] Buhu L, Manea L, Avram D and Aniculaesei G 2007 *1st International Conference – Intelligent Textiles and Mass Customisation* (Casablanca, Morocco) **1** p 565
[24] Buhu A, Manea L and Buhu L 2007, *Cortep, Buletinul Institutului Politehnic Iaşi, Section Textile and Leather, Tomul LIII (LVII) Fasc. 5* **1** p 121
[25] Manea L and Buhu L 2007 Cortep, *Buletinul Institutului Politehnic Iaşi, Section Textile and Leather Tomul LIII (LVII) Fasc. 5 **1** p 103*
[26] Manea L, Aniculaesei G, Leon A L, Hanganu L C and Potop G 2008 Buletinul Institutului Politehnic Iaşi, *Section Textile and Leather* Tomul *LIV(LVIII) Fasc.3* p. 427
[27] Manea L R, Bertea A, Nechita E, Popescu C V and Sandu I 2016 *Rev. Chim. (Bucharest)* **67** (7) 1284
[28] Manea L and Buhu A 2007 *Buletinul Institutului Politehnic din Iaşi, Section Textile and Leather Tomul LIII (LVII) Fasc. 5* **1** 109
[29] Pieptanariu M, Amariei N and Manea L 2006 *International Scientific Conference Unitech* Technical University Gabrovo (Gabrovo, Bulgaria) **2** p 316
[30] Manea L, Aniculaesei G, Leon A L and Hanganu L C 2008 *Buletinul Institutului Politehnic din Iaşi, Section Mechatronics and Precision Engineering Tomul LIV (LVIII) Fasc.4 **1*** p 59
[31] Stanescu I, Manea L R, Bertea A, Bertea A P and Sandu I C A 2016 *Rev. Chim. (Bucharest)* **67** (10) 2082
[32] Manea L R, Hristian L, Ostafe M M, Apostol L L and Sandu I 2016 *Rev. Chim. (Bucharest)* **67** (9) 1758
[33] Devaux E, Koncar V, Kim B, Campagne C, Roux C, Rochery M and Saihi D 2007 *Trans. Inst. Meas. Control* **29** 355
[34] Aniołczyk H, Koprowska J, Mamrot P and Lichawska 2004 *J Fibres & Textiles in Eastern Europe* **12** 47
[35] Cho G, Jeong K, Paik M J, Kwun Y and Sung M 2011 *IEEE Sens. J.* **11** 3183
[36] Popa A, Bucevschi A, Pustianu M, Manea L R and Sandu I 2016 *Mat. Plast.* **53**(2) 316
[37] Hristian L, Sandu A V, Manea L R, Tulbure E A and Earar K 2015 *Rev. Chim. (Bucharest)* **66** (3) 342
[38] Hristian L, Ostafe M M, Manea L R and Leon A L 2016 *IOP Conf. Series: Mater. Sci. Eng.* **145** 022014
[39] Hristian L, Ostafe M M, Manea L R and Leon A L 2016 *IOP Conf. Series: Mater. Sci.. Eng.* **145** 032004
[40] Manea L R, Hristian L, Leon A L and Popa A 2016 *IOP Conf. Series: Mater. Sci.Eng.* **145** 032007
[41] Manea L R, Hristian L, Leon A L and Popa A 2016 *IOP Conf. Series: Mater. Sci. Eng.* **145** 032006

Acknowledgments
This work was supported by a grant of the Romanian national Authority for Scientific Research and innovation, UEFISCDI, project number PN III 46 BG 2016 within PNCDI III

Electrospun Based Polyaniline Sensors – A Review

M L Rozemarie[1]*, B Andrei[1], H Liliana[1], R Cramariuc[2] and O Cramariuc[3]
[1]"Gheorghe Asachi" Technical University Iasi, Faculty of Textile and Leather Engineering and Industrial Management, 29 Dimitrie
[2]" Centre of Competence in Electrostatics and Electrotechnologies, 25, Av. Radu Beller, Bucharest, Romania
[3]Tampere University of Technology, Department of Physics, P. O. Box 692, FI-33101 Tampere, Finland

E-mail: lili191065@yahoo.com

Abstract. Polyaniline is a polymer with outstanding electronic conductivity and good response to molecules at room temperature. This review focuses on the many uses of electrospun polyaniline nanofibers as chemical sensor with high sensitivity and functionality, providing a concise report of the present status of polyaniline gas sensing applications. Latest developments in this area are reported and the capability of PANI sensors to detect various gases is emphasized. Difficulties that may obstruct practical uses of these sensors are also summarized.

1. Introduction

Many conducting polymers, such as polyaniline (PANI), poly (paraphenylen), polythiophene, poly (paraphenylenvinylen), polypyrrole, poly (3,4-ethylene-dioxythiophene), have been used in gas sensing from more then 30 years [1-5]. The conducting polymer based sensors prove to have many improved characteristics in comparison with the sensors usually based on metal oxides, and their capability to operate at room temperature is one of the most important [6-9]. Polyaniline, with a unique structure that allows it to exist in five oxidation states (figure 1), raise a special interest due to the ease of its synthesis and its versatility as sensor [10].

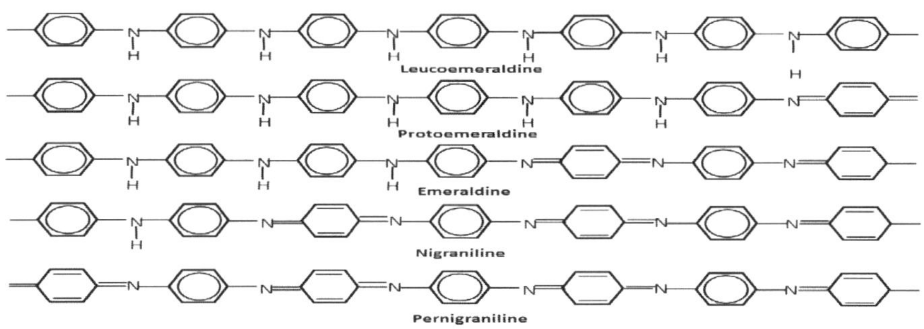

Figure 1. The five oxidation states of polyaniline.

None of the oxidation states of PANI are conductive, but protonated forms of the moderately oxidized states, especially emeraldine, are conductive [11-14]. The protonic acid doping process of PANI is illustrated in figure 2.

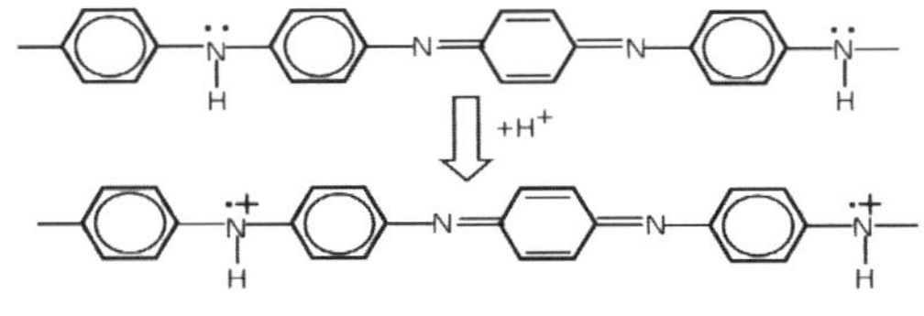

Figure 2. Protonic acid doping process of polyaniline.

2. Polyaniline as pH sensor

Polyaniline, as all the conductive polymers, has groups which can be protonated or deprotonated. Deprotonation leads to a decrease of charge carriers along the polymer chains, followed by variations in the electrical properties and modification of redox properties of PANI [1,15-18]. Polyaniline is suitable for pH sensing in aqueous medium, because conductivity decreases rapidly with increase in pH at a given potential. Optical pH sensors, based on polyaniline films prepared by chemical oxidation at room temperature, showed improved stability. The effect of pH on the change in electronic spectrum of polyaniline polymer can be explained on the basis of different degree of protonation of the imine nitrogen atoms in the polymer chain. The pH of solution could be determined by monitoring either absorption at a fixed wavelength or maximum absorption wavelength of the film. The sensor can be kept exposed in the air for over one month without any deterioration in sensor performance [19]. A polyaniline based amperometric pH sensor has been obtained using a polyaniline film coated pencil graphite electrode which is connected in series between the working and counter electrodes of a potentiostat, and immersed in the solution together with a reference electrode. When an external potential is applied, the resulting current varies with the pH solution, which provides the basis for the amperometric pH sensor [20].

3. Polyaniline for inorganic gaseous analytes detection

Gases interact with polyaniline in two ways: by a chemical reaction, such as nitrogen dioxide, sulphur dioxide, ozone, hydrogen, hydrazine, hydrochloric acid, ammonia, hydrogen sulphide, or by adsorption, such as chloroform, dichloromethane, acetonitrile, many alcohols, benzene, acetone [1].

3.1. PANI based ammonia sensors

Ammonia is a colourless gas with a suffocating odour. In case of inhalation, ingestion (solution) or skin and/or eye contact (solution/liquid), ammonia causes various health-related issues, such as dyspnea (breathing difficulty), irritation of eyes, nose and throat, chest pain and pulmonary edema. This is why The Occupational Safety and Health Administration (OSHA) have specified a threshold limit value for NH_3 in the workplace of 50 ppm [21]. Due to the dangerous effects of ammonia on human health and environment, it is absolutely necessary to monitor its trace level. To detect ammonia, electrospun nanofibres of PANI-PEO mixture which was doped with camphorsulfonic acid 10 (HCSA) were deposited on gold electrodes. The obtained sensor reacts with ammonia and alters its electrical resistance (figure 3), displaying a detection threshold of 0.5 ppm of NH_3. In figure 3, the normalized resistance [(R / Ro) - 1] is plotted as a function of analyte concentration [22]. The primary mechanism for detecting is the PANI deprotonation by ammonia, which results in an increase in resistance [23]. A field effect transistor (FET) based on polyaniline nanofibers obtained by electrospinning acts as a sensor for ammonia [24], presenting a reversible resistance change of 7% to 1 ppm NH_3 with a threshold voltage of 10 V. The diagram of the sensor is shown in figure 4.

A hybrid material based on nanostructured polyaniline and titanium dioxide [PANI-TiO_2] proved particularly effective in the detection of ammonia [25].

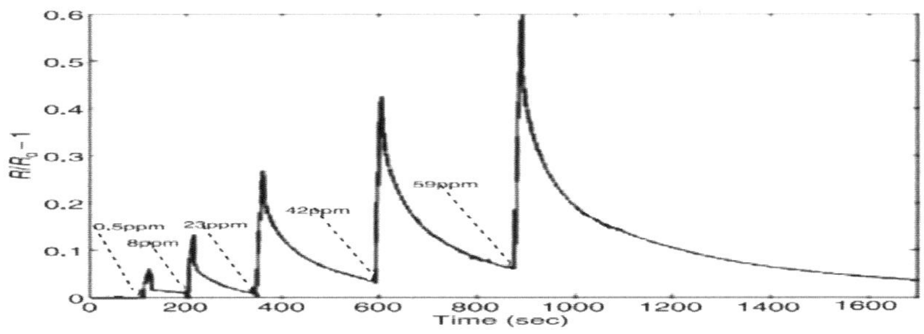

Figure 3. Real-time response of a PANI/PEO sensor to different concentrations of NH_3 gas.

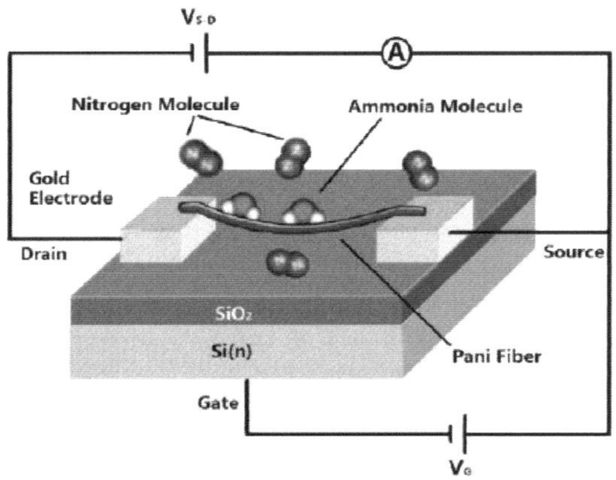

Figure 4. The diagram of the single polyaniline nanofiber field effect transistor sensor [24].

Its high sensitivity for NH_3 gas as compared to the other gases (CH_3OH, C_2H_5OH, NO_2, H_2S) was attributed by the authors to the high affinity for electrons of ammonia compared to the other gases. In order to overcome the difficulties of processing PANI, continuous + camphor-10-sulfonic acid [HCSA] doped PANI nanofibers were successfully produced by coaxial electrospinning with subsequent removal of the polymer shell and the system behaves as a sensor for both ammonia and nitrogen dioxide, showing high sensitivity and fast response time. However, resistive response for the two gases is quite different, being up to six orders of magnitude for NO_2 and less than two orders of magnitude for ammonia [26].

Electrospun PANI and PANI / ZnO nanofibers were used for the detection of NH_3 and HCl vapour at room temperature, and it was observed a decrease in resistance of the sensor on exposure to HCl vapour and an increasing resistance when exposed to NH_3 vapour. It has been found that the PANI / ZnO based sensor shows high response and sensitivity at room temperature, with very good reproducibility, in comparison with the sensor of pure PANI [27]. Contact sensors for the detection of ammonia have been obtained from mixtures of PANI electrospun nanofibres doped with HCSA and PEO that were sprayed over with gold microelectrodes. It has been observed that increasing the PANI content in the polymer mixture the sensitivity of the sensor increases, but the melting of the nanofibres significantly reduces the detection performance. Measuring conductivity vs. temperature showed that the electrons in the nano-fiber-based composite materials PANI / PEO are transported in a three-dimensional matrix [28]. Nanofibres from mixtures of PANI / poly-3-hydroxybutyrate (PHB) have been deposited by electrospinning on chemiresistors in order to investigate the effect of humidity on the ability of the sensors to detect ammonia and VOCs. It was found that the specific structure of the coating nanofibers and the physico-chemical properties of PANI / PHB allow the detection of ammonia and volatile organic compounds at high percentages of moisture, with low response time [6].

In_2O_3 / PANI nanofibers based sensors had a higher ammonia detection sensitivity than pure PANI at room temperature. The results indicated that In_2O_3 / PANI electrospun nanofibres based sensor has excellent selectivity, good repeatability and reversibility. The increased performance of the sensor for detecting ammonia can be attributed to the formation of p-n junctions between In_2O_3 and PANI [29].

A polyaniline nanofibers based sensor was tested at 100 ppm HCl and NH_3 vapour respectively. After exposure to HCl vapour the resistance of the polyaniline nanofibres decreases as a result of the doping process, and the ammonia exposure led to the de-doping process which increases the resistance of the polymer. The results showed that the sensor based on nanofibers has a much quicker response than the film-based sensor both for doping and de-doping, due to the morphology of the porous layer of nanofibres with small diameter which causes a rapid diffusion of the gas molecules [30].

3.2. PANI based nitrogen oxides (NOx)sensors

The most significant nitrogen oxides are nitrogen monoxide (NO) and nitrogen dioxide (NO_2). NO_2, which is released mainly from the combustion of fossil fuels, causes harmful effects on human health and the environment. As NO_2 is an irritant and an oxidant, which produces inflammation and oedema of the lungs if inhaled in high concentrations [31], the United States Environmental Protection Agency sets a one hour NO_2 standard at the level of 100 ppb.

Mixtures of PANI / poly (vinyl pyrrolidone) containing urease have proved to be suitable as NO_2 sensors [32,33]. Even for the basic form of polyaniline it can be measured an increase in the conductivity of the matrix in the presence of low levels of NO_2 (1-7 ppm). The answer to NO_2 in the presence of relative humidity (25% and 40%) for mixtures containing 20% wt./vol. and 80% wt./vol. PANI is substantially significant, and the amplitude of the response is an order of magnitude higher in the first case [32]. The equation for the relative sensitivity is given by:

$$S = \frac{|\Delta R|}{R_o} \cdot 100 \tag{1}$$

where ΔR is the change in resistance when exposed to the analysed gas and Ro is the baseline resistance (prior to exposure). For 5 ppm NO_2 exposures, the relative sensitivity for blends with 20% wt./vol PANI is 45% and only 9.5% for 80% wt./vol PANI mixtures. During the exposure NO_2 oxidizes PANI, which leads to an increase in the conductivity of PANI. PANI concentrations tested in these experiments ranged between 50% - 80% and 20% wt / vol [32]. These results contradict earlier research [34], which indicated that an increased of the share of PANI is favourable to increasing sensor sensitivity for NO_2 detection.

A chemiresistive sensor produced via potentiostatic deposition of a thin film of PANI on a n-octadecyl triclorosilan microelectrode was exposed to NH_3 and NO_2 and proved to have a sensitivity of 21% / ppm for NH_3 and 3760% / ppm for NO_2 and good reproducibility (sensitivity was defined as the slope of the linear range of the calibration curve). By adsorption of NH_3 on PANI the formation of ammonium ion is induced, producing the de-doping of emeraldine base to emeraldine which leads to a decrease in sensor conductivity. The detection limit for the ammonia, defined as three times the standard deviation divided by the slope of the control, was about 35 ppm [35].

A SnO_2-ZnO/PANI composite thick film obtained by coating method was used for detecting NO_2, showing high stability to NO_2 (35 ppm) at 180°C. The sensor response decreases at temperatures higher then 180°C. Despite the fact that the sensor response of analytes like NH_3, H_2, C_2H_5OH and CO was below 3%, the sensor showed a very high response of 368.9 on NO_2 [36]. A double-layered film SnO_2/PANI showed high selectivity and high response to NO_2 gas even at low concentration, with sensor response of 1-13% from the concentration range of 5-55 ppm of NO_2 and high reproducibility up to four cycles [37].

3.3. PANI based hydrogen sensors

Hydrogen, an odourless and colourless gas, is very explosive in a wide concentration range (4-75%) and has high permeability through many materials, motives enough for the development of hydrogen sensors to detect any leakage [38]. There are studies showing that hydrogen interacts directly with doped PANI nanofibers to induce a slight change in their conductivity [39, 40]. Direct uptake of hydrogen into PANI nanofibres was also observed using a quartz crystal microbalance (about 3% by mass of nanofibres). A plausible mechanism of hydrogen / doped PANI interaction involves the action of hydrogen on the charged nitrogen atoms, followed by the dissociation of the hydrogen with the formation of new NH links to the amino nitrogen of the PANI chain. Subsequently, polyaniline is returned at the initial doping state (emeraldine salt form) by charge transfer between adjacent atoms, with hydrogen release. Palladium nanoparticles doped polyaniline electrospun nanofibers were used to obtain a chemiresistor to detect hydrogen, registering a 1.8% variation of the resistance under a hydrogen concentration of 0.3% [41]. A solution that was obtained by adding SnCl2□2H2O to a solution of polyaniline in dimethylformamide (DMF) and ethanol under vigorous stirring and mixing it with polyvinylpyrrolidone was electrospun to obtain nanofibers that proved to have enhanced hydrogen detection properties at very low temperature compared to SnO_2 nanofibres [42]. Composite Al-SnO_2 / PANI electrospun nanofibres were used to detect hydrogen, yielding better results than in the case of a sensor based exclusively on the Al-SnO_2. The sensor has high sensitivity (~ 275%) in H_2 (1000 ppm) at 48°C [43]. A PANI/TiO_2:SnO_2 nanocomposite deposited onto an epoxy glass substrate was used to detect hydrogen at 27°C with a sensitivity of 1.25 (0.8 vol% H_2) [44]. A chitosan/PANI composite was used to detect hydrogen and showed high sensitivity and high selectivity, and the sensor response to concentration between 0.3 to 4% was linear [45].

3.4. PANI based hydrogen disulphide sensors

Hydrogen sulfide (H_2S) is a flammable and extremely hazardous gas; even low concentrations of H_2S can cause eyes and respiratory system irritation, while high concentrations may cause shock, inability to breathe and even death. Threshold limit for H_2S in the workplace established by OSHA is 20 ppm, and therefore monitoring is very important [2].

A sensitive and selective chemiresistive sensor for detecting hydrogen sulphide at room temperature consists of polyaniline nanofibers electrochemically functionalized with gold nanoparticles using the cyclic voltammetry technique.

Hydrogen sulphide is a weak acid and therefore does not interact significantly with PANI and its electrical conductivity is not significantly changed, but when the PANI nanofibers network is functionalized with gold nanoparticles (70-120 nm in size) by electrodeposition, the hybrid sensor demonstrates an excellent response even at low concentrations of 0.1 ppb, a dynamic range of from 0.1 to 100 ppb and very good selectivity and reproducibility (figure 5) [46].

PANI / Ag films (about 0.5 M $AgNO_3$) were analysed in terms of their chemiresistive gas detection properties; when exposed to concentrations of 10 ppm of different gases, the sensors have produced a suitable response only for H_2S.

There is an increase of the current through exposure instead of a decrease that would be expected given the electron donor nature of the H_2S (reducing gas). The reason for this effect is the dissociation of H_2S on the metal surface, resulting H^+ and HS^-. While the anion compensates the N^+ charge in the polymer chain of the PANI, H^+, whose mobility is much higher than that of the anion, provides the observed conductivity.

Further evidence of this mechanism arises from the fact that, in the presence of high concentrations of Ag, the PANI matrix, which determines the change of conductivity upon exposure to H_2S, is significantly reduced. Features such as a detection limit of 1 ppm and fast response time make this composite to be a good candidate for H_2S gas detection [47].

PANI nanofibers functionalized with gold nanoparticles are also useful in the detection of volatile sulphur compounds in human breath. Sensors made by a redox reaction between $HAuCl_4$ and PANI in emeraldine form show a good response to gases such as H_2S and CH_3SH [48].

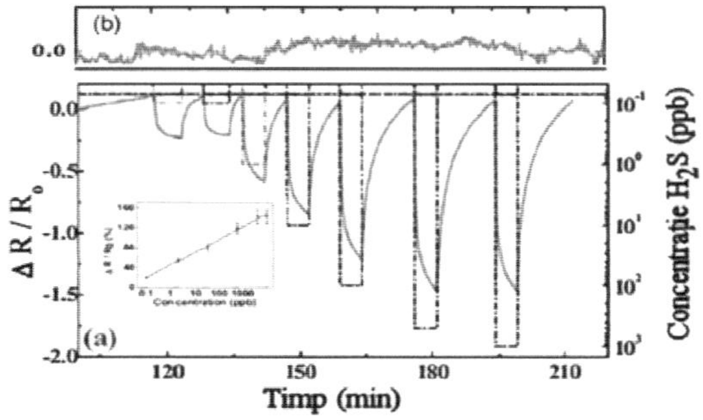

Figure 5. Response and recovery transients (solid line) of gold nanoparticles functionalized PANI nanowire network based chemiresistive sensor toward 0.1 ppb, 1 ppb, 10 ppb, 100 ppb, 500 ppb, and 1 ppm (dashed line) concentrations of H_2S gas.

3.5. PANI based hydrochloric acid sensors
Hydrochloric acid (HCl) is very dangerous for both humans and the environment. Exposure to concentrated HCl may cause circulatory collapse or asphyxia, while exposure to low concentrations solutions may lead to many health problems ranging from conjunctivitis to pulmonary oedema. As the airborne acceptable exposure limit for HCl in 8 h work day is 5 ppm, there is definite need for developing HCl sensors. [2] Nanocomposites of polyaniline and formaldehyde copolymer with a metal complex of Fe-Al obtained using thermal vacuum evaporation deposition techniques proved to be selective, fast and sensitive HCl gas sensor, which detects HCl (0.2-20 ppm) in 8-10 s and works at room temperature [49].

3.6. PANI based carbon oxides sensors
The main carbon oxides are the monoxide and the dioxide. The first is a hazardous and poisonous gas with a PEL (Permissible Exposure Limit) of 35 ppm (10-h ceiling limit) [50], while the latter is the main greenhouse gas responsible of global warming and atmosphere change, with exposure limits of 10,000 ppm (8-h Time-weighted average), and therefore there is a important need for developing sensors that can detect them.

A chemiresistive gas sensor based on polyaniline (PANI) nanofibers for CO detection was produced by a template-free electrochemical polymerization of aniline on an interdigitated electrode, a process that leads to the formation of a horizontally oriented, monolayered PANI nanofiber network. The nanosensor had an excellent responding ability on CO with a concentration as low as 1 ppm. The responding mechanism was credited to the different binding sites of CO and H_2 with PANI: the first with the unprotonated amine nitrogen atoms and the latter with the protonated amine nitrogen atoms [51]. PANI/Co_3O_4 nanocomposite sensors proved to be highly selective to CO at room temperature, with a high response of 0.81 obtained for 75 ppm CO concentration and a response time of 40 s [52].

Sensors made by vacuum-deposited PANI-Fe:Al (80:20) nanocomposite thin films produced rapid and selective detection of CO in the range 0.006-0.3 ppm at room temperature, with high sensitivity of the order of 400-600, and response times of 10 s [53] .

Sulphonated polyaniline (SPAN) was tested for sensing carbon dioxide at room temperature, showing significant conductivity change below pH 7. The impedance of SPAN films decrease when exposed to CO_2 concentrations higher then 20,000 ppm. To increase the sensor sensitivity for CO_2 SPAN was blended with polyvinyl alcohol [54]. A PANI/TiO_2 nanocomposite thin film based chemiresistive sensor developed a response for 1000 ppm CO_2 concentration of 5% at 35°C. The response was 1% with increasing temperature to 60°C. The response and recovery time characteristics of the sensor were found to be 70 s and 80 s, respectively [55].

A CO_2 room temperature gas sensor was produced using thick films of chemically synthesized cerium (Ce) doped PANI screen-printed on a glass substrate. It was found that the response decreased with an increase in the molar concentration of CeO_2, while the resistance of all Ce-doped PANI films increased with an increase in CO_2 gas concentration, supposedly because O^-_2 ions form weak bonds with π-electron clouds of PANI removing electrons from the bulk, subsequently increasing the barrier height and the resistivity [56].

3.7. PANI based sulphur dioxide sensors
Sulphur dioxide (SO_2) is a poisonous gas that affects the respiratory system; the exposure limit for SO_2 is of 5 ppm [57]. Consequently, its monitoring is much needed. Films of polyaniline doped with HCl and CSA have been used for SO_2–N_2 mixtures detecting. It was found that the electrical conductivity of PANI/HCl films is greater than that of PANI/CSA films due to the more closely packed crystalline domain which reflects its greater charge carrier mobility. The lowest detectable concentration is 200 ppm and the responses to SO_2 are reversible [58].

A polyaniline–tungsten oxide hybrid nanocomposite was used for sulphur dioxide monitoring at room temperature. The sensing response of the sensor is significantly bigger as compared to that of polyaniline nanofibres for 10 ppm SO_2 at room temperature, and the reason for this behaviour might the better porosity, the branched structure and the formation of heterojunctions. [59].

4. Polyaniline as sensor for volatile organic compounds (VOCs)
Volatile organic compounds are organic substances that have a high vapour pressure at room temperature, such as benzene, formaldehyde, methylene chloride, perchloroethylene and many others. Many of the VOCs endanger human health as well as the environment and contribute to climate change and destruction of the ozone layer [60]

Formaldehyde, which is used in many manufactured products, is one of the more toxic volatile organic compounds [61]. A sensor fabricated by using PANI intercalated MoO_3 thin films, (PANI) x MoO_3, on $LaAlO_3$ substrate exhibits distinct sensitivity to formaldehyde (HCHO) and acetaldehyde (CH_3CHO) vapours [62] PANI nanofibre based sensors (d = 300 nm) detect hydrazine which rapidly reacts with the emeraldine base and converts it to leucoemeraldină, a process influenced by humidity [41] and temperature [63]. This is why these variables have to be maintained constant (laboratory experiments) or there influence has to be compensated (operating devices).

Chloroform vapours can affect the central nervous system and damage the liver and kidneys, and this is why the recommended exposure limit is of only 2 ppm (for 60 min) [64]. A copper/PANI proved to be effective for $CHCl_3$ detection in the range of 10-100 ppm, with sensitivity values varying in the range of 1.5-3.5 [65]. A PANI/Pd nanocomposite was used for sensing methanol, a widely used substance that is flammable, explosive and toxic to human beings even in small concentrations. The short-term exposure limit for methanol is of 800 ppm. The sensors revealed a high response for 2000 ppm methanol. [66] PANI doped with camphor sulfonic acid presented a good response to alcohol vapours by undergoing a change in the electrical resistance, which decreases in the presence of small chain alcohols but increases in the case of long chain alcohols [67].

A polyaniline (PANI)/SnO_2 hybrid material was employed for sensing ethanol and acetone. The sensor is ineffective to ethanol and acetone at 30°C, but it becomes sensitive to these vapours when operated at 60 and 90°C, with an acceptable recovery time [68].

5. PANI sensors for liquefied petroleum gas (LPG)
LPG is highly inflammable and consequently its detection at room temperature is crucial. PANI based nanocomposite proved to be efficient for LPG sensing at room temperature, with excellent stability and short response and recovery times [69]. A PANI/g-Fe_2O_3 nanocomposite sensor was used for detecting LPG room temperature, exhibiting a response of 1.3 for 200 ppm LPG. The mechanism for the detection of LPG is suggested to be the result of an increase in the depletion depth due to the adsorption of gas molecules at the depletion region of the p-n heterojunction [70].

A sensitive and selective LPG sensor was fabrication based on electrospun nanofibers of PANI/ZnO nanocomposites [71]. The sensitivity of the sensor increases from 1.11% to 7.33% at 36°C, but at further increase in temperature from 36°C to 90°C the sensitivity decreased to 1.25%. Polyaniline/Barium zirconate ($BaZrO_3$) composites have been used for LPG sensing. The LPG sensitivity of about 1% at 40,000 ppm was obtained for 50 wt% $BaZrO_3$ in PANI [72-74]. A PANI/Cu_2ZnSnS_4 thin film based heterostructure was used for detecting LPG at room temperature. The maximum gas response of 44% was obtained at 0.06 vol% of LPG, but the LPG response decreased to 12% at the relative humidity of 90%. The PANI/CZTS heterojunction showed good stability and fast response and recovery time periods [74-76].

6. Conclusions

Polyaniline, alone or associated with different polymers or mineral substances, has many applications as gas sensor, due to its capacity to convert a chemical interaction (a variation in its protonation or oxidation state) into an electrical signal. Polyaniline based sensors are effective at room temperature in detecting many mineral or organic hazardous substances, showing good selectivity and stability and fast response and recovery time. This review analyses the most recent progresses in the field of PANI hybrid nanocomposites for gas/vapour sensors, pointing up their huge versatility and highlighting the most important features and possibilities for further development.

References

[1] Lange U, Roznyatovskaya N V and Mirsky V M 2008 *Analytica chimica acta* **614** 1
[2] Panday S 2016 *Journal of Science: Advanced Materials and Devices* **1** 431
[3] Adhikari B and Majumdar S 2004 *Prog. Polym. Sci.* **29** 699
[4] Manea L R, Bertea A, Nechita E, Popescu C V and Sandu I 2016 *Rev. Chim. (Bucharest)* **67** (7) 1284
[5] Manea L R, Chirita M, Hristian L and Sandu I 2016 *Mat. Plast.* **53** (3) 361
[6] Macagnano A, Perri V, Zampetti E, Bearzotti A and De Cesare F 2016 *Sensors and Actuators B Chem.* **232** 16
[7] Manea L R, Danu M C and Sandu I 2015 *Rev. Chim. (Bucharest)* **66** (6) 868
[8] Manea L R, Cramariuc B, Caunii V, Sandu I 2015 *Mat. Plast.* **52** (1) 82
[9] Manea L R, Cramariuc B, Scarlet R, Cramariuc R, Sandu I and Popescu V 2015 *Mat. Plast.* **52** (2) 180
[10] Kunzo P, Lobotka P, Micusik M and Kovacova E 2012 *Sens. Actuators B Chem.* **171-172** 838
[11] Manea L R, Hristian L, Ostafe M M, Apostol L L and Sandu I 2016 *Rev. Chim. (Bucharest)* **67** (9) 1758
[12] Hristian L, Ostafe M M, Manea L R, Sandu I G, Apostol L L and Sandu I 2016 *Rev. Chim. (Bucharest)* **67** (9) 1717
[13] Manea L R, Hristian L, Leon A L and Popa A 2016 *IOP Conf. Series: Mater. Sci. Eng.* **145** 032007
[14] Hristian L Sandu A V, Manea L R, Tulbure E A and Earar K 2015 *Rev. Chim. (Bucharest)* **66** (3) 342
[15] Manea L R, Hristian L, Leon A L and Popa A 2016 *IOP Conf. Series: Mater. Sci. Eng.* **145** 032006
[16] Manea L R and Sandu I 2015 *Rev. Chim. (Bucharest)* **66** (12) 1968
[17] Manea L R, Scarlet R, Amariei N, Nechita E and Sandu I G 2015 *Rev. Chim. (Bucharest)* **66** (4) 542
[18] Manea L R, Scarlet R, Leon A L and Sandu I 2015 *Rev. Chim. (Bucharest)* **66** (5) 640
[19] Gupta N, Sharma S, Ahmad Mir I and D. Kumar 2006 *Journal of Scientific & Industrial Research* **65** 549
[20] Gao W and Song J 2009 *Electroanalysis* **21** 973
[21] Fedoruk M J, Bronstein R, Kerger B D 2005 *J. Expo. Anal. Environ. Epidomiol.* **15** 534

[22] Liu H Q, Kameoka J, Czaplewski D A and Craighead H G 2004 *Nano Letters* **4** 671
[23] Lim J H, Phiboolsirichit N, Mubeen S, Deshusser M A, Mulehandani A and Myung N V 2010 *Nanotechnology* **21** (1)
[24] Chen D, Lei S and Chen Y 2011 *Sensors* **11** 6509
[25] Pawar S G, Patil S L, Chougule M A, Achary S N and Patil V B 2011 *Inter J Polym Mater.* **60** 244
[26] Zhang Y, Kim J J, Chen D, Tuller H L and Rutlegde G C 2014 *Adv. Funct. Mater.* **24** 4005
[27] Kondawar S, Patil P and Agrawal S 2014 *Adv. Mat. Lett.* **5** 389
[28] Li C, Chartuprayoon N, Bosze W, Low K, Lee K H, Nam J and Myung J N 2014 *Electroanalysis* **26** 711
[29] Nie Q, Pang Z, Lu H, Cai Y and Wei Q 2016 *Beilstein J. Nanotechnol.* **7** 1312
[30] Song E and Choi W J 2015 *Nanomaterials* **3** 498
[31] Purser D, Maynard R and Wakefield J 2016 *Toxicology, Survival and Health Hazards of Combustion Products*, The Royal Society of Chemistry, Cambridge, UK 419
[32] Bishop A and Gouma P 2005 *Reviews on Advanced Materials Science* **10** 209
[33] Bishop-Haynes A and Gouma P 2007 *Materials and Manufacturing Processes* **22** 764
[34] Ghosh P, Siddhantaa S and Chakrabartia A 1999 *European Polymer Journal* **35** 699
[35] Srinives S, Sarkar T and Mulchandani A 2013 *Electroanalysis* **25** 1439
[36] Xu H, Chen X, Zhang J, Wang J, Cao B and Cui D 2013 *Sens. Actuators B Chem.* **176** 166
[37] Xu H, Ju D, Li W, Gong H, Zhang J, Wang J and Cao B 2016 *Sens. Actuators B Chem.* **224** 654
[38] Wang Z, Li Z, Jiang T, Xu X and Wang C 2013 *Appl. Mater. Interfaces* **5** 2013
[39] Virji S, Kaner R and Weiller B 2006 *J. Phys. Chem. B* **110** 22266
[40] Fowler J, Virji S, Kaner R and Weiller B 2009 *J. Phys. Chem. C* **113** 6444
[41] Chen D, Guo X, Wang Z, Wang P, Chen Y and Lin L 2011a *Proceedings of the IEEE International Conference on Micro Electro Mechanical Systems (MEMS)* Cancun Mexico
[42] Sharma H J, Jamkar D V, Kondawar S.B 2013 *International Journal of Science and Research* **2** 96
[43] Sharma H J, Jamkar D V, Kondawar S B, 2015 *Procedia Materials Science* **10** 186
[44] Nasirian S and Moghaddam H M 2015 *Appl. Surf. Sci.* **328** 395
[45] Li W, Jang D M, An S Y, Kim D, Hong S – K and Kim H 2011 *Sens. Actuators B Chem* **160** 1020
[46] Shirsat M D, Baugar M A, Deshusses M A, Myung N V and Mulchandani A 2009 *Appl. Phys. Lett.* **94** 083502:1
[47] Mekki A, Joshi N, Singh A, Salmi Z, Jha P and Decorse P 2014 *Org. Electron.* **15** 71
[48] Lin C, Hayashi K and Toko K 2012 *Sens. Actuators B* **161** 504
[49] Mishra S C K, Mathur P, Yadav M, Tiwari M K, Garg S C and Tripathi P 2004 *Polymer* **45** 8623
[50] Risavi B L, Wadas J R J, Thomas C, Kupas D F 2013 *J. Emerg. Med.* **44** 637
[51] Liu C, Noda Z, Sasaki K and Hayashi K 2012 *International Journal of Hydrogen Energy* **37** 13529
[52] Sen T, Shimpi N G and Mishra S 2016 J. *Appl. Polym. Sci.* **133** 44115
[53] Mishra S C K, Mathur P and Srivastava B K 2004 *Sens. Actuators A* **114** 30
[54] Doan T, Ramaneti R, Baggerman J, van der Bent J, Marcelis A, Tong H and van Rijn C 2012 *Sensors and Actuators B Chem.* **168** 123
[55] Nimkar S H, Kondawar S B, More P S 2015 *Int. J. Res. Biosci. Agric. Technol.* **12** 167
[56] Nemade K and Waghuley S 2014 *Walailak J. Sci. Tech.* **11** 763
[57] Das S, Chakraborty S, Parkash O, Kumar D, Bandyopadhyay S, Samudrala S K, Sen A, Maiti H S 2008 *Talanta* **75** 385
[58] Kiattibutr P, Tarachiwin L, Ruangchuay L, Sirivat A and Schwank J 2002 *Reactive and Functional Polymers* **53** 29
[59] Chaudhary V and Kaur A 2015 *RSC Advances* **5** 73535

[60] Slater J M, Watt E J, Freeman N J, May I P and Weir D J 1992 *Analyst* **117** 1265
[61] Wang Y-H, Lee C-Y, Lin C-H and Fu L-M 2008 *Microsystems Technology* **14** 995
[62] Wang J, Matsubara I, Murayama N, Woosuck S and Izu N 2006 *Thin Solid Films* **514** 329
[63] Pinto H J, Ramosa I, Rojas R, Wang P C and Johnson A 2008 *Sensors and Actuators B Chem.* **129** 621
[64] Perillo P and Rodríguez D 2014 *Sens. Actuators B* **193** 263
[65] Sharma S, Nirkhe C, Pethkar S and Athawale A A 2002 *Sens. Actuators B Chem.* **85** 131
[66] Athawale A A, Bhagwat S V and Katre P P 2006 *Sens. Actuators B Chem.* **114** 263
[67] Athawale A A and Kulkarni MV 2000 *Sens. Actuators B Chem.* **67** 173
[68] Geng L, Zhao Y, Huang X, Wang S, Zhang S and Wu S 2007 *Sens. Actuators B Chem.* **120** 568
[69] Dhawale D S, Salunkhe R R, Patil U M, Gurav K V, More A M and Lokhande C D *Sens. Actuators B Chem* **134** 988
[70] Sen T, Shimpi N G, Mishra S and Sharma R P 2014 *Sens. Actuators, B Chem.* **190** 120
[71] Patil P T, Anwane R S and Kondawar S B 2015 *Procedia Mater. Sci.* **10** 195
[72] Khened B S, Machappa T, Pradeep M V N, Prasad A M V N and Sasikala M 2016 *Mater. Today Proc.* **3** 369
[73] Manea L R, Bertea A, Nechita E, Popescu C V and Sandu I 2016 *Rev. Chim. (Bucharest)* **67** (8) 1607
[74] Popa A, Bucevschi A, Pustianu M, Manea L R and Sandu I 2016 *Mat. Plast.* **53** (2) 316
[75] Hristian L, Chirita M, Manea L R and Sandu I 2016 *Mat. Plast.* **53** (4) 739
[76] Stanescu I, Manea L R, Bertea A, Bertea A P and Sandu I C A 2016 *Rev. Chim. (Bucharest)* **67**, (10) 2082

Acknowledgments
This work was supported by a grant of the Romanian national Authority for Scientific Research and innovation, UEFISCDI, project number PN III 46 BG 2016 within PNCDI III

IOP Conf. Series: Materials Science and Engineering **209** (2017) 012064 doi:10.1088/1757-899X/209/1/012064

The Effect of Alkaline Activator Ratio on the Compressive Strength of Fly Ash-Based Geopolymer Paste

A V Lăzărescu[1], H Szilagyi[2], C Baeră[3] and A Ioani[1]
[1] Technical University of Cluj-Napoca, Romania
[2] N.I.R.D. "URBAN-INCERC", Cluj-Napoca Branch, Romania
[3] N.I.R.D. "URBAN-INCERC", Timişoara Branch, Romania

E-mail: Adrian.Lazarescu@mecon.utcluj.ro

Abstract. Alkaline activation of fly ash is a particular procedure in which ash resulting from a power plant combined with a specific alkaline activator creates a solid material when dried at a certain temperature. In order to obtain desirable compressive strengths, the mix design of fly ash based geopolymer pastes should be explored comprehensively. To determine the preliminary compressive strength for fly ash based geopolymer paste using Romanian material source, various ratios of Na_2SiO_3 solution/ NaOH solution were produced, keeping the fly ash/alkaline activator ratio constant. All the mixes were then cured at 70°C for 24 hours and tested at 2 and 7 days, respectively. The aim of this paper is to present the preliminary compressive strength results for producing fly ash based geopolymer paste using Romanian material sources, the effect of alkaline activators ratio on the compressive strength and studying the directions for future research.

1. Introduction

Concrete is the most versatile, durable and trustworthy construction material and also the most used material in the construction industry. Because of this, huge quantities of Portland cement are needed and it is very well known that the cement industry is considered to be the second carbon dioxide generator. Environmental problems associated with the production of ordinary Portland cement are very well known and are closely monitored in terms of carbon dioxide amount released into the atmosphere during its production. Since industry is growing at an accelerated pace, the involvement of new, alternative, modern building materials in the industry is more than necessary.

New materials, known as geopolymers, were introduced in 1987 by Davidovits to describe a group of mineral binders. This binders could be produced by the alkali-activation of source materials of geological origin or by-product materials rich in silicon (Si) and aluminium (Al) [1,2]. The microstructure of the geopolymer materials is amorphous although the chemical composition is similar to natural zeolitic materials [3].

An alkali-activated material (or a geopolymer material) is mainly composed from one or several mineral components and one or more activators [4]. In this case, fly ash, which is a solid aluminosilicate material is used as source material in order to produce alkali-activated fly ash-based geopolymer paste. As an industrial waste, in Romania, fly ash is mainly stored, this representing one of the most common method of waste disposal and leads to land, atmosphere and ground pollution, only few quantities of it being used in different industries. The huge available amount of fly ash unused in Romania creates new perspectives on the global existing research about the production of the fly ash-based geopolymer materials.

The most common alkaline solution used to produce geopolymer concrete is a mixture of sodium silicate solution (Na_2SiO_3 solution or also known as waterglass) and sodium hydroxide solution (NaOH solution) [5]. Mixing the fly ash with the alkaline activator creates a binder that is generally known as alkali-activated fly ash-based geopolymer paste. Sand can be added to the mixture for producing geopolymer mortars and also aggregates in order to produce geopolymer concrete.

In order to obtain a material with good mechanical properties the source materials used in the production of fly ash-based geopolymer paste need to be carefully analyzed and it is necessary to fully understand the mix design of the geopolymer paste [6]. One of the most important parameters affecting the mechanical strength of the geopolymer materials is the Na_2SiO_3 solution / NaOH solution ratio [7]. Usually, researchers that investigate the production and development of alkali-activated fly ash-based geopolymers propose Na_2SiO_3/NaOH solution ratios in the range of 0.4 to 2.5 [8] with ratios of fly ash/alkaline activator between 0.5 to 3.0 [5]. Another important parameter affecting the mechanical strength of the geopolymer material is the NaOH solution concentration, and in order to produce a good geopolymer material this should be between 8M and 16M [9].

Based on the previous studies in the literature, the parameters affecting the production of alkali-activated fly ash-based geopolymer paste are the concentration of the sodium hydroxide solution, the curing temperature, the Na_2SiO_3/NaOH ratio and the fly ash/alkaline activator ratio [10,11]. The chemical composition of the source material is also an important parameter affecting the mechanical properties of the geopolymer material [12]. Previous studies also shown that using an alkaline activator composed of sodium hydroxide solution (NaOH solution) and sodium silicate solution (Na_2SiO_3 solution) leads to better mechanical properties (including compressive strengths) than using only NaOH solution as an activator [13,14]. The Na_2SiO_3 solution percentage in the alkaline activator also has an important effect since the solution favors the polymerization process adding more silicon (Si) atoms to the product and thus resulting better mechanical strength [15].

The aim of this current research is to produce alkali-activated fly ash-based geopolymer paste using Romanian local source materials and to investigate the effect of the Na_2SiO_3/NaOH solution ratio on the compressive strength of the geopolymer paste. Preliminary results will be presented in this study, further research on alkali-activated fly ash-based geopolymer paste produced using source materials from Romania still being conducted and will open new perspectives on the global existing research.

2. Materials and Methods – Experimental method

In this chapter will be presented the materials and the methods used for producing alkali-activated fly ash-based geopolymer paste using Romanian local raw materials, the mix design, the molding methods, the curing treatment proposed and the testing methods for obtaining the compressive strength results for the geopolymer paste.

2.1. Materials

The low-calcium fly ash that was used in this study is obtained from the Mintia power plant, Hunedoara county, Romania and the chemical composition was established through X-ray fluorescence analysis (XRF analysis) and it is given in table 1. Since the CaO content of the fly ash was less than 10% [16] and $SiO_2 + Al_2O_3 + Fe_2O_3 > 70\%$, $SO_3 < 5\%$ and L.O.I. < 6% [17], the fly ash used in this study complies with the ASTM requirements, classifying it as Class F fly ash.

Table 1. XRF analysis data for Mintia fly ash batch

Oxides	%
SiO_2	53.61
Al_2O_3	26.16
Fe_2O_3	7.58
CaO	2.42

MgO	1.49
SO_3	0.26
Na_2O	0.59
K_2O	2.60
P_2O_5	0.12
TiO_2	1.04
Mn_2O_3	0.08
L.O.I.	3.57

The alkaline activator used in this study was a combination between sodium silicate solution (Na_2SiO_3 solution) and sodium hydroxide solution (NaOH solution). The chemical composition of the sodium silicate solution is SiO_2=30%, Na_2O=14% and H_2O=56%. The sodium hydroxide solution was prepared dissolving the NaOH flakes, 98% purity, into water in order to obtain the desired concentration of the solution. For all the mixes in this study the NaOH solution concentration was set to 8M. To obtain a 8M NaOH solution, 320g of NaOH flakes were dissolved in water, for one liter of solution (40g x 8 = 320g NaOH/liter, where 40g is the molecular weight of NaOH). It was measured that for one kilogram of solution, the mass of NaOH was 262g (1 kg of NaOH solution, 8M = 26.2% NaOH + 73.8% water).

2.2. *Mix design, moulding and curing treatment*

Preliminary results obtained on geopolymer paste using Romanian local raw materials led to a fly ash/alkaline activator ratio of 2.0, because it has shown good workability of the alkali-activated fly ash-based geopolymer paste. The Na_2SiO_3/NaOH solutions ratio was set between 0.5 and 2.5, also according to the literature, in order to study the effect of the alkaline activator ratio on the compressive strength of the geopolymer paste. details of the mixtures of alkali-activated fly ash based geopolymer paste are given in table 2. In order to study the effect of the percentage of Na_2SiO_3 solution content of the alkaline activator on the compressive strength of alkali-activated fly ash-based geopolymer paste, for every Na_2SiO_3/NaOH ratio the percentage of waterglass was obtained.

After the preparation of the alkaline activator, 24 hours prior to mixing (by combining the sodium silicate solution with the sodium hydroxide solution), the fly ash and the activator were mixed together for ten minutes, until a homogeneous mixture was obtained. The alkali-activated geopolymer paste was then placed in 40mm x 40mm x 160mm molds and heat cured at 70°C for 24 hours. During the heat treatment a glass was put on top of the molds in order to prevent the quick release of the water from the mixtures. After demolding, the samples were kept in the climatic chamber at (20±1)°C and (60±5)% humidity until the day of the compressive strength tests (2, respectively 7 days age).

Table 2. Mixture details of alkali-activated fly ash-based geopolymer paste

Mix	Fly ash/ Alkaline liquid	Fly ash [g]	Alkaline liquid [g]	Na_2SiO_3/ NaOH	Na_2SiO_3 [g]	NaOH 8M [g]
Mix 1	2	1000	500	0.5	166.6	334.4
Mix 2				1.0	250	250
Mix 3				1.5	300	200
Mix 4				2.0	333.3	166.7
Mix 5				2.5	357.15	142.85

2.3. *Testing method*

To obtain the compressive strength results, for all the alkali-activated fly ash based geopolymer pastes, minimum three samples were tested to determinate the average value of the compressive strength.

Tests were conducted according to SR EN 196-1:2006, the standard for ordinary Portland cement concrete. The samples were tested at 2, respectively 7 days age.

3. Results and discussions

The influence of the Na_2SiO_3/ NaOH solution ratio on the compressive strength of the alkali-activated fly ash-based geopolymer paste, measured at 7 days age of the batches, is shown in figure 1. It can be easily seen that the compressive strength of the geopolymer paste increased from 7.40 MPa, for the mixes with the Na_2SiO_3/NaOH solution ratio 0.5, to a maximum of 25.30 MPa, for the geopolymer paste samples with the ratio 2.5. The increasing values of the compressive strength of the samples depend mainly on the chemical process in which the geopolymerization takes place. Increasing the Na_2SiO_3/NaOH solution ratio leads to more SiO2 species therefore more Si-O-Si bond are formed, creating a stronger material [7].

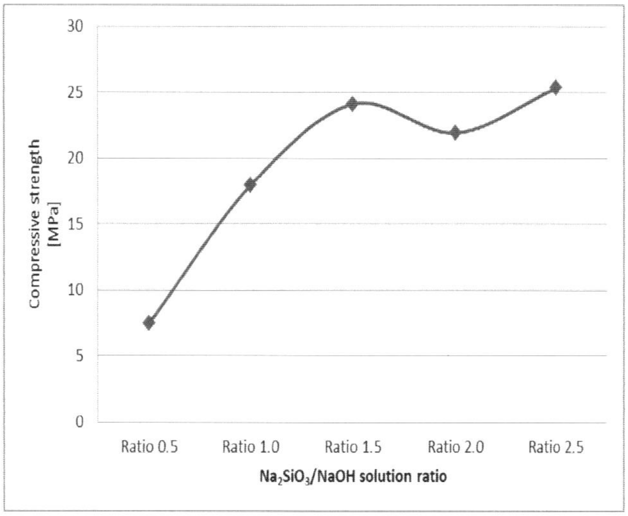

Figure 1. Influence of the Na2SiO3/NaOH solution ratio on the compressive strength of the geopolymer paste.

In order to obtain a good geopolymerization process the molar constituents of the mixtures have to be well known (table 3). Previous research have indicated that the effect of the Na_2O/SiO_2, SiO_2/Al_2O_3 and H_2O/Na_2O ratios have an important effect on the compressive strength of the alkali-activated fly ash-based geopolymer pastes [14].

Table 3. Molar constituents of the mixtures.

Mix/ Ratio	*Na_2O/ SiO_2*	*SiO_2/ Al_2O_3*	*H_2O/ Na_2O*
Mix 1 Ratio 0.5	0.16	3.801	12.742
Mix 2 Ratio 1.0	0.146	3.964	12.734
Mix 3 Ratio 1.5	0.137	4.061	12.729
Mix 4 Ratio 2.0	0.132	4.126	12.725
Mix 5 Ratio 2.5	0.128	4.172	12.722

Studies indicate that the mechanical strength of the alkali-activated fly ash-based geopolymer paste increases with the decrease of the Na_2O/SiO_2 ratio [18]. Compressive strength results obtained on geopolymer paste produced using Romanian local raw materials (figure 2) are agreed with the results held in the literature [9]. Water also has an important influence on the mechanical strength of the geopolymer paste because it helps both the destruction of the solid particles in fly ash and to achieve the necessary workability. Despite previous research [7] where the compressive strength of the alkali-activated fly ash-based geopolymer paste increased as the H_2O/Na_2O ratio increased, for the present study it can be seen (table 3) that although the ratio slightly decreased there was an important increase in the compressive strength of the samples.

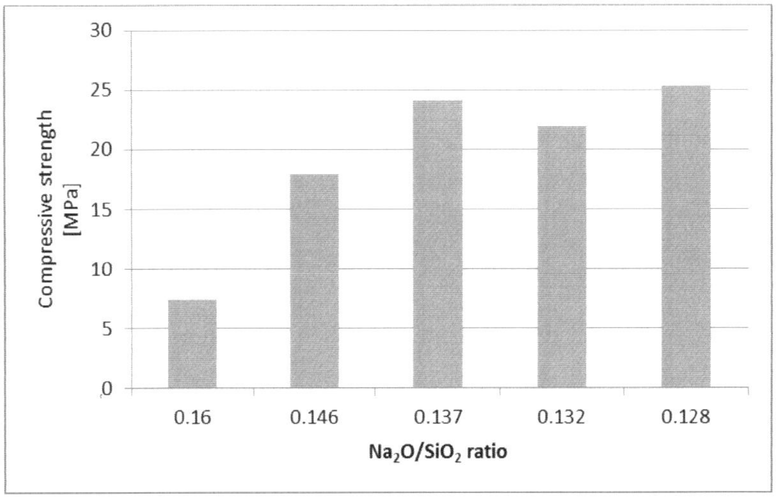

Figure 2. Effect of the molar Na_2O/SiO_2 ratio on the compressive strength of geopolymer paste.

4. Conclusions

By keeping the fly ash/alkaline liquid mixing ratio constant and changing the $Na_2SiO_3/NaOH$ solution ratio it was possible to study the effect of the alkaline activator ratio influence on the compressive strength of the alkali-activated fly ash-based geopolymer paste produced using Romanian local source materials and to investigate the geopolymerization process. Preliminary results show that increasing the sodium silicate solution content in the alkaline activator significantly increases the compressive strength of the geopolymer paste. The sodium silicate solution provides extra silicon (Si) species, which play an important role in the geopolymerization process.

Based on the presented results, it can be seen that the mix design of the alkali-activated fly ash-based geopolymer paste is an important factor in obtaining the proposed mechanical properties of the material. Different parameters affecting the compressive strength of the geopolymer paste need to be understood and carefully studied. The effect of alkaline activator ratio on the compressive strength is one of the most important factors affecting the material.

Results obtained on alkali-activated fly ash-based geopolymer paste produced using Romanian local source materials create opportunities for further research to continue. The aim of this research was both to produce alkali-activated fly ash-based geopolymer paste using Romanian source materials and to study the effect of the alkaline activator ratio on the compressive strength. Further research is still in progress and the directions are to produce more alkali-activated fly ash-based geopolymer pastes, geopolymer mortars and geopolymer concrete and study the influence of different parameters on the compressive strength of this material.

References

[1] Davidovits J 1994 High-Alkali Cements for 21st Century Conceretes. In Concrete Technology, Past, Present and Future *Proceeding of V. Mohan Malhotra Symposium, Editor: P. Kumar Metha* ACI SP-144 383-397

[2] Davidovits J 1988 Geopolymer chemistry and properties. *In: Proceeding of 1988 geopolymer conference* **1** 25-48

[3] Rangan B V 2008 Fly Ash-Based Geopolymer Concrete *Your Building Administrator* 2

[4] Silverstrim T, Martin J and Rostami H 1988 Geopolymeric Fly Ash Cement. *In: Proceedings of 1988 geopolymer conference* **1** 107-108

[5] Abdullah M, Kamarudin H, Nizar I K, Sandu A V, Binhussain M, Zarina Y and Rafiza A R 2013 *In: Rev. Chim* **64**(4) 382-387

[6] Lăzărescu A V 2017 *Proiectarea şi testarea reţetelor de beton geopolimer cu cenuşă de termocentrală PhD. Research Report* Technical University of Cluj-Napoca

[7] Mustafa A M B, Omar A K A, Kareem A and Myint S 2009 Study On The Effect Of Alkaline Activators Ratio In Preparation Of Fly Ash-Based Geopolymer *RAMM & ASMP 2009* Conference Paper

[8] Hardjito D, Wallah S E, Sumajouw D M J and Rangan B V 2004 *ACI Material Journal* **101** 467-472

[9] Hardjito D and Rangan B V 2005 Develompent and properties of low-calcium fly ash-based geopolymer concrete. *Research report GC1, Faculty of Engineering Curtin Univesity of Technology, Perth, Australia*

[10] Davidovits J 1999 Chemistry of geopolymeric sysetem, terminology *Geopolymer '99 International Conference, France* 9-44

[11] Gokhale C 2001 *The immobilization of inorganic waste through geopolymerization* University of Stellenbosch

[12] Lloyd R R and Van Deventer J S J 2005 The microstructure of geopolymers synthesized from industrial waste *1st International Conference on Engineering for Waste Treatment (WastEng), Albi, France*

[13] Palomo A, Grutzek M W and Blanco M T 1999 *Cem Concr Res* **29** 1323-1329

[14] Fernandez-Jimenez A and Palomo A 2003 *Fuel* **82** 2259-2265

[15] Criado M, Palomo A and Fernandez-Jimenez A 2005 *Fuel* **84** 2048-205.

[16] Davidovits J 2008 *Hand book of Geopolymer Chemistry and Applications* **2** 277-278

[17] Davidovits J 1994 Properties of geopolymer cements *Alkaline Cements and Concretes, Kiev, Ucraine* 131-149.

[18] Kirschner A and Harmuth H 2004 *Cerma Silic* **48** 117-120

Acknowledgments

The current investigation is carried out in the framework of the PhD. Research Program of the first author at the Technical University of Cluj-Napoca. The experimental work has been carried out in the laboratories of N.I.R.D. URBAN-INCERC, Cluj-Napoca Branch, Romania.

IOP Conf. Series: Materials Science and Engineering **209** (2017) 012065 doi:10.1088/1757-899X/209/1/012065

Investigation of Bucket Wheel Excavator Lattice Structure Internal Stress in Harsh Environment through a Remote Measurement System

M Risteiu[1], R Dobra[1], I Andras[2], M Roventa[2] and A Lorincz[3]
[1] Computer Science and Engineering Department, "1 Decembrie 1918" University of Alba Iulia, Gabriel Bethlen Str. No.5, 510009 Alba Iulia, Romania
[2] Machinery Department, University of Petrosani, University Street, No. 11, Petrosani Romania
[3] Automation Department, University of Petrosani, University Street, No. 11, Petrosani Romania

E-mail: mristeiu@uab.ro

Abstract. The paper shows the results of a lab model for strain gauges based measuring system for multiple measuring heads of the mechanical stress in lattice structures of the bucket wheel excavator for open pit mines- harsh environment. The system is designed around a microcontroller system. Because of specific working conditions, the measuring system sends data to a processing system (a PC with Matlab software), we have implemented a secure communication solution based on ISM standard, by using NRF24L01 module. The transceiver contains a fully integrated frequency synthesizer based on crystal oscillator, and a Enhanced ShockBurst™ protocol engine. The proposed solution has a current consumption around 9.0 mA at an output power of -6dBm and 12.3mA in RX mode. Built-in Power Down and Standby modes makes power saving easily realizable for our solution battery powered. The stress from structures is taken by specific strain gauges adapted to low frequency vibrations. We are using a precision 24-bit analog-to-digital converter (ADC) designed for weigh scales and industrial control applications to interface directly with a bridge sensor-instrumentation device, with low drift voltage, low noise, common mode rejection signal, frequency and temperature stability. As backup implementation for measurements a high speed storage implementation is used.

1. Introduction

The investigation of the bucket wheel excavator is done focusing on two different aspects: energy management in terms of efficiency, and excavator stability.

On the other hand, the two aspects are related each other because, because the stability issue influences the consumption of electrical energy, and/or the bucket wheel spare parts consumption.

The bucket wheel excavator is a structure complicated from mechanical and electrical point of view. Main parts of the excavator are mechanical connected each other through lattice frames and supporting cables (figure 1).

Figure 1. A bucket wheel excavator in the marching position.

As it can be seen, the lattice structure transmits the mechanical efforts in entire structure, when there is an excitation. Along its structure, the lattice is stressed differently in different points of suspension.

The purpose of this paper is to investigate the mechanical stress in site and harsh environment. The method of the investigation starts from two traditional solutions (analogic strain gauges measurement and transmission).

2. Infrastructure and Methods

For the stress monitoring, literature doesn't recommend a specific strain gauge, but the prototyping purpose, we have used Kyowa products with gauge characteristics: Uniaxial, 120 Ω, 37.00 mm x 5.20 mm, 11 ppm/°C, two wires

2.1. Stress measurement considerations

When a material receives a tensile force F, it has a stress σ that corresponds to the applied force. In proportion to the stress, the cross-section contracts and the length elongates by ΔL from the length L the material had before receiving the tensile force (see figure 2).

The ratio of the elongation to the original length is called a tensile strain and is expressed as follows:

$$\varepsilon = \frac{\Delta L}{L}$$, with: ε is Strain, L is original length, ΔL is elongation (1)

According with measuring experience, the tensile strain is expressed with a numeric value plus $x10^{-6}$ strain, με or μm/m.

The relation between stress and the strain initiated in a material by an applied force is expressed as follows based on Hooke's law:

σ= E ε, σ: Stress, E: Elastic modulus, ε: Strain.

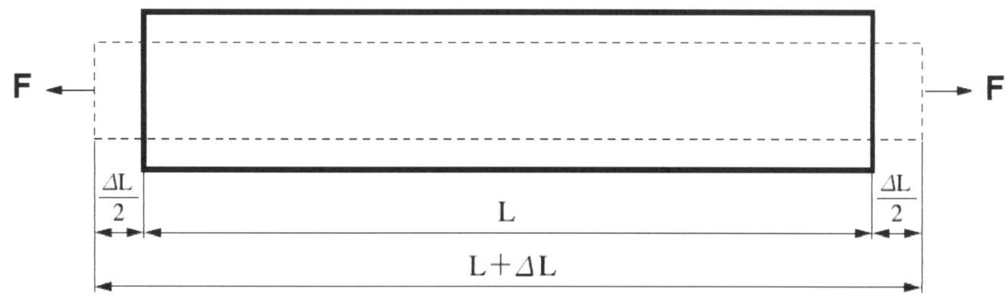

Figure 2. The mechanical model of the stressed elongation.

Stress is thus obtained by multiplying strain by the elastic modulus. When a material receives a tensile force, it elongates in the axial direction while contracting in the transverse direction. Elongation in the axial direction is called longitudinal strain and contraction in the transverse direction, transverse strain [11]. The absolute value of the ratio between the longitudinal strain and transverse strain is called Poisson's ratio, which is expressed as follows:

$$\nu = \left|\frac{\varepsilon 1}{\varepsilon 2}\right|$$, with ν is Poisson's ratio, ε1 is longitudinal strain, ε2 is transverse strain (2)

Poisson's ratio differs depending on the material. For reference, major industrial materials have the following mechanical properties including Poisson's ratio.

For the experimental setup, we detail here the mechanical properties of the used material- Carbon steel (C0.1 - 0.25%) with:

- Young's Modulus E (GPa)- 205;
- Shearing Modulus G (GPa)- 78;
- Tensile Strength (MPa)- 363- 441;
- Poisson's Ratio- 0.23- 0.3

Each metal has its specific resistance. An external tensile force (compressive force) increases (decreases) the resistance by elongating (contracting) it. Suppose the original resistance is R and a strain-initiated change in resistance is ΔR. Then, the following relation is concluded:

$$\frac{\Delta R}{R} = Ks\frac{\Delta L}{L} = Ks\varepsilon$$, where, Ks is a gage factor, the coefficient expressing strain gage sensitivity.

General-purpose strain gages use copper-nickel or nickel-chrome alloy for the resistive element, and the gage factor provided by these alloys is approximately 2.

2.1.1. Hardware implementation

Next picture shows standard referencing of the strain gauge used [5]:

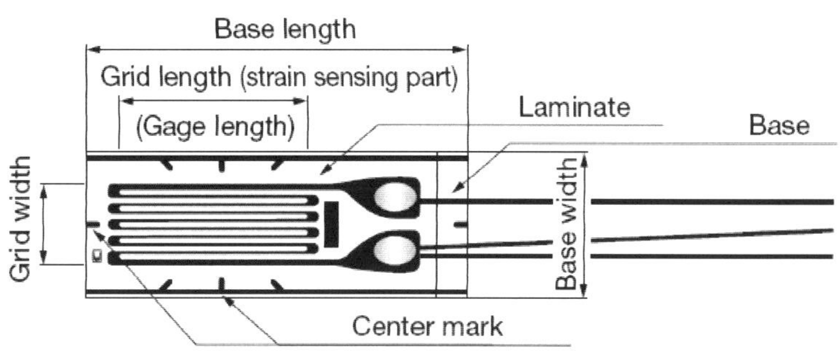

Figure 3. Typical structure of the KFG-2-120-C15-11 strain gauge.

IOP Conf. Series: Materials Science and Engineering **209** (2017) 012065 doi:10.1088/1757-899X/209/1/012065

Positioning the strain gauge in site:

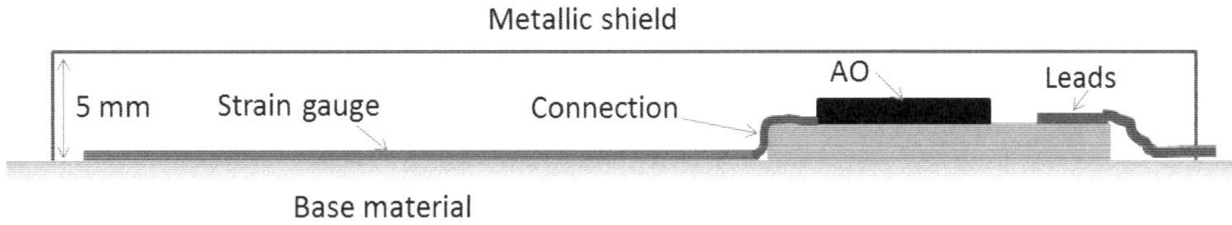

Figure 4. The designed positioning and shielding of the strain gauge.

One measuring head consists in a structure like next figure:

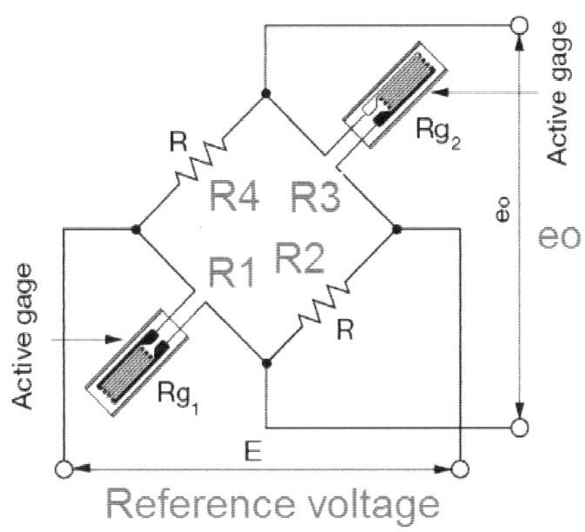

Figure 5. Bridge connected typical architecture of the two active gauges (RG1 for measuring, RG2 for compensation).

With the mathematical model:

$$eo = \frac{R1R3 - R2R4}{(R1+R2)(R3+R4)}E = \frac{1}{4}\frac{\Delta R}{R}E = \frac{1}{4}Ks\varepsilon E \tag{3}$$

Then we decided to use an instrumentation amplifier (see AO from upper figure) who is providing a high quality reference voltage, and it takes both gauges signal and processes them.

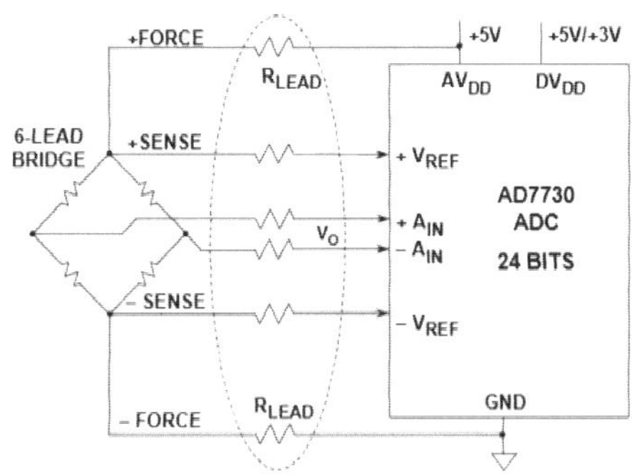

Figure 6. The proposed interface for one measuring head (with two active gauges).

2.2. Layout design considerations

The output of the AO (who includes analog digital converter) is then led to tha processing device. The block diagram of the measuring system is shown next.

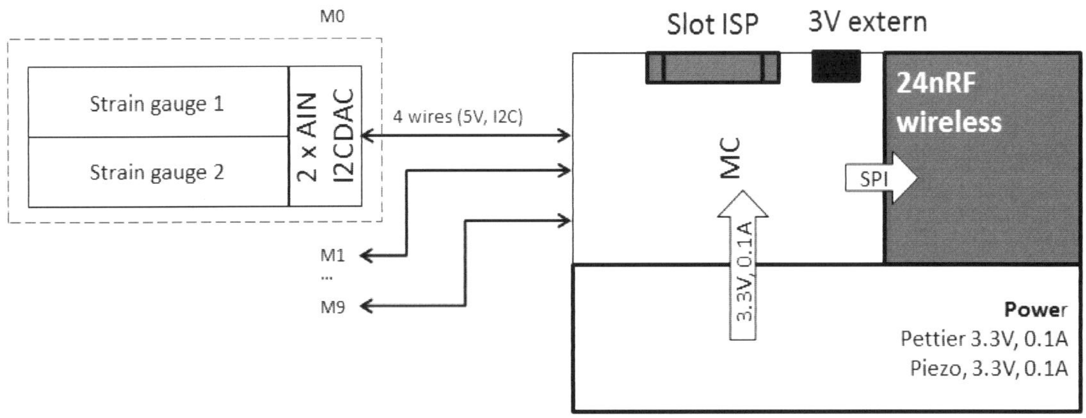

Figure 7. The proposed architecture of the designed measuring system.

One's the data is acquired from ADC (figure 6) it is provided to a microcontroller-based architecture that executes some processes locally, then sends values of elongations outside of the lattice structure (up to 60 m away), to a coordination device.

In our experimental approach, we have used 10 measuring heads (for 10 measuring points in lattice).

The values of all 10 measured elongations are sent synchronously for latter processing.

The dimensional layout of the system looks like next figure:

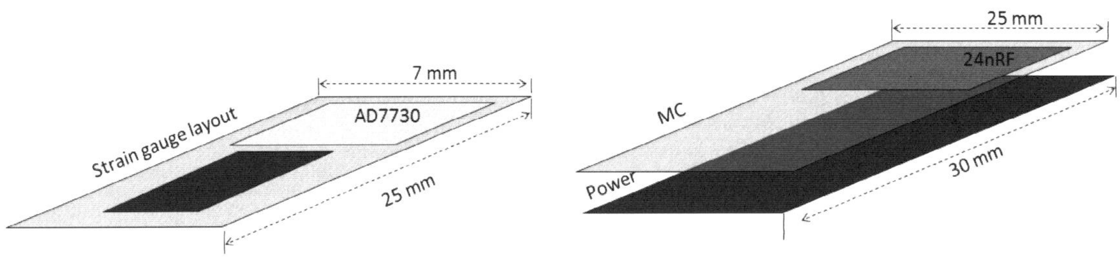

Figure 8. The dimensional layout required for harsh environment protection.

2.3. Measuring system architecture considerations

Because this research activity is a part of a research project, that requires a larger measurement infrastructure, the measuring system architecture is shown in figure 9.

The measurement module (supplied locally) measure and process some data, prepare and send to another module (called coordinator) that is connected via USB to a PC (figure 9) where runs an LaBVIEW application that display real time values, historical data (figure 10).

This application is able also to control in PWM signals some hydraulically motor (as expansion) – se next figure.

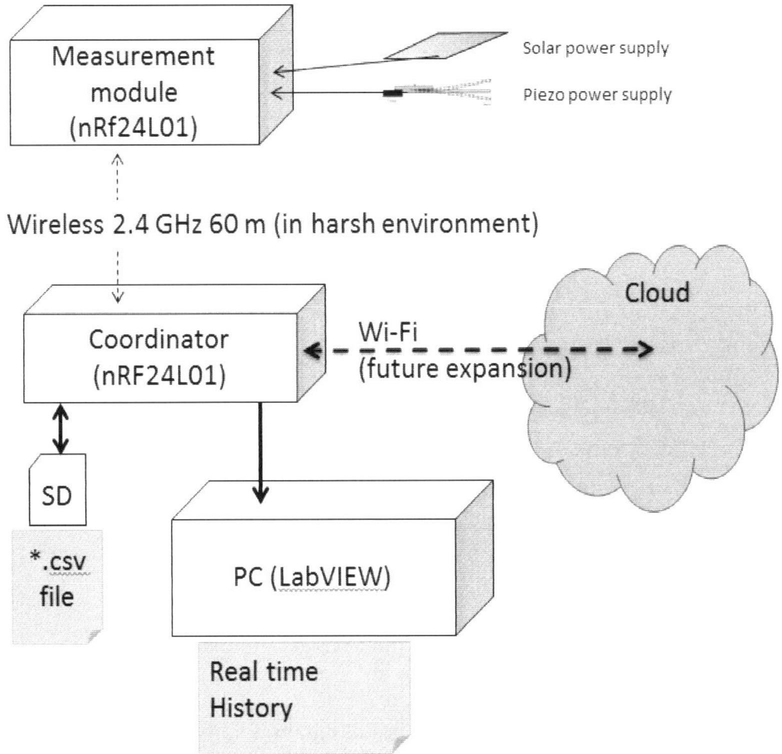

Figure 9. The architecture of the open and extensible system for complex and heterogeneous measurements.

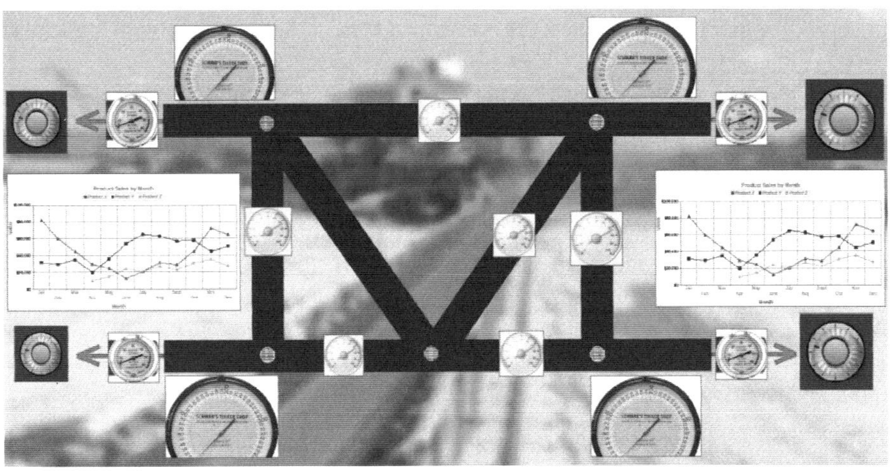

Figure 10. The software interface for real time data monitoring and control.

The intermediary device, called coordinator, is able (by software) to receive simultaneously data from more than one measuring modules.

3. Results and discussions

The application acquires data and processes it according with well-known and typical procedures. It has been calibrated to display relative elongations (μm/m).

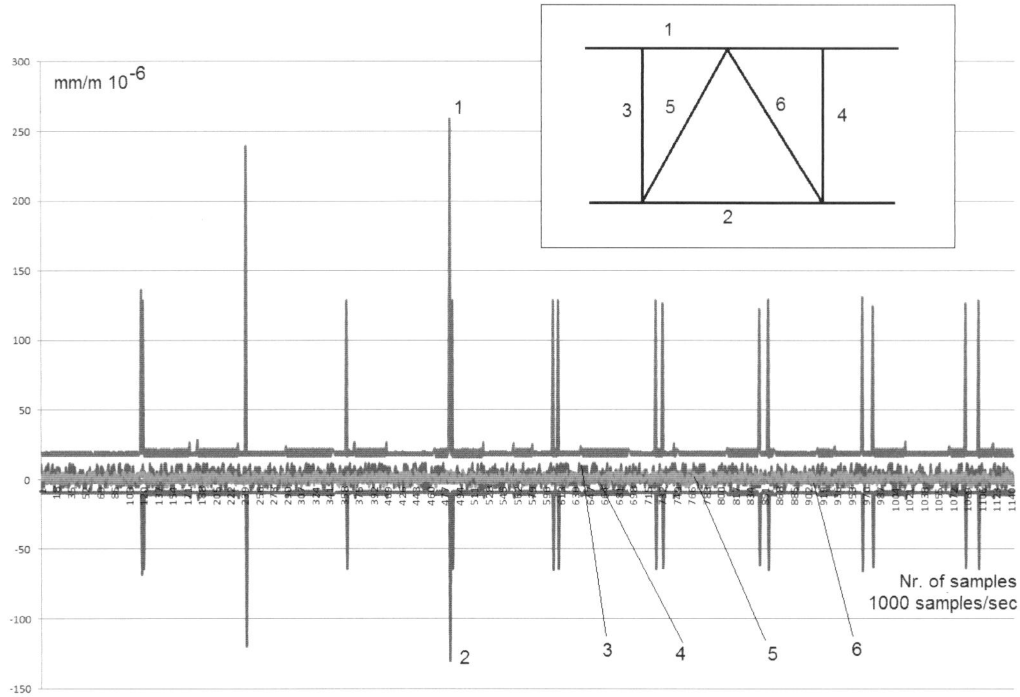

Figure 11. The synchronous display of six measured elongation values (1-6 are measurement heads).

During measuring procedures, two significant measurements have been detected:

- Simultaneously distribution of the elongation in the complex contour of the frame (figure 11).
- Comparison measurements between top and bottom horizontal parts of the frames (heads 1 and 2);

First, in the overlapped graphs, two of them are distinguished- high amplitude and fixed repetition. It is required a spectral analysis of these measured values. All the other are a least 10 times lower.

For the second part we have used a typical function from LabVIEW [5]:

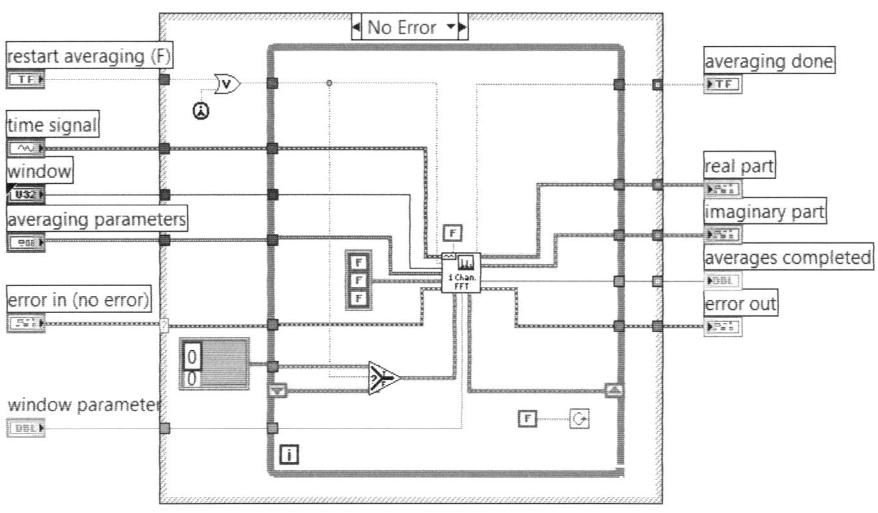

Figure 12. The designed subVI of the application- diagram window.

One of the most recognized facility of the LabVIEW functions is the easy access to the parameters configuration. According with [6,7], the accuracy of the FFT processing is influenced by the sampling rate, the number of samples taken into account for each FFT period, and the "envelope' of the interpolating function of the main frequency.

With the dedicated front panel:

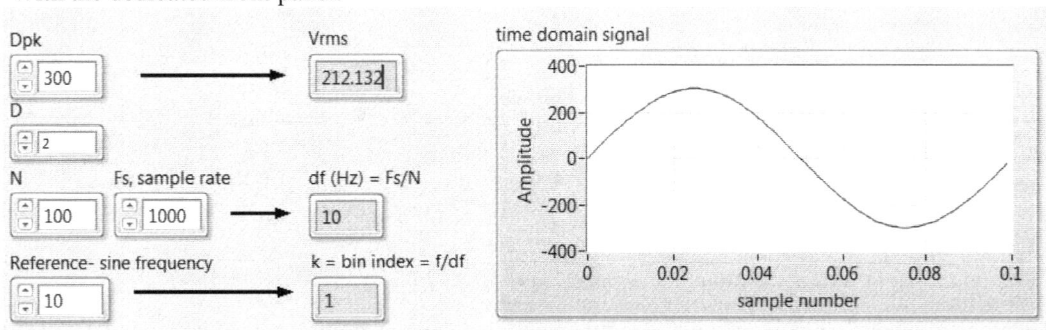

Figure 13. The designed subVI of the application- front panel window.

Then we "insert" the values from the measuring heads 1, 2, and after FFT filtering and processing we obtain:

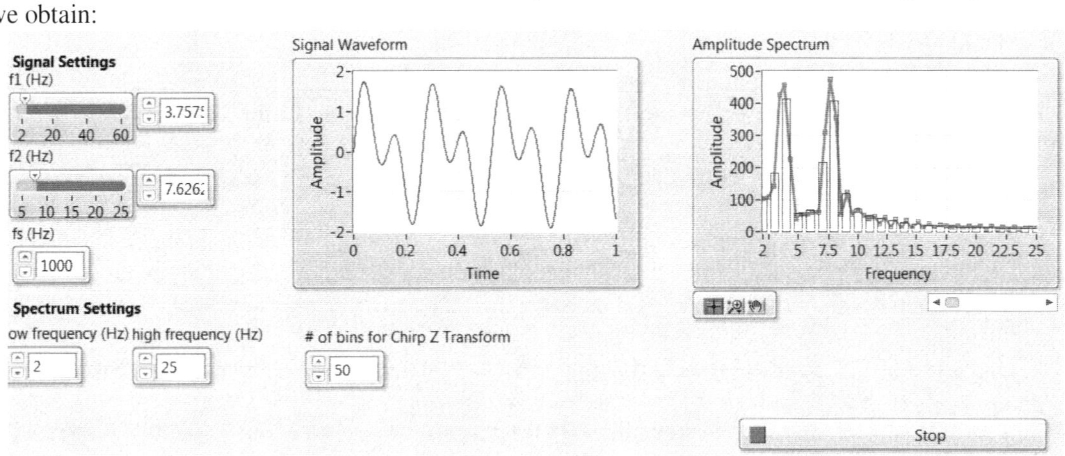

Figure 14. The designed FFT application- front panel window.

There are two main frequencies (4 Hz, and 8 Hz).

The frequency of 4 Hz represents the transmitted tension when a bucket enter into the coal (they wheel has 16 buckets, with a speed rotation 15 rot/min). According with our processing method, the second frequency is double on the "excitation" frequency. It looks like a second harmonics, with almost similar amplitude. On the other hand, the effects of cutting are transmitted along the frame as a mechanical oscillation. It is propagating along the frame and it reflects it back to the bucket wheel, with the behavior of elastic rope. It has been described complete in [8,10], where they propose an investigation of this behavior when metallic and elastic structure is very long- which is also our case. In them conclusions, in typical conditions, there is a sum of oscillating energy.

3.1 Discussion. Future work

Even it is a huge work, it is important to create a model for this elastic structure, than to test it by measurements in different frames the elongation of the sensor 1 and 2. The purpose of this investigation is to find the worst case elongation.

By creating the elastic model of the system, a better investigation might be done by vibration monitoring.

The measurements might include also the suspension cables.

The compression stress in lattice's frame should be also investigated- thinking that in the frames are welded.

4. Conclusions

In this paper we have investigated elongations, than stress in frames that composed the lattice structure of the bucket wheel excavator.

We have identified, in real site investigation, that the stress generated by the elongation is significant in top and bottom parts of the lattice frames.

The elongation, in dynamic process, has an oscillating behavior. It should be tested according with this variation. The measurement of bucket wheel excavator lattice structure internal stress was performed using. KFG-2-120-C15-11 strain gauge.

They are still parameters that must be taken into account for advanced processing.

References

[1] *** ITAtech Report n°3-V2 2015 – *Guidelines for Remote Measurements Monitoring Systems – Layout*: Longrine – Avignon – France

[2] *** Move into the future with reliable measurement 2016 *KYOWA technical report*

[3] Bandyopadhyay L K and Chaulya S K 2015 *Wireless communication in underground mines* 2015

[4] *** Kyowa strain gauges application note 5

[5] Lin W 2015 FAST Fourier Transform (FFT) and Digital Filtering Using LabVIEW *Application Note*

[6] Tomar S K and Arora A 2007 Erratum to "Reflection and transmission of elastic waves at an elastic/porous solid saturated by two immiscible fluids" [Int. J. Solids Struct. 43 (2006) 1991–2013] *International Journal of Solids and Structures* **44**(17) 5796-5800

[7] Bošnjak S, Petković Z, Zrnić N, Pantelić M and Obradović A 2010 Failure analysis and redesign of the bucket wheel excavator two-wheel bogie *Engineering Failure Analysis* **17**(2) 473-485

[8] Gottvald J 2010 The calculation and measurement of the natural frequencies of the bucket wheel excavator SchRs 1320/4x30 *Transport* **25**(3) 269-277

[9] Savković M et al. 2011 Analysis of the axle fracture of the bucket wheel excavator *Engineering Failure Analysis* **18**(1) 433-441.

[10] Bošnjak S, Pantelić M, Zrnić N, Gnjatović N and Đorđević M 2011 Failure analysis and reconstruction design of the slewing platform mantle of the bucket wheel excavator O&K SchRs 630 *Engineering Failure Analysis* **18**(2) 658-669

[11] Zoller C, Kovacs I and Ridzi M C 2007 Interpolation software to mea Length of cutting arch, mesure mechanical forces on the bucket wheels of the excavators *2nd International Conference ICQME*

Acknowledgments

The paper is based on research activity that is the subject of the invention, *Installation for in-situ measuring and recording the forces acting upon the teeth of the cutting element in rock shattering machines*, Patent number RO 123483 B1

The paper is a part of an European Union research project founded by the Coal and Steel Founding, RFCS agreement No. RFCR-CT-2015-00003 (BEWEXMIN)

IOP Conf. Series: Materials Science and Engineering **209** (2017) 012066 doi:10.1088/1757-899X/209/1/012066

Designing a Validation Method for Remote Measurements Dedicated To Investigation of Bucket Wheel Excavator Lattice Structure Internal Stress in Harsh Environment

M Risteiu[1*], R Dobra[1], I Andras[2], M Roventa[2] and A Lorincz[3]

[1] Computer Science and Engineering Department, "1 Decembrie 1918" University of Alba Iulia, Gabriel Bethlen Str. No.5, 510009 Alba Iulia, Romania

[2] Machinery Department, University of Petrosani, University Street, No. 11, Petrosani Romania

[3] Automation Department, University of Petrosani, University Street, No. 11, Petrosani Romania

E-mail: mristeiu@uab.ro

Abstract. The paper is a summarizing the validation stage of the strain gauges based remote measurements in bucket wheel lattice structures. The validation process is required when multiple and synchronized measurements are required in harsh environment. The authors propose virtual models with specific interaction facilities demanded in dynamic measurement systems (like FFT parameterization). The provided data from the remote measurement system (8 heads, discrete values) are processed in time and frequency for completing the range of values, for data synchronization for realistic strain distribution for proposed lattice structures. The proposed validation method is also acting for cleaning the inter-channel influences in measurements.

1. Introduction

The specific strength cutting of coal and of surface rocks, from the mine quarries, varies to great values in the mining field. For this reason, the splintering forces on the cutting teeth also have significant variations in the different mining fields. The size and direction of the forces from the cutting teeth also depends on the geometry of the teeth, their place and the direction of mounting in the bucket of the bucketed wheel [1,3].

Due to the complexity of the excavation process and of cinematic of the cutting tool, the accurately determination of the forces in the cutting teeth can be realized only through an experimental method in situ, reason for which it has been searched to conceive and to realize dynamometric teeth identical to the regular cutting teeth, which to provide a dynamometric bucket, and an electronic installation for measuring data into locations established previously and suited for it.

Because the measured information and recorded is represented by the specific displacements from the applying points of the SG onto the teeth, the relations between them and the components of the resulting points have been established for each tooth (normal, lateral, tangent) [8,9].

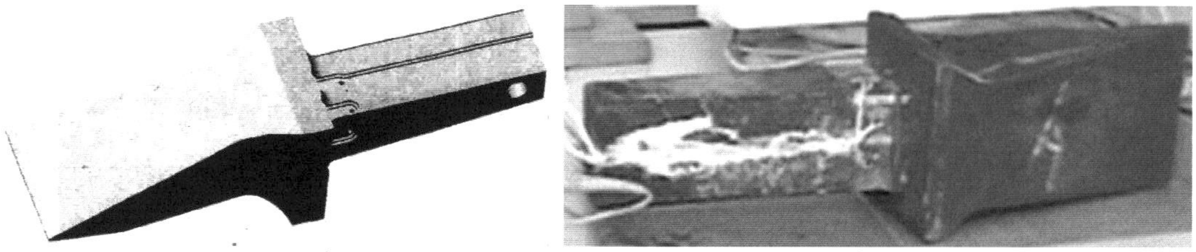

a) b)

Figure 1. Bucket wheel excavator tooth: a) The tooth with the places where the strain gage will be placed, b) The tooth with soldered strain gage.

Due to the complexity of the excavation process and of cinematic of the cutting tool, the accurately determination of the forces in the cutting teeth can be realized only through an experimental method in situ, reason for which it has been searched to conceive and to realize dynamometric teeth identical to the regular cutting teeth, which to provide a dynamometric bucket, and an electronic installation for measuring data into locations established previously and suited for it [1,3,7].

2. Bucket wheel excavator stress intensity factor

The specific strength cutting of coal and of surface rocks, from the mine quarries, varies to great values in the mining field. For this reason, the splintering forces on the cutting teeth also have significant variations in the different mining fields. The size and direction of the forces from the cutting teeth also depends on the geometry of the teeth, their place and the direction of mounting in the bucket of the bucketed wheel [2,5].

2.1. Stress intensity factor considerations

To establish the program to gather and modify the data obtained from the measurements it is necessary to establish the relations between the forces F_{x1}, F_{y1}, F_{z1} with random values, which act onto the dynamometric tooth in the splintering process, his geometrical parameters and the specific linear displacements. A crack intensity factor is referring to three different types of loading that cause displacements of the crack plane, tends to open the crack. The approach I is valid when the force is applied normal to the crack plane (tends to open the crack). The approach II refers to in-plane shear and causes the two crack surface to slide against each other. The approach III is valid when out-of-plane shear is applied and tends to tear the two crack surface apart.

First approach:

$$\sigma_x = \frac{K_1}{\sqrt{2\pi r}}\cos\frac{\theta}{2}\left(1-\sin\frac{\theta}{2}\sin\frac{3\theta}{2}\right)$$

$$\sigma_x = \frac{K_2}{\sqrt{2\pi r}}\cos\frac{\theta}{2}\left(1-\sin\frac{\theta}{2}\sin\frac{3\theta}{2}\right)$$

$$\tau_{xy} = \frac{K_1}{\sqrt{2\pi r}}\sin\frac{\theta}{2}\cos\frac{\theta}{2}\cos\frac{3\theta}{2} \qquad (1)$$

Second approach:

$$\sigma_x = -\frac{K_2}{\sqrt{2\pi r}}\sin\frac{\theta}{2}\left(1-\cos\frac{\theta}{2}\cos\frac{3\theta}{2}\right)$$

$$\sigma_y = \frac{K_2}{\sqrt{2\pi r}}\sin\frac{\theta}{2}\cos\frac{\theta}{2}\cos\frac{3\theta}{2}$$

$$\sigma_{xy} = \frac{K_2}{\sqrt{2\pi r}} \cos\frac{\theta}{2}\left(1 - \sin\frac{\theta}{2}\sin\frac{3\theta}{2}\right) \tag{2}$$

This last approach of deformation called out-of-plane shear mode, does not occur in the plane elastic problem. The stress intensity factors for all three types of approaches are denoted as K_1, K_2 and K_3.

The parameters $\tau_{xy} = \tau_{yz} = 0$ for plane stress and plane strain. The parameters $\sigma_z = 0$ for plane stress and $\sigma_z = (\sigma_x + \sigma_y)$ for plane strain [4].

Third approach:

$$\sigma_{xz} = \frac{K_m}{\sqrt{2\pi r}} \sin\frac{\theta}{2}$$
$$\sigma_{yz} = \frac{K_m}{\sqrt{2\pi r}} \cos\frac{\theta}{2} \tag{3}$$

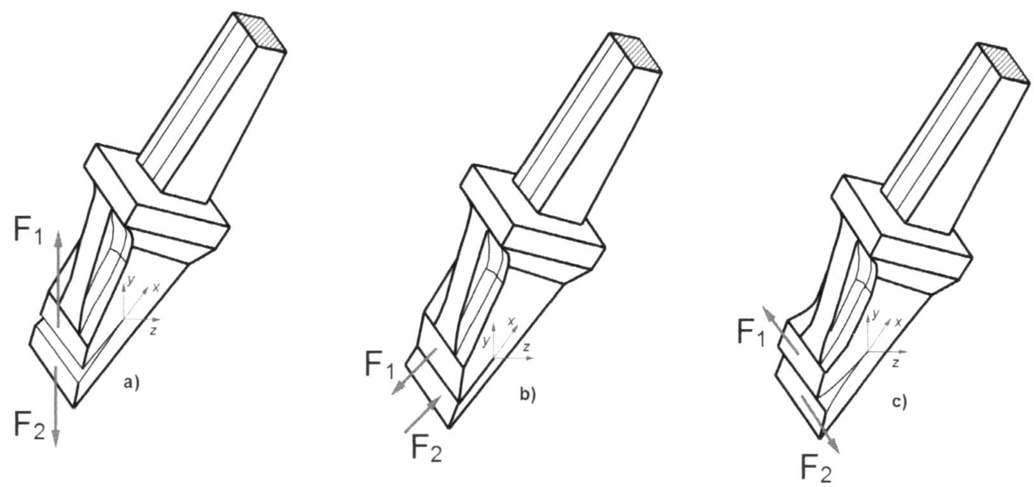

Figure 2. Deformation of the crack for bucket wheel excavator tooth.

Figure 3. Dynamometric bucket mounted onto RS 1300 type excavators.

According to figure 3 the specific deformations the forces F_{x1}, F_{y1}, F_{z1}, applied to the center of the edge of the tooth and the geometrical characteristics of the section, the following relations are obtained [6,8]:

$$\begin{cases} E\cdot\varepsilon_1 = F_{x1}\left(-\frac{1}{A}+\frac{e_1\cdot y_1}{I_z}\right)+F_{y1}\frac{l\cdot y_1}{I_z}-F_{z1}\frac{l\cdot z_1}{I_y} \\ E\cdot\varepsilon_2 = F_{x1}\left(-\frac{1}{A}+\frac{e_1\cdot y_1}{I_z}\right)+F_{y1}\frac{l\cdot y_2}{I_z}-F_{z1}\frac{l\cdot z_2}{I_y} \\ E\cdot\varepsilon_3 = -F_{x1}\left(\frac{1}{A}+\frac{e_2\cdot z_3}{I_y}\right)+F_{y1}\frac{l\cdot y_3}{I_z}-F_{z1}\frac{l\cdot z_3}{I_y} \end{cases} \tag{4}$$

From the solving of the equations system it yields:

$$\begin{cases} F_{x1} = E\dfrac{(\varepsilon_2 y_3-\varepsilon_3 y_2)+\dfrac{(\varepsilon_1-\varepsilon_2)(z_2 y_3-z_3 y_2)}{z_2-z_1}}{y_3\left(-\dfrac{1}{A}+\dfrac{e_1 y_1}{I_z}\right)+y_2\left(\dfrac{1}{A}+\dfrac{e_2 z_3}{I_y}\right)} \\ F_{y1} = \dfrac{EI_z}{l y_3}\left[\varepsilon_2+\dfrac{(\varepsilon_1-\varepsilon_2)z_2}{z_2-z_1}-\dfrac{(\varepsilon_2 y_3-\varepsilon_3 y_2)+\dfrac{(\varepsilon_1-\varepsilon_2)(z_2 y_3-z_3 y_2)}{z_1-z_2}}{y_3\left(-\dfrac{1}{A}+\dfrac{e_1 y_1}{I_z}\right)+y_2\left(\dfrac{1}{A}+\dfrac{e_2 z_3}{I_y}\right)}\right] \\ F_{z1} = EI_y\dfrac{\varepsilon_1-\varepsilon_2}{l(z_2-z_1)} \end{cases} \tag{5}$$

With these values the upper relations become:

$$\begin{cases} F_{x1} = 3020\cdot 10^4(-39{,}4584\varepsilon_1+39{,}6684\varepsilon_2-35{,}21\varepsilon_3) \\ F_{y1} = 5683\cdot 10^4(8{,}7668\varepsilon_1+1{,}2396\varepsilon_2-1{,}3072\varepsilon_3) \\ F_{z1} = 6109\cdot 10^4(\varepsilon_1-\varepsilon_2) \end{cases} \tag{6}$$

It is specified that between the forces F_{x1} F_{y1} F_{z1} and the forces on the tooth resulted from the cinematic of the working tool, the forces F_x, F_y (tangent and normal to the trajectory resulted from the rotating movement of the bucketed wheel) and F_z (resulted from the horizontal tipping of the bucketed wheel by the excavator arm for cutting radial) there are the following relations:

$$\begin{cases} F_{x1} = -F_y \sin 52^\circ + F_x \sin 38^\circ \\ F_{y1} = F_x \cos 38^\circ + F_y \cos 52^\circ \\ F_{z1} = F_z \end{cases} \tag{7}$$

2.1.1. Hardware implementation

The bucket wheel excavator lattice structure internal stress is measured with the developed programmable system.

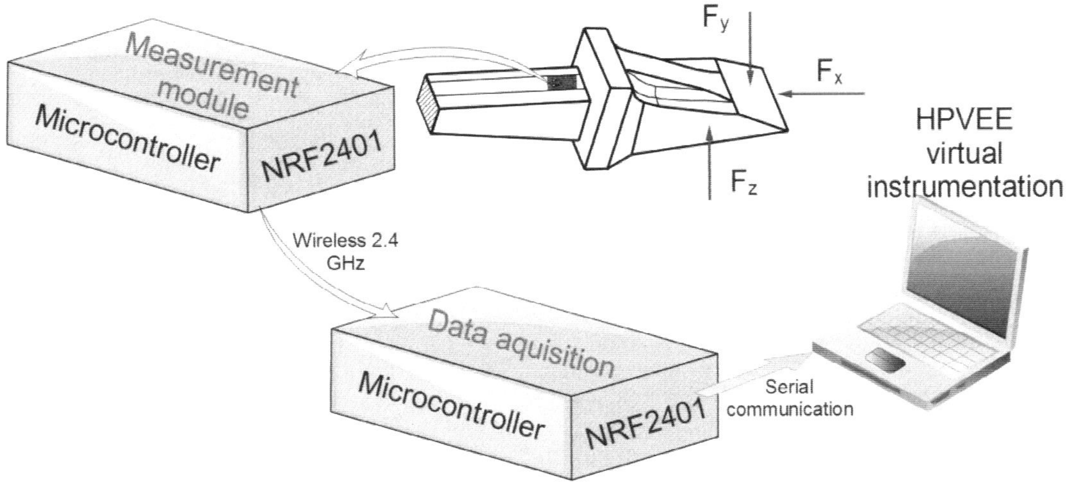

Figure 4. Remote measurements system for investigation of bucket wheel excavator.

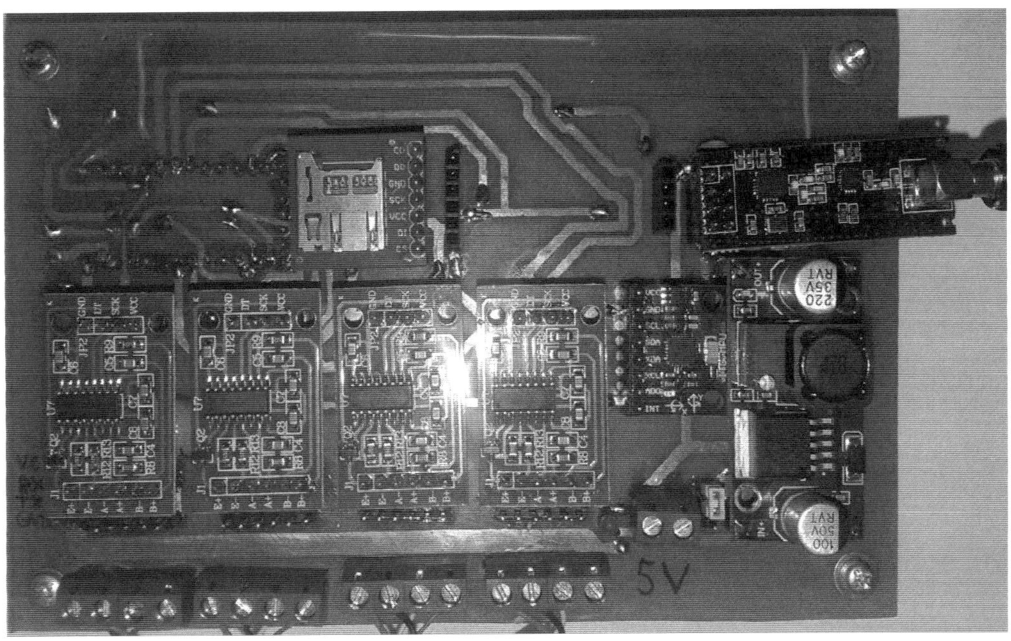

Figure 5. Testing measuring module based on Teensy microcontroller.

IOP Conf. Series: Materials Science and Engineering **209** (2017) 012066 doi:10.1088/1757-899X/209/1/012066

The nRF24L01 module is an RF module that works on the 2,4 GHz band and is perfect for wireless communication in a industrial application because it will penetrate even thick concrete walls. The nRF24L01 has a function to automatically check if the transmitted data is received at the other end. There are a couple of different versions of the NRF-family chips and they all seem to work in a similar way. These modules has an impressive range, with some versions that manages up to 1000 m (free sight) communication and up to 2000 m with a bi-quad antenna.

2.1.2. Software implementation

To start working with the Teensy, all you need to do is plug in your USB cable to your computer and your Teensy board. There are two options for programming the Teensy boards - your favorite C compiler or the Arduino IDE.

```
#include <SPI.h>
#include <SD.h>
#include "RF24.h"
#include "HX711.h"
#include<Wire.h>

#define csn 10                          //nRF24 --> csn
#define cs 6                            //microSD --> cs

const int MPU_addr=0x68;

//HX711 scale(DT,SCK);   //pinout example

HX711 scale1(A8, A9);    // #1  DT-->A8   SCK--> A9 parameter
HX711 scale2(A0, A1);    // #2  DT-->A0   SCK--> A1 parameter
HX711 scale3(3, 4);      // #3  DT-->3    SCK--> 4 parameter '
HX711 scale4(A6, A7);    // #4  DT-->A6   SCK--> A7 parameter

int nr_crt=1;

int T1 = 0;                  //T1= data from  HX711  #1
int T2 = 0;                  //T2= data from  HX711  #2
int T3 = 0;                  //T3= data from  HX711  #3
int T4 = 0;                  //T4= data from  HX711  #4

int16_t AcX=0;
int16_t AcY=0;
int16_t AcZ=0;

int16_t GyX=0;
int16_t GyY=0;
int16_t GyZ=0;

File datafile;               //creating the file for SDcard
RF24 radio(2, 10);           //initializing nRF24

byte addresses[][6] = {"0"};

struct pachet                         //creating data package
     {

       int nr_crt = 1;

       int T1 = 0;
       int T2 = 0;
       int T3 = 0;
       int T4 = 0;

       int16_t AcX=0;
       int16_t AcY=0;
       int16_t AcZ=0;

       int16_t GyX=0;
       int16_t GyY=0;
       int16_t GyZ=0;

       int16_t Temp=0;
       //char text[200]="empty";
       char text[200] = "NR_crt\t  T1,\t   T2,\t     T3,\t

     };

typedef struct pachet Pachet;
Pachet data;

void setup ()
{

  pinMode(cs, OUTPUT);
  pinMode(csn, OUTPUT);
  pinMode(2, OUTPUT);

  Wire.begin();
  Wire.beginTransmission(MPU_addr);
  Wire.write(0x6B);  // PWR_MGMT_1 register
  Wire.write(0);
  Wire.endTransmission(true);

  Serial.begin(9600);
  Serial.print("CLEARDATA");
  Serial.println("LABEL, NR_crt,T1,T2,T3,T4,AcX,AcY,AcZ,GyX,GyY,GyZ,Temp micros");
```

```
void setup ()
{

  pinMode(cs, OUTPUT);
  pinMode(csn, OUTPUT);
  pinMode(2, OUTPUT);

  Wire.begin();
  Wire.beginTransmission(MPU_addr);
  Wire.write(0x6B);  // PWR_MGMT_1 register
  Wire.write(0);
  Wire.endTransmission(true);

  Serial.begin(9600);
  Serial.print("CLEARDATA");
  Serial.println("LABEL, NR_crt,T1,T2,T3,T4,AcX,AcY,AcZ,GyX,GyY,GyZ,Temp micros");

  Serial.println("INITIALIZING SD CARD...");               //Verifing SDcard
  delay(300);
  if (!SD.begin(cs))

    Serial.println("SD CARD NOT DETECTED :(");

  else
    Serial.println("SD CARD READY FOR WRITING :)");



  datafile = SD.open("data.csv", FILE_WRITE);                 //writing data_file to SDcard

void loop()
{
  Wire.beginTransmission(MPU_addr);
  Wire.write(0x3B);  // starting with register 0x3B (ACCEL_XOUT_H)
  Wire.endTransmission(false);
  Wire.requestFrom(MPU_addr,14,true);  // request a total of 14 registers

  AcX=Wire.read()<<8|Wire.read();  // 0x3B (ACCEL_XOUT_H) & 0x3C (ACCEL_XOUT_L)
  AcY=Wire.read()<<8|Wire.read();  // 0x3D (ACCEL_YOUT_H) & 0x3E (ACCEL_YOUT_L)
  AcZ=Wire.read()<<8|Wire.read();  // 0x3F (ACCEL_ZOUT_H) & 0x40 (ACCEL_ZOUT_L)
  Tmp=Wire.read()<<8|Wire.read();  // 0x41 (TEMP_OUT_H) & 0x42 (TEMP_OUT_L)
  GyX=Wire.read()<<8|Wire.read();  // 0x43 (GYRO_XOUT_H) & 0x44 (GYRO_XOUT_L)
  GyY=Wire.read()<<8|Wire.read();  // 0x45 (GYRO_YOUT_H) & 0x46 (GYRO_YOUT_L)
  GyZ=Wire.read()<<8|Wire.read();  // 0x47 (GYRO_ZOUT_H) & 0x48 (GYRO_ZOUT_L)
  Temp=Tmp/340.00+36.53;           //equation for temperature in degrees C from datasheet

  T1=scale1.get_units(10);
  T2=scale2.get_units(10);
  T3=scale3.get_units(10);
  T4=scale4.get_units(10);
```

Monitoring and processing system of mechanical effort with high performance microcontroller is done with the aid of weight sensors which are connected to a data acquisition board .The objective of this project is to create an smart electronic weighing system programmed by MC, this device has capacity to specify the measured weight, weight differences over time and weight thresholds. Data taken from these input devices will be transmitted with an antenna, via wireless, can be stored on a card or can be read directly. Weighing sensors or tensometric stamps reprocessed by an HX711 instrumentation module. In the case of the weight sensor reading module this must be programmed, programming is done on Serial and can be done with an Arduino programmed as a programmer.

3. Results and discussions

The virtual mechanical strain bridge instrument consists on the Data acquisition module presented previously, connected to the Personal computer, the needed software for hardware functioning (Windows XP and the application strain bridge instrument program, developed with Keysight VEE Pro software).

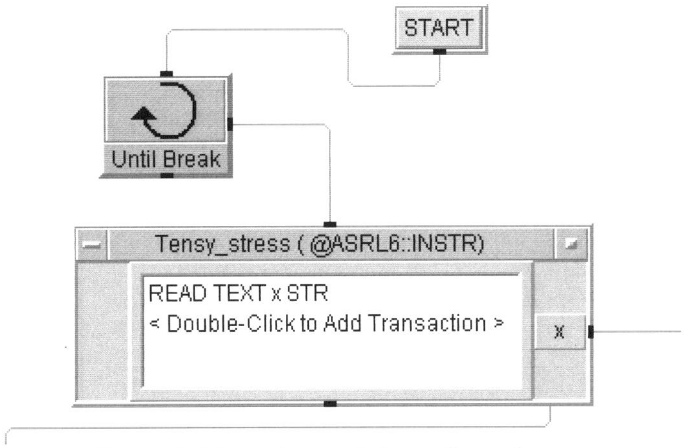

Figure 6. Measuring module configuration.

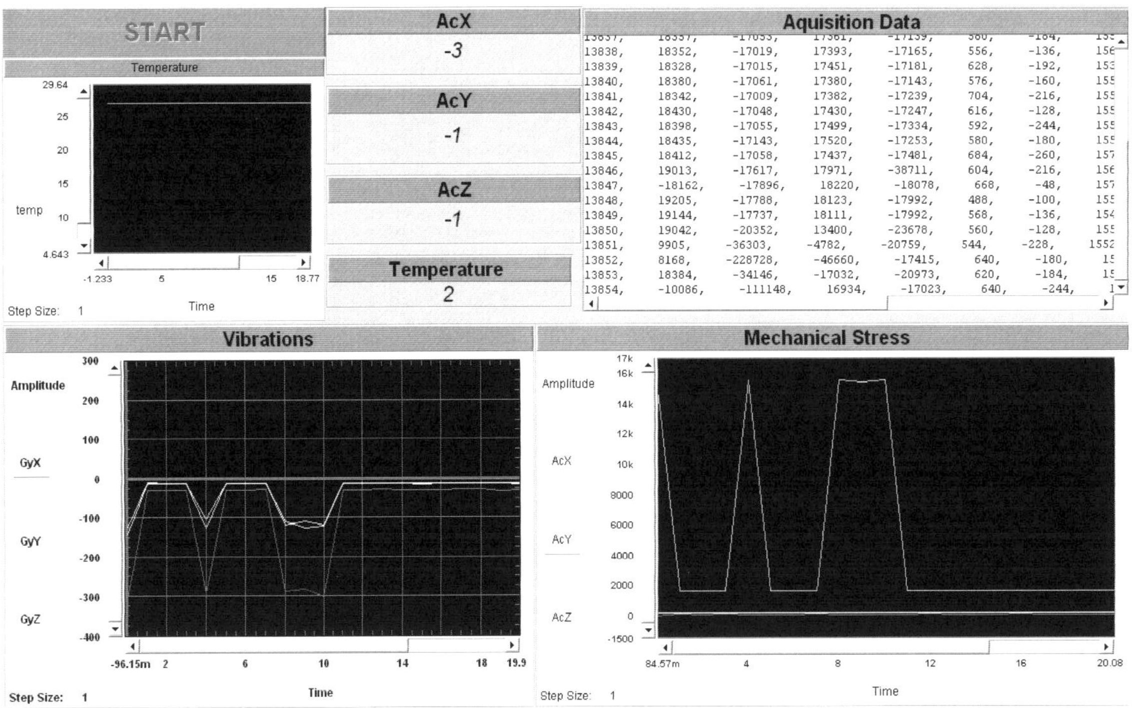

Figure 7. The graphical panel of the virtual mechanical stress measurements.

The software solution for the virtual bridge instrument was developed using the Visual Programming Environment for Data Acquisition, HP VEE and Matlab Script, Hewlett Packard .

Using the MatlabScript software, each bridge equation was specifically expressed and a graphical program was developed (figure 7). The main instruments are the acquisition data where the values from the measuring module.

Using the magnitude spectrum graphical instrument the frequency of the signals are displayed (like FFT parameterization).

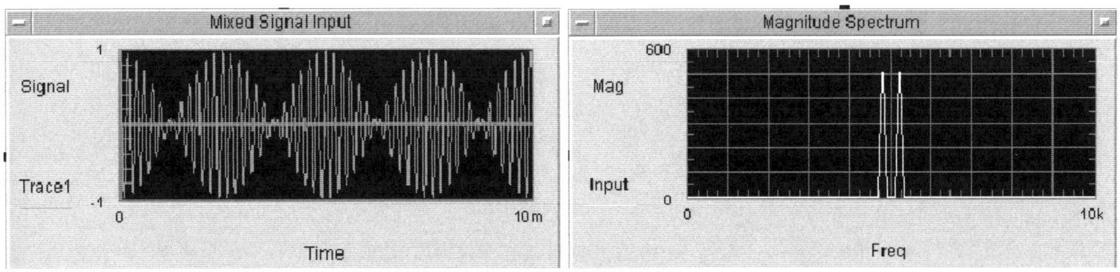

Figure 8. Spectral FFR analysis.

Based on the results of such measurements performed by the statistical processing, the relevant characteristics have led to the mechanical cutting of lignite.

4. Conclusions

The results of research performed regarding the cutting tools of the bucket wheel excavators; in order to reduce the energy consumption and the teeth wear during the lignite and overburden rock are presented.

On the basis of the methodology of analytical determination of force and energetic characteristics, starting from the results of experimental results in laboratory and taking into account the technical characteristics of the excavators

It has also been established that the dynamometric tooth to have the same geometrical parameters and to work in the same conditions as the real teeth. Following the analysis of the bucketed wheels and the way that the teeth is fixed into the holders it has resulted that the strain gage should be placed onto the tail of the tooth, in the free area between the tooth's shoulder and it's holder, in places covered with metallic plates to protect from mechanical hits, and environmental factors.

References

[1] Dumitrescu I, Nan M S, Kovacs I, Jula D, Mihăilescu S and Praporgescu G 2007 Design of an asymmetrical bucket for EsRc 1400 X 30/7 X 630 rotor excavators in agreement with the natural conditions displayed by Jilţ quarries *Analele Universităţii din Petroşani* **9 (XXXVI)** 127-132

[2] Kovacs I, Nan M S, Jula D and Tomuş O B 2007 Contributions to the study of the working mode of the rotor excavators in operation at Jilt sud and Jilt nord lignite open casts *Analele Universităţii din Petroşani* **9 (XXXVI)** 197-209

[3] Nan M S, Kovacs I, Dumitrescu I, Dimirache G, Burlan D and Radu D 2007 Study regarding the possibilities to improve the parameters of teeth on the rotor excavators in operation at Jilt sud and Jilt nord open casts *Analele Universităţii din Petroşani* **9 (XXXVI)** 63-72

[4] Ridzi M C and Zoller C 2007 Messung der Kräfte am Zahn eines Tagebaukohlebaggers *Messtechnische Briefe* **2** 16-24

[5] Ridzi M, Dumitrescu I, Zoller C and Radu D 2005 Aspects concerning the conceiving realization of dynamometric teeth of electrical measurement sketch of the displacement forces from the excavator teeth *Annals of The University of Petrosani* **7 (XXXIV)** 130-136

[6] Raaz V 2000 Optimierung der maschinen- mid verfahrenrechnischen parameter von

schaufelradbaggem für einen abbau von harteren materialen ím tagebau. *Braunkohle in Europa: Innovationen für die Zukunfl*
[7] Zoller C, Kovacs I and Ridzi M C 2007 Interpolation software to mea length of cutting arch , mesure mechanical forces on the bucket wheels of the excavators *2nd International Conference ICQME* 11-14
[8] Zoller C and Ridzi M 2005 Using Hp Vee 6 software to measure mechanical strains on the bucket wheels from the excavators *Annals of The University of Petrosani* **7 (XXXIV)** 101-108
[9] Zoller C and Simaschevici H 2003 Virtual mechanical strain bridge instrument *3rd International Conference Research And Development In Mechanical Industry RaDMI*

Acknowledgments
The paper is based on research activity that is the subject of the invention: Installation for in-situ measuring and recording the forces acting upon the teeth of the cutting element in rock shattering machines, Patent number RO 123483 B1

The paper is a part of an European Union research project founded by the Coal and Steel Founding, RFCS agreement No. RFCR-CT-2015-00003 (BEWEXMIN)

IOP Conf. Series: Materials Science and Engineering **209** (2017) 012067 doi:10.1088/1757-899X/209/1/012067

Designing Pattern Recognition-Based Method for Fast Visual Inspection of the Bucket Wheel Excavator Lattice Structure

M Risteiu[1*], A Lorincz[2], R Dobra[1], P Dasic[3], I Andras[4] and M Roventa[4]
[1] Computer Science and Engineering Department, "1 Decembrie 1918" University of Alba Iulia, Gabriel Bethlen Str. No.5, 510009 Alba Iulia, Romania
[2] Automation Department, University of Petrosani, University Street, No. 11, Petrosani Romania
[3] High Technical Mechanical School of Professional Studies, Str. Radoja Krstića 19, 37240 Trstenik, Serbia
[4] Machinery Department, University of Petrosani, University Street, No. 11, Petrosani Romania

E-mail: mristeiu@uab.ro

Abstract. The proposed paper shows some experimental results of a research in metallic structures inspection by using a high definition camera controller by high processing capabilities. The dedicated ARM Cortex-M4 initializes the ARM Cortex-M0 system for image acquiring. Then, by programming options, we are action for patterns (abnormal situations like metal cracks, or discontinuities) types and tuning, for enabling overexposure highlighting and adjusting camera brightness/exposure, to adjust minimum brightness, and to adjust the pattern's teach threshold. The proposed system has been tested in normal lighting conditions from the typical site.

1. Introduction

The investigation of the bucket wheel excavator health is one of the most important activities into a complex industry like energy production from coal. Everyone can understand this when the full chain is analyzed.

The mined coal must be correlated with the energy demand and the energetic content of the coal. This coal is then transported by conveyer belts and/or rail to the main deposit of the power plant. The amount of the coal in this deposit must be correlated with the demanded coal to be fired, and the capacity of the coal to solidify under its own weight. So, again, the mass flow of the coal is influenced by the excavator production [1,5].

When the coal fired thermal power plant works as a part of the National Energetic Dispatching System, any deviations from the energetic delivery contract (quantity and quality of the energy) are penalized. These penalties are influencing the entire energetic chain- including the excavator.

The bucket wheel excavator is a structure complicated from mechanical and electrical point of view. Main parts of the excavator are mechanical connected each other through lattice frames and supporting cables (figure 1).

Figure 1. A bucket wheel excavator in the marching position.

Figure 2. The lattices are sustaining entire structure. Some of them are welded.

As it can be seen, the lattice structure transmits the mechanical efforts in entire structure, when there is an excitation. Along its structure, the lattice is stressed differently in different points of suspension.

The purpose of this paper is to investigate in the fast way the health of the structure by implementing a pattern recognition method over pictures taken from the focused parts of the excavator. We use a high computational system (built in around a Single Board Computer- SBC) with camera that is processing the taken pictures for a fast diagnose, after some learning and calibration process. In this paper we are focused on electrical wended mechanical assembles- that are parts of excavator lattices.

2. Infrastructure and Methods

For the fast inspection of the stress in lattice structures, we are using hardware adequate: SBC- Raspberry Pi Model B 756-8308 Motherboard, a microcontroller- based camera, packed as an embedded system- Pixy.

The major advantage of the SBC is related to its processing power capacity (CPU is an ARM 1176JZF, 700MHz- RISC, RAM- 512MB, many communication ports- video, USB, WiFi), together is an important interface with external processes [9].

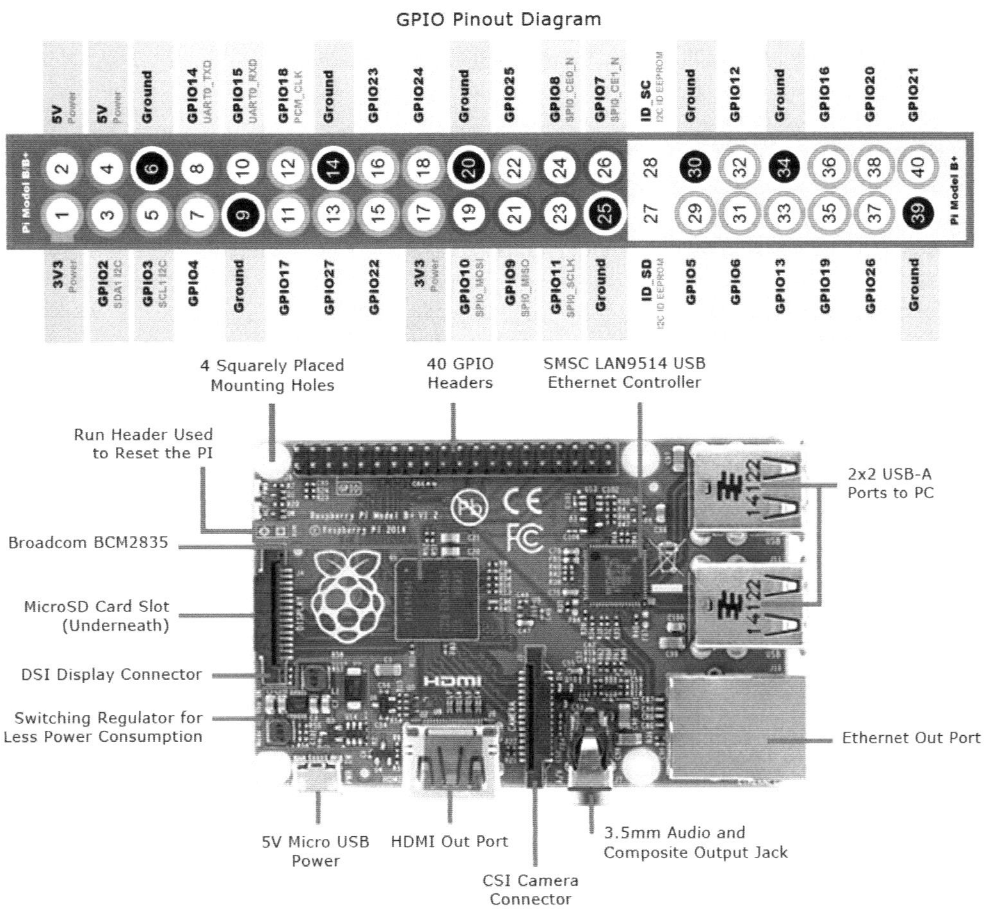

Figure 3. From its reference guide- the full GPIO map [8].

The inspection camera usually is designed around a high density CMOS image sensor. When it is integrated into an embedded system, the output of it is already a high quality processing result. The Pixy works with an NXP microcontroller: an M4 cortex at a frequency of 204MHz and a M0 cortex all at the frequency of 204 MHz, are quite powerful, running on 32 bits both.

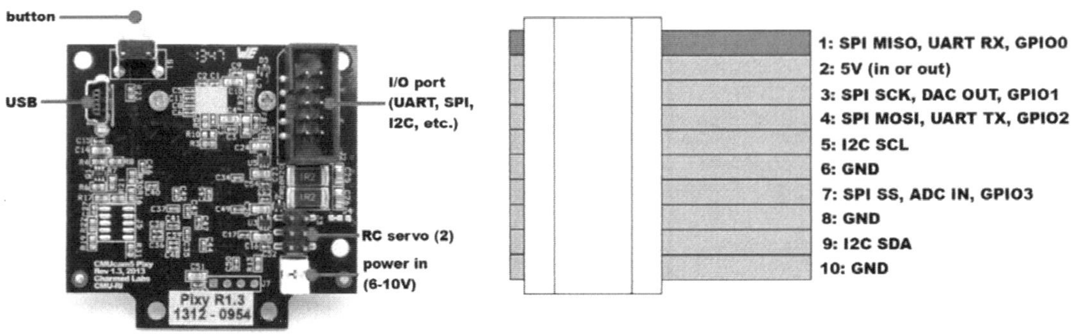

a) b)

Figure 4. The camera-based embedded system proposed for fast inspection- product view (a), pinout map (b) [3].

The embedded system with its memory 1MB (flash), where it stores its code, protocol, patterns offers some special features, and in the same time some constrains like:

- Preferable patterns should be with a color code (it cannot really distinguish the shades of gray). Theoretically through the lens it has, it may even focus very well.
- Creating some contrast through thermal imaging rooms, the higher the temperature on a body, the color is of intense yellow and the low temperature is blue.
- With memorized patterns, it creates blocks, in a significant way.

The sensor used is a CMOS WXGA (1-megapixel) HD Sensor with OmniPixel3-HS Technology OV9715. It is able to ensure 720p HD video performance at 30 frames per second (fps), with high sensitivity - 3 x 3 μm pixels and low-light sensitivity of 3.3 V/lux-sec. The OV9715 provides imaging in virtually every lighting condition from bright daylight to nearly complete darkness.

OV9715 sensor provides full-frame, sub-sampled or windowed 8-bit/10-bit images in RAW RGB format via the digital video port and with complete user control over image quality, formatting and output data transfer. The OV9715 offers a chief ray angle (CRA) of 0°

2.1. Fast inspection of the welding stress considerations

Let's analyze next figure form the quality of the welding point of view.

The most important parameters that are taken into account for quality of the welding analysis are [9]:

- Distribution of the weld material;
- The quantity of the waste (the slag after cooling);
- Porosity of the weld surface;
- Tightness should be constant and with a constant curve texture;
- Strength of the welding is required by design

All this parameters are influenced, during the arch life, by the length of the arch, by the speed of the welding head, and by the arch current [9]. The proposes three measurable parameters: the 3D trace of the weld (a); the quality of welding (eventually the speed of the weld generation) (b); the thickness of the welding cord (c).

IOP Conf. Series: Materials Science and Engineering **209** (2017) 012067 doi:10.1088/1757-899X/209/1/012067

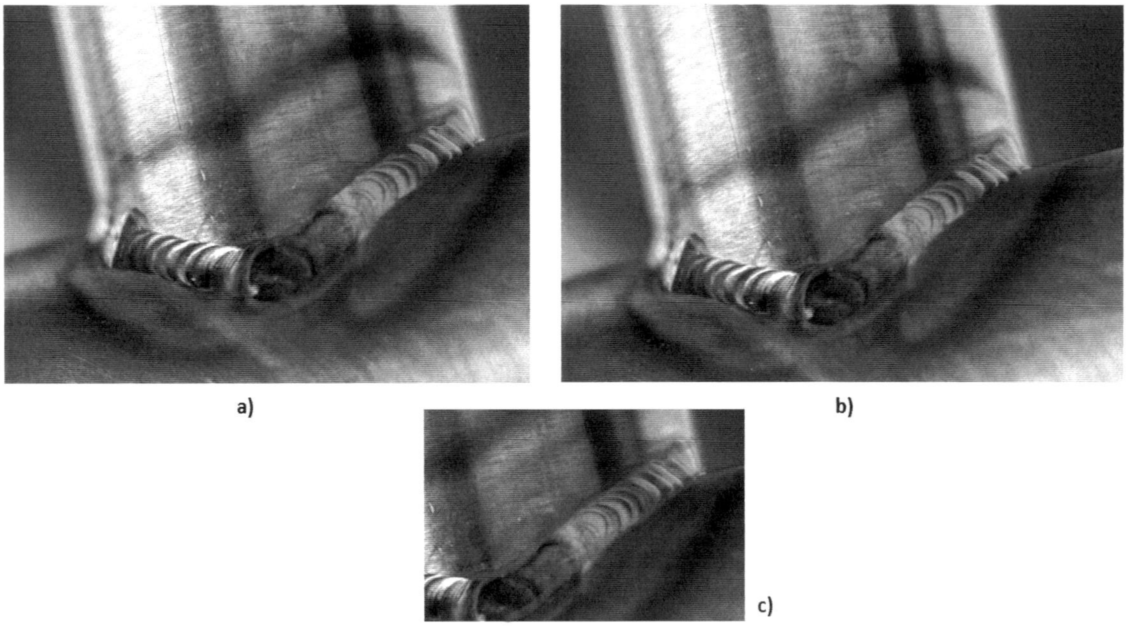

Figure 5. The evaluation of the quality of welding in a 3D exposal of the metallic assembles.

The first idea of visual inspection has started with a Matlab function designed for extracting contours, then evaluating the welding contour. The typical function is [2]:

figure;

imcontour(I,3)

In order to obtain focused contour (expected continuous contour) (figured in c) part of the figure), some extra processing are required (defining patterns according with many aspects- textures, positions). These actions are difficult to setup.

Another meet tool is developed by OpenGL programming environment [3,4,7]. We propose to analyze the *matchTemplate* function. They are developed in C++, Python. The typical pseudocode is shown for different programming environments:

C++: void matchTemplate(InputArray image, InputArray templ, OutputArray result, int method)

Python: cv2.matchTemplate(image, templ, method[, result]) → result

C: void cvMatchTemplate(const CvArr image, const CvArr* templ, CvArr* result, int method)*

Python: cv.MatchTemplate(image, templ, result, method) → None

The environment programming parameters are:

image – Image where the search is running. It must be 8-bit or 32-bit floating-point.

templ – Searched template. It must be not greater than the source image and have the same data type.

result – Map of comparison results. It must be single-channel 32-bit floating-point. If image is W \times H and templ is w \times h , then result is (W-w+1) \times (H-h+1) .

method – Parameter specifying the comparison method (see below).

The function slides through image, compares the overlapped patches of size w \times h against templ using the specified method and stores the comparison results in result. Here are the formulae for the available comparison methods (I denotes image, T template, R result). The summation is done over template and/or the image patch: x' = 0...w-1, y' = 0...h-1 [4,7].

What does mean *templ*? It means defining a template for the acquired figure, template that covers many uncontrolled conditions.

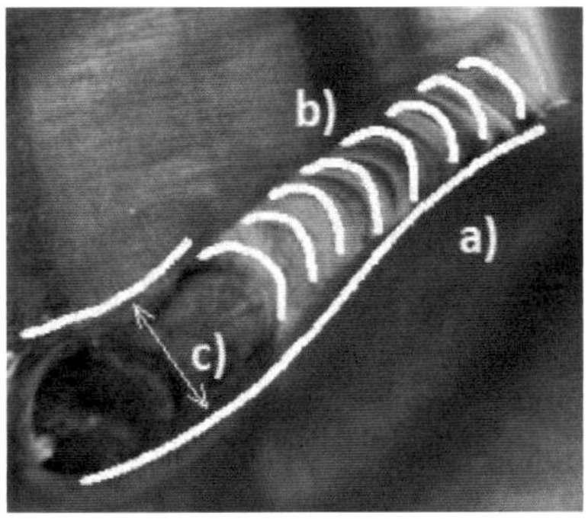

Figure 6. The defined characteristics that compose the *templ* parameter.

These parameters consist in branch of paths that are analyzed in an acquired picture inspection. These parameters are not influenced by position, contrast, and noise.

The literature offers more methods to calculate matches [4,7]. For our experiment we have used the basic method: *method=CV_TM_SQDIFF*. It uses next formula:

$$R(x,y) = \sum_{x',y'} (T(x',y') - I(x+x',y+y'))^2$$

After the function finishes the comparison, the best matches can be found as global minimums (when *CV_TM_SQDIFF* was used) or maximums (when *CV_TM_CCORR*) using the *minMaxLoc()* function.

$$R(x,y) = \sum_{x',y'} (T(x',y') I(x+x',y+y'))^2$$

In case of a color image, template summation in the numerator and each sum in the denominator is done over all of the channels and separate mean values are used for each channel (Red, Green, Blue, Hue). That is, the function can take a color template and a color image. The result will still be a single-channel image, which is easier to analyze.

2.2. Implementation pattern recognition using OpenGL and C++ for fast inspection of the welding cordon

The implementation uses two steps:

- extract focused contour;
- matches the processed contours with the templates with technical meanings.

Extracting contours of images is done through some typical steps:

```
/// Load source image and convert it to gray
src = imread( argv[1], 1 );
/// Convert image to gray and blur it
cvtColor( src, src_gray, CV_BGR2GRAY );
blur( src_gray, src_gray, Size(3,3) );
/// Create Window
char* source_window = "Source";
namedWindow( source_window, CV_WINDOW_AUTOSIZE );
imshow( source_window, src );
createTrackbar( " Canny thresh:", "Source", &thresh, max_thresh, thresh_callback );
thresh_callback( 0, 0 );
Mat canny_output;
vector<vector<Point> > contours;
```

IOP Conf. Series: Materials Science and Engineering **209** (2017) 012067 doi:10.1088/1757-899X/209/1/012067

```
vector<Vec4i> hierarchy;
/// Detect edges using canny
Canny( src_gray, canny_output, thresh, thresh*2, 3 );
/// Find contours
findContours( canny_output, contours, hierarchy, CV_RETR_TREE, CV_CHAIN_APPROX_SIMPLE, Point(0, 0) );
/// Draw contours
Mat drawing = Mat::zeros( canny_output.size(), CV_8UC3 );
for( int i = 0; i< contours.size(); i++ )
  {
   Scalar color = Scalar( rng.uniform(0, 255), rng.uniform(0,255), rng.uniform(0,255) );
   drawContours( drawing, contours, i, color, 2, 8, hierarchy, 0, Point() );
  }
```

For the second part of this implementation, the software is doing some actions:

- Loads an input image and a image patch (template);
- Perform a template matching procedure by using the OpenGL function matchTemplate with any of the method described before.
- Normalize the output of the matching procedure;
- Localize the location with higher matching probability;
- Draw a rectangle around the area corresponding to the highest match.

The typical pseudocode used is shown next:

```
using namespace std;
using namespace cv;
/// Global Variables
Mat img; Mat templ; Mat result;
char* image_window = "sudura_viteza_mare";
char* result_window = "sudura viteza mare window";
int match_method;
int max_Trackbar = 5;
/// Function Headers
void MatchingMethod( int, void* );
```

Then:

```
/// Do the Matching and Normalize
  matchTemplate( img, templ, result, match_method );
  normalize( result, result, 0, 1, NORM_MINMAX, -1, Mat() );
  /// Localizing the best match with minMaxLoc
  double minVal; double maxVal; Point minLoc; Point maxLoc;
  Point matchLoc;
  minMaxLoc( result, &minVal, &maxVal, &minLoc, &maxLoc, Mat() );
  /// For SQDIFF and SQDIFF_NORMED, the best matches are lower values. For all the other methods, the higher the better
  if( match_method == CV_TM_SQDIFF || match_method == CV_TM_SQDIFF_NORMED )
    { matchLoc = minLoc; }
  else
    { matchLoc = maxLoc; }

  /// Show me what you got
  rectangle( img_display, matchLoc, Point( matchLoc.x + templ.cols , matchLoc.y + templ.rows ), Scalar::all(0), 2, 8, 0 );
  rectangle( result, matchLoc, Point( matchLoc.x + templ.cols , matchLoc.y + templ.rows ), Scalar::all(0), 2, 8, 0 );
```

```
imshow( image_window, img_display );
imshow( result_window, result );
```

Then, the technical inspection means to evaluate/ report the results.

3. Results and discussions

The preliminary tests have been done first to train our system when different welding regimes has been established:

3.1. Testing the length of the welding

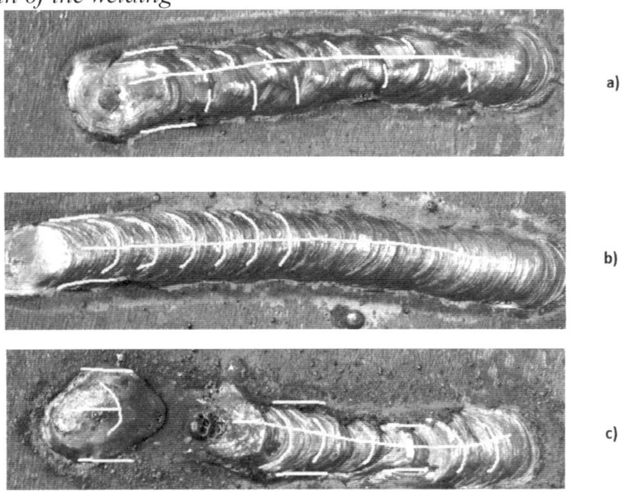

Figure 7. The patterns recognized by the *templ* parameter, according with arch length.

By running different tests (position, light, distance, zooming options), the length of the welding arch defines different patterns:

- the continuity of the weld is disturbed in short arch length (a);
- the trace and thickness of the weld is distorted (c).

3.2. Testing the speed of the welding head

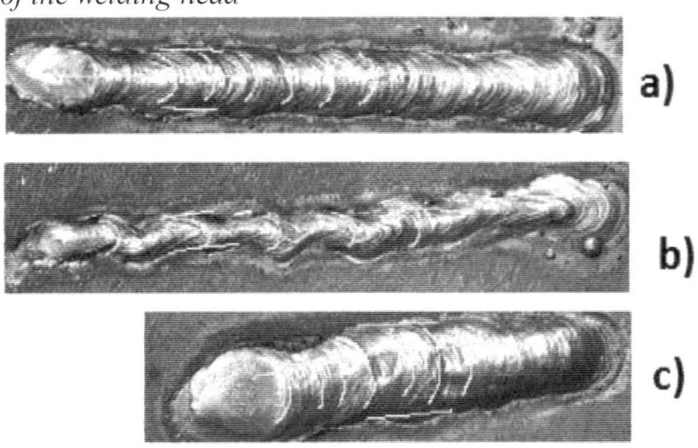

Figure 8. The patterns recognized by the *templ* parameter, according with the speed of the welding head.

By running different tests (position, light, distance, zooming options), the length of the welding arch defines different patterns:

- the thickness and the trace of the weld is disturbed(b);
- the thickness and the continuity of the weld is distorted (c).

3.3. Testing the current through the welding arch

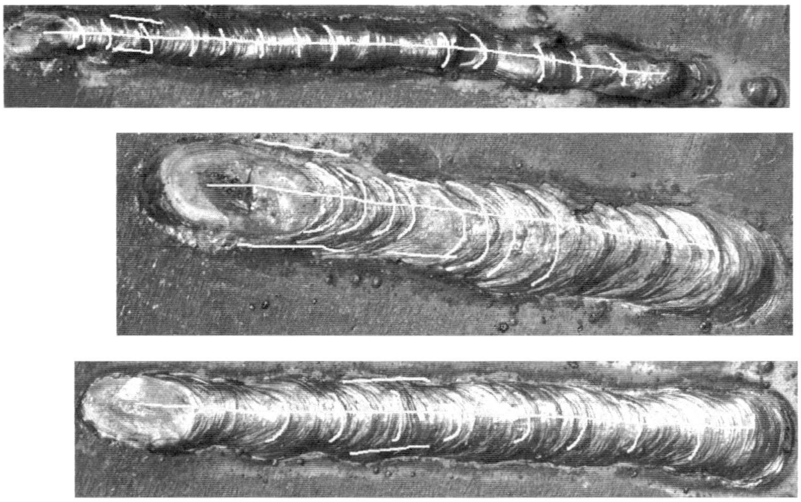

a)

b)

c)

Figure 9. The patterns recognized by the *templ* parameter, according with the current of the welding arch.

By running different tests (position, light, distance, zooming options), the length of the welding arch defines different patterns:

- the thickness and the trace of the weld is not constant (b);
- the thickness is not (c).

Then, by combining different pictures we have been required to adjust to zooming facility in order to create detect the cause of some faults in different weld. This is required in order to have a calibration database for calibrating the system for weakness inspection during the in site testing.

4. Conclusions

In this paper we have investigated different types of welds cordons according with welding regime. The tests have been done based on patterns implemented in high computing systems.

The investigation camera is powerful, with acceptable results (89% of results' success).

We consider that an important improvement (increasing the results' success) can be done by:

- Investigating, in a continuous weld contour a weakness area like a cleave
- The inspection will be done by adding zooming facility (increasing the density of the acquired dots into one pictures),
- By adding light contrast to the picture by illuminating with laser source.

References

[1] Bishop C M 2014 *Pattern Recognition and Machine Learning 3rd Edition* ISBN-10: 0-387-31073-8

[2] Stenberg E L T 2011 Quality inspection and fatigue assessment of welded structures in cooperation with volvo construction equipment AB *Master of Science Thesis in Lightweight Structures Dept of Aeronautical and Vehicle Engineering Stockholm, Sweden*

[3] Theodoridis S 2010 *Introduction in Pattern Recognition in Matlab*, ISBN 978-0-12-374486-9

[4] *** http://cmucam.org/projects/cmucam5/wiki/Porting_Guide 2017
[5] *** Edge Detection and Image Overlay with Impaired Frame 2017
[6] *** Move into the future with reliable measurement 2017 *KYOWA technical report 2016*
[7] *** Kyowa strain gauges application note 5
[8] *** OpenCV, pattern matching, ttp://docs.opencv.org/2.4/doc/tutorials/imgproc/histograms/template_matching/template_matching.html
[9] *** https://raspberrypi.stackexchange.com/questions/40318/raspberry-pi-2-can-gpio-pins-29-40-be-used-gpio-gen-input-output-configurable-in

Acknowledgments
The paper is based on research activity that is the subject of the invention brevet Installation for in-situ measuring and recording the forces acting upon the teeth of the cutting element in rock shattering machines, Patent number RO 123483 B1

The paper is a part of an European Union research project founded by the Coal and Steel Founding, RFCS agreement No. RFCR-CT-2015-00003 (BEWEXMIN)

High Accuracy Investigation of Microwave Absorption in Polymer Electrical Components on Motherboard of Computers

P Dašić[1*], C Hutanu[1*], V Jevremović[1], R Dobra[2], M Risteiu[2] and I Ileana[2]
[1] High Technical Mechanical School of Professional Studies, Str. Radoja Krstića 19, 37240 Trstenik, Serbia
[2] "1 Decembrie 1918" University, Faculty of Sciences and Applied in Electronics, 15-17 Unirii str, 510009, Alba Iulia, Romania

E-mail: dasicp58@gmail.com, hutanu_constantin@yahoo.co.uk

Abstract. Electronic operating at high frequencies can have problems with emission of high frequency noise. Once put inside an enclosure, the energy will add in phase at certain frequencies to cause resonances which will hinder the performance of the device. These absorbers are based upon open celled foam impregnated with a carbon coating. It is quite possible that in the near future, microprocessors would be to work on a frequency located in 5 to 10 GHz. In these circumstances it is useful to know how and how much of the electromagnetic field emitted by a microprocessor, it is absorbed by the circuit elements in the immediate vicinity of the microprocessor. The aim of this contribution is to demonstrate throughout high-level experimental analysis how the main electric parameters of polymer materials, which build the printed circuits and the one of electric capacitors and resistors, depend on the frequencies on which they work from the microwave range.

1. Introduction

In computers, various circuits are already operating at frequencies above 3 GHz, so we can say that these electronic circuits become microwave generating sources. They are absorbed by a very large number of materials, and once microwaves penetrate, electromagnetic energy is converted into heat. The effect generated primarily depends on its physical properties so that a material that has large dielectric losses inside it will heat up more than one with low dielectric losses [1]. When the high frequency waves are absorbed, they form so-called "hot spots". The phenomenon of maximum absorption occurs in the surfaces and volume of the metallic materials. This property of metallic materials represents a barrier to the propagation of the microwave field. Since most of the electronic components in the circuits that make up the structure of a computer are in polymeric capsules, it is necessary to know the level of microwave absorption in this type of materials [2]. In both cases, the absorption of the microwave field in metallic materials as well as the absorption into dielectric materials, the effect is the same, ie local heating. Excessive heating of electronic components due to microwave absorption can cause an accelerated aging of the structures from which the electronic components are made or abnormal functioning. In this article we studied the level of absorption of the microwave field in the metallic materials that constitute the electrolytic capacitor, in the polymeric materials that constitute the capsule of other electronic components, as well as in the insulating material which constitutes the base of the electric circuits. Following a thorough study and a large number of measurement variants, we built a measuring stand characterized by a very high accuracy of the measurements.

2. The measurement method of microwave absorption in polymer electrical components

The measurement method is of differential type and the measuring material is positioned in the space between two Horn antennas. A horn antenna facilitates the transition of TE_{10} waves emerging from the waveguide into free space. We chose the differential measure method to be able to eliminate the measurement errors caused by the voltage fluctuations of the source feeding the Gunn diode oscillator. Since we are also investigating general reflection conditions in our measurement assembly, we may possibly have to deal with returning waves that are of the same power magnitude as those generated by the microwave generator, i.e. the Gunn diode. For that reason we have to reckon with reactions affecting the oscillations generated by the Gunn oscillator. For that reason an isolator (one-way waveguide) is connected between the oscillator and the slotted measuring line. The isolator provides for good transmission power (in the arrow's direction), meaning very little wave attenuation in one direction, while providing very high attenuation in the other. Typical values of this kind of field displacement isolator for the X-band from 8.5 GHz up to 9.6 GHz amount to a minimum of 30 dB reverse attenuation and maximum 0.5 dB for forward attenuation. As such the reflected or returning wave is subjected to severe attenuation by the isolator and consequently is unable to affect the oscillator. In order, to not disturb the normal operation of the oscillator that generates the microwave field by returning the reflected wave, the measuring line is provided with an isolator which attenuates with 54 dB the eventual wave reflected by the sample positioned between the Horn antennas. The isolator is a special high-frequency ferrite component, in which the propagation of the wave is influence. The magnetic materials used here are ferrite with high electrical resistivity. High resistance is a necessity to keep attenuation losses as low as possible. A ferrite plate is used at the waveguide edge where the magnetic field vector circulates. This is located in the static magnetic field B of a permanent magnet. The direction of this magnetic field is parallel to the electric field lines of the TE10-mode, i.e perpendicular to the waveguide's wide side. The magnetic moments in the ferrite carry out a precession motion around the direction of the permanent magnetic field with the frequency $f = \varsigma\mu B/h$. Here f means frequency, ς is the Landé factor, μ is Bohr's magneton, B is the magnetic field strength and h Planck's constant (of action). The ferrite plate can absorb circularly polarised electromagnetic waves of the frequency specified by the equation above. However, the absorption only appears for one rotation direction, the one predetermined by the direction of the static magnetic field.

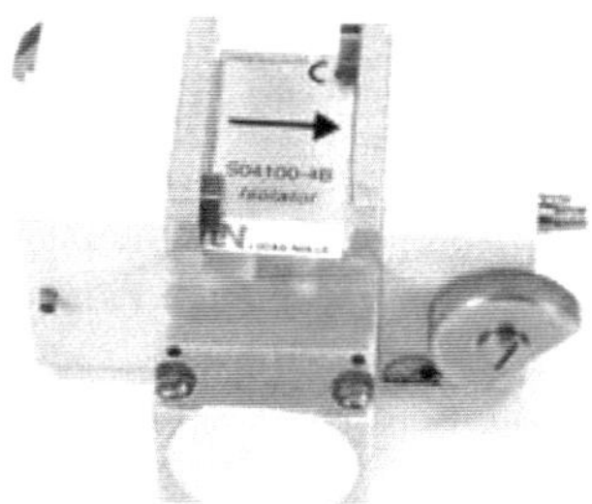

Figure 1. The isolator SO4100-4B (Frequency range: 8.5 to 9.6 GHz).

The condition imposed and monitored in the measurement activity is that, irrespective of the frequency of the oscillator generating the microwave field, the intensity of the microwave field in the space in which the sample to be placed, is to be the same. For this purpose, the measuring line is constructed with a two-branch bifurcation. The microwave field produced by the Gunn diode is divided by a waveguide section that has the property of roughly dividing the incident wave. The role of each of the two branches after bifurcation, is:

- One of the branches guides the microwave field to a separate detector connected to a virtual measuring device.
- The second branch guides the microwave field to the Horn antenna system.

The microwave field interacting with the sample is taken over by the second Horn antenna in the system and directed by a rectangular guide segment toward a detector device. The electrical voltage at the output of the detector device is taken over by an LNC connected via the SMA cable with a X-Band Measurement Interface. This device is provides for an operating frequency range between 8.3 - 9.9 GHz in the X-band range.

3. The experimental analysis of polymer materials

The experimental analysis bench is composed by four categories:

- Main branch having the structure composed of: Gunn oscillator, ferrite isolator and cross directional coupler SO4100-4R (Input Direct Line);
- The first branch that has the structure of the following components: R100/N adapter (waveguide-coax transition) SO4100-4H, LNC "Low Noise Converter", SMA cable and X-Band Measurement Interface;
- The second branch (emissive branch) that has the structure of the following components: Variable attenuator SO4100-4D, Slotted line, N socket/SMA plug adapter, Coaxial measuring probe for slotted guide, LNC, dual BNC/DIN cable, SMA cable, X-Band Measurement Interface, waveguide adapter and Horn antenna;
- The third branch (the receiving branch) that has the structure of the following components: Horn antenna, two waveguide adapters, coaxial adapter, power sensor, dual BNC/DIN cable and power meter .

3.1. Description and role of the main components

Gunn oscillator: An X-band oscillator with mechanical frequency adjustment is used to generate high-frequency electromagnetic waves. Frequency adjustment is performed using a micrometer screw. Operating voltage: 8 to 10 V DC, power: + 17 dBm (50mW), frequency range: 8.5 to 9.6 GHz, connection : BNC socket.

Directional coupler SO4100-4R: Cross couplers serve to separately measure the wave travelling in forward and reverse propagation in a transmission line. This can be carried out simultaneously for the receiving direction. The cross coupler consists of a main arm and a side arm situated at a right angle above this, in which, as indicated in the figure below, a portion of the power travelling in the main arm is decoupled. The decoupling is normally carried out using apertures or holes in the waveguide wall shared by the two arms, as specified in the image below (cross-shaped holes).

Figure 2. *Directional coupler SO4100-4R* (Frequency range: 8.5 to 9.6 GHz).

LNC receiver: operates just like a receiver for satellite television reception. The basic circuit is that

IOP Conf. Series: Materials Science and Engineering **209** (2017) 012068 doi:10.1088/1757-899X/209/1/012068

of a superheterodyne receiver with low-noise input stage, microwave mixer, IF amplifier and rectifier. The X-band LNC in conjunction with an internal oscillator of 10 GHz and mixer generates receives the X-band signal to be measured with a frequency around 9 GHz and generates a proportional signal with an intermediate frequency (IF) in the range of 1 GHz. This IF signal is supplied by the LNC to the receiver card of the UniTrain-I Experimenter via an SMA cable and there it is further processed and rectified. The LNC is also supplied with operating voltage via the SMA cable. Input sensitivity > -75 dBm, volume range ≥ 50 dB, 16 dB gain.

Variable attenuator: are designed to absorb a certain portion of the incoming microwave power and conduct the rest. The attenuation strength of a variable attenuator can be affected both by its length as well as by the position of resistive material inside the waveguide. Here in the middle of the waveguide's width, i.e. the location where the strength of the electric field is maximum, an absorbing plate is immersed penetrating into the interior of the waveguide.

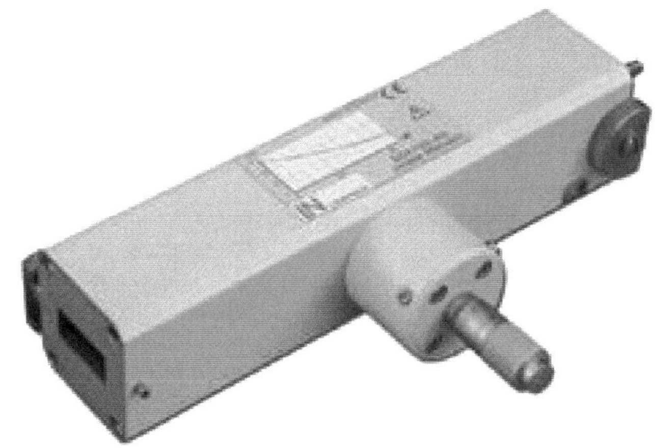

Figure 3. *Variable attenuator* (Frequency range: 8.5 to 9.6 GHz).

X-Band Measurement Interface: The "X-Band Measurement Interface" experiment card is designed as the measurement and control interface for all waveguide experiments. The card is equipped with an integrated short-circuit protected power supply for the Gunn diode. Operating voltage +5V± 0.1V, adjustable DC operating voltage for Gunn diode 0...10 V, logarithmic receiver with dynamic response up to 50dB, 16-bit data, bandwidth 500 MHz, sensitivity -70 dBm, frequency measurement 8.3 ... 9.9 GHz.

3.2. Generating and propagating the microwave field in the measuring stand

The microwave field generated by the oscillator Gunn will pass through the insulator with negligible attenuation. When passing through the directional coupler (Cross coupler), the microwave field will be divided into two approximately equal intensity beams. The first microwave beam will cross the adapter (waveguide-coax transition) and will be taken over by the LNC and transmitted via the SMA cable to X-Band Measurement Interface. The second microwave beam will cross the variable attenuator, it will come and go the slotted line and it will reach the first Horn antenna in the antenna system where the test sample is placed. The micrometric screw of the variable attenuator will be adjusted for each microwave frequency so that the levels of the electrical signals taken over by the probes and transmitted through the two LNCs are the same. In this way, we have the certainty that for each frequency of the microwave field generated by the Gunn oscillator, the intensity of the microwave field that will interact with the measurement sample placed between the Horn antennas, will have the same value [3]. Under these conditions, the microwave field attenuation measurements after its interaction with the sample to be measured will be accurate, having the same incident reference value of the incident beam [4]. The emissive side of the measuring stand is shown in figure 4. The elements that make up the measuring cell (the Horn antenna system and the detector element) are shown in figure 5. The whole measuring stand, consisting of the main branch, the first branch, the second branch, as well as the branch of the measurement and measurement of the attenuation after passing the microwave beam through the measurement sample, is shown in figure 6. Positioning of the sample to be measured inside the Horn antenna system is done by means of dielectric wires made of materials that do not absorb the energy of the microwave field from the incident beam.

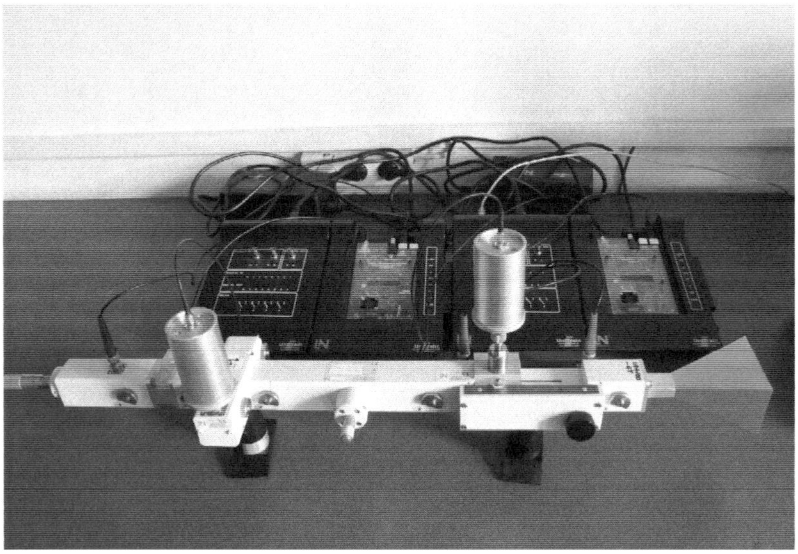

Figure 4. The emissive side of the measuring stand.

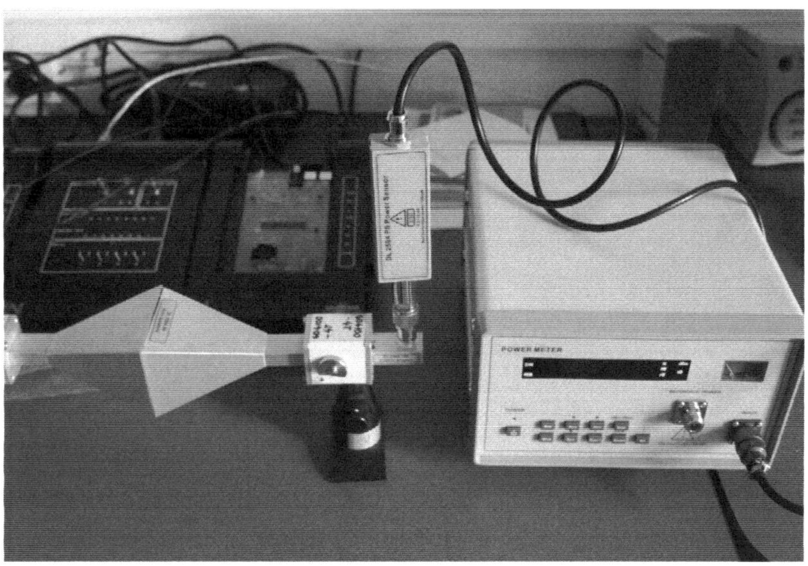

Figure 5. The measuring cell and attenuation measuring system.

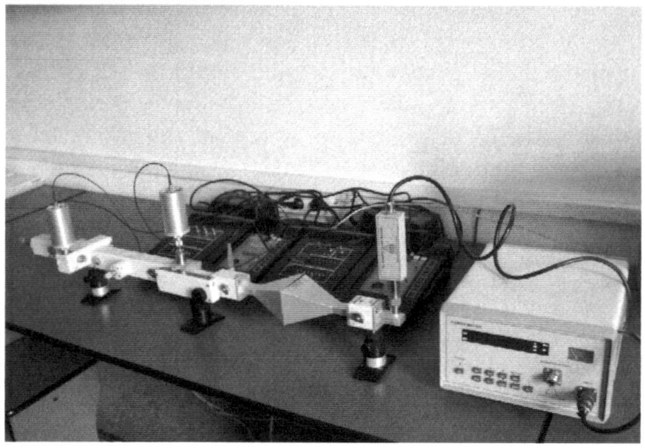

Figure 6. The whole measuring stand.

4. Experimental Results

Measured probes were two different types of non-polarized ceramic capacitors and two electrical resistors, which were taken from an electrical circuit in the close proximity of a PC computer's microprocessor. The values of the measured electrical quantities are shown in table 1 to table 4.

Table 1. Measured parameter values (sample no.1).

No.	Frequency (GHz)	*Level (dBm)*	*Attenuation correction (mm)*	*Component Type (P_0=0.312 mW)*	*Power level (mW)*
1	9.00	-22.5	0.00	Capacitor (C1)	0.293
2	9.05	-22.5	0.06	Capacitor (C1)	0.294
3	9.10	-22.5	0.02	Capacitor (C1)	0.293
4	9.15	-22.5	0.00	Capacitor (C1)	0.292
5	9.20	-22.5	0.07	Capacitor (C1)	0.290
6	9.25	-22.5	0.04	Capacitor (C1)	0.287
7	9.30	-22.5	0.01	Capacitor (C1)	0.275
8	9.35	-22.5	0.00	Capacitor (C1)	0.283
9	9.40	-22.5	0.00	Capacitor (C1)	0.286
10	9.45	-22.5	0.04	Capacitor (C1)	0.290

Table 2. Measured parameter values (sample no.2).

No.	Frequency (GHz)	*Level (dBm)*	*Attenuation correction (mm)*	*Component Type (P_0=0.312 mW)*	*Power level (mW)*
1	9.00	-22.5	0.01	Capacitor (C2)	0.290
2	9.05	-22.5	0.03	Capacitor (C2)	0.290
3	9.10	-22.5	0.00	Capacitor (C2)	0.285
4	9.15	-22.5	0.00	Capacitor (C2)	0.286
5	9.20	-22.5	0.03	Capacitor (C2)	0.283
6	9.25	-22.5	0.01	Capacitor (C2)	0.277
7	9.30	-22.5	0.01	Capacitor (C2)	0.269
8	9.35	-22.5	0.00	Capacitor (C2)	0.273
9	9.40	-22.5	0.05	Capacitor (C2)	0.276
10	9.45	-22.5	0.04	Capacitor (C2)	0.284

IOP Conf. Series: Materials Science and Engineering **209** (2017) 012068 doi:10.1088/1757-899X/209/1/012068

Table 3. Measured parameter values (sample no.3).

No.	Frequency (GHz)	*Level (dBm)*	*Attenuation correction (mm)*	*Component Type (P_0=0.312 mW)*	*Power level (mW)*
1	9.00	-22.5	0.00	Resistor (R1)	0.293
2	9.05	-22.5	0.00	Resistor (R1)	0.296
3	9.10	-22.5	0.03	Resistor (R1)	0.295
4	9.15	-22.5	0.03	Resistor (R1)	0.296
5	9.20	-22.5	0.01	Resistor (R1)	0.2.95
6	9.25	-22.5	0.05	Resistor (R1)	0.297
7	9.30	-22.5	0.00	Resistor (R1)	0.299
8	9.35	-22.5	0.00	Resistor (R1)	0.299
9	9.40	-22.5	0.06	Resistor (R1)	0.301
10	9.45	-22.5	0.07	Resistor (R1)	0.301

Table 4. Measured parameter values (sample no.4).

No.	Frequency (GHz)	*Level (dBm)*	*Attenuation correction (mm)*	*Component Type (P_0=0.312 mW)*	*Power level (mW)*
1	9.00	-22.5	0.03	Resistor (R2)	0.283
2	9.05	-22.5	0.01	Resistor (R2)	0.286
3	9.10	-22.5	0.00	Resistor (R2)	0.288
4	9.15	-22.5	0.02	Resistor (R2)	0.288
5	9.20	-22.5	0.04	Resistor (R2)	0.293
6	9.25	-22.5	0.00	Resistor (R2)	0.297
7	9.30	-22.5	0.02	Resistor (R2)	0.299
8	9.35	-22.5	0.00	Resistor (R2)	0.303
9	9.40	-22.5	0.01	Resistor (R2)	0.302
10	9.45	-22.5	0.00	Resistor (R2)	0.303

5. Conclusions

From the analysis of the experimental data obtained at the power measurements absorbed by the studied samples, the following conclusions can be drawn:

- In the case of electric capacitors, a microwave field absorption is observed near the frequency of 9.30 Ghz;
- In the case of electric resistors, there is a progressive decrease in the absorption of the microwave field when it increases the frequency of the electromagnetic field with which the probe interacts in the space between the two Horn antennas;

The situation encountered in the behavior of electric capacitors, we can say that the dielectric assembly inside the capsule as well as the capsule itself shows an increased absorption near the frequency of 9.30 GHz. Since the measurement time was short and due to the fact that the introduction of a temperature sensor into the interaction space of the sample with the microwave field would significantly disturb the measured values, it was avoided to measure temperature variations [5]. In a more detailed analysis, we will analyse the interaction of the microwave field separately for the dielectric in the capsule, and then an analysis of the microwave field interaction only for the electric capacitor capsule. This future study will help determine which of the two components has a major interaction with the microwave field [6]. A primary conclusion is that these types of electrical capacitors would not have been recommended to be found in the electronic circuits of the PC type if they would operate at frequencies close to the reference value of 9.30 GHz.

In case of the study of the microwave field absorption phenomenon in the volume of electric resistors, it can be said that this phenomenon is possible to meet at frequencies lower than 9.00 GHz. In the near future, we are planning to study the behaviour of resistors in microwave fields whose frequency ranges from 7.50 to 9.00 GHz.

References

[1] Jebbor N, Bri S, Sánchez A M and Chaibi M 2012 Experimental complex permittivity determination of low-loss dielectric materials at microwave frequency band *International J. of advanced scientific and technical research* **2**(5) 129–134

[2] Application Note Agilent 1449-1/2/3/4 Fundamentals of RF and microwave power measurements *Agilent Application Note 150-1 Spectrum Analysis. Amplitude and Frequency Modulations*

[3] Terhzaz J, Ammor H and al. 2006 Determination of the complex permittivity of dielectric materials at microwave frequency using rectangular waveguide measurements and Newton-Raphson method *EDP Sciences Materiaux & Techniques* **94** 227–233

[4] Elmajid H, Terhzaz J, Ammor H, Chaïbi M and Mediavilla A 2015 A New Method to Determine the Complex Permittivity and Complex Permeability of Dielectric Materials at X-Band Frequencies *International J. of Microwave and Optical Technology* **10** (1) 35–39

[5] Adriano L.P, Mirabel C.R, Joaquim J.B 2015 Experimental measurements and numerical simulation of permittivity and permeability of Teflon in X band *J. Aerosp.Technol. Manag. São José dos Campos.* **3** (1) 59–64

[6] Murata K, Hanawa A and Nozaki R 2005 Broadband complex permittivity measurement techniques of materials with thin configuration at microwave frequencies *J. Applied Phys* **98**

IOP Conf. Series: Materials Science and Engineering **209** (2017) 012069 doi:10.1088/1757-899X/209/1/012069

Designing and Implementation a Lab Testing Method for Power Cables Insulation Resistance According with STAS 10411-89, SR EN ISO/CEI/17025/2005

R Dobra[1*], D Pasculescu[2], G. Marc[1], M Risteiu[1] and A Antonov[3]
[1]Department of Computer Science, "1 Decembrie 1918" University Alba-Iulia, Gabriel Bethlen Str., No.5, 510009 Alba Iulia, Romania
[2]Department of Computers and Power Engineering, University of Petrosani, 20 Universitatii str, Petrosani, România
[3] Electrical and Mechanical Risks Laboratory, National Research and Development Institute for Labor Protection "Alexandru Darabont", Bucharest, Blvd. Ghencea no. 35A, district 6, Romania

E-mail: remusdobra@upet.ro

Abstract. Insulation resistance measurement is one of the most important tests required by standards and regulations in terms of electrical safety. Why these tests are is to prevent possible accidents caused by electric shock, damage to equipment or outbreak of fire in normal operating conditions of electrical cables. The insulation resistance experiment refers to the testing of electrical cable insulation, which has a measured resistance that must be below the imposed regulations. Using a microcontroller system data regarding the insulation resistance of the power cables is acquired and with SCADA software the test results are displayed.

1. Introduction

Measurement of insulation resistance is one of the most important tests required by standards and norms in terms of electrical safety. The reason these tests are done is to prevent accidents caused by electric shock, damage to equipment or fire in normal operation of electrical cables. In time resistance testing, test voltage remains constant, and resistance values increase over time if insulation is good. The Fluke 1550 C has an internal timer that allows a constant test voltage (500 V, 1 kV, 2.5 kV or 5 kV) to be applied for up to 99 minutes, although 10 minutes or less is The typical value [1, 3].

Standard time resistance tests include polarization and dielectric absorption index tests. Despite the "pretentious" terms, the tests themselves are quite simple. The dielectric absorption test compares the values of two resistance readings, such as a reading of 60 seconds and one to 30 seconds. The higher the rate the better the insulation. In general, a ratio greater than 1.4 to 1.6 indicates good insulation. The polarization index test is the same as the 1 minute and 10 minute test. The resistance at 10 minutes is divided by the resistance of 1 minute. Indicators from 2 to 4 or more indicate good insulation. The polarization index test can be performed on new equipment to establish a base value and then periodically to predict when the equipment starts to fail. If the index decreases over time, it means that isolation deteriorates.

2. Technical safety requirements for mine electric cables and their accessories

Taking into consideration the mechanical stresses in the mining industry (dynamic and static crushing, simultaneous torsion and flexion, alternate torsion and flexion), the flexible mine electric cables with natural or synthetic rubber insulation, as well as the mine armored cables with polyethylene or PVC insulation and PVC sheath, intended for use in electric installations and equipment for powering, signaling, telecommunications and telemechanic, in the underground and quarries mine workings have to assure a strength appropriate to these stresses. If for the electric and non-electric equipment or components and protective components used in areas with explosion hazards specific types of protection had been designed according to the dangerous zone the equipment is intended to be used and to the chosen category of equipment, in the case of electric cables especially in large electric networks where particularly dangerous short-circuits may occur, since short-circuits develop heat, electric arcs, and metallic droplets that may represent an imminent danger of occurrence of unwanted events: fires or explosions and environment pollution [3,5].

There are two different approaches related to mechanical influences of insulation characteristics of the electrical power cables:

- Structural integrity analysis
- Dielectric strengths of the insulation materials

E. David, in its paper [2] proves and describes the elastic motion of the power cable insulation materials when it is "activated" by the mechanical stress. It is generally accepted that electrical performances of the insulation materials that composes the power cables are influenced by the mechanical stress.

On the other hand, the literature [1] offers values for dielectric strength of insulating materials, relative to nitrogen, unless units of kV/mm are indicated.

Both aspects are subject of our paper research: to prove the variation of the material insulation capabilities when different mechanical efforts are applied to the power cables.

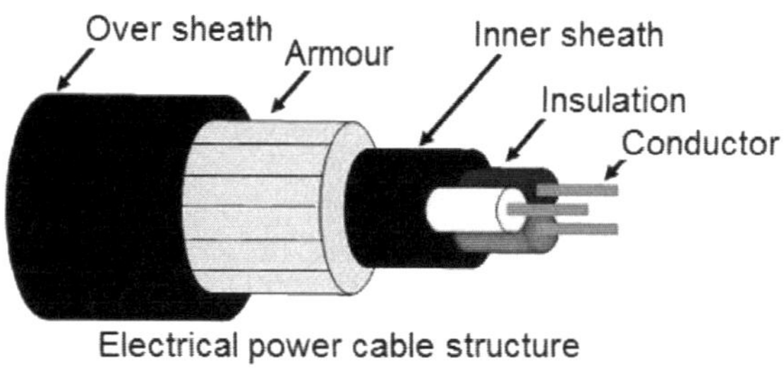

Figure 1. A bucket wheel excavator in the marching position.

3. The standardization testing method for cables insulation resistance

When tested for the resistance to flame propagation, in accordance with SR EN 60332-1-1:2005 and SR EN 60332-1-2:2005, the insulated conductor or the cable successfully passes the resistance to flame propagation or increased resistance to flame propagation if the distance between the lower edge of the upper support and the limit of the carbonized zone is higher than 50 mm, and in the case when the carbonized area spreads down on a length higher than 540 mm starting from the lower limit of the upper support, then the test is considered to be failed and as consequence the cables are not allowed to be used in the underground of firedamp mines, non-firedamp mines or quarries. In order to perform these tests, in the ENExEMEIP laboratory had been conceived and carried out test stands with software interface for control of the test cycles of static and dynamic stresses [5,6,7].

According with invoked standards [3,4] three mechanical tested are proceeded:

Squash test

There is a squashing force two (F) that acts over an impressive area (I) (see next figure): is carried out by means of a rig provided with a ram weighing 20 ± 1kg, falling upon the test sample, which is placed on a 80 mm diameter CopperZinc cylinder

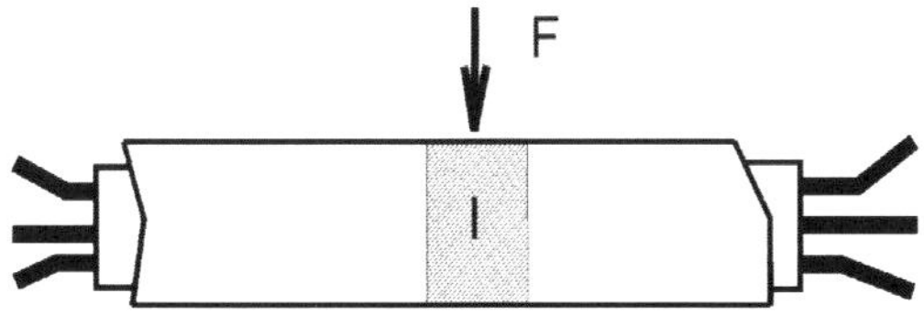

Figure 2. The static crushing test

Torsion test

There are two contrary torsion (T, -T) torques that act over an impressive area (I) (see next figure):

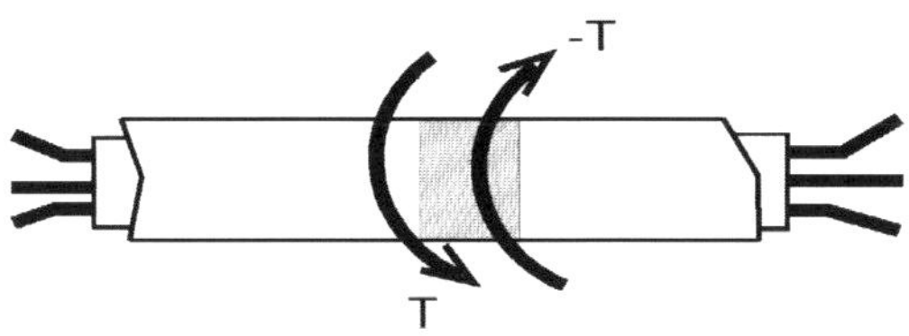

Figure 3. The simultaneous torsion and flexion test.

Shear test

There are two contrary forces (F, -F) that act over an impressive area (I) (see next figure)

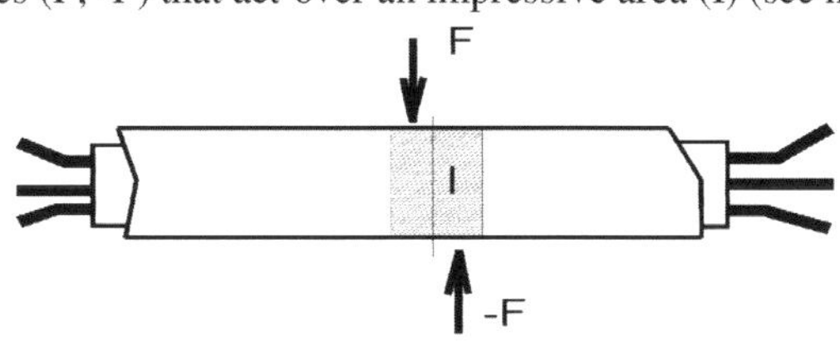

Figure 4. The dynamic crushing test.

4. Test of mine electric cables and their accessories

The simultaneous torsion and flexion test is performed with the apparatus and according to the main assembly diagram shown in figure 5. For the needs of ensuring the rotary and translation motions for the cable samples submitted to this test the rig in figure 4 had been designed. The rig ensures through its translation device, a number of 7...8 coming and going cycles with the speed of 20...30 m/min, and through its torsion device at each cycle 3 rotations (6π radians) on one side and 3 rotations (6π radians) on the other side are generated.

The distance between sample the locking devices, in their wide spaced position shall be 2,5 m, and the stroke length of the translation motion shall be 1,5 m. The test samples locked in these locking devices is connected to the control apparatus. The system monitors and acquisition the following parameters:

- the insulation resistance in order to detect the insulation flaws
- the currents on the three phases in order to detect short-circuits or splitting of conductors inside the cable
- the number of strikes in case of dynamic crushing or the number of twists in case of torsion test
- the pressure applied in case of static crushing test.

IOP Conf. Series: Materials Science and Engineering **209** (2017) 012069 doi:10.1088/1757-899X/209/1/012069

The current on its three phases R, S, T is measured by means of current transformers. This current in case of normal operation without faults is identical for each of the three phases. The current increases when a short-circuit occurs and decreases when a phase is split.

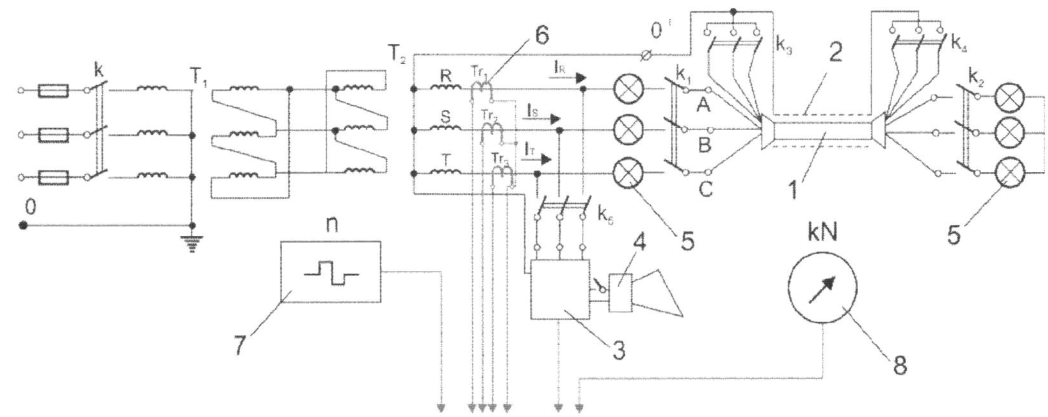

Figure 5. Laboratory testing method for power cables.

A IRDH 375 isometric model was used at the test bench. It is designed to monitor the insulation resistance of the main circuits (AC , AC / DC 0 ÷ 793V and DC 0 ÷ 650). Standard time resistance tests include polarization and dielectric absorption index tests. Despite the "pretentious" terms, the tests themselves are quite simple [8].

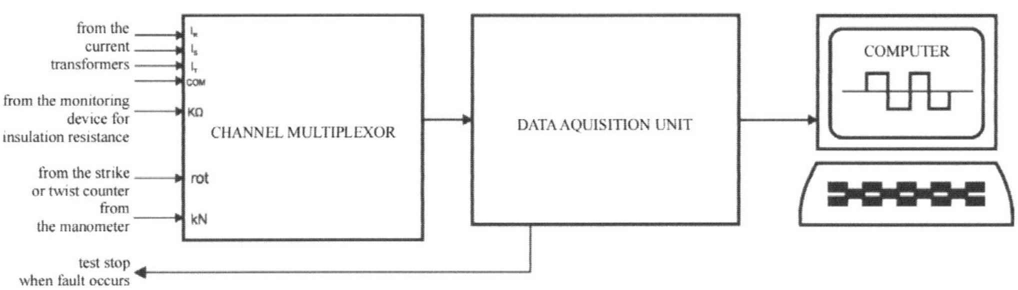

Figure 6. The control system for the testing cycles in the dynamic and static crushing tests.

The dielectric absorption test compares the values of two resistance readings, such as a reading of 60 seconds and one to 30 seconds. The higher the rate the better the insulation. In general, a ratio greater than 1.4 to 1.6 indicates good insulation. The polarization index test is the same as the 1 minute and 10 minute test. The resistance at 10 minutes is divided by the resistance of 1 minute. Indicators from 2 to 4 or more indicate good insulation.

The polarization index test can be performed on new equipment to establish a base value and then periodically to predict when the equipment starts to fail. If the index decreases over time, it means that isolation deteriorates. Note that the polarization index is a ratio, not an absolute value. Taking the reports from two tests done under the same conditions, cancel the temperature and humidity variables.

5. Software testing method for power cables insulation resistance

The insulation resistance is permanently monitored and in case an insulation fault occurs, a signal is transmitted to the data acquisition unit, in order to stop the test. The electric cable network in underground are carried out with mine electric cables that, together with the assembly of electric equipment and device in the electromechanical installations, have the role of ensuring protection against electrocution, fire and/or explosion hazards.

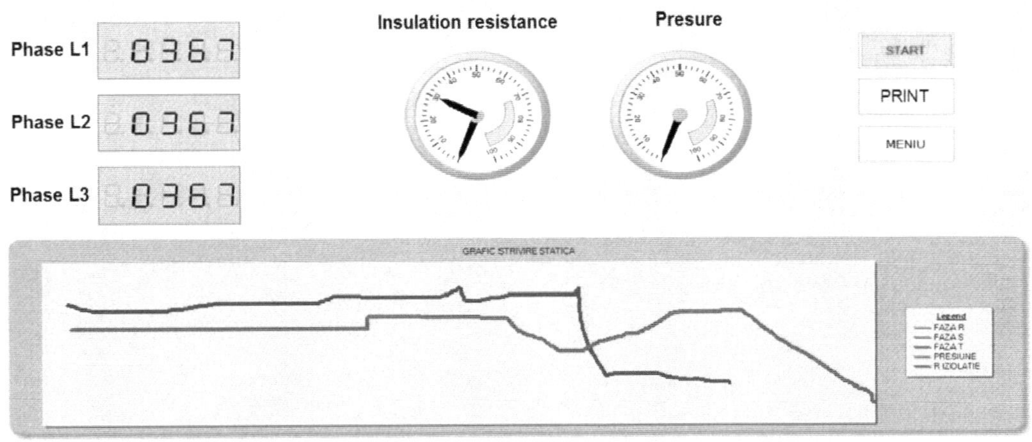

Figure 7. Data acquisition software for testing electrical power cables.

On this interface are monitored currents on the three phases of the test conductor, the insulation resistance and the pressure applied by the crushing piston on the conductor, at the bottom there is a graph showing all parameters monitored along with the time at which the cable testing was performed.

An electric signal is taken over from a digital numerator and then recorded together with the other data for setting out the number of rotations or twists up to occurrence of fault. The pressure applied to the dynamic crushing test stand is displayed and monitored by a digital manometer that transmits the data to the data acquisition unit, through the multiplexor.

The program is made in the WinTR software, it can be configured as a web server facilitating the access from the Ethernet network of another operator that can track the execution of the test in real time, the server is redundant, can send messages to the address Email, GSM or Skype network, the customer, operator or service staff can receive messages about the test result, problems with the operation of the equipment.

In laboratory was test some different power cables. One example is presented below:

Cable type: 3x35+3x16/3E

The cable consists of: 3 phase active conductors with 36,04 mm2 section isolated in rubber; the earthing conductor (special grounding) with 37.08 mm2 section is distributed concentrically around each active phase; the red rubber outer jacket is embossed: CMCCGCEf - 3x35 + 3x16 / 3E - 6kV ICME ECAB S.A. 2003

Dimensional measurements:

- Average outside diameter of the cable sample: 52,26 mm;
- Average thickness of outer shell 7,02 mm;
- Protective conductor (special earthing) is formed 63 tiles of copper tin with ϕ 0.5 mm and the total section 37.08 mm2;
- Phase active conductors are made up of 7 bundles of 39 tinned copper wires with ϕ 0.41 mm with 36.04 mm2 section insulated into rubber of the same color each: brown;
- Average insulation thickness 4.63 mm

From the last point of application of the force, a cable end was made to more than 200 mm and the first series (j = 1) of consecutive kicks was applied to the occurrence of insulation faults, phases or short circuits. In 4 different places spaced at approx. 200 mm was applied a series of strokes. The results for the 5 requests are as follows:

- strain in the first hitting place (j =1): number of hits until an insulation fault occurs: n11 = 144; number of hits until a phase is interrupted: n21 = 204;
- strain in the second hitting place (j =2): number of hits until an insulation fault occurs n12 = 150 ; number of hits until a phase is interrupted n22 = 180 ;
- strain in the third hitting place (j =3): number of hits until an insulation fault occurs n13 = 120; number of hits until a phase is interrupted n23 = 162 ;
- strain in the fourth hitting place (j =4): number of hits until an insulation fault occurs n14 = 147 ; number of hits until a phase is interrupted n24 = 202 ;
- strain in the fifth hitting place (j =5): number of hits until an insulation fault occurs n15 = 149 ; number of hits until a phase is interrupted n25 = 198 ;

Disconnection preventive factor:
$$k = \frac{\sum_{j=1}^{5} n_{1j}}{\sum_{j=1}^{5} n_{2j}} = \frac{710}{946} = 0{,}751 \qquad (1)$$

6. Conclusions

The electric cable network in underground are carried out with mine electric cables that, together with the assembly of electric equipment and device in the electromechanical installations, have the role of ensuring protection against electrocution, fire and/or explosion hazards.

The safety specific technical requirements for mine cable are not regulated at international level, such as each country applies its own national regulations / standards. In Romania the requirements in STAS 10 411 apply. In order to successfully pass the above mentioned tests, the electric power cables as well as the signaling and control cables used in underground mine workings, along with being manufactured with materials adequate from a point of view of their mechanical, thermal, electric characteristics and flame, the cables shall have a proper constructional form with a special protective conductor, designed for connection to earth in the general network.

References

[1] HG nr. 1049 / 2006 Regarding the minimum requirements for ensuring the safety and health of workers in surface or underground mining
[2] David E, Parpal J L and Crine J P 1994 - Influence of mechanical strain and stress on the electrical performance of XLPE cable insulation *Proceedings of IEEE International Symposium on Electrical Insulation Proceeding*
[3] STAS 10411-89: Electrical cables for mining purposes. Test methods
[4] SR 11388, 2000: Common test methods for electric cables an conductors
[5] SR HD 629.1 S2, 2006: Test requirements on accessories for use on power cables of rated voltage from 3,6/6(7,2) kV up to 20,8/36(42) kV. Part 1: Cables with extruded insulation
[6] SR EN 60332 - 1-2: 2005, Tests on electric and optical fiber cables under fire conditions. Part 1-2: Test for vertical flame propagation for a single insulated wire or cable. Procedure for 1kW pre-mixed flame
[7] SR EN 60332 - 1-1: 2005 - Tests on electric and optical fiber cables under fire conditions. Part 1-1: Test for vertical flame propagation for a single insulated wire or cable. Apparatus
[8] SR EN 60332 - 2-1: 2005: Tests on electric and optical fiber cables under fire conditions. Part 2-1: Test for vertical flame propagation for a single small insulated wire or cable

IOP Conf. Series: Materials Science and Engineering **209** (2017) 012070 doi:10.1088/1757-899X/209/1/012070

Decreasing the Functioning Consumptions of Plastics Injection Moulding Machines

V Vasilache[1*], M Vasilache[1] and A Vasilache[1]
[1] S. C. VASILACHE s.n.c., 25/2 Măgurei street, Târgu-Mureş, Romania

E-mail: themannextdoor2007@yahoo.com

Abstract. A share of about 75% in the cost of a plastic moulded part is held by the energy consumption of the injection machine. It is the heating system of the plasticising unit which is responsible for the energophague character of the process. The transfer rate from the heating elements to the plasticising cylinder depends hardly on the geometry of the system. A new heating system is designed, replacing the classical systems which are applied on the exterior of the cylinder with an "engrooved system". Proposed heating system leads to decreasing of energy consumption up to 30 % and maintenance costs up to 10 %. A supplementary possibility to decrease the maintenance costs is to modify the geometry of the injection torpedo. The proposed geometry eliminates the possibility of any breaking or gripping related to the injection torpedo.

1. Introduction

The field of plastic parts production is targeted by all economical-industrial strategies. This sector is an energophague one and an important consumer of water resources. Decreasing of energetic resources consumption is one of the goals of the Romanian National Strategy for Sustainable Development. The output of the Project implementation consists of subassemblies designed to be put on the existing equipment for plastics transforming in our country, in order to decrease the energetic consumptions of them.

Both duration (over 3 years), and consistency of the activities by which the authors address the problem and propose a series of solutions make this set of works to be different by a certain simple work of engineering design.

Granting an invention patent involves for a technical solution: novelty on world scale, technical progress and industrial applicability. *OSIM granting 2 invention patents* [1], [2] to the technical solutions addressed in our Project certifies the *technical – scientific priority* of the solutions on a national and international scale, becoming a clear indicator of the performance level of the results.

Studies on experimental models, digital simulations were performed, experimental stands were erected, prototypes were fabricated and "in situ" determinations to different plastics transformers economical agents were performed in order to attain the results which allow technical - economical conclusions related to the stated theme. All lasted for more than three years.

Documentation of state of art had a very precise purpose, representing a solid argument in the authors orientation related to what is yet to be done and what is proposed to be developed as technical, scientific and with economic effects novelty.

The screw (the part which transmits mechanical solicitations to the polymer mass) geometry is very important upon injection phenomenon which (due to polymer chains breakings, apparition of micro-cracks with free radicals aside of them, re-shaping of molecular spheroids) is not only a mechanical process but a chemical one too, some certain properties having the possibility to totally change. A screw geometry which escaped from calculation rigors or was developed in the idea that the only

IOP Conf. Series: Materials Science and Engineering **209** (2017) 012070 doi:10.1088/1757-899X/209/1/012070

purpose of the rotating screw is to transport the plastic grains, may lead to completely unexpected secondary effects. The screw geometry must hence be adapted to the structure and properties of the respective polymer.

The injection head or torpedo is fixed on top of the screw. This component is one of the most solicited, working in condition like:

- temperature: up to 300°C
- pressure: up to abt. 2500 bar
- torsion stress
- maximal wear.

The polymer melt is injected through the injection nozzle after exiting through the torpedo orifices (the torpedo works like a one-way valve, blocking the backwards flow of the plasticized polymer stream).

Most of the torpedoes from the top of the screws have a part of „closing bushing" type and one of „Top" type, which, both, have a surface of „slope" type mated one to the other. The „ top" is screwed on the frontal extremity of the feeding screw. The „closing bushing" type part has an abt. 2.5 ÷ 3 mm thick wall, mated in an adjustment of H7/f6 with the plasticising cylinder which has an abt. 32.5 ÷ 42.5 mm thick wall. Heat is transmitted from exterior to interior of the cylinder. The wall of the "bushing" gets faster to a certain dimensional value than the interior of the cylinder. Thus, a small tight of the "bushing" appears. At a certain moment within the functioning cycle, the screw executes an axial retreat movement in the plasticising cylinder (altogether with the „top" which is screwed in the screw). The slope surface of the „Top" pushes on the corresponding surface (also a slope surface) of the „bushing" in order to put in axial movement the „bushing" too. The axial force decomposes, a radial component resulting too. This component emphasises the grip of the „bushing" and, at a certain moment, this grip becomes so strong that the rod of the conical top breaks. The solution proposed by the authors consists in replacing the constructive componence „conical top – closing bushing" with another one, which eliminates the existence of the "slope" type surfaces.

As regarding the solutions for the heating system, it is found that all the existent solutions for the small machines, with max. 3 heating zones, consist from ceramic sleeves with resistive wire [3], from aluminium casted sleeves with tubular resistances introduced in some channels or from resistive wire which is coiled on a mica or mineral core[4]. All these solutions have an interface zone between the resistive element and the housing and a second interface zone between the housing and the exterior of the plasticising cylinder. The fewer the interface zones are, the highest is the efficiency of conductive transmission of heat. The solution proposed by the authors consists in eliminating one of the two interface zones, by means of directly „grooving" some resistive elements in longitudinal channels which are milled on the exterior surface of the plasticising cylinder.

2. Experimental procedures

2.1 Work methods

The following methods, from point of view of research and/or design process, were used for running the Project:

- direct research, at natural scale (in situ);
- research by means of experimental proceedings, by means of laboratory studies on physical models;
- research by means of mathematical, numerical proceedings.

2.2 Constructive definition of the injection unit subject to simulation / experiments

An injection torpedo is always provided between the feeding screw and the injection nozzle, as a back-flow valve (in case of thermoplastic materials).

The newest solution, illustrated in figure 1, shows the final variant of the construction of the torpedo [1]. The torpedo body (1) has a piston type zone (a) which mates an adjustment with the plasticising cylinder (not figured). Four equidistant inclined holes (A) related to the longitudinal axle of the body are provided between the piston zone and the threaded end of the torpedo. These holes communicate in an axial hole (B) where there is a ball (3) and a pin (4). When the screw rotates, the

IOP Conf. Series: Materials Science and Engineering **209** (2017) 012070 doi:10.1088/1757-899X/209/1/012070

melted material is pushed in the zone between the end of the thread helix and the piston zone. Here, the melted stream is divided in four streams through holes (A) and re-unifies in hole (B) where the ball is pushed until it tampons the pin. The melted stream is divided again, around the ball, and it is re-united after overpassing the ball. The screw advances with a great force in the moment of injection and, due to the piston zone, a corresponding pressure is created in front of him.

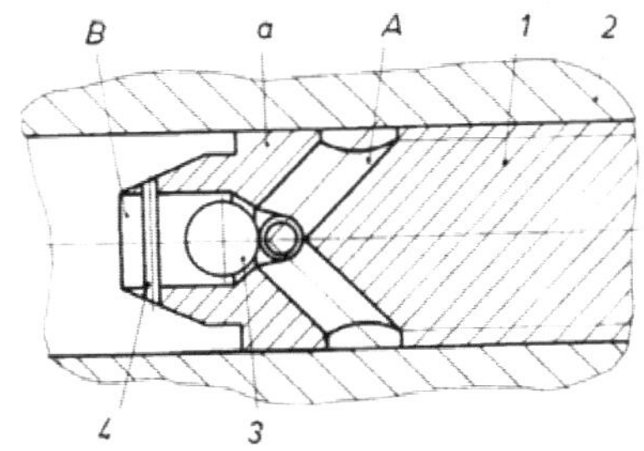

Figure 1. The injection TORPEDO proposed by the authors (final variant) [1,7].

It may be considered, simplifying, for our scope, that the plasticised fluid is a Newtonian fluid. The introduced errors do not influence significantly the problem. If we may consider, so, the pressure as being constant over all the fluid mass and, because the injection nozzle diameter is much smaller than the ball diameter, a force appears pushing the ball to the smaller diameter step. The melted polymer retreat is thus blocked and the melt is forced to evacuate through the only un-blocked orifice – the orifice of the injection nozzle. The new construction achieves a superior homogenization of the plasticised mass, does not introduce de-balancing and does not bring the danger of auto-grip.

Dimensional finalising of the injection torpedo geometry was done by means of Finite Element Method (FEM) and by means of photoelastometry method.

The injection unit is the most important part of the plastics injection moulding machine. This assembly is also the most energophague among all the assemblies of the machine, which is the reason for what it is the target of our actions to reduce the energetic consumptions of the plastics injection moulding machines. Our concerns on this direction were materialised in developing a new heating system for the injection unit, which system is subject of patenting procedure [5].

We gave up to the constructive solution of the well known heating sleeves and we have adopted a new system (figure 2, figure 3).

The injection cylinder (1) is provided with a number of longitudinal channels (A) on the circumference. An equal number of tubular rectilinear resistances (2) with the number of desired heating zones is introduced in each of the channels. The resistances of each zone are connected in derivation by means of some metallic parts (3), after they had been tightened by some colliers (4). Each zone is insulated with mineral insulation, an individual housing engulfing (5) the whole assembly; the interior surface of the housing is polished and behaves like a thermal reflector.

The new construction, with smaller overall dimensions than the classical one, brings spectacular results related to energetic consumptions.

The tubular heating resistances are an important element of the new heating system (figure 4).

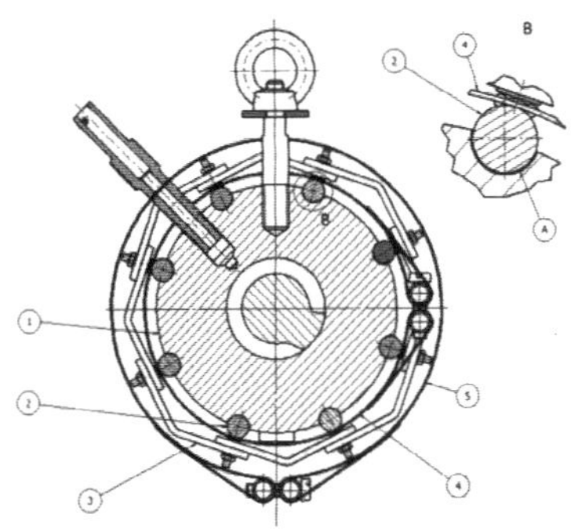

Figure 2. Transversal section through the injection unit which is equipped with the new solution [5].

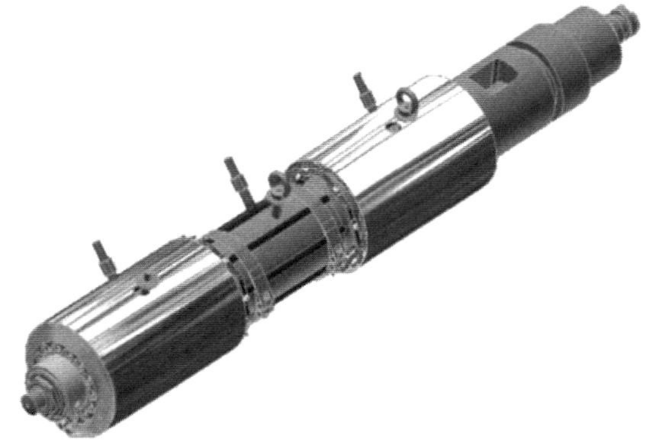

Figure 3. The heating system, mounted on an injection unit of a 80 tons machine (the middle zone reflector and insulation are replaced) [9].

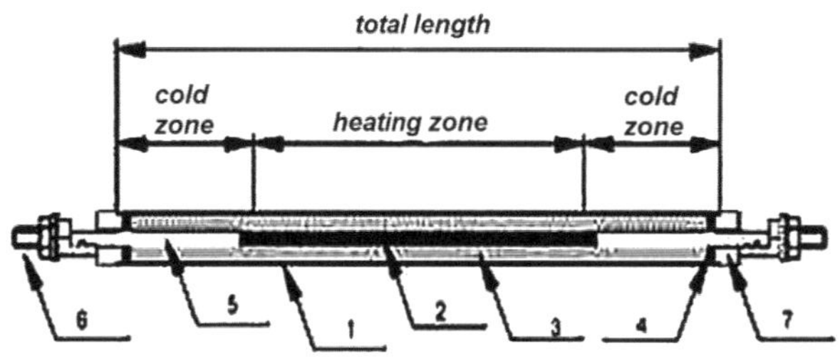

Figure 4. Principled construction of a tubular electrical resistance [authors' creation].

2.3 Contributions to the conception of the new solutions

Two problems were followed within the study of the injection unit: geometry of the screw top (i.e. the injection torpedo) and the unit heating.

The experiments related the two assemblies: the injection torpedo and the heating system of the injection unit.

2.3.1. The injection Torpedo – proposed variant. MEF simulation was run for torpedoes with diameter Ø25, Ø30, Ø35, Ø40, Ø45.

Firstly, the parts and the assembly were 3D modelled. The torpedo material is: 38MoCrAl09 SR EN 10083 – 1 + A1 : 2002.

Structural analyse of the parts was then performed. Functioning conditions were as in table 1:

Table 1. Functioning conditions of the injection torpedoes.

	U.M.	Torpedo Ø25	Torpedo Ø30	Torpedo Ø35	Torpedo Ø40	Torpedo Ø45
Press. in front of hole ØA	MPa	163	155	214	164	130
Press. behind holes ØF	MPa	0.2	0.2	0.2	0.2	0.2

Structural results obtained for each simulation are:

- equivalent stress (minimal value, maximal value, coloured contouring of value on part surface) [Mpa]
- maximal principal stress (minimal value, maximal value, coloured contouring of value on part surface) [Mpa]
- minimal principal stress (minimal value, maximal value, coloured contouring of value on part surface) [Mpa]
- deformation (minimal value, maximal value, coloured contouring of value on part surface) [mm]
- safety coefficient (minimal value, maximal value, coloured contouring of value on part surface)

Photoelastometry research was performed for Ø35 torpedo. The study was effectuated on models moulded from epoxy resin DINOX 010, with cold strengthening, using the method of « icing » the tensions. The method consists in stressing the model concomitant with its heating to the temperature of up to 100° C and slow cooling, under stress [6]. In this way tensions are « iced » in the model and their configuration is not changed even if the model is cut.

Pressure application upon active model was realized by means of a hydraulic stand (figure 5) consisting of a metallic cylinder (1) into which the model (2) is introduced. A thermocouple (3) is introduced for monitoring the temperature during the model solicitation. Manometer (6) measures the pressure delivered by a gear pump (4) which is driven by an electric motor (5). A safety valve (7) limits the set pressure. The solicitation cylinder is installed in a thermostat oven (8). Pressure and temperature are kept constant during the test.

After stress application and cooling, the models are cut into slices that contain the zones of interest: central hole zone, inclined holes zone and joining area of the piston zone. The obtained slices were studied in a polariscope, obtaining the respective isochromates. The photo-elastic constant of the material is determined by means of calibration and then the value of the difference between the principal stresses alongside the isochromates is established. A report was issued at the end of the study, containing the isobars for σ_2 and the isobars for σ_1.

The stand for hydro-mechanical tests is a construction which allows testing the samples.

The stand (figure 5, figure 6) was intended to be multifunctional, in order to be used for some other ulterior tests. It consists of the following subassemblies: Chassis with protection covers; Hydraulic installation; Electric motor; Vertical device; Horizontal device; Electric driving box; Electric heating box.

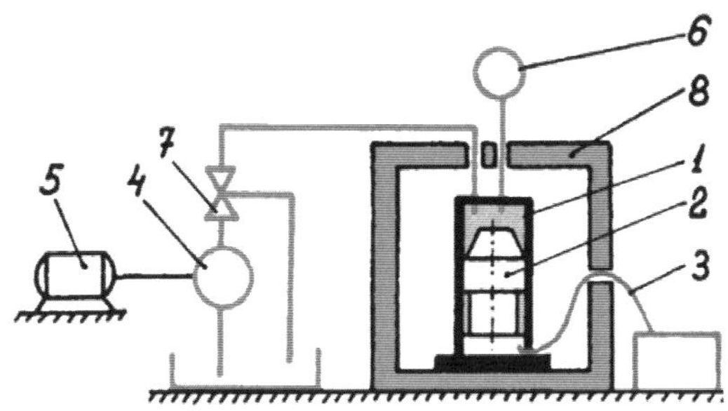

Figure 5. Principle diagram of the installation for stressing the active model [8].

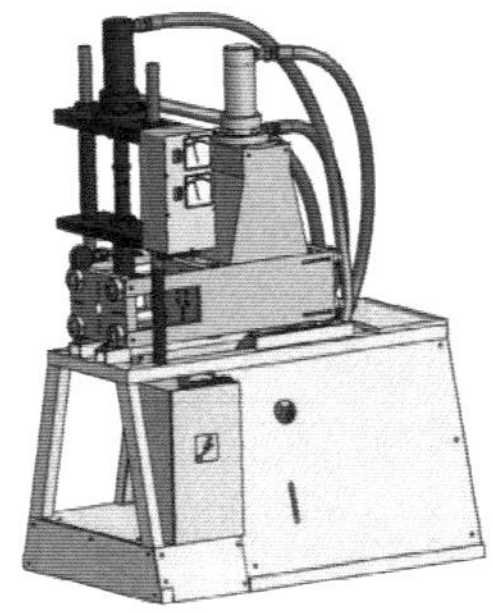

Figure 6. 3D view of the stand for hydro-mechanic tests [authors' creation].

The stand for researching the mixing in the plasticised mass (figure 7) represents, mainly, a mould for injecting some calibration parts. The calibration parts are long type parts – in fact some cones with thin walls.

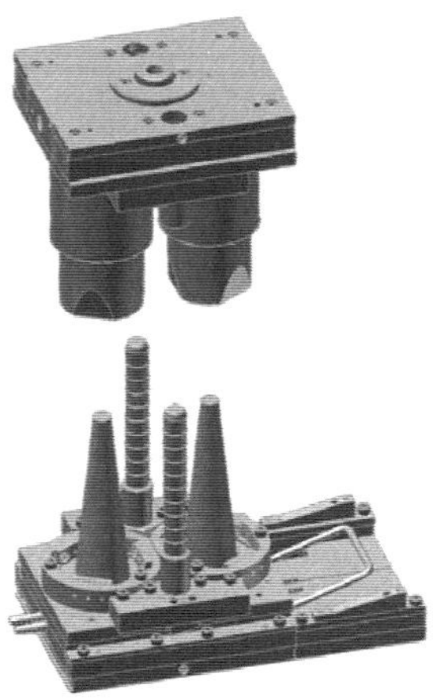

Figure 7. The stand for research the mixing (the mould) in open position [authors' creation].

The research of the effect upon mixing of the melted plastic material by introducing the new torpedo was performed by mounting the classical solution at the end of the feeding screw and injecting a part from white or colourless virgin material, then injecting the same part from the same material but with some pigment added. It was counted the number of parts that were necessary to obtain a uniform coloured part, with no white stains.

The pigmented material was cleaned out from the injection unit, the new torpedo replaced the classical torpedo at the front end of the feeding screw and the same injections were repeated, using a different coloured pigment. The injections related to both torpedoes were performed on the same machine, with the same mould, attention being given to maintaining the same settings of the injection machine and to use material from the same bag. A report was issued at the end of the tests, containing the necessary number of injections for obtaining a uniform coloured sample, with no colour stains.

2.3.2. The Heating System – proposed variant. Experimental research was performed for plasticising units with feeding screw diameter of 30 mm, 40 mm and 45 mm, in classical variant and in new variant, the experimental models being executed at 1:1 scale. The method consisted in mounting on the injection moulding machine the experimental model of the plasticising unit and measuring the consumption of electrical energy. The cylinder material is: 38MoCrAl09 SR EN 10083 – 1 + A1 : 2002. Functioning conditions: 300°C. Three different size machines were used, covering the domain of the small capacity injection machines: 250 kN, 500 kN and 800 kN.

Measurements were performed by means of an installation comprising a cylinder of the plasticising unit, three heating sleeves R (both classical variant, and new variant), three thermocouples of the machine (one for each heating sleeve), three temperature regulators of the machine (one for each heating sleeve), three electrical energy metering devices, three current metering instruments A and three voltage metering instruments V (one for each heating sleeve). The electrical energy metering devices were mounted in a stand aside the injection machine, while the other components were mounted on the injection moulding machine.

A report was issued at the end of the tests, containing among other registrations, the effective electricity consumption on the heating sleeve during each coupling of the thermoregulator.

3. Results and discussions

3.1. Project approach

The tests Protocol was issued within Activity 7 “Issuing tests Protocols” and it was internally advised.

As to the Project, two categories of tests are envisaged:

- digital experiments, by means of FEM simulation (research of mechanical behaviour), effectuated within Activity 4 “FEM simulation”
- experiments on the stand (testing mechanical behaviour and evaluation of the effect of introducing the new torpedo on mixing of the melt plastic material), effectuated within Activity 11 “Testing the prototypes on stands and comparison of experimental results to the numerical ones”

The purpose of the experiments was to establish a data base with behaviour results which can allow, then, the establishing of the final geometry of the two subassemblies. Experiments were performed for the injection Torpedo in order to determinate the mechanical behaviour of the new solution and to determinate the effect of introducing the new solution upon the mixing of the plasticised material.

Experiments were performed for the heating system of the injection unit, in order to determinate the effect of introducing the new solution upon the electrical energy consumption.

The photos were taken by the authors, using the following equipment:

1. Digital photographic camera SONY MAVICA FD91 – 1 pcs.
2. Photographical reflector 1000 W – 2 pcs.
3. Photographical tripod – 3 pcs.
4. Feeding cable with 3 sockets – 1 pcs.

IOP Conf. Series: Materials Science and Engineering **209** (2017) 012070 doi:10.1088/1757-899X/209/1/012070

3.2. Experiments upon the injection torpedo

3.2.1. Digital experiments. A specialised software was used, dedicated to mechanical engineers: INVENTOR 2009.

Hardware used in performing the tasks:

- Graphic station DELL M6400 Covette:
- processor INTEL Core 2 Duo (2.8 GHz, 6MB L2 Cache, 1066 MHz FSB)
- memory RAM 2048 MB (2x1024), 1066 MHz DDR3 Dual Channel Memory
- video 512 MB NVIDIA Quadro FX2700M (dedicated memory 512 MB)
- Hard disc 2x500 GB, 7200 rpm
- 8x DVD+/-RW
- monitor LCD 17 inch, 1920x1200 resolution
- printer XEROX 7500 Phaser: A3, colour, laser.

The Ø25 (figure 8), Ø30, Ø35, Ø40, Ø45 torpedoes were 3D modelled.

Figure 8. 3D Model of Ø25 Torpedo [8].

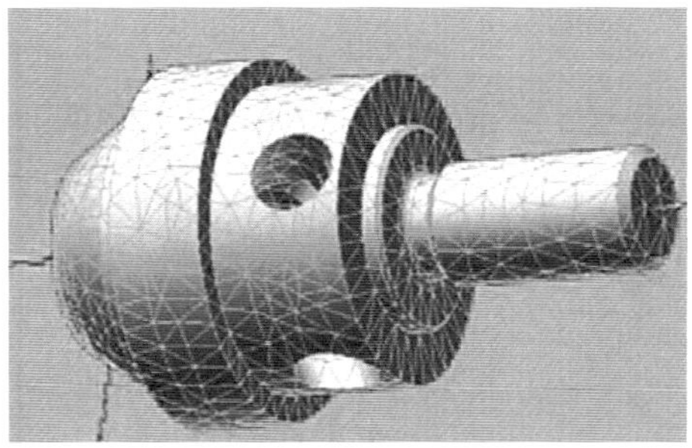

Figure 9. Discretization of the part Body [8].

FEM simulations of charge functioning were performed for 36 variants for each of these 5 dimensions for the part Torpedo Body. Discretization was done using triangular elements. An example of discretization network appears in figure 9.

The 180 digital simulations (36 variants for each of the 5 dimensions of torpedoes) have generated 180 Reports.

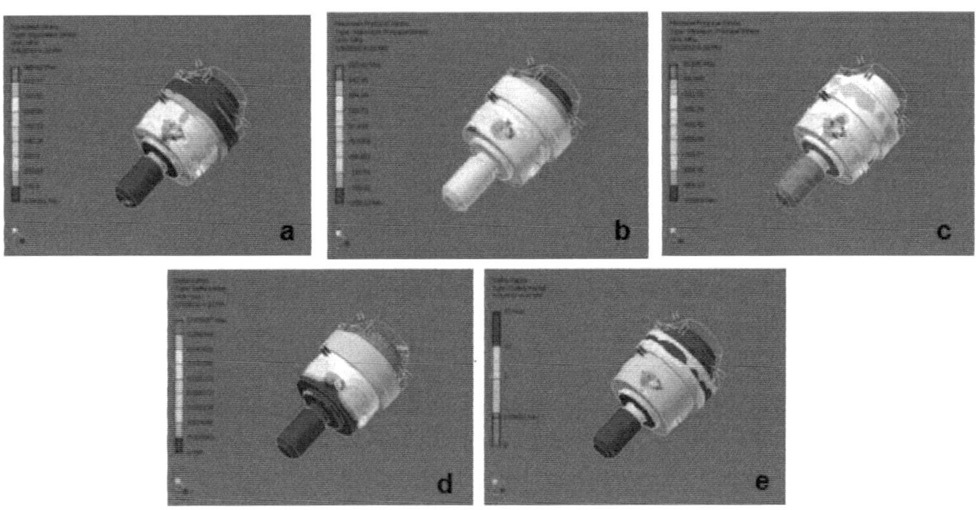

Figure 10. Solicitation diagrams: a – equivalent stress; b - Maximum Principal Stress; c - Minimum Principal Stress, d- Deformation, e – Safety Factor [10].

3.2.2. Experiments using the photoelastometry method. The model (figure 11) was moulded in a mould especially made for this experiment. The mould has 4 retractable cores (for the 4 holes Ø9), and for the zone Ø31 of the model it has two dismountable half-rings.

The model volume is of 43800 mm^3.

Necessary materials and equipment for manufacturing the model appear in table 2:

Table 2. Necessary materials and equipment for manufacturing the sample of Torpedo.

Material	Properties
Epoxy Resin DINOX 010 - P	Density $\rho_R = 1,18 \div 1,25$ [g/cm^3] [6]
TETA	triethylenetetramine
Demoulding agent	(oil wax in CCl_4)
Balloon 250 ml	Provided with agitating system and thermostat
Mould	

In figure 12 appears the hydro-mechanic stand dedicated to stressing the models.

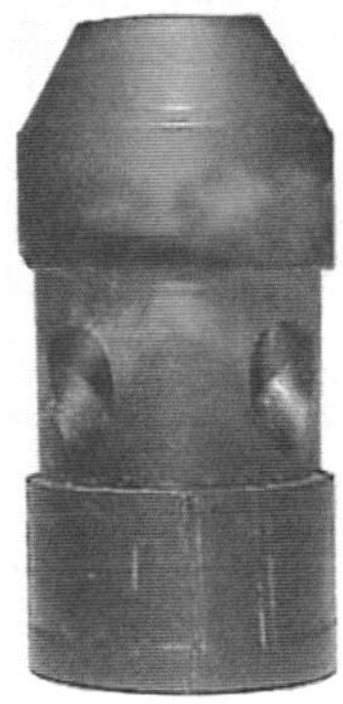

Figure 11. Model of Ø 35 torpedo before solicitation [8].

Figure 12. The hydro-mechanic stand [8].

After solicitation and cooling, the models were cut by using a diamond disk and slices were obtained which contained the interest zones of the model: the central hole zone, the zone of inclined holes and of joining of the piston zone. The slices were studied in a polariscope in yellow sodium light, obtaining the related isochromates. The photo-elastic constant of the material was determined by means of calibration and then the value of the difference of the principal stresses alongside the isochromates was determined.

Figure 13 and figure 14 catch aspects of the models after solicitation.

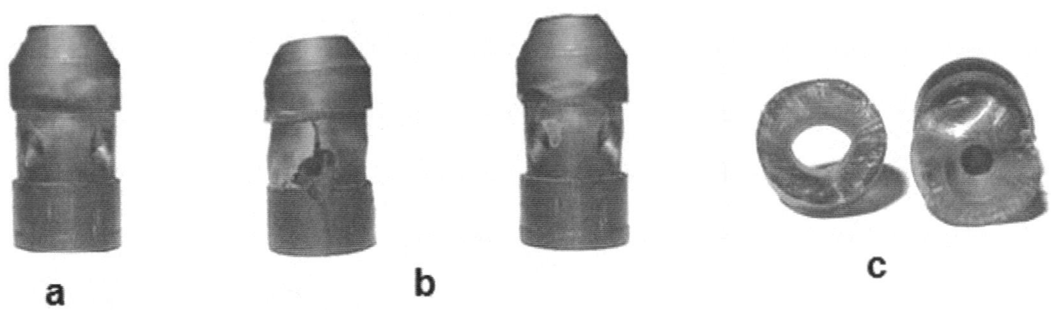

Figure 13. Model of Ø 35 torpedo before solicitation (a) and after solicitation (b and c) [8].

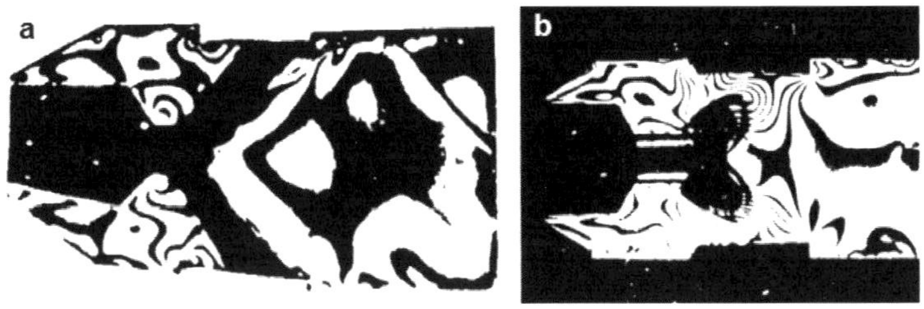

Figure 14. Isobars for Torpedo Body 35 (a) with details (b) [8].

3.2.3. Experiments in real functioning conditions. Testing the mechanical behaviour and evaluation of effect of new Torpedo upon mixing of the plasticised material in the plasticising cylinder were the goals of the experiment.

An especially designed and manufactured stand (a special injection mould) was used, manufactured within Activity 10 „Acquisition of research services (Outsourced manufacturing of stands for tests).

Tests were performed on an 80 tf thermoplastic plastic injection moulding machine (ARBURG Allrounder 420M/800-225) and on a 160 tf one (NEGRI BOSSI V160-610).

The machine is equipped, firstly, with the classical torpedo (figure 15).

Secondly, the machine is equipped with the new injection torpedo (figure 15). Details relating the technical solution of the new injection torpedo may be found in [1].

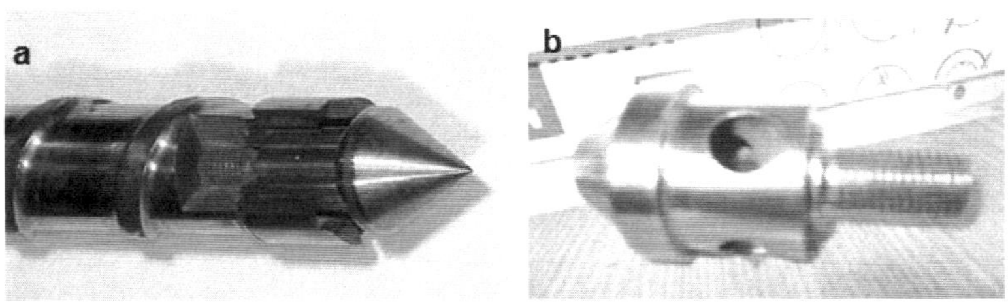

Figure 15. Photograph of the torpedo: a - Classical injection torpedo (sectioned for visibility of details), b - The new injection torpedo [9].

The testing sample must be the longest possible (the ratio between the length and the maximum diameter of the sample must be at least 3), in order to better observe the pigment traces and to analyse the mixing of the plasticised material that passed through the torpedo. The conical shape facilitates visualisation of the possible colour traces. Consequently, the sample had a conical shape, with thin walls of abt. 2 mm (for fast cooling and reducing the possibility that in the interior of the material mass to be residual traces of colour). Reinforcements on the inner and outer face of the sample were designed in order also to facilitate the visualisation of the colour traces. A smooth area was designed on the inner face, with no reinforcement, of abt. 50x40 mm for applying after injecting the sample, a self-adhesive label for identification of the sample.

A witness sample had been moulded, white coloured, from low density polyethylene (figure 16). The material had been dried, before moulding, during 1 hour at 65°C.

Blue pigment had been then added in the feeding hopper and so many samples were moulded using the classical torpedo until the sample became uniformly coloured (figure 16).

The feeding hopper was then emptied and the blue pigment was very well cleaned. The plasticising/injecting unit was emptied. White material was fed in the hopper and it was circulated through the plasticising/injecting unit until all the blue traces disappeared.

Red pigment had been then added in the feeding hopper and so many samples were moulded using the new torpedo until the sample became uniformly coloured (figure 16).

Moulding with both technical solutions of torpedo were performed on the same machine, with the same mould, attention being paid to keeping the same settings of the machine and to using material from the same bag. The time interval between the two sets of testing was not longer than the necessary period for cleaning the pigmented material from the unit and the hopper and feeding with new material. A report was issued at end of each set of testing containing the results of the experiment.

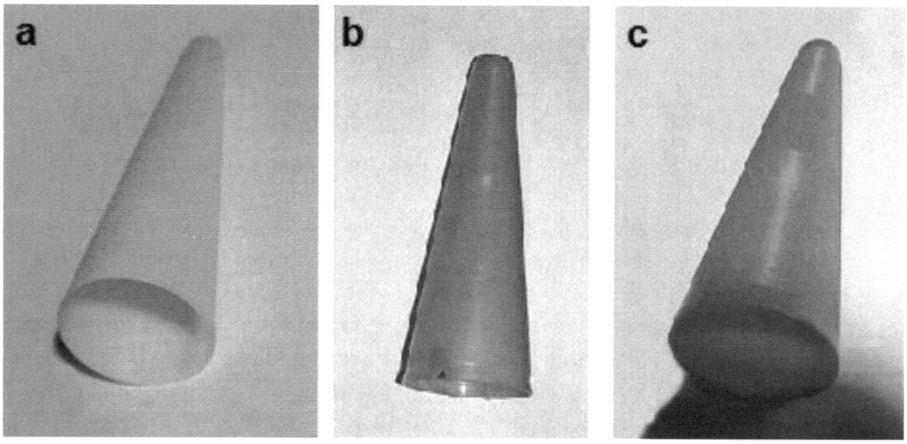

Figure 16. Photograph of the obtained samples: a - Witness sample, white coloured, b - Sample moulded with classical torpedo, c - Sample moulded with new torpedo [authors' creation].

3.3. Experiments upon the Heating System

As related from the testing protocol, a single category of experiments were performed:

- experiments in conditions of real functioning

The equipment:

- three plastics injection moulding machines, of 25 tf, of 50 tf and of 80 tf, covering thus the domain of the small capacity machines. The machines were endowed from manufacturer with thermocouples for measuring the temperatures of the heating zones and with thermoregulators for thermostating these zones.
- a measuring stand, comprising three electricity metering instruments, three ampermeters and three voltmeters
- a digital display clock

Temperature was set for each zone at 300°C.

The times that the thermoregulators were connected to the heating sleeves and the related electricity consumptions were recorded in the test report.

Each measurement lasted for minimum five hours.

Sets of measures were performed on the three already named machines, on each machine with both the classical (ceramic sleeves) and the new heating system.

4. Discussions

4.1. Injection Torpedo

4.1.1. Results and interpretations related to the digital experiments. We have digitally simulated 180 constructive variants of injection Torpedoes, for diameters Ø25, Ø30, Ø35, Ø40, Ø45 (36 variants for each dimension). The maximum value and its localisation for equivalent stress, for maximum principal stress, for minimum principal stress, for deformation, for safety coefficient were the purpose for each simulated variant.

It comes out from the Analyse Reports that the equivalent stress has its maximum value in the area of intersecting of the inclined orifices from the torpedo body, as it can be seen in figure 17 (the red coloured area).

IOP Conf. Series: Materials Science and Engineering **209** (2017) 012070 doi:10.1088/1757-899X/209/1/012070

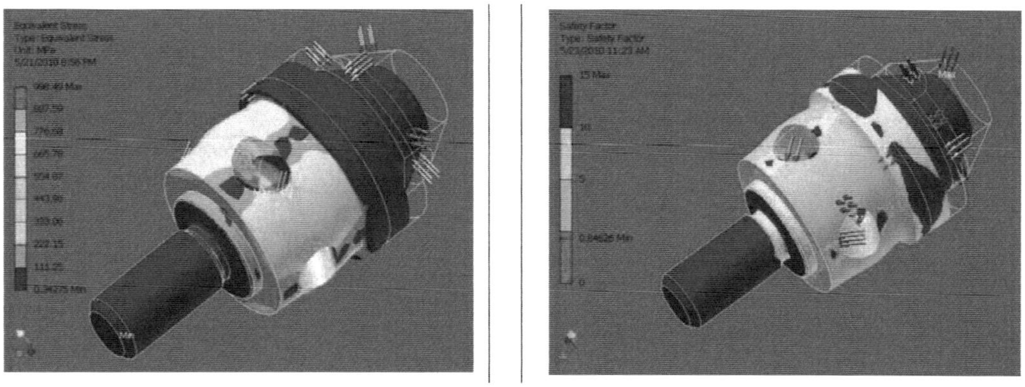

Figure 17. Location of the maximum value of equivalent stress (a) versus location of the minimum value of safety coefficient [10].

The Analyse Reports reveal that the minimum value for the safety coefficient is located in the area of intersecting of the inclined orifices from the torpedo body, as it can be seen in figure 17 (the red coloured area).

OBSERVATION: most of the safety coefficients are less than 1.

INTERMEDIARY CONCLUZION: The geometries had to be modified for the Ø 30 mm, Ø 35 mm, Ø 40 mm and Ø 45 mm torpedoes. Solutions had to be found for solving the stresses in the area of intersecting of the inclined orifices and the axial hole from the torpedo body (most of the graphical representations indicated this zone to be the critical zone).

The visualization of the variation trend for the studied solicitations (equivalent stress, maximum principal stress, minimum principal stress, deformation and safety coefficient) was necessary in order to make a right decision related to how the torpedo geometry could be modified for obtaining an optimization of its behaviour under load relating it to the variable elements of the geometry (d, B, R_t and R_s).

The respective variation diagrams were obtained by studying the variation of the maximum value for the equivalent stress, for the maximum principal stress, for the minimum principal stress, for the deformation and for the safety coefficient as a function of d and B for $R_t \in \{0,1,2.5,5\}$ and $R_s \in \{0,1.5,2\}$. Such a diagram is illustrated in figure 18.

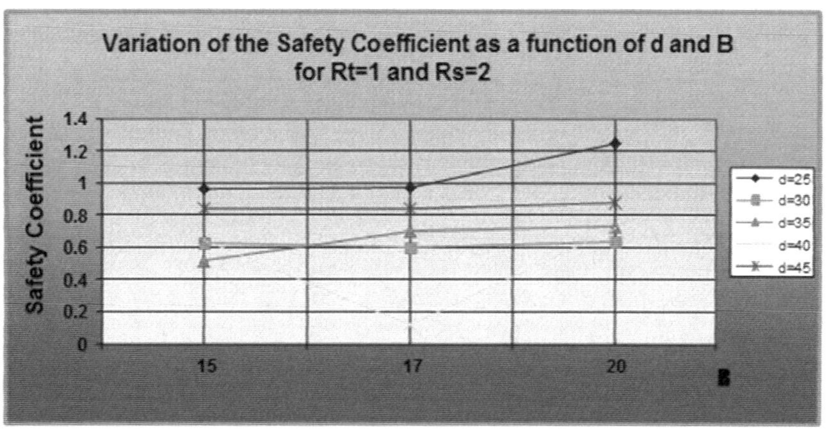

Figure 18. Variation of the safety coefficient as a function of d and B for Rt=1 and Rs=2 [authors'creation].

The variation function of deformation related to dimension B of the torpedo body reaches a maximum point for the Ø35 mm torpedo and a minimum one for the Ø40 mm torpedo for the same dimension B = 17 mm.

The value of the function is greater for dimension 20 mm than for dimension 15 mm for all diameters. The variation is extremely small only for diameter Ø30 mm.

INTERMEDIARY CONCLUZION: Dimensions d and B do not have an important influence on the variation of the maximal value of the maximum principal stress.

Some other dimensions than B, R_S, R_T must be modified in order to obtain safety coefficients greater than 1 at all the torpedoes. The angle between the inclined holes had been modified from 60° to 50°, as well as dimension E, dimension F and only for torpedoes Ø40 and Ø45 dimension I

4.1.2. Results and interpretations of photoelastometry experiments. The variation diagram had been drawn related to the principal stresses which were obtained by means of photoelastometry and FEM (Figure 19).

The experimental results show, as a rule, greater values for stresses than the FEM values.

The photoelastometry method was used in order to check the work hypothesis that had been used in our numeric calculations and for a fast highlight of the areas with stress concentrators.

The photoelastometry (red colour) versus FEM (blue colour) difference σ_1 - σ_2 was put on the horizontal axis in figure 19, since the radial distance of the determination points (green colour) was put on the vertical axis. The graphic representation is overlapped on the axial section of the torpedo body (the same scale was used for the axial section and the vertical axis) in order to illustrate the most suggestive possible the stress localization in the torpedo body.

It is noticeable that stress increase quasi-exponential in the area of intersection between the axial hole and the inclined orifices. The same thing is illustrated in figure 10, where the red coloured areas are located in the same zones.

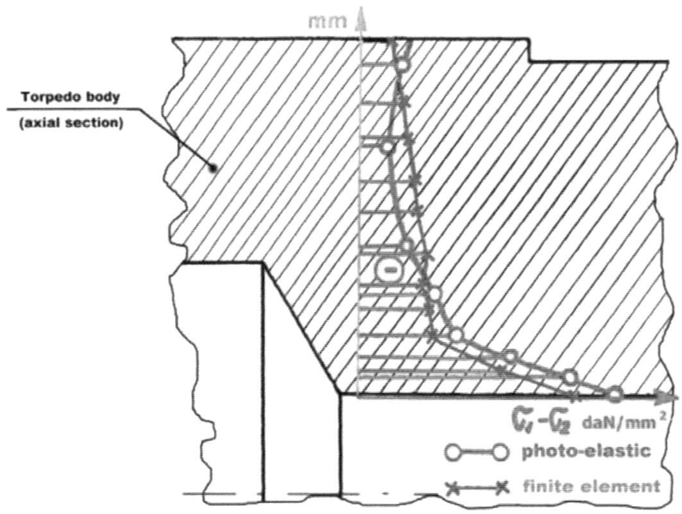

Figure 19. Comparison between the values σ_1 - σ_2 which were obtained by means of photoelastometry and by FEM [11].

Models have been manufactured and tested after corroboration of results of the two investigation methods - numerical and experimental – with the following dimensional modifications (figure 20):

- in zone I joining radius of 2.5 mm;
- in zone II a rectilinear transit between the central hole and the inclined holes;
- in zone III joining radius of 1.5 mm

The analyse of the longitudinally cut slices from the tested model marked out a decrease of stress concentration in the already mentioned areas and, also, stress decrease on the path II - III (stress-breaking zone at initial model) figure 13.

The great number of isochromates (which are the geometric place of the points with an equal difference of principal stresses) in the holes zone is a result of the existence of an only one compression stress on longitudinal direction. The principal stresses within the figure plane at the top of the part are sensitively close as magnitude and of the same sign – so, their difference being very small. (figure 14).

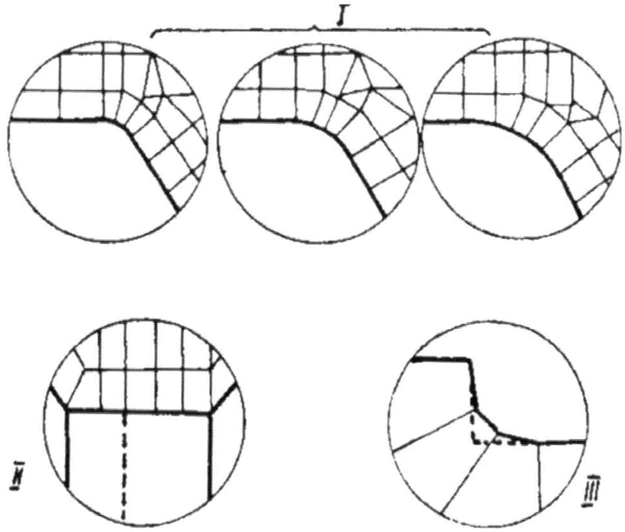

Figure 20. Zones with dimensional modifications [11].

4.1.3. Results and interpretations after experiments in real functioning conditions
Results of the tests are illustrated in table 3:

Table 3. Centralization of results of researches upon mixing.

Nr. of experiment	Torpedo type	Pigment	Necessary number of injections for obtaining an uniform colour
1	classical	-	1
2	classical	blue	9
3	new	-	1
4	new	red	6

It may be observed that fewer injections are necessary with the new torpedo than with the classical one in order to obtain a uniform colour of the moulded part. This thing reveals a better mixing of the plastic melt by using the new torpedo than the classical one.

4.2. The Heating System

4.2.1. Results and interpretations after experiments in real functioning conditions. Tests were performed using three plastic injection moulding machines: of 25 tf, of 50 tf and of 80tf. Both the classical and the new heating system were tested on each machine.

By analysing consumptions of the 25 tf, 50 tf and 80 tf machines that were equipped with the classical heating system, we have noticed that temperature allocation on each heating sleeve corresponds to the allocation of consumption.

It is easy to compare for each machine dimension the consumption of the classical heating system versus the new one, by studying the diagrams which we have issued within the Project. The result of the comparison, illustrated in table 4, is clearly favourable to equip the machines with the new heating system.

Table 4. Synthesis of electricity consumption.

Measuring conditions		Classical system	New system
1. Machine type		25 tons Moulding machine (MI63/25TP)	
Consumption after one hour [W]		3654	2267
Consumption after two hours [W]		5489	2808
Consumption after three hours [W]		7062	3286
Consumption after five hours [W]		9439	4359
Difference after five hours	kW	5.080	
	%	53.819	
1. Machine type		50 tons Moulding machine (MI100/50TP)	
Consumption after one hour [W]		3051	2463
Consumption after two hours [W]		4347	3133
Consumption after three hours [W]		5601	3736
Consumption after five hours [W]		8148	5405
Difference after five hours	kW	2.743	
	%	33.665	
1. Machine type		80 tons Moulding machine (MI250/80TP)	
Consumption after one hour [W]		3797	3684
Consumption after two hours [W]		5497	4481
Consumption after three hours [W]		7600	5471
Consumption after five hours [W]		10903	7545
Difference after five hours	kW	3.358	
	%	30.799	

It is to be noticed, for all the studied machines, that the difference classic/new is smallest for the heating sleeve No. 2. This situation is complying to reality behaviour, since heating sleeve No. 2 is located between sleeve No. 1 and sleeve No. 3, borrowing lateral heating from both sides.

5. Conclusions

We have studied within our Project aspects related to the „heart" of the plastics injection moulding machine: the plasticising/injection unit. The studied aspects refer to increase the reliability of the plasticising/injection unit and to decrease the electrical energy consumption.

The Project approached the practical problem of issuing some new constructive solutions for the plasticising/injection unit, solutions which had been patented by the authors (RO-99019, RO-99128 and patent request 00641), relating respectively to a dozing – injecting Torpedo and to a Heating System for the plasticising/injection units for thermoplastics. The following objectives were aimed by the Project:

- Study of mechanical behaviour of a new dozing – injecting Torpedo
- Study of mixing of the plasticised material, through the new torpedo as well as through the classical one
- Study of electrical energy consumption, with the new heating system as well as with the classical one

These objectives were fulfilled by means of performing digital simulations, photoelastometry experiments, experiments with some prototypes on different specially created stands, experiments and measurements performed on machines in functioning condition.

A starting point consisted in the great number of defections of the existing dozing – injecting Torpedoes. Repairing of such a defection consumes a working shift time (8 hours), counting the cooling time, emptying the unit, the effective repair, re-heating and re-filling the unit.

The second starting point consisted in the great consumption of electrical energy of the plastics injection moulding machine, the energetic consumption sharing abt. 75% in the cost price calculation for the moulded parts.

We have theoretically and practically demonstrated within the Project that adopting of some other geometry for the dozing – injecting Torpedo leads to eliminating the reliability problems of the classical torpedo, as well as that adopting of some other conception for the Heating System leads to a significant decrease of the electrical energy consumption.

We may state, finally, that the desired objective was attained: we have found new geometries for some essential sub-assemblies of the plastics injection moulding machines, their adopting leading to increasing the reliability of the machine and to decreasing its energetic consumption.

References

[1] Vasilache V and Vasilache M 1990 *Torpila pentru dozat - injectat* Patent No. 99019/ Romania
[2] Vasilache V and Vasilache M 1990 *Sistem de încălzire pentru unitățile de injectat mase plastic termoplaste* Patent No. 99128/ Romania
[3] Vasilache V, Grama L and Vasilache M 2009 Decreasing the Energetic Consumption at the Plastics Injection Molding Machines *Nonconventional Technologies Review* **1**
[4] Vasilache V, Vasilache M and Vasilache A 2013 *Mașini de injectat material plastic* Ed. Universității Petru Maior Tg-Mureș ISBN978-606-581-066-2
[5] Vasilache V, Vasilache M and Vasilache A 2015 Sistem de încălzire *Buletinul Oficial de Proprietate Industrială – secțiunea Brevete de Invenție* **3**
[6] Theocaris P S, Boleanțu L, Buga M ș.a. 1977 *Analiza experimentală a tensiunilor*, ET, București
[7] Grama L and Vasilache V 2009 *Optimization of the Plasticizing Unit Construction*
[8] Vasilache V – *Contributions to constructive improvement of the small capacity plastics injection moulding machines for thermoplastics*
[9] Șoaită D, Vasilache V, ș. a. 2015 Referat de testări pentru prototipuri: Torpila pentru injectare, Sistem de încălzire "*Reducerea Consumurilor de Funcționare la Mașinile de Injectat Materiale Plastice*" Proiect de cercetare finanțat de UE în cadrul POSCCE Tg-Mureș
[10] Șoaită D, Vasilache V, ș. a. 2014 Referat privind simularea prin metoda elementului finit "*Reducerea Consumurilor de Funcționare la Mașinile de Injectat Materiale Plastice*" Proiect de cercetare finanțat de UE în cadrul POSCCE, Tg-Mureș
[11] Grama L and Vasilache V 2009 Improvements of the geometry of the injection torpedo *XXIII MicroCAD International Scientific Conference*

Acknowledgements

This Project was supported by a grant of the Romanian National Authority for Scientific Research, project RECOPLAS (POS-CCE, O.2.3.3., ID 1292, SMIS 38814, ctr. No 405/01.08.2012).

IOP Conf. Series: Materials Science and Engineering **209** (2017) 012071 doi:10.1088/1757-899X/209/1/012071

Mechatronic Hydraulic Drive with Regulator, Based on Artificial Neural Network

Y Burennikov, L Kozlov*, V Pyliavets and O Piontkevych
Vinnytsia National Technical University, 95 Khmelnytske shose, Vinnytsia, Ukraine

*E-mail: osna2030@gmail.com

Abstract. Mechatronic hydraulic drives, based on variable pump, proportional hydraulics and controllers find wide application in technological machines and testing equipment. Mechatronic hydraulic drives provide necessary parameters of actuating elements motion with the possibility of their correction in case of external loads change. This enables to improve the quality of working operations, increase the capacity of machines. The scheme of mechatronic hydraulic drive, based on the pump, hydraulic cylinder, proportional valve with electrohydraulic control and programmable controller is suggested. Algorithm for the control of mechatronic hydraulic drive to provide necessary pressure change law in hydraulic cylinder is developed. For the realization of control algorithm in the controller artificial neural networks are used. Mathematical model of mechatronic hydraulic drive, enabling to create the training base for adjustment of artificial neural networks of the regulator is developed.

1. Introduction

The study of operating processes of metal processing allows to determine the optimum processing modes [1-3]. It improves the quality of detail processing, reduces the number of defects. The capabilities of the technological equipment for metal treatment under pressure relating to the ensuring of optimum modes significantly extend in case of the use of mechatronic hydraulic drives with proportional program control [4-7]. Currently, the development and introduction of mechatronic hydraulic drives in technological machines for metal treatment under pressure corresponds to the global trends of design of such machines, and the development of structure and determination of effective algorithms of mechatronic hydraulic drive control is a recent scientific and technical problem [4].

The objective of this paper is to develop a scheme of a mechatronic hydraulic drive with regulator that provides the possibility of implementation of the necessary load laws of a workpiece in the process of their deformation.

2. The scheme of a mechatronic hydraulic drive

The scheme of a mechatronic hydraulic drive is presented in figure 1. The hydraulic drive includes an adjustable pump 1, a distributor 2, a hydraulic cylinder 3 and a controller 4. The adjustable pump is equipped with a regulator 19 that includes a spool valve 8 with a spring 9, throttles 14, 16 and a valve 7 with an electromagnet 6. A faceplate 13 of an adjustable pump 1 is controlled by servo-cylinders 11 and 12. The hydraulic fluid is delivered to servo-cylinder 11 from a regulator 19 through the damper 15. The controller 4 through the amplifier 17 and 18 is connected to the distributor 2 and 5 through the amplifier 5 to the regulator 19. The signals given by the controller 4 modes determine the hydraulic drive operating regime and load changes laws of workpieces in the process of processing. The regulator 19 is connected through the hydraulic line 20 with the distributor 2. The hydraulic line 10 is connected to the controller 4 through the pressure sensor 21.

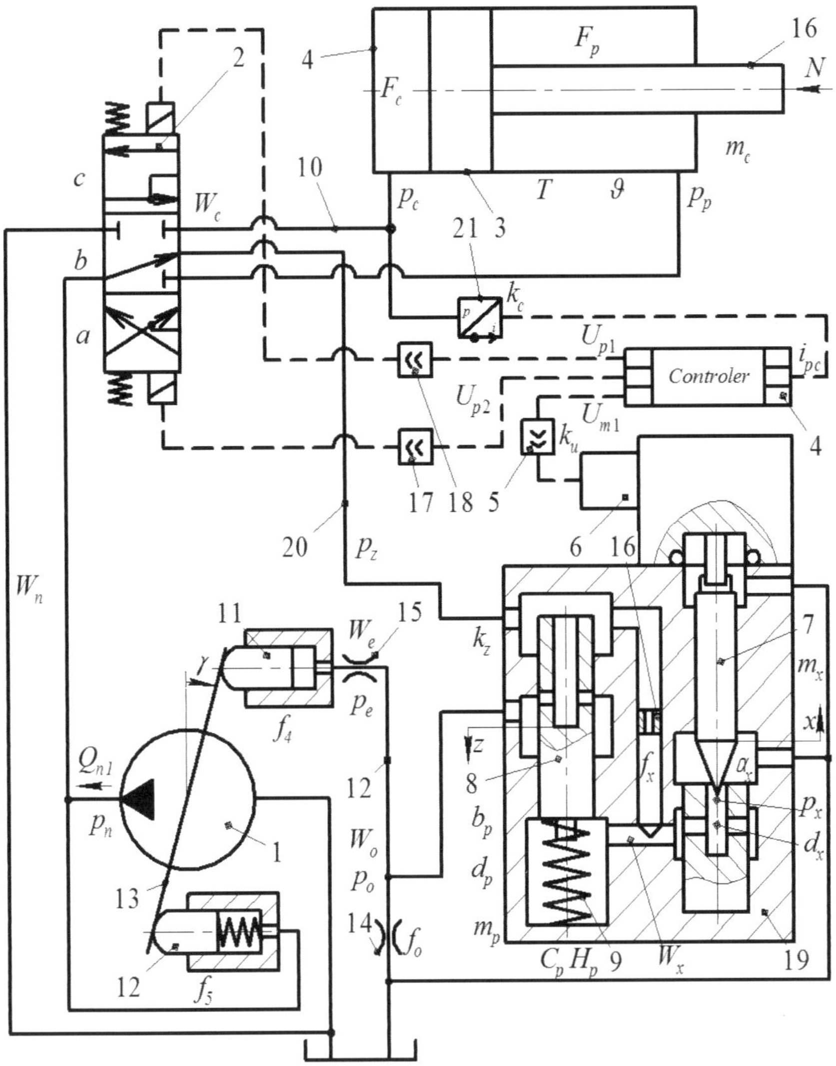

Figure 1. The scheme of a mechatronic hydraulic drive.

The hydraulic drive operates in such a way. When finding a distributor in a position (*b*) the adjustable pump 1 is disconnected from the hydraulic cylinder, but delivers a hydraulic fluid through the hydraulic line 20 to the regulator 19. The controller 4 thus delivers a zero signal to the electromagnet 6. The hydraulic fluid from the hydraulic line 20 will enter through the throttle 16 and valve work window to the drain, and simultaneously through the spool valve work window 8 and damper 15 to servo-cylinder 11. A faceplate 13 of an adjustable pump 1, which is currently under servo-cylinders 11 and 12 operation will unfold downward the angle γ reduction and the pump output flow will be minor. Delivery pressure of the adjustable pump 1 will be determined by adjusting the spring 9 and will be 1.0 MPa, and the pump output flow in this mode compensates only a small flow of hydraulic fluid that occurs in the regulator 19. Thus when finding a distributor 2 in a positions (*b*) pump discharge mode is provided, in which capacity loss in the hydraulic drive is minor.

When switching the distributor into position (*a*) hydraulic fluid from the adjustable pump 1 through the distributor 2 will flow into a hydraulic cylinder 3 head end. The hydraulic cylinder 3 rod will move from left to right until it stops at the work piece to be deformed. Drain of the hydraulic fluid from the hydraulic cylinder rod end is also through distributor 2. Therewith the controller 4 generates a signal U_{m1}, which through the amplifier 5 is delivered to the electromagnet 6. The electromagnet 6 influences the valve 7 and creates pressure p_x in the spring chamber of the spool valve 9. Therewith the value of pressure p_x is proportional to the signal U_{m1}. Pressure p_x influences the spool valve 8 and with the spring 9 forms pressure p_x above the upper end of the spool valve. Pressure p_x will determine the magnitude of pressure p_z, which occurs in the hydraulic cylinder head end. Thus, the value of pressure p_z, which determines the force *N* on hydraulic cylinder 3 rod 16 will be proportional to the signal U_{m1} value at the output of the controller 4.

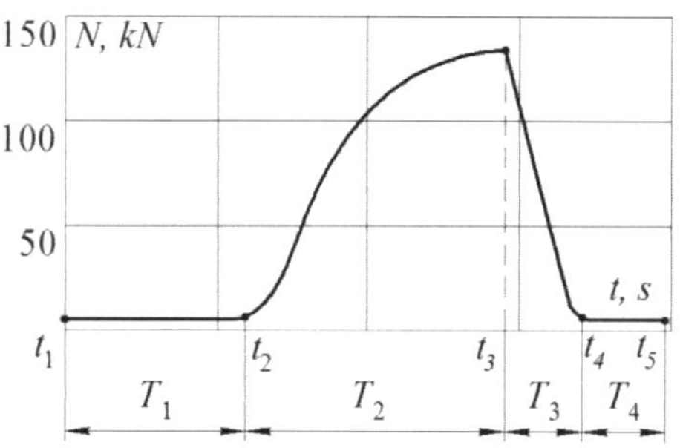

$T_1=t_2–t_1$ – period of tool advance;
$T_2=t_3–t_2$ – period of workpiece loading;
$T_3=t_4–t_3$ – period of workpiece discharging;
$T_4=t_5–t_4$ – period of tool returning.

Figure 2. The workpiece load law

Setting the law of variation of signal U_{m1} over time may provide the necessary law of variation of pressure p_z in the hydraulic cylinder 3 head end and accordingly the law of variation of the value of force *N* on the rod [4]. Figure 2 shows the dependence of the force *N* on the hydraulic cylinder 3 rod from the time during operation of the mechatronic hydraulic drive in the mode when the distributor 2 is in position (*a*). This mode is called load conditions and has three specific periods. During $T_1 = t_2–t_1$ the controller 4 generates a signal $U_{m1} = 0$ and in this period hydraulic cylinder rod 16 with the tool fixed on it is fed to the work piece. Force p_z in the hydraulic cylinder 3 head end is thus determined by adjusting the spring 9 of the regulator 19. In time t_2 period by a signal from the pressure sensor controller 21 starts changing signal U_{m1}, by law, which is determined by the control algorithm and is optimal for this type of processing and type of workpiece [4]. Workpiece loading takes place during the period $T_2 = t_3–t_2$. At time t_3 the law of variation of signal U_{m1} changes and force *N* changes accordingly, influencing the workpiece. Unloading of the workpiece takes place during $T_3 = t_4–t_3$. At time t_4 distributor 2 switches into position (*c*). The hydraulic fluid from the adjustable pump 1 through the distributor 2 will be delivered to the hydraulic cylinder 3 head end and the rod will move from right to left deflecting the tool from the workpiece. In the period $T_4 = t_5–t_4$ the signal U_{m1} at the output of the controller is zero, and the pressure p_{pz} in the head end will be determined by adjusting the regulator 19 spring 9. In the period t_5-t_4 mode, called "tool returning" will be implemented.
Formation of required law during processing is provided by the controller. The controller functions as a regulator, a block diagram of which is shown in figure 3. The regulator includes shift units 2 and 5, and artificial neural networks 3 and 4. The signal p_c from the pressure sensor transfers to the input regulator. Depending on the operating regime of mechatronic, hydraulic drive regulator generates the following signals: U_{p1}, U_{p2} – voltage fed to the electromagnets of the distributor; U_{m1} – voltage fed to the electromagnet of the pump regulator.

Shift units 2 and 5 maintain operation of the regulator at such mechatronic hydraulic drive operating modes:
1. Pump discharge mode $U_{p1}=U_{p2}=0$; $U_{m1}=0$
2. Valve hydraulic-cylinder rod travel with the tool

2.1. Accelerated tool advance U_{p1}=0; U_{p2}=max; U_{m1}=0
2.2. Workpiece loading U_{p1}=0; U_{p2}=max; $U_{m1}=F_1(t)$
2.3. Workpiece discharging U_{p1}=0; U_{p2}=max; $U_{m1}=F_2(t)$
3. Tool returning U_{p1}=max; U_{p2}=0; U_{m1}=0
$F_1(t)$, $F_2(t)$ are indicated – transfer functions of neural networks 3 and 4.

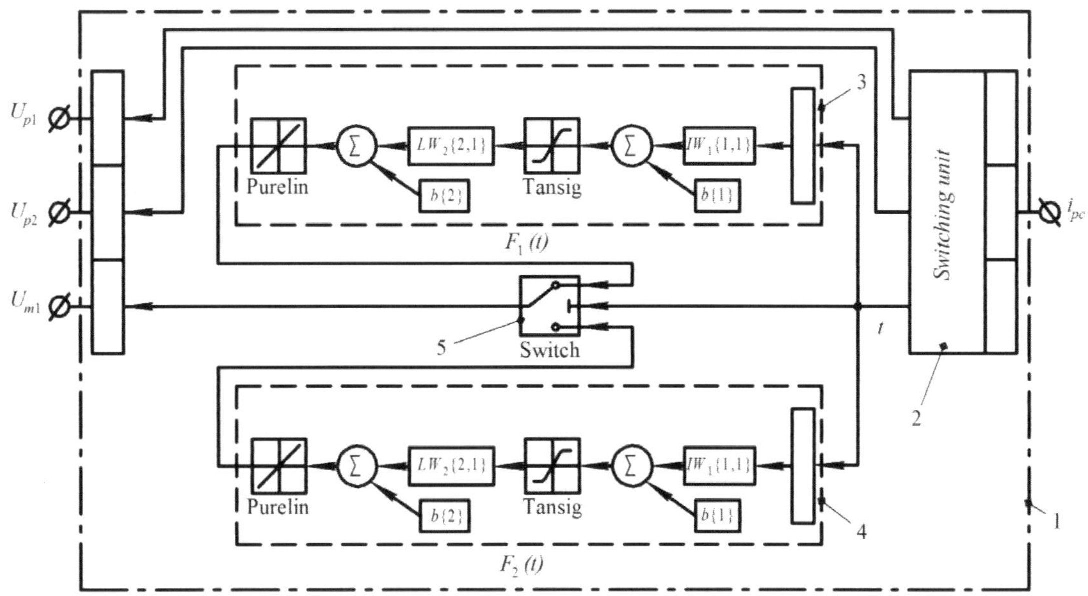

Figure 3. The block diagram of the controller, based on artificial neural network.

Necessary laws of load changes on the workpiece that are determined by the nature of pressure p_z change during the processing are formed by neural networks. Load law on the workpiece during the period $T_2=t_3-t_2$ is determined by the neural network 4, during the period $T_4=t_5-t_4$ by the neural network 4.

Checking of operability and efficiency of mechatronic hydraulic drive at the design stage may be performed on the basis of simulative mathematical modeling.

3. Mathematical model and simulation results of mechatronic hydraulic drive

Mathematical model of mechatronic hydraulic drive comprises of moment (1) equation acting on the pump faceplate 13 (figure 1), force (2) equation acting on the valve 7, force (3) equation acting on the spool 8, force (4) equation acting on the 3 hydraulic-cylinder rod 16 and flow continuity (5) equation for hydraulic lines between the regulated pump 1, hydraulic cylinder 3 and regulator 19, flow continuity (6) equation for hydraulic lines between the spool 8, a damper 15 and the throttle valve 4, flow continuity (7) equation for hydraulic line between the throttle valve 16 and the valve working window 7, flow continuity (8) equation for hydraulic lines between the damper 15 and servo-cylinder 11, voltage drop (9) equation in the circuit of the electromagnet 6, the dependence (10) of compliance factor b_n, which describes the total compression of the hydraulic fluid and rubber-steel pipelines, dependence (11), which determines the adjusted modulus the hydraulic fluid given the presence of the gas phase. Structure blocks that implement transfer functions $F_1(t)$, $F_2(t)$ of neural networks are also included into mathematical model.

The mathematical model of the mechatronic hydraulic drive is designed at such basic assumptions and simplifications. The focused hydraulic drive parameters are considered; the temperature during working hours does not change; wave processes in hydraulic lines were not considered; flow coefficients through the throttle, working windows of a spool 8 valve 7 are constant; pressure loss in hydraulic lines and in working windows of the distributor 2 were not inspected; hydraulic drive operation mode is non-cavitating; dry fiction force in hydraulic cylinder does not depend on the rod travel speed; operation of an electromagnet amplifier of the pump control valve is measured with the help of the proportional link.

$$I\frac{d^2\gamma}{dt^2}=p_c\cdot f_5\cdot l-p_e\cdot f_4\cdot l-\frac{\pi\cdot\rho\cdot v_k\cdot d_4\cdot l_4}{\varepsilon_n}\cdot\frac{d\gamma}{dt}\cdot\cos\gamma+m_0+m_1\cdot Q_{n1}+$$
$$+m_2\cdot p_{n1}+m_3\cdot Q_{n1}^2+m_4\cdot p_{n1}^2+m_5\cdot p_{n1}\cdot Q_{n1}+m_f(\omega_n) \quad (1)$$

$$m_x\frac{d^2x}{dt^2}=p_x\cdot\frac{\pi\cdot d_x^2}{4}-k_m\cdot i_{m1}-\left(\frac{\pi\cdot\rho\cdot v_k\cdot d_x\cdot l_x}{\varepsilon_x}\right)\cdot\frac{dx}{dt} \quad (2)$$

$$m_p\frac{d^2z}{dt^2}=p_c\frac{\pi\cdot d_p^2}{4}-p_x\frac{\pi\cdot d_p^2}{4}-C_p\cdot(H_p+z)-\frac{\pi\cdot\rho\cdot v_k\cdot d_p\cdot l_p}{\varepsilon_p}\cdot\frac{dz}{dt} \quad (3)$$

$$m_c\frac{d\upsilon}{dt}=p_c\cdot F_c-\pi\cdot D_{cz}\cdot(q_0+k_q\cdot p_c)-N \quad (4)$$

$$F_7\cdot d_8\cdot k_1\cdot n_n\cdot tg\gamma-k_{n1}\cdot p_{n1}=\mu\cdot k_z\cdot z\cdot\sqrt{\frac{2\cdot|p_{n1}-p_0|}{\rho}}\cdot sign(p_{n1}-p_0)+\beta_n\cdot W_c\frac{dp_c}{dt} \quad (5)$$

$$\mu\cdot k_z\cdot z\sqrt{\frac{2\cdot|p_c-p_0|}{\rho}}\cdot sign(p_c-p_0)=\mu\cdot f_0\cdot\sqrt{\frac{2\cdot p_0}{\rho}}+\beta_p\cdot W_0\frac{dp_0}{dt}+\mu\cdot f_e\cdot\sqrt{\frac{2\cdot|p_0-p_e|}{\rho}}\cdot sign(p_0-p_e) \quad (6)$$

$$\mu\cdot f_x\cdot\sqrt{\frac{2\cdot|p_c-p_x|}{\rho}}\cdot sign(p_c-p_x)=\mu\cdot\pi\cdot d_x\cdot x\cdot\sin\left(\frac{\alpha_x}{2}\right)\cdot\sqrt{\frac{2\cdot p_x}{\rho}}+\beta_p\cdot W_x\frac{dp_x}{dt} \quad (7)$$

$$\mu\cdot f_e\cdot\sqrt{\frac{2\cdot|p_0-p_e|}{\rho}}\cdot sign(p_0-p_e)=\beta_p\cdot W_e\frac{dp_e}{dt}-\exp(A_b\cdot p_e+B_b\cdot t^\circ+C_b\cdot\varepsilon_e+D_b)-f_4\cdot l\cdot\frac{d\gamma}{dt}\cdot\cos\gamma \quad (8)$$

$$L_e\frac{di_{m1}}{dt}+i_{m1}\cdot R_e=\begin{cases}0 & by \quad t_1\le t<t_2\\ k_u\cdot F_1(t) & by \quad t_2\le t<t_3\\ k_u\cdot F_2(t) & by \quad t_3\le t<t_4\\ 0 & by \quad t_4\le t<t_5\end{cases} \quad (9)$$

$$\beta_n=\frac{1}{E_p}+\frac{d_{mp}}{\delta_{mp}\cdot E_{mp}(p)} \quad (10)$$

$$E_p=\frac{1}{\beta_p}=E_{p0}\frac{W_f/W_a+1}{W_f/W_a+(E_{p0}\cdot p_0)/p^2} \quad (11)$$

where p_c, p_p, p_e, p_z, p_0 – pressure at the in-outlet of the hydraulic cylinder 3 (figure 1), in the pump control system 1; z – coordinate position of the spool regulator 8; γ –rotation angle of the adjustable pump faceplate 1; f_0 – throttle area 14 in the pump control system; F_c, F_p, f_4, f_5, f_e – piston surface area of the hydraulic cylinder 3, of servo-cylinders of the regulated pump, of the damper of pump servo-cylinder; D_c, d_p, d_7, d_8, d_x – diameters of hydraulic cylinder 3, of slide valve 8 of regulator, feeds plungers of the adjustable pump and the contact circle of the pump feeds with the faceplate, working window of the valve 7; k_q, k_m, k_n, k_c, k_u – coefficients specific friction force in the hydraulic cylinder 3, efforts proportionality of the electromagnet, of pump leaks 1, pressure transducer reinforcement 21 and reinforcers 5; L_e, R_e – coefficient of induction and active resistance of the winding 6; i_{m1}, i_{pc} – currents in the winding of the valve 17, in the outlet of the pressure transducer 21; N, T – load force on the throttle 16 and friction in the hydraulic cylinder 3; μ – flow coefficient through the throttle and spool elements; ρ – hydraulic fluid density; l_p, l_4, l – spool contact length 8, pump servo-cylinder 11, arm of action of the servo-cylinders of the regulated pump 1; I – moment of the inertia of the faceplate 13 of the pump; m_p – spool mass 8; W_0, W_c – hydraulic lines between the regulator 8, throttle 14 and damper 15, between the pump 1 and hydraulic cylinder 3; n_n – rate speed of the pump shelf assembly 1; k_1 – quantity of the feed plunger 1; q_0 – specific friction force in hydraulic cylinder 3; ε_p, ε_n – gaps between the spool 8, servo-cylinder of the pump 11 and their corpuses; $F_1(t)$, $F_2(t)$ – transfer functions, which are implemented by neural networks; m_0, m_1, m_2, m_3 – coefficients of dependency of the resistance moment on the faceplate 13 of the pump of flow and pressure; A_b, B_b, C_b, D_b – coefficient in flow formula from servo-cylinder control camera 11; β_p – reduced compliance coefficient of the gas-liquid mixture; β_n – reduces compliance coefficient of the rubber-metal pipeworks and gas-liquid mixture; E_{p0}, E_p, $E_{pm}(p)$ – elasticity modulus of the working liquid, reduced elasticity modulus of the gas-liquid mixture and rubber-metal pipeworks; δ_{mp} – thickness of the pipework wall; W_f – liquid volume in the gas-liquid mixture at pressure p; W_a – gas volume in the gas-liquid mixture at atmosphere pressure.

Neural networks used in a regulator of mechatronic hydraulic drive are generated with the help of the application Neural Network Toolbox that is a part of the software package MATLAB [8].
Classical neural networks with a direct signal propagation are used. These neural networks perform the function:

$$net{=}newff\,(PR[S1,\ S2..SN1]\ TF1,\ TF2..TFN1,\ BTF,\ BLF,\ PF) \tag{12}$$

where PR – signal values in the input and output of the neural network, taken form the database, $S1$, $S2..SN1$ – quantity of neurons in the hidden layers of the neural networks; $TF1$, $TF2..TFN1$ – activation functions for the hidden layers; BTF – learning neural networks function; BLF – weight coefficients and bias adjustment feature; PF – inaccuracy function.

Function parameters *newff*: activation function for the hidden layer neurons – *tansig*, for the output neuron – *purelin*, neural network learning function by the classical scheme of the back propagation of error – *traingd*, function of the weight coefficients and bias settings– *learngdm*, inaccuracy function – *mse* are used in the process of neural network generation and learning.

During the training process for the termination process, the following parameters are controlled: *epochs* – the maximum number of learning cycles, *goal* – functional training threshold value, *max-fail* – the maximum permissible level of error exceeding of the reference sample in comparison with the training.

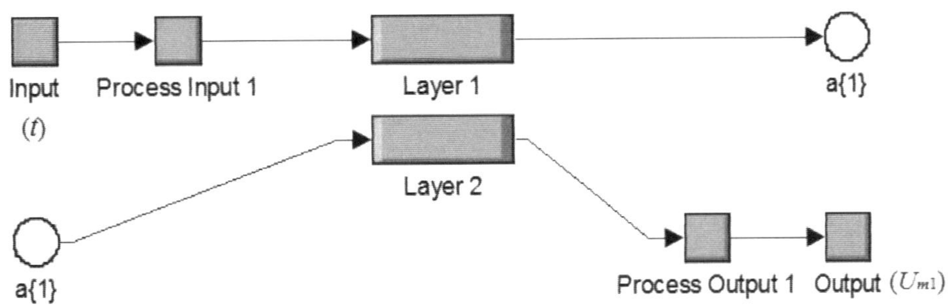

Figure 4. Structure of neural network that tools the workpiece load law.

Neural network, providing workpiece load has 6 hidden layer neurons. The structure of the neural network is formed in MATLAB-Simulink environment and is presented in figure 4. Figure 5 shows the structure of the hidden layer neurons of the neural network.

For the neural network instruction training bases built on the results of work are used [4]. Neural network training base that provides work piece loading consisted of 27 pairs of data, and the neural network base that provides work piece unloading consisted of 22 pairs of data.

In the course of the neural network training, providing work piece loading 25 iterations were held. Weight coefficients and neural network bias were determined.

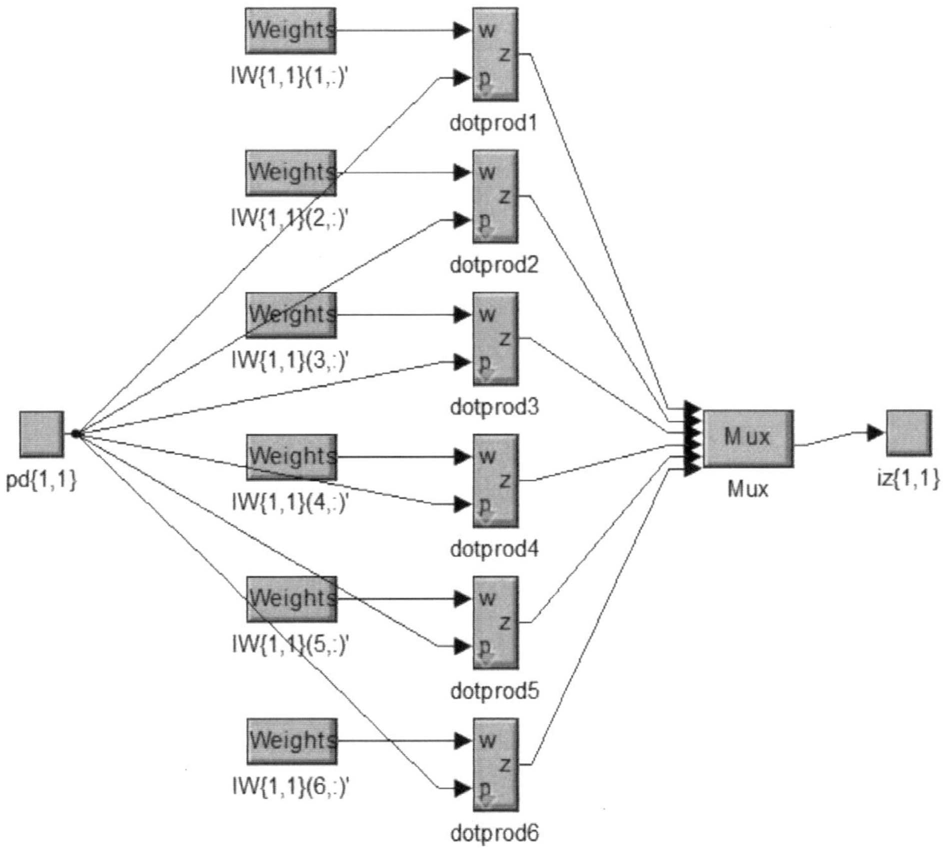

Figure 5. Structure of neurons hidden layer in a neural network that tools the workpiece load law.

Weight coefficients vector of neurons hidden layer of this neural network: $iw_1\{1,1\}$ = (- 8,619; 8,010; -8,415; -11,631; 5,880; -5,190).
Neurons hidden layer displacement vector: $b\{1\}$ = (8.107, -3.439, 0.321, -2.158, 2.676, -7.290).Inputs vector for the neuron output layer: $LW_2\{2,1\}$ = (0,009; 0,033; -0,127; -0,125; 0,677; -2,062).Output neuron displacement: $b\{2\}$ = (- 2.052).

The maximum deviation value result of the neural network of the corresponding value of the training base is 2.2%.

Neural network that provides unloading of the workpiece has 7 neurons in the hidden layer. In the course of training of this neural network 12 iterations are held.

Weight coefficients and neural network bias were determined. Weight coefficients vector of the neural network is the following: $iw_1\{1,1\}$ = (- 9,642; -9,793; 9,678; 9,777; -9,737; 9,69; -9,803).
Vector displacement of neurons hidden layer: $b\{1\}$ = (9.931, 6.497, -3.468, 0.086, -3.173, 6.558, -9.571).

Vector displacement of the neuron output layer: $LW_2\{2,1\}$ = (0,025; 0,022; -0,163; -0,186; 0,278; -0,233; 0,165).

Output neuron bias: $b\{2\}$ = (0.035).

The maximum deviation value result of the neural network of the corresponding value of the training base of 2.4%.

4. Conclusion

Represented scheme of the mechatronic hydraulic drive with regulator based on artificial neural networks enables loading and unloading of workpieces with necessary laws during processing. The use of neural networks enables implementation of complex nonlinear load laws and minimize labor costs to prepare regulatory control algorithms owing to existing developed mathematical apparatus of artificial neural networks and distributed software. Designed mathematical model of mechatronic hydraulic drive is the basis for determining its operability and performance in the design phase.

References

[1] Kotsiubivska K I, Burennikov Y A, Rusu I and Sivak I O 2009 Radial extrusion of workpieces using backpressure *Buletinul Institutului Politehnic Din Iasi "Stiinta si Ingineria Materialelor"* **LV** (LIX) 195 – 202

[2] Коцюбивская Е И, Сивак И О, Алиева Л И and Куценко С. В. 2008 Оценка деформируемости заготовок при радиальном выдавливании с противодавлением *Обработка металлов давлением* 29 – 33

[3] Сивак И О and Коцюбивская Е И 2007 Пластичность металлов при объёмном напряжённом состоянии *Удосконалення процесів і обладнання обробки тиском в металургії і машинобудуванні* 73 – 76

[4] Сивак І О, Бурєнніков Ю А and Козлов Л Г та інші 2010 Підвищення ресурсу пластичності металів шляхом оптимального управління параметрами процесів засобами гідроприводів з пропорційним керуванням *Вінницький національний технічній університет (Звіт про науково-дослідну роботу)* 138

[5] Губарев О П and Ганпанцурова О С 2016 Адаптація логіки керування пневматичним виконавчим модулем мехатронної системи *Вісник Національного технічного університету "Харківський політехнічний інститут"* **41**(1213) 32 –38

[6] Kozlov L G, Burennikov Yu A and Kovalchuk V A 2009 Dynamic characteristics of the proportionally –controlled pressure valve *Buletinul Institutului Politehnic Din Iasi "Stiinta si Ingineria Materialelor"* **LV** (LIX) 203 – 209

[7] Козлов Л Г, Ковальчук В А and Кушпіта Н В 2009 Динамічні характеристики гідропривода з пропорційним керуванням величини тиску *Вісник Хмельницького національного технічного університету* **4** 28 – 33

[8] Дьяконов В and Круглов В 2001 Математические пакеты расширения MATLAB *СПб: Питер* 480

Influence of the Process Parameters on the Properties of Diamax Deposits Obtained by Flame Thermal Spray

M A Baciu[1], C Nanu[2], G I Sandu[3], B F Toma[4], C Bejinariu[3], A Cazac[3] and S L Toma[3*]

[1] "Aerostar" SA Bacau, Str. Condorilor, No. 9, 600302, Romania
[2] Clinical Emergency Hospitals, P. Neamt, 610003, Romania
[3] Technical University "Gheorghe Asachi", Fac Mat. Sci& Engineering Iasi, 700050, Romania
[4] University of Medicine & Pharm. "Grigore T Popa", Fac. of Medicine, Iasi, 700115, Romania

E-mail: stl_toma@yahoo.com

Abstract. The paper aims to determine the influence of the process parameters, namely: C_2H_2 gas flow rate and inclination angle of the spraying gun on physico and mechanical properties of the hardalloyed layers of Diamax 10999 Eutalloy, on steel support - obtained by flame thermal spray process. For this purpose, the two technological parameters varied on three levels and in each case were evaluated the deposits properties. Investigations conducted by electronical microscopy SEM, X-ray, micro-hardness and by adherence evaluation and of the deposits porosity allowed the establishment of the performant deposit. Thus it was found that at the decreasing of the spraying distance,the deposit porosity decreases; in layer appear the phenomena of overheating, issue that determine the adherence reducing in average of 22%, and also the modification of chemical composition. The results recorded have afforded the obtaining of an optimum domain of variation of the process parameters.

1. Introduction

The thermal spray process in oxygase flame, at low pressures, is one of the oldest spraying processes, being considered also the cheapest process for layers obtaining, which use for melting the input material the heat developed by burning of a gaseous fuel mixture (usually acetylene - C_2H_2 in combination with oxygen - O_2), [1-3]. The process uses, as added materials, metals (pure metal or alloys based on Al, Zn, Fe, Ni, Co, etc.) as non-metallic materials in the form of powders of different micrometric sizes or wire, [4,5]. Flame-spray coatings have often been used to protect against corrosion and/or wear protection of various tracks, to dimensional reconstruction, to obtain new properties or to obtain surfaces with a decorative aspect, [6]. Studies, in the field of thermal spraying, [7,8] have demonstrated that a pulverized particle is strongly attached to the substrates of a substrate if its velocity and temperature is sufficiently high, [9, 11]. In thermal flame spraying, the temperature and particle velocity at the moment of impact with the surface of the substrate is dependent on the C_2H_2/O_2 ratio, the impact angle and respectevelly on the spray distance, [7,8].Obviously the modification of the spray jet temperature - by increasing it, will have a positive effect on the temperature of the particles and implicitly on the deposition properties.

The Diamax deposits, which mainly containing tungsten (W) and nickel (Ni), are widespread due to their combination of unique properties: hardness and resilience, offered by tungsten carbide (WC) - the compound which is formed by spraying and provides excellent wear resistance. Generally, the metal matrix composite coatings on the basis of WC (MMC) [12-13], use special deposition processes that require high temperatures which induce secondary effects, such as the formation of undesirable brittle phases (W_2C and the amorphous binder) [14-15] and increase the porosity of deposits, [16].

The present researches are focused on the study of the process parameters influence, respectivelly: of C_2H_2 flow rate and the angle of inclination of the spray gun against the surface of the substrate (α), on the physical, chemical and mechanical properties of the Diamax 10999 Castolin deposits, produced by flame spray.

2. Experimental procedures

2.1. Materials and process parameters

In the researches, it was deposed Diamax 10999 Eutalloy, produced by Castolin Switzerland - containing 15% W - 30% W - as 45% Ni, 15% Cr, 5% Co and Fe - which enssure the wear resistence at high temperatures, [23]. For the the deposits obtaining, a SuperJet -S-Eutalloy type torch, manufactured by Castolin Switzerland, was used.

The deposits were carried out on a frontal flat surface of some samples, Ø40×50mm – EN 582, made of C15-DIN 17210-1652 steel, with the following composition C—0.14, Mn—0.43, Cr—0.3, Ni—0.3, Si—0.15, P—0.04, and S—0.04 (wt.%)).

The frontal flat surface of the cylindrical samples was cleaned by grit-blasting of Al_2O_3 (grade 24), which are driven by a compressed air jet, at a pressure of 4bar. The parameters of the flame spraying process are presented in Table 1.

The coating process used was the manual one which uses the support for the spray gun. Therefore, the amount of material deposited is a characteristic of the system. However, the test conditions of each passes may be considered in the range of 50 to 80 μm of thickness The final coating varied between 250–480 μm of thickness.

Table 1 - Deposition parameters.

Parameters	Value
C_2H_2 gas flow rate (l/min)	20/22/24
O_2 gas flow rate (l/min)	38
Spray angle (^{0})	60/75/90
Spraying distance (mm)	80
Number of spraying passes	6
Scan speed of the gun, (m/s)	0.16

2.2. Characterization methods

The coatings obtained by flame spraying were examined with stereomicroscope Optika SZM 2 — made in Italy, for the purpose of excluding the coatings which contain defects.

The microstructural investigations realized in the cross-section of the samples, were done with a optical microscope Optika coupled with a scanning electronic SEM microscope, of the Vega Tescan LMH II type, (made by Czech Republic). The image analysis was carried out with software ImagePro, (product by Media Cybernetics, Bethesda, MD, USA) and allowed the estimate the WC content, the determining the porosity of the cross section of the polished coatings and the mean free path between the reinforcing particles. About ten images of each coating system were used to determine the average

WC content (in vol.%). For determination the mean free path between the reinforcing particles was used the count/size function in the ImagePro software and about twelve images of each coating. On the images was drawn a random line and was recorded the number of times that the line intersected WC particles. Because ImagePro presents the length of the line in pixels, was used a line to measure the scale bar to convert pixels in to micrometres. The mean free path was determined used following equation: $\lambda = \frac{1-V_p}{N_l}$, were: λ is mean free path, V_p is volume fraction of the reinforcing particles and N_l – is the number of particle intersected per unit length of test line.

The micro-hardness HV was determined using digital microdurimeter CV - 400D, manufactured by CV Instruments - made by Japan. Measurements were made using a load of 300 g for 10 s. On each sample were carried out twelve random measurements, accordance standard E384, [18]. Indentations were spaced at a distance of at least four times the diagonal of the previous indent as per ASTM C1327 to avoid strain-hardening effects and possible cracking of the WC particles [19].

Qualitative analysis phase was performed using X-ray diffractometer, X'PERT PRO MRD type, product by PANalytical, Holland, which works in 2θ angle range from 0° to 90° and a High Score Plus software.

The adherence of the obtained deposits was determined by tensile test of the coatings deposed on the frontal flat surface of the samples (for the determination of tensile adherence strength R_H) in accordance with EN 582. Twelve tests were performed on each sample using an Instron machine (Instron Corp., Canton, MA) at a crosshead speed of 0.075 cm/min.

The wear resistance of the coated bars was evaluated using an RFT-III (Kyowa R/D Ltd., Japan) reciprocating friction and wear tester at room temperature without lubrication. Tests were performed under loads of 200N for 5h, at rotation speed 200rpm and at room temperature of 23^{0}C and relative humidity of 26%. Three tests were performed for each condition for repeatability and the averages of the three test results are given in the paper

3. Results and discussions

Figure 2 shows the microstructure of the deposit were made from mixtures of powders with chemically stable, based on the W and Ni. In the SEM images, it is observed that in each layer there are the following distinct phases: a light phase containing W bonded in the form of carbides (WC) and a darker phase with a high Ni content. Black areas were identified as pores. It can be seen that the degree of homogeneity of the structure increases with the increase in C_2H_2 flow. This is explained by the fact that as the gas flow increases, it increases the temperature of the spray jet and maintains the molten state of the particles at the moment of impact with the surface of the substrate.

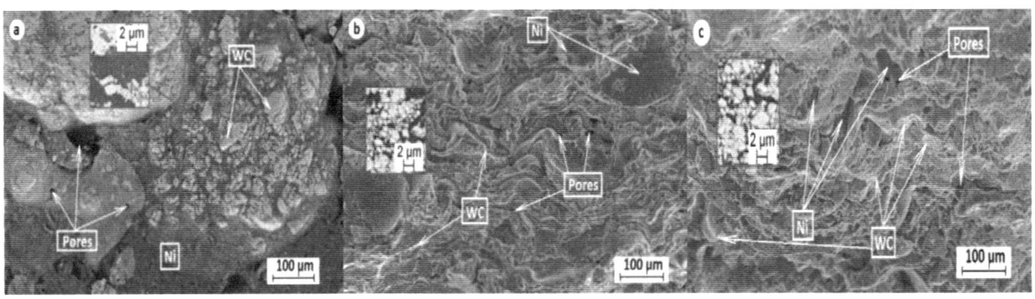

Figure 1. SEM micrographics of Diamax 10999 Castolin coatings deposited using flame spraying, inclination angle α = 90^0, at C_2H_2 gas flow rate: 20 l/min (a); 22 l/min (b) and 24 l/min (c).

Table 2 shows that the porosity of WC deposits decreased as the inclination angle (α) increased. This phenomenon was also noticeable by P.C. Du and colab [17].

Table 2. Porosity values of Diamax 10999 coatings obtained in different technological conditions.

Nr Crt	Materials	Inclination angle, α [^{0}]	C_2H_2 gas flow rate, [l/min]	Porosity, [%]	Mean free Path, [μm]
1	Diamax 10999 Castolin	60	20	18.8	84.3
2			22	15.6	62.4
3			24	11.3	37.5
4		75	20	13.6	46.8
5			22	8.8	28.4
6			24	7.2	24.6
7		90	20	12.6	32.4
8			22	5.3	21.5
9			24	4.5	9.2

The amount of Diamax deposition obtained by flame spraying was determined by X-ray diffraction (XRD) analysis. In Figure 4 there are presented XRD profiles of the deposits made at different flows of C_2H_2. It can be seen that XRD profiles contain mainly the quantities of phase: WC and Ni. The WC and Ni drops are preponderant at angles of 360, 480 and respectively 400. Figure 4 shows that when the C_2H_2 flow increases, there is a relative increase in the primary peak of the WC, compared to the peak corresponding to the Ni content in the layer. It can be asserted that by increasing the temperature of the gas jet it is allowed to fix of a larger amount of WC in the layer, aspect which contributes to the increase of the wear resistance of the Diamax 10999 layers.

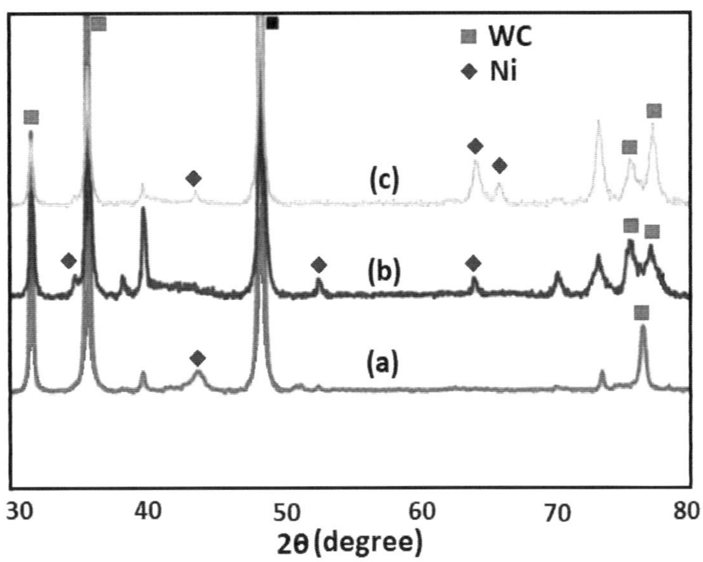

Figure 2. XRD profile of the Diamax 10999 Castolin coatings obtained using α = 90^0 and C_2H_2 gas flow rate: 20 l/min (a); 22 l/min (b) and 24 l/min (c).

Vickers' microhardness indentation determinations performed on Diamax 10999 deposits allowed a first estimate of how will behave the depositions at erosion. Due to the used indentation load, relatively low (300gf), the depths produced were located both in areas with a high contain of WC and in the areas of matrix existence of Ni, which led to large variations of HV microhardness. Figure 3 is represented the effect of the C_2H_2 flow rate on the microhardness of the coating layer. Figure 3 also shows that as the C_2H_2 flow rate increased, the microhardness HV also increased. Diamax powders produced the hardest coatings with a hardness of 509 HV0.3 when a flow rate of 24 l/min of C_2H_2 was used. This is to be expected because the hardness of the layer should increase when the particles hit the molten substrate and spread evenly therein. The microdurality values recorded in areas with a high Ni content were relatively low, around 112 $HV_{0,3}$, being lower than in the areas rich in WC. The microhardness values recorded in areas with a high Ni content were relatively low, around 112 $HV_{0,3}$, being lower than in the areas rich in WC. In addition, the high porosity of the Ni matrix, deposited by flame spray, allows an easy fixation of the hard particles (WC), due to the decrease of the resistance at deformation of the previous formed layer. This phenomenon was also observed in other studies involving WC, [20]. Although microhardness is not the only indicator of wear resistance of a coating, it is the simplest way to characterize it.

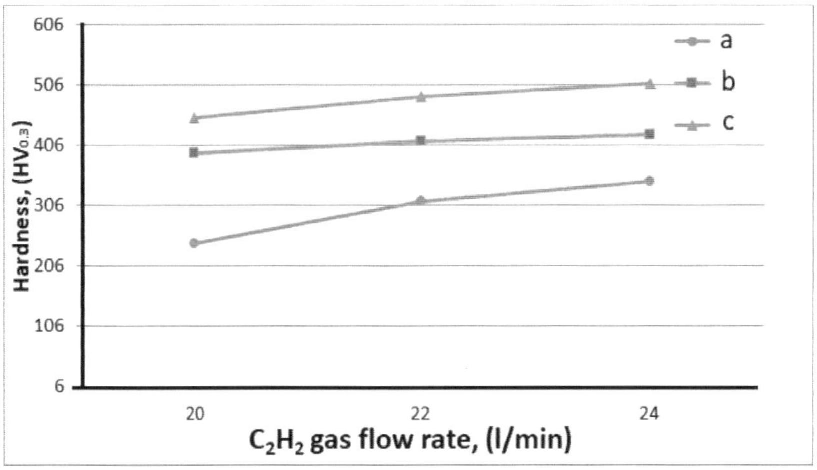

Figure 3. The variation of the micro hardness of Diamax 10999 Castolin coatings obtained using different C_2H_2 gas flow rate and spraying angles: $\alpha = 60^0$ (a); $\alpha = 75^0$ (b); and $\alpha = 90^0$ (c).

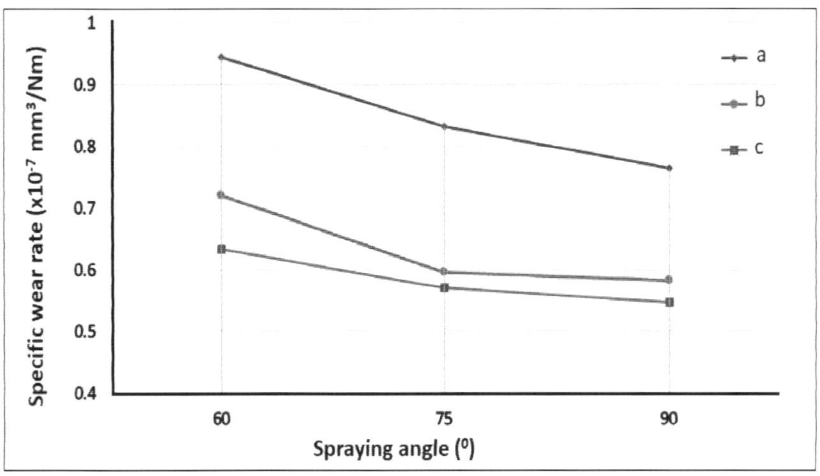

Figure 4. Specific wear rate of Diamax 10999 Castolin coatings deposited at the different spray angles and C2H2 gas flow rate: 20 l/min (a); 22 l/min (b) and 24 l/min (c).

Figure 4 represents the wear velocity, specific to the coatings obtained at different spray angles. All the layers obtained have a relatively low wear rate (below 10^{-7} mm^3/Nm), although a slight decrease in the wear rate is observed, with about 12% by increasing the spraying angle. These conclusions have been obtained also by other researchers and can be explained both by increasing the deposition porosity, [21] and by the low deposition rate, [22] - recorded at low angle of inclination.

The C_2H_2 flow also influences significantly the adherence of the Diamax 10999 deposits, as shown in Figure 5. It was found that the values of the stresses R_H increase by more than 12%, with the increase of the compressed air flow from 20l/min to 24l/min. This can be explained by the fact that the increasing flow results in the increase of the jet temperature consequently, of the sprayed particles temperature, which contributes to the increasing number of particles attached to the substrate. It can be stated that the increase of sprayed particles velocity and temperature by the increase of jet temperature was achieved - appearance also encountered by other researchers, [4]. Also, it can be observed that with the decrease of the spray angle, the adhesion of the deposition decreases. Thus, breaking strength - R_H, decreases by over 18% when decreasing the spray angle from 90^0 to 60^0 .This aspect can be explained by the fact that by the inclination of the spray gun, the spray jet degrades the surface previously created by overheating.

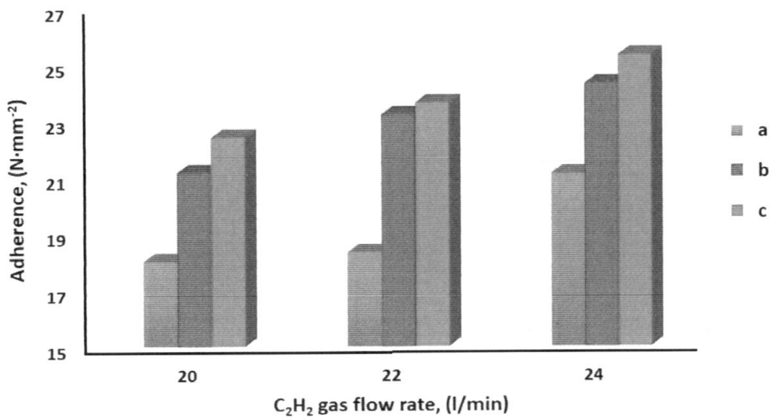

Figure 5. The variation of the adherence of Diamax 10999 Castolin coatings obtained using different C2H2 gas flow rate and spraying angles: α = 600 (a); α = 750 (b); and α = 900 (c).

4. Conclusions

The Diamax depositions, made by oxy-acetylene flame spraying, using a cold spray installation, at low pressure and various process parameters, were investigated to identify the optimal process parameters needed to obtain wear adherent layers and resistant. Thus, it was found that:

- By increasing the flow rate of C_2H_2 homogeneous layers are obtained - Figure 1, the quantity of pores in the layer decreases - Table 2, increases the adherence of the coating to the substrate - figure 5.
- The X-ray analysis showed that by increasing the flow rate of C_2H_2 in the layer, a larger quantity of WC is fixed, thus recording a significantly growth of the depositions microhardness - Figure 2
- The decrease of the spraying angle caused an increase of the wear rate by approximately 12% - v. Figure 4 and the decrease by more than 17% of the adherence of the Diamax deposits.

References

[1] Toma S T, Bejinariu C, Baciu R, Radu S 2013 The effect of fron. nozzle geom. and gas press.on the steel coat. prop. obtain. by wire arc spraying *Surf & Coat. Technol.* **220** 266–270

[2] L. Pawlowski 2008 *in: 2nd ed., The Science and Engineering of Thermal Spray Coatings, Ed. Wiley, Chichester, England*

[3] V. Nedeff, E. Mosnegutu, M. Panainte, M. Ristea, St. Toma, M. Agop 2012 Dynamics in the boundary layer of a flat particle *Powder Technology* **221** 312–317

[4] Toma S T 2013 The infl. of jet gas temp on the characteristics of steel coating obtained by wire arc spraying *Surf & Coat. Technol.* **220** 261–265

[5] Guilemany J M, Paco J M, Nutting J, Miguel J R 1999 *Metall. Mater. Trans. A* **30** 1913

[6] Irimiciuc S A, Agop M, Nica P, Gurlui S, Mihaileanu D, Toma S and Focsa C 2014 Dispersive effects in laser ablation plasmas *Jpn. J. Appl. Phys* **53**

[7] Tillmann W, Abdulgader M, Anjami N and Hagen L 2015 Study. the effect of the air-cap config. in twin-wire arc-spraying process on the obtained flowcharac. usi design of experiment oriented fluid simulation *J. Therm. Spray Technol.* **24(1-2)** 46-54

[8] Jean-Baptiste D, Hanlin L, Christian C and Jean-Marie M 2015 Infl.of gas flow param.s and nozzle design on sec. atomiz. in a rotating twin-wire arc spray system *Proceedings of the International Thermal Spray Conference* **1** 486-492

[9] König, J, Lahres M and Methner O 2014 Quality Designed Twin Wire Arc Spraying of Aluminum Bores *Journal of Thermal Spray Technology* **24** (1-2) 63-74

[10] Matz M M and Aumiller M 2014 Prac. Compar. of Cylin. Nozzle and DeLaval Nozzle for Wire Arc Spraying *Journ. of Thermal Spray Technology* **23** (8) 1470-1477.

[11] Cimpoeşu N, Trincă L C, Dascălu G, Stanciu S, Gurlui S O and Mareci D 2016 Electrochem. charact. of a new biodegradable femnsi alloy coated with hydroxyapatite-zirconia by pld technique journal of chemistry, Article ID 9520972 http://dx.doi.org/10.1155/2016/9520972

[12] Berger L M, Saaro S, Naumann T, Wiener M, Weihnacht V, Thiele S and Suchánek J 2008 *Surf. Coat. Technol.* **202** 4417

[13] Celik E, Culha O, Uyulgan B, Azem N F, Ozdemir I and Turk A 2006 *Surf. Coat. Technol.* **200** 4320

[14] Verdon C, Karimi A and Martin J L 1998 *Mater.Sci. Eng. A* **246** 11

[15] Stewart D A, Shipway P H, McCartney D G 1999 *Wear* **225–229** 789

[16] Izquierdo J, Bolat G, Cimpoesu N, Trinca LC, Mareci D and Souto R S 2016 Electrochemical charact. of pulsed layer deposited hydr.-zirc. layers on Ti-21Nb-15Ta-6Zr alloy for biomedical application *Applied Surface Science* **385** 368-378

[17] Du P C, Zhu X P, Meng Y, Feng H, Wang Q F, and Lei M K 2017 Water-lubric. trib. behavior of WC-Ni coat. deposited by off-angle HVOF spraying *Surface & Coatings Technology* **309** 663–670

[18] ASTM E384, 2011 Standard Test Method for Knoop and Vickers Hardness of Materials *West Conshohocken* USA

[19] Melendez N M and McDonald A G 2013 Development of WC-based metal matrix composite coatings using low-pressure cold gas dynamic spraying *Surface & Coatings Technology* **214** 101–109

[20] Jafari M, Enayati M H, Salehi M, Nahvi S M and Park C G 2013 Microstruct. and mechan. characterizations of a novel HVOF-sprayed WC-Co coating deposited from electroless Ni-P coated WC-12Co powders *Mater. Sci. Eng. A* **578** 46–53

[21] Engqvist H G, Botton A, Ederyd S, Phaneuf M, Fondelius J and Axén N 2000 Wear phenomena on WC-based face seal rings *Int. J. Refract. Met. Hard Mater.* **18** 39–46

[22] Schneider D S and Stephens L S 2006 An experimental study on the impact of interface temperature on thermally induced wear transitions in dry sliding *J. Tribol.* **128** 460–468

[23] *** 2016 Eutectic Castolin, Material safety data sheet

IOP Conf. Series: Materials Science and Engineering **209** (2017) 012073 doi:10.1088/1757-899X/209/1/012073

Associated Polymers, Solvents and Doping Agents to Make Polyaniline Electrospinnable

A Bertea[1*], L R Manea[1], A Bertea[1] and L Hristian[1]
[1]"Gheorghe Asachi" Technical University Iasi, Faculty of Textile and Leather Engineering and Industrial Management, 29 Dimitrie

E-mail: eurotexmed@yahoo.com

Abstract. Polyaniline (PANI) is a conductive polymer that has both metal (electrical, electronic, optical and magnetic properties) and polymer characteristics (low density, low-cost and resistance to chemicals). Polyaniline becomes a conductor by treatment with a dopant that acts by extracting electrons (oxidation) or by inserting electrons (reduction). The reduced solubility of PANI in all common solvents restricts its capacity to be electrospun into uniform fibers. The present paper reviews the methods to increase the solubility of PANI by blending it with other polymers and doping it with organic acids, highlighting the best polymer/solvent couples and doping agents.

1. Introduction

Conducting polymers, which have been discovered in 1976 [1], have many useful properties such as mechanical flexibility, ease of processing and, above all, adjustable electrical conductivity [2]. Amongst these polymers, polyaniline (PANI) has been studied for many applications [3] such as chemical sensing [4-6], logic circuit components [7], electromagnetic shielding [8], and anticorrosion coating [9] due to its uncomplicated synthesis, room temperature process, high conductivity and relative environment stability [10]. Polyaniline has the general polymeric structure shown in fig. 1. Polyaniline contrasts from the other inherently conducting polymers, such as polypyrroles and polythiophenes, in that it has three readily accessible oxidation states, which range from the fully reduced leucoemeraldine state (y = 1) to the half oxidized emeraldine (y = 0.5) form to the fully oxidized pernigraniline state (y = 0) (fig. 1). Polyaniline shows the highest conductivity in the emeraldine state [11].

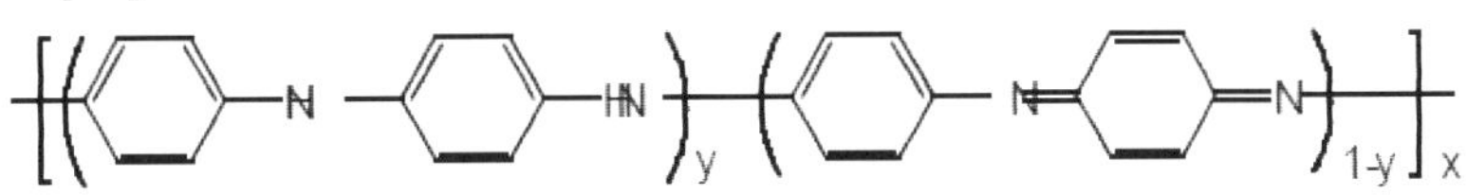

Figure 1. y = 1 (leucoemeraldine) 0.5 (emeraldine) and 0 (pernigraniline).

The electrical properties of polyaniline can be easily adjusted from insulating to conducting range by a doping/de-doping procedure, when the electrical properties are changed by the partial oxidation or reduction of the polymer chain [12]. The electrical conductivity can be increased by many orders of magnitude from 10^{-10}–10^{-5} to 10^{2}–10^{5} $S\times cm^{-1}$ upon doping, which covers the whole insulator-semiconductor-metal range [13].

2. Alternatives to obtain the electrospinning solution/dispersion (associated polymers and solvents)

Despite the great potential of PANI, its processing remains a difficult problem to handle, due to its insolubility in ordinary organic solvents. Because of this, PAN is hardly processed alone by electrospinning, and various other polymers are mixed with PANI in the spinning solution [14-18]. Approximately 1% solubility was observed in N-methyl, 2-pyrrolidone (NMP), which also acts as a plasticizer [19-23]. It was also found that 2-hydroxyethyl ammonium formate is a versatile solvent as reaction medium in preparing organo-soluble polyaniline with reasonable molecular weights by oxidative coupling polymerization. Low viscosity and superficial tension of PANI.HCSA (polyaniline doped with camphorsulfonic acid) in a chloroform solution prevents electrospinning, which becomes possible only by the addition of another polymer such as poly (ethylene oxide) (PEO) or poly (methyl methacrylate) (PMMA) to the chloroform solution [24-27]. The use of PEO leads to superior uniformity of the spinning solution, while when using PMMA phase separation phenomena occur. In addition, PANI / PEO nanofibers are more conductive [28]. Mixtures of polyaniline with polyethylene oxide (PEO - Mw 6000,000) and polymethyl methacrylate (PMMA - Mw 996,000) [8, 28] were also tested. Chloroform was used as solvent [29], and the undissolved part has been removed by filtration. The used solutions were 1% wt% PANI.HCSA / 1 wt% PEO or 1% wt% PANI.HCSA / 5 wt% PMMA, both solutions lacking phase separation tendency. At these concentrations the electrospinning polymer jet was uniform and stable. A study on the use of the electrospinning technique for PANI-poly (styrene sulfonic acid) (SPS), poly (vinyl butinol) (PVB) and polyethylene oxide (PEO) blends shows that the introduction of polyethylene oxide into the electrospinning solution alters the hydrophilicity of the obtained nanofibers, enhancing the response to humidity, improves the sensitivity and decreases hysteresis. All nanofiber sensors obtained this way show an impedance variation of about three orders of magnitude in the 20-90% RH range with high sensitivity. For example, the nanofiber response from PANI-SPAS / PEO / PVB (mass ratio 20/3/14) shows a rapid response, with a response time of $t_{90\%}$ for adsorption and desorption processes of 8 s and 6 s respectively [30]. Nanofibers from mixtures of PANI with poly (styrene sulfonic acid) solved in N, N-dimethyl formamide exhibited a highly porous structure, and their impedance varied rapidly and reversibly in an exponential fashion with the amount of adsorbed water [31]. Poly (3-hydroxybutyrate) (PHB) - PANI mixtures were electrospun using chloroform and 2,2,2-trifluoroethanol (TFE) as solvents. In the case of chloroform, the nanofibers produced were resistant but had a heterogeneous surface, with a wide scattering of diameters and spongy surface structure; the use of TFE leads to a significant improvement in the properties of nanofibers, which became smooth, uniform and thin, but with a lesser mechanical strength. An interesting effect is the improvement of PANI conformation, from the coil-like shape to a shape much closer to linearity, resulting both from the use of TFE and from the interaction between PHB and PANI. Changing the solvent also leads to increased fiber hydrophilicity as well as improved thermal resistance [32]. A polylactic acid (PLA) and PANI mixture with low polylactic acid content (about 1%) demonstrated solubility in chloroform and was electrospun to give nanofibers with diameters in the range of 10 nm - 300 nm [33].

3. Doping agents

Doping and returning to the de-doped form plays a key role in the sensing mechanism of conductive polymers [34]. By controlling the pH of the dopant acid solution, the doping agents can be added in any desired amount until all imine nitrogen atoms (half of the total) are doped. Moreover, dopants can be removed by the interaction of the emeraldine salt form with common bases, such as ammonium hydroxide. The ability to switch between conductive and insulating forms makes PANI a polymer sensitive to acidic / basic and reducing / oxidizing compounds such as ammonia (NH_3), nitrogen dioxide (NO_2), hydrogen (H_2), some volatile organic compounds [35-38]. Relatively numerous doping agents have been tested in the production of polyaniline with applications in the field of sensors. A study of the potential of some of the most common mineral acids as polyaniline doping agents has highlighted the fact that perchloric acid has the highest effectiveness, followed by sulfuric acid, nitric acid, hydrochloric acid and hydroiodic acid, as it can be seen in fig. 2 [39].

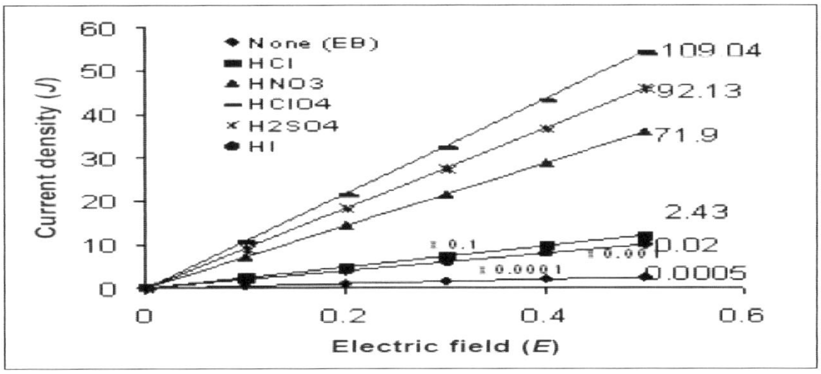

Figure 2. Scaled plots of current density versus electric field for emeraldine salt with different dopants showing the slopes as the conductivity in S/cm.

A relatively frequently used doping agent is 10-camphor sulfonic acid (HCSA), an organic acid that confers a certain solubility and high conductivity to polyaniline [8, 11, 28, 38, 40]. The doping procedure involves grinding 1 g of PANI base, 1.32 g of HCSA and 30 mL of m-cresol in a homogenizer and shaking at an initial temperature of 50 ° C for about 45 minutes followed by drying for 24 hours to obtain a fine powder [38-40]. In another procedure, camphorsulfonic acid is first fully dissolved in chloroform, and then an adequate amount of polyaniline base is dispersed in the solution. This is sonicated for 5 min, kept under magnet stirring for 6 to 8 hours and filtered before mixing with polyethylene oxide [41]. A schematic representation of HCSA-doped polyaniline is shown in fig. 3 [42].

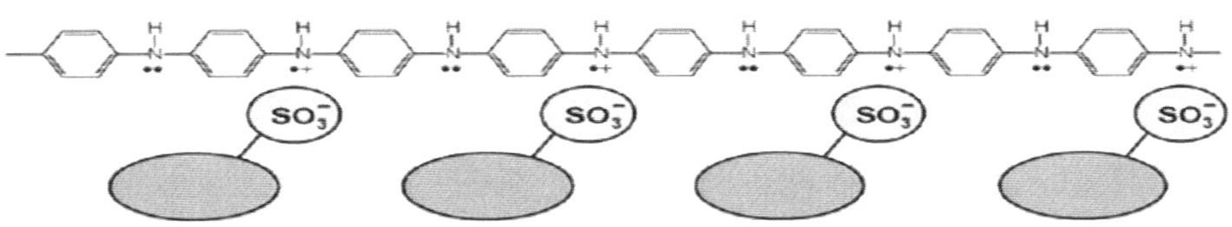

Figure 3. HCSA doped polyaniline.

The way HCSA doping affects the conductive characteristics of polyaniline is illustrated in 4. Also HCSA was used to dope polyaniline that was mixed with poly (ethylene oxide) in the proportions of 1/10 to 1/2, observing an increase in the conductivity when the PANI content was increasing [43]. Polyaniline doped with HCSA was dissolved in a mixture of chloroform / ethanol (9: 1 by volume) and the resulting solution was mixed with poly (vinyl pyrrolidone) (PVP) (1.5 PVP / PAN) and polyethylene oxide (PEO) (0.62 PEO / PANI), the two polymer solutions being then electrospun to obtain sensors. It was found that PANI-PEO was the most porous layer with good electrical performance and high sensitivity to ammonia, but fragile when subjected to environmental stress such as high humidity and temperature, while PANI-PVP proved to be a very good sensor for NO_2, but it was strongly influenced by humidity even at low values [44].

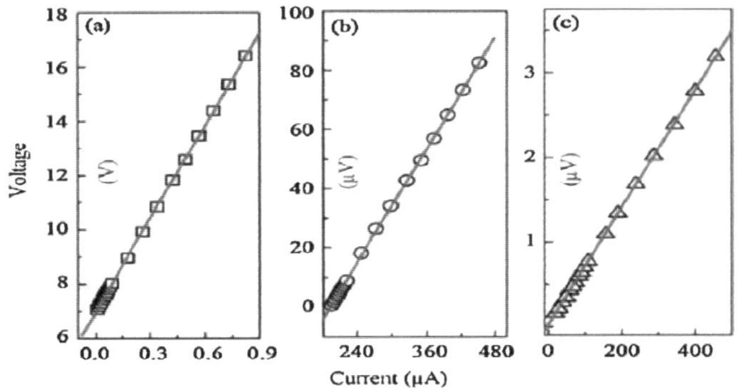

Figure 4. Current–voltage characteristics of (a) PANI base, (b) PANI (HCl doped) and (c) PANI (HCSA doped) [5].

Dodecylbenzene sulfonic acid (DBSA) appeared as a promising dopant and surfactant and is frequently applied in various protocols for soluble PANI synthesis [45]. A few examples are polymerization in solution [46] emulsification / reverse emulsion polymerization [47] and re-doping of emeraldine base (EB-PANI) [48]. The solubility of PANI doped with dodecylbenzene sulfonic acid (DBSA) or camphor sulfonic acid in chloroform and xylene, has been reported. The long chain of DBSA is supposed to confer solubility to the synthesized polymer in ordinary organic solvents [47]. To study the role of DBSA as dopant, blends of vinylidene polyfluoride (PVDF) / emeraldine base (EB) and PVDF / PANI doped with DBSA, respectively, were electrospun. It has been noted that both emeraldine base and emeraldine salt affect the solution properties and have a significant influence on the process, the morphology and diameter of the fiber. Electrospinning of the solution containing doped PANI is easier to perform compared to a solution containing PANI salt because both the viscosity and the ionic conductivity of the solution are significantly modified by the addition of doped PANI. In order to obtain uniform fibers, the applied tension should be adjusted when PANI salt is used to obtain small diameter nanofibers [49]. A comparative study of three dopants: hydrochloric acid, sulfuric acid and formic acid highlighted the superior efficacy of the latter, providing the highest conductivity [50]. Doped polyaniline was obtained by aniline oxidation polymerization using ammonium peroxide sulphate $(NH_4)_2S_2O_8)$ as dopant and the obtained polymer was electrospun after mixing with polyvinyl alcohol and polyvinyl alcohol - $AgNO_3$. The electrospinning of 0.036 M PANI solution in N-methyl-2-pyrrolidone resulted in the formation of micro to nanoparticles, while electrospinning a PANI / PVA solution altered the electrospinning behavior and resulted in the formation of fine composite electrospun fibers. When the composition is further modified by the addition of $AgNO_3$, three-component electrospun composite fibers are obtained [51]. Enzymes, such as urease, were also used as dopants for PANI [52-58]. The urease is fixed in the PANI-PVP matrix, which results in an increase in the sensitivity of the nanofibers to ammonia.

4. Conclusions

The paper summarizes the types of polymer that a currently associated with PANI to improve its processability, the concept of doping and the types of dopants that are most frequently used. Various organic/inorganic acids are analyzed as effective dopants to improve the conducting properties of polyaniline, proving that the electronic and optical properties of polyaniline can be simply controlled and tailored by the oxidizing/reducing agents and acid/base doping added during the polymerization process. The nature of the solvent used to make PANI spinnable depends decisively by the nature of the dopant and that of the associated polymer that are used, and some of the most efficient solvents are m-cresol and xylene which solve PANI doped with dodecylbenzene sulfonic acid or camphor sulfonic acid.

References

[1] Olejnik R, Babar D G, Slobodian P and Matyas J 2015 *Mater. Sci. and Eng.* **108** 1
[2] Lu X, Wang C and Wei Y 2009 *Small* **5** 2349
[3] Zhan C, Yu G, Lu Y, Wang L Wujcik E and Wei S 2017 *J. Mater. Chem. C.* **7** 1569
[4] Zhang H D, Tang C-C, Long Y-Z, Zhang J-C, Huang R, Lib J-J and Gub C-Z 2014 *Sensor Actuat A-PHYS* **219** 123
[5] Fratoddi I, Vendittia I, Cametti C and Russo M V 2015 *Sensor Actuat. B-Chemical* **220** 534
[6] Gremlica M, Olejnik R, Babar D G, Slobodian P 2014 Nanocon Proceedings Available online: http://nanocon2014.tanger.cz/files/proceedings/20/reports/3131.pdf (Accessed on: 14 March 2017)
[7] Tseng R J, Huang J, Ouyang J, Kaner R B and Yang Y 2005 *Nano. Lett.* **5** 1077
[8] Chen D, Lei S and Chen Y, 2011 *Sensors* **11** 6509
[9] Gvozdenović M, Jugović B, Jambrec D, Stevanović J and Grgur B 2012 *Zaštita Materijala* **53** 353
[10] Long Y Z, Chen Z J, Shen J Y, Zhang Z M, Zhang L J, Huang K, Wan M X, Jin A Z, Gu C Z and Duvail J L 2006 *Nanotechnology* **17** 903
[11] Wallace G G, Spinks G M and Teasdale P R 2003 *Conductive electroactive polymers: intelligent materials systems CRC Press* Boca Raton Florida
[12] Ameen S, Akhtar M S, Kim Y S and Shin H-S 2011 *Fabrication, Doping and Characterization of Polyaniline and Metal Oxides: Dye Sensitized Solar Cells*, in L. Kosyachenko (Ed.) *Solar Cells - Dye-Sensitized Devices* Intech Rijeka Croatia p 95
[13] Abd Razak S I, Wahab I F, Fadil F, Dahli F N, Khudzari A Z and Adeli H 2015 *AMSE* **2015** 1
[14] Manea L R, Bertea A, Nechita E, Popescu C V and Sandu I 2016 *Rev. Chim. (Bucharest)* **67** (7) 1284
[15] Manea L R, Chirita M, Hristian L and Sandu I 2016 *Mat. Plast.* **53** (3) 361
[16] Manea L R, Danu M C and Sandu I 2015 *Rev. Chim. (Bucharest)* **66** (6) 868
[17] Manea L R, Cramariuc C, Caunii V and Sandu I 2015 *Mat. Plast.* **52** (1) 82
[18] Manea L R, Cramariuc C, Scarlet R, Cramariuc R, Sandu I and Popescu V 2015 *Mat. Plast.* **52** (2) 180
[19] Bıcak N, Senkal B F and Sezer E 2005 *Synthetic Met.* **155** 105
[20] Manea L R and Sandu I 2015 *Rev. Chim. (Bucharest)* **66** (12) 1968
[21] Manea L R, Scarlet R, Amariei N, Nechita E and Sandu I G 2015 *Rev. Chim. (Bucharest)* **66** (4) 542
[22] Manea L R, Scarlet R, Leon A L and Sandu I 2015 *Rev. Chim. (Bucharest)* **66** (5) 640
[23] Manea L R, Nechita E and Sandu I 2015 *Rev. Chim. (Bucharest)* **66** (11) 1841
[24] Calin M A, Khenoussi N, Schacher L, Adolphe D, Manea L R, Gradinaru I, Zetu I and Stratulat S 2013 *Mat. Plast.* **50** (4) 257
[25] Manea L R, Hristian L, Ostafe M M, Apostol L L and Sandu I 2016 *Rev. Chim. (Bucharest)* **67** (9) 1758
[26] Hristian L, Ostafe M M, Manea L R, Sandu I G, Apostol L L and Sandu I 2016 *Rev. Chim. (Bucharest)* **67** (9) 1717
[27] Manea L R, Hristian L, Leon A L and Popa A 2016 *IOP Conf. Series: Mater. Sci. Eng.* **145** 032007
[28] Attout A, Yunus S and Bertrand P 2008 *Polym. Eng. Sci.* **48** 1661
[29] Norris I D, Shaker M M, Ko F K and MacDiarmid A G 2000 *Synthetic Met.* **114** 109
[30] Lin Q, Li Y and Yang M 2012 *Sensors and Actuat. B-Chemical* **161** 967
[31] Lin Q, Li Y and Yang M 2012 *Sensors and Actuat. B-Chemical* **171– 172** 309
[32] Fryczkowski R, Gorczowska M, Fryczkowska B, Janicki J 2013 *Synthetic Met.* **166** 14
[33] Serrano W, Melendez A, Ramos I and Pinto N J 2014 *Polymer* **55** 5727
[34] Bai H and Shi G 2007 *Sensors* **7** 267
[35] Virji S, Kojima R, Fowler J D, Villanueva J G, Kaner R B and Weiller B H 2009 *Nano Res.* **2**

135
[36] Yan X B, Han Z J, Yang Y and Tay B K 2007 *Sensors and Actuat. B. Chemical* **123** 107
[37] Zhang T, Mubeen S, Yoo B, Myung N V and Deshusses M A *Nanotechnology* **20** 1
[38] Babu V J, Vempati S, Ramakrishna S 2013 *Mat. Sci. and App.* **4** 1
[39] Catedral M D, Tapia A K G, Sarmago R V, Tamayo J P and del Rosario E J 2010 *Science Diliman* **16** 41
[40] Gustavsson J M, Innis P C, Heb J, Wallace G G and Tallman D E 2009 *Electrochim. Acta* **54** 1483
[41] Moutsatsou P, Coopman K, Smith M B and Georgiadou S 2015 *Polymer* **77** 143
[42] Pinto N J, Ramosa I, Rojas R, Wang P C, Johnson A T 2008 *Sensors and Actuat. B-Chemical* **129** 621
[43] Li C, Chartuprayoon N, Bosze W, Low K, Lee K H, Nam J and Myung N V 2014 *Electroanalysis* **26** 711
[44] Macagnano A, Zampetti E, Pantalei S, De Cesare F, Bearzotti A and Persaud K C 2011 *Thin Solid Films* **520** 978
[45] Bhadra S, Khastgir D, Singha N K and Lee J H 2009 *Prog. Polym. Sci.* **34** 783
[46] Barra M E, Leyva M M and Gorelova B G 2001 *J. Appl. Polym. Sci.* **80** 556
[47] Bilal S, Gul S and Ali Shah K A A 2012 *Synth. Met.* **162** 2259
[48] Titelman G I, Siegmann A, Narkis M and Wei Y 1998 *J. Appl. Polym. Sci.* **69** 2205
[49] Merlini C, Pegoretti A, Medeiros Araujo T, Ramoa S D A S, Schreiner W H and de Oliveira Barra G M 2016 *Synth. Met.* **213** 34
[50] Melad O and Jarur M 2016 *Chemistry & Chem Tech.* **10** 41
[51] Shahi M, Moghimi A, Naderizadeh B and Maddahb B 2011 *Scientia Iranica C* **18** 1327
[52] Bishop A and Gouma P 2005 *Rev. Adv. Mater. Sci.* **10** 209
[53] Hristian L, Sandu A V, Manea L R, Tulbure E A and Earar K 2015 *Rev. Chim. (Bucharest)* **66** (3) 342
[54] Hristian L, Bordeianu D L, Iurea P, Sandu I and Earar K 2014 *Mat. Plast.* **51** (4) 405
[55] Manea L R, Bertea A, Nechita E, Popescu C V and Sandu I 2016 *Rev. Chim. (Bucharest)* **67** (8) 1607
[56] Popa A, Bucevschi A, Pustianu M, Manea L R and Sandu I 2016 *Mat. Plast.* **53** (2) 316
[57] Hristian L, Chirita M, Manea L R and Sandu I 2016 *Mat. Plast.* **53** (4) 739
[58] Stanescu I, Manea L R, Bertea A, Bertea A P and Sandu I C A 2016 *Rev. Chim. (Bucharest)* **67** (10) 2082

Acknowledgments
This work was supported by a grant of the Romanian national Authority for Scientific Research and innovation, UEFISCDI, project number PN III 46 BG 2016 within PNCDI III

The Influence of Process Parameters on the Characteristics of Electrospun 3D Nanostructures

A Bertea[1*], L R Manea[1], A Popa[2] and A Bertea[1]
[1]"Gheorghe Asachi" Technical University of Iasi- Romania, Faculty of Textiles, Leather and Industrial Management;
[2]"Aurel Vlaicu" University of Arad, Romania, Engineering Faculty, Textile Department

E-mail: eurotexmed@yahoo.com

Abstract. Electrospinning is a fast developing technique that employs electrostatic repulsive forces to produce ultrafine fibres with application in fields like environment protection, medicine, sensors and many others. The characteristics of the polymer jet and the properties of the electrospun nanofibres are highly influenced by technological and environmental parameters. This paper offers a report on the main processing parameters that may influence the characteristics of the obtained nanofibres. The influence of flow rate, spinneret to collector distance and applied voltage on maximum fibre length, average fibre diameter, diameter uniformity and nanofibre quality is reviewed.

1. Introduction

Electrospinning is an economical and uncomplicated method to produce polymer fibres at low basis weight and at a small fibre diameter and pore size [1-5]. Electrospinning is different from traditional wet/dry fibre spinning; if in the traditional methods mechanical pulling forces lead to fibres in the micrometre range, in electrospinning electrical pulling forces enable the production of nanofibres. A representation a typical electrospinning setup is shown in figure 1.

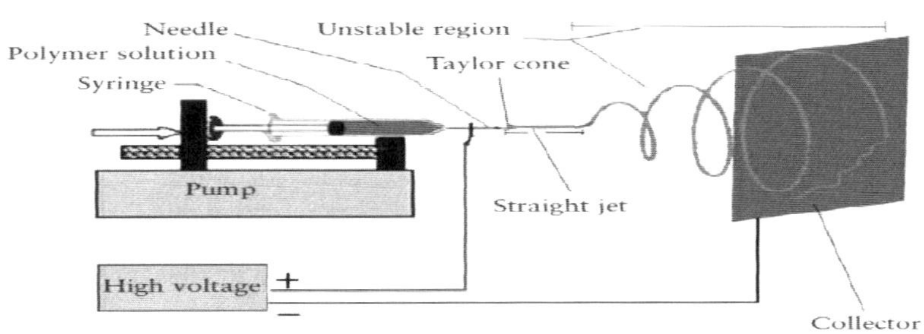

Figure 1. Schematic electrospinning setup [1].

High voltage is linked to the end of a capillary that contains the polymer solution, forcing the extinction of the hemispherical surface of the solution from the end of the capillary and creating a Taylor cone. When the electric field is increased, the repulsive electrostatic force overwhelms the surface tension and the charged strand emerges out of the end of the Taylor cone [6-9]. The polymer jet elongates and the solvent evaporates, forming fibres that are deposited on the collector (figure 2).

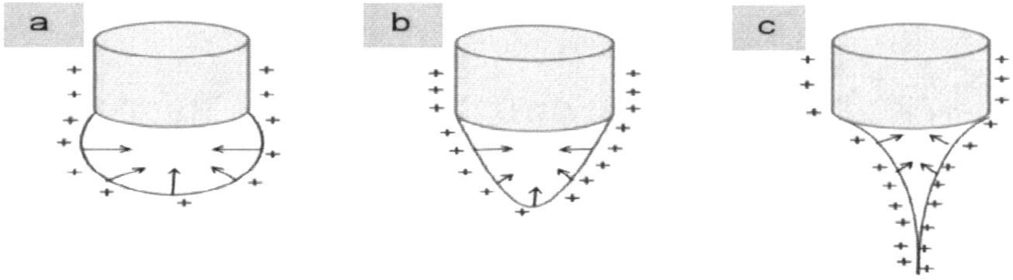

Figure 2. The mechanism of the effect of charges on the polymeric droplets [10].

The process can be tuned by controlling three main groups of parameters: solution properties, processing conditions and ambient conditions. Process variables, such as fluid flow rate, spinneret to collector distance and electric field strength have a significant influence on the properties of the obtained nanofibres [11-14].

2. Fluid flow rate

Flow rate is considered to be as one of the main parameters that control the diameter of the fibre and its distribution, the initial shape of the droplet, the trajectory of the jet, the persistence of Taylor cone and deposition area [15-18]. The polymer solution flow rate within the syringe is an important process parameter; generally a lower polymer solution flow rate is recommended to have enough time for polarization [19-22]. If the feed rate is very high, exceeding a critical value, which varies with the polymer system, the delivery rate of the polymer solution jet to the capillary tip surpasses the rate at which the solution is detached from the tip by electric forces [1], larger droplets are formed, and polymer fibres with a high density of beads are formed, because of the reduced solidification time of the polymer filament before reaching the collector and of low tension forces.

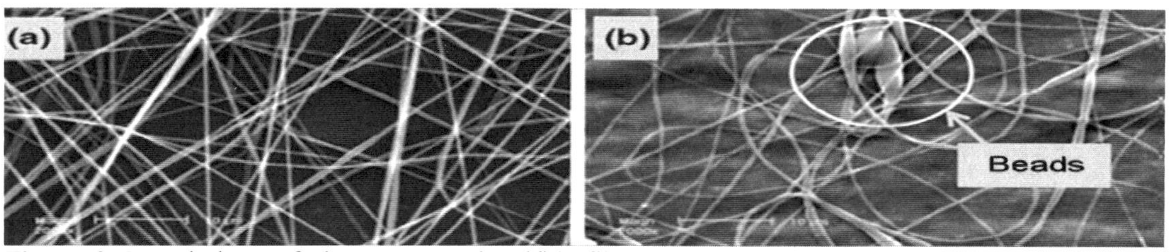

Figure 3. Morphology of electrospun polysulfide fibres at different flow rates – a) 40 mL/h; b) 66 mL/h [23].

The way the shape of the polymer cone changes with increasing flow rate is shown in figure 4. In a study on the influence of flow rate on the polystyrene fibre morphology and size it was observed that bead formation occurs when the flow rate is higher than 0.1 mL/min, and the pore size and diameter of fibres increases [26]. An investigation on the effect of the flow rate on morphologies of the polysulfone fibres from 20% polysulfone/ N,N-dimethylacetamide solution at 10 kV showed that bead fibres with thicker diameters were obtained by the flow rate of 0.66 mL/h [27].

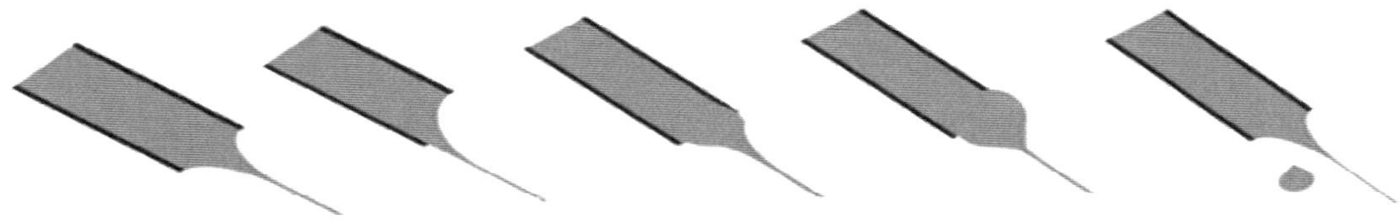

Figure 4. Formation of various jets with increasing flow rate [24, 25].

Similar results have been reported for nylon 6, where bead fibres formed when the feeding rate exceeded 4 mL/min [28]. To increase the productivity of electrospinning, the feed rate can be increased as the applied stress increases, which usually leads to smaller diameter of the nanofibres [29], but when the flow rate is too high, as larger volume of solution is drawn from the needle tip, which needs a longer time to dry, there is the possibility that fibres merge together and make webs instead of fibres [15].

Furthermore, when the flow rate increases, fibre diameter distribution becomes wider. Since increases and decreases in the polymer solution flow rate influence the nanofibre formation and diameter, it is preferable to maintain a minimum flow rate which ensure a balance between the leaving polymeric solution and its replacement with a new one during jet formation [10].

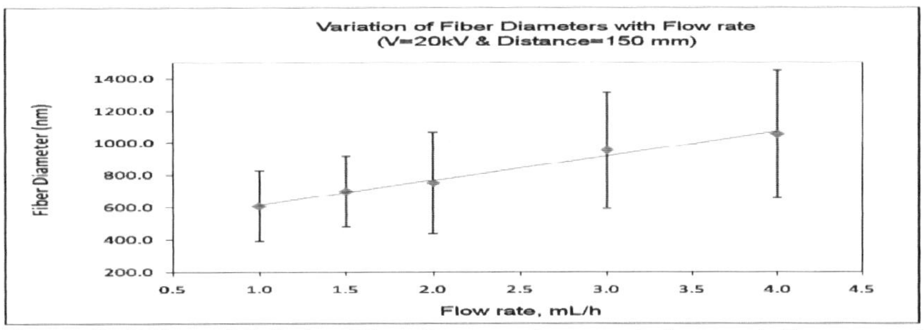

Figure 5. Effect of Solution Flow Rate on the morphology and diameters of nanofibres - Variation of average fibre diameters with flow rate [30].

Choosing an appropriate flow rate leads to a limitation of the formation of any defects such as blobs, splitting and branched fibres. In addition, a stable and constant flow rate is required [31]

3. Spinneret to collector distance

By proper setting of the distance between the metallic needle tip and the collector it can be controlled the morphology and the diameter of the electrospun fibres. If the distance is too small, the fibre will not have enough time to solidify before it reaches the collector, while if the distance is too high, it is possible to obtain fibre with droplets at the surface [19,32]. The solvent removal capacity being an essential factor in obtaining high quality fibres, it is advisable to carefully set the distance to the collector. It has been shown that a small change in this distance significantly influences the characteristics of the fibres. Furthermore, as the distance between the charged nozzle and the collector increases, the level of the electric field between the two decreases, forming fewer charged ions, allowing bending instabilities and whipping action to elongate and decreases the diameter of the polymer jet [31, 33]. Most studies concluded that defective and large-diameter nanofibres are obtained when the spinneret to collector distance is small, while the diameter of the nanofibre is reduced as the distance was increased [34]. Electrospun poly (vinyledene fluoride) nanofibres obtained from a 28 wt% solution at 12 kV showed a decrease in diameter from 397 nm to 314 nm when the distance between the nozzle and the collector increased from 15 cm to 16 cm (figure 6), as well as improved uniformity [35]. A study of the influence of the spinneret to collector distance D on the diameter of the fibres electrospun from 12% polyetherimide solution (PEI), using as solvents a mixture of dimethylacetamide /tetrahydrofuran 1:1 came to the conclusion that the smallest values of the diameter are obtained for a distance of 45 mm [36].

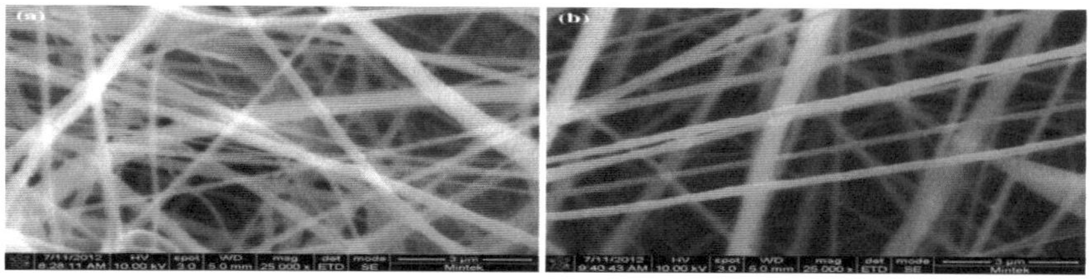

Figure 6. SEM micrographs of poly(vinyledene fluoride) fibres at spinning distance of: a. 15 cm; b. 16 cm [35].

Usually the appropriate distance between the metallic needle tip and collector varies with the polymer system, but there were cases when no effect on the morphology of the nanofibre occurred when changing the distance between the metallic needle and collector [37].

4. Applied voltage

The applied electrical potential is one of the most important parameters of the electrospinning process because it influences directly both the dynamics of the fluid flow and the morphological characteristics of the electrospun fibres, but its degree of significances varies with the polymer solution concentration and the distance between the nozzle and the collector [38]. Only when the applied voltage exceeds a threshold value, the charged polymer jets are ejected from the Taylor cone [39]. The size of the threshold voltage is dependent on the nature of the polymer and the polymer-solvent system [10]. The applied voltage is the most contradictory parameter in affecting nanofibres morphology. There are studies showing that an increase in applied voltage leads to the formation of smaller diameter nanofibres due to the extent of the polymer solution in correlation with the increased rejection forces in the polymer jet [40, 41], but also studies which have demonstrated that there is no essential influence of the electric field on the diameter of electrophilic polyethylene oxide (PEO) nanofibres [42, 43]. There are studies that highlight the fact that higher voltage facilitate the formation of high diameter fibres, such as poly (vinyl alcohol) (PVA) / water [37]. Therefore, the applied voltage influences the diameter of the fibre, but the level of significance varies depending on the nature and concentration of the polymer solution and the distance between the electrode and the collector [44]. It is generally accepted that an increase in the applied voltage leads to an increase in the deposition rate, therefore there is a greater probability of defect formation [40, 45]. The length and diameter of the electrospun fibre decrease with the increase in applied voltage without any change in pore size [46, 47]. For nanofibres obtained with low electrical voltage a uniform morphology with lesser defects and drops is obtained [48-53]. The study on a mixture of PANI-HCSA / PEO highlighted the fact that higher voltage leads to thinner nanofibres, but also a greater diversity of diameters and a wider distribution of diameters, as it can be seen in figure 7 [29].

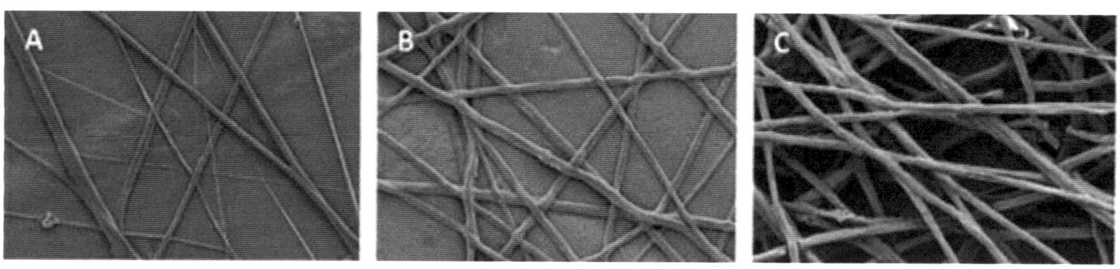

Figure 7. Applied voltage influence: a) 13.5 kV; b) 9.2 kV; c) 5 kV [29].

5. Conclusions

This paper offers an overview of the electrospinning process parameters (applied electric field strength, distance between the spinneret and collector and fluid flow rate) that influence the nanofibres production. All these parameters significantly affect the morphology of the electrospun nanofibres, and by controlling them it is possible to produce nanofibres for specific functions. The fluid flow rate influence the droplet size and its initiating shape at the capillary tip, the trajectory of the jet and the maintenance of Taylor cone, the fibre diameter distribution and the morphology of the nanofibres. The correct distance between the nozzle and collector varies with the polymer system, but most frequently a slight decrease in the average fibre diameter is noticed by the distance increases. The applied voltage has a contradictory influence on nanofibres morphology, as there are studies that found that average fibre diameters is directly proportional to the applied voltage, but many studied that reported a inverse proportionality.

6. References

[1] Haghi A and Zaycov G 2012 *Advances in Nanofibre Research 2*, Smithers Rapra Technology Ltd, Shropshire, UK
[2] Manea L R, Bertea A, Nechita E, Popescu C V and Sandu I 2016 *Rev. Chim. (Bucharest)* **67** (7) 1284
[3] Manea L R, Chirita M, Hristian L and Sandu I 2016 *Mat. Plast.* **53** (3) 361
[4] Manea L R, Danu M C and Sandu I 2015 *Rev. Chim. (Bucharest)* **66** (6) 868
[5] Manea L R and Sandu I 2015 *Rev. Chim. (Bucharest)* **66** (12) 1968
[6] Manea L R Scarlet R, Leon A L and Sandu I 2015 *Rev. Chim. (Bucharest)* **66** (5) 640
[7] Manea L R, Nechita E and Sandu I 2015 *Rev. Chim. (Bucharest)* **66** (11) 1841
[8] Manea L R, Cramariuc B, Caunii V and Sandu I 2015 *Mat. Plast. (Bucharest)* **52** (1) 82
[9] Manea L R Cramariuc B, Scarlet R, Cramariuc R, Sandu I and Popescu V 2015 *Mat. Plast. (Bucharest)* **52** (2) 180
[10] Haider A, Haider S and Kang I-K. 2015 *Arab. J. Chem.* **6** 1
[11] Calin M A, Khenoussi N, Schacher L, Adolphe D, Manea L R, Gradinaru I, Zetu I and Stratulat S 2013 *Mat. Plast. (Bucharest)* **50** (4) 257
[12] Manea L R, Hristian L, Ostafe M M, Apostol L L and Sandu I 2016 *Rev. Chim. (Bucharest)* **67** (9) 1758
[13] Hristian L, Ostafe M M, Manea L R, Sandu I G, Apostol L L and Sandu I 2016 *Rev. Chim. (Bucharest)* **67** (9) 1717
[14] Manea L R, Hristian L, Leon A L and Popa A 2016 *IOP Conf. Series: Mater. Sci.Eng.* **145** 032007
[15] Zargham S, Bazgir S, Tavakoli A, Rashidi A S and Damerchely R 2012 *J. Eng. Fiber. Fab.* **7** 42
[16] Hristian L, Sandu A V, Manea L R, Tulbure EA and Earar K 2015 *Rev. Chim. (Bucharest)* **66** (3) 342
[17] Hristian L, Bordeianu D L, Iurea P, Sandu I and Earar K 2014 Mat. Plast. (Bucharest) **51** (4) 405
[18] Manea L R, Hristian L, Leon A L and Popa A 2016 *IOP Conf. Series: Mater. Sci.Eng.* **145** 032006
[19] Afshari M *Electrospun Nanofibres* 2017 Woodhead Publishing
[20] Scarlet R, Manea L R, Sandu I, Martinova L, Cramariuc O and Sandu I G 2012 *Rev. Chim.* **63** (7) 688
[21] Scarlet R, Manea L R Sandu I, Cramariuc B and Sandu A V 2012 *Rev. Chim.* **63** (8) 777
[22] Manea L R, Scarlet R and Sandu I 2015 *Rev. Chim.* **66** (10) 1622
[23] Wang C, Zhang W, Huang Z H, Yan E Y and Su Y H 2006 *Pigment Resin Tech.* **35** (2) 278
[24] Shamim Z, Saeed B, Amir T, Abo Saied R and Rogheih D 2012 *J. Eng. Fabrics Fibers* **7** (2) 42
[25] Laudenslager M J and Sigmund W M 2012 *Electrospinning* Bhushan B (Ed.) *Encyclopedia of*

Nanotechnology Springer Publishers pp 769–775
[26] Megelski S, Stephens J S, Chase D B and Rabolt J F 2002 *Macromolecules* **35** 8456
[27] Yuan X, Zhang Y, Dong C and Sheng J 2004 *Polym. Int.* **53** 1704
[28] Chowdhury M and Stylios G 2010 *Int. J. Basic Appl. Sci.* **10** 10
[29] Moutsatsou P, Coopman K, Smith M B and Georgiadou S 2015 *Polymer* **77** 143
[30] Khan Z and Kafiah F M 2013 Conference on Membrane Science and Technology, Available online: https://www.academia.edu/16840322/Preparation_of_Polysulfone_Electrospun_Nanofibres_Effect_of_Electrospinnng_and_Solution_Parameters.pdf (Accessed on: 14 February 2017)
[31] Zafar M, Najeeb S, Khurshid Z., Vazirzadeh M, Zohai S, Najeeb B and Sefat F 2016 *Materials* **9** 73
[32] Ki C S, Baek D H, Gang K D, Lee K H, Um I C and Park Y H 2005 *Polymer* **46** 5094
[33] Yalcinkaya F, Yalcinkaya B and Jirsak O 2016 *Dependent and Independent Parameters of Needleless Electrospinning*, in S. Haider, A. Haider (Eds), *Electrospinning - Material, Techniques and Biomedical Applications* Intech Rijeka Croatia p 67
[34] Wang T and Kumar S 2006 *J. Appl. Polym. Sci.* **102** 1023
[35] Matabola K P and Moutloali R. M. 2013 *J. Mater. Sci.* **48** 547
[36] Manea L R Bertea A P, Nechita E, Popescu C V and Sandu I 2016 *Rev. Chim (Bucharest)* **67** (7) 1284
[37] Zhang C, Yuan X, Wu L, Han Y and Sheng J 2005 Eur. Polym. J. **41** 423
[38] Yördem O S, Papila M and Menceloglu Y Z. 2008 *Mater. Des.* **29** 34
[39] Li Z and Wang C 2013 *One-Dimensional Nanostructures Electrospinning Technique and Unique Nanofibres* Springer Science&Business Media London
[40] Sill T J and von Recum H. A. 2008 *Biomaterials* **29** 1989
[41] Supaphol P, Mit-Uppatham C and M. Nithitanakul. 2005 *J Polym. Sc.i B Polym. Phys.* **43** 3699
[42] Chowdhury M, Stylios G 2010 *Int. J. Basic Appl. Sci.* **10** 116
[43] Reneker D H and Chun I 1996 *Nanotechnology* **7** 216
[44] Zhou F L, Gong R H and Porat I 2009 *Polym. Eng. Sci.* **49** 2475
[45] Ramakrishna S, Fujihara K, Teo W E, Lim T C and Ma Z. 2005 *An Introduction to Electrospinning and Nanofibres* World Scientific Publishing Co. Pte. Ltd., Singapore
[46] Beachley V and Wen X 2009 *Mat. Sci. Eng. C* **29** 663
[47] Bhardwaj N and Kundu S C 2010 *Biotechnol. Adv.* **28** 325
[48] Tan S H, Inai R, Kotaki M and Ramakrishn S 2005 *Polymer* **46** 6128
[49] Manea L R, Bertea A., Nechita E, Popescu C V and Sandu I 2016 *Rev. Chim. (Bucharest)* **67** (8) 1607
[50] Popa A, Bucevschi A, Pustianu M, Manea L R and Sandu I, 2016 *Mat. Plast. (Bucharest)* **53** (2) 316
[51] Hristian L, Chirita M, Manea L R and Sandu I 2016 *Mat. Plast. (Bucharest)* **53** (4) 739
[52] Hristian L, Ostafe M M, Manea L R and Leon A L *2016 IOP Conf. Series: Mater. Sci.Eng.* **145** 032004
[53] Hristian L, Ostafe M M, Manea L R and Leon A L 2016 *IOP Conf. Series: Mater. Sci.Eng.* **145** 022014

Acknowledgments
This work was supported by a grant of the Romanian national Authority for Scientific Research and innovation, UEFISCDI, project number PN III 46 BG 2016 within PNCDI III.

IOP Conf. Series: Materials Science and Engineering **209** (2017) 012075 doi:10.1088/1757-899X/209/1/012075

Study of Mechanical Properties of Wool Type Fabrics using ANCOVA Regression Model

L Hristian[1*], M M Ostafe[1], L R Manea[1] and L L Apostol[1]
[1] "Gheorghe Asachi" Technical University of Iasi, Faculty of Textile, Leather & Industrial Management, Department of Engineering and Design of Textile Products, Blvd. Mangeron, No.28, Iasi, Romania

E-mail: hristian@tex.tuiasi.ro

Abstract. The work has achieved a study on the variation of tensile strength for the four groups of wool fabric type, depending on the fiber composition, the tensile strength of the warp yarns and the weft yarns technological density using ANCOVA regression model. ANCOVA checks the correlation between a dependent variable and the covariate independent variables and removes the variability from the dependent variable that can be accounted for by the covariates. Analysis of covariance models combines analysis of variance with regression analysis techniques. Regarding design, ANCOVA models explain the dependent variable by combining categorical (qualitative) independent variables with continuous (quantitative) variables. There are special extensions to ANCOVA calculations to estimate parameters for both categorical and continuous variables. However ANCOVA models can also be calculated using multiple regression analysis using a design matrix with a mix of dummy-coded qualitative and quantitative variables.

1. Introduction

Analysis of covariance models combine analysis of variance with techniques from regression analysis. With respect to the design, ANCOVA models explain the dependent variable by combining categorical (qualitative) independent variables with continuous (quantitative) variables [1-3]. There are special extensions to ANOVA calculations to estimate parameters for both categorical and continuous variables. ANCOVA models can however, also be calculated using multiple regression analysis using a design matrix with a mix of dummy-coded qualitative and quantitative variables [4,5]. In the latter approach, ANCOVA is considered as a special case of the General Linear Model (GLM) framework [6]. The ANCOVA is most useful in that it explains an ANOVA's within-group variance, and controls confounding factors. Firstly, the analysis of variance splits the total variance of the dependent variable into: variance explained by the independent variable (also called between groups variance) and unexplained variance (also called within group variance).

The ANCOVA looks at the unexplained variance and tries to explain some of it with the covariate(s). Thus it increases the power of the ANOVA by explaining more variability in the model [7-9]. This also makes the ANCOVA the model of choice when analyzing semi-partial correlations in an experiment, instead of the partial correlation analysis which requires random data [10-12]. The one-way ANCOVA (analysis of covariance) can be thought of as an extension of the one-way ANOVA to incorporate a covariate. Like the one-way ANOVA, the one-way ANCOVA is used to determine whether there are any significant differences between two or more independent (unrelated) groups on a dependent variable [12-14]. However, whereas the ANOVA looks for differences in the group means, the ANCOVA looks for differences in adjusted means (i.e., adjusted for the covariate) [15-17]. As such, compared to the one-way ANOVA, the one-way ANCOVA has the additional benefit of allowing you to "statistically control" for a third variable (sometimes known as a "confounding variable"), which you believe will affect your results [18-20]. ANCOVA is a model that relies on linear regression wherein the dependent variable must be linear to the independent variable. The origins of MANCOVA as well as ANOVA stem from agriculture where the main variables are concerned with crop yields.

On the other hand, regression is also a statistical tool which is available in many variants. Its variants include the linear regression model, simple linear regression, logistic regression, nonlinear regression, nonparametric regression, robust regression, and stepwise regression [21-23]. Regression is the relationship of a dependent variable and independent variable to each other. In this model, there is one dependent variable and one or more independent variables. There is also an effort to understand the change of the values of the dependent variable due to changes in one of the independent variants. In this situation, the other independent variants remain fixed [23-26].

2. Experimental part

2.1. Materials and methods

The study was conducted on woven materials made of combed wool type yarns used for manufacturing outwear clothing, on 24 articles. The variation limits of the composition and structural characteristics for the tested woven materials are indicated in table 1.

Table 1. Variation limits of composition and structural characteristics.

Variant code	Fibrous composition	Yarn count balanced	Nm_{warp}	Nm_{weft}	Fabric structure
A3			52/2	52/2	D2/2
A4			52/2	52/2	plain
A7		$Nm_{warp}=Nm_{weft}$	48/2	48/2	D2/2
A10	100% Wool		48/2	48/2	crepe
A1			40/2	24/1	D2/2
A8		$Nm_{warp}\neq Nm_{weft}$	48/2	30/1	D2/1
A23			56/2	37/1	plain
B1			64/2	64/2	P6/6
B3			60/2	60/2	D2/2
B12		$Nm_{warp}=Nm_{weft}$	52/2	52/2	Plain
B14	45%Wool+55%PES		48/2	48/2	D2/2
B7			52/2	52/1	D2/1
B11		$Nm_{warp}\neq Nm_{weft}$	52/2	30/1	D2/1
B15			64/2	37/1	Plain
C2			60/2	60/2	Plain
C3		$Nm_{warp}=Nm_{weft}$	52/2	52/2	D2/2
C7			60/2	60/2	D2/1
C1	45%Wool+52%PES+3%D		56/2	37/1	D2/1
C4		$Nm_{warp}\neq Nm_{weft}$	52/2	52/1	D2/1
C5			56/2	37/1	D2/1
D1			64/2	64/2	D2/1
D3		$Nm_{warp}=Nm_{weft}$	64/2	64/2	Plain
D7	60%PES+40%CELO		60/2	60/2	D2/2
D10			52/2	52/2	D2/1

The tensile testing was performed using an H 1K-S UTM Tinius Olsen (Hounsfield) testing machine, with a 1 kN load cell. The tests were done accordingly to standard (SR EN ISO 13934-1:2002) on both directions – weft and warp [26].

3. Results and discussions

3.1. Collection, systematization and processing of experimental data

The Based on the experimental data, the following variables were included in the ANCOVA regression model:
- dependent variable (Y) is tensile strength variance, Pr_{warp} [daN];
- nominal independent variable is group of woven fabrics depending on fiber composition *(100% Wool; 60% PES+40%CELO; 45% Wool + 55% PES; 45%Wool + 52% PES + 3% Dorlastan)* – three dummy variables D_1, D_2 respective D_3;
- quantitative independent variables (X_1 respective X_2) are warp yarns tensile strength Pr_{warp} (daN) and weft yarns technological density D_{weft} (yarns/10cm).

3.2. Hypothesis formulation

H_0: There are no significant differences between tensile strength values of woven fabrics depending of fiber composition, warp yarns tensile strength Pr_{warp} (daN) and weft yarns technological density D_{weft} (yarns/10cm);
H_1: There are significant differences between tensile strength values of woven fabrics depending of fiber composition, warp yarns tensile strength Pr_{warp} (daN) and weft yarns technological density D_{weft} (yarns/10cm).

3.3. Formulation of the regression model

The nominal independent variable "fiber composition" has 4 categories, so it can be built three dummy variables called alternative variables. Fabrics to which we relate (D1, D2, D3 = 0) will be made of 100% wool. Therefore, all the performances will be made in comparison with this group of fabrics.

The transformation in the dummy variables is summarized in table 2.

Table 2 Transformation in the dummy variables is.

Group	D1	D2	D3	The fiber composition
B	1	0	0	45% wool + 55%Pes
C	0	1	0	45% wool + 52%Pes + 3%Dorlastan
D	0	0	1	60%Pes+40%Celo
A	0	0	0	100% wool

The ANCOVA model with three dummy variables is defined as relation 1:

$$Y = a_0 + a_1 D_1 + a_2 D_2 + a_3 D_3 + b1\ X1 + \varepsilon \quad (1)$$

The regression, as a conditioned mean, has the following forms:

$$M\,(Y/D) = a_0 + b1\ X1;\ D1, D2, D3 = 0 \quad (2)$$

for tensile strength variance of 100% wool woven fabrics;

$$M\,(Y/D) = (a_0 + a_1) + b1\ X1;\ D1 = 1;\ D2, D3 = 0 \quad (3)$$

for variance of tensile strength of 45% Wool + 55% PES woven fabrics;

$$M\,(Y/D) = (a_0 + a_2) + b1\ X1;\ D2 = 1;\ D1, D3 = 0 \quad (4)$$

for tensile strength variance of 45% Wool + 52% PES + 3% Dorlastan woven fabrics;

$$M\,(Y/D) = (a_0 + a_3) + b1\ X1;\ D3 = 1;\ D1, D2 = 0 \quad (5)$$

for tensile strength variance of 60% Pes+40%Celo woven fabrics.

ANCOVA model coefficients are calculated in Table 3.

The estimators and the estimates of the model are defined similar to previous models.

Model estimates:

$a_0 = 29{,}637$
$a_1 = 45{,}874$
$a_2 = 45{,}522$
$a_3 = -25{,}955$
$b_1 = 0{,}014$

The estimated regression model has the form:

$$Y = 29{,}637 + 45{,}874D1 + 45{,}522D2 - 25{,}955D3 + 0{,}014D_{weft}$$

Table 3 ANCOVA model Coefficints[a].

Model		Unstandardized Coefficients		Standardized Coefficients	t	Sig.
		B	Std. Error	Beta		
1	(Constant)	29,637	9,543		3,106	0,006
	D1	45,874	4,160	0,701	11,027	0,000
	D2	45,522	4,337	0,663	10,497	0,000
	D3	-25,955	4,706	-0,325	-5,515	0,000
	Dweft (yarn/10cm)	0,014	0,039	0,020	0,365	0,719
2	(Constant)	32,960	2,771		11,896	0,000
	D1	46,283	3,918	0,707	11,812	0,000
	D2	45,957	4,078	0,669	11,268	0,000
	D3	-26,058	4,595	-0,327	-5,671	0,000

a. Dependent Variable: Pr_{warp} (daN)

3.4. Interpretation

a) $a_0 = 29{,}637$ means the average estimated value for the tensile strength of fabrics made of *100% wool* while the tensile strength of the warp yarns and the weft yarns technological density should be coded with 0 (X 1 = 0).

b) $a_0 + a_1 = 29{,}637 + 45{,}874 = 75{,}511$ means the average estimated value for the tensile strength of the fabrics made of *45% wool +55% PES* while the tensile strength of the warp yarns and the weft yarns technological density should be coded with 0 (X 1 = 0).

c) $a_0 + a_2 = 29{,}637 + 45{,}522 = 75{,}159$ means the average estimated value for the tensile strength of the fabrics made of *45% wool+52% PES+3% Dorlastan* while the tensile strength of the warp yarns and the weft yarns technological density should be coded with 0 (X_1=0).

d) $a_0 + a_3 = 29{,}637 - 25{,}955 = 3{,}682$ means the average estimated value for the tensile strength of the fabrics made of *100% PES* while the tensile strength of the warp yarns and the weft yarns technological density should be coded with 0 (X_1=0).

e) b_1 shows the variation in tensile strength of woolen fabrics as long as the tensile strength of the warp yarns and technological density of the weft yarns increase themselves with a unit.

It results from the coefficients table that sig.is < 0.05 (with the exception of sig.of D_{weft} (yarns/10 cm), the value being 0.719).

Although sig.for D_{weft} (yarns/10cm) exceeds the value of 0.05 (which shows that the density of the weft yarns does not significantly influence the dependent variable) has not been ejected during the Backward processing method (table 4).

Regression is the method for forecasting and prediction of a continuous outcome. It is the method to use for the continuous outcome and is based on one or more continuous predictor variables.

Interpreting the sig value it can be concluded that there are significant differences between the values of the tensile strength of the fabrics according to the composition of the fiber, the tensile strength of the warp yarns Pr_{warp} (daN) (it is rejected the null hypothesis Ho).

Table 4. Variables Entered/Removed[a].

Model	Variables Entered	Variables Removed	Method
1	D3, Dweft (yarn/10cm), D2, D1[b]		Enter
2		Dweft (yarn/10cm)	Backward (criterion: Probability of F-to-remove >= ,100)

a. Dependent Variable: Pr_{warp} (daN)
b. All requested variables entered.

The technological density of weft yarns D_{weft} (yarns/10cm) has no significant change in tensile strength for the wool fabric type.
Ho: between the values of the tensile strength of the fabrics there are no significant differences depending on the fiber composition, the tensile strength of the warp yarns Pr_{warp} (N) and technological density of weft yarns D_{weft} (yarns /10cm).
H1: between the values of the tensile strength of the fabrics there are significant differences depending on the fiber composition, the tensile strength of the warp yarns Pr_{warp} (daN) and technological density of weft yarns D_{weft} (yarns/10cm).

3.5. Hypotheses on errors

3.5.1. M (ε) = 0 (zero average error). Statistical assumptions about errors in this study can be formulated as follows:
Ho: M (ε) = 0
H1: M (ε) ≠ 0
Verification of these assumptions are made by applying t-test Student for testing the error (Unstandardized Residual), and the results are shown in table 5.

Table 5. Student t-test for testing of errors mean.

	Test Value = 0					
	t	df	Sig. (2-tailed)	Mean Difference	95% Confidence Interval of the Difference	
					Lower	Upper
Unstandardized Residual	0,000	23	1,000	0E-8	-2,886564	2,886564

From table 5 it can be shown that the value of Sig. = 1 (> 0.05), so Ho is accepted, the average errors is Zero, M (ε) = 0.

3.5.2. V (εi) = σ^2 (homoscedasticity hypothesis). To test this hypothesis a non-parametric test of correlation between the estimated errors and the dependent variables is applied and it is calculated the Spearman correlation coefficient and Student test for this coefficient (Table 6). In statistics, Spearman's rank correlation coefficient is a nonparametric measure of rank correlation (statistical dependence between the ranking of two variables). The Spearman correlation between two variables is equal to the Pearson correlation between the rank values of those two variables; while Pearson's correlation assesses linear relationships, Spearman's correlation assesses monotonic relationships (whether linear or not). If there are no repeated data values, a perfect Spearman correlation of +1 or −1 occurs when each of the variables is a perfect monotone function of the other.

Hypothesis statistics:
H0: the correlation coefficient is insignificantly larger than zero (the null hypothesis of the Student t-test is accepted);
H1: the correlation coefficient is significantly larger than zero (the null hypothesis of the Student t-test is rejected).

Table 6. Spearman test to verify the homoscedasticity hypothesis.

			D1	D2	D3	Pr_{warp} (daN)	D_{weft} (yarn/10cm)	Unstandardized Residual
Spearman's rho	D1	Correlation Coefficient	1,000	-0,370	-0,287	0,536**	0,226	0,113
		Sig. (2-tailed)	0.000	0,075	0,174	0,007	0,289	0,600
	D2	Correlation Coefficient	-0,370	1,000	-0,258	0,431*	0,202	-0,056
		Sig. (2-tailed)	0,075	.	0,223	0,036	0,344	0,796
	D3	Correlation Coefficient	-,287	-0,258	1,000	-0,646**	-0,227	-0,048
		Sig. (2-tailed)	,174	0,223	.	0,001	0,287	0,822
	Pwarp (daN)	Correlation Coefficient	0,536**	0,431*	-0,646**	1,000	0,393	0,475*
		Sig. (2-tailed)	0,007	0,036	0,001	.	0,057	0,019
	Dweft (yarn/10cm)	Correlation Coefficient	0,226	0,202	-0,227	0,393	1,000	0,152
		Sig. (2-tailed)	0,289	0,344	0,287	0,057	.	0,477
	Unstandardized Residual	Correlation Coefficient	0,113	-0,056	-0,048	0,475*	0,152	1,000
		Sig. (2-tailed)	0,600	0,796	0,822	0,019	0,477	.

**. Correlation is significant at the 0.01 level (2-tailed).
*. Correlation is significant at the 0.05 level (2-tailed).

Values of Sig. t for D1 correlations - expected errors (0.600) D2 - expected errors (0.796) D3 - expected errors (0.822), D_{weft} (yarns/10 cm) - expected errors (0.477), Pr_{warp} (daN) - expected errors (0.019) are > 0.05, so the null hypothesis of the Student test is rejected, so the model is homoscedastic.

3.5.3. $\varepsilon i \sim N(0, \sigma^2)$ - normality hypothesis. Testing normality of the distribution of errors is the Kolmogorov-Smirnov non-parametric test (table 7).

Table 7. Testing normality hypothesis. One-Sample Kolmogorov-Smirnov.

		Unstandardized Residual
N		24
Normal Parameters[a,b]	Mean	0E-7
	Std. Deviation	6,8359
Most Extreme Differences	Absolute	0,170
	Positive	0,168
	Negative	-0,170
Kolmogorov-Smirnov Z		0,832
Asymp. Sig. (2-tailed)		0,493

a. Test distribution is Normal; b. Calculated from data.

Value Sig = 0.493 (more than the value of 0.05), so it supports the hypothesis of normality (Ho)

3.5.4. cov (εi, εi) –testing of errors autocorrelation. Hypothesis statistics:
Ho: ρ = 0 (errors are not autocorelated)
H1: ρ ≠ 0 (errors are autocorelated)
Verification is done with Durbin Watson test (table 8).

Table 8. Durbin Watson test for testing of errors autocorrelation.

Model	R	R Square	Adjusted R Square	Std. Error of the Estimate	Durbin-Watson
1	0,974[b]	0,949	0,942	7,33073	2,168

Dependent Variable: Pr_{warp} (daN)

Value of 2.168 is compared to the calculated value of the test (dl, du). It is noted that the value obtained is within the range (du, 4 - du), which leads to the acceptance of the null hypothesis (no autocorrelation of errors).

3.6. Testing collinearity of independent variables

Table 9. Testing collinearity of independent variables. Coefficients.

Model		Unstandardized Coefficients		Standardized Coefficients	t	Sig.	95,0% Confidence Interval for B		Collinearity Statistics	
		B	Std. Error	Beta			Lower Bound	Upper Bound	Tolerance	VIF
1	(Constant)	29,637	9,543		3,106	0,006	9,663	49,611		
	Dweft (yarn/10cm)	0,014	0,039	0,020	0,365	0,719	-0,068	0,096	0,854	1,171
	D1	45,874	4,160	0,701	11,02	0,000	37,166	54,581	0,655	1,528
	D2	45,522	4,337	0,663	10,49	0,000	36,445	54,599	0,664	1,507
	D3	-25,955	4,706	-0,325	-5,515	0,000	-35,805	-16,105	0,761	1,314

The indicator VIF has a high value of between (1,171; 1,528) indicating that there is collinearity between the independent variables used in the model.

4. Conclusions

By applying and interpreting regression model ANCOVA sig value can conclude that there are significant differences between the variance of tensile strength of woven fabrics depending on fiber composition, tensile strength of warp yarns Pr_{warp} (daN), the null hypothesis Ho is rejected.

Technological weft density D_{weft} (yarns/10cm) does not significantly influence the tensile strength variation of woollen woven fabrics.

The ANCOVA model permits us to evaluate the homogeneous character of a population by separating and testing of the effects caused by the considered factors.

If the test data obtained after trials on the elements of a sample taken from one homogeneous population are divided in distinct groups, then the mean values of the groups do not differ significantly between themselves.

In the case of a non-homogeneous population, the deviations of the individual values as compare to the mean value are not anymore accidental and when dividing the test data in distinct groups the mean values are different between themselves significantly because of some causes having a systematic action.

ANCOVA model included as independent variables, both dummy or alternative variables (fiber composition) and numerical variables (tensile strength of warp and weft yarn density technology) and as the dependent variable, tensile strength woven fabrics.

References

[1] Zhao J, Cook R J and Wu C 2015 *Canadian Journal of Statistics* **43** (2) 240
[2] Stanescu I, Manea L R, Bertea A, Bertea A P and Sandu I 2016 *Rev. Chim. (Bucharest)* **67** (11) 2276
[3] Manea L R, Scarlet R, Leon A L and Sandu I 2015 *Rev. Chim. (Bucharest)* **66** (5) 640
[4] Boruvka A and Cook R J 2016 *Biostatistics* **17** (2) 350
[5] Hristian L, Sandu A V, Manea L R, Tulbure E A and Earar K 2015 *Rev. Chim. (Bucharest)* **66** (3) 342
[6] Hristian L, Chirita M, Manea L R and Sandu I 2016 *Mat. Plast.* **53** (4) 738
[7] Aalen O O, Cook R J and Roysland K 2015 *Lifetime Data Analysis* **21** (4) 579
[8] Manea L R, Bertea A, Nechita E, Popescu C V and Sandu I 2016 *Rev. Chim. (Bucharest)* **67** (8) 1607
[9] Zeng L, Cook RJ, Wen L and Boruvka A 2015 *Statistics in Medicine* **34** (24) 3181
[10] Popa A, Bucevschi A, Pustianu M, Manea L R and Sandu I 2016 *Mat. Plast.* **53** (2) 316
[11] Hristian L, Ostafe M M, Manea L R and Leon A L 2016 *IOP Conf. Series: Mater. Sci. Eng.* **145** 022014
[12] Bertea A, Manea L R, Bertea A P and Sandu I 2016 *Rev. Chim. (Bucharest)* **67** (12) 2446
[13] Mcisaac M A and Cook R J 2015 *Statistics in Medicine* **34** (21) 2899
[14] Zeng L, Cook R J, Wen L and Boruvka A 2015 *Statistics in Medicine* **34** (24) 3181
[15] Manea L R, Danu M C and Sandu I 2015 *Rev. Chim. (Bucharest)* **66** (6) 868
[16] Zhong Y and Cook RJ 2015 *Statistics in Medicine* **34** (6) 901
[17] Stanescu I, Manea L R, Bertea A, Bertea A P and Sandu I C A 2016 *Rev. Chim. (Bucharest)* **67** (10) 2082
[18] Manea L R, Hristian L, Ostafe M M, Apostol L L and Sandu I 2016 *Rev. Chim. (Bucharest)* **67** (9) 1758
[19] Wu Y and Cook R J 2015 *Biometrics* **71** (3) 782
[20] Manea L R, Chirita M, Hristian L, Popa A, Sandu I 2016 *Mat. Plast.* **53** (3) 361
[21] Sutradhar R and Cook R J, 2016 *Journal of Medical Statistics and Informatics* **4** 1
[22] Manea L R, Bertea A, Nechita E, Popescu C V and Sandu I 2016 *Rev. Chim. (Bucharest)* **67** (7) 1284
[23] Hristian L, Ostafe M M, Manea L R, Sandu I G, Apostol L L and Sandu I 2016 *Rev. Chim. (Bucharest)* **67** (9) 1717
[24] Ostafe M M, Avram D and Hristian L 2016 *Buletinul Agir* **3** 40
[25] Hristian L, Vilcu C, Bordeianu D L, Ostafe M M and Apostol L L 2016 *Buletinul Agir* **3** 31
[26] Standards SR EN ISO 13934-1: 2002

IOP Conf. Series: Materials Science and Engineering **209** (2017) 012076 doi:10.1088/1757-899X/209/1/012076

Thermal Analysis of a Structural Solution for Sustainable, Modular and Prefabricated Buildings

D N Isopescu[1], S G Maxineasa[1] and O Neculai[1*]
[1] Department of Civil and Industrial Engineering, Faculty of Civil Engineering and Building Services, Technical University „Gheorghe Asachi" of Iaşi, Romania

E-mail: oana.neculai@tuiasi.ro

Abstract. In the construction field, the design principles for an efficient and operational use of buildings and a minimal impact on the environment are essential aspects of sustainable development. In this regard, several aspects must be taken into consideration, such as: durability, easy maintenance, flexibility in interior design, and reduced energy consumption. Decreasing energy consumption in buildings during the service life (heating / cooling / drinking water / electricity) can mean lower costs, but also a lower impact on the environment. The paper presents the thermal analysis for a GF+1F height structure, consisting of several identical, adjacent and / or overlapped metallic cubic modules. The spaces inside this cubes ensemble solve the functionality of a family home building. The good carrying capacity, the rapidity of execution, the superior degree of thermal insulation and the minimum losses of material in execution were the main advantages provided by this structural solution. Regarding the thermal comfort for the users of this constructive system, the thermal analysis showed that the internal temperatures are constant and uniform, without cold surfaces or temperature fluctuations. In addition, humidity is controlled and there is no risk of condensation.

1. Introduction

The measures that can be taken to improve energy efficiency in all economy sectors are mainly represented by:

- introducing products and technologies with high energy efficiency;
- promoting the use of renewable energy sources by end-users;
- reducing the environmental impact of industrial activities, and of producing, transporting, distributing and consumption of all forms of used energy;
- applying modern principles of energy management, modern energy control and management systems for monitoring, continued assessment of energy efficiency and energy consumption forecasting;
- granting financial and fiscal incentives, under the law, for investments based on energy efficiency principles;
- developing the market for alternative energy services.

The construction sector accounts for 40% of the total energy consumption in the European Union (EU). Reducing energy consumption in this domain is, therefore, a priority in energy efficiency targets. The buildings' energy efficiency, governed by two fundamental principles, almost zero energy consumption (NZEB), and durability (appropriate carrying capacity over the entire service life) implies:

- using renewable energy sources;
- constructing buildings with prefabricated, modular and standardized systems, to meet the quality requirements and achieve a balance between the energy needed to construct a building and the amount of energy saved during its service life;
- protecting the environment both during the construction and use of buildings, and through the use of low-toxicity, recyclable and renewable materials.

The construction sector is closely linked to the energy efficiency field by applying the provisions of the Energy Efficiency Directive 2012/27/EU. The provisions of this Directive require that in the coming years the EU member states to respect the policy of the European Union and thus, by the end of year 2020, to satisfy the proposed targets for reducing conventional energy consumption. According to Directive 2010/31/EU of the European Parliament on buildings energy performance, starting with year 2020, newly built houses in the EU must have almost zero energy consumption (NZEB).

Energy efficiency in constructions is a principle embodied in various forms: energy-independent buildings, passive / active buildings, eco-friendly buildings (eco-buildings) and green buildings. Passive buildings are of current interest; these are buildings with low energy consumption, their passivity being realized by the balance between the energy consumed from general systems and the energy produced by their own systems.

The design of passive constructions is based on the advantages offered by the site, the interior comfort requirements, as well as the superior performance of the existing construction products on the local market, in order to diminish energy consumption. Taking advantage of the site's opportunities, designers can incorporate energy efficient strategies and principles into the structural design, aiming to use renewable energy systems.

Green buildings (also known as green constructions or sustainable buildings) refer to the structural concept and use of processes that are responsible for environmental protection and are resource-efficient throughout the life cycle of a building: from design, construction, operation, maintenance, renovation to demolition.

Modulation creates the premises for buildings' industrialization. By optimizing the work, modulation opens the way to the widespread use of prefabricated elements and can lead to significant decreases in material consumption. The use of prefabricated elements is one of the most important elements in the constructions' industrialization process, with positive influences in ensuring quality and meeting sustainability requirements.

The paper presents the thermal analysis of a building with a GF+1F height regime, [1], made of several identical cubic modules, with metallic structure, placed adjacent and / or overlapped. The building was designed to meet the cumulative requirements of the following concept definitions: passive building, green building, modulation- prefabrication and durability. The good carrying capacity, the rapidity of execution, the superior degree of thermal insulation, as well as the minimum material loss on site were the main advantages provided by this structural solution.

2. The constructive solution description

The presented GF+1F building starts with the multiplication of a cubic module, having the bearing structure made of steel profiles with square tubular cross-sections for columns, and with rectangular tubular cross-sections for beams, [1-5].

The cube side is 4.00 m, Figure 1. At least 4 cubic modules are considered necessary to make a functional model for a 3-room apartment, and a supplementary cubic module is required for the staircase. A proposed functional model is illustrated in a three-dimensional representation in figure 2 [1].

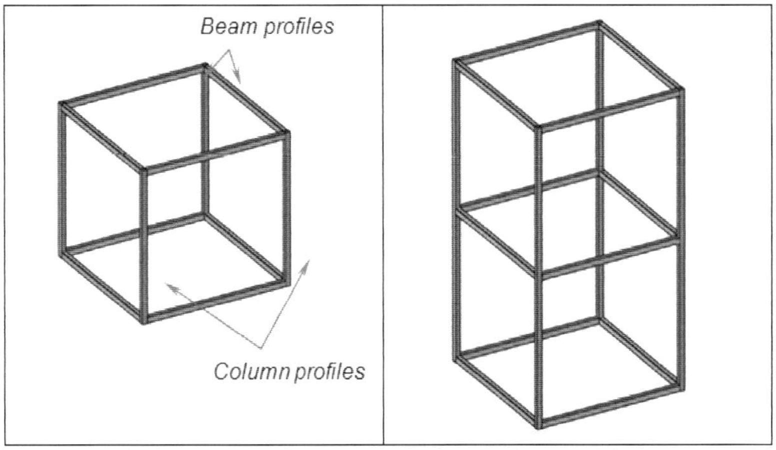

Figure 1. The cubic module and the vertical assembly of two modules, [1].

The cube's faces, forming the walls, are made of various hybrid construction products, which, besides ensuring the perimeter closings, incorporate systems for heating and ventilation based on energy from renewable sources existing at the site. In order to stiffen the structure, the floors are made of profiled steel sheets and reinforced concrete, cast in position on the construction site. The roof is of the terrace type, offering multiple possibilities: a green terrace which, by summing up its positive, ecological and economical characteristics, brings a relevant contribution to achieving the future-oriented sustainable architecture principles, or an uncirculated terrace**,** as a placement location for the energy production equipment from renewable sources.

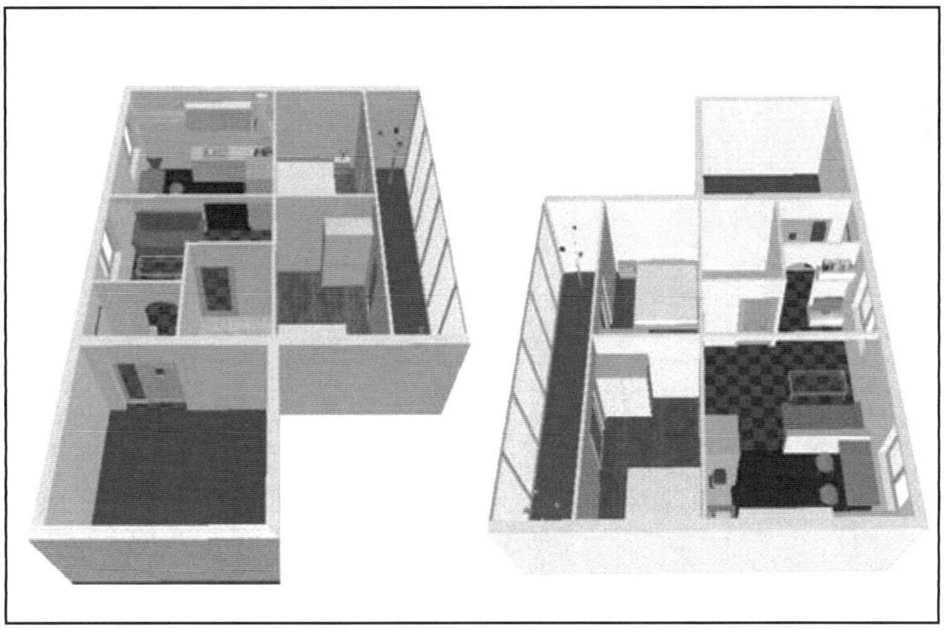

Figure 2. Three-dimensional image for a proposed functional model, [1].

The connection between the foundation, and the column and beams is achieved by using a metallic element fixed in the concrete raft foundation. The beams are slid into the U profiles, the columns are inserted in the square section profiles, and then all the elements are fastened with bolts, as depicted in Figure 3. Therefore, a semi-rigid joint is created between the foundation, and the column and beams.

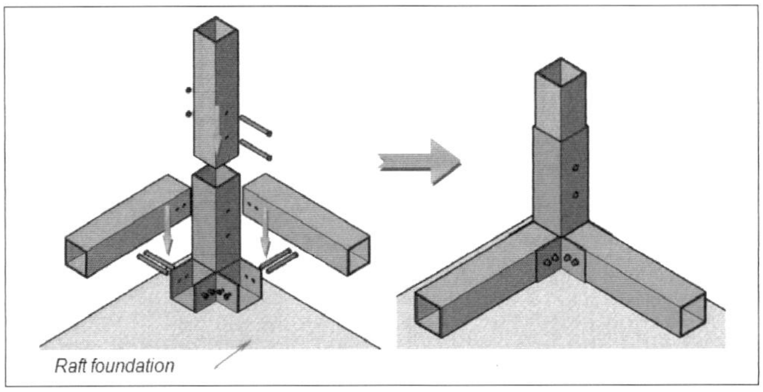

Figure 3. Joint between foundation – column – beams [1].

At the superstructure, the joints between the columns and beams are realized using a metallic element made of four profiles inserted into the columns and intersecting beams, as shown in Figure 4. All the elements are fastened with bolts. For this joint type, installing hybrid sandwich panels for perimeter closings or internal compartments is easy and fast, Figure 5.

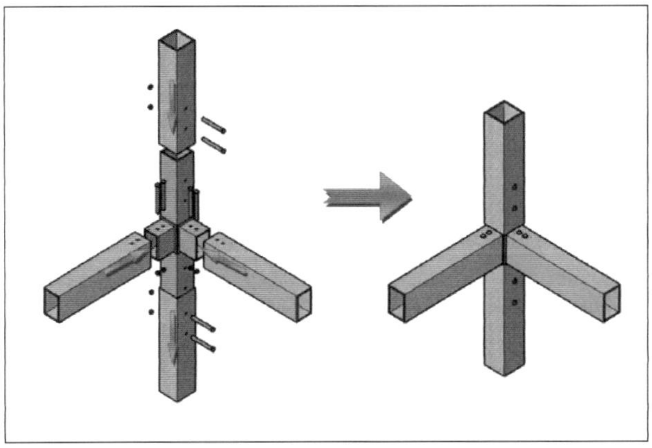

Figure 4. Joint between columns – beams [1].

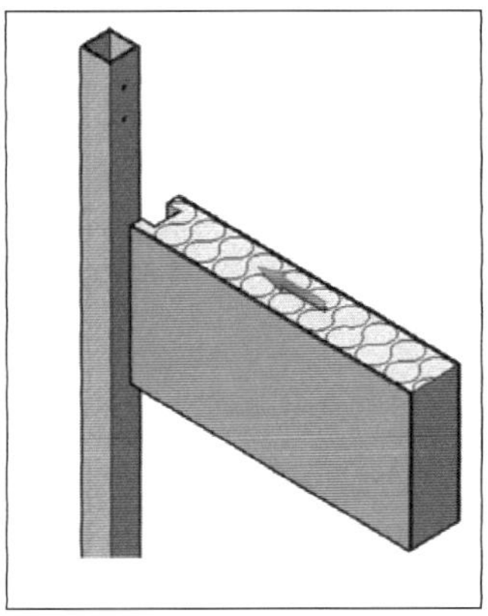

Figure 5. Sandwich panels mounting for exterior walls [1].

The mixed-structure slabs are made of reinforced concrete and folded metallic sheets having a double purpose: adding strength and also ensuring the co-operation of the vertical bearing elements under horizontal actions from wind or earthquake, Fig 6.b. The structure of the slab is shown in Figure 6.a.

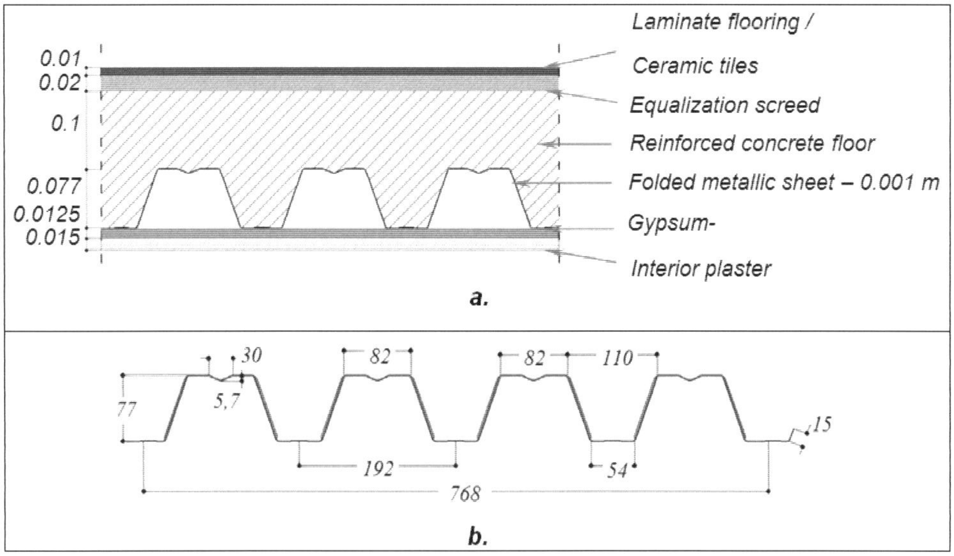

Figure 6. Intermediary slab: a. layers, [m]; b. folded metallic sheet detail, [mm].

3. Thermal analysis of the building

The thermal analysis of the building has been carried out in conformity with the 2010 modified form of the Romanian Code C107/2005 "Normative regarding the calculation of thermal properties of building construction elements", [6, 7]. The thermal performances have been assessed by considering an outside design temperature of Te=-21^{0}C and an inside design temperature of Ti=20^{0}C.

3.1. Calculating the normalized global thermal insulation coefficient

The normalized global thermal insulation coefficient, G_N, has been determined by taking into account the number of building levels, N, and the ratio between the envelope area, A, and the building volume, V. Table 1 presents the considered geometrical characteristics of the analysed building.

Table 1. Geometrical characteristics of the envelope.

Building element	Area
Exterior longitudinal wall 1 (matt part + joinery)	179.67 m^2
Exterior longitudinal wall 2 (matt part + joinery)	179.67 m^2
Exterior transversal wall 1 (matt part + joinery)	67.71 m^2
Exterior transversal wall 2 (matt part + joinery)	67.71 m^2
Exterior longitudinal wall 1 (matt part)	162.93 m^2
Exterior longitudinal wall 2 (matt part)	162.93 m^2
Exterior transversal wall 1 (matt part)	67.71 m^2
Exterior transversal wall 2 (matt part)	67.71 m^2
Exterior joinery – corresponding to longitudinal wall 1 (windows + doors)	16.74 m^2

Building element	Area
Exterior joinery – corresponding to longitudinal wall 2 (windows + doors)	16.74 m^2
Exterior joinery – corresponding to transversal wall 1 (windows + doors)	0 m^2
Exterior joinery – corresponding to transversal wall 2 (windows + doors)	0 m^2
Roof terrace	161.85 m^2
Slab on grade	161.85 m^2
Envelope area	818.46 m^2
Heated volume	1294.77 m^3

Taking into account that the building under study comprises of two levels and therefore $N = 2$ and $A/V = 0.63$, according to the stipulations from Annex 2 of the C107/1-2010 national standard, [6, 7], the G_N coefficient equals 0.51. In order to fulfil a proper level of global thermal insulation, the following condition must be satisfied, considering G as the global thermal insulation coefficient:

$$G \leq G_N \ [W/m^3K]$$

3.2. Evaluating the global thermal insulation coefficient, G
In this section the following analysis phases have been performed:

- evaluating the unidirectional thermal resistances;
- numerical modelling and analysis of the thermal bridges specific to the considered building;
- determining the corrected specific thermal resistances of the envelope elements.

3.2.1. Determining the unidirectional thermal resistances, R. In order to evaluate the unidirectional thermal resistances, R, the expression (1) has been used. The resulted values are presented in Table 2.

$$R = R_i + R_e + \sum_k R_k = \frac{1}{\alpha_i} + \frac{1}{\alpha_e} + \sum_k \frac{\delta_k}{\lambda_k} \qquad (1)$$

where:

α_i - convective heat transfer coefficient of the inner surface of the element, [W/m²K];

α_e- convective heat transfer coefficient of the outer surface of the element, [W/m²K];

δ_k- thickness of the layer „k", [m];

λ_k - thermal conductivity of the layer „k", [W/mK].

Table 2. Specific unidirectional thermal resistances of the envelope elements.

Construction element	R [m^2K/W]
Exterior wall	6.25
Slab on grade	7.84
Roof terrace	8.42

IOP Conf. Series: Materials Science and Engineering **209** (2017) 012076 doi:10.1088/1757-899X/209/1/012076

3.2.2. Numerical modelling of the thermal bridges and evaluating the linear thermal transfer coefficients, Ψ. The RDM 6 software has been used with the goal of analysing the thermal bridges characteristic to the considered construction. During the assessment, 3 vertical and 7 horizontal thermal bridges have been examined, and the resulted values for the linear thermal transfer coefficients, Ψ, are presented in Table 3.

Table 3. Linear thermal bridges coefficients.

No.	Analysed thermal bridge	Value of Ψ, [W/(m·K)]
1.	"outer corner" vertical thermal bridge 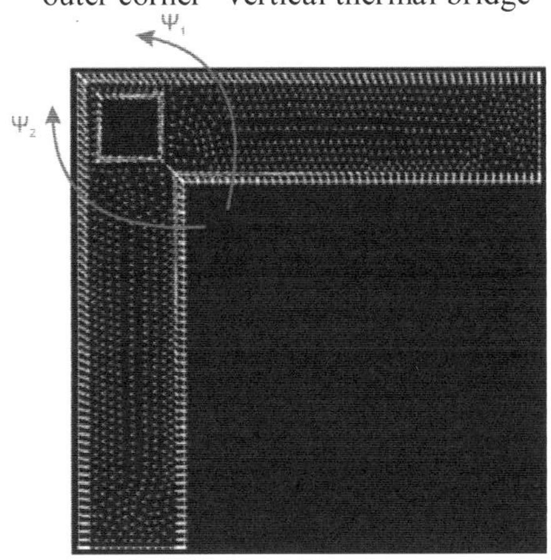	$\Psi_1=\Psi_2=0.03$
2.	"exterior and interior walls intersection" (between apartments) vertical thermal bridge 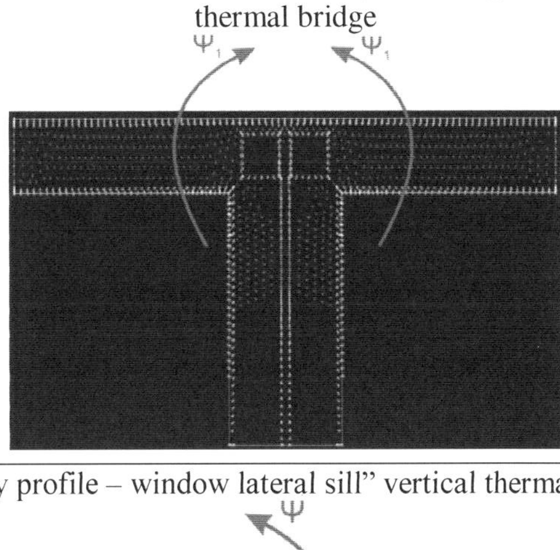	$\Psi_1=0.09$
3.	"joinery profile – window lateral sill" vertical thermal bridge 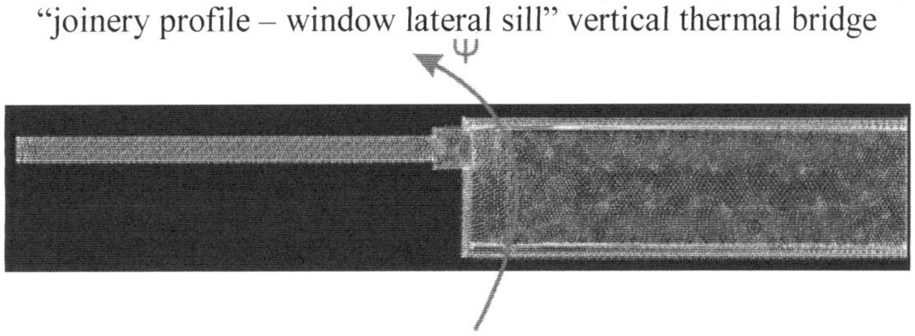	$\Psi_1=0.07$

IOP Conf. Series: Materials Science and Engineering **209** (2017) 012076 doi:10.1088/1757-899X/209/1/012076

No.	Analysed thermal bridge	Value of Ψ, [W/(m·K)]
4.	"attic" horizontal thermal bridge 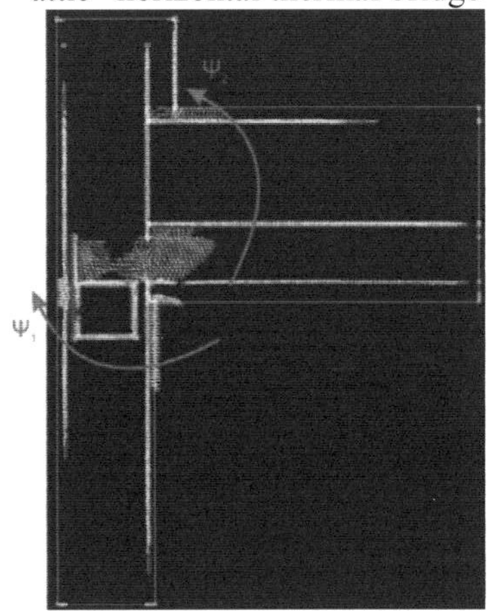	Ψ_1=0.33 Ψ_2=0.21
5.	"exterior wall and slab over ground floor intersection" horizontal thermal bridge 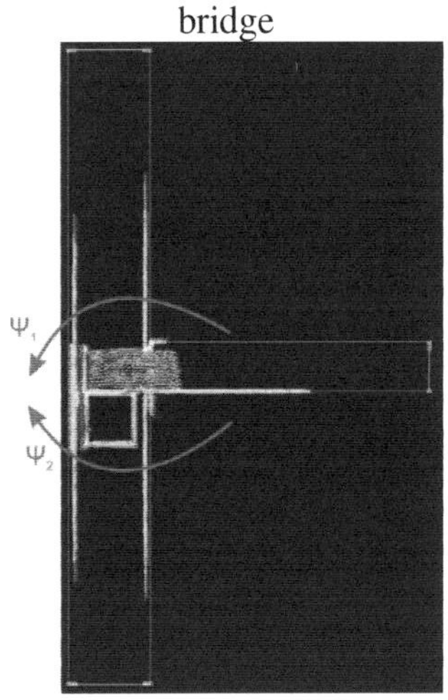	Ψ_1=0.12 Ψ_2=0.19
6.	"exterior wall and slab on grade intersection" horizontal thermal bridge 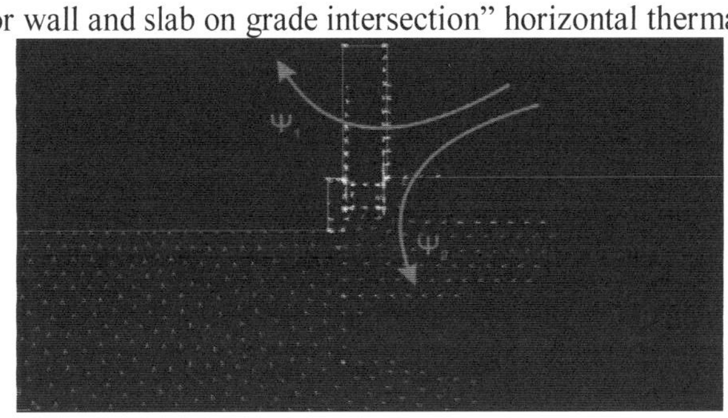	Ψ_1=0.31 Ψ_2=0.35

IOP Conf. Series: Materials Science and Engineering **209** (2017) 012076 doi:10.1088/1757-899X/209/1/012076

No.	Analysed thermal bridge	Value of Ψ, [W/(m·K)]
7.	"interior wall and slab on grade intersection" (between apartments) horizontal thermal bridge 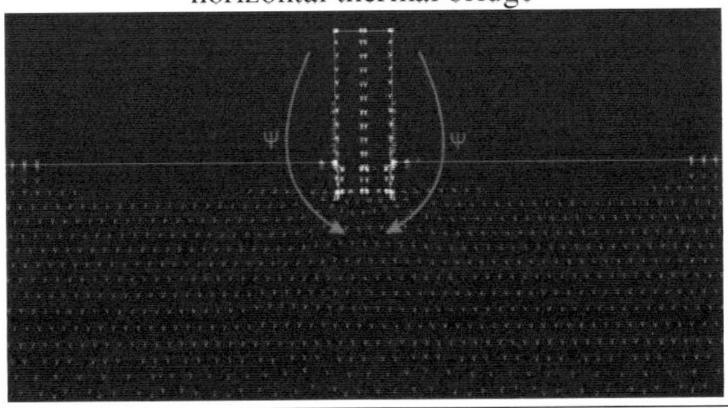	Ψ=0.08
8.	"slab on grade and metallic profiles intersection" horizontal thermal bridge 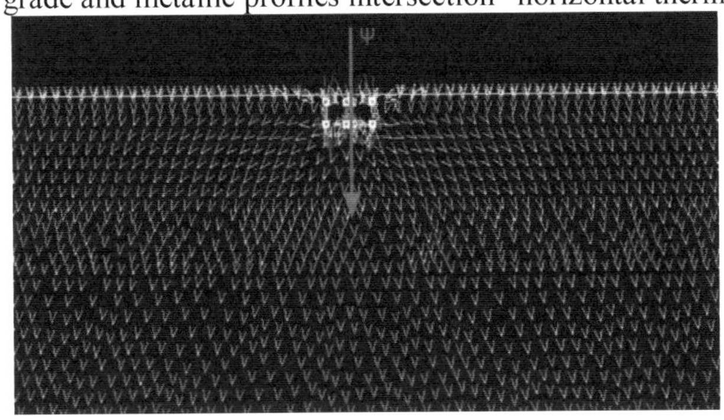	Ψ_1=0.16
9.	"joinery profile – window lintel" horizontal thermal bridge 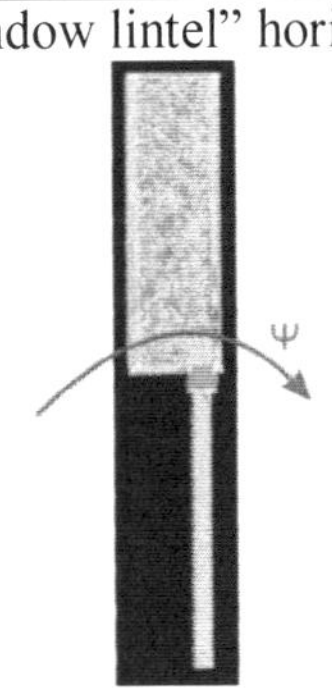	Ψ=0.07
10.	"joinery profile – window sill" horizontal thermal bridge 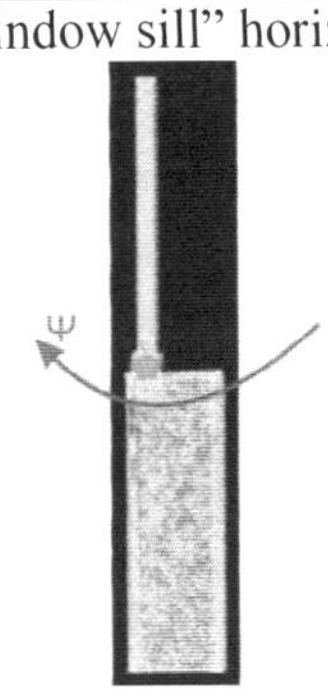	Ψ=0.07

3.2.3. Evaluating the specific thermal resistances, R'. The specific thermal resistances corrected with the value of the envelope elements thermal bridges coefficients are presented in Table 4. The resulted values have been attained by using the expression (2), where l is the length of the thermal bridge, and U' is the thermal transmittance of the element.

$$U' = \frac{1}{R} + \frac{\Sigma \psi l}{A} + \frac{\Sigma \lambda}{A}; \quad R' = \frac{1}{U'} \tag{2}$$

Table 4. Values of the specific thermal resistance.

Element	Thermal bridge	Ψ [W/mK]	l [m]	Ψ·l [W/K]	R [m^2K/W]	U' [W/m^2K]	R' [m^2K/W]
Longitudinal wall 1	Outer corner	0.03	16	0.48			
	Exterior and interior walls intersection	0.09	16	1.44			
	Joinery profile – window lateral sill	0.07	14.40	1.01			
	Attic	0.33	20.72	6.84			
	Exterior wall and slab over ground floor intersection	0.12	20.72	2.49			
		0.19	20.72	3.94			
	Exterior wall and slab on grade intersection	0.31	20.72	6.42			
	Joinery profile – window lintel	0.07	7.90	0.55			
	Joinery profile – window sill	0.07	7.90	0.55			
Longitudinal wall 1 – total		A = 162.93		23.72	6.25	0.31	3.23
Longitudinal wall 2	Outer corner	0.03	16	0.48			
	Exterior and interior walls intersection	0.09	16	1.44			
	Joinery profile – window lateral sill	0.07	14.40	1.01			
	Attic	0.33	20.72	6.84			
	Exterior wall and slab over ground floor intersection	0.12	20.72	2.49			
		0.19	20.72	3.94			
	Exterior wall and slab on grade intersection	0.31	20.72	6.42			
	Joinery profile – window lintel	0.07	7.90	0.55			
	Joinery profile – window sill	0.07	7.90	0.55			
Longitudinal wall 2 – total		A = 162.93		23.72	6.25	0.31	3.23
Transversal wall 1	Outer corner	0.03	16	0.48			
	Attic	0.33	7.81	2.58			
	Exterior wall and slab	0.12	7.81	0.94			

Element	Thermal bridge	Ψ [W/mK]	l [m]	Ψ·l [W/K]	R [m²K/W]	U' [W/m²K]	R' [m²K/W]
	over ground floor intersection	0.19	7.81	1.48			
	Exterior wall and slab on grade intersection	0.31	7.81	2.42			
Transversal wall 1 – total		A = 67.71		7.90	6.25	0.27	3.70
Transversal wall 2	Outer corner	0.03	16	0.48			
	Attic	0.33	7.81	2.58			
	Exterior wall and slab over ground floor intersection	0.12	7.81	0.94			
		0.19	7.81	1.48			
	Exterior wall and slab on grade intersection	0.31	7.81	2.42			
Transversal wall 2 – total		A = 67.71		7.90	6.25	0.27	3.70
Roof terrace	Attic	0.21	57.07	11.98			
Roof terrace – total		A = 161.85		11.98	8.42	0.19	5.26
Slab on grade	Exterior wall and slab on grade intersection	0.35	57.06	19.97			
	Interior wall and slab on grade intersection	0.08	24	1.92			
	Slab on grade and metallic profiles intersection	0.16	40	6.40			
Slab on grade – total		A = 161.85		28.29	7.84	0.21	4.76

3.2.4. Evaluating the overall effective thermal insulation coefficient, G. Table 5 presents the values of the global thermal insulation coefficient, G, that was determined by using the expression (3).

$$G = \frac{\Sigma\left(\frac{A_j\ \tau_j}{R'_j}\right)}{V} + 0.34\,n = \frac{\Sigma\left(\frac{A_j\ \frac{t_i - t_j}{t_i - t_e}}{R'_j}\right)}{V} + 0.34\,n \tag{3}$$

where:

t_i - conventional inside design temperature [°C],

t_e - conventional outside design temperature [°C],

t_j - temperature at the outer side of the envelope [°C], which can be the outside design temperature or the temperature from the unheated spaces/rooms that are placed outside the building envelope;

τ - exterior temperatures correction factor;

A_j - the area of the building element "j" [m²], having the specific thermal resistance R';

n - ventilation rate [h^{-1}];

0.34 - the ratio between the mass caloric capacity and the apparent density of air.

Table 5. The global thermal insulation coefficient.

Building element	A [m^2]	U' [m^2K/W]	R' [m^2K/W]	τ	$\frac{A \cdot \tau}{R'}$ [1/KW]
Exterior longitudinal wall 1	162.93	0.31	3.23	1	50.44
Exterior longitudinal wall 2	162.93	0.31	3.23	1	50.44
Exterior transversal wall 1	67.71	0.27	3.70	1	18.30
Exterior transversal wall 2	67.71	0.27	3.70	1	18.30
Exterior joinery L1	16.74	1.30	0.77	1	21.74
Exterior joinery L2	16.74	1.30	0.77	1	21.74
Exterior joinery T1	0	1.30	0	1	0
Exterior joinery T2	0	1.30	0	1	0
Roof terrace	161.85	0.19	5.26	1	30.76
Slab on grade	161.85	0.21	4.76	0.50	17
Heated volume V = 1294,77[m^3]				$\sum \frac{A \cdot \tau}{R'} = 228.72$	
n = 0,5 [h^{-1}] = ventilation rate					
G = 0.35[W/(m^3K)]					
G = 0.35 < G_N = 0.51					

4. Conclusions

Structural steel can be regarded as an innovative construction material. In the last years, in the current context of global sustainable development, the steel manufacturing industry has made tremendous progress towards minimizing the environmental impact of their products. Due to the high mechanical proprieties and the development of a large number of structural elements' forms, steel represents one of the most common material choices for resolving different civil engineering issues. Generally, structural steel is used for constructing long span bridges and skyscrapers, but nowadays, this material is used more and more to assemble various residential buildings.

The article presents an analysis of a structure which satisfies the following general requests: a good load carrying capacity, rapid assembly on the construction site, and an adequate management of natural resources consumption in the pre-operation stage. All the above mentioned can constitute decisive facts for future users, but one of the most important is represented by the thermal behaviour of the building, and it is this particular detail that is put under evaluation in this analysis.

The considered steel construction can be regarded as an advanced and innovative one, offering a high level of structural modularity and mobility. The study has shown that the proposed solution offers a high level of thermal resistance. Analysing the numerical modelling and the undertaken evaluations, we can conclude that the construction global thermal insulation coefficient is considerably lower than the one imposed by the Romanian national standard. This fact has direct influence over the level of the energy consumed in order to satisfy a proper level of indoor climate conditions. Taking into account that the Romanian construction sector uses an impressive amount of energy obtained from non-renewable resources, minimizing consumption levels will have an important positive effect over the environment.

The good carrying capacity, the rapidity of execution, the superior degree of thermal insulation, as well as the minimum material loss on site are the main advantages of this structural solution, and beside this fact, they demonstrate that the proposed solution might be included in the list of solutions efficient energetic and of protective environmentally.

References

[1] Steel cubic module structure for prefabricated constructions, Contract "Gheorghe Asachi" Technical University of Iasi, Faculty of Civil Engineering and Building Services – S.C. NEXT STUDIO S.R.L., 2016

[2] SR EN 1993-1-1:2006, Eurocode 3: Design of steel structures. Part 1-1: General rules and rules for buildings, Romania Standardization Association, Bucharest, Romania, 2006

[3] SR EN 1993-1-1/AC, Eurocode 3: Design of steel structures. Part 1-1: General rules and rules for buildings, Romania Standardization Association, Bucharest, Romania, 2006

[4] SR EN 1993-1-1/NA, Eurocode 3: Design of steel structures. Part 1-1: General rules and rules for buildings, National Annex, Romania Standardization Association, Bucharest, Romania, 2008

[5] SR EN 1998–1: 2004, Eurocode 8: Design of structures for earthquake resistance – Part 1: General rules, seismic actions and rules for buildings Romania Standardization Association, Bucharest, Romania, 2004

[6] C107/1-2010, Normative regarding the calculation of thermic properties of building construction elements, 2010 modified form

[7] Government Ordinance no. 386/2016 regarding the modification and completion of C107-2005, Normative regarding the calculation of thermic properties of building construction elements, issued by the Ministry of Transport, Construction and Tourism - Government Ordinance no. 2.055/2005

Acknowledgements

This paper was realized in the framework of the project EFECON – ECO-INNOVATIVE PRODUCTS AND TECHNOLOGIES FOR ENERGY EFFICIENCY IN CONSTRUCTION, POC/71/1/4 - Knowledge Transfer Partnership, Cod MySMIS: 105524, ID: P_40_295, Project co-financed by the European Regional Development Fund.

Furthermore, the authors acknowledge the contribution and thank SC NEXT STUDIO SRL which provided insight that greatly assisted the research.

The Analysis of an Internal Combustion Engine Breakdown - Case Study

I Agape[1*], L Gaiginschi[1], G Ianuş[2] and N Cimpoeşu[3]
[1] Department of Mechanical Engineering and Automotive, Technical University "Gh.Asachi", Iaşi
[2] Department of Mechanical Engineering, mecatronics and Robotics, Technical University "Gh.Asachi", Iaşi
[3] Department of Material Science, Technical University "Gh.Asachi", Iaşi

E-mail: iulian.agape@tuiasi.ro

Abstract. The paper concerns with the general assembly conditions and circumstances that could cause fatal damage to an internal combustion engine during operation. In the first part, the focus is on identifying the majority of dysfunctions that may result in the connecting rod destruction. Possible causes of the connecting rod destruction that led to engine failure are presented sequentially, starting with those of conceptual nature (the right choice of material, constructive solution and cross-sectional design from a geometric-dimensional point of view), then passing through issues related to operation (the severity of thermal and mechanical operation - and lubrication regimes) and last but not least, the effects of accumulated fatigue and potential structural defects of the material. This review of possible causes takes into account the particularities induced by the ignition engines and by the type of equipped machinery. The synthesis contains a case study on a 6-L medium displacement diesel engine in which the damage cause is accurately identified and which has been investigated with a scanning electron microscope (SEM).

1. Introduction

The general issue of internal combustion engines failure is very complex and comprehensive. In principle, an analysis of such failure must take into account the way the constructive/functional requirements are met by the engine's mobile organs, the particularities of the operating regime of the equipment (specified by speed, load, the corresponding thermal regime and engine cooling conditions), lubrication regimes of the engine friction couples and the quality of the used lubricant.

It is also necessary to take into account the wear accumulated during the engine operating, as well as possible internal discrete defects of the engine's parts. The systemic approach to this complex of conceptual, functional and intrinsic problems involves, on the one hand, the multidisciplinary engineering analysis and the polyvalent training of the operator team and, on the other hand, requires specific investigation equipment. This paper is in fact an application of analysis techniques in a particular case [1,2].

The subject of this case study is a 4-stroke compression ignition engine with 6-cylinder in-line, turbo-compressor medium-turbocharged and liquid-cooled. The engine is an older generation of the Case IH family and has equipped a grain-harvesting combine. Its cylinder capacity is 8.7 [liters] and runs at 230 [HP] at a nominal speed of 2300 [rpm]. The engine compression ratio is 18.

This engine failed during operation, the engine running in full load. At the time of the damage, the engine was about 1200 hours of operation since the last major repair and about 100 hours since the last oil change.

2. Experimental investigations to identify the cause of engine failure

The engine is of the direct injection type and uses the Meurer principle for self-ignition of the mixture. Consequently, in spite of overcharging, the load of such an engine is medium; the engine is fitted with only two valves per cylinder and is medium-speed. The use of the pellicular burning principle gives the engine a certain functional insensitivity to fuel quality.

Internal combustion engines with piston which equips agricultural aggregates have as functional particularity very high thermal regimes, due to low speeds and consequently low relative speed between the engine and the ambient air flow - on one hand - and the high loading regimes specific to the operation, on the other hand. Under these circumstances it is possible that such an engine may fail as a result of compromising engine lubrication functions, as a consequence of the extremely severe thermal regime applied. It is well known that under such severe conditions the viscosity of the lubricant, the strength of the lubricating film under the shear stress, and even the flows circulated in the engine lubrication system decrease, and the lubricating medium may degrade as a consequence of oxidation reactions accelerating.

In the first phase, the lubricant problem was removed from the list of possible causes of engine failure. A degree of oxidative degradation of 47 [%] was determined for the lubricant by the SKF technique of measuring the dielectric constant of the lubricant. It is appreciated that the degradation of oil is acceptable considering its accumulated use time and the seasonal specificity of the operation of the equipped machine.

The engine was damaged as a result of the failure of the connecting rod from engine cylinder no. 1, positioned in the cylinder line at the end of the engine, opposite to the power outlet.

Given the destructive effects of the damage in the context of the general casualties of engine breakdown and their causes, the experimental investigations to identify the possible causes of the failure in the present case have focused on the engine's connecting rod from cylinder 1 and the sensing points of it, respectively the two bolts and the connecting rod small end sections.

Such an analysis should take into account some constructive/functional solutions used by the engine manufacturer to design the connecting rod. Thus, the piston pin type is full floating. The full-floating pin is free to rotate in the connecting rod and in the bosses, while plugs or snap-ring locks prevent it from working out against the sides of the cylinder. The solution is also common in the case of medium power engines with aluminum alloy pistons.

This solution takes into account that the piston pin material often has a lower thermal expansion coefficient than that of the piston material. Starting the engine is ensured by providing a functional clearance between the piston pin and the rod bushing. At the operating regime, this clearance tends to decrease, due to the increased dilatation of the bushing, but has no dangerous effects. To ensure sufficient lubrication in the piston pin bearings, oil access holes are provided both in the piston bosses and the connecting rod small end eye, corresponding to the oil hole in rod bushing.

The rod bushing for the cylinder 1 connecting rod is deeply deformed (its initial cylindrical shape degenerates into a quasi-flattened one), it has numerous blows, tears caused by shock. It is noted the break-through of the bushing in the direction of the generator, in an area disposed approximately diametrically opposed to the corresponding oil hole. By analyzing the inner working surface of the bush, it is found that it is not affected by pitting (wear of surface fatigue contact), there are no dislocations of the anti-friction layer and no adhesive wear marks, that emphasize that the rod bushing has fulfilled its functional role and has not initiated the dangerous situation that eventually led to the destruction of the engine.

The rod cap shows superficial blows on its outer surfaces. There are no significant changes to the geometric-dimensional configuration of the cap. The holes for fixing the cap to the body have been found free from the bolt remains.

The rod body is broken by tearing into two parts, the rupture being produced approximately halfway of the beam.

The upper part (towards the connecting rod small end) is strongly deformed and lacks a piece of the connecting rod small end eye. A fragment of the piece emitted from the rod eye was found in the engine crank when dismantling the engine, and was analyzed separately. The lower part of the rod body is also strongly deformed in the oscillation plane, the tear section indicating that the rod has been loaded at uniform bending stresses in the above-specified plane, above the breaking limit.

Comparing the recovered sections of the bolt with other bolts from the other cylinders, it is noted the elongation of this bolt remain (by about 0.48 [mm] or 0.63 [%] at the beginning of the thread), and thinning it (with 0.11 [mm] or 1.2 [%] in the middle of the third section) indicating unequivocally that rupture of this bolt occurred under predominantly stretching effort.

Applying the free lubrication solution of the bearing formed by the connecting rod small end bushing and the piston pin by using the oil access hole has the great disadvantage that the bore is an important stress concentrator by decreasing the active section of the connecting rod eye under the bending load.

In the case of the present engine the free lubricating solution is taken to the extreme, in the sense that the oil access hole has along its length two sections of approximately equal lengths; the section disposed to the outer fiber is essentially an enlargement, in order to capture an increased flow of oil. This enlargement towards the outside of the oil access hole accentuates the character of the orifice stress concentrator. Taking into account these observations, the analysis of the end sections of the cut-off zone of the connecting rod eye becomes a priority.

Last but not least, the solution of connecting rod screw discharge from lateral shear stresses in the separation plane between the cap and the rod body is of an old type. The screws are provided with median sections (adjacent to the plane of separation between the cap and the rod body) which are mounted without clearance in the corresponding cap and rod body locations. The solution is disadvantageous because it allows additional loading of shear stressing on screws and it is not very reliable over time.

Regarding the mechanism of connecting rod failure from the engine cylinder 1, two scenarios can be developed:

A - The destruction of the connecting rod was initiated by the failure of the connecting rod small end eye's dangerous sections, thus the movement of the rod was not anymore guided by the evolution of the piston in the cylinder. Subsequently, in a second phase, rupture of the bolts and of the rod body took place, as a result of the stresses induced by the impact of the rod body with the side walls of the engine block.

B -The destruction of the connecting rod was caused initially by the failure of the bolts and later of the rod's small end and body, as a result of the impact of the body against the side walls of the engine block. In this case, the movement of the rod big end was no longer synchronous with the rotation of the crankshaft throw corresponding to the cylinder 1. The rupture of the connecting rod small end eye occurred in a second phase of destruction

The investigations were deepened on the basis of electron microscopy analysis for the fracture sections of the two bolts and the rupture section of the rod's small end eye containing the oil hole. This last section was identified at the ruptured piece of the rod's eye found in the inferior crankcase of the damaged engine.

For analysis, a high-performance electronic microscope [3-5] (the TESCAN brand equipped with BEKER electronic cannon) was used, which allows, besides macro- and microscopic analysis of metal surfaces, also the analysis of the chemical composition at well defined points on the analyzed surfaces, with the detection of substances deposited on surfaces by assessing the energy potentials of relatively new chemical bonds, usually of a metallic nature.

Figures 1 and 2 show the photographic results of the electron microscope investigation of the cut-off part from the connecting rod small end eye, at the level of the surface that contains in its median area the oil hole. It is noted that the surface positioned to the inner fiber (the contact area with the rod bushing) has preserved its cylindrical shape after rupture. The breaking surface captured in the photograms is positioned along the direction of the rod's eye generic cylinder generator.

Figure 1. The rupture section of the rod's eye.

Figure 2. Detail from the rupture section of the rod's eye.

Results interpretation should take into account the connecting rod's material, respectively steel alloyed with manganese and chrome. In the photograms one can notice the existence in the breaking surface of two distinct areas:

- a more flat but smaller area with a "fresh" appearance of metallic luster arranged to the outer fiber (zone 2, as shown in figure 3);
- another area, larger, disposed to the inner fiber of the rod's eye (zone 1, as shown in Figure 3). This area, remarkable in terms of breaking surface (fibrous, drawing in "lines", "faults" or "waves") reveals the fatigue breaking of the material.

This type of steels rupture highlights the progressive propagation (according to the drawing lines) of an initial microfissure, with the evolution of rupture over time during the operation, initiated by a possible endogenous factor. Such a rupture can be initiated even by a microscopic defect of the crystalline structure of the steel and occurs at exploitation of the analyzed component at fatigue under high loading of the resistant section. This kind of rupture can be initiated and propagated reaching catastrophic break even after several hundred hours of cyclical fatigue.

The flat and metallic luster appearance zone 2 occurs in the load cycles immediately preceding the catastrophic breakage. It is even possible that the specific appearance of the surface is partly due to the "beating" (micro-vibrations) of the surface with a conjugated metal surface over the last few functional cycles of the engine in the area of total rupture, when, although the section under analysis is pierced by tearing, the embedded section of the connecting rod's eye is still sufficiently strong to guide the piston pin. These findings are clear arguments in favor of scenario "A" of engine destruction. Also, the microfissure could not be initiated by mounting the piston pin in the rod bushing because - as we have already mentioned, the solution used in this engine is with the piston pin freely mounted in the eye of the rod's small end.

A chemical analysis of the surface composition was carried out for the rupture area, revealing the following: first of all, the basic elementary composition of the steel from which the connecting rod is made (oxygen, carbon, iron, manganese and chromium). Secondly, the chemical analysis emphasized the fixing of four metallic elements (aluminum, sodium, magnesium and calcium) and an organic polymeric compound (silicone polysiloxane) originating from insoluble substances contained in the engine lubricating oil, which penetrated the microfissure during the operation of the engine

Figure 3. The cut-off piece from the rod's small end.

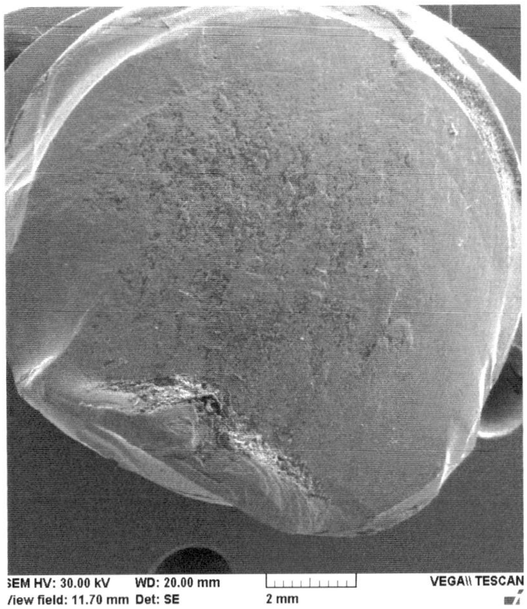

Figure 4. The breakage section of the bolt 2.

This last aspect highlights the different "ages" of the two areas of rupture. The formation of metallic bonds with high energy potential, needed to fix the metal atoms (aluminum, sodium, magnesium, calcium) to the base metallic material of the connecting rod does not occur instantly, and shows that the rupture in zone 1was made before the rupture in zone 2 (which did not have the time required for contamination with various metals).

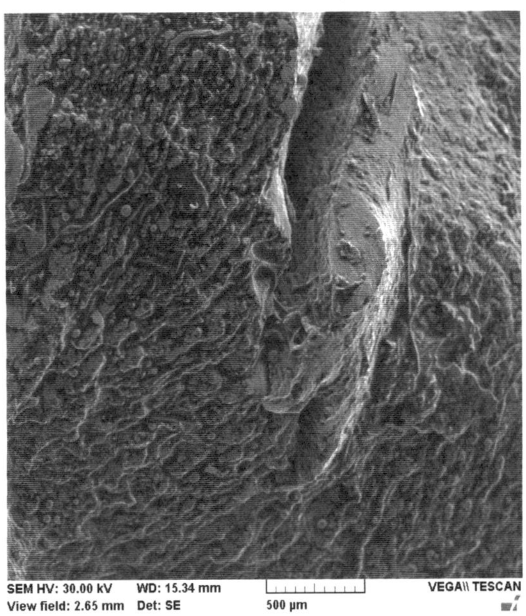

Figure 5. The rupture surface of the bolt 1.

IOP Conf. Series: Materials Science and Engineering **209** (2017) 012077 doi:10.1088/1757-899X/209/1/012077

The chemical analysis of the rupture surface validates the theory of rupture in time, by fatigue of this section, with the obvious propagation of the rupture from the inner fiber to the outer one of the rod's small end eye.

Figure 4 shows the breakage section of the bolt 2, with the typical appearance of rupture by dynamic loading with shock, under a combined effort of eccentric tightening with bending, favored by course thread tension concentrators.

Figure 5 shows the rupture surface of the bolt 1 and which, in qualitative terms, detaches itself from the rupture of the bolt 2. The irregular surface appearance with peaks of material alternating with material squeezes and excavations reveals the occurrence of rupture by dynamic loading with shock under pure bending effort. In this case, the presence on the bolt body of threaded tension concentrators plays an insignificant role.

3. Conclusions

Engine failure was caused by the fatigue rupture of the connecting rod small end's eye from the engine's cylinder no. 1 along the direction of the generator, in the section of the rod's eye that contained the oil hole, and which favored breaking by concentrating the stresses.

The rupture section was strongly stressed by bending in the oscillation plan dynamically, in asymmetric alternating fatigue cycle. Weakness of the breakage section was generated by a structural intimate defect which led by exploitation over time to the initiation of a microfissure on the inner fiber of the rod's eye. This microfissure progressively propagated to the outer fiber as the engine was operated until the whole section was penetrated, culminating in catastrophic breakdown.

References

[1] Basshuysen R and Schäfer F 2015 Handbuch Verbrennungsmotor *Grundlagen, Komponenten, Systeme, Perspektiven, Springer Vieweg*

[2] ASM Handbook 2002 Volume 11: Failure Analysis and Prevention *American Society of Metals-International*

[3] Perju M C, Nejneru C, Vizureanu P and Axinte M 2013 *Metalurgia International* **18**(5) 174-177

[4] Bârcă E S, Plaiasu A G, Abrudeanu M, Istrate B, Luca D, Munteanu C 2015 Journal *of Optoelectronics and Ddvanced Materials* **17**(9-10) 1522-1527

[5] Tugui C A, Vizureanu P, Iftimie N, Steigmann R 2016 Some aspects over the quality of thin films deposited on special steels used in hydraulic blades, *IOP Conference Series-Materials Science and Engineering* **147** 012040

IOP Conf. Series: Materials Science and Engineering **209** (2017) 012078 doi:10.1088/1757-899X/209/1/012078

Evaluation of Resin-Resin Interface in Direct Composite Restoration Repair

S Stoleriu*, S Andrian, G Pancu, I Nica and G Iovan
Odontology, Periodontology and Fixed Prosthodontics Department, "Grigore T. Popa" University of Medicine and Pharmacy, Faculty of Dental Medicine, Iași, Romania

E-mail: stoleriu_simona@yahoo.com

Abstract. The aim of this study was to evaluate the resin-resin interface when a universal bonding agent was used in two different strategies in direct restoration repair. Two composite resins (a micro-filled hybrid and a nano-filled hybrid) as old restorations that have to be repair, a universal bonding agent and a micro-filled hybrid composite resin (different then that aged) as new material for repair were chosen for the study. Non-aged samples were used as control and aged samples were used as study groups. The universal bonding agent was applied in etch-and-rinse and in self-etch strategies. The interface between old and new composite resins was evaluated by SEM and the microleakage was assessed by scoring the dye penetration. Very good adaptation of the two different composite resins placed in direct contact in non-aged samples was recorded. No gaps or defects were visible and strong resin-resin contact was observed. After aging, enlargement of resin-resin junction were observed in most of the samples and a increased dye penetration was recorded irrespective of the strategy (etch-and-rinse or self-etch) used for bonding agent application.

1. Introduction

Since their use in dentistry in early 60s, composite resins have been the most used material for final direct restorations due to their continuous improvement of esthetic quality, mechanical properties and ability to bond to dental tissues mediated by adhesive systems [1]. Unfortunately, in oral environment, composite resins are prone to deterioration and degradation [2,3]. For many years the treatment of failed restorations consisted in complete removal of fillings. This approach leaded to excessive removal of sound enamel and dentine due to the practitioner's need to „do something", while trying to discriminate between the perfect matched composite resin and the dental tissues and providing freshly new cut enamel for bonding the new material [4,5]. The repetitive cycle of replacing old restorative materials would lead to larger cavities [6-8], increased risk to harm the pulp during the removal of the filling and increased cost of the procedure.

The minimally invasive concept was introduced in dentistry in early 70s with the major goal to preserve as much as possible the tooth structures. This approach focuses on early diagnosis of caries disease, remineralization of incipient lesions and reduction of cariogenic bacteria, minimum preparation in cavitated caries lesions, and repair rather than replacement of defective restorations [9-13]. Composite repair is considered a conservative procedure, which is increasingly accepted and practiced [14,15]. It is estimated nowadays that 50 to 71 percent of the dentist activity is represented by the repair of existing restorations [10]. Repair is indicated in fillings fracture [16], material discoloration or worn areas [17], poor anatomic form, secondary caries, tooth fracture and tooth pain/sensitivity. This procedure consists in adding a fresh new layer of material in contact with an old restorative material from which a part had been removed by fracture, by wear or by the practitioner. The success of the repair relies on interfacial coupling and long-term retention between the surfaces involved in repair.

Factors like composite surface contamination [18], polishing [5] and ageing [19] are responsible for impairing of direct bonding. Chemical compatibility (related to organic matrix and polymerisation method) between the two composite resins could also affect bonding, since the brand of pre-existing composite resins is often unknown [20,21]. The age of the old restoration and the elapsed time since the repair are also important factors that can affect the bonding [22]. In oral environment, the bond between old and new composite resins is prone to degradation promoted by the direct contact with moisture and by salivary enzymatic activity [22].

Different methods for surface treatment and various intermediary agents were tested in previous studies in order to obtain an ideal adhesion between the two different materials used in the repair procedure [17,23,24]. It was demonstrated that an intermediate adhesive agent increases the repair bond significantly due to surface wetting and chemical bond with the new composite [25-29]. All dental adhesive systems have three different components: etching agent, primer and bonding agent. Clinically, acid-etching might be applied in a separate step (etch-and-rinse technique) or it might be provided by acidic functional monomers (self-etch technique). The most recent novelty in dental adhesion was the introduction of universal or multi-mode bonding agents. Often these materials consist in a mixture of all three components in the same bottle. According to the manufacturer, they can be applied either in etch-and-rinse as in self-etch procedure.

Clinically, good bond strength is reflected by the interface ability to prevent the microleakage [30]. The leakage is responsible for bond degradation, marginal staining of the restoration, biofilm accumulation and the risk of caries lesions onset. Previous studies that investigated microleakage of the composite repair interface showed differences between the adhesive systems investigated, but no data are available regarding universal bonding agents [31].

The aim of this study was to evaluate the resin-resin interface when a universal bonding agent was used in two different strategies in direct restoration repair.

2. Experimental procedures

2.1. Obtaining the composite resin specimens

Two different composite resins were chosen for this study: a micro-filled hybrid (Filtek Z250, 3M ESPE- MH1) and a nano-filled hybrid (Premise, Kerr NH). Forty quadrangular specimens of each material were obtained by placing unpolymerized composite resin into molds having 5 mm in length, 5 mm in width and 2 mm in high. One layer of 2 mm was placed using a buccal spatula and a condenser, and was polymerized for 40 seconds using a LED light unit (LED B, Guilin Woodpwcker Medical Instrument Co., Ltd, China) having the light intensity of 850-1000mW\cm^2 and the wavelength of 420-480 nm. Before the last increment polymerization, a translucent Mylar strip and a 2mm thick glass slab will be place on top of the surface layer in order to obtain a flat, smooth surface and to prevent the formation of the oxygen inhibited layer. Twenty specimens of each material were used to obtain control samples and twenty specimens were aged by storing in citric acid for one week.

2.2. Composite resin repair and samples preparation

One composite resin (Z-mack, Zermack- MH2) was chosen to be place in contact with the non-aged and aged composite resins as a repair material. For bonding the micro-filled hybrid resin to aged resins, an universal bonding agent (G Premio Bond, GC Corporation- UBA) was used in two different strategies: etch-and-rinse (strategy 1), and self-etch (strategy 2). The layout of the groups is presented in table 1.

Table 1. The layout of the groups.

	non-aged MH1	non-aged NH	aged MH1	aged NH	UBA strategy 1	UBA strategy 2	MH2
Group 1	X				X		X
Group 2	X					X	X
Group 3		X			X		X
Group 4		X				X	X
Group 5			X		X		X
Group 6			X			X	X
Group 7				X	X		X
Group 8				X		X	X

In strategy 1, 35% phosphoric acid etchant gel (3M-ESPE, St. Paul, MN, USA) was applied for 15 seconds on one of 2mmx5mm surface of the resin specimens using the tip of the syringe, then the acid was removed using the water spray from the dental unit and gently dry using the air spray. For bonding procedure, the application of UBA was made according to the producer instructions. Using disposable applicator brushes, the universal adhesive agents was applied by scrubbing on the resin surface for 20 second, gently air dry for 5 seconds using medium air pressure and then lightcured for 20 seconds.

In order to obtain the samples, the repair composite was applied in contact with non-aged or aged specimens. For that non-aged and aged composite specimens were reassembled in 10 mm length, 5 mm width and 2 mm high molds, with the surface where the universal adhesive agent was applied toward the middle of the mold. MH2 was inserted in one increment of 2mm, using the same procedure as for MH1 and NH. Different shade was chosen for repair material for a better assessment of the repair interface. The samples were then removed from the molds and stored in distilled water for 24 hours.

2.3. Microleakage assessment by dye penetration and SEM evaluation of the samples interface

The samples from each group were used for evaluation of adhesive efficiency by microleakage assessment and for SEM evaluation of the interface. All the extern surfaces of the samples will be coated with two layers of water resistant nail varnish, except the 10mm x5mm surface that was finished. The samples were immersed in 2% methylene blue buffered dye solution (pH=7) for 4 hours (Cavalcanti, 2004).

After that the samples were transversely sectioned using diamond discs (Komet Dental, Brasseler GmbH&Co, Germany), under watercooling. The sections were examined using a optical microscope (Carl-Zeiss AXIO Imager A1m) coupled with a high resolution digital camera at 50x magnification. The dye penetration was assessed according to the following scores: 0- no dye penetration, 1- dye penetration on less then a half of the interface, 2- dye penetration on more then a half of the interface, but less then whole interface, 3- complete dye penetration of the interface.

A scanning electron microscope (VEGA II LSH, Czech Republic) was used to analyze the aspect of the repair interface. The morphology of UB layer in groups 1-8, and the microgaps formation between UB and MH1, NH, and MH2 were performed at x100, x300, and x1000 magnification.

3. Results and discussions

Some SEM aspects of resin-resin interface in groups 1-8 are presented in figure 1. Very good adaptation of the two different composite resins placed in direct contact in all control groups was recorded. No gaps or defects were visible and strong resin-resin contact was observed. In groups 4-8, enlargement of resin-resin junction were observed in most of the samples.

Table 2. Dye penetration scores in control and study groups.

		No of samples					
	In groups	score 0	score 1	score 2	score 3	score 4	Mean score
Group 1	10	10	-	-	-	-	0
Group 2	10	10	-	-	-	-	0
Group 3	10	10	-	-	-	-	0
Group 4	10	10	-	-	-	-	0
Group 5	10	1	6	1	2	-	1.4
Group 6	10	1	7	1	1	-	1,2
Group 7	10	1	6	2	-	1	1,4
Group 8	10	1	7	1	1	-	1,2

In figure 1 some samples scored with 1 according to dye penetration are exemplified (groups 5 and 8). The dye penetration scores in all eight groups are presented in table 2. In groups 1-4 there was no sample scored more than 0 for dye penetration. In groups 5-8, increased values of scores were registered when compared to control.

Group 1

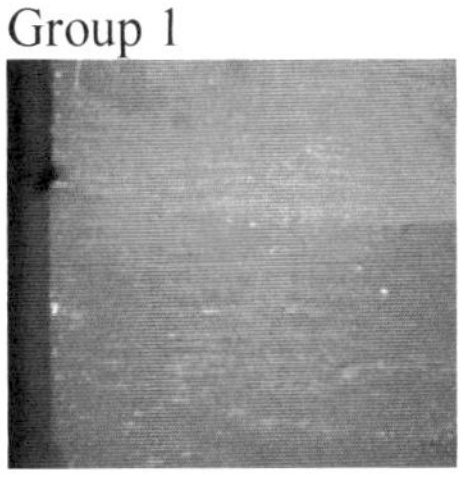

Dye penetration-
Score 0

SEM X 1000

Group 2

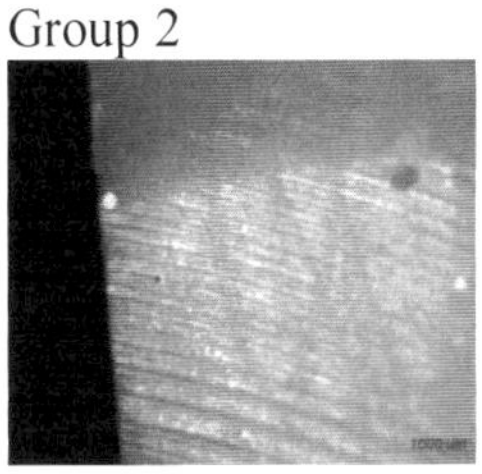

Dye penetration-
Score 0

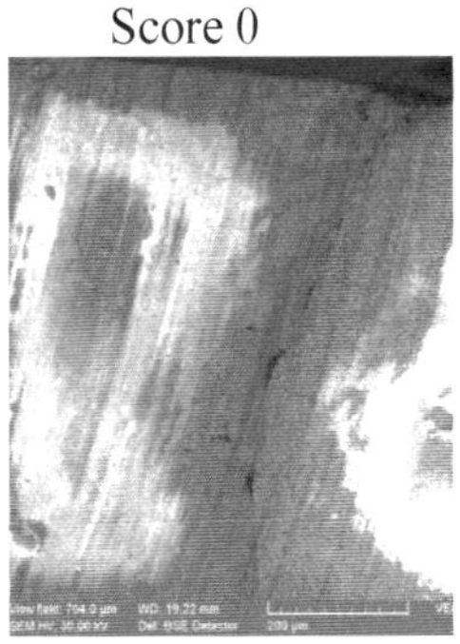

SEM X 1000

Group 3

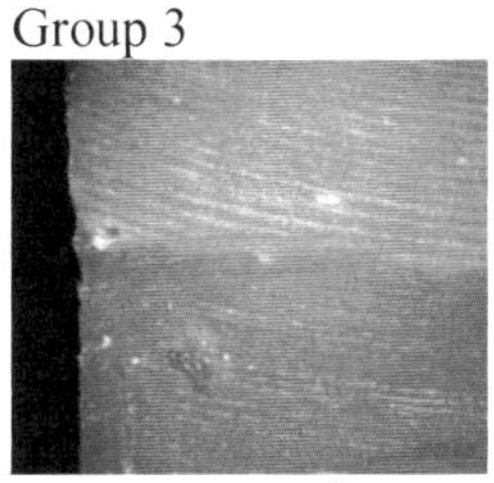

Dye penetration-
Score 0

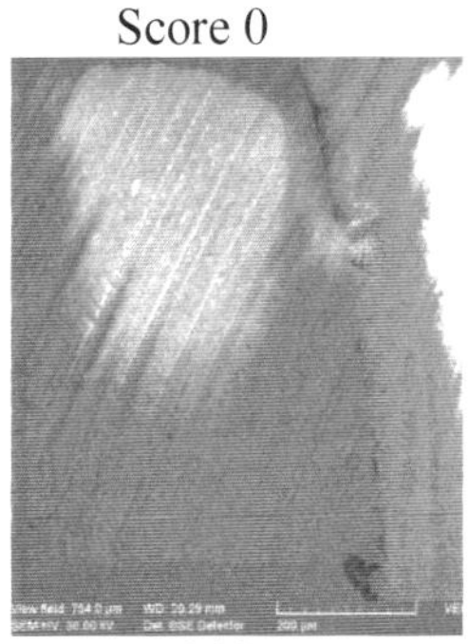

SEM X 300

Group 4

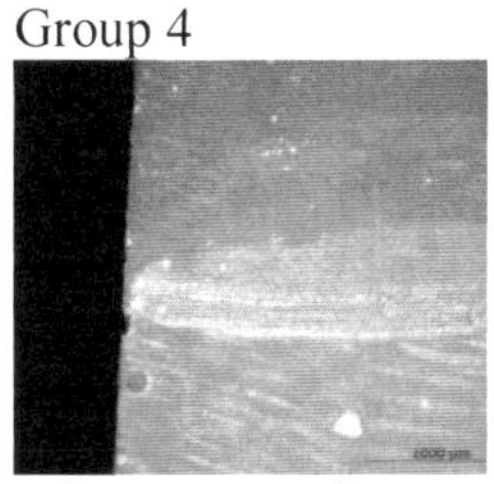

Dye penetration-
Score 0

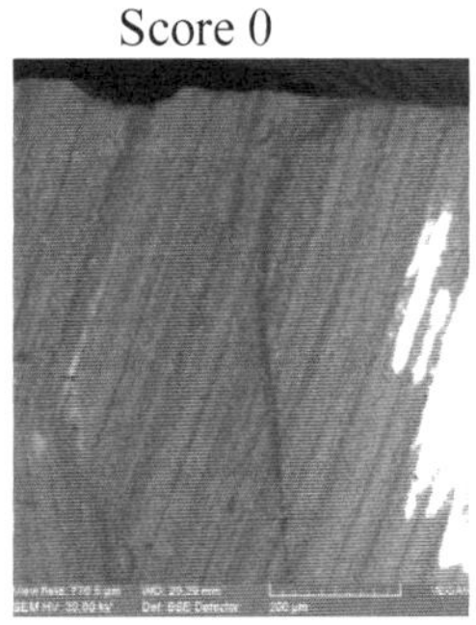

SEM X 300

Group 5

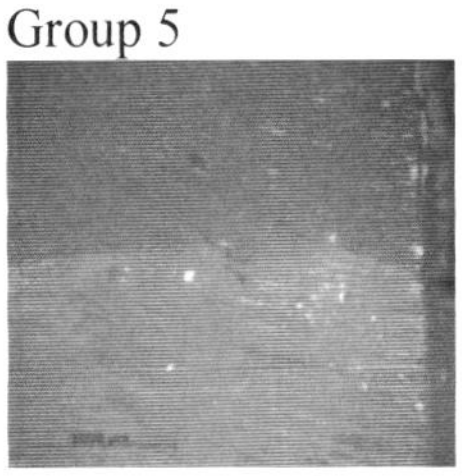

Dye penetration-
Score 1

Group 6

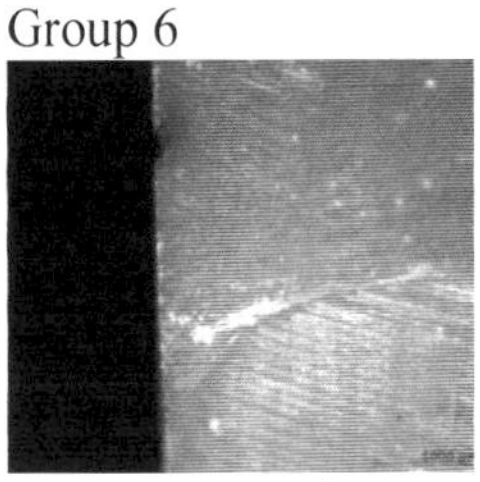

Dye penetration-
Score 0

Group 7

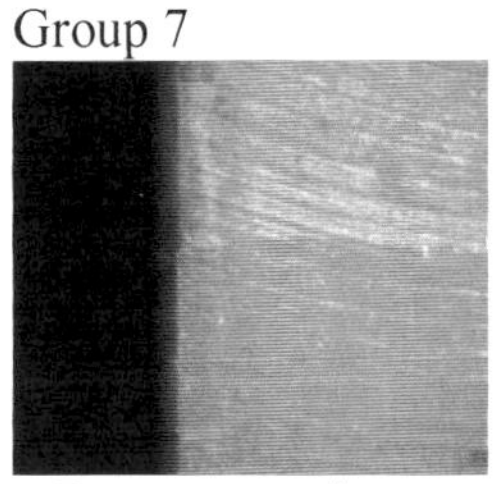

Dye penetration-
Score 0

Group 8

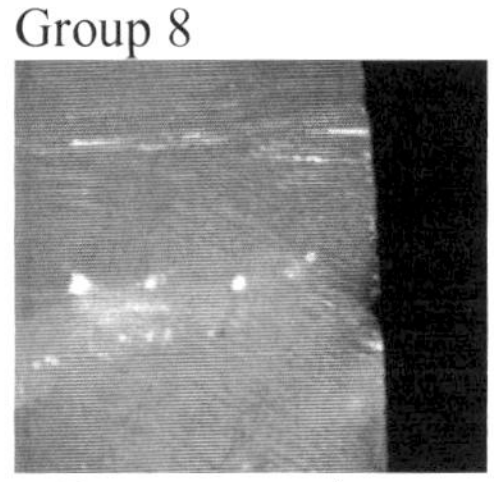

Dye penetration-
Score 1

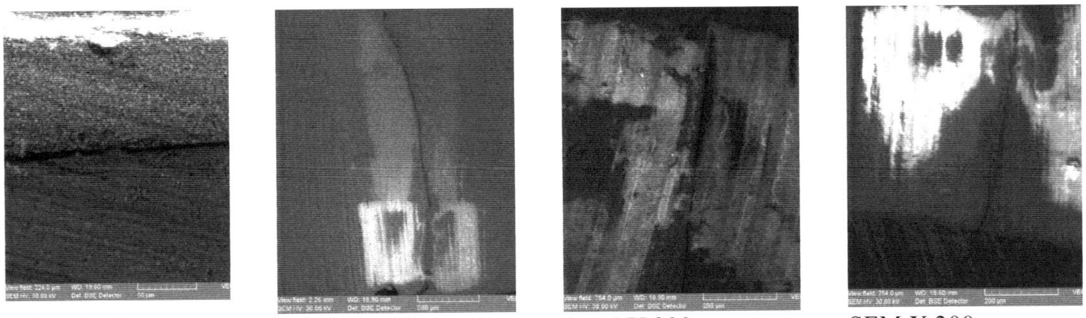

SEM X 300 SEM X 1 00 SEM X 300 SEM X 300

Figure 1. Aspects of dye penetration and SEM aspect of the interface in control and study groups.

It was demonstrated that microleakage evaluation is a suitable method to assess the performance of adhesive systems especially when few clinical data are available [31-35]. In order to obtain the composite resins aging, different storage media and techniques were used in previous studies: immersion in artificial saliva, saline solution, deionized water, water [36], citric acid [37] or thermocycling. The idea of using citric acid for aging was to mimic the action of acidic beverages or food. Studies that analysed the effect of different storage media showed similar effects on filler exposure when using citric acid, water storage [5] and termocycling. Despite termocycling and water storage, citric acid leads to resin matrix degradation [28] and a lesser surface roughness. The increased surface roughness after immersion in water might be explained by swelling instead of matrix degradation [28]. Resin-to-resin bonding after aging was evaluated in previous studies [38].

In composite resin repair a big challenge is represented by the adhesion between a new composite resin and a pre-existing one, which is unknown most of the time. Generally, when placing two composite layers in contact, the adhesion is obtained through the oxygen-inhibited un-polymerized resin layer [39]. Aged composite resins have lesser or no un-reacted monomers, so pour adhesion is expected at the interface with the newer resin. In our study increased values of dye penetration were recorded at resin-resin interface after both composite resins aging as a result of the less efficient adhesion.

Previous investigations showed that microleakage at the resin-resin interface is not correlated to the type of pre-existing composite resin [31]. Almost similar dye penetration was recorded in present study when nano-filled hybrid and micro-filled hybrid were used as pre-existing materials.

In order to obtain a good composite repair procedure, micromechanical adhesion is desirable. For that, increasing the surface area of the pre-existing resin or application of an intermediate layer (like bonging agent) was recommended [5, 40, 41]. For self-etch bonding systems, the acidic primer seems to provide a clean surface [42] and for etch and rinse bonding systems phosphoric acid is responsible for this action. When phosphoric acid was used in etch and rinse bonding systems, it was demonstrated that no significantly effect on improving the bonding in composite repair procedure [29, 43]. Previous studies demonstrated that both etch and rinse and two-step self-etching systems lead to similar results in composite repair. Universal adhesives or multi-mode adhesives contain all the bonding components in one bottle and they can be applied either in etch-and-rinse or self-etch approaches. In our study the universal bonding agent applied in both strategies determined comparable results when used as intermediate layer in composite repair.

Still conflicting results were reported when evaluating the multiple variables involved in bonding at the resin-to-resin interface. The literature reveals a lack of common opinion regarding the behaviour of different bonding agents used in direct composite resins repair and very little information regarding the efficiency of universal bonding agents. The few studies that have been published aimed only to evaluate the bond of universal adhesive agents to dental hard tissues. In our study the use of universal adhesive system in both strategies for composite repair conducted to similar result regarding the adhesion efficiency. The results of our study are preliminary, future studies being necessary to investigate more the bonding capacity of universal adhesive agents used in composite resin repair.

4. Conclusions

Very good adaptation of micro-filled hybrid composite resin used to repair the non-aged micro-filled hybrid and non-aged nano-filled hybrid composite resin was recorded. The direct contact between both non-aged composite resin and the new composite resin when using universal bonding agent as intermediate layer did not leaded to gaps formation or visible defects. After aging, enlargement of resin-resin junction and a increased microleakage was recorded, irrespective of the strategy (etch-and-rinse or self -etch) used for bonding agent application.

References

[1]. Roulet J F, Roulet J F, Wilson N H F, and Fuzzi M 2001 *Advances in Operative Dentistry*. vol 1, Quintessence Publishing Co Inc
[2]. Soderholm K J, Zigan M, Ragan M, Fischlschweiger W, and Bergman M. 1984 *J. Dent. Res.* **63** 1248–1254
[3]. Finer Y, and Santerre J P 2004 *J. Dent. Res.* **83**: 22-26
[4]. Söderholm K-J M, and Roberts MJ 1991 *Scand. J. Dent. Res.* **99** 2 173- 80
[5]. Shahdad S A, and Kennedy J G 1998 *J. Dent.* **26** 8 685-94
[6]. Krejici I, Lieber C M, and Lutz F 1995 *Dent. Mater.* **11** 34-40
[7]. Gordan V V 2000 *J. Esthet. Dent.* **12** 209-215
[8]. Gordan V V, Mondragon E, and Shen C 2002 *Quintessence Int*. **33** 273-278
[9]. Murdoch-Kinch C A, and McLean M E 2003 *J. Am. Dent. Ass.* **134** 87-95
[10]. Tyas M J, Anusavice K J, Frencken J E, and Mount G J 2000 *Int. Dent. J.* **50** 1 1-12
[11]. Mjör I A, and Gordan V V 2002 *Oper. Dent.* **27** 528-534
[12]. Staehle H J. Minimally invasive restorative treatment. J Adhes Dent 1999; 1: 267-284
[13]. Chalmers J M 2006 *J. Can. Dent. Ass.* **72** 435-440
[14]. Gordan V V, Mjör I A, Blum I R, and Wilson N 2003 *J. Am. Dent. Ass.* **134** 317-323
[15]. Blum I R, Schriever A, Heidemann D, Mjör I A, and Wilson N H F 2003 *Eur. J. Dent. Educ.* **7** 41-48
[16]. Chalkley Y, and Chan D C N 1986 *J. Prosthet. Dent.* **56** 4 441-444
[17]. Brosh, T, Pilo R, Bichacho N, and Blutstein R 1997 *J. Prosthet. Dent.* **77 2** 122-126
[18]. Eiriksonn S O, Pereira P N R, Swift E J Jr, Heymann H O, Sigurdsson A 2004 *Dent. Mater*. **20** 37-44
[19]. Mitsaki-Matsou H, Karanika-Kouma A, Papadoyiannis Y, and Theodoridou-Pahine S 1991 *Quintessence Int*. **22** 475-481
[20]. Kao E C, Pryor H G, and Johnston W M 1988 *J. Prosthet. Dent.* **60** 328-333
[21]. Gregory W A, Pounder B, and Bakus E 1990 *J. Prosthet. Dent*. **64** 664-668
[22]. Finer Y, and Santerre J P 2004 *J. Dent. Res.* 2004 **83** 22-26
[23]. Lucena-Martín C, González-López S, and Navajas-Rodrígues de Mondelo J M 2001 *J. Prosthet. Dent.* **86** 5 481-8
[24]. Öztas N, Alaçan A, and Bardakcy Y 2003 *Oper. Dent.* **28** 2 149-54
[25]. Padipatvuthikul P, and Mair L 2007 *Dent. Mater. J.* **23** 4 519-525
[26]. Yesilyurt C, Kusgoz A, Bayram M, and Ulker M *2009 J. Esthet. Rest. Dent.* **21** 4 251-260
[27]. Junior S, Ferracane J, and Bona A 2009 *Den.t Mater. J.* 2009 **25** 4 442-451
[28]. Rinastiti M, Özcan M, Siswomihardjo W, and Busscher H 2010 *J. Dent.* **38** 1 29-38
[29]. Papacchini F, DallÓca S, Chieffi N, Goracci C, Sadek F T, Suh B I, Tay F R, and Ferrari M 2007 *J. Adhes. Dent*. 2007 **9** 25–31
[30]. Hadavi F, Hey J H, Ambrose E R, and Elbadrawy H E 1993 *Oper. Dent*. **18** 1 2-7
[31]. Celik C, Cehreli SB, and Arhun N. 2015 *Eur. J. Dent.* 2015; **9** 1 92–99
[32]. Aschenbrenner C M, Lang R, Handel G, and Behr M. 2012 *Clin. Oral Investig*. **16** 191–200

[33]. Bagis Y H, Baltacioglu I H, and Kahyaogullari S 2009 *Oper. Dent.* **34** 578–85
[34]. Iovan G, Stoleriu S, Pancu G, Topoliceanu C, Ciobanu M C, and Andrian S 2014 *Romanian Journal of Oral Rehabilitation* **6** 2 18-23
[35]. Iovan G, Stoleriu S, Andrian S, Dia V, Căruntu I D 2004 *Rev Med Chir Soc Med Nat Iaşi* **108** 4 894-898
[36]. Yap A U J, Sau C W, Lye K W 1999 *Oper. Dent.* **24** 371–376
[37]. Gregory W A, Pounder B, and Bakus E 1990 *J. Prosthet. Dent.* **64** 664-668
[38]. Rasmussen S 1996 *Int J Adhesion and Adhesives* **16** 147–154
[39]. Ozcan M, Kojima A N, Pekkan G, Mesquita A M, and Bottino M A 2014, *Int. J. Adh. Adhes.* **49** 97–102
[40]. Costa T R, Ferreira S Q, Klein-Júnior C A, Loguercio A D, and Reis A 2010 *Oper. Dent.* **35** 231–7
[41]. Cehreli S B, Arhun N, and Celik C. 2010 *Oper. Dent* **35** 337–44
[42]. Cavalcanti A N, De Lima A F, Peris A R, Mitsui F H, and Marchi G M 2007 *J. Esthet. Restor. Dent.***19** 90–8
[43]. Hannig C, Laubach S, Hahn P, and Attin T. 2006 *J. Adhes. Dent.* **8** 35–40

IOP Conf. Series: Materials Science and Engineering **209** (2017) 012079 doi:10.1088/1757-899X/209/1/012079

SEM Evaluation of Surrounding Enamel after Finishing of Composite Restorations– Preliminary Results

G Iovan[1], S Stoleriu[1], S Solomon[1], A Ghiorghe[1], A V Sandu[2] and S Andrian[1]
[1]Department of Odontology, Periodontology and Fixed Prosthodontics, Faculty of Dental Medicine, "Grigore T. Popa" University of Medicine and Pharmacy, Iaşi, Romania
[2]Faculty of Materials Science and Engineering, "Gheorghe Asachi" Technical University of Iaşi, Romania

E-mail: stoleriu_simona@yahoo.com

Abstract. The purpose of this study was to analyze the surface characteristics of the enamel adjacent to composite resin after finishing the restoration with different diamond and tungsten carbide burs. The topography of enamel was observed by using a scanning electron microscope. Finishing with extra-/ultra-fine carbide burs, and extra-fine diamond burs resulted in smooth surfaces. In few areas some superficial scratches with no clinical relevance were observed. Deep grooves were observed on the surface of enamel when fine diamond burs were used. Finishing of composite restorations with coarse burs should be avoided when there is a high risk of touching and scratching adjacent enamel during the procedure.

1. Introduction

Most composite restorations need to be finished in order to adapt to adjacent dental tissues. Finishing aims to remove excessive material from the margins and improve the contour of the restoration.

Different finishing instruments are available at the moment, among the most frequently used being the diamond and carbide burs. During finishing, both restoration and adjacent tooth may be subjected to excessive abrasion by over-instrumentation. However limited information is available regarding the iatrogenic potential of this procedure on the enamel adjacent to the restoration.

Most of the information is derived from the orthodontic data related to enamel injury after bracket removal. Despite the minor contradictions related to the variability of the tested instruments, technical protocols and research methods, most of the studies concluded that both diamond and tungsten carbide burs are fast and effective instruments for resin removal but they remove also a layer of enamel and roughen its surface[1-4].

As regarding restorations, although there are numerous studies on different methods of finishing and polishing the surface of restorations[5-7], which polishing technique is least damaging for the tooth surface is still a mater of debate. Sparse data is available[3, 8] although for successful treatment, preserving the integrity of adjacent dental tissues should be equally important as achieving a smooth restoration.

The purpose of this study was to analyze the surface characteristics of the enamel adjacent to the composite resin after finishing the restoration with different types of burs.

2. Experimental procedures

In this study 16 extracted human molars were used. The teeth had been stored in refrigerated distilled water and used for maximum 1 month following extraction.

Buccal and lingual V class cavities were prepared using a cylindrical diamond bur with water spray and high speed. The dimensions of of the cavities were approximately 1.5mm depth, 4mm wide and 2mm high. All the margins of the cavities were prepared butt-joint in enamel and finished with fine-grit diamond bur. The cavities were cleaned with water and lightly air-dried using cotton pellets. The teeth were restore with a microhybrid composite - G-aenial Posterior (shade A3) and a self-etching one component adhesive - G-aenial^{TM} Bond (GC Corporation, Tokyo, Japan) using a bulk-technique.

The composite resin was polymerized against a Mylar matrix. In order to minimize the extension of the material at the margins, the matrix was tightened and held by fingers beyond the margins of the cavities. Due to the normal convexities of the dental crown, composite overflows were present on the adjacent enamel in all specimens. Half of each restoration was finished using one of the tested rotary instrument, simulating the clinical procedure of removing the excessive material extending over the margins of the cavities.

The buccal restorations were finished with tungsten carbide burs and the lingual restorations were finished with diamond burs, for approximately 10s. All the preparations were performed by the same operator. 4 buccal restorations were used for testing each type of the carbide burs. 2 of the corresponding lingual restorations were used for testing a fine diamond bur and the other two restorations for testing a superfine diamond bur. The characteristics of the tested instruments are listed in table 1.

Table 1. Technical characteristics of the finishing instruments.

Instrument		Band	Grit/ Rotation per minute	Code/Manufacturer
Diamond	Fine Tapered round-end	No color band	40μm 450,000 r.p.m.	C10F ISO 199/016 Top Diamco Deltadent, Italy
	Extrafine Tapered round-end	Yellow	20-30μm 300,000 r.p.m	TR-25EF ISO 199/016 MANI Inc, Japan
Tungsten Carbide	Extra-fine Straight-cut Tapered, safe end	Yellow	20 blades 160,000 r.p.m.	H 135SF-014-FG ISO 500314166041014 NTI Kahla GmbH, Germany
	Ultra-fine Straight-cut Tapered, safe end	White	30 blades 160,000 r.p.m.	H 135UF-014-FG ISO 500314166031014 NTI Kahla GmbH, Germany
	One-step finishing Straight cross-cut Tapered, safe end	Blue	16 blades 20,000 r.p.m.	H135S-014-FG ISO 500314166072014 NTI Kahla GmbH, Germany

The prepared teeth were split in an axial mesio-distal direction. The surface of the enamel at the joint with composite in both halves of the restorations (one half resulted from using only Mylar matrices and one half after finishing with a tested bur) were observed by scanning electron microscopy using a VEGA II LSH (Tescan Czech Republic). The integrally computer-assisted microscope has a cannon of electrons with tungsten filament, which may reach a resolution from 3nm to 30KV, and a magnification power up to 1.000.000X in the resolution mode, an accelerating voltage between 200 V and 30 kV, scanning rates between 200 ns and 10 ms per pixel.

Five of the specimens evaluated as being less abraded on adjacent enamel (one for each type of bur) were cut in an buccal-lingual direction and the section were also examined with the microscope.

Before each examination, the surface of the specimen was etched for 10s with phosphoric acid 37% and than washed with distilled water and dried. The acid etching was used in order to remove the smear layer that could change the appearance of the underneath enamel.

3. Results and discussions

In our study both diamond and carbide burs resulted in abrading the enamel surrounding the restoration, although all the tested instruments were specifically indicated by the manifacturer for finishing of composite restorations. The enamel injuries were in most cases superficial and inconsistent. As expected, the severity of the lesion was related to the efficacy of the bur to cut/abrade the substrate.

The tested diamond burs included a fine grit (40μm) and an extra-fine grit (20μm).

For the fine diamond bur, large scratches were observed on both enamel and composite resin even when 100X magnification was used (figure 1 a, b). Some of the scratches were visible with magnification loupes (3X). The fine diamond bur was included in the study because it is clinically widely used for effective removal of gross overhangs of restorations. The heavy abrasion was the result of the dimensions of the diamond grit. Obviously when using these type of burs for finishing composite restorations there is a high risk of overinstrumentation and damaging the interface with the enamel and even the surrounding enamel. Therefore such instruments should be used only for gross excess and care should be taken to avoid the contact with adjacent enamel. Once the excessive composite material is thinned, the fine burs should be changed with less aggressive instruments.

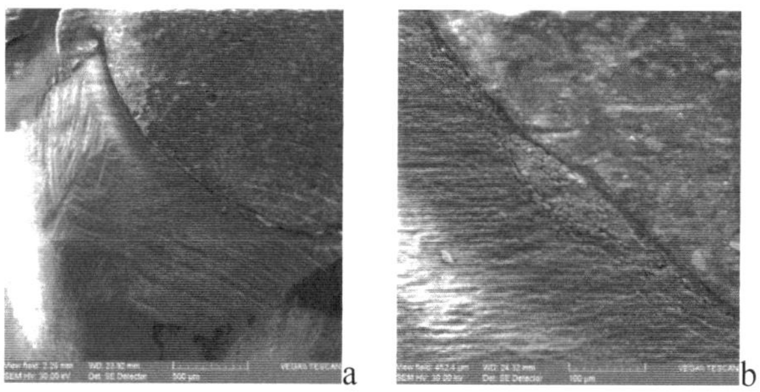

Figure 1. SEM images of enamel and composite surfaces in sample finished with fine diamond burs (a. 100X, b. 500X).

Using an extra-fine diamond bur decreased the risk of severe injuries. The scratches were superficial. For 100X magnification, the specimens finished with the extrafine diamond burs did not show any significant differences between the surfaces of enamel adjacent to unpolished and polished composite. However, in terms of composite surface it was noticed a different appearance between the two halves of the restorations. On the finished surface, the glass fillers looked much larger probably because of the grinding action of the diamond bur which removed the superficial layer of resin and exposed the mineral particles (figure 2a). When higher magnification was used, fine scratches were noted on the surface of enamel adjacent to the polished composite (figure 2b).

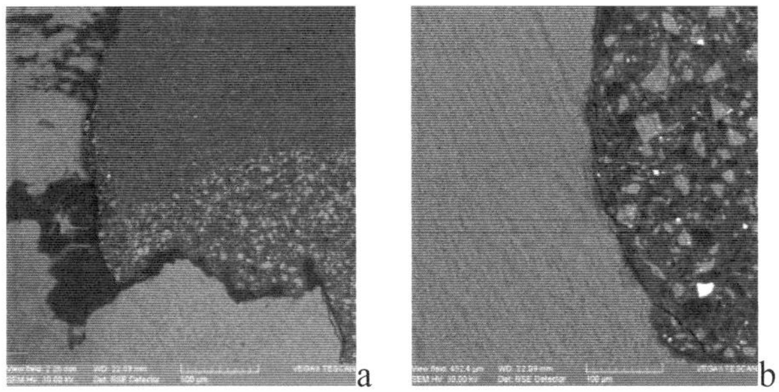

Figure 2. SEM images of enamel and composite surfaces in sample finished with extra-fine diamond bur (a. 100X, b. 500X).

The tested carbide burs were straight-cut with 20 blades (super-fine) and 30 blades (ultra-fine). These burs can be used in multiple-step procedure, but we decided to use them separately in order to distinguish between their iatrogenic potential.

In contrast with diamonds, carbide-finishing burs perform a cutting rather than a grinding action. Carbide finishing burs were designed to selectively cut the composite resin and conserve the enamel[9]. However conventional carbide burs may scratch the enamel due to the shape and sharpness of their blades. Working parts with nonrounded angles may damage enamel surfaces[10]. This applies to both burs with nonrounded flat ends and those with pointed cutting tips.

In our study when the carbide burs were used for polishing, the scratches were superficial and inconstant. Wide, deep grooves were absent.

For one sample finished with the extra-fine carbide there was a clear line of demarcation between normal enamel and the area where perikymata was flatten because of the cutting action of the bur (figure 3a). Within this area the bur "smoothed" the enamel without "scratching" it. In higher magnification very fine trails seemed to be present on the surface of enamel (figure 3b).

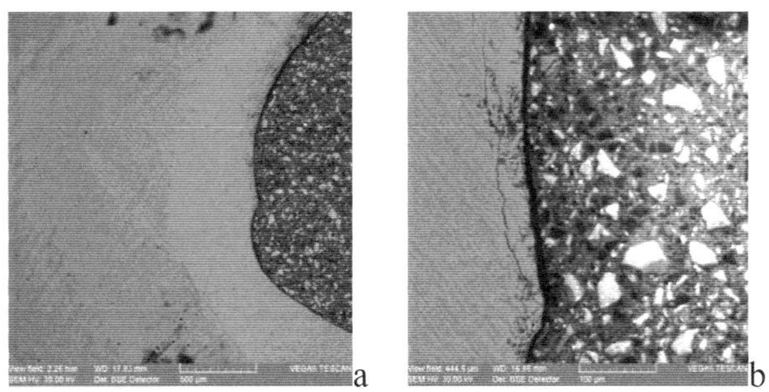

Figure 3. SEM images of enamel and composite surfaces in sample finished with extra-fine carbide bur (a. 100X, b. 500X).

Polishing with the ultra-fine carbide burs resulted in enamel margins with little or no scratches at all. The surfaces of enamel adjacent to the finished composite did not show any trace of instrumentation in lower magnification (figure 4a). In higher magnification some areas with shallow irregularities could be observed on the surfaces of enamel (figure 4b).

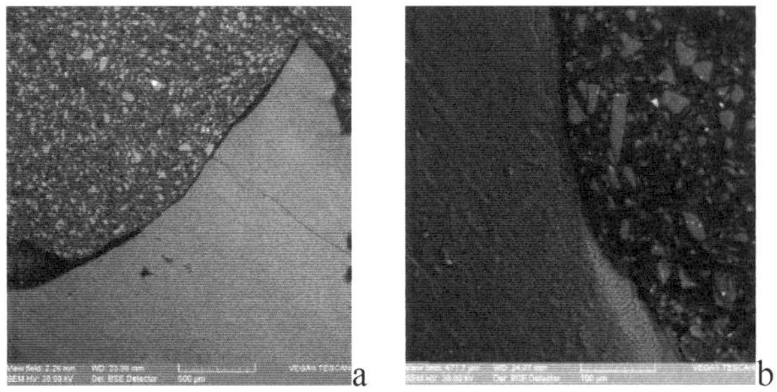

Figure 4. SEM images of enamel and composite surfaces in sample finished with ultra-fine carbide bur (a. 100X, b. 500X).

The third tested instrument was also a carbide bur designed to be use in single-step procedure. This bur has the same shape but fewer blades with an additional cross cut which creates a roughness depth of around 1µm.

The color code (blue) and the number of blades (16) suggest a high cutting efficiency, which usually results in rough surfaces. However the surface of surrounding enamel was smooth. No significant differences could be noted between the surfaces of enamel adjacent to the areas where Mylar matrix was used and the polished surface (figure 5a). In two specimens when higher magnification was use, a clear delineation was noticed between the two areas of enamel with several trails on the surface of enamel although no deep grooves were observed.

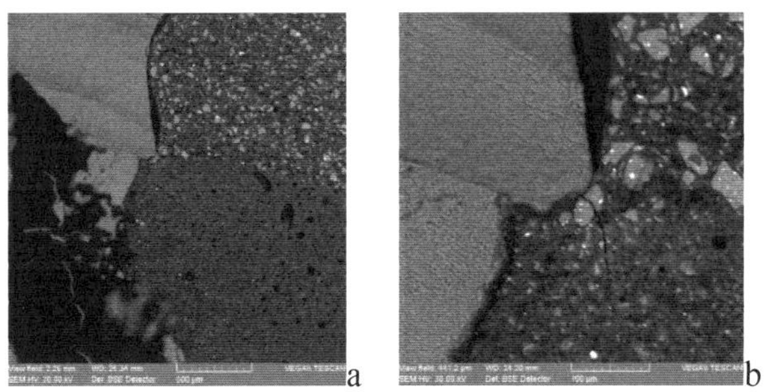

Figure 5. SEM images of enamel and composite surfaces in sample finished with cross-cut carbide bur (a. 100X, b. 500X).

When axial sections were assessed, a superficial loss of the adjacent enamel was observed at the margins of all the restoration after finishing (figure 6).

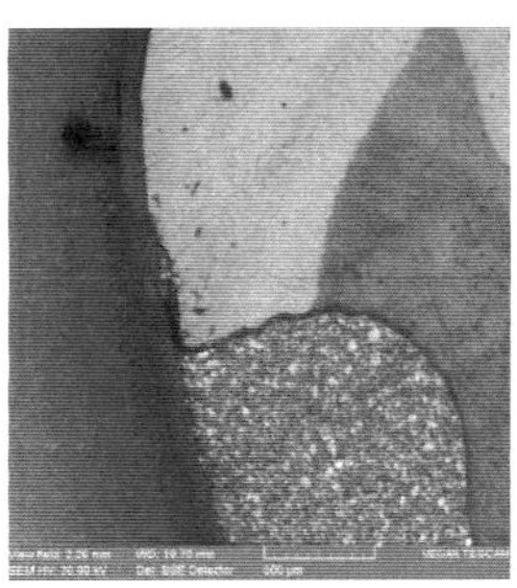

Figure 6. SEM image of the interface between composite and the enamel for finishing with ultra-fine carbide bur.

In our study both diamond and carbide burs resulted in cutting enamel at the joint with the composite resin. The enamel damage was superficial and inconsistent. These results are supported by a previous study which found iatrogenic abrasion of surrounding enamel during composite shaping and finishing procedures for both diamond and tungsten carbide burs [11].

The severity of the lesion was related to the type of the instrument and most important to the size of the grit or cutting blade. Best results were obtained by using extrafine diamond burs and ultrafine carbide burs. This is consistent with the results of previous studies [1,3]. In case of the cross-cut carbide bur the architecture of the blades seems to prevent the severe injury of the enamel and still being efficient in cutting the composite resin. The recommended rotation speed is low, which might contribute also to make the bur less harmful for the enamel.

Most of the data about the impact of polishing on enamel is provided by numerous studies related to resin removal after bracket debonding. In spite of the numerous studies, choosing between diamond and carbide burs have been controversial. Faria-Junior and coworkers found that polishing with an oxide disc system resulted in less enamel roughness than did the multilaminated carbide bur system [12], while Cehreli and coworkers found better results when using the 16-blade tungsten carbide bur than using ultrafine diamond bur [13]. Other studies observed that tungsten carbide drill with 30 blades during 30 s after bracket debonding resulted in higher surface variance for marks as well as deep and large scratches. This method however caused regular thickness wear, which is in accordance of our findings [14]. Similarly Vidor and coworkers found that the use of the tungsten drill (30 blades) in unidirectional movements can be recommended to remove large volumes of resin remnants although it produced grooves on the enamel surface [15]. Ryf and coworkers found that clean up with carbide burs resulted in higher variability of surface alteration [16]. Ulusoy found that 12-fluted carbide burs resulted in an irregular enamel surface in SEM evaluation, showing horizontal scars with a consistent pattern while 30-fluted burs resulted in less scarring of the enamel surface [17].

When considering adhesive removal after bracket debonding the enamel damage might be related not only to finishing procedure but also to the stress induced by the debonding of brackets and by the orthodontic treatment itself. Therefore the results of these studies cannot be extrapolated to restorative dentistry. Moreover the heterogeneity of the tested instruments, the technique used (high speed, low speed, water-spray), the time of finishing make comparison almost impossible between the results of different studies.

There are plenty of studies related to the efficacy of different finishing and polishing systems on the surface of the composite restorations as smoothness of the restoration is considered very important for the long-term success of the restoration [18]. Very few studies evaluate the impact of these procedures on adjacent dental surfaces. It was suggested that although resin remnants removal procedures cause deterioration of enamel surface, no clinical problem would be caused because the enamel surface layer contained a fluoride-rich layer of 50μm depth [19]. However the roughness of any surface exposed to oral environment would increase the retention of the bacterial biofilm. The threshold for plaque accumulation on the surface of restorative materials is considered to be 0.2μm [5,6]. The threshold for enamel is most probably different, still rough enamel at the margins of the restorations would significantly increase the risk for caries not because of the higher susceptibility of the abraded enamel but mostly because of the retention of bacterial biofilm.

Most of the studies found that finishing with both diamond and carbide burs result in enamel damage [4]. These damages can be assigned to the higher hardness of both diamond and tungsten carbide in comparison with enamel. In spite of these results the finishing burs are still widely used to remove composite resin because the advantages are significant. They are more effective than other rotary instruments and they have a versatile design allowing their access even in difficult-to reach areas. Whether the damage produced to enamel is clinically relevant or not depends on factors related not only to the instrument, but also to the technique and patient.

Our results indicate that coarse diamond burs abrade enamel severely. On the other hand both diamond and carbide ultrafine burs seem to be acceptable solutions. Considering these preliminary results, more research is necessary in order to assess the impact of finishing procedures on the long-term effects of these iatrogenic maneuvers. Also the dentin-composite interface should be assessed after finishing and polishing procedures since dentin is less resistant than enamel and the cervical areas of restorations are more prone to failure due to marginal defect and secondary caries.

4. Conclusions

Finishing of composite restorations with both diamond and carbide burs resulted in abrasion of adjacent enamel.

Finishing with extra-/ultra-fine carbide burs and extra-fine diamond burs resulted in smooth surfaces. SEM images showed only inconstant superficial scratches of the enamel adjacent to composite restoration.

Severe injury of enamel was observed only when fine diamond burs were used. Coarse rotary instruments should be avoided when there is a high risk of touching adjacent enamel during the finishing of composite restorations.

References

[1] Schmidlin PR, Gohring TN 2004 *Oper. Dent.* **29** 80-6
[2] Alessandri Bonetti G Zanarini M Incerti Parenti S Lattuca M Marchionni S Gatto MR 2011 *Am. J. Orthod. Dentofacial. Orthop.* **140** 696-702
[3] FerrarisF, Conti A 2014 *Int. J. Esthet. Dent.* **9** 70-89
[4] Janiszewska-Olszowska J, Szatkiewitcz T, Tomkowski R, Tandecka K, Grocholewicz K 2014 *Med. Sci. Monit.* **20** 1991-2001
[5] Edemir U, Sancakli HS, Yildiz E 2012 *Eur. J. Dent.* **6** 198-205
[6] Avsar A, Yuzbasioglu E, Sarac D 2015 *Adv. Clin. Exp. Med.* **24** 881-890
[7] Iovan G, Stoleriu S, Nica I, Solomon S, Munteanu A, Andrian A 2016 *Mat.Plast.* **53** 755
[8] Maresca C, Pimenta LA, Heymann HO, Ziemiecki TL, Ritter AV 2010 *J. Esthet. Dent.* **22** 104-112
[9] Glazer HS 2009 *Dent. Today.* **28** 122, 124-5
[10] Radlanski RJ 2001 *J. Orofac. Orthop.* **62** 229-304
[11] Mitchell CA, Pintado MR, Douglas WH 2002 *J. Prosthet. Dent.* **88** 320-8
[12] Faria-Júnior EM, Guiraldo RD, Berger SB, Correr AB, Correr-Sobrinho L, Conteras EF, Lopes MB 2015 *Am. J. Orthod. Dentofacial. Orthop.***147** 324-9
[13] Cehreli ZC, Lashimipathy M, Yazici R 2008 *Dent. Traumatol.* **24** 177-182
[14] Cardoso LAM, Valdrighi HC, Filho MV, Correr AB 2014 *Dental. Press. J. Orthod.* **19** 105
[15] Vidor MM, Felix RP, Marchioro EM, Hahn L 2015 *Dental. Press. J. Orthod.* **20** 61
[16] Ryf S, Flury S, Palaniappan S, Lussi A, van Meerbeek B, Zimmerli B 2012 *Eur. J. Orthod.* **34** 25-32
[17] Ulusoy C 2009 *J. Appl. Oral. Sci.*2009 **17** 209-215
[18] Ghiorghe A C, Iovan G, Topoliceanu C, Sandu AV, Andrian S 2013 *Rev. Chim. (Bucharest)* **64** 1436-1440
[19] Kim SS, Park WK, Son WS, Ahn HS, Ro JK, Kim YD 2007 Am. J. Orthod. Dentofacial. Orthop. **132** 71-6

Approaching on Colorimetric Change of Porous Calcareous Rocks Exposed in Urban Environmental Conditions from Iasi – Romania

V Pelin[1*], O Rusu[2*], I Sandu[3], V Vasilache[4], S Gurlui[2], A V Sandu[5,6], M M Cazacu[2,7] and I G Sandu[5,6]

[1] Department of Environmental Science, Faculty of Geography-Geology, Doctoral School of Chemistry and Life Science and Earth, Alexandru Ioan Cuza University of Iasi, Carol I Blvd, no. 20A, Iasi, Romania
[2]Atmosphere Optics, Spectroscopy and Lasers Laboratory - LOASL, Faculty of Physics, Alexandru Ioan Cuza University of Iasi, Carol I Blvd, no. 11, Iasi, Romania
[3] Interdisciplinary Platform ARHEOINVEST, Alexandru Ioan Cuza University of Iasi, Carol I Blvd, no. 11, build G, Iasi, Romania
[4] Department of Interdisciplinary Research – Field Sciences, Alexandru Ioan Cuza University of Iasi, Lascar Catargi Blvd, no. 54, Iasi, Romania.
[5] Faculty of Materials Science and Engineering, Gheorghe Asachi Technical University of Iasi, Dimitrie Mangeron Street, no. 51A, Iasi, Romania
[6] Romanian Inventors Forum, Sf. Petru Movila Street, no. 3, Iasi, Romania
[7] Department of Physics, Gheorghe Asachi Technical University of Iasi, Dimitrie Mangeron Blvd, no. 59A, Iasi, Romania

E-mail: vasilepelin@yahoo.com; onuka23is@yahoo.com

Abstract. According to the scientific literature, the pollution phenomenon is strongly related by the urban activity from the last decades, with direct effects on the state of conservation of the stone constructions also. This paper presents a preliminary study on the colorimetric evolution of the lithic surfaces exposed under strongly traffic influence from the urban microclimate conditions. The analysed lithic surfaces are similar with the building stone from the structure of an historical monument (from 19th century), such as the *Stone Bridge* in Iasi - Romania, located in the immediate vicinity of the roadside loop with the same name. The colour change monitoring for the above-mentioned geomaterials aims at anticipating the effects of postponing the decongestion of car traffic and implicitly initiating the assessment of the effects of pollution over this historic monument, which is in an advanced state of deterioration and degradation.

1. Introduction

In the recent decades, the urban environment has undergone major changes, including cities such as Iasi (Romania), where the interaction between the atmosphere and historic stone monuments leads to the alteration of the lithic surfaces and the appearance of a new superficial layers [1] that ultimately affect the geomaterials [2] used to implement these objectives. Thus, taking into account the intensive use in local constructions of porous sedimentary indigenous lithic materials specific to the Iasi area [3], it is necessary to perform interdisciplinary studies, to monitor the colour changes of the surfaces affected by the atmospheric pollution by gravimetric deposition of airborne particles [4], given that for the scientific preservation of historical monuments the use of the same geomaterials as the original ones must be taken into account in all restoration interventions[5,6] but in an entirely different environment than the initial state [7,8].

The paper also presents the evaluation of the color changes of the lithic surfaces as a way to interpret the degree of pollution and its dynamics over time [9].

2. Materials and method

The lithic material exposed to urban environmental conditions is a sedimentary and oolithical calcareous rock (Fig. 1), form during the Sarmatian geological age, and was sampled from Paun – Repedea village, Iasi County.

CIE $L^*a^*b^*$ colorimetric investigations were performed using a Lovibond® RT 300 (Reflectance Tintometer D65/10°) spectrophotometer, monitoring the chromatic deviation at the same measurement points in each specific work step.

Scanning Electron Microscopy (SEM) coupled with energy dispersive X-ray (EDX) allowed the recording of the micrographic analyzed area and recording the spectrum based on which to determine elemental composition for lithic materials. A scanning electron microscope model Vega II LSH – Tescan®, coupled with an EDX detector, model Quantax QX2 – Bruker - Roentec® was used (Fig. 1).

Colorimetric investigations were carried out in constant laboratory conditions, at a temperature of 21°C and relative humidity (RH) of 60%.

Figure 1. SEM image (100xbse) on oolithical calcareous sample surface.

3. Experimental part

The current urban environmental exposure area is located in the city of Iasi (Romania) in the immediate vicinity of a 19th century historical monument, called *Stone Bridge* (*Podul de Piatra*), which is located over the Bahlui River and borders South of a major and crowded road junction, bearing the same name: *Stone Bridge* (figure 2a and 2b). Thus, the sampling area is located between the historic building and the eastern part of the intersection, with the following geographical coordinates: 47.158035 N, 27.575451 E (figure 2c and 3).

For exposure to urban environmental conditions, six samples of calcareous rock, numbered P1 to P6, were cut using roughly rectangular shapes with at least two planar surfaces to facilitate fastening on support and to have a uniformity of the gravimetric deposition areas of the airborne particles, while allowing for colorimetric investigations before and after each exposure (figure 4).

Exposure of stone samples was performed in two distinct periods, as follows: Oct. 10 ÷ Nov. 09, 2016 (*31 days*) and Nov. 16, 2016 ÷ Feb. 16, 2017 (*93 days*). Taking into account that 19 days of precipitation were recorded in the first 31 days of exposure [10], the samples were withdrawn, stored and dried under laboratory conditions for 96 hours, after which the colorimetric measurements were

IOP Conf. Series: Materials Science and Engineering **209** (2017) 012080 doi:10.1088/1757-899X/209/1/012080

carried out. Also, after the second exposure within 93 days, the samples were withdrawn and dried under the same laboratory conditions for a new set of colorimetric measurements (table 1).

Figure 2. *Podul de Piatră – Stone Bridge* Intersection from Iasi, Romania: a - older *(historical)* bridge; b – newer bridge; c – geographical position for exposure area.

Figure 3. Exposure place, near road junction, with geographical coordinate: 47.158035 N, 27.575451 E.

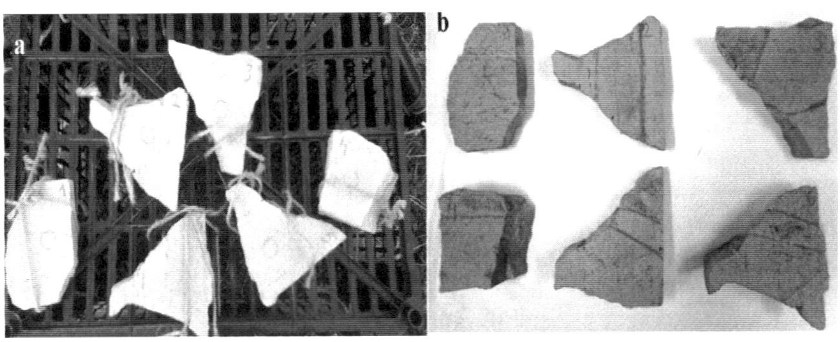

Figure 4. Lithical samples: a – fresh calcareous rocks, before exposure period; b – after 124 exposure days (right image is obtained in laboratory conditions).

Table 1. Colorimetric change after the two exposure periods.

Lithical samples	ΔE^*_{ab}	
	after 31 exposure days	after 124 exposure days
P1	0.91	9.29
P2	2.14	9.80
P3	2.32	10.97
P4	2.15	7.51
P5	2.62	**11.39**
P6	1.11	**7.26**

Concerning the colorimetric evolution, the colour change was measured for each coordinate (L^*, a^* and b^*), as compared to its initially value, on the same sample and in the same point, whereas the total change of colour (ΔE^*_{ab}) was calculated in accordance with the following equation [11-13]

$\Delta E^*ab = \sqrt{(\Delta L^*)^2 + (\Delta a^*)^2 + (\Delta b^*)^2}$, in which:

- ΔL^* represents the change in light intensity in the respective point, at different time intervals, as compared to the initial value ($\Delta L^* = L^*_n - L^*_{initial}$), where L^* represent the lightness to darkness coordinate and varying from 0 (black) to 100 (white);
- Δa^* represents the chromatic modification of the coordinates of axis a^* ($+a^*$ indicating red and $-a^*$ green), from the same point, at different time intervals, as compared to its initial value ($\Delta a^* = a^*_n - a^*_{initial}$);
- Δb^* represents the chromatic modification of the coordinates of axis b^* ($+b^*$ means yellow and $-b^*$ blue) while respecting the same method of calculation ($\Delta b^* = b^*_n - b^*_{initial}$).

4. Results and discussions

According to literature and analyzing the results presented in Table 1, an excess of the minimum value for total color change ($\Delta E^*_{ab} > 5$) [11] for each of the six samples is observed after 124 exposure days, which confirms the meeting of preliminary conditions of alteration of the lithic surfaces [4,7,8].

Thus, for the preliminary assessment of the atmospheric pollution conditions, elemental analysis by SEM-EDX technique was performed by two microprobes, taken from the sample with the minimum value of the color change ($\Delta E^*_{ab} = 7.26$, sample P6) and respectively from the sample with the maximum value ($\Delta E^*_{ab} = 11.39$, sample P5), the results obtained are presented in Figure 5 and table 2.

Also, analyzing the elemental compositions (table 2), the following aspects can be observed:

- increasing the presence of Si and Al elements on the surface samples (outside) as compared to the non-exposed inner part (inside) indicates the deposition of aluminosilicates from the street dust (both due to of peri-urban soils erosion and the increase of buildings progress with direct implications for the street dust form);
- the sulphur appearance on the sample surfaces, exposed to urban conditions, indicates the presence precursor conditions of $CaSO_4$ as a first effect of pollution[14,15], especially due to the biomass combustion and using Diesel engines;
- the occurrence of sodium and chlorine on the outer surface of the P5 sample indicates the uses of sodium chloride to defrosting the adjacent streets to the exposure area. The fact that these two elements do not appear on the surface of the sample P6 can be the explanation of the higher value of the total colour change, where: $\Delta E^*_{ab} = 11.39$ for P5 vs. $\Delta E^*_{ab} = 7.26$ for P6.

During the 124 days of exposure, there were 63 days of specific precipitation to the cold season in the temperate continental climate [10]. Besides the elemental chemical changes signaled on the exterior lithic surfaces, these conditions contribute to the colorimetric evolution previously presented. Also, a major role in the transport and subsequent deposition of atmospheric particles are the winds from this topoclimate, which is predominantly from the West - Northwest direction (WNW) to the South – East (SE), according to field measurements [10,16] (figure 6).

Table 2. SEM-EDX - elemental composition in percentage (inside vs. outside) for lithical micro-samples P5 and P6, after 124 exposure days.

Elements	Sample P5 (ΔE^*_{ab} = 11.39)				Sample P6 (ΔE^*_{ab} = 7.26)			
	% atomic		% weight		% atomic		% weight	
	inside	outside	inside	outside	inside	outside	inside	outside
Calcium	32.08	21.48	48.96	35.55	29.79	32.09	46.20	46.82
Silicon	12.90	14.14	13.80	16.41	12.33	19.47	13.39	19.91
Aluminum	0.97	1.65	0.99	1.84	1.23	1.76	1.29	1.72
Carbon	0.24	0.75	0.11	0.37	0.26	0.13	0.12	0.06
Iron	0.70	0.76	1.49	1.75	0.65	0.79	1.42	1.59
Manganese	0.37	0.31	0.77	0.70	0.30	0.00	0.61	0.00
Potassium	1.57	1.13	2.34	1.83	1.50	2.47	2.27	3.52
Magnesium	0.65	1.02	0.60	1.02	0.93	0.74	0.87	0.66
Sodium	0.60	1.84	0.52	1.74	0.87	0.94	0.77	0.79
Oxygen	49.93	55.52	30.42	36.68	51.31	40.85	31.76	23.79
Sulfur	0.00	0.59	0.00	0.79	0.00	0.33	0.00	0.39
Chlorine	0.00	0.53	0.00	0.77	0.00	0.00	0.00	0.00
Titanium	0.00	0.28	0.00	0.55	0.46	0.43	0.85	0.75
Phosphorus	0.00	0.00	0.00	0.00	0.37	0.00	0.45	0.00
Total	100	100	100	100	100	100	100	100

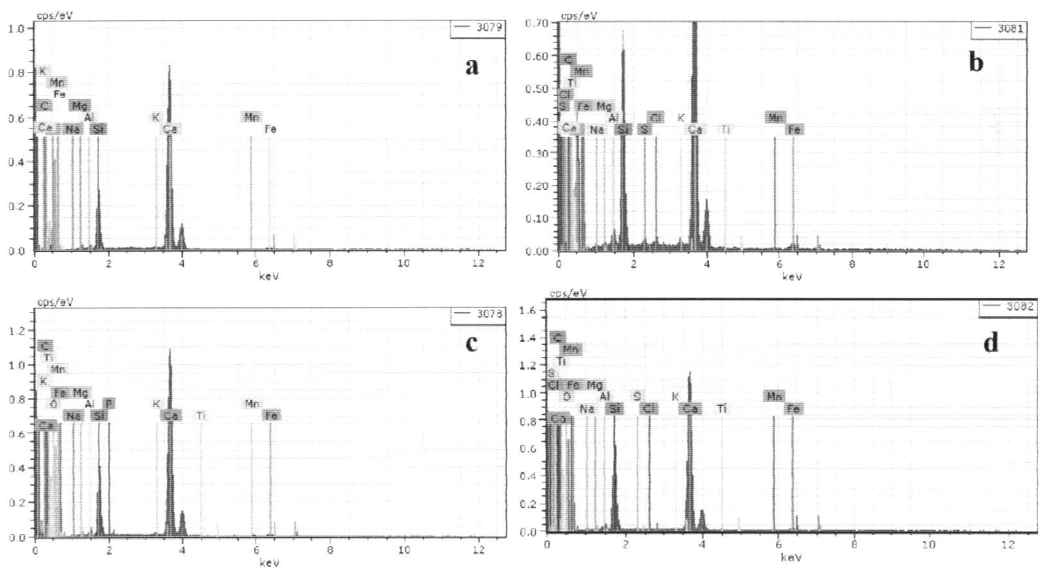

Figure 5. EDX spectral analysis for lithical micro-samples, after 124 exposure days: a – P5 inside; b – P5 outside; c – P6 inside; d – outside.

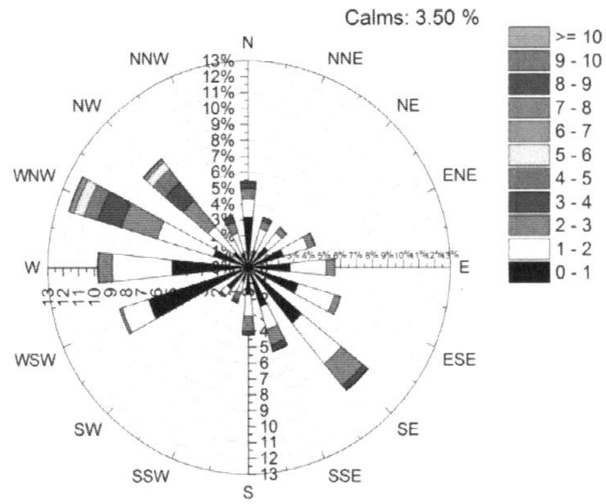

Figure 6. Wind rose for exposure periods, in *Stone Bridge* Intersection – Iasi, Romania.

5. Conclusions

Taking into consideration the fact that, starting with September 25, 2014, the Commission of the European Union calls on Romania to take measures to reduce air pollution, including in the city of Iasi [17] it is necessary to study all aspects of this issue due to the multiple unwanted effects, including historical monuments. Thus, the interdisciplinary researches carried out in the *Podul de Piatra* (*Stone Bridge*) Intersection [16] confirm the various forms or degrees of pollution with major implications on the Iasi community. Also, the colorimetric evolution - during only four months - of the lithic surfaces exposed in the current urban environment, characterized mainly by the increase of the road traffic, is an additional argument for the permanent monitoring and management of the respective topoclimate [18].

References

[1] Sabbioni C 1995 *Sci Total Environ* **167** 49-55
[2] Fookes P G 1991 *Q J Eng Geol* **24 (1)** 3-15
[3] Grasu C, Branzila M, G.C. Miclaus G C and Bobos I 2002 *Sarmatian System of Foreland Basin of the Eastern Carpathians* Technical Publishing, Bucharest, Romania
[4] Zereini F and Wiseman C L S 2010 *Urban Airborne Particulate Matter* Springer - Verlag Berlin Heidelberg, Germany
[5] Sandu I, Branzila M and Sandu I G 2009 *The Scientific Conservation of Stone Monuments* Alexandru Ioan Cuza University Publishing, Iasi, Romania
[6] Jokilehto J 1998 *Conserv Manage Archa* **2** 229-233
[7] Camuffo D 2014 *Microclimate for Cultural Heritage* (second edition) Elsevier, San Diego CA, United States of America
[8] Crabbe H and Hamilton R 2009 *The Effects of Air Pollution on Cultural Heritage* Springer - Verlag Berlin Heidelberg, Germany
[9] Grossi D, Del Lama E A, Garcia-Talegon J, Iñigo A C and Santiago Vicente-Tavera S 2015 *Int J Conserv Sci* **6(3)** 313-322
[10] *** http://www.meteoromania.ro/anm2/
[11] Schanda J 2007 *Colorimetry. Understanding the CIE System* Wiley-Interscience Hoboken, New Jersey, United States of America
[12] Pelin V, Sandu I, Munteanu M, Iurcovschi C T, Gurlui S, Sandu A V, Vasilache V, Branzila M and Sandu I G 2016 *IOP Conf Ser-Mat Sci* **133** 012061
[13] Pelin V, Sandu I, Gurlui S, Brinzila M, Vasilache V and Sandu I G 2016 *Rev Chim-Bucharest* **67(12)** 2568-2572
[14] Lal Gauri K, Holdren G C 1981 *Environ Sci Technol* **15(4)** 386-390

[15] Cardell C, Delalieux F, Roumpopoulos K, Moropoulou A, Auger F and Van Grieken R 2003 *Constr Build Mater* **17** 165-179
[16] Breaban I G and Paiu M 2016 *Int Multi Sci Geoco* **II** 79-86
[17] ***Memo/14/537/25/09/2014 European Commission (http://europa.eu/rapid/press-release_MEMO-14-537_en.htm)
[18] Ozga I, Ghedini N, Bonazza A, Morselli L and Sabbioni C 2009 *Wit Trans Ecol Envir* **123** 259-269

Acknowledgments
We thank Mrs. Gina Tiron and Mr. Adrian Timofte, from the National Meteorological Administration Romania, for their support in obtaining the weather data used in this article.

Using Quality Management Methods in Knowledge-Based Organizations. An Approach to the Application of the Taguchi Method to the Process of Pressing Tappets into Anchors

M A Țîțu[1*], A B Pop[2] and Ș Țîțu[3]
[1] Lucian Blaga University of Sibiu, Victoriei Street no. 10, code 550024, Sibiu, România
[2] SC TechnoCAD SA, Vasile Alecsandri no. 72, code 430351, Baia Mare, România
[3] Iuliu Hațieganu University of Medicine and Pharmacy, Victor Babeș Street no. 8, code 400012 Cluj-Napoca, România

E-mail: mihail.titu@ulbsibiu.ro

Abstract. This paper presents a study on the modelling and optimization of certain variables by using the Taguchi Method with a view to modelling and optimizing the process of pressing tappets into anchors, process conducted in an organization that promotes knowledge-based management. The paper promotes practical concepts of the Taguchi Method and describes the way in which the objective functions are obtained and used during the modelling and optimization of the process of pressing tappets into the anchors.

1. Introduction. An Overview

G. Taguchi proposes a systematic and efficient method to achieve an experiment, leading to an optimum formula in terms of performance and costs. Taguchi's method uses an orthogonal array which allows division of the experimental plan area and also to select the orthogonal array for experimentation, which facilitates a large number of variables with a small number of experiments.

Experimental determinations are influenced by the variability of environmental conditions - temperature, humidity, dust, etc., by the variability of the characteristics of the raw materials and components used and also by the workers different operating methods [1]. G. Taguchi calls this variability - noise factors [2]; the strategy he uses is not to control these factors (which sometimes can be expensive), but instead to minimize their impact.

In this research paper, the Taguchi's method will be used to present a study on the modelling and optimization of certain variables of the process of pressing tappets into anchors.

So, after inserting the tappets [3] (ident no 44075) into their anchors (ident no 43983) there are several nonconformities: deformed tappets (concentricity 0.12, higher than 0.12mm) and tappet length, non-compliant level 20.2+/-0.1 (higher than 20.3mm). This operation is performed by SCHMIDT Press Control 1001, figure 1.

The press is connected to a monitor in which specialized programs are entered for each type of component, with all documentation requirements. For each product the force and the press stroke are introduced in a graph, figure 2.

When the component is changed, the program has to be changed as well but not before saving the previous program. The program displays the graphs of the previous 100 processed components, and

IOP Conf. Series: Materials Science and Engineering **209** (2017) 012081 doi:10.1088/1757-899X/209/1/012081

the left panel on the screen displays the quantity of the last order as well as the percentage of defects. The evolution of reuse, expressed in PPM, is presented above in figure 3.

Figure 1. The SCHMIDT Press Control 1001.

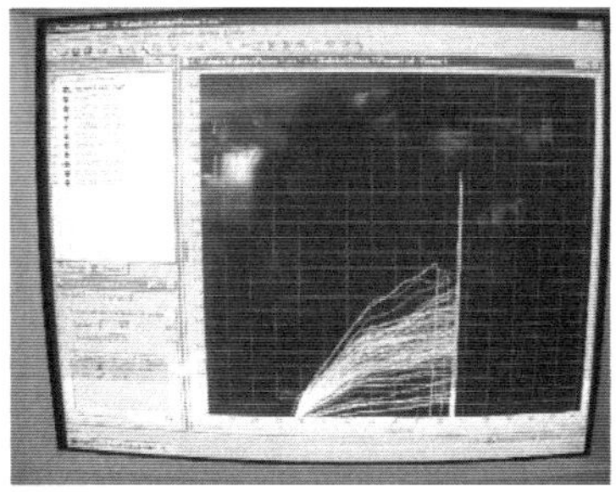

Figure 2. The Monitor of the SCHMIDT Press Control 1001.

Although, in certain weeks, an outage is detected, in all 11 weeks analyzed the evolution of PPM is decreasing; in the last 4 weeks, the level of refuse is around 4%. However, one must optimize the process so that it reaches 0.5%, the level accepted by the customer.

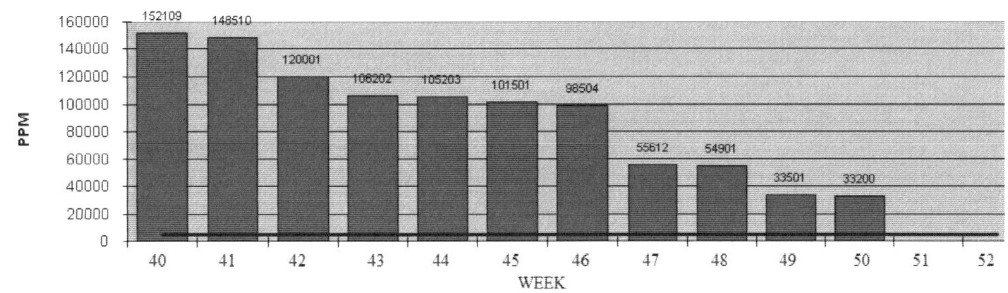

Figure 3. Internal PPM.

Nonconformities are presented in figure 4: the most frequent nonconformities are level 20.2+/-0.1 too high and © 0.12 too high, i.e. 80% of all nonconformities.

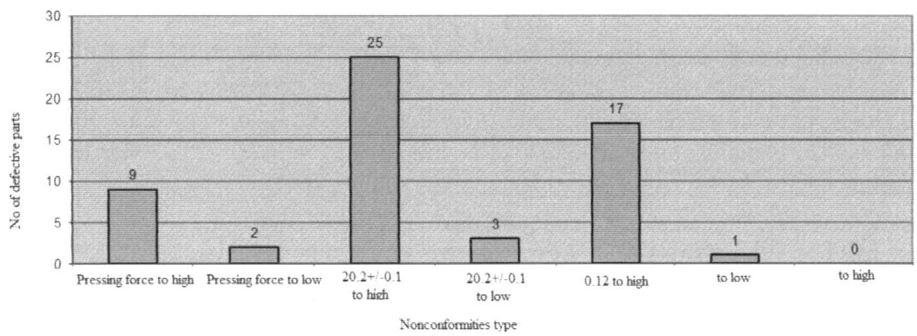

Figure 4. Types of defects.

2. The Selection of Features which Require Optimization

In order to be able to choose features that require optimization a brainstorming session was held, with several participants: quality manager, quality analyst, production manager, production-line supervisor, engineer, maintenance technician. Due to the fact that the two major defects, level 20.2+/-0.1 and concentricity value lower than 0.12, cannot be permanently quantified, they are considered qualitative criteria. Qualitative criteria to be monitored:

a) Level 20.2+/-0.1 and concentricity < 0.12;
b) Other defects;

All criteria must be minimized.

The selected factors are shown in the table below, on two levels. Next the Taguchi Method will be used [4, 5, 6, 7].

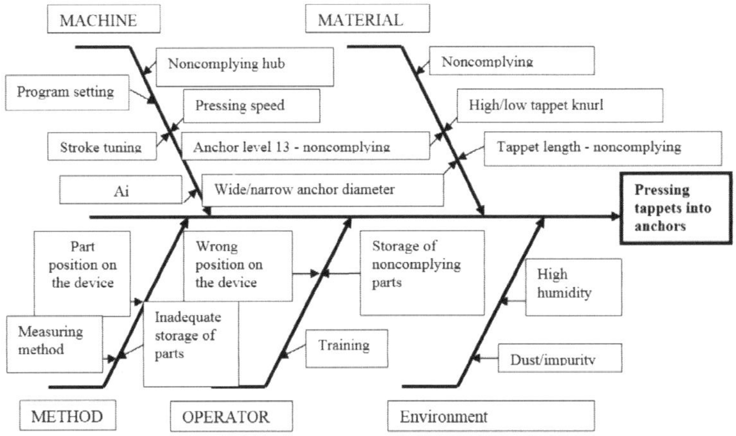

Figure 5. Ishikawa Diagram.

Table 1. Controlled factors and their levels.

Factor	Controlled Factor	Level 1	Level 2
A	Hubs	current	new
B	Pressing race	0,3kN	>0,3kN
C	Tappet lubrication	No lubrication	With lubrication
D	Pressing pressure	4 bar	6.6 bar
E	Press stroke	29.2 mm	32.7 mm
F	Component pairing	random	set

3. Developing experiments and the matrix

For the selected case, there are 64 (2^6) trials; an experiment matrix L64 must be developed, which is compatible with the objectives of the chosen experiment, but because of the fact that each level of each factor is combined with which level of the other factors one matrix with 8 trials is enough, of a factorial fractioned plan, in which the first column will include the number of trials, in the last column the calculation of the Ac controlled factors interaction and in the central columns the levels of the controlled factors.

Table 2. The factorial fractioned plan for 6 factors on 2 levels.

Trial No	Controlled Factor						**AC**
	A	**B**	**C**	**D**	**E**	**F**	
1	1	1	1	1	1	1	1
2	1	2	1	1	2	2	1
3	1	1	2	2	2	1	2
4	1	2	2	2	1	1	2
5	2	1	1	2	1	2	2
6	2	2	1	2	2	2	2
7	2	1	2	1	2	1	1
8	2	2	2	1	1	2	1

For each of the trials nonconformities were sampled every two hours, arranged according to the two groups of criteria which need to be minimized:

a) Level 20.2+/-0.1 and concentricity < 0.12;

b) Other defects;

a) The answers for the first criteria (Level 20.2+/-0.1 and concentricity < 0.12) are shown in the table 3.

Table 3. Answers for the first analysed criterion.

Trial no	Answer				Average y	Standard deviation s	S/N dB
	Interval 1 quantity	Interval 2 quantity	Interval 3 quantity	Interval 4 quantity			
1	3	1	2	5	2.75	1.48	-9.89
2	2	2	5	2	2.75	1.30	-9.66
3	2	2	1	1	1.50	0.50	-3.98
4	2	1	1	2	1.50	0.50	-3.98
5	0	0	1	0	0.25	0.40	6.53
6	0	1	0	0	0.25	0.40	6.53
7	1	1	2	0	1.00	0.71	-1.77
8	2	1	3	0	1.50	1.12	-5.45
					≈1.44		≈ -2.71

First the average is calculated for every line, then the standard deviation, according to the formula:

$$s = \sqrt{\frac{\sum(X - \bar{X})^2}{n}} \quad (1)$$

Where: s – standard deviation

X – Individual value;

$\bar{X}$ – Average of values;

n – Number of samples.

The ratio signal/noise is required so that the process might be optimized; in the case of a criterion that needs to be minimized it is obtained according to the formula below:

$$\frac{S}{N} = -10\log(s^2 + y^2)\,(dB) \quad (2)$$

b) The answers for the second criterion (other defects) are show in the table 4.

The comparison of the average results for each level, which is equal to the arithmetic mean of the results of the ratio S/N for each trial where the controlled factor is on the same level, such as:

A on level 1 (A1), average result:

$$A1 = \frac{[(-9.89) + (-9.66) + (-3.98) + (-3.98)]}{4} = -6.88\,(dB) \quad (3)$$

A on level 2 (A2), average result:

$$A2 = \frac{[6.53 + 6.53 + (-1.77) + (-5.45)]}{4} = 1.46\,(dB) \quad (4)$$

Table 4. Answers for the second analysed criterion.

Trial no	Answer				Average y	Standard deviation s	S/N dB
	Interval 1 quantity	Interval 2 quantity	Interval 3 quantity	Interval 4 quantity			
1	1	1	0	1	0.75	0.43	1.26

2	0	1	1	0	0.50	0.50	3.01
3	0	0	1	0	0.25	0.43	6.07
4	1	0	0	0	0.25	0.43	6.07
5	0	0	0	1	0.25	0.43	6.07
6	0	1	0	0	0.25	0.43	6.07
7	1	1	0	0	0.50	0.50	3.01
8	2	1	0	0	0.75	0.83	-0.97
					≈0.44		≈3.82

Average results S/N for each factor are displayed in the table 5:

Table 5. Average results of the ratio S/N for each of the controlled factors depending on the analysed criterion and level.

Average result for the ratio S/N for the first criterion which needs to be minimized		No factors	Average result for the ratio S/N for the second criterion which needs to be minimized	
Level 1	Level 2		Level 1	Level 2
-6.88	1.46	A	4.10	3.55
-2.28	-3.14	B	4.10	3.55
-1.62	-3.79	C	4.10	3.55
-6.69	1.27	D	1.58	6.07
-3.20	-2.22	E	3.11	4.54
-4.91	-0.51	F	4.10	3.55

The comparison of average results for each level is equal to the arithmetic mean of the results of all averages of measured values where the controlled factor was on the same level such as:

A on level 1 (A1), average result:

$$A1 = \frac{[2.75 + 2.75 + 1.50 + 1.50]}{4} = 2.13\ (dB) \tag{5}$$

A on level 2 (A2), average result:

$$A2 = \frac{[0.25 + 0.25 + 1.00 + 1.50]}{4} = 0.75\ (dB) \tag{6}$$

Average results of the values measured for each factors are shown in the table 6.

The average effect, calculated as difference between medium results of each controlled factor on the same level and the total average of answers (T), such as:

$$EA1 = A1 - T = (-6.88) - (-2.71) = -4.17\ (dB) \tag{7}$$

$$A2 = A2 - T = 1.46 - (-2.71) = 4.17\ (dB) \tag{8}$$

Table 6. Average results of the measured values for each of the controlled factors depending on the analysed criterion and level.

Average result for measured values for the first criterion that needs to be minimized		No factors	Average result for measured values for the second criterion that needs to be minimized	
Level 1	Level 2		Level 1	Level 2
2.13	0.75	A	0.44	0.44
1.38	1.50	B	0.44	0.44
1.50	1.38	C	0.44	0.44
2.00	0.88	D	0.63	0.25
1.50	1.38	E	0.50	0.38
1.69	1.19	F	0.44	0.44

The results of the average effect for each factor and for the first analyzed criterion are shown in the table 7.

Table 7. Effects of the first quality criterion.

EFFECT on the S/N ratio		No factors	EFFECT Measured Value	
Level 1	**Level 2**		**Level 1**	**Level 2**
-4.17	4.17	A	0.69	-0.69
0.43	-0.43	B	-0.06	0.06
1.09	-1.08	C	0.06	-0.06
-3.98	3.98	D	0.56	-0.56
-0.49	0.49	E	0.06	-0.06
-2.20	2.20	F	0.25	-0.25

If the value of the interaction between factors A and C was 0, then the two line would be parallel.

The answer of the relative S/N ratio on the interaction of factor A1C1 for the first criterion is the sum of all answers of S/N ratio when both factors are on level 1 in that case:

$$A1C1 = \frac{[(-9.89) + (-9.66)]}{2} = -9.78\ (dB) \tag{9}$$

Calculations are similar for all the other factors.The interaction is calculated considering the general average of S/N answers and the average of all measured values, for the first analyzed criterion:

$$IA1C1 = A1C1 - T - EA1 - EC1 = (-9.78) - (-2.71) - (-4.17) - 1.09 = -3.99\ (dB) \tag{10}$$

All the calculations for the other factors are similar, and the data is shown in the table 8.

Table 8. Interaction effects for the first analysed criterion.

Interaction	A1C1	A1C2	A2C1	A2C2
Effect on the S/N ratio	-3.99	3.98	3.98	-3.99
Effect on the measured value	0.56	-0.57	-0.56	0.56

The results of the average effect for each factor and for the second analyzed criterion are displayed in the table 9.

Table 9. Interaction effects for the second analysed quality criterion.

EFFECT on the S/N ratio		No factors	EFFECT Measured Value	
Level 1	Level 2		Level 1	Level 2
0.28	-0.27	A	0.00	0.00
0.28	-0.27	B	0.00	0.00
0.28	-0.27	C	0.00	0.00
-2.24	2.25	D	0.19	-0.19
-0.71	0.72	E	0.06	-0.06
0.28	-0.27	F	0.00	0.00

The interaction of the Ac factors depending on the general average of S/N answers and the measured values for the second analyzed criterion are shown in the table below:

Table 10. Interaction effects for the SECOND analysed criterion.

Interaction	A1C1	A1C2	A2C1	A2C2
Effect on the S/N ratio	-2.24	2.24	2.24	-2.26
Effect on the measured value	0.19	-0.19	-0.19	0.19

4. Results

The two types of results can be best compared by using the graphs in figure 6. The higher the value of the S/N ratio, the lower the loss, i.e. the higher the production performance. The result are to be compared by means of the following figures: 7; 8, 9 and 10: Factor A (the hub type) is the most important one, it has the highest S/N ratio result, for the first criterion; thus for the first criterion A2 should be selected, and for the second A1, i.e. the highest values. Factor D (pressure) which acts directly on nonconformities, exerts an important influence on both the first and the second criterion, but mostly on the second one. For both criteria, D2 should be selected. For factor F (component pairing) the same things might be stated but it has a lower effect than previous factors. For the first criterion F2 should be selected and for the second, F1. Factor C (tappet lubrication) reduced dispersion and improves productivity. A1C2 combination, for both the first and second criterion, reduces the dispersion of the process results. For both criteria, C1 should be selected. Factor E (press stroke) and factor B (pressing force) exert low influence on the first criterion, and for the second criterion factor E has a slightly higher influence on result dispersion. For both criteria, B1 and E2 must be selected.

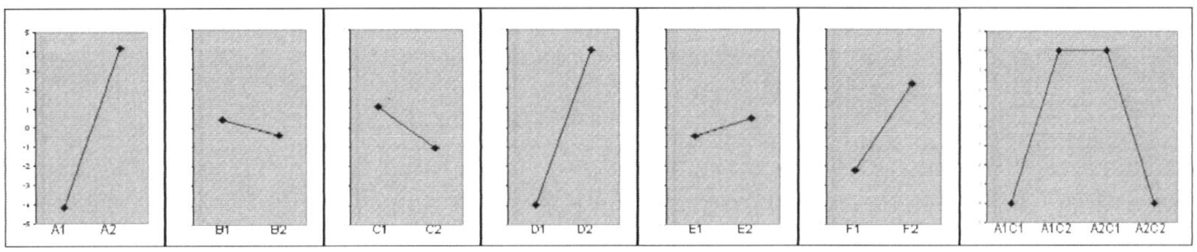

A1=-4.17	B1=0.43	C1=1.09	D1=-3.98	E1=-0.49	F1=-2.2	A1C1=-3.99	A2C1=3.98
A2=4.17	B2=-0.43	C2=-1.08	D2=3.98	E2=0.49	F2=2.2	A1C2=3.99	A2C2=-3.98

Figure 6. The effects of the first quality criterion on the signal-noise ratio (S/N) depending on the factors selected for the analysis.

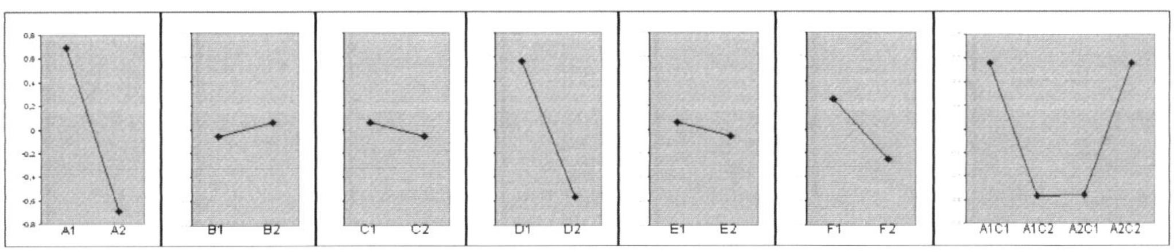

A1=0.69	B1=-0.06	C1=0.06	D1=0.56	E1=0.06	F1=0.25	A1C1=0.56	A2C1=-0.56
A2=-0.69	B2=0.06	C2=-0.06	D2=-0.56	E2=-0.06	F2=-0.25	A1C2=-0.57	A2C2=0.56

Figure 7. The effects of the second quality criterion on the signal-noise ratio (S/N) depending on the factors selected for the analysis first quality criterion.

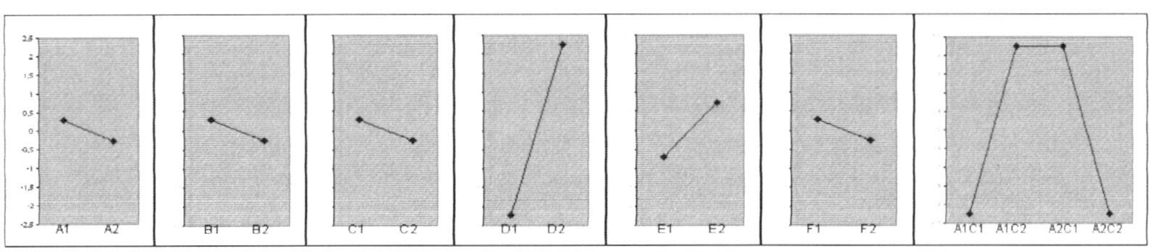

A1=0.28	B1=0.28	C1=0.28	D1=-2.24	E1=-0.71	F1=0.28	A1C1=-2.24	A2C1=2.24
A2=-0.27	B2=-0.27	C2=-0.27	D2=2.25	E2=0.72	F2=-0.27	A1C2=2.24	A2C2=-2.26

Figure 8. The effects of the first quality criterion on the value measured depending on the factors selected for analysis.

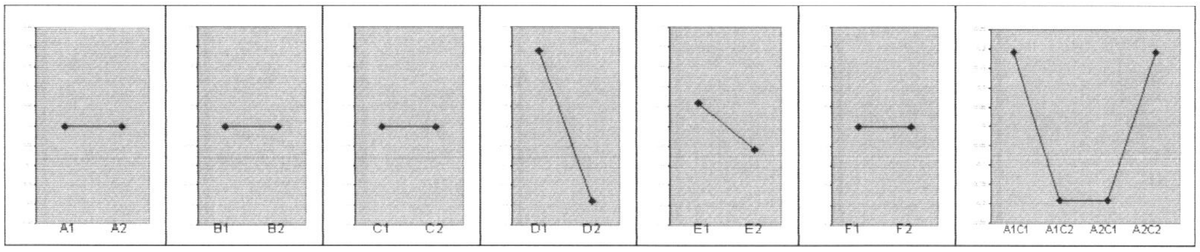

A1=0	B1=0	C1=0	D1=0.19	E1=0.06	F1=0	A1C1=0.19	A2C1=-0.19
A2=0	B2=0	C2=0	D2=-0.19	E2=-0.06	F2=-0	A1C2=-0.19	A2C2=0.19

Figure 9. The effects of the second quality criterion on the value measured depending on the factors selected for analysis.

5. The levels of the global optimization factors for the S/N factors

The table below (Table 11) displays the criteria which must be optimized, containing the level of each factor which optimizes this criterion and the highest level of the S/N ratio result:

Table 11. The S/N contribution corresponding to the optimal level of each factor.

		Features to be optimized		
No.	Factor description	Optimal level/ contribution	Criterion I	Criterion II
A	Hubs	Optimal Level	2	1
		CONTRIBUTION	4.17	0.28
B	Pressing force	Optimal Level	1	1
		CONTRIBUTION	0.43	0.28
C	Tappet lubrication	Optimal Level	1	1
		CONTRIBUTION	1.09	0.28
D	Pressing pressure	Optimal Level	2	2
		CONTRIBUTION	3.98	2.25
E	Part stroke	Optimal Level	2	2
		CONTRIBUTION	0.49	0.72
F	Component pairing	Optimal Level	2	1
		CONTRIBUTION	2.20	0.28

The table shows that factors A and F produce influence conflicts between the two quality criteria which must be optimized; the two factors require an arbitration so as to choose the level of each of the previous factors. Calculation are performed for the first optimization of the S/N ratio. For factor A, the sum of S/N ratio concentration is:

$$A1: -4.17 + 0.28 = -3.89 \quad (11)$$

$$A2: 4.17 - 0.27 = 3.89 \quad (12)$$

The sum of S/N ratio concentration for A2 is the highest, and thus level 2 will be selected (in the case of the two-level factors, the sums A1 and A2 are always opposed algebraic values).

$$B1: 0.43 + 0.28 = 0.71 \quad (13)$$

$$B2: -0.43 - 0.27 = 0.27 \quad \textit{Select level } 1 \quad (14)$$

$$C1: 1.09 + 0.28 = 1.37 \quad (15)$$

$$C2: -1.08 - 0.27 = -1.35 \quad \textit{Select level } 1 \quad (16)$$

$$D1: -3.98 - 2.24 = -6.22 \quad (17)$$

$$D2: 3.98 + 2.25 = 6.22 \quad \textit{Select level } 2 \quad (18)$$

$$E1: -0.49 - 0.71 = -1.20 \quad (19)$$

$$E2: 0.49 + 0.72 = 1.21 \quad \textit{Select level } 2 \quad (20)$$

$$F1: -2.20 + 0.28 = -1.92 \quad (21)$$

$$F2: 2.20 - 0.27 = -1.93 \quad \textit{Select level 2} \tag{22}$$

The selections are presented in Table 12 below.

Table 12. Experimental Values and Effects obtained.

The first optimization /N			S/N Contribution and effect upon the measured value			
No	Factor description	Level	Criterion I		Criterion II	
A	Hubs	2	4.17		-0.27	
				-0.69		0.00
B	Pressing force	1	0.43		0.28	
				-0.06		0.00
C	Tappet lubrication	1	1.09		0.28	
				0.06		0.00
D	Pressing pressure	2	3.98		2.25	
				-0.56		-0.19
E	Part stroke	2	0.49		0.72	
				-0.06		-0.06
F	Component pairing	2	2.20		-0.27	
				-0.25		0.00
S/N ratio theoretically result, in dB			2.06		0.50	
Average S/N, in dB			-2.71		3.82	
Measured value theoretically resulted				-0.26		-0.04

The resulted configuration: **A2B1C1D2E2F2**. Therefore: The S/N ratio for Criterion I – level 20.2+/-0.1, is improved, the theoretical S/N ratio being 2.06, and the value theoretically measures -0.26, is lower than the average value. The S/N ratio for Criterion II – other defects is also improved, the S/N ratio theoretically resulted being 0.50, and the theoretically measured value -0.04, is very close to zero. This first optimization of the S/N ratio reveals the final variant for the levels of the analyzed factors and for the performance of the validation trial.

6. The validation of experiments. Conclusions and original contributions.

The obtained results are very close to those expected, the value of the S/N ratio is higher than the theoretical value both for the first and the second criteria.

For series production the following working procedure shall be implemented:

The problems caused by the ne hub shall be monitored with a view to its maintenance and for the feasibility of the process.The current hub is no longer utilized. The pressing force shall be set at 0.3kN.The tappets shall be lubricated before pressing.

The pressing pressure shall be of 6.6 bar. The press stroke shall be set at 32.7 mm.

The component pairing shall be set in advance. This last factor should be analyzed because the component pairing is not at all time efficient, and thus solutions should be found with a view to optimizing time consumption.

Table 13. The experiment validation.

Trial no.	Answer criterion I				Average	Standard deviation	S/N in dB
	1	2	3	4			
1	0	0	1	0	0.25	0.43	6.07
Trial no.	Answer criterion I				Average	Standard deviation	S/N in dB
	1	2	3	4			
1	0	0	0	1	0.25	0.43	6.07

IOP Conf. Series: Materials Science and Engineering **209** (2017) 012081 doi:10.1088/1757-899X/209/1/012081

The validation experiment was performed again for the validation of the previously set levels, and the results are displayed in the table below (table 13).

One trial was performed for each of the criteria, the defect samples being taken as in the case analyzed above every two hours.

References

[1] Pop A B 2015 *Applied Mechanics and Materials* **809-810** 123-128

[2] Țîțu M A, Oprean C and Boroiu A 2011 Experimental research applied to increase product and service quality (in Romanian) *AGIR Publishing House, București*

[3] Florea A and Florea R 2013 Mecanisme si organe de masini Sibiu *Editura Universitatii Lucian Blaga*

[4] Țîțu M A and Pop A B 2016 Contribution on Taguchi's Method Application on the Surface Roughness Analysis in End Milling Process on 7136 Aluminium Alloy *20th Innovative Manufacturing Engineering and Energy Conference (IManEE 2016) IOP Conf. Series: Materials Science and Engineering* **16**

[5] Medan N and Bănică M 2016 Taguchi versus full factorial design to determine the influence of process parameters on the impact forces produced by water jets used in sewer *cleaning 20th Innovative Manufacturing Engineering and Energy Conference (IManEE 2016) IOP Conf. Series: Materials Science and Engineering* **161** 012016

[6] Mishra A K R, Sheokand R and Srivastava R K 2012 *International Journal of Scientific and Research Publications* **2**(10)

[7] Murthy M, Babu KM and Kumar R S *2014* Optimization of Machinability Parameters of Al6061 using Taguchi Technique *International Journal of Current Engineering and Technology* **3**

Use of Incineration Solid Waste Bottom Ash as Cement Mixture in Cement Production

N H Jun[1], M M A B Abdullah[1, 2], T S Jin[2], A A Kadir[3,1], C A Tugui[4] and A V Sandu[1,4]

[1]Center of Excellence Geopolymer and Green Technology (CEGeoGTech), School of Materials Engineering, Universiti Malaysia Perlis (UniMAP), Perlis, Malaysia.
[2]Faculty of Engineering Technology, Universiti Malaysia Perlis (UniMAP), Kangar, Perlis.
[3]Department of Water and Environmental Engineering, Faculty of Civil and Environmental Engineering, Universiti Tun Hussein Onn Malaysia, Johor, Malaysia
[4]Faculty of Materials Science and Engineering, Gheorghe Asachi Technical University of Iasi, Romania

E-mail: hooi.jun.ng88@gmail.com[1], mustafa_albakri.unimap.edu.my[1,2], sjtan@unimap.edu.my[2]

Abstract. Incineration solid waste bottom ash was use to examine the suitability as a substitution in cement production. This study enveloped an innovative technology option for designing new equivalent cement that contains incineration solid waste bottom ash. The compressive strength of the samples was determined at 7, 14, 28 and 90 days. The result was compared to control cement with cement mixture containing incineration waste bottom ash where the result proved that bottom ash cement mixture able achieve its equivalent performance compared to control cement which meeting the requirement of the standards according to EN 196-1. The pozzolanic activity index of bottom ash cement mixture reached 0.92 at 28 days and 0.95 at 90 and this values can be concluded as a pozzolanic material with positive pozzolanic activity. Calcium hydroxide in Portland cement decreasing with the increasing replacement of bottom ash where the reaction occur between $Ca(OH)_2$ and active SiO_2.

1. Introduction

Today incineration solid waste (ISW) such as blast furnace slag, fly ash, silica fume, limestone, and natural pozzolans are widely used in construction field. The use of ISW helps to utilize by products of industrial manufacturing processes and reduced the amount of cement required while convert waste materials into value added products [1]. Each of the ISW possesses different properties and presents potential use for cement and concrete industry by reacts differently in the presence of water while reduce the dependency on Portland cement and also replace part of the clinker to enhance the performance of the hydrated cement [2,3]. Many ISW have been successfully used as cementing materials where this application would be the future trend in cement industry [4].

The amount of ISW bottom ash (BA) mainly come from major portion of the waste combustion products that generated from combustion furnace and accounting for 20-30 wt.% of the original waste

[5][6][7]. ISWBA is a coarse, fused, glassy texture, and granular material make it an ideal substitute for natural aggregates. It also often used as low cost replacement for sand in concrete blocks and based in road construction [8]. Anyway, the quality and chemical composition may vary due to the coal used and technology applied especially the temperature during the combustion process. Furthermore, disposing the ash efficiently and economically is very challenging else these ashes gone to the landfills causing wastes of land resources and environmental pollution problems.

In cement industry point of view, use of ISWBA in large scale is beneficial to sustainable development and environmental friendly which could partially reduce its carbon footprint and reduce high energy of cement manufacturing. Studies of the utilization of ISWBA to produce ecofriendly Portland cement are gaining attention in relation to recycle and reuse of these materials [9]. This paper shows most significant results on analyzing and evaluating the material with mechanical properties of cement containing ISWBA. Little attention has been given to the possibility of ISWBA to be utilized in Portland cement. Optimization of ISWBA content into cement mixture to produce the desired strength according to standard is needed and the result shown in this study is optimistic and possible to promote the use of ISWBA in large quantities and could technically apply in cement industry. The compressive strength, pozzolanic strength activity index, and phase analysis of ISWBA cement mixture are reported and compared with controlled Portland cement in this paper.

2. Materials and Methods

The bottom ash that used in this study was located in Manjung Coal-Fired Power Station, Lumut, Perak, Malaysia and the plant is composed of three identical incineration rotating speed lines for a total capacity of 2100MW per day. The cement used in all the mixtures was from Cement Industries of Malaysia Berhad (CIMA). In order to mix equally with Portland cement, ISWBA was undergo milling to less than 150um to increase the surface area and enhancing chemical reactivity.

Cement was replace by ISWBA with simple volume replacement where five different mixture including a reference mortar with 10%, 20%, 30% and 40% replacement ratio. The water to cement ratio was fixed at ratio 0.5 in all the samples where the details of mixes proportion are shown in Table 1. By the way, there is a plain cement sample without ISWBA that serves as control sample. Cement mortar samples was casted and mix until homogeneous by molded in 50 x 50 x 50mm^3 molds. The samples were demolded after 24 hours to ensure the sufficient strength for demolding and then put in curing tank until the curing age reach at 7, 14, 28, and 90 days.

The best possible performance in term of optimizing the mixture composition can be through different ISWBA composition and choose the best combination by comparing the desired characteristic and properties of cement. The method of choosing the optimization mixture in this study is based on the minimum requirement of standard EN 196-1.

Table 1. Mix proportions of different curing day.

ISWBA content (%)	w/c ratio	Water (kg/m^3)	Cement (kg/m^3)	ISWBA (kg/m^3)
0	0.5	270	540	0
10	0.5	270	486	54
20	0.5	270	432	108
30	0.5	270	378	162
40	0.5	270	324	216

3. Results and discussions

3.1. Compressive strength

Compressive strength is the critical parameter to determine the design process and influence the cost of cement industry. In this study, the compressive strength were conducted accordance to ASTM C109 where the compressive strength increased with curing time for all the series. With the progress of curing ages, compressive strength of ISWBA cement mixture increase at a faster rate after 7 days of curing. Fig. 1 shows all the series of mixtures and the control mixture generally produced highest compressive strength at age of 7, 14, 28, and 90. The compressive strength of 10% substitution of ISWBA into Portland cement mortar (33.62MPa, 35.24MPa) was lower than control Portland cement (36.52MPa, 37.23MPa) but the difference of compressive strength was less distinct on 28 days of curing where the value was comparable for 90 days of curing. The compressive strength development with addition of 40% ISWBA at 7, 14, 28, and 90 days experienced a decline in compressive strength by 59.4%, 63.9%, 30.4%, 25.5% respectively compared to control Portland cement.

It is believe that the addition of ISWBA in Portland cement may altered the nature of hydration phases. The compressive strength of 10% ISWBA reaching 92.1% in the 28 days curing was slightly higher than commercial cement according to EN 196-1 and classified as class 32.5 or 32.5R Portland cement. This finding proved that ISWBA suitable to substitute in Portland cement as cementitious material. Besides that, fine particles of ISWBA absorbed largest portion of cement paste to ensure the texture and compactness of the mixture through providing nucleation sites for the hydration of C_3S and C_2S while accelerating the hydration process and enhances the inter particle friction on the flow of the mortar [10]. Furthermore, it also enter the spaces that exist between the particles of cement to improve the packing and contribute to the formation of C-S-H which it is the main contributor compound of strength development [11].

Partial replacement of ISWBA into Portland cement reduced the proportion of primary hydration reactant ($(CaO)_3.SiO_2$, C_3S) and dicalcium silicate ($(CaO)_2.SiO_2$, C_2S), the reduction of C_3S and C_2S led to insufficient formation of Ca/Si C-S-H which causing retardant of hydration process [12] and the result is expected to display lower mechanical strength since the reduction of C-S-H compounds of ISWBA cement mixture samples. This result indicates that the addition of ISWBA was not able to promote the mechanical strength but only bring dilution effect to the hydrate system and retard the cementitious reaction. Hence, it can be concluded that more cement in the composition leads to higher early strength and the best composition is 10% ISWBA replacement.

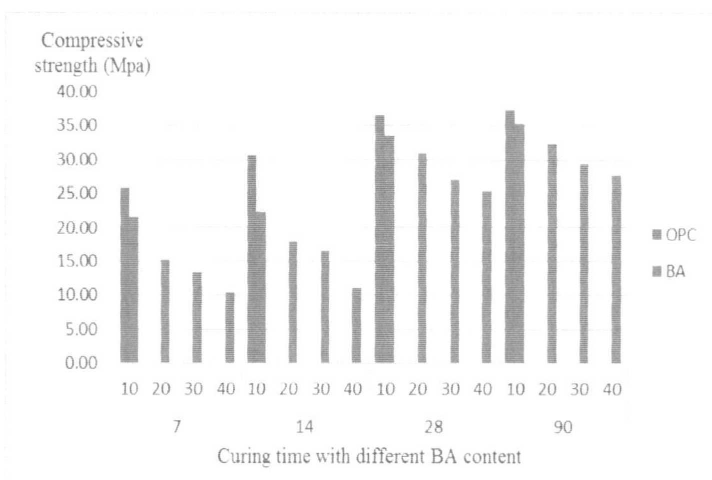

Figure 1. Compressive strength of Portland cement and various bottom ash content with different curing time.

3.2. Pozzolanic Activity Index (PAI)

Pozzolanic activity index defined as percentage of compressive strength compared to the control where the index that greater than 0.75 at 28 days and 0.85 at 90 days consider as a positive pozzolanic activity according to ASTM C618 and ASTM C311. A test of pozzolanic activity is considered a suitable means of evaluation the strength contribution potential of a mineral admixture [13].

The summation of silica (SiO_2), aluminium oxide (Al_2O_3), and iron oxide (Fe_2O_3) is more than 70% can be assumed as pozzolanic materials with pozzolanic activity [4,14] whereas these compounds are the major components in ISWBA [15]. In this study, fine particles of ISWBA react with calcium hydroxide and produce C-S-H gel. After long curing period, the reactive silica in ISWBA reacts with alkali calcium hydroxide that produced by the hydration process of cement and forms calcium silicates and aluminate hydrates, these are the new reaction products that exhibited binding character as shown in Equation 1. This could be explained from the PAI value which the 90 days of curing is higher 2.7% than 28 days indicated that PAI increases slightly with age as can be seen in Fig. 2. Thus, the pozzolanic reaction can contribute to strength development as a function of curing time where the higher the value of PAI indicates that the good reaction between ISWBA and Portland cement.

$$Ca(OH)_2 \rightarrow Ca^{2+} + 2OH^-$$

$$Ca^{2+} + 2OH^- + SiO_2 \rightarrow C\text{-}S\text{-}H$$

$$Ca^{2+} + 2OH^- + Al_2O_3 \rightarrow C\text{-}A\text{-}H \quad (1)$$

However, the pozzolanic properties of ISWBA was very weak and causes the mechanical properties always lower than Portland cement for the mixture of ISWBA that investigated in this study. The reduction in mechanical strength was due to cement dilution which the addition of ISWBA might not counterbalanced to the mechanical properties of Portland cement. Anyway, pozzolanic activity in the ISWBA make it possible to be used as replacement material for Portland cement and its addition could have a beneficial role in the development of hydrated cement paste.

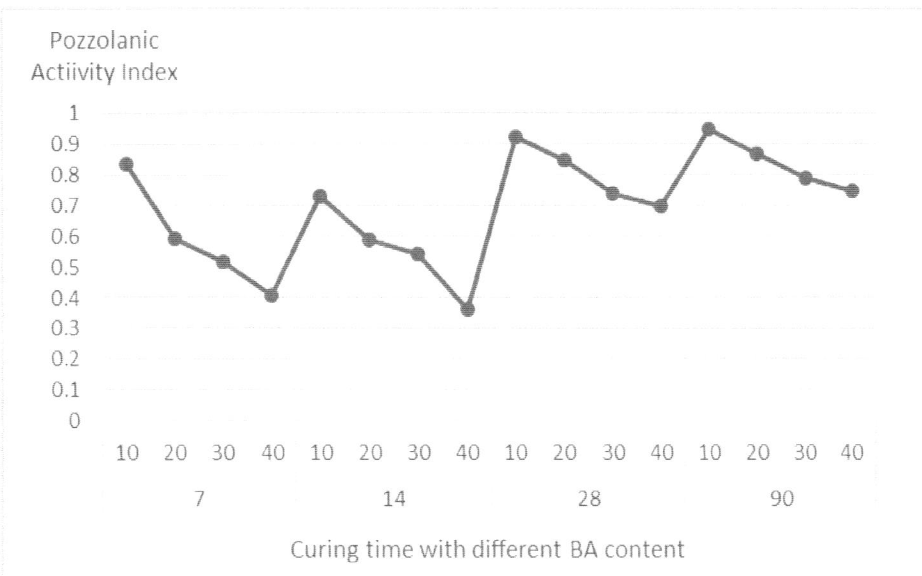

Figure 2. Pozzolanic activity index of ISWBA in Portland cement mortar.

3.3. Phase analysis

The XRD analysis was carried out with 2θ scanning ranging between 10° to 70° by 1s counting time after 90 days of curing and presented in Fig. 3. The Portland cement phase was compared with ISWBA cement mixture. According to the result shown in Fig. 3, portlandite, calcite, dicalcium silicate, and quartz was observed as main hydration products. For 10 and 20% of ISWBA replacement, the amount of portlandite produced was lower because the amount of Portland cement was reduced. Hence the hydration of calcium silicate phases also reduced subsequently. The relative decrease of $CaCO_3$ peaks and increases of C-S-H peaks intensity even at higher level of cement was due to the reaction occur with Ca(OH)2. Whereas for 30% and 40% ISWBA replacement mixtures, there was some silica that did not react with Portland cement.

There are phase composition change in the 10-40% in ISWBA cement mixture was observed in the XRD spectrum indicates that the absolute intensities of portlandite peaks of ISWBA cement mixture was lower than control Portland cement. Assuming the peak intensity of portlandite is equal to the amount of the Portland cement, the lower peak intensity of portlandite indicates that the influence of quantities of ISWBA is significant. Portlandite produced is the product of cement hydration and also the result of pozzolanic reactions of ISWBA and it forms phases of C-S-H with the silica from ISWBA.

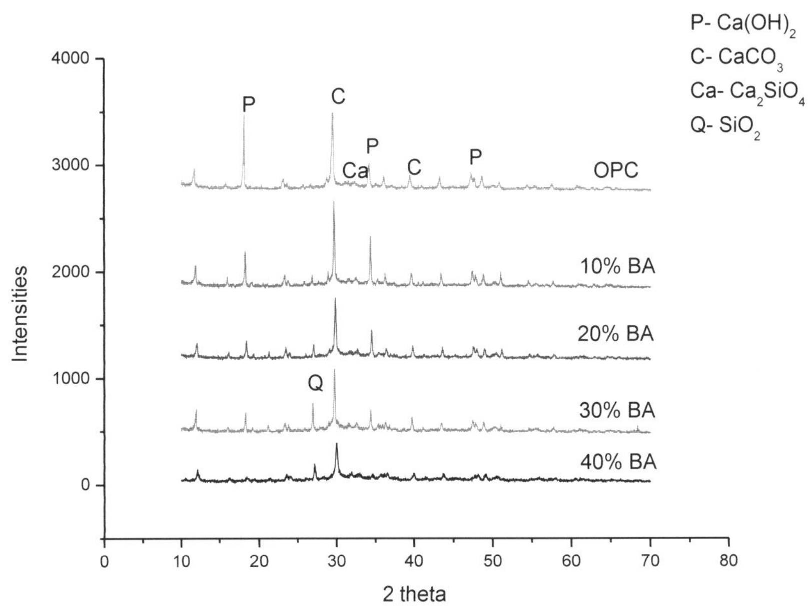

Figure 3. Phase analysis of ISWBA cement mixture.

4. Conclusion

The result of this study leads to good improvement by using large replacement of ISWBA through simple and beneficial way to industrial application.

It can be concluded that up to 40% replacement level can be used in Portland cement and there is no significant negative effect on strength properties at any curing age. The compressive strength of ISWBA cement mixture are gradually decreased with the addition amount of ISWBA. At 7 and 14 days, all mixes showed the compressive strength test lower than control, but as the curing age increased to 28 and 90 days, all mixes showed almost comparable strength result to control. The compressive strength is meeting the requirement for commercial cement of class 32.5 or 32.5R class as per EN 197-1.

The pozzolanic activity of ISWBA is relatively lower and it might delay the hydration process of cement at early curing age. The pozzolanic activity index reached 0.92 at 28 days and 0.95 at 90 days for 10% replacement ISWBA, such values allow and confirm the use of ISWBA in Portland cement and can be concluded as a pozzolanic material with positive pozzolanic activity. For phase analysis, main hydration products such as portlandite, calcite, dicalcium silicate, and quartz was observed.

Calcium hydroxide (Ca(OH)2) in Portland cement decreasing with the increasing replacement of ISWBA where the reaction occur between Ca(OH)2 and active SiO2 in ISWBA. The peak intensity of portlandite is equal to the amount of the Portland cement, the lower peak intensity of portlandite indicates that the influence of quantities of ISWBA is significant.

References

[1] Lothenbach B, Scrivener K and Hooton R D 2011 *Cem. Concr. Res.* **41** 1244-1256
[2] Pourkhorshidi A R, Najimi M, Parhizkar T, Jafarpour F and B. Hillemeier 2010 *Cem. Concr. Compos.* **32** 794–800
[3] Dembovska L, Bajare D, Pundiene I and Vitola L 2017 *Procedia Eng.* **172** 202–210
[4] Lessard J-M, Omran A, Tagnit-Hamou A and Gagne R 2017 *Constr. Build. Mater.* **132** 565–577
[5] Filipponi P, Polettini A, Pomi R and Sirini P 2003 *Waste Manag.* **23** 145–56
[6] Polettini A, Pomi R and Carcani G 2005 *Resour. Conserv. Recycl.* **43** 403–418
[7] Baite E, Messan A, Hannawi K, Tsobnang F and Prince W 2016 *Constr. Build. Mater.* **125** 919–926
[8] Cheriaf M, Rocha J C and Pera J 1999 *Cem. Concr. Res.* **29** 1387–1391
[9] Valle-Zermeño R, Gomez-Manrique J, Giro-Paloma J, Formosa J and Chimenos J M 2017 *Sci. Total Environ.* **581** 897–905
[10] Rafieizonooz M, Mirza J, Salim M R, Hussin M W and Khankhaje E 2016 *Constr. Build. Mater.* **116** 15–24
[11] Toutanji H, Delatte N, Aggoun S, Duval R and Danson A 2004 *Cem. Concr. Res.* **34** 311–319
[12] Ercikdi B, Cihangir F, Kesimal A, Deveci H and Alp I 2009 *J. Hazard. Mater.* **168** 848–856
[13] Cheng A 2012 *Mater. Des.* **36** 859–864
[14] Polettini A, Pomi R and Fortuna E 2009 *J. Hazard. Mater.* **162** 1292–1299
[15] Chimenos J, Segarra M, Fernandez M and Espiell F 1999 *J. Hazard. Mater.* **64** 211-222

Acknowledgments

The authors would like to thanks to CEGeoGTech @ UniMAP and School of Materials Engineering, University Malaysia Perlis (UniMAP). Author also wish to acknowledge assistance or encouragement from colleagues, special work by technical staff and support from supervisor Prof. Madya Dr. Mustafa Al Bakri.

IOP Conf. Series: Materials Science and Engineering **209** (2017) 012083 doi:10.1088/1757-899X/209/1/012083

Coagulation-Flocculation Process in Landfill Leachate Treatment: Focus on Coagulants and Coagulants Aid

M A Kamaruddin[1,2], M M A Abdullah[3,2], M S Yusoff[4], R Alrozi[5] and O Neculai[6]
[1]Environmental Technology Division, School of Industrial Technology, Universiti Sains Malaysia, 11800 Penang, Malaysia
[2]Center of Excellence Geopolymer and Green Technology (CEGeoGTEch), Universiti Malaysia Perlis (UniMAP), Perlis, Malaysia
[3]Faculty of Engineering Technology, Universiti Malaysia Perlis (UniMAP), Perlis, Malaysia
[4]School of Civil Engineering, Universiti Sains Malaysia, 14300 Nibong Tebal, Penang, Malaysia
[5]Faculty of Chemical Engineering, Universiti Teknologi MARA, Penang, Malaysia E-mail: anuarkamaruddin@usm.my
[6] Department of Civil and Industrial Engineering, Faculty of Civil Engineering and Building Services, Technical University „Gheorghe Asachi" of Iaşi, Romania

E-mail: anuarkamaruddin@usm.my

Abstract. In physico-chemical treatment, the separation of suspended particles from the liquid phase is usually accomplished by coagulation, flocculation and sedimentation. Coagulation-flocculation processes have been widely used as alternative treatment to remove leachate pollutants such as BOD, COD, TSS, heavy metals, colour, and nitrogen compounds prior to other treatment methods. It is often coupled with treatment methods like biological process, chemical oxidation, adsorption or filtration to achieve desirable effluent quality. In spite of being economical, the dewatering and disposal of the precipitated sludge could be laborious and time-consuming. In this manuscript, brief discussions on coagulant and coagulants aid in landfill leachate treatment is discussed with respect to their mechanism.

1. Introduction

Coagulation is generally defined as destabilization of a colloidal suspension or solution. On the other hand, flocculation is the process whereby the destabilized particles are propagated to gather and agglomerate into larger flocs. Although coagulation and flocculation are two distinct processes, they have been used interchangeably due to the confusion resulted from water treatment industry itself [1].

The same goes for the terms 'coagulant' and 'flocculant'. However, it is pertinent to assert that 'coagulant' is often used to describe any substances that can induce both coagulation and flocculation [2].

1.1. Colloids in Landfill Leachate

Dispersed solids in landfill leachate comprise non-sedimentary suspending materials with negligible sedimentation velocity; and colloids are the major constituents of these non-sedimentary particles. Colloids are generally defined as solid particles with a size between 1 nm and 1 mm [3,4,5]. Some

IOP Conf. Series: Materials Science and Engineering **209** (2017) 012083 doi:10.1088/1757-899X/209/1/012083

studies had even revealed that particles with a size up to 10 cm might act as colloids as well [6].

Matura and co-researchers had noted that the colloids in landfill leachate were mostly inorganic and mainly consisted of carbonates and clays. In addition, colloids can be classified as hydrophobic (resistant to water bonding) or hydrophilic (affinity for water bonding). Owing to the attraction of hydrophilic colloids for water, they are difficult to be removed from dispersion. However, this classification does not work for colloids in wastewater as it is hard to differentiate between the two [1].

1.2. Colloid and Contaminant Mobilization

Colloids are important in the context of speciation, fate, transport, biogeochemistry, bioavailability and toxicity of contaminants in the environment [7]. For landfill leachate particularly, several researchers had asserted similar conclusions [8,9]. Heavy metals are bound to colloids by adsorption, precipitation (and surface precipitation) and ion exchange [11]. Owing to the great affinity of colloids for heavy metals, around 70 to 90% of metals in landfill leachate are in colloidal or organic complexes (Baun & Christensen, 2004). Excluding highly mobile elements like arsenic, selenium and rubidium, most trace elements bind at least 20% individually to colloids and some metals like aluminium, iron and lead may go up to nearly 100% as shown in table 1.

Table 1. Binding of elements to colloidal particles in landfill leachate.

Element	Binding to colloids (%)
As, Se and Rb	0
Ca, Co, Cr, Ni, Cu, Zn, Sr, Ba, V and Sb	≥20
Al, Fe and Pb	~100

2. Coagulants and Coagulant Aids

2.1. Coagulants

Coagulants are chemicals added into water or wastewater to perform coagulation. They can be classified broadly into three types i.e. metallic salts (or hydrolysing metallic salts), polymeric metallic salts (or pre-hydrolysed/pre-polymerised metallic salts) and polymers. Polymeric coagulants encompass both natural and synthetic polymers. Detailed classification of coagulants is depicted in Figure 1. Amongst them, microbial flocculants or bioflocculants are not produced or applied in industrial scale, albeit literatures regarding their bench scale coagulation efficiencies are scanty in recent years. This is due to their low yield and high production cost [12].

Moreover, their coagulation mechanisms are inherently complicated. Therefore, bioflocculants are not reviewed in depth in this work. The same goes for other less applicable coagulants. They are excluded since they are less favoured in water treatment industry. In addition, general requirements for coagulants in water treatment are: (1) trivalent metallic cations or polymers that have been proven efficient as coagulants; (2) nontoxic and do not cause any detrimental physiological effects on human health; and (3) insoluble or low solubility in the pH ranges common in water treatment to minimise chemical residual in effluent [13]. Owing to that, coagulant for particular wastewater treatment has to be selected carefully based on their feasibilities. Advantages and drawbacks of commonly used coagulants are shortlisted in table 2.

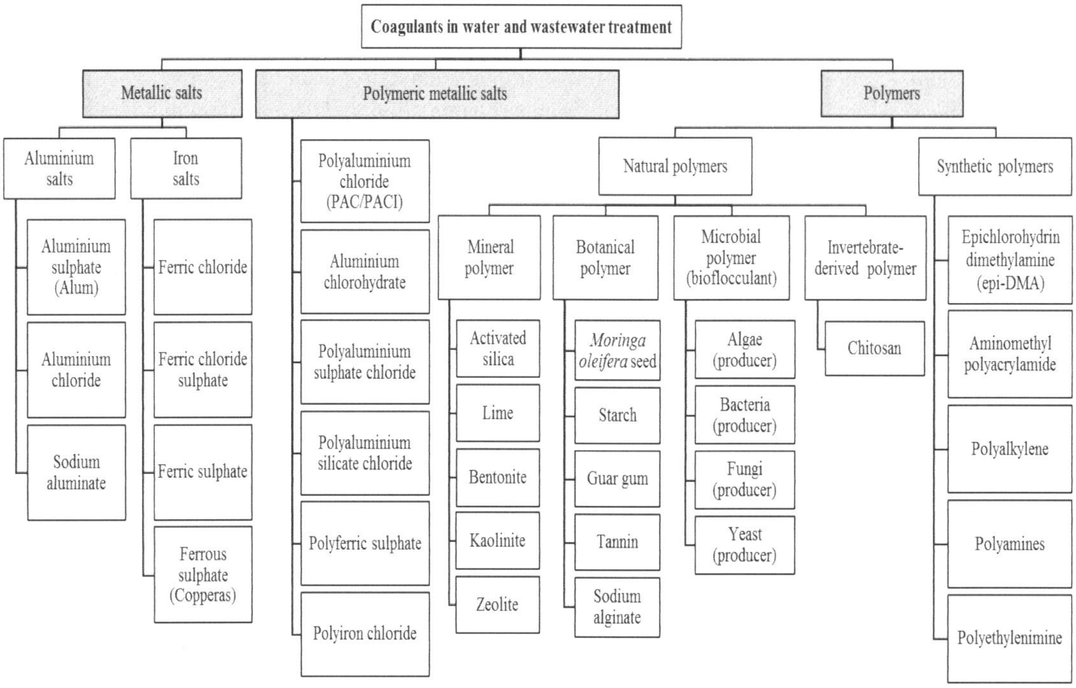

Figure 1. Classification of coagulants in leachate treatment.

Table 2. Degree of colloid stability as a function of zeta potential.

Zeta potential (mV)	Colloid stability
from 0 to ±5	Rapid coagulation or flocculation
from ±10 to ±30	Incipient instability
from ±30 to ±40	Moderate stability
from ±40 to ±60	Good stability
more than ±61	Excellent stability

2.2. Coagulant Aids

Coagulant aids are applied in tandem with primary coagulants (mostly metallic coagulants) to optimize coagulation process. They accelerate clarification, strengthen the flocs, improve sludge settleability, reduce coagulant dosage and treatment cost, provide a wider effective pH range and elevate contaminant removal efficiency [14]. Shammas [11] had postulated that there are four types of typical coagulant aids i.e. alkalinity additives, pH adjusters, particulate additives and polymers.

Despite most literatures only consider particulate additives and polymers as coagulant aids (by means of post-orthokinetic agglomeration), alkalinity additives and pH adjusters are also included in this review because they provide conducive environments for optimum coagulation-flocculation (by means of orthokinetic agglomeration). Indeed, coagulant aids are also commonly classified into two types i.e. particle binders and nucleating agents based on their action mechanisms. For particle binders, they promote flocculation like primary coagulants while the nucleating agents act as nucleation sites to facilitate floc formation.

Most of the commonly used coagulant aids are polyeelctrolytes, especially those synthetic polymers like polyacrylamides, polydiallyldimethyl ammonium chloride (poly-DADMAC) and polystyrene sulfonate. Polyelectrolytes are polymers with ionisable functional groups such as

IOP Conf. Series: Materials Science and Engineering **209** (2017) 012083 doi:10.1088/1757-899X/209/1/012083

carboxyl, amino or sulphonic groups. They can be cationic (with positive functional groups), anionic (with negative functional groups) or amphoteric (with both positive and negative functional groups). Polymers without any ionisable functional group are nonionic. Comparison of these typical coagulant aids is summarised in Table 3.

2.3. Polyaluminium Chloride (PAC)

Polyaluminium chloride (PAC) is a highly effective polymerised inorganic coagulant. It is the dominant variety of polymeric inorganic salts or so-called the New Age coagulants in wastewater treatment. Indeed, PAC is a generic name for various salts with a general formula, $Al_nCl_{(3n-m)}(OH)_m$ that can be formed by polymerising aluminium chloride under different proprietary reaction conditions. They include aluminium chlorohydrate, basic aluminium chloride, polybasic aluminium chloride, aluminium hydroxide and aluminium oxychloride. They are sold in both aqueous and dry powder forms. Among them, aluminium chlorohydrate (ACH), $Al_2Cl(OH)_5$ is the most common PAC in the market. Comparison between these two PACs is shown in **Error! Reference source not found.** 2.

Knowledge on Al speciation is crucial for PAC coagulation-floculation optimisation. When PAC is introduced into water, aluminium (Al) hydrolyses into diverse monomeric and polymeric forms, depending upon pH [16,17,18,19], counter-ions, Al salt concentration [20], temperature [21] and hardness [22]. Though Al hydrolysis reactions at different pH values is not yet ascertained, several studies in the past have shed some light on their predominant species in different pH ranges (Figure 3). Al^{3+} $[Al(H_2O)_6{}^{3+}]$, Al_{13} $[AlO_4Al_{12}(OH)_{24}(H_2O)_{12}{}^{7+}]$ and aluminate ions $[Al(OH)_4]^-$ seem to be the dominant species at pH <4, ≈5 and >7, respectively. In the presence of counter-ions like humic acid, fulvic acid, sulphates, phosphates and fluorides, speciation and precipitation of Al will vary. In light of space limitation, other less significant alluded factors are left out of this discussion.

Amongst all hydrolysed Al species, metastable pre-hydrolysed polynuclear Al_{13} species (a.k.a. tridecameric Al or Keggin Al_{13}) is believed to be the main active species for coagulation. Its superiority over other Al species especially Al^{3+} (key Al species from alum and $AlCl_3$ hydrolysis) is supported by the fact that it has the greatest charge neutralisation ability, higher structural stability and nanosized molecular diameter. However, this theory is now challenged by the discovery of Al_{30} $\{[(AlO_4)_2Al_{28}(OH)_{56}(H_2O)_{26}]^{18+}\}$ and its characterisation in recent years [4,9,11,12]. Its preponderance over Al_{13} has been postulated since it is bigger (1 nm diameter and 2 nm length) more positively charged, more thermal resistant and stable in hydrolysis process especially at low pH. Despite the coagulation mechanism of PAC is still under debate, it is suggested that adsorption, charge neutralisation, sweep flocculation and bridging do occur during coagulation. [8] had revealed that precipitate formation of PAC increased coagulation performance substantially; contrariwise, the charge neutralisation was insignificant after certain PAC dosage. Studies had proven that it is effective in removing suspended matters, dissolved organic matter, turbidity, COD, colour and humic acid.

Table 3. Comparison of coagulant aids in water and wastewater treatment.

Type	Chemical	Advantage	Disadvantage
Synthetic cationic polymers	Polydiallyldimethyl ammonium chloride (poly-DADMAC), polydimethyl aminomethyl polyacrylamide, polyvinylbenzyl & trimethyl ammonium chloride	Produces denser shear-resistant sludge; For the use with metallic coagulants.	Increases complexity of coagulation-flocculation stage but improves process performance and economy when used properly. Remarks Application of polyacrylamides may

Synthetic neutral polymers	Polyacrylamides & polyethylene oxide		release residual acrylamide into the environment. It is possibly genotoxic and neurotoxic carcinogenic. Its biodegradation could take up to a few weeks' time. Therefore, polyacrylamide usage is banned or limited in several countries.
Synthetic anionic polymer	Anionic polymer A110 (Kemira), hydrolysed polyacrylamides, polystyrene sulfonate, polyacrylic acid & polyacrylates	Produces larger shear-resistant flocs by promoting bridging.	
Natural polymers	Sodium alginate	Particularly suited for the use with ferric salts, but if used properly can be effective when used with alum	Less expensive, but less efficient than synthetic polymers
	Moringa oleifera seed extract, starch, chitosan, guar gum, tannin	Cheaper additives for increasing settling velocity and reducing coagulant dosage.	
Inorganic coagulant aids	Aluminium chloride	For the use with organic polymer coagulants	Increases complexity of coagulation-flocculation, but improves process performance and economy when used properly

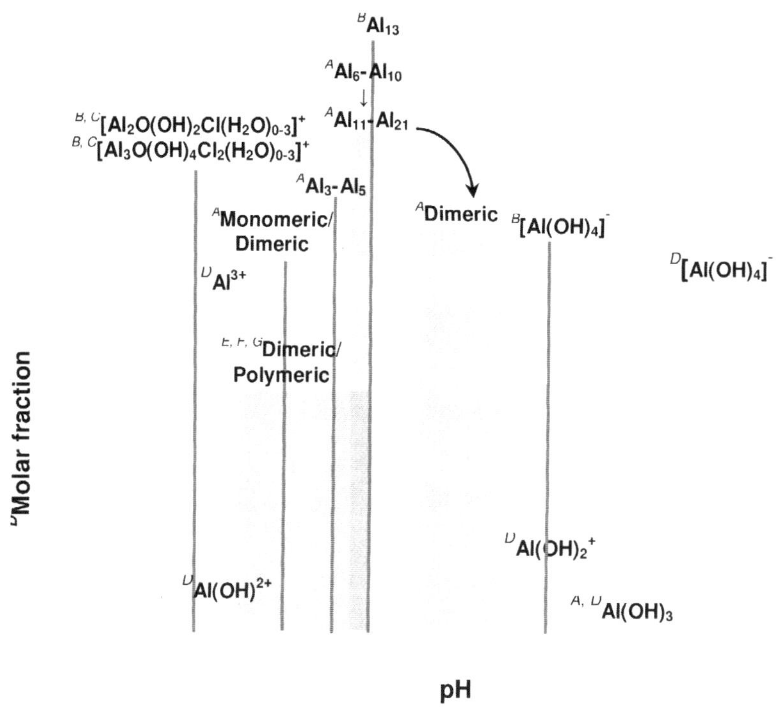

Figure 2. Predominance of Al species at different pH values.

PAC was developed in Japan in 1960s and has been commercially applied in water and wastewater treatment in Japan, the former USSR and Europe since the '70s, followed by America in the 1980s, rivalling against traditional metallic coagulants like alum and ferric chloride. In recent years, utilisation of PAC has surged and substitutes traditional metallic coagulants gradually due to its high coagulation efficiency, rapid precipitation, less dosage, less effect on pH, lower alkalinity consumption, lesser sludge production, low cost and wider availability compared to traditional coagulants. Now, it is extensively utilised in water and wastewater treatment across numerous countries, namely Japan, France, Germany and North America.

Despite of its prominence in landfill leachate treatment, the application of PAC may introduce toxic Al residuals into the aquatic environment. This had been verified by [18] using optimum PAC-treated synthetic and surface water samples. Residuals like very toxic Al_{13} and toxic Al^{3+} are of utmost concern. In soil environment, Al_{13} can be formed at pH 3.6-5.7 with peak dominance at pH 4.1-4.8, and subsequently persist over wide pH range (pH 3.6-10.6). Other organic Al-complexes and monomeric Al species like sulphate, phosphate and silicate Al species are far less toxic. Prolonged exposure of Al may induce bioaccumulation in human brain, bones and liver [22]. In fact, low molecular weight Al species can even penetrate biological membrane, threatening any living organism.

3. Conclusions

Despite having recent advancement of landfill leachate treatment technology, it is well perceived that coagulation and flocculation is still predominant for the treatment of landfill leachate is widely practiced by the many landfill operator. Accordingly, it is assumed that pre-hydrolysed coagulants such as PACl, PFCl, PFS and PAFCl can be considered as the superior coagulants due to better colour removal and high pH affinity towards leachate. In fact, Ferrous sulfate can also considered as a better coagulant over other metallic salts. Having said that, emerging of starch-based coagulants in leachate treatment is seen as promising approach which can reduce dependent of pure coagulant which can effectively improve sludge production during flocculation process. However, until now, the applicability of the discussed coagulants are yet to be documented and extensively discussed. Further studies required to evaluate their application for landfill leachate treatment with respect to broad range of its parameters that need to be observed in following to Department of Environment permissible discharge limit.

References

[1] Bratby J 2006 Coagulation and flocculation in water and wastewater treatment. *IWA Publishing*

[2] Leopold P and Freese S D 2009 A simple guide to the chemistry, selection and use of chemicals for water and wastewater treatment (WRC Report No. TT 405/09). *Republic of South Africa: Water Science cc*

[3] Letterman R D, Amirtharajah A and O'Melia C R 1999 Coagulation and flocculation. In American Water Works Association (AWWA) (Ed.), *Water quality and treatment: A handbook of community water suppliers*

[4] Dempsey B A 2006 Coagulation characteristics and reactions. *Interface science in drinking water treatment: Fundamentals and applications* 5-24 doi:10.1016/S1573-4285(06)80071-2.

[5] Aziz H A, Alias S, Adlan M N, Faridah, Asaari A H and Zahari M S 2007 Colour removal from landfill leachate by coagulation and flocculation processes *Bioresource Technology*, **98**(1) 218-220 doi:10.1016/j.biortech.2005.11.013

[6] Al-Shamrani A A, James A and Xiao H 2002 Destabilisation of oil-water emulsions and separation by dissolved air flotation *Water Research* **36**(6) 1503-1512. doi:10.1016/S0043-1354(01)00347-5

[7] IUPAC 2016. IUPAC Compendium of Chemical Technology (Gold Book), electronic version release 2.2. Retrieved 6th May, 2016, from http://goldbook.iupac.org
[8] Baun D L and Christensen T H 2004 Speciation of heavy metals in landfill leachate: A review *Waste Manage. Res* **22(1)** 3-23 doi:10.1177/0734242X04042146
[9] Christian P, Von der Kammer F, Baalousha M and Hofmann T 2008 Nanoparticles: structure, properties, preparation and behaviour in environmental media *Ecotoxicology,* **17**(5) 326-343 doi:10.1007/s10646-008-0213-1
[10] Wigginton N S, Haus K L and Hochella M F 2007 Aquatic environmental nanoparticles *Journal of Environmental Monitoring* **9**(12) 1306-1316 doi:10.1039/B712709J.
[11] Chanudet V and Filella M 2008 Size and composition of inorganic colloids in a peri-alpine, glacial flour-rich lake. *Geochimica et Cosmochimica Acta* **72**(5) 1466-1479. doi:10.1016/j.gca.2008.01.002
[12] Gustafsson O and Gschwend P M 1997. Aquatic colloids: Concepts, definitions, and current challenges *Limnology and Oceanography* **42**(3) 519-528
[13] Matura M, Ettler V and Klementová M 2011. Transmission electron microscopy investigation of colloids and particles from landfill leachates. *Waste Management & Research,* **0**(0), 1-12. doi:10.1177/0734242X11408382
[14] Matura M, Ettler V, Ježek J, Mihaljevič M, Šebek O, Sýkora V and Klementová M 2010. Association of trace elements with colloidal fractions in leachates from closed and active municipal solid waste landfills *Journal of Hazardous Materials* **183**(1-3) 541-548 doi:10.1016/j.jhazmat.2010.07.058
[15] Filella M, Chanudet V, Philippo S and Quentel F 2009 Particle size and mineralogical composition of inorganic colloids in waters draining the adit of an abandoned mine, *Applied Geochemistry* **24**(1) 52-61 doi:10.1016/j.apgeochem.2008.11.010
[16] Doucet F J, Lead J R and Santschi P H 2007 Colloid-trace element interactions in aquatic systems. Environmental colloids and particles: Behaviour, separation and characterisation *IUPAC series. Chichester* 95-157
[17] Bin G, Cao X, Dong Y, Luo Y and Ma L Q 2011. Colloid deposition and release in soils and their association with heavy metals *Critical Reviews in Environmental Science and Technology* **41**(4) 336-372. doi:10.1080/10643380902871464
[18] Gregory J and Duan J 2001 Hydrolyzing metal salts as coagulants *Pure and Applied Chemistry* **73**(12) 2017-2026 doi:10.1351/pac200173122017
[19] Kaszuba M, Corbett J, Watson F M and Jones A 2010. High-concentration zeta potential measurements using light-scattering technique *Philosophical Transactions of The Royal Society A-Mathematical Physical & Engineering Sciences* **368**(1927) 4439-4451. doi:10.1098/rsta.2010.0175
[20] Lyklema J 2005 *Fundamentals of interface and colloid science* **V** Soft colloid. Amsterdam: Elsevier
[21] Delgado A V, Gonzalez-Caballero F, Hunter R J, Koopal L K and Lyklema J 2005. Measurement and interpretation of electrokinetic phenomena (IUPAC Technical Report). *Pure and Applied Chemistry* **77**(10) 1753-1805. International Union of Pure and Applied Chemistry (IUPAC). doi:10.1351/pac200577101753
[22] Alley E R 2007. Water quality control handbook (2nd Edition). WEF & McGraw-Hill

Acknowledgments

This research is fully funded projects associated with Solid Waste Management Cluster (SWAM), Grant No. 1001/CKT/870023 administered by Science and Engineering Research Center (SERC) Universiti Sains Malaysia. Authors gratefully acknowledge contribution in this paper whom directly or indirectly towards the outcome of the present research

Nickel (Ni) Microalloying Additions in Sn-Cu Lead-free Solder. Short Review

M A A Mohd Salleh[1], I G Sandu[1,2], M M A Abdullah[1,3], I Sandu[1,4] and N A Saleh[1]
[1]Centre of Excellence Geopolymer and Green Technology, School of Materials Engineering, Universiti Malaysia Perlis (UniMAP), Perlis, Malaysia
[2]Faculty of Materials Science and Engineering, Gheorghe Asachi Technical University of Iasi, Romania
[3]Faculty of Engineering Technology, Universiti Malaysia Perlis (UniMAP), Perlis, Malaysia
[4]Laboratory of Scientific Investigation, Arheoinvest Platform, Alexandru Ioan Cuza University of Iasi, Romania

E-mail: arifanuar@unimap.edu.my

Abstract. In this digital-age era, solder plays important role in electronic packaging industries. As interconnects material, solder provide an electrical and mechanical support to the electronics devices. Solder usually consist of two or more addition of microalloying. By microalloying addition, the solidification structure can be modified. This paper reviews the addition of Ni as microalloying in Sn-Cu lead free solder. Small additions of Ni resulted with an improvement of solder in microstructure and in intermetallic compounds. The stabilization of hexagonal structure of Cu_6Sn_5 in lead-free solder alloys occurred due to present of Ni.

1. Introduction

In decades, soldering act as main process of electronic packaging industry. It is a process of metallurgical joining between two metals which are used solder as a filler under its melting point [1]. Usually, the metallurgical joining process temperature used is below 400°C [2]. The physical interaction between the metals make electronic packaging industry growth rapidly and more advances in electrical connection and give good mechanical support [3]. The good performance and quality of solder create a multi-function of electronic devices that have been used in this century such as cellular phones and digital cameras [2].

Tin-lead solders have been used widely for several years ago in electronic technology to make an improvement in the performance. The presence of lead in the electronics device can cause problems to human and the environment because it has higher toxicity. Due to environmental and health concern, the European Union makes a general instruction remove the used of lead in electronic packaging industry that has been issued by Waste of Electrical and Electronic Equipment (WEEE) and the Directive of the restriction of the Use of Certain Hazardous Substances (RoHS) after July 2006 [3].

Nowadays, lead-free solders alloy take a led towards the green technologies to replace the conventional tin-lead solder alloy [4–6]. Kang [1] summarized that there a few requirements needed by lead-free solders alloy exchange in terms of wettability, melting temperature, thermal fatigue resistance and cost. Currently, Sn-Ag-Cu alloy is one of the most excellent solder composition to select as replacement due to their performance [1]. However, the increasing cost of silver makes Sn-

Cu-Ni alloy is targeted in electronic packaging industry [7]. Sn-Cu-Ni alloy has been discovered due to limitation of Sn-Cu in terms of poor wettability, low resistance to thermal fatigue and high melting temperature.

The additions of nickel are proven to improve the flow characteristic and increase the amount of eutectic microstructure of the solder alloys [8]. Recently, Sn-Cu-Ni or known as SN100C is patented by Nihon Superior have been chosen as matrix based alloy in electronic packaging industries [9]. The additional of third or fourth elements in solder alloys has been known to improve the mechanical, physical and thermal properties of the solders. Usually the additional of microalloying element such as indium, titanium, zinc, bismuth, nickel and aluminum are used by the researcher because of their excellent properties [10]. With the addition of microalloying, the better solder alloys have been introduced and used in application of electronics device.

According to Reichencker [11], the growth of intermetallic compounds (IMCs) between two metals reduces the strength of solder joint due to brittle nature of these compound [11]. The crack growth in the intermetallic compounds (IMCs) also give limitation of the compound [12]. To increase the performance of the solder, the addition of third or fourth elements are needed in the solder such as zinc, bismuth and phosphorus [13].

2. Fabrication Method

There are several of fabrication technique in lead-free development. This technique usually is divided into two groups which are casting and powder metallurgy method [14]. Both of the method is commonly used in electronic industry by giving different properties of soldering alloy. Figure 1 shows the densification process for both method in bulk solder by showing that the casting method is depending on the cooling rate of molten metal while for powder metallurgy, heating rate during sintering cause the particle growth in soldering alloy [14].

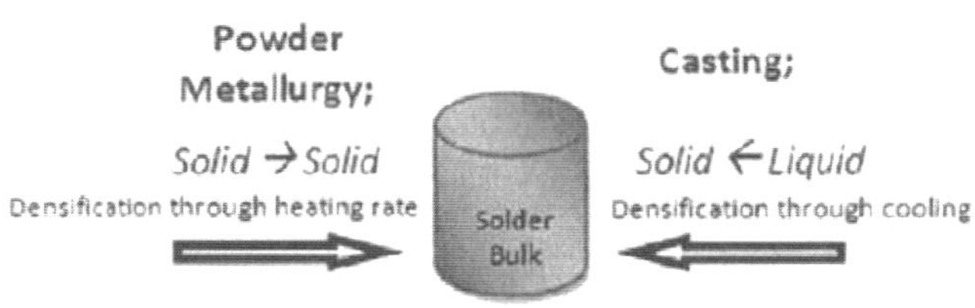

Figure 1. Densification process for casting and powder metallurgy method in bulk solder [14].

Powder metallurgy is also recently popular in electronic packaging applications [14]. In this method, sintering play role important part to reveal the end properties [15]. For sintering method, the direction of heating is transfer from outside to inside of compacted powder and usually it gives poor microstructure characteristic to the solder. To recover the limitation faced, advance of microwave-assistance for rapid sintering is introduced and forms uniform heating. It is also lower in cost and energy efficient [15]. The microstructure of powder metallurgy technique for Sn-0.7Cu is shown in figure 2 by using optical microscope which contains pores [14].

Casting is one of technique used by melting the alloy above their melting point. The technique used by melting the alloys and poured into a mold that has desired shaped and then it through a solidification process. By using this method, solidification of molten solder is faster but the grain growth only focuses by heating.

The rapid of solidification can be done with the help of medium such water or air at low temperature.

Figure 2. Microstructure image for powder metallurgy [14].

Figure 3 shows the microstructure image for casting method for Sn-0.7Cu by using optical microscope which consists of finer β-tin and eutectic region (Sn + Cu_6Sn_5) [14].

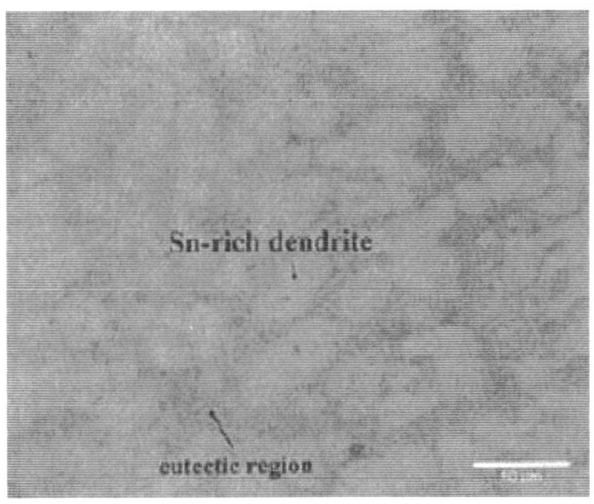

Figure 3. Microstructure image for casting method by using optical microscope [14].

3. Effects of Ni to microstructure in Sn-Cu solders

In a recent year, several researchers have been developed improvement of Sn-Cu solder alloy. Because of poor wettability, high melting temperature and low resistance to thermal fatigue, some of the alloying elements are added into Sn–Cu solder alloys such as Ni and rare earths [8]. The additional of Ni in Sn-0.7Cu solder was studied to improve flow characteristics and there several reasons are reported which is (i) it includes the composition of the commercial solder SN100C, (ii) it is important to start with a hypoeutectic alloy but it must close to the binary eutectic composition and (iii) it can make comparisons with the binary system Sn-Cu [8,16].

The addition of Ni in Sn-0.7Cu also can increase the amount of eutectic microstructure and it can be seen in Figure 4 [8]. Due to application in lead free solder, the melting temperature used for Sn-Cu-Ni solder alloy is around 240°C [7]. Figure 5 showed the system of Sn-Cu-Ni has been studied by Ghosh [7] especially in the sections from Ni_3Sn to Cu_3Sn and from Ni_3Sn_2 to Cu_6Sn_5 [7]. The increasing content of Ni in Sn-Cu-Ni makes Cu_6Sn_5 phases to become more rounded and lower the volume fraction for Sn-dendrites [8].

A.K. Dahle [16] investigated that the Ni addition in Sn-Cu are shifted from the hypoeutectic binary composition to ternary eutectic which is can improves the fluidity and reduce the freezing range.

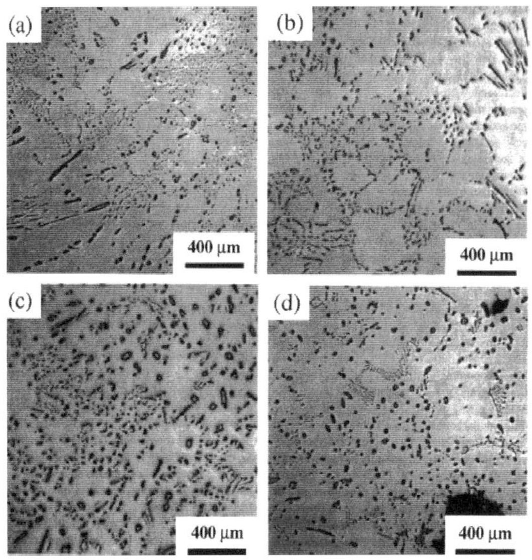

Figure 4. Example the additional of nickel in Sn–0.7Cu (a) Pure (b) 100 ppm Ni, (c) 600 ppm Ni and (d) 1000 ppm Ni [8].

Due to present of Ni, the eutectic are weakly irregular in the binary system while strongly irregular in the ternary alloys [16].

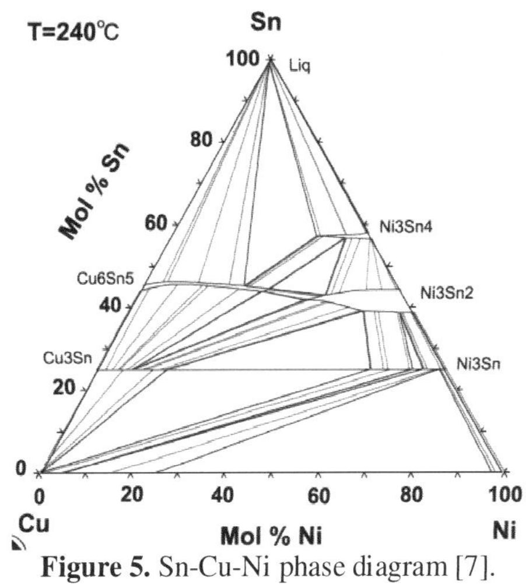

Figure 5. Sn-Cu-Ni phase diagram [7].

According to Stuart [17], the uniformity and stability of microstructure are identified for lead-free solder alloys in achieving of mechanical properties balance. The additions of microalloying can give refinement to Cu_6Sn_5 and give improvement the solder reliability [17]. By addition of Ni into Sn-0.7Cu solder, indirectly it gives improvement in solder revolution, and this formulation (Sn-Cu-Ni) was known as SN100C and patented by Nihon Superior Co. Ltd. Todays, Sn-Cu-Ni appears as new composite solder with a good compromise between cost and quality.

4. Effects of Ni to Intermetallic Compound of Sn-Cu solders

The important factors which developed by influencing the solder joint reliability are the intermetallic compounds (IMCs) layer [18]. Formation of intermetallic compounds (IMCs) is one of mechanisms to connect between the solder and substrate. Harris [12] who investigates the presence of the intermetallic Ag_3Sn and Cu_6Sn_5 in the tin solid solution of Sn-3.5Ag and Sn-0.7Cu state that it will improves hardness and resistance to fatigue [12]. The brittle nature of the intermetallic layer was confirmed by Reichencker [13] has excessive IMCs growth which decreases the strength of solder joint [13]. The addition of third or fourth elements is one of way to give excellent performance of the interfacial IMCs layers which help to increase the solder alloy properties.

Usually addition of another element in solder has effect between the solder and substrate reaction that can make the IMCs growth rate instable, it also can change the physical properties of the phase formed in solder and form additional layer reaction at the interface of IMCs. Nogita [19] claims that the high temperature hexagonal phase are stabilize with the present of Ni in the intermetallic. In intermetallic compound layer between the solder alloy and Cu substrate, it can be seen a visible crack which is known to be Cu_6Sn_5 and apparent significantly fewer cracks are present in the Cu_6Sn_5 layer when Ni are added into Sn-Cu solder. Figure 6 shows the cross sectioned ball grid alloy on Cu substrate by Sn-Cu solder with Ni and without Ni [19].

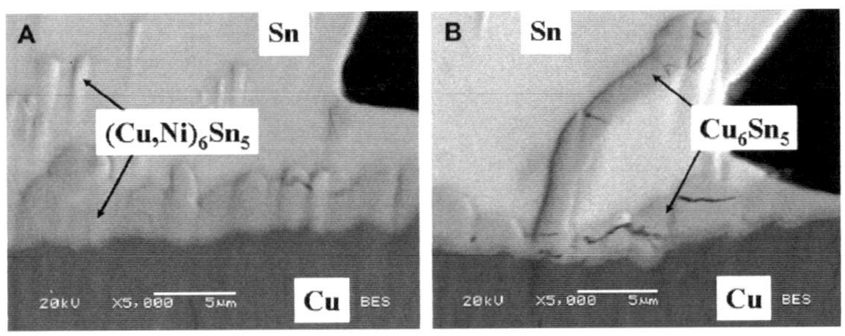

Figure 6. The cross sectioned ball grid alloy on Cu substrate (a) Sn-0.7Cu-0.05 Ni solder and (b) Sn-0.7Cu solder.

Cu_6Sn_5 is a relatively brittle phase after thermal cycling are stated by Stuart [17]. Due to volumetric changes experienced during polymorphic transformation between the hexagonal and monoclinic Cu_6Sn_5 crystallographic, the cracking are occurred at equilibrium condition [17]. However, the creep and mechanical properties of Cu_6Sn_5 by the present of Ni at the elevated temperature is still unknown [20].

5. Conclusions

This study is focused on the microstructure and intermetallic compounds of lead-free solder alloy due to the additions of the Ni as a microalloying element. The previous studied was show the effect of Ni in microstructure and intermetallic compounds that give improvement in the solder alloy. With the Ni addition in Sn-Cu, the microstructure is shifted from the hypoeutectic binary composition to ternary eutectic which is can improves the fluidity and reduce the freezing range. While the high temperature hexagonal phase are stabilize with the present of Ni in the intermetallic compounds.

References

[1] Efzan E and Marini A 2012 *International Journal of Engineering.* **1**(1) 2305– 8269
[2] Sona M and Prabhu K 2013 *Journal of Materials Science: Materials in Electronics* **24**(9) 3149-3169
[3] Ma H and Suhling J 2009 *Journal of Materials Science* **44**(5) 1141-1158
[4] Somidin F, Mohd Salleh M and Ahmad K 2012 *Advanced Materials Research* **620** 105-111

[5] Musa S, Mohd Salleh M and Norainiza S 2013 *Advanced Materials Research* **795** 518-521
[6] Mohd Salleh M, McDonald S, Gourlay C, Yasuda H and Nogita K 2016 *Materials & Design*, **108** 418-428
[7] Schmetterer C, Flandorfer H, Luef C, Kodentsov A and Ipser H 2008 *Journal of Electronic Materials* **38**(1) 10-24
[8] Mu D, McDonald S, Read J, Huang H and Nogita K 2016 *Current Opinion in Solid State and Materials Science* **20**(2) 55-76
[9] Ramli M, Saud N, Mohd Salleh M, Nazree Derman M, Mohd Said R and Nasir N 2014 *Materials Science Forum* **803** 228-232
[10] Sun L and Zhang L 2015 *Advances in Materials Science and Engineering* **2015** 1-16
[11] Parent J, Chung D and Bernstein I 1988 *Journal of Materials Science* **23**(7) 2564-2572
[12] Siewert T A, Madeni J C and Liu S,2003 *Proceedings of the APEX Conference on Electronics Manufacturing*, Anaheim, California
[13] Yahya I, Ghani N, Salleh M, Hamid H, Ahmad Z and Mayappan R 2012 *2012 35th IEEE/CPMT International Electronics Manufacturing Technology Conference (IEMT)*
[14] Salleh M A A M, Al Bakri A M M, Somidin F, Sandu A V, Saud N, Kamaruddin H,
[15] McDonald S D and Nogita K 2013 *Revista de Chimie (Bucharest)* **64**(7) 725-728
[16] Nai S, Kuma J, Alam M, Zhong X, Babaghorbani P and Gupta M 2009 *Journal of Materials Engineering and Performance* **19**(3) 335-341
[17] Felberbaum M, Ventura T, Rappaz M and Dahle A 2011 *JOM* **63**(10) 52-55
[18] McDonald S, Nogita K, Read J, Ventura T and Nishimura T 2012 *Journal of Electronic Materials* **42**(2) 256-262
[19] Zeng G, Xue S, Zhang L, Gao L, Dai W and Luo J 2010 *Journal of Materials Science: Materials in Electronics* **21**(5) 421-440
[20] Nogita K 2010 *Intermetallics* **18**(1) 145-149
[21] Mu D, Huang H, McDonald S and Nogita K 2012 *Journal of Electronic Materials* **42**(2) 304-311
[22] Achitei D, Galusca D G, Vizureanu P, Carabet R, Cimpoesu N 2009 *Metalurgia International* **14** 45-48

Acknowledgments
The authors gratefully acknowledge the Ministry of Higher Education Malaysia, Universiti Malaysia Perlis (UniMAP) and Nihon Superior Co. Ltd.

IOP Conf. Series: Materials Science and Engineering **209** (2017) 012085 doi:10.1088/1757-899X/209/1/012085

Review of Geopolymer Behaviour in Thermal Environment

K Zulkifly[1], H C Yong[1], M M A B Abdullah[2,1], L Y Ming[1], D Panias[3] and K Sakkas[3]

[1]Center of Excellence Geopolymer and Green Technology (CEGeoGTech), School of Materials Engineering, Universiti Malaysia Perlis (UniMAP), Perlis, Malaysia

[2]Faculty of Engineering Technology, Universiti Malaysia Perlis (UniMAP), Perlis, Malaysia

[3]School of Mining and Metallurgical Engineering, National Technical University of Athens, Athens, Greece

E-mail: khairunnisa.rus2014@gmail.com & cyheah@unimap.edu.my

Abstract. This paper presents a review of the thermal resistance of geopolymers towards the fire and elevated temperature. Geopolymers are one of the new class materials, emerging in far- reaching potential in fire and thermal resistance areas. The review includes thethermal performance of geopolymers before and after the thermal exposure. An overview on the recent progress of the geopolymers in high-temperature application is discussed in detail to keep pace with the development in this field. The characterization of geopolymers before and after the thermal treatment in term of microstructural, and crystallographic are also reviewed.

1. Introduction

The exponential increase of articles on geopolymers provides a better understanding on the applicability of geopolymers as a prominent alternative to the traditional cement which leads to a revolution in civil engineering. Geopolymer, is a state-of-the-art inorganic material that has been found displaying enhanced performances in thermal stability and fire resistance. In the urgent necessity to enhance the thermal or fire resistance, geopolymers present an enormous potential product such as geopolymer concrete, coating, lining and composite matrix to combat the problems.

There are several specific properties of geopolymers which lead to particular suitability in the high temperature application such as tunnel fire resistance, furnace lining, wall panel and thermal insulation. Davidovits [1] had emphasised the ceramic- like properties henceforth the formation of several crystalline aluminosilicate phases with the potentially high melting point up to 1100°C. These physiochemical characteristics believed as pivotal to preserve the structural integrity upon heating at high temperatures. It is in the sense that the chemical composition of the raw materials itself, admixtures, aggregates and alkaline used determine the production of geopolymers with a longer life cycle.

Be that as it may, it must be emphasised that even though geopolymers are incombustible materials, it still undergoes allowable strength deteriorations with increasing exposing temperature. Any construction materials, mainly ordinary Portland cement (OPC) will end its life cycle or undergo a significant strength loss at elevated temperatures (500-550°C) due to dehydration of OPC systems with formation of calcium oxide. During the rehydration, calcium hydroxide is formed, which is a high-end process leading to the formation of thermal cracking henceforth it degrades in strength irreversibly [2, 3].

Thermal expansion, thermal conductivity, strength retention, spalling and melting point are the critical macroscopic characteristics when assessing the suitability of the high-temperature performance of geopolymers. Hlavacek et al. [4] recognised there are high shrinkage and sintering effect associated with the elevated temperature resistance. In the microstructural point of view, the thermal analysis of the geopolymer for instances the thermodynamics, dehydration, phase stability and morphology are those of the major concern. Thus, this paper summarises the thermal performances as well as the effect of extent heating rates to the structural evolution and phase stability of geopolymeric materials.

2. Evolution of geopolymer during thermal exposure

The physical evolution of geopolymer during thermal exposure are of the major concern used to assess the behaviour after firing geopolymers sample. Geopolymers molecular structures particularly MK-based geopolymers are stable even when exposed at temperatures above 700°C is in many cases far superior to Portland cements [5-8]. Temperature condition change in term of the exposure time and heating rates are one of the ultimate factors that affecting the molecular structure henceforth entirely concerned in the material deterioration. But then, the benchmarks used to assess fire resistance could be defined differently according to their applications.

Considering the fire endurance assessment of geopolymers, the evolution of geopolymer during thermal exposure is of interest. On the basis, geopolymers undergo phase changes, namely thermal shrinkage, crystallisation, thermal conductivity and mechanical strength when exposed to fires. The responses of geopolymers towards the elevated temperature hereafter lead to microstructural stresses. This causes a major decline in the mechanical properties of the geopolymers. Microstructural variations as a results of high-temperature exposure can be analysed through differential thermogravimetry (DTG)/TG, Fourier transform infrared (FTIR), and X-ray diffraction analyser (XRD).

According to the study conducted by Barbosa and MacKenzie [6], geopolymers based on potassium sialate displayed extremely superior thermal stability. These geopolymers displayed high degree of thermal stability with no sign of melting up to 1000°C. Beyond these, few data have been compiled to have an insight into the microstructure change of such geopolymer concrete and mortar after thermal exposure.

2.1. Visual appearance

Serious damage of cracking and spalling of concrete structure are the major drawback of OPC concrete when it was subjected to rapid heating rates. Additional thermal stresses induced the explosive spalling commonly causes severe cracking within concrete's structure. According to Hertz and Kristian [9], spalling of concrete occurs due to the high internal stresses caused by water evaporation or temperature gradients inside the concrete. Referring to (figure 1), after 119min of ISO 834 fire, sudden explosive spalling leading to a loss of 60 mm of concrete consequently lower strength [10].

Figure 1. Explosive spalling that occur to High Performance Concrete (HPC) [10].

The violent breaking off of OPC concrete fragments at high temperatures provide as an essential input in the development of sustainable construction materials which is geopolymer. Sakkas et al. [11] reported that FeNi slag- based geoplymers shows little no crack on the surface after subjected to a propone flame with temperature higher than 1100°C for 120 min. Referring to X-ray diffractograms, there is an oxidation of magnetite to maghemite (γ-Fe_2O_3) when thermally treated at 500°C, whereas the red colour is formed when it is treated to 1000°C which transforms maghemite to hematite. Simultaneously, new aluminosilicate phases such as albite and olivine were formed [12, 13].

Physical evolution of fly ash geopolymer after firing was perceived to have a reddish in colour with diversifying degree (figure 2 and figure 3). The degree of red tinge was observed to be corresponding to the iron content in each different type of fly ash. The higher amount of iron oxide which is 16.4% (figure 2) cause remarkable change in colour than fly ash that consist 4.1% of iron oxide (figure 3). Even so, the inferior amount of iron oxide prevent macro cracking due to dimensional stability and preferred particle connectivity associated with sintering [14].

The improvement in the interconnectivity through sintering could increase the mechanical strength; in some cases can be more than five times the strength at ambient temperature [15, 16]. According to Rickard et al. [17], sintering can be defined as the "heal" of crack induced during the dehydration phase (phase in which water escaping caused the structural damage) and hence improve the mechanical properties of the materials. However, severe macro crack was observed for fly ash geopolymer (figure 2) and the crack healed as the temperature is was increase to 900°C and 1000°C. This was attributed to the dimensional instability caused by the heterogeneous melting [14].

Saker et al. [18] reported that fly ash based geopolymer shows a significant colour change when subjected to ISO 834 [19] fire curve in gas fired furnace. In companion to OPC concrete when subjected to same temperature profile as geopolymer, it did not display much change in colour, but greyness of the cement faded. Colour change in OPC is due to the higher reduction of moisture. Whereas the reddish tinge from the geopolymer concrete at 1000°C was because of the high iron oxide content on the fly ash and oxidised [18, 20].

Susceptible surface cracking and spalling was observed on OPC after the high temperature ranges starting 600°C. It is mainly due to the large amount of differential strain and thermal gradient which cause the pore pressure to exceed the tensile stress of the structure element. The relative less cracking and spalling of geopolymer concrete allowed it to be considered for use in high- temperature application such as furnace.

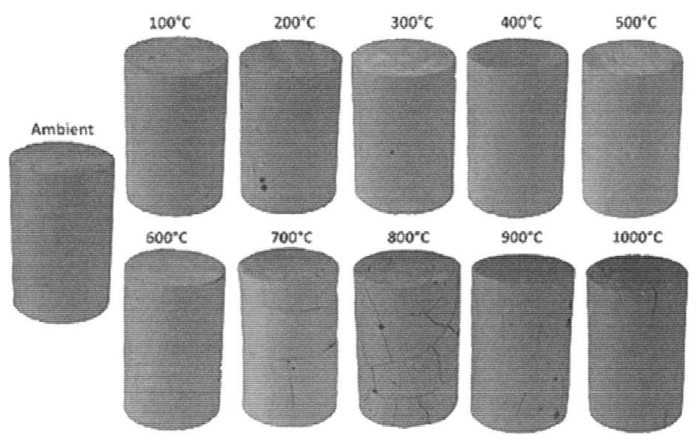

Figure 2. Fly ash geopolymer (16.4% iron oxide) fired under various temperatures [14].

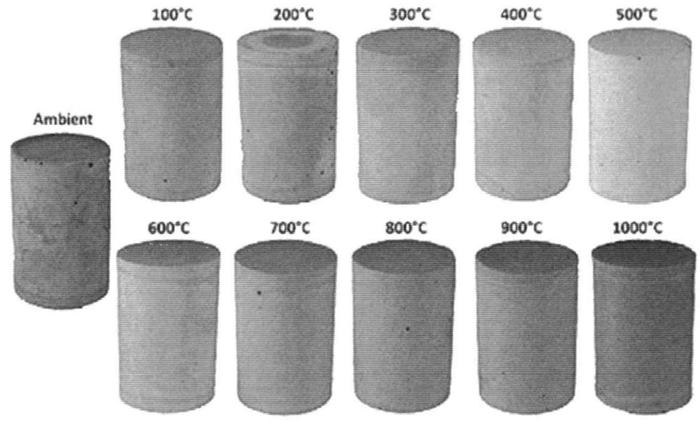

Figure 3. Fly ash geopolymer (4.1% iron oxide) fired under various temperatures [14].

On the other hand, after heating to 800°C and 1000°C the geopolymer mortar prepared by using calcined ore- dressing tailing of bauxite and ground granulated blast furnace slag showed visible cracks of 0.25±0.1 mm [21]. Similar cracks were observed for fly ash geopolymers containing high concentration of iron oxide and α- quartz [22]. Visual inspection indicated that larger sample generates more cracks compared to smaller one caused by the thermal differential during heating. The differential of oxidised and non- oxidised regions contributed to the cracking in the samples.

According to Zhang et al. [23], the breeding of cracking was contributed by the inhomogeneous distribution of binders and aggregates in the geopolymer concrete specimens. The preparation of specimen caused the floating upward of geopolymer binder and sinking downward of coarse aggregates. The exorbitant inhomogeneity between both unintended layers in the geopolymer concrete specimens, exacerbated the problem of crack. Therefore, according to Perera et al. [24], it was practical to cure under controlled relative humidity with mild heating between 40°C to 60°C to obtain crack free state.

2.2. Thermal shrinkage

Thermal shrinkage or expansion can be the basic response that causes failure in a fire. It is evaluated through the volumetric changes of geopolymers that occurs during given high temperature. Some studies proved that thermal incompatibility between geopolymer and aggregates renders them to thermal expansion. Thermal shrinkage is caused by the evaporation of water and hydroxyl groups OH which consequently weakens the geopolymer's structure system. Rahier et al.[25] studied the synthesis of dehydroxylated clay suspended in alkali sodium silicate solution at low-temperature curing denoted as "low temperature inorganic polymer glass (LTIPG). The LTIPG material shrunk about 5% attributed to the evaporation of water at 250°C after polymerization. However, the evaporation of water can occur at room temperature as well.

Rovnanik et al.[26] claimed that the shrinkage alkali-activated slag (AAS) is comparable to the shrinkage of OPC with a maximum of only 6.6% at 1200°C. It was reported that the increase in shrinkage for AAS is 5% between 40°C to 1000°C sequentially dominates the mass loss and the decrease in bulk density. This was attributed to the moisture loss and the liberation of water from the structure as a result of the thermal shrinkage which unfortunately weakens the geopolymer structure. The dissimilarity of density of structure, as well as the large number of micro-cracks in conjunction to the increased in the total porosity rapidly changed the morphology of AAS paste when the material is heated up to 800°C (figure 4).

On the basis, thermal shrinkage can be categorised into four characteristic regions which are the structural resilience, dehydration, dehydroxylation and sintering. Pore parameter of mortar before and after heating, closely affecting the thermal properties of geopolymers. The expansion and shrinkage as results of the thermal exposure were caused by the dehydration, dehydroxylation, phase transformation of quartz and viscous sintering. Ye et al.[21] acquiesced in the results that the porosity accumulate momentously after exposing to 400°C and reaches maximum at 1000°C. However, at 1200°C, there was almost 2% reduction in porosity due to the viscous flow and surface tension joint action.

Previous studies [21, 22, 27-29] observed that the dependence of bulk density under thermal exposure is not erratic, it actually undergoes several obstructing processes. First is the weight loss which derives from the density loss. Next is the shrinkage in the temperature ranging from ambient temperature to 800°C, followed by the sintering at the higher temperature that causes yield in the density. Later, is the expansion of quartz, engendering in a density loss. Also, the amount of shrinkage due to firing can be measured by using Vernier callipers as an average shrinkage using at least three samples.

Recent work by Duxson et al. [29] elucidated the physical evolution of geopolymeric materials derived from metakaolin and synthesised with sodium, mixed- alkali and potassium activating solutions (Na- NaK- K) during thermal exposure. According to the results, K- geopolymer implied prominent influence to the thermal shrinkage with the increasing of Si/Al ratios. This was attributed to the densification caused by reduction in porosity during dehydroxylation and sintering. Besides, according to previous literature geopolymers prepared using K-containing activators with a high Al/Si ratio remained their thermal stability up to 1300- 1400°C [6]. It is pronounced that the geopolymer specimen experienced recrystallization to feldspar leucite and kalsilite at 1000°C. This statement was further supported by Barbosa et al. [6], who stated that potassium geopolymers with low Al/Si ratio show significant large-grained in consistency and porous when heated >1200°C. The reduction of shrinkage in length and diameter when the geopolymer prepared using K activators was also concurred by Bakharev and Tatiana [28].

2.3. Crystallisation

The structural changes of fly ash geopolymers under thermal exposure were quite complex due to the influence of secondary phases in precursor for instances quartz, mullite, unburnt carbon, magnetite and other various iron species.

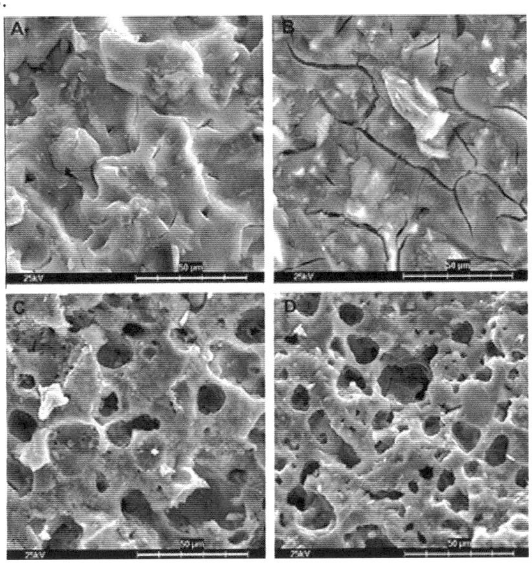

Figure 4. The micrographs of AAS paste treated at a room temperature (A) and after exposure to 600°C (B), 800°C (C) and 1200°C (D) [26].

The high-temperature exposure increased the predisposition towards the liberation and oxidation of iron species which accounted for the formation of stable crystalline phase [22]. Heah et al. [30] and Bakharev [28] studied the thermal properties of fly ash geopolymers, and they reported that the increase in iron oxide species, such as hematite, after high-temperature exposure. Comparatively, for metakaolin geopolymers, the crystallisation of dense sodium aluminosilicate phases namely nepheline was formed when exposed to temperature 800°C [31].

Figure 5 presents the phase composition of fly ash geopolymers with iron oxide content of 16.4%. This fly ash geopolymers showed relatively broad humps of amorphous with the crystalline components originated from fly ash. Analysis of XRD pattern indicated a slightly increased concentration of hematite after heating to 500°C. This was the typical characteristic of aluminosilicates matrix after the geopolymerization at high-temperature.

The drop of amorphous phases from 66% to 43% at 1000°C might due to the crystallisation into nepheline ($NaAlSiO_4$) and anortoclase ($NaAlSi_3O_8$). Mineral phases of quartz, hematite and maghemite were likely reduced in intensity. These phases have higher melting points than 1000°C (studied temperature range) which is mullite, 1830°C [32] and quartz (β- cristobalite), 1713°C .This reflected that mullite and quartz are the unique sintering agents [33].

2.4. Mechanical strength at elevated temperature

There is considerable on- going interest in adopting geopolymers in construction industry especially concerning the residual strength of the material after exposure to high temperature or fire. In general, the mechanical strength and microstructural analysis of all geopolymers have very distinct properties both before and after exposed to heating. Rickard et al.[8, 14, 16, 17, 34] supported that the main mechanism for the mechanical strength gain or losses is anticipated to dehydration damage, dimensional stability, sintering and melting. The mechanism that corporated to the reduction in mechanical strength is accord well with the formation of cracking too.

An earlier qualitative study by Perera et al. [24] showed that the strength of geopolymers decrease linearly with the increase in porosity. Porosity of geopolymers was highly dependent on the amount of water extant in the original precursors followed by the weight loss due to the curing conditions. According to Bakharev [28], geopolymer mateials prepared with sodium containing activators deteriorate rapidly in strength as it reaches 800°C, which is connected to the intense amount of build-up pore in the structures.

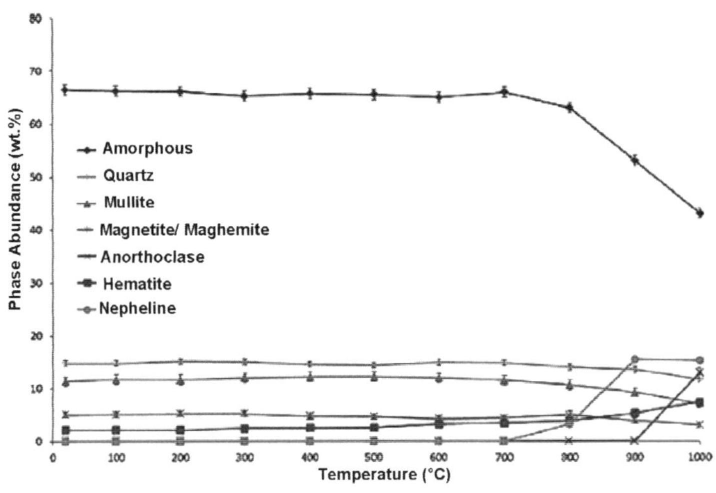

Figure 5. Fly ash geopolymer (16.4% iron oxide) fired under various temperatures [14].

According to Bakhharev [35], fly ash geopolymers with sodium containing activator had very small average pore size of about 4.5 nm and caused these materials to have low permeability. The high stresses in the pore wall developed movement of uncombined water and therefore greater degradation with rising temperature contributed from the collapse of fine pores and extent of shrinkage. As stated above, the loss of water after exposure to temperatures up to 1200°C led to more pores (figure 6) in the structure and consequently lessen the strength [36].

This trend was supported by Duxson et al. [37] whereby the microstructure of the material that contains a less dense gel and nanometer size pore improves the thermal performance and finally increases the compressive strength. This can be achieved by increasing the silicate concentration in the activating solution, successively leads to a massive increase in densification of the material up to temperature between 600°C to 900°C [8].

The high- strength geopolymers exhibited poor thermal properties, whereas the relatively low initial strength geopolymers exhibited impressive thermal properties. The unexposed fly ash geopolymer with early strength of 100 MPa exhibited severe macro cracking when heated to 600°C but retained its minimum strength of 2 MPa. Attractively, beyond 700°C, the strength improved to 15 MPa until reaching 1000°C. The SEM image revealed that at elevated temperature exposure the sintering of geopolymers produced a more homogenous and exceptional connection between the microstructure [14].

The exposure at elevated temperature caused changes in the phase composition and hence the mechanical strength of the geopolymers. This was supported by Bernal et al. [38] for slag geopolymers synthesised with high- modulus activators. When the slag geopolymer was exposed to the temperature of 1200°C, the compressive strength increased by 40% compared to the unexposed geopolymers. This was attributed to the formation of crystalline phases such as akermatite and gehlenite.

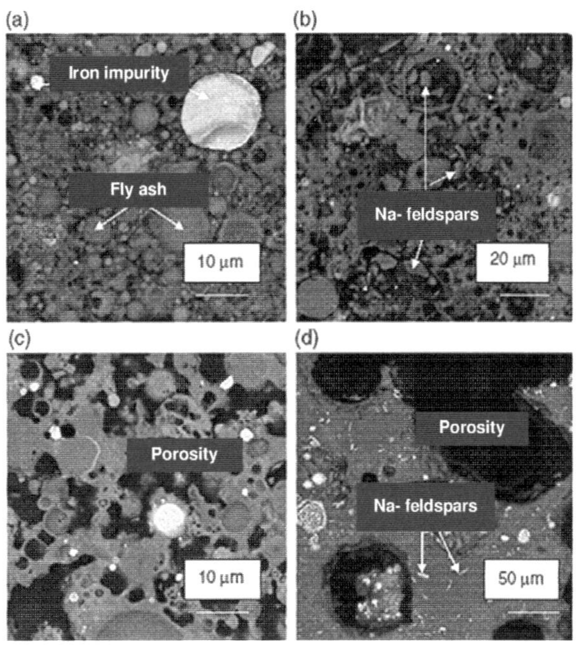

Figure 6. SEM micrograph of fly ash geopolymer using Na alkaline activators for unexposed sample (a), exposed to elevated temperature, 800°C (b), 1000°C (c) and 1200°C (d).

3. Conclusions

In this paper, the thermal behaviour of geopolymers was reviewed. With the global scarcity towards development of new materials with improved thermophysical properties and the necessary issues of reutilising of industrial by- products, alkali –activated binders so- called geopolymers, materials with very high fire resistance ratings is necessary for the long-term basis. Increased demands and reduced supply of Portland cements as common construction materials require industries to participate in employing geopolymers through the dissolution and polycondensation of aluminoslicate sources in highly alkaline solution. As the literature has shown, geopolymers is an alternative fire resistance construction materials that can be adopted. Finally, geopolymerization methods are encouraged to be realised applying either in existing structures or as novel technologies.

References

[1] Davidovits J 2013 Geopolymer cement. A review *Geopolymer Institute, Technical papers* **21** 1-11
[2] Arioz O 2007 Effects of elevated temperatures on properties of concrete.*Fire safety J.* **42** 516-22
[3] Alarcon-Ruiz L, Platret G, Massieu E and Ehrlacher A 2005 The use of thermal analysis in assessing the effect of temperature on a cement paste *Cem. Concr. Res..* **35** 609-13
[4] Hlaváček P, Šmilauer V, Škvára F, Kopecký L and Šulc R 2015 Inorganic foams made from alkali-activated fly ash: Mechanical, chemical and physical properties *J. Eur. Ceram. Soc.* **35** 703-09
[5] Rahier H, Van Mele B and Wastiels J 1996 Low-temperature synthesized aluminosilicate glasses *J.Mater.Sci.***31** 80-85
[6] Barbosa V F and MacKenzie K J 2003 Synthesis and thermal behaviour of potassium sialate geopolymers *Mater.Letters* **57** 1477-82
[7] Barbosa V F and MacKenzie K J 2003 *Mater. Res. Bulletin* **38** 319-31
[8] Duxson P, Lukey G C and van Deventer J S 2007 Physical evolution of Na-geopolymer derived from metakaolin up to 1000 C *J.Mater. Sci.* **42** 3044-54
[9] Hertz K D 2003 Limits of spalling of fire-exposed concrete *Fire safety J.* **38** 103-16
[10] Klingsch E 2014 Explosive spalling of concrete in fire in the Department of Civil, Environmental and Geomatic Engineering Doctoral and Habilitation Theses (Institute of Structural Engineering- Zurich)
[11] Sakkas K, Nomikos P, Sofianos A and Panias D 2013 Inorganic polymeric materials for passive fire protection of underground constructions. *Fire Mater.* **37** 140-50.
[12] Sidhu P 1988 Transformation of trace element-substituted maghemite to hematite *Clays and Clay Miner.* **36** 31-38
[13] Przepiera A, Przepiera K, *J.Therm. Anal. Calorim.* **40** 1131-38
[14] Rickard W D, Kealley C S and Riessen A 2015 *J. Amer.Ceram. Soc.* **98** 929-39
[15] Kuenzel C, Grover L M, Vandeperre L, Boccaccini A R and Cheeseman C R 2013 Production of nepheline/quartz ceramics from geopolymer mortars. *J. Eur. Ceram.Soc.* **33** 251-58
[16] Rickard W D, Temuujin J and van Riessen A 2012 *J.Non-Cryst. Solids.* **358** 1830-39.
[17] Rickard W D and Van Riessen A 2014. *Cem. Conc.Compos.* **48** 75-82
[18] Sarker P K, Kelly S and Yao Z *Mater. Design* **63** 584-92
[19] Standard I 834: 1994 Fire resistance tests-Elements of building construction. I*nter. Organiz. Standar. Geneva*
[20] Zhao R and Sanjayan J G 2011 Geopolymer and Portland cement concretes in simulated fire. *Magaz.Concr. Res.* **63** 163-73
[21] Ye J, Zhang W and Shi D 2014 Effect of elevated temperature on the properties of geopolymer synthesized from calcined ore-dressing tailing of bauxite and ground-granulated blast furnace slag. *Constr. and Build. Mater.* **69** 41-48

[22] Rickard W D, Riessen A and Walls P 2010 Thermal Character of Geopolymers Synthesized from Class F Fly Ash Containing High Concentrations of Iron and α□Quartz. *Inter. J. App. Ceram.Tech.* **7** 81-88

[23] Zhang H Y, Kodur V, Qi S L, Cao L and Wu B 2014 Development of metakaolin–fly ash based geopolymers for fire resistance applications. *Constr. Build. Mater.* **55** 38-45

[24] Perera D S, Uchida O, Vance E R and Finnie K S 2007 Influence of curing schedule on the integrity of geopolymers. *J. Mater. Sci.* **42** 3099-106

[25] Rahier H, Mele B and Wastiels J 1996 Low-temperature synthesized aluminosilicate glasses. *J. Mater. Sci.* **31** 80-85

[26] Rovnaník P, Bayer P and Rovnaníková P Characterization of alkali activated slag paste after exposure to high temperatures. *Constr. Build. Mater.* **47** 1479-87

[27] Duan P, Yan C, Zhou W and Luo W 2015 Thermal Behavior of Portland Cement and Fly Ash-Metakaolin-Based Geopolymer Cement Pastes. *Arab. J. Sci. Eng.* **40** 2261–69

[28] Bakharev T 2006 Thermal behaviour of geopolymers prepared using class F fly ash and elevated temperature curing. *Cem. Concr. Res.* **36** 1134-47

[29] Duxson P, Lukey G C and Deventer J S 2006 Thermal evolution of metakaolin geopolymers: Part 1–Physical evolution. *J. Non-Cryst. Solids.* **352** 5541-55

[30] Cheng-Yong H, Yun-Ming L, Abdullah M M A B, and Hussin K 2017 Thermal Resistance Variations of Fly Ash Geopolymers: Foaming Responses. *Sci. Rep.* **7** 45355

[31] Duxson P, Lukey G C and Deventer J S 2006 Evolution of gel structure during thermal processing of Na-geopolymer gels. *Langmuir* **22**(21) 8750-57

[32] Schneider H, Schreuer J and Hildmann B 2008 Structure and properties of mullite—a review. *J. Eur. Ceram. Soc.* **28** 329-44

[33] Kovalchuk G. and Krivenko P 2009 Producing fire-and heat-resistant geopolymers. *Geopolymers: Structures, Processing, Properties and Industrial Applications* 229-68

[34] Xu H, Li Q, Shen L, Wang W and Zhai J 2010 Synthesis of thermostable geopolymer from circulating fluidized bed combustion (CFBC) bottom ashes. *J. Hazard. Mater.* **175** 198-204

[35] Bakharev T 2005 Resistance of geopolymer materials to acid attack. *Cem.Conc. Res.* **35** 658-70

[36] Temuujin J, Minjigmaa A, Rickard W and Van Riessen A 2012 Thermal properties of spray-coated geopolymer-type compositions. *J. Therm. Anal. Calorim.* **107** 287-92

[37] Duxson P, Provis J L, Lukey G C, Mallicoat S W, Kriven W M and Van Deventer J S 2005 Understanding the relationship between geopolymer composition, microstructure and mechanical properties. *Colloids and Surfaces A: Physicochem. Eng. Aspec.* **269** 47-58

[38] Bernal S A, Rodríguez E D, de Gutiérrez R M, Gordillo M and Provis J L 2011 Mechanical and thermal characterisation of geopolymers based on silicate-activated metakaolin/slag blends. *J. Mater. Sci.* **46** 5477

Acknowledgements
The authors would like to acknowledge the funding support from "Partnership for Research in Geopolymer Concrete" (PRI-GeoC-689857) sponsored by the European Union and Fundamental Research Grant Scheme (FRGS) (9003-00544 and 9003-00540) by Ministry of Education, Malaysia.

Hybrid TiO_2-Gigantochloa Albociliata Charcoal in Dye Sensitized Solar Cell

L F A Talip[1,2], M M Ramli[1,2*], S S M Isa[1,2], D S C Halin[2], N S Mazlan[1,2], N A M Anhar[1,2], N A Danial[1,2] and M R Muda[1,2]
[1] School of Microelectronic Engineering, University Malaysia Perlis, 01000 Malaysia
[2]Center of Excellence Geopolymer and Green Technology, School of Materials Engineering, Universiti Malaysia Perlis (UniMAP), Perlis, Malaysia

E-mail: mmahyiddin@unimap.edu.my

Abstract. The Dye Sensitized Solar cell (DSSC) is an alternative to the silicon solar cell because it is low cost and easy to fabricate. In previous work, Remazol Orange (RO) was used as a dye sensitizer in DSSC but the efficiency is still low, 0.13%. In order to increase the device performance, TiO_2 thin film as the working electrode is hybridized with high conducting and absorption material which is bamboo charcoal powder (BCP). It is founded that the nanoparticle size of TiO_2-BCP composite was smaller compared to pristine TiO_2. The ratio of TiO_2 and BCP did not give any significant effect towards the particle size. The efficiency of RO DSSC was highly improved by 84.6% at higher carbonization temperature, 1100 °C compared to 500 °C during pyrolysis process due to its capability in absorbing more dye as it has larger specific area.

1. Introduction

Dye Sensitized Solar Cell (DSSC) is one of photovoltaic device that aroused world's attention due to its low cost production and high efficiency. There are many efforts that have been applied in order to enhance the efficiency of DSSC. One of the recent techniques is by hybridizing all parts with carbon based materials like carbon nanotubes [1], graphene [2] and even carbon nanotubes-graphene hybrid system [3]. Carbon based material such as carbon nanotubes and graphene are highly favoured due to their extraordinary structure, conductivity and other properties. However these materials are quite expensive.

Bamboo charcoal is a newly developed material with high conductivity, high absorption and other extraordinary properties. Bamboo is an abundant plant in tropical countries including Malaysia, thus it offers a low cost alternative towards other carbon family. The most noticeable one is a micro-porous material with excellent adsorption property due to its large surface area where it could absorb more dye when the electrode (TiO_2-BCP) was immersed in the dye solution [4].

In the previous work, an alternative to Ruthenium as dye sensitizer was studied and investigated as this material was very expensive and not environmental friendly [5]. The results found that a water based dye, Remazol Orange potential to be used as a dye sensitizer with efficiency of 0.2% at the condition of 2.5 mM immersed for 24 Hours. In this project, TiO_2 as the nanomaterial for working electrode was hybridized with BCP in order to improve Remazol Orange DSSC device performance as illustrated in figure 1. Herein, the structural investigation on the hybrid system and also the efficiency of the DSSC were investigated.

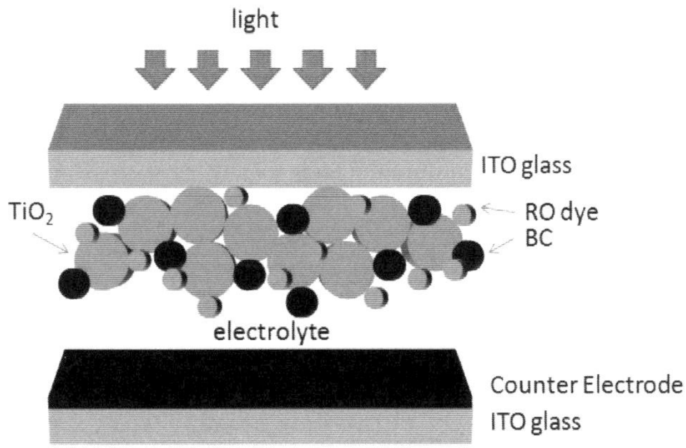

Figure 1. DSSC structure.

2. Materials and Methods

2.1. Bamboo Charcoal Powder Preparation

There are two types of BCP using in this experiment which are the commercial sample and in-house sample. For commercial sample, the BCP was purchased from the JitraGrow Resources (carbonized at 500°C). While for in-house sample, the BCP was prepared using furnace at the temperature of 1100°C. For carbonization process, firstly, the bamboo was chopped into small segments. The carbonization process of bamboo was as followed: (1) the temperature was set to 300°C within 10 minutes and retained for 5 minutes before increased to 600°C and retained for 15 minutes and (2) the temperature was ramped up to 1100°C within 30 minutes and this temperature retained for one hour. Then, the time was set for 2 hours to let it ramped down to the room temperature.

2.2. Remazol Orange Dye Preparation

Remazol Orange (RO) dye is a water based dye which was used as the dye sensitizer in this experiment. It is easily to get as it is widely used in textile industry. RO was purchased from the AR Alatan Sdn. Bhd and was diluted before being used. The powder of RO has a molarity weight at 617.54 g/mol. The solution is prepared by mixing 23 mg powder and 15 mL DI water to form 2.5 mM concentration. The calculation was show as Equation 1.

Molar mass (mg):

$$[(\text{Molecular weight (g/mol)})\ (\text{concentration (mMolar)})\ (\text{volume (mL)})] \qquad (1)$$

2.3. TiO_2-bamboo charcoal powder hybrid

The preparation procedures for a working electrode of TiO_2 (P25)-BCP composite were as followed: (1) preparation of the stabilizer by mixing the 0.1 mL of high concentration of acetic acid solution ($C_2H_4O_2$) with 50 mL DI water to produce 0.035 molar solution; (2) preparation of the suspension by mixing TiO_2-BCP composite particles (1:0.08 g) with stabilizer (15 mL) and Triton X-100 (0.05 mL); (3) deposition of suspension on top of the ITO glass substrate using Doctor Blade method; (4) annealing of this substrate at 450°C for 30 minutes on hot plate and (5) immersion of TiO_2-BCP thin film into the solution of RO for 24 hours in order to produce the working electrode.

2.4. Assembling and testing the DSSC

The counter electrode was produced using a black deposition from the soot of combustion of flame. The working electrode was sandwiched with soothed conductive glass by facing each other the active sides of anode and cathode and held together using binder clips as shown in figure 2. Then, a drop of iodine electrolyte was injected between the two electrodes. The PV Cell Testing LS1000 Solar Simulator was used to illuminate the DSSC and a digital source meter (Keithley 2450) was used to obtain the I_{sc} and V_{oc}. From the graph, the efficiency of the DSSC can be calculated by using Equation 2 and 3:

Figure 2. The assemble of dye sensitized solar cell (DSSC) using RO as a dye.

$$Efficiency, \eta = \frac{Jsc \quad x \quad Voc \quad x \quad FF}{Pin} \tag{2}$$

where

$$FF = \frac{P\max}{Isc \quad x \quad Voc} \tag{3}$$

2.5. Structural investigation
The structural and morphology of the TiO_2-BCP composite were analyzed using Atomic Force Microscopy (AFM) and Field Emission Scanning Electron Microscope (FESEM).

3. Results and discussions

3.1. Characterizations of BC and TiO_2-BCP composite
Figure 3 (a) shows the bamboo which was carbonized at 500°C for commercial sample before being grinded to form powder as in figure 3 (b). According to the Kuromoto et al., bamboo which was carbonized at temperature above 500°C has higher adsorption capacity where the degradation of hemicellulose and cellulose of bamboo occur at this stage [6]. This process is important to provide porous structure in the BC. The adsorptive properties of the BC depend on the microporosities of the BC. Figure 3 (c and d) show the comparison of TiO_2 powder before and after being mixed with BCP powder.

The mass ratio of TiO_2-BCP was varied in order to obtain the best optimized ratio. It was observed that the white colour of TiO_2 powder turned to black pestle after mixed with BCP and it can be concluded that the mass of BCP increased the colour of TiO_2-BCP composite. Figure 4 shows the FESEM images of TiO_2 and TiO_2-BCP (prepared with commercial sample) composite after annealed at 450°C where the slurry was homogenously dispersed on the ITO substrate. It was also observed that the surface of the TiO_2-BCP thin film was quite rough compared to the TiO_2 sample. It can be seen that the TiO_2 nanoparticle was clearly in spherical shape. After hybrid with the BCP, it was observed that the TiO_2 nanoparticle become stick together with the BC and the size of the TiO_2-BCP becomes smaller.

The surface morphology of this thin film was further observed using AFM as shown in figure 5. Generally TiO_2 consist of three types of structure which are anatase, ructile and brookite and this structure formation depends on the annealing temperature.

IOP Conf. Series: Materials Science and Engineering **209** (2017) 012086 doi:10.1088/1757-899X/209/1/012086

Figure 3. Images of (a) bamboo charcoal, (b) bamboo charcoal powder, (c) TiO_2 (P-25) and (d) TiO_2 (P-25)-BCP composite.

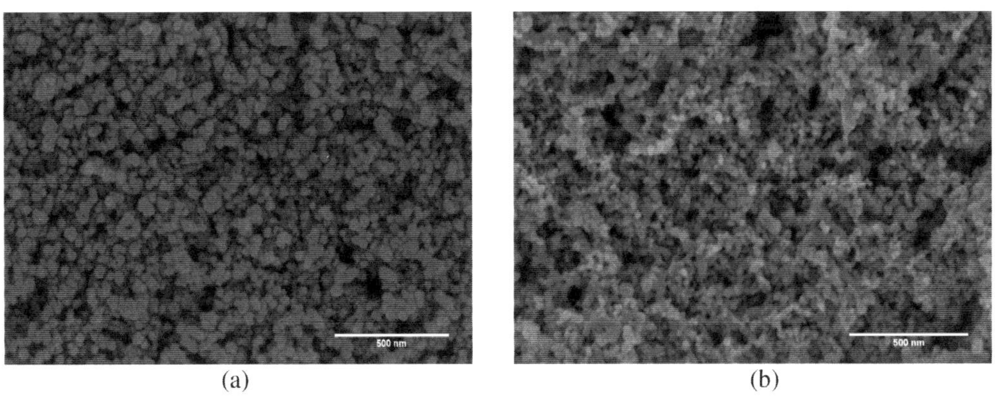

Figure 4. FESEM images of (a) TiO_2 (P-25) and (b) TiO_2 (P-25)-BCP composite

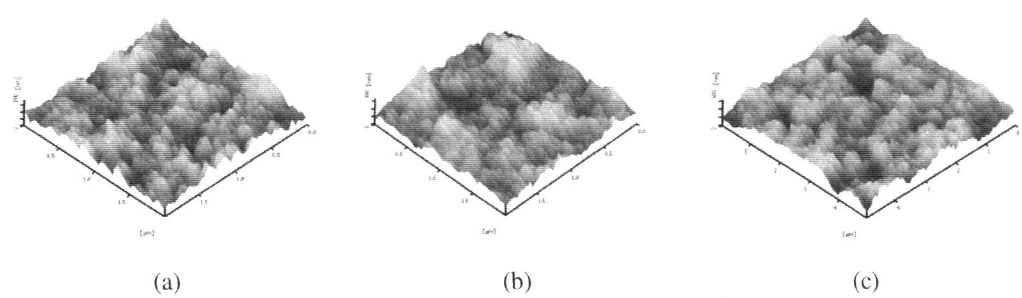

Figure 5 AFM images of TiO : BCP at different ratio (a) 1:0.02; (b) 1:0.04 ; (c) 1:0.08.

The annealing process was done at 450°C after being optimized in the previous work [5], where anatase structure which is more suitable for photocatalytic process was successfully obtained. From the images, it found that the particles of the TiO_2 are in irregular shape and it is expected to agglomerate and becomes larger as the temperature increased [5]. Table 1 shows the comparison of grain size for deposited TiO_2-BCP thin film at different mass ratio of 1:0.02, 1:0.06, 1:0.08 g respectively. The results show that TiO_2 film was very uniform and the crystal structure was quite small. The average diameter was measured between 65.70 nm to 71.58 nm. It shows that the grain sizes are not affected by the ratio of the composite. However, there was a big different of the It shows that the presence of BCP has inhibited the growing size of the TiO_2.These findings were consistent for all samples (commercial and in-house BCP).The average grain size between the pure TiO_2 and the TiO_2-BCP composite was decreased by 52%. This finding proved the result of FESEM images which were observed in figure 5 where the size of the TiO_2 becomes smaller after hybrid with the BCP.

Table 1. Grain size for deposited TiO_2 thin film.

Mass Ratio	Averages Grain Size, nm
TiO_2 only	131
1: 0.02	71.58
1: 0.04	67.01
1: 0.08	65.70

3.2. Effect of BC carbonization temperature towards efficiency of the DSSC.
Figure 6 shows the current voltage characteristic (IV) graph for different carbonization temperature of BC towards efficiency of the DSSC. The carbonization temperature of in-house BCP at 1100°C shows the highest device efficiency which was 1% followed by 500°C (commercial sample) at 0.23 % and RO DSSC with efficiency of 0.13% as presented in table 2. The performance of RO DSSC was highly improved by 84.6% using hybrid TiO_2-BC composite. It was expected that the pyrolysis of bamboo during carbonization process resulted a high quality porous carbonaceous structure.

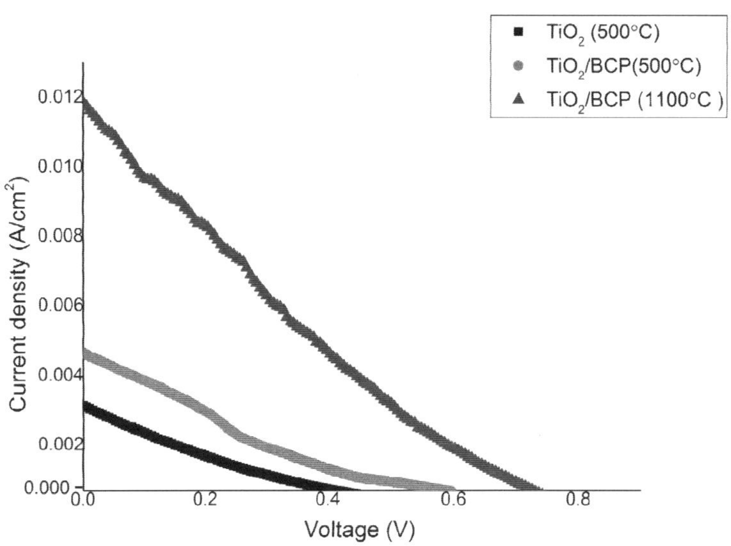

Figure 6. The IV graph of different carbonization temperature of BC towards efficiency of the DSSC.

This porous structure was obtained when the parenchyma cell wall of bamboo charcoal become thinner. The formation of crevice in the parenchyma cell wall is also one of the major factors to promote the formation of this porous structure, thus increase the surface area of the charcoal [7]. It was proved that higher carbonization temperature (> 800°C) has improved the porous structure, hence produce high conducting and adsorption material, and thus increase the device efficiency.

Table 2. The efficiency of DSSC.

Sample	V_{OC}	I_{SC}	J_{SC}	P_{MAX}	FF	η %
TiO_2	0.54	0.003138	0.001569	0.000258	0.152538	0.13
TiO_2/BCP(500 °C)	0.6	0.004642	0.002321	0.00045	0.161608	0.23
TiO_2/BCP(1100 °C)	0.8	0.011866	0.005933	0.001919	0.202121	1.00

4. Conclusions

In this work, the efficiency of RO DSSC was improved by using TiO_2/BCP composite as the working electrode. Structural investigations at annealing temperature of 450°C were done using FESEM and AFM where the nanoparticle size of the composite was smaller compared to pristine TiO_2. It was also observed that the mass ratio of TiO_2 and BCP did not affect the size of nanoparticle. These findings were consistent for all samples; commercial and in-house prepared at 500°C and 1100°C respectively. The DSSC performance was investigated using solar cell simulator. From the result, the efficiency of the DSSC was increased by 84.6% as the carbonization temperature increased from 0.13% up to 0.23% for 500°C and 1% for 1100°C. It was expected that, carbonization process at higher temperature promotes porous structure which increase the surface area of the charcoal, hence increase the absorption of the dye.

References

[1] Yang Z, Li L, Lin H, Luo Y, He R, Qiu L, Ren J and Peng H 2012 Penetrated and aligned carbon nanotubes for counter electrodes of highly efficient dye-sensitized solar cells *Chem. Phys. Lett.* **549** 82–5

[2] Nemala S S, Kartikay P, Prathapani S, Bohm H L M, Bhargava P, Bohm S and Mallick S 2017 Liquid phase high shear exfoliated graphene nanoplatelets as counter electrode material for dye-sensitized solar cells *J. Colloid Interface Sci.* **499** 9–16

[3] Ma J, Li C, Yu F and Chen J 2014 3D single-walled carbon nanotube/graphene aerogels as Pt-free transparent counter electrodes for high efficiency dye-sensitized solar cells *ChemSusChem* **7** 3304–11

[4] Chou C, Chen C, Lin S, Lu W and Wu P 2015 Preparation of TiO 2 / bamboo-charcoal-powder composite particles and their applications in dye-sensitized solar cells *Adv. Powder Technol.* **26** 711–7

[5] Isa S S M, Ramli M M, Hambali N A M A, Muda M R, Kasjoo S R, Isa M M, Nor N I M, Ahmad N, Khalid N, Selamat N M, Anhar N A M and Sensitizer A D 2016 Remazol Orange Dye Sensitized Solar Cell 543–6

[6] Asada T, Ishihara S, Yamane T, Toba A, Yamada A and Oikawa K 2002 Science of Bamboo Charcoal: Study on Carbonizing Temperature of Bamboo Charcoal and Removal Capability of Harmful Gases *J. Heal. Sci.* **48** 473–9

[7] Lin S H, Hsu L Y, Chou C S, Jhang J W and Wu P 2014 Carbonization process of moso bamboo (Phyllostachys pubescens) charcoal and its governing thermodynamics, *J. Anal. Appl. Pyrolysis.* **107** 9–16

IOP Conf. Series: Materials Science and Engineering **209** (2017) 012087 doi:10.1088/1757-899X/209/1/012087

Potential of Starch Nanocomposites for Biomedical Applications

N H Zakaria[1,2,*], N Muhammad[1,2] and M M A B Abdullah[3,2]
[1] School of Manufacturing Engineering, Universiti Malaysia Perlis, Perlis, Malaysia
[2] Center of Excellence Geopolymer & Green Technology, School of Materials Engineering, Universiti Malaysia Perlis, Perlis, Malaysia
[3] Faculty of Engineering Technology, Uniciti Alam Campus, Universiti Malaysia Perlis, Perlis, Malaysia

E-mail: [a]nurulhusna.zakaria@gmail.com, [b]noorhafiza@unimap.edu.my, [c]mustafa_albakri@unimap.edu.my

Abstract. In recent years, the development of biodegradable materials from renewable sources based on polymeric biomaterials have grown rapidly due to increase environmental concerns and the shortage of petroleum sources. In this regard, naturally renewable polymers such as starch has shown great potential as environmental friendly materials. Besides, the unique properties of starch such as biodegradable and non-toxic, biocompatible and solubility make them useful for a various biomedical applications. Regardless of their unique properties, starch materials are known to have limitations in term of poor processability, low mechanical properties, poor long term stability and high water sensitivity. In order to overcome these limitations, the incorporation of nano size fillers into starch materials (nanocomposites) has been introduced. This review aims to give an overview about structure and characteristics of starch, modification of starch by nanocomposites and their potential for biomedical applications.

1. Introduction

Synthetic polymeric materials have been extensively used in every field of human activity. However, the concern related with their usage are potential pollution due to non-degradable properties, no renewability and high costs [1, 2]. Thus, intense research towards development of new environmental friendly materials based on biopolymers has been conducted. Natural biopolymers are produced by living organisms such as animals, plants, algae and microorganism. Among of these natural polymers, the most abundant polysaccharides known as starch has received considerable attention as biodegradable materials.

The unique physiochemical and functional characteristics of starch, vary from different plant sources like maize, wheat, cassava and potato. Starch has been famously used in food, pharmaceutical and biomedical applications because of its biocompatibility, total degradability without toxic residues, low cost, wide availability, renewability and has thermoplastic behavior [3–5]. In addition, starch has already known to have great potential in biomedical fields including tissue engineering scaffolds, bone cements, drug delivery systems and stent [6–9].

Starch feasible to process into thermoplastic materials through the use of various industrial techniques, in which similar to those commonly used for synthetic thermoplastic such as injection molding, extrusion, and thermo molding to produce porous or dense materials [10]. Nonetheless, thermoplastic starch (TPS) products are water sensitive and would lead to poorer mechanical and physical properties, due to their hydrophilic nature. In order to improve the mechanical properties and water resistance, starch can be altered by several methods such as blending with synthetic or natural polymers to form composite materials. Recently, special attention has been focused on the use of nano sized fillers into starch matrix to form nanocomposites. The application of nanocomposites could yield significant enhanced mechanical properties of starch based materials as compared to conventional micro composites. This is because the mutual effect between filler at nanometer size permits them to form molecular bridges in polymer matrix.

In this paper, a review on the research progress for the development of nanocomposites based on starch will be presented. The review focus on the influences of the incorporation of starch, layered silicate, and hydroxyapatite nanofillers on the properties of the composites.

2. Structural and Properties of Starch

Starch is a polymeric carbohydrate, which stored in plant tissues as insoluble semicrystalline granules that vary in shape, size and structure, relying on the sources. Native starch is composed primarily of two homopolymers (glucans), which are amylose (a linear polymer) and amylopectin (a highly branched polymer) as shown in (figure 1), as well as minor components such as lipids and proteins. The structure of the starch granule depends on the distribution of the amylose and amylopectin, while the ratios between amylose and amylopectin varies depending on the starch source because of differences in geographic origin and culture conditions. Table 1 shows the amylose and amylopectin content of starches from different plantl sources.

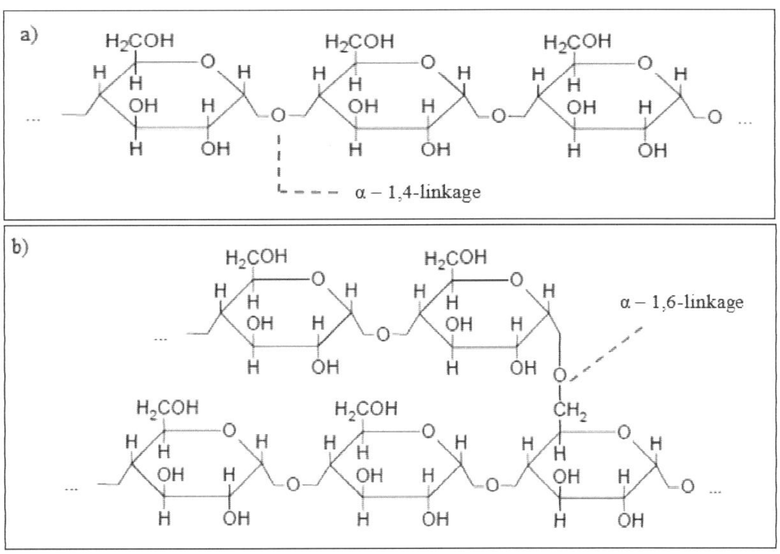

Figure 1. Chemical structures of a) amylose and b) amylopectin [11].

Amylose is a linear structure primarily based on α-D (1, 4) repeating units with a molecular weight of 10^5 - 10^6 and can have a degree of polymerisation (DP) as high as 600. The large number of hydroxyl groups in the structure have proven hydrophilic properties of the polymer, resulting a material with high affinity for moisture. On the other hand, amylopectin is highly multiple branched polymer consists of short α-1, 4 chains linked by α-1, 6 bond containing 10 – 60 glucose units and side chains with 15 – 45 glucose units. The molecular weight of amylopectin ranging from 10^7 – 10^9 [12].

Starch consists three types of crystallinity, which are the A type primarily from cereal starches like maize, wheat and rice; while the B type are from tubers like potato, cassava and sagoo; and lastly the C type contain both the polymorphic A and B type crystallinity patterns, which typically found in bean and other root starches.

Starch properties, such as water absorption, gelatinization and pasting, retrogradation, and susceptibility to enzymatic attack, are depend on the amylose/amylopectin content and granule size [13]. Furthermore, the functional properties of starches are also influence by other factors like chemical modification, system composition, pH, and ionic strength of the media.

Table 1. Amylose and amylopectin content of starch from different sources [14].

Type of starch	Amylose (%)	Amylopectin (%)
Amylomaize	48-77	23-52
Banana	17-24	76-83
Corn	17-25	75-83
High-amylose corn	55-70	30-45
Potato	17-24	76-83
Rice	15-35	65-85
Sorghum	25	75
Cassava	19-22	28-81
Wheat	20-25	75-80
Waxy	<1	>99
Yam	9-15	85-91

3. Starch for Biomedical Applications

The unique physiochemical and functional characteristics of natural starches such as good biocompatibility, biodegradability, non-toxic and degradation as requirement make them useful for a wide range of biomedical applications [15]. Several starch based biodegradable polymers have been widely studied over the last few years, particularly in bone tissue engineering, drug delivery systems and hydrogels.

As an example, studies investigated by Boesel et al. [16] showed that starch based biodegradable bone cements are able to provide a temporary structural support and disappear gradually thereafter, thus allow the ingrowth of new bone for complete healing process. In addition, starch based biodegradable polymers for bone tissue engineering scaffold have been reported. According to Gomes et al. [17], an ideal scaffolds must be design using a biomaterial which possess sufficient degradation rate in which compatible with the formation of new tissue. Thus, the selection of starch for scaffold application is applicable. Other potential attraction of starch based biodegradable polymers are reported in drug delivery systems. Drug delivery system using biodegradable starch is advantageous in deliverability process without need for surgical removal [11].

The successful of the above studies indicates that starch has potential as biomaterials to apply in medical field. Nevertheless, current starch based biodegradable polymers have obviously inferior mechanical properties, thus restraining the capability to be used in the large number of biomedical applications. Consequently, a development of nanocomposites starch material has been introduced to eliminate the limitations of biodegradable starch alone.

4. Starch Based Nanocomposites

As previously mentioned, most of native starches are limited in their direct application due to its inferior mechanical properties, poor long term stability and high water sensitivity [18]. Thus, in order to overcome these limitations, the integration of various nanofillers like particles, fibres and layers into starch matrix to form nanocomposites have received considerable attention. The integration of nanofiller size has greatly influences the final properties of the nanocomposites as compared to conventional micro composites. This is because the mutual effect between filler at nanometer size permits them to form molecular bridges in starch matrix [19], hence resulting an enhancement in nanocomposites properties, such as mechanical properties, thermal stability, moisture resistance, oxygen barrier property and biodegradation rate. Additionally, nanocomposites offer an extensive advantages over conventional composites, as their use of relatively small amounts of fillers (less than 15%) able to achieve desired properties without adding excessive weight and can help improve optical transparency of the composite, since the nanofillers are much smaller than the light wavelength and do not scatter light [20].

Nevertheless, the properties of the nanocomposites material, rely not only on the properties of its constituents, but also on the interfacial bonding between matrix and the filler. Good dispersion of the nanofillers within the starch matrix will cause the strong interface adhesion and thus allowing the formation of a rigid molecular network which can enhance diverse properties of the nanocomposites. Furthermore, there are some aspects that influence the enhancement of starch matrix characteristics, which are plasticizer/additive use during processing, starch source, natural starch modification, existence of other polymer and nano reinforcement aspect ratio [21].

4.1. Starch/Layer Silicate

Development of biopolymer including starch reinforcing nanoparticles such as layer silicate has been one of the newest innovative steps in the nanocomposites. The addition of low percentage layer silicate into starch matrix has proven to be a favourable option in enhancing mechanical and barrier properties of the biopolymer. There are four possible dispersion state of layer silicate in starch matrix which are, phase separated, intercalated, exfoliated and disordered intercalated (partially exfoliated) [22]. Montmorillonite (MMT) is the most widely studied type of layer silicate in starch based nanocomposite because it is environmental friendly, availability at lower cost and processability [23].

Several studies have reported that starch nanocomposite reinforced layer silicate are commonly show an improvement in properties, particularly in tensile strength, Young's modulus, thermal stability, moisture resistance, oxygen barrier property and biodegradation rate. The potential application of these materials was further investigated by Ardakani et al. [1] using solution casting method to prepare nanocomposite films from potato, corn, wheat, MMT, CMMT (modified by citric acid), 30B-MMT and glycerol. The addition of potato starch, CMMT and glycerol was found to improve mechanical properties of the films at optimum conditions. Another research conducted by Muller et al. [24], reported bionanocomposite films prepared by extrusion method using cassava starch, bentonite clay and glycerol. The nanocomposite films presented an outstanding improvements in mechanical properties and reduced the water vapour permeability when compared with native starch films. Gao et al. [25] studied starch/clay nanocomposites using various clay materials with different hydrophilicities by the film blowing method. The properties of the starch nanoclay film were significantly influenced by the hydrophilicity of the clay, in which medium range hydrophilicity favoured the nanocomposites formation. Furthermore, investigation by Avella et al. [26] showed a good increase in modulus and tensile strength for starch-nanocomposites prepared from MMT and potato starch.

The success of the above studies indicates that layer silicate show potential in improving the mechanical and barrier properties of starch-based biomaterials as well as exhibited other desired functions in various applications.

4.2. Starch/hydroxyapatite

Various embodiments of the present nanotechnology include starch/hydroxyapatite (HA) nanocomposites. HA has been used widely in biomedical applications, particularly for contact with bone tissue due to resemblance properties to mineral bone. HA could be found in natural or synthetic form and it possesses favourable properties such as, biocompatible and osteoconductive (cell regeneration process) [24, 25]. Owing to these properties, HA able to promote fast cell regeneration without intermediate connective tissues.

A number of studies [26–29] have been proven that HA promotes faster cell regeneration without intermediate connective tissue, and it is currently used in clinical applications. However, HA is challenging to shape due to its brittleness and lack of flexibility. Similarly, starch materials have some problems associated with mechanical properties. Thus, combination of hydroxyapatite with starch materials may offer the solution for the brittleness issue. Sadjadi et al. [31] claimed that the polar nature of starch encourage a good adhesion between starch and hydroxyapatite. From the research conducted by Sadjadi, the presence of starch has significant effect on the final morphology of HA due to the interaction of OH- groups from starch and Ca^{2+} from HA. In another study, Ayza et.al [12] suggested that the starch acted as the pore generator in membranes and other porous ceramics. This porous criteria is crucial in bone tissue engineering as porous architecture directly related to the scaffolds mechanical strength, cell migration, and this would ensure bone oxygenation and angiogenesis. Additionally, nano sized HA may have other special properties due to small size and huge specific surface area. Recent research suggested that, nano sized HA has reported significant increase in osteoblast adhesion as compared to micro sized ceramic materials [31, 32].

5. Conclusions

Based on the review, starch biomaterial was found to be one of the most promising candidates for various industry applications. The excellent biocompatibility, total degradability without toxic residues, low cost, wide availability and renewability properties of starch would open many applications in biomedical fields, such as bone tissue engineering and drug delivery systems. In addition, starch biomaterials can be process into thermoplastic materials through the use of various industrial techniques, in which similar to those widely used for synthetic thermoplastic. Recently, special attention has been given on nanotechnology by reinforce nano sized fillers into starch matrix to develop novel nanocomposite materials. The application of nanocomposites could significantly enhanced the properties of starch based materials, which would be strongly required for biomedical applications.

References

[1] Majdzadeh-Ardakani K, Navarchian A H and Sadeghi F 2010 *Carbohydr. Poly* **79**(3) 547–554
[2] Chung Y L, Ansari S, Estevez L, Hayrapetyan S, Giannelis E P and Lai H M 2010 *Carbohydr. Polym* **79**(2) 391–396
[3] Tanetrungroj Y and Prachayawarakorn J 2015 *Songklanakarin J. Sci. Techno* **37**(2) 193–199
[4] Ochubiojo E and Rodrigues A 2012 *Sci. Heal. Soc. Asp. Food Ind* 488.
[5] Sanyang M L, Sapuan S M, Jawaid M, Ishak M R and Sahari J 2015 *Polymers (Basel)* **7** 1106–` 1124
[6] Chaunier L 2014 *Laryngoscope* **124** 875–881
[7] Torres F and Arce D 2015 *Biodegrad. Polym. Nanocomposites* 73–94
[8] Marques A P, Reis R L and Hunt J A 2002 *Biomaterials* **23**(6) 1471–1478
[9] Velasquez D, Pavon-djavid G, Chaunier L, Meddahi-pellé A and Lourdin D 2015 *Carbohydr. Poly* **124** 180–187
[10] María Guadalupe Lomelí Ramíreza S I, Satyanarayanab K G and Graciela Bolzon de Munizc T S F S Valcineide Tanobec 2011 *Carbohydr. Polym* **86**(4) 1712–1722
[11] Lu D R, Xiao C M and Xu S J 2009 *eXPRESS Polym. Lett* **3**(6) 366–375
[12] Lorente-Ayza S M M M, Orts M J, Pérez-Herranz V 2015 *J. Eur. Ceram. Soc* **35**(8) 2333–2341.
[13] Bajer D, Kaczmarek H and Bajer K 2013 *Carbohydr. Polym* **98**(1) 477–482
[14] Garc N L, Fam L, Accorso N B D and Goyanes S 2015 *Eco-Friendly Polymer Nanocomposites* 17–77.
[15] Fernando G. Torres M W, Troncoso O P and Vega J 2015 *Polym. from Renew. Resour* **6**(3) 91–

103
[16] Boesel R L R L F, Mano J F 2004 *J. Mater. Sci. Mater. Med* **15** 73–83
[17] Gomes M E, Godinho J S, Tchalamov D, Cunha A M and Reis R L 2002 *Mater. Sci. Eng. C*, **20** 19–26
[18] Souza A C, Benze R, Ferrão E S, Ditchfield C, Coelho A C V and Tadini C C 2012 *LWT - Food Sci. Technol* **46**(1) 110–117
[19] Motwani M S *et al.* 2011 *Biotechnolgy Appl. Biochem* **58**(1) 2–13
[20] Elham K A, Bifari N, Khan S B, Alamry K A, Abdullah M A 2016 *Curr. Pharm. Des* **22** 3007–3019
[21] Xie F, Pollet E, Halley P J and Avérous L 2013 *Prog. Polym. Sci* **38**(10–11) 1590–1628
[22] Kumar P, Sandeep K P, Alavi S, Truong V D and Gorga R E 2010 *J. Food Eng* **100**(3) 480–489
[23] Le Corre D D A, Bras J 2010 *Biomacromolecules* **11** 1139–1153
[24] Carmen F Y, Müller M O, Laurindo J B 2011 *Ind. Crops Prod* **33**(3) 605–610
[25] Gao W, Dong H, Hou H and Zhang H 2012 *Carbohydr. Polym* **88**(1) 321–328
[26] Avella M, De Vlieger J J, Errico M E, Fischer S, Vacca P and Volpe M G 2005 *Food Chem* **93**(3) 467–474
[27] Armentano I, Dottori M, Fortunati E, Mattioli S and Kenny J M 2010 *Polym. Degrad.Stab* **95**(11) 2126–2146
[28] Paul D R and Robeson L M 2008 *Polymer (Guildf)* **49**(15) 3187–3204
[29] Roslana M R, Mohd N, Chengc E M and Mamatd N 2016 *Int. J. Mech. mechatronics Eng* **16**(1) 36–41
[30] Liu T J W H 2010 *Int. J. Nanomedicine* **5** 299–313
[31] Sadjadi M S, Meskinfam M and Jazdarreh H 2010 *Int. J. Nano. Dim* **1**(1) 57– 63
[32] Bianco A, Cacciotti I, Lombardi M, Montanaro L and Gusmano G 2007 *J. Therm. Anal. Calorim* **88**(1) 237–243
[33] Minh D P, Martínez M G, Nzihou A and Sharrock P 2013 *J. Therm. Anal. Calorim* **112**(3) 1145–1155

A Study on the Use of Mortar Utama Cement Type 420 as Concrete Admixture

Asrullah[1*] and A Mulyadi[2]
[1,2]Engineering Faculty, Palembang University, Dharmapala Street 1A Palembang 30136 South Sumatra Indonesia

[*]E-mail : asrull66@yahoo.co.id

Abstract- This research was conducted at laboratory scale in the form of experiment for the purpose to know the value of concrete strength by using Mortar Utama Cement Type 420 as concrete mixing additive. The concrete mixing method being used was SNI 03-2834-2000. The concrete quality being used was K 250, K 300 and K 350. Additions of Mortar Utama Cement were for 5%, 10%, 15% and 20% of cement weight. The strength was tested on the 3rd, 14th, 21st and 28th days of concrete making. According to the test result, it can be concluded that the highest strength of concrete for K 250 with 5% addition was 275.09 Kg/cm^2, for K 300 with 5% addition was 325.32 kg/cm^2, while for K 350 with 5% addition was 368.48 kg/cm^2. Additions of Mortar Utama cement type 420 were able to influence the strength of concrete with simple linear regression model for K 250: Y = -2.005X + 272.7 with R^2 = 0.757, for K 300: Y = -3.061X + 328.3 with R^2 = 0.731, and for K 350:Y = -3.114X + 362.5 with R^2 = 0.785.

1. Introduction

Concrete is part of construction structure with important role in constructing. Concrete is a mix of cement, water, fine aggregate (sand), and coarse aggregate (gravel/split), added with various additional materials [1].

Mortar Utama Cement is manufactured by PT Cipta Mortar Utama, which is the first and the greatest national company to manufacture ready-to-use instant cement (premixed mortar) using modern technology known by Mortar Utama brand. The main material of Mortar Utama cement is selected sand, cement PC 1 of the best quality of Holcim, premium filler and Europe additive.

Cement PC1 (Portland Cement 1) is cement type one of the best quality with high strength, filler contained inside it to improve its density and to reduce its mixing porosity, while the additive contained inside it is water additive useful to improve its consistency, adhesive capacity, water retention, durability and strength [2]. The use of additive for concrete mix gets more common in Indonesia. In general, additive used in concrete can be divided into two types: chemical natured additive and mineral natured additive.

Admixture is added during mixing and or during casting (placing), thus it is used more in improving casting performance. On the other hand, mineral natured additive is added during mixing, more cement-like work, used in improving its strength performance [3]. Mortar Utama cement has various types appropriate to its usage. This research uses an additive of Mortar Utama cement Type 420. Using additive of Mortar Utama cement Type 420, it is expected to improve the strength of concrete.

2. Materials and Methods

2.1. Materials

The research was carried out at the Laboratory of Main Station of the Office of Public Work of South Sumatera, Palembang, Indonesia. The materials used in this research were : Portland Cement, Holcim Cement brand, Coarse aggregate in the form of split stone from Lahat, South Sumatra, Fine aggregate in the form of sand from Sekayu, tap water from PDAM (Palembang City Water Utility company) Tirta Musi, and Admixture in the form of Mortar Utama Cement type 420 manufactured by PT Cipta Mortar Utama.

2.2. Methods

Methods of research included: 1) Examination of the characteristics of fine and coarse aggregates, namely testing of specific gravity and water absorption of coarse aggregate, testing of specific gravity and water absorption of fine aggregate, testing of analysis on fine and coarse aggregates, testing of water content of the aggregates; 2) Carrying out of normal concrete mix with additions of Mortar Utama Cement type 420 for 5%, 10%, 15% and 20% of cement weight used for each characteristic of concrete. 3) Concrete slump test. The test object was made using a mold

15x15x15 cm with qualities of concrete being made were K 250, K 300 and K350 respectively. Concrete strength test was conducted on the 3rd, 14th, 21st and 28th days of concrete making.

3. Results and discussions

3.1. Test Result of Fine and Coarse Aggregate

The results are presented in Table 1, it can be stated that: the filter analysis is in zone I and qualifies as a concrete mixture [4].

Table 1. Value of Characteristics of Fine and Coarse Aggregates.

Type of Test	Value		
	Sand	Split ¾ inch	Split 1 ½ inch
1. Filter Analysis			
36.10 (1½")	-	-	100
25.40 (1")	-	100	53.63
19.10 (3/4")	-	87.98	16.50
12.70 (1/2")	-	49.32	2.43
9.25 (3/8")	100	19.59	0.00
# 4	98.04	0.00	-
# 8	89.72	-	-
# 16	72.19	-	-
# 30	22.15	-	-
# 50	1.04	-	-
# 100	0.22	-	-
# 200	0.12	-	-
2. Fineness modulus:	3.17	6.92	7.83
3. Content Weight : Loose	1.462	1.337	1.330
Solid	1.551	1.415	1.419
4. Basic specific gravity of dry saturated surface	2.54	2.57	2.58
5. Absorption (%)	1.19	2.26	2.10
6. Organic dirt	No : 3	-	-
7. Through filter No.200 Mud Content	0.60 %	-	-
8. Wear (%)	-	-	23. 24 %

Color standard organic dirt no 3 meets the required maximum no. 4, for the standard organic content of the color no. 5 the need for fine aggregate washing before use [5]. The mud content of 0.60%, categorized as eligible from the maximum permissible sludge content of 5% [6]. Rough aggregate wear of 23.24%, meets normal standards permitted maximum 25% [6]. The fine aggregate fineness modulus of 3.17 qualifies between 2.5-38 and the rough aggregate fineness modulus of 6.92 and 7.83 qualifies between 6-8 [6]. From the available data, it follows that overall fine aggregate and coarse aggregate characteristics are eligible for use as concrete mixtures.

3.2. Concrete Mix Composition

The results presented in Table 2 are the composition of concrete mixture without addition of main mortar type 420 cement or concrete mixture with addition of main mortar type 420. Value of water factor of cement (w/c) Which can be calculated for concrete without the addition of the main mortar type 420 cement is 0.50 for the characteristics of K 250, K 300 and K 350 concrete. Along with the addition of 420.4%, 15% and 20% main mortar cement %, There was a decrease of cement water factor (w/c) respectively: 0.48, 0.46, 044 and 0.42 for all types of concrete characteristics. The value of the cement water factor (w/c) required in the range 0.50-0.45 [7]. The decrease in the value of the cement water factor (w/c) is due to an increase in the amount of cement in the concrete mixture while the amount of water remains.

Table 2. Material Needed for Each Characteristic of Concrete.

No	Type of Material	Need for Material 1 M^3		
		K 250	K 300	K350
1	Cement	402 kg	451 kg	500 kg
2	Sand	598 kg	599 kg	596 kg
3	Split size ¾ inch (19.00mm)	493 kg	565 kg	480 kg
4	Split size 1 1/2 inch (37.50mm)	669 kg	548 kg	579 kg
5	Water	205 liters	225 liters	250 liters
6	Mortar Utama Cement Type 420 5%	20.1 kg	22.55 kg	25 kg
7	Mortar Utama Cement Type 420 10%	40.2 kg	45.1 kg	50 kg
8	Mortar Utama Cement Type 420 15%	60.3 kg	67.65 kg	75 kg
9	Mortar Utama Cement Type 420 20%	80.4 kg	90.20 kg	100 kg

3.3. Slump Value

Slump value test was conducted to each concrete mix. The result presented in table 3 showed that with the increase of main mortar type 420 inside the concrete mixture, then the slump value is decreased. For K 250 of 82 mm it decreases successively to 75 mm, 70 mm, 65 mm and 58 mm. This condition applies also to K 300 from 79 mm to 72 mm, 66 mm, 62 mm and 59 mm. To K 350 from 72 mm to 68 mm, 64 mm, 61 mm and 58 mm. For the addition of main mortar type 420 by 20% does not meet the specified requirement that the slump value should be in the range of 100-60 mm [7]. According to Asrullah [8] the decrease in slump value is due to an increase in the number of main mortar cement type 420 in concrete mix while the amount of water remains, so that the concrete mixture becomes thick and tend not homogeneous.

Table 3. Slump Value of Some Variations.

Type of Concrete	Slump Value (mm)		
	K 250	K 300	K 350
Non-Added Concrete (0%) of Mortar Utama cement Type 420	82	79	72
Concrete added with 5% of Mortar Utama cement Type 420	75	72	68
Concrete added with 10% of Mortar Utama cement Type 420	70	66	64
Concrete added with 15% of Mortar Utama cement type 420	65	62	61
Concrete added with 20% of Mortar Utama cement type 420	58	59	58

3.4. Test Result of Strength

The results presented in Table 4 show that the standard deviation (S) value for the 28-day-old concrete is fliktuated and unstable. This is due to in each of the compressive strength of the concrete on the auxiliary characteristics there is a difference resulting in an inconsistent standard deviation (S). The results in Table 5 show that concrete without the addition of the main mortar type 420 (0%) meets the predetermined standard, ie for K 250 the compressive strength value is 261,69 kg/cm2, K 300 the compressive strength value is 311,35 kg/cm^2, K 350 value of compressive strength 351,28 kg / cm^2. The value of compressive strength of concrete with the addition of main mortar type 420 is increased by the addition of main mortar type 420 by 5% of cement weight .ie K 250 value of compressive strength 275,09 kg /

cm^2, K 300 value of compressive strength 325,32 kg / cm^2, K 350 value of compressive strength 368,48 kg / cm^2. While the addition of main mortar type 420 by 10%, 15% and 20% decreased.

Table 4. Standard Deviation(S) of Concrete on the 28th Day.

Type of Concrete	Standard Deviation (S) (kg/cm^2)		
	K 250	K 300	K 350
Non-Added Concrete (0%) Mortar Utama cement Type 420,	0.69	0.56	2.49
Concrete added with 5% of Mortar Utama cement Type 420	3.24	2.20	6.92
Concrete added with 10% of Mortar Utama cement Type 420	3.63	2.85	3.34
Concrete added with 15% of Mortar Utama cement type 420	6.42	4.25	5.89
Concrete added with 20% of Mortar Utama cement type 420	0.45	1.28	6.79

Table 5. Concrete Strength on the 28th day.

Type of Concrete	Strength of Concrete Characteristic (kg/cm^2)		
	K 250	K 300	K 350
Non-Added Concrete (0%) Mortar Utama cement Type 420,	261.69	311.35	351.28
Concrete added with 5% of Mortar Utama cement Type 420	275.09	325.32	368.48
Concrete added with 10% of Mortar Utama cement Type 420	257.73	309.65	323.47
Concrete added with 15% of Mortar Utama cement type 420	239.29	290.22	312.64
Concrete added with 20% of Mortar Utama cement type 420	229.95	252.37	301.35

According to Asrullah [9] this is because there is a big change to the water factor of cement (w/c) accompanied by decrease of slump value which implication there is decrease of concrete compressive value. Even if there is an increase of concrete compressive strength value on the addition of 5% of main mortar type 420 this is due to the change of cement water factor (w/c) and slump value is still in fairness, given the composition of the main mortar type 420 containing additives that have a role in Increase the value of concrete compressive strength.

3.5. Simple Linear Regression Relationship Model

A simple linear regression analysis describes the relationship of a free variable (addition of Primary Cement type 420) and an independent variable (Concrete strength). In this research, the free variable (X) is addition of Mortar Main cement type 420, while the independent variable (Y) is concrete strength. The regression model in figure 1 is a simple linear regression for K 250 concrete which is the relationship between the value of concrete compressive strength with the addition of main mortar type 420 with the equation Y = -2.005X + 272,7. From this equation it is seen that the minus value [10] in the regression coefficient shows the dependent variable number (Y) based on the independent variable (X). This means that if the addition of Main Mortar Type 420 cement increases, it will cause a decline in the value of compressive strength of the concrete. Furthermore, according to Asrullah [11] the coefficient of determination R^2 is worth 0.757, this figure is close to the value of 1, it means to have a strong correlation between the addition of Main Mortar type 420 cement and the compressive strength of concrete. In addition, the coefficient shows that the effect of the addition of Mortar Utama Type 420 cement to the value of mortar compressive strength of 75.70%.

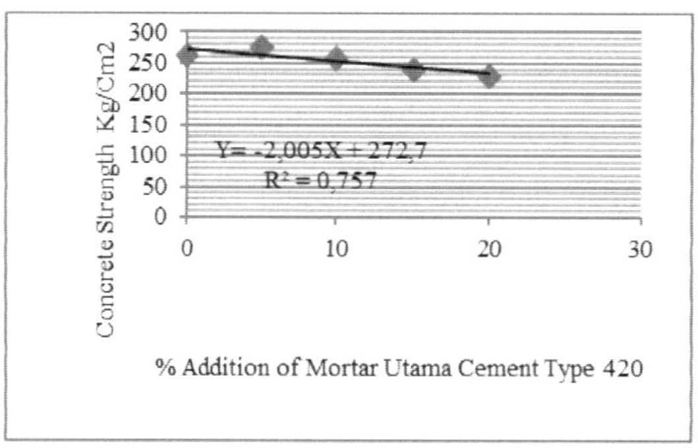

Figure 1. Relationship between the Strength of Concrete K 250 and Addition of Mortar Utama Cement Type 420.

The regression model in figure 2 is a simple linear regression for K 300 concrete which is the relationship between the value of concrete compressive strength with the addition of the main mortar type 420 with the equation Y = -3.061X + 328.3. From this equation it is seen that the minus value [10] in the regression coefficient shows the dependent variable number (Y) based on the independent variable (X). This means that if the addition of Main Mortar Type 420 cement increases, it will cause a decline in the value of compressive strength of the concrete. Furthermore, according to Asrullah [11] the coefficient of determination R^2 is worth 0.731, this number is close to the value of 1, it means to have a strong correlation between the addition of Main Mortar type 420 cement and concrete compressive strength. In addition, the coefficient shows that the effect of the addition of Mortar Utama Type 420 mortar to the value of mortar compressive strength of 73.10%.

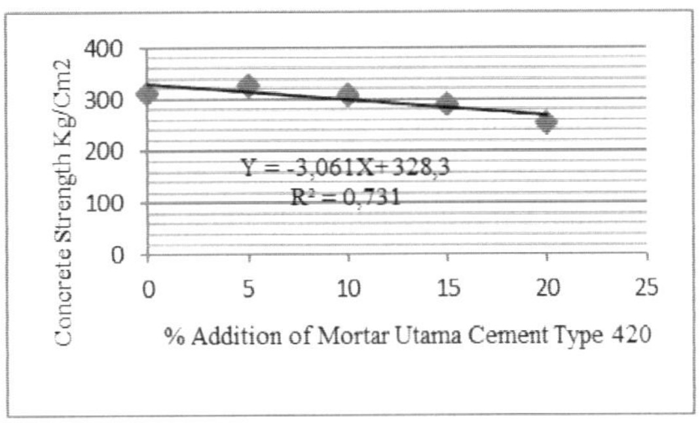

Figure 2. Relationship between the Strength of Concrete K 300 and Addition of Mortar Utama Cement Type 420.

The regression model in figure 3 is a simple linear regression of concrete K 350 which is the relationship between the value of concrete compressive strength with the addition of main mortar type 420 with the equation Y = -3.114X + 362.5. From this equation it is seen that the minus value [10] in the regression coefficient shows the dependent variable number (Y) based on the independent variable (X). This means that if the addition of Main Mortar Type 420 cement increases, it will cause a decline in the value of compressive strength of the concrete. Furthermore according to Asrullah [11] the coefficient of determination R^2 is worth 0.7851, this number is close to the value of 1, it means to have a strong correlation between the addition of Main Mortar type 420 cement and concrete compressive strength. In

IOP Conf. Series: Materials Science and Engineering **209** (2017) 012088 doi:10.1088/1757-899X/209/1/012088

addition, the coefficient shows that the effect of the addition of Main Mortar Type 420 cement to the value of mortar compressive strength of 78.51%

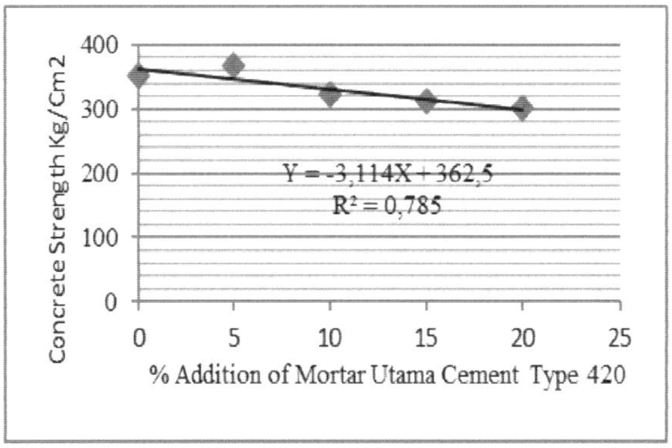

Figure 3. Relationship between the Strength of Concrete K 350 and Addition of Mortar Utama Cement Type 420.

The relationship between addition of Mortar Main cement type 420 (free variable) and slump value (independent variable) is presented in figures 4,5,6. The regression model in figure 4 is a simple linear regression of K 250 which is the relationship between the slump value of the concrete with the addition of the main mortar type 420 with the equation Y = -1,16X + 81.6. From this equation it is seen that the minus value [10] in the regression coefficient shows the dependent variable number (Y) based on the independent variable (X). This means that if the addition of Main Mortar Type 420 cement increases, it will cause a decrease in slump value. Furthermore, according to Asrullah [11] the coefficient of determination R^2 is worth 0.995, this figure is close to the value of 1, it means to have a strong correlation between the addition of Main Mortar Type 420 cement and concrete slump value. In addition, the coefficient shows that the effect of the addition of Mortar Utama Type 420 mortar to the value of mortar compressive strength of 99.50%.

The regression model in figure 5 is a simple linear regression of K 300 which is the relationship between the slump value of concrete with the addition of the main mortar type 420 with the equation Y = -X + 77.6. From this equation it is seen that the minus value [10] in the regression coefficient shows the dependent variable number (Y) based on the independent variable (X).

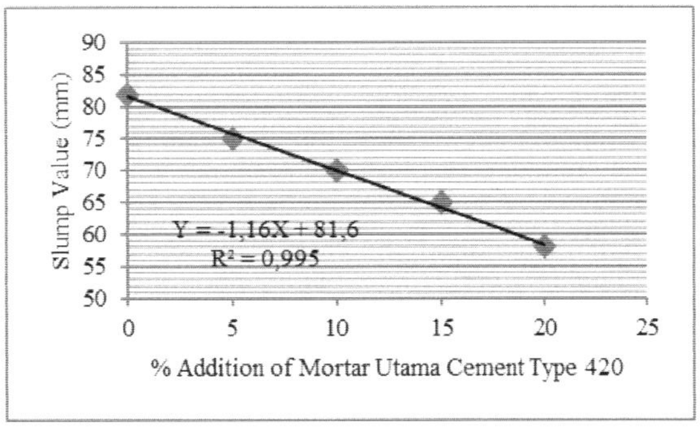

Figure 4. Relationship of Slump Value of Concrete K 250 and Addition of Mortar Utama Cement Type 420.

This means that if the addition of Main Mortar Type 420 cement increases, it will cause a decrease in slump value. Furthermore according to Asrullah [11] the coefficient of determination R^2 is worth 0.972, this number is close to the value of 1, it means to have a strong correlation between the addition of Main Mortar type 420 cement and concrete slump value. In addition, the coefficient shows that the effect of the addition of Mortar Main Type 420 cement to the value of mortar compressive strength of 97.20%.

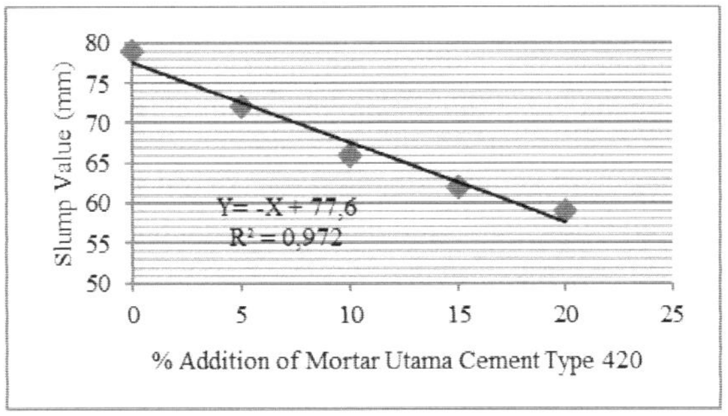

Figure 5. Relationship between Slump Value of Concrete K 300 and Addition of Mortar Utama Cement Type 420.

The regression model in figure 6 is a simple linear regression K 350 which is the relationship between the slump value of concrete with the addition of the main mortar type 420 with the equation Y = -0.7X + 71.6.

From this equation it is seen that the minus value [10] in the regression coefficient shows the dependent variable number (Y) based on the independent variable (X). This means that if the addition of Main Mortar Type 420 cement increases, it will cause a decrease in slump value.

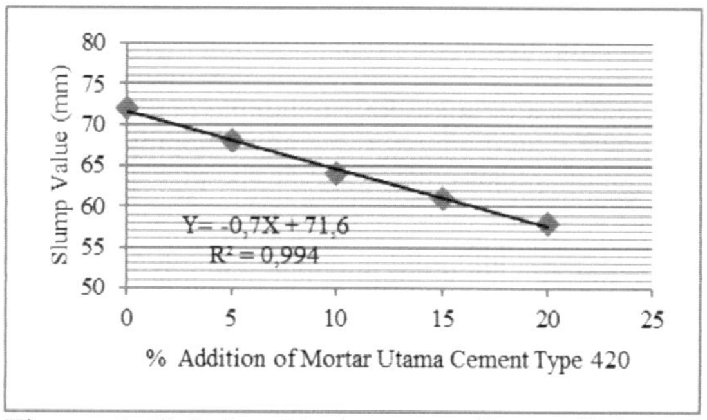

Figure 6. Relationship between Slump Value of Concrete K 350 and Addition of Mortar Utama Cement Type 420.

Furthermore according to Asrullah [11] the coefficient of determination R^2 is worth 0.9943, this number is close to the value of 1, it means to have strong correlation between addition of Main Mortar Type 420 cement and concrete slump value. In addition, the coefficient shows that the effect of the addition of Mortar Utama Type 420 mortar to the value of mortar compressive strength of 99.43%.

4. Conclusions

From the result of this research, we conclude that:

1. The highest strength of concrete K 250 was achieved by adding 5% of Mortar Utama Cement Type 420 with strength value 275.09 kg/cm^2, Y = -2.005X + 272.7 with R^2 = 0.757.
2. The highest strength of concrete K 300 was achieved by adding 5% of Mortar Utama Cement Type 420 with strength value 325.32 kg/cm^2, Y = -3.061X + 328.3 with R^2 = 0.731.
3. The highest strength of concrete K 350 was achieved by adding 5% of Mortar Utama Cement Type 420 with strength value 368.48 kg/cm^2, Y = -3.114X + 362.5 with R^2 = 0,785

References

[1] Asroni A 2010 Beam and Reinforced Beam Plates *Graha Ilmu Yogyakarta, First Edition.*

[2] Utama Pt C M 2012 *MM 2100 Industrial Town*
[3] ***http://andykasipil.blogspot.co.id/2012/01/materials for concrete. html.
[4] Mulyono T 2003 *Concrete Technology*
[5] SNI 03-2816-1992 *Method of Testing Organic Inside Sand for Sand Mixture of Mortar and Concrete,* National Standardization Agency.
[6] Ministry of Public 1971 *Rules of Reinforced Concrete Indonesia.*
[7] SNI 03-2834-2000 *Methods of Mix Making of Normal Concrete* National Standardization Agency.
[8] Asrullah 2015 Concrete Strength Analysis K 250 With Addition of Mortar Utama Cement Type 420 With Rough Aggregate Size (Split) Uniform. *J. of Civil Engineering Program Study* **5(10)** 9-20.
[9] Asrullah 2016 Analysis of Addition of Mortar Utama Cement Type 420 Against Concrete Strength K 250 Using Split Different Size *J. of Civil Engineering Program Study* **5(11)** 9-19.
[10] Ahmad I A 2009 Analysis of Effect Temperature Against Concrete Strength *J. of Civil Engineering Civil Engineering Study Program* **16 (2)** 63-69.
[11] Asrullah 2016 The effect of Adding Cement Mortar Utama (MU) Type 420 Against Concrete Strength K 350 *Journal of Civil Engineering Program Study* **6(12)** 16-29.

Acknowledgments
We would like to thank Mr. Zulkifli S Mukti, SH, MH, Rector of Palembang University, Palembang, South Sumatra, Indonesia, whose help and assistance made possible the completion of research and the writing of this paper.

Laboratory Study of Methane Flux from Acid Sulphate Soil in South Kalimantan

W Annisa[1*], D Cahyana[1], H Syahbuddin[2] and A Rachman[3]
[1] Indonesian Swampland Agricultural Research Institute, Banjarbaru, Indonesia
[2] Indonesian Agroclimatic and Hidrological Research Institute, Bogor, Indonesia
[3] Indonesian Soil Research Institute, Bogor, Indonesia

E-mail: annisa_balittra@yahoo.com

Abstract. Addition of organic matter in waterlogged conditions will enhance methanogenesis process that produces greenhouse gases. Fresh organic material is considered reactive because it contains carbons that is subject to decompose, therefore, when it exposed to acid sulphate soil, both in natural condition (aeration required) and intensive (aeration not required) will lower the value of redox potential. This experiment aimed to determine the flux of methane (CH_4) from various locally available organic materials applied to acid sulphate soil. The experiment was arranged in factorial design with two factors. The first factor was the source of organic matter, i.e. fresh rice straw, fresh purun, fresh cattle manure, composted rice straw, composted purun and composted cattle manure, and control. The second factor was the management of organic matter i.e. placed on the soil surface with no tillage and mixed with soil during tillage. The results showed that application of fresh organic matter into inundated acid sulphate soil increased CH_4 fluxes up to 23.78 μg CH_4 g^{-1} d^{-1} which was higher than from composted organic matter (4.327 μg $CH_4.g^{-1}.d^{-1}$). Methane flux due to organic matter management was significantly negatively (p=0.001) correlated with soil redox potential (Eh) with R2 of - 0.76. Organic matter placed on the soil surface with no tillage produced methane flux ranged from 0.33 to 20.78 g CH_4 g^{-1} d^{-1}, which was lower than methane flux produced from organic matter mixed with soil during tillage (0.38 to 27.27 g CH_4 g^{-1} d^{-1}). Composting organic matter before application and mixing them with the soil through tillage are highly recommended to reduce greenhouse gas emissions from cultivated acid sulphate soils.

1. Introduction

Organic carbon in the soil is a source of greenhouse gas emissions. Higher amount of organic carbon found under flooded paddy soil than under rain fed because of slower organic breakdown takes place under anaerobic condition, however, continuous inundation of paddy field may stimulate the formation of CH_4 [1]. Changchun Song et al reported that higher CH_4 emissions (12.80 mg m^{-2} d^{-1}) under continuously inundated condition in Sanjiang Plain, China than under seasonally inundated condition (8.56 mg m^{-2} day^{-1}) [2]. Setyanto & Abubakar (2006) found that maintaining standing water at 5 cm height in paddy field released methane emissions around 254 kg CH_4 ha^{-1} $season^{-1}$, which is much higher than 136 kg CH_4 ha^{-1} $season^{-1}$ emitted from the intermittent irrigation treatment [3]. The availability of organic substrates and low redox potential under continuous flooded paddy fields are a good environment for methanogenic bacteria to grow, therefore, its existence will be dominating than methanotrophic bacteria resulting in higher methane flux[4].

Another function of organic materials in acid sulphate soil is to retain reductive conditions which cause in the release of Fe^{2+} ions to increase the concentration of Fe^{2+} in the soil solution resulted in the suppression of pyrite oxidation. In addition, high quality organic materials may also chelate toxic substances in the soil. The lignin content of organic materials applied into the acid sulphate soil affects the capability of organic matter as an electron donor in the reduction process. In general, the higher the lignin content of the organic material the lower the capability of organic matter as an electron donor. Banjarese farmers have a traditional manner in undertaking land preparation for rice cultivation by managing fresh organic materials in flooded condition through a series of practices known as Tajak-Puntal-Balik-Hampar (trowel-puntal-turn over-spread). This technique for managing organic materials is commonly practiced in acid sulphate soil area in South Kalimantan.

Methane is a greenhouse gas that develops in reductive conditions of acid sulphate soil due to inundation and addition of organic matter. Methanogenic bacteria as a producer of CH_4 exists in anaerobic conditions with a redox potential of <-200 mV [5]. Annisa *et al* found that adding organic matter into acid sulphate soils planted with rice lower the redox potential up to <-250 mV. Two processes, methane production and consumption, are responsible for microbial methane emissions from aquatic environments such as rice field [6,7]. According to the thermodynamic theory, methane is formed after the majority of ferric iron reduced to ferrous iron or in other words oxidants (oxygen> nitrate> sulphate and iron (III) oxides). This experiment aimed to determine the flux of methane (CH_4) from various sources of locally available organic matter under two different managements in acid sulphate soils.

2. Materials and Methods

2.1. Site description

The site was hydrological affected by flood water tides, therefore it is designated as overflow type B. The soil is classified as Typic Sulphaquent based on Soil Survey Staff [8]. The monthly temperature ranged between 33.6-36.8°C (figure 1). The difference between the highest and the lowest temperatures is less than 5°C.

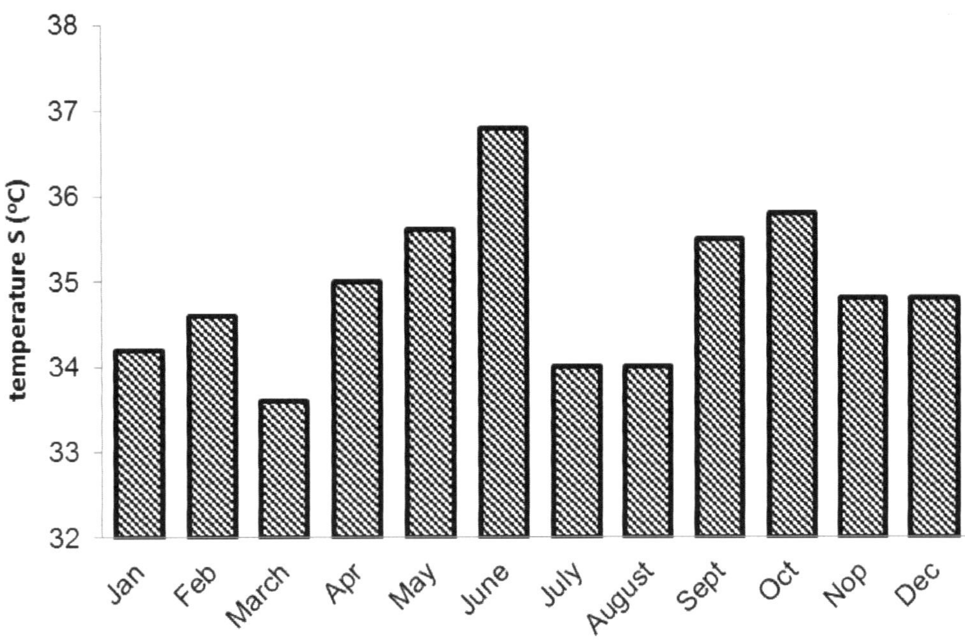

Figure 1. The monthly temperatures of the Mandastana region, near the site, in 2012.

2.2. Soil and organic matter characteristics

The soil used in this study has no accumulation of Fe (III) oxides in soil layers, the oxalate extractable Fe content were low and decreased with depth (Table 1). Iron content were 0.14-1.66%, which is smaller than those in a normal acid sulphate soil of 2–3%. The wet climate helps keep the soil wet, thus hampering the formation of Fe oxides [9]. The high organic nature of the soil also plays an important role in inhibiting the formation of Fe-oxides in this soil, keeping redox potential low most of the time.[10]

Table 1. Soil chemical properties of Acid Sulphate Soil used in laboratory experiment.

Parameter	Depth (cm)			
	0-20	20-40	40-60	>60
pH H_20	5.12	5.13	4.41	3.49
C-org (%)	9.50	8.33	8.19	7.64
Total N (%)	0.28	0.20	0.23	0.20
Fe^{2+} (mg.kg^{-1})	673.55	784.30	1314.1	3521.1
SO_4^{2-}(mg.kg^{-1})	1026.2	1349.4	1722.1	3880.3
FeS_2 (%)	0.14	0.25	0.60	1.66

The organic materials used in this experiment were locally available and commonly used by farmers in acid sulphate soil include: rice straw, purun and cattle manure. Each of these organic materials have different characteristics as shown in Table 2. Fresh rice straw had the highest C/N ration of 92.71 followed by fresh purun (66.01), fresh cattle manure (36.47), composted purun (31.99), composted rice straw (28.26), and the least was composted cattle manure (20.81). Composting of those organic materials increased the nutrient P from 12.09% to 12.59%; K from 0.43 to 1.39% and Fe from 0.23 to 3.41%.

Table 2. Chemical properties of organic material.

	Organic Material Type					
	S_f	P_f	C_f	RS_c	PR_c	CM_c
C organic (%)	50.62	47.13	33.13	41.15	41.20	32.93
N (%)	0.546	0.714	0.910	1.456	1.288	1.582
Ratio C/N	92.71	66.01	36.47	28.26	31.99	20.81
P (%)	0.093	0.197	0.114	0.214	0.207	0.590
K(%)	0.899	0.689	0.432	1.390	1.131	0.588
Fe (%)	0.228	1.385	0.278	0.707	3.409	0.549

S_f = fresh rice straw; P_f= fresh *purun*; C_f= fresh cattle manure; RS_c= composted rice straw; PR_c = composted *purun*; CM_c = composted cattle manure.

2.3. Experimental design

The study was conducted from April to June 2012 at the Soil and Plant Laboratory of Indonesian Swampland Agricultural Research Institute (ISARI) in Banjarbaru, South Kalimantan. Soil samples were taken at 0-20 cm soil depth (rhizosphere zone) using a PVC pot with a 10 cm diameter and 35cm height from Tanjung Harapan village, Alalak Sub District, Barito Kuala Regency, South Kalimantan (030 10'S; 1140 36'E). The soil samples were then immediately transported to the laboratory for experimental preparation.

The experiment was arranged in a factorial randomized block design with 2 factors and 3 replications. The first factor was seven sources of organic matter e.g (1) without organic matter (control), (2) fresh rice straw (Sf), (3) fresh purun (Eleocharis dulcis; Pf), (4) fresh cattle manure (Cf), (5) composted rice straw (RSc), (6) composted purun (PRc), and (7) composted cattle manure (CMc).

The second factor was two organic matter management e.g (1) organic matter placed on the soil surface with no tillage (NT) and (2) organic matter mixed with soil during tillage (TM).

The soil samples were then incubated for 2, 4, 6, and 8 weeks in a room temperature before gas emission were sampled. The gas samples were taken periodically using a syringe and analyzed.

Methane flux, pH and Eh measured 4 times, i.e: 2, 4, 6 and 8 week after incubation. The top part of the tube was covered with PVC cap to prevent gas exchange during gas sampling periods.

The bottom part of the tube had hole (1 cm in diameter) for water drainage during decomposition periods.

Methane and carbon dioxide concentration in the syringe were immediately determined using Varian 4900 Gas Chromatograph (GC) with a flame ionization detector and helium as a gas carrier. While soil redox-potential (Eh) was measured using field electrode. Emission of methane and carbon dioxide were calculated using the equation:

$$E = Kx \frac{Vhs}{B} \frac{Wm}{Vm} \frac{273.2}{273.2 + T} \quad (1)$$

Where:

E	= CH_4 flux (kg ha^{-1})
Kx	= CH_4 concentration at GC
Vhs	= headspace volume (ml)
B	= soil weight (g)
Wm	= molecular weight of CH_4 and CO_2
m	= volume of 1 mole of gas at standard temperature and air pressure
T	= average temperature inside the chamber during gas sampling (°C)

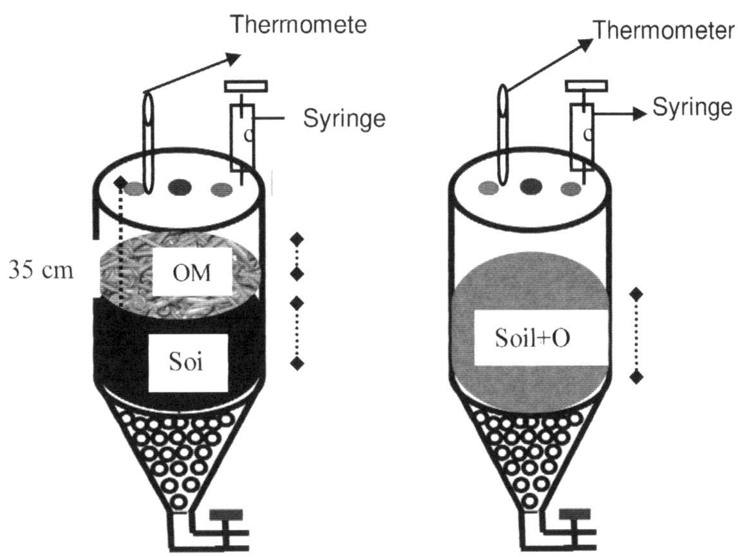

(a) Organic matter placed on the soil surface (b) organic matter mixed with soil

Figure 2. Schematic diagram of laboratory experiment.

2.4. Data Analysis

Analysis of variance (ANOVA) was used to measure the significance among treatments and then mean comparison was calculated with Duncan's Multiple Range Test (DMRT). Statistical analyzes were performed using SAS software for Windows ver. 9.0.

3. Results and discussions

3.1. Methane Flux in Acid Sulphate Soil

Banjarese farmers commonly apply fresh organic matter as green manure under anaerobic condition. This practice may have released a significant amount of methane as the fresh organic matters undergo decomposition process under saturated condition. Our study indicated that applying fresh organic matter into the inundated acid sulphate soil had significantly higher methane emission than composted organic matter. The fresh organic matter produced 23.78 μg CH_4 $g^{-1}d^{-1}$ while the composted organic matter produced only 4.327 μg CH_4 g^{-1} d^{-1} (figure 3).

The decomposing process of organic matter in flooded soils depends not only on the quality of organic substrates but also on the amount of electron acceptor and electron acceptor consumption rate associated with DOC (dissolved organic carbon). [11] The results from this study also confirmed that the type and maturity of organic matter applied into acid sulphate soil affected the amount of CH_4 emitted, which is in accordance with Reddy and DeLaune who found that around 70% of carbon lost from wetland is in the form of methane.

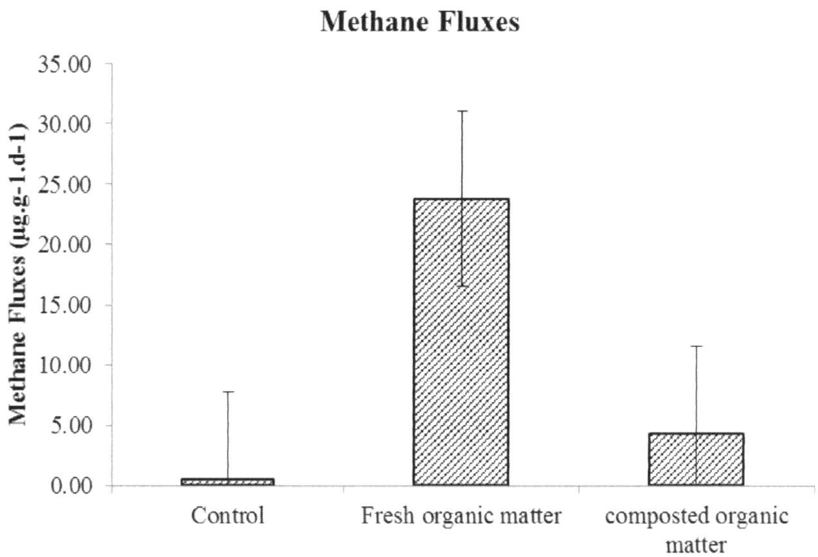

Figure 3. Methane flux from fresh and composted organic matters.

Another advantage of applying composted organic matter over fresh organic matter is the higher readily available nutrients for plant uptake. The chemical analysis of initial organic matter showed that composted organic matter had higher nutrient content than fresh organic matter. The six sources of organic matter used in this study had various C/N ratios ranked as follows fresh rice straw > fresh purun > fresh cattle manure > composted purun > composted rice straw > composted cattle manure. The advance of decomposition process of organic material applied were determined using absorption data from FTIR at a wave number of 2020-2920 cm^{-1}, low absorption indicates a less decomposition process is taken place. The lowest absorption was found from composted cattle manure followed by composted rice straw, fresh cattle manure, composted purun, fresh purun, and the highest was from fresh rice straw. In general, the absorption was found higher from fresh organic matter than the decomposed organic matter indicating a more active decomposition process in the fresh organic matter

IOP Conf. Series: Materials Science and Engineering **209** (2017) 012089 doi:10.1088/1757-899X/209/1/012089

than in the composted one. During the composition process, it was also observed that the content of aliphatic methyl was decreasing. Therefore, the alipathic methyl content of organic matter can be used as an indicator of the maturity of compost in addition to the C/N ratio.

Incubation of organic matter in the acid sulphate soil produced different pattern of methane flux both under NT and TM treatments. High initial methane flux during the first 4 weeks of incubation for NT then decreased with time of incubation. In contrast, the TM showed an initial low methane flux, reached the peak at 6 weeks of incubation then decreased (figure 4). Methane flux from NT ranged from 0.33 to 20.78 g CH_4 g^{-1} d^{-1} while 0.38 to 27.27 g CH_4 g^{-1} d^{-1} from TM (figure 5). However, in general the total amount of methane flux under TM was higher as compared to NT. The reason for this is because by mixing organic matters with soil during tillage will directly expose the organic matters to microorganisms to degrade organic matter resulting in faster decomposition rate. While organic matters in the NT will undergo more reductive condition that slower decomposition rate. Cheng-Fang et al reported that CH_4 emission from paddy field with organic materials placed on the soil surface emits lower CH_4 emission than from organic materials in wetland environment [12]. Figures 4 and 5 also imply that methane flux from fresh rice straw increased significantly after 2 weeks of incubation and significantly different with the others. Kongchum reported that the application of straw into soil affects the production of methan, therefore, it is critical to decompose rice straw before applying into acid sulphate soil as a source of soil organic matter to minimize methane emission [13].

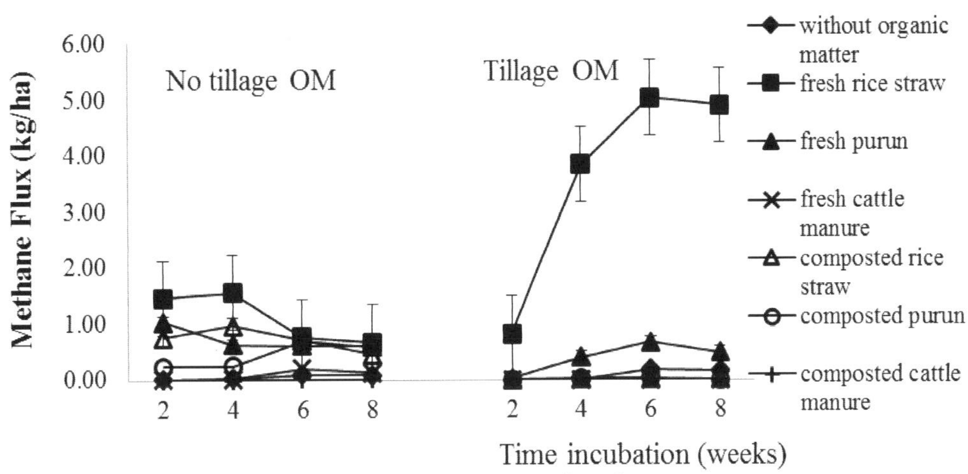

No Tillage OM = No tillage with organic matter placed on the soil surface
Tillage OM= Organic matter is mixed with soil with tillage

Figure 4. Methane flux from organic matter application at acid sulphate soils.

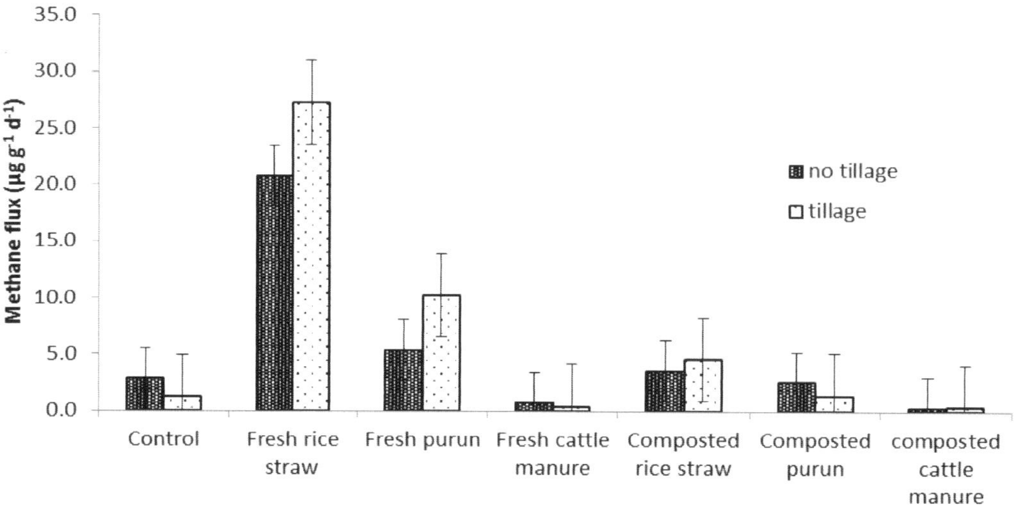

Figure 5. Methane flux from different sources of organic matter and tillage in acid sulphate soils.

3.2. Redox Potensial

Soil microbial respiration associated with oxidation-reduction processes affects the value of redox potential (Eh) and it can be used as an indicator of biogeochemical processes in the soil. Figure 6 shows that applying of fresh organic matters into the soil lower the redox potential compared with both without organic material (control) and composted organic matter applications. The lowest redox potential (-188.51 mV) was found under application of fresh rice straw followed by fresh purun (-161.47 mV), composted rice straw (-134.09 mV), control (-121.89 mV), composted purun (-119.70 mV), fresh cattle manure (-99.45 mV), and the least was from composted cattle manure (-75.61 mV) after 8 weeks of incubation. The kind and quality of organic materials applied which provide a substrate for microbial metabolic processes determine the level of CH_4 emissions. Fresh organic matter with C/N ratio of 36-90 indicated faster reduction in soil redox potential as compared with composted of organic matter with C/N ratio of <31[14].

Redox potential is highly correlated with methane flux, especially in waterlogged condition. Figure 7 shows the exponential correlation between redox potential and methane emission with coefficient of determination (R2) of 0.76 (p=0.001). Methane flux increased with the decrease in redox potensial and increased significantly at redox potential <-150 mV. Previous studies indicated that optimum CH_4 production occurs in the range of Eh -150 mV to -300 mV where methanogenic bacteria as a producer of CH_4 works optimally with Eh $\leq$ -150 mV [15,16].

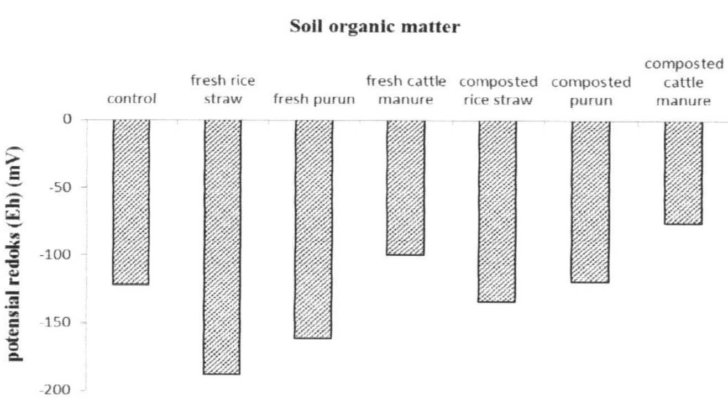

Figure 6. Soil Potensial Redoks (Eh) from various kind and form of

organic matters applied to acid sulphate soils.

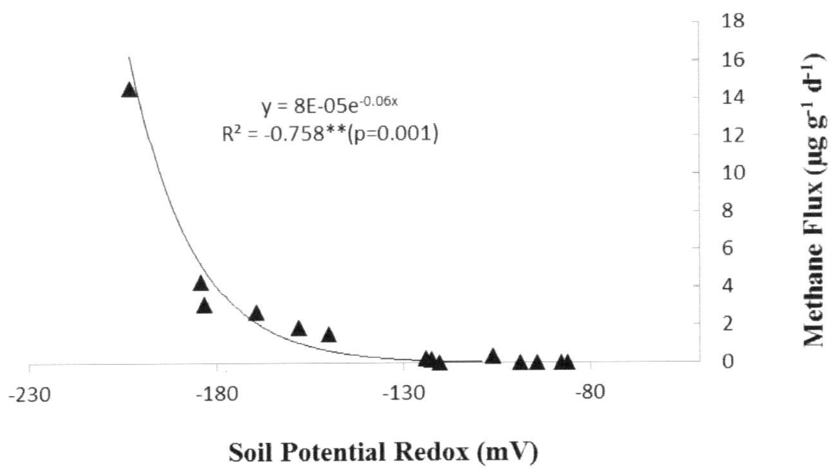

Figure 7. Correlation Soil redox potential (Eh) with methane flux.

3.3. Correlation between Fe^{2+} and SO_4^{2-}

Methane is the end product of the organic matter decomposition in waterlogged conditions. Higher CH_4 formed by the addition of fresh organic matter associated with a quality of organic material. Previous studies indicated that factors that influence the decomposition of organic materials in wetlands is the quality of the organic material indicated by N content, the ratios of C to N, lignin to N, and C to P [17].

Application of fresh or composted organic matter into acid sulphate soils may increase soil pH since organic matter as an electron donor in the reduction process consumes $H^{(+)}$ thus increasing soil pH. In this study, we found that application of composted cattle manure increased soil pH up to 5.65 with the following decrease in Fe-soluble (0.47 $cmol^{(+)}\ kg^{-1}$) and SO_4^{2-} (0.61 $cmol^{(+)}\ kg^{-1}$). Those conditions increased the reduction process of iron hydroxide (oxidative) to iron reductive.

The organic materials can suppress the reductive dissolution of iron with chelating process. Under reductive condition, thermodynamics do not explain sufficiently the reduction of Fe^{3+} because there is another non-enzymatic mechanism associated with the reduction of Fe^{3+}. Therefore, in addition to organic matter, sulphate is also determine the intensity of Fe^{3+} as it is used as a primary electron acceptor by microbes in the process of Fe^{3+} reduction. Figure 8 shows a high correlation between Fe^{2+} and SO_4^{2-} with a coefficient of determination (R2) of 0.84 which indicates that each equivalent anion (SO_4^{2-}) is balanced by an equivalent of the cation (Fe^{2+}). In Thermodynamics, sulphate reduction occurs in conditions where soil redox potential (Eh) lower than the reduction of Fe^{3+} (Eh <-100 mV), however, previous studies indicated that the Fe^{3+} can be reduced by sulfur with the help of bacteria. Reddy and DeLaune, (2008) studied that some microorganisms such as *Thiobacillus thioxidans*, *T. ferroxidans*, *Sulfolobus spp* utilize Fe^{3+} as an electron acceptor with sulfur as an electron donor.

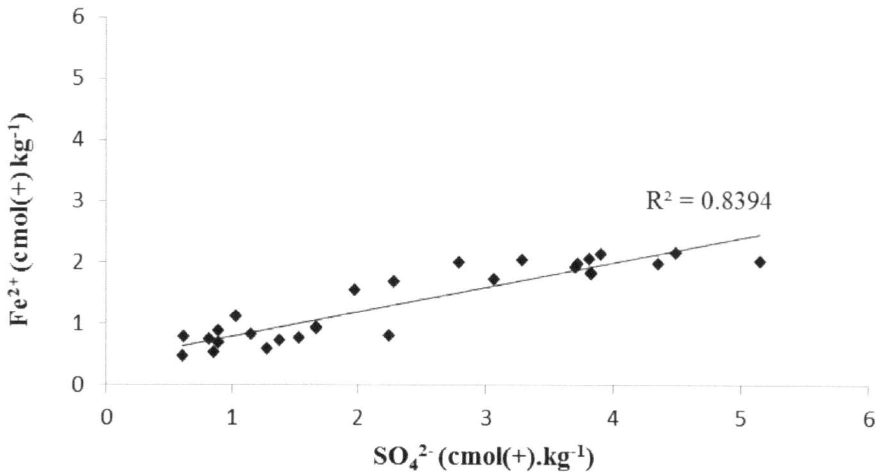

Figure 8. Correlation between Fe^{2+} and SO_4^{2-}.

4. Conclusions

Methane flux in both no tillage with organic matters placed of soil surface and soil organic matter mixed with soil during tillage was significantly negatively (p=0.001) correlated with soil redox potential (Eh) with a correlation of determination (R2) of 0.76. Application of fresh organic matter in the inundated condition produced CH_4 fluxes as high as 23.78 µg CH_4 g^{-1} d^{-1} which was significantly higher than CH_4 fluxes of 4.327 µg CH_4 g^{-1} d^{-1} from composted organic matter. The lowest soil redox potential of -188.51 mV was found from fresh rice straw which was 2.5 times lower as compared with redox potential from composted cattle manure (-75.61 mV). The low redox potential under fresh rice straw treatment is responsible for the high methane fluxes. Applying organic materials on the soil surface without tillage produced slightly lower methane fluxes of 0.33 to 20.78 g CH_4 g^{-1} d^{-1} as organic materials mixed soil during tillage (0.38 to 27.27 g CH_4 g^{-1} d^{-1}). It is critical to decompose organic materials before applying to cultivated acid sulphate soils to minimize methane emission.

References

[1] Esmizade Z, Landi A and Gilani A 2010 Evaluating the amount of carbonic greenhouse gasses (GHGes) emission from rice paddies *World Congress of Soil Science, Soil Solutions for a Changing World* 68-70

[2] Changchun S, Yan B, Wang Y, Lou Y and Zhao Z 2003 Fluxes of carbon dioxide and methane ` from swamp and impact factors in Sanjiang Plain, China *Chinese Science Bulletin* **48** (24) 2749-2753

[3] Watanabe I, Hashimoto T and Shimoyama A 1997 Methane-oxidizing activities and methanotrophic populations associated with wetland rice plants *Biol. Fertil. Soils* **24** 261-265

[4] Setyanto P and Abubakar R 2006 Evaluation of methane emission and potential mitigation from flooded rice field. *Jurnal Litbang Pertanian* **25** (4) 139-148

[5] Minamikawa K and Sakai N 2006 The effect of water management based on soil redox potential on methane emission from two kinds of paddy soils in Japan. *Agric. Ecosyst. Environ.* **107** 397-407

[6] Annisa W, Purwanto B H and Shiddieq D 2011 Effect of rice straw and purun at various level of decomposition to iron concentrations in acid sulphate soils. *Soil and Climate Journal. Special Edition Swamp*

[7] Rudd J W M and Taylor C D 1980 Methane cycling in aquatic environments. *Adv. Aquat. Microbiol* **2** 77–150

[8] Soil Survey Staff 2010 Keys to Soil Taxonomy. Eleventh Edition Handbook *Natural Resources Conservation Service-USDA* **336** 346

[9] Konsten C J M, Van Breemen N, Suping S, Aribawa I B and Groenenberg J E 1994 *Effects* of flooding on pH of rice producing, acid sulphate soils in Indonesia. *Soil Sci. Soc. Am. J.* **58** 871– 883

[10] Muhrizal S, Shamshuddin J, Fauziah M.A. and Husni H 2006 Changes in iron-poor acid sulphate soil upon submergence *Geoderma* **131** 110-122

[11] Reddy K R and Delaune R D 2008 The Biogeochemistry of Wetland *Science and Application.* CRC Press

[12] Cheng-Fang Z D, Zhi-Kui K, Sheng Z Z, Ping W J, Li C M 2012 Effects of Tillage and Nitrogen Fertilizers on CH4 and CO2 Emissions and Soil Organic Carbon in Paddy Fields of Central China. *Journal pone. 0034642* doi: 10.1371

[13] Kongchum 2005 Effect of plant residue and water management practices on soil redox chemistry, methane emission, and rice productivity. *A Dissertation.* The Department of Agronomy & Environmental Management. Louisiana State University and Agricultural and Mechanical College

[14] Roger P A 2001 Production, oxidation, emission and consumption of methane by soils: A review. *Eur. J. Soil Biol.* **37** 25-50

[15] Hou A X, Chen G X, Wang Z P, Van Cleemput O and Patrick W H 2000 Methane and Nitrous Oxide Emissions from a Rice Field in Relation to Soil Redox and Microbiological Processes. *Soil Sci. Soc. Am. J.* **64** 2180-2186

[16] Setyanto P, Rosenani A B, Boer R, Fauziah C I and Khanif M J 2004 The effect of rice cultivars on methane emission from irrigated rice field *Indon. J. Agric. Sci* **5**(1) 20-31

[17] Richardson J L and Vepraskas M J 2004 *Wetland Soils. Genesis, Hydrology, Landscapes and Classification.* Lewis Publisher

IOP Conf. Series: Materials Science and Engineering **209** (2017) 012090 doi:10.1088/1757-899X/209/1/012090

PIV Study of Aeration Efficient of Stepped Spillway System

M A Abas[1 *], R Jamil[2], M R Rozainy, M A Zainol[3], M N Adlan[4] and C W Keong[5]
[1] School of Mechanical Engineering, Universiti Sains Malaysia, Engineering Campus, 14300 Nibong Tebal Malaysia.
[2,3,4] School of Civil Engineering, Universiti Sains Malaysia, Engineering Campus, 14300 Nibong Tebal Malaysia.
[5] School of Mechanical Engineering, Universiti Sains Malaysia, Engineering Campus, 14300 Nibong Tebal Malaysia.

E-mail: ceremy@usm.my

Abstract. This paper investigates the three-dimensional (3D) simulation of Cascade aerator system using Lattice Boltzmann simulation and laboratory experiment was carried out to investigate the flow, aeration and cavitation in the spillway. Different configurations of stepped spillway are designed in this project in order to investigate the relationship between the configurations of stepped spillway and cavitation in the flow. The aeration in the stepped spillway will also be investigated. The experimental result will be compared with the simulated result at the end of this project. The figure of flow pattern at the 3rd step in simulation and experiment for Set 1 and Set 2 are look similar between LBM simulation and the experiment findings. This will provide a better understanding of the cavitation, aeration and flow in different configurations of the stepped spillway. In addition the occurrence of negative pressure region in the stepped spillway, increases the possibility of cavitation to occur. The cavitation will damage the structure of the stepped spillway. Furthermore, it also founds that increasing in barrier thickness of the stepped spillway will improve the aeration efficiency and reduce the cavitation in stepped spillway.

1. Introduction

In the late 1980s, Lattice Gas Automata method is invented. It able to solve the Navier- Stokes equations of fluid motion. The evolution of lattice gas model proceeds in two steps that take place during each time step, which are collision and streaming steps. For the streaming step, the particles move to new sites relative to their previous positions and velocities. In contrast, the collision step means that the particles collide and scatter according to collision rules specified in the formulation. In the lattice gas, all particles momentum conservation of the particles is reduced to conservation of the vector sum of the velocities. However, the Lattice Gas Automata is difficult to solve complex collision in 3D and it consists of statistical noise. Therefore, the Lattice Boltzmann Method LBM is introduced. The LBM is able to solve the statistical noise.

Lattice Boltzmann Method, LBM is a computational simulation method used for simulation of fluid flow. The fundamental idea of the LBM is to create a simplified kinetic model that incorporates the essential physics of microscopic processes. Thus, the macroscopic averaged properties obey the desired macroscopic equations [1]. Besides, LBM also is a simulation technique that utlizes the distribution function to replace tagging of each particle as applied in molecular dynamic simulations. The distribution function acts as a representative for collection of particles. This is the so-called mesoscale formulation. In addition, LBM is able to recover the Navier-Stokes equation with a proper choice of collision operator [2].

The main idea of Boltzmann is to bridge the gap between micro-scale and macro-scale. LBM has advantages of both macroscopic and microscopic approaches. LBM is easy to treat multi-phase and multi-component flows because it does not need to trace the interfaces between different phases. In additional, LBM can be applied to parallel processes computing. Thus, LBM can save the computer resources significantly. Figure 1 shows the technique of simulation.

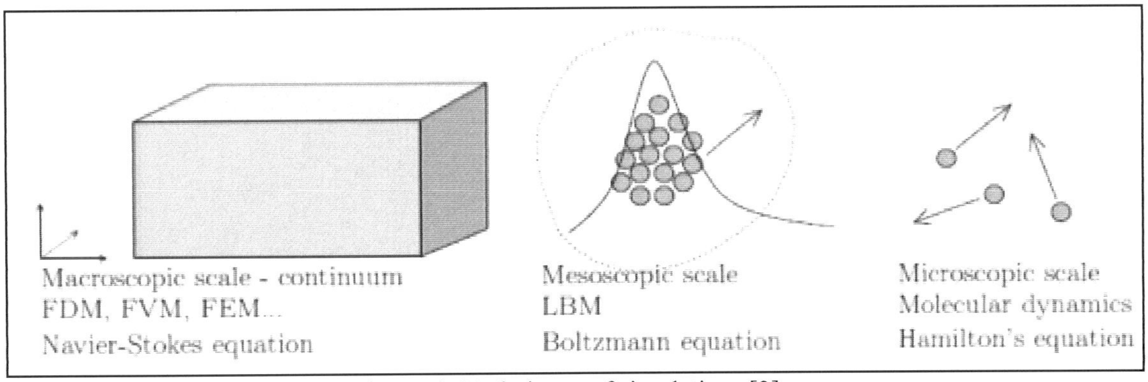

Figure 1. Techniques of simulations [2].

In this project, Lattice Boltzmann Method Simulation and laboratory experiment are carried out to investigate the flow, aeration and cavitation in the spillway. Different configurations of stepped spillway are designed in this project in order to investigate the relationship between the configurations of stepped spillway and cavitation in the flow. The aeration in the stepped spillway will also be investigated. The experimental result will be compared with the simulated result at the end of this project. This will provide a better understanding of the cavitation, aeration and flow in different configurations of the stepped spillway.

In the laboratory experiment, different configurations of stepped spillway are built. The 3mm thickness of Perspex is selected in this project. The Perspex is cut to certain dimension based on the configuration of the stepped spillway. Then, the Perspex plates are connected in order to build the stepped spillway. The model of the stepped spillway is shown in the figure 2. After the stepped spillway is built, the experiment will be carried out. Firstly, the water is poured into the water tank. Then, the water will flow down the stepped spillway. The velocity and pressure on the barrier of the step edges located in the non-aerated and aerated flow region are determined in order to investigate the aeration. Besides, the flow of the stepped spillway is observed.

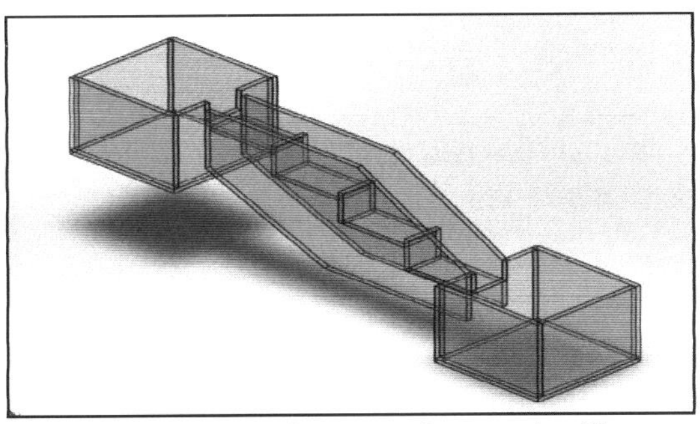

Figure 2. Schematic diagram of stepped spillway.

The Lattice Boltzmann Method (LBM) based solver is selected to simulate the flow in the stepped spillway. The LBM solver has been found effective for simulation of fluid with high accuracy on the computed solution [1,2]. Afterwards, the Paraview software is used to post-process the resultsand then analyzed.

2. Multiphase Equation

Multiphase system in water will consist of both liquid and vapour of water. Multiphase is selected because the cavitation and aeration are investigated in this project. The cavitation and aeration involve both the liquid and vapour of the water. Thus, multiphase is selected in LBM. Besides, since only water is used a medium in this project so single component is selected.

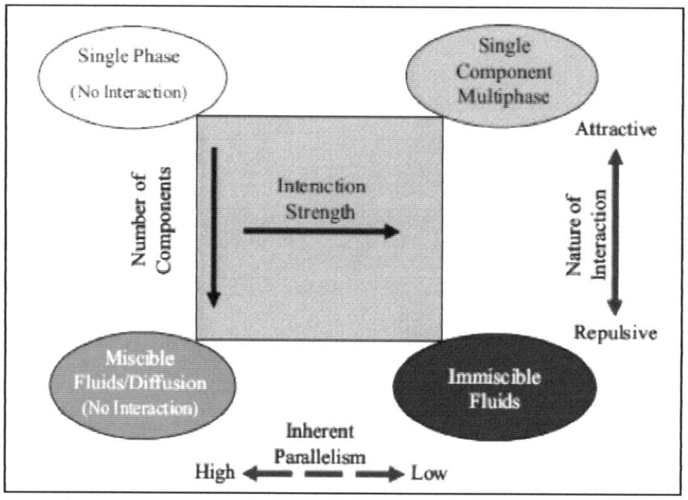

Figure 3. Conceptual framework for LBM models [4].

2.1. Cavitation

Cavitation can be defined as a phenomenon of catastrophic transition from liquid to vapor. If the cavitation is happened at the limit of pure liquid's tensile strength, the cavitation will be called as homogeneous cavitation. In contrast, if the preexisting bubbles in the structure of the liquid cause the cavitation, the cavitation will be called as heterogeneous cavitation. The total energy to create a vapour bubble can be written as:

$$\Delta E = 2\,\pi r \sigma + \pi r 2 \Delta P \tag{1}$$

Where σ is surface tension, ΔP is pressure and r = bubble radius. When the pressure is negative, it will show the possibility of cavitation. [6] Thus, when the negative value of the pressure is increased, the possibility of the cavitation is increased too.

2.1.1. Pressure in LBM. The inter-particle forces from equation of state (EOS) [4,7] are used to described the pressure, P:

$$\boldsymbol{P} = \boldsymbol{\rho R T} + \frac{\boldsymbol{GRT}}{2}[\psi(\rho)]^2 \tag{2}$$

where T is the temperature, R is the universal gas constant, G is the interaction strength and ψ is the interaction potential. ψ can be expressed as:

IOP Conf. Series: Materials Science and Engineering **209** (2017) 012090 doi:10.1088/1757-899X/209/1/012090

$$\psi(\rho) = \psi_0 \, exp\left(-\frac{\rho_0}{\rho}\right) \tag{3}$$

2.1.2. D3Q19

2.1.2.1 Boundary Conditions. There are three types boundaries in the LBM method:
1) Periodic boundary condition
2) No-slip boundary condition
3) Free slip boundary condition

Periodic boundary condition is the phenomena where the f leave the domain and it i will re-enter on the opposite side. It happened during the steaming phase. Periodic boundary condition can be applied to the bulk phenomena. Besides, no slip boundary condition is the phenomena that no slip and bounce back is occurred. It means that the component is in rest at the boundary. The free slip boundary condition is applied to compute the wall that does not contain any friction. Thus, the tangential velocity will be remained the same.

The velocity vector of the D3Q19 consists x, y and z components. The velocity vector is specified from which density/pressure. Density/pressure is computed on the basis of conditions inside the domain.

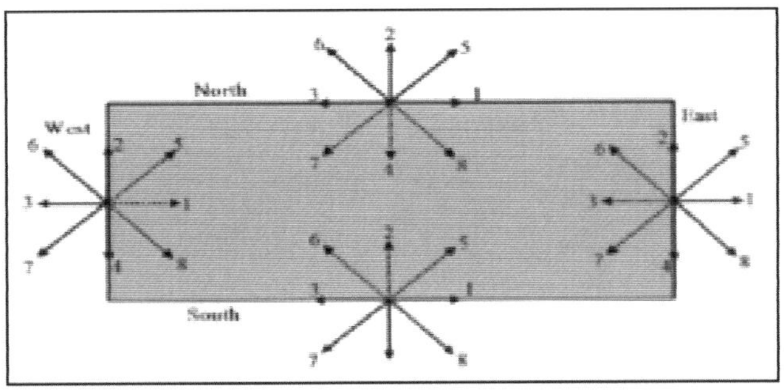

Figure 4. No-slip boundary condition.

2.2. Aeration Equation

In cascade aerator, the water falls down the steps. It is utilizing the potential energy of water in order to create interfaces for efficient gas transfer. Besides, the splashing of the water will create the turbulence in the water. The transfer of gases will be occurred when the entrained air is dispersed in the form of bubbles throughout the receiving body of water. The amount of the air entrained will be depended on velocity of the water.

The equation of the power dissipation rate in the stepped spillway will be expressed as:

$$P = \rho g Q \Delta H \tag{4}$$

Where ρ = density of water, g =gravity, Q = flow rate of water and ΔH = velocity head.

Thus, the equation of the power dissipation rate per unit width can be expressed as

$$\omega = \frac{P}{m} \tag{5}$$

Where m is the width of the step in the stepped spillway. The positive linear relationship between the aeration efficiency and energy dissipation rate is expressed as the equation below [5].

$$E20 = 0.0015\ \omega + 0.01 \tag{6}$$

the velocity head can be expressed as;

$$\Delta H = \frac{V^2}{2g} \tag{7}$$

in which V is the velocity of the water.

3. Results and discussions

3.1. Validation

The configurations of the stepped spillway are shown in table 1. There are two sets of experiments and three set of simulation in this project. The relationship between the configuration of stepped spillway and the aeration in the water of the stepped spillway is investigated in this project using both LBM simulation and PIV experiment. Additionally, the cavitation phenomenon in the stepped spillway is also investigated in this project. The velocity results are determined from both the simulation and experiment in order to study the aeration phenomena in the stepped spillways the pressure results are also determined using the simulation to investigate the cavitation phenomena.

Table 1. Dimension of the prototype of stepped spillway for Set 1, Set 2 and Set 3.

Set	Dimension of Prototype of Stepped Spillway (mm)		
	Length of Step (include Barrier)	Height of Step (include barrier)	Thickness of Barrier
1	130	70	3
2	130	70	9
3	260	70	3

The velocity result for the simulation is extracted using Paraview software. The velocity results that obtained from the Paraview are shown in three directions (x-direction, y- direction and z-direction). The total magnitude of the velocity can be obtained using formula of;

$$\sqrt{(x - direction)^2 + (y - direction)^2 + (z - direction)^2} \tag{8}$$

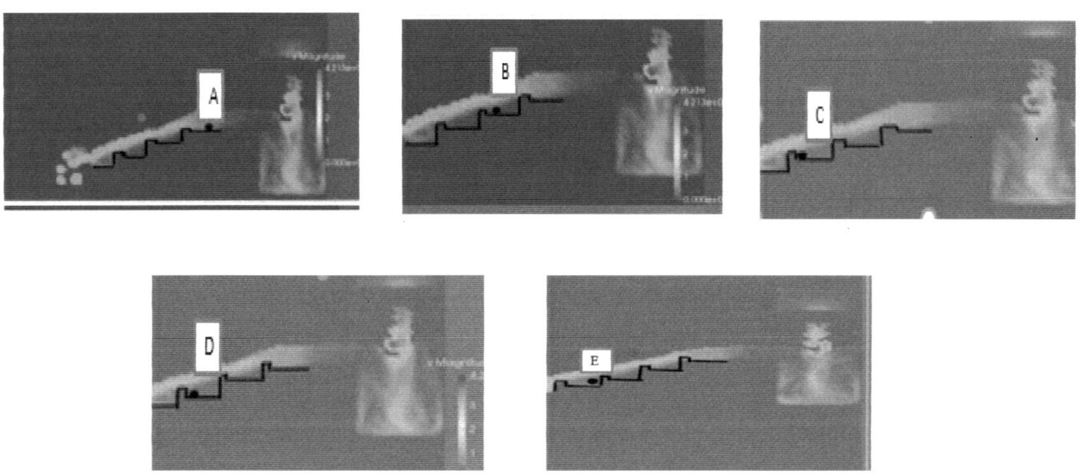

Figure 5. Point A, B, C, D and E at (1st, 2nd, 3rd and 4th) of the stepped spillway (Black point)

IOP Conf. Series: Materials Science and Engineering **209** (2017) 012090 doi:10.1088/1757-899X/209/1/012090

The velocity results are obtained from four different points of the stepped spillway to validate the simulation and experimental findings. One point was selected at the 1st step (highest step), one point at the 2nd step and two points at the 3rd step. The locations of these points are shown in figure 5. The graph of the velocity results for Set 1 and Set 2 are plotted as shown in figure 6 and figure 7.

Based on the figure 6 and figure 7, it was shown that the flow velocity of the water increases along the main flow direction of the water. The velocity is increased due to the slop change along the main flow direction of the water. The highest velocity is located at the point D of the stepped spillway. From figure 6 and figure 7, the result of simulation and experiment for Set 1 and Set 2 show good conformity in the velocity results obtained. Both the simulation and experiment velocity result show the same trend as demonstrated figure 6 and figure 7 respectively. The percentage error between the simulation and experiment result is less than 11 percent. Given the high conformity of results between the simulation and experiment velocity, this proof that LBM simulation's capability to determine the velocity of the water on the stepped spillway with accuracy. The use of pure simulation will eliminate the need for an expensive experimental procedure of stepped spillway design in the future.

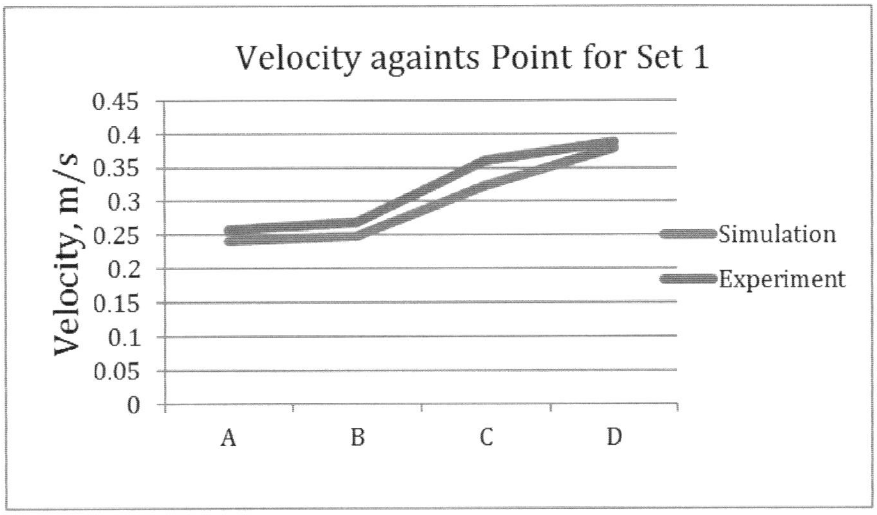

Figure 6. Graph of Velocity against Point for Set 1.

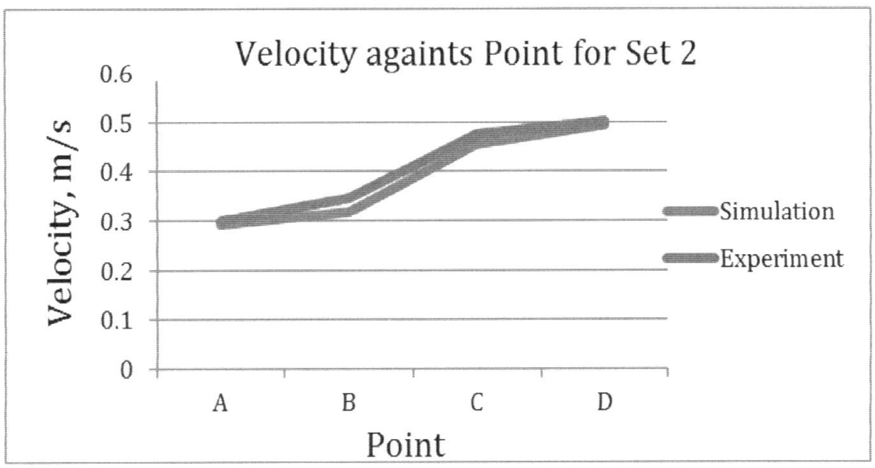

Figure 7. Graph of Velocity against Point for Set 2.

Table 2. Percentage error of velocity between simulation and experiment.

Point	Percentage Error (%)	
	Set 1	Set 2
A	6.20	2.66
B	8.15	8.096
C	10.53	4.00
D	2.83	1.20

The figure of flow pattern at the 3rd step in simulation and experiment for Set 1 and Set 2 are shown in figure 8 and figure 9. It founds that the flow pattern of the LBM simulation look similar to the experiment findings. In Figure 8, it was shown that some of the water would flow back to the reserved direction when the water is hit the 3rd step's barrier.

In figure 9, it shows that the recirculation phenomena happened at the location where near to the 3rd step's barrier. Some air is trapped into the recirculation causing small aeration be developed. Most of the aeration can be seen occuring at the final step of the stepped spillway. This observation demonstrates the high conformity of both simulation and experimental data in the flow pattern for Set 1and Set 2. Again, this will proof that the LBM simulation can be used to accurately predict the result for the stepped spillway design.

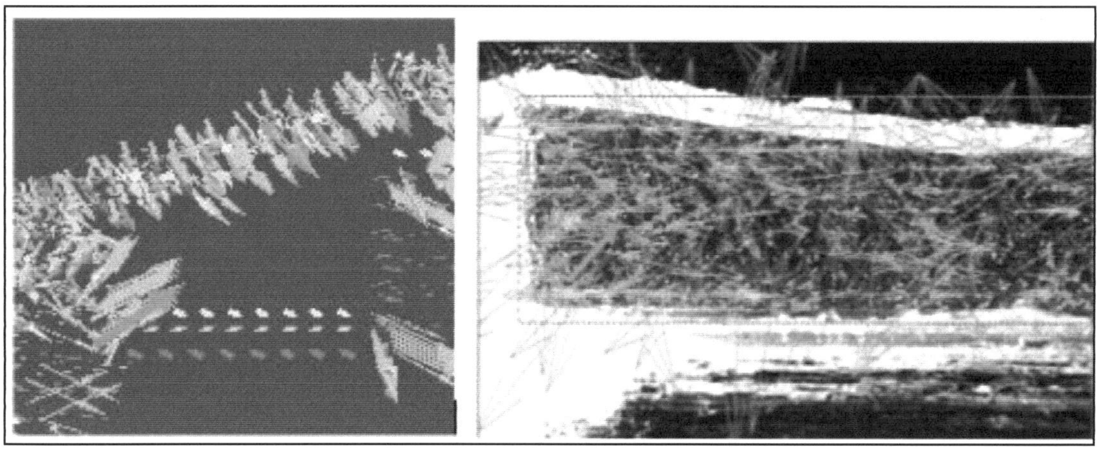

Figure 8. Flow pattern in 3rd step of the stepped spillway for Set 1.

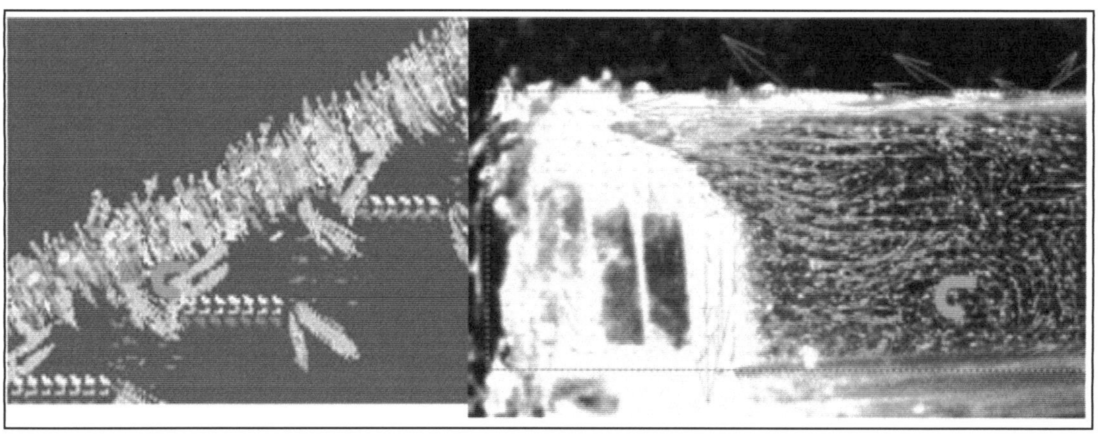

Figure 9. Flow pattern in 3rd step of the stepped spillway for Set 2.

3.2. Aeration Efficiency

Aeration process is used to remove the constituents in water in this project. Aeration will raise the dissolved oxygen content of the water. The turbulence of aeration process will cause the scrubbing process, which will help to remove the dissolved gases from the water. Then, the dissolved gas will escape into the surrounding air. The recirculating proses in the water will help to trap the air in the water.

The aeration efficiency is expressed as shown in Equation (9).

$$E20 = 0.0015\ \omega + 0.01 \tag{9}$$

This equation is created by the positive linear relationship between the aeration efficiency and the energy dissipation rate. [5] From the equation (9), it shows that the aeration efficiency is dependent on the energy dissipation rate. When the energy dissipation rate is increased, the aeration efficiency is increased. It means that the velocity of the water flow will affect the aeration efficiency. The faster the water velocity, the higher the aeration efficiency. Thus, the better aeration result will be obtained if the water velocity is higher.

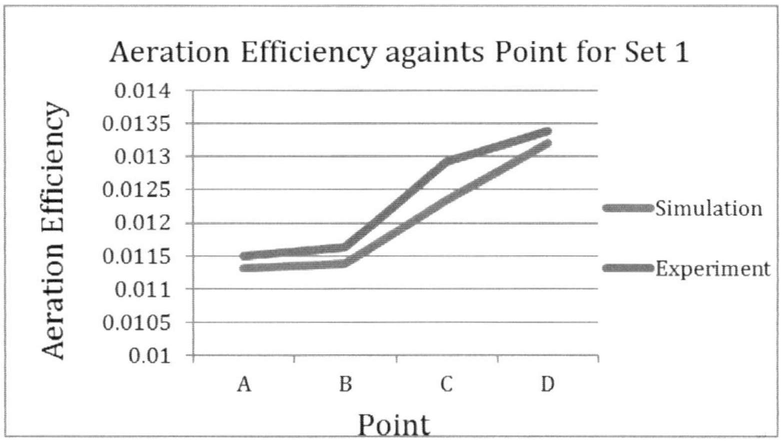

Figure 10. Bar chart of aeration efficiency against point for Set 1.

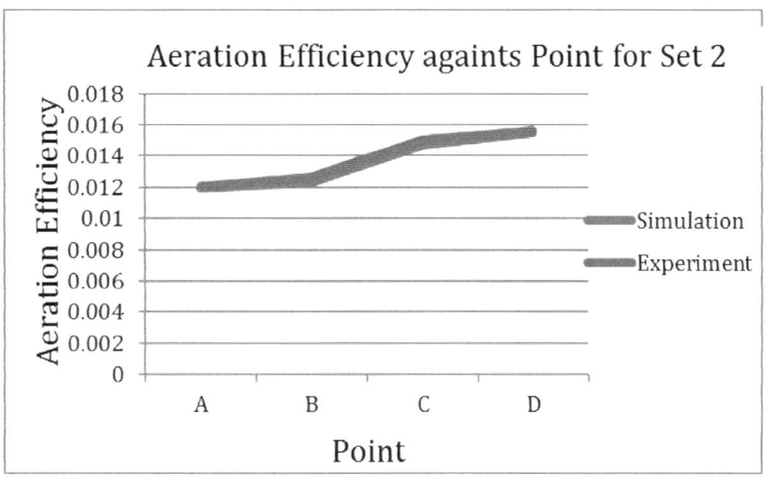

Figure 11. Bar chart of aeration efficiency against point for Set 2.

The aeration efficiency along the points is tabulated and plotted in Table 5 and 6. Based on the figure 10 and figure 11, the aeration efficiency increases along the points. The location point D has the highest aeration efficiency for both Set 1 and Set 2. Thus, when the velocity increases, the energy dissipation rate and aeration efficiency also increase. Set 2 has a better aeration efficiency than Set 1.

4. Conclusions

The velocity of the water will gradually increase along the main flow of the water over the stepped spillway. Additionally, it is also shown that the pressure decreased along the stepped spillway due to the increase in velocity of the water. As the pressure of the water is negative there is a great possibility for the cavitation to occur at the vicinity of the region. In this project, it was shown that the velocity of the water would influence the energy dissipation rate per unit width and the aeration efficiency for the stepped spillway. It was found also that as the velocity increases, the energy dissipation rate per unit width would also increase.

According to equation (3), it was found that as the energy dissipation rate per unit width is increased, the aeration efficiency of the stepped spillway would be increased. The maximum aeration efficiency is located at the 4th step of the stepped spillway. From the water flow pattern figure, it was found that there exist substantial air trapped region at the 4th step of the stepped spillway. This is the main reason the higher accumulation of aeration efficiency at the 4th step of the stepped spillway.

References

[1] Chen S, and Doolen 1988 G Lattice Boltzmann method for fluid flows *Annu. Rev. Fluid Mech* **30** 329-364
[2] Mohamad A A 2011 Lattice Boltzmann method *Springer-Verlag London*
[3] Melling A 1997 Tracer particles and seeding for particle image velocimetry *Meas. Sci. Technol.* **8** 1406–1416
[4] Mohamad A A "Lattice Boltzmann Method" *Fundamentals and Engineering Applications with Computer Codes.*
[5] Daneshfaraz R , Zogi 2013"Investigation of Cavitation in Stepped Spillway of Siah-Bishe Dam by Using Flow-3d Model," *International Research Journal of Applied and Basic Sciences*
[6] Sukop M C and Or D 2004 Lattice Boltzmann method for modeling liquid-vapor interface configurations in porous media *Water Resour. Res.* **40**
[7] Gabbana A "Accelarating the D3Q19 Lattice Boltzmann Model with Open ACC and MPT."

Acknowledgments

The authors would like to acknowledge Ministry of Education Malaysia for providing LRGS grant No. 203/PKT/6726001 – Riverbank Filtration for Drinking water Source Abstraction to fund this research.

IOP Conf. Series: Materials Science and Engineering **209** (2017) 012091 doi:10.1088/1757-899X/209/1/012091

Electrospinning Polyaniline for Sensors

N Amariei[1], L R Manea[1*], A P Bertea[1], R Cramariuc[2], A Bertea[1] and O Cramariuc[3]
[1]"Gheorghe Asachi" Technical University of Iasi- Romania;
[2]Centre of Competence in Electrostatics and Electrotechnologies, 25, Av. Radu Beller, Bucharest, Romania
[3] Tampere University of Technology, Department of Physics, P. O. Box 692, FI-33101 Tampere, Finland

E-mail: lili191065@yahoo.com

Abstract. Polyaniline is a conductive polymer (organic metal) with multiple uses as a sensor for mineral acids, ammonia, VOCs etc. The yield of the electrochemical reactions being proportional to the surface area, it is expected that polyaniline nanofibers to provide great opportunities for applications in electronic nanodevices. This article reviews electrospinning configurations (classic, coaxial, with 2 collectors, with rotary collector, with liquid collector etc.) used to obtain 3D structures of electrospun polyaniline, alone or mixed with other polymers, which are used in sensors.

1. Introduction

Polyaniline is an extremely versatile, highly conductive polymer with many uses. Depending on these, the electrospinning systems that can be used vary significantly, especially with respect to the nature of the collector.

2. Electrospinning alternatives for producing PANI nanofibers

There is a relatively large number of configurations used for the electrospinning of polyaniline (PANI), alone or mixed with other polymers. Thus, in order to electrospinn a PANI / polyvinylpyrrolidone (PVP) solution to obtain NO_2 sensors, Bishop uses a configuration comprising a DC voltage source, a programmable syringe pump and an aluminum collector plate. The voltage applied by the power supply is big enough to overcome the superficial tension of solution drops formed at the end of the syringe, causing the formation of a polymer blend jet. The electrostatic forces between the collector plate and the tip of the needle help driving the fibers to the collector. Nanofibers of PVP-PANI and PVP-PANI-Urate are formed at the application of 15 kV and 20 kV voltages, with flow rates ranging from 5 □L / min - 50 □L / min and a distance between the needle and collector of 5- 6 cm [1]. Nanofibers used for ammonia detection were obtained by placing about 0.5 mL of PANI / poly(ethylene oxide) (PEO) solution in a hypodermic syringe and connecting the needle to 0.8 kV with respect to the grounded counter-electrode located at approximately 2 mm from the needle tip (figure 1).

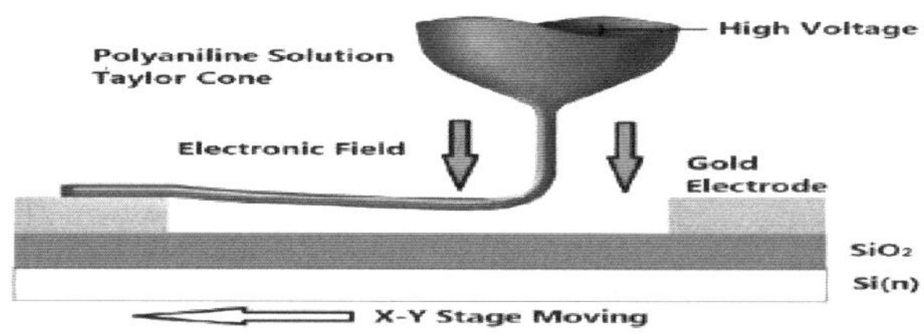

Figure 1. PANI/PEO near field electrospinning and device schematic [2].

The substrate is displaced in a computer controlled trajectory at a speed of 30 mm/s. By applying different trajectories and velocities, the amount and position of the fibers can be controlled [2]. In order to overcome the difficulty of PANI processing into fibers, the production of PANI doped with +camphor-10-sulfonic acid [HCSA] was achieved by coaxial electrospinning followed by the removal of the shell polymer, (figure 2) [3].

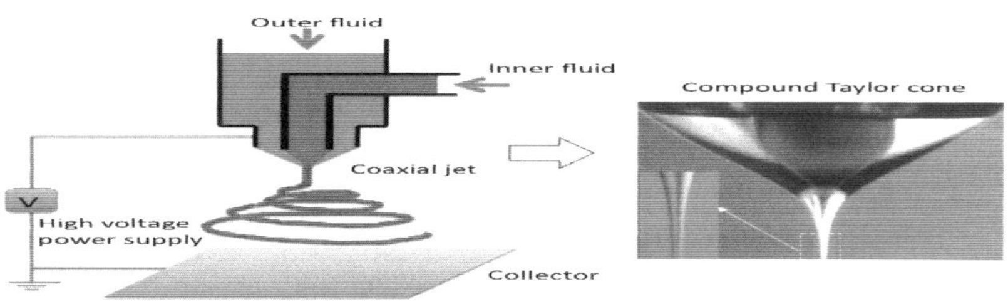

Figure 2. Basic setup for coaxial electrospinning of core-shell nanofibers [3].

The core was 2% by weight PANI, mixed with various amounts of HCSA, dissolved in a 5: 1 mixture of chloroform and dimethylformamide (DMF), and the shell was 15% by weight poly(methyl methacrylate) (PMMA) in DMF. The system behaves as a sensor for ammonia and nitrogen dioxide, both showing high sensitivity and rapid response time [4]. Co-polymer PANI / poly (vinyl chloride - acrylonitrile) nanofibers were obtained using an aluminum foil wrapped on a rotating drum with a 12 cm distance between the spinning nozzle and the collector, a flow rate of 1.0 mL/h, 15 kV voltage and 55% -65% humidity [5].

3. Electrospinning of 3D structures

A general problem that arises in nanofiber electrospinning is the predominantly two-dimensional character of the obtained structures, although a porous 3D-type structure would be more favorable in many areas of use, including sensors. Several methods have been explored to increase the porosity of the electrospun fibrous matrices.

A common strategy involves electrospinning of mixtures of polymer solution and readily removable materials (such as water-soluble salts or polymers) which are then removed after the process: this method involves risks related to the stability of the nanofibers after eliminating the add-on substances [6-8].

Using a chemical expansion agent (azodicarbonamide) in the electrospinning solution increases the pore size (in the case of poly-caprolactone up to 50-70 mm), creating 3D structures [9]. An alternative approach is the use of ultrasonication, which can increase the pore size of the electrospun structures [10], without producing a true 3D structure, especially in terms of thickness.

It has been attempted to obtain 3D structures by replacing the traditional collector with a spherical collector with a series of needles [11], when reproducible results are difficult to obtain due to the lack of precise control over parameters such as the spherical collector dimensions, needle length, distance etc. Another variant tested for this purpose involves the use of liquid collector systems, which produce high porous 3D fibrous skeletons [12].

4. Electrospinning with nanofibers orientation

As a result of the bending instability, the path of an electrospun polymer jet is extremely chaotic, so that nanofibers are typically collected randomly overlayed in network / nonwoven forms [13] that can be used for certain applications, such as filtration / separation. However, the unilateral alignment of electrospun nanofibers can be crucial for other applications, such as microelectronic devices manufacturing [14,15].

An experimental static system with an alternating electrostatic field between two collectors can be used to improve the control of the orientation and positioning of electrospun fibers of PANI / PEO and PANI / PMMA, allowing the alignment of nanofibers on the substrate between collectors (figure 3) [16].

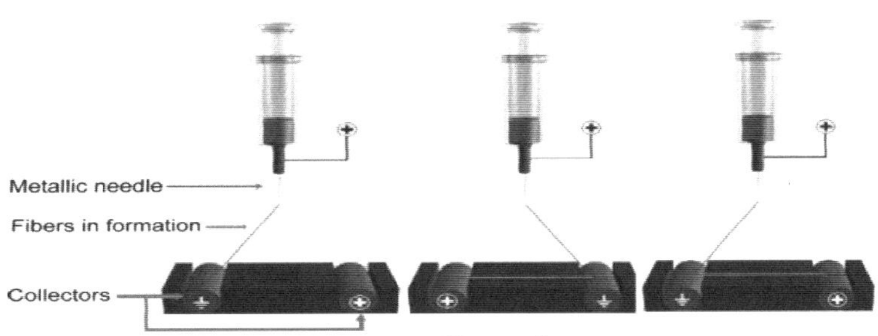

Figure 1. Electrospun fibers alignment set up.

Whereas one collector is grounded, the other is kept at 6 kV and this electric state is alternated between the two collectors using high voltage reed relays, forcing the electrostatically charged polymer to bounce from one collector to the other producing aligned fibers.

Inspired by parallel plates collector and rotary collectors, a rotating drum from copper wire was used to collect aligned fibers [17].

Another electrospinning technique that allows the regular orientation of the obtained nanofibers used a grounded collector consisting of a glass Petri dish, a metal plate being placed inside the bath and covered with distilled water up to a depth of 5 mm. A thin copper wire was passed over the edge of the bath to connect the metal plate to the ground and a motorized pickup roller with a rotation speed equivalent to 0.05 m / sec takes the disordered nanofiber mixture formed on the water surface and deposits it in an orderly manner, as can be seen in figure 4 [18]. Moreover, a dynamic substrate liquid system has been developed to support and assemble the electrospun nanofibers [19].

In a similar approach, a dynamic fluid system in the form of a vortex surface with a rotating drum is used to prepare well-aligned continuous yarns from nanofibers [20]. Another technique of manufacturing uniaxially aligned electrospun fibers uses two needles with opposite voltages that spray simultaneously and the spun fibers with opposite charges attract each other combining into a yarn, which is wound by a cylinder collector rotating at high speed (figure 5) [21].

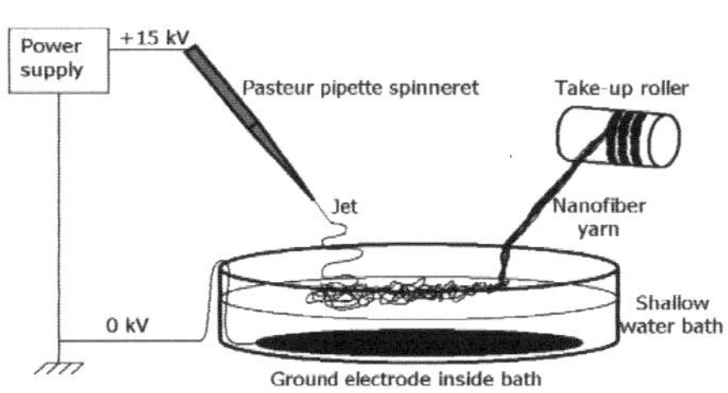

Figure 2. Water bath grounded collector electrode electrospinning.

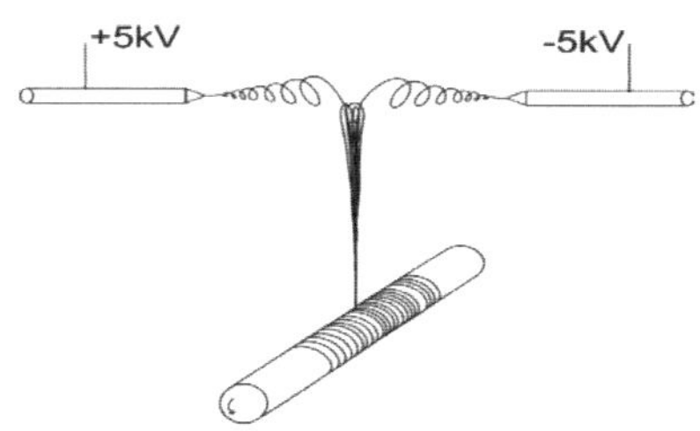

Figure 3. Schematic two needles electrospinning setup.

Rotary drum collectors, unlike stationary or rotatable collectors, provide increased evaporation time and, implicitly, increased ordering, even if the thickness of the nanofibrous deposition is limited and the nanofibers can't be aligned at any time. Other difficulties are related to the fact that not all materials can be electrospun with rotary collectors. Rigid materials, with limited fluidity, lead to fibers that break at lower rotational speed. The rotating collector method is time-consuming and energy-intensive and the degree of fiber alignment decreases with increasing spinning time and fiber layer thickness [17].

The degree of alignment of electrospun nanofibers can be controlled by a magnetic field if the polymer solution contains a small amount of magnetic material, as the magnetized nanofibers are stretched and paralleled (figure 6) [23, 24]. Highly ordered nanofibers can be obtained by using an auxiliary electric field to control nanofiber deposition, stabilizing the unsteady jets by controlling the shape and intensity of the macroscopic electric field between the anode and the cathode (Figure 7) [25-28].

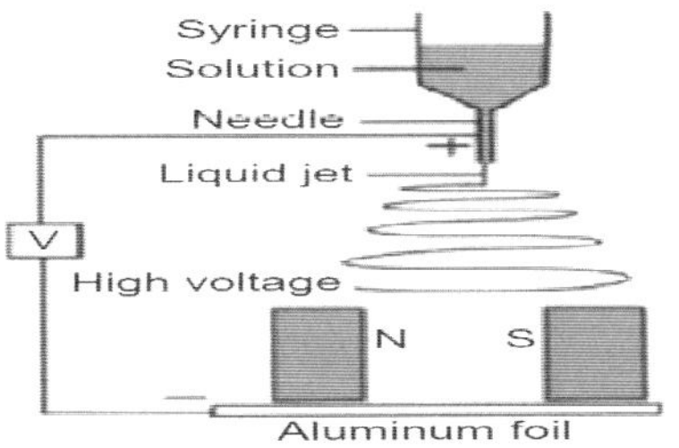

Figure 4. Magnetic electrospinning.

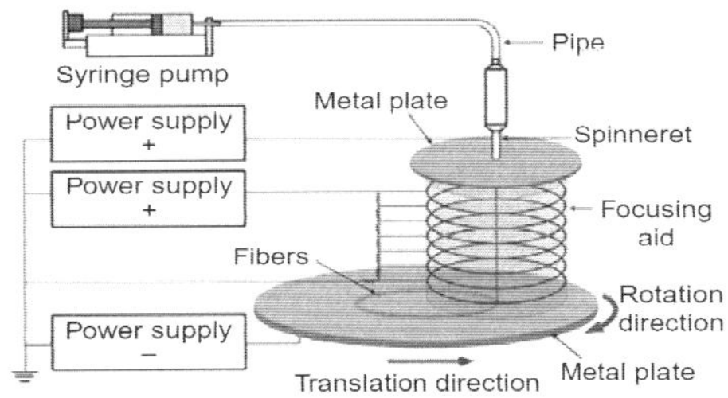

Figure 5. Schematic diagram of the focusing electric-field-assisted electrospinning setup.

Well aligned nanofibers can be obtained by using a modified electrospinning equipment with a rotating nozzle (centrifugal electrospinning), when the forces contributing to stretching the solution droplet into fibers are a combination of centrifugal force and electrostatic force [29-35]. Another method uses a collector obtained by coupling a T-shaped electrode and a rectangular electrode, with the use of a gas stream to force the nanofibers to point to the positive side of the rectangular electrode [36-43]. The Near Field Electrospinning (NFES) method uses a reduced distance between the spinning tip and the collector electrodes (500 □m to 3 millimeters) to manipulate the electrospun jet into aligned filament structures, (figure 8) [40, 44-54].

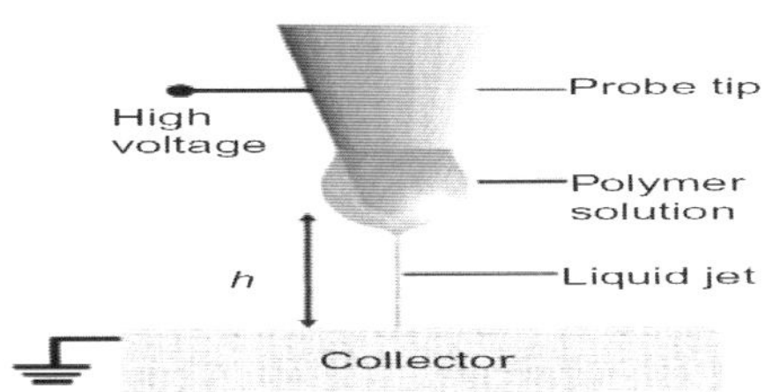

Figure 8. Schematic representation of near field electrospinning.

5. Conclusions

Many electrospinning set-ups (classic, coaxial, with 2 collectors, with rotary collector, with liquid collector etc.) are used to obtain aligned nanofibers or 3D structures of electrospun polyaniline, alone or mixed with other polymers, which are used in sensors. The paper reviews the most significant achievements in these directions.

References

[1] Bishop A and Gouma P *Reviews on Advanced Materials Science* 2005 **10** 209
[2] Chen D, Lei S and Chen Y *Sensors* 2011 **11** 6509
[3] Li F and Zhao Y S Y 2010 *Core-Shell Nanofibers: Nano Channel and Capsule by Coaxial Electrospinning*, in Kumar, A. (Ed.), *Nanofibers* Intech, Vukovar Croatia

[4] Zhang H D, Tang C C, Long Y Z, Zhang J C, Huang R, Li J J and Gu C Z 2014 *Sensors and Actuators A* **219** 123
[5] Li X Q, Liu W W, Liu S, Li M, Li Y G and Ge M 2014 *Chinese Chem Lett* **25** 83
[6] Nam J, Huang Y, Agarwal S and Lannutti J 2007 *Tissue Eng.* **13** 2249
[7] Ekaputra A K, Prestwich G D and Cool S M 2008 *Biomacromolecules* **9** 2097
[8] Guimaraes A, Martins A, Pinho E D, Faria S, Reis R L and Neves N M 2010 *Nanomedicine* **5** 539
[9] Nedjari S, Hebraud A and Schlatter G 2015 *Organized Assembly of Electrospun Nanofibers: From 1D to 3D*, in Mitchell, G. (Ed), *Electrospinning - Principles, Practice and Possibilities* (The Royal Society of Chemistry Cambridge UK)
[10] Lee J B, Jeong S I, Bae M S, Yang D H, Heo D N, Kim C H, Alsberg E and Kwon I K 2011 *Tissue Eng. Part A* **17** 2695
[11] Blakeney B A, Tambralli A, Anderson J M, Andukuri A, Lim D J, Dean D R and Jun H W 2011 *Biomaterials* **32** 1583
[12] Chen H, Peng Y, Wu S and Tan L P 2016 *Materials* **9** 272
[13] Ding Y, Hou H, Zhao Y, Zhu Z and Fong H 2016 *Prog. Polym. Sci.* **61** 67
[14] Dan Y, Evoy S and Johnson C 2008 *Nanowire Research Progress* **3** 95
[15] Yuan H, Zhou Q and Zhang Y 2016 *Improving fiber alignment during electrospinning*, in M. Afshari, (Ed.), *Electrospun Nanofibers* (Woodhead Publishing Duxford UK)
[16] Attout A, Yunus S and Bertrand P 2008 *Polymer engineering and science* **48** 1661
[17] Katta P, Alessandro M, Ramsier R and Chase G 2004 *Nano Lett.* **4** 2215
[18] Smit E, Buttner U and Sanderson R 2005 *Polymer* **46** 2419
[19] Rao P S, Sathyanarayana D N and Palaniappan S 2002 *Macromolecules* **35** 4988
[20] Yousefzadeh M, Latifi M and Teo W-E 2011 *Polym. Eng. Sci.* **51** 323
[21] Pan H, Li L, Hu L and Cui X 2006 *Polymer* **47** 4901
[22] Katta P, Alessandro M, Ramsier R D, G. G. 2004 *Nano Letters* **4** 2215
[23] Ajao J A, Abiona A A, Chigome S and Fasasi A 2010 *J. Mater. Sci.* **45** 2324
[24] Liu Y, Zhang X, Xia Y and Yang H 2010 *Adv. Maters.* **22** 2454
[25] Deitzel J, Kleinmeyer J, Hirvonen J and Tan N B 2001 *Polymer* **42** 8163
[26] Deitzel J, Kleinmeyer J, Harris D and Tan N B 2001 *Polymer* **42** 261
[27] Carnell L S, Siochi E J and Holloway N M 2008 *Macromolecules* **41** 5345
[28] Cui X, Li X L and Xu F 2011 *Appl. Phys.* A **103** 167
[29] Badrossamay M R, Mcllwee H A, Goss J A and Parker K. 2010 *Nano Lett.* **10** 2257
[30] Khamforoush M and Mahjob M 2011 *Mater. Lett.* **65** 453
[31] Calin M A, Khenoussi N, Schacher L, Adolphe D, Manea L R, Gradinaru I, Zetu I and Stratulat S 2013 *Materiale Plastice* **50** (4) 257
[32] Scarlet R, Manea L R, Sandu I, Cramariuc B and Sandu A V 2012 *Rev. Chim.* **63** (8) 777
[33] Scarlet R, Manea L R, Sandu I, Martinova L, Cramariuc O and Sandu I G 2012 *Rev. Chim.* **63** (7) 688

[34] Manea L R, Scarlet R and Sandu I 2015 *Rev. Chim.* **66** (10) 1622
[35] Popescu V, Manea L R, Popescu G 2009 *Management of Technological Changes* ed Rusu C. (Alexandroupolis, Greece, 3-5 September 2009) (Proceedings of the 6th International Conference on the Management of Technological Changes, Book 2) pp 769-772
[36] Zhou W, Li Z, Zhang Q, Liu Y and Wei F 2007 *J. Nanosci. Nanotechnol.* **7** 2667
[37] Manea L R, Hristian L, Leon A L and Popa A 2016 *IOP Conf. Series: Mater. Sci. Eng.* **145** 032006
[38] Hristian L, Ostafe M M, Manea L R and Leon A L *2016 IOP Conf. Series: Mater. Sci. Eng.* **145** 032004
[39] Hristian L, Ostafe M M, Manea L R and Leon A L 2016 *IOP Conf. Series: Mater. Sci.Eng.* **145** 022014
[40] Kameoka J and Craighead H G 2003 *Appl. Phys. Lett.* **83** 371
[41] Chang C, Limkrailassiri K and Lin L 2008 *Appl. Phys. Lett.* **93** 123111
[42] Manea L R, Hristian L, Leon A L and Popa A 2016 *IOP Conf. Series: Mater. Sci. Eng.* **145** 032007
[43] Earar K, Matei M N, Sandu A V, Hristian L, Bejinariu C and Sandu I G 2015 *Revista de Materiale Plastice* **52** (1) 98
[44] Maftei D, Asaftei I V, Sandu I, Manea L R, Birsa L M and Earar K 2015 *Rev. Chim.* **65** (5) 673
[45] Manea L R, Cramariuc B, Caunii V and Sandu I 2015 *Materiale Plastice* **52** (1) 82
[46] Popescu V, Manea L R and Amariei N 2009 *Materiale Plastice* **46** (1) 95
[47] Manea L R, Scarlet R, Amariei N, Nechita E and Sandu I G 2015 *Rev. Chim.* **66** (4) 542
[48] Asaftei I V, Sandu I G, Birsa L M, Manea L R and Earar K 2015 *Rev. Chim.* **66** (3) 336
[49] Hristian L, Sandu A V, Manea L R, Tulbure E A and Earar K 2015 *Rev. Chim.* **66** (3) 342
[50] Hristian L, Bordeianu, D L, Iurea P, Sandu I and Earar K 2014 *Materiale Plastice* **51** (4) 405
[51] Manea L R, Danu M C and Sandu I 2015 *Rev. Chim.* **66** (6) 868
[52] Manea L R, Nechita E, Danu M C and Agop M 2015 *J. Comput. Theor. Nanosci.* **12** (11) 4693
[53] Leon A L, Manea L R, 2008 ITC&DC: 4th International Textile Clothing & Design Conference *Magic World of Textiles* ed Z Dragcevic (Dubrovnik, Croatia, 5-8 October 2008) (Faculty f Textile technology, University of Zagreb sections) E Book of Proceedings pp 803-806
[54] Manea L R, Stanescu I, Nechita E and Agop M 2015 J. Comput. Theor. Nanosci. **12** (11) 4373

Acknowledgments
This work was supported by a grant of the Romanian national Authority for Scientific Research and innovation, UEFISCDI, project number PN III 46 BG 2016 within PNCDI III

IOP Conf. Series: Materials Science and Engineering **209** (2017) 012092 doi:10.1088/1757-899X/209/1/012092

The Influence of Polymer Solution on the Properties of Electrospun 3D Nanostructures

N Amariei[1], L R Manea[1*], A P Bertea[1], A Bertea[1] and A Popa[2]
[1]"Gheorghe Asachi" Technical University Iasi
[2]"Aurel Vlaicu" University of Arad, Romania, Engineering Faculty, Textile Department

E-mail: lili191065@yahoo.com

Abstract. The remarkable versatility of electrospinning is the result of the fact that many factors influence the process results. The characteristics of the electrospun nanofibres depend crucially, besides technological parameters, on the nature of the polymer solution. In the present article it is overviewed the way the properties of the polymer solution, such as concentration, viscosity, conductivity, surface tension, volatility characteristics of the used solvent, influence the main features of the obtained nanofibres.

1. Introduction

Electrospinning is the cheapest and the most straightforward way to produce extremely thin polymer fibers with diameters down to the nanometer scale, which have very high surface area. In electrospinning the mechanical force that is applied to the end of a jet in conventional fiber spinning is substituted by the electric body force that acts on elements of charged fluid [1-5]. A schematic diagram of the electrospinning process is shown in Figure 1.

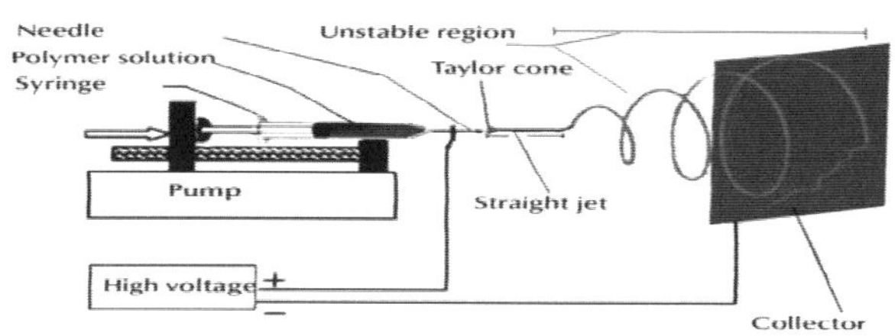

Figure 1. Electrospinning setup [1].

A significant quality of the electrospinning process is its versatility that allows the production of nanofibers of various compositions with controlled structures and multiscale and multifunctional assembly possibilities. Using electrospinning process one can produce a large diversity of materials with enhanced and tuneable properties for a wide range of applications [6-10]. In the beginning, electrospinning has been exclusively applied on the production of nanofibers of natural and synthetic polymers and carbon, with applications in filtration, textile and biomedical systems [11-13], but in the last years this process was extended to fiber formation based also on metal oxides, metals, and composite/hybrid systems [14-17].

2. Main polymer solution characteristics that influence on the properties of electrospun nanofibers

2. 1. Viscosity

The surface tension and viscoelastic properties of the polymer solution are among the key parameters of electrospinning process [18-20]. These characteristics can be controlled by appropriate choice of the polymer concentration, which determines whether the electrospinning process can take place and has a dominant effect on the fiber diameter, as well as its morphology [21, 22]. The viscosity of the solution is usually considered to be the dominant parameter which decides fiber diameter [23] and if the solutions will electrospin at all. The viscosity of the electrospun polymer solution should be situated within a properly delineated range, outside which electrospinning can not be realised. A too low viscosity may result in interruption of polymeric filaments and droplets of polymer (electrospray), while a too high viscosity makes it impossible to extrude the polymer. The minimum required viscosity threshold corresponds to a certain polymer concentration in the solution to be electrospun and varies depending on the molecular weight of the polymer and the nature of the used solvent. The importance of electrostatic viscosity of the polymer solution is illustrated in figure 2.

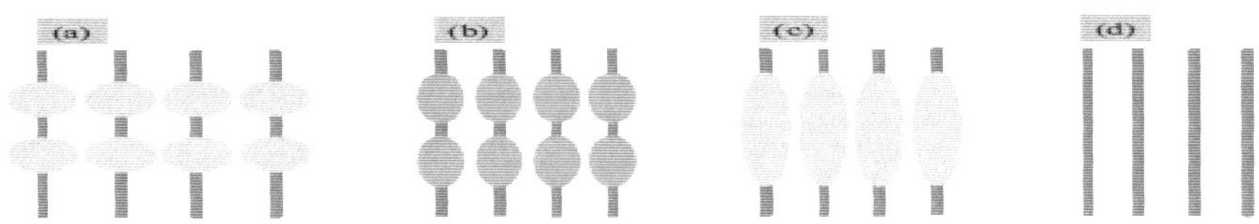

Figure 2. The viscosity increases from a) to d): a) drops; b) drops slightly elongated; c) elongated drop; d) nanofibers [24].

In general, a spinning solution with a viscosity of 1-20 poise values and values of surface tension of 35-55 dyn/cm^2 is considered to be a solution suitable for electrospinning. At viscosities of more than 20 poise, electrospinning is prevented because of polymer flow instability due to the high cohesion of the solution. When the solution viscosity is increased, the polymer beads are larger, the average distance between them is longer, and the diameter of the obtained fibers is increased. The formation of such beads occurs because the fluid flow breaks into droplets when the viscosity of the spinning solution is less than 1 poise [25]. As the solution concentration increases, the shape of the beads changes from spherical to spindlelike [1]. The lower the polymer concentration, the smaller the diameter of the electrospun nanofibers, but the quantitative ratio between the concentration of the polymer solution and the diameter of obtained nanofibers is mostly variable.

Nevertheless, reducing the diameter of nanofibers by lowering the concentration of the polymer solution is limited by the possibility of processing the solution through electrospinning, as well as the possibility of reducing the diameter in post-processing treatments [26]. A change of the surface morphology of individual fibers, obtained by producing of pits, pores or protuberances on their surface or modifying the circular shape of their cross-section by regular or irregular patterning of the surface provides an alternative or complementary method to increase the surface area of the nanofibers [27].

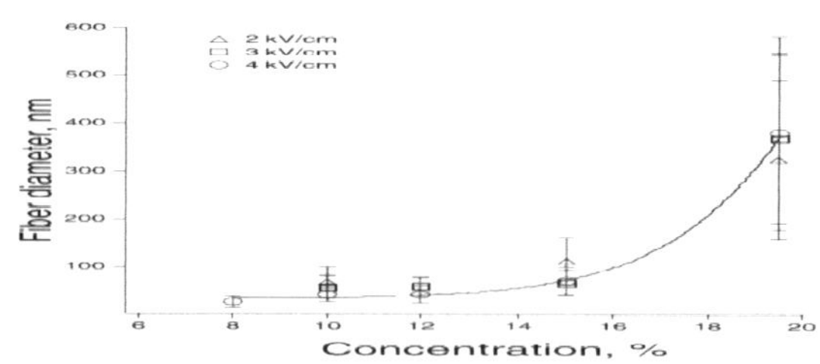

Figure 3. The relationship between fiber diameter and the polymer solution concentration at different values of the electric field [28].

The close relationship between the viscosity and the concentration of the polymer solution to be electrospun has been studied for polymers such as poly (lactic-co-glycolic acid) (PLGA) [29], polyethylene oxide (PEO) [30, 31], polyvinyl alcohol (PVA) [32-34], polymethyl methacrylate (PMMA) [35], polystyrene (PS) [36], poly-L-lactic acid (PLLA) [23], gelatine [28] and dextran [24]. A study on the relationship between viscosity and concentration on electrospinning of seven linear homopolymers of poly(methylmethacrylate) with different molecular weights using dimethylformamide (DMF) as solvent showed that the electrospinning of the polymethyl metacrylate (PMMA) of low molecular weight at concentrations less then six times bigger then the overlap concentration (concentration of polymer at which the viscosity of solution changes abruptly) only produced beads, while high molecular weight PMMA produced fibers with some beads at concentrations in this range [34].

An innovative approach to reduce the viscosity of the polymer solution is to use low-frequency vibration of the solution during the process [37, 38]. The vibration facilitates provisional untangling of the polymer chains by the disruption van der Waals interactions between the polymer chains, which leads to a reduction in the solution viscosity. The viscosity of a PMMA solution could be reduced by an order of magnitude when a vibration with a frequency of 300 rad/s was applied to the solution [37, 38]. The vibration of the tip of the capillary (about 400 kHz) during electrospinning of a poly (butylene succinate) (PBS) / $CHCl_3$ solution led to the generation of nanofibers thinner in comparison with the situation in which no vibration were applied [39]. At higher concentrations of the polymer solution electrospinning can be realised only with the application of vibration. Another advantage of this technique is the possibility to electrospin suspensions and coagulated material which otherwise cannot be electrospun [27].

2.2. Conductivity

The electrospinning process fundamentally requires electrical charge transfer from the electrode to the polymer droplet located at the ending point of the injection needle, hence a minimum electrical conductivity in solution is essential to the electrospinning; solutions devoid of conductivity can not be electrospun.

The conductivity indicates the number of electric charges found on the surface of the polymer solution to be electrospun, and controls its tendency to form nanofibers and not nanodrops.

When dissolving a polymer in a solvent, the conductivity of the solution increases due to the availability of ionic species in the polymer (mostly being impurities or additives). However, with increasing concentration of the polymer in the solution, the electrical conductivity may decrease [10]. If the polymer itself has ionic capabilities of polyelectrolyte type, the conductivity of the solution will be much higher (compared to the solutions of the polymer without electric charges) and significantly dependents on the concentration [28].

A correction of the lack of conductivity of the polymer solution can be obtained by adding electrolyte, when the increased number of electrical charge cause an increase in the elongation capacity of the solution, favouring the formation of smooth fibers, with smaller diameter [29] .

Inorganic salts such as NaCl (0.01 M) [21, 30, 31] or ionic organic compounds such as pyridinium formate (PF) [10], palladium diacetate [32], trialchilbenzil ammonium chloride [33, 34] can be used to improve the conductivity of the polymer solution.

In this case the size of the ions affects the morphology of nanofibers; the obtained diameters are smaller as the size of the ions is smaller, as a result of their increased mobility, which leads to a higher tensile force of the jet of electrospun polymer, favouring the uniformity of the fibers and the reduction of the number of flaws [35].

2. 3. Surface tension

The surface tension is the property of the liquid that shows the strong cohesiveness of the liquid molecules, indicating the dissimilarity of the phases at the interface [36].

The surface tension, which is strongly influenced by the nature of the solvent from the polymer solution, is a very important factor in electrospinning. This is why the correct selection of the solvent is critical not only to obtain a homogeneous solution of the polymer, but to obtain the right surface tension. Basically, if all the other conditions are set, the surface tension establishes the upper and lower limits of the range in which electrospinning can be achieved [37-42]. Lower surface tension of the spinning solution aids electrospinning to occur at lesser electric field

[1, 43-47]. In the absence of any electric field, the polymer solution is retained at the tip of the capillary by the surface tension of the liquid. When applying the electric field, the polymer jet is formed as a result of the action of the electric charges that overcome the surface tension of the solution, and therefore a decrease in the surface tension of the polymer solution induces instability to the jet and increases the breaking tendency of the extruded filament, with the formation of drops [48-53]. The surface tension can be adjusted by altering the polymer / solvent ratio in the electrospinning solution [54-60. Another method of reducing the surface tension is to add surfactant to the solution, a procedure which ensures more uniformity to the fiber. However, not necessarily a lower surface tension of the solution will always be suitable for electrospinning [36, 61-65].

3. Conclusions

Nanofibers can be manufactured by interfacial polymerization and electrospinning, but electrospining it the technique that offers a large versatility in what regards polymer compositions and the possibility to control the structure of the nanofibers.

This paper presents new research in the field electrospinning as a process used to produce nanofibers and offers a report on the main characteristics of the polymer solution that may effect the properties of the obtained nanofibers. The influence of viscosity, conductivity and surface tension is reviewed.

References

[1] Haghi A K 2012 *Electrospinning of nanofibers in textiles* CRC Press, Boca Raton Finland

[2] Calin M A, Khenoussi N, Schacher L, Adolphe D, Manea L R, Gradinaru I, Zetu I and Stratulat S 2013 *Mat. Plast.* **50** (4) 257

[3] Scarlet R, Manea L R, Sandu I, Cramariuc B and Sandu A V 2012 *Rev. Chim.* **63** (8) 777

[4] Scarlet R, Manea L R, Sandu I, Martinova L, Cramariuc O and Sandu I G 2012 *Rev. Chim.* **63** (7) 688

[5] Manea L R, Scarlet R and Sandu I 2015 *Rev. Chim.* **66** (10) 1622

[6] Cavaliere S 2016 *Electrospinning for advanced energy and environmental applications* (CRC Press, Boca Raton, Finland)

[7] Popescu V, Manea L R and Popescu G 2009 *Management of Technological Changes* ed Rusu C. (Alexandroupolis, Greece, 3-5 September 2009) (Proceedings of the 6th International Conference on the Management of Technological Changes, Book 2) pp 769-772

[8] Zhou W, Li Z, Zhang Q, Liu Y and Wei F 2007 *J. Nanosci. Nanotechnol.* **7** 2667

[9] Manea L R, Hristian L, Leon A L and Popa A 2016 *IOP Conf. Series: Mater. Sci. Eng.* **145** 032006

[10] Hristian L, Ostafe M M, Manea L R and Leon A L *2016 IOP Conf. Series: Mater. Sci. Eng.* **145** 032004

[11] Huang L, Nagapudi K., Apkarian R P and Chaikof EL 2001 *J Biomater Sci Polym* Ed. **12** 979

[12] Frenot A and Chronakis I 2003 *Curr.pin. Colloid Interface Sci.* **8** 64

[13] Hristian L, Ostafe M M, Manea L R and Leon A L 2016 *IOP Conf. Series:*

Mater. Sci.Eng. **145** 022014
[14] Greiner A and Wendorff J H 2007 *Angew. Chem. Int. Ed. Engl.* **46** 5670
[15] Manea L R, Scarlet R, Amariei N, Nechita E and Sandu I G 2015 *Rev. Chim.* **66** (4) 542
[16] Asaftei I V, Sandu I G, Birsa L M, Manea L R and Earar K 2015 *Rev. Chim.* **66** (3) 336
[17] Hristian L, Sandu A V, Manea L R, Tulbure E A and Earar K 2015 *Rev. Chim.* **66** (3) 342
[18] Attout A, Yunus S and Bertrand P 2008 *Polymer eng. and science* **48** 1661
[19] Ginestra P, Ceretti E and Fiorentino A 2016 *A. Proceedings of The Second CIRP Conference on Biomanufacturing* **49** 8
[20] Demir M M, Yilgor I, Yilgor E and Erman B 2002 *Polymer* **43** 3303
[21] Zong X H, Kim K, Fang D, Ran S F, Hsiao B S and Chu B 2002 *Polymer* **43** 4403
[22] Jun Z, Hou H, Schaper A, Wendorff J H and Greiner A 2003 *e-Polym* **9** 1
[23] Luzio A, Canesi EV, Bertarelli C and Caironi M 2014 *Materials* **7** 906
[24] Abd Razak SI, Wahab I F, Fadil F, Dahli F, Khudzari A and Adeli H 2015 *Advances in Materials Science and Engineering* **2015** 1
[25] Zong X H, Bien H, Chung C-Y, Yin L H, Fang D F, Hsiao B S, Chu B and Entcheva E 2005 *Biomaterials* **26** 5330
[26] Andrady A L 2008 *Science and technology of polymer nanofibers* (John Wiley & Sons, Inc., Hoboken New Jersey)
[27] Sukigara S, Gandhi M, Ayutsede J, Micklus M and Ko F 2003 *Polymer* **44** 5721
[28] Ki C S, Baek D H, Gang K D, Lee K H, Um I C and Park Y H 2005 *Polymer* **46** 5094
[29] Huang Z.-M, Zhang Y-Z, Kotaki M and Ramakrishna S 2003 *Compos. Sci.Technol.* **63** 2223
[30] Son W K. Youk J H, Lee T S and Park W H 2004 *Polymer* **45** 2916
[31] Ding B, Kim H Y, Lee S C, Shao C L, Lee D R, Park S J, Kwang G B and Choi C J 2002 *J Polym Sci B Polym Phys* **40** 1261
[32] Lee J S, Choi KH, Ghim H D, Kim S S, Chun D H, Kim H Y and Lyoo W S, 2004 *J Appl Polym Sci.* **93** 1638
[33] Kim J R., Choi S W, Jo S M, Lee W S and Kim B C 2005 *Journal of the Electrochemical Society* **152** A295
[34] Gupta P, Elkins C, Long T E and Wilkes G L 2005 *Polymer* **46** 4799
[35] Stanger J, Tucker N and Staiger M 2005 *Electrospinning* (Rapra Technology, Report 190)
[36] Jiang H L, Fang D F, Hsiao B S, Chu B and Chen W L 2004 Biomacromolecules **5** 326
[37] He J-H, Wan Y-Q and Yu J-Y 2004 *International Journal of Nonlinear Sciences and Numerical Simulation* **5** 253
[38] Wan Y-Q, He J-H, Wu Y and Yu J-Y 2006 *Materials Letters* **60** 3296
[39] Wan Y-Q, He J-H, Wu Y and Yu J-Y 2007 *International Journal of Electrospun Nanofibers and Applications* **1** 17
[40] McKee M G, Hunley M T, Layman J M and Long T E 2006 *Macromolecules* **39** 575

[41] Haghi A K and Zaikov G E 2011 *Development of nanotechnology in textiles* (Nova Science Publishers, Inc., New York)
[42] Lee C H, Shin H J, Cho I H, Kang Y-M, Kim I A, Park K-D and Shin J-W 2005 *Biomaterials* **26** 1261
[43] Wannatong L, Sirivat A and Supaphol P 2004 *Polymer International* **53** 1851
[44] Yu J H, Fridrikh S V and Rutledge G C 2004 Advanced Materials **16** 1562
[45] Zeng J, Xu X Y, Chen X S, Liang Q Z., Bian X C, Yang L X and Jing X B 2003 Journal of Controlled Release **92** 227
[46] Choi J S, Lee S W, Jeong L, Bae S H, Min B C, Youk J H and Park W H 2004 *International Journal of Biological Macromolecules* **34** 249
[47] Bhardwaj N, Kundu S C 2010 Biotechnology Advances **28** 325
[48] Afshari M 2017 *Electrospun Nanofibers* (Woodhead Publishing Cambridge)
[49] Li Z, Wang C 2013 *One-Dimensional nanostructures Electrospinning Technique and Unique Nanofibers* (Springer Berlin)
[50] De Vrieze S, van Camp T, Nelvig A, Hagstrom B, Westbroek P and De Clerck K, 2009 *J Mater Sci* **44** 1357
[51] Pillay V, Dott C, Choonara Y, Tyagi C, Tomar L, Kumar P, du Toit L and Ndesendo V 2013 *Hindawi Publishing Corporation, Journal of Nanomaterials* **2013** 1
[52] Arumugam G K, Khan S and Heiden PA 2009 *Macromolecular Materials and Engineering* **294** 45
[53] Manea L R, Bertea A P, Nechita E, Popescu C V and Sandu I 2016 *Rev.Chim.* **67** 8
[54] Manea L R, Hristian L, Leon A L and Popa A 2016 *IOP Conf. Series: Mater. Sci. Eng.* **145** 032007
[55] Maftei D, Asaftei I V, Sandu I, Manea L R, Birsa L M and Earar K 2015 *Rev. Chim.* **65** (5) 673
[56] Manea L R, Cramariuc B, Caunii V and Sandu I 2015 *Mat. Plast.* **52** (1) 82
[57] Popescu V, Manea L R and Amariei N 2009 *Mat. Plast.* **46** (1) 95
[58] Manea L R, Danu M C and Sandu I 2015 *Rev. Chim.* **66** (6) 868
[59] Leon A L, Manea L R, 2008 ITC&DC: 4th International Textile Clothing & Design Conference *Magic World of Textiles* ed Z Dragcevic (Dubrovnik, Croatia, 5-8 October 2008) (Faculty f Textile technology, University of Zagreb sections) E Book of Proceedings pp 803-806
[60] Manea L R, Bertea A, Nechita E, Popescu C V and Sandu I 2016 *Rev. Chim. (Bucharest)* **67** (7) 1284
[61] Hristian L, Ostafe M M, Manea L R, Sandu I G, Apostol L L and Sandu I 2016 *Rev. Chim. (Bucharest)* **67** (9) 1717
[62] Manea L R, Chirita M, Hristian L and Sandu I 2016 *Mat. Plast.* **53** (3) 361
[63] Popescu V, Radu C D and Manea L R 2010 *Industria Textila* **61** (1) 23
[64] Popescu V, Manea L R, Curteza A and Vasluianu E 2011 *Tekstil* **60** (7) 306
[65] Manea L R, Cramariuc B, Scarlet R, Cramariuc R, Sandu I and Popescu V 2015 *Mat. Plast.* **52** (2) 180

Designing Predictive Diagnose Method for Insulation Resistance Degradation of the Electrical Power Cables from Neutral Insulated Power Networks

R Dobra[1], D Pasculescu[2], M Risteiu[1], G Buica[3] and V Jevremović[4]
[1] Computer Science and Engineering Department, "1 Decembrie 1918" University of Alba Iulia, Gabriel Bethlen Str. No.5, 510009 Alba Iulia, Romania
[2] Electrical Department, University of Petrosani, University Street, No. 11, Petrosani Romania
[3] Electrical and Mechanical Risks Laboratory, National Research and Development Institute for Labor Protection "Alexandru Darabont", Bucharest, Blvd. Ghencea no. 35A, district 6, Romania
[4]High Technical Mechanical School of Professional Studies, Str. Radoja Krstića 19, 37240 Trstenik, Serbia

E-mail: remusdobra@upet.ro

Abstract. This paper describe some possibilities to minimize voltages switching-off risks from the mining power networks, in case of insulated resistance faults by using a predictive diagnose method. The cables from the neutral insulated power networks (underground mining) are designed to provide a flexible electrical connection between portable or mobile equipment and a point of supply, including main feeder cable for continuous miners, pump cable, and power supply cable. An electronic protection for insulated resistance of mining power cables can be made using this predictive strategy. The main role of electronic relays for insulation resistance degradation of the electrical power cables, from neutral insulated power networks, is to provide a permanent measurement of the insulated resistance between phases and ground, in order to switch-off voltage when the resistance value is below a standard value. The automat system of protection is able to signalize the failure and the human operator will be early informed about the switch-off power and will have time to take proper measures to fix the failure. This logic for fast and automat switch-off voltage without aprioristic announcement is suitable for the electrical installations, realizing so a protection against fires and explosion. It is presented an algorithm and an anticipative relay for insulated resistance control from three-phase low voltage installations with insulated neutral connection.

1. Introduction

In underground mining are found different cable needs for both fixed and mobile installations in mines where electrification is the main source of power. A primary goal in the development of different compounds for cable insulations is to obtain physical and electrical characteristics that are stable at elevated temperatures in either wet or dry environments. From an engineering and design viewpoint, high temperature resistance is highly desirable and increases the safety factor during periods of emergency. Insulation stability during an emergency overload is of extreme importance [1].

Operating temperatures must be kept in the standard rage because if the current load increases the degradation of insulations and coverings will accelerate. Thermal environment can affect the degradation of insulation at elevated levels since the speed of a chemical reaction is doubled with a 10°C rise in temperature [2]. The continuous exposure of cable to heavy environmental conditions is a major concern for cable engineers. All polymer-type coverings undergo degradation over time. Environment, installation and chemical composition of the polymer significantly influence longevity. Figure 1 shows a type of underground mining cables and what are the materials of which it is built.

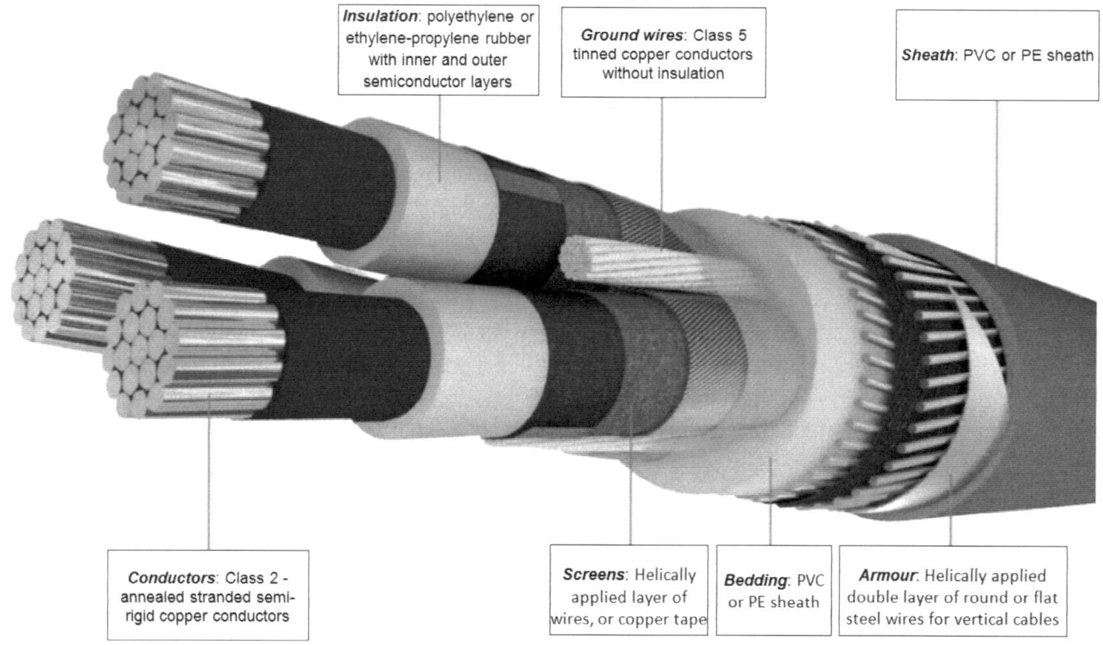

Figure 1. Materials composition of the electrical power cables from neutral insulated power networks.

Many cable failures are the direct result of excessive tension. A cable that has been "stretched" no longer has the balanced construction. Tension on the conductors subjects the individual wires in the strand to compression and shear. These thin wires are damaged and will break more easily when bent or flexed. In some cases of mechanical damage, the failure is instant, and the cause can be assigned on the spot. Many times, however, the cable components are only "injured" and become latent failures [3]. For the electrical systems (regarding the increase of their reliability and particularly for the increase of the safety in their usage) solid improvements of protection subsystems are needed, for the classical type with contacts and switches, as well as for the second generation type, with discrete components, semiconductors and integrated logical or analogical circuits [4].

There are two possible ways of achieving these conceptual improvements in the field of electric and electronic protection, by using micro programmable, dedicated electronic circuits, such as microprocessors, micro controllers, etc and by implicating the electronic computer, hence the control and command virtual systems.

2. Anticipative Strategy for Insulation Resistance Degradation

There are electronic protection devices against resistance degradation of the electrical power cables from neutral insulated power networks and they are realized for normal electrical atmosphere, but also for hazardous electrical areas. This are destined to be used in tree-phase electrical network with insolated neutral connection.

The control of insulated resistance is made as fallow:

- *Preventive*, is made when the network is not supplied with voltage and act before the principal circuit breaker therefore the relay realized a blocking command to the switchgear when the value of the insulation resistance is under impose limit.
- *Permanent*, is made when the network is supplied with voltage and act after closing the principal switchgear therefore if insulation resistance is diminished the relay signalized and give automate disconnect command respectively.

The functionality principle of this device consist of applied a continuous voltage cu minus polarity for the controlled electrical installation and the resulting current is measured because this current is dependent on the insulated resistance value.

2.1. Predictive diagnose method for insulation resistance degradation

In order to avoid unexpected switching-off voltage, when failure of electrical insolated resistance appear, is presented in figure 2 a predictive protection strategy for the insulated resistance control relay. If a degradation process of insolated resistance in tacking place this function is like an integral temporal function and can be identify on the certain time interval with a temporal discreet value. The idea is to calculate the degradation speed ($\Delta R/\Delta t$) in the beginning degradation process of insolated resistance because the accepted limit value is known and so we can determine the time reserve.

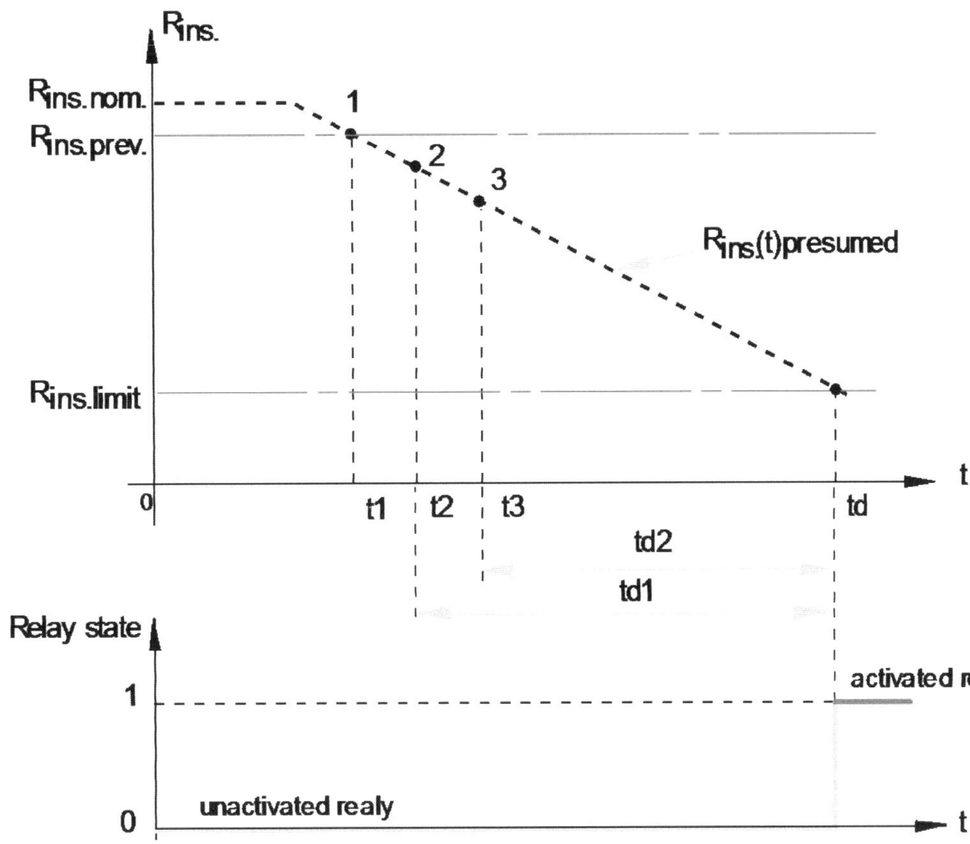

Figure 2. The predictive diagnose method for insulation resistance degradation.

This reserve of time can help the energetic operator to maintain electrical installations supplied with voltage. This method is functional if the degradation characteristic is not changed very much.

2.2. Predictive flowchart for insulation resistance degradation

The relay functionality flowchart shown in picture 2 sequentially run through a few steps. At *ET0* all the acquisition system ports are initialized in order to signalize the functionality of the relay, after this was voltage supplied.

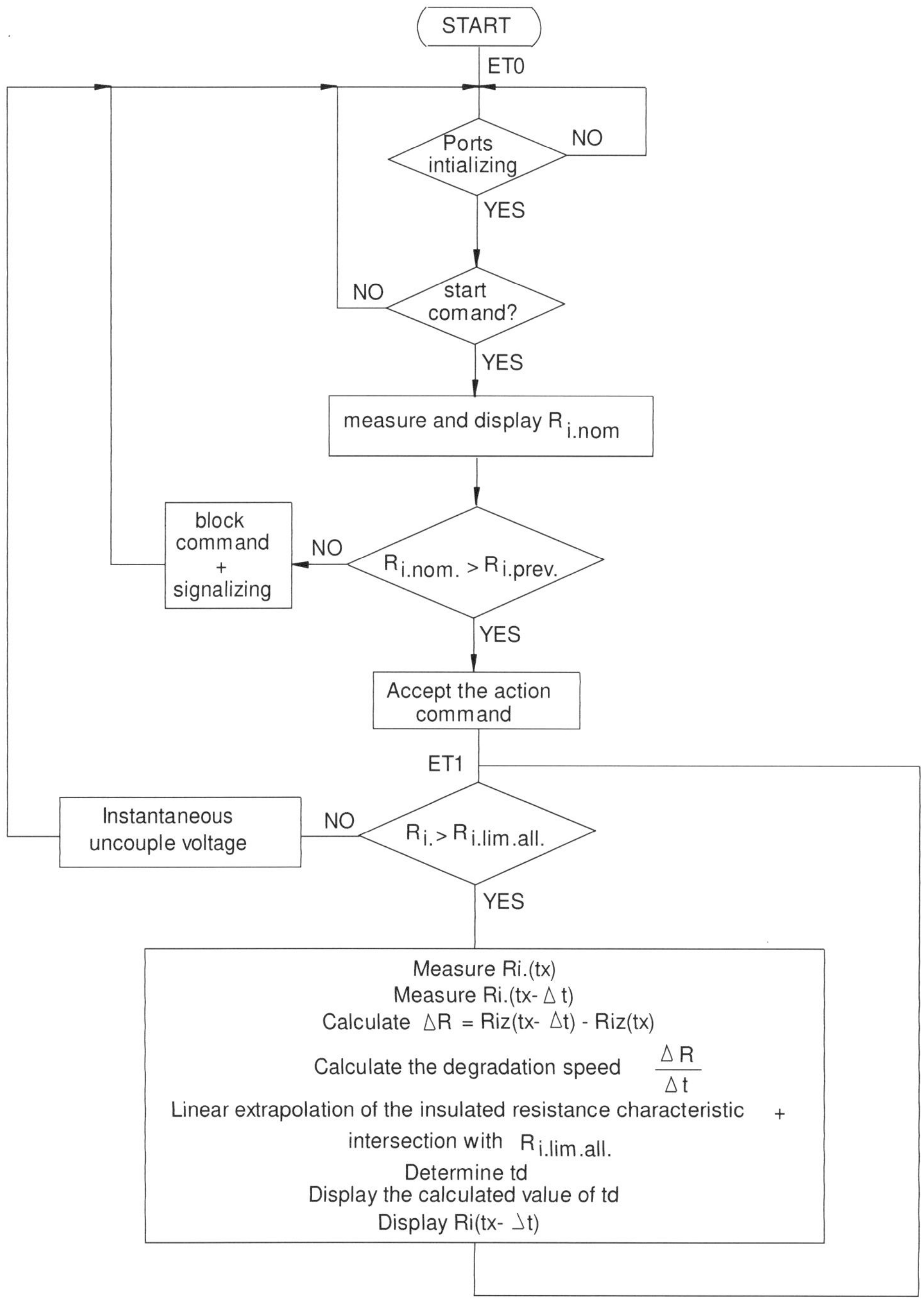

Figure 3. Predictive flowchart for insulated resistance control.

Measure the voltage over resistance r realizing insulated resistance conversion, Riz, which is again measured and displayed.

The measured value of the nominal insulated resistance (Ri.nom.) in compared with the preventive insulated resistance (Ri.prev.) threshold value, which has a standardization value [5]:

- If the rated value of insulated resistance *Ri.nom* is greater than Ri.prev the relay allows the switching devices (K) to close and in this point the consumers (Zs) will receive electric voltage, jump to ET1;
- If *Ri.nom* is smaller than Ri.prev the relay obstructs a possible voltage supply command to consumers. The relay signalizes the deficiency and command an ET0 jump.
- If insulated resistance (Ri) is smaller then allowed limit insulated resistance (Ri.lim.all.) the relay give a command to switching devices (K) which instantaneous uncouple voltage. After that the relay give the signalizing signal command an ET1 jump;
- If Ri is greater than Ri.lim.all. then the resistance Ri(tx) is measured on the certain moment of time tx; after a certain time Δt the insulated resistance Ri(tx - Δt) is measure again; the next step is to calculate the difference ΔR = Ri(tx) - Ri(tx - Δt); the insolated resistance degradation speed is calculated; the insolated resistance characteristic is linear extrapolating and will be intersected with the insolated resistance characteristic which is standardization allowed Ri limit allowed; now the microcontroller can determine the time duration td which is the prevision time to safe voltage maintaining; the next step is to display the calculated value of the safety time duration td and also the microcontroller will display the last insulated resistance value Ri(tx - Δt); the last step of the algorithm is the jump command to ET1 and after that the cycle is repeating following the describe logic.

3. Predictive diagnose electronic protection

Technical realization is possible by implementing electronic systems like microcontrollers, microprocessors or using a PC. In order to understand more easily is synthesized an electronic insulated resistance relay with virtual anticipation of the insulated resistance and using a PC, a data acquisition board and an adequate program.

This strategy creates an advantage because the switching-off voltage events can be prevented. Basically, by applying this predictive strategy the time reserve of the safe functionality of the energetic system will be more easily to monitories because the insulated resistance remains in the safety limits that are technically accepted.

The insulating resistance protection type switch is created for automatic, permanent and preventive control of the status of isolation resistances in low voltage, triphased circuits, with an isolated null to the ground. This switch uses information about the operational current, injected in the controlled network, which proves to be a measure of the state of isolation resistances. The switch has the capacity to control isolation faults between each phase and ground, on the one hand, and between the null of the power transformer and ground on the other hand.

The state of the isolation resistance is an electro security global indicator, which can outline accidental touches of electric circuits at dangerous potentials.

The electronic protection switch is made to accomplish the following two specific functions:

- it does not allow tension connection on a certain part of the circuit, if the isolation resistance does not compel with the minimal legal values, for a certain guard tension of the network;
- it automatically and instantaneously cuts the tension, when the isolation resistance to the ground decreases under a legally imposed minimal value.

This virtual anticipation relay, presented in figure 4 and 5, is used for insulated resistance control and the currents absorbed by the consumers (1) who are supplied from a electrical transformer (2) with a insolated neutral connection to the ground trough the impedance Z_{iz0}. The anticipative insulated resistance relay is integrated inside a computer (3) hardware, which has a data acquisition board (4) on the PCI port (5) or on the USB (6) input. The connection between the anticipative relay and the measuring interface (7) is made trough an shielded cable (8) or a USB cable (9). The insulated resistance control is made by the current injection and the system unbalance. The continuous current is infected in the system trough a voltage operative source (10).

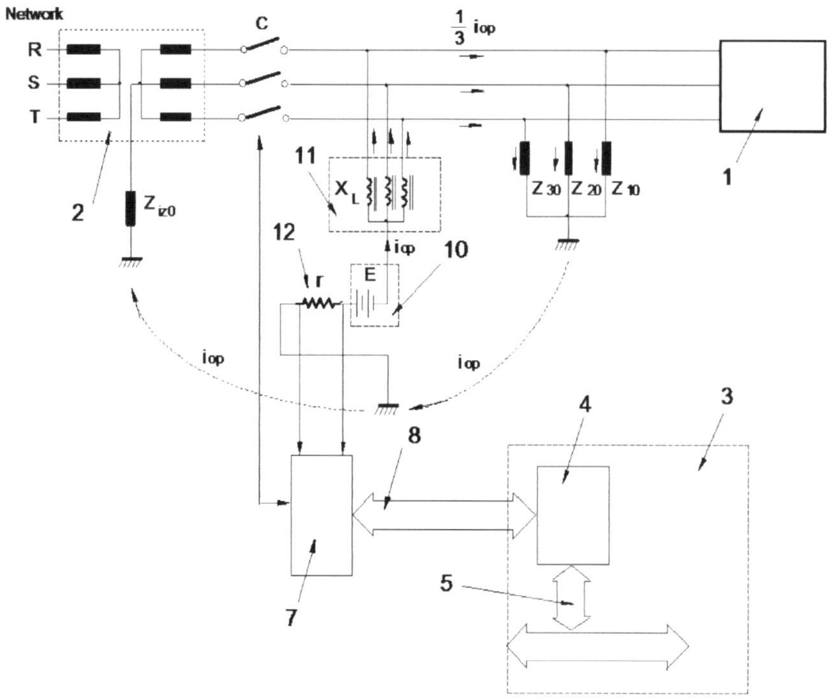

Figure 4. The predictive insulation resistance protection using a measuring interface.

By developing the implementation principle of virtual protection systems in real electro energetic systems, we open new opportunities, only limited by the imagination of the application architect, in order to realize intelligent protection systems, with direct consequences over the increase of the viability of these systems and over their degree of security.

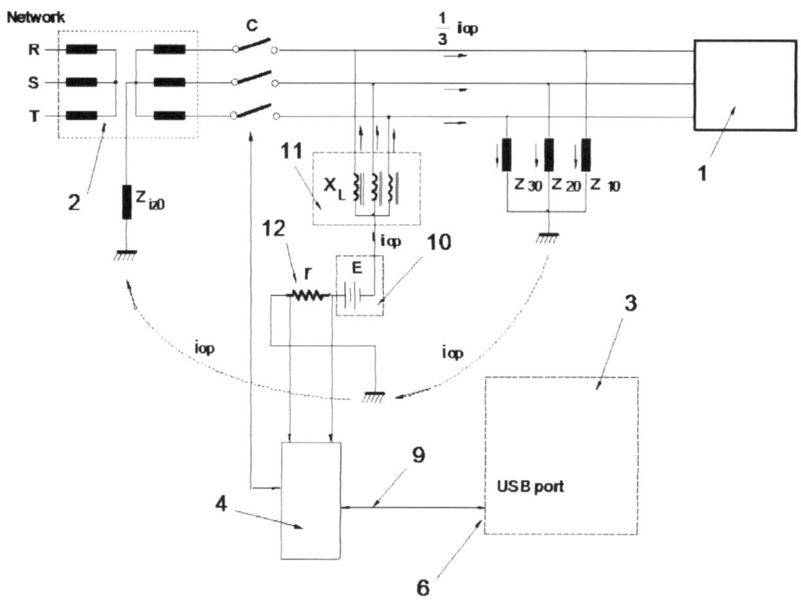

Figure 5. The predictive insulation resistance protection using a data acquisition board.

3.1. Virtual instrumentation for simulating the insulating resistance degradation

In order to understand test the method is synthesized an electronic insulated resistance relay with virtual anticipation of the insulated resistance and using a PC, a data acquisition board and an adequate program. To express electrical insulating properties, volume resistivity or dielectric strength is widely used as an index. The volume resistivity is expressed in terms of electrical resistance calculated per unit volume of the test piece. The dielectric strength is expressed in terms of the magnitude of voltage withstood without destruction of the test piece when a specified amount of voltage is applied for a specified period of time. In both cases, a higher value means better electrical characteristics.

The virtual instrumentation for real time measurement and processing of various physical values is the result of the development of microelectronic technologies, implemented in the computer technology. Through the hardware and the software which are specific to the electronic computers – PC, using specific interfaces for data acquisitions, based on specific emulation programming of signal analysis and measurement instruments, a conceptual leap has been realized regarding the measurement, the control and the computerized command of processes. Practically, the electronic computer is becoming the virtual, like is presented in figure 6, universal instrument, which can be transformed into any kind of measurement, analysis, command and fine tuning equipment, through specific programming, called for by users.

In order for the shown system to become operational, it is necessary to install, in the process dedicated PC, a function driver of the data acquisition system and the software which is adapted to the application, meaning that to the graphic simulation programme we need to attach and set the graphic instrument for data acquisition, from the programmed application tool bar [6].

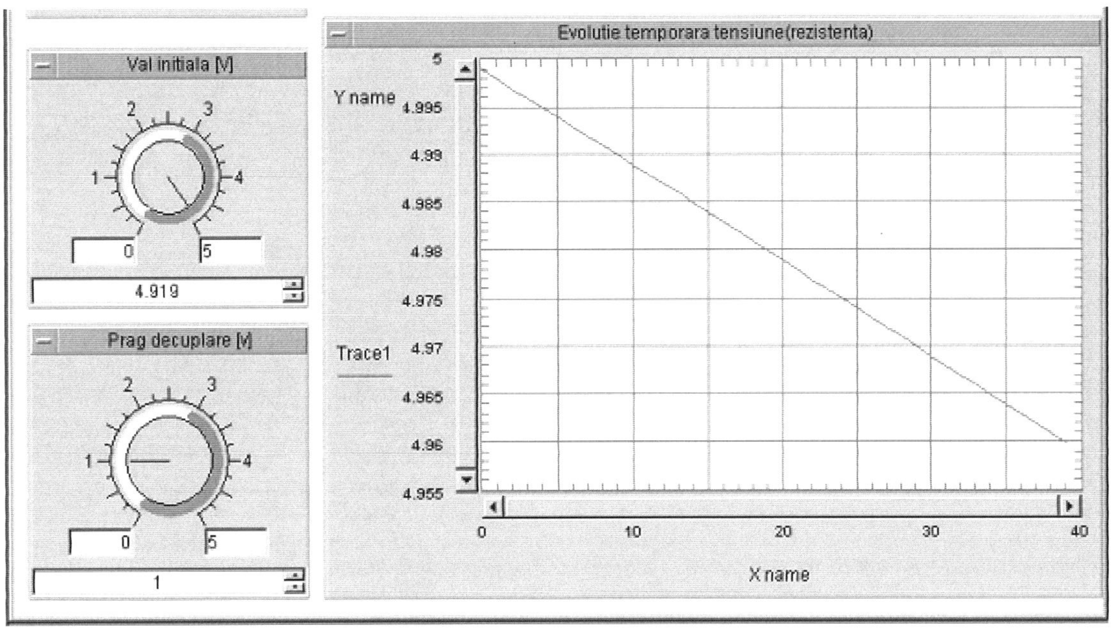

Figure 6. Predictive flowchart for insulated resistance control.

4. Conclusions

This paper presents a predictive strategy for insulated resistance control. Trough this method, the control of the insulated resistance is made with monitoring of the safety functionality time reserve from the energetic system. In this case the insulated resistance parameter remains in the standardizations limits. Conceiving a relay based on this strategy shown in this paper we have the possibility to take anticipative measures that can prevent voltage uncoupling and all the events that come after this.

IOP Conf. Series: Materials Science and Engineering **209** (2017) 012093 doi:10.1088/1757-899X/209/1/012093

The implemented functionality algorithm consists on new conception and applications of a new strategy to control the insulated resistance from electrical network in the hazardous areas. The implemented algorithm creates a possibility to take some anticipative measures in order to prevent undesirable voltage disconnecting.

References

[1] Amyot N, David E and Lee S Y 2002 *IEEE Transactions on dielectrics and electrical insulation* **9**(3) 458 – 466

[2] Bascom E C, Von Dollen and al. 1994 Computerized underground cable fault location expertise *Transmission and Distribution Conference, Proceedings of the 1994 IEEE Power Engineering Society*

[3] Hyvönen P, Oyegoke B and Aro M 2001 Advanced diagnostics test and measurement methods for power cable systems on-site *Literature review with discussion. Report TKK SJT-49. Helsinki University of Technology* Finland

[4] Noda Taku 2008 *IEEJ Transactions on Electrical and Electronic Engineering* **3**(5) 549-559.

[5] Zoller C, Chirigiu G and Dobra R 2012 Method for anticipative automatic control of insulation resistors in low voltage three-phase networks with neutral insulated from the earth *Patent number RO00123449, OSIM Bucuresti*

[6] Zoller C L, Dobra R, Chirigiu G and Şerban H 2007 *Revista Minelor* **5**(191) 6 – 9

Evaluation of Electrical Characteristics of Protective Equipment - a Prerequisite for Ensuring Safety and Health of Workers at Work

G Buică[1], C Beiu[1], A Antonov[2], R Dobra[3] and D Păsculescu[4]
[1] Electrical and Mechanical Risks Laboratory, National Research and Development Institute for Labor Protection "Alexandru Darabont", Bucharest, Blvd. Ghencea no. 35A, district 6, Romania
[2] Certification Body ICSPM-CS, National Research and Development Institute for Labor Protection "Alexandru Darabont", Bucharest, Blvd. Ghencea no. 35A, district 6, Romania
[3] Department of Electrical and Electronic Engineering, Faculty of Exact Sciences and Engineering, University "1 Decembrie 1918" Alba-Iulia, Region Alba, Romania
[4] Department of Computers and Power Engineering, University of Petrosani, 20 Universitatii str, Petrosani, România

E-mail: gbuica@protectiamuncii.ro

Abstract. The protecting electrical equipment in use are subject to various factors generated by the use, maintenance, storage and working environment, which may change the characteristics of protection against electric shock. The study presents the results of research on the behaviour over time of protective characteristics of insulating covers of material of work equipment in use, in order to determine the type and periodicity of safety tests. There were tested and evaluated safety equipment with plastic and insulating rubber covers used in operations of verifying functionality, safety and maintenance of machinery used in manufacturing industries and specific services from electric, energy and food sector.

1. Introduction

Protective equipment used in electrical installations must be manufactured, imported, marketed and used only when complying with the applicable national health and safety requirements, technical regulations and good practice codes corresponding to the level of technical progress, and the conformity assessment guidelines.

Safety and health work legislation applicable to these categories of protective equipment, defined as work equipment (WE), must cumulatively meet the safety requirements of Law 245/2004 and the following conditions:

a) these products must comply with the safety and health provisions in use established by EO no.1146 / 2006 concerning the minimum safety and health requirements for the use of work equipment by workers at work;
b) these products must guarantee compliance with the technical conditions laid down in the specific Romanian standards, applicable to the certain product categories, such as: SR EN 60900: 2013 for low voltage electrical insulating hand tools, SR EN 61111: 2010 for electrical

insulating matting, SR EN 61478/2004, SR EN 61478 / A1-2004, etc.) in order to ensure compliance with the minimum safety and health requirements in Annex no.2 of EO no.1146 / 2006.

In the applicable Romanian legislation, the minimum conditions of admissibility of these product categories are ensured if complying with the Romanian standards applicable to protective equipment, depending on use conditions and the manufacturing materials, in compliance with the instructions provided by the Technical Instructions and the provisions related to the safety and health at work, established by the Law no. 319/2006 and the EO no.1146/2006, art. 4, paragraph (1), letter i) [5,9].

The research aimed to:

- Establish the necessary technical conditions for the identification of risks, evaluation and technical verification of the protective equipment used in the electrical installations, in order to prevent the professional risks generated by the external influences (normal use environments, petroleum products and mineral oils or potentially explosive oils) ;
- Identify the necessary technical and safety requirements for the verification and technical inspection of these categories of protective equipment;
- Identify the necessary technical conditions for the identification of the electrical and mechanical risks, which allows the evaluation of the behaviour in time of insulating material characteristics of protective covers for protective equipment in use, in order to determine the types and the periodicity of carrying out the safety test at the employer level;
- Establish the necessary technical conditions for the conformity assessment in order, if necessary, to bring into conformity the protective equipment in use.

The research consisted of monitoring the behavior over time of the protective characteristics of the insulating material and insulating matting, for new protective products in order to place on the market, as well as for protective equipment in use, to determine the types and periodicity of safety tests.

2. Identified risk factors that influence the behavior of electrical insulating equipment in use

In electrical installations there are a variety of types of protection equipment in use, each of them suitable for certain areas of use and types of electrical installations (electrical insulating mats, electrical insulating hand tools, voltage detectors / indicators / testers, electrical insulating ladders, electrical insulating platforms, etc.). Live working protective equipment used in electrical installations is designed to protect both workers in work areas and unauthorized persons in the vicinity for electrical shock. These protective equipment must provide the designated safety function for use under the intended conditions of use, so that not to endanger exposed workers.

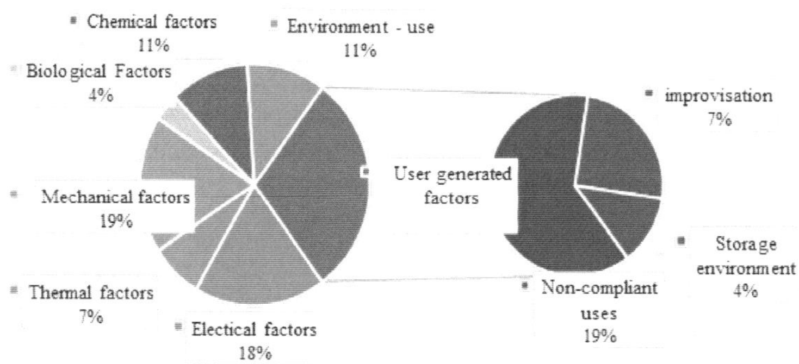

Figure 1. The identified risk share, generated by the use, maintenance, storage and environment.

Live working protective equipment used in electrical installations must guarantee, at the design and manufacturing stage, an adequate level of safety in order to comply with the occupational safety and health requirements for protection against electric shock and protection due to external influence [1,4,6,7]. During use, the protective equipment have been found to suffer from electrical, mechanical, thermal, chemical, damage that may alter their functionality, performance characteristics and, in particular, insulating characteristics.

Also, it was found that the user has a bad influence over insulating characteristics by faulty use, storage under environmental conditions that alter the protective characteristics and the lack of cleanliness.

Figure 1 shows the share of identified risk factors generated by the use, maintenance, storage and environment that can alter the protection characteristics against electric shock.

It was found that the highest share, 30%, was due to the users' risk factors (lack of knowledge about the use, storage and maintenance), and followed by the mechanical factors with 19% and the electric factors with 15% and environmental factors with 11%.

3. Assessing the safety characteristics of electrical insulating materials

By ensuring that the occupational health and safety conditions for the prevention of work accidents and occupational diseases, at the employer's level, are achieved by using safe protective equipment attesting compliance with legislative and technical provisions.

The study was carried out on specimens of new protective equipment in order to assess the conformity safety characteristics of electrical insulating materials for the purpose of placing on the market and on specimens of protective equipment in use to test the safety level, to keep them in use or to withdraw from use, if the insulating material does not provide the designated electrical insulating characteristics.

Protective equipment tested and evaluated from occupational safety and health point of view, were made of insulating materials such as insulating rubber or electrically insulating covers in one or two layers, which present mechanical strength under both normal working conditions and in extreme temperatures working conditions between -20°C and + 70°C and dielectric properties ensuring the user's protection against electric shock for live working or near live installations with a nominal voltage up to 1000 V ac. and 1500 V cc.

Many defects that may occur for the protective equipment it is of interest in the case of live working, the defects consisting in the deterioration, puncture or contamination of the insulating material cover of the protective equipment and the influence of the environment in which it is used: the influence of lubricants, raw materials from production such as water, sugar, oil and food aromas) [2,3].

In this respect, the technical and protective features against electric shock were taken into account, both the nature of the material from which the protective equipment was made and the influence of the various environmental factors in use, established by the manufacturer or the environmental conditions in which they can be used by workers in electric and food domain.

The use of certain categories of equipment to protect against electric shock is determined by the material manufacturing recipe, based on components that provide the specific characteristics of the mixture (e.g. insulating rubber designed for normal environments, is not resistant to mineral oils and other oil products).

Other categories of equipment designed to protect against electric shock can also be used in oil environments, for example in the case of insulating rubber, based on components that provide the specific characteristics, can be resistant to mineral oils and petroleum products.

The monitoring, diagnosis and evaluation of the behavior of the protective features of the electrical insulating covers of the protective equipment against electric shock has been carried out since commissioning until 10 years in use.

Also, with the provision of new protective equipment by the manufacturers, were carried out the monitoring, diagnosis and evaluation of protection characteristics of the insulating material, made on the basis of different recipes, which determine different behaviors of the technical characteristics and safety in use.

New protective equipment has been tested and evaluated, initially conditioned on immersion in water, according to the test methods set out in the standards.

During the monitoring of electrical insulating covers of equipment against electric shock, the following elements were analyzed: the existence of the component elements, the technical status (mechanical, thermal, electrical, chemical damage), the electrical insulating protection characteristic, as well as the existence and durability of the marking.

Given that some visual test may be subjective by their nature, new witnesses were retained for the purpose of comparing the results.

In case of explosion hazard, the use of insulating materials for developing explosion-proof equipment raises higher issues due to the microclimate conditions which may influence decisively their quality, with regard to the insulation's degradation, thus leading to a higher risk for the electrical equipment to break down and to generate possible failures and/or explosions [8].

4. Results and discussions

The study was carried out on specimens of new and used protective equipment in the specific activities of manufacturing industries and services specific to the electric and food domain.

Both visual examinations and destructive tests were performed to verify the dielectric characteristics and the performance level achieved by the material recipe established at the various manufacturers.

It was followed for four years, the behaviour of technical and protective characteristics of new protective equipment for use in normal and with oil environments, the samples of insulated material were stored and not sold in parallel with new specimens made by the same manufacturer at the time of the research study.

It was followed for a period of ten years, the behavior of the technical and protective characteristics of new protective equipment intended for use in normal and with oil environments, the behavior of dielectric characteristics, from three manufacturers, each one using own recipes and different components.

Electrical insulating matting are classified in accordance with the requirements of SR EN 61111: 2010 as follows: by an electrical class (Class 0, Class 1, Class 2, Class 3, Class 4) and, where applicable, by adding the suffix "C" in the electric class, meaning resistant to very low temperature. The Electrical insulating mattings were tested in accordance with the requirements of standard SR EN 61111: 2010 [10,11,13].

Figure 2 a and b shows the leakage current values for 4 batches of electrical insulating mats from three manufacturers and having different insulating rubber compound recipes. From the tests performed, the basic elements of the composition of the rubber compound influence the insulating properties of the dielectric and mechanical resistance; some of the mats (recipe 1K) not pass the tests for safety.

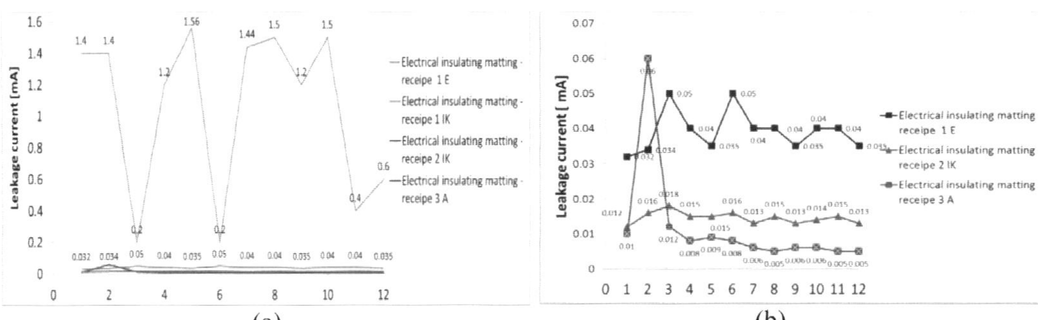

(a) (b)

Figure 2. a Leakage current - new electrical insulating mattings; b - Leakage current - new electrical insulating mattings – details for recipe 1 E, 2 IK, 3A.

Analyzing the leakage current values for the electrical insulating mat - recipe 1K, it was found that the manufacturing technology was not respected.

Figure 3 shows the values of the leakage current for one batch of 10 electrical insulating mattings from the same manufacturer, tested before and after four years of storage.

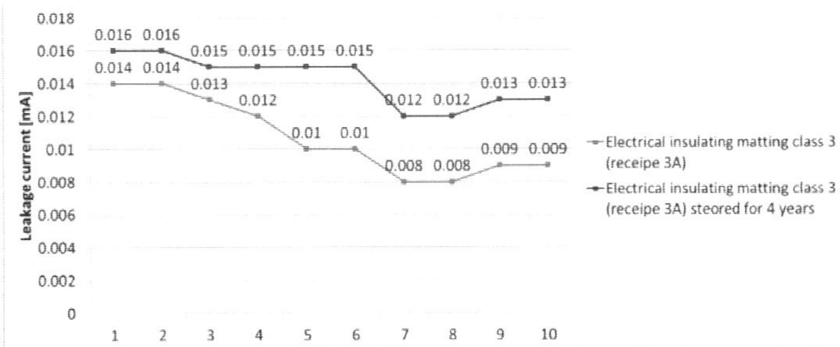

Figure 3. Leakage current new electrical insulating mattings, depending on thickness and voltage range applied for two batches made at four years.

From the performed tests, the recipe established by the manufacturer, regardless of the date of the manufacturing, in the conditions of storage in dust-free environments, guarantees the preservation of the dielectric characteristics even if the product has not been used [2,3]. The composition of the mixture established by the manufacturers, without modification of the recipe and manufacturing process, influences the dielectric characteristic and the mechanical resistance, validated be safety tests on electrical insulating matting (recipe 1E, recipe 2K and recipe 3A).

Figure 4 shows the values of leakage current for a batch of 4 electric insulating mattings having 10 m in length, in use for 10 years (electrical power station).

The test voltage for this batch of electrical insulating matting in use for 10 years (in electrical power station), was 20 kV, for electric class 2, tested under normal environmental conditions. The recipe for these electrical insulating mattings is 3A. It was found that after 10 years of use the protection characteristics were not altered. In the case of the electrical insulating mat 3, the high leakage currents were due to the mechanical stress generated by the workers' traffic in the control room of the power station.

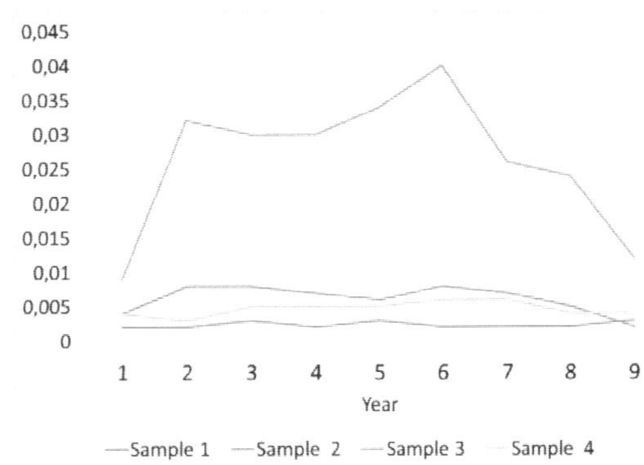

Figure 4. Leakage current - electrical insulating matting in use for 10 years.

IOP Conf. Series: Materials Science and Engineering **209** (2017) 012094 doi:10.1088/1757-899X/209/1/012094

Tests were carried out on other batches of electrical insulating mattings in use, following the behavior of electrical insulating mat in wear, contact with oils, dust, water and mold. It has been found that, over time, the insulating material loses its electrically insulating properties due to wear, contact with oils, dust, etc., the recorded results show the increase in leakage current due to the inappropriate use and maintenance of these products.

In this respect, manufacturers of electrical insulating rubber equipment must ensure that, when placed on the market for use, compliance with the applicable safety and health requirements, they must fully respect the declared, related to recipe (technical, functional, constructive and safety characteristics) related to thickness and color, including manufacturing technology [2, 3].

The study also analyzed the behavior of hand tools used by workers for five years, which fall under the category of electrically insulating tools for live working with rated voltage up to 1000Vac and 1500 V dc, equipment manufactured and designed not to endanger the user or the electrical installations, if they are properly used.

The test voltage was 10 kV, in accordance with the methods specified in SR EN 60900: 2010, for testing of new types of new various categories insulating hand tools (keys, pliers, screwdrivers, hex keys, and extension keys).

From the tests made, it resulted that the insulating material covers influences the dielectric characteristics and the mechanical resistance, the new tested insulating hand tools passing the safety tests.

Figure 5 shows the leakage current values for 6 sets of 10 new insulating hand tools. Verification of the dielectric characteristics was performed after the hand tools were mechanically tested, immersed in water and at extreme temperatures.

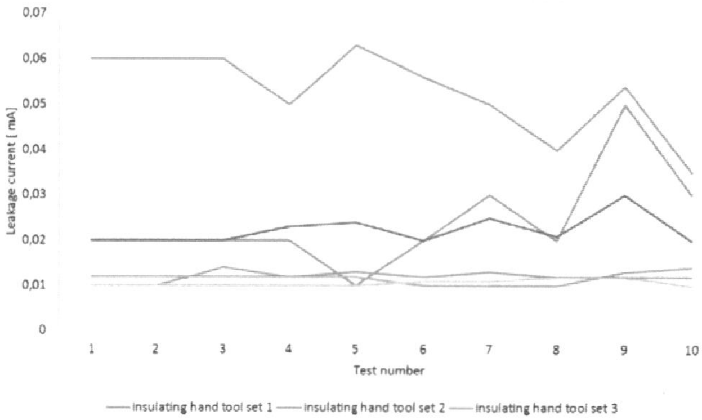

Figure 5. Leakage currents - new insulating hand tools.

The insulating sheath of hand tools must have two layers of insulating material with different colors, the insulating inner sheath being a rigid sheath with very high insulating characteristics, and the outer sheath, that protects the inner sheath, to absorb the mechanical shock, improving ease of use.

If the outer sheath is damage, the insulating hand tools must be replaced (out of use), a situation that occurs in the case of long time use.

Figure 6 shows the leakage current values for 6 sets of 10 insulating hand tools. Tool sets 1, 2, 3 and 4 were used in maintenance and set 5 and 6 in electrical industry.

From 40 samples used in the manufacturing industry, 5 samples punctured the insulation (not recorded in the chart). From batch 2, only one tool has been approved for use. Also, analyzing together with users, the biggest values of leakage currents for a series of tools, were found that the tools have been subjected to repeated mechanical shocks.

To validate the results, a batch of new tools was subjected to the mechanical shocks described by the users and verified that the materials of the insulating coatings maintain their dielectric characteristics.

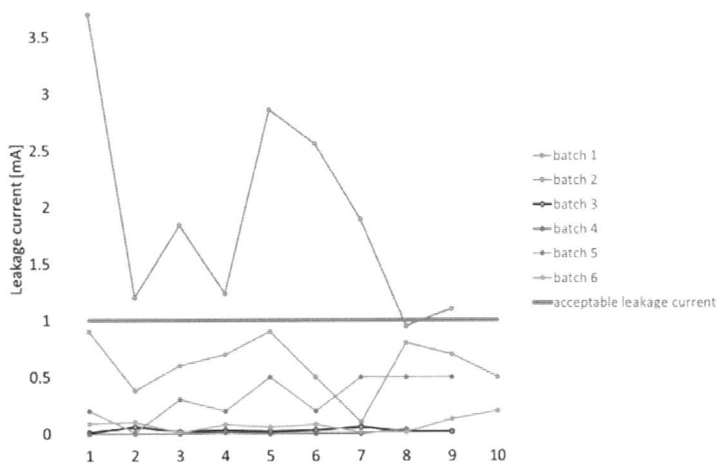

Figure 6. Leakage current values for 6 sets of 10 insulating hand tools in use.

The protective equipment must be tested by the user, both for safety status, in order to identify the defects by visual inspection (e.g. missing pieces, cracks, pinches, burrs, foreign inclusions, etc.) and the technical status, mainly related to their maintenance.

According to the provisions of Romanian legislation and technical regulations, protective equipment must be tested before use, commissioning, periodically in use and after repairs or alterations / upgrades [1, 3, 7].

It is recommended that the dielectric characteristics of the protective equipment to be tested once a year and, as the case may be, whenever it is found, after visual examination, that the insulating material of the equipment is not reliable.

The study presents the results of the research on the behavior over time of the protective characteristics for insulating material or insulating rubber of new and used work equipment, in order to determine the types and the periodicity of the safety tests, There are no explicit provisions at national level regarding the period during which these tests to be carried out.

5. Conclusions

By ensuring a safety and health level for workers, in order to prevent the risks of injury and professional disease, an objective underlined by the general objectives of the EUROPA 2020 Strategy, the research study aims to develop a verification methodology of electrical insulating, taking into account that during use, the electrical insulating equipment is subjected to various factors generated by the way of use, maintenance, storage and the working environment, which can alter the protection characteristics against electric shock.

Regular testing of protective equipment used in electrical installations is a mandatory safety requirement in order to guarantee and certify the fulfillment of the WE safety characteristics for workers both for sectors underlined by this study (electrical and food sector) and for national economy sectors.

This requirement must be ensured by the WE manufacturers used in the electrical installations, in order to notify the periodicity of carrying out the safety tests in the User Manual / Technical Instructions of the WE.

The results of periodic test are an important information in the management of occupational risks, in order to select and acquire effective equipment with high performance, long duration and to establish the most appropriate technical diagnostic and inspection/control of the safety level for protective equipment.

References

[1] Antonov A et al. 2014 *Environmental Engineering and Management Journal* **13(6)** 1361-1354.
[2] Buică G, Dobra R, Păsculescu D and Tătar A *IOP Conference Series: Materials Science and Engineering* **133** 1.
[3] Buică G 2010 Contributions of diagnostic methods of safty and health at work in electrical installations *PhD Thesis, Universitas, Petroşani* 70 – 149
[4] INCDPM 2013 Guide for the application of GD No.1146/2006 regarding the minimum safety and health requirements for the use of work equipment by workers (in Romanian) *National Institute of Research and Development for Occupational Safety* http://www.inpm.ro/files/publicatii/2013-05.01-ghid-t.pdf
[5] INCDPM 2014 Risk management within professional equipment maintenance activities to prevent work accidents and occupational diseases - a prerequisite for increasing the competitiveness of employers on the market, research study.
[6] Kurihara, Satoshi, et al. 2003 Construction of remote monitoring system for separative measurement of leakage current of outdoor insulators, Properties and Applications of Dielectric Materials *Proceedings of the 7th International Conference IEEE* **1**
[7] Law no. 319/2006 on the safe and healthy at work.
[8] Pasculescu V M, Vlasin N I et. Al. 2016 Considerations on impact resistance tests for electrical equipment used in explosive atmospheres *Proceedings of the 16th International Multidisciplinary Scientific Geoconference SGEM 2016, Section Science and Technologies in Geology, Exploration and Mining* **II** 181-188
[9] RERA Order no. 23/2011 for the approval of the Technical Norm for establishing the requirements for the execution of live working in electrical installations, code NTE 010/11/00, http://www.anre.ro/ro/energie-electrica/legislatie/norme-tehnice/normative-tehnice-energetice-requirements.
[10] SR EN 60060-1: 2011 High-voltage test techniques. Part 1: General definitions and test.
[11] SR EN 60060-2: 2011 High-voltage test techniques – Part 2: Measuring systems.
[12] SR EN 60900:2005 Live working. Hand tools for use up to 1000V a.c and 1500V d.c.
[13] SR EN 61111:2010 Live working. Electrical insulating matting.
[14] SR ISO 2230:2008 Rubber products. Guidelines for storage.

Hierarchical modeling of professional skills in the field of castings manufacture engineering

V Samuilă[1], V F Soporan[1], G Conțiu[1], S Pădurețu[1], T R Lehene[1] and M M Vescan[1]

[1]Technical University of Cluj-Napoca, Department of Environmental Engineering Entrepreneurship and Sustainable Development, Romania

vfsoporan@gmail.com

Abstract. The paper presents a method of hierarchizing professional skills in the manufacturing of molded parts (castings) by using and adapting the FAHP algorithm (Fuzzy Analitical Hierarchy Process). Assessments are made regarding the peculiarities of the professional training process, specifying the activities to be carried out and the competences necessary for their development. The contribution of the design of the method extends to the design of the hierarchy system architecture, the linguistic determination of the importance of each characteristic, the construction of the fuzzy ordering matrices for each stage of the process, the determination of the share of the characteristics for each hierarchy step and establishing the hierarchy of the characteristics taking into account the influences of the others, grouped at the level of the steps and within the global matrix. The research carried out represents the support for generating an instrument of hierarchy of professional competencies that can be used in various professional and institutional contexts. Case study on the hierarchy of professional skills in the manufacturing of molded parts engineering.
Keywords: Materials engineering, castings manufacture professional skills, hierarchy, AHP method, standard occupational curriculum.

1 Introduction

The paper is determined by the fact that the manufacturing processes of the molded parts (castings) are particularly complex and in permanent consolidation due to the mutations favored by the dynamics of the fields of use, the diversification of the casting alloys, the innovations to the technological process and the appearance of the new restrictions related to the required quality of the products and the existence of the necessary conditions to ensure ecoresponsibility. In this context, it is appreciated the need to adapt the educational and training forms to the new realities and evolution trends.

The objective of the paper is to structure, from the point of view of the interdisciplinary scientific research, a tool for modeling and hierarchizing of the competences that must be included at the educational level for the specialization related to the manufacturing of castings within the specific training in the field of materials engineering.

The problem to be solved consists in the structuring of the professional competencies in terms of the correlation of the educational plan with the requirements of the profession in the field of engineering regarding casting manufacturing.

From the methodological point of view, the form of structure analysis consists of the following steps: establishment of the hierarchy tool and elements necessary for its use; structuring the data characterizing the professional activities specific to the field; conceptualizing and implementing the hierarchy tool; formalizing input data within it; processing input data through the proposed algorithm; use of results in the curriculum structure.

In the paper are presented some contributions in the process of substantiation of the professional competences in the manufacturing of castings: the adaptation of the FAHP method for its use as a ranking tool of professional skills; analyzing and determining by institutional means the data characterizing specific professional activities; the construction and algorithmization of the hierarchy specific to the FAHP method; establishing on a statistical basis the linguistic assessments to be introduced at the level of the hierarchy; establishing under various conditions the hierarchy of professional competences required in exercising of the profession; the use of the data obtained at the level of the occupational standard and the construction of the training plan of the specialization "castings manufacture engineering".

The results of the research activity provide the basis for a methodology for determining the professional competency hierarchy in order to structure the occupational standards and the support for the construction of the educational plan, with an example in the manufacturing of the molded parts.

At the level of the specialized literature we notice the classification of the specific notions, related mainly to the traceability and knowledge management in order to use them. The construction of competency references was done mainly by using empirical methods, specific to public authorities, having inherent responsibilities in this field [1]. An analysis of the possibilities of using the FAHP algorithm in the modeling of work situations in competency characterization is presented by Farouk, Bonjour and Dulmet [2]. An important support in structuring the research theme was determined by the experience gained in the PhD thesis elaborated by Potra Florina Liliana [3] and Pop Alin Lenut [4], the discussions and lectures of Professor Potra Teodor [5] and the writing of a synthesis on the need to professionalize the environmental engineer [6]. The paper included the principles of hierarchy presented at the level of juridical-engineering convergence. [7]

At the same time, the activities carried out at the level of casting manufacturing were taken from the analysis of the domain from the perspective of the circular economy. [8]

The qualitative and quantitative assessments used were based on the report 50th Census of world casting production. [9]

In determining the characteristics, the assessments made in the portrait of the engineer at "L'Institut Mines-Telecom" were taken into account. [10]

2 Considerations on the particularities of the professional training process in the manufacture of castings

Considering the institutional validation methodology of professional training, it is appreciated that the process of professional training takes place by following several stages: establishing the activities; establishing competencies; developing the occupational standard; validation of the occupational standard; institutionalization of the educational plan; authorization of training activities; carrying out training activities, according to the established institutional framework; individual certification of training. In the context of this paper, by taking into consideration the regulatory framework, it is appreciated that the establishment of the activities carried out in the casting industry must consider the following elements: technological establishment and conditioning; technical and technological specificity of the fields served by the production of castings; technological specificity of the participants in the process.

IOP Conf. Series: Materials Science and Engineering **209** (2017) 012095 doi:10.1088/1757-899X/209/1/012095

Based on these considerations, in the first stage of the procedure, a questionnaire asked the specialists in the field of analysis to specify the activities currently carried out by engineers in this field for the different phases of the entire manufacturing process. Using the answers given in a narrow context, that of the representatives of the professional association and of the specialists in the field of the university professional training, the ones considered as being essential in the engineering praxis were retained.

In assessing the competencies required to carry out an engineering activity, these are organized according to the following groups: the group of transversal competences (CT); the group pf general professional competence (CPG) and the specific professional skills group (CPS).

The following types of questionnaires were used for determining and assessing the importance of competencies: a questionnaire for determining the importance of transversal skills in castings manufacturing; a questionnaire to determine the importance of general professional skills at the manufacturing stage of castings; a questionnaire to determine the importance of specific professional skills at the manufacturing stage of castings. The interrogated specialists were recruited from the following categories: specialists working in the field of castings manufacturing in a technical position; specialists working in the castings manufacturing in a managerial or administrative or commercial position; specialists working in the field of education and training; specialists working at the research and design level; specialists working in other areas, specifying the area. The results of the investigation were the basis for structuring the features to be used in the design of the architecture of the competency system.

3 Applying the FAHP method in the analysis of the influences on the engineering training system in the casting industry

The stages of the hierarchy method refer to the following aspects: designing the architecture of the process hierarchy system; linguistic determination of the importance of each feature in the hierarchy process; constructing fuzzy ordering matrices for each stage of hierarchy; determining the share of the characteristics for each hierarchy; determining the share of the characteristics taking into account the influences of each level of hierarchy; establishing the hierarchy of the influence of the characteristics in the process analyzed according to the calculated weight, taking into account the influence of each hierarchical stage; observations and discussions regarding the results obtained.

3.1 Designing the architecture of the skills system in castings manufacturing

The design of the architecture of the system of competencies in the manufacturing engineering of castings includes the following stages: establishment of the characteristic elements of the competencies; the establishment of levels of competence characteristics and grouping criteria for each level; assigning the characteristic elements for the first stage of hierarchy and establishing the groups of the first ranking hierarchy; grouping the structures of the first stage of hierarchy at the second stage of hierarchy; the establishment of the global hierarchy structure.

For the present case, the system to be architecturally structured comprises 81 characteristic features $(ECC)_i, i \in \{1 \dots 81\}$, the first stage of hierarchy comprises 25 groups of detailed competences $(1GCD)_j, j \in \{1 \dots 25\}$, and the second stage of hierarchy contains 3 skill groups $(2GC)_k, k \in \{1,2,3\}$, which constitute components of the global hierarchy group. The overall global hierarchy structure will be composed of the groups determined on the second level of hierarchy, i.e. the representation of transversal skills, general professional skills and specific professional skills.

Therefore, the architecture resulting in the analysis is shown in **Figure 1**.

3.2 Determining the importance of each feature within the hierarchy of professional skills specific to the manufacture of castings

Determining the importance of each characteristic within the hierarchy of professional skills specific to the manufacture of castings includes the following steps: specifying the objective and the criteria according to which the hierarchy is made. The objective of the analysis is to determine a multi-step hierarchy of characteristics, in view of establishing a hierarchy of "characteristic features of

IOP Conf. Series: Materials Science and Engineering **209** (2017) 012095 doi:10.1088/1757-899X/209/1/012095

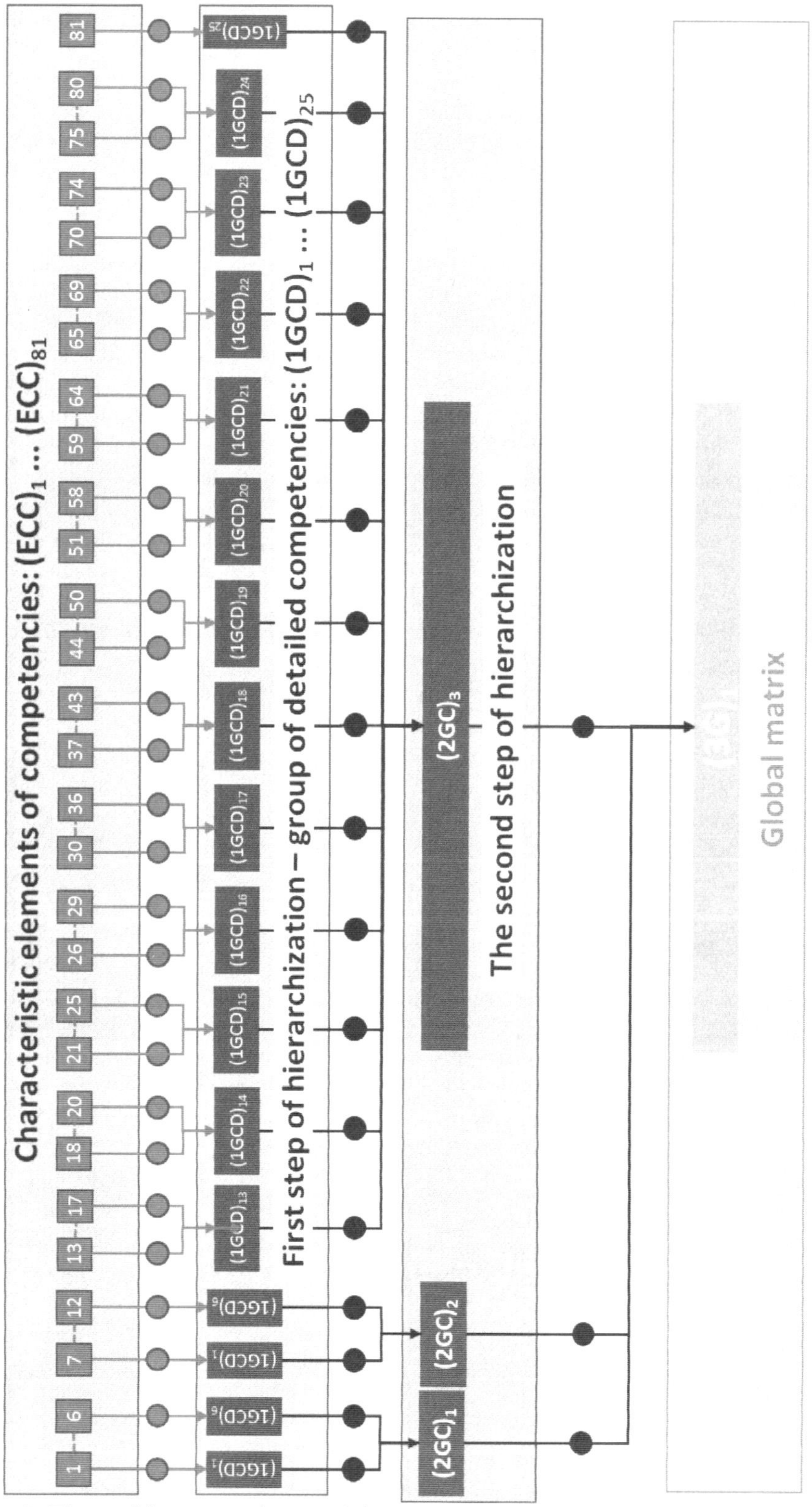

Figure 1. The architecture scheme of the system of professional competencies through FAHP

competences" in "manufacturing engineering of castings". The objective is subdued to the motivation to achieve the circular economy and the criteria represent the means to achieve it. The importance, against the set objective, is expressed in five levels of appreciation: very important, important, medium, low and very low; the establishment of a scoring system through a correlation between the previously defined levels of importance and the fuzzy numbers. For the analysis carried out, a number of 5 levels of quantification of importance and a scoring correlation through fuzzy numbers were established as follows: 11 - very important, 9 - important, 7 - medium, 5 - low and 3 - very low; the assignment of fuzzy numbers for each parameter in the analysis according to the specified criteria. For the case analized the attribution of characteristic fuzzy numbers $(ECC)_i$, detailed competence groups $(1GCD)_j$ and competence grouping $(2GC)_k$ are presented in Table 1.

Table 1 Fuzzy numbers assigned to the components of the analyzed architecture

Architecture component	Fuzzy number assigned	Architecture component	Fuzzy number assigned	Architecture component	Fuzzy number assigned	Architecture component	Fuzzy number assigned	Architecture component	Fuzzy number assigned
$(ECC)_1$	$\tilde{9}_\alpha$	$(ECC)_{23}$	$1\tilde{1}_\alpha$	$(ECC)_{45}$	$\tilde{9}_\alpha$	$(ECC)_{67}$	$1\tilde{1}_\alpha$	$(1GCD)_8$	$1\tilde{1}_\alpha$
$(ECC)_2$	$\tilde{7}_\alpha$	$(ECC)_{24}$	$1\tilde{1}_\alpha$	$(ECC)_{46}$	$\tilde{9}_\alpha$	$(ECC)_{68}$	$1\tilde{1}_\alpha$	$(1GCD)_9$	$1\tilde{1}_\alpha$
$(ECC)_3$	$\tilde{7}_\alpha$	$(ECC)_{25}$	$\tilde{9}_\alpha$	$(ECC)_{47}$	$\tilde{9}_\alpha$	$(ECC)_{69}$	$\tilde{9}_\alpha$	$(1GCD)_{10}$	$\tilde{9}_\alpha$
$(ECC)_4$	$1\tilde{1}_\alpha$	$(ECC)_{26}$	$1\tilde{1}_\alpha$	$(ECC)_{48}$	$\tilde{9}_\alpha$	$(ECC)_{70}$	$\tilde{9}_\alpha$	$(1GCD)_{11}$	$1\tilde{1}_\alpha$
$(ECC)_5$	$1\tilde{1}_\alpha$	$(ECC)_{27}$	$\tilde{9}_\alpha$	$(ECC)_{49}$	$\tilde{9}_\alpha$	$(ECC)_{71}$	$\tilde{9}_\alpha$	$(1GCD)_{12}$	$\tilde{7}_\alpha$
$(ECC)_6$	$\tilde{7}_\alpha$	$(ECC)_{28}$	$\tilde{9}_\alpha$	$(ECC)_{50}$	$1\tilde{1}_\alpha$	$(ECC)_{72}$	$\tilde{9}_\alpha$	$(1GCD)_{13}$	$\tilde{7}_\alpha$
$(ECC)_7$	$1\tilde{1}_\alpha$	$(ECC)_{29}$	$\tilde{9}_\alpha$	$(ECC)_{51}$	$\tilde{9}_\alpha$	$(ECC)_{73}$	$\tilde{9}_\alpha$	$(1GCD)_{14}$	$\tilde{9}_\alpha$
$(ECC)_8$	$1\tilde{1}_\alpha$	$(ECC)_{30}$	$1\tilde{1}_\alpha$	$(ECC)_{52}$	$\tilde{9}_\alpha$	$(ECC)_{74}$	$\tilde{9}_\alpha$	$(1GCD)_{15}$	$1\tilde{1}_\alpha$
$(ECC)_9$	$1\tilde{1}_\alpha$	$(ECC)_{31}$	$\tilde{9}_\alpha$	$(ECC)_{53}$	$\tilde{9}_\alpha$	$(ECC)_{75}$	$\tilde{9}_\alpha$	$(1GCD)_{16}$	$\tilde{5}_\alpha$
$(ECC)_{10}$	$\tilde{9}_\alpha$	$(ECC)_{32}$	$\tilde{9}_\alpha$	$(ECC)_{54}$	$\tilde{7}_\alpha$	$(ECC)_{76}$	$1\tilde{1}_\alpha$	$(1GCD)_{17}$	$1\tilde{1}_\alpha$
$(ECC)_{11}$	$1\tilde{1}_\alpha$	$(ECC)_{33}$	$1\tilde{1}_\alpha$	$(ECC)_{55}$	$1\tilde{1}_\alpha$	$(ECC)_{77}$	$1\tilde{1}_\alpha$	$(1GCD)_{18}$	$\tilde{9}_\alpha$
$(ECC)_{12}$	$\tilde{7}_\alpha$	$(ECC)_{34}$	$\tilde{9}_\alpha$	$(ECC)_{56}$	$\tilde{9}_\alpha$	$(ECC)_{78}$	$\tilde{9}_\alpha$	$(1GCD)_{19}$	$1\tilde{1}_\alpha$
$(ECC)_{13}$	$\tilde{9}_\alpha$	$(ECC)_{35}$	$\tilde{9}_\alpha$	$(ECC)_{57}$	$1\tilde{1}_\alpha$	$(ECC)_{79}$	$\tilde{9}_\alpha$	$(1GCD)_{20}$	$\tilde{9}_\alpha$
$(ECC)_{14}$	$1\tilde{1}_\alpha$	$(ECC)_{36}$	$\tilde{9}_\alpha$	$(ECC)_{58}$	$\tilde{9}_\alpha$	$(ECC)_{80}$	$\tilde{9}_\alpha$	$(1GCD)_{21}$	$\tilde{9}_\alpha$
$(ECC)_{15}$	$1\tilde{1}_\alpha$	$(ECC)_{37}$	$\tilde{9}_\alpha$	$(ECC)_{59}$	$1\tilde{1}_\alpha$	$(ECC)_{81}$	$\tilde{7}_\alpha$	$(1GCD)_{22}$	$1\tilde{1}_\alpha$
$(ECC)_{16}$	$1\tilde{1}_\alpha$	$(ECC)_{38}$	$\tilde{9}_\alpha$	$(ECC)_{60}$	$\tilde{9}_\alpha$	$(1GCD)_1$	$\tilde{9}_\alpha$	$(1GCD)_{23}$	$\tilde{7}_\alpha$
$(ECC)_{17}$	$1\tilde{1}_\alpha$	$(ECC)_{39}$	$\tilde{9}_\alpha$	$(ECC)_{61}$	$\tilde{9}_\alpha$	$(1GCD)_2$	$\tilde{7}_\alpha$	$(1GCD)_{24}$	$\tilde{9}_\alpha$
$(ECC)_{18}$	$\tilde{9}_\alpha$	$(ECC)_{40}$	$1\tilde{1}_\alpha$	$(ECC)_{62}$	$1\tilde{1}_\alpha$	$(1GCD)_3$	$\tilde{7}_\alpha$	$(1GCD)_{25}$	$\tilde{7}_\alpha$
$(ECC)_{19}$	$1\tilde{1}_\alpha$	$(ECC)_{41}$	$1\tilde{1}_\alpha$	$(ECC)_{63}$	$1\tilde{1}_\alpha$	$(1GCD)_4$	$1\tilde{1}_\alpha$	$(2GC)_1$	$\tilde{5}_\alpha$
$(ECC)_{20}$	$\tilde{9}_\alpha$	$(ECC)_{42}$	$\tilde{9}_\alpha$	$(ECC)_{64}$	$\tilde{9}_\alpha$	$(1GCD)_5$	$1\tilde{1}_\alpha$	$(2GC)_2$	$\tilde{7}_\alpha$
$(ECC)_{21}$	$\tilde{9}_\alpha$	$(ECC)_{43}$	$\tilde{9}_\alpha$	$(ECC)_{65}$	$\tilde{9}_\alpha$	$(1GCD)_6$	$\tilde{7}_\alpha$	$(2GC)_3$	$1\tilde{1}_\alpha$
$(ECC)_{22}$	$\tilde{9}_\alpha$	$(ECC)_{44}$	$\tilde{9}_\alpha$	$(ECC)_{66}$	$\tilde{9}_\alpha$	$(1GCD)_7$	$1\tilde{1}_\alpha$		

3.3 Construction of fuzzy ordination matrices for each stage of the hierarchy

In the AHP algorithm, the fuzzy ordering matrices attached to each level are created, which in order to constitute they need at least two components. The analyzed structure corresponds to the one presented in Figure 1, where 81 characteristic features of competences appear, 25 groups on the first level of ranking and 3 groups on the second level of ranking.

Expressing the ordering matrices is done by means of triangular cut fuzzy numbers.

$\tilde{3}_\alpha, \tilde{5}_\alpha, \tilde{7}_\alpha, \tilde{9}_\alpha$ and $\widetilde{11}_\alpha$, corresponding to the assessments of the importance of the feature or structure analyzed, as follows: $\tilde{3}_\alpha$ – very low importance; $\tilde{5}_\alpha$ – low importance; $\tilde{7}_\alpha$ – medium importance; $\tilde{9}_\alpha$ – big importance and $\widetilde{11}_\alpha$ – very big inportance. The use of this structure, that of the triangular cut fuzzy numbers, is motivated by the fact that this typology is considered as generating reasonable results. Therefore, they are:

$\tilde{3}_\alpha$= (1-β)(1+2α)+β(5-2α); $\tilde{5}_\alpha$= (1-β)(3+2α)+β(7-2α); $\tilde{7}_\alpha$= (1-β)(5+2α)+β(9-2α);

$\tilde{9}_\alpha$= (1-β)(7+2α)+β(11-2α); $\widetilde{11}_\alpha$= (1-β)(9+2α)+β(9-2α)

and their inverses are given by:

$$\tilde{3}_\alpha^{-1} = \frac{1-\beta}{5-2\alpha} + \frac{\beta}{1+2\alpha};\ \tilde{5}_\alpha^{-1} = \frac{1-\beta}{7-2\alpha} + \frac{\beta}{3+2\alpha};\ \tilde{7}_\alpha^{-1} = \frac{1-\beta}{9-2\alpha} + \frac{\beta}{5+2\alpha};\ \tilde{9}_\alpha^{-1} = \frac{1-\beta}{11-2\alpha} + \frac{\beta}{7+2\alpha}$$

$$\widetilde{11}_\alpha^{-1} = \frac{1-\beta}{13-2\alpha} + \frac{\beta}{9+2\alpha}$$

The problem is solved with moderate appreciation in the situation of α = 0,5 and β = 0,75; the fuzzy number and their inverse have the following values: $\tilde{3}_\alpha$= 3,5000; $\tilde{5}_\alpha$=5,5000; $\tilde{7}_\alpha$=7,5000; $\tilde{9}_\alpha$ = 9,000; $\widetilde{11}_\alpha$ = 11,5000; $\tilde{3}_\alpha^{-1}$=0,4375; $\tilde{5}_\alpha^{-1}$ = 0,2292; $\tilde{7}_\alpha^{-1}$ = 0,1563; $\tilde{9}_\alpha^{-1}$ = 0,1188; $\widetilde{11}_\alpha^{-1}$ = 0,0958.
For example, some fuzzy ordination matrices determined in the procedure used are presented:

$(MOF1)_{15} = (1GCD)_{15}$ =

	$(ECC)_{21} = \tilde{9}_\alpha$	$(ECC)_{22} = \tilde{9}_\alpha$	$(ECC)_{23} = \widetilde{11}_\alpha$	$(ECC)_{24} = \widetilde{11}_\alpha$	$(ECC)_{25} = \tilde{9}_\alpha$
$(ECC)_{21} = \tilde{9}_\alpha$	1	1	$\widetilde{11}_\alpha^{-1}$	$\widetilde{11}_\alpha^{-1}$	1
$(ECC)_{22} = \tilde{9}_\alpha$	1	1	$\widetilde{11}_\alpha^{-1}$	$\widetilde{11}_\alpha^{-1}$	1
$(ECC)_{23} = \widetilde{11}_\alpha$	$\widetilde{11}_\alpha$	$\widetilde{11}_\alpha$	1	1	$\widetilde{11}_\alpha$
$(ECC)_{24} = \widetilde{11}_\alpha$	$\widetilde{11}_\alpha$	$\widetilde{11}_\alpha$	1	1	$\widetilde{11}_\alpha$
$(ECC)_{25} = \tilde{9}_\alpha$	1	1	$\widetilde{11}_\alpha^{-1}$	$\widetilde{11}_\alpha^{-1}$	1

$(MOF1)_{18} = (1GCD)_{18} =$

	$(ECC)_{37} = \tilde{9}_\alpha$	$(ECC)_{38} = \tilde{9}_\alpha$	$(ECC)_{39} = \tilde{9}_\alpha$	$(ECC)_{40} = \widetilde{11}_\alpha$	$(ECC)_{41} = \widetilde{11}_\alpha$	$(ECC)_{42} = \tilde{9}_\alpha$	$(ECC)_{43} = \tilde{9}_\alpha$
$(ECC)_{37} = \tilde{9}_\alpha$	1	1	1	$\widetilde{11}_\alpha^{-1}$	$\widetilde{11}_\alpha^{-1}$	1	1
$(ECC)_{38} = \tilde{9}_\alpha$	1	1	1	$\widetilde{11}_\alpha^{-1}$	$\widetilde{11}_\alpha^{-1}$	1	1
$(ECC)_{39} = \tilde{9}_\alpha$	1	1	1	$\widetilde{11}_\alpha^{-1}$	$\widetilde{11}_\alpha^{-1}$	1	1
$(ECC)_{40} = 1\tilde{1}_\alpha$	$\widetilde{11}_\alpha$	$\widetilde{11}_\alpha$	$\widetilde{11}_\alpha$	1	1	$\widetilde{11}_\alpha$	$\widetilde{11}_\alpha$
$(ECC)_{41} = 1\tilde{1}_\alpha$	$\widetilde{11}_\alpha$	$\widetilde{11}_\alpha$	$\widetilde{11}_\alpha$	1	1	$\widetilde{11}_\alpha$	$\widetilde{11}_\alpha$
$(ECC)_{42} = \tilde{9}_\alpha$	1	1	1	$\widetilde{11}_\alpha^{-1}$	$\widetilde{11}_\alpha^{-1}$	1	1
$(ECC)_{43} = \tilde{9}_\alpha$	1	1	1	$\widetilde{11}_\alpha^{-1}$	$\widetilde{11}_\alpha^{-1}$	1	1

$(MOF2)_1 = (2GC)_1 =$

	$(1GCD)_1 = \tilde{9}_\alpha$	$(1GCD)_2 = \tilde{7}_\alpha$	$(1GCD)_3 = \tilde{7}_\alpha$	$(1GCD)_4 = \widetilde{11}_\alpha$	$(1GCD)_5 = \widetilde{11}_\alpha$	$(1GCD)_6 = \tilde{7}_\alpha$
$(1GCD)_1=\tilde{9}_\alpha$	1	$\tilde{9}_\alpha$	$\tilde{9}_\alpha$	$\widetilde{11}_\alpha^{-1}$	$\widetilde{11}_\alpha^{-1}$	$\tilde{9}_\alpha$
$(1GCD)_2=\tilde{7}_\alpha$	$\tilde{9}_\alpha^{-1}$	1	1	$\widetilde{11}_\alpha^{-1}$	$\widetilde{11}_\alpha^{-1}$	1
$(1GCD)_3=\tilde{7}_\alpha$	$\tilde{9}_\alpha^{-1}$	1	1	$\widetilde{11}_\alpha^{-1}$	$\widetilde{11}_\alpha^{-1}$	1
$(1GCD)_4=\widetilde{11}_\alpha$	$\widetilde{11}_\alpha$	$\widetilde{11}_\alpha$	$\widetilde{11}_\alpha$	1	1	$\widetilde{11}_\alpha$
$(1GCD)_5=\widetilde{11}_\alpha$	$\widetilde{11}_\alpha$	$\widetilde{11}_\alpha$	$\widetilde{11}_\alpha$	1	1	$\widetilde{11}_\alpha$
$(1GCD)_6=\tilde{7}_\alpha$	$\tilde{9}_\alpha^{-1}$	1	1	$\widetilde{11}_\alpha^{-1}$	$\widetilde{11}_\alpha^{-1}$	1

The global fuzzy ordering matrix has the following form:

G =

	$(2GC)_1 = \tilde{5}_\alpha$	$(2GC)_2 = \tilde{7}_\alpha$	$(2GC)_3 = \widetilde{11}_\alpha$
$(2GC)_1 = \tilde{5}_\alpha$	1	$\tilde{7}_\alpha^{-1}$	$\widetilde{11}_\alpha^{-1}$
$(2GC)_2 = \tilde{7}_\alpha$	$\tilde{7}_\alpha$	1	$\widetilde{11}_\alpha^{-1}$
$(2GC)_3 = \widetilde{11}_\alpha$	$\widetilde{11}_\alpha$	$\widetilde{11}_\alpha$	1

3.4 Determination of shares for each stage of hierarchy

After fully loading the fuzzy ordering matrices, the fuzzy FAHP algorithm continues with steps that are strictly dependent on the problem being considered. Consequently, the own value theory and the vectors corresponding to the fuzzy ordination matrices, $(MOF1)_i$ și $(MOF2)_j$. will be used. Thus, for each characteristic its share will be determined, expressed by a number between 0 and 1, which shows its importance in all the characteristics that occur in the analyzed situation. These shares are no longer independent, they are consequences of the interactions between the features. Therefore, the contribution of fuzzy theory to solving problems, such as the one analyzed, in which linguistic input data is used, and several interdependent inputs lead to a single conclusion is evident.

3.5 Determining theshares of the characteristics considering the influences of each stage of the hierarchy

The calculation of the share of a characteristic at the level of the influences of the two stages of the hierarchy and of the global one is done with the formula:

Share $(ECC)_i$ = Share $(ECC)_i^{stage\ 1}$x Share $(ECC)_i^{stage\ 2}$x Share $(ECC)_i^{global\ matrix}$

Table 2 presents the shares of the 81 characteristics by taking into account the influences of each stage of the hierarchy.

Table 2. Selected values of the shares of the characteristics by taking into account the influences of the hierarchical stages

Name of the characteristic	Symbol	Share	Position in the hierarchy	Importance/ group
CT04: The ability to work in an international context	$(ECC)_4$	0.016146171	13	CFI – Group I
CT05. The ability to self-evaluate, to make professional choices.	$(ECC)_5$	0.016146171	14	CFI – Group I
CPG1: Knowledge and understanding of a broad field of knowledge in the field of fundamental sciences as well as the ability to analyze and synthesize what is associated with them from the perspective of casting production.	$(ECC)_7$	0.033747852	8	CFI – Group I
CPG2: The ability to mobilize resources from a scientific and technical field related to the given specialty.	$(ECC)_8$	0.033747852	9	CFI – Group I
CPG3: Using engineer-specific methods and tools: identifying and solving problems, collecting and	$(ECC)_9$	0.033747852	10	CFI – Group I

IOP Conf. Series: Materials Science and Engineering **209** (2017) 012095 doi:10.1088/1757-899X/209/1/012095

interpreting data, using computer tools.				
CPG5:. Ability to take into account the environmental protection objectives.	$(ECC)_{11}$	0.033747852	11	CFI – Group I
CPS02.02: Performing a comparative study (quality, cost, availability, ...) of compatible processes.	$(ECC)_{19}$	0.025033269	12	CFI – Group I
CPS03.03: Writing technological sheets.	$(ECC)_{23}$	0.070178487	2	CFI – Group I
CPS03.04: Standardization of operations.	$(ECC)_{24}$	0.070178487	3	CFI – Group I
CPS04.01: Validation of the production system and verification tests.	$(ECC)_{26}$	0.004100396	44	CIM – Group I
CPS05.01: Analyzing the real load of the production sector	$(ECC)_{30}$	0.064971121	6	CFI – Group I
CPS07.07: Identification of faults and installations anomalies and coordination of preventive maintenance actions of the sector.	$(ECC)_{50}$	0.103192165	1	CFI – Group I
CPS05.04: Sorting the tasks of different production sectors by optimizing all criteria and taking into account different constraints.	$(ECC)_{33}$	0.064971121	7	CFI – Group I
CPS10.03: Identifying and implementing prevention and protection means and procedures for personal and collective safety.	$(ECC)_{67}$	0.070178487	4	CFI – Group I
CPS10.04: Ensuring and implementing measures related to safety and environmental protection.	$(ECC)_{68}$	0.070178487	5	CFI – Group I
CPS07.01: Team appointment and organization of the service.	$(ECC)_{44}$	0.009409941	21	CI – Group II
CPS07.02: Adjusting the distribution of staff according to the tasks to be performed, their skills, qualifications and availability.	$(ECC)_{45}$	0.009409941	22	CI – Group II
CPS07.03: Decipher the existing operation indicators.	$(ECC)_{46}$	0.009409941	23	CI – Group II
CPG4: Using experimental methods in a context of research and innovation.	$(ECC)_{10}$	0.005884519	42	CIM – Group III
CT01: Developing the spirit of entrepreneurship and the ability to take into account economic objectives.	$(ECC)_{1}$	0.00456452	43	CIM – Group III
CPS08.01: Performing control of problems with deadlines, quality, safety and environmental protection.	$(ECC)_{51}$	0.001349475	57	CIS – Group IV
CPS08.02: Performing periodic inspection of machinery.	$(ECC)_{52}$	0.001349475	58	CIS – Group IV
CT02: Ability to take into account social and ethical objectives.	$(ECC)_{2}$	0.000893768	71	CIFS – Group V
CTAC Competence other than those previously mentioned in the transversal skills	$(ECC)_{6}$	0.000893768	73	CIFS – Group V

The influence of characteristics on the skills required for the “casting engineer” profession is shown in Figure 2.

The influence of characteristics on the skills required for the “casting engineer” profession is shown in Figure 2. Another analysis that can be done by evaluating the global shares is the one related to mentioning the order of importance of the groups formed on the first stage of hierarchy. The overall share of the group is determined by summing the global shares of the features present within it. Using this procedure, Figure 3 shows the global shares of the groups present on the first stage of hierarchy.

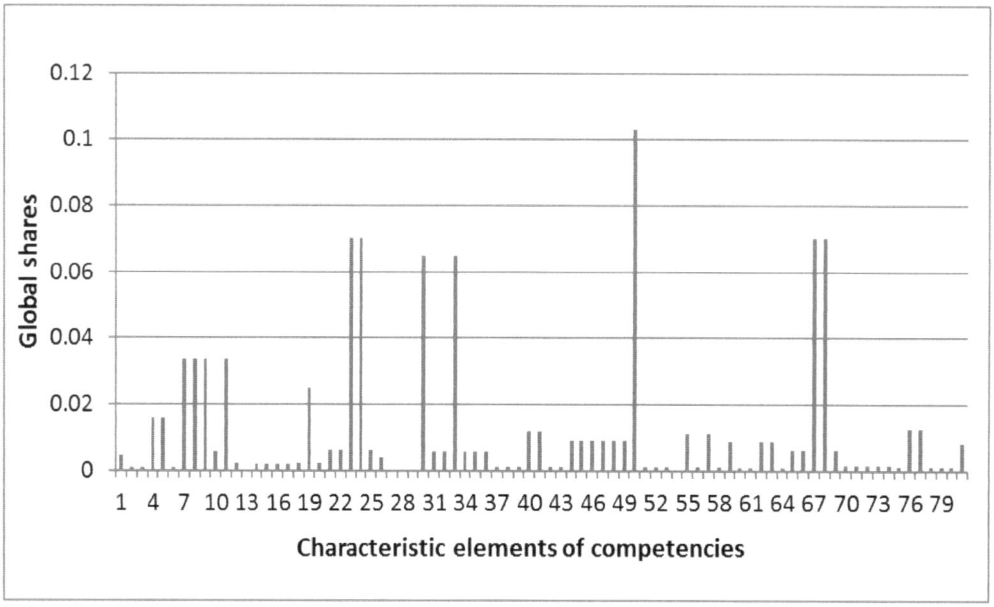

Figure 2. Global shares values and characteristic elements of competencies

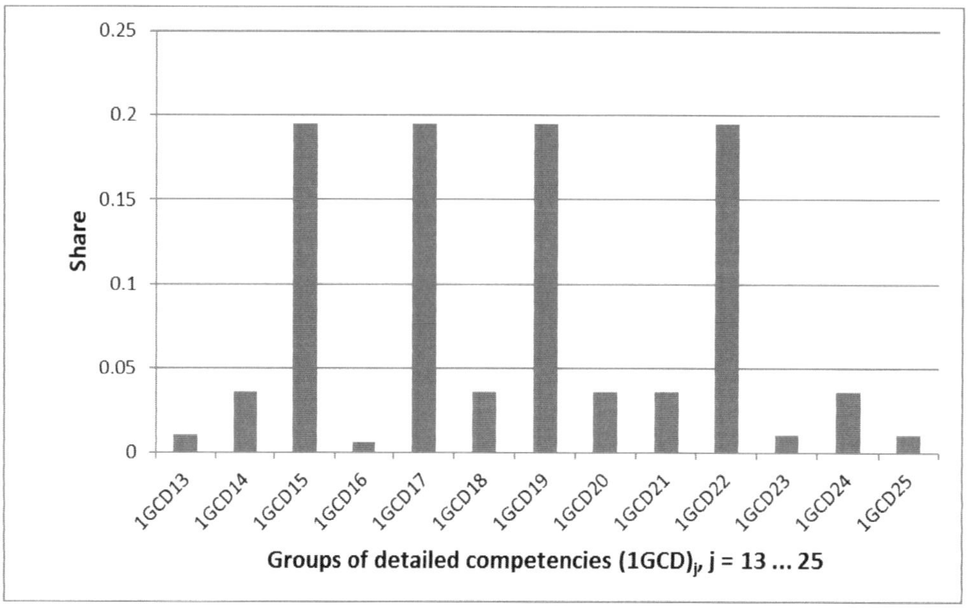

Figure 3. Shares of the groups on the first stage of hierarchization

4 Interpretation of the results obtained

Considering the results presented in Table 2, we find that from the point of view of the value of the global share, determined mathematically, we have to deal with five groups of importance (by returning to the linguistic appreciation):

- The group of very important characteristics (CFI) – Group I;
- The group of important characteristics (CI) – Group II;
- The group of medium importance characteristics (CIM) – Group III;

- The group of low importance characteristics (CIS) – Group IV;
- The group of very low importance characteristics (CIFS) – Group V.

In the group of very important features, of which 14 characteristics are part, the following are ranked according to the hierarchy criteria of the second step:

- 2 characteristics of the transversal competences (CT4 and CT5) of the 6 present in the analysis, with a share of 14.28% of the total of the very important characteristics;
- 4 characteristics of the general professional competences class (CPG1, CPG2, CPG3 and CPG5) of the 6 present in the analysis, with a share of 28.56% of the very important characteristics;
- 8 characteristics of the specific professional competences class (CPS03.03, CPS03.04, CPS02.02, CPS05.01, CPS05.04, CPS07.07, CPS10.03, CPS10.04) of the 69 present in the analysis, with a share of 57.12% of all the very important features.

From the previous presentation, it is noted that within the very important features, the essential role belongs to the transversal competences (CT) and general professional ones (GPs) their share exceeds 42.84 %. Therefore, it is appreciated that in the curriculum the disciplines that support the mentioned competencies must be properly represented.

In an expression of linguistic assessment, the priority competences to be insisted on, in the educational processes resulting from the analysis, have the following expression: CT04: The ability to work in an international context; CT05. The ability to self-evaluate, to make professional choices; CPG1: Knowledge and understanding of a broad field of knowledge in the field of fundamental sciences, as well as the ability to analyze and synthesize what is associated with them from the perspective of manufacturing molded parts; CPG3: Using engineer-specific methods and tools: identifying and solving problems, collecting and interpreting data, using computer tools; CPG5: Ability to consider environmental objectives; CPS03.03: Writing technological sheets; CPS03.04: Standardization of operations; CPS02.02: Performing a comparative study (quality, cost, availability, ...) of compatible processes; CPS05.01: Analyzing the real load of the production sector; CPS05.04: Sorting the tasks of different production sectors by optimizing all criteria and taking into account different constraints; CPS07.07: Identification of installation faults and anomalies and coordination of preventive maintenance actions of the sector; CPS10.03: Identifying and implementing means and procedures for prevention and protection of personal and collective security; CPS10.04: Ensuring and implementing measures related to safety and the protection of the environment.

From the point of view of the specific professional competencies, their order of priority, determined according to the methodology developed in the research, shows as follows: 1. CPS07.07: Identification of defects and anomalies of the installations and coordination of preventive maintenance actions of the sector; 2. CPS03.03: Writing technological sheets; 3. CPS03.04: Normalization of operations; 4. CPS10.03: Identification and implementation of prevention and protection means and procedures for personal and collective safety; 5. CPS10.04: Ensuring and implementing measures related to safety and environmental protection; 6. CPS05.01: Analyzing the real load of the production sector; 7. CPS05.04: Sorting the tasks of the different production sectors by optimizing all the criteria and taking into account different constraints; 8. CPS02.02: Performing a comparative study (quality, cost, availability, ...) of compatible processes.

5 Conclusions

In conclusion of the presentation, which reflects the researches undertaken, it is appreciated that starting from the determination the hierarchy of professional skills in the manufacturing engineering of castings it is possible to structure a tool that puts into operation the following methodological steps: 1. Establishing the activities that are carried out in the field to be analyzed; 2. Determining the professional competencies that are necessary for carrying out the analyzed activities by means of specialized questionnaires addressed to the specialists present on the different levels of action of the field; 3. Studying and analyzing the institutional documents regulating the vocational training in the analyzed field (global documents, European documents, national documents); 4. Experiences of good practice in the field of training of the analyzed domain (global situation, European situation, national situation); 5. Determining the characteristics of the competencies that will be subjected to the hierarchy process; 6. Establishing the general criteria for assessing the characteristics; 7. Establishment of the linguistic assessment scale or levels; 8. Linguistic assessment of the characteristics according to the general criterion or the established criteria; 9. Establishing and justifying criteria for the allocation of attributes within the hierarchical stages and the formalization of classes through the characteristics; 10. Linguistic appreciation of classes established according to the general criterion; 11. Establishment and justification of criteria for the allocation of classes established on the higher ranking hierarchy; 12. The linguistic appreciation of the structures built on the upper levels according to the established general criterion; 13. Continuing the allocation process up to the level of the last hierarchy; 14. Linguistic assessment of the last formations grouped according to the general criterion or established criteria; 15. Allocation of fuzzy numbers according to linguistic assessments at the level of architectural features and classes. 16. Formation of the fuzzy ordering matrices at the level of the classes constituted by comparing the fuzzy numbers assigned to the characteristics; 17. Calculation of local shares on established hierarchy levels; 18. Calculation of the global shares for each component characteristic; 19. Determination of the importance groups of the characteristics analyzed according to the decreasing values of the global shares: the group of very important characteristics; the group of important characteristics; the group of medium importance characteristics; the group of low importance characteristics; the group of very low importance characteristics; 20. Interpretation of results through the hierarchy of competences required to conduct a profession.

References

[1] CNFPT 2001 *Référentiels d'emploi – activités – compétences, Guide méthodologique* (Paris: Centre national de la fonction publique territoriale, direction du développement des compétences territoriales Paris)

[2] Belkadi F, Bonjour E and Dulmet M 2009 De la modelisation des situations de travail a la caracterisation des competences: Une approche par logique floue, *Journal Europeen des Systemes Automatises (JESA) Lavoisier* **43** 35-71

[3] Ureche F L 2015 *Cercetări privind optimizarea parametrilor procesului tehnologic de prelucrare a materialelor prin electroeroziune electrică* (Cluj-Napoca: Universitatea Tehnică din Cluj-Napoca Teză de doctorat)

[4] Pop A L 2015 *Contribuții asupra analizei convergențelor tehnico-juridice la nivelul gestiunii deșeurilor municipale din perspectiva economiei circulare* (Cluj-Napoca: Universitatea Tehnică din Cluj-Napoca Teză de doctorat)

[5] Potra T 2009 *Noțiuni preliminar asupra elementelor de logică fuzzy și algoritmul FAHP Notițe de curs* (Cluj-Napoca: Universitatea Tehnică din Cluj-Napoca)

[6] Soporan V F 2016 Nevoia de profesionalizare instituționalizată în ingineria mediului *Univers ingineresc* **7** 4 – 6

[7] Pop A L and Soporan V F 2015 Architecture hierarchy systems in engineering and legal convergences arhitectura sistemelor de ierarhizare in convergențele juridico-inginerești *Acta Technica Napocensis - Series: Environmental Engineering and Sustainable Development Entrepreneurship – EESDE* **2** 43 – 48

[8] Soporan V F, Crișan M, Lehene T and Pop A L, 2016 *IOP Conf. Ser.: Mater. Sci. Eng.* **133** 012063

[9] 2016 50TH Census of world casting production *Modern Casting* **12** 25 – 29.

[10] SocioVision 2014 *L'Institut Mines-Télécom présente Portrait de l'ingenieur 2030* (Paris: L'Institut Mines-Télécom)

IOP Conf. Series: Materials Science and Engineering **209** (2017) 012096 doi:10.1088/1757-899X/209/1/012096

Engineering of the institutionalization of the circular economy at the level of casting production

M M Vescan[1], V F Soporan[1], D M Crişan[1], T R Lehene[1], S Pădureţu[1] and V Samuila[1]

[1]Technical University of Cluj-Napoca, Department of Environmental Engineering Entrepreneurship and Sustainable Development, Romania

vfsoporan@gmail.com

Abstract. This paper is motivated by the necessity of introducing the principles of circular economy at the level of different social – economic activities, and from this point of view one of the fields with a special potential is that of the manufacture of castings. Objective: to connect to the organizing and application of the methodology of the circular economy principles. The proposed method is an innovating one, being connected to the use of institutionalization engineering.

Formulating the subject: The subject formulated to be solved aims at the introduction of new approaches, defined through institutionalization engineering, which proposes to set the correlation of actions between the specifics of the circular economy and the specific elements of the manufacture of castings. Research method: An institutional structuring operation was imposed for the optimization of the research method, in which different versions interact at the following levels: the level of public policies, the level of the regulatory framework, the level of technical solutions and the level of financing solutions and financial instruments. The determination of the optimal solution established in a dynamic context, favorable for the requirements of the different actors present within the process, appeals to the elements of critical thinking, specific for the engineer's actions. Achievement of the research activity: The research activity structures a methodology of quantifying the contributions of each stage of the manufacturing process for castings at the fulfilling of the specific conditions of the circular economy, indicating the critical areas of action for more efficient actions of the circular economy, according to the market economy requirements, where there is a potential of implementing the technical solutions by quantizing the financial solutions and the opportunity of using the financial instruments. The major contribution of the research: The proposed methodology, with examples at the level of castings manufacture, sets the bases of a new field of action of the engineering thinking, namely, that of circular economy institutionalization functioning. Conclusions of the research activity: The proposed methodology represents the bases of establishing a new instrument of action at the level of institutionalized functioning of the circular economy.

1 Introduction

The paper is motivated by the necessity of introducing the principles of the circular economy at the level of the different socio-economic activities, and one of the areas with a great potential in this respect is the production of molded parts. The objective is related to the structure of a methodology for applying the principles of the circular economy, and the proposed method is a novel one, being related to the use of institutionalization engineering. The problem to be solved is to introduce a new approach, defined by the institutionalization engineering, which aims to establish the correlation of the actions between the specifics of the circular economy and the specific elements of the casting production. In the optimization of the methodology of analysis, a structure of institutional functioning has been imposed, in which the following layers interact in different variants: the level of public policies, the level of the regulatory framework, the level of the technical solutions and the level of financing solutions and the financial instruments, the schematic presentation is made in Figure 1.

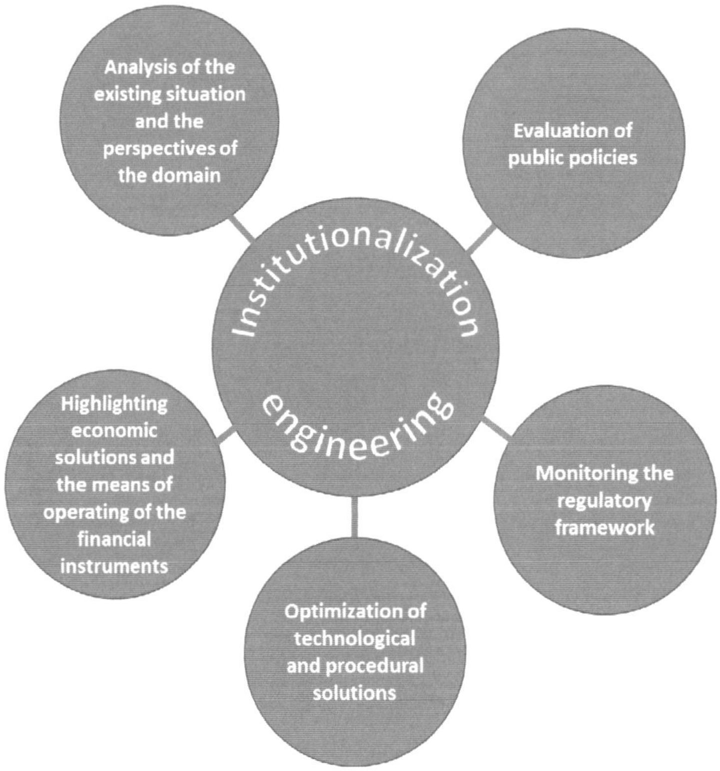

Figure 1. Schematic of the engineering actions of the institutionalization of casting production

The determination of the optimal solution set in a dynamic context, favorable to the requirements of the various actors present in the process, calls for critical thinking elements specific to the engineering actions. The research activity structures a methodology to quantify the contributions of each stage of the casting process in order to meet the conditions specific to the functioning of the circular economy, indicating the critical areas for action to streamline the circular economy in line with the requirements of the market economy where there is potential for implementation. Technical solutions are assessed by quantifying financial solutions and the opportunity to use financial instruments. The proposed methodology, exemplified in the manufacturing of molded parts, sets the basis for a new field of action of engineering thinking, that of the engineering of the institutionalization of the circular economy. The openness offered by the proposed methodology is the basis for the creation of a new instrument of action at the level of the institutionalized functioning of the circular economy. From the examination of the specialized literature we can see the existence of some concerns in the field of reporting the production of molded parts to the requirements of the circular economy. It is worth noting, in this context, the methodology for assessing the tendencies in the manufacturing of molded parts from the perspective of the circular economy [1]. At the same time, at the level of the practical actions, the report of studies and researches on the consolidation of the circular economy through projects of increased exploitation of the industrial products by the participation of six Centers Techniques Industriels from France, which reminds that starting within the year 1997, started the incorporation of the spent foundry mixture into building materials [2]. Institutionally, there is a tendency to materialize procedures for determining the material cycle in an economy [3]. Examples of economic and social structures are presented at the level of the trajectories of enterprises in achieving the objectives of the circular economy [4]. An institutional synthesis of the institutional developments for the promotion of the circular economy highlights the constituent elements of the circular economy that have the vocation to be directly integrated into government actions [5]. Also, the circular economy institutionalization is characterized by the annual report of the circular economy in the competition for resources [6] and the typological construction of the regional programs of the circular economy for mobilizing resources and minimizing the losses through the innovative regional economy [7]. Historically, given the authors' concerns and concerns, we want to present two Romanian landmarks related to certain stages of the economy. Thus, in the "Wire Industry Survey SA" published in 1938, it is formulated in a modem manner for those times, comparable to those present at present, “If iron and steel have gained more and more utilization, facilitating light and solid constructions, while the spread of known warehouses to date is limited. The need for iron and steel increases to a greater extent than the reuse of old iron can help us. Other means of saving have been sought, being found in: suitable methods for executing a thing with minimal material expense; Constructive forms appropriate to the demands and qualities of the materials; Raising steel by adding other metals to increase its strength; Protection against acids and humidity. ... The exhibition (Dusseldorf Exhibition of 1937) aims to show that old things have an appreciable economic value, and they can serve to manufacture new objects, and thus have a raw material character. ... Thus, in the case of scrap iron, it shows how to sort, cut large pieces, press in packs, which are then taken to the high furnace where they are mixed with iron ore to be rebuilt in cast iron or sent to the Martin or Electric steel for the production of steel, saving up to 80 % of the ore. The process is also applied for a long time and in our country (Romania), with the electric furnace in operation at the Turda Plain plant, it brings its contribution, using all those scrap iron that remain after the lamination of the wires and its other sections." [8]. At theoretical level [9] This prefigures smartly and realistically "the need to establish the institutional framework given by the defined convergences of public policies, juristic regulations, technical solutions and economic and financial ones in the sense of ensuring sustainable development.” [10].

2 General considerations regarding the institutionalization of the circular economy.

Below are some elements of Professor Vasile Soporan's synthesis on the institutionalization of the circular economy [11]. It presents the following elements of the analysis made for the institutionalization of the circular economy: triggering mechanisms for the development of the circular economy; Motivation for "circular economy institutionalization"; The values at which the circular economy is reported; The conditionalities envisaged from the perspective of the circular economy; The concept of circular economy; The fields of action of the circular economy; Specific approaches to "circular economy institutionalization"; The questions necessary to substantiate the "engineering of the circular economy institutionalization".

The triggering of the mechanisms for the development of the circular economy took into account the following: institutional reporting to the continuity of the three states: the past, the present and the future; The existence or perception of discontinuities by their emphasis by Marc Halevy [12] presents the following forms: ecological discontinuities, informational discontinuities, organizational discontinuities, economic discontinuities and discontinuities of the meaning of life; Individual awareness of needs and development materializes in the community space, in its specific forms, at the philosophical, theological-religious, political, economic level, from the perspective of the rejection of the order of nature triggered by the political order, the social order, the religious order and the military order, Some economic-social structure, trying to optimize the balance of division and hierarchy between the close and the enemy.

The motivation for "circular economy institutionalization" is based on the functioning of the whole by the synergistic cumulation of actions designed and developed cumulatively at the level of actions of four levels already mentioned in the presentation.

Circular economy reporting takes into account that the material, product or service has an expression at the following values: the functional value; material value, energy value, environmental value and social value. Correct assessment should take into account the considerations considered in the design of a shaped material, product or service: conditionalities of functionality, safety conditions, machining conditions, environmental conditioning, durability conditioning and conditioning disposal at the end of the life cycle.

The concept of circular economy used is based on Vincez Aurez et Laurent Georgeault's wording: "The circular economy is a principle of economic organization that aims at systematically reducing the amount of raw materials and energy on the whole of a product or service life cycle and at all levels of organization of a society, in order to ensure the protection of biodiversity and the proper development of individuals. At an institutional level, the circular economy consists of three areas of action with seven orientation axes: D1 - waste management (A11 - waste recycling axis); D2 - economic supply field (A21 - sustainable provisioning axis, A22 – eco-concept axis, A23 - industrial and territorial ecology axis, A24 - functional economy axis); D3 - the daily consumption area (A31 - the length of service life, A32 - the responsible consumption generation axis) "[13].

The specific approaches to the "engineering of the circular economy institutionalization" are proposed to be carried out as follows: addressing the issue (specific status and issues, state of the public policies, regulatory framework, technical and procedural solutions, financial solutions and financial instruments); Institutionalization of processes (defining the actors of institutionally regulated processes, institutions, institutionalized instruments, institutional actions, taxation and institutionalized financing); The technical solution of the problem; Financial settlement of the problem.

3 Particularities of the production of molded parts from the perspective of the circular economy

From the perspective of the circular economy, the particularities of casting production concern the following aspects, presented in the seminar "Situation of the production of molded parts in Romania and the needs for structuring the vocational training" [14]: the evolution of production of castings at global level (2000-2015); The structure of the production of molded parts, with specifications in the year 2015; The economic status of casting producers; The specific investment made by molded cast makers, exemplifying the situation in 2015; The average production of industrial establishments, with examples at the level of the first 10 producers of castings and Romania; The share of the casting production in the specific technologies for putting into operation the material for machine building; Global institutional conclusions on the issues faced by molded cast producers; The motivation for the sustainable production of molded parts from the perspective of circular economy; The conditionality of the research activities at the manufacturing stage of castings.

Evolution of casting production is presented using the World Casting Production annual reports published by WFO and published in December of each year in Modern Casting [15] and a cumulative synthetic report of WFO [17].

From the analysis of the mentioned WFO reports, the global institutional conclusions on the issues faced by the producers of castings in the last half century show: the transition from the "linear economy" to the "circular economy"; The emergence of conditions for the use of material resources; Conditions for the use of energy resources; Imposing conditioning on health and the environment; The need for intelligent use of complex computerized processes or adaptation to the requirements of the "digital agenda"; The influence of sophisticated complex processes at the global market (instability by de-globalization); Diversification of use areas for castings made by casting; Increasing characteristics and requirements in operation; Increasing competition from materials and manufacturing technologies for shaped products.

The motivation for the sustainable production of castings from the perspective of the cyclical economy is determined by the quantitative importance of molded parts, which exceeded 100 million tons, the significant consumption of resources needed for their production, the impact processes generate on the environment and the fact that hazardous substances are used to a significant extent, and generated waste also falls into this category. From the perspective of the presented ones, it is clear that this branch of industry is suitable for analysis from the perspective of the circular economy. The position is supported by the fact that the serviced fields are of a special complexity, aiming at a deployment from the automotive industry to nuclear power. Therefore, the European document "Towards a Circular Economy: zero waste program for Europe" [15] finds an appropriate field of application in the casting industry, at least through the design and innovation approaches for a circular economy: "Reducing the amount of material needed to perform a particular service (easy load); Extending product life (sustainability); Reducing the use of energy and materials in the production and use phases (efficiency); Reducing the use of hazardous materials or difficult to recycle in products and production processes (substitution); The creation of outlets for secondary (recycled) raw materials (based on standard, public procurement, etc.); Designing products that are easier to maintain, repair, upgrade, re-fabricate or recycle (eco-design); The development of services that consumers need in this regard (maintenance / repair services etc); Stimulating and supporting waste reduction and quality separation activities by consumers; Stimulating separation and collection systems that minimize the costs of recycling and reuse; Facilitating the grouping of activities to prevent the transformation of by-products (industrial symbiosis) into waste and encouraging the extension and improvement of the consumer choice through rental, loan or exchange of services as an alternative to the holding of products, At the same time, the interests of consumers in terms of costs, protection, information, contractual clauses, insurance aspects etc. "[16].

It is therefore appreciated that the research activities at the manufacturing of casting parts also concern the issue of the development of the circular economy through the conditions that impinge upon it. Thus, the field of casting production, even if it is one of the oldest technological processes for putting metal materials in their functional form, and not only them, requires a permanent connection to the conditions of the socio-economic activities. The major needs of innovation in this area can be grouped into two levels: the level of technical and technological innovation and the level of change in the way markets are approached. From a pragmatic point of view, at the level of actors present in the production of cast pieces, the rules of survival and development make for the substantial collaborative dynamics between the information and the actions carried out.

Starting from the objective of the paper and in view of the activities carried out within the Technical University of Cluj-Napoca, in the last years, from a multidisciplinary perspective, we consider the following topics addressed in the context of the eco-responsibility needs: research on the design and functioning of “Collaborative Entrepreneurial Platform” in casting industry as a virtual action tool; production of molded parts in terms of digital communication requirements in the virtual space; research on the development of the methodology for assessing the tendencies in the production of castings from the perspective of the circular economy; contributions to the determination of the “unnecessary part” generated in the manufacturing processes of the molded parts and determination of the solutions for increasing its usefulness from the perspective of the circular economy; research on the optimization and digitization of institutional waste audit procedures under the conditions of casting production; research on the use of "Life Cycle Analysis - ACV" in the hierarchy of manufacturing processes of molded parts at the level of provision of conditions related to the provision of sustainable development; researches on the use of "critical thinking" in the manufacturing of molded parts; optimization of engineering training systems in the manufacturing of molded parts through an AHP (Analytical Hierarchy Process) hierarchy at the level of transversal and professional skills; rewriting casting manufacturing processes from the perspective of using fast prototyping and 3D printing capabilities. Grouped in one major theme, the issue mentioned at the research level could be titled: "Engineering of the Institutionalization of the Economy in the Manufacturing of Molded Pieces".

4 Considerations on the Model Approach for the Manufacturing of Castings from the Circular Economy perspective

Following the analysis of the casting approach variants, a method has been used that is structured on the schematic representations of Figure 1.

The criteria for evaluating public policies and the regulatory framework from a cyclical economy perspective take into account the incidents that can be created at the economic, environmental, social and administrative levels. The economic incidence is based on the following criteria: the financial viability of measures to promote the circular economy; the possibility of economic development; the possibility of partnering. Impacts related to environmental protection are assessed in terms of the following criteria: the possibility of reducing the environmental impact; the possibility of waste reduction at source; creating a global environmental advantage. Social incidents are appreciated by the following criteria: acceptance of solutions offered to the public and stakeholders; the absence of risks with negative side-effects other than those related to environmental protection. Administrative impacts are measurable according to the following criteria: conformity assurance; ease of office; ease of administration; the measurable character of the processes through administrative formulas.

The evaluation of economic instruments that promote the circular economy at the manufacturing stage of castings is based on a set of indicators of appreciation of the effect of production on the environment: energy efficiency; the content of recycled materials at the process level; recyclability of the product at the end of its life cycle; water resource economy; conserving resources through process deployment; GHG emission at production level; measures to prevent waste in the manufacturing process; the percentage of renewable materials used in manufacturing; the negative effects of the manufacturing process on workers, animals, plants, air, water and soil; the content of toxic material in the casting process; the quantities of packaging used in the manufacturing process; transport distances carried out in the manufacturing and dispatching of cast parts; facilitating repairs and refurbishment of castings; managing the castings at the end of the life cycle. From the perspective of the significance of the indicators presented, the economic instruments used for the promotion of the circular economy have the following expression: green (ecological) supply; additional usage fees for the resources used; consignment systems for technological infrastructure usable in manufacturing processes; waste management incentive systems for the production of molded parts from a circular economy perspective; recycling of the solid waste generated; application of the principle the user pays; obtaining credits for greenhouse gas reduction situations; application of the wider producer responsibility.

The analysis of the technological and procedural processes used from the perspective of the circular economy and the functioning of the economic instruments for the promotion of the circular economy represent the most important area in the evaluation of the resources that can be used to achieve the objectives of the circular economy. The stages of the manufacturing process under consideration are: the analysis of the specific information that influences the field (market analysis, the analysis of the regulations influencing the production of castings, the analysis of the processes evolution and of the technological infrastructure); analysis of production preparation from the point of view of the available technologies and technological infrastructure (analysis of constructive design, analysis of technological design, analysis of the technological manufacturing preparation); analysis of the casting process (analysis of direct activities, analysis of indirect activities); analysis of the results of the manufacturing process (the stage of the analysis of the useful part of the manufacturing process - the cast parts delivered to the unproductive parts resulting from the manufacturing process, the unpleasant parts resulting from the manufacturing process); life cycle exploitation and disposal (re-use of molded parts, recycling of waste parts, waste management); analysis of other phases that contribute to casting (analysis of the production and distribution of utilities, analysis of other phases of the process); process analysis at the level of the whole from the perspective of compliance with environmental conditions and sustainable development (pollution intensity, quantification of specific energy consumption per unit of product,

IOP Conf. Series: Materials Science and Engineering **209** (2017) 012096 doi:10.1088/1757-899X/209/1/012096

other consumption per product unit, waste generated per unit of unit, consumption of recycled material introduced on the production process at the product unit, consumption of recycled material introduced into the production process on the product stream, energy consumption recovered in the casting process). Within this activity, the stages of the manufacturing process of casting together with the measures envisaged to achieve the objectives of the circular economy are presented in Table 1.

Table 1. The stages of the casting process and the measures projected to achieve the circular economy (selection of points in the analysis carried out).

No.	Step or stage of the technological process	Specific measures projected in terms of circular economy
1	1. Analysis of influencing specific information field; 1.1. Market analysis and influencing factors; 1.1.1. Analyze developments beneficiaries castings	Quantification main user industry castings and competing technological procedures, such as mechanical construction welded or processed commodity plastics.
2	1. Analysis of influencing specific information field; 1.3. Analysis of the evolution of processes and technological infrastructure; 1.3.1. Making process analysis and modification	Reducing energy consumption and materials in the process of developing specific product unit. Increasing the use of recycled materials specific process for obtaining the castings. Avoiding placing in the batch composition of the elements that hinder the recycling process of castings at the end of their life cycle. Avoiding substances that generate hazardous waste generated at the level that causes a reduction in the possible use specific circular economy.
3	1. Analysis of influencing specific information field; 1.3. Analysis of the evolution of processes and technological infrastructure;1.3.3. Analysis of the casting process	Loss of material and energy decrease occurring during the casting process through the use of technology and infrastructure. Organization of the technological process by developing opportunities for recovery and reintroduction technological flow generated losses.
4	2. Analysis of manufacturing preparation in terms of technology and technological infrastructure available; 2.1. constructive concept analysis; 2.1.1. Management tools for computer aided design	Development information tools that generate constructive forms-based functional and management of databases on constructive changes made in technological contexts data.
5	2. Analysis of manufacturing preparation in terms of technology and technological infrastructure available; 2.2. Analysis of technological design; 2.2.1. Solidification by correlation analysis and optimization feeder system, the cooling of constructive typologies designed	Development of software for computer aided design of solidification process, which, based on models created to perform a specific consumption optimization in the construction of the assembly molding technology.
6	2. Analysis of manufacturing preparation in terms of technology and technological infrastructure available; 2.2. Analysis of technological design; 2.2.2. Analysis of the cooling process and design optimization from the point of view of stresses which are formed in the casting pieces	Development of software for computer aided design of castings cooling process, which, based on models created to achieve constructive forms of optimization in order to reduce the destructive effects of stress that develops in the alloy mass.
7	2. Analysis of manufacturing preparation in terms of technology and technological infrastructure available; 2.2. Analysis of technological design; 2.2.3. Optimization of the technological constructivism alloy	Development of operational design software, computer aided process for obtaining castings, usable in the offices of study and analysis technology with current practice in the analyze domain

	composition, the casting characteristics of the shape and conditions of casting	
8	2. Analysis of manufacturing preparation in terms of technology and technological infrastructure available; 2.3. Technological preparation of production analysis; 2.3.1. Selection, analysis and optimization of process flow steps	Developing technological analyzes aimed at enhancing technology flows, whole, and in terms of aspects related to the context of developing circular economy, instead of the dominant linear development lately. The analysis in this respect, it is necessary to achieve each stage of the technological process. The analysis of development in the context of circular economy are essential.
9	2. Analysis of manufacturing preparation in terms of technology and technological infrastructure available; 2.3. Technological preparation of production analysis; 2.3.2. Choice or imposition technological infrastructure needed for the process flow steps	Development and analysis of infrastructure technologies used in terms of capacity development to protect the environment, conditions of employment and sustainable development. In this context, the quantification of the effects on the development of circular economy.
10	5. Analysis of exploitation and disposal at end of life; 5.1. Reuse of casting pieces; 5.1.1. Management components with the potential to be recycled	Organizing recycling flows in the castings at end of life on typology and alloys constituting, aiming to avoid the inclusion of parts which by their composition affects the ability to reintroduce the productive flows, especially in the materials increased sensitivity to compositional and structural changes.
11	5. Analysis of exploitation and disposal at end of life; 5.1. Reuse of casting pieces; 5.1.1. Management components with the potential to be recycled	Organizing recycling flows in the castings at end of life on typology and alloys constituting, aiming to avoid the inclusion of parts which by their composition affects the ability to reintroduce the productive flows, especially in the materials increased sensitivity to compositional and structural changes.
12	5. Analysis of exploitation and disposal at end of life; 5.1. Reuse of casting pieces; 5.1.2. Developing technologies reconditioning castings	Development of recycling technologies massive and complex parts to increase their lifespan.
13	5. Analysis of exploitation and disposal at end of life; 5.1. Reuse of casting pieces; 5.1.3. Development alloys typing parts and castings to ensure interchangeability and linking overall constructive on this criterion	Conducting an active database on typing and alloy parts to ensure their interchangeability conditions at constructive solutions used.
14	5. Analysis of exploitation and disposal at end of life; 5.2. Recycling of worn parts; 5.2.1. Establishment of collection systems based on flow assurance criteria ensuring recycling	The development of circular economy principles collection systems for the purposes of efficient use of resources with potential for use in the production of castings.
15	5. Analysis of exploitation and disposal at end of life; 5.2. Recycling of worn parts; 5.2.2. The development of circular economy principles collection systems for the purposes of efficient use of resources with potential for use in the production of castings.	Develop training techniques recycled materials production of castings, especially for sorting, crushing and quality assurance charging materials and establishment of training centers castings materials industry.
16	5. Analysis of exploitation and disposal at end of life; 5.2. Recycling of worn parts; 5.2.3. Market promotion of recycled materials for competitive production of castings	Using institutional regulations and financial instruments to stimulate the market for recycled materials used or generated castings industry.
17	5. Analysis of exploitation and disposal at end of life; 5.3. Waste management; .3.1. Charging heredity material management;	Development activities quantify the impact that it may have to introduce new elements in the alloy

		formation, in addition to improving the assessment characteristics.
18	5. Analysis of exploitation and disposal at end of life; 5.3. Waste management; 5.3.2. Application of the principles of extended producer responsibility in relation castings	Using the principles of extended producer responsibility in the global management of production processes castings, meaning thinking from the very beginning of their elimination at the end of their life cycle.
19	5. Analysis of exploitation and disposal at end of life; 5.3. Waste management; 5.3.3. Management of historical industrial dumps and landfills specific manufacturing processes tuned parts	Development of material recovery potential of industrial dumps and new landfills generated in manufacturing of castings.
20	6. Analysis of other phases contributing to the cast pieces; 6.1. Analyze production and distribution utilities; 6.1.1. Analysis of water resources	Managing the environmental principles of circular economy and water resources used.
21	6. Analysis of other phases contributing to the cast pieces; 6.1. Analyze production and distribution utilities; 6.1.2. Analysis of energy	Managing the environmental principles of circular economy and energy resources to increase their efficiency.
22	6. Analysis of other phases contributing to the cast pieces; 6.1. Analyze production and distribution utilities; 6.1.3. Analysis of production and distribution of compressed air	Managing the environmental principles of circular economy and the activities of production and distribution of compressed air.

5 The need to design the collaborative entrepreneurial platform for the durable manufacture of "PAC – FDPT" castings

Given the complexity of the issues that arise in the transition of the manufacture of castings from the linear economy to the objectives of the circular economy, the fundamental proposal that has been developed since 2013 [17] is related to the construction of an entrepreneurial platform to solve the issue proposed and set forth.

The entrepreneurial approach of global factors of influence to the requirements of the manufacturing castings from the perspective of small-sized companies and the conditions imposed by an increasingly globalised market. The platform is a portal that can bring together the well-defined partners of a community, but also the outsiders, who try out the private offers and requests for information, advice, design, processing, services, promotion, sales, etc. within a virtual space, in a given social and economic context, aiming at concluding some transactions.

The "PAC – FDPT" missions are expressed as follows: direct virtual business relationship between the members of the platform; strengthening the presence of small-sized companies on the global market; supporting a strategic concept such as the development of a specific market of materials and the economic development of a particular geographical area; serving an extended and associative economic entity among several economic stakeholders organised into clusters, development centres or agencies; analysing the offers and requests of the participants and the economic stakeholders; ensuring their structuring on different criteria: formal qualification, competence, specific activity, medium-term development intent; demand and supply consolidation, which involves: aggregating data on the demand and supply of internal markets of organisations, assessing the critical level from which they are relevant, aggregating the relevant supply and demand by material / symbolic relationships between the participants who generated it; participation in the virtual economy, meaning that once consolidated, the demand and supply are exposed to the global virtual marketplace with a view to concluding transactions with external partners; business intermediation for the benefit of its members and partners; mediating demand and supply, by ensuring interconnection with the participants within the virtual pole established; operation according to the "business incubator" principle by starting up businesses (start-ups) through assistance, advice, training; supporting the launch of innovative projects and their development with the help of the virtual environment; managing the creative potential and the science of making it available at the level of the social network created, at the level of mentors, trainers and trainees; developing a stronger demand and offer for a much stronger global market presence; achieving an effective platform to support the Romanian inclusion and consolidation markets and their alignment with the requirements of global markets.

The "PAC – FPT" architecture features the following elements: "PAC - FPT – PS" static platform and "PAC - FPT – PD" dynamic platform. The attributes of the static platform "PAC-FPT-PS" are shown in Figure 2, and the components of the dynamic "PAC-FPT-PD" platform in Figure 3.

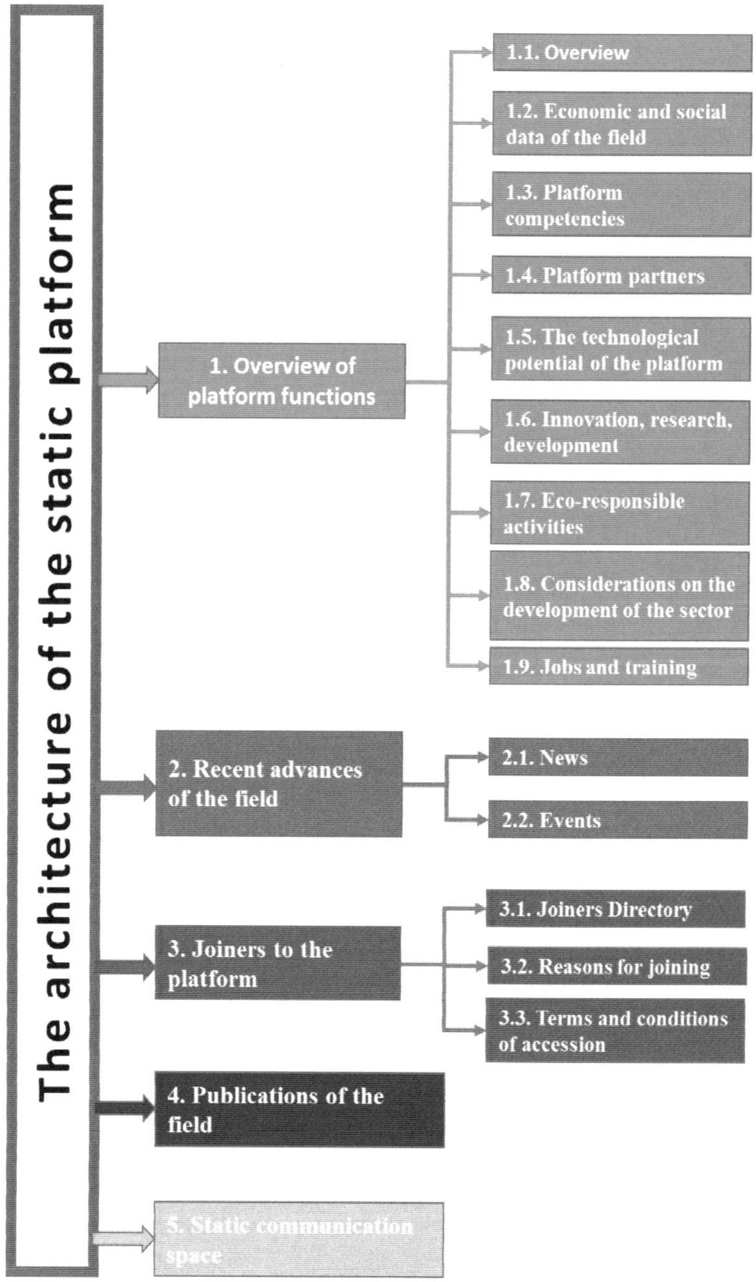

Figure 2. Schematic of static platform attributes "PAC-FPT-PS"

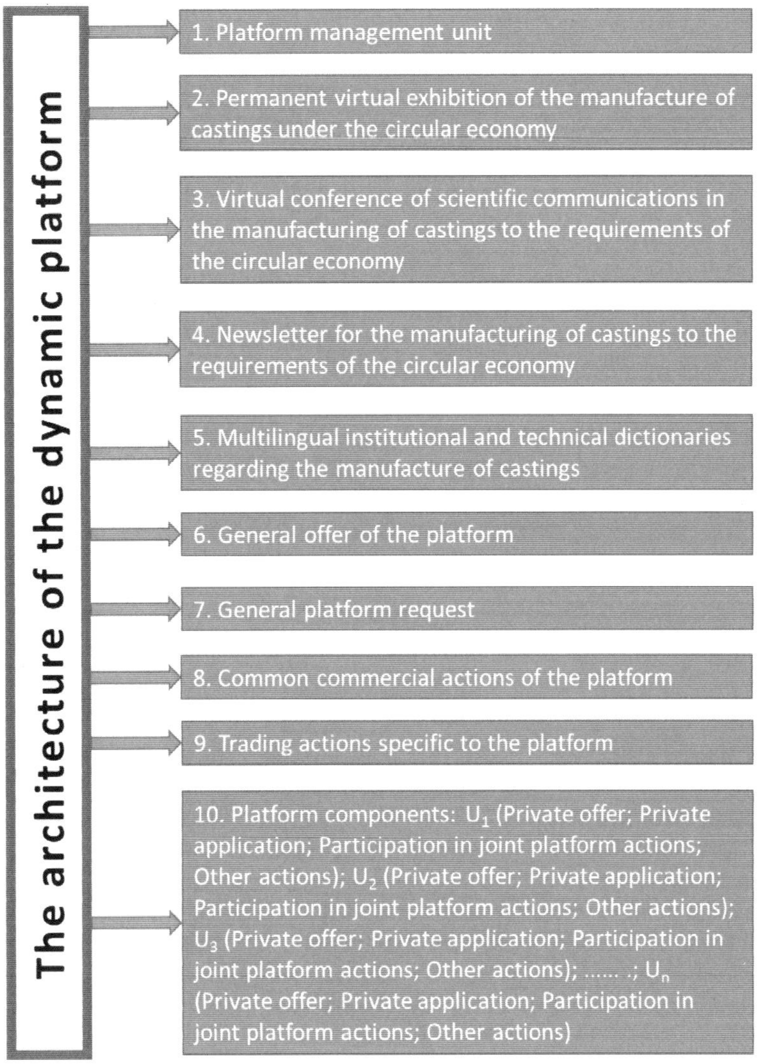

Figure 3. Schematic of dynamic platform components

The "PAC - FPT – PS" instruments aim the following aspects: 1. Theoretical aspects regarding the manufacturing of castings under the circular economy; 2. The current situation of the manufacturing of castings with respect to the observance of the rules of the circular economy; 3. Mechanisms of functioning of the castings market in terms of circular economy; 4. Circular economy tools adapted to the manufacture of castings; 5. Public policies regarding the manufacture of castings under the circular economy; 6. Regulatory framework for the manufacture of castings under the circular economy; 7. Technical and technological solutions for ensuring the manufacture of castings under the circular economy; 8. Funding solutions and financial instruments for ensuring the manufacture of castings under the circular economy; 9. Examples of good practice and branches for the manufacture of castings under the cyclical economy; 10. The stakeholders of the manufacturing process of castings under the circular economy (level of action: local, regional, national, European and global): 10.1. Castings manufacturers: 10.1.1. Ferrous alloys: 10.1.1.1. Cast iron (a. Grey and alloy cast iron; b. Nodular graphite cast iron; c. Malleable cast iron; d. Special cast iron and other types of cast iron); 10.1.1.2. Steels (Carbon steels; Alloyed steels; Special steels and other types of steels); 10.1.2. Non-ferrous alloys: 10.1.2.1. Copper-based alloys; 10.1.2.2. Aluminium alloys; 10.1.2.3. Magnesium alloys; 10.1.2.4. Zinc alloys; 10.1.2.5. Other types of non-ferrous alloys); 10.2. Suppliers: 10.2.1. Suppliers of raw materials for the casting industry; 10.2.2. Suppliers of equipment for the casting industry; 10.2.3. Providers of service and computer products for the casting industry; 10.2.4. Consultancy and technical and managerial assistance service providers for the manufacture of castings; 10.2.5. Service providers specific to the circular economy; 10.2.6. Providers of educational and training services; 10.2.7. Intermediary service providers in the manufacture of castings; 10.2.8. Media, advertising and virtual media service providers; 10.3. Units of scientific and technological research in the manufacture of castings adapted to the requirements of the circular economy; 10.4. Socio-professional and scientific associations; 10.5. Public regulation, implementation and monitoring authorities; 10.6. Banking units and investment funds for the manufacture of castings under the circular economy; 10.7. Units for managing institutional funds (local, national, European) that relate to the manufacture of castings under the circular economy; 10.8. Beneficiaries of castings manufacturing; 11. Events in the manufacturing of castings with respect to the requirements of the circular economy: 11.1. Fairs and exhibitions relating to the manufacture of castings; 11.2. Congresses and conferences; 11.3. Seminars; 11.4. Other events; 12. The issue of the manufacture of castings from the perspective of the circular economy; 13. Publications on the manufacture of castings from the perspective of the circular economy: 13.1. Specialty books; 13.2. Specialty magazines; 13.3. Doctoral theses; 13.4. Institutional reports; 13.5. Product catalogues; 13.6. Collection of public policies and regulations; 13.7. Virtual media for information dissemination.

6 Conclusions

The present paper emphasizes the importance of the transition process of the castings manufacturing from the modalities of carrying out the transition from ”linear economy” to the ”circular economy”. At the same time, this paper, which aims to continue the work presented in recent years, presents a methodology of ensuring that processes are carried out at the level of the objectives of the circular economy. Within this framework, from the perspective of the circular economy, the following stages are followed (the process is generalized, with the potential to develop in other productive domains as well): analysis of the domain specific information according to the principles of economic intelligence; analysis of manufacturing preparation in terms of technology and technological infrastructure available; analysis of the castings manufacturing process; analysis of the results of the manufacturing process; analysis of the exploitation of castings and their elimination at the end of the life cycle; analysis of the other phases that contribute to casting; process analysis at general level from the perspective of compliance with environmental conditions and sustainable development. Considering the complexity of the productive field, related to the market requirements from the perspective of the circular economy, the importance of the position of the actors present in the process is emphasized: the uses of castings, the beneficiary of castings, their producer and the supplier of the materials and the energy resources used. The method of analysis regarding the achievement of the objectives of the circular economy included encompassing and quantifying the effects of the process from the perspective of sustainable development and environmental protection through the evaluation indicators: consumption of material per unit of product, energy consumption per unit of product; the effects on the product unit created on the environment; generation of by-products per unit of product; generation of by-products on the total unused portion produced; generation of waste per unit of product; generation of recycled materials per unit of product and on the generated unused part; the share of recycled materials on the life cycle of castings with direct use on the life cycle of cast parts with direct use on the casting stream; the share of recycled materials on other technological flows; the share of recycled materials generated by other technological fluxes used in the casting process; the share of energy recovered at the level of the technological flow with direct use; the share of recovered energy delivered to other beneficiaries; the social and environmental effects that they create. Regarding the perspective of the procedures and indicators, the importance of introducing digitization at the level of casting manufacturing through the collaborative entrepreneurial platform is emphasized.

7 References

[1] Soporan V F, Crişan M, Lehene T and Pop A L, 2016 *IOP Conf. Ser.: Mater. Sci. Eng.* **133** 012063

[2] Petitpain M 2014 *Economie circulaire: Projet valorification croisees de co-produits industriels* (Epernon: Cerib. Etudes et Recherches)

[3] Ghewy X, Paquier J L, Gregoire P, Gicquiaux C and Louis H 2013 *Le cycle des matieres dans l'economie francais* (Paris: Commissariat general au developpement durable Service de l'observation et des statistique)

[4] AFEP 2017 *Trajectoires economie circulaire. 33 entreprises se mobilisent avec 100 engagements* (Paris : Association francaise des entreprises privees – afep)

[5] Perret B 2014 *L'economie circulaire, etat des lieux et perspectives. Raport n° 009548-06* (Paris: CGEDD – Conseil General de l'Environnement et du Developrpement Durable)

[6] Valerian F 2014 *L'economie circulaire, ou la competition pour les ressources. Etude annuelle 2014* (Paris: Conseil General de l'Economie, Ministere de l'Economie de l'Industrie et du Numerique)

[7] be circular be.brussels 2016 Programme regional en economie circulaire 2016-2020. Mobiliser les ressources et minimiser les richesses perdue: pour une economie regionale inivante. Be circular be.brussels, Mars 2016.

[8] Diamant I 1938 *Buletinul de studii Industria Sârmei S.A. octombrie 1938* (Cluj: Industria Sarmei SA)

[9] Georgescu-Roegen N 1995 *Le decroissance. Entropie – Ecologie – Economie* (Paris: Edition sang de la terre)

[10] Soporan VF 2015 *Universitatea, ecoantreprenoriatul și dezvoltarea comunităților* (Cluj-Napoca: Pro Ecologic Concept 188-189)

[11] Soporan V F 2016 *Ingineria instituționalizării economiei circulare – notițe de curs* (Cluj-Napoca: Univeristatea Tehnică din Cluj-Napoca)

[12] Halevy M 2015 *Les grandes mutations économiques: comment s'y adapter* (Bordeaux: Centre Entreprise et Communication Avancee Université Hommes-Entreprises Sens et Croissance 5-10)

[13] Aurez V and Georgeault L 2016 *Economie circulaire. Systeme economique et finitude de ressources* (Louvain-la- Neuve: De Boeck Superieur 115)

[14] Soporan V F 2017 *Situația fabricației pieselor turnate și instituționalizarea formării profesionale în România și la nivel global* (Cluj-Napoca: Universitatea Tehnică din Cluj-Napoca Seminarul "Situația fabricației pieselor turnate în România și nevoile de structurare a formării profesionale")

[15] Comisia Europeană 2014 *Comunicare a Comisiei către Parlamentul European, Consiliu, Comitetul Economic și Social și Comitetul regiunilor: Spre o economie circulară: un program "deșeuri zero" pentru Europa"*, (Bruxelles: COM(2014) 398 final/2)

[16] Comisia Europeană 2014 *Comunicare a Comisiei către Parlamentul European, Consiliu, Comitetul Economic și Social și Comitetul regiunilor: Spre o economie circulară: un program "deșeuri zero" pentru Europa"* (Bruxelles: COM(2014) 398 final/2 4-5)

[17] Vescan M M 2016 *Studii și cercetări cu privire la comunicarea virtuală specifică fabricației pieselor turnate din perspectiva economiei circulare* (Cluj-Napoca: Universitatea Tehnicdin Cluj-Napoca Referat teză de doctorat)

IOP Conf. Series: Materials Science and Engineering **209** (2017) 012097 doi:10.1088/1757-899X/209/1/012097

Greening Solutions Applicable in the Tailing Ponds Tăusani and Bosneag from Moldova Nouă

I F Burlacu*, G Deak, M C Raischi, A Daescu, S Zamfir, B Uritescu, C Cirstinoiu and M V Olteanu
National Institute for Research and Development in Environmental Protection, Adresse : Splaiul Independenței, no. 294, district 6, Bucharest, postal code 060031, Romania

E-mail: iasmina.burlacu@incdpm.ro

Abstract. This study aims to propose solutions for greening of the tailings ponds resulted from mining activities with transboundary impacts. As case study, are proposed for greening the Boşneag and Tăuşani tailing ponds because they pollute Moldova Nouă, Danube and towns on the Serbian side of the Danube with particles in suspension. We analyzed four scenarios of modeling dispersion of particles in suspension (copper and other heavy metals) from the Tăuşani and Boşneag tailing ponds in the theoretical background where pollution has cross-border nature and require studying the transport of pollutants over a long distance from the source and modeling dispersion of particles in suspension in the atmosphere, these were performed using TAPM model, able to simulate the aspects mentioned. After running the software for modeling the dispersion of particles, was revealed that the pollution generated from the pollution sources taked into consideration is very high and significantly affects quality of life on considerable areas both in Romania and Serbia, thus amplifying the need to implement greening solutions of the analyzed area. Following the results obtained are presented three alternatives solutions for greening the area studied, aiming at minimizing the impact on the environmental and population.

1. Introduction

Romania has a long history of mining of useful mineral substances, from which the most important to remember are: precious metals (gold and silver), copper, lead, zinc, uranium, manganese, salt and coal. Thanks to the long exploitation period and the extraction technologies from the mining areas and from the affiliated industrial areas of mineral processing, the environment was affected in a very aggressive manner, leading to an important burden over the soil, water, air and the ecosystems. The problems have extended across the borders due the migration of pollutants through the hydrographic network or due to the wind that carried them, therefore we face a cross border pollution situation [1].

In carrying out of this paper, the tailing ponds Tăuşani – Boşneag were chosen, related to the former mining enterprise from the Moldova Nouă, placed on the left bank of the Danube, on the Moldova Nouă town's territory, between Moldova Veche town and Coronini town, near the wet area, Ostrovul from the Natural Park Iron Gates. The mining enterprise Moldova Nouă started operating in 1965 through underground exploitation of copper mineral resources with over 0.3 Cu, in Florimunda, Suvarov, The Great Valley, Gărâna, Vărad areas, with a maximum processing of 1.6 million tons/year. Since 1974, the exploitation in the mining pit of copper banatites has started, with a maximum level reached in 1988 of 3.3 tons/year. Since the closure of the activity in 2008, an estimated total of sterile material around 128.000 tons accumulated on the tailing ponds [2].

This area was chosen to be analyzed because it represents a major pollution source, with thin particles being carried in the atmosphere by the local wind (Cosava), from the tailing ponds resulted due to the copper processing. We wish to emphasize the necessity of the greening of this area, because the impact of the tailing ponds is reflected over the environment factors, also representing a situation where cross border pollution is present [3,4,5].

Over a total surface of approximately 150 ha, after the decantation of the water coming from the copper preparation plant and after the evaporation of these waters, a dry deposit has resulted, composed of thin powders, with an average thickness of 1.50 – 2 m. The surface that represents the subject of this paper is of 122.2 ha and is presented in figure 1 and the rest of the surfaces are considered to be rehabilitated [5].

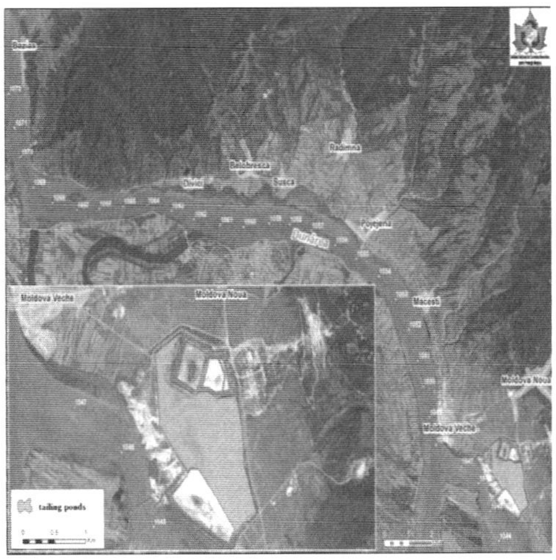

Figure 1. Localization of the studied area.

2. Materials and methods

Because the analyzed area is characterized by a complex field and the pollution has a cross border character, it requires the study of the transportation of the specific pollutant over a wide distance from the source. The dispersion modeling of the powders into the atmosphere has been made using the TAPM model, capable of stimulating the already mentioned aspects.

The TAPM model (The Air Pollution Model, developed by CSIRO – Australia) is capable to stimulate the emissions provided from all types of sources: punctual, linear and surface, also a great number of sources (e.g. all the sources for which there are detailed data at a national level). Also, the model can stimulate the significant spatial variability of the meteorological conditions, induced by the complexity of the topographic surface (the proximity of the sea, tall, complex mountain areas) and also the radiation differences, considering the advection and diffusion of the pollutants in a non – stationary and inhomogeneous field. Thanks to the calculation procedure with included domains through which the calculations are refined, the model can be applied to a mesoscale and also to a local scale [6].

2.1. Entry data

Meteorological data used as entry data for the model are provided from a synoptic scale analysis model (LAPS) and they consist of modeled data in intervals consisting of 6 hours in a geographic network – longitude/latitude with a resolution of 0.75 degrees (approx. 75 km) which covers the Northern Hemisphere.

Field data are provided from the US Geological Survey, Earth Resources Observation Systems (EROS), Data centre Distributed Active Archive Centre (EDC DAAC), with a resolution of 30 seconds (approx. 1 km).

It have been made 4 calculations scenarios, with emission sources of different sizes, to stimulate the existing situation and also the reduction of the concentrations and of the affected areas due to the pollutant concentrations, as the tailing ponds which represent the pollution sources are being rehabilitated, the purpose of making these stimulations being the necessity of rehabilitating this pollution source as fast as possible and for that, infringement procedures against Romania were released, but another reason is to calculate with how much the pollution could be reduced, during the rehabilitation process, when different amounts of the surface of the source would already be rehabilitated, through different greening methods. Four powder stimulations have been made for the total amount of suspension powders (TSP) and also for the fractions PM10 and PM 2.5, but we chose to present the results for the total suspension powders (TSP) being the most conclusive. The modeling domain was of 30 x 30 km, the grid calculation for the pollution having a resolution of 250 m, being used a total amount of 13689 receptors. The thresholds of the atmospheric pollutants established through the legislation in force are presented in table 1.

Table 1. The maximum admitted concentrations for the TSP, according to STAS 12574 – 87.

No. crt.	Chemical compound / Pollutant	Limit for protection	Mediation period	Maximum admitted concentration (μg/m^3)
1	TSP	Population	24 h	150
		Population	Year	75

3. Results and discussions

3.1. The first scenario (S100)

In the case of the first studied scenario (S100), a source with a surface of 122.2 ha was considered (considered the actual situation, presented in the figure 2), in the subsequent scenarios the surface of this source being reduced by 25% - the 2nd scenario (S75), by 50% - the 3rd scenario (S50%) and by 75% - the 4th scenario (S25). In the figure 2 is represented the distribution of the monthly average concentrations of total suspension powders in the case of the 1st scenario [5].

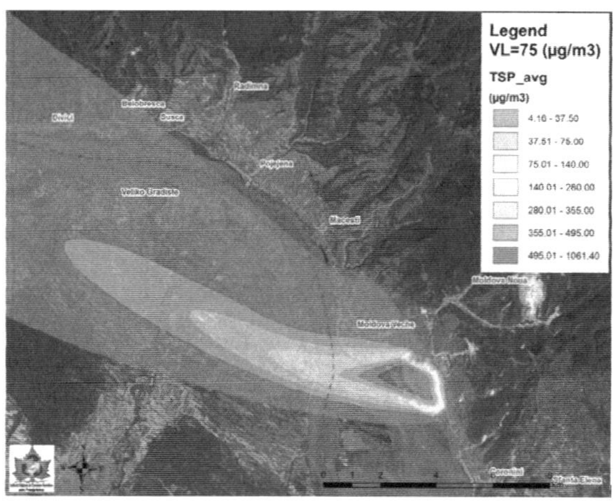

Figure 2. The distribution of the monthly average concentrations of TSP – S100.

To evaluate the air quality and the areas affected by suspension powders, the results of the mathematical modeling of the dispersion of the pollutants into the atmosphere was compared with the limits of the maximum concentrations from the 104/2011 law, regarding the quality of air. After the map analysis, great exceeding of the imposed limits can be noticed for all the fractions of the analyzed suspension particles. Very high concentrations, obtained as a result of the modeling, which exceed the limit value, can be noticed not only in the proximity of the pollution source, but also on distances exceeding 10 km in the dominant wind direction for the total suspension particles (TSP) [5].

3.2. The second scenario (S75)

This scenario implies that 25% of the total emission surface was rehabilitated, the rest of 75% remaining in the most unfavourable conditions for the quality of air (dry, with a great potential of pollution with suspension powders). In the figure 3 is represented the distribution of the monthly average concentrations of total suspension powders in second scenario.

The analysis of the modeling results of the dispersion of the atmosphere pollutants in the second scenario indicates a great similarity of the spatial distribution of the maximum daily concentrations of total suspension particles (TSP) and a very slight decrease of the maximum obtained concentrations, but the exceeding of the threshold values are of the same order of magnitude as in the case of the first scenario and extended over approx. the same distance in the wind direction (around 10 – 13 km).

3.3. The third scenario (S50)

The third scenario, implies that a half of the total emission surface has been rehabilitated, the rest of 50% remaining in the most unfavorable conditions for the air quality (dry, with a great potential of suspension powders pollution). In the figure 4 is represented the distribution of the monthly average concentrations of total suspension powders.

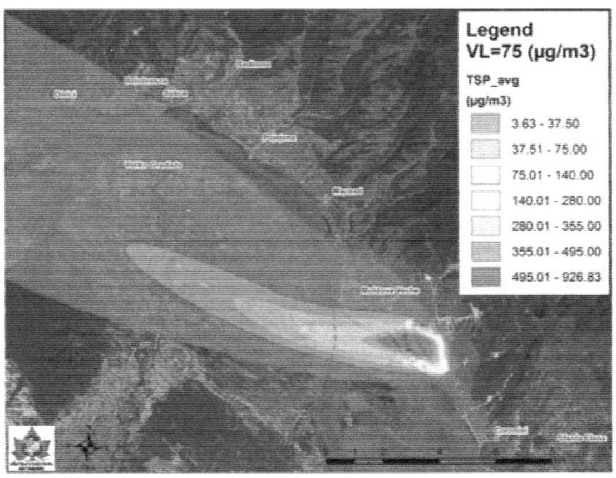

Figure 3. The distribution of the monthly average concentrations of TSP – S75.

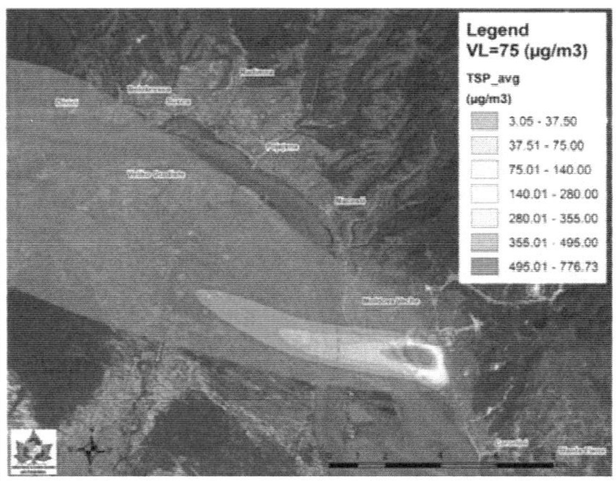

Figure 4. The distribution of the monthly average concentrations of TSP – S50.

The obtained results from the mathematical modeling in the 3rd scenario for the TSP indicate a reduction of the area in which concentrations which exceed the limit value of 150 $\mu g/m^3$ can be found, the distance in the direction of the wind where these values were obtained being of approx. 8 km, but concentrations which can reach half of the limit value have been obtained on the same surface as in the 1st scenario (the situation being considered the present one) [5].

3.4. The fourth scenario (S25)

In this scenario it was implied that 75% of the total emission surface has been rehabilitated, the rest of 25% remaining in the most unfavourable conditions for the air quality (dry, with a great potential for

IOP Conf. Series: Materials Science and Engineering **209** (2017) 012097 doi:10.1088/1757-899X/209/1/012097

pollution with suspension powders). In the figure 5 is represented the monthly average concentrations of total suspension powders.

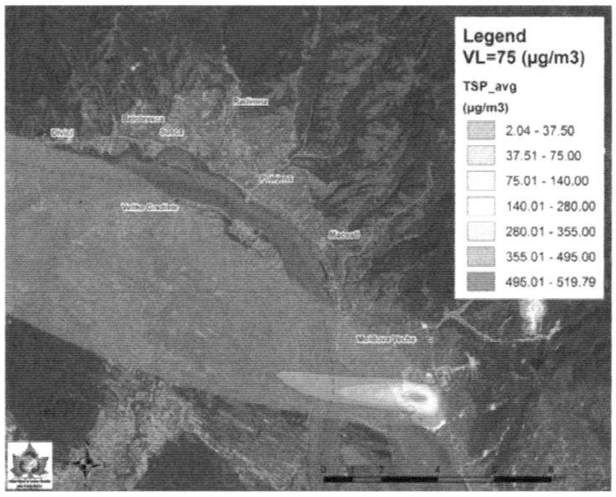

Figure 5. The distribution of the monthly average concentrations of TSP- S25.

TSP concentrations of which the values exceed the limit threshold of 150 µg/m^3 have been obtained after the mathematical modeling in the 4th scenario over a distance (in the dominant wind direction) of approx. 9-10 km. The reduction of the surface on which exceeding of the limit imposed by law are obtained is evident but the maximum concentrations, though reduced compared to the 1st scenario, exceed the limit value with an order of magnitude. After the rolling of the modelling software of the powder dispersion it resulted the fact that the generated pollution by the considered pollution sources is very high and it significantly affects the quality of life on areas on the Romanian territory as well as Serbia, this way amplifying the necessity of the implementation of a greening solutions of the analyzed area [5].

3.5. Greening solutions

For greening the area it is necessary to be made in advanced, a study regarding the stability of the tailing ponds, which must have geotechnical and electrometrical measurements done, laboratory investigations, topographic measurements, to be able to establish the characteristics of the tailing ponds [5]. In relation to the obtained results as following the investigations of the geometrical and geomechanical characteristics of the tailing ponds, these will lead to the choosing of the best greening solutions. Also, an important factor regarding the choosing of a greening solution is the economical factor that has to do with the costs that will be necessary for the implementation technical greening/closing tailing ponds project. The proposed options for the greening of the studied area are presented in table 2.

Table 2. Proposed options of rehabilitation for the studied area.

Option I	Option I: „*in situ*" stabilization - dry remedial
Option II	Option 2: capitalizing the useful area of the mixture deposited in the lake and the relocation of the sterile material resulted in the mining enclosure
Option III	Option 3: in which the 1st and the 2nd option are combatted

3.6. Option I – „in situ" stabilization – dry remedial

In the context of Tăuşani and Boşneag tailing ponds, one of the greening methods that have been proposed in the present study is the „*dry option*". Through its implementation, the decrease of the impact on the environment of these tailing ponds is being pursued and also the fitting into the landscape of the studied emplacement.

The steps of the work regarding the greening through „*in situ*" stabilization – with the placement of a soil cover – "dry option" are the following :

- covering with a layer of **bentonite geocomposites** (0,002- 0,003m);
- covering with a layer to collecting and evacuate the rain water, the cover being composed of a gravel layer and an absorbent draining and collecting system (0.25 – 0.5 m) (0,25-0,5m);
- covering of the emplacement with a layer of **vegetal soil**, between 0.3 – 1 m which will be on with perennial herbs will be planted, these are presented in the figure 6.

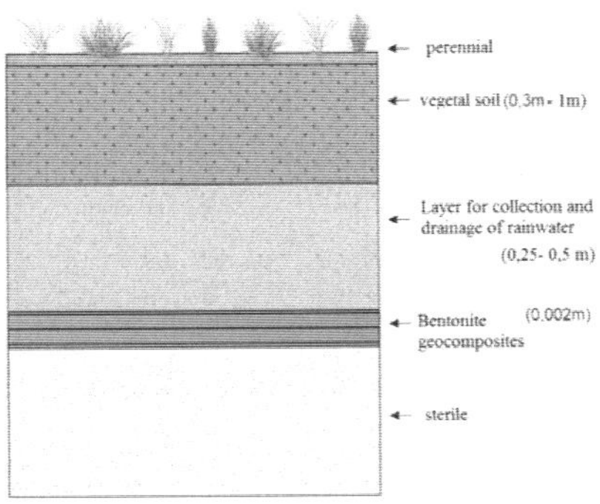

Figure 6. Scheme of closure and greening of the studied area.

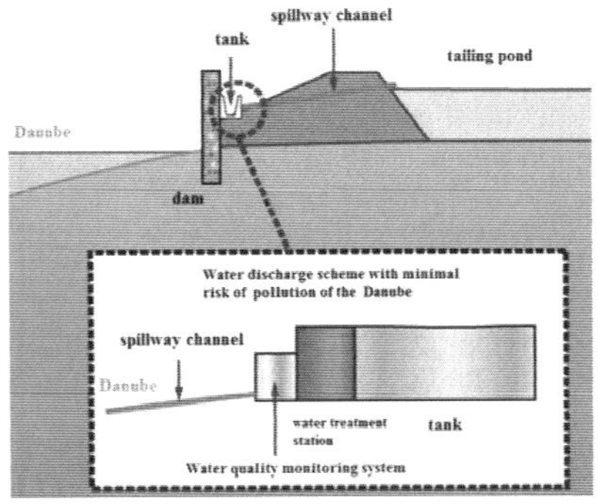

Figure 7. Representation of tailing ponds, Treatment Water Tank Collector and Spillway Channel.

Following the analysis of the decision matrix on the geosynthetic sealing of the tailing ponds, *bentonite geocomposites* have been chosen because they present: competitive cost; impermeability; good resistance to frost /thaw and wet / dry; good behavior on slopes and differentiated deposits; easy installation under adverse climatic and field conditions; replacing compacted clay or geomembranes at ponds, lagoons, canals, irrigation works, marshland, reedbeds; overlays are easily achieved using bentonite paste; have a sealing function, are compatible with cathodic protection. Covering of the area studied with *bentonite geocomposite* of approximately 0,002m-0,003m (consisting of 3 layers of high density polypropylene/geomembrane geotextile layer, bentonite layer in the form of powder or granules, polypropylene nonwoven geotextile coating) having the role of sealing the tailings ponds [7,8,9].

Followed by a *permeable layer for collecting and evacuating rainwater*, consisting of a gravel layer and a system of drainage and collecting drains of approx. 0.25-0.5m. If the water quality in the collection tanks does not comply with the NTPA 001/2005 water quality norms, in order to be discharged into the Danube, it is necessary to construct a treatment plant, where the water is brought to the appropriate parameters in order to be able to be discharged into the Danube through the spillway shown in figure 7. The last stage of the ecological process of the Tăuşani and Boşneag tailing ponds is covering the site with a layer of *vegetal soil* (0.3m-1m) and its grazing with perennial species [10].

3.7. Option II

In accordance with Order 95/2005 regarding the establishing acceptance criteria and preliminary procedures for the acceptance of waste storage and the national list of waste accepted in each class of landfill for waste from the excavation of metal ores, identified with the code 01 01 01 it isn't recommended to apply a recovery method.

As a result of the laboratory analyzes carried out within the National Institute for Research and Development in Environmental Protection (INCDPM), the material samples from the tailings ponds resulted the values presented in table 3. These analyzes were carried out within the project *" Studies for achieving shore protection of wetland area Divici-Pojejena "* carried out by INCDPM.

Table 3. Results of the chemical analyzis of the steril samples from the Tăuşani-Boşneag tailing ponds.

Sampling depth	Sampling date	Cu	Cr	Cd	Ni	Zn	Pb	Hg
				[mg/kg]				
0-15 cm	16.07.2014	1589	25.69	2.86	3.99	904	1.16	0.31
15-30 cm		948	34.47	1.99	13.81	505	27.82	0.30

The high concentration of copper from the samples taken and analyzed from the tailings ponds shown that the possibility of utilizing the copper recovery . According to laboratory analyzes, the percentage of copper that can be extracted is about 75% -80% (representing about 12597 tonnes) of the amount of copper present in tailings ponds [11].

It is proposed that the technological processes of capitalization of the useful part by separation from the sterile part be done in the processing plant number 1 of the economic operator in the area, by building a new production capacity, located in the proximity of the Tăuşani and Boşneag tailing ponds and subsequently applying greening option 1 [12,13].

The application of Option 2, in addition to Option 1, brings an economic contribution by capitalizing on existing metals that can be recovered from tailings ponds.

3.8. Option III

Option III combines option I and option II, proposing the utilization of useful parts from the sterile site, relocation of the site to the old mining site and subsequent application of *"in situ stabilization"* - dry remediation. Thus, by applying options II and III, it is possible to solve the problem related to the storage of hazardous wastes directly on the soil, which is a significant a source of pollution. Depending on the existing funds, the application of these solutions will lead to the proper placement of hazardous waste and to minimizing the risks by reducing the impact on the soil, while improving the aspect related to the landscape. Also using the useful part of the tailings ponds would integrate important quantities of cupric concentrates into the economic circuit.

From a technical point of view, we consider the application of option III to be more feasible because it combines the capitalization of the Tăuşani and Boşneag tailing ponds in the economic circle through the use of clean environmental technologies that don't produce pollution, relocation of the site and subsequently the application of a *in situ* dry remediation that minimizes the impact of the dispersion of particulate matter on the environment, on the Danube River and cross-border pollution, as well as landscaping of the area.

4. Conclusions

Due to the mathematical modeling of the dispersion of pollutants into the atmosphere, results the need to take urgent measures for greening the study area and to maintain a high humidity of the material located on the surface of the tailings ponds exposed to wind erosion.

In accordance with the four simulated scenarios it can be stated that even if 75% of the surface of the Tăuşani - Boşneag tailing ponds would be ecologized (with no further measures to reduce the emission of dust from the non - certified surface), the concentrations of suspended particulate matter transported by the wind would exceed the limits imposed by legislation to protect human health and vegetation. The pollution generated by the analyzed sources is high and significantly affecting the quality of life on considerable areas both on the territory of Romania and on the Serbian side, the cross-border impact being obvious.

Of the three proposed greening options, technically, we consider the application of option III to be more reliable, as it allows the capitalization of the useful part from the sterile material, relocating the site to the old mining area and then applying the *in situ* stabilization- a dry remediation that contributes to minimizing the impact of suspended particulate matter on the environment,on the Danube River and the Serbian side and also contribute to landscaping the area. It is important to take into account the cost-effectiveness of such an operation and it is necessary to develop a cost-benefit analysis and also performing site stability studies based on geotechnical, electrometric, laboratory, topographical measurements, in order to determine the characteristics of the tailings ponds. The methodology used in the mathematical modeling scenarios as well as the proposed greening options can also be used as a reference for the greening of other tailing ponds.

This paper was carried out within and with the support of the scientific researchers of the National Institute for Research and Development in Environmental Protection through the project "Researches on establishing the best industrial waste treatment techniques applicable in Romania in accordance with the EU requirements" within Core Program.

References

[1] INCDPM 2004 *Ecological restoration and rendering of urban and industrial landfills in the economic circuit. Solutions on and adjacent to these lands, industrial and/or commercial buildings, leisure areas, dwellings, etc.*

[2] Beutură D, Rogobete G, Beutură S, Bertici R, Tudor C and Ţimbota O 2005 Estimation of the impact of mining on the soil, water and air in Moldova Nouă, *Caraş Severin County Office of Pedological and Agrochemical Studies Timişoara* "Politehnica" University of Timişoara - Pedogenic Factors and Processes in the Temperate Zone 4 S* 231-239

[3] INCDPM, NIBIO and INCDDD 2015- 2016 *Mapping and Assessment of the Ecosystem Services in Divici- Pojejena Wetland and Identification of their Contribution to the Economic Sectors* (WETECOS)

[4] Matei M, Laslo L, Ciobotaru N, Musat C, Boboc M, Raischi M, Deák G 2016 *International Journal of Environmental Science* **1**(1) 265-271

[5] INCDPM Core Program 2015 *Researches on greening the tailing ponds and dumps resulting from mining activities with cross- border impact*

[6] Gillette D A 1980 Major contributors of natural primary continental aerosols: Source mechanisms. In Aerosols, Anthropogenic and Natural Sources and Transport *The New York Academy of Sciences* 348-358

[7] INCDPM Core Program 2014 Researches on the use of geosynthetic materials for ecological restoration and rendering in the economic circuit of industrial non-hazardous landfills - Phase 1 - *Study on the durability of geotextiles for hydrotechnical constructions*; Phase 2- *Analysis of the impact of geotextiles used in hydro-technical constructions on biodiversity on the Danube and preventive solutions for ecological recovery* 09060132

[8] INCDPM 2015 *Long-term perfomance of geotextiles used for Danube River banks protection*, Conference Trabzon Turkey

[9] INCDPM Core Program 2013 *Researches on the criteria identification for the selection of geomembranes used for the closure of non-hazardous industrial landfills in order to their ecological rehabilitation* 09060132.1

[10] Matei M, Raischi M, Ciobotaru N, Laslo L, Boboc M, Zamfir S A, Deák G 2017 *Journal of environmental protection and ecology* **18(1)** 235-245

[11] INCDPM 2014-2015 Studies for achieving shore protection of wetland area Divici-Pojejena (Europe-Aid/134910/4/SER/RO, CODE 1344)

[12] INCDPM 2003 *Preparation of the Environmental Balances for level I and II at the Ovidiu-Constanţa Ecological Waste and Industrial Waste Depot*

[13] INCDPM 2004 *Technical Expert on the equivalence of the solution adopted by the designer for the Sludge and Sludge Technological Sludge Storage Facility, DUCTIL STEEL S.A*

Efficient Use of AUVs in the Maritime Environment

G Slămnoiu[1], O Radu, V Roşca, C Pascu, G Surdu, E Curcă, R G Damian and A Rădulescu
Research Center for Navy, 4 Ştefăniţă Vodă str, 900402, Constanţa, România

E-mail: staff@ccsfn.ro

Abstract. Autonomous underwater vehicle (AUV) a.k.a. underwater drones are subsea vehicles which operate in the underwater environment independently of direct human input. There is a growing interest in underwater data collection by using autonomous underwater vehicles within the oceanographic research community. In this paper, the Iver 2 AUV is examined to accomplish accurate side-scan data while executing well planned missions. Therefore, this papers goal is to establish the optimal use conditions for the AUV that RCN has so that we maximize the detection probability of sea bottom objects that can be risk factors and at the same time to cover a surface as large as possible during a single mission.

1. Introduction

The capability of protecting marine areas and important objectives situated near the shore against dangers associated with surface or underwater threats is critical for providing security in marine activities. Significant categories of threats are ships, small crafts and divers – these can either transport terrorists or various dangerous substances and materials in the protected area or can deposit such items on the sea bottom in the area, or can even conduct terrorist action themselves.

This is the reason that the mission profile for a SIROLC-type system (System for detection, localization, tracking and identification of risk factors addressing important strategic objectives in littoral areas) should focus on two main objectives. The first one is to detect, locate and identify, through regular inspections and post-mission analysis the presence on the sea bottom or near it of objects that can become dangerous to the shipping activities in the area and for the protected objective.

For this purpose SIROLC will have in its configuration an AUV-type subsystem for underwater surveillance [1] and which is fitted with specific sensors for orientation and determination of its own position and for determining the conductivity and salinity parameters, needed for calculating the sound speed profile at depth (information needed for increasing the performance of sonar systems). For observing the objects situated on the sea bottom the AUV subsystem is fitted with a high resolution side scanning sonar.

A second objective is that of signaling the presence near the protected area of small ships or divers. There are various systems that can detect and track targets that move on surface or underwater, the most often used being of the active type (sonar and radar) [1].

The AUV has a main role in the missions of detection, localization, tracking and identification of risk factors situated on the sea bottom as it can be deployed in relatively large areas with a high coverage degree with spatial and temporal resolutions that can not be achieved through other means. At the same time, the AUV efficiency is very high when compared to the costs and duration of measurements [3].

After analyzing the answers provided by the potential beneficiaries to a requirements form, the operational requirements for the SIROLC system have been defined and a part of these also cover the AUV subsystem:

- surveillance of the underwater environment in the harbor area and other coastal objectives, ship anchorage ranges and mandatory pass points;
- warning about presence of underwater danger (explosive devices – mines, improvised devices, etc.) in the interest areas.

The main operational characteristics of the AUV of type IVER 2 that RCN has are:

- autonomy: min. 8 h / 2 – 2.5 kts (depending on sensor package used)
- speed: max. 4 kts
- maximum immersion depth: 100 m

The fitted side scanning sonar has the following characteristics:

- frequency: 450 kHz center (430 – 470 kHz)
- pulse duration: 400 μs
- signal processing: pulse compression (CHIRP pulse)
- hydro-acoustic antennas: 2 pieces, width of the directivity characteristic (VxO): 60° x 0,5° (-3 dB);

2. Factors that affect the AUV capabilities of discovering underwater targets

The AUV capabilities of discovering underwater targets are affected by the following factors:

- side scanning sonar performance: angle and distance resolutions, frequency, type of signal used, emission power, etc.
- environmental conditions: profile of the sound speed on depth, ambient noise;
- sea bottom type;
- AUV evolution parameters as defined through the mission profile (path, speed, depth), and the oscillations of the platform (due to the influence of the environment or to the mission profile – points of direction change);
- own noise level at different evolution speed;
- target characteristics: target strength depending on dimensions, geometry, etc.;
- operator ability: experience in analysis of sonar images.

2.1. Side scanning sonar performance

Accounting for the side scanning sonar characteristics (signal processing with pulse compression, using pulses with linear frequency modulation) we can calculate the resolution in range (perpendicular to the path traveled) with the formula [4]:

$$\Delta r_y = c/(2\times B) \tag{1}$$

where B is the bandwidth of the pulse (40 kHz in our case).

Considering an average value for the speed of sound in a marine environment of 1500 m/s, this gets us a range resolution of Δr_y = 18,75 mm.

Due to the reduced width of the directivity characteristic in a horizontal plane, this theoretical range resolution, parallel to the direction of movement, is extremely good. Also, considering the horizontal swath width of the directivity characteristic and the distance from AUV to target, we have determined the following values for the range resolution:

Table 1. Range resolution, parallel with the AUV movement path.

r [m]	10	20	40	60	80	100
Δr_x [m]	0.14	0.28	0.56	0.84	1.12	1.40

IOP Conf. Series: Materials Science and Engineering **209** (2017) 012098 doi:10.1088/1757-899X/209/1/012098

We can affirm that:

- at small distances from the AUV, the range resolution perpendicular on the path traveled is reduced and the one parallel with the traveled path is very good;
- at large distances from the AUV, the range resolution perpendicular to the path traveled is good but the one parallel to the traveled path is reduced.

Considering the dimensions of the sonar antennas and the maximum range of these (adjustable between approx. 10 and 150 meters) we have to define a maximum upper speed limit for the AUV so that during the pulse repetition interval there should be enough time for the emitted signal to reach a possible target, be reflected and reach back to the sonar antenna [5]. This condition is represented by the following equation:

$$v_{max} = L \cdot f_p = \frac{L \cdot c}{2 \cdot R} \tag{2}$$

where:

- v_{max} = maximum AUV speed;
- L = antenna length;
- f_p = pulse repetition frequency;
- c = speed of sound in the environment (assumed 1500 m/s);
- R = maximum sonar range.

We have calculated the maximum AUV speeds for different sonar ranges with the current antennas.

Table 2. Maximum AUV speed vs. side scanning sonar range.

R [m]	10	20	30	50	100	150
v_{max} [m/s]	32.4	16.2	10.8	6.48	3.24	2.16

Considering that the maximum speed that the AUV can reach is about 2.057 m/s we conclude that in any navigation conditions the underwater targets, even those situated at the extreme limit of the detection range, will be "illuminated" and seen by the side scanning sonar.

In figure 1 below we present the results of a simulation of the propagation loss for the side scanning sonar, for a single side of the directivity characteristic, at a depth of 30 meters and using real data for the sound velocity profile (SVP).

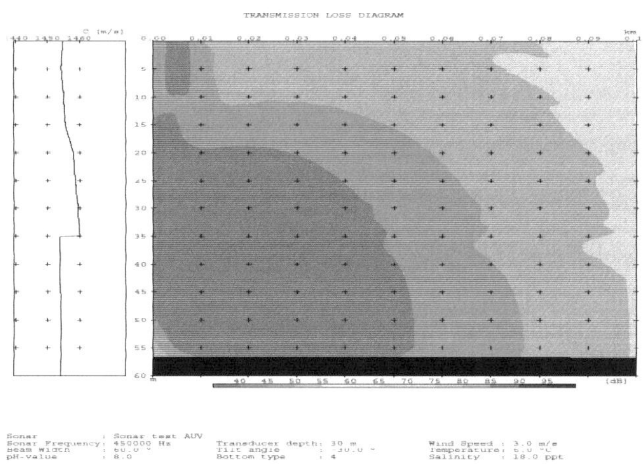

Figure 1. Propagation loss (TL), f = 450 kHz, real SVP Black Sea, across Mangalia, February (winter).

We find that the working frequency of the side scanning sonar is low enough so that it reduces propagation loss.

From the global analysis of the side scanning sonar characteristics we can conclude that this one has good performance (very good resolution, relatively low working frequency, etc.).

2.2. Influence of the AUV mission parameters

In regular use, the recordings provided by the side scanning sonar can also contain distortions caused by the AUV oscillations while it travels the mission path (these are presented in figure 2) and also due to variations of the vehicle speed.

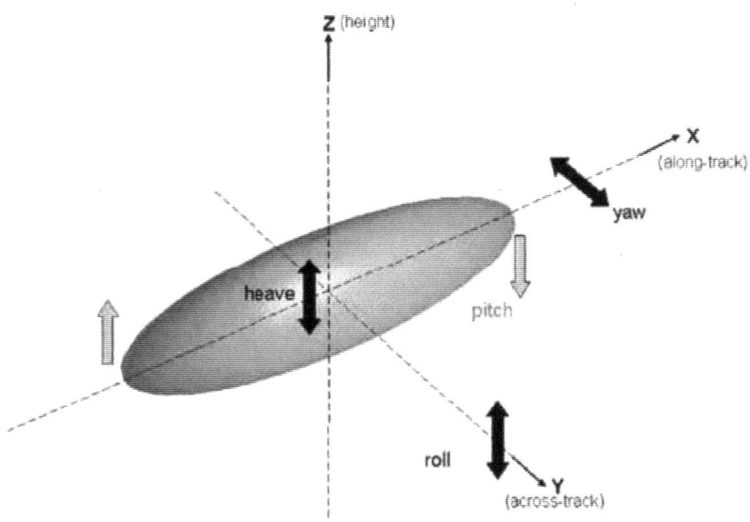

Figure 2. Variations of the AUV attitude that can degrade the side scanning sonar performance [5].

During experiments conducted in 2015 and 2016 we have recorded data about roll, pitch, depth and track variation - a graphical representations of these are in figure 3, figure 4 and figure 5.

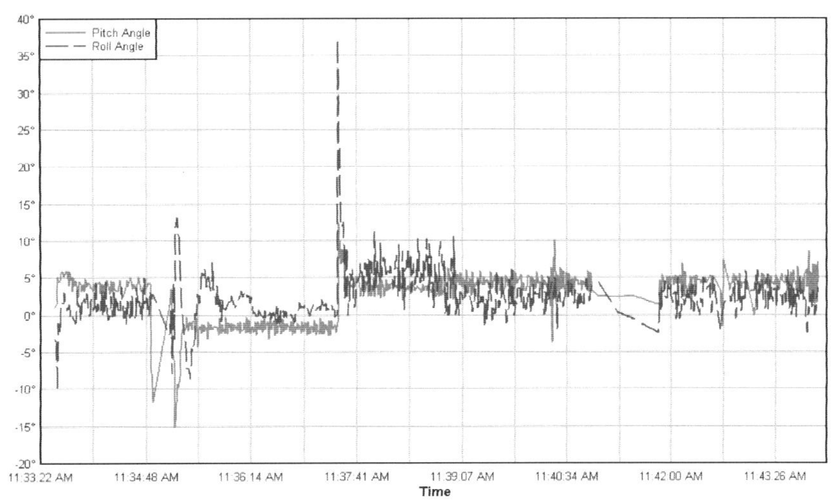

Figure 3. Oscillations of the pitch (red) and roll (blue) of the AUV while in immersion.

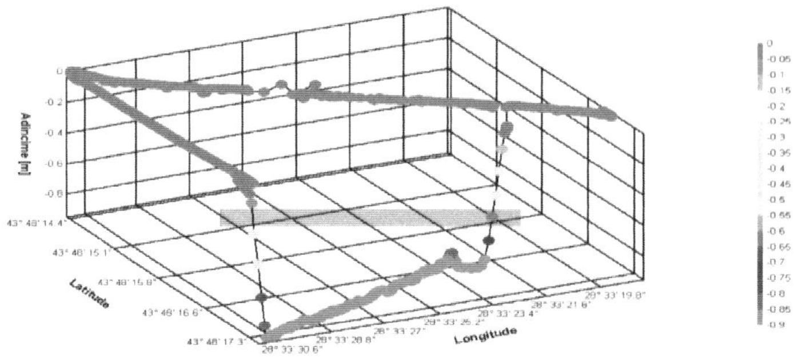

Figure 4. Vertical oscillations (depth) of the AUV while sailing at surface and immersed.

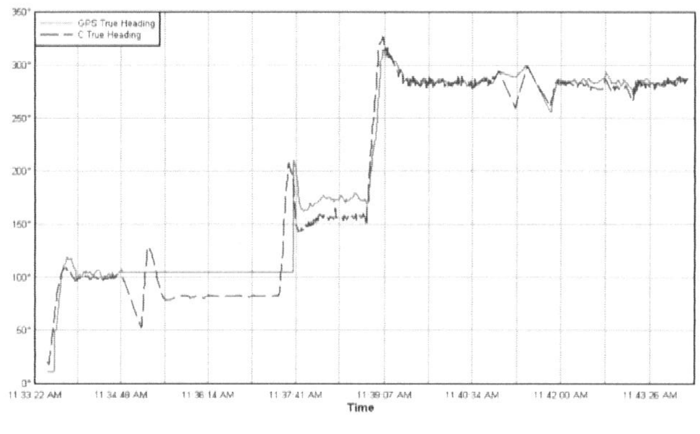

Figure 5. Variations of the AUV direction (yaw) while sailing at surface and immersed.

From the analysis of the data about the oscillations of this platform and without insisting too much on the characteristics of the area where the AUV has collected the direction or depth changes data, the following conclusions can be drawn:

- in the case of surface navigation the roll and pitch are quite large in amplitude as they are influenced significantly by the sea state and the direction of movement (the attack angle of the waves);
- in the case of underwater navigation the roll of the platform is larger than the pitch, this is mainly because the AUV attempts to correct its attitude in the presence of underwater currents, especially at slow movement speeds of around 2 kts;
- the variations of depth and path, especially when moving underwater, are reduced.

In missions where it is required to move only at surface, the vehicle can follow accurately the programmed path, using data from GPS, magnetic compass and accelerometers to compensate for the possible deviations caused by waves, currents and so on.

In underwater missions, as the vehicle does not have a Doppler velocity loch (DVL) or an inertial navigation system, compensating for the path deviations caused by the currents could not be done accurately. Later on, this causes errors in geo-referencing the images obtained from the side-scanning sonar.

Diving/ surfacing of the underwater platform can be done very suddenly, without large variations of the geographic position and this allows for a relatively easy mission planning but for obtaining relevant data that is as close to reality as possible the sonar recordings must be started only after the AUV stabilizes at the scheduled depth.

2.3. Acoustic characterization of underwater targets

For cylindrical bodies, the target strength is calculated with the formula [4] below:

$$TS = 10\lg\left[\frac{rL^2}{2\lambda}\left(\frac{\sin(k\cdot L\cdot \sin\theta)}{k\cdot L\cdot \sin\theta}\right)^2 \cdot \cos^2\theta\right] \tag{3}$$

where:

- r – cylinder radius;
- L – cylinder length;
- λ – wave length (ratio between the sound speed underwater and the sonar working frequency);
- k – wave number (2π/λ);
- θ – incidence angle of the probing signal, considered relative to the normal vector on the cylinder surface generator.

For a marine mine (MMA – a cylinder of 1.96 meters in length and diameter of 0.45 meters) and for a heavy torpedo (caliber 533 mm and length of 7.8 meters), when considering a sonar that works on a central frequency of 450 Khz, the target strength data is presented in figure 6.

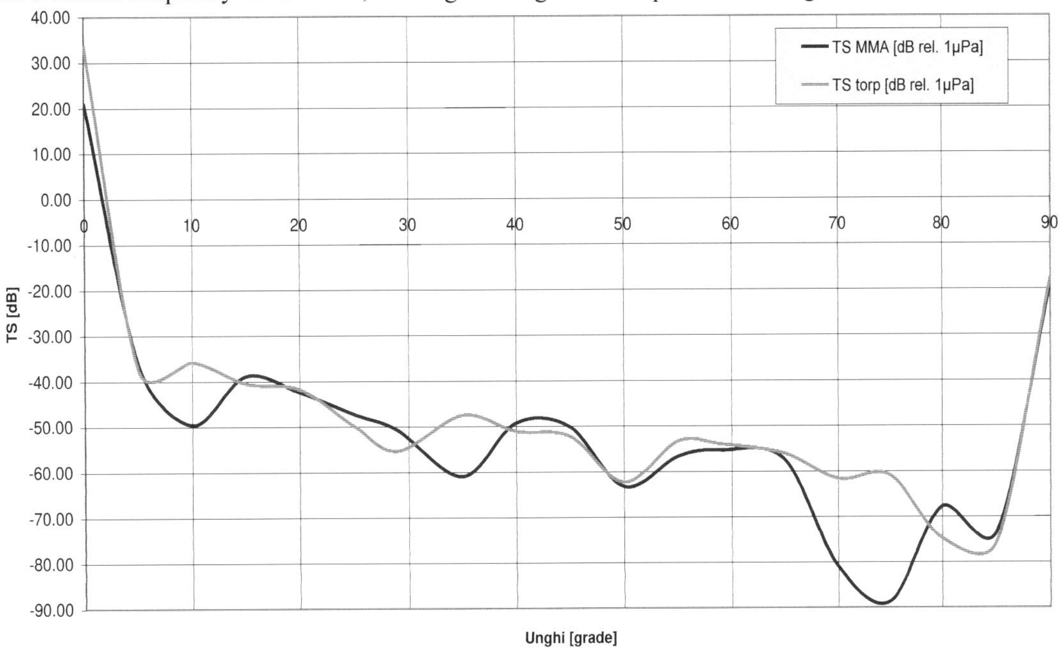

Figure 6. Target strength for a sea bottom mine (TS MMA) and for a heavy torpedo (TS torp).

It can be noticed that the TS varies across a large range of values which will have quite a large influence on the sonar discovery range. For the theoretical evaluation of the discovery capabilities of underwater targets by the side scanning sonar fitted to the autonomous underwater platform we'll consider targets with a TS = -20 and -45 dB rel. 1 µPa at 1 m, values considered adequate enough for the qualitative and quantitative evaluation of the system performance.

For the duration of the evaluation the AUV will operate in immersion, at depths of at most 5 – 20% of the depth of the area to explore. We'll only consider only modifications to the reflective properties of the sea bottom (rocky or muddy bottom) because the environmental noise level – generated by shipping traffic or sea state – has a very low influence on the general performance of the sonar in the frequency band that is used.

Theoretical evaluation of the discovery capabilities of underwater targets by the side scanning sonar fitted to the mobile autonomous underwater platform has been conducted for the scenarios presented in the following table.

Table 3. Scenarios in which we evaluate the performance of the side scanning sonar from the AUV.

Row	Target strength [dB]	Sonar immersion [m]	Water Depth [m]	Sound Velocity Profile	Specific characteristics
1.	-20	30	50	February	Sea state – SS 0 Bottom type – mixed rock and mud
2.	-20	40	50	July	Sea state – SS 0 Bottom type – mixed rock and mud
3.	-20	40	50	July	Sea state – SS 0 Bottom type – mixed rock and mud Bottom slope +10%
4.	-45	40	50	February	Sea state – SS 0 Bottom type – mud
5.	-45	40	50	July	Sea state – SS 0 Bottom type – mixed rock and mud Bottom slope +10%

Following the simulations run, we have evaluated the following relevant characteristics:

- acoustic rays distribution mode;
- propagation loss;
- target detection probability.

Results of evaluations for target detection probability for the scenarios presented in table 3 above are presented in figures 7 to 11.

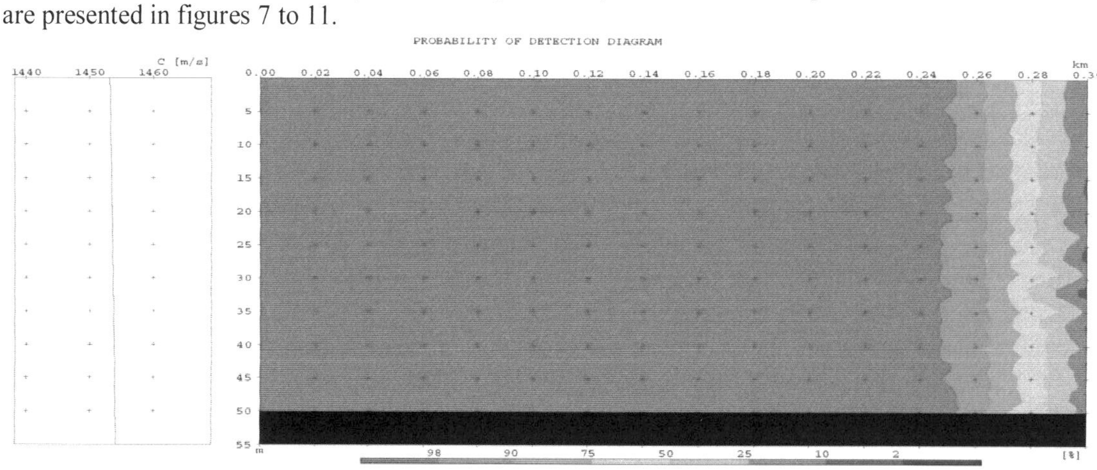

Figure 7. Detection probability distribution – scenario 1.

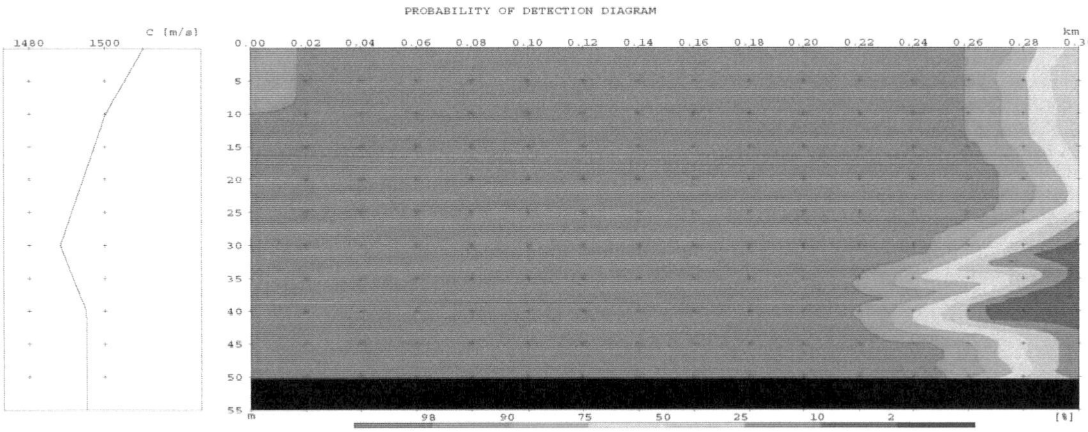

Figure 8. Detection probability distribution – scenario 2.

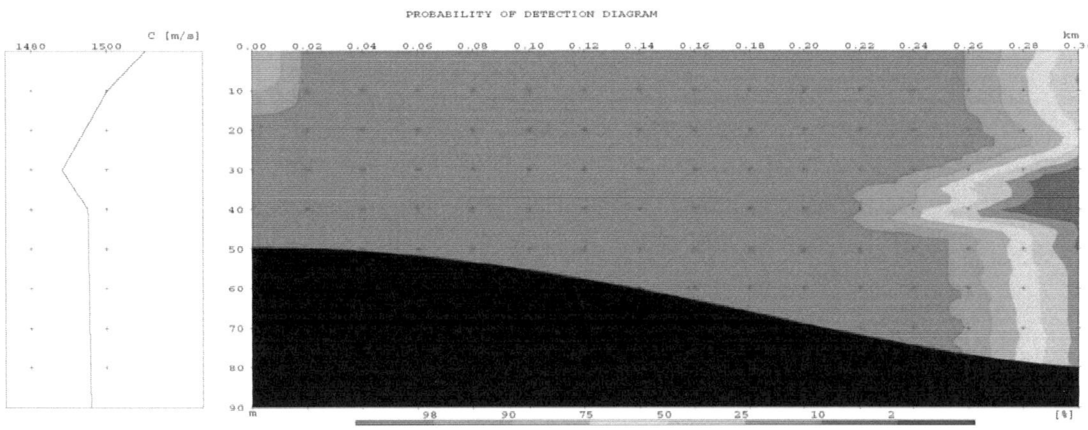

Figure 9. Detection probability distribution – scenario 3.

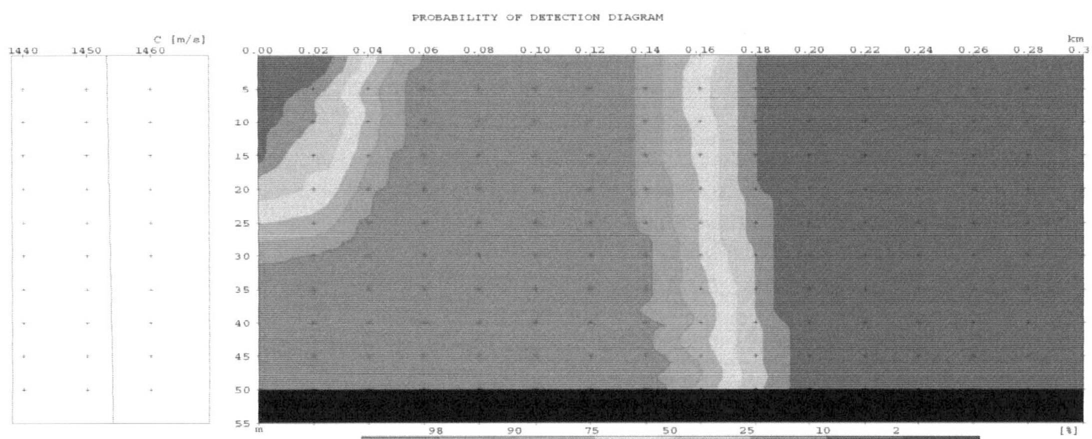

Figure 10. Detection probability distribution – scenario 4.

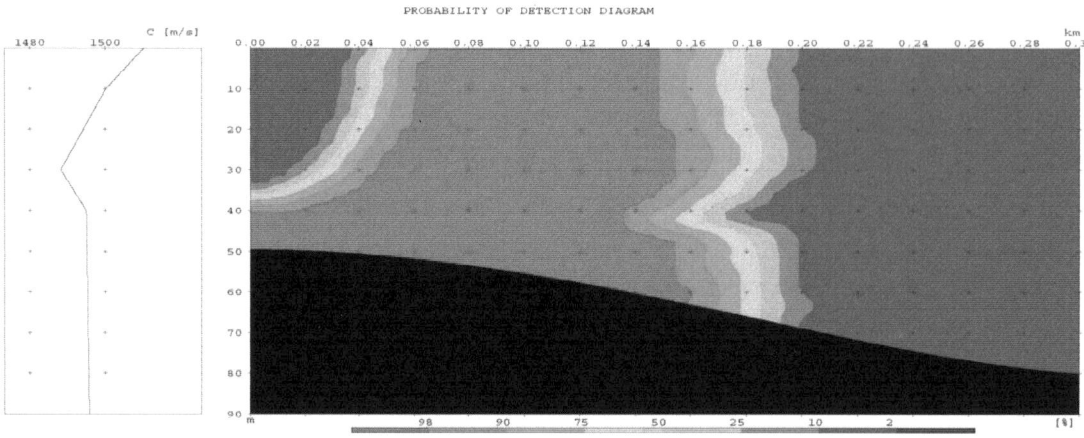

Figure 11. Detection probability distribution – scenario 5.

According to the results of the simulations we can affirm that:

- for a target strength with high values (above -20 dB rel. 1 μPa at 1 m) the probability of detection is over 98% at the maximum sonar range (adjustable between 10 – 200 m), no matter what the environment conditions are (sound speed profile specific for winter/summer, sea bottom type – muddy or rocky, water basin depth 20 – 50 m) and the AUV activity parameters (immersion between 10 to 45 meters);
- adaptation of the AUV immersion depending of the sound speed profile and the profile of the water basin bottom can lead to an increase of the target detection probability within the limit of the maximum sonar range;
- for targets with a low target strength (depending on the target viewing angle, dimensions, geometry or materials used for fabrication - under dB -20 dB rel. 1 μPa at 1 m) discovery can be done with a reasonable probability (> 90%) only at small distances, only when the AUV adapts its immersion very precisely and the maximum sonar range limit is adjusted for the sound speed profile and the sea bottom profile.

3. Conclusions

The major advantages provided by installing the side scanning sonar on the AUV are:

- excellent manoeuvrability;
- flexible development platform that can accept a wide range of sensor types and other instruments dedicated for specific missions;
- relatively long mission length;
- operating personnel can stay out the action range of dangerous items;
- significant economy of resources;
- possibility to conduct covert ops.

Experiences conducted in the previous stages have demonstrated that missions that require the platform to function on surface, it can follow accurately the programmed path, using GPS, magnetic compass and accelerometers to compensate for any path deviations due to waves, currents, etc.

The capabilities of the AUV fitted with a side scanning sonar are mostly influenced by:

- target strength (TS): the more this one has lower values, the higher that the probability of detection is at low distances only;
- knowing the precise profile of the sound speed with depth, sea bottom type and geometry: the sailing depth of the underwater platform and the maximum discovery range of the side scanning sonar depend on those two parameters in order to obtain a coverage rate as high as possible for the exploration area at a high detection probability

For determining the distance to target and target dimensions, knowing the precise sound speed in the current environment has a major importance. For this, assuming an error rate of 1% in the sound speed value, the calculations for targets and distances to the AUV path (position) are equally affected by the same 1% error.

In the context of mission definition, the main factors that affect the discovery capabilities of underwater targets are:

- quality of charts used for mission planning: the resolution of the chart determines the positioning precision of the waypoints, and the chart scale and detail level (depths of the area, sea bottom type, shipwrecks, etc) influences the mission depth plan for the AUV and thus the dimensions of the "blind spot" in the sonar data recordings, false targets, reflections from rock areas, and so on.
- planning the path of the autonomous underwater platform through immersion depth planning (usually 8% – 20% of the distance scale), overlaying the sonar swath image data, stabilization of the underwater platform on path before beginning to explore the interest area, and so on.

In missions at depth, as the vehicle does not have a Doppler velocity loch (DVL), compensating for the path deviations due to sea currents cannot be done accurately. This causes errors in geo-referencing the images obtained from the side scanning sonar. In order to decrease these errors the mission profile must be defined so that the platform returns to surface often and runs position correction calculations.

References

[1] Hanganu M and Slămnoiu G 2010 *Platformele de foraj marin românești și securitatea acestora* (București: Editura U.N.Ap. „Carol I") 166-167 170-183
[2] Button R W, Kamp J, Curtin B and Dryden T 2009 A Survey of Missions for Unmanned Undersea Vehicles (Santa Monica, CA: RAND Corporation) 69 - 102
[3] Carfield S and Hillenbrand C 2002 Defence applications for unmanned underwater vehicles *Technology and applications of autonomous underwater vehicles* 161-178
[4] Waite A D 2002 *Sonar for Practising Engineers* **70-74** 192-200
[5] Blondel P 2009 *The handbook of sidescan sonar* **20** 56

Acknowledgements

This paper has been created through the Program of Priority Areas - PN II, carried out with the support of MEN - UEFISCDI, within the framework of the project "System for detection, localization, tracking and identification of risk factors for strategic objectives on seaside areas-SIROLC", contract no. 302/2014.

IOP Conf. Series: Materials Science and Engineering **209** (2017) 012099 doi:10.1088/1757-899X/209/1/012099

Risk Factors Detection for Strategic Importance Objectives in Littoral Areas

G Slămnoiu[1], O Radu, V Roşca, C Pascu, G Surdu, E Curcă, R G Damian and A Rădulescu
Research Center for Navy, 4 Ştefăniţă Vodă str, 900402, Constanţa, România

E-mail: staff@ccsfn.ro

Abstract. With the invention and development of underwater explosive devices the need to neutralize them has also appeared, both for enemy and for own devices once conflicts are finished. The fight against active underwater explosive devices is a very complicated action that requires a very careful approach.Also, in the current context, strategic importance objectives located in the littoral areas can also become targets for divers or fast boats (suicidal actions).The system for detection, localization, tracking and identification of risk factors for strategic importance objectives in littoral areas has as one of its components an AUV and a hydro-acoustic sub-system for determining the 'fingerprints' of potential targets.The overall system will provide support for main missions such as underwater environment surveillance (detection, monitoring) in harbor areas and around other coast objectives, ship anchorage areas, mandatory pass points and also provide warnings about the presence of underwater and surface dangers in the interest areas.

1. Introduction

Risk factors, for the purposes of this paper, consist in those internal, or external elements, situations or conditions that may, by their nature, affect the security of the country, resulting in adverse effects or attainment of our fundamental interests.

In recent years, a new category of risks - the non-classical asymmetric ones - which can consist of armed and non-armed actions, has been manifested in the last years, aiming to affect the national security by causing direct or indirect consequences on the economic and social life of the country. Among the risks of this type can be listed:

- transnational and international political terrorism, including its biological and informational forms;
- actions that can address the safety of domestic and international transport systems;
- individual or collective actions of unlawful access of information systems.

Among the above, international terrorism manifests itself in an unprecedented acute form, with multiple effects on state security and international stability in general.

Ports are generally vulnerable to terrorist attacks due to their size, general water and land accessibility, heavy traffic of materials and people, positioning in densely populated areas, etc. As ports have a high density of transport routes (roads, railways and waterways), they are much more open to terrorist attacks than other areas. The concentration of passengers, goods, property and business in the harbor areas are as many potential targets.

Vessels and drilling platforms are a symbol. Above the company, the state itself is targeted by a terrorist attack. The crews, the passengers or the ship's destination are also irrelevant. A marine drilling and extraction platform is one of the most appealing targets and can be attacked by terrorists in three environments (water, air or underwater) and its damage / destruction is strongly mediated due to the economic impact and environmentally friendly.

In recent years, due to technical progress, accumulated experience, and the internationalization of the terrorist phenomenon, attacks on naval targets have become more frequent and well organized, ranging from attacking civilian ships with no defense capabilities to attacking large-scale combat ships. The experience of asymmetric actions such as those in the Gulf (1991, 2003), Yugoslavia (1999), Afghanistan (2002) demonstrated the vulnerability to cyber attacks or the possibilities of countering simple means on sophisticated means and research - Capable of operating virtually all electromagnetic spectrum.

On November 3, 2003, the Romanian Parliament adopted a law approving Government Emergency Ordinance no. 80/2003 for the acceptance by Romania of the amendments to the annex to the International Convention for the Safety of Life at Sea, 1974, and the International Code for the Security of Ships and Port Facility (ISPS Code) adopted at the International Maritime Organization Conference in London, Between 9 and 13 December 2002. There is no express provision under the ISPS Code on the protection of port facilities and ships in ports against terrorist or diversionary attacks launched from underwater [1].

2. Potential scenarios of terrorist actions against strategic coastal objectives

Table 1 presents a number of representative features that could be used to draw up scenarios of possible attacks [2].

Table 1. Characteristics of some possible terrorist attack scenarios.

Elements	Examples	
Authors	• Transnational Terrorist Groups • Islamic non-affiliated • Foreign nationalists	• Disgruntled employees • Others
Objectives	• Mass victims • Destruction of port, ship, drilling and extraction platform, etc.	• Undermining economic activities • Destruction of the environment
Locations	• ports • Stationary or deployment areas • Mandatory crossing points	• Roads and passes • Natural Reserves or Tourist Areas
Targets	• Military or commercial ships • Oil platforms • Ferries / cruise ships	• Port areas • Waterways
Tactics	• Explosible in suicide ships • Crash / hitting ships • Missiles launched from ships • Marine mines and / or improvised explosive devices	• Diversion divers • Underwater vehicles • remote / autonomous • Intentional sinking of merchant ships

Based on the information presented in table 1, it is possible to create numerous unique scenarios of attack, logic and operational veracity, based on different combinations of authors, objectives, locations, targets and tactics. In this way, plausible scenarios will be created and can not be avoided with limited anti-terrorist resources.

In order to be able to define operational features of a system of protection against such attackers, in addition to good ktsowledge of environmental conditions (propagation, position of natural obstacles, fund profile, statistical properties of clutter, etc.) Also considering those characteristics of the threat related to the movement and its "fingerprint" from the perspective of the sensor used in the detection and identification process [3]. Table 2 presents such data from the perspective of hydroacoustic detection (active or passive).

Table 2. Characteristics of targets from detection perspective using active sonar systems [1].

Attacker/weapon	Speed [kts]	Action depth [m]	Target strength [dB rel. 1 μPa/1 m]	Generated noise level
Diver	0,5 - 1	1 - 20	-25	Low
Torpedo	10 - 50	> 20	-20	High
Fast ship	5 - 40	0,1 – 0,5	?	High, but strongly oscillating
Mine	-	5 - 10	-15	-
AUV / ROV	2 - 5	5 – maximum zone depth	-20	Low
Mini submarine	2 - 5	5 - 30	-10	Low
Unforeseen attackers (e.g. dolphins)	1 - 35	1 - 50	-15 to -30	Medium - High

We note that potential aggressors with high availability have evolving and specific characteristics from the point of view of sonar detection dispersed in a relatively wide range of values so that the process of establishing some technical and operational requirements becomes extreme difficult. Consider, for example, the case of divers:

- Equipment can be purchased relatively easily;
- The production of explosive charges or their acquisition on obscure channels should not be considered a difficulty;
- May be launched by pleasure boats, beaches nearby, objective vessels, commercial vessels, etc.;
- The specific characteristics vary significantly depending on the sonar system (working frequency, modulation type, etc.), the breathing circuit used (closed or open) and so on;
- Evolves in an environment whose features make detection difficult: noise due to extremely high traffic, shallow waters, high masking possibilities, etc.

At 1 kts (≈ 0.5 m / s), the diver travels a distance of 500 m in about 17 minutes. These 500 m are declarative detection distances for active sonar systems for discovering divers, ensuring a detection probability of more than 90%. If we take into account the time needed to detect, identify and classify the target, establish the trajectory, assess its degree of danger, and so on, it can be noticed that the time required for the intervention of the counter-action team remains for about 8-10 minutes, which will involve the use of ultra-fast means of travel, localization and "instantaneous" neutralization of the aggressor. This can be aggravated by the fact that the target strength for an open-circuit diver can reach -27 dB (at 100 kHz), the air bubbles trailing in a ship's path mask the target for a significant time, and the presence of various objects in the surveillance area, in addition to allowing divers to hide, significantly impedes the operator's activity [4].

From the analysis of the terrorist actions that targeted merchant ships and port facilities to the present, the following were noted:

- Attacks with the use of small and fast ships loaded with explosives are the easiest to organize and fulfill. Such an attack on a ship, harbor facilities or drilling platforms involves first of all an attempt to approach the target with one or more craft until the arsenic reaction to neutralize the threat is late.

- Any organization, however poor or poorly funded, may engage in "amateur divers", but this is not the only risk. Strong organizations can afford to train professional teams of underwater terrorists. Diving can be equipped with me or improvised explosive devices that they can place on the ship's body, on various submersible elements of marine platforms or port infrastructure.

On the basis of the conclusions presented, we consider that the detection of threats to the strategic objectives at the seaside should focus, but not be limited to, the two types of targets. In this vision, the following research directions have been identified:

- The use of physical barriers to delay the aggressor;
- The use of passive detection methods and non-acoustic sensors, as well as the development of robust data fusion algorithms to increase the probability of detection and simultaneously decrease the probability of false alarms;
- Conducting in-depth research and creating specific footprint databases in order to increase detection and target classification capabilities simultaneously with the decrease of the interval between the first alarms alignment and the moment of aggression identification / classification.

3. Detection of risk factors against strategic objectives at the seaside

The passive detection of small and fast craft or self-directed divers may be based on the identification of their "fingerprints" in the marine noise recordings, and this identification of "fingerprints" is limited to finding hidden periodicity in a generally uncorrelated noise. Periodicity is provided by the propeller, propeller, engine, or respiratory cycle shaft in the case of the diver.

It has been mentioned in many specialized papers [5], [6], [7] that this periodicity of the acoustic fingerprint of the ship or diver can be observed in the hydrophone signal spectrogram and that the analysis of the cycle frequency in the output signal of the hydrophone allows detection of the ship or diver and estimation of its respiration rate.

As shown in [8] a method of processing signals with very good results for the detection and estimation of the rotation frequency of the propeller and the rate of respiration is the method using the DEMON algorithm (Detection of Modulation On Noise) and for a hydro-acoustic system The use of a DEMON algorithm according to the scheme illustrated in the following figure is proposed for the determination of the potential target fingerprints (HDAPT).

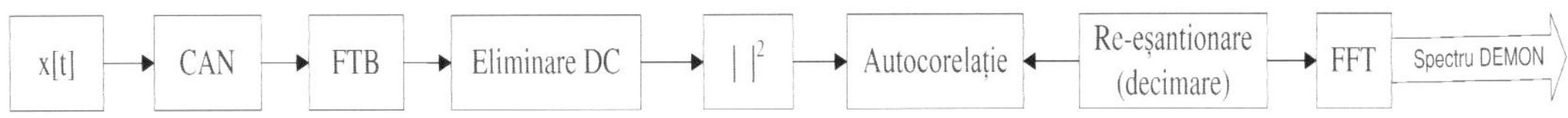

Figure 1. Block diagram - DEMON HDAPT [8].

For the implementation of the HDAPT algorithm and the display of results in clear, intelligible formats for the operator, two software applications specific to the detection of ships and self-directed divers have been developed. The signal source, representing the noise of the moving ship or the breath of the diver, can be in analogue format (connection to the audio inputs of the computing system) or in digital format (ASCII, * .wav, * .mp3 etc.).

In case of vessel detection, the DEMON diagram is displayed in the main menu (figure 2). The operator can identify on the DEMON diagram whether in the ambient noise picking from the hydrophone there are signals with specific characteristics of the noise produced by the propeller propeller. Based on the data obtained by applying the HDAPT algorithm, the application calculates and displays the ship's specific characteristics and displacement characteristics: the rotational speed of the portable spindle (rot / min) and the number of propeller blades.

Using a variety of ship noise records, it was found that, in general, under normal weather and normal traffic conditions, it is necessary to apply a medium band pass filter, resulting in higher output

(Figure 2, right). In situations where the natural noise level is significant, the results obtained are superior if a low frequency range is used for the band pass filter (figure 2, left).

In addition to the results of the HDAPT algorithm, the signals acquired after filtering JF (left) and MF (right) are also displayed on the screen.

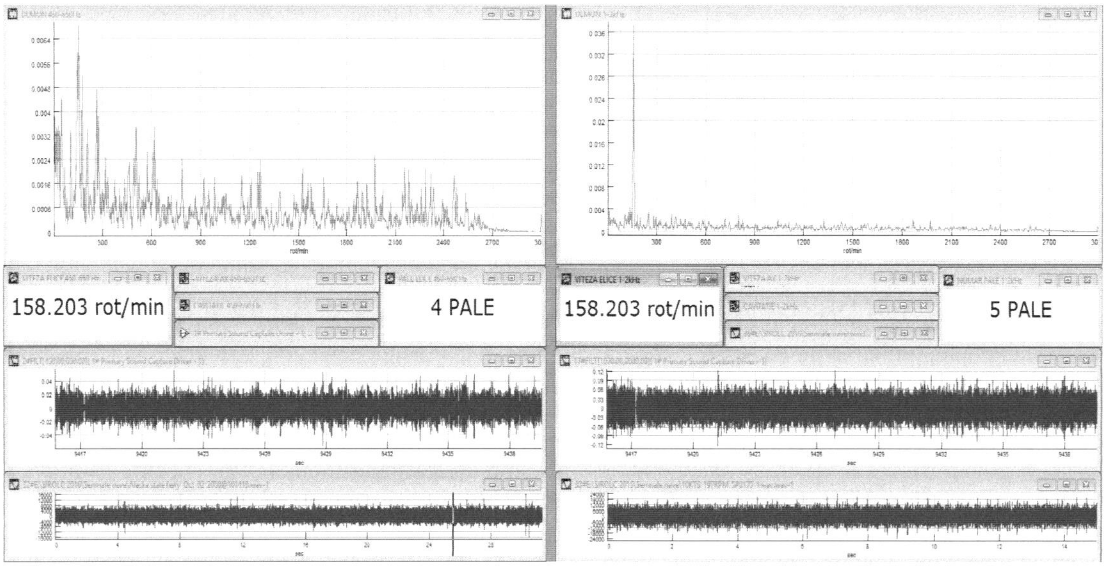

Figure 2. Graphical operator interface for detection of Alaska State Ferry.

For the detection of a diver in the algorithm, parts of the hydrophone signal are processed, representing about 4 to 5 relatively stable breathing periods. Considering the limits for a normal respiration between 10 and 30 breaths per minute (0.15 ... 0.5 Hz) [9], it results that a rhythm of 10 breaths per minute, it takes about 30 seconds, while when the diver is moving fast (with over 30 breaths per minute) the interval required for detection would be reduced to one-third. On the other hand, the higher the number of periods of breathing, the more the algorithm applied is less sensitive to relative rhythm variations and ambient noise. For these reasons, a diver detection software application was designed and developed that simultaneously processes 10, 20 and 30 second intervals from the hydrophone signal.

Figure 3 shows the graphical interface of the operator, observing the results of the hydrophone noise processing on three horizontal sections (bottom-up - for 10, 20 and 30 s intervals):

- The DEMON diagram;
- Spectrogram;
- The calculated breathing rate value;
- The oscillogram of the acquired noise.

On the right of each section is a circular instrument, which automatically signals (the needle in the red area) the diver's detection. If the ambient noise is due to natural phenomena (wind, rain, etc.), even if the noise level covers the noise produced by the breath of the diver, the application calculates correctly all the parameters because there are no impulsive noise characteristics.

In observation by the operator, the main indication of the presence of a diver near the hydrophone is the presence on the spectrogram of the vertical bands specific to the acoustic energy emitted by the diver's diving apparatus during inhalation of the air [9].

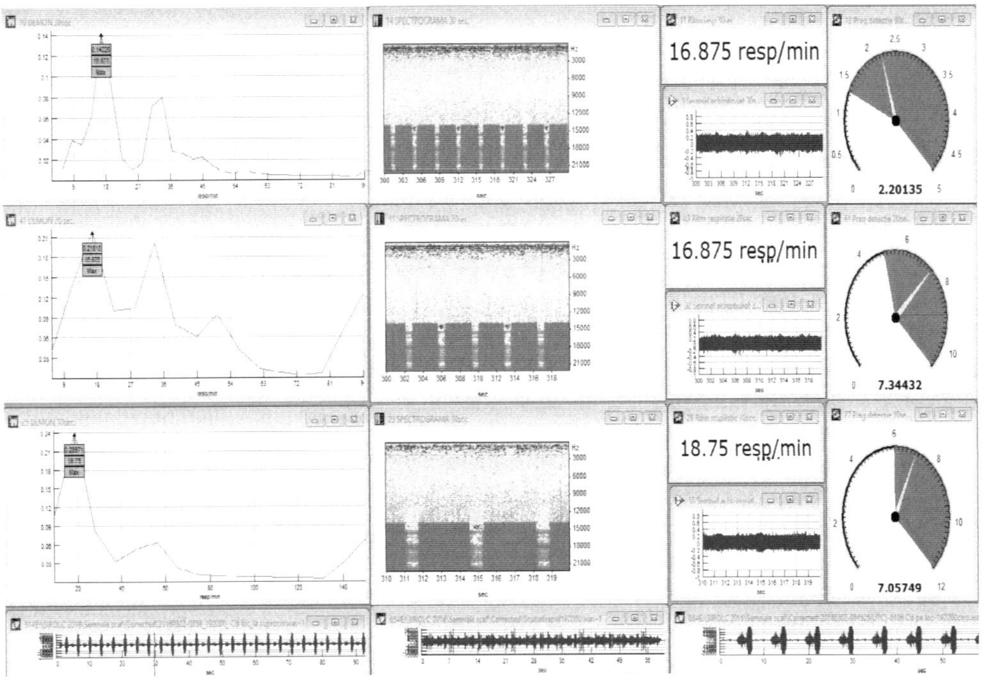

Figure 3. Operator Graphic Interface for Divers Detection (strong wind conditions).

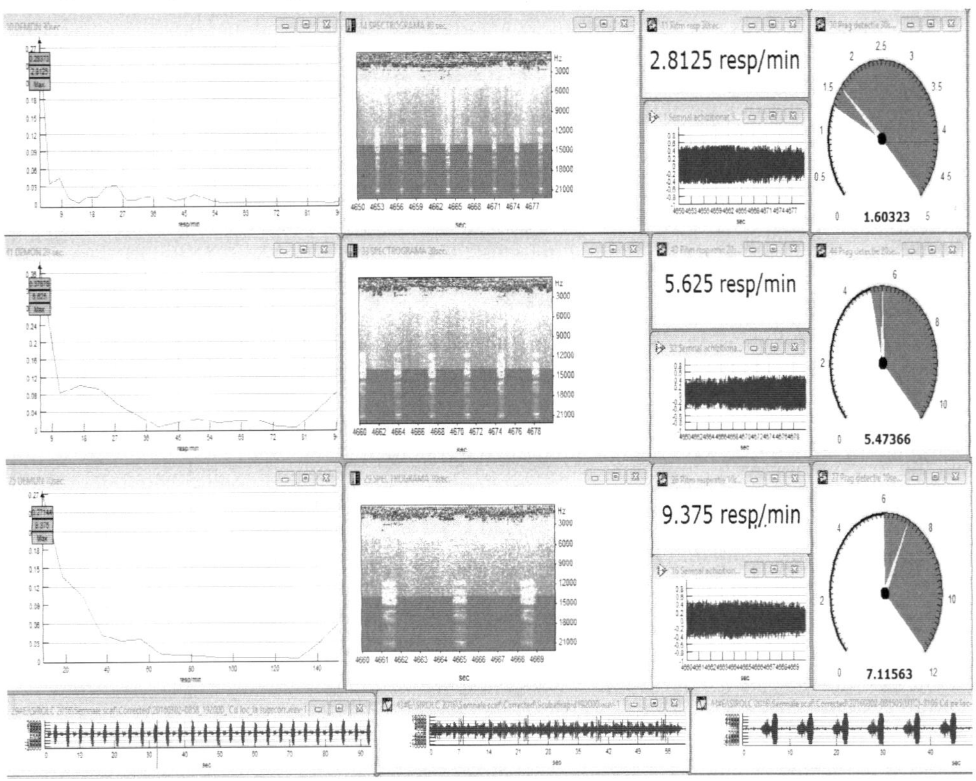

Figure 4. Detection of divers in strong wind conditions and simultaneous evolution of a helicopter and a ski-jet in the area.

As with small craft detection software, for testing and training, the operator can select and observe the results for overhead diver noise and other noises (eg from a helicopter, ski-jet, etc.) pre-recorded.As can be seen in Figure 4, due to the high level and the impulsive character (but with less repetition) of helicopter and ski-jet noise, the DEMON diagram moves to the left, the calculated periods are lower, and the final result is significantly affected (the noise produced by the diver is the same as in Figure 3).

4. Conclusions

There are no safety and security solutions that guarantee 100% protection of strategic objectives in seaside areas, but a set of measures can be implemented to help mitigate risks. In this respect, the detection of risk factors is an imperative requirement, and the detection of threats such as fast surface craft and autonomous divers are a first step.The HDAPT signal processing algorithm (DEMON) has produced good results for detecting the threats mentioned under normal operating conditions, and results are similar to those obtained by applying algorithms of the same type presented in the literature.When used with other sources of information (AUVs, surface sensors - radar, AIS etc.) of hydroacoustic threat detection systems incorporating the applications described in the paper and some information fusion solutions, the level of Protection of strategic objectives in coastal areas will increase significantly.In the future, it is intended to test the software applications developed using a wide range of noise recordings (for boats, ships and divers) performed under different hydro-meteorological conditions and presence of disturbing factors (naval, air traffic, etc.).

References

[1] Hanganu M and Slămnoiu G 2010 *Platformele de foraj marin româneşti şi securitatea acestora* (Bucureşti: Editura U.N.Ap. „Carol I") 37, 50-51, 124-129, 170-177

[2] Parfomak P W and Frittelli J 2007 *Maritime Security: Potential Terrorist Attacks and Protection Priorities*: Congressinal Research Service The Library of Congress) 7

[3] Kessel R T 2007 Protection in ports: countering underwater intruders *Undersea Defence Technology Europe* (Naples, IT: UDT Europe) 4

[4] Sharma N S, Yakubovskiy A M and Zimmerman M J 2013 SCUBA Diver Detection and Classification in Active and Passive Sonars - a Unified Approach *2013 IEEE International Conference on Technologies for Homeland Security (HST)* 193-194

[5] Donskoy D M, Sedunov N A, Sedunov A N and Tsionskiy M A 2008, Variability of SCUBA diver's acoustic emission *Proc. SPIE 6945*, Optics and Photonics in Global Homeland Security IV 694515 doi: 10.1117/12.783500

[6] Labat V and Dare D 2014, Analyse de signaux acoustiques marins: identification de frequences caracteristiques via la methode DEMON, *Congres Francais d'Acoustique (12; 2014; Poitiers)* Poitiers France 345-350 HAL Id: hal-01087837

[7] Chung K W, Sutin A, Sedunov A and Bruno M 2011, DEMON Acoustic Ship Signature Measurements in an Urban Harbor, *Advances in Acoustics and Vibration* vol. 2011 Article ID 952798, doi:10.1155/2011/952798

[8] Slămnoiu G, Radu O, Roşca V, Pascu C, Damian R, Surdu G, Curcă E and Rădulescu A 2016 *IOP Conf. Ser.: Mater. Sci. and Eng.* **145** 082013

[9] Donskoy D M, Sedunov N A, Sedunov A N and Tsionskiy M A 2008, Variability of SCUBA diver's acoustic emission, *Proc. SPIE 6945,694515 Optics and Photonics in Global Homeland Security IV* 1-11

Acknowledgements

This paper has been created through the Program of Priority Areas - PN II, carried out with the support of MEN - UEFISCDI, within the framework of the project "System for detection, localization, tracking and identification of risk factors for strategic objectives on seaside areas-SIROLC", contract no. 302/2014.

IOP Conf. Series: Materials Science and Engineering **209** (2017) 012100 doi:10.1088/1757-899X/209/1/012100

Assessment and Rehabilitation Issues Concerning Existing 70's Structural Stock

E Sabareanu [1,2]
[1] CCI Department, Politehnica University Timisoara, 2 T. Lalescu Str., Timisoara, Romania
[2] SC SELLM SRL, 23 Gh. Lazar str, 300344Timisoara, Romania

E-mail: office@sellm.ro

Abstract. The last 30 years were very demanding in terms of norms and standards change concerning the structural calculus for buildings, leaving a large stock of structures erected during 70-90 decades in a weak position concerning seismic loads and loads level for live loads, wind and snow. In the same time, taking into account that a large amount of buildings are in service all over the country, they cannot be demolished, but suitable rehabilitation methods should be proposed, structural durability being achieved. The paper proposes some rehabilitation methods suitable in terms of structural safety and cost optimization for diaphragm reinforced concrete structures, with an example on an existing multi storey building.

1. Introduction

The towns' development during the historical ages involves difficult changes in architectural and structural conception [1]. Usually, the change between two different approaches for these aspects appears gradually, and follows the civilisation steps of changes. Because our evolution has grown very fast, the cities must adapt and change with the same speed. This aspect leaded to a peculiar situation: the central town areas are populated with old buildings from different historical stages which does not comply with modern structural safety requirements [2] [3], are important for the town history and also not comply with current architectural requirements for use of interior spaces [4]. Examples of these aspects are given in literature for existing buildings of different types: from historical building from XVIII Century [2] to rather modern buildings from the XX Century with industrial purpose [5], [6] or hospitals and administration purpose [7]. Taking into account the large distribution of changes occurred in buildings structures, load type and intensity, the literature proposes lines of priorities of the upgrade process [8], [9], and identifies ways of assessment for these structures taking into account the stability [10], [11] and strength [12], [13], mainly from seismic point of view [14], [15].

The building considered in the present paper can be classified after the literature in the category of buildings with large durability [16] that can be considered for rehabilitation in order to use its entire life span at maximum capacity. It is a complex of fours independent reinforced concrete structures that formed a spa complex that includes a hotel, situated in one of the most valued Romanian wellness resorts.

IOP Conf. Series: Materials Science and Engineering **209** (2017) 012100 doi:10.1088/1757-899X/209/1/012100

2. General aspects concerning the building status

The building in discussion is placed in South Carpathians Mountains and was designed and built during 1970 decade. The assembly is consisting in four sections, for which a general view can be seen in Figure 1. Besides D sector, all sectors have three basements used for wellness and treatment, the superstructure level being different. The highest sector is B – where hotel facilities were provided – that has 17 floors.

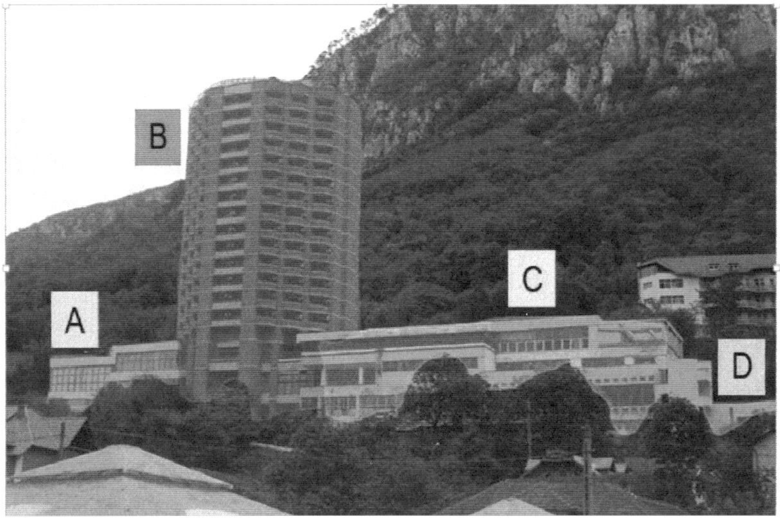

Figure 1. General view.

2.1. Structural system

All the structures are cast in place reinforced concrete structures, in mixt solution – spatial frames / diaphragms in different proportion chosen in regard with the number of levels and horizontal print for each section.

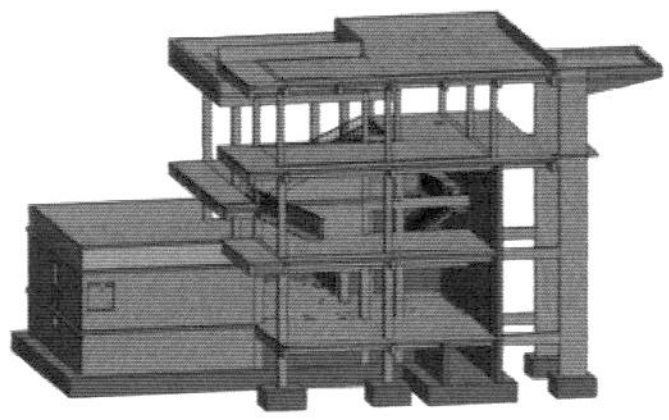

Figure 2. Section A structure

Figure 3. Section C structure.

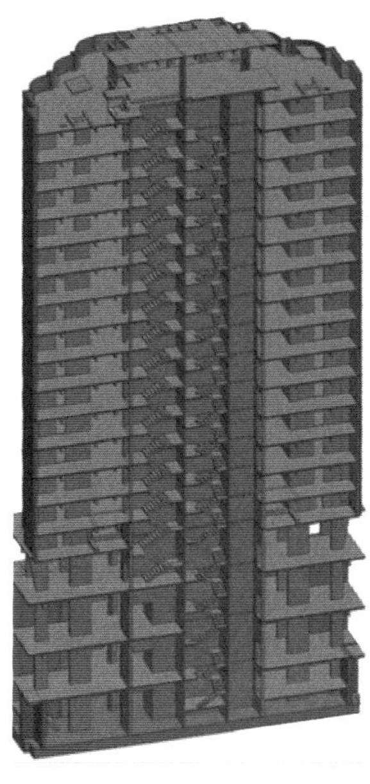

Figure 4. Section B structure.

As figures 2, 3 and 4 shows, for the small heights sectors, the spatial frames prevail, the diaphragms being used only on the structure perimeter. The situation differs for section B, where the concrete walls are mainly used. The particularity here is that the columns are prevalent at the first four ground floors, and concrete walls at the rest of the floors. An important aspect to be mentioned is the joint between sectors, which is a small dimension joint, provided only for settlement.

The foundations are slab foundations 70 cm thick for the main part of the structure, and there are some isolated foundations for façade columns support.

The materials used are concrete C12/15 for infrastructure, and superstructure above +20.00m, and C16/20 for the rest of the superstructure. The reinforcement is made of PC52 longitudinal rebars and OB 37 for stirrups at all elements.

The coherence between the design and site situation was checked using nondestructive tests based of vibration change at transmission in different environments [17]. In the areas where the result were not conclusive, carrots were taken for laboratory tests. Following the tests, the compliance of design requirements for materials resistance was 80% fulfilled. The geometry and reinforcement content was 99% respected.

2.2. Structural integrity

Generally, the structure shows a good behaviour and integrity. Following the detailed visual analysis of the structural elements, it was observed that they did not show generalized degradation or pronounced cracks. Some small degradation of concrete cover layer (Figure 6), small segregation areas and some small carbonation areas were found at sector A (Figure 5).

Figure 5. Carbonated concrete beams and exposed reinforcement.

Figure 6. Visible rebars on columns due to inadequate concrete cover.

Water infiltrations are also present in some areas, and technical floor show ware marks. Finally, it can be said that the structures shown a good behavior in time, and it is suitable for rehabilitation and further use.

3. Considerations regarding the suitability of the structure to the current resistance and stability requirements

Structure design was finished during 1976, when the following conditions were stated.

- The seismic design norms prior to the 1977 earthquake were P13-63 and P3-70, associated with the seismic map given by STAS 2923-63, which frames a large number of localities in the macro seismic zone 6 for which, according to the norms in Force, it was not required to design the structures to the effect of seismic forces, regardless of the material from which they were made. This provision led to the situation that the entire 1963-1977 built-up fund in the 6th zone (among which Baile Herculane where the discussed building is situated), is not calculated at the action of seismic forces.

- The level of normalized loads was lower than the ones required by present norms used for seismic rehabilitation purpose P100-1/2006 and P100-3/2008, according to which the examined building is in the seismic area with the ground acceleration ag = 0.16g.
- The partial safety factors considered in the calculation of the calculations were much lower than those in the Eurocodes. A comparative study of these issues is outlined in table 1.

Table 1. Seismic zoning, partial safety factors and standard loads.

1976 –building design	2016 –assessment	Difference [%]
SEISMIC ZONES		
Normative P13-63, and STAS 2926-63: It was not considered a seismic area	Normative P100/1-2006: Grade 7.5 on the Richter scale:- Design acceleration to earthquake 0.16g	
SAFETY FACTORS		
- Own weight permanent elements (structural elements)		
1.1	1.35	22.7%
- Weight of insulation, finishing		
1.3	1.35	3.8%
- Live loads		
1.3	1.5	12.5%
LOADS		
- Live load for Hotels		
150 x 1.3 = 210 daN/m^2	150 x 1.5 = 450 daN/m^2	13.3%
- live loads for halls		
200 x 1.3 = 260 daN/m^2	200x 1.5 = 300 daN/m^2	13.3%
- live loads for traffic areas		
400 x 1.3 = 520 daN/m^2	400 x 1.5 = 600 daN/m^2	13.3%
- live loads for roofs		
75 x 1.4 = 105 daN/m^2	75x1.5=112.5 daN/ m^2	7.1%
MATERIALS		
- minimum concrete class existing in structures C12/15	- minimum concrete class by norms for M ductility class C20/25	

The first issue to be discussed was the response of overall structure to the seismic requirement of present norms. From this point, the following considerations can be made:

- Structure geometry:
  - This requirement is considered respected because the construction of all sectors is regular both on horizontal and vertical direction. For the B sector, the centre of rigidity and the centre of the masses coincide, the building being symmetrical in both directions, a major advantage being not required at general torsion.

   - The structural system is continuous and ensures a clear, direct and uninterrupted transmission of the seismic forces, regardless of their direction, to the ground.
   - It can be considered that the slabs insures the lateral stiffness at each level and thus the seismic forces are transmitted relatively uniformly to all diaphragms and columns
   - There are no significant discontinuities in the route of the seismic forces discharge.
   - The structural regularity in plan and elevation of the sections is satisfied.
- Structural redundancy: there is a high degree of redundancy in the structure, since the failure of one diaphragm does not affect the entire building. As a ductility level, the structure is considered to have a medium ductility because it can form plastic hinges only at the base of the diaphragms and these have the required level of reinforcement for medium ductility elements (such as bulb-shaped ends).
- Because at the design date the site was not considered a seismic area, there are no seismic joints between the sectors.

In the light of the findings considering the state of the structure and the change in norms and utility of the structure, the following conclusions can be considered.

The structures respect the original project in a large proportion. However, the concrete classes in the project used in the structure of resistance do not fully meet the requirements imposed by the real norms. According to the non-destructive tests, there were differences between the concrete classes in the project and those from the expert elements. The actual concrete classes were used to calculate the resistance structure.

The degree of fulfilment of seismic conditions for the detailed qualitative assessment of structure is 64% from the required bearing capacity at seismic loads, meaning a construction which, under the effect of the design earthquake, may present structural degradations that do not significantly affect Structural safety but where non-structural degradations may be important. On the other side, the structures are separated by a 2.5cm compression joint that is not sufficient to allow specific level shifts caused by seismic action so buildings can collide.

4. Considered rehabilitation

The main issue that had to be corrected was the behaviour at seismic loads. The seismic joints were re-designed, so that their final size to summing the level displacements of the adjacent buildings to which a safety space of 5cm is added, obtaining a 12.5cm total length. Because the vicinity was insured with cantilevers, both slabs and beams were cut 5 cm each near the settlement joint in order to increase its size at all levels. In this way, the lateral drift for each floor is solved for the new seismic conditions.

The bearing capacity of all elements had also to be insured. Measures to increase it was taken for each type of structural element. The main beams were consolidate by adding reinforcement both on upper and lower sides, the columns received additional reinforcement on all sides. The diaphragms did not require an increase of strength due to their massiveness. Foundations check both as pressure on the ground and as reinforcement. The pressure at the base of the foundation was verified based on the conventional pressure presented in the geotechnical study and the capacity of the land is not exceeded.

A problem was to insure the lateral displacement of the first ground levels, weaker due to lack of diaphragms compared with other floors. Since it is not technologically possible to consolidate all columns, a study has been carried out in several consolidation variants. Considering the technological conditions of execution, the solution was to introduce three diaphragms at the level of the ground floor in positions with maximum influence on lateral structural movement.

Degradations produced by a bad execution (insufficient coating) were repaired by removing the segregated concrete until a solid, rugged support was found. After dust clearing, the exposed reinforcements were treated with passive solutions and a new coating of mono-component sulphate-resistant thixotropic single-component thixotropic mortar was used.

Degradation caused by environmental factors (degraded cover layer, exposed reinforcements) were treated also by cleaning the affected surfaces and replace the lost concrete with the same type of mortar.

5. Conclusions

The case presented shows that structures in the 1970's have a high durability and their life span is not yet consumed. It is easy to notice that the structures that have a high density of reinforced concrete diaphragms in both directions also fulfil the conditions of lateral displacement and seismic loads imposed by the current norms.

Thus, we can assume that the rehabilitation of these structures is a simplified process compared to similar reinforce concrete framestructures, implying only punctual rehabilitation in areas with diaphragm deficiency.

In conclusion, if the structures are properly maintained, the rehabilitation process is economically viable and the building's durability is greatly improved.

References

[1] Narita A-M, Gurza V, Oprita R, Keller A, Apostol I, Mosoarca M and Bocan C 2016 *Pollack Periodica* **11** 15-26
[2] Lute M 2015 *15th International Multidisciplinary Scientific Geoconf. (SGEM 2015) (Albena, Bulgaria, 18-24 June 2015)* **2** 569-574
[3] Dan S, Iures L and Badea C 2013 *13th International Multidisciplinary Scientific Geoconf. (SGEM 2013) (Albena, Bulgaria, 16-22 June 2013)* 451-456
[4] Gioncu V and Mazzolani FM 2010 Final Conference on COST Action C26: Urban Habitat *Constructions under Catastrophic Events (Naples, Italy, 16-18 Sep 2010)* 87-92
[5] Lute M and Marincu A 2016 *16th International Multidisciplinary Scientific Geoconf. (SGEM 2016)* (Albena, Bulgaria, 30 Jun-06 Jul 2016) **II** 579-586
[6] Lute M 2016 *IOP Conf. Ser.: Mater.Sci.Eng.* **133** 012022
[7] Lute M 2016 IOP Conf. Ser.: Mater.Sci.Eng. **133** 012033
[8] Kolz E 2001 Burge M SEI 11 202-206
[9] Iacob F 2010 *Phys. Lett. A* **374** 1332-1335
[10] Tesfamariam S, Loeppky JL and Bazabeh MA 2017 *Struct. Infrastruct. Eng.* **13** 554-566
[11] Iacob F and Lute M 2015 *J. Math. Phys.* **56** 56
[12] Yamin LE, Hurtado A, Rincon R, Dorado JF and Reyes JC 2017 *Eng. Struct.* **138** 308-323
[13] Lavan O, Wilkinson PJ 2017 *J. Struct. Eng.* **143** 04016205
[14] Shu Z, Li S, Gao M and Yuan Z 2017 *Earthquake Eng.* **12** 297-307
[15] Ozmen H.B and Inel M 2016 *EARTHQ STRUCT* **10** 1315-1330
[16] Puskas A and Moga L 2015 *SDP* **10**(2) 165-176
[17] Iacob F 2014 *Cent. Eur. J. Phys.* **12**(9) 628-636

IOP Conf. Series: Materials Science and Engineering **209** (2017) 012101 doi:10.1088/1757-899X/209/1/012101

Laboratory Performance Testing of Two Types of Geotextiles used in Danube Hydrotechnical Works

F D Dumitru, M A Moncea*, A M Panait, M Olteanu and L Laslo
National Institute for Research and Development in Environmental Protection, 294 Splaiul Independentei, 6th District, 060031, Bucharest, ROMANIA

E-mail: moncea.andreea@gmail.com

Abstract. The products included in the geosynthetic category (geotextiles, geogrids, geonets, geomembranes, geosynthetic clay liners, geofoam, geocells (cellular confinement) and geocomposites) have multiple civil engineering uses, being effectively used as drains and filters in civil and environmental works. The paper aims to test the performance of the geotextiles used for the ecological restoration of riverbank defences on the Danube River, between Calarasi and Braila. For this purpose, the analysed geosynthetics were subjected to hydrolysis and oxidation degradation in laboratory conditions. To evaluate the effect of the two degradation mechanisms, the specimens were subjected to tensile tests at room temperature. The three analysed parameters were tensile strength, elongation and failure mode. The results showed that the tensile strength values for the samples subjected to oxidation and hydrolysis are lower than the ones corresponding to the reference samples, while the elongation values determined after mechanical testing showed that hydrolysis influences the fibre flexibility. The failure mode of the analysed geotextiles highlighted the tendency of the samples to fail either in the calibrated area or at the grip, indicating that in use the geotextile will break in the strained region.

1. Introduction

Geosynthetics are synthetic products in the form of sheets, strips or textile fibres, woven or non-woven, in which at least one component is a polymer (synthetic or natural) [1]. These materials have been developed mainly because of the need to increase the durability of applications subjected to severe exploitation conditions [2], being suitable for many applications in hydraulic engineering (dams, basins, erosion / coastal protection, etc.) [3].

Considering the long-term performance needed for this type of materials, durability tests are required, taking into account the degradation mechanisms that affect them (ultraviolet light degradation, radiation degradation, chemical degradation, hydrolysis degradation, swelling, extraction, delamination, oxidative degradation and biological degradation), which can act independently or synergistically [4]. Therefore, geotextiles degradation may occur as a result of interactions with environmental factors under temperature conditions and mechanical stresses. The most important factors that favour chemical degradation mechanisms are the pH of the environment, solar radiation, temperature, and mechanical stress [5].

Hydrolysis degradation of the geotextiles is a process by which polyethylene terephthalate (PET) decomposes by its reaction with water, the reaction being reversible [6]:

$$H[\text{-OOC-}\varnothing\text{-COOCH}_2\text{CH}_2^{-}]_n\text{OH} + (2n\text{-}1)\,H_2O \rightarrow n\,(\,\text{HOOC} - \varnothing - \text{COOH}) + n\,(\text{HOCH}_2\text{CH}_2\text{OH}) \quad (1)$$

However, if the polyester fibre is degraded the reaction becomes non-reversible.

The behaviour of polyethylene subjected to hydrolysis degradation by various chemicals was considerably studied [7,8,9]. There are two types of hydrolysis degradation that can affect the geosynthetics, namely internal or external degradation [10] by damaging the polymer chain, either throughout the cross-section of the polyethylene (internal hydrolysis) or at the surface (external hydrolysis), leading to a lower molecular weight.

The other degradation mechanism studied, oxidation, is a reactive process by which the elements of a material lose electrons when exposed to oxygen and its valence is correspondingly increased, having a significant impact on geotextile material, determining its fragility, surface cracking, discoloration and most importantly molecular weight loss, which directly results in a loss of mechanical strength [4,11]. The oxidative stability of polypropylene and polyethylene geotextiles can be determined through an oven-ageing test [12].

Therefore, the aim of this study is to determine the durability of the geotextiles used for the ecological restoration of riverbank defences on the Danube River, considering that hydrolysis and oxidation degradation are two the mechanisms that will mostly affect the in-situ performance of the geosynthetics. In order to assess the degree of degradation induced by these two mechanisms, the performance mechanical properties of the two geotextiles (Terrafix and Secutex) will be evaluated.

2. Materials and methods

The two geotextiles used for laboratory determinations are Terrafix and Secutex, produced by NAUE GmbH & Co. KG. The geotextiles were tested after being used in hydraulic works for the defense of Danube's banks in order to reduce erosion in the areas: Bala Branch, Epurașu Island and Caleia Branch.

Terrafix geotextile (figure 1a), used on the riverbank to protect against erosion, is made of non-woven, multilayer, coloured polypropylene fibre and white polyester, sand ballast.

The second geotextile, Secutex was used on the crest, as it is a nonwoven geotextile made from short fibres which can be used in different projects, among its applications being separation, filtration, protection and drainage.

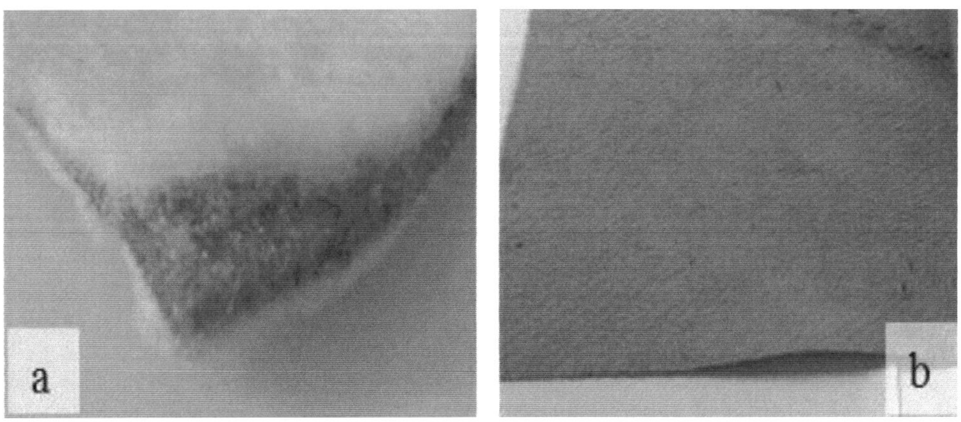

Figure 1. The analysed geosynthetics: a) Terrafix and b) Secutex.

In order to determine the performance of geotextiles in environments which can affect the mechanical properties, samples from the two materials were subjected to different temperature and pH conditions, considering the durability tests performed on them, as it can be seen in figure 2.

These determinations were established based on the requirements of the EN 14030 and EN ISO 13438 standards, being modified accordingly to laboratory conditions. The testing conditions (time, temperature, pH, reagents) are centralized in table 1.

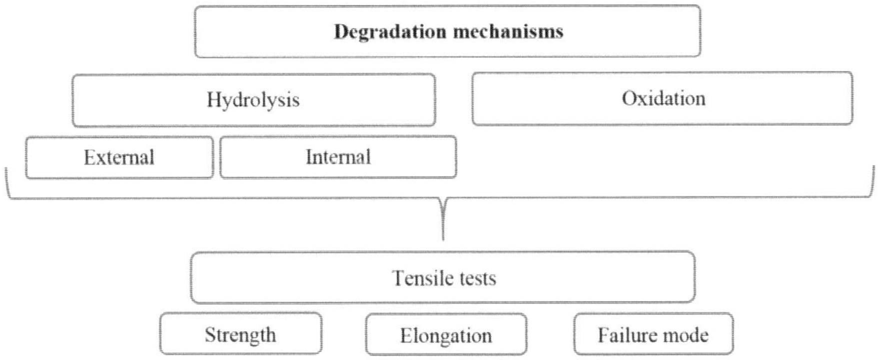

Figure 2. Experimental design of the study.

Table 1. Laboratory conditions for durability tests performed on the two analysed geotextiles (Terrafix and Secutex).

Determination		Laboratory conditions	
Hydrolysis	external	Maintaining period	30 days
		Temperature, °C	22
		pH	12
		Reagents	-
	internal	Maintaining period	30 days
		Temperature, °C	22
		pH	5
		Reagents	Calcium hydroxide
Oxidation degradation		Maintaining period	14 days
		Temperature, °C	Cicles of 110 °C/8 hours + 37 °C/16 hours
		pH	-
		Reagents	-

Durability tests were completed by determinations of tensile tests carried out in a universal testing machine MP 0.5t, model 97 with mechanical clamping jaws, at a constant speed of about 1.3 mm / min at room temperature. The tensile strength was determined using the following relationship [13]:

$$\alpha_f = \frac{F_f}{W_s} \tag{2}$$

where: α_f - tensile strength, N/m;
F_f – maximum breaking load, N
W_s – specified width of test samples, m

Also, to determine the elongation was used the following relationship:

$$Elongation = \frac{l_i - l_f}{l_i} \times 100 \ (\%) \tag{3}$$

where: l_i - initial length of the specimen, mm;
l_f - final length of the specimen, after the tensile test, mm.

All the tests were performed three times in order to ensure reproducibility and the results were determined as the average between the 3 values (the difference between these values does not exceed ± 10% from their mean value).

After the tensile tests, the failure mode of the samples was analysed, considering features like failure type, area and location.

3. Results and discussion

3.1. Hydrolysis

For the evaluation of the internal hydrolysis, the specimens were kept in an acid environment (an admixture of water and HCl was mixed with a soil, maintaining it continuously moist), while for the external hydrolysis the specimens were kept in an alkaline soil, which contained hidrated powder $Ca(OH)_2$. Figure 3 shows the tensile strengths of the geotextile materials subjected to hydrolysis degradation for 30 days. The internal hydrolytic degradation caused the mechanical strengths of the Terrafix samples to decrease from 26 kN/m (reference value) to 23,42 kN/m (a decrease of ~10%), while the external hydrolysis lead to a decrease of only 5% of the tensile strength values. However, the Secutex geosynthetic presented a more significant decrease of the mechanical resistance values. The degradation through internal hydrolysis decreased the tensile strength of the samples with an average of 35.5% (14.25 kN/m compared with the reference value of 22.1 kN/m), while after external hydrolysis the samples showed a reduction of about 29% (15.75 kN/m from the reference value equal to 22.1 kN/m).

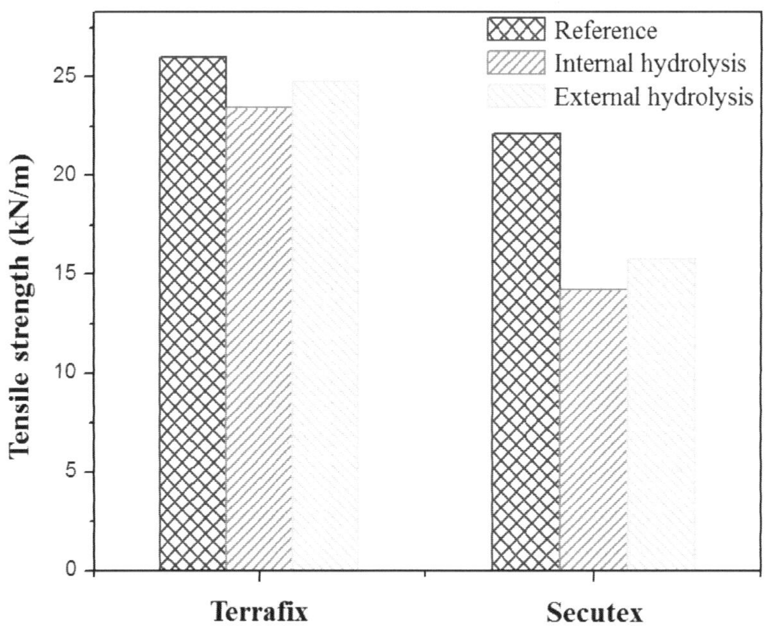

Figure 3. Tensile strength of the two analyzed geosynthetics after hydrolysis degradation.

It can be seen that the internal hydrolysis is more damaging than the external hydrolysis. The decrease of tensile strength can be explained by the fact that the hydrolysis processes depend on the phenomena of convection and diffusion transport of hydroxyl ions (OH^-) to the polymer structure, and in addition, the moisture acts as a catalyst in this chemical process [6,10].

The other analysed parameter, elongation, indicated higher values in the case of specimens subjected to internal hydrolysis tests (figure 4), with an average increase of ~56% for the Terrafix specimens, which according to Saathoff et. al [14,15] reduces the degradation which can occur during installation and allows a structural flexibility. The Secutex samples presented elongation values of ~24%. This behaviour can be explained as a result of the chemical reactions involved in this process, which influences the flexibility of polymeric fibres.

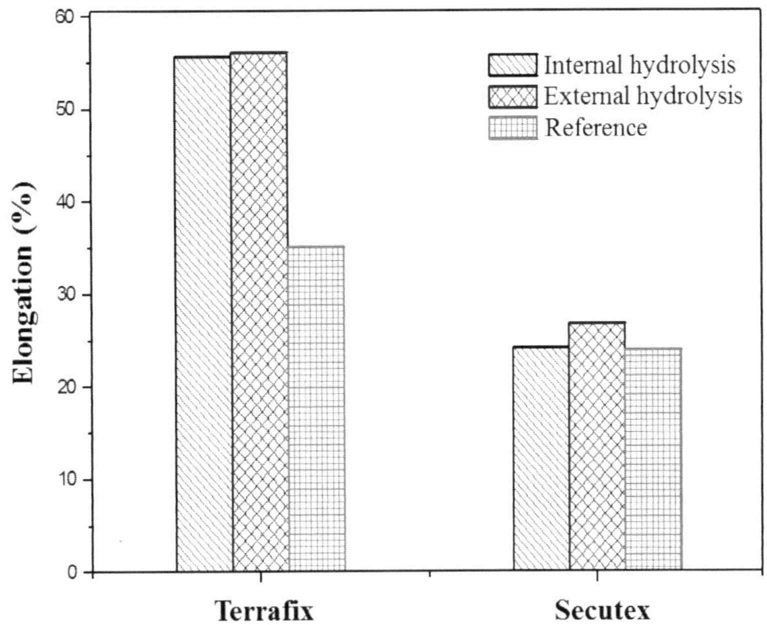

Figure 4. Elongation values of Terrafix and Secutex geotextiles after hydrolysis.

Regarding the other analysed parameter of the mechanical testing, failure mode, the Terrafix geotextile presented the breakage in the middle of the lateral calibrated area, for the specimens subjected to internal hydrolysis, while the ones with external hydrolysis degradation failed at the top/bottom of the grip, which indicates a higher pressure of the top/bottom clamping jaw. The Secutex samples subjected to internal hydrolysis showed a tendency to fail explosively at the bottom grip/tab, which could be explained by a proper distribution of stresses in the loaded strip.

The geotextile specimens damaged through external hydrolysis showed a tendency to fail angular in the middle of the calibrated area.

3.2. Oxidation degradation

The oxidation resistance of the analysed geotextiles (Terrafix and Secutex), assessed by determining the tensile strength of the materials, is shown in figure 5. It can be seen that the tensile strength decreases in the case of specimens subjected to the oxidation test by approximately 10% in the case of Terrafix and 47% for the Secutex geotextile.

The action of oxygen on the polymeric fibres causes the breaking of the polymer chains, and consequently the decrease of the mechanical resistances. The different behaviour of the analysed geosynthetics is determined by the nature of the polymer, degree of crystallization, fibre thickness, etc.

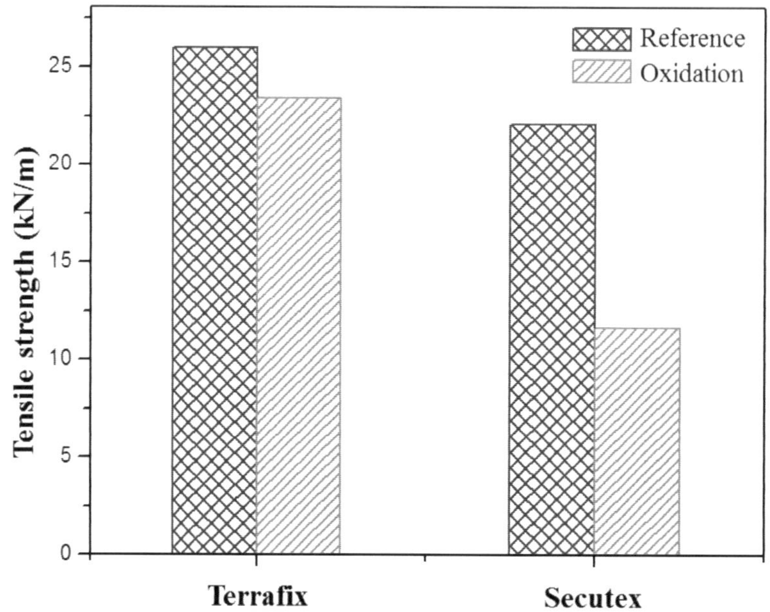

Figure 5. Tensile strength of the geotextiles after oxidation degradation.

The calculated elongation of the specimens subjected to the oxidation degradation is plotted in figure 6. The results show that for the Terrafix geotextile, the elongation values are superior to the reference. This behaviour is due to the fact that this geosynthetic is a composite material made of both polyethylene and polyester fibres, therefore the oxidizing medium can act differently on the fibres, considering that the ester fibres contain in their structure carboxyl groups (-COO^-). For the Secutex geotextile, which is made entirely of polyolefin fibres (polypropylene and polyethylene), the determined elongation was lower than the reference value.

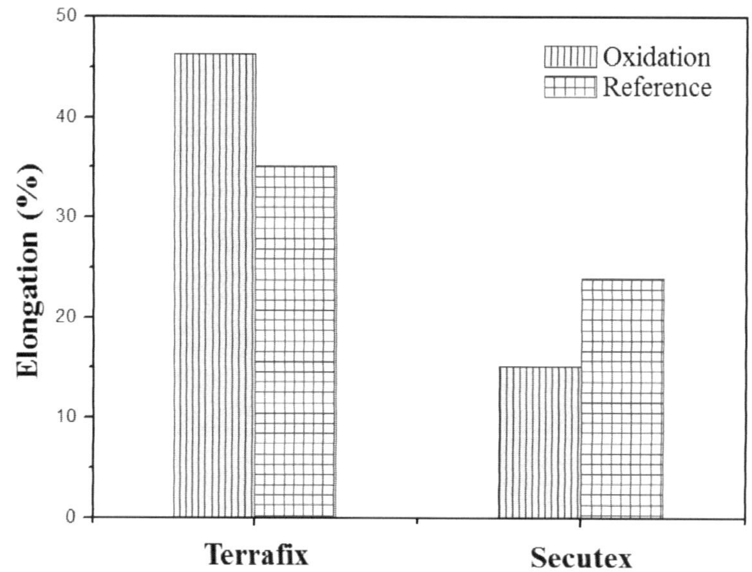

Figure 6. Elongation values of the analysed geotextiles after oxidation degradation.

The failure modes of the two geotextiles subjected to oxidation degradation showed that the Terrafix specimens tended to fail at the top/bottom of the grip, indicating a higher pressure of the clamping jaws, while the Secutex samples showed a tendency to fail at the top/bottom of the calibrated area.

4. Conclusions

The main conclusion of this work are the following:

- The performance testing of the two analysed geotextiles (Terrafix and Secutex) showed that the internal hydrolysis produces more damage to the geotextiles than the external hydrolysis.
- The Secutex geosynthetic presented a higher tendency to degrate under the influence of oxidation and hydrolysis. This behaviour could be explained considering that most reactions occur at the surface of the polymeric material, therefore the reduced thickness of this material makes it less resistant.
- The elongation values were higher after the hydrolysis degradation than after oxidation degradation, due to the chemical reactions involved in the process, which influence the flexibility of synthetic fibers.
- Considering that the Terrafix geotextile failure trend has been at the gripping area suggests that during functioning, the material may be at risk of yielding in the area where the load is applied. The Secutex geotextile samples tendend to fail in the calibrated area.

References

[1] Radchenko V G and Semenkov V M 1992 *Hydrotechnical Construction* **26** 685
[2] Han J and Guo J 2017 *Frontiers of Structural and Civil Engineering* **11** 56
[3] Heibaum M, Fourie A, Girard H, Karunaratne G P, Lafleur J and Palmeira E M 2006 *Proceedings of 8th International Conference on Geosynthetics*, ed. J Kuwano & J Koseki (Rotterdam, Netherlands: Millpress) 76-120.
[4] The Textile Institute 2007 *Geosynthetics in civil engineering*, ed. R.W. Sarsby (Cambridge, England: Woodhead Publishing Limited)
[5] Ingold T S and Miller K S 1990 *Geotextiles Handbook*, London, England: Thomas Telford Ltd.
[6] Greenwood J H, Schroeder H F and Voskamp W 2012 *Durability of Geosynthetics*, Gouda, The Netherlands: Stichting CURNET
[7] East G C and Rahman M 1999 *Polymer* **40** 2281
[8] Ito M and Nagai K 2008 *Polymer Degradation and Stability* **93** 1723
[9] Doll W W, Goodrum R and Sprague C J 1995 *Proceedings of Geosynthetics '95*, St. Paul, U.S.A.: Industrial Fabrics Association International 975-988.
[10] Hsuan Y G, Koerner R M and Lord A E 1993 *Geosynthetic Soil Reinforcement Testing Procedures*, ASTM STP 1190, ed. S.C. Jonathan Cheng(Philadelphia, U.S.A.: ASTM International) pp. 228–244.
[11] Vink P and Fontijn H F N 2000 *Geotextiles and Geomembranes* **18** 333
[12] International Organization for Standardization 2004 *ISO 13438:2004 - Geotextiles and geotextile-related products. Screening test method for determining the resistance to oxidation*, Geneva, Switzerland: ISO
[13] ASTM 2011 *D4595-11: Standard Test Method for Tensile Properties of Geotextiles by the Wide-Width Strip Method*, West Conshohocken, U.S.A.: ASTM International
[14] Saathoff F and Witte J 1994 *Geosynthetics World* **5** 1
[15] Saathoff F, Oumeraci H and Restall S 2007 *Geotextiles and Geomembranes* **25** 251

IOP Conf. Series: Materials Science and Engineering **209** (2017) 012102 doi:10.1088/1757-899X/209/1/012102

Testing Method for External Cladding Systems - Incerc Romania

A Simion[1] and H Dragne[2,3]
[1] Laboratory of testing and research for building fire safety from National Institute INCERC Bucharest, National Research and Development Institute URBAN-INCERC, Bucharest, Romania
[2] Laboratory of testing and research for building fire safety from National Institute INCERC Bucharest, National Research and Development Institute URBAN-INCERC, Bucharest, Romania
[3]PhD, Technical University of Civil Engineering, Bucharest, Romania

E-mail: simion_i_adrian@yahoo.com

Abstract. This research presents a new testing method in a natural scale for external cladding systems tested on buildings with minimum than 3 floors [1]. The testing method is unique in Romania and it is similar about many fire testing current methods from European Union states. Also, presents the fire propagation and the effect of fire smoke on the building façade composed of thermal insulation. Laboratory of testing and research for building fire safety from National Institute INCERC Bucharest, provides a test method for determining the fire performance characteristics of non-loadbearing external cladding systems and external wall insulation systems when applied to the face of a building and exposed to an external fire under controlled conditions [2]. The fire exposure is representative of an external fire source or a fully-developed (post-flashover) fire in a room, venting through an opening such as a window aperture that exposes the cladding to the effects of external flames, or an external fire source. On the future, fire tests will be experimented for answer demande a number of high-profile fires where the external facade of tall buildings provided a route for vertical fire spread.

1. Introduction

The test method simulates a fully developed fire in a room abutting the face of a building and venting through an aperture. In this test, the external cladding system is applied to a vertical external masonry surface simulating the face of a building in the form of a main face together with a perpendicular side wing, with the cladding system attached in the manner specified by the test sponsor. At the base of the main vertical masonry wall an opening is provided through which the fire can vent. Extend of damage caused to the external cladding system is evaluated, particulary the ability of the external cladding system to resist the propagation of the fire upwards or penetration through the system. Any falling debris and fire penetration was recorded. The test is concerned not only with the reaction to fire properties of the materials, as to now in Romania, but also with system performance in terms of overall integrity and stability. The large-scale test method covered by this test, determine the comparative burning characteristics of exterior wall assemblies by evaluating fire spread over the external surface, fire spread internally within the system under test and mechanical response, that is, the degree of distortion and local or global collapse [3].

2. Experimental procedures

The test facility is illustrated schematically in figure 1.

The parameters recorded until the test was: temperature 18 °C, humidity 81.6 % and wind speed 0.4 m/s. The thermal insulation of the facade was made of fireproof polystyrene with a thickness of 10 cm over which an exterior and protection decor layer has been applied (figure 2).

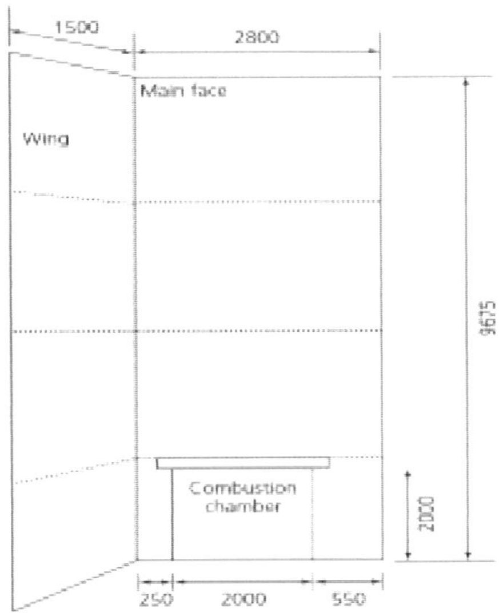

Figure 1. Large-scale cladding test facility (dimensions in mm) [4].

Figure 2. Building façade composed of thermal insulation.

Inside the chamber was a stack of wood with a thermal load of 450 Mj (figure 3).

Figure 3. Stack of wood.

Thermocouples was placed at two levels to measure both internal and external fire spread (figure 4, 5).

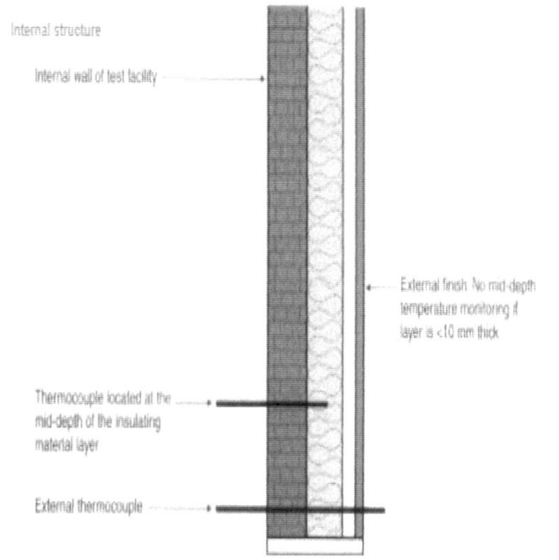

Figure 4. Location of termocouples at level 2.

Figure 5. Thermocouples at level 2.

After ignition of the stack of wood (figure6), start time for measuring fire spread occurs when the temperature recorded by any external thermocouple at level 1 reaches 200 C above the ambient temperature value and remains above that level for a period of at least 30 seconds (figure 7).

Figure 6. Ignition stack of wood.

Figure 7. Temperature recorder.

Failure due to external fire spread will have occurred if the temperature rise of any of the external thermocouples at level 2 exceeds 600 C for a period of at least 30 seconds within 15 minutes of the start of the recording period.

Failure due to internal fire spread will have occurred if the temperature rise of any of the internal thermocouples at level 2 exceeds 600 C for a period of at least 30 seconds within 15 minutes of the start of the recording period. Details of any system collapse, spalling or delamination was reported (figure 8, 9, 10).

Figure 8. The moment of fire propagation process in height.

Figure 9. Generalized fire of the thermal system on the front with the combustion chamber.

Figure 10. General combustion of the wing termal system.

3. Results and discussions

The fire resistance test of the thermal insulation system lasted 1 hour. After finish the resistance fire test, the records values of temperatures have been processed and interpreted. Damage is recorded in the following areas: flame spread on surface, flame spread in cavities or insulation area of façade damaged or detached. In the first 15 minutes since ignition the fire, no thermocouple on the second floor recorded temperatures at least 600 °C. The maximum temperature in the combustion chamber was about 1000 °C and the maximum face temperature above 800 °C. The complete firing time of the polystyrene insulation thermal rehabilitations systems, was 40 minutes from the ignition of the fire. During the fire test, they fell burned droplets and a large amount of black-grey smoke was released in atmosphere.

4. Conclusions

This test method was developed to address systems installed to masonry structures. Many European states have similar tests for façade systems recognising the need to understand the system interactions [5]. The INCERC test method, provides a means of assessing performance against a realistic scenario in terms of fire load and provides information on the performance of the system and not just the exposed surface. This research presents the fire propagation and the effect of fire smoke on the façade building. The results obtained from the test provide the data, needed to evaluate the fire protection performance of the external cladding systems. Is most important as the fire performance of external walls, covers fire resistance to restrict fire spread across a site boundary, the combustibility of the outer surface to minimise the possibility of ignition from an external source and subsequent fire spread up the external facade. Following this test it results that it is necessary for external walls which form elements of structure then the provisions for fire resistance will apply. However, the external envelope of a building should not provide a medium for fire spread. The use of combustible materials in the cladding system in tandem with unstopped cavities is a potential route for rapid fire spread. In this moment, in Romania is not obligate the testing of the fire resistance for external cladding systems on buildings. As an alternative means of compliance it is now possible to meet the fire performance criteria of the external cladding system of the masonry face of a building given which could hope contribute to elaborate the new provisions of fire legislation in Romania [6].

References

[1] Lalu O, Anghel I, Codescu S and Branisteanu B 2016 Experimental research on fire behavior analysis for polystyrene insulation thermal rehabilitations systems *Romanian Journal of Civil Engineering* **7** 208

[2] Lalu O 2016 Research on developing sustainable solution for thermal rehabilitation of the building envelope. Behaviour in case of fire, fire safety solutions *Technical University of Construction* 4-10

[3] Lannon T 2011 Structural fire engineering *ICE Publishing* 32-35

[4] BS 8414-1 2002 Fire performance of external cladding systems applied to the face of the building *BSI* 4-6

[5] Coldwell S and Baker T 2003 BR-135 - Fire performance of external thermal insulation for walls of multi-storey buildings *BRE Bookshoop* 10-14

[6] Lalu O and Bubulete L 2017 Termosistem pentru placari exterioare utilizat pentru anveloparea constructiilor. Raspuns la actiunea focului *Construction Magazine* **133** 8

Steel – Concrete Materials Performance in Composite Joints Configuration

M Pop, O Corbu and P Pernes
Technical University of Cluj – Napoca, Civil Engineering Faculty, 25 Baritiu str. 620002, Romania

E-mail: maria.pop@dst.utcluj.ro

Abstract. In many buildings there is a need to combine reinforced concrete, steel or composite members. The paper aims to an experimental program focused on the behaviour of the materials steel and concrete what makes up for a composite joint configuration. Material tests were performed prior each type of testing. The main purpose is to investigate the main parameters that affect the response and the contribution of the two materials. The tests were performed at the TUCN – Laboratory of Civil Engineering Faculty, Romania. The results of the tests on the materials were used for preliminary experimental analysis and presented in the paper.

1. Introduction

Multi-storey buildings have been a preferred investor option for centuries and a challenge for engineers involved in materializing these projects. Due to the growth of the service sector, tourism and other age-specific branches on the one hand and due to acute managerial influences, such as the cost-benefit ratio, on the other hand. An optimal structural model for their realization is the steel-concrete composite structure solution. Structures composed of steel and concrete have had a significant development since their conception, both materials having structural potential and good fire resistance and good cooperation between them. Although the two materials are different, they complement each other with comparable thermal expansion. The materials used must have adequate strength and deformability, fundamental performance of the materials to withstand the various actions. Over time, it has resistance to damage or wear. Another important aspect is the minimal influence on the environment both during manufacture and in operation [1-6].

2. The design materials

Working in the structure of the two materials can refer to the whole building, resulting in mixed steel - concrete structures or sub - assemblies, resulting in composite steel - concrete elements:

- concrete
- steel

Multi-story structures in composite steel-concrete aim to make the most of the advantages of each of the components in order to obtain preponderant technical and economic performances. In figure 1 are showed a composite structures The International Finance Center.

Steel-concrete composite structures can achieve material savings ranging from 15 to 50%, depending on the type of structure and execution, compared to steel structures and also lower weights compared to reinforced concrete structures.

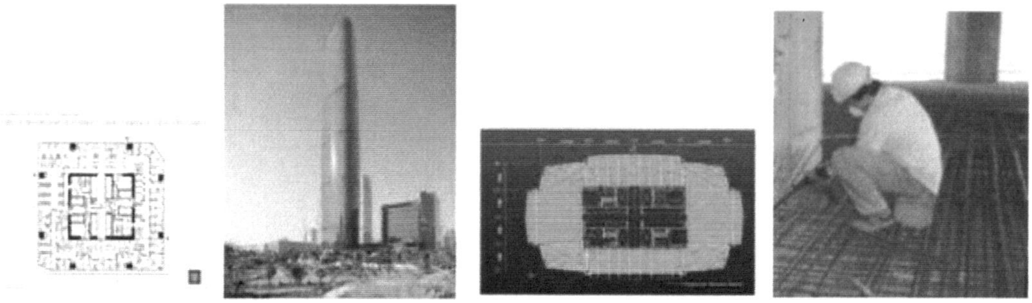

Figure 1. Composite Structures.

Compressive strength of concrete depends on many factors such as water-cement ratio, cement strength, quality of concrete material, and quality control during production of concrete.

Concrete strength was determined on three specimens showed in figure 2: cube 150mm dimensions, samples were taken at the site were the concrete casting was made and tested after 7 days and 28 days.

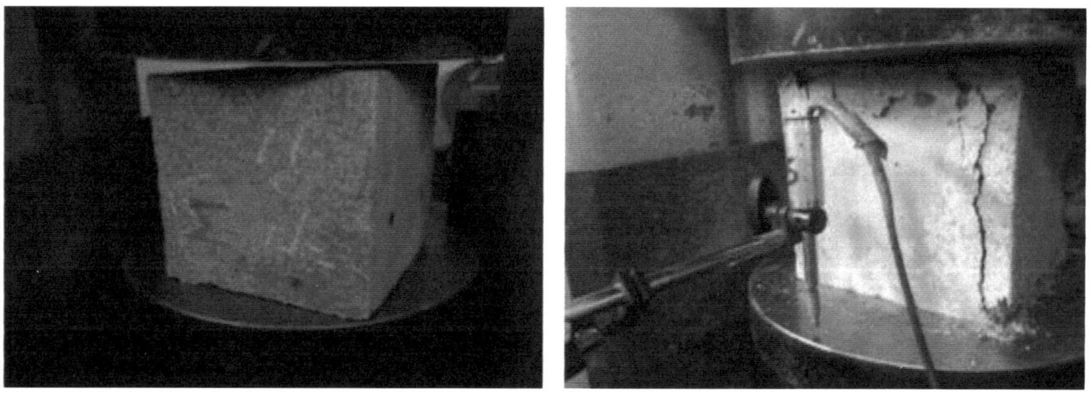

Figure 2. Concrete Strenght of the tested specimens.

Compressive strength of concrete depends on many factors such as water-cement ratio, cement strength, quality of concrete material, and quality control during production of concrete.
Due to the testing of the concrete materials (see table 1), the results showed similar behaviour and the concrete, the proper class was achieved in all specimens.

Table 1. Concrete characteristic values.

Testing	$f_{cm,cube}$	Unit
7days	31,31	N/mm^2
28days	32,25	N/mm^2

Due to the testing of the concrete materials (see table 2), the results showed similar behaviour and the concrete, the proper class was achieved in all specimens.

Steel resistance is an important feature of structural steel, depending on its chemical composition, such as the percentage of carbon or manganese, the heat treatment used, and all the processes during the profiles. The resistance to steel flow can be determined experimentally by stretching the steel [7-

14]. The steel materials were tested also according to the tensile testing procedure ECCS [11] .The specimens were tested to determine the characteristics physical and mechanical. All the tested specimens had the same material properties.

Ultimate tensile strength is measured by the maximum stress that a material can withstand while being stretched or pulled before breaking. In the study of strength of materials, tensile strength, compressive strength, and shear strength can be analysed independently.

Three coupons were fabricated from each type of steel elements and tested at room temperature with the INSTRON universal testing machine at the UTCN Civil Engineering Laboratory, Romania. universal testing machine and the elongations were measured using a video extensometer module. Coupon specimen dimensions are showned in figure 3.

Figure 3. Steel copuons and traction test press.

In order to determine the physical-mechanical characteristics of the structural steel, the tensile test was performed for the metal profiles. The tensile strength was determined by the tensile testing of the metal profiles: IPE 270 Europrofiles, with samples taken from the web and the flanges of the steel profiles.

The results of the tensile test on the tree coupons tested are given in figure 4. The ultimate elongation shows the steel elements have a good ductility.

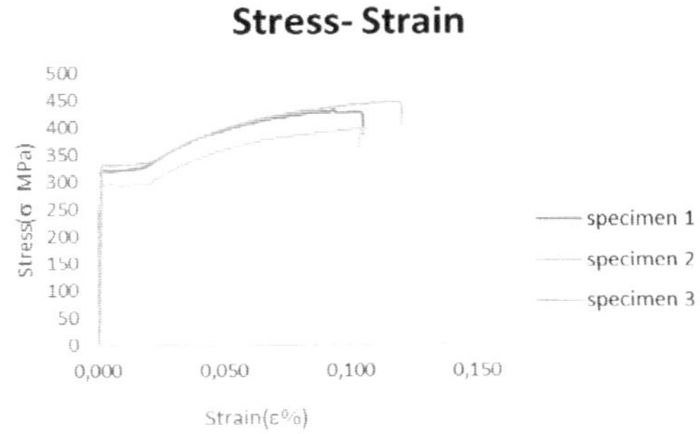

Figure 4. Stress–Strain curve.

In figure 4 the stress-strain curve (σ-ε) of the steel is illustrated. It can be noticed that the initial steel has a stress-strain, defined by the f_y flow velocity and the deformation ε_y.

The aim of the test of the steel traction test is to determine the ultimate strength and elastic limits by determining the tensile and compressive strengths.

3. Conclusions

The tested materials in the experimental program have adequate strength and deformability, fundamental performance of the materials to withstand the various actions. Another important aspect is the minimal influence on the environment both during manufacture and in operation.

The research recommended for composite steel-concrete elements that it is necessary to use a minimum concrete class C20/25.

The adhesion in the structure of the two materials can refer to the whole building, resulting in mixed steel - concrete structures or sub - assemblies, resulting in composite steel - concrete elements.

References

[1] Chira N Campian C 2010 *Rehabilitation of an industrial building in three different solutions* (IABSE Symposium) 702-710

[2] ECCS 1990 *Design of Composite Joints for Buildings*

[3] Iuhos V Campian C 2016 *Specific issues on the design and expertise of steel columns for overhead lines* (Elsevier Procedia Technology Journal)

[4] Gioncu F Mazzolani M 2002 *Ductility of seismic resistant steel structures* (London)

[5] Gundy P Thomas R Bennetts T 1980 *Beam to Column Moment Connections* **106** 313-330

[6] Ck I. Bradford M Chen W 2009 *Second order inelastic analysis of composite framed structures based on the refined plastic hinge method* 31

[7] EN European structural steel standard 2004 *Hot rolled products of structural steel* 31

[8] SR EN 12620: 2002+A1 2008 *Aggregates for concrete*

[9] SR EN 4 2004 *Design of composite steel and concrete structures*

[10] ECCS 2004 *European Structural Steel Standard: Hot rolled products of structural steels*

[11] EN 10002-1 2004 *European Standard Metallic materials – Tensile testing Part 1; Method of test at ambient temperature*

[12] Campian C Sav V Chira N Chira A 2012 Strength and ductility of concrete encased composite columns (*Stessa, Proceedings of the 7th International Conference on behavior of steel structures in seismic areas*, Santiago, Chile)

[13] Heyman J 1999 *The Science of Structural Engineering*

[14] Mathe A Popa A Campian C 2011 The influence of Semi-Rigid Connections upon the Performance of Steel Structures Seismically Excited (*Acta Technica, Civil Engineering and Arhitecture Romania*) 189-203

Acknowledgments

The research was supported with the help of the Concrete Laboratory, Civil Engineering Faculty of Cluj-Napoca.

IOP Conf. Series: Materials Science and Engineering **209** (2017) 012104 doi:10.1088/1757-899X/209/1/012104

Experimental Researches on the Durability Indicators and the Physiological Comfort of Fabrics using the Principal Component Analysis (PCA) Method

L Hristian[1*], M M Ostafe[1], L R Manea[1] and L L Apostol[1]
[1] "Gheorghe Asachi" Technical University of Iasi, Faculty of Textile, Leather & Industrial Management, Department of Engineering and Design of Textile Products, Blvd. Mangeron, No.28, Iasi, Romania

E-mail: hristian@tex.tuiasi.ro

Abstract. The work pursued the distribution of combed wool fabrics destined to manufacturing of external articles of clothing in terms of the values of durability and physiological comfort indices, using the mathematical model of Principal Component Analysis (PCA). Principal Components Analysis (PCA) applied in this study is a descriptive method of the multivariate analysis/multi-dimensional data, and aims to reduce, under control, the number of variables (columns) of the matrix data as much as possible to two or three. Therefore, based on the information about each group/assortment of fabrics, it is desired that, instead of nine inter-correlated variables, to have only two or three new variables called components. The PCA target is to extract the smallest number of components which recover the most of the total information contained in the initial data.

1. Introduction

Analysis of the Principal Components (PCA) is a descriptive method of data multi-dimensional analysis. The PCA objectives are: synthesis of the initial information contained in a big size table; visualization of the similarities and differences between individuals from the standpoint of the considered variables [1-3]. Principal Components Analysis (PCA) is a useful statistical technique that has found application in fields such as face recognition and image compression, and is a common technique for finding patterns in data of high dimension [4-6]. This technique introduces a mathematical concept that covers standard deviation, covariance, eigenvectors and eigenvalues [7-9]. Principal component analysis is a dimension-reduction technique. It can be used as an exploratory data analysis tool, but is also useful for constructing predictive models, as in principal components analysis regression (also known as PCA regression or PCR) [10-12]. For data with a very large number of variables, the Principal Components platform provides an estimation method called the Wide method that enables you to calculate principal components in short computing times. These principal components can then be used in PCA regression [13-15]. At the beginning, the methods of multivariate analysis of data were applied in fields of humanities: biology, psychology, sociology [16-19]. Later, there were applied in many other areas, such as: psihometria, genetics, medicine, economics and industry [18-21]. Outstanding contributions to theoretical and practical development of these methods had, in particular, Galton and K. F. Pearson, by the studies in the field of regression and correlation analysis, Spearman and H. C. Hotteling by factor analysis and research of the principal components analysis. At present, the more specialized softwares enable a multitude of multivariate analysis methods of the data for the statistics processing of the research results: SPSS, Statistics, SAS, ADDAD [20-22]. The main asset of PCA is its ability to identify genetic structures in very large datasets within negligible computational time, and the absence of any assumption about the underlying population genetic model [21-26].

2. Experimental part

2.1. Materials and methods

In this study, there were analyzed four articles from five groups of worsted type fabric, made of different compositions, representing the number of individuals like:
Group A (100% Wool) - encoded items: A1; A2; A3; A4;
Group B (45% Wool + 55% Pes) - encoded items : B1; B2; B3; B4;
Group C (45% Pes + 55%Wool) - encoded items: C1; C2; C3; C4;
Group D (44%Pes + 54% Wool +2%Dorlastan) - encoded items: D1; D2; D3; D4;
Group E (60% Pes + 40% Celo) - encoded items: E1; E2; E3; E4

In making the fabrics from groups A, B, C and E, they were used Sirospun yarns which are twisted appearance yarns, called Jasper or yarn from double roving fibers. At group D fabrics, the weft threads are elastomeric yarns with core made of filament and their sheaths are made of fibers obtained also by the Siorospun process, through simultaneous supplying of the drawing frame with the roving and the filament core.The joining of the two components takes place in front of the delivery rollers so that by simultaneous twisting of them it is made the yarn structure with the filament core covered by the fibers. Using two rovings which will be rolled separately and a filament core at one of the sides, it is obtained a twisted yarn from a spun yarn and a core yarn . Most polymers and elastomers used for industrial or commercial applications are composites, which contain solid fillers.

The values of indicators of physiological comfort and durability were obtained by standard methods and processed in SPSS. The variables included in the analysis are: tear strength, Pr (daN); volumetric filling coefficient, Cvu; tenacity, t (cN/tex); the thickness; g (mm); mass, M (g/m^2); air permeability, Pa (m^3/min.mm); the degree of compactness, Kt (%); rigidity, R (mg.cm); porosity, Pz (%). This technique aims to reduce the number of controlled variables (columns) of the matrix data, as far as possible at two or three. Thus, based on information about each group/types of fabrics, it is desirable that instead of nine interrelated variables to have only two or three new variables, called components. The PCA goal is to extract the smallest number of components to recover as much of the total information contained in the initial data indenter.

3. Results and discussions

After processing the data in SPSS programme, applying PCA,the following results were obtained on statistical variables: descriptive statistics indicators (Descriptive Statistics); the correlation matrix; calculated values for both $\chi 2$ test statistics and KMO statistics; the variance of the variables; eigenvalues and variance explained by each factorial axis; variable coordinates on factorial axes; contributions of the variables at the inertia of factorial axes and graphics.

3.1. Descriptive Statistics Indices

The statistical parameters calculated for each variable are shown in table 1. (Descriptive Statistics Output). Statistics is concerned with the scientific method by which information is collected, organised, analysed and interpreted for the purpose of description and decision making. Methods which help us describe the key features of data from interval and ratio scale measurement, fall into two groups, these are measures of central tendency and measures of dispersion. The measures of central tendency help us to get a feel for how alike individual measurements are to one another. The measures of dispersion help us to appreciate how spreads out the individual scores are from one another.
Analyzing the data from table 1, which contains information about each independently analyzed variable, it can be observed that:

- the variable Pr (daN), the breaking strength is characterized by its 64.843 average and it 521.950 variance; the maximum force at breaking it was obtained for article E4 made of 100% Pes, Pr = 106.1 daN and the minimum value for article A1, made of 100% wool, Pr = 31.4 daN;
- the variable Cvu, volumetric filling coefficient is characterized by 39.334 average and 160,363 variance; the maximum volumetric filling coefficient was obtained to article D4, made of 44%Pes + 54%Wool +2%Dorlastan, Cvu = 65,32, and the minimum value was obtained to article A1 made of 100% wool, Cvu=21,34%;coating.

Table 1. Descriptive Statistics.

Variabile	N	Minimum	Maximum	Mean	Std. Deviation	Variance
Pr(daN)	20	31,4	106,1	64,843	22,8462	521,950
Cvu	20	21,34	65,32	39,334	12,66344	160,363
g(mm)	20	0,2872	0,6975	0,42753	0,144938	0,021
M(g/m^2)	20	170,3	238,7	202,555	18,0991	327,578
t(cN/tex)	20	5,9	18,2	12,620	3,5847	12,850
Pa(m^3/mm.min)	20	14,2	33,9	22,570	6,3101	39,817
Kt(%)	20	75,23	98,29	87,5275	6,47263	41,895
R(mg.cm)	20	48,22	99,10	71,7220	15,42990	238,082
Pz(%)	20	58,50	79,21	67,4185	5,17203	26,750
Valid N (listwise)	20					

Analogous to this, the other variables are analyzed too. Following this independent analysis of every variable it is observed that the homogeneous variable is that related to thickness of the fabrics. The same trend is observed for the variable fabric porosity.

3.2. Correlation Matrix (output of Correlation Matrix)

The correlation matrix shows the correlation coefficient values of the variables considered in pairs. It is a square matrix symmetrical about the main diagonal (equal to one, because a variable is perfectly correlated with itself). The shape of the correlation matrix is shown in table 2, after the data have been standardized.

The analysis of the correlation coefficients of the matrix allows the assessment of the application of the PCA. High values of coefficients (greater than + 0.5 and less than - 0.5) show that there are significant statistically links between the considered variables (direct connection if the coefficient values are positive, the reverse link if the coefficient values are negative). For example, from Table 2 it is observed that there are:

- significant statistical connections (direct link) between:

- Pr (daN) and Pa (m^3/mm.min);
- M (g/m^2) and Kt (%); Pz (%);
- Pa (m^3/mm.min) and Pr (daN);
- Kt (%) and M (g/m^2);
- Pz (%) and M (g/m^2).

- significant statistical connections (indirect link) between :

- Pr (daN) and R (mg.cm);
- Cvu and Pa (m^3/mm.min);
- g (mm) and Pa (m3/mm.min); Pz (%);
- M (g/m^2) and R (mg.cm);
- Pa (m^3/mm.min) and Cvu, g (mm); R (mg.cm);
- R (mg.cm) and Pr (daN); M (g/m2); Pa (m3/mm.min); Kt (%);
- Pz (%) and g (mm).table 2.

Table 2. Correlation Matrix.

		Pr (daN)	Cvu	g (mm)	M (g/m^2)	Pa (m^3/mm. min)	Kt (%)	R (mg.cm)	Pz (%)
Correlation	Pr(daN)	1,000	0,105	-0,580	0,238	0,815	0,192	-0,439	0,304
	Cvu	0,105	1,000	-0,548	0,123	-0,358	0,408	0,138	0,222
	g(mm)	-0,580	-0,548	1,000	0,076	-0,041	0,052	0,097	-0,136
	M(g/m^2)	0,238	0,123	0,076	1,000	0,244	0,594	-0,162	0,684
	Pa(m^3/mm.min	0,815	-0,358	-0,041	0,244	1,000	0,112	-0,489	0,230
	Kt(%)	0,192	0,408	0,052	0,594	0,112	1,000	-0,069	0,252
	R(mg.cm)	-0,439	0,138	0,097	-0,162	-0,489	-0,069	1,000	-0,506
	Pz(%)	0,304	0,222	-0,136	0,684	0,230	0,252	-0,506	1,000
Sig. (1-tailed)	Pr(daN)		0,330	0,004	0,156	0,000	0,209	0,026	0,096
	Cvu	0,330		0,006	0,303	0,061	0,037	0,281	0,174
	g(mm)	0,004	0,006		0,374	0,431	0,413	0,342	0,284
	M(g/m^2)	0,156	0,303	0,374		0,150	0,003	0,247	0,000
	Pa(m^3/mm.min	0,000	0,061	0,431	0,150		0,319	0,014	0,164
	Kt(%)	0,209	0,037	0,413	0,003	0,319		0,386	0,142
	R(mg.cm)	0,026	0,281	0,342	0,247	0,014	0,386		0,011
	Pz(%)	0,096	0,174	0,284	0,000	0,164	0,142	0,011	

[a]Determinant = 0,001

In this case, PCA can be applied. A feature of the correlation matrix is that the number of correlation coefficients increases greatly when the number of variables (k) included in the analysis increases regardless of the volume of statistics community.

The number of the correlation coefficients is: k (k-1)/2. For the experimental data which are showing values for nine variables, the number of the correlation coefficients is 36 (Table 3). This significant increase of the correlation coefficients shows the impossibility of explanation of the links between the variables only by analysis of the values found in the correlation matrix. In the Correlation Matrix output it is shown, also, the value of the correlation matrix determinant. It can take values in the range [0, 1] and it shows the strenght of the correlations between the variables. The determinant of the analyzed database has the value: Determinant = 0,001 which shows that there are strong statistical relations between the statistical variables (Correlation coefficient values are greater than 0.5). In this case the application of PCA can be achieved.

3.3. $\chi 2$ and KMO statistics (output for KMO and Barlett's)

To test the hypothesis of independence between the statistical variables, SPSS provides calculated values of the corresponding test statistics. Statistics test (output KMO and Bartlett's) is used to test whether the correlation matrix is an identity (unit) matrix, so if there is a statistics connection between the statistical variables.

For this, it makes the following statistical hypotheses:

Ho: the hypothesis of independence (correlation matrix is an identity matrix);

H1: the hypothesis of dependence

To test these hypothesis, the SPSS program provides both the calculated value of the statistic $\chi2$ = 142,577and the probability value associated with the calculated test statistic (Sig = 0,000 lower than 0,05) thus it is rejected the Ho hypothesis that allowed the existence of the independence of the variables (table 3). Therefore, it can be assumed with a 95% probability that between the variables there are statistical links (connections).

Table 3 KMO and Bartlett's Test.

Kaiser-Meyer-Olkin Measure of Sampling Adequacy.		0,467
	Approx. Chi-Square	142,577
Bartlett's Test of Sphericity	df	36
	Sig.	0,000

3.4. The Variance variables (output of Communalities)
Standardizing of the variables yields new variables of zero average and one variance. The variances of statistics variables are presented in the Communalities output. The variance of values after removing of the factors are calculated on the basis of the results from the output matrix component (table 4).

For example, for the volumetric filling coefficient **Cvu** it is obtained $\sigma_1^2 = 0,882$. Low values of the variance of the variables after extractionof the factors (Extraction column) shows that those variables can be eliminated from the proper analysis because they are not correlated to the factorial axes.

Table 4. Communalities.

Variable	Initial	Extraction
Pr(daN)	1,000	0,711
Cvu	1,000	0,882
t(cN/tex)	1,000	0,852
g(mm)	1,000	0,550
M(g/m^2)	1,000	0,346
Pa(m^3/mm.min)	1,000	0,812
Kt(%)	1,000	0,258
R(mg.cm)	1,000	0,562
Pz(%)	1,000	0,499

Extraction Method: Principal Component Analysis

3.5. The Eigenvalues λ_k associated with each factorial axis and the total variance explained by each factorial axle (output of Total Variance Explained)
The eigenvalues of the correlation matrix are shown in the Total Variance Explained output, initial Eigenvalues column (Table 5). In Table 5 it is shown that the eigenvalues of the correlation matrix are:

$\lambda_1 = 3,532$; $\lambda_2 = 1,941$; $\lambda_3 = 1,651$; $\lambda_4 = 0,895$; $\lambda_5 = 0,545$; $\lambda_6 = 0,243$; $\lambda_7 = 0,107$; $\lambda_8 = 0,079$; $\lambda_9 = 0,007$;

The eigenvalues correspond to inertia explained by the factorial axes. Their sum is the total inertia of the cloud of points equal to the number of statistical variables of the original data table or the amount of elements from the main diagonal of the correlation matrix.

Table 5. Total Variance Explained.

Component	Initial Eigenvalues			Extraction Sums of Squared Loadings		
	Total	% of Variance	Cumulative %	Total	% of Variance	Cumulative %
1	3,532	39,245	39,245	3,532	39,245	39,245
2	1,941	21,564	60,808	1,941	21,564	60,808
3	1,651	18,348	79,156	1,651	18,348	79,156
4	0,895	9,945	89,102			
5	0,545	6,059	95,161			
6	0,243	2,696	97,857			
7	0,107	1,185	99,041			
8	0,079	0,877	99,919			
9	0,007	0,081	100,000			

Based on the values from Table 6 it may result:

$$\sum_{k=1}^{k} \lambda_k = In \qquad (1)$$

where: In is the total inertia of the cloud of points.

Based on the values from Table 6 it may result:

$$\sum_{k=1}^{k} \lambda_k = 3{,}532 + 1{,}941 + 1{,}651 + 0{,}895 + 0{,}545 + 0{,}243 + 0{,}107 + 0{,}079 + 0{,}007 = 9$$

Factorial variance explained by each axis is calculated from the relationship:

$$\%Varian'a = \frac{\lambda_k}{In} \qquad (2)$$

The first factorial axis explains 3,532/9=39,244% of the total variance of the cloud points.

The first two factorial axes together explain 60,808% of total variance.

The first three factorial axes together explain 79,156%.

The number of factorial axes which are to be interpreted in the PCA was chosen according to two criteria:

- the Kaiser criterion which implies the choice of that number of factorial axes for which the corresponding of own values are higher than one. According to this criterion, three factorial axes are chosen, corresponding to their own values:

($\lambda_1 = 3{,}532$; $\lambda_2 = 1{,}941$; $\lambda_3 = 1{,}651$) >1.

These axes explain the biggest differences between statistical units in terms of the considered variables.

- Cattel's criterion involves the graphical representation of its own values (the Scree Plot chart) and the pursuit of a sudden fall of inertia explained by them. There are chosen the axes that precede this sudden change in the slope of the chart of the eigenvalues (Figure 1).

From the diagram shown in Figure 1, three factorial axes are selected for interpretation. These axes explains the biggest differences between statistical units in terms of the considered variables.

3.6. The eigenvectors

The eigenvectors associated with eigenvalues of the correlation matrix allow obtaining the coefficients of the variables in the linear equation of the main axes. Eigenvectors coordinates do not appear in the results of PCA but serve to calculate the coordinates of the variables in the principal axes.

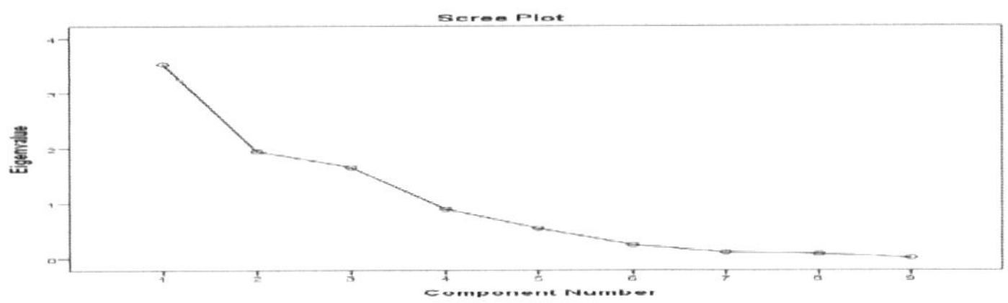

Figure 1. Graphical representation of the eigenvalues of the correlation matrix.

3.7. The coordinate of X_j variable on k factorial axis (output of the Component Matrix)

The values of the coordinates of the variables on the factorial axes show the correlation of the value of coefficients, between the variables X_j and that factorial axis,and they are calculated according to the relation:

$$G_k(j) = \sqrt{\lambda_k}.u_{kj} \quad (3)$$

where: λ_k is the own value appropriate to the k factorial axis;

u_{kj} are the coordinates of the eigenvector associated with the eigenvalue λ_k.

The coordinate values of the variables are presented in Table 6.

The values in Table 6 show the position of the variables on the factorial axes. For example, the variable **Pr** (daN) has a positive coordinate position on the first factorial axis (0.802) and a negative coordinate on the second factorial axis (-0.259); data will be plotted in the positive quadrant values of the first factorial axis and in the negative values quadrant of the second factorial axis. The high values of coordinate variables on the factorial axes listed that they are strongly correlated with that factorial.

For example, the variables **t** (cN/tex) and **g** (mm) are connected to the first factorial axis which indicates that these variables explain the significant differences between statistical units (there are significant differences between statistical units in terms of values obtained for these variables).

Table 6. The coordinates of the variables in the first three factorial axes.

Variable	Component		
	1	2	3
Pr(daN)	0,802	-0,259	-0,407
Cvu	0,406	0,847	0,047
t(cN/tex)	0,864	0,325	-0,184
g(mm)	-0,560	-0,487	0,608
$M(g/m^2)$	0,581	-0,094	0,707
$Pa(m^3/mm.min)$	0,527	-0,731	-0,198
Kt(%)	0,470	0,192	0,617
R(mg.cm)	-0,580	0,475	0,104
Pz(%)	0,700	-0,092	0,388

Extraction Method: Principal Component Analysis.

However, between these two variables there is a reverse link because they have opposite sign coordinates and the statistical units with high values for the variable **t** (cN/tex) have lower values for the variable **g** (mm). The coordinates of variables in the factorial axes represent the coefficients of the linear equation of the relationship between variables. For example, for the data in Table 7, the first axis is a new variable defined by the linear combination of the initial variables, having the form:

F = 0,802 Pr + 0,406 Cvu + 0,864 t - 0,560 g + 0,581 M + 0,527 Pa + 0,470 Kt - 0,580 R + 0,700 Pz

To identify the variables explaining the second factorial axis, choose those variables from Table 7 (Component 2 column) that have high coordinate values. Note that the formation of the second factorial axle is explained only by the variable **g** (mm).
From the analysis of the values of Table 7, it is noted that the variables **g** (mm), **M** (g/m^2), and **Kt** (%) explain most the difference between the statistical units for the third factorial axle (column of the 3rd component), with values higher than 0.5.

3.8. The contribution of X_j variable at inertia of k axis (output for the Component Score Coefficient Matrix)

The contribution of a variable to the formation of a factorial axis is calculated with the relationship

$$CTR_k(j) = \frac{G_k(j)}{\lambda_k} \tag{4}$$

where: $G_k(j)$ is coordinate of the variable on k factorial axis;

λ_k is the appropriate own value corresponding to the k factorial axis.

Similarly to the interpretation of the values in Table 7, the high values of the contributions show the significant importance of the respective variable in differentiating the considered statistical units.

Table 7. Contributions of the variables to the inertia of the first three factorial axis.

Variable	Component		
	1	2	3
Pr(daN)	0,227	-0,133	-0,247
Cvu	0,115	0,436	0,028
t(cN/tex)	0,245	0,168	-0,111
g(mm)	-0,159	-0,251	0,368
M(g/m2)	0,164	-0,049	0,428
Pa(m3/mm.min)	0,149	-0,376	-0,120
Kt(%)	0,133	0,099	0,374
R(mg.cm)	-0,164	0,245	0,063
Pz(%)	0,198	-0,047	0,235

Extraction Method: Principal Component Analysis

For the data from the Table 7, all variables contribute to the formation of the first factorial axis. For the second factorial axis , the variable that contributes to itself formation is *Cvu*; for the third factorial axis, the variables that contribute to this axis formation are: *g* (mm) and *M* (g/m^2).

3.9. The graphical representation

The representation of the variable points in the first two factorial axes is shown in figure 2.

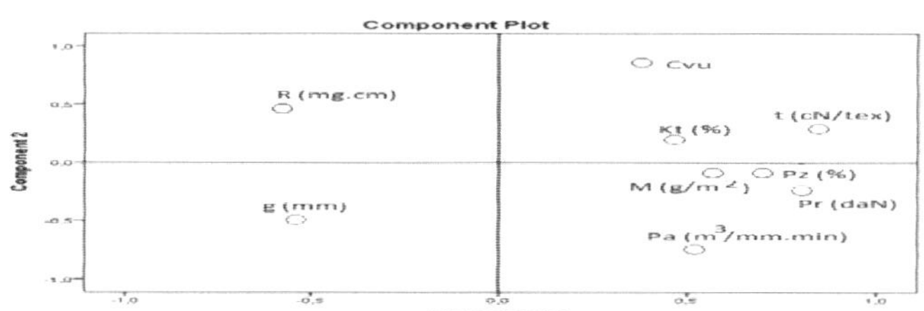

Figure 2. Position of the variables on the first two factorial axes.

The first factorial axis represented on the horizontal shows that between the variables *M* (g/m^2), *Pz* (%), *Kt* (%), *Pr* (daN) and *t* (cN/tex) there is a strong direct connection and between the variables *M* (g/m^2), *Pz* (%) and the variable *R* (mg.cm) there is a reverse link. The representative variables for the second factorial axis, represented on the vertical, are *Cvu* and *Pa* (m^3/mm.min): between them there is a reverse link.

For the interpretation of the third factorial axis, it is possible representing of the variable points in three-dimensional space (figure 3). Because it is difficult to interpret the variables in the graph of figure 3, the coordinate values of variables on the third factorial axis were analyzed according to the values obtained in table 7. The graphical representation of the variety of fabrics from those five groups, on the first two factorial axes, is shown in Figure 4.

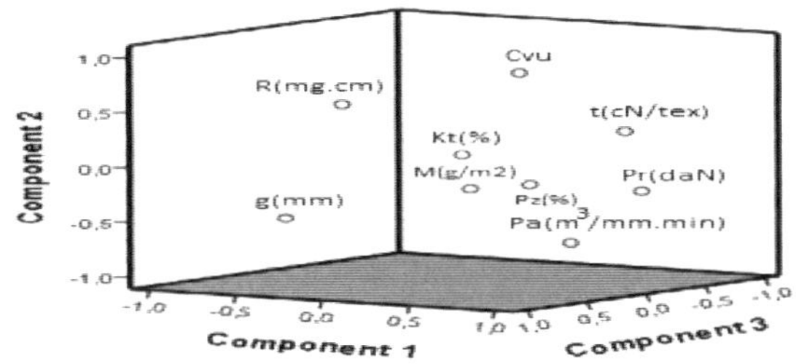

Figure 3. The representation of the variables in the first three-factorial axis system.

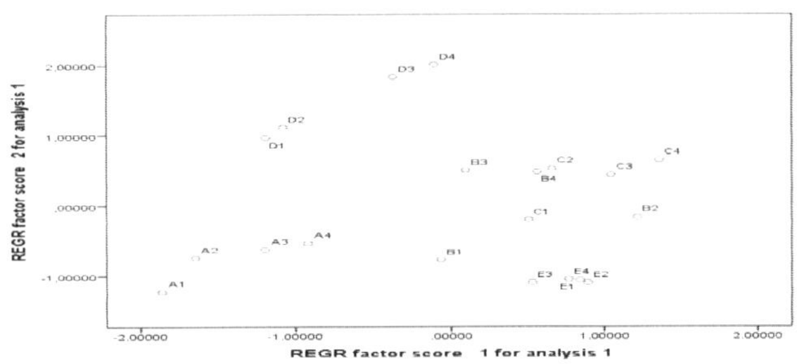

Figure. 4. Position of the assortments of fabrics on the first two factorial axes.

4. Conclusions

The first factorial axis highlights two groups of types of fabrics: The first group is made up of the articles B2 and C4 (with an extreme position to the right of the graph in figure 4) and the second group of articles A1 and A2 (with an extreme position to the left of the graph). The items from the first group are characterized by higher values than average fort he following characteristics: breaking force, tenacity, weight, air permeability, degree of compactness and porosity of woven fabrics, because the percentage of polyester fibers in the composition of yarns, as opposed to the items of the second group, the articles A1 and A2, where the yarns are only made of wool fibers. The second factorial axis highlightes the D4 item, which is characterized by high values of the degree of compactness, stiffness and porosity of the fabric as opposed to articles E2 and E1 where where the volumetric filling coefficient, thickness, weight, degree of compactness, stiffness and porosity of the fabrics are below the average. By applying the PCA mathematical model it was obtained the controlled decrease of the number of variables (columns) of the data matrix, at both components, which contain information about each group/types of fabrics.

References

[1] Schmitt E, Rato T, Ketelaere B D, Reis M and Hubert M 2016 *J. of Chemomet.* **30** (4) 163
[2] Popa A, Bucevschi A, Pustianu M, Manea L R and Sandu I 2016 *Mat. Plast* **53** (2) 316
[3] Hristian L, Sandu A V, Manea L R, Tulbure EA and Earar K 2015 *Rev. Chim. (Bucharest)* **66** (3) 342
[4] Andrecut M 2009 *Journal of Comput. Biology* **16** (11) 1593
[5] Manea L R, Bertea A, Nechita E, Popescu C V and Sandu I 2016 *Rev. Chim. (Bucharest)* **67** (8) 1607
[6] Hristian L, Ostafe M M, Manea L R and Leon A L *2016 IOP Conf. Series: Mater. Sci. Eng.* **145** 032004
[7] Xl Qi, T Meyer, TR Stanford and C Constantinidis 2012 *The Journal of Neuroscience* **32** 6161
[8] Hristian L, Ostafe M M, Manea L R and Leon A L 2016 *IOP Conf. Series: Mater. Sci.Eng.* **145** 022014
[9] Bertea A, Manea L R, Bertea A P and Sandu I 2016 *Rev. Chim.* **67** (12) 2446
[10] Zhong Y and Cook RJ 2016 *Biostatistics* **17** (3) 437
[11] Hristian L, Chirita M, Manea L R and Sandu I 2016 *Mat. Plast.* **53** (4) 738
[12] Abdi H and Williams L J 2010 *Wiley Interdisciplinary Reviews: Computational Statistics* **2** (4) 433
[13] Manea L R, Danu M C and Sandu I 2015 *Rev. Chim. (Bucharest)* **66** (6) 868
[14] Stanescu I, Manea L R, Bertea A, Bertea A P and Sandu I 2016 *Rev. Chim. (Bucharest)* **67** (11) 2276
[15] Llaneras F and Picó J 2010 *Journal Biomed Biotechnology* **34** (753904) 1084
[16] Stanescu I, Manea L R, Bertea A, Bertea A P and Sandu I C A 2016 *Rev. Chim. (Bucharest)* **67** (10) 2082
[17] Manea L R, Hristian L, Ostafe M M, Apostol L L and Sandu I 2016 *Rev. Chim. (Bucharest)* **67** (9) 1758
[18] Rato T J and Reis M S 2013 *Ind. & Eng. Chemist. Research* **52** 13685
[19] Manea L R, Scarlet R, Leon A L and Sandu I 2015 *Rev. Chim. (Bucharest)* **66** (5) 640
[20] Manea L R, Chirita M, Hristian L, Popa A and Sandu I 2016*Mat. Plast.* **53** (3) 361
[21] Mertiny P and Ellyin F 2002 *Composites Part A: Applied Scien and Manufact.* **33** (12) 1615
[22] Manea L R, Bertea A, Nechita E, Popescu C V and Sandu I 2016 *Rev. Chim. (Bucharest)* **67** (7) 1284
[23] Vanhatalo E and Kulahci M 2016 *Quality and Reliabil. Eng. Internat.* **32** (4) 1483
[24] Shen H and Cook RJ 2015 *Biometrical Journal* **57** (5) 725
[25] Manea L R, Hristian L, Leon A L and Popa A 2016 *IOP Conf. Series: Mater. Sci.Eng.* **145** 032007
[26] Manea L R, Hristian L, Leon A L and Popa A *IOP Conf. Series: Mater. Sci.Eng.* **145** 032006

Using Medical-Device Wearable to Improve Hemodialysis Patient's Live and Access the Holistic Health

W L Chen[1], C-C Wu[2,*] and C D Kan[3,*]
[1]Department of Engineering and Maintenance, Kaohsiung Veterans General Hospital, Kaohsiung, Taiwan.
[2]Department of Mechanical Engineering, Chienkuo Technology University, Changhua City, Taiwan.
[3]Division of Cardiovascular Surgery, Department of Surgery National Cheng Kung University Hospital, College of Medicine, National Cheng University Tainan

E-mail: lynnchen.k@gmail.com, wcj@ctu.edu.tw, * kcd56@mail.ncku.edu.tw

Abstract. The increasing incidence of end-stage renal disease (ESRD) is the major burden to health budgets and a threat to public health worldwide. For many years, Taiwan has been ranked the first in the world for the number of hemodialysis patients. For solving the above-mentioned circumstance, we demonstrate the project, here, which goal is to construct the holistic health for hemodialysis patient. The project is to design a wearable medicine-device which can simultaneously measure and monitor the vital sign, including heart rate (HR), pulse oximetry (SPO2), continuous non-invasive blood pressure (c-NIBP), and total body water (TBW), of hemodialysis patient. By aid of the device we design, hemodialysis patients will get better health care than before. This device comprises three techniques. The first is named "Using phonoangiography technique to early detect the dysfunction of arteriovenous access by arteriovenous access (AVA) stenosis detector". The stenosis detector based on autoregressive model was employed to simultaneously estimate the status of AVA life cycle and to tract changes in frequency spectra. It helps hemodialysis patients to early detect the dysfunction of AVA and alarms them to make a return visit. The second technique is named "Physiological detecting device for wearable medical device and encoding algorithm development". The feature of the second technique is to optimize the prognosis by analyzing physiological signals, including water content index, pulse oximetry, and blood pressure in the meanwhile. The third technique is named "Intelligent and smart tourniquet". This technique aims to preclude AVA dysfunction caused by inappropriate hemostasis.

1. Introduction

The increasing incidence of end-stage renal disease (ESRD) is the major burden to health budgets and a threat to public health worldwide. For many years, Taiwan has been ranked the first in the world for the number of hemodialysis patients. According to the United States Renal Data System 2015 Annual Data Report, 2015, the prevalent and incidence rates of ESRD in Taiwan are 3,138 (per million population, pmp) and 458 respectively. By 2015, total annual cost of the National Health Insurance (NHI) program in Taiwan is up to NT$34,200,000,000 due to the 82,000 ESRD patients. For solving the above-mentioned circumstance, we demonstrate the project, here, which goal is to construct the holistic health for hemodialysis patient. The project is to design a wearable medicine-device which can simultaneously measure and monitor the vital sign, including heart rate (HR), pulse oximetry (SPO_2), continuous non-invasive blood pressure (c-NIBP), and total body water (TBW), of hemodialysis patient. By aid of the device we design, hemodialysis patients will get better health care than before. This device comprises three techniques. The first technique is named "Using phonoangiography technique to early detect the dysfunction of arteriovenous access by arteriovenous access (AVA) stenosis detector". The stenosis detector based on autoregressive model was employed to simultaneously estimate the status of AVA life cycle and to tract changes in frequency spectra. It helps hemodialysis patients to early detect the dysfunction of AVA and alarms them to make a return visit.

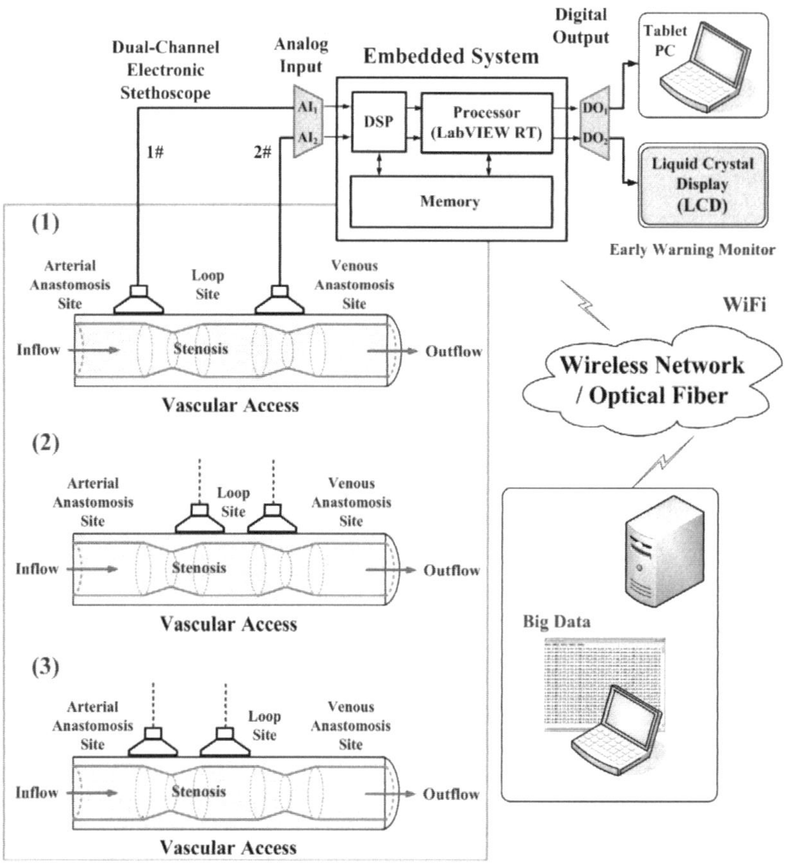

Figure 1. Block diagram of the rapid screening system.

Dialysis vascular access stenosis and failure arise from thrombosis to inflow or/and outflow stenosis, due to long-term traumatic puncture for dialysis treatment. The interior of the vascular access exhibits pathological changes, including the formation of intimal hyperplasia and changes in aneurysmal deformability. Any narrowing of the vascular access causes vibration and turbulent flow and also produces bruits murmur sounds [1-3]. AVG has a higher patency rate than an arteriovenous fistula. Ultrasound detection is a useful and noninvasive technique with a highly accurate detection rate for DOS estimation. Through ultrasonic image examinations, > 50% in the vascular accesses results in hemodynamic or clinical abnormality; and > 70% stenosis in a vascular access, percutaneous transluminal angioplasty or surgical treatment is required to dilate the stenotic lesion. However, its techniques and decision-making depend on the operation skill of the physician. In addition, auscultation is a real-time and noninvasive screening method to detect the local conditions of hemodynamic motions during dialysis therapy. However, only one measurement results in lack of information, because frequency sounds are dependent on the stenosis sites, multiple stensosis, monitoring sites, and DOS. Thus, this study proposes the multi-site auscultation method is carried out to increase the accurate estimations in inflow site, mid site, and outflow site, as shown in figure 1.

Previous studies [4-6] show PAG signals can indicate the hemodynamic conditions, such as flow velocities and flow rates. PAG signals can also be decomposed to time-frequency or frequency-based features and have reported that the changes in frequencies and amplitudes. The characteristic frequencies indicate the spectral peaks between 25Hz and 800Hz. Thus, the characteristic frequencies and frequency spectra can be applied to separate the normal condition from the single and multiple stenosis with different acoustic data length and sampling frequency.Through big data collection, the sound wave patterns can be captured in different data length of the streaming acoustic data along the dialysis vascular access. Based on phonoangiography technique [6], the Burg AR method [7] is used to extract key parameters in the frequency spectra using various combinations of stenostic conditions in an AVG biophysical model, as seen in figure 1 The multiple regression method [8] is employed to find the positive correlation between the key parameters and the DOSs. These big spectral data are further used to establish training data for intelligent machine learning model. In this study, radial-based functions perform the similarity degree measurement for pattern recognition. Then, color relation analysis (CRA) [9] is employed to separate the normal condition from stenostic conditions. The in vitro experimental results will show higher hit rates, true-positive rates, and true-negative rates.

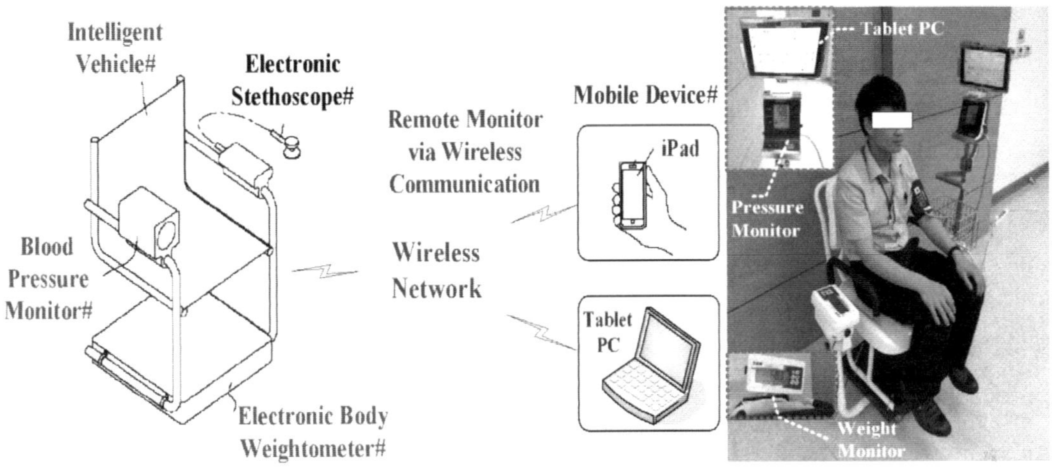

Figure 2. An intelligent healthcare chair for physiological measurements in predialysis healthcare.

The second technique is named "Physiological detecting device for wearable medical device and encoding algorithm development". The feature of the second technique is to optimize the prognosis by analyzing physiological signals, including water content index, pulse oximetry, and blood pressure in the meanwhile. Maintaining adequate dry weight and fluid volume balance is an important issue for dialysis patients. Malnutrition and sodium intake are the primary factors that cause fluid volume imbalance and changes in body weights. Inadequate dry weight control results in higher levels of blood pressures and is related to various complications, such as volume overload, hypertension, congestive symptoms, and cardiovascular diseases. Moreover, inadequate fluid removal provokes hypotension during dialysis treatment. Thus, we propose an early warning tool based on fuzzy color reason analysis (CRA) in predialysis healthcare for hypervolemia screening. The anthropometric method is a rapid, non-invasive, and simple technique for estimating the total body water (TBW). In this study, Watson standard formula is employed to estimate cross-sectional standard of TBW with the patient characteristics, including gender, age, height, and weight. Contrast with the experienced anthropometric formulas, Watson formula has $<$ 2% of margin errors and provides a criterion as a reference manner to estimate the TBW in patient's normal dry weight. In addition, inadequate dry weight and TBW controls will lead to higher blood pressures. The systolic blood pressure (SBP) is also an indicator to evaluate pre-hypertension of 120 – 139 mmHg and hypertension of $\geq$ 140 mmHg. Therefore, the levels of two indicators, TBW and SBP, are parameterized with fuzzy membership grades to describe the normal and the specific ranges of undervolemia and hypervolemia. A CRA utilizes a hue-saturation-value color model to design a color perceptual manner for separating normal condition from hypervolemia or undervolemia. Normalized hue angle and saturation value provide a promising visual representation with color codes to realize the patients' diagnosis. Dialysis patients with hypertension demonstrated that the proposed model can be used in clinical applications. In addition, a healthcare chair is carried out to measure blood pressure and weight in predialysis. The proposed assistant tool integrates an electronic pressure monitor, an electronic weight monitor, and fuzzy CRA is also intended to be established in an intelligent vehicle via a WiFi wireless local area network for cloud computing.

The third technique is named "Intelligent and smart tourniquet". This technique aims to preclude AVA dysfunction caused by inappropriate hemostasis.

2. Methodology

2.1. PAG Signal Acquisition and Feature Extraction

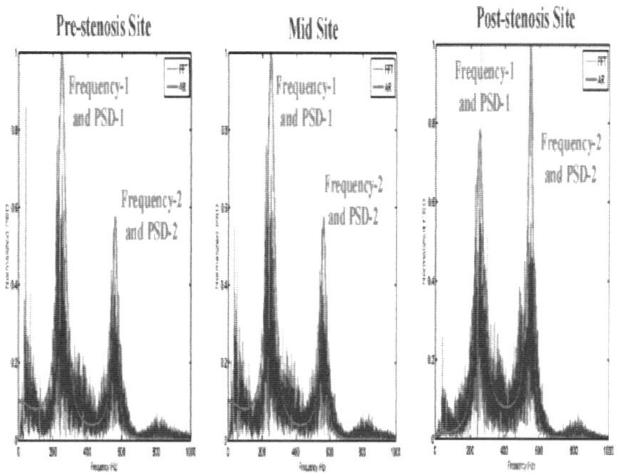

Figure 3. Characteristic frequencies at pre-stenosis site, mid site, and post-stenosis site.

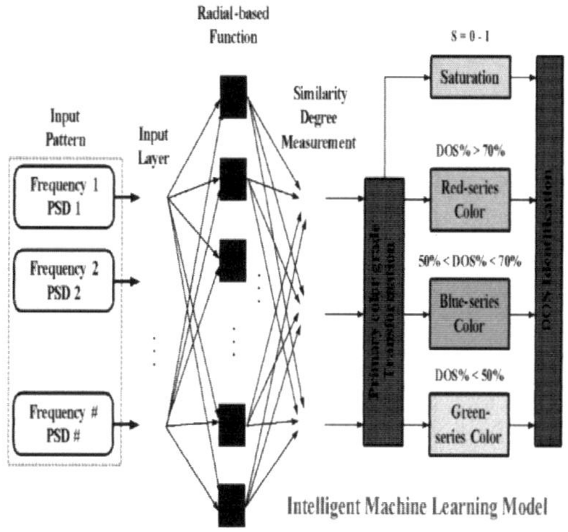

Figure 4. The proposed intelligent machine learning model: radial-based color relation analysis.

As shown in figure 3, fast Fourier transform (FFT) algorithm was used with discrete Fourier transform (DFT) to transform the time-domain PAG signal into discrete frequency-domain representation, as follows [10]:

$$x[n] = PAG[n],\ \Omega = \frac{2\pi n}{N},\ n = 0, 1, \ldots, N\text{-}1 \quad (1)$$

$$y[n'] = \sum_{n=0}^{N-1} x[n] e^{-j\Omega n},\ n' = 0, 1, \ldots, N\text{-}1 \quad (2)$$

where $x[n]$ is a discrete time PAG signal (sampling data of original PAG signal, $PAG[n]$); n and n' are the sampling points; and $y[n']$ is a periodic and extended from frequency, $f = 0$, to sampling frequency, f_s. The DFT was defined in the region between 0 and f_s. The first half of the frequency range (from 0 to the Nyquist frequency, f_s / 2) was sufficient to identify the component frequencies in the FFT data. After the DFT process, a PAG signal can be decomposed into a number of discrete frequencies, $y[n']$, in the specific range from 0 to 1,000 Hz. The frequency spectrum was normalized with the maximum amplitude, $\sqrt{}[n']$, as

$$\Phi[n'] = [\varphi_1,\ \varphi_2,\ \cdots,\ \varphi_{n'},\ \cdots,\ \varphi_p] = \frac{abs(y)}{\max[abs(y)]},\ n' = 1, 2, 3, \ldots, p \quad (3)$$

Then, the Burg AR method [11-13] was employed to smoothen the frequency spectrum over a continuous range. To obtain the reliable characteristic frequencies, the AR model with AR order was determined to minimize the sum of the residual energies using the Levinson–Durbin recursion algorithm [14-15]. In this study, stenosis combinations are considered at the pre-stenosis site and post-stenosis site, as shown in figure 4, and all the 171 combinations of data sets are used to characterize the primary components of frequency features. PAG signals with different window lengths, $n = 0, 1, 2, \ldots, N - 1$, and the choice of AR order were suggested to achieve a good evaluation analysis by a clinician, including (1) matching the time-limited observation (10-s record for a consumer electronic stethoscope), (2) reducing spectral broadening and spectral leakage, (3) distinguishing the characteristic features in the specific ranges, and (4) enhancing the frequency resolution within the observed window lengths. Two characteristic frequencies could be obtained to determine the spectral peaks between 200 – 300 Hz and 500 – 600 Hz in two marginal distributions, symbolized as frequency-1 and frequeny-2. Therefore, distinct peaks of power spectral densities (PSDs) can be found as key features for the stenosis screening at the pre-stenosis site and post-stenosis site, symbolized as four combinations (pre-PSD-1, pre-frequency-1), (pre-PSD-2, pre-frequency-2), (post-PSD-1, post-frequency-1), and (post-PSD-2, post-frequency-2). nflow and outflow stenoses.

3. Implementation of Proposed Screening Model

The concept of the fuzzy CRA reason was derived from the fuzzification operations and HSV color model to describe perceptual color relationships for hypervolemia screening. Fuzzification operations can map mathematical input variables into specific gray grades using Gaussian, *Z* sigmoidal, and *S* sigmoidal membership functions (MFs). The MF parameters were static or could be changed dynamically based on different HD patients. The proposed screening model could be implemented in an embedded system, as shown in figure 5.

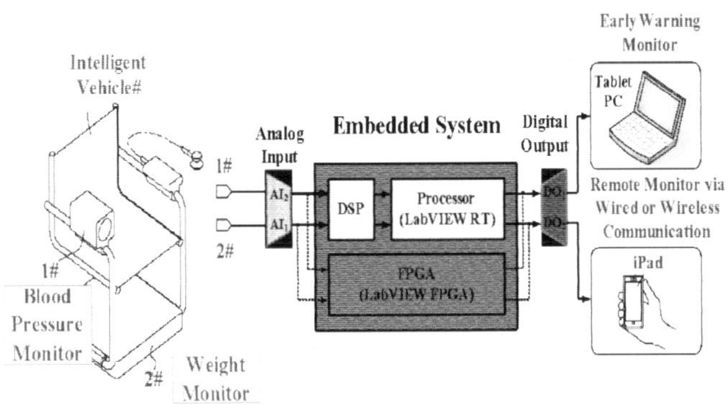

Figure 5. The proposed fuzzy CRA screening Model and the implementation of screening model in an embedded system.

Its model uses straightforward mathematical computations to achieve the inference procedures for real-time applications. An embedded system (National Instruments™ myRIO-1900, Austin, Texas, U.S.A) can be applied to establish a prototype screening algorithm within a short design cycle. We also integrated electronic body weightometer and blood pressure monitor for weight and blood pressure measurements using a healthcare vehicle/chair. The vehicle was equipped with commercial sensors and was designed to carry out wireless communication for unconstrained physiological monitoring. A WiFi (IEEE 802.11 Standard) wireless local area network was used for linking mobile devices (smart phones, personal digital assistant, or iPad) and portable computers support (laptop), while being operated on 2.4 GHz industrial, science, and medical frequency bands. It was designed for portable devices with low power consumption and short-range communications in common household and mobile appliances.

4. Case Study in Hypertensive HD Patients

The experimental data from hypertensive HD patients were used to verify the proposed screening model. Data of 12 participating subjects, aged 37 – 67 years, were shown in tables 1 and 2. For these case studies, Watson, Hume, Sahlgrenska formulas were used to calculate the normality TBWs and estimated TBWs. The normal dry weight was prescribed for each patient by the nephrologists. Then the TBWrat was parameterized to rapidly indicate the changes in fluid volume balance, as TBWrat > 1, TBWrat ≈ 1, or TBWrat < 1. In fuzzification operations, the membership grades of TBWrat and SBP were converted to gray grades and three primary color grades using the gray-grade intensity adjustment.

The CRA utilized HSV color model to map gray grades into color codes, green series color, blue series color, and red series color, for the estimation of patients' diagnosis. For instance, Subject 7# (male, aged 54.9 years) had hypertension and diabetes, a case in which changes in dry weight of +2.0 kg could result in higher SBPs and fluid volume imbalance. Inadequate control of fluid volume or inability to maintain an appropriate dry weight for chronic HD patients was identified as a key factor to cause excess mortality. These problems could lead to chronic volume overload with hypertension and left ventricular hypertrophy, while subsequently causing cardiovascular symptoms and also increasing the patency rate of complications. In addition, some patients removed extra weights to achieve appropriate dry weight, resulting in uncomfortable symptoms. Therefore, it is important to screen early in predialysis stage (3 times a week) and then perform dialysis for volume control or administer antihypertensive drugs.

Subject 7# satisfied with an index, H_C = 0.9992 (359.7064°) for red series color, and S = 0.9794; thus, this case study can be agreed as a patient with increasing weight that led to "hypervolemia." In addition, for Subject 9# (male, aged 66.1 years), the inference results indicated "normal condition," satisfying with an index, H_C = 0.5369 (193.2840°) for blue series color.

However, the saturation index, S = 0.4485, was less than 0.5, due to the subject having slight decreases in weight and prehypertension. This observation led to a suggestion of maintaining the appropriate BMI and dry weight. In contrast with Subject 11# (male, aged 66.9 years), he had changes in weight of −1.7 kg, and the inference results indicated "undervolemia," satisfying with an index, H_C = 0.2950 (106.2000°) for green series color, and saturation index, S = 0.5303. For color code in imaging version, the screening results provided a promising suggestion to gradually control fluid volume and appropriate BMI for HD healthcares. This finding confirmed that the proposed screening model could detect fluid volume imbalance in its early stages in HD patients, six with hypervolemia, two with normal condition, and four with undervolemia. Experimental results for 12 HD subjects are shown in tables 1 and 2.

Table 1. The experimental results for TBW estimation.

Patient No.	A (year)	Baseline W (kg)	Change in W (kg)	H (m)	Normality TBW, TBW_{nor}			Estimated TBW			Error (%), (TBW- TBW_{nor}) / TBW_{nor}		
					(1)	(2)	(3)	(4)	(5)	(6)	(4)-(1)/(1)	(5)-(2)/(2)	(6)-(3)/(3)
1	37.0	92.6	+1.7	1.73	48.6	47.1	47.6	49.2	47.6	48.2	+1.17	+1.07	+1.31
2	43.7	94.2	+1.3	1.75	48.2	46.9	47.4	48.6	47.3	47.8	+0.90	+0.80	+1.00
3	46.1	61.2	+1.7	1.72	37.1	37.6	35.9	37.7	38.1	36.5	+1.54	+1.34	+1.73
4	52.0	75.9	+1.7	1.71	41.3	41.7	40.9	41.9	42.2	41.5	+1.38	+1.21	+1.52
5	53.6	87.3	+1.9	1.77	45.7	46.3	46.6	46.3	46.8	47.3	+1.39	+1.21	+1.49
6	54.2	62.5	-2.4	1.72	36.8	38.0	36.3	36.0	37.3	35.4	-2.19	-1.87	-2.42
7	54.9	81.0	+2.0	1.72	42.9	43.4	43.1	43.6	44.0	43.8	+1.57	+1.36	+1.70
8	56.5	78.6	-1.5	1.69	41.7	42.3	41.7	41.2	41.9	41.2	-1.20	-1.05	-1.32
9	66.1	67.8	-0.4	1.70	37.7	40.1	39.0	37.6	40.0	38.9	-0.36	-0.29	-0.38
10	66.7	64.3	-0.7	1.76	36.6	39.3	38.0	36.4	39.1	37.7	-0.64	-0.53	-0.67
11	66.9	58.6	-1.7	1.69	33.9	36.2	34.1	33.3	35.6	33.4	-1.68	-1.39	-1.83
12	67.0	59.2	-1.5	1.68	33.9	36.1	34.0	33.4	35.7	33.4	-1.48	-1.22	-1.61

Note: 1. Symbols, (1) and (4), mean Watson formulas; 2. Symbols, (2) and (5), mean Hume formulas; 3. Symbols, (3) and (6), mean Sahlgrenska formulas

Table 2. The experimental results for hypertensive HD patients.

Patient No.	A (year)	Baseline W (kg)	Change in W (kg)	H (m)	SBP (mmHg)	$TBW_{rat}=TBW/TBW_{nor}$			Hypertension Diabetes	Fuzzy CRA (H_C / S)		
						(1)	(2)	(3)		(1)	(2)	(3)
1	37.0	92.6	+1.7	1.73	168.8	1.0117	1.0107	1.0131	√	0.9998	0.9998	0.9998
										0.9855	0.9887	0.9873
2	43.7	94.2	+1.3	1.75	177.2	1.0090	1.0082	1.0100	√	0.9999	1.0000	1.0000
										0.9890	0.9855	0.9893
3	46.1	61.2	+1.7	1.72	145.6	1.0153	1.0133	1.0173	√	0.9828	0.9817	0.9834
										0.8629	0.8245	0.8766
4	52.0	75.9	+1.7	1.71	154.6	1.0138	1.0120	1.0151	√	0.9973	0.9973	0.9973
										0.9578	0.9546	0.9630
5	53.6	87.3	+1.9	1.77	163.6	1.0139	1.0121	1.0149	√	0.9995	0.9995	0.9995
										0.9830	0.9820	0.9835
6	54.2	62.5	-2.4	1.72	143.3	0.9780	0.9813	0.9758	√	0.2581	0.2586	0.2555
										0.4691	0.3562	0.5417
7	54.9	81.0	+2.0	1.72	160.6	1.0157	1.0136	1.0169	√	0.9992	0.9992	0.9992
										0.9794	0.9780	0.9803
8	56.5	78.6	-1.5	1.69	158.8	0.9879	0.9895	0.9868	√	0.1667	0.1667	0.1667
										0.9062	0.9130	0.9004
9	66.1	67.8	-0.4	1.70	130.0	0.9964	0.9971	0.9962	X	0.5369	0.5179	0.5153
										0.4485	0.6444	0.6583
10	66.7	64.3	-0.7	1.76	140.0	0.9936	0.9947	0.9933	X	0.5769	0.6368	0.5639
										0.1536	0.1112	0.1672
11	66.9	58.6	-1.7	1.69	140.0	0.9831	0.9860	0.9817	√	0.2950	0.3085	0.2905
										0.5303	0.4191	0.5722
12	67.0	59.2	-1.5	1.68	140.4	0.9852	0.9877	0.9838	√	0.3002	0.3131	0.2931
										0.4401	0.3445	0.5484

Note: 1. Symbol (1) means Watson formulas; 2. Symbol (2) means Hume formulas; 3. Symbol (3) means Sahlgrenska formulas

5. Discussion

Cardiovascular disease is the most common cause of mortality in HD patients, and hypertension is also a significant factor for cardiovascular disease pathogenesis. Hypervolemia, hypernatremia, and hyperkalemia are important risk factors for their morbidities and mortalities. In previous studies [3, 4, 16-18], the relationship between hypervolemia and malnutrition was regarded to be the key indicator of predisposition to hypertension in HD patients, further preventing the progression of cardiovascular event rate. During dialysis treatment, inadequate fluid removal and blood volume ultrafiltration control or fluid volume imbalance provoked hypotension, associated with clinical symptoms for nursing decreases SBP by > 20 mmHg or decreases mean arterial pressure by 10 mmHg [19].

Hence, anthropometric formulas, such as Watson, Hume, Sahlgrenska, and Lee formulas [20], provide well-known standard methods to estimate TBW. This method might cause large systematic errors, while TBW varied from the average, obese, and overhydrated patients. In particular, in obese patients, the estimated results led to large errors, in average or close-to-average weights, and the Watson, Hume, and Sahlgrenska formulas provided a promising reference with narrow margins of error [2]. For prescribed the normal dry weight, estimated results with a total average error of < 2% for 12 subjects using three formulas are shown in Table 1. The robustness range of normal condition was ± 2% of changes in TBW and changes in SBP from 110 to 135 mmHg. That is, the Watson formula has been verified and can be applied to estimate TBW. This indicates that a rapidly, safe, and simple method can be used from a cross-sectional standard for bedside applications in predialysis stage. In this study, the objective was to establish an intelligent vehicle with a warning tool for predialysis healthcare. For physiologica measurements to screen hypervolemia, these findings can provide a promising suggestion to make changes in drink / food and to control extra body weights.

6. Conclusion

A strategy to monitor and control the fluid volume status and hypertension is an important clinical issue in HD patients. Dietary sodium restriction and fluid volume control have been performed to improve malnutrition. A promising method that can provide an accurate assessment to achieve and maintain a stable BMI and dry weight is needed. For prescribed the normal dry weight, Watson formulas for male and female subjects have been validated to estimate cross-sectional TBW with an average error of <1.2% to indicate the fluid volume imbalance. The proposed fuzzy CRA with the TBW_{rat} and SBP is used to separate the normal condition from hypervolemia or undervolemia. The fuzzy CRA has a flexibility inference mechanism and no iterative computations to update model parameters. The recognition coefficient, ζ, monotonously increases to enhance better contrast in classification applications, while has capability of self-regulation in the primary color grades. Hence, this simple technique can be easily implemented in a tablet PC and an intelligent vehicle via wireless connection, which only requires few patient characteristics, such as *S*, *A*, *W*, and *H* parameters, as shown by the data in the dashboard in Figure 8. This individualized tool can enhance the priority in data read, cloud computing, and cloud storage for patient demands. In addition, antihypertensive medication is a directed manner to control blood pressure, further preventing the progression of congestive heart failure and improving cardiovascular outcomes. In routine examinations, TBW and blood pressure screenings can be used to evaluate individualized characteristics for drink, food, and pharmacologic controls. We may have a cross-sectional reference to objectively direct dry weight management. In addition, this promising model is an individualized tool for dry weight maintenance in predialysis healthcare.

References

[1] Kurata M 1982 *Numerical Analysis for Semiconductor Devices* (Lexington, MA: Heath)
[2] Selberherr S 1984 *Analysis and Simulation of Semiconductor Devices* (Berlin: Springer)
[3] Sze S M 1969 *Physics of Semiconductor Devices* (New York: Wiley–Interscience)
[4] Dorman L I 1975 *Variations of Galactic Cosmic Rays* (Moscow: Moscow State University Press) p 103
[5] Caplar R and Kulisic P 1973 *Proc. Int. Conf. on Nuclear Physics (Munich)* **1** 517
[6] Cheng G X 2001 *Raman and Brillouin Scattering—Principles and Applications* (Beijing: Scientific)
[7] Szytula A and Leciejewicz J 1989 *Handbook on the Physics and Chemistry of Rare Earths* **12** 133
[8] Kuhn T 1998 Density matrix theory of coherent ultrafast dynamics *Theory of Transport Properties of Semiconductor Nanostructures* chapter **6** 173–214
[9] Kuhn T, Binder E, Rossi F, Lohner A, Rick K, Leisching P, Leitenstorfer A, Elsaesser T and Stolz W 1994 Coherent excitonic and free-carrier dynamics in bulk GaAs and heterostructures *Coherent Optical Interactions in Semiconductors: Proc. NATO Advanced Research Workgroup (Cambridge, UK, 11–14 August 1993) (NATO Advanced Study Institute, Series B: Physics* **330***)* 33–62
[10] Agarwal R 2010 *Hypertension* **56** 512-517
[11] Matthew R W 2010 *Hypertension* **56** 341-343
[12] Pogue V, Rahman M, Lipkowitz M, Toto R, Miller E, Faulkner M, Rostand S, Hiremath L, Sika M, Kendrick C, Hu B, Green T, Appel L and Phillips R A 2009 *Hyperension* **53** 20-27
[13] Agarwal R, Flynn J, Pogue V, Rahman M, Reisin E and Matthew R 2014 *J. Am. Soc. Nephrol.* **25** 1630-1646
[14] Inci A, Kursat S, Kutsal D A, Ulman C, and Veysel Y 2015 *Clinical Nephrology and Urology science* **2** 1-8
[15] Lee S W, Song J H, Kim G A, Lee K J and Kim M J 2001 *Nephrol. Dial. Transplant.* **16** 91-97
[16] Vallozzi L, Van Torre P, Hertleer C, Rogier H, Moeneclaey M and Verhaevert J 2010 *IEEE Transactions on Antennas and Propagation* **58 (4)** 1357-1368
[17] Shoji T, Tsubakihara Y, Fujii M and Imai E 2004 *Kidney International* **66** 1212-1220
[18] Ohashi Y, Otani T, Tai R, Okada T, Tanaka K, Tanaka Y, Sakai K and Aikawa A 2012 *Kidney Blood Pressure Research* **36** 231-241
[19] Watson P E, Watson I D and R D Batt R D 1980 *Am. J. Clin. Nutr.* **33(1)** 27-39

[20] Hwang H S, Hong Y A, Yoon H E, Chang Y K, Kim S Y, Kim Y O, Jin D C, Kim S H, Kim Y L, Kim Y S, Kang S W, Kim N H and Yang C W 2016 *Medicine* **95(7)** 1-8

Acknowledgments
This work is supported in part by the Ministry of Science and Technology, Taiwan, under contract number: MOST 105-2218-E-075B-001 and MOST 105-2221-E-006-087-MY2, duration: March 1, 2016 ~ February 28, 2017 and is also supported in part by the research grant of Kaohsiung Veterans General Hospital, under contract no. VGHKS 105-070.

Hazard, Vulnerability and Capacity Mapping for Landslides Risk Analysis using Geographic Information System (GIS)

D A P Sari[1], S Innaqa[2] and Safrilah[3]
[1] Environmental Engineering Program Study, Universitas Bakrie, Jakarta, Indonesia
[2] Disaster Management Program Study, Indonesia Defence University, IPSC Sentul, Indonesia
[3] Civil Engineering Program Study, Universitas Bakrie, Jakarta, Indonesia

E-mail: deffi.sari@bakrie.ac.id, deffiayu@gmail.com

Abstract. This research analyzed the levels of disaster risk in the Citeureup sub-District, Bogor Regency, West Java, based on its potential hazard, vulnerability and capacity, using map to represent the results, then Miles and Huberman analytical techniques was used to analyze the qualitative interviews. The analysis conducted in this study is based on the concept of disaster risk by Wisner. The result shows that the Citeureup sub-District has medium-low risk of landslides. Of the 14 villages, three villages have a moderate risk level, namely Hambalang, Tajur, and Tangkil, or 49.58% of the total land area. Eleven villages have a low level of risk, namely Pasir Mukti, Sanja, Tarikolot, Gunung Sari, Puspasari, East Karang Asem, Citeureup, Leuwinutug, Sukahati, West Karang Asem West and Puspanegara, or 48.68% of the total land area, for high-risk areas only around 1.74%, which is part of Hambalang village. The analysis using Geographic Information System (GIS) prove that areas with a high risk potential does not necessarily have a high level of risk. The capacity of the community plays an important role to minimize the risk of a region. Disaster risk reduction strategy is done by creating a safe condition, which intensified the movement of disaster risk reduction.

1. Introduction

One of the goals of the Republic of Indonesia as stated in the Act of 1945 is to protect the people and the country of Indonesia from all threats. This time, the threat to the country's defense is dominated by non-military threats are considered more complex and multidimensional than military threats. One form of tangible non-military threats facing Indonesia today is a natural disaster. At a certain scale, the disaster could have an impact on national stability and a threat to the development [1].

Bogor Regency in West Java has vulnerability to landslides. During the past 10 years, landslides have dominated the disastrous events contained in Bogor, as many as 116 events with thousands of human casualties and damage to facilities and settlements [2]. Bogor Regency morphological conditions are mostly high plains, hills and mountains, and relatively high rainfall. The morphological condition of Bogor Regency is mostly in the form of plateau, hills and mountains, and has high rainfall. Land formation of some areas of Bogor Regency can be seen from the image data DEM (Digital Elevation Model) processed from satellite images as shown in Figure 1.

To minimize the impact of landslides, disaster risk reduction must be done with careful consideration of the each area's characteristics. Assessment and risk mapping is considered a part of pre-disaster stages to reduce the impact of potential disasters and as part of efforts for preparation to face the disaster [3].

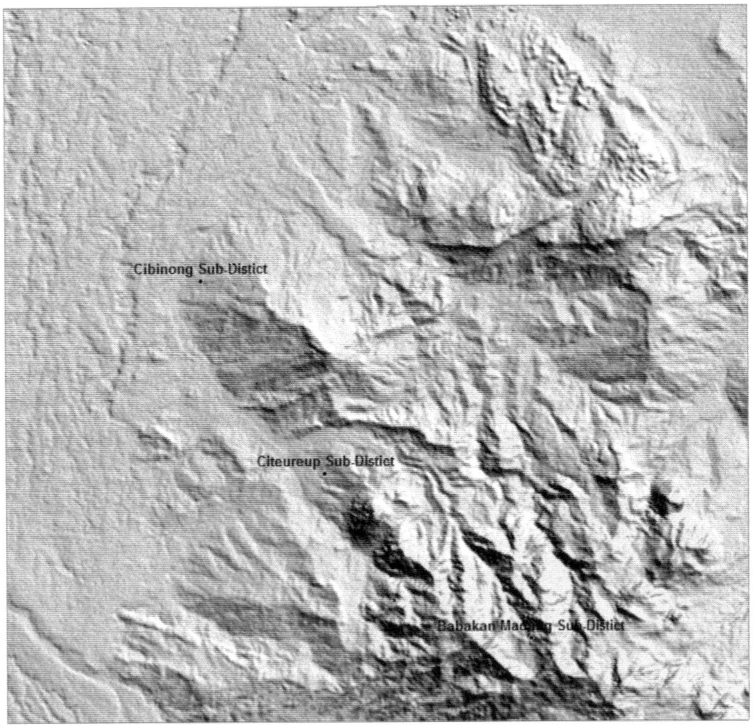

Figure 1. Digital Elevation Model (DEM) part of Bogor Regency
Source: SRTM 30 DEM from www.earthexplorer.usgs.gov.

Based on the background and the description above, the formulation of the problem in the form of risk assessment in the context of landslide disaster risk reduction in Citeureup sub-district is described in research goal which is to analyze the level of risk of landslides in Citeureup sub-district based on its hazards, vulnerability and capacity using Geographic Information System (GIS).

2. Materials and methods

The method used in this study is a qualitative method. However, descriptive quantitative approach is done in weighting or scoring of quantitative data.

First, scoring techniques are used in the determination of threat/hazard, vulnerability, and capacity parameters. The results are poured in the form of maps in the form of overlay of these three components using software Geographic Information System (GIS). The final risk value is derived from the arithmetic operations of all three components, namely Risk (R) = Threat / Hazard (H) x {Vulnerability (V) / Capacity (C)}.

Analysis of mapping used to represent the results of the research, then Miles and Huberman analytical techniques used to analyze the qualitative interviews.

The analysis conducted in this study is based on the concept of disaster risk by Wisner *et.al.* Disaster risk assessment is analyzed with the help of Geographic Information System (GIS) applications, as well as parameter scoring techniques used in the assessment of landslide threats/hazards, physical, social, and economic vulnerabilities, and capacity of the community. Parameters, indicators, weights and values used to assess the level of risk in this study are presented in table 1.

IOP Conf. Series: Materials Science and Engineering **209** (2017) 012106 doi:10.1088/1757-899X/209/1/012106

Table 1. Variables, parameters, indicators, the weight and value of threats, vulnerabilities and capacities.

Value	Weight	Indicator	Parameter	Variable
1	0.3	Dry (<1000 mm / year)	Rainfall	Threat
2		Dry (1001 to 2000 mm / year)		
3		Medium (2001-3000 mm / year)		
4		Wet (3000-3001 mm / year)		
5		Very Wet (> 4000 mm / year)		
1	0.4	0-8%	Slope	
2		8-15%		
3		15-25%		
4		25-45%		
5		>45 %		
1	0.1	Alluvium	Geology	
2		Pratersier		
3		Miocene facies sediments		
4		Result of Volcanic Old Quaternary		
5		Results Volcano Young Quaternary		
1	0.2	Forests	Land Use	
2		Mixed Forest		
3		Gardens, Mixed Garden		
4		Paddy field, Farm, Bush / Thicket		
5		Settlement, Industry		
1	0.4	<30 buildings/ha	Home Density	Vulnerability (physical)
3		30-80 building/ha		
5		>80 buildings/ha		
1	0.3	<31 units of	Total Common Facilities	
3		32-47 units of		
5		>48 units		
1	0.3	<30%	Built Area	
3		30-60%		
5		> 60%		
1	0.6	<15 inhabitants/ha	Density	Vulnerability (social)
3		of 15-25 inhabitants/ha		
5		>25 inhabitants/ha		
1	0.2	<20	Poverty Ratio	
3		20-40		
5		>40		
1	0.2	<18 inhabitants	Total Disability	
3		18-36 inhabitants		
5		>36 inhabitants		
1	0.4	Private/goverment employees	Primary Source of Income	Vulnerability (economy)
3		Processing industry, trade		
5		Agriculture, miner/quarrying		
1	0, 3	<50 units	Total Industry	
3		50-100 units		
5		>100 units		
1	0.3	<30%	Productive Land Area	
3		30-50%		
5		>50%		
1	0.2	Not present	Early Warning System	Capacity
5		Present		
1	0.1	Not present	Evacuation Route	
5		Present		
1	0.1	Not Available	Alarm	
5		Available		
1	0.1	Not available	Availability of the Equipment	
3		Not adequate		
5		Complete		
1	0.3	Never	Disaster Education/Training	
3		Somewhat often		
5		Often		
1	0.2	Not present	Mutual Cooperation	
5		Present		

3. Results and discussions

3.1. Land Slide Risk Assessment

Landslides are defined as the displacement of slope-forming materials, those are debris avalanche, rocks, soil, or mixed material moving down or out of the slope. This incident is influenced by several factors, including slopes that are bare and the condition of soil and rocks are fragile. The main trigger of landslide is the high intensity of rainfall. In addition, human activity can also be a driver of landslides, namely development in protected areas, plantation activities or fields in cliff areas.

Risks as potential losses that can be death, life threats, material losses, and disruption of socio-economic activities of communities caused by disasters. Risk analysis aims to identify areas based on their risk level of disaster [4]. Assessment results become a reference in the formulation to cope with the negative impact of the disaster.

Risk is a function of a combination of hazard and vulnerability, and its relationship is expressed as a result of multiplication between hazard and vulnerability [5]. Disaster risk is formed at least by two factors, the place where the disaster is located, usually called a place that has the potential for disaster (hazard) and the level of vulnerability (vulnerability). Mathematically, the risks can be formulated as follows:

$$\text{Risk} = f\ \{\text{Hazard, Vulnerability}\}$$

In addition to hazard and vulnerability factors, there are other factors that can play a role in increasing or decreasing disaster risk, ie capacity (capacity). Capacity is defined as a system, community or community potentially exposed to danger to adapt or change to achieve or maintain an acceptable level of function and structure [5]. This is determined by the extent to which such a social system is able to organize itself to enhance the capacity to learn from past disasters for the better protection of the future and to enhance risk reduction measures.

Risk assessment is one part of the disaster risk assessment. The risk assessment itself is part of disaster risk management. Disaster risk management is one of the phases in pre-disaster, defined as a process of identifying disaster risk types, assessments, and monitoring of results, conducted repeatedly and periodically with the objective of minimizing disaster risk [6]. According to the Asian Disaster Reduction and Response Network/ADRRN (2010), disaster risk management aims to avoid, mitigate, or divert harmful impacts from threats / hazards through prevention, mitigation and preparedness activities and efforts.

In the context of disaster prevention and development policy decisions that can pose a risk of disaster, mapping plays an important role in every stage of disaster management, and is very helpful in disaster risk reduction [7]. Landslide risk map produced can describe areas with low risk, medium, and high. The risk level is calculated by scoring the parameters of threats, vulnerabilities, and capacities, which produces maps the level of threat, vulnerability, and capacity. From the results of the mapping is done categorization of risk using the matrix multiplication method VCA *(Vulnerable Capacity* Assessment). To determine the level of risk, result of scoring vulnerability-capacity beforehand calculated and classified using VCA table to get the class of Vulnerability/Capacity (V/C). The result can be seen in the matrix Table 2 below.

3.2. Water absorption properties porous geopolymer

The features and properties of the porous geopolymer, for example, porosity, pore size distribution, pore morphology, and pore connectivity (commonly identified as the relationship between open and closed porosity), depend strongly on the composition and processing method. Research study shows that the chemical compositions of the geopolymers (in terms of their H_2O/Al_2O_3 and Na_2O/Al_2O_3 ratios) were systematically varied, allowing the porosity and water retention properties to be adjusted [6].

Figure 2 shows that geopolymers synthesized with higher H_2O/Al_2O_3 ratios are more porous, their larger pore size and higher pore volume giving them good water absorption and water retention properties but decreasing their mechanical strength. The limited mechanical strength of this type of geopolymers may have restriction on some applications. By contrast, the geopolymers synthesized with lower H_2O/Al_2O_3 ratios are denser, with smaller pore sizes and lower pore volumes, resulting in better water retention and mechanical properties than materials with higher H_2O/Al_2O_3 ratios. Thus, these two types of geopolymers are suitable for different water retention applications [6].

Table 2. Matrix vulnerability-capacity landslides in Citeureup District.

V/C		Capacity		
		High	Medium	Low
Vulnerability	Low			
	Medium		Tangkil, Hambalang, Tajur, Sanja, West Karang Asem, Puspanegara, Puspasari, Citeureup, East Karang Asem, Leuwinutug	
	High		Pasir Mukti, Sukahati, Tarikolot, Gunung Sari	

Based on the results of the matrix multiplication, Tangkil, Hambalang, Tajur, Leuwinutug, Sanja, West Karang Asem, Puspanegara, Puspasari, Citeureup, and East Karang Asem has a medium vulnerability and capacity, and including in medium V/C. Pasir Mukti, Sukahati, Tarikolot, and Gunung Sari are including as areas with high vulnerability and medium capacity, and belongs to high V/C.

Pasir Mukti, Sukahati, Tarikolot, and Gunung Sari has a high vulnerability, influenced by the number of vulnerable people, ie persons with disabilities, commonly found in the region. Furthermore, there are many community-based economic activities in that region, shown by the number of factories located in Pasir Mukti, Sukahati, Tarikolot, and Gunung Sari.

Furthermore, from V/C calculation, additional of threats was added. This process is performed by using a calculator filed on the QGIS device. The result can be seen in the matrix table 3 below.

Table 3. Matrix hazard-vulnerability to landslides in Citeureup district.

Risk Matrix		V/C		
		High	Medium	Low
Threat/Danger	Low	Pasir Mukti, Sukahati, Tarikolot, Gunung Sari	Sanja, West Karang Asem, Puspanegara, Puspasari, Sanja, Leuwinutug, East Karang Asem	
	Medium		Tangkil, Hambalang, Tajur	
	High			

Risk Matrix of landslides in Citeureup sub-district shows that rural areas/villages are in the low-medium threat zone, which means that those villages are belong to landslide risk of low to moderate class. Only three villages namely Hambalang, Tajur, and Tangkil are in the moderate risk class, while Pasir Mukti, Sukahati, Tarikolot, Gunung Sari, Sanja, Karang Asem West, Puspanegara, Puspasari, Sanja, Leuwinutug, and East Karang Asem East are in low risk class.

If an area has low vulnerability, although that area is located in high hazard zones, it will make the area has low risk of disaster. Due to low vulnerabilities that exist in the region, so that the disaster will not affect the physical, social, and economic condition of the region significantly.

Result from interviews indicate a high awareness of the potential for landslides or land movements in the Citeureup sub-district. Some areas are located in the hills, and in the area around the river which is prone to landslides and subsidence. However, this condition does not exist in every region in Citeureup. Only a few villages or locations that have prone points to landslides, namely Hambalang and Tajur. Both villages are located in the Hambalang hills and is the highest area in Citeureup sub-district. The region has the morphology of hills with a slope >15%. The level of landslides risk in Citeureup district can be seen in figure 2.

Landslides incident were happened several times in the area of research. However, those incident did not cause casualties or collapsed buildings. Landslides occur every year in Citeureup sub-District, with different location of landslides. Although the intensity of the landslide is low and did not cause casualties, but the landslides in 2007 has led Hambalang villagers to be relocated because of the discovery of cracks extending around settlements.

Danger or threat is a natural phenomenon or man-made which has the potential to threaten human life, damage environment and loss of property. Referring to the definition of danger according to Awotona [8], that the danger can happen anywhere and anytime, and trigger disasters when meeting with vulnerable conditions [8]. The impact is generally in the form of loss and damage to humans, which is caused by the inability of man-made systems in protecting themselves and assets/possessions he had.

Based on the results of research conducted by Dianardi *et.al* (2012), morphology or the outside structure of rocks that making up the earth's surface is the cause of landslide in Hambalang Hills region [9]. Tajur and Hambalang villages with a medium-high landslide potential, has a height of 500 meters with a maximum slope of 25°. This causes low slope stability and resulted in susceptible ground movement. Rock composer which have fine-grained texture causes high absorption of runoff and low adhesion to the ground, so the potential for land subsidence is high.

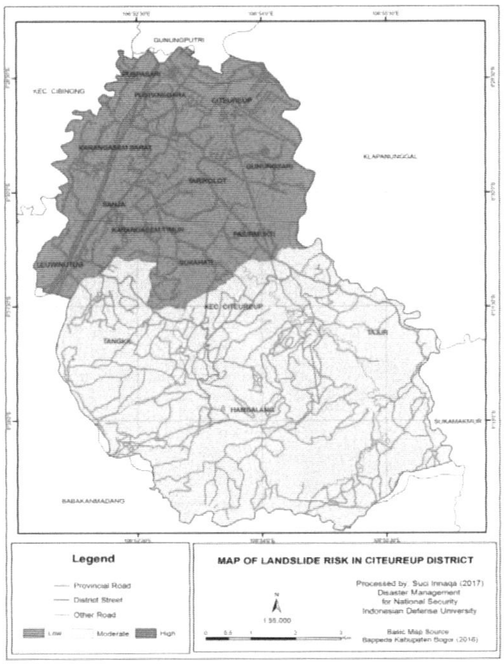

Figure 2. Map of Landslide Risk in Citeureup District. Note: Red (High Risk), Yellow (Moderate Risk), Green (Low Risk).

In addition to the slope, land use is one of the criteria that influence the occurrence of landslides.

Land which there is a building or tree crops with strong roots actually reduce the potential for landslides than vacant land or just overgrown plants without strong roots like vegetables. The area with the mountainous topography or have high slope has high potential for ground movement or landslides. This is in accordance with the results of the study that describes the highest village area (Hambalang, and part of Tajur) that includes in high class category for potential ground movement. This condition is inevitable, because the condition of a region is *given* to the region itself. Later, vulnerabilities and capacities of a those region which would make the threat/danger become a disaster or not. Map of threat generated in this study are shown in figure 3 below.

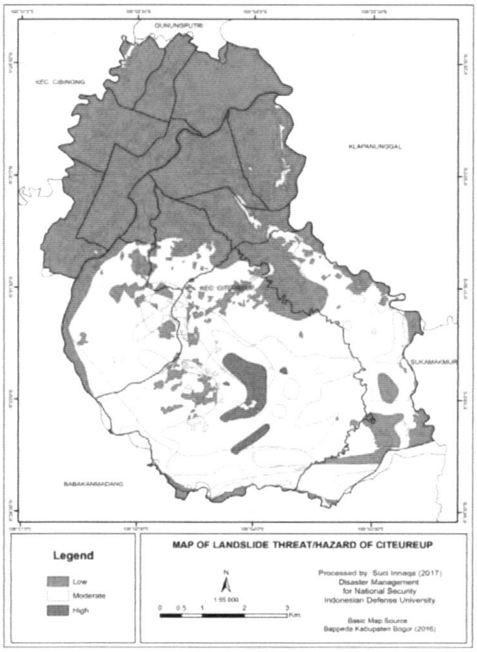

Figure 3. Map of landslide threat/danger of Citeureup District. Note: Red (High Threat), Yellow (Moderat Threat), Green (Low Threat).

Based on the research result, tabel 4 shows the level of lanslides hazard in Citeureup District. Based on the scores of hazard parameters, the hazard level in Kecamatan Citeureup is dominated by low-moderate hazards. The total area of low hazard area is 3,325.17ha (48,68%), medium hazard area is 3,386.43 (49.58%), and the area in high hazrd is only 118,96 (1.74%). A total of 11 villages are located in low hazard areas, while three villages, Hambalang, Tangkil and Tajur are in moderate hazard areas. High landslide hazard areas are found only in a small part of Hambalang Village.

3.3. Physical, Social, Economic Vulnerability and Capacity related to Landslides Disaster

Efforts to reduce disaster can be done either by reducing the level of vulnerability. Identification and vulnerability assessments should be done to reduce the impact of disaster. Wisner *et. al* explain to understand the vulnerability using Pressure Release Model (PAR). PAR consists of two parts, represented by two diagrams, Pressure Model and Release Model. To discuss vulnerabilities, researchers used a model to analyze the pressure in understanding the root causes of vulnerability in Citeureup District. Vulnerabilities is known as one of the factors that influence a natural event to turn to a disaster. Disasters will happen when hazards meet vulnerable conditions. In this discussion, the physical vulnerability describe the level of infrastructure damage that occurs during a disaster.

Tabel 4. Landslides hazard level in Citeureup District.

Village	Population (person)	Area (ha)	Hazard					
			Low	%	Medium	%	High	%
Tangkil	813	776.42	183.83	23.68	592.59	76.32		
Hambalang	11097	2295.33	221.10	9.63	1955.27	85.18	118.96	5.18
Tajur	12527	1110.45	426.44	38.40	684.01	61.60		
Pasir Mukti	10343	188.67	172.93	91.66	15.74	8.34		
Sukahati	9616	309.32	193.49	62.55	115.83	37.45		
Leuwinutug	14308	223.17	222.47	99.69	0.70	0.31		
Sanja	15185	292.21	292.21	100.00				
Karang Asem Barat	31521	276.60	276.60	100.00				
Karang Asem Timur	11184	92.96	92.38	99.37	0.59	0.63		
Tarikolot	20655	293.07	290.65	99.17	2.42	0.83		
Gunung Sari	13108	324.51	311.43	95.97	13.08	4.03		
Citeureup	17280	367.69	367.69	100.00				
Puspanegara	18304	139.41	139.41	100.00				
Puspasari	14833	140.74	134.55	95.60	6.20	4.40		
TOTAL	**200774**	**6830.55**	**3325.17**	**48.68**	**3386.43**	**49.58**	**118.96**	**1.74**

Citeureup sub-District can be divided into two characteristic regions, regions with urban characteristics (including West Karang Asem and Puspanegara), and characterized by rural, namely the village of Tarikolot, Sanja, Leuwinutug, East Karang Asem, Citeureup, Sukahati, Tajur, Tangkil, Hambalang, Pasir Mukti, Puspasari, and Gunung Sari. Physical vulnerability scoring result shows that West Karang Asem, Puspanegara, and Puspasari are in high physical vulnerability, the village of Karang Asem Sanja and East Karang Asem are in moderate physical vulnerability, and nine other villages are in the lower levels of physical vulnerability. Area with urban characteristics have a more complex activity, which is more densely populated, that lead to the construction of public facilities nearby. According to Wisner *et.al*, a fragile physical environment can lead to a condition of an area to be unsafe. Hazardous locations and the condition of the infrastructure and buildings that are not protected is the cause of the development of vulnerability. Although the roots of the problem and the dynamic pressure has been overcome, but the unsafe conditions that triggered the danger by potential soil movement prone to make the area subject to landslides.

3.3.1. Physical Vulnerability. Physical vulnerability often associated with spatial planning perspective, which is associated with the arrangement of settlements, the network of infrastructure and tools that support the social economic, and strategic areas the territory. Spatial planning is a form of intervention from the development activities that are multidimensional and allows various forms of disaster risk reduction activities to be integrated. In determining the form of the activities carried out, of course, depend on the type of disaster and the purpose of the activity. For example, the Citeureup District has done a landslide risk reduction by holding crops that can bind water. The aim is to reduce the absorption of water directly to the soil so that runoff water will not seep into the ground. Betung bamboo tree were planted in several locations including areas that prone to landslides. This is an appropriate efforts in the environment so that disaster risk reduction strategies can be realized. Additionally, attempts to do next is to build a sewer to drain the water flow so that runoff water will not seep into the soil in the area that is friable and has a hollow structure.

The value of physical vulnerability is obtained from the scores of three parameters derived from the Village Potential Data of 2014 issued by Statistics Indonesia (*Badan Pusat Statistik*-BPS). The first parameter used in assessing physical vulnerability is house density. According to BPS Bogor District (2016), the total number of houses in Citeureup sub-district is 54,304 houses, mostly in West Karang

Asem Village, which is 8543 houses and the lowest in Tangkil Village, 226 houses. However, East Karang Asem village has the highest house density, which is 34.50 houses per hectare, and Tangkil Village has the lowest house density of 0.29 houses per hectare.

The second parameter to assess physical vulnerability is the number of public facilities. Based on the Village Potential Data, Hambalang Village has the largest total of public buildings, with 54 buildings dominated by religious facilities, while the village with the smallest number of facilities is Tangkil which has only six public facilities buildings. Most educational facilities are located in Puspanegara and Karang Asem Barat villages. Religious facilities are commonly found in Hambalang Village, namely musholla or langgar (BPS, 2014).

The third parameter is the percentage of land area built in Citeureup sub-district. Puspasari village has the largest widened area of 87.02%. The smallest built-up land area is in Hambalang Village, which is only 7.51%.

3.3.2. Social Vulnerability. Social vulnerability illustrates the vulnerability to the safety of people if a disaster occurs. Based on calculations that have been done, there are four villages that have high social vulnerability, namely Pasir Mukti, Sukahati, Tarikolot, and Citeureup. Hambalang village, Tajur, and Tangkil have low social vulnerability, while seven villages/other villages are in the medium level of vulnerability. The population density has a role in determining a region included in vulnerable category or not. The higher the density the higher the condition followed by social activities in it. Although the highest population density is in the Village Puspanegara, East Karang Asem, West Karang Asem and Puspasari, but the four regions are in the middle class of social vulnerability. In addition to overcrowding, there are other indicators in the form of people with disabilities and poor people who are part of the vulnerable groups.

Based on PAR models, there are few constituent parts that affect the vulnerability: the root causes, dynamic pressures and unsafe conditions. The root cause of vulnerability is always located on the problems of economic, demographic, and political processes, which may affect the distribution and allocation of resources in the community.

Based on this research, areas that have a high population density may not necessarily be in high social vulnerability criteria. This is because the other indicators such as vulnerable groups that trigger the vulnerability of an area when viewed from the side of the social community. Vulnerable population requires more attention, especially the population of disabilities. This is due to population disabilities have limitations in performing self-rescue efforts, and limited participation in complicity life situations. Therefore, it is necessary for the proper handling of individual persons with disabilities so that they will have the same response with other residents who do not have disability issues.

3.3.3. Economic Vulnerability. Economic vulnerability shows the condition of the vulnerability of the resource and economic activity of society. Citeureup sub-district's economic activity is divided into two, namely relying on the industrial sector and agro-tourism. Agrowisata concentrated in hilly areas like Hambalang and Tajur, while for the industry is concentrated in the West Karang Asem and Tarikolot. Economic vulnerability seen from productive land area, number of industries, and the source of the main income. The region has productive land area, primarily located in rural areas, are extremely vulnerable in the event of a disaster. Productive land in the form of farms, fields, and plantations become damaged and economic activity disrupted. Wisner *et.al* states, economic vulnerability has basically two concepts, namely the vulnerability of macroeconomic and micro-economic vulnerability. From the standpoint of looking at macro economy more widely, such as the vulnerability of the economic growth of a country when disaster strikes, this may impact on the market or economic activity. From the standpoint of microeconomics, economic vulnerability focuses on the impact of shocks to the individual or household level, where household income is the main thing that is most influential. The decline in household income levels make a household is unable to meet basic needs.

IOP Conf. Series: Materials Science and Engineering **209** (2017) 012106 doi:10.1088/1757-899X/209/1/012106

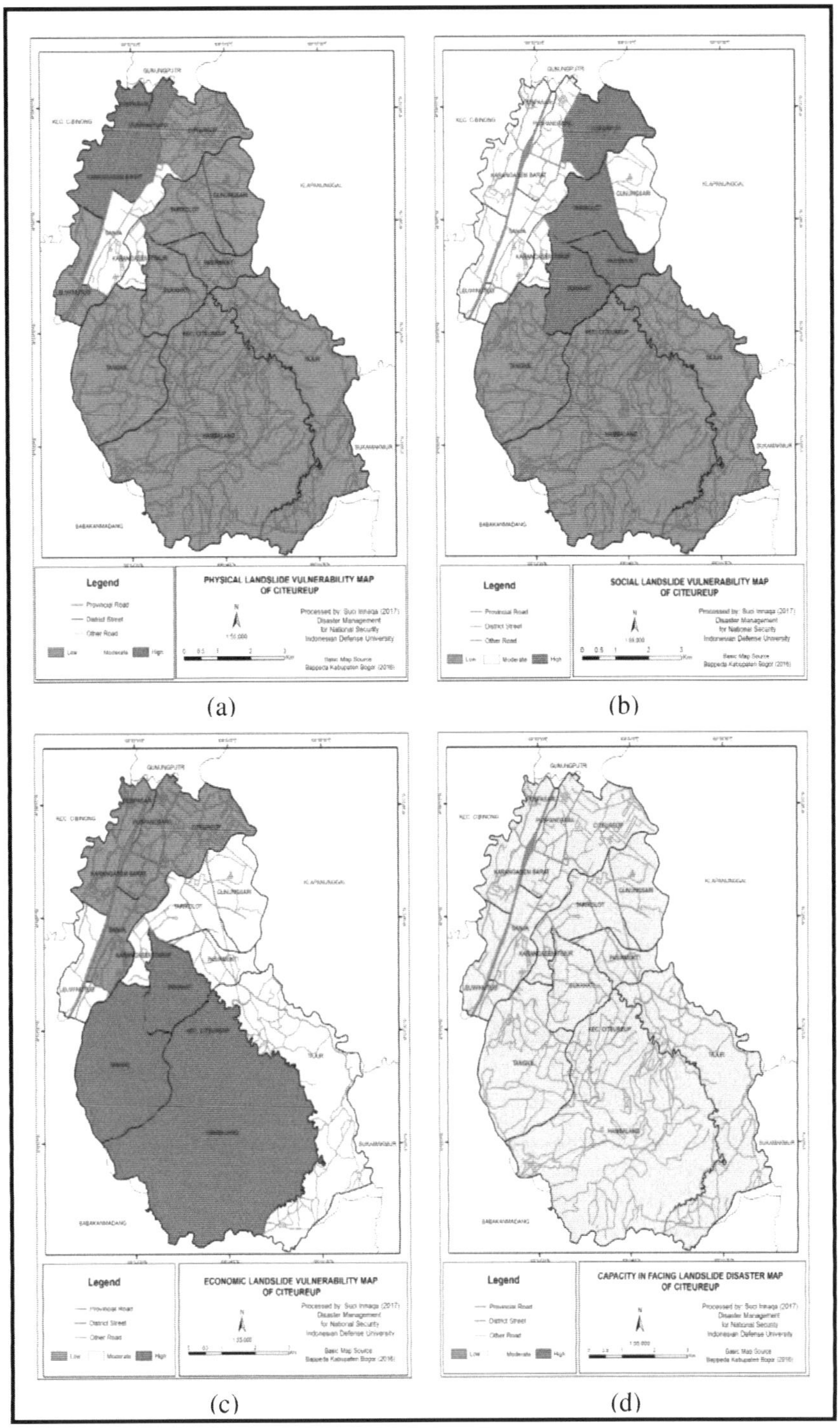

Figure 4. Physical Landslide Vulnerability Map (a), Social Vulnerability Map(b), Economic Vulnerability Map(c), and Capacity in Facing Landslide Disaster-Map in Citeureup sub-district(d). Note: Red (High), Yellow (Moderate), Green (Low).

The impact will lead to the creation of conditions of household poverty. Most villagers Hambalang and Tajur build their livelihoods in the agricultural sector that produces crops. Village/other urban livelihoods depend largely on the industry, working as laborers or as an employee.

Fragile local economic conditions resulting from livelihood at risk, as well as a low income can increase vulnerability. Therefore, efforts should be done to strengthen the livelihoods of the rural for example with income diversification opportunities, such as the creative economic activities so that each family have other income sources besides the main income.

3.3.4. Capacity. Capacity can be defined as the ability of the regions and communities in the face of disaster. There is a reciprocal relationship between vulnerability and capacity, especially in access to resources (assets), both natural psychological, social or financial. Improved access to resources indicate an increased capacity to face and recover from hazard events, nor vice versa. Increasing the capacity of the area can be done through several strategies, one of them by relying on community participation and development orientation.

The study assessed the physical capacity of the existence of early warning systems, evacuation routes, signs for danger signs, and availability of equipment, while social rated capacity of educational disaster institutions and cooperativeness. Based on the results of scoring each indicator of the capacity, the whole villages in the Citeureup sub-district included in the medium capacity class. That is, the capacity in the Citeureup sub-district is good enough. Based on the findings, the physical capacity in the Citeureup sub-district can be said to have a low value. This is due to the lack of early warning systems, evacuation routes, and signs for danger signs. This is because of the Citeureup sub-district has not been included in the priority procurement and construction of an early warning system of Local Disaster Management Agency *(Badan Penanggulangan Bencana Daerah-BPBD)* Bogor Regency. The reason is the landslide that occurred in Citeureup sub-district has not reached the level of severe catastrophic events, despite having a region with high potential. Similarly, the presence of evacuation routes and danger signs mark does not exist in the Citeureup sub-district, although the government has drawn up districts with the initiative as a step preventive evacuation path. However, the capacity of a high social value, due to the introduction and dissemination on disaster reduction has been done, even in formal education. Moreover, the existence of village institutions such as youth clubs, village-level disaster task force, as well as other institutions to be a forum of mutual cooperation in the motion disaster risk reduction.

Distribution of physical, social, economic vulnerabilities and capacities in the Citeureup sub-district can be seen in figure 4. The figure shows that area of economic vulnerability is the largest compare to social and physical vulnerability. Red color shows areas with high vulnerability. However, overall capacity of Citeureup sub-district falls into medium category (yellow color). When overall capacity of an area is high, it could compensate the high vulnerability and decrease the risk in facing disaster.

4. Conclusions

The conclusion of this research is that Citeureup sub-district has low-level landslide risk, although it has high potential for land movement. This shows that areas with high landslide potential do not necessarily have high risk level. Villages with moderate risk are Hambalang, Tajur, and Tangkil villages, while low-risk villages include Pasir Mukti, Sanja, Tarikolot, Gunung Sari, Puspasari, Karang Asem Timur, Citeureup, Leuwinutug, Sukahati, West Karang Asem West and Puspanegara. The capacity of the community plays an important role to minimize the risk of a region. Disaster risk reduction strategy is done by creating a safe condition, which intensifies the movement of disaster risk reduction.

References

[1] Wisner B, Blaikie P, Cannon T and Davis I 2004 *At Risk: Natural Hazards, People's Vulnerability and Disaster. Second Edition.* London and New York:Routlede

[2] BNPB 2016 *Data dan Informasi Bencana (DIBI)*, [online], http://dibi.bnpb.go.id/

[3] Hadmoko D, Lavigne F, Sartohadi J, dan Winaryo P 2010 Landslide Hazard an Risk Assessment and Their Apllication in Risk Management and Landuse Planning in Eastern Flank of Menoreh Mountains *Journal of Natural Hazard* **54** 623–642

[4] Mardiatno D, Marfai M A, Rahmawati K, Tanjung R, Sianturi R and Yossi S 2012 *Assessment of Multirisc of flood and sea water raise at sub-district of north of Pekalongan*. Yogyakarta: MPPDAS Fakulty of Geography, UGM, Yogyakarta, Indonesia.

[5] United Nations ISDR 2004 *Living with Risk: A global review of disaster reduction initiatives* **1**. Geneva: United Nations ISDR

[6] Asian Disaster Reduction and Response Network (ADRRN) 2010 *Terminology of Reducing the Disaster Risk*. Bangkok: ADRRN & UNISDR.

[7] Oktari D 2015 *Mapping Effectiveness Using Unmanned Aircrafts for Rapid Assessment in Banjir Bandang Hazard Hazard on Way Ela Natural Dam in Central Maluku District, Maluku Province*: PhD Thesis. Faculty of Defense Management, Disaster Management Study Program. Bogor: University of Defense.

[8] Awotona A 1997 *Reconstruction After Disaster: Issues and Practices*. USA: Ashgate Publishing Company.

[9] Dianardi et al. 2012 *Geological Condition of Hambalang Area and Surrounding District of Citeureup and Cileungsi*. Faculty of Geological Engineering. Bandung: Padjadjaran University.

Acknowledgments

The authors would like to thank Prof. Dr. Sobar Sutisna, M.Surv and Brigjen (Purn.) Makmur Supriyatno, B.Sc, S.Pd, M.Pd from Indonesia Defense University who have been supervising this research and to the University of Bakrie, Jakarta Indonesia for sponsoring the joint research.

IOP Conf. Series: Materials Science and Engineering **209** (2017) 012107 doi:10.1088/1757-899X/209/1/012107

Application of Artificial Neural Network to Predict the use of Runway at Juanda International Airport

J C P Putra[1,*] and Safrilah[1]

[1]Department of Civil Engineering, Universitas Bakrie, Jl. HR Rasuna Said , Kav. C-22, Jakarta Selatan, Indonesia, 12940

E-mail: jouvan.chandra@bakrie.ac.id

Abstract. Artificial neural network approaches are useful to solve many complicated problems. It solves a number of problems in various areas such as engineering, medicine, business, manufacturing, etc. This paper presents an application of artificial neural network to predict a runway capacity at Juanda International Airport. An artificial neural network model of backpropagation and multi-layer perceptron is adopted to this research to learning process of runway capacity at Juanda International Airport. The results indicate that the training data is successfully recognizing the certain pattern of runway use at Juanda International Airport. Whereas, testing data indicate vice versa. Finally, it can be concluded that the approach of uniformity data and network architecture is the critical part to determine the accuracy of prediction results.

1. Introduction

Indonesia is a big country that has a numbers of islands. Consequently, the easiest way to transport is by using air transportation. This leads to airport become an important part of economic growth due to its role for connecting cities and nations [1]. However, increasing of air traffic demand without improvement of airport infrastructure is the main problem that effect to severe traffic management. According to the International Airport Transportation (IATA) [2], during 2013 to 2014 there was increasing about 5.1% regarding the demand of air travel. This is equivalent to more than 3 billion passengers. In addition, Juanda International Airport in Indonesia is serving the city of Surabaya, East Java and surrounding areas. It is located in District Sedati, Sidoarjo. According to Umilia [3], the number of passengers using the service of Juanda International Airport is in the range of 30000 to 35000 people per day and it makes Juanda International Airport become the second busiest airport in Indonesia after Soekarno-Hatta International Airport due to its aircraft movement and passengers. Therefore, predicting the airport capacity in terms of its runway use can be very useful to be considered as one of strategies to avoid the congestion of air traffic management.

The numerous methods have been previously developed regarding to this problem were deal with probability, statistics, and operations research [4]–[6]. Khanmohammadi *et al.*, [4] introduced a new type of multilevel input layer ANN that can handle nominal variables and is interpretable in a sense that one can easily see the relationship between different input variables and output variables. Marinelli *et al.*, [5] proposed a method based on the Bee Colony Optimization (BCO) to find an optimal flight gate assignment for a given schedule. This approach is expected to solve the problem in the gate scheduling as a key of airport operation. Additionally, Gianazza [6] had combined tree search methods with neural network to forecast air traffic workload. However, the consideration of using ANN in those problems can be beneficial in solving nonlinear problems [4].

Based on the study above, it had motivated the author to conduct a research about predicting the use of runway at airport by using artificial neural network.

2. Artificial Neural Networks

Artificial neural networks (ANN) are the processing system that imitates the concept of biological nerve system. According to Ata [7], ANN is consisted of a group of units called "neurons" that are analogous to nerve neurons and its connected with each other by weights. In order to determine the weights, the ANN use inductive learning process. This process is deal with the numbers of data that will be trained. Once it done, a new pattern may be come out to them for prediction or classification [8]. Subsequently, the main issue of using ANN is how to handle the nominal variables due to converting nominal variable to numerical variable introduce order to the variable which is not desired [4]. Additionally, the process of data conversion from initial data to normalization data need some of adjustments prior to be inputted to hidden layer as shown in figure 1.

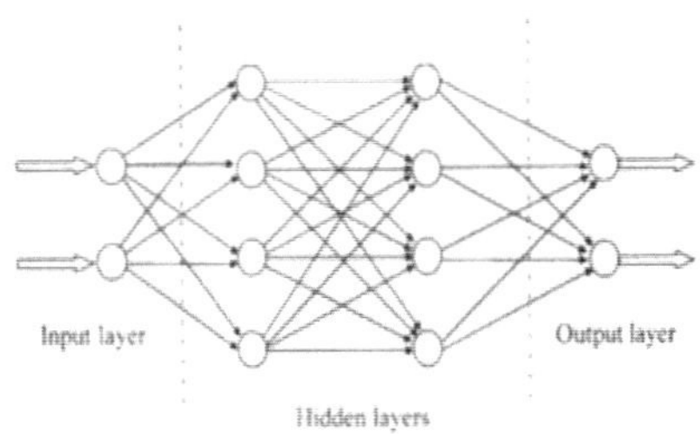

Figure 1. Schematic diagram of a multilayer feed-forward neural network [9].

Based on figure 1, it described that artificial neural network is consisted of three steps which are input layer, some hidden layers, and an output layer. Each single neuron is connected to other neurons by using adaptable synaptic weights [9]. Generally, knowledge is assigned as a set of connection weights. Besides, training is a process of modifying some connection weight properly using a suitable learning method. A learning mode is used in network and it need to prepare the input data in order to generate the desired output. Subsequently, figure 2 shows how information is processed using single node.

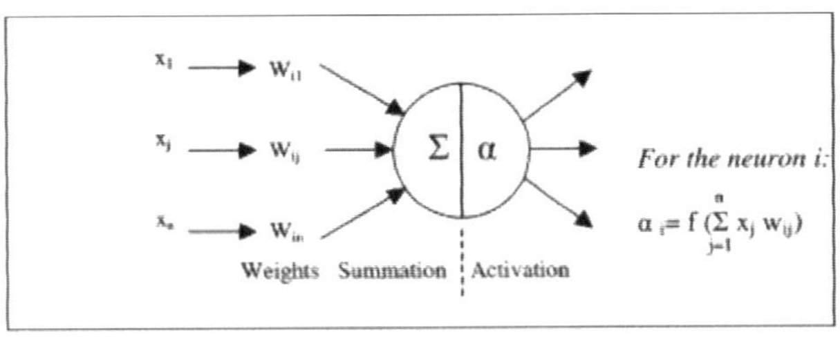

Figure 2. Information processing in neural network [9].

According to Kalogirou et al., [9] the node receives weighted activation of other nodes through its incoming connections. For the first step is the summation of entire neurons. Then the result is come off by activation function. Finally, to activate value for each of outgoing connections is achieved by multiplying it with the specific weight and transferred to the next node [9].

3. Methodology

The use of the runway at Juanda International Airport during January to March 2016 are considered as our research. It retrieved from the AirNav Indonesia-Surabaya. Subsequently, the following preparations are made up prior to analysis the data:

3.1. Input Data

Prior to train the data, it needs to prepare the initial data. The following step is to divide the initial data into two categories which are training data and testing data. A number of training data will be used for network training, whereas testing data will be used once the network training done. This is aimed to evaluate whether the output data can recognize certain patterns which is not previously learned.

The data of the use of runway at Juanda International Airport are used as training data and testing data regarding to this research. Prior to use that data as input, it needs to conduct normalization process or pre-processing regarding the data. The normalization process is needed to convert the data within the range of 0 to 1. In addition, this method is commonly known as an approach to activate the function of binary sigmoid. Although the normalization of data is expected in the range of 0 to 1, yet practically it will be normalized in the range of 0.1 to 0.9. The following equation is used to normalize the data:

$$X' = \frac{0.8(X - b)}{(a - b)} + 0.1 \quad (1)$$

Where;
X' : Normalized data
X : Initial data
a : Maximum value of initial data
b : Minimum value of initial data

However, the initial data for January to March 2016 have a different dimension in which the initial data for January and March have 31 data. While, the initial data for February only has 29 data. Therefore, the initial data for January and March are adjusted to 29 data. In addition, the approach of how to input the data in ANN is described as in table 1.

Table 1. Input Data Pattern.

Pattern	Input Data	Target
1	Data at the date of 1 to date of 29	Data at the date of 30
2	Data at the date of 2 to the date of 30	Data at the date of 31
3	Data at the date of 3 to the date of 31	Data at the date of 32
•	•	•
•	•	•
•	•	•
29	Data at the date of 29 to the date of 57	Data at the date of 58

3.2. Determination of Network Architecture

The selection of the number of layers and pattern of each layer is the principal of network architecture. Based on feedback neural network architecture, this research uses only one artificial neural networks multilayer which is 29 – 10 – 1 pattern. This means that 29 data for input data, 10 data as hidden layer, and only 1 data used as a target.

3.3. Learning and Training Process

A learning process in artificial neural network aims to adjust the weighting factor in order to obtain the desired weight. In this research, back propagation approaches as a learning process is adopted. Furthermore, there are several steps of algorithms preparation for backpropagation learning, namely:

1. The initial weight initialization
2. Forward propagation steps in order to find an error by using activation function such as:
   - Input layer to hidden layer 1 and it continued to hidden layer 2

     Sigmoid function, y: $f(x) = \frac{1}{1 + \sigma x}$ (2)
   - Hidden layer 2 to output layer

     Identity function: $f(x) = x$ (3)
3. Backward propagation is a training process of weighting value of artificial neural network. This is carried out by decreasing total error system to entire data through weighting correction using either gradient descent method or adaptive learning rate.

3.4. Testing Process

Testing of artificial neural network is carried out to identify whether the trained network is capable to recognize the certain patterns. To evaluate the trained network, there are several approaches used regarding to this research, namely

1. Mean Square Error (MSE): MSE is an estimator that measures the average of the squares of the errors or deviations. It is a risk functions, corresponding to the expected value of the squared error loss or quadratic loss. The difference occurs because of randomness or because the estimator doesn't account for information that could produce a more accurate estimate [10]. Subsequently, it values always non-negative and closer to zero are better.
2. Regression: regression is a statistical process for estimating the relationship among variables. It includes many techniques for modelling and analyzing several variables

3.5. Codes preparation for training data and testing data

This step is carried out in order to train the normalization data and test the output data. It is aimed to assess whether the data capable to recognise certain patterns. The codes for training data and testing data are shown in figure 3 and figure 4 respectively.

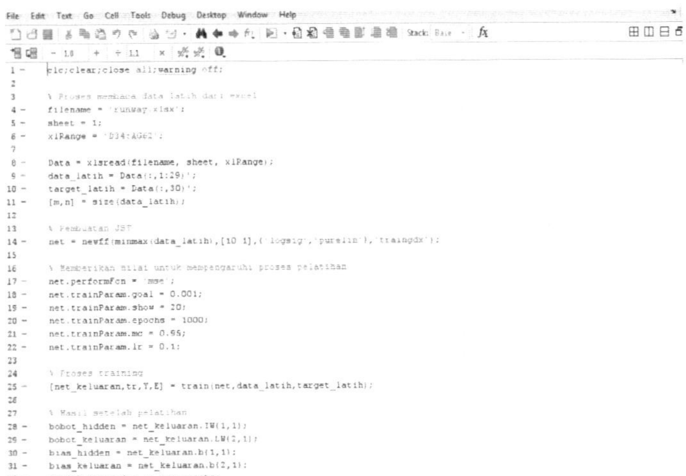

```
clc;clear;close all;warning off;

% Proses membaca data latih dari excel
filename = 'runway.xlsx';
sheet = 1;
xlRange = 'D34:AG62';

Data = xlsread(filename, sheet, xlRange);
data_latih = Data(:,1:29)';
target_latih = Data(:,30)';
[m,n] = size(data_latih);

% Pembuatan JST
net = newff(minmax(data_latih),[10 1],{'logsig','purelin'},'traingdx');

% Memberikan nilai untuk mempengaruhi proses pelatihan
net.performFcn = 'mse';
net.trainParam.goal = 0.001;
net.trainParam.show = 20;
net.trainParam.epochs = 1000;
net.trainParam.mc = 0.95;
net.trainParam.lr = 0.1;

% Proses training
[net_keluaran,tr,Y,E] = train(net,data_latih,target_latih);

% Hasil setelah pelatihan
bobot_hidden = net_keluaran.IW{1,1};
bobot_keluaran = net_keluaran.LW{2,1};
bias_hidden = net_keluaran.b{1,1};
bias_keluaran = net_keluaran.b{2,1};
jumlah_iterasi = tr.num_epochs;
```

Figure 3. Developed code for training normalization data.

Figure 3 shows the code for training normalization data where the number of initial input data is 29 and the initial target data is the data of number 30. Besides, the code involves 1000 iterations of epochs.

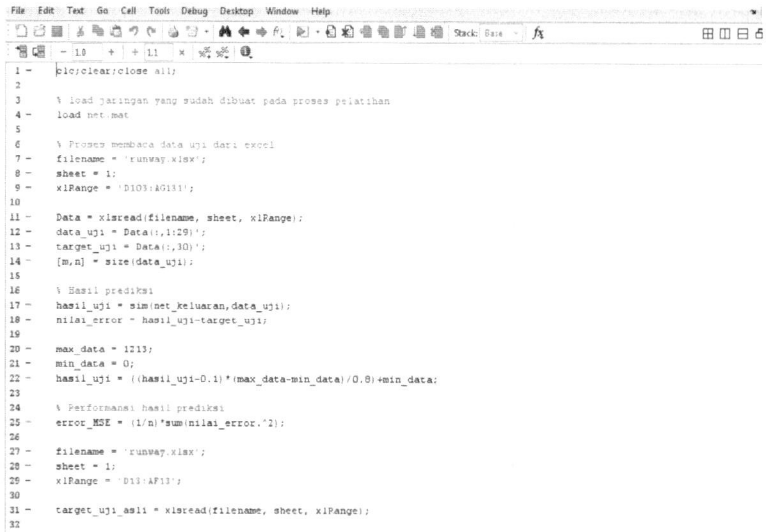

Figure 4. Developed code for testing output data.

Figure. 4 shows the number of testing output data is 29 data with the target of data is the data of number 30. This means the dimension of testing output data is [29 x 29] with the maximum output data is 1213.

4. Results

The results are presented to show both training network outcome and testing network outcome in following subsection:

4.1. Training network outcome

Once the code for training network has made up, then the process of training network is shown as in figure 5.

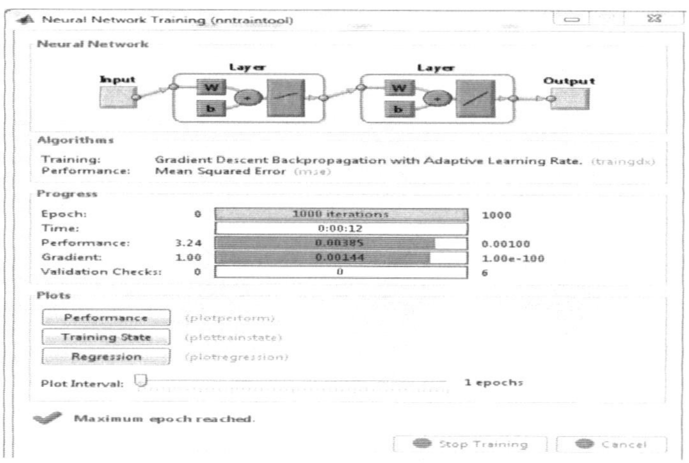

Figure 5. Neural Network Training Tool for Trained Networking.

Based on figure 5, it shows that neural network for training tool complete 1000 iterations within 12 sec. Consequently, the following approach needs to estimate the average of the squares of the deviation. It shows in figure 6.

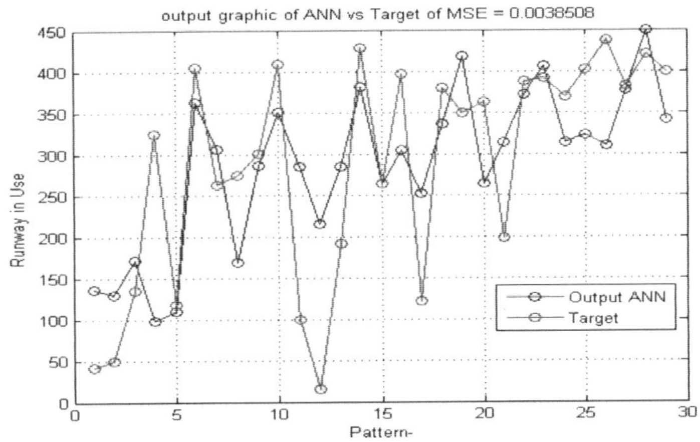

Figure 6. MSE of Trained Networking.

Based on figure 6, the value of MSE is 0.003858. This value shows that the estimation is quite good. Additionally, it also supported by the behaviour of target that is capable to imitate the pattern of target. Subsequently, the following step is to estimate the relationship between input data and target whether these two variables are really connected. It used regression approach as shown in figure 7 to complete it.

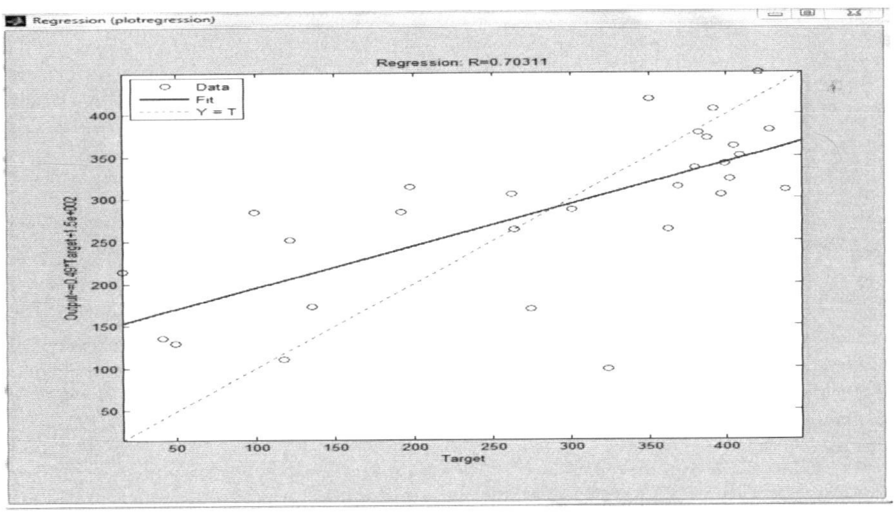

Figure 7. Regression of Trained networking.

The regression results as depicted in figure 7 shows that the relationship between input data and target are quite good as the result shows the value of 0.70311. It means that these two variables between input data and target are having positive relation, if one of them is decreasing, then the other one will follow it and vice versa.

4.2. Testing network outcome

This step is needed to test the network in order to identify the capability of trained network to recognize certain patterns properly. Once the code for testing network done, then the value of MSE can be calculated and the result is shown as in figure 8.

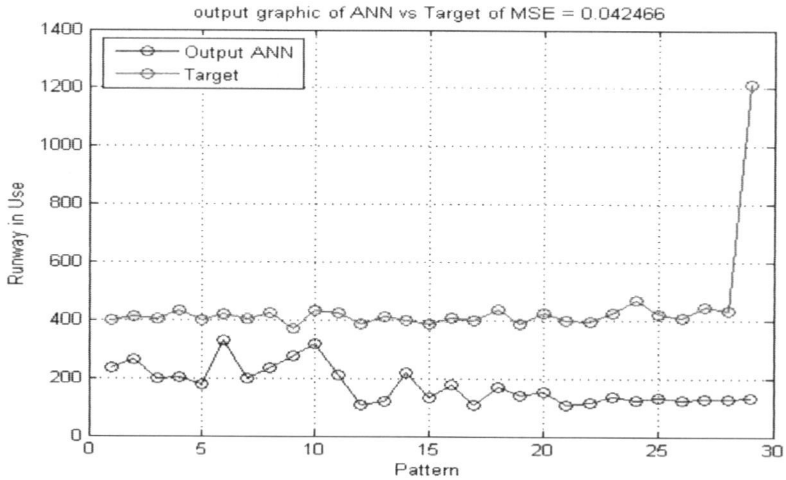

Figure 8. Regression of tested networking.

Figure 8 shows that the pattern between target data and input data is not precise with the value obtained of 0.042466. Consequently, this pattern will affect to the regression results as depicted in figure 8.

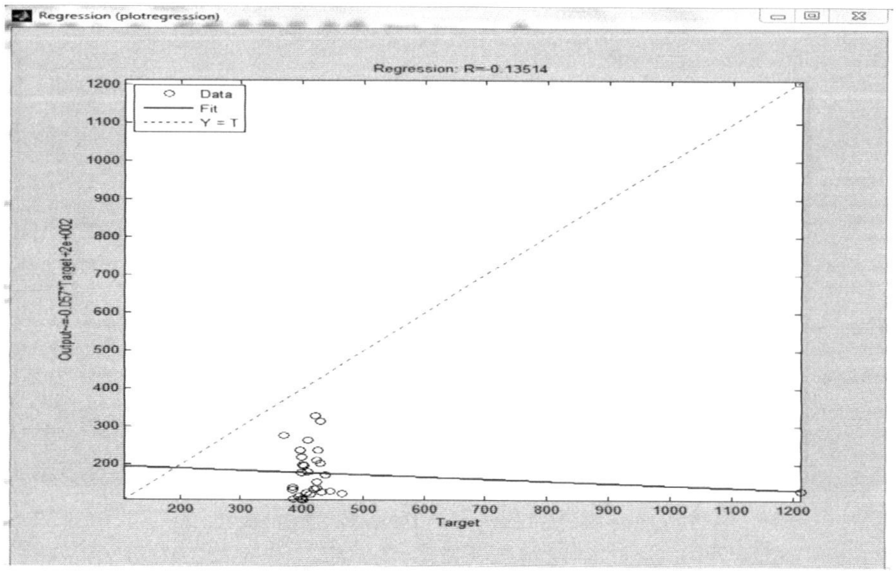

Figure 9. Regression of Tested Networking.

The figure 9 shows the pattern that is resulted between trained network and tested network. From these two variables are obtained the result value of 0.13514. This means that the relationship between trained network and tested network are in weak condition. Consequently, each of these variables having a weak correlation.

This result is also affected by the process of adjusting the data dimension which is changed from the dimension of [31 x 31] to the dimension of [29 x 29]. Besides, according to Khanmohammadi et al., [4] one of factors that contribute to error in predicting using artificial neural networks is process of converting nominal variables into numerical variables. Therefore, it will increases the complexity of datasets that leads to more difficult for interpreting the result of neural network model due to being a black box model

5. Conclusions

Based on the results of this research, several findings can be drawn as follows:

To uniformity the data, it needs a certain approach. Unless, the trained network will not capable to recognize the pattern of data.

It needs to model a number of network architectures to obtain an optimum model which is indicated by its regression value.

Generally, the model that developed using artificial neural network has capability to imitate the fluctuation of runway capacity of Juanda International Airport during January 2016 to March 2016.

References

[1] Alodhaibi S, Burdett R L and Yarlagadda P K D V 2017 *Procedia Eng.* **174** 1100–1109
[2] IATA 2004 *International Air Transport Association*
[3] Umilia E 2014 *Procedia - Soc. Behav. Sci.* **135** 172–177
[4] Khanmohammadi S, Tutun S and Kucuk Y 2016 A *Procedia Comput. Sci.* **95** 237–244
[5] Marinelli M, Dell'Orco M and Sassanelli D 2015 *Transp. Res. Procedia* **5** 211–220
[6] Gianazza D 2010 *Artif. Intell.* **174**(7–8) 530–549
[7] Ata R 2015 *Renew. Sustain. Energy Rev.* **49** 534–562
[8] Kalogirou S 2001 *Renew. Sustain. Energy Rev.* **5**(4) 373–401
[9] Kalogirou S A 2003 *Artificial intelligence for the modeling and control of combustion processes: A review* **29**(6)
[10] Lehmann E L and Casella G 1998 Theory of Point Estimation, Second Edition Springer *Texts in Statistics* **41**(3)

Review Study on Runway Capacity Parameters and Improvement

Safrilah[1] and J C P Putra[2]
[1]Civil Engineering Department, Universitas Bakrie, , HR. Rasuna Said Road, C22 South Jakarta 12920, Indonesia

Email : safrilah@bakrie.ac.id

Abstract. The demand of air travel continues to increase over time, due to its short travel time, reliability and safety. Problems then arise when airport capacity, especially airside (mainly runway) capacity cannot cope with the demand. Some airports build the expensive additional infrastructure, while some others believe that manage on system is more efficient and effective. The study gathering information from various source about parameters related to runway capacity so that the improvement made in the future will solve right on target. To accommodate wide number of factors, the study classify the parameters into five categories in which operation/procedure related parameters play an important role (52%). To facilitate future research on runway capacity, the study also tabulates methods used by various scholars to improve runway capacity

1. Introduction

Domestic air travel growth in Indonesia is dealing with 16.25% increase in passenger's number from previous year [1] in which Juanda International Airport in Surabaya (SUB) contributes 9.78% from total national passenger. As one of the biggest airport in Indonesia, Juanda faces the classic capacity problem, they have to continue accommodate demand rate of runway use (both for departure and arrival) which is increasing over time.

Furthermore, Angkasa Pura I as the airport management operator that responsible to manage Juanda states that the airport need for urgent actions to address its overcapacity with their significant growth in passenger number (13% per year). According to that, the existing and planning actions to increase Juanda's capacity are operate an additional terminal (Terminal 2), preparing masterplan for another additional terminal (Terminal 3) and two other runways. The solutions mentioned before require additional infrastructures build and hence generate huge additional cost and land acquisition problems.

However, with a closer look to the actual data shows in Figure 1, Juanda capacity problems didn't appear in the whole slot time. The runway fails to serve demanded movements typically happen only in morning peak hours, and the rest movements satisfied by the existing runway. The current management system and coordination between infrastructures, technology and the stakeholder seems not working properly, and the easiest way come in first is to expand physical area as mentioned before.

In contrary to the solving actions done by Juanda, several researches from scholars, institutions, government agencies studied the more efficient and effective way to optimize runway capacity and minimizing the desire to build another costly infrastructure.

According to Abeyratne [3], Secretariat Study Group on Traffic Peaks at International Airports proposed a couple of solution in order to manage airport congestion more efficiently, which were looking for any possible solution to satisfy the peak time and improve the traffic flow management quality.

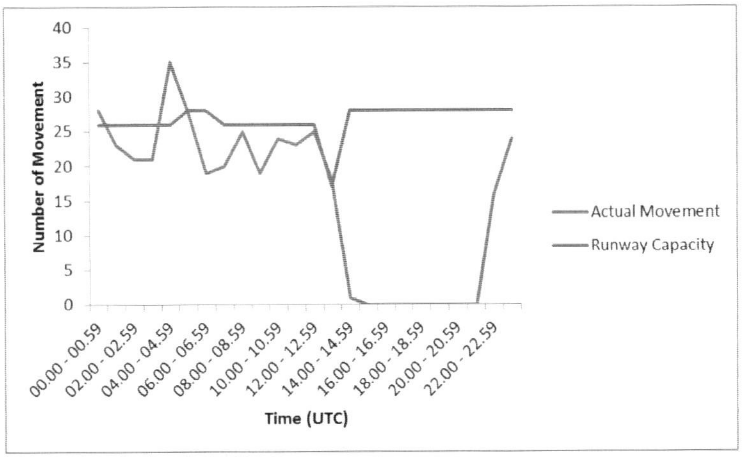

Figure 1. Daily Movement vs Declared Capacity on Juanda Airport [2].

The other approach are: solving aircraft landing and takeoff problems and simulate model that satisfy different stakeholder simultaneously [4], and study on wake turbulence on each aircraft type and sequencing aircraft to reduce effect on wake turbulence and hence increasing capacity [5].

Different solutions proposed and models structured by scholars and institutions vary depends on which runway capacity factors have been looking on through and each airport particular consideration. The study will provides data and result from the previous research and perform a review on most influencing factor on runway capacity based on the data. The output generated expected to be used as reference in future studies in runway optimization.

2. Research Design

The paper designed in four steps: first, gathering definition of runway capacity from scholars perception; second, identifying parameters related to runway capacity and shortlisted into different categories; third, reviewing on previous research methods and analysis, and fourth, drawing conclusion from the research.

3. Definition of Runway Capacity

Runway capacity has a tight a relationship with airport performance to serve demand for air travel, and the inability of runway capacity to accommodate demand turns into one significant factor of delay. The problems related to runway capacity resolved with improving capacity number by additional runway or/and system improvement.

The correct perception of runway capacity and its significant factors needs to address precise solution of runway capacity problems. Definition on how scholars interpret general runway capacity tabulated in table 1 and may vary as a result of operation system, physical condition of airport and external factors.

Runway capacity is often divided into practical hour capacity, maximum throughput capacity, sustained capacity and declared capacity [6][7]. Practical hour capacity is defined as average movement number of aircraft on a runway per hour with accepted delay (4 minutes per operation); maximum throughput capacity is aircraft movement number with continuous demand; sustained capacity is average number of movements per hour that can be maintained for several hours; and declared capacity is capacity per hour, independently determined by the schedule coordination in each airport.

Table 1. Definition of Runway Capacity from Scholars.

No	Source	Definition
1	Pavlin *et al*, 2006 [8]	Maximum service rate and rely only on the physical and operational characteristic of runway and aircraft type.
2	Horonjeff, 2010 [9]	Maximum number of operation accommodated by facilities for some time period
3	Ashford *et al*, 2011 [10]	Maximum number of continuous operation of aircraft performed during certain time interval on a specific runway configuration under given weather condition and acceptable level of delay
4	OTA, 2012 [11]	Number of air operation that airport and ATC system could accommodate in a unit of time
5	Kolos-Lakatos *et al*, 2013 [12]	Maximum possible number accommodated within certain time period for continuous demand for service
6	Airnav Indonesia, 2014 [13]	Number of aircraft movements (departure and arrival) accommodated in certain time period based on configuration of runway, types of aircraft, percent touch and go, separation standard applied, supporting facilities and weather condition
7	Sawali *et al*, 2016 [14]	Ability of a runway to manage movements of aircraft (arrival and departure) in given unit of time

4. Parameters Identification and Classification

As mentioned above, runway capacity may vary as a result of wide range of causative factors. Popular factor influencing runway capacity including: mix of aircraft types, runway occupancy time and separation between movements which then adapted by Airnav Indonesia to be implemented in Indonesian airports, while some of scholars derived each factor to get detailed significant factor.

To overcome the great number of factors/parameters related to runway capacity, this paper classify each parameters into five categories, adapted from main components of air transportation on dynamic capacity-demand balance research which are: operations/procedures, human factors, infrastructure/geometry, aircraft performance, flight plans acording to Zhao *et al.* [15] and modified into: operations/procedures (flight plans categorized as operation), human factors, infrastructure/geometry, aircraft performance, external factors (several source mentioned external factors such as weather condition and noise).

Table 2 is constructed to get better and systematic classification system.

Table 2. Classification of Runway Capacity Factors.

Category	Description
Operation/procedures	Factors related to anything that makes activities functioned as addressed
Geometry of airside facilities	Factors related to physical characteristic of airside facilities
Aircraft performance	Factors related to aircraft characteristic and its derivative
Human factors	Factor related to pilot, ATC officer and human related
External	Factors excluding 4 categories above and unmanageable factors

Based on the classification system, the paper then collects parameter data from two kinds of sources: research result done by scholars and practical implementation in the form of case studies. table 3 shows how each scholar has different result on significant parameter in influencing number of runway capacity.

Table 3. Parameter Identification from Various Scholars.

No	Source	Factor/Parameter	Category
1	Airnav Indonesia, 2014 [13]	Aircraft mix	Operation/procedures
		Runway occupancy time	Operation/procedures
		Separation between movements	Operation/procedures
2	Poldy, 1982 [16]	Separation standards	Operation/procedures
		Aircraft characteristic	Aircraft performance
		Runway configuration	Geometry of airside facilities
		Movement mix	Operation/procedures
		Air traffic control strategies	Human factors
3	FAA, 1983 [17]	Ceiling and visibility	External factors
		Runway use configuration	Geometry of airside facilities
		Aircraft mix	Operation/procedures
		Percent arrival	Operation/procedures
		Percent touch and go	Operation/procedures
		Exit taxiway location	Geometry of airside facilities
4	Horonjeff, 2010 [9]	Runway geometry, including configuration, number, spacing and direction (orientation);	Geometry of airside facilities
		Configuration of exit taxiways, including number and location;	Geometry of airside facilities
		Gate number, size and arrangement in apron area;	Geometry of airside facilities
		Runway Occupancy Time (ROT) for both of departure and arrival;	Operation/procedures
		Aircraft types and combination;	Operation/procedures
		Weather condition;	External factors
		Noise policy;	Aircraft performance
		Ratio between arrival and departure;	Operation/procedures
		Operator's strategy under each weather and noise condition;	Human factors
		Wake vortex separation;	Operation/procedures
		Navigation tool and procedures availability.	Operation/procedures
5	Pavlin *et al*, 2006 [8]	Apron number and dimension;	Geometry of airside facilities
		Aircraft types using the facilities;	Operation/procedures
		Weather condition;	External factors
		Noise regulation;	Aircraft performance
		Procedures of ATC (Air Traffic Controller);	Operation/procedures
		Runway Occupancy Time (ROT).	Operation/procedures
6	Kolos-Lakatos *et al*, 2013 [12]	Wake vortex separation;	Operation/procedures
		Runway Occupancy	Operation/procedures
		Environment and noise	External factors
7	Sakif *et a*l, 2015 [18]	Separation;	Operation/procedures
		Aircraft mix;	Operation/procedures
		Exit location configuration.	Geometry of airside facilities
8	Buttler, 2008 [5]	In trail separation	Operation/procedures
		lateral separation	Operation/procedures

Table 3. Parameter Identification from Various Scholars.

No	Source	Factor/Parameter	Category
		Sequencing and separation of departing and landing aircraft on intersect runway	Operation/procedures
		Sequencing and separation of departing and landing aircraft on single runway	Operation/procedures
		The sequencing of aircraft approaching airports located in close proximity to one another	Operation/procedures
9	Sawali *et al*, 2016 [14]	Runway occupancy time	Operation/procedures
		fleet mix	Operation/procedures
		separation	Operation/procedures
		reaction time of the pilot	Human factors
10	OTA, 2012 [11]	Aircraft performance characteristic	Aircraft performance
		Wake vortex turbulence	Aircraft performance
		Weather condition;	External factors
		Airfield and airspace configuration	Geometry of airside facilities
		Aircraft noise	Aircraft performance
		ATC equipment and procedures	Operation/procedures
		Demand consideration	Operation/procedures
11	Ashford *et al*, 2011 [10]	Meteorological condition	External factors
		Airfield layout, runway configuration and operational strategy	Geometry of airside facilities
		Aircraft arrival and departures ratio	Operation/procedures
		Aircraft fleet mix	Operation/procedures
		Runway occupancy time	Operation/procedures
		ATC related factors	Operation/procedures
12	Bubalo *et al*, 2011 [19]	Local topography	External factors
		Prevailing weather condition	External factors
		Layout, configuration, availability of runway(s)	Geometry of airside facilities
		Runway exits	Geometry of airside facilities
		Taxiways	Geometry of airside facilities
		Apron area	Geometry of airside facilities
		Aircraft parking stands	Geometry of airside facilities
		ATC capabilities	Operation/procedures

5. Previous Study Method and Analysis

The different approach to deal with runway optimization based on various interests and condition of each airport system. Airnav Indonesia apply DORATASK method which calculating runway capacity based on actual time (take-off and landing time) and manage to improve capacity by reducing runway occupancy time, similar with research of Pavlin *et al.* [8]; while Horonjeff *et al.* [9], Kolos-Lakatos *et al.* [12], Buttler [5] and Sawali *et al.* [14] considering separation between aircraft plays an important role on runway capacity with deriving separation problems into different approach. The detail of method and stage taken by different scholars and airport presents on table 4.

Table 4. Method and Analysis of Previous Study.

No	Source	Method	Poin of Consideration	Stages	
1	Airnav Indonesia, 2014 [13]	DORATASK Method	DORATASK method requires primary and secondary data, low cost (no demands on specific software), applicable, easily socialized, accurate (most of data is a primary data). DORATASK Method consisting 3 stages and 16 steps	Runway Physical Capacity Calculation	a. Collecting Runway Occupancy Time (ROT) with field survey b. Calculating Arithmetical Mean ROT (AMROT) c. Calculating mix index for runway use d. Calculating Mean Runway Occupancy (MROT) e. Calculating Physical Capacity Runway (PCR) f. Calculating Aerodrome Physical Capacity (APC)
				Theoretical Runway Capacity Calculation	a. Collecting flight time for each type of aircraft between Outer Marker (OM) and threshold (THR) and calculating Mean

IOP Conf. Series: Materials Science and Engineering **209** (2017) 012108 doi:10.1088/1757-899X/209/1/012108

No	Source	Method	Poin of Consideration	Stages
				Flight Time (MT) b. Calculating average approach speed c. Determine Safety Separation (SS) d. Determine total separation between 2 consecutive landing (TS) e. Weighted time between 2 consecutive landing (MTTS) f. Determine number of landings in one hour interval (P) g. Determine number of takeoffs in one hour interval (D) h. Determine Theoretical Runway Capacity (TRC)
			Runway Declared Capacity Runway (DCR) Calculation	$DCR = \frac{UP_A.TRC_S + UP_B.TRC_B + \cdots.}{UP_A + UP_R + \cdots}$
2	Poldy, 1982 [16]	Intermovement Time (IMT) Matrix	Constructing pair probability matrix	$\begin{matrix} p_{ij} & A & D \\ A & t_{AA} & t_{AD} \\ D & t_{DA} & t_{DD} \end{matrix}$
			Calculating interrmovement time for a sequence of movement during a busy period	$T = \sum_{ij} p_{ij} t_{ij}$

IOP Conf. Series: Materials Science and Engineering **209** (2017) 012108 doi:10.1088/1757-899X/209/1/012108

No	Source	Method	Poin of Consideration	Stages
3	FAA, 1983 [17]	FAA Method	FAA Method using secondary data with verification process, low cost, easily socialized, short period of survey	Calculating capacity $C = 3600/T$ a. Runway-use configuration selection b. Identify figure number for capacity using figure released by FAA c. Determine percentage of Aircraft class C and D and calculate the mix index d. Determine percent arrivals e. Determine hourly capacity base (C*) f. Determine the percentage of touch and go operation during VFR operation and determine touch and go factor (T) g. Determine the location of exit taxiways and determine the exit factor (E) h. Calculate the hourly capacity of the runway component by: **Hourly capacity = $C^{*}.T.E$**
4	Horonjeff, 2010 [9]	Time-Space Concept	Spacing between aircraft considering as the most significant factor among others, hence time-space diagram used in time space concept is very useful and easy to understand	a. Two aircrafts may not conduct an operation on the runway at the same time. b. Arriving aircraft has a priority over departing aircraft c. Departures is allowed if the runway is clear and the subsequent arrival is at least a certain distance from the runway threshold

No	Source	Method	Poin of Consideration	Stages
				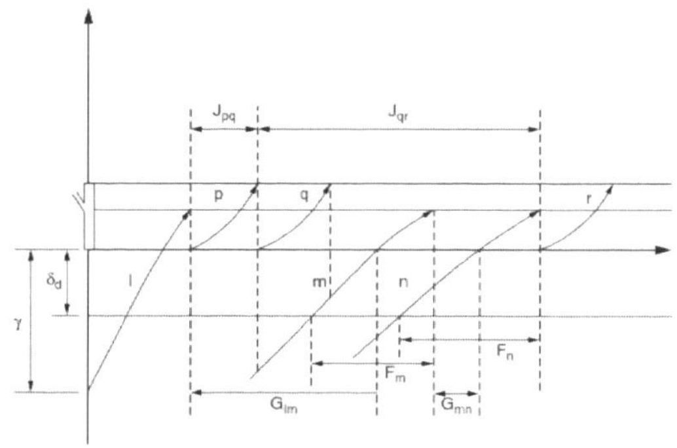 **Figure 2 Time Space Diagram Concept for Mixed Operation**
5	Pavlin *et al*, 2006 [8]	Reducing ROT concept	ROT considering as one of the most important factors related to runway capacity (increasing up to 15% capacity)	a. Calculating Runway Occupancy Time Arrival (ROTA) = time between threshold crossing and aircraft tail vacating the runway b. Calculating Runway Occupancy Time Departure (ROTD) = time between crossing the holding stop bar and main gear lifting off the runway
6	Kolos-Lakatos *et al*, 2013 [12]	Probabilistic queuing model, RECAT (Re-categorization) Traffic Mix model	Re-Categorization of aircraft based on wake turbulence after reduced separation minima considering effective in increasing capacity	a. Categorizing aircraft in six groups, range from A (largest commercially operating aircraft generating most severe wake turbulence) to F (small business jets and turboprops with weak vortices b. Considering separation standard between aircraft category

No	Source	Method	Poin of Consideration	Stages

Follower (NM)

Leader	A	B	C	D	E	F
A	2.5/3	5	6	7	7	8
B	2.5/3	3	4	5	5	7
C	2.5/3	2.5/3	2.5/3	3.5	3.5	6
D	2.5/3	2.5/3	2.5/3	2.5/3	2.5/3	5
E	2.5/3	2.5/3	2.5/3	2.5/3	2.5/3	4
F	2.5/3	2.5/3	2.5/3	2.5/3	2.5/3	2.5/3

c. Calculating minimum time separation between leading and following aircraft T_{ij}

$$T_{ij} = max\left[\frac{r+s_{ij}}{v_j} - \frac{r}{v_i}, ROTi\right] \text{when } v_i > v_j$$

r: is the length of the final approach.
i : is the leading airplane's weight group.
j : is the following aircraft's group.
Sij : is the minimum required separation between the leading aircraft type and the following aircraft type.
Tij : is the minimum time interval between successive arrivals for aircraft i and aircraft j.
vi : is the final approach velocity for the aircraft i.
vj : is the final approach velocity for the aircraft j.

d. Considering buffering time 19 sec

e. Calculating probability arrival pair

$$P_{ij} = 0.5x\ (percentage\ of\ grup)$$

f. Calculating expected time an arrival pair takes to land

$$E(t) = \sum P_{ij}\,(T_{ij} + b)$$

g. Calculating RECAT capacity

IOP Conf. Series: Materials Science and Engineering **209** (2017) 012108 doi:10.1088/1757-899X/209/1/012108

No	Source	Method	Poin of Consideration	Stages
				$\mu = \frac{1\,hr}{E(t)} x3600 sec/hour$
7	Sakif *et al*, 2015 [18]	Probabilistic model		$T_{ij} = max\left[\frac{r+s_{ij}}{v_j} - \frac{r}{v_i}, ROTi\right]$ when $v_i > v_j$ r: is the length of the final approach. i : is the leading airplane's weight group. j : is the following aircraft's group. Sij : is the minimum required separation between the leading aircraft type and the following aircraft type. Tij : is the minimum time interval between successive arrivals for aircraft i and aircraft j. vi : is the final approach velocity for the aircraft i. vj : is the final approach velocity for the aircraft j
8	Buttler, 2008 [5]	NextGen Concept	NextGen represents automation in routine aspect of air traffic control, it will safely reduce separation between aircraft and permit more efficient routing of planes	

<table>
<tr><th>No</th><th>Source</th><th>Method</th><th>Poin of Consideration</th><th>Stages</th></tr>
<tr><td>9</td><td>Sawali et al, 2016 [14]</td><td>Sequencing method</td><td>Sequence between two consecutive flights considering as the ost critical factor in runway capacity</td><td>a. Data input : flight number, aircraft departure, type of operation, scheduled time and runway use
b. Rule A : applicable for consecutive departure on the same runway
Rule B : applicable to all sequence with succeeding flight is arrival operation
Rule S : Any other sequence outside Rule A and Rule B
c. Input for algorithm
<table>
<tr><th>Input Parameters</th><th>Description</th></tr>
<tr><td>Flight Number (Flight No.)</td><td>Each is mentioned during command exchange by the controller.</td></tr>
<tr><td>Aircraft Type (Type)</td><td>This indicates the make and model of an aircraft.</td></tr>
<tr><td>Time Data Points</td><td></td></tr>
<tr><td>Entry</td><td>Time of actual entry or hover on runway entry threshold</td></tr>
<tr><td>Mid 1</td><td>For arrivals, this is the time of touchdown; for departures, this is the start of roll time</td></tr>
<tr><td>Mid 2</td><td>For arrivals, this is left blank; for departures, this is the time of wheels up</td></tr>
<tr><td>Exit</td><td>For arrivals, this is the time of full exit to a taxiway; for departures, this is the time of hover on runway exit threshold</td></tr>
<tr><td>Operation (D/A)</td><td>A departing aircraft is labeled "D"; an arriving aircraft is labeled "A".</td></tr>
<tr><td>Runway (RWY)</td><td>The runway that is used by an aircraft during actual operations.</td></tr>
<tr><td>Sequence</td><td>The sequencing of air traffic controllers in Manila follows a first-come first-served basis, regardless of aircraft category. This sequence is used to determine current maximum capacity.</td></tr>
</table>
d. Output of the algorithm
<table>
<tr><th>Input Parameters</th><th>Description</th></tr>
<tr><td>Sequence Number (No.)</td><td>The numbering will follow the initial order of the input schedule.</td></tr>
<tr><td>Aircraft</td><td>The same details from the inputs will also be printed in the results.</td></tr>
<tr><td>Runway Occupancy Time (ROT)</td><td>The calculated Runway Occupancy Time for a flight is the sum of the assigned separation time (SEP), and its Rolling Time (RT_).</td></tr>
<tr><td>Actual Time (ATD/ATA)</td><td>The time each flight leaves the runway, by hovering above the threshold for departures, or by exiting through a taxiway for arrivals, will be calculated using the algorithm for observed flights.</td></tr>
</table>
e. Manual optimization</td></tr>
<tr><td>10</td><td>Ashford et al, 2011 [10]</td><td>Optimization model</td><td>Combine both analytical and empirical analysis, reliable and realistic</td><td>$\min \sum_{i=1}^{N} F_i(X_{i+1}, Y_{i+1})$</td></tr>
</table>

No	Source	Method	Poin of Consideration	Stages
				subject to $X_{i+1} = \max(0, X_i + a_i - u_i) \quad i \in I$ $Y_{i+1} = \max(0, Y_i + d_i - v_i) \quad i \in I$ $X_1 = X_0 \geq 0;\ Y_1 = Y_0 \geq 0$ (given initial condition $0 \leq v_i \leq \phi_i(u_i) \quad \phi_i(u) \in \Phi \quad i \in I$ $0 \leq u_i \leq B_i \quad i \in I$ Where N = number of sequenced capacity curves u_i = arrival capacity at ith time slot v_i = departure capacity at ith time slot F = nondecreasing scalar function X_i = demand for departures at ith time slot Y_i = demand for arrivals at ith time slot a_i = demand for arrivals at ith time slot d_i = demand for departures at ith time slot $\phi_i(u)$ = arrival/departure capacity curve, which determines capacity at ith time slot Φ = set of capacity curves that represent all runway configurations at all weather conditions
11	Bubalo *et al*, 2011 [19]	SIMMOD Simulation	Possible to analyze and calculate airside capacity with complex and dynamic factors	Input : combined flight schedule Processed automatically with Visual SIMMOD Software package

6. Conclusions

Review on this study was based on review on runway capacity among scholars. Air transportation is an exclusive area in transportation world, the research conducted mainly for regulation need and held by government. The parameters related to runway capacity were gathered from various source, research paper, reports and books. The whole parameters collected were classified into five different categories: aircraft performance, geometry of airside facilities, operation/procedures, human factors and external factor to simplify further research.

The pie chart on figure 5 shows that operation/procedure related factors contribute the most on runway capacity (52 %). Derivation of operation/procedure factors will discuss separately.

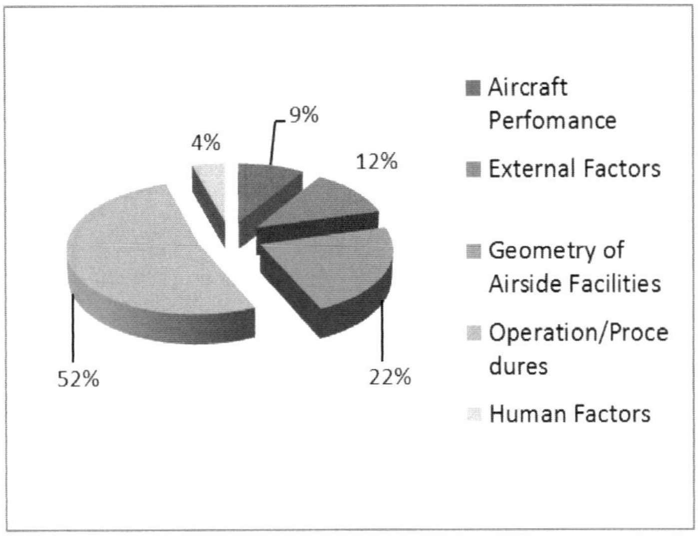

Figure 3. Parameters Related to Runway Capacity.

References

[1] Directorate General of Civil Aviation, 2015 *Air Transport Traffic*. http://hubud.dephub.go.id/?id/news/detail/2374

[2] Airnav Indonesia 2016 *ADC Monthly Report 2016*. Directorate General of Civil Aviation

[3] Abeyratne R I R 2000 Management of airport congestion through slot allocation *J. Air Transp. Manag.* **6**(1) 29–41

[4] Potts C N, Mesgarpour M and Bennel J A 2009 *A Review Of Airport Runway Optimization*. Eurocontrol

[5] Butler V 2006 *Increasing Airport Capacity Without Increasing Airport Size.* Reason Foundation

[6] Odoni A R 2002 *Airfield Capacity in A*irport Systems: Planning, Design and Management. Mc Graw Hill Education

[7] Sherry L 2008 *RUNWAY OPERATIONS: Computing Runway Arrival Capacity*, G. M. U. Center for Air Transportation System Reseach
[8] Pavlin S, Zuzic M and Pavicic S 2006 Runway occupancy time as element of runway capacity. *Traffic Plan. Rev.* **18**(4) 293–299
[9] Horonjeff R, Mckelvey F X, Sproule W J and Young S B 2010 *Planning and Design of Airports 5th Edition*. McGraw Hill Education
[10] Ashford N J, Mumayiz S A and Wright P H 2011 *Airport engineering: Planning, design, and development of 21st-century airports 4th Edition*. John Wiley & Sons, Inc
[11] Office of Technology Assessment (OTA), 2012 *Airport and Air Traffic Control System: Airport Capacity Alternatives* in Airport and Air Traffic Control System. U.S. Government Printing Office
[12] T. Kolos-Lakatos and R. Hansman, 2013 T*he Influence of Runway Occupancy Time and Wake Vortex Separation Requirements on Runway Throughput.* MIT International Center for Air Transportation
[13] Airnav Indonesia, 2014 *Manual for Runway Capacity Calculation.* Direktorat Safety and Standard Airnav Indonesia
[14] F. C. A. Sawali, H. S. O. Palmiano, and L. V. U. Vega, 2016 *Evaluation of the Runway Capacity of the Ninoy Aquino International Airport*. 23rd annual conference of the transportation science society of the Philippines Quezon City, Philippines
[15] W. Zhao, R. Curran, and H. A. Abbass, 2015 Dynamic Capacity-Demand Balance in Air Transport Systems Using Co-*evolutionary Computational Red Teaming*, in *Air Transport and Operations*, pp. 1–11
[16] F. Poldy, 1982 *BTE Publication Summary: Airport Runway Capacity*. Watson Ferguson and Co, Brisbane. [17] Federal Aviation Administration, "Airport Capacity and Delay," 1983.
[18] S. Sakif, S. S. Position, and E. Mohammadia, 2015 *Modelling, Optimazation of Runway Occupane Time.* Proceedings of the ITRN 2015
[19] B. Bubalo and J. R. Daduna, 2011 *Airport capacity and demand calculations by simulation-the case of Berlin-Brandenburg International Airport," NETNOMICS Econ. Res. Electron. Netw.*, vol. 12, no. 3, pp. 161–181

IOP Conf. Series: Materials Science and Engineering **209** (2017) 012109 doi:10.1088/1757-899X/209/1/012109

Growth and Production of Some Variety Corn (Zea mays L.). Planted under the Canopy of Palm Oil 12 Years Old in Swamp Land

Syafrullah[1], N Marlina[2*], S E Rahim[2], R I S Aminah[1], Midranisiah[3], Rosmiah[1] and F Sakalena[4]

[1] Agriculture Faculty, University of Muhammadiyah Palembang, Palembang, South Sumatra, Indonesia
[2] Agriculture Faculty, University of Palembang, Palembang, South Sumatra, Indonesia
[3] STIPER Sriwigama Palembang, Palembang, South Sumatra, Indonesia
[4] Agriculture Faculty, University of Baturaja, Ogan Komering Ulu, South Sumatra. Indonesia

Email: marlina002@yahoo.com

Abstract. This research was conducted in wetlands Semambu Village, District of North Indralaya, Ogan Ilir, South Sumatra Province, Indonesia, which lasted from July 2015 to February 2016. The observation of a microclimate indicate that the average intensity of light outside the auspices of the plot 1968.9 $m^2s\ mol^{-1}$, under waranet 1502.40 $mol^{-1}\ m^2s$, below paranet 721.99 $mol^{-1}\ m^2s^{-1}$ and under waranet 439.25 $\mu mol\ m^2s^{-1}$ - equivalent to the light interception 1 or 100%, 76%, 37% and 22%. Results of soil chemical analysis that the soil has a low fertility study (H_20 pH of 3.32, organic C 4.47%, total N 0.35%, Bray P 13.30 ppm, K-ea 0.26 me / 100g, CEC 19.6 rne / 100g and Al-ea 3.28 me / 100g). Tests on 22 genotypes of maize grown with light interception 100%, 76%. 37% and 22%, by calculating tolerance index based on the weight of dry seed cob-1 was found four genotypes of maize tend to be tolerant to low-intensity light that were genotype B 41, Pioneer 27, Sukmaraga and Sugihan. The test results of corn planted in beds shade with light interception 100%, 76%, 37% and 22% for groups of maize tolerant and sensitive, followed by application of urea 0 kg ha^{-1}, 100 kg ha^{-1} 200 kg ha^{-1}, 300 kg ha^{-1} and 400 kg ha^{-1} indicate that maize and 41 and Pioneer 27 by Urea 300 kg ha^{-1} gives better results than other varieties at different intensities of light oil palm age of 12 years with applications Urea fertilizer 300 kg ha^{-1}, indicating that the B 41 and Pioneer 27 tends to give better results compared with other varieties. The application of a polyculture system palm-maize can produce 1000 kg of dry grain of corn in a 1 ha of oil palm cultivation.

1. Introduction

The Republic of Indonesia today is faced with various problems that are global, national, regional and even sub-sub-regional. These issues must be given a conceptual response comprehensive and totality. One of them is the development of the agricultural sector. Indonesia faced with the problem of realization of national food security and improving the welfare of farmers.

Therefore, the Ministry of Agriculture will give priority to the development and optimization of specific technologies that are technically relevant to the condition of the land, and land development suboptimal. One effort that can be done is suboptimal land development. namely swamp (lowland, wetland), which are very widespread in Indonesia [1].

Vast swamp land in Indonesia is estimated at 33 million ha, consisting of tidal land and 20 million ha of lowland swamp 13 million ha [2]. Furthermore, Subagyo [3], explains that the lowland swamp land in addition to rice production centers, it has also become a regional maize production. This is evidenced by the results of research Rosmiah *et al*, [3] and Marlina *et al*, [4], that the production of corn and sweet corn can be increased in the swampy wetlands with biological organic fertilizer. Maize cultivation in the swamp in which is still conventional become essential to most places. Corn generally grown in paddy fields after rice in monoculture, land surjan continuously with monoculture and in intercropping surjan land with plantations or horticultural crops (such as cocoa, citrus). Corn productivity in wetlands is very diverse. The results of the research in Karang Agung, South Sumatra Province showed yield of corn in the wetlands around 3.5-5.5 ton dry seed ha-1, depending on the typology of land, varieties, and the input of ameliorant and fertilizers [5].

In the national economy, corn contributor (GDP) is the second largest after rice in food crops. Donations corn to Gross Domestic Product (GDP) continues to increase every year, even during the economic crisis. In 2000 the contribution of maize in the national economy reached Rp 9.4 trillion and in 2003 increased to Rp 18.2 trillion. These conditions indicate the magnitude of the role of corn in spurring the growth of food crops and the national economy in general.

The productivity of oil palm growers are also extremely low at only 5 tons - of fresh fruit bunches ha^{-1} yr^{-1}, which is far from the existing potential is 20-30 tons of fresh fruit bunches per hectare per year. Oil palm plantations belonging to the people of an area of 3.01 million hectares, or 40.15% of the total national oil palm plantations [6] . That area could potentially big enough to be developed through a pattern of polyculture with maize as a crop sidelines.

To overcome the various problems faced by farmers wetlands. with principle to coordinate the desire of farmers to have a garden of palm oil. while maintaining the function of wetlands as buffers national food availability. One effort that can be developed through plant-based oil palm polyculture cropping with crops. Paddy fields or non-paddy fields have turned into oil palm monocultures immature still cultivated rice cultivation in the first planting season and maize planting season II. Thus the intensification of polyculture in palm oil plantation area can help contribute rice and maize production from Lebak swamp. The existence of oil palm on land that has been over the function will take place until one cycle is 20-25 years. Therefore there is a time lag between 10-15 years of a cycle of oil palm with the potential to do the cultivation of maize. Polyculture aims to improve the biological and economic results from a field in it is so associated with the type of crops grown and the population in one unified system of cultivation or planting patterns [7][8]. Furthermore Purnomo [9] explains that the introduction of maize as a crop is plagued by low production stream. Low productivity is related to the intensity of light at the bottom of the stand which is low compared to the open. in addition to be competition between plants for nutrients and water.

As the plant among palm. maize varieties appropriate election will determine the level of productivity that can be achieved. Atman [10] concluded that the use of varieties of composite or hybrid varieties recommended in corn cultivation under the canopy of coconut trees. Mohr and Schopher [11] stated that the plant's ability to tolerate shade is determined by the genetic properties of plants. Nevertheless. the appearance of a variety depends on the interaction of genetic and environmental factors [12]. Research by Rochman et al. [13] revealed that on in wetlands is still no specific varieties appropriate for use farmers as a whole and evenly.

Information on plant responses to light with several treatments have sufficient level of shade available. but information on maize varieties tolerant of low light. especially corn are cultivated with oil palm is still limited. Although corn as C_4 plants have different responses to light compared with C_3 plants because of the different characteristics of photosynthesis, need to be tested because it is possible that there are varieties of corn that is tolerant of low light has the potential to be developed as intercrops under palm canopy has produced.

In recent decades wetlands as buffers national food production is threatened by the emergence of the phenomenon of the transition function of wetlands into oil palm and rubber plantations or non-agricultural [13]. According to Minister of Agriculture [14] over the wetland function reaches 100,000 ha per year either by plantations and non-plantation. This phenomenon is of course a major threat to national food security and the government's efforts in achieving self-sufficient food. Agricultural development wetlands have become a mainstay for increased production of food crops. With proper management through the application of appropriate science and technology, wetlands have great prospects for development into productive agricultural land, especially in order to preserve food self-sufficiency, diversification of production, income generation and employment, as well as agribusiness development area [15].

2. Materials and methods

Land availability optimal depletion of food due to land conversion, has prompted the government to shift food production to sub-optimal land, including for wetlands, both in swampy bogs and in the tidal marsh.

The pattern of palm plantations is also growing rapidly in a swamp area with conventional cropping systems (monoculture), the number of oil palm plantations on these systems varies between 130-148 plants per hectare so that they have a growing space (space between canopy) which allows for growing crops between rows of corn as a staple crop plants sidelines, especially when oil palm age of 3-6 years with using corn crop varieties tolerant or adaptive low light (shaded). Palm frond that has been completely closed space between the plants would cause the intensity of light is the main constraint development patterns of poly-corn oil.

The corn crop is grown in lowland swamp as the plants will suffer from stress light, and a variety of stress-induced marginal land conditions, resulted in a decrease in the rate of metabolism. These stresses will look at the symptoms of physiological, morphological and anatomical plants resulting in low production of biological crops. Information about the various plant responses to low light (shaded) is sufficient available, but the information of the corn is tolerant of low light is still very limited. One of the technologies that will enhance the production of maize as a crop stream is through varieties tolerant of low light, followed by the input of technology-specific fertilization.

Besides the low light intensities as the cause of low production of corn, swampy bog fertility factor is also a determinant of productivity. Various efforts to improve the quality of wetlands has also done many of them by giving ameliorant (pH soil improvement materials such as limestone), the addition of organic material. organic fertilizers and inorganic fertilizers. Information on this matter has been sufficiently available, but the corn crop fertilization action information specific locations as plant sidelines are not yet available.

The addition of nitrogen nutrients through fertilization is believed to increase the rate of photosynthesis through increasing the chlorophyll content of leaves and increased levels of the enzyme ribulosebiphosphate carboxylase / oxygenase (Rubisco). Adequacy of Rubisco could be expected to improve the adaptability of plants through (1) the mechanism of increasing the efficiency of light absorption ("efficiency of light trapping"), and (2) the mechanism of low light tolerance (tolerance of light deficit).

The mechanism of adaptation to light reflected from changes in leaf characteristics such as leaf area index (LAI), weight specific leaf (BDS), the content of chlorophyll a and chlorophyll b and the ratio of chlorophyll a and chlorophyll b, reducing the reflection of light, a reduction transinisi light and reduction khloroflas non pigmented. With the addition of nutrient nitrogen through proper fertilization and specific conditions of the swamp is expected to overcome the negative effects of low light intensity. To achieve the goal of optimizing the biological production of corn plants that was developed at the space in between trees of oil palm plantations in wetlands need to do a series of studies initiated from the inventory of the condition of the field and light interception in the canopy of palm, testing the tolerance of corn varieties to low-light conditions, efforts to increase production through nitrogen fertilization in corn shaded conditions and testing different varieties of corn that were given extra nitrogen in the space between rows of oil palm plantations have produced.

Plot trials made with a size of 2 m x 1.5 m by 8 plots in oil palm canopy 12 years old, the distance between subplots within 50 cm and between replications 8 meters (distance canopy of palm). Number of plots within one repetition is 8 plots. Two weeks before planting given Dolomite 6 tonnes / ha, three days before planting Furadan 3 G sprinkled evenly over the surface of the plot as much as 15 kg ha-1. Seeds planted by ditugal as much as 1-3 seeds per planting hole by using a spacing of 75 cm x 25 cm and a week after growing thinning with one plant per hole. Manure in the form of chicken manure with a rate of 2.5 ton / ha given at planting. NPK fertilizers, such as urea 400 kg / ha, SP-36 225 kg ha-1 and 125 kg KCl / ha. Fertilization is first given 45 SP-36 kg, 25 kg KCl / ha and urea 20% of the dose given appropriate treatment at the plant age 10 day after planting (DAP). Fertilization both SP-36 90 kg. 50 kg KCI / ha and 40% urea according to treatment at 30 DAP. Fertilization third given both SP-36 90 kg, 50 kg KCl / ha and 40% urea according to treatment at age 50 DAP. The fertilizer is given the array within 5-10 cm side row of plants along with hoarding and mechanical weed control.

Final plant height observations done at age 65 DAP, together with sampling for observation leaves leaf area. specific leaf area. Observations of chlorophyll a and b and total leaf chlorophyll. Observation of leaf dry weight using oven with a temperature of 70 ° C for 48 hours. Measurement of leaf area, then, by measuring the length and width of leaf samples, calibrated using leaf statue meter to determine the constant of leaf area index. Next, the total leaf area leaf area is determined by multiplying the sample with the average number of leaves of the plant samples at the last observation. Specific leaf area determination by comparing the sample divided leaf area leaf dry weight of the sample, (Dry weight of leaves using an oven temperature of 70 ° C for 48 hours). Root dry weight at the end of the study by taking the roots washed and then dried in an oven (at 70 ° C for 48 hours). Observation components of the production is done when the corn cob was physiologically mature. and the wind dried cob weight measurement.

3. Results and discussions

3.1. Results

Microclimate conditions under the canopy of palm oil during the study as shown in table 1, shows that the average light intensity M2S 430 mol-1, an average temperature of 34°C and humidity is 66.24%.

Table 1. Microclimate conditions under the canopy of palm ages 0-12 years.

Climatic components	Observation I	II	III	IV	Average
Light intensity (μmol m^2s^{-1})	487.67	397.33	412	425	430.50
Temperature (^oC)	33.77	34.43	34.87	35.20	34.57
Humidity (%)	78.78	76.21	73.5	76.47	75.23

Results of analysis of variance to the observed variables indicate that the treatment effect is significant on plant height, number of chlorophyll a and b, the number of leaves, leaf length, leaf area, segment length, cob without the husk, the weight of dry seed cob^{-1}, the weight of dry seed ha^{-1}, except the leaf width, treatment effect was not significant (table 2).

Further test results with the smallest difference test. in Table. 2. the group CIR tolerant maize varieties. show that varieties B 41 which has the highest plant height (184.0 cm) and leaves the longest (80.33 cm) was statistically different from the other varieties.

Corn Pioneer 27 have the highest number of leaves (13.16 strands). the highest weight of cob without husk (89.58 g) and the weight of the highest shelled cob-1 (99.12 g). Sukmaraga shows the widest leaf width (7.83 cm) and the longest segment (12.66 cm).

Table 2. Results of analysis of variance varieties influence on the observed variables.

Variable	Variety	CV (%)
Plant height	**	8.72
Number of leaves	**	10.86
Length of leaves	**	10.1
Width of leaves	Ns	11.75
Length of steam	**	12.86
Dry weigh of cob no husk	**	12.86
Dry weight seed cob^{-1}	**	14.37

Sugihan being the lowest for all the observed variables. Meanwhile. the corn-sensitive group CIR. dry seed weight of cob-1 obtained at the highest local varieties of maize. Tg.lago (67.25 %).

Table 3. Effect of different varieties of maize to the observed variables.

Variety	Plant height	Number of leaves	Length of leaves	Width of leaves	Length of segment	BTTK	BPPKT
B41	184.00 B	13.00 B	80.33 B	7.06 A	12.58 B	82.66 BC	83.62 B
Pioneer 27	179.58 B	13.16 BC	76.83 A	7.50 B	10.66 A	89.58 BC	89.12 B
SMRG	187.75 B	11.16 A	79.75 B	7.83 BC	12.66 B	75.16 B	83.68 B
Sugihan	176.25 B	11.91 B	74.08 A	7.22 A	11.50 B	78.91 B	89.37 B
L164	168.58 A	10.25 A	75.16 A	6.37 A	11.08 A	56.54 A	54.00 A
NT105	154.00 A	10.25 A	69.75 A	6.18 A	12.91 B	67.83 A	57.12 A
LMR	160.83 A	10.91 A	66.91 A	6.16 A	9.25 A	62.62 A	63.25 A
TG LAGO	163.18 A	10.08 A	79.91 B	6.56 A	11.75 B	63.58 A	67.25 A
BNT 0.05	22.03	1.81	11.19	1.18	2.18	13.44	16.04

3.2. Discussion

Generally shade formed by the canopy of palm oil ranges from 50% -80% depending on the age of the palm. Low-intensity light effects that will continually affect the growth of corn, since the phase of seed germination to process pematengan response to various corn varieties, fertilizers Urea with the same doses, it produces two groups of plant height. CIR-tolerant varieties have plant height were not significantly different. The same thing happened to a group of plants that are sensitive CIR. The condition is thought to be due CIR tolerant maize plants still have a better metabolism, thus increasing plant height.

Corn B 41 and Pioneer 27 have a number of leaves more than other varieties. This is allegedly due to the optimal utilization of nitrogen by two of these varieties, although in limited light intensity conditions. Utilization of nitrogen through the mechanism of PEP-karbosilase can stimulate the accumulation photosintat so that the number of leaves more. The same thing happened to the length and width of leaves. B 41 segment length of about 1-2 cm longer than the sensitive group of corn CIF, except for NT 105. The length of this segment is in response CIR tolerant plants to increase plant height due to increased activity on the network meristim auxin.

In terms of the result of dry seed cob^{-1} crop^{-1} and weighs cob without husk varieties of Pioneer 27 and B 41 to give hope to be developed as a crop stream on gawangan palm (polyculture oil palm plantations-corn) with the ability of the results of dry seed ha^{-1} reaches 5:25 tons for Pioneer 27 and 4.96 tons respectively. B 41 and Pioneer 27 are varieties with high stem growth with the number of leaves, leaf width and length greater than the other varieties. So, it is thought to have the ability to capture more light and with the availability of sufficient nitrogen and supported by micro-environmental conditions with optimum temperature for growth of corn with the air humidity is high enough, it is supposedly enables optimal rate of photosynthesis will result fotosintat better. Fotosintat results will be stored in the form of grain or cob. This statement is supported by Kusumasari *et al.* [15] which states that a sturdy root system and is capable of high production in a variety of soil conditions and can be classified as corn that is tolerant of low light intensity and B 41 is a strain of corn that efficient nutrient [16].

4. Conclusions

4.1. Conclusion
From this study, it can be concluded that the maize varieties of B 41 and Pioneer 27 (group tolerant CIR) can grow and develop quite well on below canopy of oil palm age of 12 years, that the adoption of polyculture palm-corn with levels of occupational planting 0.7 and the area to be planted 0.4 ha, and that from one ha of oil palm planted monoculture with genotypes of B 41 and Pioneer 27 can be generated approximately one ton of dry shelled corn.

4.2. Recommendations
Further research is needed to determine optimal spacing of oil palm (wide distance canopy) which allows the interception of light optimal for the growth of maize as polyculture in oil palm plantations (intercrops) because in this study observation was not conducted for optimal LAI palm.

References

[1] Lakitan, B and N. Gofar, 2013 *Policy technological innovations for sustainable land management suboptimal.* National Seminar Papers Suboptimal land. Palembang, 20- 21 September 2013

[2] Wetlands and Coastal Directorate, Department of Public Works, 2009 *Potential and challenges Development of swamps in Indonesia.* Papers on Swamp Management Workshop Seminar in support of National Food Security. Jakarta. Coordination for Infrastructure and Regional Development. Coordinating Ministry of Economic Affairs

[3] Subagyo, A. 2006. *Land Swamp Lebak in Didi Ardis S et al. Characteristics and Swamp Land Management.* Great Hall Research and Development of Agricultural Land Resources. Bogor. Pages 99-116

[4] Rosmiah, Gusmiatun, R. I.S. Aminah, Minwal and N. Marlina, 2017 *Effectiveness of Biological Organic Fertilizer plus Inorganic Fertilizer on the Growth and Production Plant Variety some Hybrid Corn (Zea mays L.) and Weed in Lowland Swamp Area.* International Journal of engeenering Reseach and Science and Tecnology, 6 (1): 1-1

[5] Marlina, N., N. Amir., R.I.S. Aminah., L. Nisfuriah, G. A. Nasser., Y. Purwanti and Asmawati, 2017 *Organic and Inorganic Fertilizers Application on NPK Uptake and Production of Sweet Corn in Inceptisol Soil of Lowland Swamp Area.* Matec Web of 97.01106 ETIC Conference 2016: 1-11

[6] Ismail, G.I., 1993 Eight Years of Agricultural Research in the Swamp Land 1985-1993

[7] Dir. Gen. Of Estate, 2010 *Statistics of Plantation.* Agriculture department. The Directorate General of Plantation Indonesia. Jakarta

[8] Abdurachman and E.E. Ananto, 2000 *The concept of Sustainable Agriculture Development in Swamp Land to support the resilience of the Food and Agricultural Development.* National Seminar on Agricultural Research and Development in Land Swamp. Bogor, 25 to 27 July 2000.23

[9] Pumomo, D, 2005 *Maize Varietal response to Simple Irrigation.* Agrosains. 7: 86-93

[10] Atman, 2007 *Plant between coconut with corn.* Tambua Scientific Journal, 2, 187-193

[11] Mohr H, Schopfer P. 1995. *Plant physiology.* 629pp. Berlin: Springer-Verlag

[12] Sujiprihati, M., Gratitude and R. Yunisanti, 2006 *Stability Analysis of Results of seven sweet corn population using methods aditive play multicative interaction effect.* Bul. Agron: **34** 93-97

[13] Rochman, Ahmad, Muhammad Noor and Rina Yanti, 2010 *Prospects and Strategy Development of Agribusiness Cultivation of corn crops in Land Swamp: Constraints and Challenges.* Proceedings of the National Cereals Week 2010

[14] Sudana, W. 2005. *Potential and Prospects of Land Swamp as a Source of Agricultural Production.* Agricultural Policy Analysis (3): 141-151

[15] Kusumasari, A.C., B. Prayudi and A. Supriyo, 2013 *Effect of shelter on the corn P27 performance at Teak wood plantation.* National Seminar at Agriculture Faculty, University of Trunojoyo, Madura, East Java, Indonesia

[16] Moelyohadi, Y., M.U. Aaron, Munandar, R. Biological and N. Gofar, 2012 *Utilization of Different Types of Biological Fertilizer on Crops Maize (Zea mays L.) Efficient Hara in Marginal Dry Land.* Suboptimal Land journal. ISSN 2252-6188 (1): 31-39

IOP Conf. Series: Materials Science and Engineering **209** (2017) 012110 doi:10.1088/1757-899X/209/1/012110

Highly Polluted Wastewaters Treatment by Improved Dissolved Air Flotation Technology

I C Moga[1*], C I Covaliu[2], M G Matache[3] and B I Doroftei[1]
[1]SC DFR Systems SRL, 46 Drumul Taberei Street, ap. 23, Bucharest, Romania
[2]Department of Biotechnical Systems, University Politehnica of Bucharest, 313 Splaiul Independenței Street, București, Romania
[3]National Institute of Research-Development for Machines and Installations Designed to Agriculture and Food Industry, 6 Ion Ionescu de la Brad Street, Bucharest, Romania

E-mail: corinamoga@yahoo.com

Abstract. Numerous investigations are oriented towards the development of new wastewater treatment technologies, having high efficiencies for removing even low concentrations of pollutants found in water. These efforts were determined by the destroyer impact of the pollutants to the environment and human's health. For this reason this paper presents our study concerning an improved dissolved air flotation technology for wastewater treatment. There is described a dissolved air flotation (DAF) installation composed by two equipments: pressurized capsule and lamellar settling. Also, there are presented some advantages of using nanoparticles as flotation collectors.

1. Introduction

Flotation represents a separation process (physical and chemical) of pollutant (solid matter and/or fats) by establishing a contact in different phases: the floating pollutant, the liquid phase and air. Nowadays, this is an important separation technique, frequently used around wold, and its applications varies from selective separation of minerals to ions and even microorganisms [1,2].

Artificial flotation represents a process in which air is forced introduced into the wastewater. The air bubbles rising to the surface adhere to the suspended particles or emulsified fats presented in the liquid and lead pollutants to the wastewater surface. The large interface contact surface produced by the formation of air bubbles in the water mass requires the necessity of an energy input proportional to their superficial energy - "potential energy". This process is based on the adheration capacity of the dispersed particles, solid or liquid at the gas/air bubbles, forming conglomerates that using the ascending force of the bubble rise to the water surface.

Dissolved air flotation (DAF) was also worldwide used during the last fifty years for the removal of oils, greases and suspended solids. Nowadays, DAF systems are frequently applied in wastewater treatment field, product recovery, and thickening of biological solids in industries such as textile and leather, paper and cellulose, food processing etc. Even though the technology advances and the range of applications for DAF is expanded, the engineers and process designers frequently use outdated and insufficient design data for DAF systems for industrial pretreatment.

The DAF technology is generally applied in industrial wastewater treatment plants (WWTPs), where the amount of pollutants is above the average (municipal wastewaters). The research team members introduced a modified DAF unit in a laboratory leachate wastewater treatment plant. The efficiency of wastewater treatment depends on pollutant types and concentrations.

2. Experimental procedures

The dissolved air flotation unit consists of two main equipment: a pressurized capsule and a lamellar settler. One of the most important element of DAF is the pressurized capsule in which the water comes into contact with the compressed air. Air consumption in the transfer process is (20 – 100) Ndm^3/m^2 water, that is (1.5 - 5.0)% of the water volume. Inside the pressurized capsule which is a cylindrical chamber having at the ends 2 caps are introduced water and air under pressure. Under pressure conditions (6 bar) the air in dissolved into the water mass. The detention time and the contact surface area are important in order to increase the amount of dissolved air, with benefit to the flotation process (an increase of the specific flow rate and of the saturation coefficient). Achieving a very fine dispersion of the water may increase the contact area.

At the top of the capsule is located the circuit for water feeding with the help of 4 sprinklers as fine dispersed droplets and not as jet. At the bottom of the capsule is positioned the air supply figure 1. The researchers found that introducing moving plastic elements inside the capsule will lead to obtaining a longer time contact between air bubbles introduced through the circular pipe and the water.

Figure 1. The pressurized capsule for a DAF unit and an example of a plastic element.

The longitudinal settler consists in two compartments placed in series: a lamellar zone and a technical room (figure 2). At the usual lamellar clarifiers the clogging phenomenon can appear between the settling plates. The flotation process inside the clarifier can overcome the clogging. The air-water mixture enters into the clarifier with the help of a transport system that consists of pipes and three funnels. Funnels are diffusers. So, the air bubbles - water mixture does not "wash" the slab foundation. As the air is depressurized, small air bubbles are formed and the bubbles rise to the water surface. On their way up the air bubbles gather suspended particles, leading them to the surface, from where are targeted, as foam, to skimmer and discharged outside the system.

At the bottom of settler the sludge is deposited and the treated water is discharged through a pipe into the emissary. An efficient settling for the entire clarifier length is obtained by using well designed plates systems, mounted obliquely (52^0). The stability of the liquid and sludge retention is given by the rectangular cross section of the settling and the interior construction. Through a system of connected suction pipes is collected the settled sludge from the bottom of lamellar clarifier. Through a hydrociclone this sludge is pumped by the sludge pump placed in the technical chamber. Periodically, dense mineralized sludge is discharged into dewatering textile bags from where it is dehydrated and manually removed.

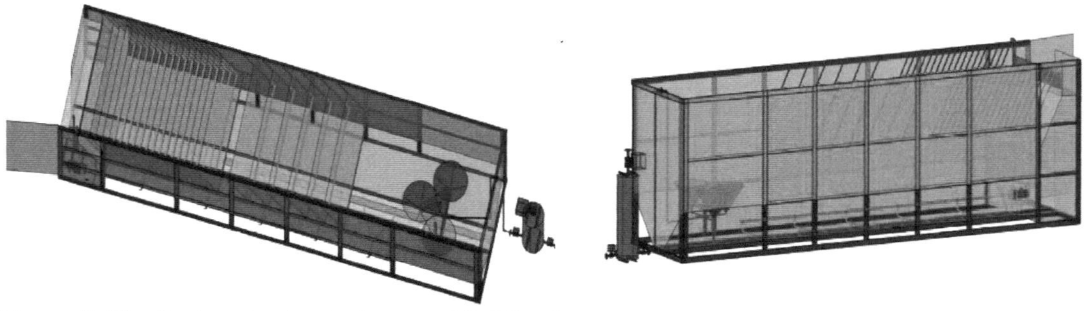

Figure 2. The final settler for the improved DAF unit.

3. Results and discussions

Figure 3 presents the laboratory scheme and installation for the DAF unit. The water recirculation inside the DAF system was considered, as shown in figure 3. The laboratory DAF facility consist of a lamellar clarifier, a pressurized capsule, a pressure clean/treated water pump, a valve and a foam removal equipment. The air is introduced inside the pressure capsule with the help of a compressor.

By treated water recirculation a clear advantage appears - air saturation of part of the treated water returns into DAF. The laboratory facility was tested in an industrial (food processing) wastewater treatment plant. The DAF unit was installed to remove the very fine chopped poultry feathers, which have passed the grills and all the previous wastewater treatment stages.

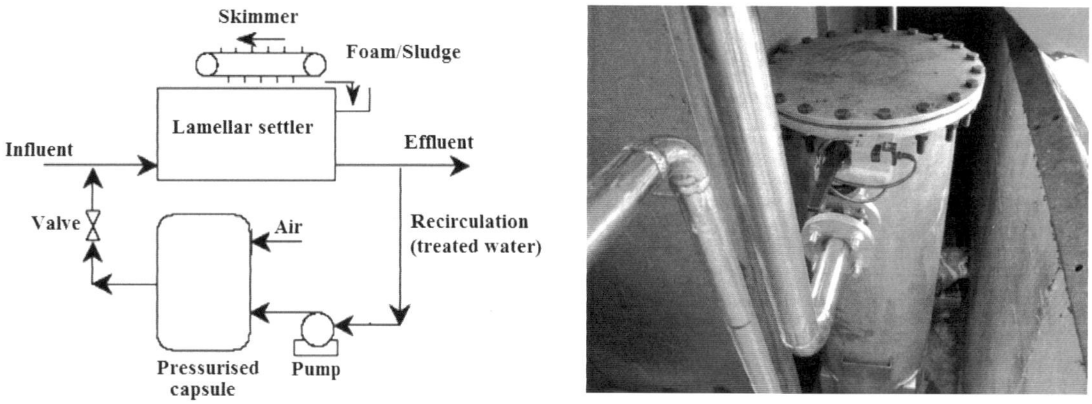

Figure 3. Experimental laboratory DAF unit installed in an industrial wastewater treatment plant.

For a correct designing of the lamellar settler, mathematical modelling and numerical simulations were realized by the researchers. The role of the numerical simulations was to determine the profiles of pollutants inside the settler. In order to build the mathematical model for the determination of the pollutant profile inside the lamellar settler, the general equation of dispersion was considered and the obtained results are shown in Figure 4. As it can be observed, in the front part or the settler, where the air-water mixture is introduced, is the highest concentration of pollutants. In the flotation zone, the most part of pollutants is removed. In the second part of the lamellar clarifier the water treatment process is completed and the pollutant concentration (suspended solids mostly represented by fine chopped feathers) decreases from 300 mg/l to 25 mg/l.

During the experimental measurement bubbles were formed; the bubbles were produced by supersaturating the water with air introduced under pressure (6 bars), followed by a depression (using a valve and 3 diffuserd) to normal atmospheric values. The formed bubbles had a small diameter and the chopped feathers adhered to the bubbles and formed a foam layer at the water surface.

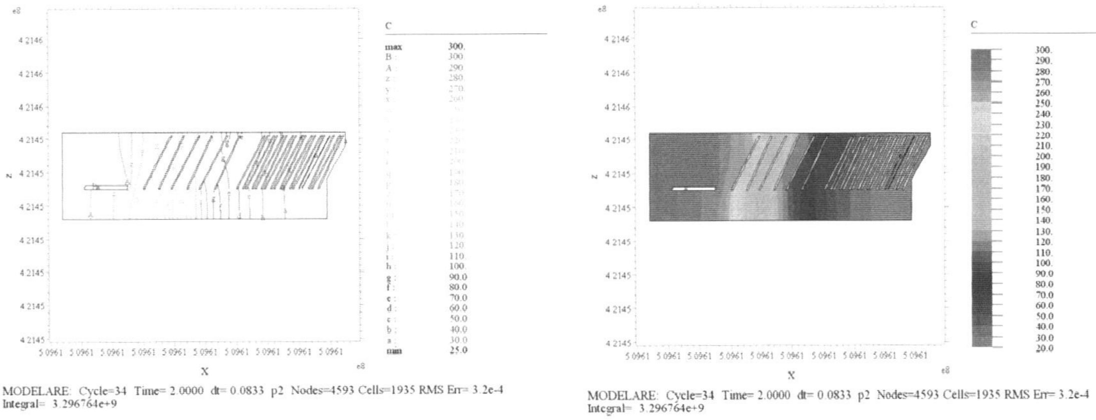

Figure 4. Pollutant (suspended solids) profile inside the lamellar settler – numerical simulations.

The installation was designed for a capsule pressure of 6-8 bars and a retention time of (1 - 3) min in the pressured capsule, where the water contacts the compressed air in the presence of the plastic elements. The depressure valve together with the 3 diffusers, placed inside the clarifier, plays an important role in the appearance of air bubbles in the wastewater mass.

Foaming is a process of surface separation of the impurities present in the foam generated by the diffusion of air into the body of water.

Formation of the bubble – suspended solid flocs depends on the depth of the flotation tank, and in the experimental laboratory installation a depth of 2.5 m was realized. The water retention time in the flotation tank was 1.5 hours.

The improved DAF unit removed 82% of the suspended solids.

Some of the innovative aspects and the advantages of the dissolved air flotation unit are:

- higher efficiency of the process obtained through the plastic moving elements inside the capsule that "tease" trail of bubbles to the surface;
- existing a large air-water surface contact offered by the using of sprinklers for water dispersion in very fine droplets within the capsule;
- facile installation of the pressurised capsule even in existing clarifiers;
- providing higher treatment efficiency without using any bio-products or consumables for enhancing the biological degradation processes - completely organic process;
- smaller footprint required as compared to conventional longitudinal settlers and flocculation;
- shorter flocculation times, as compared to gravity separation, because a smaller floc particle size is required;
- modular construction.

Recently, flotation reagents have an essential role in the pollutant removal boath as sludge or foam. During the developing of industrial applications for flotation, the major advantage resulted from acquiring very effective flotation reagents. Depending on their role the flotation reagents are divided into collectors, frothers, modifiers, depressants and flocculants.

The purpose of the collector is to selectively form a hydrophobic layer on a given pollutant surface without binding to the pollutant particles in flotation slurry; thus allowing the hydrophobized particles to attach to air bubbles, which can be recovered in the froth product.

In the scientific literature is described a new technology where as collectors can be hydrophobic nanoparticles adsorb onto much larger, hydrophilic particle surfaces to facilitate the attachment to air bubbles in flotation, Figure 5. Also, it is showed that the role of nanoparticles is to facilitate particle-bubble attachment and/or to minimize detachment. The goal of the researchers is to consider the influence of nanoparticle parameters on the various stages of particle flotation for identifying the important role of nanoparticles and to optimize nanoparticle properties for this application.

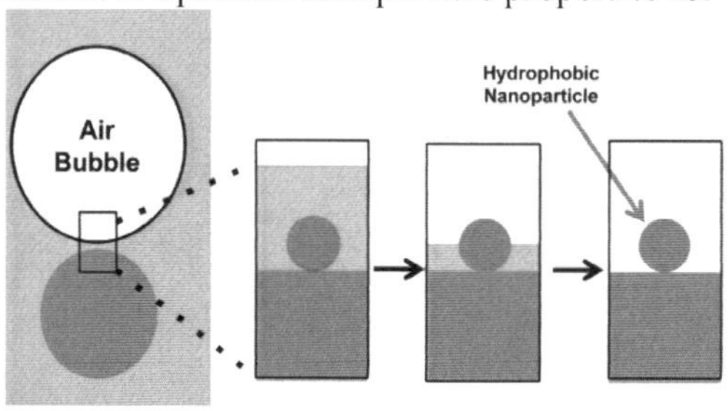

Figure 5. Thinning and rupture of the liquid film between a mineral particle with adsorbed nanoparticles and an air bubble [4].

Moreover it was found that smaller and more hydrophobic nanoparticles are the most efficient flotation collectors. The forces required to pull a nanoparticle-coated sphere from the air/water interface of a bubble into the water was determined via micromechanical measurements. The maximum pull-off force ranged from 0.0086 μN for a clean 55 μm glass bead to 1.9 μN for a bead bearing adsorbed 46 nm diameter polystyrene spheres [3].

4. Conclusions

Some of the advantages of the dissolved air flotation unit such as high efficiency of the wastewater process, facile installation sustain the further research in this field. Appling nanoparticles in DAF process could represent an efficient and innovative way to increase the DAF efficiency.

The research team, during the implementation of project no. PN-III-P2-2.1-PTE-2016-0183 will develop and test a pilot DAF installation. A nanomaterial dosing systems will be included in the treatment stage to obtain a more stable foam and a better pollutant removal efficiency. In order to validate the new proposed technology will be designed and realised wastewater treatment stage with a capacity of 24 m^3/day.

References

[1] Heiskanen K, Kirjavainen V, Laapas H 1991 *Int. J. Miner. Process.* **33**(1-4) 263-274
[2] Edwards C, Kipkie W, Agar G 1980 *Int. J. Miner. Process.* **7**(1) 33-42
[3] Senior G, Thomas S 2005 *Int. J. Miner. Process.* **78**(1) 49-61
[4] Yang S, Pelton R 2011 Nanoparticle Flotation Collectors II: The Role of Nanoparticle Hydrophobicity, Ontario, Canada

Acknowledgements

This work was supported by a grant of the Romanian National Authority for Scientific Research and Innovation, CNCS/CCCDI – UEFISCDI, project number PN-III-P2-2.1-PTE-2016-0183 (25/2016 “Improved flotation treatment technology for heavily loaded wastewaters”), within PNCDI III.